BIOLOG

HOW LIFE WORKS

BIOLOGY HOW LIFE WORKS

Volume 1: Cells, Genetics, and Evolution

Chapters 1–24

SECOND EDITION

James Morris
BRANDEIS UNIVERSITY

Daniel Hartl
HARVARD UNIVERSITY

Andrew Knoll
HARVARD UNIVERSITY

Robert Lue
HARVARD UNIVERSITY

Melissa Michael
UNIVERSITY OF ILLINOIS AT URBANA-CHAMPAIGN

ANDREW BERRY, ANDREW BIEWENER,
BRIAN FARRELL, N. MICHELE HOLBROOK
HARVARD UNIVERSITY

A Macmillan Education Imprint

PUBLISHER Kate Ahr Parker
ACQUISITIONS EDITOR Beth Cole
LEAD DEVELOPMENTAL EDITOR Lisa Samols
SENIOR DEVELOPMENTAL EDITOR Susan Moran
DEVELOPMENTAL EDITOR Erica Champion
EDITORIAL ASSISTANTS Jane Taylor, Alexandra Garrett, Abigail Fagan
REVIEW COORDINATOR Donna Brodman
PROJECT MANAGER Karen Misler
DIRECTOR OF MARKET DEVELOPMENT Lindsey Jaroszewicz
EXECUTIVE MARKETING MANAGER Will Moore
ART AND MEDIA DIRECTOR Robert Lue, Harvard University
EXECUTIVE MEDIA EDITOR Amanda Nietzel
SENIOR DEVELOPMENT EDITOR FOR TEACHING & LEARNING STRATEGIES Elaine Palucki
SENIOR MEDIA PRODUCER Chris Efstratiou
PROJECT EDITOR Robert Errera
MANUSCRIPT EDITOR Nancy Brooks
DIRECTOR OF DESIGN, CONTENT MANAGEMENT Diana Blume
DESIGN Tom Carling, Carling Design Inc.
ART MANAGER Matt McAdams
ILLUSTRATIONS Imagineering
CREATIVE DIRECTOR Mark Mykytiuk, Imagineering
PHOTO EDITOR Christine Buese
PHOTO RESEARCHERS Lisa Passmore and Richard Fox
PRODUCTION MANAGER Paul Rohloff
COMPOSITION MPS Limited
PRINTING AND BINDING King Printing Co., Inc.
COVER IMAGE: Thorsten Henn/Getty Images

Library of Congress Control Number: 2015950976
ISBN-13: 978-1-319-04882-2
ISBN-10: 1-319-04882-X

Printed in the United States of America

Third printing

Macmillan Learning
W. H. Freeman and Company
One New York Plaza
Suite 4500
New York, NY 10004-1562
www.macmillanhighered.com

DEDICATION

To all who are curious about life and how it works

ABOUT THE AUTHORS

JAMES R. MORRIS is Professor of Biology at Brandeis University. He teaches a wide variety of courses for majors and non-majors, including introductory biology, evolution, genetics and genomics, epigenetics, comparative vertebrate anatomy, and a first-year seminar on Darwin's *On the Origin of Species*. He is the recipient of numerous teaching awards from Brandeis and Harvard. His research focuses on the rapidly growing field of epigenetics, making use of the fruit fly *Drosophila melanogaster* as a model organism. He currently pursues this research with undergraduates in order to give them the opportunity to do genuine, laboratory-based research early in their scientific careers. Dr. Morris received a PhD in genetics from Harvard University and an MD from Harvard Medical School. He was a Junior Fellow in the Society of Fellows at Harvard University, and a National Academies Education Fellow and Mentor in the Life Sciences. He also writes short essays on science, medicine, and teaching at his Science Whys blog (http://blogs.brandeis.edu/sciencewhys).

DANIEL L. HARTL is Higgins Professor of Biology in the Department of Organismic and Evolutionary Biology at Harvard University and Professor of Immunology and Infectious Diseases at the Harvard Chan School of Public Health. He has taught highly popular courses in genetics and evolution at both the introductory and advanced levels. His lab studies molecular evolutionary genetics and population genetics and genomics. Dr. Hartl is the recipient of the Samuel Weiner Outstanding Scholar Award as well as the Gold Medal of the Stazione Zoologica Anton Dohrn, Naples. He is a member of the National Academy of Sciences and the American Academy of Arts and Sciences. He has served as President of the Genetics Society of America and President of the Society for Molecular Biology and Evolution. Dr. Hartl's PhD is from the University of Wisconsin, and he did postdoctoral studies at the University of California, Berkeley. Before joining the Harvard faculty, he served on the faculties of the University of Minnesota, Purdue University, and Washington University Medical School. In addition to publishing more than 400 scientific articles, Dr. Hartl has authored or coauthored 30 books.

ANDREW H. KNOLL is Fisher Professor of Natural History in the Department of Organismic and Evolutionary Biology at Harvard University. He is also Professor of Earth and Planetary Sciences. Dr. Knoll teaches introductory courses in both departments. His research focuses on the early evolution of life, Precambrian environmental history, and the interconnections between the two. He has also worked extensively on the early evolution of animals, mass extinction, and plant evolution. He currently serves on the science team for NASA's mission to Mars. Dr. Knoll received the Phi Beta Kappa Book Award in Science for *Life on a Young Planet*. Other honors include the Paleontological Society Medal and Wollaston Medal of the Geological Society, London. He is a member of the National Academy of Sciences and a foreign member of the Royal Society of London. He received his PhD from Harvard University and then taught at Oberlin College before returning to Harvard.

ROBERT A. LUE is Professor of Molecular and Cellular Biology at Harvard University and the Richard L. Menschel Faculty Director of the Derek Bok Center for Teaching and Learning. Dr. Lue has a longstanding commitment to interdisciplinary teaching and research, and chaired the faculty committee that developed the first integrated science foundation in the country to serve science majors as well as pre-medical students. The founding director of Life Sciences Education at Harvard, Dr. Lue led a complete redesign of the introductory curriculum, redefining how the university can more effectively foster new generations of scientists as well as science-literate citizens. Dr. Lue has also developed award-winning multimedia, including the animation "The Inner Life of the Cell." He has coauthored undergraduate biology textbooks and chaired education conferences on college biology for the National Academies and the National Science Foundation and on diversity in science for the Howard Hughes Medical Institute and the National Institutes of Health. In 2012, Dr. Lue's extensive work on using technology to enhance learning took a new direction when he became faculty director of university-wide online education initiative HarvardX; he now helps to shape Harvard's engagement in online learning to reinforce its commitment to teaching excellence. Dr. Lue earned his PhD from Harvard University.

MELISSA MICHAEL is Director for Core Curriculum and Assistant Director for Undergraduate Instruction for the School of Molecular and Cellular Biology at the University of Illinois at Urbana-Champaign. A cell biologist, she primarily focuses on the continuing development of the School's undergraduate curricula. She is currently engaged in several projects aimed at improving instruction and assessment at the course and program levels. Her research focuses primarily on how creative assessment strategies affect student learning outcomes, and how outcomes in large-enrollment courses can be improved through the use of formative assessment in active classrooms.

ANDREW BERRY is Lecturer in the Department of Organismic and Evolutionary Biology and an undergraduate advisor in the Life Sciences at Harvard University. With research interests in evolutionary biology and history of science, he teaches courses that either focus on one of the areas or combine the two. He has written two books: *Infinite Tropics*, a collection of the writings of Alfred Russel Wallace, and, with James D. Watson, *DNA: The Secret of Life*, which is part history, part exploration of the controversies surrounding DNA-based technologies.

ANDREW A. BIEWENER is Charles P. Lyman Professor of Biology in the Department of Organismic and Evolutionary Biology at Harvard University and Director of the Concord Field Station. He teaches both introductory and advanced courses in anatomy, physiology, and biomechanics. His research focuses on the comparative biomechanics and neuromuscular control of mammalian and avian locomotion, with relevance to biorobotics. He is currently Deputy Editor-in-Chief for the Journal of Experimental Biology. He also served as President of the American Society of Biomechanics.

BRIAN D. FARRELL is Director of the David Rockefeller Center for Latin American Studies and Professor of Organismic and Evolutionary Biology and Curator in Entomology at the Museum of Comparative Zoology at Harvard University. He is an authority on coevolution between insects and plants and a specialist on the biology of beetles. He is the author of many scientific papers and book chapters on the evolution of ecological interactions between plants, beetles, and other insects in the tropics and temperate zone. Professor Farrell also spearheads initiatives to repatriate digital information from scientific specimens of insects in museums to their tropical countries of origin. In 2011–2012, he was a Fulbright Scholar to the Universidad Autónoma de Santo Domingo in the Dominican Republic. Professor Farrell received a BA degree in Zoology and Botany from the University of Vermont and MS and PhD degrees from the University of Maryland.

N. MICHELE HOLBROOK is Charles Bullard Professor of Forestry in the Department of Organismic and Evolutionary Biology at Harvard University. She teaches an introductory course on biodiversity as well as advanced courses in plant biology. She studies the physics and physiology of vascular transport in plants with the goal of understanding how constraints on the movement of water and solutes between soil and leaves influences ecological and evolutionary processes.

ASSESSMENT AUTHORS

JEAN HEITZ is a Distinguished Faculty Associate at the University of Wisconsin in Madison, WI. She has worked with the two-semester introductory sequence for biological sciences majors for over 30 years. Her primary roles include developing both interactive discussion/recitation activities designed to uncover and modify misconceptions in biology and open-ended process-oriented labs designed to give students a more authentic experience with science. The lab experience includes engaging all second-semester students in independent research, either mentored research or a library-based meta-analysis of an open question in the literature. She is also the advisor to the Peer Learning Association and is actively involved in TA training. She has taught a graduate course in "Teaching College Biology," has presented active-learning workshops at a number of national and international meetings, and has published a variety of lab modules, workbooks, and articles related to biology education.

MARK HENS is Associate Professor of Biology at the University of North Carolina Greensboro, where he has taught introductory biology since 1996. He is a National Academies Education Mentor in the Life Sciences and is the director of his department's Introductory Biology Program. In this role, he guided the development of a comprehensive set of assessable student learning outcomes for the two-semester introductory biology course required of all science majors at UNCG. In various leadership roles in general education, both on his campus and statewide, he was instrumental in crafting a common set of assessable student learning outcomes for all natural science courses for which students receive general education credit on the sixteen campuses of the University of North Carolina system.

JOHN MERRILL is Director of the Biological Sciences Program in the College of Natural Science at Michigan State University. This program administers the core biology course sequence required for all science majors. He is a National Academies Education Mentor in the Life Sciences. In recent years he has focused his research on teaching and learning with emphasis on classroom interventions and enhanced assessment. A particularly active area is the NSF-funded development of computer tools for automatic scoring of students' open-ended responses to conceptual assessment questions, with the goal of making it feasible to use open-response questions in large-enrollment classes.

RANDALL PHILLIS is Associate Professor of Biology at the University of Massachusetts Amherst. He has taught in the majors introductory biology course at this institution for 19 years and is a National Academies Education Mentor in the Life Sciences. With help from the Pew Center for Academic Transformation (1999), he has been instrumental in transforming the introductory biology course to an active learning format that makes use of classroom communication systems. He also participates in an NSF-funded project to design model-based reasoning assessment tools for use in class and on exams. These tools are being designed to develop and evaluate student scientific reasoning skills, with a focus on topics in introductory biology.

DEBRA PIRES is an Academic Administrator at the University of California, Los Angeles. She teaches the introductory courses in the Life Sciences Core Curriculum. She is also the Instructional Consultant for the Center for Education Innovation & Learning in the Sciences (CEILS). Many of her efforts are focused on curricular redesign of introductory biology courses. Through her work with CEILS, she coordinates faculty development workshops across several departments to facilitate pedagogical changes associated with curricular developments. Her current research focuses on what impact the experience of active learning pedagogies in lower division courses may have on student performance and concept retention in upper division courses.

ASSESSMENT CONTRIBUTORS

ELENA R. LOZOVSKY, Principal Staff Scientist, Department of Organismic and Evolutionary Biology, Harvard University

FULTON ROCKWELL, Research Associate, Department of Organismic and Evolutionary Biology, Harvard University

VISION AND STORY OF *BIOLOGY: HOW LIFE WORKS*

Dear students and instructors,

One of the most frequent questions we get about the second edition of *Biology: How Life Works* is, "Has science really changed that much in three years?"

Ongoing discoveries in biology mean that a new edition of an introductory biology textbook will certainly have some new science content. But, more importantly, our second edition is new in the sense that we had the opportunity, for the first time, to listen to students and instructors who used *HLW* in the classroom. The second edition is responsive to this group and their input has proven invaluable.

What we heard from this community is that the philosophy of *HLW* resonates with students and instructors. They appreciate a streamlined text that rigorously focuses on introductory biology, an emphasis on integration, a modern treatment of biology, and equal attention to text, assessment, and media. These elements haven't changed—they are the threads that connect the first and second editions. In fact, all of the changes of the second edition are integrated within the framework of the first edition; they are not simply add-ons.

We are particularly excited about the work we've done in assessment. In the first edition, we worked with a creative and dedicated team of assessment authors to create something wholly new: not a standard test bank, but a thoughtful, curated, well-aligned set of questions that can be used for teaching as well as testing. These questions are written at a variety of cognitive levels. In addition, they can be used in a variety of contexts, including pre-class, in-class, homework, and exam, providing a learning path for students.

Our approach was so well received that we took it a step further in the second edition. The *HLW* team is excited to have Melissa Michael, our lead assessment author in the first edition, join us as a lead author in the second edition. Her new role allows her to work more closely with the text and media, which makes for an even tighter alignment among these various components.

Instructors have told us that they especially like the activities that can be used in class to foster active learning among students. In response, the second edition includes a rich set of activities across the introductory biology curriculum. Some are short, taking just a couple of minutes to explore a specific topic or concept. Others are longer, spanning several class periods and exploring topics and concepts across many chapters.

Although students and instructors appreciated a streamlined text, we also heard that more attention was needed in ecology. In response, we added a chapter that focuses on physical processes and global ecology. This new chapter also had a ripple effect throughout the later chapters of Part 2, giving us more space to explore other ecological concepts more deeply as well.

The media in *HLW* is many layered, so that a static visual synthesis on the page becomes animated online — and even interactive — through visual synthesis maps and simulations, using a consistent visual language and supported by assessment. Media resources for this new edition have been expanded to reflect its increased emphasis on global ecology — for example, there is a new Visual Synthesis figure and online map on the flow of matter and energy in ecosystems. We also developed additional media resources that focus on viruses, cells, and tissues.

We feel that this edition is a wonderful opportunity for us to continue to develop an integrated set of resources to support instructor teaching and student learning in introductory biology. Thank you for taking the time to use it in the classroom.

Sincerely,
The *Biology: How Life Works* Author Team

RETHINKING **BIOLOGY**

The *Biology: How Life Works* team set out to create a resource for today's biology students that would reimagine how content is created and delivered. With this second edition, we've refined that vision using feedback from the many dedicated instructors and students who have become a part of the *How Life Works* community.

We remain committed to the philosophy of *How Life Works*: a streamlined, integrated, and modern approach to introductory biology.

Thematic

We wrote *How Life Works* with six themes in mind. We used these themes as a guide to make decisions about which concepts to include and how to organize them. The themes provide a framework that helps students see biology as a set of connected concepts rather than disparate facts.

- The scientific method is a deliberate way of asking and answering questions about the natural world.
- Life works according to fundamental principles of chemistry and physics.
- The fundamental unit of life is the cell.
- Evolution explains the features that organisms share and those that set them apart.
- Organisms interact with one another and with their physical environment, shaping ecological systems that sustain life.
- In the 21st century, humans have become major agents in ecology and evolution.

Selective

It is unrealistic to expect the majors course to cover everything. We envision *How Life Works* not as a reference for all of biology, but as a resource focused on foundational concepts, terms, and experiments. We explain fundamental topics carefully, with an appropriate amount of supporting detail, so that students leave an introductory biology class with a framework on which to build.

Integrated

How Life Works moves away from minimally related chapters to provide guidance on how concepts connect to one another and the bigger picture. Across the book, key concepts such as chemistry are presented in context and Cases and Visual Synthesis figures throughout the text provide a framework for connecting and assimilating information.

WHAT'S NEW IN **THE SECOND EDITION?**

Expanded ecology coverage on physical processes and global ecology provides additional emphasis on ecological concepts, while ensuring that content is integrated into the larger theme of evolution. *Learn more about the expanded ecology coverage on page xviii.*

Visual Synthesis Figures and Online Maps on the Flow of Matter and Energy through Ecosystems, Cellular Communities, and Viruses allow students to explore connections between concepts through dynamic visualizations. *Learn more about the new media on page xix.*

Lead Author Melissa Michael guides our assessment team in refining and expanding our collection of thoughtful, well-curated assessments. Dr. Michael's role ensures a tight alignment between the assessments and the media and text. *Learn more on page xvi.*

A rich collection of in-class activities provides active learning materials for instructors to use in a variety of settings. *Learn more about the new in-class activities on page xvii.*

Improved LaunchPad functionality makes it easier to search and filter within our expansive collection of assessment questions. *Learn more about new LaunchPad functionality on page xii.*

TABLE OF CONTENTS

The table of contents is arranged in a familiar way to allow its easy use in a range of introductory biology courses. On closer look, there are significant differences that aim to help biology teachers incorporate the outlooks and research of biology today. **Key differences** are identified by ● and **unique chapters** by ✪.

EVOLUTION COVERAGE: Chapter 1 introduces evolution as a major theme of the book before discussing gene expression in Chapters 3 and 4 as a foundation for later discussions of the conservation of metabolic pathways and enzyme structure (Chapters 6–8) and genetic and phenotypic variation (Chapters 14 and 15). After the chapters on the mechanisms and patterns of evolution (Chapters 21–24), we discuss the diversity of all organisms in terms of adaptations and comparative features, culminating in ecology as the ultimate illustration of evolution in action.

CHEMISTRY: Chemistry is taught in the context of biological processes, emphasizing the key principle that structure determines function.

THE CELL: The first set of chapters emphasizes three key aspects of a cell—information flow (Chapters 3 and 4), actively maintaining a constant internal environment (Chapter 5), and harnessing energy (Chapters 6–8). Placing these basic points at the start of the textbook gives them emphasis and helps students build their knowledge of biology naturally.

CASE STUDIES: Biology is best understood when presented using real and engaging examples as a framework for synthesizing information. Eight carefully positioned Cases help provide this framework. For example, the Case about your personal genome is introduced before the set of chapters on genetics and is revisited in each of these chapters where it serves to reinforce important concepts.

GENETICS: The genetics chapters start with genomes and move to inheritance to provide a modern, molecular look at genetic variation and how traits are transmitted.

UNIQUE CHAPTERS: *Biology: How Life Works*, Second Edition includes chapters that don't traditionally appear in introductory biology texts, one in almost every major subject area. These novel chapters represent shifts toward a more modern conception of certain topics in biology and are identified by ✪.

"The approach to teaching is something my colleagues and I had been waiting for in a textbook. However, the text is flexible enough to accommodate a traditional teaching style."

–Steve Uyeda, *Pima Community College*

To hear the authors talk about the table of contents in more depth, visit **biologyhowlifeworks.com**

BIOGEOCHEMICAL CYCLES: We present the carbon cycle as a bridge between the molecular and organismal parts of the book, showing how different kinds of organisms use the biochemical processes discussed in the first half of the book to create a cycle that drives life on Earth and creates ecosystems. The carbon cycle along with other biogeochemical cycles—sulfur and nitrogen—provides the conceptual backbone around which prokaryotic diversity is organized.

PLANT DEFENSE: The chapter on plant defense provides a strong ecological and case-based perspective on the strategies plants use to survive their exploitation by pathogens and herbivores.

DIVERSITY AND PHYSIOLOGY: Diversity follows physiology in order to provide a basis for understanding the groupings of organisms and to avoid presenting diversity as a list of names to memorize. When students understand how organisms function, they can understand the different groups in depth and organize them intuitively. To give instructors maximum flexibility, brief descriptions of unfamiliar organisms and the major groups of organisms have been layered in the physiology chapters, and the diversity chapters include a brief review of organismal form and function.

NEW CHAPTER 48: BIOMES AND GLOBAL ECOLOGY is part of the greatly expanded ecology coverage on physical processes and global ecology. The new coverage broadens connections between ecological concepts, and is carefully integrated into the larger theme of evolution.

RETHINKING THE TEXTBOOK **THROUGH LAUNCHPAD**

Ordinarily, textbooks are developed by first writing chapters, then making decisions about art and images, and finally assembling a test bank and ancillary media. *Biology: How Life Works* develops the text, visual program, and assessment at the same time. These three threads are tied to the same set of core concepts, share a common language, and use the same visual palette, which ensures a seamless learning experience for students throughout the course.

The text, visuals, and assessments come together most effectively through LaunchPad, Macmillan's integrated learning management system. In LaunchPad, students and instructors can access all components of *Biology: How Life Works*.

LaunchPad resources for *How Life Works* are flexible and aligned. Instructors have the ability to select the visuals, assessments, and activities that best suit their classroom and students. All resources are aligned to one another as well as to the text to ensure effectiveness in helping students build skills and develop knowledge necessary for a foundation in biology.

NEW IN LAUNCHPAD FOR BIOLOGY: *HOW LIFE WORKS*, SECOND EDITION

Functionality to search the question database and filter questions for a number of variables including Core Concept, difficulty level, Bloom's level, and class setting allows instructors to make best use of the robust assessment assets of *How Life Works*.

New question types include sequenced questions and multiple true-false. *Learn more on page xviii.*

Metadata tags for each question show at a glance information, including instructional guidance for select questions.

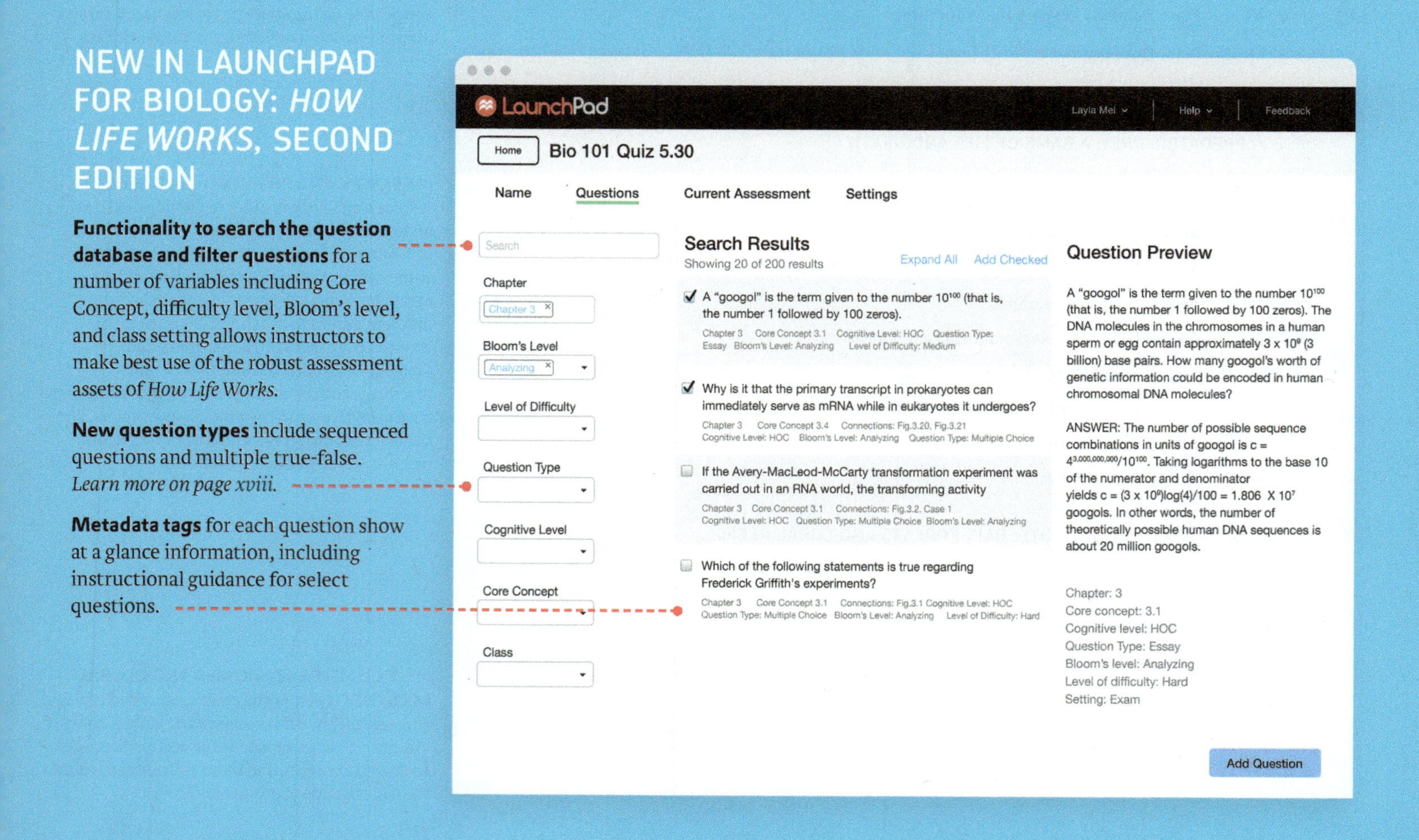

LAUNCHPAD

Where content counts.
Where service matters.
Where students learn.

LAUNCHPADWORKS.COM

Powerful, Simple, and Inviting

LaunchPad for *How Life Works* includes:

The complete *Biology: How Life Works* interactive e-Book

Carefully curated multimedia visuals and assessments, assignable by the instructor and easily accessible by students.

LearningCurve adaptive quizzing that puts "testing to learn" into action, with individualized question sets and feedback for each student based on his or her correct and incorrect responses. All the questions are tied back to the e-Book to encourage students to use the resources at hand.

Pre-built units that are easy to adapt and augment to fit your course.

A Gradebook that provides clear feedback to students and instructors on performance in the course as a whole and individual assignment.

LMS integration allows LaunchPad to be easily integrated into your school's learning management system so your Gradebook and roster are always in sync.

LEARNINGCURVE

LEARNINGCURVEWORKS.COM

Students agree that LearningCurve is extremely helpful in their studies.

Macmillan's LearningCurve adaptive quizzing is part of almost every LaunchPad and has been enthusiastically embraced by students and instructors alike. LearningCurve provides specific feedback for every question and includes links to relevant sections of the e-Book. Questions are written exclusively for the LaunchPad in which they are offered and the question banks are robust and varied.

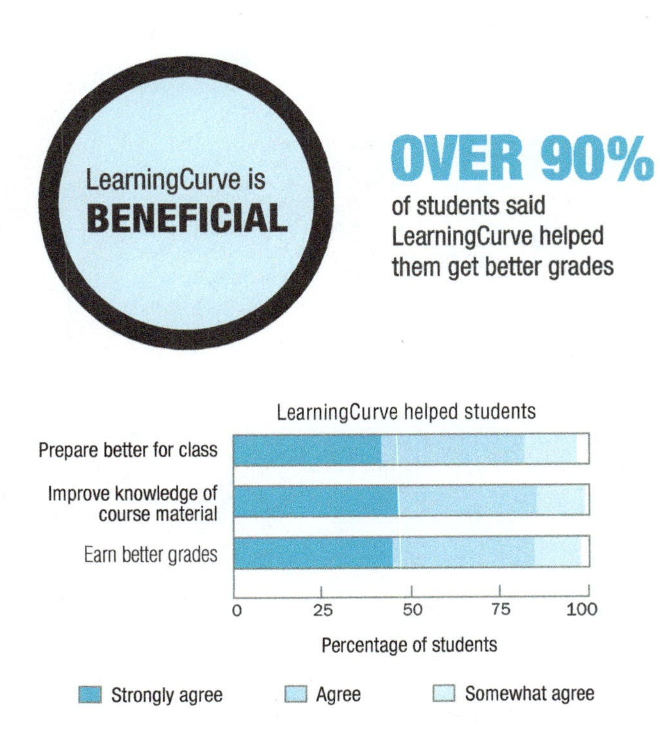

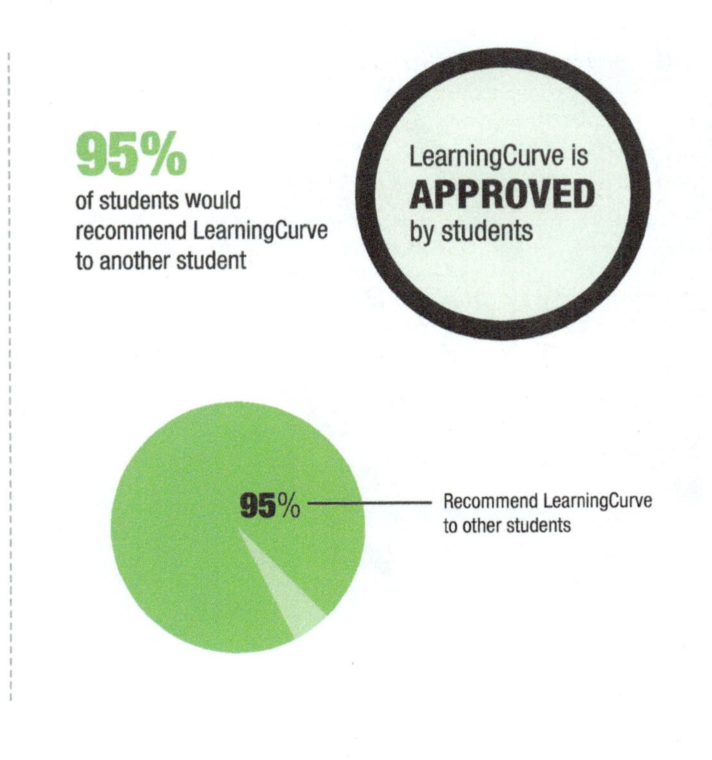

RETHINKING THE VISUAL PROGRAM

The art in the text of *How Life Works* and the associated media in LaunchPad were developed in coordination with the text and assessments to present an integrated and engaging visual experience for students.

Two of the biggest challenges introductory biology students face are connecting concepts across chapters and building a contextual picture, or visual framework, of a complex process. To help students think like biologists, we provide **Visual Synthesis** figures at twelve key points in the book. These figures bring together multiple images students have already seen into a visual summary, helping students see how individual concepts connect to tell a single story.

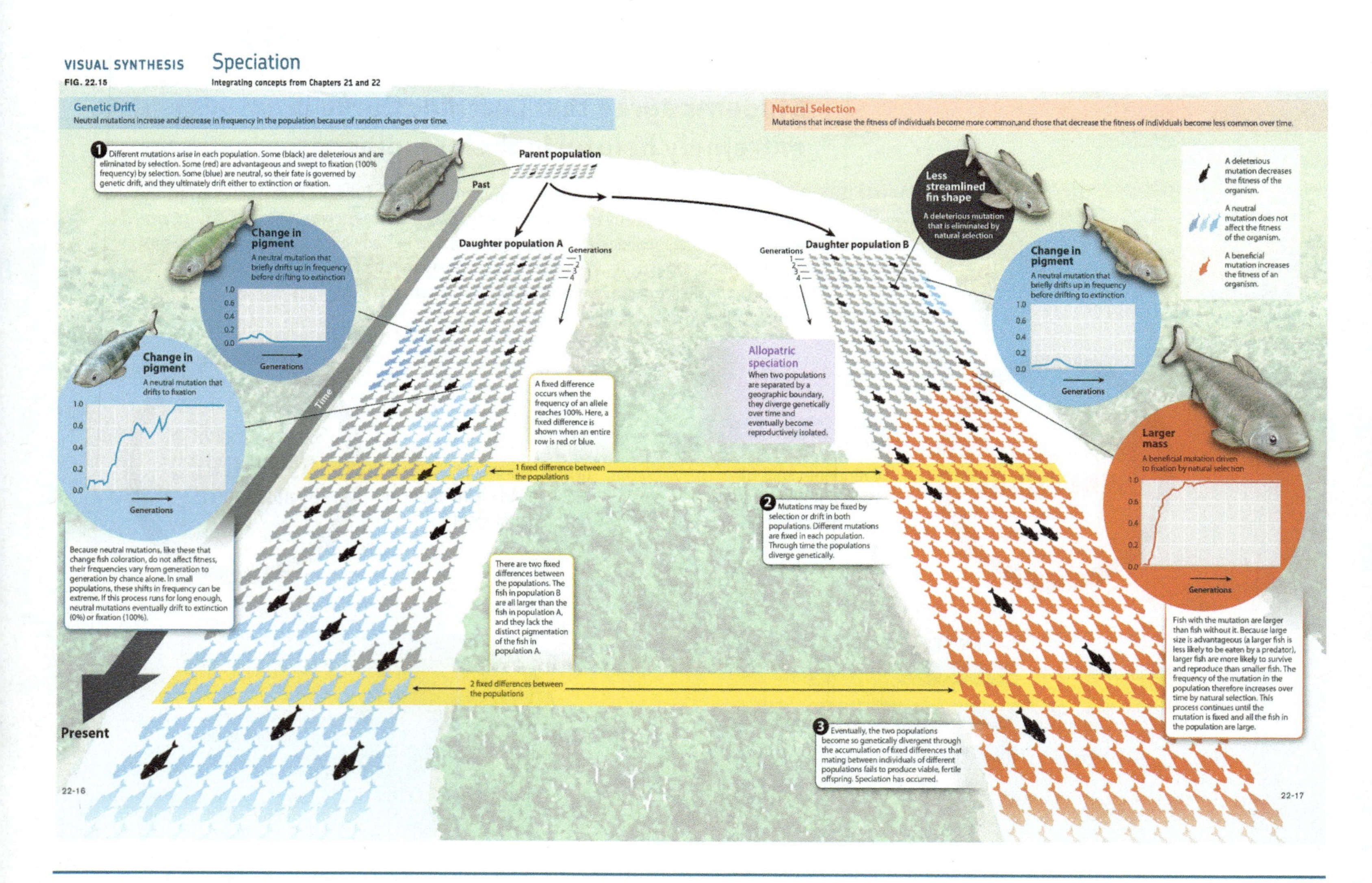

This Visual Synthesis figure on Speciation brings together multiple concepts from the chapters on evolution.

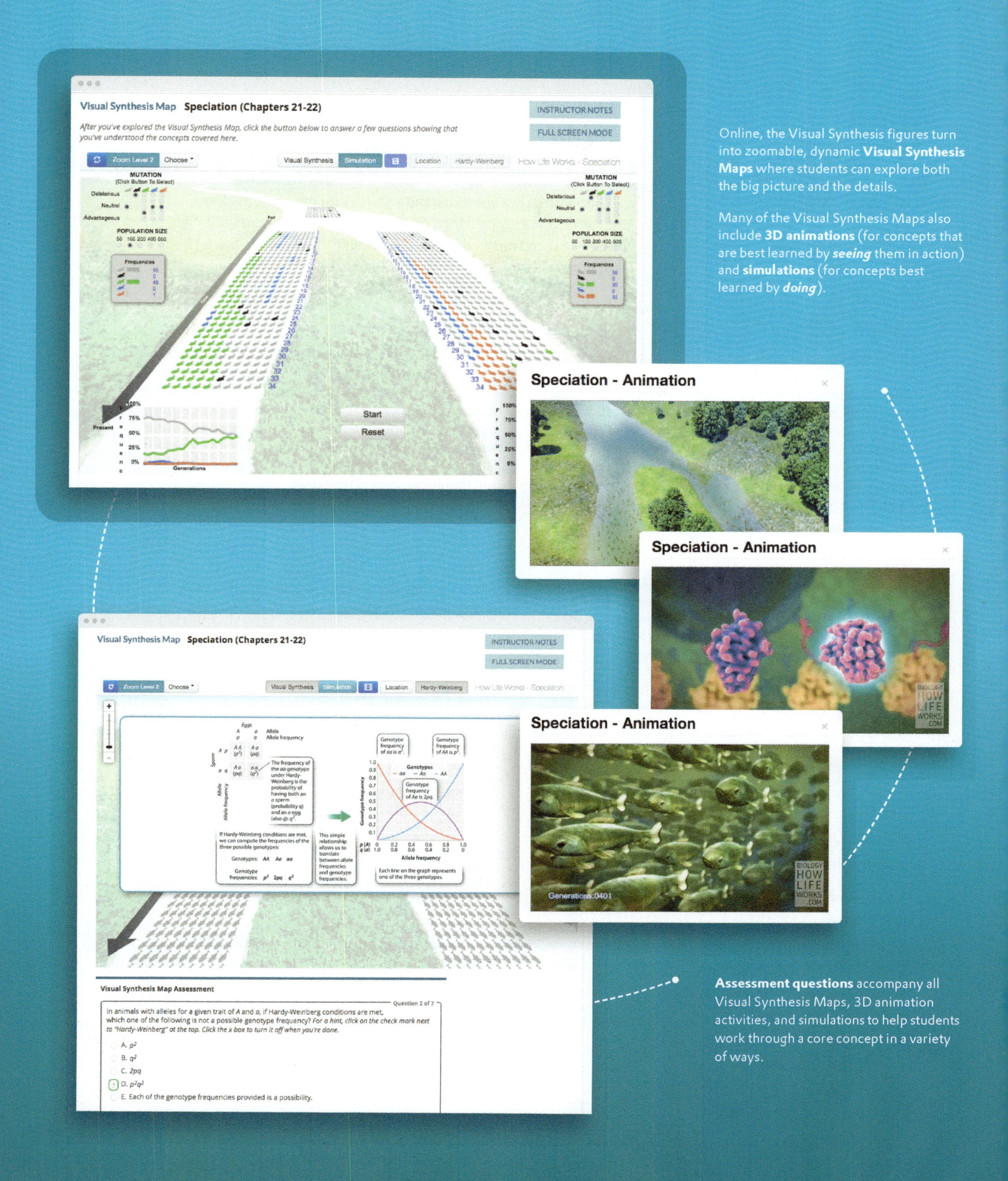

Online, the Visual Synthesis figures turn into zoomable, dynamic **Visual Synthesis Maps** where students can explore both the big picture and the details.

Many of the Visual Synthesis Maps also include **3D animations** (for concepts that are best learned by ***seeing*** them in action) and **simulations** (for concepts best learned by ***doing***).

Assessment questions accompany all Visual Synthesis Maps, 3D animation activities, and simulations to help students work through a core concept in a variety of ways.

RETHINKING ASSESSMENT

Well-designed assessment is a tremendous tool for instructors in gauging student understanding, actively teaching students, and preparing students for exams. The *Biology: How Life Works* assessment author team applied decades of experience researching and implementing assessment practices to create a variety of questions and activities for pre-class, in-class, homework, and exam settings. All assessment items are carefully aligned with the text and media and have the flexibility to meet the needs of instructors with any experience level, classroom size, or teaching style.

Alignment

If the questions, exercises, and activities in the course aren't aligned with course objectives and materials, practice with these resources may not help students succeed in their exam or in future biology courses. Each *How Life Works* assessment item is carefully aligned to the goals and content of the text, and to the assessment items used in other parts of the course. Students are guided through a learning path that provides them with repeated and increasingly challenging practice with the important concepts illustrated in the text and media.

Flexibility

The *How Life Works* assessment authors teach in a variety of classroom sizes and styles, and recognize that there is a wide diversity of course goals and circumstances. Each set of materials, from in-class activities to exam questions, includes a spectrum of options for instructors. All the assessment items are housed in the LaunchPad platform, which is designed to allow instructors to assign and organize assessment items to suit the unique needs of their course and their students.

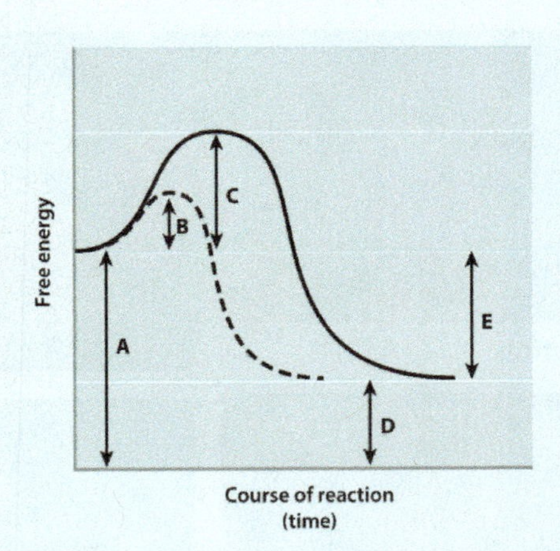

Sample assessment questions, including a sequenced question based on a graph.

Answer the following questions about the reactions shown in the graph.

Which arrow indicates the activation energy of the catalyzed reaction?

❑ A | ■ **B** | ❑ C | ❑ D | ❑ E

Which arrow indicates the activation energy of the uncatalyzed reaction?

❑ A | ❑ B | ■ **C** | ❑ D | ❑ E

Which arrow represents the free energy of the substrate?

■ **A** | ❑ B | ❑ C | ❑ D | ❑ E

Which arrow indicates the free energy of the products?

❑ A | ❑ B | ❑ C | ■ **D** | ❑ E

Which arrow indicates the change in free energy (ΔG) of the reaction?

❑ A | ❑ B | ❑ C | ❑ D | ■ **E**

Is the reaction illustrated by the solid line endergonic or exergonic?

a.) endergonic **b.) exergonic**

Is the reaction illustrated by the dashed line endergonic or exergonic?

a.) endergonic **b.) exergonic**

Which of the following reactions would you predict could be coupled to ATP synthesis from ADP + Pi? Select all that apply.

a.) creatine phosphate + H_2O → creatine + Pi, ΔG -10.3 kcal/mol

b.) phosphoenolpyruvate + H_2O → pyruvate + Pi, ΔG -14.8 kcal/mol

c.) glucose 6-phosphate + H_2O → glucose + Pi, ΔG -3.3 kcal/mol

d.) glucose 1-phosphate + H_2O → glucose + Pi, ΔG -5.0 kcal/mol

e.) glutamic acid + NH_3 → glutamine, ΔG +3.4 kcal/mol

The emperor penguins of Antarctica live on a diet of fish and crustaceans obtained from the cold Antarctic seawaters. During their annual breeding cycle, however, they migrate across the frozen continent to their breeding grounds 50 miles away from the sea (and 50 miles away from their source of food). For over two months the male emperor penguins care for and incubate the eggs while the females return to the sea to feed. During this time a male penguin can lose up to 50% of its biomass (by dry weight). Where did this biomass go?

a.) It was converted to CO_2 and H_2O and then released.

b.) It was converted to heat and then released.

c.) It was converted to ATP molecules.

RETHINKING **ACTIVITIES**

Active learning exercises are an important component of the learning pathway and provide students with hands-on exploration of challenging topics and misconceptions. The second edition of *Biology: How Life Works* includes a new collection of over 40 in-class activities crafted to address the concepts that students find most challenging.

The activities collection was designed to cover a range of classroom sizes and complexity levels, and many can be easily adapted to suit the available time and preferred teaching style. Each activity includes a detailed activity guide for instructors. The activity guide introduces the activity, outlines learning objectives, and provides guidance on how to implement and customize the activity.

Many of the assessment questions and activities throughout *How Life Works* incorporate experimental thinking and data analysis. In addition, two activity types round out the assessment collection by providing applied practice with the data sets and examples from the text. Through ***Working with Data*** activities, students explore and analyze the experiment from a *How Do We Know?* figure from the text. ***Mirror Experiment*** activities introduce students to a new scientific study that relates to, or "mirrors," one of the *How Do We Know?* experiments and ask them to apply what they have learned about data analysis to this new scenario.

List of Materials

1. Activity Guide (file name: 1 Activity Guide Ch11 Double Double Double)
2. In-class Presentation (file name: 2 In-class Presentation Ch11 Double Double Double)
3. Exam Questions (file name: 3 Exam Qs Ch11 Double Double Double)

Description

Students are asked to identify corresponding depictions of chromosomes and DNA pre- and post-replication. The questions take the form shown in this sample, in which students must select one of the DNA representations that most closely depicts the chromosome labeled with the question mark.

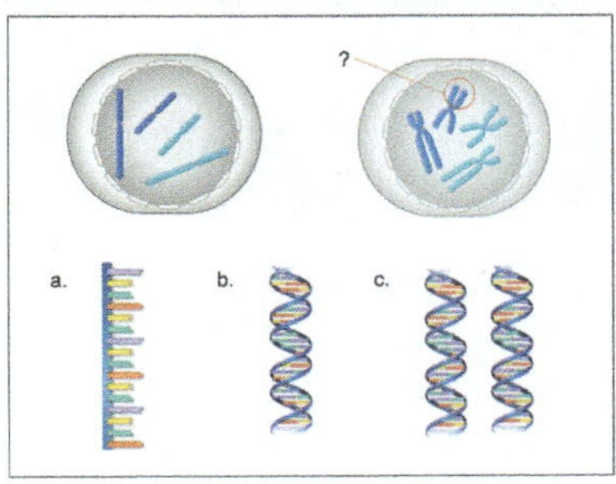

The first clicker question, showing only the mitosis figures, will verify that the students recognize these image types. This might be skipped if the activity comes soon after covering mitosis.

This sequence works very well if students are encouraged to discuss the questions with each other, but with essentially no other instructor guidance or input. Furthermore, clicker results are **not** shown until the final clicker is completed.

Class size and timing

This activity can be used in essentially any class size since it is based on clicker questions. It can be quite brief, 5-10 minutes, even with summary discussion.

Connection to Vision and Change:

This activity is aimed at the Core Concept 3. Information Flow, Exchange, and Storage.

Excerpts of activities from Chapters 11 and 36

Mirror Experiment Activity 36.20

The experiment described below explored the same concepts as the one described in Figure 36.20 in the textbook. Read the description of the experiment and answer the questions below the description to practice interpreting data and understanding experimental design.

Mirror Experiment activities practice skills described in the brief Experiment and Data Analysis Primers, which can be found by clicking on the "Resources" button on the upper right of your LaunchPad homepage. Certain questions in this activity draw on concepts described in the **Experimental Design** and **Data and data Presentation** primers. Click on the "Key Terms" buttons to see definitions of terms used in the question, and click on the "Primer Section" button to pull up a relevant section from the primer.

Experiment

Background

As you have learned, the somatosensory cortex is responsible for processing "touch" stimuli. If someone were tickling the bottom of your foot, mechanoreceptors in the skin of your foot would fire action potentials. These signals would (ultimately) be relayed to the somatosensory cortex portion of your brain, and then your motor cortex. Following this chain of events, you might jerk your foot away from the tickler.

If you were to take a cross section of the somatosensory cortex, you would find that neurons are arranged in six distinct layers; the first layer would be composed of superficial neurons located near the brain surface, and the sixth layer would be composed of the "deepest" neurons (that is, those closest to the white matter). How are neurons that respond to touch stimuli organized in the somatosensory cortex? Do neurons in the six different layers of the somatosensory cortex respond to different types of stimuli?

Hypothesis

Vernon Mountcastle hypothesized that researchers could create a diagram of the somatosensory cortex by tracking which neurons responded to different types of touch stimuli.

Experiment

Mountcastle exposed cats to two types of stimuli: (1) cutaneous or superficial stimuli, which included touching hairs or touching the skin; and (2) deep stimuli, which included bending and extending joints or touching the connective tissue surrounding muscles. He was able to track which neurons in the somatosensory cortex fired action potentials in response to these two types of stimuli, and measured their firing rates (Figure 1).

Results

Mountcastle determined that neurons involved in processing the same type of stimuli are organized in "vertical columns." These columns are composed of cells belonging to different layers of the somatosensory cortex stacked one on top of another. These results demonstrated that just because a neuron responds to deep stimuli does not mean that this neuron will be found deep within the brain; similarly, a neuron that responds to cutaneous stimuli will not necessarily be located near the brain surface. In addition to identifying vertical columns of neurons, Mountcastle

Question 1 of 8

Mountcastle noted that within the somatosensory cortex, neurons that respond to the same type of stimulus are arranged in vertical columns. One way Mountcastle discovered this vertical arrangement was by inserting a measuring device (to record action potentials) into the somatosensory cortex at different angles. This approach is depicted in Figure 2 below. Which of the following statements is true regarding measuring devices that were inserted into the brain at angles of 45° and 90°?

- A. The measuring device inserted into the brain at a 90° angle will likely encounter neurons belonging to the same vertical column.
- B. The measuring device inserted into the brain at a 45° angle will likely encounter neurons that belong to different vertical columns.
- C. All of the neurons encountered by the measuring device inserted into the brain at a 90° angle will likely respond to the same type of stimuli.
- D. Neurons encountered by the measuring device inserted into the brain at a 45° angle will likely respond to different types of stimuli.
- E. All of the answer options are true.

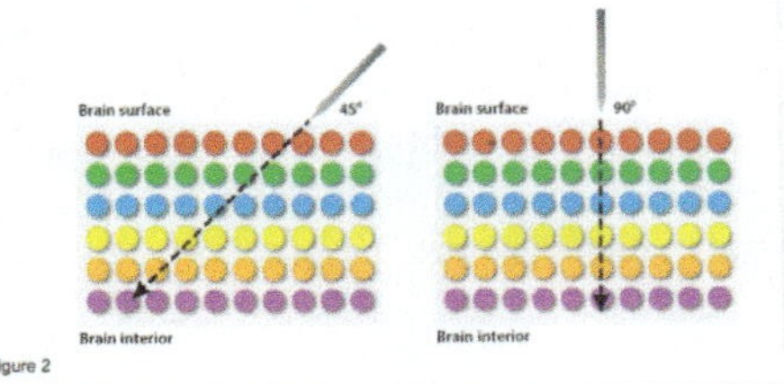

Figure 2

Data from Mountcastle, V. B., 1957. Modality and topographic properties of single neurons of cat's somatic sensory cortex. *J Neurophysiol.* 20, 408-34.

Submit

WHAT'S NEW IN THE SECOND EDITION?

From the start, *Biology: How Life Works* was envisioned not as a reference book for all of biology, but as a resource focused on foundational concepts, terms, and experiments, all placed in a framework that motivates student interest through a coherent and authentic presentation of current science. In preparing this edition, we carefully considered the latest breakthroughs and incremental, but nevertheless significant, changes across the fields of biology. We also reached out to adopters, instructors not using our book, and primary literature to determine what concepts and details are relevant, important, and necessary additions. Our integrated approach to text, media, and assessment means that all changes are carefully reflected in each of these areas.

MAJOR CHANGES AND UPDATES

We've greatly expanded the coverage of ecology in the second edition of *Biology: How Life Works*.

A new ecology chapter, Chapter 48: Biomes and Global Ecology, takes a broad look at ecology on the largest scale. It begins with how and why climates are distributed as they are around the world and introduces Earth's major biomes. Biomes crystallize the relationships among ecology, evolution, and physical environment—landscapes look different in different parts of the world because of the morphological and physiological adaptations that plants and animals have made to different climates. The chapter is distinguished by extensive discussion of biomes in the aquatic realm, especially in the oceans.

Chapter 47: Species Interactions, Communities, and Ecosystems **includes expanded coverage of the ways species interact with one another in communities.** This chapter now has more detail on facilitation, herbivory, and biodiversity.

Chapter 49: The Anthropocene: Humans as a Planetary Force **includes new discussions exploring how human activities affect ecology**. The chapter now examines fracking and its effects on the carbon cycle, habitat loss and its effects on biodiversity, and the overexploitation of resources and its effects on community ecology. The chapter ends with a new section on conservation biology that explores how conservationists are working to preserve natural habitats.

A new Visual Synthesis figure on the flow of matter and energy through ecosystems illustrates, explores, and physically situates the relationships among concepts from Chapters 25, 26, 47, 48, and 49. In LaunchPad, students and instructors can interact with an accompanying dynamic, zoomable, and interactive Visual Synthesis Map based on this figure.

Our new collection of over 40 in-class activities provides tools for instructors to engage their students in active learning. In-class activities are designed to address difficult concepts, and can be used with a variety of classroom sizes and teaching styles. Each activity includes a detailed activity guide for instructors.

We've expanded our collection of high-quality assessment questions by adding over 1000 new questions. New questions are particularly focused on higher-order thinking, including questions based on figures or data, and questions that ask students to consider how perturbing a system would affect outcomes. As in the first edition, questions are carefully aligned with core concepts from the text. New and revised assessment questions also accompany Visual Synthesis Maps, simulations, animations, and other visual media, to more effectively probe student understanding of the media tools they've explored.

The second edition also includes several new question types. **Sequenced questions** ask students several, individually scored questions about a single scenario or system. These questions often build on one another to guide students from lower-order thinking to higher-order thinking about the same concept. **Multiple true–false questions** ask students several, individually scored true–false questions about a single scenario or system.

Improved functionality in LaunchPad allows instructors to search the question database and filter questions by a number of variables, including core concept, difficulty level, Bloom's level, and class setting. Metadata tags for each question show additional information at a glance, including instructional guidance for select questions.

NEW MEDIA

Cell Communities Visual Synthesis Map to accompany the printed Visual Synthesis figure

Virus Visual Synthesis Map to accompany the printed Visual Synthesis figure

New Visual Synthesis figure and map on the Flow of Matter and Energy in Ecosystems

Virus Video featuring author Rob Lue

Cell Membrane simulation

New Animations

Chapter 9: Basic Principles of Cell Signaling

Chapter 9: G protein-coupled Receptor Signaling

Chapter 9: Signal Amplification

Chapter 10: Dynamic Nature of Microtubules

Chapter 10: Motor Proteins

Chapter 10: Dynamic Nature of Actin Filaments

Chapter 19: Lac Operon

Chapter 20: ABC Model of Floral Development

Chapter 40: Glucose Absorption in the Small Intestine

Chapter 42: Gastrulation

NEW TOPICS AND OTHER REVISIONS

The following is a detailed list of content changes in this edition. These range from the very small (nucleotides shown at physiological pH) to quite substantial (an entire new chapter in the ecology section). Especially important changes are indicated with an asterisk (✪).

✪ New coverage of functional groups (Chapter 2)

Nucleotides now shown at physiological pH (Chapter 3)

Amino acids now shown at physiological pH (Chapter 4)

The story of the evolution of photosynthesis now brought together in a single major section at the end of Chapter 8 (section 8.5)

✪ Chapters 9 and 10 streamlined to better match our mission statement

A new discussion of cellular response and what determines it (Chapter 9)

✪ New inclusion of the trombone model of DNA replication (Chapter 12)

✪ Addition of CRISPR technology (Chapter 12)

Expanded coverage of retrotransposons and reverse transcriptase (Chapter 13)

A new *How Do We Know?* figure explaining Mendel's experimental results (Chapter 16)

New coverage of the mechanism of *X*-inactivation (Chapter 19)

An expanded discussion of non-random mating and inbreeding depression (Chapter 21)

✪ Addition of the effect of mass extinctions on species diversity (Chapter 23)

Updated discussion of the relationship between Neanderthals and *Homo sapiens*, as well as Denisovans (Chapter 24)

Significantly revised link between the carbon cycle, biodiversity, and ecology (Chapter 25)

New branching order of the eukaryote tree to reflect new research in the past three years (Chapter 27 and onward)

A new paragraph on ciliates (Chapter 27)

A new explanation of protist diversity (Chapter 27)

✪ A new discussion of plant nutrients with a table (Chapter 29)

✪ An enhanced discussion of seeds, including the development of the embryo and dispersal structures (Chapter 30)

New coverage of the genetic advantages of alternation of generations, and how it allows inbreeding (Chapter 30)

Addition of apomixis (Chapter 30)

The section on the role of plant sensory systems in the timing of plant reproduction moved from Chapter 30 to Chapter 31

Completely revised explanation of the basis for angiosperm diversity (Chapter 33)

Brief descriptions of unfamiliar organisms and the major groups of organisms layered in the animal physiology chapters to make it easier to teach physiology before diversity (Chapters 35-42)

✪ Brief review of organismal form and function in the plant and animal diversity chapters (Chapters 33 and 44), allowing these chapters to be used on their own or before the physiology chapters and giving instructors maximum flexibility

✪ A new section on the composition of blood (Chapter 39)

New diagrams of hormone feedback loops in the menstrual cycle (Chapter 42)

A new introduction to the immune system (Chapter 43)

✪ A new discussion of nematodes (Chapter 44)

Introduction of a newly discovered species, *Dendrogramma enigmatica* (Chapter 44)

A simplified population growth equation (Chapter 46)

A new discussion of facilitation (Chapter 47)

An expanded discussion of herbivory (Chapter 47)

A new example of microbial symbionts (Chapter 47)

A new discussion of biodiversity and its importance (Chapter 47)

✪ An entirely new chapter on physical processes that underlie different biomes (Chapter 48)

- Differential solar energy around the globe and seasonality
- Wind and ocean currents
- Effects of circulation and topography on rainfall
- Expanded discussion of terrestrial biomes
- Freshwater and marine biomes
- Integration of concepts of biogeochemical cycles from Chapters 25 and 26 with ecological concepts
- Global patterns of primary production
- Global biodiversity

✪ A new exploration of the effect of fracking on the carbon cycle (Chapter 49)

✪ New coverage of habitat loss and biodiversity (Chapter 49)

✪ New coverage of overexploitation of resources and its effects on community ecology (Chapter 49)

✪ A new Core Concept and discussion of conservation biology (Chapter 49)

PRAISE FOR *HOW LIFE WORKS*

I have taught botany and then Biology II for over 20 years and have been very frustrated when I have realized how little knowledge students retained. Since we have gone to this textbook, I find that the questions students are asking in class are much more probing than those in the past, and the students seem much more engaged in the topics. I am hopeful that this approach will help our students be deeper thinkers and better scientists.

– **GLORIA CADDELL**, University of Central Oklahoma

One of the things that really sold me on this text was the LaunchPad system: easy to use; intuitive navigation; really good questions that match the sophistication of the text; love the LearningCurve activities; use most of the animations in my lectures!

– **SARA CARLSON**, University of Akron

This is the best set of questions I've ever seen in a textbook. They are thorough and the right mix of challenging the student with requiring memorization of important details.

– **KURT ELLIOT**, Northwest Vista College

We have all seen an improvement in our students' understanding of the material this year, the first year that we used the Morris text.

– **ANUPAMA SESHAN**, Emmanuel College

I like the figures, especially the 3D ones — we focus on "perceptual ability" training in our classes and figures that encourage students to think about cells in 3D are excellent!

– **KIRKWOOD LAND**, University of the Pacific

These chapters all seem to draw students through the course by referencing what they have learned previously and then adding new information. This makes the course seem like a complete story instead of a series of encyclopedia entries to be learned in isolation.

– **TIM KROFT**, Auburn University at Montgomery

We used this book last year and overall felt that it represented a major improvement from our previous text.

– **PETER ARMBRUSTER**, Georgetown University

Good questions are just as important as a good textbook. The available variety of assessment tools was very important for our adoption of this text.

– **MATTHEW BREWER, Georgia State University**

If the whole book reads like this I would love to use it! This is the way I like to teach! I want students to understand rather than memorize and this chapter seems aimed at this.

– **JENNIFER SCHRAMM, Chemeketa Community College**

The artwork seems very clear-cut and geared to giving the students a very specific piece of information with a very simple example. This should greatly help students with forming a visual image of the various subjects.

– **CHRIS PETRIE, Eastern Florida State College**

The writing style is excellent, it makes a great narrative and incorporates key scientific experiments into the explanation of photosynthesis.

– **DIANNE JENNING, Virginia Commonwealth University**

I think *HLW* does a better job of presenting introductory material than our current text, which tends to overwhelm students.

– **LAURA HILL, University of Vermont**

With the quick checks and the experiments the first chapter already has the learners thinking about experiments and critical analysis.

– **JOHN KOONTZ, University of Tennessee Knoxville**

This book moves teaching away from merely understanding all of the bold terms in a textbook in order to spit them back on a multiple-choice test. I can use this text in order to prepare my students to understand and learn the general principles and concepts in biology and how those concepts translate across different levels of biology. I would not trade this textbook for any other book on the market.

– **PAUL MOORE, Bowling Green State University**

ACKNOWLEDGMENTS

Biology: How Life Works is not only a book. Instead, it is an integrated set of resources to support student learning and instructor teaching in introductory biology. As a result, we work closely with an entire community of authors, publishers, instructors, reviewers, and students. We would like to thank this dedicated group.

First and foremost, we thank the thousands of students we have collectively taught. Their curiosity, intelligence, and enthusiasm have been sources of motivation for all of us.

Our teachers and mentors have provided us with models of patience, creativity, and inquisitiveness that we strive to bring into our own teaching and research. They encourage us to be life-long learners, teachers, and scholars.

We feel very lucky to be a partner with W. H. Freeman and Macmillan Learning. From the start, they have embraced our project, giving us the space and room to achieve something unique, while at the same time providing guidance, support, and input from the broader community of instructors and students.

Beth Cole, our acquisitions editor, deserves thanks for taking on the second edition and becoming our leader. She keeps a watchful eye on important trends in science, education, and technology, carefully listens to what we want to do, and helps us put our aspirations in a larger context.

Lead developmental editor Lisa Samols continues to have just the right touch—the ability to listen as well as offer intelligent suggestions, serious with a touch of humor, quiet but persistent. Senior developmental editor Susan Moran has an eye for detail and the uncanny ability to read the manuscript like a student. Developmental editor Erica Champion brings intelligence and thoughtfulness to her edits.

Karen Misler kept us all on schedule in a clear and firm but always understanding and compassionate way.

Lindsey Jaroszewicz, our market development manager, is remarkable for her energy and enthusiasm, her attention to detail, and her creativity in ways to reach out to instructors and students. Will Moore, our marketing manager, refined the story of *How Life Works* 2e and works tirelessly with our sales teams to bring the second edition to instructors and students everywhere.

We thank Robert Errera for coordinating the move from manuscript to the page, and Nancy Brooks for helping to even out the prose. We also thank Diana Blume, our design director, Tom Carling, our text and cover designer, and Sheridan Sellers, our compositor. Together, they managed the look and feel of the book, coming up with creative solutions for page layout.

In digital media, we thank Amanda Nietzel for her editorial insight in making pedagogically useful media tools, and Keri deManigold and Chris Efstratiou for managing and coordinating the media and websites. They each took on this project with dedication, persistence, enthusiasm, and attention to detail that we deeply appreciate.

We are extremely grateful to Elaine Palucki for her insight into teaching and learning strategies, Donna Brodman for coordinating the many reviewers, and Jane Taylor, Alexandra Garrett, and Abigail Fagan for their consistent and tireless support.

Imagineering under the patient and intelligent guidance of Mark Mykytuik provided creative, insightful art to complement, support, and reinforce the text. We also thank our illustration coordinator, Matt McAdams, for skillfully guiding our collaboration with Imagineering. Christine Buese, our photo editor, and Lisa Passmore and Richard Fox, our photo researchers, provided us with a steady stream of stunning photos, and never gave up on those hard-to-find shots. Paul Rohloff, our production manager, ensured that the journey from manuscript to printing was seamless.

We would also like to acknowledge Kate Parker, Publisher of Sciences, Chuck Linsmeier, Vice President of Editorial, Susan Winslow, Managing Director, and Ken Michaels, Chief Executive Officer, for their support of *How Life Works* and our unique approach.

We also sincerely thank Erin Betters, Jere A. Boudell, Donna Koslowsky, and Jon Stoltzfus for thoughtful and insightful contributions to the assessment materials.

We are extremely grateful for all of the hard work and expertise of the sales representatives, regional managers, and regional sales specialists. We have enjoyed meeting and working with this dedicated sales staff, who are the ones that ultimately put the book in the hands of instructors.

Countless reviewers made invaluable contributions to this book and deserve special thanks. From catching mistakes to suggesting new and innovative ways to organize the content, they provided substantial input to the book. They brought their collective years of teaching to the project, and their suggestions are tangible in every chapter.

Finally, we would like to thank our families. None of this would have been possible without their support, inspiration, and encouragement.

Contributors, First Edition

Thank you to all the instructors who worked in collaboration with the authors and assessment authors to write Biology: How Life Works *assessments, activities, and exercises.*

Allison Alvarado, University of California, Los Angeles
Peter Armbruster, Georgetown University
Zane Barlow-Coleman, formerly of University of Massachusetts, Amherst
James Bottesch, Brevard Community College
Jessamina Blum, Yale University
Jere Boudell, Clayton State University
David Bos, Purdue University
Laura Ciaccia West, Yale University*
Laura DiCaprio, Ohio University
Tod Duncan, University of Colorado, Denver
Cindy Giffen, University of Wisconsin, Madison
Paul Greenwood, Colby College
Stanley Guffey, The University of Tennessee, Knoxville
Alison Hill, Duke University
Meg Horton, University of North Carolina at Greensboro
Kerry Kilburn, Old Dominion University
Jo Kurdziel, University of Michigan
David Lampe, Duquesne University
Brenda Leady, University of Toledo
Sara Marlatt, Yale University*
Kelly McLaughlin, Tufts University
Brad Mehrtens, University of Illinois at Urbana-Champaign
Nancy Morvillo, Florida Southern College
Jennifer Nauen, University of Delaware
Kavita Oommen, Georgia State University
Patricia Phelps, Austin Community College
Melissa Reedy, University of Illinois at Urbana-Champaign
Lindsay Rush, Yale University*
Sukanya Subramanian, Collin College
Michelle Withers, West Virginia University

*Graduate student, Yale University Scientific Teaching Fellow

Reviewers, Class Testers, and Focus Group Participants

Thank you to all the instructors who reviewed and/or class tested chapters, art, assessment questions, and other Biology: How Life Works *materials.*

First Edition

Thomas Abbott, University of Connecticut
Tamarah Adair, Baylor University
Sandra Adams, Montclair State University
Jonathon Akin, University of Connecticut
Eddie Alford, Arizona State University
Chris Allen, College of the Mainland
Sylvester Allred, Northern Arizona University
Shivanthi Anandan, Drexel University
Andrew Andres, University of Nevada, Las Vegas
Michael Angilletta, Arizona State University
Jonathan Armbruster, Auburn University
Jessica Armenta, Lone Star College System
Brian Ashburner, University of Toledo
Andrea Aspbury, Texas State University
Nevin Aspinwall, Saint Louis University
Felicitas Avendano, Grand View University
Yael Avissar, Rhode Island College
Ricardo Azpiroz, Richland College
Jessica Baack, Southwestern Illinois College
Charles Baer, University of Florida
Brian Bagatto, University of Akron
Alan L. Baker, University of New Hampshire
Ellen Baker, Santa Monica College
Mitchell Balish, Miami University
Teri Balser, University of Florida
Paul Bates, University of Minnesota, Duluth
Michel Baudry, University of Southern California
Jerome Baudry, The University of Tennessee, Knoxville
Mike Beach, Southern Polytechnic State University
Andrew Beall, University of North Florida
Gregory Beaulieu, University of Victoria
John Bell, Brigham Young University
Michael Bell, Richland College
Rebecca Bellone, University of Tampa
Anne Bergey, Truman State University
Laura Bermingham, University of Vermont
Aimee Bernard, University of Colorado, Denver
Annalisa Berta, San Diego State University
Joydeep Bhattacharjee, University of Louisiana, Monroe
Arlene Billock, University of Louisiana, Lafayette
Daniel Blackburn, Trinity College
Mark Blackmore, Valdosta State University
Justin Blau, New York University
Andrew Blaustein, Oregon State University
Mary Bober, Santa Monica College
Robert Bohanan, University of Wisconsin, Madison
Jim Bonacum, University of Illinois at Springfield
Laurie Bonneau, Trinity College
David Bos, Purdue University
James Bottesch, Brevard Community College
Jere Boudell, Clayton State University
Nancy Boury, Iowa State University
Matthew Brewer, Georgia State University
Mirjana Brockett, Georgia Institute of Technology
Andrew Brower, Middle Tennessee State University
Heather Bruns, Ball State University
Jill Buettner, Richland College
Stephen Burnett, Clayton State University
Steve Bush, Coastal Carolina University
David Byres, Florida State College at Jacksonville
James Campanella, Montclair State University
Darlene Campbell, Cornell University
Jennifer Campbell, North Carolina State University
John Campbell, Northwest College
David Canning, Murray State University
Richard Cardullo, University of California, Riverside
Sara Carlson, University of Akron
Jeff Carmichael, University of North Dakota
Dale Casamatta, University of North Florida

Anne Casper, Eastern Michigan University
David Champlin, University of Southern Maine
Rebekah Chapman, Georgia State University
Samantha Chapman, Villanova University
Mark Chappell, University of California, Riverside
P. Bryant Chase, Florida State University
Young Cho, Eastern New Mexico University
Tim Christensen, East Carolina University
Steven Clark, University of Michigan
Ethan Clotfelter, Amherst College
Catharina Coenen, Allegheny College
Mary Colavito, Santa Monica College
Craig Coleman, Brigham Young University
Alex Collier, Armstrong Atlantic State University
Sharon Collinge, University of Colorado, Boulder
Jay Comeaux, McNeese State University
Reid Compton, University of Maryland
Ronald Cooper, University of California, Los Angeles
Victoria Corbin, University of Kansas
Asaph Cousins, Washington State University
Will Crampton, University of Central Florida
Kathryn Craven, Armstrong Atlantic State University
Scott Crousillac, Louisiana State University
Kelly Cude, College of the Canyons
Stanley Cunningham, Arizona State University
Karen Curto, University of Pittsburgh
Bruce Cushing, The University of Akron
Rebekka Darner, University of Florida
James Dawson, Pittsburg State University
Elizabeth De Stasio, Lawrence University
Jennifer Dechaine, Central Washington University
James Demastes, University of Northern Iowa
D. Michael Denbow, Virginia Polytechnic Institute and State University
Joseph Dent, McGill University
Terry Derting, Murray State University
Jean DeSaix, University of North Carolina at Chapel Hill
Donald Deters, Bowling Green State University
Hudson DeYoe, The University of Texas, Pan American
Leif Deyrup, University of the Cumberlands
Laura DiCaprio, Ohio University
Jesse Dillon, California State University, Long Beach
Frank Dirrigl, The University of Texas, Pan American
Kevin Dixon, Florida State University
Elaine Dodge Lynch, Memorial University of Newfoundland
Hartmut Doebel, George Washington University
Jennifer Doll, Loyola University, Chicago
Logan Donaldson, York University
Blaise Dondji, Central Washington University
Christine Donmoyer, Allegheny College
James Dooley, Adelphi University
Jennifer Doudna, University of California, Berkeley
John DuBois, Middle Tennessee State University
Richard Duhrkopf, Baylor University
Kamal Dulai, University of California, Merced
Arthur Dunham, University of Pennsylvania
Mary Durant, Lone Star College System
Roland Dute, Auburn University
Andy Dyer, University of South Carolina, Aiken
William Edwards, Niagara University
John Elder, Valdosta State University
William Eldred, Boston University
David Eldridge, Baylor University
Inge Eley, Hudson Valley Community College
Lisa Elfring, University of Arizona
Richard Elinson, Duquesne University
Kurt Elliott, Northwest Vista College
Miles Engell, North Carolina State University
Susan Erster, Stony Brook University
Joseph Esdin, University of California, Los Angeles
Jean Everett, College of Charleston
Brent Ewers, University of Wyoming
Melanie Fierro, Florida State College at Jacksonville
Michael Fine, Virginia Commonwealth University
Jonathan Fingerut, St. Joseph's University
Ryan Fisher, Salem State University
David Fitch, New York University
Paul Fitzgerald, Northern Virginia Community College
Jason Flores, University of North Carolina at Charlotte
Matthias Foellmer, Adelphi University
Barbara Frase, Bradley University
Caitlin Gabor, Texas State University
Michael Gaines, University of Miami
Jane Gallagher, The City College of New York, The City University of New York
Kathryn Gardner, Boston University
J. Yvette Gardner, Clayton State University
Gillian Gass, Dalhousie University
Jason Gee, East Carolina University
Topher Gee, University of North Carolina at Charlotte
Vaughn Gehle, Southwest Minnesota State University
Tom Gehring, Central Michigan University
John Geiser, Western Michigan University
Alex Georgakilas, East Carolina University
Peter Germroth, Hillsborough Community College
Arundhati Ghosh, University of Pittsburgh
Carol Gibbons Kroeker, University of Calgary
Phil Gibson, University of Oklahoma
Cindee Giffen, University of Wisconsin, Madison
Matthew Gilg, University of North Florida
Sharon Gillies, University of the Fraser Valley
Leonard Ginsberg, Western Michigan University
Florence Gleason, University of Minnesota
Russ Goddard, Valdosta State University
Miriam Golbert, College of the Canyons
Jessica Goldstein, Barnard College, Columbia University
Steven Gorsich, Central Michigan University
Sandra Grebe, Lone Star College System
Robert Greene, Niagara University
Ann Grens, Indiana University, South Bend
Theresa Grove, Valdosta State University
Stanley Guffey, The University of Tennessee, Knoxville
Nancy Guild, University of Colorado, Boulder
Lonnie Guralnick, Roger Williams University
Laura Hake, Boston College
Kimberly Hammond, University of California, Riverside
Paul Hapeman, University of Florida
Luke Harmon, University of Idaho
Sally Harmych, University of Toledo
Jacob Harney, University of Hartford
Sherry Harrel, Eastern Kentucky University
Dale Harrington, Caldwell Community College and Technical Institute
J. Scott Harrison, Georgia Southern University
Diane Hartman, Baylor University

Mary Haskins, Rockhurst University
Bernard Hauser, University of Florida
David Haymer, University of Hawaii
David Hearn, Towson University
Marshal Hedin, San Diego State University
Paul Heideman, College of William and Mary
Gary Heisermann, Salem State University
Brian Helmuth, University of South Carolina
Christopher Herlihy, Middle Tennessee State University
Albert Herrera, University of Southern California
Brad Hersh, Allegheny College
David Hicks, The University of Texas at Brownsville
Karen Hicks, Kenyon College
Alison Hill, Duke University
Kendra Hill, South Dakota State University
Jay Hodgson, Armstrong Atlantic State University
John Hoffman, Arcadia University
Jill Holliday, University of Florida
Sara Hoot, University of Wisconsin, Milwaukee
Margaret Horton, University of North Carolina at Greensboro
Lynne Houck, Oregon State University
Kelly Howe, University of New Mexico
William Huddleston, University of Calgary
Jodi Huggenvik, Southern Illinois University
Melissa Hughes, College of Charleston
Randy Hunt, Indiana University Southeast
Tony Huntley, Saddleback College
Brian Hyatt, Bethel College
Jeba Inbarasu, Metropolitan Community College
Colin Jackson, The University of Mississippi
Eric Jellen, Brigham Young University
Dianne Jennings, Virginia Commonwealth University
Scott Johnson, Wake Technical Community College
Mark Johnston, Dalhousie University
Susan Jorstad, University of Arizona
Stephen Juris, Central Michigan University
Julie Kang, University of Northern Iowa
Jonghoon Kang, Valdosta State University
George Karleskint, St. Louis Community College at Meramec
David Karowe, Western Michigan University
Judy Kaufman, Monroe Community College
Nancy Kaufmann, University of Pittsburgh
Ramneet Kaur, Georgia Gwinett College
John Kauwe, Brigham Young University
Elena Keeling, California Polytechnic State University
Jill Keeney, Juniata College
Tamara Kelly, York University
Chris Kennedy, Simon Fraser University
Bretton Kent, University of Maryland
Jake Kerby, University of South Dakota
Jeffrey Kiggins, Monroe Community College
Scott Kight, Montclair State University
Stephen Kilpatrick, University of Pittsburgh, Johnstown
Kelly Kissane, University of Nevada, Reno
David Kittlesen, University of Virginia
Jennifer Kneafsey, Tulsa Community College
Jennifer Knight, University of Colorado, Boulder
Ross Koning, Eastern Connecticut State University
David Kooyman, Brigham Young University
Olga Kopp, Utah Valley University
Anna Koshy, Houston Community College
Todd Kostman, University of Wisconsin, Oshkosh
Peter Kourtev, Central Michigan University
William Kroll, Loyola University, Chicago
Dave Kubien, University of New Brunswick
Allen Kurta, Eastern Michigan University
Ellen Lamb, University of North Carolina at Greensboro
Troy Ladine, East Texas Baptist University
David Lampe, Duquesne University
Evan Lampert, Gainesville State College
James Langeland, Kalamazoo College
John Latto, University of California, Santa Barbara
Brenda Leady, University of Toledo
Jennifer Leavey, Georgia Institute of Technology
Hugh Lefcort, Gonzaga University
Brenda Leicht, University of Iowa
Craig Lending, The College at Brockport, The State University of New York
Nathan Lents, John Jay College of Criminal Justice, The City University of New York
Michael Leonardo, Coe College
Army Lester, Kennesaw State University
Cynthia Littlejohn, University of Southern Mississippi
Zhiming Liu, Eastern New Mexico University
Jonathan Lochamy, Georgia Perimeter College
Suzanne Long, Monroe Community College
Julia Loreth, University of North Carolina at Greensboro
Jennifer Louten, Southern Polytechnic State University
Janet Loxterman, Idaho State University
Ford Lux, Metropolitan State College of Denver
José-Luis Machado, Swarthmore College
C. Smoot Major, University of South Alabama
Charles Mallery, University of Miami
Mark Maloney, Spelman College
Carroll Mann, Florida State College at Jacksonville
Carol Mapes, Kutztown University of Pennsylvania
Nilo Marin, Broward College
Diane Marshall, University of New Mexico
Heather Masonjones, University of Tampa
Scott Mateer, Armstrong Atlantic State University
Luciano Matzkin, The University of Alabama in Huntsville
Robert Maxwell, Georgia State University
Meghan May, Towson University
Michael McGinnis, Spelman College
Kathleen McGuire, San Diego State University
Maureen McHale, Truman State University
Shannon McQuaig, St. Petersburg College
Susan McRae, East Carolina University
Lori McRae, University of Tampa
Mark Meade, Jacksonville State University
Brad Mehrtens, University of Illinois at Urbana-Champaign
Michael Meighan, University of California, Berkeley
Douglas Meikle, Miami University
Richard Merritt, Houston Community College
Jennifer Metzler, Ball State University
James Mickle, North Carolina State University
Brian Miller, Middle Tennessee State University
Allison Miller, Saint Louis University
Yuko Miyamoto, Elon University
Ivona Mladenovic, Simon Fraser University
Marcie Moehnke, Baylor University
Chad Montgomery, Truman State University
Jennifer Mook, Gainesville State College
Daniel Moon, University of North Florida
Jamie Moon, University of North Florida
Jeanelle Morgan, Gainesville State College
David Morgan, University of West Georgia
Julie Morris, Armstrong Atlantic State University
Becky Morrow, Duquesne University
Mark Mort, University of Kansas
Nancy Morvillo, Florida Southern College
Anthony Moss, Auburn University

Mario Mota, University of Central Florida
Alexander Motten, Duke University
Tim Mulkey, Indiana State University
John Mull, Weber State University
Michael Muller, University of Illinois at Chicago
Beth Mullin, The University of Tennessee, Knoxville
Paul Narguizian, California State University, Los Angeles
Jennifer Nauen, University of Delaware
Paul Nealen, Indiana University of Pennsylvania
Diana Nemergut, University of Colorado, Boulder
Kathryn Nette, Cuyamaca College
Jacalyn Newman, University of Pittsburgh
James Nienow, Valdosta State University
Alexey Nikitin, Grand Valley State University
Tanya Noel, York University
Fran Norflus, Clayton State University
Celia Norman, Arapahoe Community College
Eric Norstrom, DePaul University
Jorge Obeso, Miami Dade College
Kavita Oommen, Georgia State University
David Oppenheimer, University of Florida
Joseph Orkwiszewski, Villanova University
Rebecca Orr, Collin College
Don Padgett, Bridgewater State College
Joanna Padolina, Virginia Commonwealth University
One Pagan, West Chester University
Kathleen Page, Bucknell University
Daniel Papaj, University of Arizona
Pamela Pape-Lindstrom, Everett Community College
Bruce Patterson, University of Arizona, Tucson
Shelley Penrod, Lone Star College System
Roger Persell, Hunter College, The City University of New York
John Peters, College of Charleston
Chris Petrie, Brevard Community College
Patricia Phelps, Austin Community College
Steven Phelps, The University of Texas at Austin
Kristin Picardo, St. John Fisher College
Aaron Pierce, Nicholls State University
Debra Pires, University of California, Los Angeles
Thomas Pitzer, Florida International University
Nicola Plowes, Arizona State University
Crima Pogge, City College of San Francisco
Darren Pollock, Eastern New Mexico University
Kenneth Pruitt, The University of Texas at Brownsville
Sonja Pyott, University of North Carolina at Wilmington
Rajinder Ranu, Colorado State University
Philip Rea, University of Pennsylvania
Amy Reber, Georgia State University
Ahnya Redman, West Virginia University
Melissa Reedy, University of Illinois at Urbana-Champaign
Brian Ring, Valdosta State University
David Rintoul, Kansas State University
Michael Rischbieter, Presbyterian College
Laurel Roberts, University of Pittsburgh
George Robinson, The University at Albany, The State University of New York
Peggy Rolfsen, Cincinnati State Technical and Community College
Mike Rosenzweig, Virginia Polytechnic Institute and State University
Doug Rouse, University of Wisconsin, Madison
Yelena Rudayeva, Palm Beach State College
Ann Rushing, Baylor University
Shereen Sabet, La Sierra University
Rebecca Safran, University of Colorado
Peter Sakaris, Southern Polytechnic State University
Thomas Sasek, University of Louisiana, Monroe
Udo Savalli, Arizona State University
H. Jochen Schenk, California State University, Fullerton
Gregory Schmaltz, University of the Fraser Valley
Jean Schmidt, University of Pittsburgh
Andrew Schnabel, Indiana University, South Bend
Roxann Schroeder, Humboldt State University
David Schultz, University of Missouri, Columbia
Andrea Schwarzbach, The University of Texas at Brownsville
Erik Scully, Towson University
Robert Seagull, Hofstra University
Pramila Sen, Houston Community College
Alice Sessions, Austin Community College
Vijay Setaluri, University of Wisconsin
Jyotsna Sharma, The University of Texas at San Antonio
Elizabeth Sharpe-Aparicio, Blinn College
Patty Shields, University of Maryland
Cara Shillington, Eastern Michigan University
James Shinkle, Trinity University
Rebecca Shipe, University of California, Los Angeles
Marcia Shofner, University of Maryland
Laurie Shornick, Saint Louis University
Jill Sible, Virginia Polytechnic Institute and State University
Allison Silveus, Tarrant County College
Kristin Simokat, University of Idaho
Sue Simon-Westendorf, Ohio University
Sedonia Sipes, Southern Illinois University, Carbondale
John Skillman, California State University, San Bernardino
Marek Sliwinski, University of Northern Iowa
Felisa Smith, University of New Mexico
John Sollinger, Southern Oregon University
Scott Solomon, Rice University
Morvarid Soltani-Bejnood, The University of Tennessee
Vladimir Spiegelman, University of Wisconsin, Madison
Chrissy Spencer, Georgia Institute of Technology
Kathryn Spilios, Boston University
Ashley Spring, Brevard Community College
Bruce Stallsmith, The University of Alabama in Huntsville
Jennifer Stanford, Drexel University
Barbara Stegenga, University of North Carolina, Chapel Hill
Patricia Steinke, San Jacinto College, Central Campus
Asha Stephens, College of the Mainland
Robert Steven, University of Toledo
Eric Strauss, University of Wisconsin, La Crosse
Sukanya Subramanian, Collin College
Mark Sugalski, Southern Polytechnic State University
Brad Swanson, Central Michigan University
Ken Sweat, Arizona State University
David Tam, University of North Texas
Ignatius Tan, New York University
William Taylor, University of Toledo
Christine Terry, Lynchburg College
Sharon Thoma, University of Wisconsin, Madison
Pamela Thomas, University of Central Florida
Carol Thornber, University of Rhode Island
Patrick Thorpe, Grand Valley State University

Briana Timmerman, University of South Carolina
Chris Todd, University of Saskatchewan
Gail Tompkins, Wake Technical Community College
Martin Tracey, Florida International University
Randall Tracy, Worcester State University
James Traniello, Boston University
Bibit Traut, City College of San Francisco
Terry Trier, Grand Valley State University
Stephen Trumble, Baylor University
Jan Trybula, The State University of New York at Potsdam
Alexa Tullis, University of Puget Sound
Marsha Turell, Houston Community College
Mary Tyler, University of Maine
Marcel van Tuinen, University of North Carolina at Wilmington
Dirk Vanderklein, Montclair State University
Jorge Vasquez-Kool, Wake Technical Community College
William Velhagen, New York University
Dennis Venema, Trinity Western University
Laura Vogel, North Carolina State University
Jyoti Wagle, Houston Community College
Jeff Walker, University of Southern Maine
Gary Walker, Appalachian State University
Andrea Ward, Adelphi University
Fred Wasserman, Boston University
Elizabeth Waters, San Diego State University
Douglas Watson, The University of Alabama at Birmingham
Matthew Weand, Southern Polytechnic State University
Michael Weber, Carleton University
Cindy Wedig, The University of Texas, Pan American
Brad Wetherbee, University of Rhode Island
Debbie Wheeler, University of the Fraser Valley
Clay White, Lone Star College System
Lisa Whitenack, Allegheny College
Maggie Whitson, Northern Kentucky University
Stacey Wild East, Tennessee State University
Herbert Wildey, Arizona State University and Phoenix College
David Wilkes, Indiana University, South Bend
Lisa Williams, Northern Virginia Community College
Elizabeth Willott, University of Arizona
Mark Wilson, Humboldt State University
Ken Wilson, University of Saskatchewan
Bob Winning, Eastern Michigan University
Candace Winstead, California Polytechnic State University
Robert Wise, University of Wisconsin, Oshkosh
D. Reid Wiseman, College of Charleston
MaryJo Witz, Monroe Community College
David Wolfe, American River College
Kevin Woo, University of Central Florida
Denise Woodward, Penn State
Shawn Wright, Central New Mexico Community College
Grace Wyngaard, James Madison University
Aimee Wyrick, Pacific Union College
Joanna Wysocka-Diller, Auburn University
Ken Yasukawa, Beloit College
John Yoder, The University of Alabama
Kelly Young, California State University, Long Beach
James Yount, Brevard Community College
Min Zhong, Auburn University

Second Edition

Barbara J. Abraham, Hampton University
Jason Adams, College of DuPage
Sandra D. Adams, Montclair State University
Richard Adler, University of Michigan, Dearborn
Nancy Aguilar-Roca, University of California, Irvine
Shivanthi Anandan, Drexel University
Lynn Anderson-Carpenter, University of Michigan
Christine Andrews, The University of Chicago
Peter Armbruster, Georgetown University
Jessica Armenta, Austin Community College
Brian Ashburner, University of Toledo
Ann J. Auman, Pacific Lutheran University
Nicanor Austriaco, Providence College
Felicitas Avendano, Grand View University
J. P. Avery, University of North Florida
Jim Bader, Case Western Reserve University
Ellen Baker, Santa Monica College
Andrew S. Baldwin, Mesa Community College
Stephen Baron, Bridgewater College
Paul W. Bates, University of Minnesota, Duluth
Janet Batzli, University of Wisconsin, Madison
David Baum, University of Wisconsin, Madison
Kevin S. Beach, The University of Tampa
Philip Becraft, Iowa State University
Alexandra Bely, University of Maryland
Lauryn Benedict, University of Northern Colorado
Anne Bergey, Truman State University
Joydeep Bhattacharjee, University of Louisiana, Monroe
Todd Bishop, Dalhousie University
Catherine Black, Idaho State University
Andrew R. Blaustein, Oregon State University
James Bolton, Georgia Gwinnett College
Jim Bonacum, University of Illinois at Springfield
Laurie J. Bonneau, Trinity College
James Bottesch, Eastern Florida State College
Lisa Boucher, The University of Texas at Austin
Nicole Bournias-Vardiabasis, California State University, San Bernardino
Nancy Boury, Iowa State University
Matthew Brewer, Georgia State University
Christoper G. Brown, Georgia Gwinnett College
Jill Buettner, Richland College
Sharon K. Bullock, University of North Carolina at Charlotte
Lisa Burgess, Broward College
Jorge Busciglio, University of California, Irvine
Stephen Bush, Coastal Carolina University
David Byres, Florida State College at Jacksonville
Gloria Caddell, University of Central Oklahoma
Guy A. Caldwell, The University of Alabama
Kim A. Caldwell, The University of Alabama
John S. Campbell, Northwest College
Jennifer Capers, Indian River State College
Joel Carlin, Gustavus Adolphus College
Sara G. Carlson, University of Akron
Dale Casamatta, University of North Florida
Merri Lynn Casem, California State University, Fullerton
Anne Casper, Eastern Michigan University
David Champlin, University of Southern Maine
Rebekah Chapman, Georgia State University
P. Bryant Chase, Florida State University
Thomas T. Chen, Santa Monica College
Young Cho, Eastern New Mexico University
Sunita Chowrira, University of British Columbia
Tim W. Christensen, East Carolina University

Steven Clark, University of Michigan
Beth Cliffel, Triton College
Liane Cochran-Stafira, Saint Xavier University
John G. Cogan, The Ohio State University
Reid Compton, University of Maryland
Ronald H. Cooper, University of California, Los Angeles
Janice Countaway, University of Central Oklahoma
Joseph A. Covi, University of North Carolina at Wilmington
Will Crampton, University of Central Florida
Kathryn Craven, Armstrong State University
Lorelei Crerar, George Mason University
Kerry Cresawn, James Madison University
Richard J. Cristiano, Houston Community College Northwest
Cynthia K. Damer, Central Michigan University
David Dansereau, Saint Mary's University
Mark Davis, Macalester College
Elizabeth A. De Stasio, Lawrence University
Tracy Deem, Bridgewater College
Kimberley Dej, McMaster University
Terrence Delaney, University of Vermont
Tracie Delgado, Northwest University
Mark S. Demarest, University of North Texas
D. Michael Denbow, Virginia Polytechnic Institute and State University
Jonathan Dennis, Florida State University
Brandon S. Diamond, University of Miami
AnnMarie DiLorenzo, Montclair State University
Frank J. Dirrigl, Jr., University of Texas, Pan American
Christine Donmoyer, Allegheny College
Samuel Douglas, Angelina College
John D. DuBois, Middle Tennessee State University
Janet Duerr, Ohio University
Meghan Duffy, University of Michigan
Richard E. Duhrkopf, Baylor University
Jacquelyn Duke, Baylor University
Kamal Dulai, University of California, Merced
Rebecca K. Dunn, Boston College
Jacob Egge, Pacific Lutheran University
Kurt J. Elliott, Northwest Vista College
Miles Dean Engell, North Carolina State University
Susan Erster, Stony Brook University
Barbara I. Evans, Lake Superior State University
Lisa Felzien, Rockhurst University
Ralph Feuer, San Diego State University
Ginger R. Fisher, University of Northern Colorado
John Flaspohler, Concordia College
Sam Flaxman, University of Colorado, Boulder
Nancy Flood, Thompson Rivers University
Arthur Frampton, University of North Carolina at Wilmington
Caitlin Gabor, Texas State University
Tracy Galarowicz, Central Michigan University
Raul Galvan, South Texas College
Deborah Garrity, Colorado State University
Jason Mitchell Gee, East Carolina University
T.M. Gehring, Central Michigan University
John Geiser, Western Michigan University
Carol A. Gibbons Kroeker, Ambrose University
Susan A. Gibson, South Dakota State University
Cynthia J. Giffen, University of Michigan
Matthew Gilg, University of North Florida
Sharon L. Gillies, University of the Fraser Valley
Leslie Goertzen, Auburn University
Marla Gomez, Nicholls State University
Steven Gorsich, Central Michigan University
Daniel Graetzer, Northwest University
James Grant, Concordia University
Linda E. Green, Georgia Institute of Technology
Sara Gremillion, Armstrong State University
Ann Grens, Indiana University, South Bend
John L. Griffis, Joliet Junior College
Nancy A. Guild, University of Colorado, Boulder
Lonnie Guralnick, Roger Williams University
Valerie K. Haftel, Morehouse College
Margaret Hanes, Eastern Michigan University
Sally E. Harmych, University of Toledo
Sherry Harrel, Eastern Kentucky University
J. Scott Harrison, Georgia Southern University
Pat Harrison, University of the Fraser Valley
Diane Hartman, Baylor University
Wayne Hatch, Utah State University Eastern
David Haymer, University of Hawaii at Manoa
Chris Haynes, Shelton State Community College
Christiane Healey, University of Massachusetts, Amherst
David Hearn, Towson University
Marshal Hedin, San Diego State University
Triscia Hendrickson, Morehouse College
Albert A. Herrera, University of Southern California
Bradley Hersh, Allegheny College
Anna Hiatt, East Tennessee State University
Laura Hill, University of Vermont
Jay Hodgson, Armstrong State University
James Horwitz, Palm Beach State College
Sarah Hosch, Oakland University
Kelly Howe, University of New Mexico
Kimberly Hruska, Langara College
William Huddleston, University of Calgary
Carol Hurney, James Madison University
Brian A. Hyatt, Bethel University
Bradley C. Hyman, University of California, Riverside
Anne Jacobs, Allegheny College
Robert C. Jadin, Northeastern Illinois University
Rick Jellen, Brigham Young University
Dianne Jennings, Virginia Commonwealth University
L. Scott Johnson, Towson University
Russell Johnson, Colby College
Susan Jorstad, University of Arizona
Matthew Julius, St. Cloud State University
John S. K. Kauwe, Brigham Young University
Lori J. Kayes, Oregon State University
Todd Kelson, Brigham Young University, Idaho
Christopher Kennedy, Simon Fraser University
Jacob Kerby, University of South Dakota
Stephen T. Kilpatrick, University of Pittsburgh, Johnstown
Mary Kimble, Northeastern Illinois University
Denice D. King, Cleveland State Community College
David Kittlesen, University of Virginia
Ann Kleinschmidt, Allegheny College
Kathryn Kleppinger-Sparace, Tri-County Technical College
Daniel Klionsky, University of Michigan
Ned Knight, Reed College
Benedict Kolber, Duquesne University
Ross Koning, Eastern Connecticut State University

John Koontz, The University of Tennessee, Knoxville
Peter Kourtev, Central Michigan University
Elizabeth Kovar, The University of Chicago
Nadine Kriska, University of Wisconsin, Whitewater
Tim L. Kroft, Auburn University at Montgomery
William Kroll, Loyola University of Chicago
Dave Kubien, University of New Brunswick
Jason Kuehner, Emmanuel College
Josephine Kurdziel, University of Michigan
Troy A. Ladine, East Texas Baptist University
Diane M. Lahaise, Georgia Perimeter College
Janice Lai, Austin Community College, Cypress Creek
Kirk Land, University of the Pacific
James Langeland, Kalamazoo College
Neva Laurie-Berry, Pacific Lutheran University
Brenda Leady, University of Toledo
Adrienne Lee, University of California, Fullerton
Chris Levesque, John Abbott College
Bai-Lian Larry Li, University of California, Riverside
Cynthia Littlejohn, University of Southern Mississippi
Jason Locklin, Temple College
Xu Lu, University of Findlay
Patrice Ludwig, James Madison University
Ford Lux, Metropolitan State University of Denver
Morris F. Maduro, University of California, Riverside
C. Smoot Major, University of South Alabama
Barry Margulies, Towson University
Nilo Marin, Broward College
Michael Martin, John Carroll University
Heather D. Masonjones, University of Tampa
Scott C. Mateer, Armstrong State University
Robert Maxwell, Georgia State University
Joseph McCormick, Duquesne University
Lori L. McGrew, Belmont University
Peter B. McIntyre, University of Wisconsin, Madison
Iain McKinnell, Carleton University
Krystle McLaughlin, Lehigh University
Susan B. McRae, East Carolina University
Mark Meade, Jacksonville State University
Richard Merritt, Houston Community College Northwest
James E. Mickle, North Carolina State University
Chad Montgomery, Truman State University
Scott M. Moody, Ohio University
Daniel Moon, University of North Florida
Jamie Moon, University of North Florida
Jonathan Moore, Virginia Commonwealth University
Paul A. Moore, Bowling Green State University
Tsafrir Mor, Arizona State University
Jeanelle Morgan, University of North Georgia
Mark Mort, University of Kansas
Anthony Moss, Auburn University
Karen Neal, Reynolds Community College
Kimberlyn Nelson, Pennsylvania State University
Hao Nguyen, California State University, Sacramento
John Niedzwiecki, Belmont University
Alexey G. Nikitin, Grand Valley State University
Matthew Nusnbaum, Georgia State University
Robert Okazaki, Weber State University
Tiffany Oliver, Spelman College
Jennifer S. O'Neil, Houston Community College
Kavita Oommen, Georgia State University
Nathan Opolot Okia, Auburn University at Montgomery
Robin O'Quinn, Eastern Washington University
Sarah A. Orlofske, Northeastern Illinois University
Don Padgett, Bridgewater State University
Lisa Parks, North Carolina State University
Nilay Patel, California State University, Fullerton
Markus Pauly, University of California, Berkeley
Daniel M. Pavuk, Bowling Green State University
Marc Perkins, Orange Coast College
Beverly Perry, Houston Community College
John S. Peters, College of Charleston
Chris Petrie, Eastern Florida State College
John M. Pleasants, Iowa State University
Michael Plotkin, Mt. San Jacinto College
Mary Poffenroth, San Jose State University
Dan Porter, Amarillo College
Sonja Pyott, University of North Carolina at Wilmington
Mirwais Qaderi, Mount Saint Vincent University
Nick Reeves, Mt. San Jacinto College
Adam J. Reinhart, Wayland Baptist University
Stephanie Richards, Bates College
David A. Rintoul, Kansas State University
Trevor Rivers, University of Kansas
Laurel Roberts, University of Pittsburgh
Casey Roehrig, Harvard University
Jennifer Rose, University of North Georgia
Michael S. Rosenzweig, Virginia Polytechnic Institute and State University
Caleb M. Rounds, University of Massachusetts, Amherst
Yelena Rudayeva, Palm Beach State College
James E. Russell, Georgia Gwinnett College
Donald Sakaguchi, Iowa State University
Thomas Sasek, University of Louisiana at Monroe
Leslie J. Saucedo, University of Puget Sound
Udo M. Savalli, Arizona State University West Campus
Smita Savant, Houston Community College Southwest
Leena Sawant, Houston Community College Southwest
H. Jochen Schenk, California State University, Fullerton
Aaron E. Schirmer, Northeastern Illinois University
Mark Schlueter, Georgia Gwinnett College
Gregory Schmaltz, University of the Fraser Valley
Jennifer Schramm, Chemeketa Community College
Roxann Schroeder, Humboldt State University
Tim Schuh, St. Cloud State University
Kevin G. E. Scott, University of Manitoba
Erik P. Scully, Towson University
Sarah B. Selke, Three Rivers Community College
Pramila Sen, Houston Community College
Anupama Seshan, Emmanuel College
Alice Sessions, Austin Community College
Vijay Setaluri, University of Wisconsin
Timothy E. Shannon, Francis Marion University
Wallace Sharif, Morehouse College
Mark Sherrard, University of Northern Iowa
Cara Shillington, Eastern Michigan University
Amy Siegesmund, Pacific Lutheran University

Christine Simmons, Southern Illinois University, Edwardsville
S. D. Sipes, Southern Illinois University, Carbondale
John Skillman, California State University, San Bernardino
Daryl Smith, Langara College
Julie Smith, Pacific Lutheran University
Karen Smith, University of British Columbia
Leo Smith, University of Kansas
Ramona Smith-Burrell, Eastern Florida State College
Joel Snodgrass, Towson University
Alan J. Snow, University of Akron–Wayne College
Judith Solti, Houston Community College, Spring Branch
Ann Song, University of California, Fullerton
C. Kay Song, Georgia State University
Chrissy Spencer, Georgia Institute of Technology
Rachel Spicer, Connecticut College
Ashley Spring, Eastern Florida State College
Bruce Stallsmith, The University of Alabama in Huntsville
Maria L. Stanko, New Jersey Institute of Technology
Nancy Staub, Gonzaga University
Barbara Stegenga, University of North Carolina
Robert Steven, University of Toledo
Lori Stevens, University of Vermont
Mark Sturtevant, Oakland University
Elizabeth B. Sudduth, Georgia Gwinnett College
Mark Sugalski, Kennesaw State University, Southern Polytechnic State University
Fengjie Sun, Georgia Gwinnett College
Bradley J. Swanson, Central Michigan University
Brook O. Swanson, Gonzaga University
Ken Gunter Sweat, Arizona State University West Campus
Annette Tavares, University of Ontario Institute of Technology
William R. Taylor, University of Toledo
Samantha Terris Parks, Georgia State University
Jessica Theodor, University of Calgary
Sharon Thoma, University of Wisconsin, Madison
Sue Thomson, Auburn University at Montgomery
Mark Tiemeier, Cincinnati State Technical and Community College
Candace Timpte, Georgia Gwinnett College
Nicholas Tippery, University of Wisconsin, Whitewater
Chris Todd, University of Saskatchewan
Kurt A. Toenjes, Montana State University, Billings
Jeffrey L. Travis, The University at Albany, The State University of New York
Stephen J. Trumble, Baylor University
Jan Trybula, The State University of New York at Potsdam
Cathy Tugmon, Georgia Regents University
Alexa Tullis, University of Puget Sound
Marsha Turell, Houston Community College
Pat Uelmen Huey, Georgia Gwinnett College
Steven M. Uyeda, Pima Community College
Rani Vajravelu, University of Central Florida
Moira van Staaden, Bowling Green State University
Dirk Vanderklein, Montclair State University
William Velhagen, Caldwell University
Sara Via, University of Maryland
Christopher Vitek, University of Texas, Pan American
Neal J. Voelz, St. Cloud State University
Mindy Walker, Rockhurst University
Andrea Ward, Adelphi University
Jennifer Ward, University of North Carolina at Asheville
Pauline Ward, Houston Community College
Alan Wasmoen, Metropolitan Community College
Elizabeth R. Waters, San Diego State University
Matthew Weand, Southern Polytechnic State University
K. Derek Weber, Raritan Valley Community College
Andrea Weeks, George Mason University
Charles Welsh, Duquesne University
Naomi L. B. Wernick, University of Massachusetts, Lowell
Mary E. White, Southeastern Louisiana University
David Wilkes, Indiana University, South Bend
Frank Williams, Langara College
Kathy S. Williams, San Diego State University
Lisa D. Williams, Northern Virginia Community College
Christina Wills, Rockhurst University
Brian Wisenden, Minnesota State University, Moorhead
David Wolfe, American River College
Ramakrishna Wusirika, Michigan Technological University
G. Wyngaard, James Madison University
James R. Yount, Eastern Florida State College
Min Zhong, Auburn University

BIOLOGY

HOW LIFE WORKS

CONTENTS

CHAPTER 3 NUCLEIC ACIDS AND TRANSCRIPTION 49

CHAPTER 4 TRANSLATION AND PROTEIN STRUCTURE 69

CHAPTER 11 **CELL DIVISION**

Variations, Regulation, and Cancer 219

CHAPTER 12 **DNA REPLICATION AND MANIPULATION** **247**

BIOLOGY
HOW LIFE WORKS

PART 1

FROM CELLS TO ORGANISMS

"Nothing in biology makes sense except in the light of evolution."

—THEODOSIUS DOBZHANSKY

CHAPTER 1

Life

Chemical, Cellular, and Evolutionary Foundations

Core Concepts

1.1 The scientific method is a deliberate way of asking and answering questions about the natural world.

1.2 Life works according to fundamental principles of chemistry and physics.

1.3 The fundamental unit of life is the cell.

1.4 Evolution explains the features that organisms share and those that set them apart.

1.5 Organisms interact with one another and with their physical environment, shaping ecological systems that sustain life.

1.6 In the 21st century, humans have become major agents in ecology and evolution.

Giovanna Griffo/Gallery Stock.

Every day, remarkable things happen within and around you. Strolling through a local market, you come across a bin full of crisp apples, pick one up, and take a bite. Underlying this unremarkable occurrence is an extraordinary series of events. Your eyes sense the apple from a distance, and nerves carry that information to your brain, permitting identification. Biologists call this cognition, an area of biological study. Stimulated by the apple and recognizing it as ripe and tasty, your brain transmits impulses through nerves to your muscles. How we respond to external cues motivates behavior, another biological discipline. Grabbing the apple requires the coordinated activities of dozens of muscles that move your arm and hand to a precise spot. These movements are described by biomechanics, yet another area of biological research. And, as you bite down on the apple, glands in your mouth secrete saliva, starting to convert energy stored in the apple as sugar into energy that you will use to fuel your own activities. Physiology, like biomechanics, lies at the heart of biological function.

The study of cognition, behavior, biomechanics, and physiology are all ways of approaching **biology,** the science of how life works. **Biologists,** scientists who study life, have come to understand a great deal about these and other processes at levels that run from molecular mechanisms within the cell, through the integrated actions of many cells within an organ or body, to the interactions among different organisms in nature. We don't know everything about how life works—in fact, it seems as if every discovery raises new questions. But biology provides us with an organized way of understanding ourselves and the world around us.

Why study biology? The example of eating an apple was deliberately chosen because it is an everyday occurrence that we ordinarily wouldn't think about twice. The scope of modern biology, however, is vast, raising questions that can fire our imaginations, affect our health, and influence our future. How, for example, will our understanding of the human genome change the way that we fight cancer? How do bacteria in our digestive system help determine health and well-being? Will expected increases in the temperature and acidity of seawater doom coral reefs? Is there, or has there ever been, life on Mars? And, to echo the great storyteller Rudyard Kipling, why do leopards have spots, and tigers stripes?

We can describe six grand themes that connect and unite the many dimensions of life science, from molecules to the biosphere. These six themes are stated as Core Concepts for this chapter and are introduced in the following sections. Throughout the book, these themes will be visited again and again. We view them as the keys to understanding the many details in subsequent chapters and relating them to one another. Our hope is that by the time you finish this book, you will have an understanding of how life works, from the molecular machines inside cells and the metabolic pathways that cycle carbon through the biosphere to the process of evolution, which has shaped the living world that surrounds (and includes) us. You will, we hope, see the connections among these different ways of understanding life, and come away with a greater understanding of how scientists think about and ask questions about the natural world. How, in fact, do we know what we think we know about life? And we hope you will develop a basis for making informed decisions about your life, career, and the actions you take as a citizen.

1.1 THE SCIENTIFIC METHOD

How do we go about trying to understand the vastness and complexity of nature? For most scientists, studies of the natural world involve the complementary processes of observation and experimentation. **Observation** is the act of viewing the world around us. **Experimentation** is a disciplined and controlled way of asking and answering questions about the world in an unbiased manner.

Observation allows us to draw tentative explanations called hypotheses.

Observations allow us to ask focused questions about nature. Let's say you observe a hummingbird like the one pictured in **Fig. 1.1** hovering near a red flower, occasionally dipping its long beak into the bloom. What motivates this behavior? Is the bird feeding on some substance within the flower? Is it drawn to the flower by its vivid color? What benefit, if any, does the flower derive from this busy bird?

Observations such as these, and the questions they raise, allow us to propose tentative explanations, or **hypotheses.** We might, for example, hypothesize that the hummingbird is carrying pollen from one flower to the next, facilitating reproduction in the plant. Or we might hypothesize that nectar produced deep within the flower provides nutrition for the hummingbird—that the hummingbird's actions reflect the need to take in food. Both hypotheses provide a reasonable explanation of the behavior we observed, but they may or may not be correct. To find out, we have to test them.

Charles Darwin's classic book, *On the Origin of Species,* published in 1859, beautifully illustrates how we can piece together individual observations to construct a working hypothesis. In this book, Darwin discussed a wide range of observations, from pigeon breeding to fossils and from embryology to the unusual animals and plants found on islands. Darwin noted the success of animal breeders in selecting specific individuals for reproduction and thereby generating new breeds for agriculture or show. He appreciated that selective breeding is successful only if specific features of the animals can be passed from one generation to the next by inheritance. Reading economic treatises by the English clergyman Thomas Malthus, he understood that limiting environmental resources could select among the variety of different individuals in populations in much the way that breeders select among cows or pigeons.

Gathering together all these seemingly disparate pieces of information, Darwin argued that life has evolved over time by means of natural selection. Since its formulation, Darwin's initial hypothesis has been tested by experiments, many thousands of

FIG. 1.1 A hummingbird visiting a flower. This simple observation leads to questions: Why do hummingbirds pay so much attention to flowers? Why do they hover near red flowers? *Source: Charles J. Smith.*

them. Our knowledge of many biological phenomena, ranging from biodiversity to the way the human brain is wired, depends on direct observation followed by careful inferences that lead to models of how things work.

A hypothesis makes predictions that can be tested by observation and experiments.

Not just any idea qualifies as a hypothesis. Two features set hypotheses apart from other ways of attacking problems. First, a good hypothesis makes predictions about observations not yet made or experiments not yet run. Second, because hypotheses make predictions, we can test them. That is, we can devise an experiment to see whether the predictions made by the hypothesis actually occur, or we can go into the field to try to make further observations predicted by the hypothesis. A hypothesis, then, is a statement about nature that can be tested by experiments or by new observations. Hypotheses are testable because, even as they suggest an explanation for observations made previously, they make predictions about observations yet to be made.

→ **Quick Check 1** Mice that live in sand dunes commonly have light tan fur. Develop a hypothesis to explain this coloration.

Once we have a hypothesis, we can test it to see if its predictions are accurate. Returning to the hummingbird and flower, we can test the hypothesis that the bird is transporting pollen from one flower to the next, enabling the plant to reproduce. Observation provides one type of test: If we catch and examine the bird just after it visits a flower, do we find pollen stuck to its beak or feathers? If so, our hypothesis survives the test.

Note, however, that we haven't proved the case. Pollen might be stuck on the bird for a different reason—perhaps it provides food for the hummingbird. However, if the birds *didn't* carry pollen from flower to flower, we would reject the hypothesis that they facilitate pollination. In other words, a single observation or experiment can lead us to reject a hypothesis, or it can support the hypothesis, but it cannot prove that a hypothesis is correct. To move forward, then, we might make a second set of observations. Does pollen that adheres to the hummingbird rub off when the bird visits a second flower of the same species? If so, we have stronger support for our hypothesis.

We might also use observations to test a more general hypothesis about birds and flowers. Does red color generally attract birds and so facilitate pollination in a wide range of flowers? To answer this question, we might catalog the pollination of many red flowers and ask whether they are pollinated mainly by birds. Or we might go the opposite direction and catalog the flowers visited by many different birds—are they more likely to be red than chance alone might predict?

Finally, we can test the hypothesis that the birds visit the flowers primarily to obtain food, spreading pollen as a side effect of their feeding behavior. We can measure the amount of nectar in the flower before and after the bird visits and calculate how much energy has been consumed by the bird during its visit. Continued observations over the course of the day will tell us whether the birds gain the nutrition they need by drinking nectar, and whether the birds have other sources of food.

In addition to observations, in many cases we can design experiments to test hypotheses. One of the most powerful types of experiment is called a controlled experiment. In a controlled experiment, the researcher sets up several groups to be tested, keeping the conditions and setup as similar as possible from one group to the next. Then, the researcher deliberately introduces something different, known as a **variable,** into one group that he or she hypothesizes might have some sort of an effect. This is called the **test group.** In another group, the researcher does not introduce this variable. This is a **control group,** and the expectation is that no effect will occur in this group.

Controlled experiments are extremely powerful. By changing just one variable at a time, the researcher is able to determine if that variable is important. If many variables were changed at once, it would be difficult, if not impossible, to draw conclusions from the experiment because the researcher would not be able to figure out which variable caused the outcome. The control group plays a key role as well. Having a group in which no change is expected ensures that the experiment works as it is supposed to and provides a baseline against which to compare the results of the test groups.

For example, we might test the hypothesis that hummingbirds facilitate pollination by doing a controlled experiment. In this case, we could set up groups of red flowers that are all similar to one another. For one group, we could surround the flowers with a fine mesh that allows small insects access to the plant but keeps hummingbirds away. For another group, we would not use a mesh. The variable, then, is the presence of a mesh; the test group is the flowers with the mesh; and the control group is the flowers without the mesh since the variable was not introduced in this group.

Will the flowers be pollinated? If only the group without the mesh is pollinated, this result lends support to our initial hypothesis. In this case, the hypothesis becomes less tentative and more certain. If both groups are pollinated, our hypothesis is not supported, in which case we may discard it for another explanation or change it to account for the new information.

→ **Quick Check 2** Design a controlled experiment that tests the hypothesis that cigarette smoke causes lung cancer.

FIG. 1.2 The scientific method.

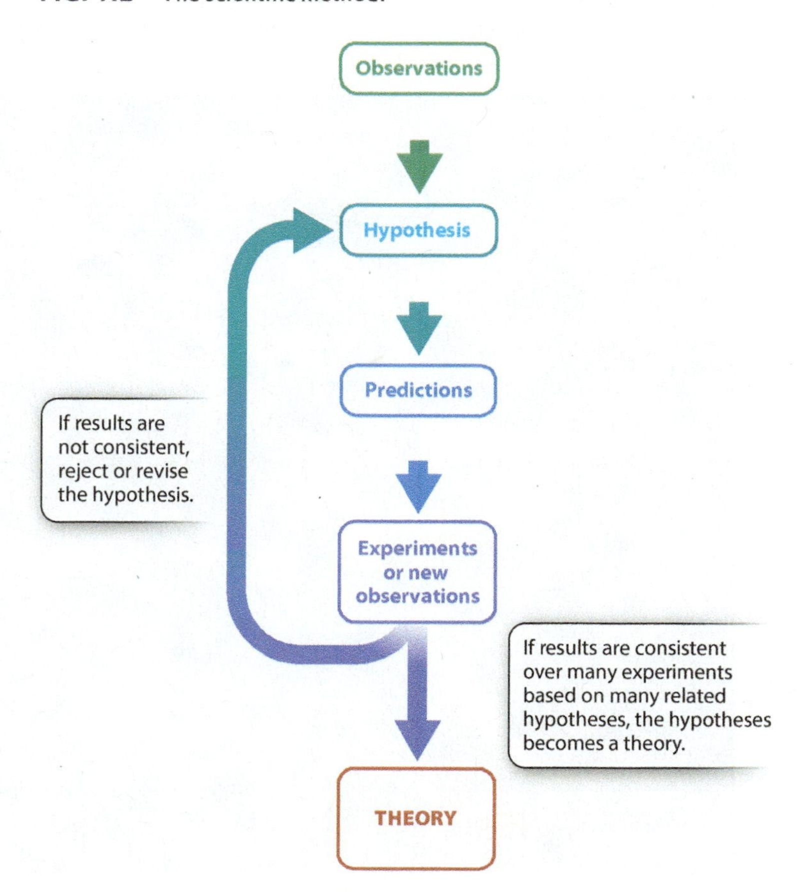

Using observations to generate a hypothesis and then making predictions based on that hypothesis that can be tested experimentally are the first two steps in the **scientific method,** outlined in **Fig. 1.2.** The scientific method is a deliberate and careful way of asking questions about the unknown. We make observations, collect field or laboratory samples, and design and carry out experiments or analyses to make sense of things we initially do not understand. The scientific method has proved to be spectacularly successful in helping us to understand the world around us. We explore several aspects of the scientific method, including experimental design, data and data presentation, probability and statistics, and scale and approximation on LaunchPad.

To emphasize the power of the scientific method, we turn to a famous riddle drawn from the fossil record (**Fig. 1.3**). Since the nineteenth century, paleontologists have known that before mammals expanded to their current ecological importance, other large animals dominated Earth. Dinosaurs evolved about 210 million years ago and disappeared abruptly 66 million years ago, along with many other species of plants, animals, and microscopic organisms. In many cases, the skeletons and shells of these creatures were buried in sediment and became fossilized. Layers of sedimentary rock therefore record the history of Earth.

Working in Italy, the American geologist Walter Alvarez collected samples from the precise point in the rock layers that corresponds to the time of the extinction. Careful chemical analysis showed that rocks at this level are unusually enriched in the element iridium. Iridium is rare in most rocks on continents and the seafloor, but is relatively common in rocks that fall from space—that is, in meteorites. From these observations, Alvarez and his colleagues developed a remarkable hypothesis: 66 million years ago, a large (11-km diameter) meteor slammed into Earth, and in the resulting environmental havoc, dinosaurs and many other species became extinct. This hypothesis makes specific predictions, described in Fig. 1.3, which turned out to be supported by further observations. Thus, observational tests support the hypothesis that nearly 150 million years of dinosaur evolution were undone in a moment.

General explanations of natural phenomena supported by many experiments and observations are called theories.

As already noted, a hypothesis may initially be tentative. Commonly, in fact, it will provide only one of several possible ways of explaining existing data. With repeated observation and experimentation, however, a good hypothesis gathers strength,

HOW DO WE KNOW?

FIG. 1.3

What caused the extinction of the dinosaurs?

BACKGROUND Dinosaurs were diverse and ecologically important for nearly 150 million years but became extinct about 66 million years ago.

OBSERVATION

Photo Source: Kirk Johnson, Denver Museum of Nature & Science.

HYPOTHESIS The impact of a large meteorite disrupted communities on land and in the sea, causing the extinction of the dinosaurs and many other species.

PREDICTIONS Independent evidence of a meteor impact should be found in rock layers corresponding to the time of the extinction and be rare or absent in older and younger beds.

FURTHER OBSERVATIONS

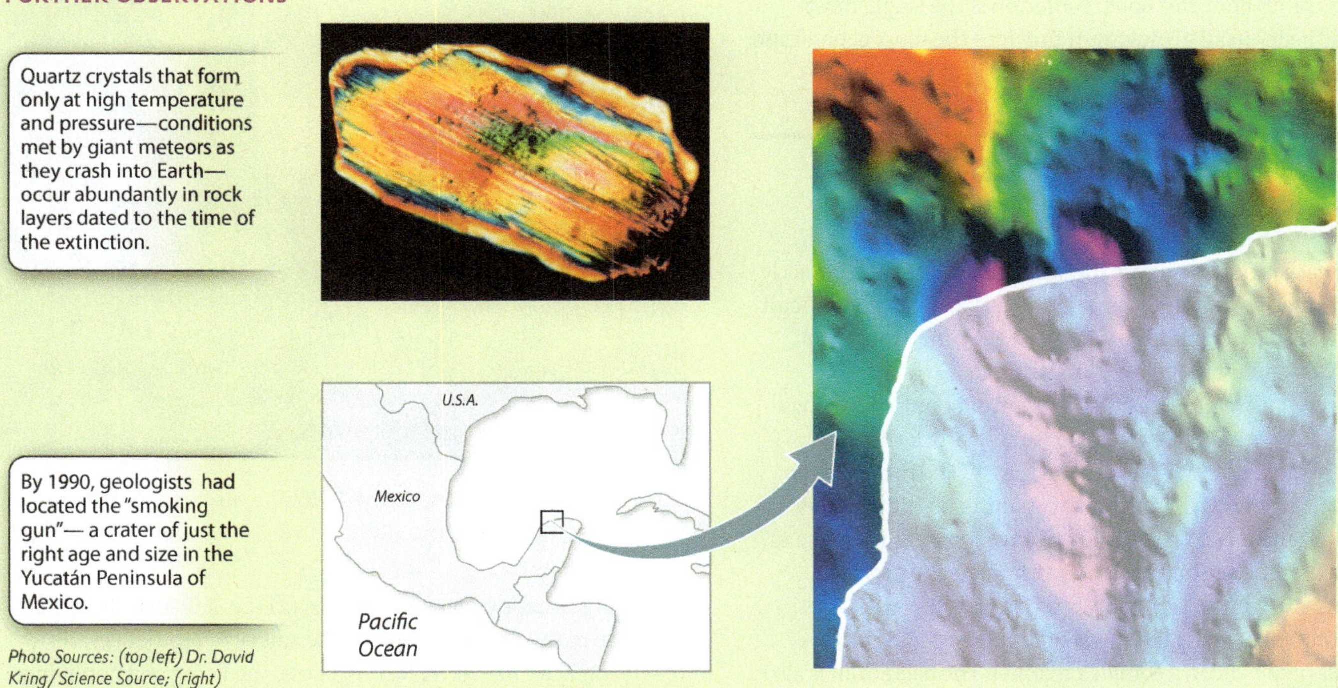

Photo Sources: (top left) Dr. David Kring/Science Source; (right) Image courtesy of V. L. Sharpton/ Lunar and Planetary Institute.

CONCLUSION A giant meteor struck Earth 66 million years ago, causing the extinction of the dinosaurs and many other species.

FOLLOW-UP WORK Researchers have documented other mass extinctions, but the event that eliminated the dinosaurs appears to be the only one associated with a meteorite impact.

SOURCE Alvarez, W. 1998. *T. rex and the Crater of Doom.* New York: Vintage Press.

and we have more and more confidence in it. When a number of related hypotheses survive repeated testing and come to be accepted as good bases for explaining what we see in nature, scientists articulate a broader explanation that accounts for all the hypotheses and the results of their tests. We call this statement a **theory,** a general explanation of the world supported by a large body of experiments and observations (see Fig. 1.2).

Note that scientists use the word "theory" in a very particular way. In general conversation, "theory" is often synonymous with "hypothesis," "idea," or "hunch"—"I've got a theory about that." But in a scientific context, the word "theory" has a specific meaning. Scientists speak in terms of theories only if hypotheses have withstood testing to the point where they provide a general explanation for many observations and experimental results. Just as a good hypothesis makes testable predictions, a good theory both generates good hypotheses and predicts their outcomes. Thus, scientists talk about the theory of gravity—a set of hypotheses you test every day by walking down the street or dropping a fork. Similarly, the theory of evolution is not one explanation among many for the unity and diversity of life. It is a set of hypotheses that has been tested for more than a century and shown to provide an extraordinarily powerful explanation of biological observations that range from the amino acid sequences of proteins to the diversity of ants in a rain forest. In fact, as we discuss throughout this book, evolution is the single most important theory in all of biology. It provides the most general and powerful explanation of how life works.

FIG. 1.4 A climber scaling a rock. Living organisms like this climber contain chemicals that are found in rocks, but only living organisms reproduce in a manner that allows for evolution over time. *Source: Scott Hailstone/Getty Images.*

1.2 CHEMICAL AND PHYSICAL PRINCIPLES

We stated earlier that biology is the study of life. But what exactly *is* life? As simple as this question seems, it is frustratingly difficult to answer. We all recognize life when we see it, but coming up with a definition is harder than it first appears.

Living organisms are clearly different from nonliving things. But just how different is an organism from the rock shown in **Fig. 1.4?** On one level, the comparison is easy: The rock is much simpler than any living organism we can think of. It has far fewer components, and it is largely static, with no apparent response to environmental change on timescales that are readily tracked.

In contrast, even an organism as relatively simple as a bacterium contains many hundreds of different chemical compounds organized in a complex manner. The bacterium is also dynamic in that it changes continuously, especially in response to the environment. Organisms reproduce, which minerals do not. And organisms do something else that rocks and minerals don't: They evolve. Indeed, the molecular biologist Gerald Joyce has defined life as a chemical system capable of undergoing Darwinian evolution.

From these simple comparisons, we can highlight four key characteristics of living organisms: (1) complexity, with precise spatial organization on several scales; (2) the ability to change in response to the environment; (3) the ability to reproduce; and (4) the capacity to evolve. Nevertheless, the living and nonliving worlds share an important attribute: Both are subject to the basic laws of chemistry and physics.

The living and nonliving worlds follow the same chemical rules and obey the same physical laws.

The chemical elements found in rocks and other nonliving things are no different from those found in living organisms. In other words, all the elements that make up living things can be found

FIG. 1.5 Composition of (a) Earth's crust and (b) cells in the human body. The Earth beneath our feet is made up of the same elements found in our feet, but in strikingly different proportions.

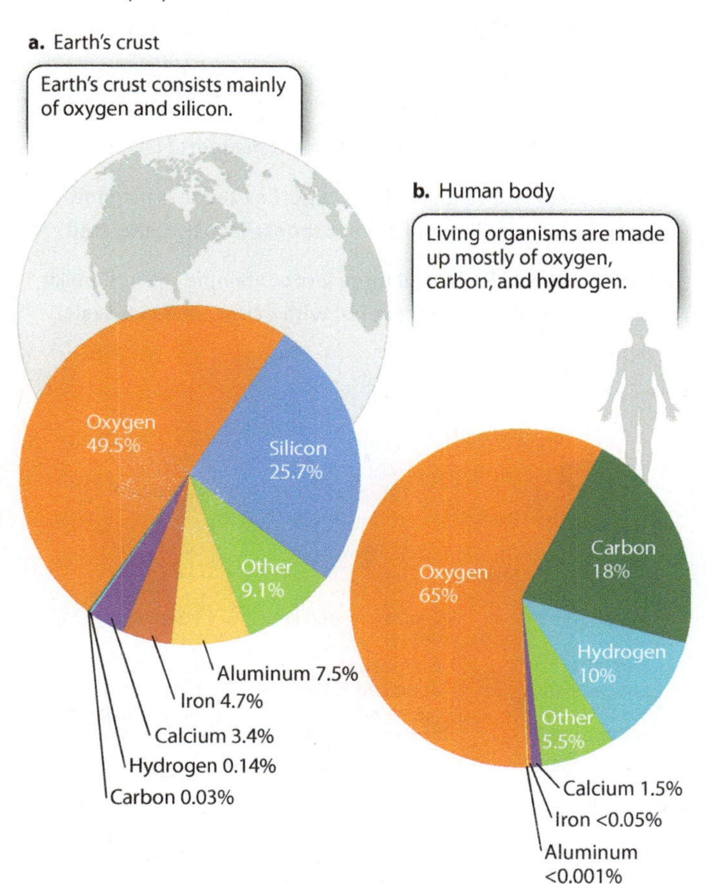

in the nonliving environment—there is nothing special about our chemical components when taken individually. That said, the *relative* abundances of elements in organisms differ greatly from those in the nonliving world. In the universe as a whole, hydrogen and helium make up more than 99% of known matter, while Earth's crust contains mostly oxygen and silicon, with significant amounts of aluminum, iron, and calcium (**Fig. 1.5a**). In organisms, by contrast, oxygen, carbon, and hydrogen are by far the most abundant elements (**Fig. 1.5b**). As discussed more fully in Chapter 2, carbon provides the chemical backbone of life. The particular properties of carbon make possible a wide diversity of molecules that, in turn, support a wide range of functions within cells.

All living organisms are subject to the physical laws of the universe. Physics helps us to understand how animals move and why trees don't fall over; it explains how redwoods conduct water upward through their trunks and how oxygen gets into the cells that line your lungs. Indeed, two laws of thermodynamics, both of which describe how energy is transformed in any system, determine how living organisms are able to do work and maintain their spatial organization.

The **first law of thermodynamics** states that energy can neither be created nor destroyed; it can only be transformed from one form into another. In other words, the total energy in the universe is constant, but the form that energy takes can change. Living organisms are energy transformers. They acquire energy from the environment and transform it into a chemical form that cells can use. All organisms obtain energy from the sun or from chemical compounds. Some of this energy is used to do work—such as moving, reproducing, and building cellular components—and the rest is dissipated as heat. The energy that is used to do work plus the heat that is generated is the total amount of energy, which is the same as the input energy (**Fig. 1.6**). In other words, the total amount of energy remains constant before and after energy transformation.

The **second law of thermodynamics** states that the degree of disorder (or the number of possible positions and motions of molecules) in the universe tends to increase. Think about a box full of marbles distributed more or less randomly; if you want to line up all the red ones or blue ones in a row, you have to do work—that is, you have to add energy. In this case, the addition of energy increases the order of the system, or, put another way, decreases its disorder. Physicists quantify the amount of disorder (or the number of possible positions and motions of molecules) in a system as the **entropy** of the system.

Living organisms are highly organized. As with lining up marbles in a row, energy is needed to maintain this organization. Given the tendency toward greater disorder, the high level of organization of even a single cell would appear to violate the second law. But it does not. The key is that a cell is not an isolated system and therefore cannot be considered on its own; it exists in an environment. So we need to take into account the whole system, the cell plus the environment that surrounds it. As energy

FIG. 1.6 Energy transformation and the first law of thermodynamics. The first law states that the total amount of energy in any system remains the same. Organisms transform energy from one form to another, but the total energy in any system is constant.

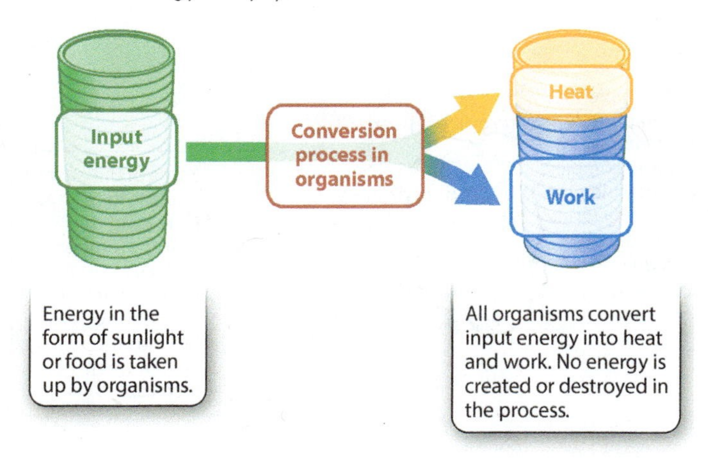

is harnessed by cells, only some is used to do work; the rest is dissipated as heat (Fig. 1.6). That is, conversion of energy from one form to another is never 100% efficient. Heat is a form of energy, so the total amount of energy is conserved, as dictated by the first law. In addition, heat corresponds to the motion of small molecules—the greater the heat, the greater the motion, and the greater the motion, the greater the degree of disorder. Therefore, the release of heat as organisms harness energy means that the total entropy for the combination of the cell and its surroundings increases, in keeping with the second law (**Fig. 1.7**).

The scientific method shows that living organisms come from other living organisms.

Life is made up of chemical components that also occur in the nonliving environment and obey the same laws of chemistry and physics. Can life spontaneously arise from these nonliving materials? We all know that living organisms come from other living organisms, but it is worth asking *how* we know this. Direct observation can be misleading here. For example, raw meat, if left out on a plate, will rot and become infested with maggots (fly larvae). It might seem as though the maggots appear spontaneously. In fact, the question of where maggots come from was a matter of vigorous debate for centuries, until application of the scientific method settled the issue. In the 1600s, the Italian physician and naturalist Francesco Redi hypothesized that maggots (and hence flies) in rotting meat come only from other flies that laid their eggs in the meat.

To test his hypothesis, Redi set up an experiment in which he placed meat in three glass jars (**Fig. 1.8**). One jar was left open, a second was covered with gauze, and the third was sealed with a cap. The jars were left in a room with flies. Note that in this experiment, the three jars were subject to the same conditions—the only difference was the opening of the jar. The open jar allowed

FIG. 1.7 Energy transformation and the second law of thermodynamics. The second law states that disorder in any system tends to increase. Entropy can decrease locally (inside a cell, for example) because the heat released increases disorder in the environment.

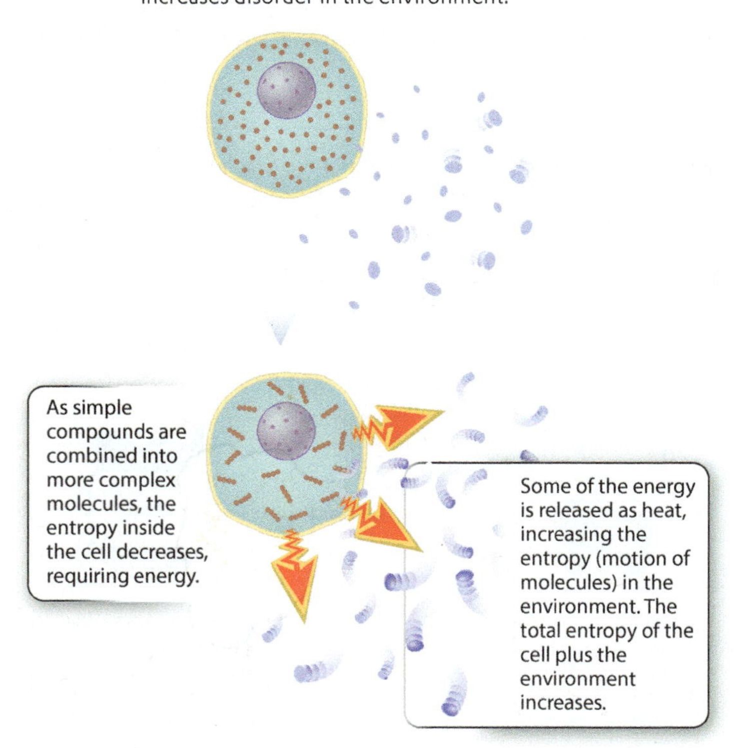

HOW DO WE KNOW?

FIG. 1.8

Can living organisms arise from nonliving matter?

BACKGROUND Until the 1600s, many people believed that rotting meat spontaneously generates maggots (fly larvae).

HYPOTHESIS Francesco Redi hypothesized that maggots come only from flies and are not spontaneously generated.

EXPERIMENT Redi used three jars containing meat. One jar was left open; one was covered with gauze; one was sealed with a cap.

RESULTS

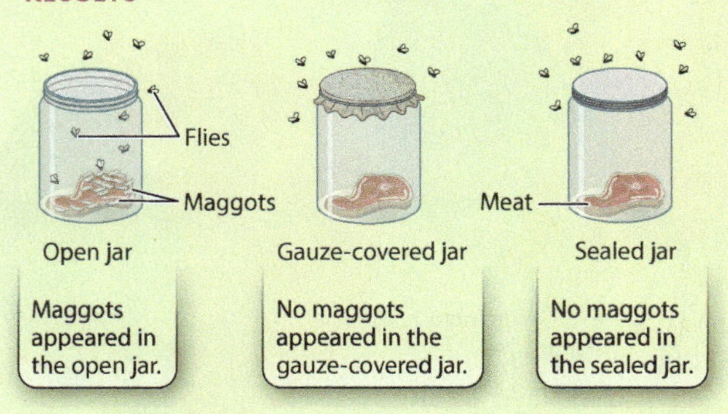

Maggots appeared in the open jar.

No maggots appeared in the gauze-covered jar.

No maggots appeared in the sealed jar.

CONCLUSION The presence of maggots in the open jar and the absence of maggots in the gauze-covered and sealed jars supported the hypothesis that maggots come from flies and allowed Redi to reject the hypothesis that maggots are spontaneously generated.

FOLLOW-UP WORK Redi's experiment argued against the idea of spontaneous generation for insects. However, it was unclear whether his results could be extended to microbes. Applying the scientific method, Louis Pasteur used a similar approach about 200 years later to investigate this question (see Fig. 1.9).

HOW DO WE KNOW?

FIG. 1.9

Can microscopic life arise from nonliving matter?

BACKGROUND Educated people in Pasteur's time knew that microbes grow well in nutrient-rich liquids like broth. It was also known that boiling would sterilize the broth, killing the microbes.

HYPOTHESIS Pasteur hypothesized that if microbes were generated spontaneously from nonliving matter, they should reappear in sterilized broth without the addition of microbes.

EXPERIMENT Pasteur used two flasks, one with a straight neck and one with a swan neck. The straight-neck flask allowed dust particles with microbes to enter. The swan-neck flask did not.

RESULTS

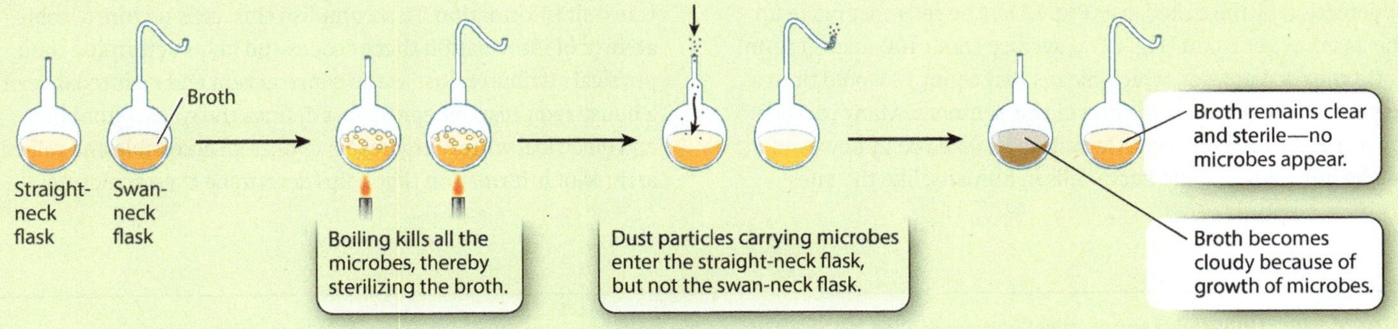

CONCLUSION The presence of microbes in the straight-neck flask and the absence of microbes in the swan-neck flask supported the hypothesis that microbes come from other microbes and are not spontaneously generated.

DISCUSSION Redi's and Pasteur's research illustrate classic attributes of well-designed experiments. Multiple treatments are set up, and nearly all conditions are the same in them all—they are constant, and therefore cannot be the cause of different outcomes of the experiment. One key feature—the variable—is changed by the experimenter from one treatment to the next. This is a place to look for explanations of different experimental outcomes.

for the passage of flies and air; the jar with the gauze allowed for the passage of air but not flies; and the sealed jar did not allow air or flies to enter. Over time, Redi observed that maggots appeared only on the meat in the open jar. No maggots appeared in the other two jars, which did not allow access to the meat by flies. These observations supported Redi's hypothesis that flies come from other flies, and did not provide support for the alternative hypothesis that maggots arise spontaneously from meat.

Redi demonstrated that living organisms come from other organisms, but some argued that his conclusion might apply only to larger organisms—microscopic life might be another matter entirely. It was not until the nineteenth century that the French chemist and biologist Louis Pasteur tested the hypothesis that microorganisms can arise by spontaneous generation (**Fig. 1.9**).

Pasteur filled two glass flasks with broth that had first been sterilized over heat—one with a straight vertical neck and the other with a curved swan neck. As in Redi's experiments, there was only one variable, in this case the shape of the neck of the flask. The straight-neck flask allowed airborne dust particles carrying microbes to fall into the sterile broth, while the swan-neck flask prevented dust from getting inside. Over time, Pasteur observed that microbes grew in the broth inside the straight-neck flask but not in the swan-neck flask. From these observations, Pasteur rejected the hypothesis that microbes arise spontaneously from sterile broth. Instead, exposure to microbes carried on airborne dust particles is necessary for microbial growth.

Redi's and Pasteur's experiments demonstrated that living organisms come from other living organisms and are not generated spontaneously from chemical components. But this raises the question of how life arose in the first place. If life comes from life, where did the first living organisms come from? Although today all organisms are produced by parental organisms, early in Earth's history this was not the case. Scientists hypothesize that life initially emerged from chemical compounds about 4 billion years ago. That is, chemical systems capable of evolution arose from chemical reactions that took place on the early Earth. We'll return to the great question of life's origin in Case 1: The First Cell and in Chapters 2 through 8.

1.3 THE CELL

The **cell** is the simplest entity that can exist as an independent unit of life. Every known living organism is either a single cell or an ensemble of a few to many cells (**Fig. 1.10**). Most bacteria (like those in Pasteur's experiment), yeasts, and the tiny algae that float in oceans and ponds spend their lives as single cells. In contrast, plants and animals contain billions to trillions of cells that function in a coordinated fashion.

Most cells are tiny, their dimensions well below the threshold of detection by the naked eye (**Fig. 1.11**). The cells that make up the layers of your skin (Fig. 1.11a) average about 100 microns (μm) or 0.1 mm in diameter, which means that about 10 would fit in a row across the period at the end of this sentence. Many bacteria are less than a micron long. Certain specialized cells, however, can be quite large. Some nerve cells in humans, like the ones pictured in Fig. 1.11b, extend slender projections known as axons for distances as great as a meter, and the cannonball-size egg of an ostrich in Fig. 1.11c is a single giant cell.

The types of cell just mentioned—bacteria, yeasts, skin cells, nerve cells, and an egg—seem very different, but all are organized along broadly similar lines. In general, all cells contain a stable blueprint of information in molecular form; they have a discrete boundary that separates the interior of the cell from its external environment; and they have the ability to harness materials and energy from the environment.

Nucleic acids store and transmit information needed for growth, function, and reproduction.

The first essential feature of a cell is its ability to store and transmit information. To accomplish this, cells require a stable archive of information that encodes and helps determine their physical attributes. Just as the construction and maintenance of a house requires a blueprint that defines the walls, plumbing, and electrical wiring, organisms require an accessible and reliable archive of information that helps determine their structure

FIG. 1.10 **Unicellular and multicellular organisms.** All living organisms are made up of cells: (a) bacteria; (b) brewer's yeast; (c) algae; (d) cheetahs; (e) humans. *Sources: a. Steve Gschmeissner/Science Source; b. Steve Gschmeissner/Science Source; c. Michael Abbey/Getty Images; d. Sven-Olof Lindblad/Science Source; e. Megapress/Alamy.*

FIG. 1.11 **Cell diversity.** Cells vary greatly in size and shape: (a) skin cells; (b) nerve cells; (c) ostrich egg. *Sources: a. Biophoto Associates/Science Source; b. Dr. Jonathan Clarke. Wellcome Images; c. Hemis/Alamy.*

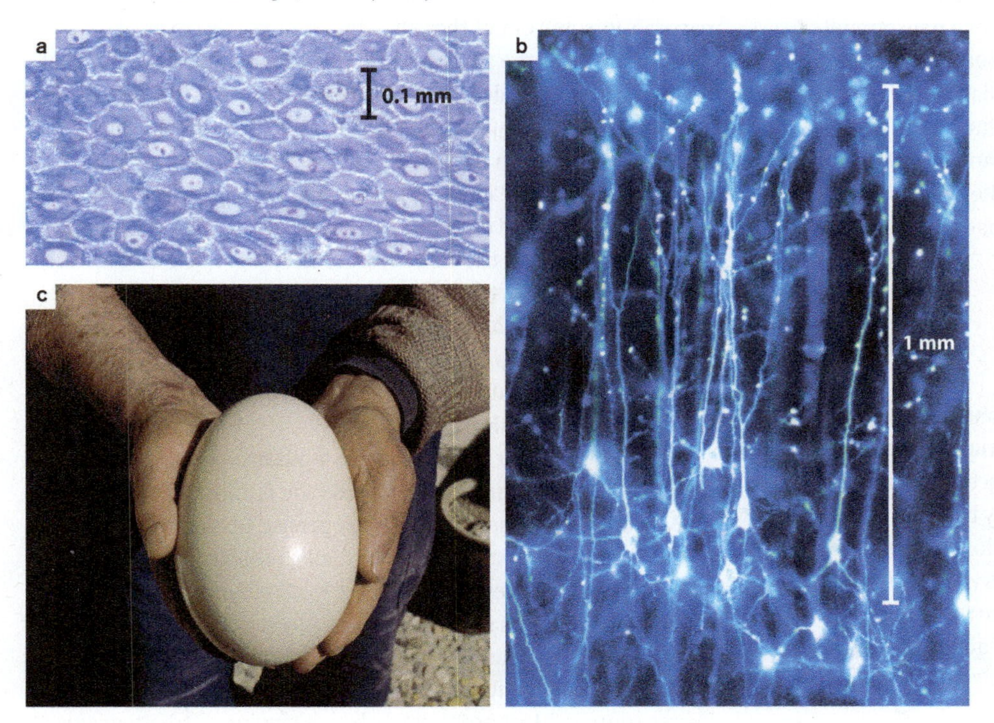

and metabolic activities. Another hallmark of life is the ability to reproduce. To reproduce, cells must be able to copy their archive of information rapidly and accurately. In all organisms, the information archive is a remarkable molecule known as **deoxyribonucleic acid,** or **DNA** (**Fig. 1.12**).

FIG. 1.12 **A molecule of DNA.** DNA is a double helix made up of varying sequences of four different subunits.

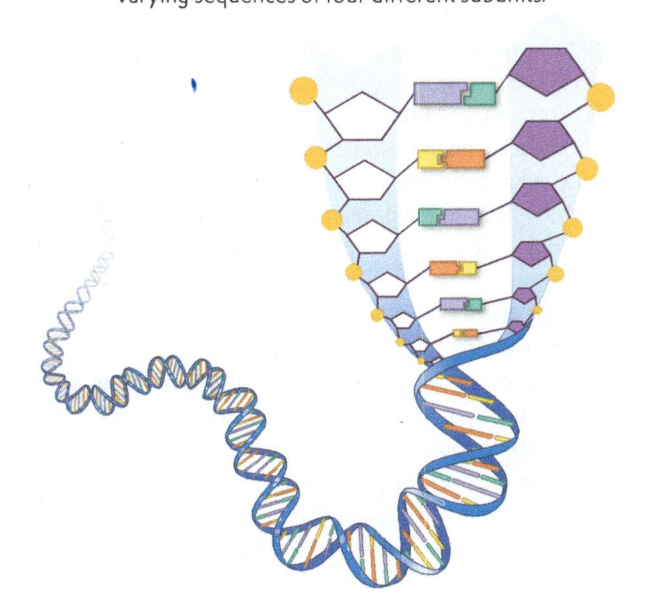

DNA is a double-stranded helix, with each strand made up of varying sequences of four different kinds of molecules connected end to end. It is the arrangement of these molecular subunits that makes DNA special; in essence, they provide a four-letter alphabet that encodes cellular information. Notably, the information encoded in DNA directs the formation of **proteins,** the key structural and functional molecules that do the work of the cell. Virtually every aspect of the cell's existence—its internal architecture, its shape, its ability to move, and its various chemical reactions—depends on proteins.

How does the information stored in DNA direct the synthesis of proteins? First, existing proteins create a copy of the DNA's information in the form of a closely related molecule called **ribonucleic acid,** or **RNA.** The synthesis of RNA from a DNA template is called **transcription,** a term that describes the copying of information from one form into another. Specialized molecular structures within the cell then "read" the RNA molecule to determine what building blocks to use to create a protein. This process, called **translation,** converts information stored in the language of nucleic acids to information in the language of proteins.

The pathway from DNA to RNA (specifically to a form of RNA called messenger RNA, or mRNA) to protein is known as the **central dogma** of molecular biology (**Fig. 1.13**). The central dogma describes the basic flow of information in a cell and, while there are exceptions, it constitutes a fundamental principle in biology. As proteins are ultimately encoded by DNA, we can define specific stretches or segments of DNA according to the proteins that they encode. This is the simplest definition of a **gene:** the DNA sequence that corresponds to a specific protein product.

DNA has another remarkable feature. In addition to storing information, it is easily copied, or **replicated,** allowing genetic

FIG. 1.13 The central dogma of molecular biology, defining the flow of information in all living organisms from DNA to RNA to protein.

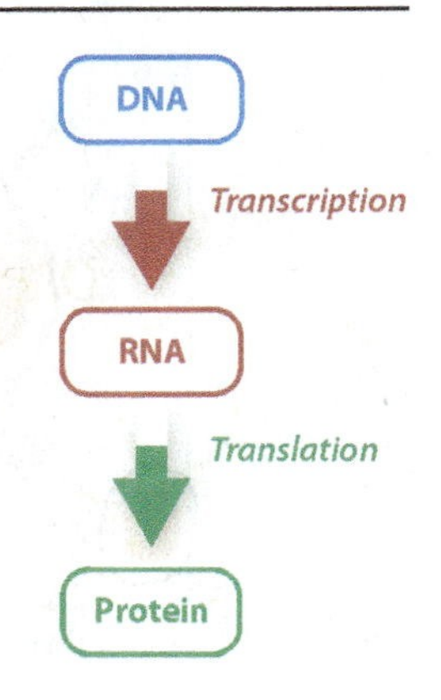

information to be passed from cell to cell or from an organism to its progeny. Each organism's DNA archive can be stably and reliably passed from generation to generation in large part because of its double-stranded helical structure. During replication, each strand of the double helix serves as a template for a new strand. Replication is necessarily precise and accurate because mistakes introduced into the cell's information archive may be lethal to the cell. That said, errors in DNA can and do occur during the process of replication, and environmental insults can damage DNA as well. Such changes are known as **mutations;** they can spell death for the cell, or they can lead to the variations that underlie the diversity of life and the process of evolution.

→ **Quick Check 3** How does the central dogma help us to understand how mutations in DNA can result in disease?

Membranes define cells and spaces within cells.

The second essential feature of all cells is a **plasma membrane** that separates the living material within the cell from the nonliving environment around it (**Fig. 1.14**). This boundary between inside and outside does not mean that cells are closed systems independent of the environment. On the contrary, there is an active and dynamic interplay between cells and their surroundings that is mediated by the plasma membrane. All cells require sustained contributions from their surroundings, both simple ions and the building blocks required to manufacture macromolecules. They also release waste products into the environment. As discussed more fully in Chapter 5, the plasma membrane controls the movement of materials into and out of the cell.

In addition to the plasma membrane, many cells have internal membranes that divide the cell into discrete compartments, each specialized for a particular function. A notable example is the **nucleus,** which houses the cell's DNA. Like the plasma membrane, the nuclear membrane selectively controls movement of molecules into and out of it. As a result, the nucleus occupies a discrete space within the cell, separate from the space outside the nucleus, called the **cytoplasm.**

Not all cells have a nucleus. In fact, cells can be grouped into two broad classes depending on whether or not they have a nucleus. Cells without a nucleus are called **prokaryotes,** and cells with a nucleus are **eukaryotes.**

The first cells that emerged about 4 billion years ago were prokaryotic. Their descendants include the familiar bacteria, found today nearly everywhere that life can persist. Some prokaryotes live in peaceful coexistence with humans, inhabiting our gut and aiding digestion. Others cause disease—salmonella, tuberculosis, and cholera are familiar examples. The success of these cells depends in part on their small size, their ability to reproduce rapidly, and their ability to obtain energy and nutrients from diverse sources. Most prokaryotes live as single-celled organisms, but some have simple multicellular forms.

Eukaryotes evolved much later, roughly 2 billion years ago, from prokaryotic ancestors. They include familiar groups such as animals, plants, and fungi, along with a wide diversity of single-celled microorganisms called protists. Eukaryotic organisms exist as single cells like yeasts or as multicellular organisms like humans. In multicellular organisms, cells may specialize to perform different functions. For example, in humans, muscle cells contract; red blood cells carry oxygen to tissues; and skin cells provide an external barrier.

The terms "prokaryotes" and "eukaryotes" are useful in drawing attention to a fundamental distinction between these two groups of cells. However, today, biologists recognize three domains of life—**Bacteria, Archaea,** and **Eukarya** (Chapters 26 and 27). Bacteria and Archaea both lack a nucleus and are therefore prokaryotes, whereas Eukarya are eukaryotic. Archaea are single-celled microorganisms, many of which flourish under seemingly hostile conditions, such as the hot springs of Yellowstone National Park.

Metabolism converts energy from the environment into a form that can be used by cells.

A third key feature of cells is the ability to harness energy from the environment. Let's go back to our introductory example of eating an apple. The apple contains sugars, which store energy in their chemical bonds. By breaking down sugar, our cells harness this energy and convert it into a form that can be used to do the work of the cell. Energy from the food we eat allows us to grow, move, communicate, and do all the other things that we do.

FIG. 1.14 **The plasma membrane.** The plasma membrane surrounds every cell and controls the exchange of material with the environment. *Sources: (top) Dr. Gopal Murti/SPL/Science Source; (bottom) Don W. Fawcett/Science Source.*

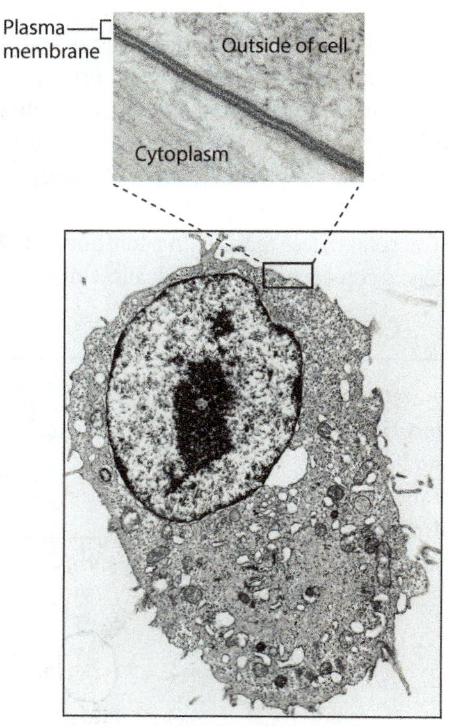

Organisms acquire energy from just two sources—the sun and chemical compounds. The term **metabolism** describes chemical reactions by which cells convert energy from one form to another and build and break down molecules. These reactions are required to sustain life. Regardless of their source of energy, all organisms use chemical reactions to break down molecules, releasing energy in the process that is stored in a chemical form called **adenosine triphosphate,** or **ATP.** This molecule enables cells to carry out all sorts of work, including growth, division, and moving substances into and out of the cell.

Many metabolic reactions are highly conserved between organisms, meaning the same reactions are found in many different organisms. This observation suggests that the reactions evolved early in the history of life and have been maintained for billions of years because of their fundamental importance to cellular biochemistry.

A virus is genetic material in need of a cell.

It's worth taking a moment to consider viruses. A virus is an agent that infects cells. It is smaller and simpler than cells. Why, then, aren't viruses the smallest unit of life? We just considered three essential features of cells—the capacity to store and transmit information, a membrane that selectively controls movement in and out, and the ability to harness energy from the environment. Viruses have a stable archive of genetic information, which can be RNA or DNA, surrounded by a protein coat and sometimes a lipid envelope. But viruses cannot harness energy from the environment. Therefore, on their own viruses cannot read and use the information contained in their genetic material, nor can they regulate the passage of substances across their protein coats or lipid envelopes the way that cells do. To replicate, they require a cell.

A virus infects a cell by binding to the cell's surface, inserting its genetic material into the cell, and, in most cases, using the cellular machinery to produce more viruses. In this way, it is often said that a virus "hijacks" a cell. The infected cell may produce more viruses, sometimes by lysis, or breakage, of the cell, and the new viruses can then infect more cells. In some cases, the genetic material of the virus integrates into the DNA of the host cell.

We discuss viruses many times throughout the book. Each species of Bacteria, Archaea, and Eukarya is susceptible to many types of virus that are specialized to infect its cells. Several hundred types of virus are known to infect humans, and the catalog is still incomplete. Useful tools in biological research, viruses have provided a model system for many problems in biology, including how genes are turned on and off and how cancer develops.

1.4 EVOLUTION

The themes introduced in the last two sections stress life's unity: Cells form the basic unit of all life; DNA, RNA, and proteins carry out the molecular functions of all cells; and metabolic reactions build and break down macromolecules. We need only look around us, however, to recognize that for all its unity, life displays a remarkable degree of diversity. We don't really know how many species share our planet, but reasonable estimates run to 10 million or more. Both the unity and the diversity of life are explained by the process of **evolution,** or change over time.

Variation in populations provides the raw material for evolution.

Described in detail, evolution by **natural selection** calls on complex mathematical formulations, but at heart its main principles are simple, indeed unavoidable. When there is variation within a population of organisms, and when that variation can be inherited (that is, when it can be passed from one generation to the next), the variants best suited for growth and reproduction in a given environment will contribute disproportionately to the next generation. As Darwin recognized, farmers have used this principle for thousands of years to select for crops with high yield or improved resistance to drought and disease. It is how people around the world have developed breeds of dog ranging from terriers to huskies (**Fig. 1.15**). And it is why antibiotic resistance is on the rise in many disease-causing microorganisms. Life has been shaped by evolution since its origin, and the capacity for Darwinian evolution may be life's most fundamental property.

FIG. 1.15 **Artificial selection.** Selection over many centuries has resulted in remarkable variations among dogs. Charles Darwin called this "selection under domestication" and noted that it resembles selection that occurs in nature. *Source: © Rob Brodman 2011.*

The apples in the bin from which you made your choice didn't all look alike. Had you picked your apple in an orchard, you would have seen that different apples on the same tree looked different—some smaller, some greener, some misshapen, a few damaged by worms. Such variation is so commonplace that we scarcely pay attention to it. Variation is observed among individuals in virtually every species of organism. Variation that can be inherited provides the raw material on which evolution acts.

The causes of variation among individuals within a species are usually grouped into two broad categories. Variation among individuals is sometimes due to differences in the environment; this is called **environmental variation.** Among apples on the same tree some may have good exposure to sunlight; some may be hidden in the shade; some were lucky enough to escape the female codling moth, whose egg develops into a caterpillar that eats its way into the fruit. These are all examples of environmental variation.

The other main cause of variation among individuals is differences in the genetic material that is transmitted from parents to offspring; this is known as **genetic variation.** Differences among individuals' DNA can lead to differences among the individuals' RNA and proteins, which affect the molecular functions of the cell and ultimately can lead to physical differences that we can observe. Genetic differences among apples produce varieties whose mature fruits differ in taste and color, such as the green Granny Smith, the yellow Golden Delicious, and the scarlet Red Delicious.

But even on a single tree, each apple contains seeds that are genetically distinct because the apple tree is a sexual organism. Bees carry pollen from the flowers of one tree and deposit it in some of the flowers of another, enabling the sperm inside pollen grains to fertilize egg cells within that single flower. All the seeds on an apple tree contain shared genes from one parent, the tree on which they developed. But they contain distinct sets of genes contributed by sperm transported in pollen from other trees. In all sexual organisms, fertilization produces unique combinations of genes, which explains in part why sisters and brothers with the same parents can be so different from one another.

Genetic variation arises ultimately from mutations. Mutations arise either from random errors during DNA replication or from environmental factors such as ultraviolet (UV) radiation, which can damage DNA. If these mutations are not corrected, they are passed on to the next generation. To put a human face on this, consider that lung cancer can result from an environmental insult, such as cigarette smoking, or from a genetic susceptibility inherited from the parents.

In nature, most mutations that harm growth and reproduction die out after a handful of generations. Those that are neither harmful nor beneficial can persist for hundreds or thousands of generations. And those that are beneficial to growth and reproduction can gradually become incorporated into the genetic makeup of every individual in the species. That is how evolution works: The genetic makeup of a population changes over time.

Evolution predicts a nested pattern of relatedness among species, depicted as a tree.

Evolutionary theory predicts that new species arise by the divergence of populations through time from a common ancestor. As a result, closely related species are likely to resemble each other more closely than they do more distantly related species. You know this to be true from common experience. All of us recognize the similarity between a chimpanzee's face and body and our own (**Fig. 1.16a**), and biologists have long known that we share more features with chimpanzees than we do with any other species.

Humans and chimpanzees, in turn, share more features with gorillas than they do with any other species. And humans, chimpanzees, and gorillas share more features with orangutans than they do with any other species. And so on. We can continue to include a widening diversity of species, successively adding monkeys, lemurs, and other primates, to construct a set of evolutionary relationships that can be depicted as a tree (**Fig. 1.16b**). In this tree, the tips or branches on the

FIG. 1.16 Phylogenetic relationships among primates. (a) Humans share many features with chimpanzees. (b) Humans and chimpanzees, in turn, share more features with gorillas than they do with other species, and so on down through the evolutionary tree of primates. Treelike patterns of nested similarities are the predicted result of evolution. *Source: AP Photo/Bela Szandelszky.*

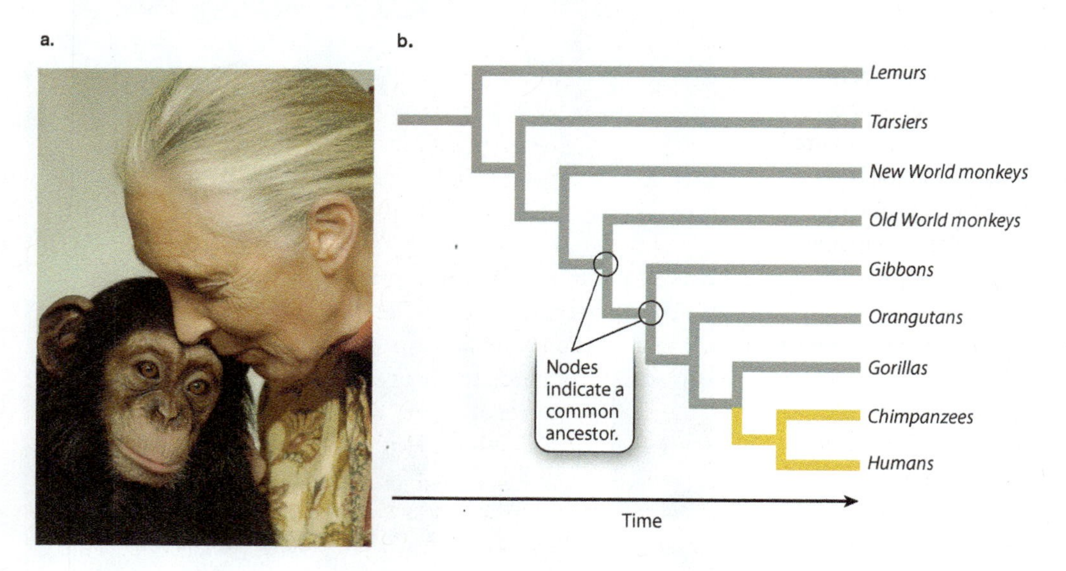

FIG. 1.17 The tree of life.

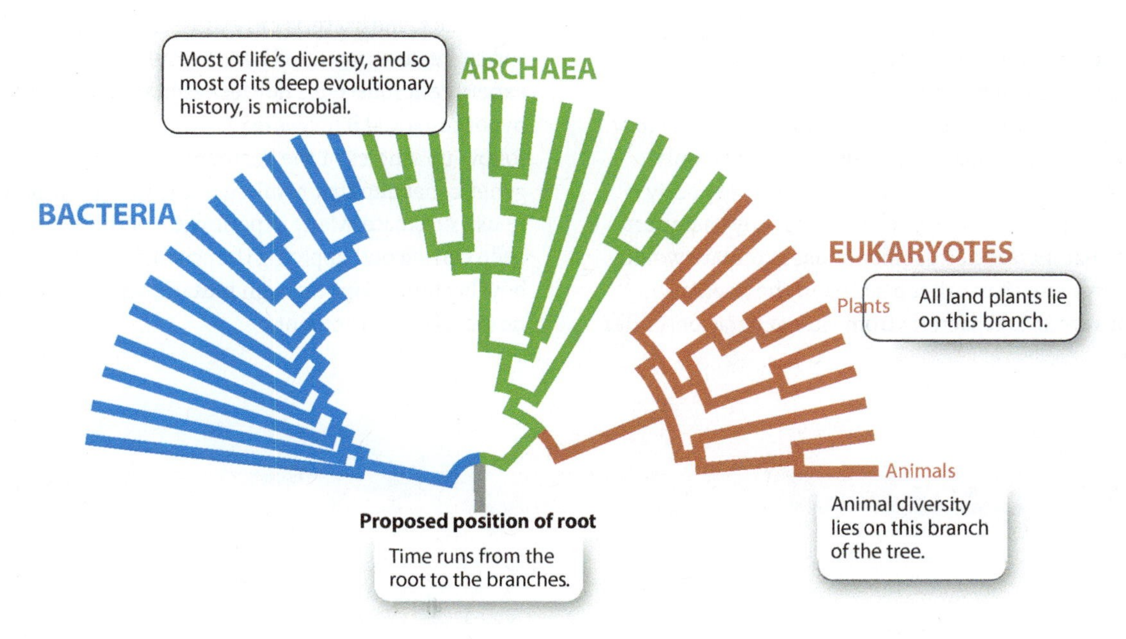

right represent different groups of organisms, nodes (where lines split) represent the most recent common ancestor, and time runs from left to right.

Evolutionary theory predicts that primates should show a nested pattern of similarity, and this is what morphological and molecular observations reveal. We can continue to add other mammals and then other vertebrate animals to our comparison, in the process generating a pattern of evolutionary relationships that forms a larger tree, with the primates confined to one limb. And using comparisons of DNA among species, we can generate still larger trees, ones that include plants as well as animals and the full diversity of microscopic organisms. Biologists call the full set of evolutionary relationships among all organisms the tree of life.

This tree, illustrated in **Fig. 1.17,** has three major branches representing the three domains mentioned earlier and is made up mostly of microorganisms. The last common ancestor of all living organisms, which form a root to the tree, is thought to lie between the branch leading to Bacteria and the branch leading to Archaea and Eukarya. The plants and animals so conspicuous in our daily existence make up only two branches on the eukaryotic limb of the tree.

The tree of life makes predictions for the order of appearance of different life-forms in the fossil record. For example, Fig. 1.16—one small branch on the greater tree—predicts that humans should appear later than monkeys, and that primates more akin to lemurs and tarsiers should appear even earlier. The greater tree also predicts that all records of animal life should be preceded by a long interval of microbial evolution. As we will see in subsequent chapters, these predictions are confirmed by the geologic record.

Shared features, then, sometimes imply inheritance from a common ancestor. In combination, molecular studies such as comparisons of DNA sequences and fossils show that the close similarity between humans and chimpanzees reflects descent from a common ancestor that lived about 6 million years ago. Their differences reflect what Darwin called "descent with modification"—evolutionary changes that have accumulated over time since the two lineages split. For example, as discussed in Chapter 24, the flat face of humans, our small teeth, and our upright posture all evolved within our ancestors after they diverged from the ancestors of chimpanzees. At a broader scale, the fundamental features shared by all organisms reflect inheritance from a common ancestor that lived billions of years ago. And the differences that characterize the many branches on the tree of life have formed through the continuing action of evolution *since* the time of our earliest ancestors.

Four decades ago, the geneticist Theodosius Dobzhansky wrote, "Nothing in biology makes sense except in the light of evolution." For this reason, evolution permeates discussions throughout this book, whether we are explaining the molecular biology of cells, how organisms function and reproduce, how species interact in nature, or the remarkable biological diversity of our planet.

Evolution can be studied by means of experiments.

Both the nested patterns of similarity among living organisms and the succession of fossils in the geologic record fit the predictions of evolutionary theory. Can we actually capture evolutionary processes in action? One way to accomplish this is in the laboratory. Bacteria are ideal for these experiments because they reproduce rapidly and can form populations with millions of individuals. Large population size means that mutations are likely to form in nearly every generation, even though the probability that any individual cell will acquire a mutation is small. (In contrast to bacteria, think about trying evolutionary experiments on elephants!)

One such experiment is illustrated in **Fig. 1.18.** The microbiologists Santiago Elena and Richard Lenski grew populations of the common intestinal bacterium *Escherichia coli* in liquid medium, with the organic acid succinate as the only source of food. In general, *E. coli* cells feed on succinate poorly if at all, leading the researchers to hypothesize that any bacterium with a mutation that increased its ability to use succinate would reproduce at a faster rate than other bacteria in the population. The key questions were: Did bacteria from later generations differ from those of earlier generations—that is, did evolution occur?—and did the bacteria evolve an improved ability to use succinate?

In fact, the bacteria did evolve an improved ability to use succinate, demonstrated by the results shown in Fig. 1.18. This experiment illustrates how experiments can be used to test hypotheses and it shows evolution in action. Furthermore, follow-up studies of the bacterial DNA identified differences in genetic makeup that resulted in the improved ability of *E. coli* to use succinate. Many experiments of this general type have been carried out, applying the scientific method to demonstrate how bacteria adapt through mutation and natural selection to any number of environments.

HOW DO WE KNOW?

FIG. 1.18

Can evolution be demonstrated in the laboratory?

BACKGROUND *Escherichia coli* is an intestinal bacterium commonly used in the laboratory. It grows poorly in liquid media where the only source of food is succinate. Santiago Elena and Richard Lenski wondered whether *E. coli* grown for 20,000 generations in succinate would evolve in ways that improved their ability to metabolize this compound.

HYPOTHESIS Any bacterium with a random mutation that increases its ability to utilize succinate will reproduce at a faster rate than other bacteria in the population. Over time, such a mutant will increase in frequency relative to other types of bacteria, thereby demonstrating evolution in a bacterial population.

EXPERIMENT Cells of *E. coli* can be frozen in liquid nitrogen, which keeps them in a sort of suspended animation in which no biological processes take place, but the cells survive. At the beginning of the experiment, the researchers froze a large number of samples of the starting bacteria ("Ancestral"). As the experiment progressed, they took samples of the bacterial populations at intervals ("Later") and grew them together with a thawed sample of the starting bacteria in succinate. They then compared the rate of growth of the ancestral bacteria with that of the bacteria taken at later time points.

RESULTS At each time interval, the cells from later time points grew more rapidly than the ancestral cells when the two populations were grown together in succinate.

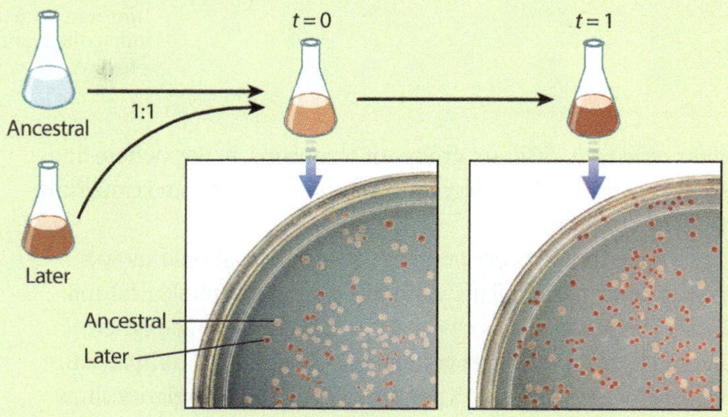

Source: Neerja Hajela, Michigan State University.

Over time, the bacteria grown in succinate showed more and more improvement in growth compared to the starting bacteria.

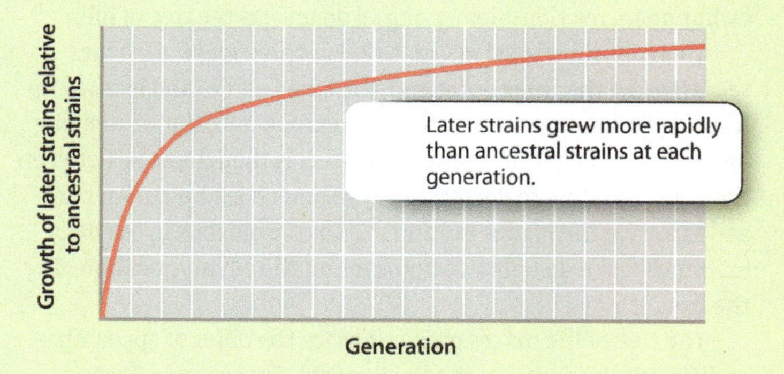

CONCLUSION Evolution occurred in the population: The bacteria evolved an improved ability to metabolize succinate.

FOLLOW-UP WORK Other experimental studies have shown that bacteria are able to evolve adaptations to a wide variety of environmental conditions.

SOURCE (top) Elena, S. F., and R. E. Lenski. 2003. "Evolution Experiments with Microorganisms." *Nature Review Genetics* 4:457–469.

Experiments in laboratory evolution help us to understand how life works, and they have an immensely important practical side as well. They allow biologists to develop new and beneficial strains of microorganisms that, for example, remove toxins from lakes and rivers. And they show how some of our worst pathogens develop resistance to drugs designed to eliminate them.

→ **Quick Check 4** How might the heavy-handed use of antibiotics result in the increase of antibiotic-resistant cells in bacterial populations?

1.5 ECOLOGICAL SYSTEMS

Watching a movie, we don't need a narrator to tell us whether the action is set in a tropical rain forest, the African savanna, or Arctic tundra. The plants in the scene make it obvious. Plants can be linked closely with environment—palm trees with the tropics, for example, or cacti with the desert—because the environmental distributions of different species reflect the sum of their biological features. Palms cannot tolerate freezing and so are confined to warmer environments; the cactus can store water in its tissues, enabling it to withstand prolonged drought. But the geographic ranges of palms and cacti reflect more than just their physical tolerances. They are also strongly influenced by interactions with other species, including other plants that compete for the limited resources available for growth, animals that feed on them or spread their pollen, and microorganisms that infest their tissues. **Ecology** is the study of how organisms interact with one another and with their physical environment in nature.

Basic features of anatomy, physiology, and behavior shape ecological systems.

Apple trees, we noted, reproduce sexually. How is this accomplished given that the trees have no moving parts? The answer is that bees carry pollen from one flower to the next, enabling sperm carried within the pollen grain to fuse with an egg cell protected within the flower (**Fig. 1.19**). The process is much like the one discussed at the beginning of the chapter in which hummingbirds carry pollen from flower to flower. In both examples, the plants complete their life cycles by exploiting close interactions with animal species. Birds and bees visit flowers, attracted, as we are, by their color and odor. Neither visits flowers with the intent to pollinate; they come in expectation of a meal. As the birds and bees nestle into flowers to collect nutritious nectar, pollen rubs off on their bodies.

But if animals help apples to fertilize their eggs, how do apple trees disperse from one site to another? Again, the plants rely on animals—the apple's flesh attracts mammals that eat the fruit, seeds and all, and spread the seeds through defecation. Humans tend to be finicky—we carefully eat only the sweet outer flesh, discarding the seedy core. Nonetheless, even as humans have selected apples for their quality of fruit, we have dispersed apple trees far beyond their natural range.

The many ways that organisms interact in nature reflect their basic functional requirements. If a plant is to grow and reproduce, for example, it must have access to light, carbon dioxide, water, and basic nutrients. All plants have these requirements, so one plant's success in gathering nutrients may mean that fewer nutrients are available for its neighbor—the plants may compete for limited resources needed for growth (**Fig. 1.20a**). Animals in the same area require organic molecules for nutrition, and some of them may eat the plant's leaves or bark (**Fig. 1.20b**). Predation also benefits some organisms at the expense of others. In combination, the physical tolerances of organisms and the ways that organisms interact with one another determine the structure and diversity of communities.

FIG. 1.19 Bee pollination. Many plants reproduce sexually by exploiting the behavior of animals. Here a honeybee pollinates an apple flower. *Source: Donald Specker/ Animals Animals – Earth Scenes.*

FIG. 1.20 Ecological relationships. (a) The trees in a rain forest all require water and nutrients. Neighboring plants compete for the limited supplies of these materials. (b) Plants provide food for many animals. Leaf-cutter ants cut slices of leaves from tropical plants and transport them to their nests, where the leaves grow fungi that the ants eat. *Sources: a. Louise Murray/Getty Images; b. Gail Shumway/Getty Images.*

In short, ecological relationships reflect the biomechanical, physiological, and behavioral traits of organisms in nature. Form and function, in turn, arise from molecular processes within cells, governed by the expression of genes.

Ecological interactions play an important role in evolution.

G. Evelyn Hutchinson, one of the founders of modern ecology, wrote a book called *The Ecological Theater and the Evolutionary Play* (1965). This wonderful title succinctly captures a key feature of biological relationships. It suggests that ecological communities provide the stage on which the play of evolution takes place.

As an example, let's look again at plants competing for resources. As a result of this competition, natural selection may favor plants that have more efficient uptake of nutrients or water. In the example of animals eating plants, natural selection may favor animals with greater jaw strength or more efficient extraction of nutrients in the digestive system. In turn, plants that avoid predation by synthesizing toxic compounds in their leaves may gain the upper hand. In each case, interactions between organisms lead to the evolution of particular traits.

To take another example from mammal–plant interactions: Selection for fleshy fruits improved the dispersal of apples because it increased the attractiveness of these seed-bearing structures to hungry mammals. Tiny yeasts make a meal on sugary fruits as well, in the process producing alcohol that deters potential competitors for the food. Already in prehistoric times, humans had learned to harness this physiological capability of yeasts, and for this reason the total abundance and distribution of *Vitis vinifera,* the wine grape, has increased dramatically through time.

1.6 THE HUMAN FOOTPRINT

The story of life has a cast of millions, with humans playing only one of many roles in an epic 4 billion years in the making. Our own species, *Homo sapiens*, has existed for only the most recent 1/200 of 1% of life's history, yet there are compelling reasons to pay special attention to ourselves. We want to understand how our own bodies work and how humans came to be: Curiosity about ourselves is after all a deeply human trait.

We also want to understand how biology can help us conquer disease and improve human welfare. Epidemics have decimated human populations throughout history. The Black Death—bubonic plague caused by the bacterium *Yersinia pestis*—is estimated to have killed half the population of Europe in the fourteenth century. Casualties from the flu pandemic of 1918 exceeded those of World War I. Even King Tut, we now know, suffered from malaria in his Egyptian palace approximately 3300 years ago. Throughout this book, we discuss how basic biological principles are helping scientists to prevent and cure the great diseases that have persisted since antiquity, as well as modern ones such as AIDS and Ebola.

Furthermore, we need to understand the evolutionary and ecological consequences of a human population that now exceeds 7 billion. All species affect the world around them. However, in the 21st century human activities have taken on special importance because our numbers and technological abilities make our footprint on Earth's ecology so large. Human activities now emit more carbon dioxide than do volcanoes, through industrial processes we convert more atmospheric nitrogen to ammonia than nature does, and we commandeer, either directly or indirectly, as much as 25% of all photosynthetic production on land. To chart our environmental future, we need to understand our role in the Earth system as a whole.

And, as we have become major players in ecology, humans have become important agents of evolution. As our population has expanded, some species have expanded along with us. We've seen how agriculture has sharply increased the abundance and distribution of grapes, and the same is true for corn, cows, and apple trees. At the same time, we have inadvertently helped other species to expand—the crowded and not always clean environments of cities provide excellent habitats for cockroaches and rats (**Fig. 1.21**).

Other species, however, are in decline, their populations reduced by hunting and fishing, changes in land use, and other

FIG. 1.21 **Species that have benefited from human activity, including (a) corn, (b) rats, and (c) cockroaches.**

Sources: a. Fred Dimmick/iStockphoto; b. Arndt Sven-Erik/age footstock; c. Nigel Cattlin/Alamy.

FIG. 1.22 Extinct and endangered species. Humans have caused many organisms to become extinct, such as (a) the dodo, (b) passenger pigeon, (c) Bali tiger, and (d) dusky seaside sparrow. Burgeoning human populations have diminished the ranges of many others, including (e) the white rhinoceros. *Sources: a. Science Source; b. G. I. Bernard/Science Source; c. Look and Learn/Bridgeman Images; d. P.W. Sykes/U.S. Fish and Wildlife Service; e. James Warwick/Science Source.*

human activities. When Europeans first arrived on the Indian Ocean island of Mauritius, large flightless birds called dodos were plentiful (**Fig. 1.22**). Within a century, the dodo was extinct. Early Europeans in North America were greeted by vast populations of passenger pigeons, more than a million in a single flock. By the early twentieth century, the species was extinct, a victim of hunting and habitat change through expanding agriculture. Other organisms both great (the Bali tiger) and small (the dusky seaside sparrow) have become extinct in recent decades. Still others are imperiled by human activities; the magnificent white rhinoceros of Africa is threatened by both habitat destruction and poaching. Whether any rhinos will exist at the end of this century will depend almost entirely on decisions we make today.

Throughout this book, we return to the practical issues of life science. How can we use the principles of biology to improve human welfare, and how can we live our lives in ways that control our impact on the world around us? The answers to the questions critically depend on understanding biology in an *integrated* fashion. While it is tempting to consider molecules, cells, organisms, and ecosystems as separate entities, they are inseparable in nature. To tackle biological problems, whether building an artificial cell, stopping the spread of infectious diseases such as HIV or malaria, feeding a growing population, or preserving endangered habitats and species, we need an integrated perspective. In decisively important ways, our future welfare depends on improving our knowledge of how life works. ■

Core Concepts Summary

1.1 The scientific method is a deliberate way of asking and answering questions about the natural world.

Observations are used to generate a hypothesis, a tentative explanation that makes predictions that can be tested. page 4

On the basis of a hypothesis, scientists design experiments and make additional observations that test the hypothesis. page 5

A controlled experiment typically involves several groups in which all the conditions are the same and one group where a variable is deliberately introduced in order to determine if that variable has an effect. page 5

If a hypothesis is supported through continued observation and experiments over long periods of time, it is elevated to a theory, a sound and broad explanation of some aspect of the world. page 6

1.2 Life works according to fundamental principles of chemistry and physics.

The living and nonliving worlds follow the same chemical rules and obey the same physical laws. page 8

Experiments by Redi in the 1600s and Pasteur in the 1800s demonstrated that organisms come from other organisms and are not spontaneously generated. page 10

Life originated on Earth about 4 billion years ago, arising from nonliving matter. page 11

1.3 The fundamental unit of life is the cell.

The cell is the simplest biological entity that can exist independently. page 12

Information in a cell is stored in the form of the nucleic acid DNA. page 12

The central dogma describes the usual flow of information in a cell, from DNA to RNA to protein. page 13

The plasma membrane is the boundary that separates the cell from its environment. page 14

Cells with a nucleus are eukaryotes; cells without a nucleus are prokaryotes. page 14

Metabolism is the set of chemical reactions in cells that build and break down macromolecules and harness energy. page 14

A virus is an infectious agent composed of a genome and protein coat that uses a host cell to replicate. page 15

1.4 Evolution explains the features that organisms share and those that set them apart.

When there is variation within a population of organisms, and when that variation can be inherited, the variants best able to grow and reproduce in a particular environment will contribute disproportionately to the next generation, leading to a change in the population over time, or evolution. page 15

Variation can be genetic or environmental. The ultimate source of genetic variation is mutation. page 16

Organisms show a nested pattern of similarity, with humans more similar to primates than other organisms, primates more similar to mammals, mammals more similar to vertebrates, and so on. page 16

Evolution can be demonstrated by laboratory experiments. page 18

1.5 Organisms interact with one another and with their physical environment, shaping ecological systems that sustain life.

Ecology is the study of how organisms interact with one another and with their physical environment in nature. page 19

These interactions are driven in part by the anatomy, physiology, and behavior of organisms, that is, the basic features of organisms shaped by evolution. page 19

1.6 In the 21st century, humans have become major agents in ecology and evolution.

Humans have existed for only the most recent 1/200 of 1% of life's 4-billion-year history. page 20

In spite of our recent arrival, our growing numbers are leaving a large ecological and evolutionary footprint. page 21

Solving biological problems requires an integrated understanding of life, with contributions from all the fields of biology, including molecular biology, cell biology, genetics, organismal biology, and ecology, as well as from chemistry, physics, and engineering. page 22

Self-Assessment

1. Name and summarize the steps in the scientific method.
2. Differentiate among a guess, a hypothesis, and a theory.
3. Describe the difference between a test group and a control group, and explain why they are important.

4. State the first and second laws of thermodynamics and describe how they apply to living organisms.
5. Describe what it means to say that a cell is life's functional unit.
6. Describe the experimental evidence that demonstrates that living organisms come from other living organisms.
7. Explain how evolution accounts for both the unity and the diversity of life.
8. Name and describe several features that determine the shape of ecological systems.
9. Name three ways that humans have affected life on Earth.
10. Summarize the six themes that are discussed in this chapter.

Log in to LaunchPad to check your answers to the Self-Assessment questions, and to access additional learning tools.

CASE 1

The First Cell

Life's Origins

Deep underground, in Mexico's Cueva de Villa Luz, the cave walls drip with slime. The rocky surfaces are teeming with colonies of mucus-producing bacteria. No sunlight reaches these organisms far beneath Earth's surface. Instead, the bacteria survive by capturing energy released as they oxidize hydrogen sulfide gas that exists within the cave. As a by-product of that reaction, the microbes produce sulfuric acid, making the slime that oozes from the cave walls—dubbed "snottites" by researchers—as corrosive as battery acid.

Snottites might be stomach turning, but they're intriguing, too. The organisms that produce snottites are called extremophiles because they live in places where humans and most other animals cannot survive. Such microorganisms may tell us something about life when Earth was young.

All cells require an archive of information, a membrane to separate the inside of the cell from its surroundings, and the ability to gather materials and harness energy from the environment.

From cave-dwelling bacteria to 100-ton blue whales, the diversity of life on Earth is astounding. Yet all of our planet's organisms, living and extinct, exist on branches of the same family tree. Bacteria that produce snottites, swordfish, humans, hydrangeas—all evolved from a single common ancestor.

When and how life originated are some of the biggest questions in biology. Earth is nearly 4.6 billion years old. Chemical evidence from 3.5-billion-year-old rocks in Australia suggests that biologically driven carbon and sulfur cycles existed at the time those rocks were formed. In the eons since, the first primitive life-forms have evolved into the millions of different species that populate the planet today.

How did the first living cell arise? Scientists generally accept that life arose from nonliving materials—a process called abiogenesis—and thousands of laboratory experiments performed over the past sixty years provide glimpses of how this might have occurred. In our modern world, the features that separate life from nonlife are relatively easy to discern. But Earth's first organisms were almost certainly much less complicated than even the simplest bacteria alive today. And before those first truly living things appeared, molecular systems presumably existed that hovered somewhere between the living and the nonliving.

All cells require an archive of information, a membrane to separate the inside of the cell from its surroundings, and the ability to gather materials and harness energy from the environment. In modern organisms, the cell's information archive is DNA, the double-stranded molecule that contains the instructions needed for cells to grow, differentiate, and reproduce. Without that molecular machinery, life as we know it would not exist.

DNA is critical, and it's complex. Among the organisms alive today, the smallest known genome belongs to the bacterium *Carsonella rudii.* Even that genome contains nearly 160,000 DNA base pairs. How could such sophisticated molecular systems have arisen?

The likely answer to that question is: step by step. Laboratory experiments have shown how precursors to nucleic acids might have come together under chemical conditions present on the young Earth. It's exceedingly unlikely that a molecule as complex as DNA was employed by the very first living cells. As you'll see in the chapters that follow, scientists have gathered evidence suggesting

Snottites deep underground in a Mexican cave. These stalactite-like slime formations are produced by bacteria that gain energy by reacting hydrogen sulfide gas with oxygen. *Source: Kenneth Ingham Photography.*

that RNA, rather than DNA, stored information in early cells and, indeed, did much more than that, catalyzing chemical reactions much as proteins do today.

While some kind of information archive was necessary for life to unfold, there is more to the story. Living things must have a barrier that separates them from their environment. All cells, whether found as single-celled bacteria or by the trillions in trees or humans, are individually encased in a cell membrane.

Once again, scientists can only guess at how the first cell membranes came about, but research shows that the molecules that make up modern membranes possess properties that may have led them to form spontaneously on the early Earth. At first, the membranes were probably quite simple—straightforward (but leaky) barriers that kept the contents of early cells separated from the world at large. Over time, as chance variations arose, those membranes that provided a better barrier were favored by natural selection. Moreover, proteins became embedded in membranes, providing gates or channels that regulated the transport of ions and small molecules into and out of the cell.

A third essential characteristic of living things is the ability to harness energy from the environment. Here, too, it's feasible that a series of natural chemical processes led to entities that could achieve this feat. Simple reactions that produced molecular by-products would have enabled more complex reactions down the road. Ultimately, that collection of reactions—combined with an archive of information and enclosed in some kind of primitive membrane—evolved into individual units that could grow, reproduce, and evolve.

Such a series of events may sound unlikely. However, some scientists argue that given the chemicals present on the early Earth, it was likely—and even inevitable—that they would come together in such a way that life would emerge. Indeed, relatively simple, naturally occurring materials such as metal ions have been shown to play a role in key cellular reactions. Billions of years after the first cells arose, some of those metal ions—such as complexes of iron and sulfur—still play a critical role in cells.

That's one reason researchers are so interested in studying organisms found today in extreme environments. The sulfur-hungry snottites in the Cueva de Villa Luz help us to understand life 1–2 billion years ago, when oxygen was less plentiful and hydrogen sulfide more abundant. Still earlier, when life arose about 4 billion years ago, the planet's atmosphere contained no oxygen—humans couldn't survive in such a world and neither could snottite bacteria. However, by studying modern extremophiles living in oxygen-free

A hydrothermal vent. Some scientists think that this type of vent provided a favorable environment for chemical reactions that led to the origin of life. *Source: Image courtesy of New Zealand American Submarine Ring of Fire 2007 Exploration, NOAA Vents Program, the Institute of Geological & Nuclear Sciences and NOAA-OE.*

environments, scientists may uncover clues about how Earth's first cells came together and functioned.

Did life arise just once? Or could it have started up and died out several times before it finally got a foothold? If, given Earth's early chemistry, life here was inevitable, could it have arisen elsewhere in the universe? The study of life's origins produces many more questions than answers—and not just for biologists. The mystery of life spans the fields of biology, chemistry, physics, and planetary science. Though the questions are vast, our understanding of life's origins is likely to come about the same way life itself arose: step by step.

? CASE 1 QUESTIONS

Special sections in Chapters 2–8 discuss the following questions related to Case 1.

1. **How did the molecules of life form?** See page 45.
2. **What was the first nucleic acid molecule, and how did it arise?** See page 59.
3. **How did the genetic code originate?** See page 83.
4. **How did the first cell membranes form?** See page 91.
5. **What naturally occurring elements might have spurred the first reactions that led to life?** See page 128.
6. **What were the earliest energy-harnessing reactions?** See page 139.
7. **How did early cells meet their energy requirements?** See page 146.
8. **How did early cells use sunlight to meet their energy requirements?** See page 170.

CHAPTER 2

The Molecules of Life

Core Concepts

2.1 The atom is the fundamental unit of matter.

2.2 Atoms can combine to form molecules linked by chemical bonds.

2.3 Water is essential for life.

2.4 Carbon is the backbone of organic molecules.

2.5 Organic molecules include proteins, nucleic acids, carbohydrates, and lipids, each of which is built from simpler units.

2.6 Life likely originated on Earth by a set of chemical reactions that gave rise to the molecules of life.

Science Photo Library/Alamy.

When biologists speak of diversity, they commonly point to the 2 million or so species named and described to date, or to the 10–100 million living species thought to exist in total. Life's diversity can also be found at a very different level of observation: in the molecules that make up each and every cell. Life depends critically on many essential functions, including storing and transmitting genetic information, establishing a boundary to separate cells from their surroundings, and harnessing energy from the environment. These functions ultimately depend on the chemical characteristics of the molecules that make up organisms.

In spite of the diversity of molecules and functions, the chemistry of life is based on just a few types of molecule, which in turn are made up of just a few elements. Of the 100 or so chemical elements, only about a dozen are found in more than trace amounts in living organisms. These elements interact with one another in only a limited number of ways. So, the question arises: How is diversity generated from a limited suite of chemicals and interactions? The answer lies in some basic features of chemistry.

2.1 PROPERTIES OF ATOMS

Since antiquity, it has been accepted that the materials of nature are made up of a small number of fundamental substances combined in various ways. Today, we call these substances **elements.** From the seventeenth century through the end of the nineteenth, elements were defined as pure substances that could not be broken down further by the methods of chemistry. In time, it was recognized that each element contains only one type of **atom,** the basic unit of matter. By 1850, about 60 elements were known, including such common ones as oxygen, copper, gold, and sodium. Today, 118 elements are known. Of these, 94 occur naturally and 24 have been created artificially in the laboratory. Elements are often indicated by a chemical symbol that consists of a one- or two-letter abbreviation of the name of the element. For example, carbon is represented by C, hydrogen by H, and helium by He.

Atoms consist of protons, neutrons, and electrons.

Elements are composed of atoms. The atom contains a dense central **nucleus** made up of positively charged particles called **protons** and electrically neutral particles called **neutrons.** A third type of particle, the negatively charged **electron,** moves around the nucleus at some distance from it. Carbon, for example, typically has six protons, six neutrons, and six electrons (**Fig. 2.1**). The number of protons, or the atomic number, specifies an atom as a particular element. An atom with one proton is hydrogen, for example, and an atom with six protons is carbon.

Together, the protons and neutrons determine the **atomic mass,** the mass of the atom. Each proton and neutron, by definition, has a mass of 1, whereas an electron has negligible mass. The number of neutrons in atoms of a particular element can vary, changing its mass. **Isotopes** are atoms of the same element that have different numbers of neutrons. For example, carbon has three isotopes: About 99% of carbon atoms have six neutrons and six protons, for an atomic mass of 12; about 1% have seven neutrons and six protons, for an atomic mass of 13; and only a very small fraction have eight neutrons and six protons, for an atomic mass of 14. The atomic mass is sometimes indicated as a superscript to the left of the chemical symbol. For instance, ^{12}C is the isotope of carbon with six neutrons and six protons.

FIG. 2.1 **A carbon atom.** Each carbon atom has six protons, six neutrons, and six electrons. The net charge of any atom is neutral because there are as many electrons as protons.

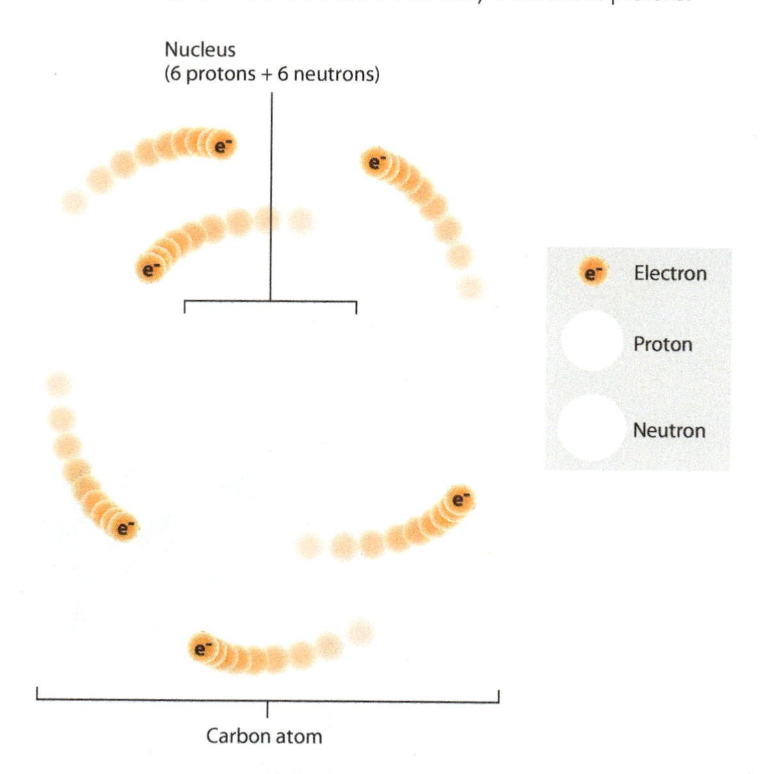

Typically, an atom has the same number of protons and electrons. Because a carbon atom has six protons and six electrons, the positive and negative charges cancel each other out and the carbon atom is electrically neutral. Some chemical processes cause an atom to either gain or lose electrons. An atom that has lost an electron is positively charged, and one that has gained an electron is negatively charged. Electrically charged atoms are called **ions.** The charge of an ion is specified as a superscript to the right of the chemical symbol. Thus, H^+ indicates a hydrogen ion that has lost an electron and is positively charged.

Electrons occupy regions of space called orbitals.

Electrons move around the nucleus, but not in the simplified way shown in Fig. 2.1. The exact path that an electron takes cannot

FIG. 2.2 Electron orbitals and energy levels (shells) for hydrogen and carbon. The orbital of an electron can be visualized as a cloud of points that is more dense where the electron is more likely to be. The hydrogen atom contains a single orbital, in a single energy level (a and c). The carbon atom has five orbitals, one in the first energy level and four in the second energy level (b and c).

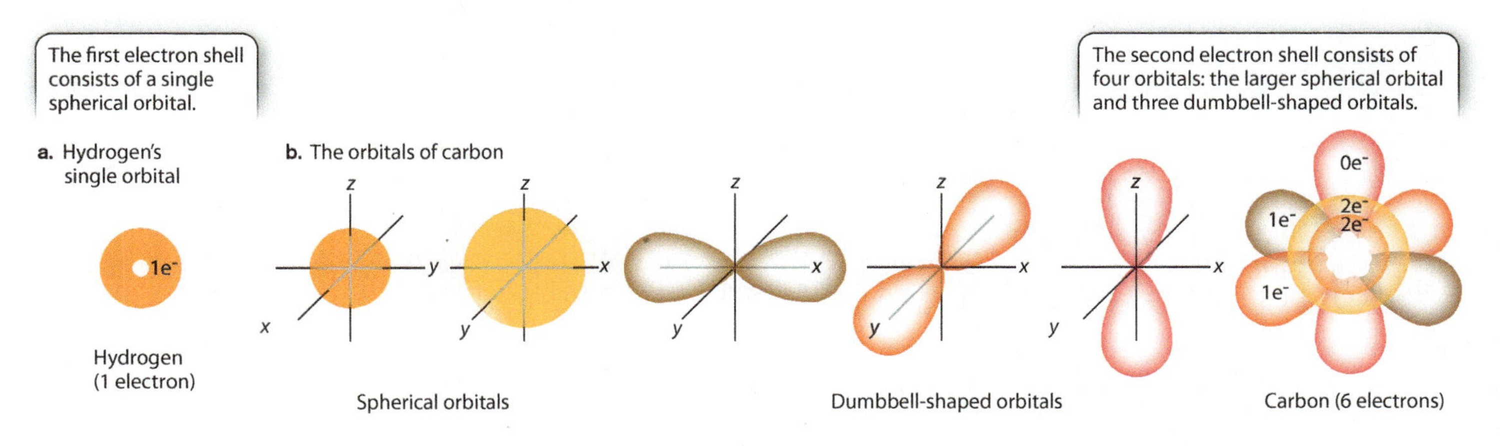

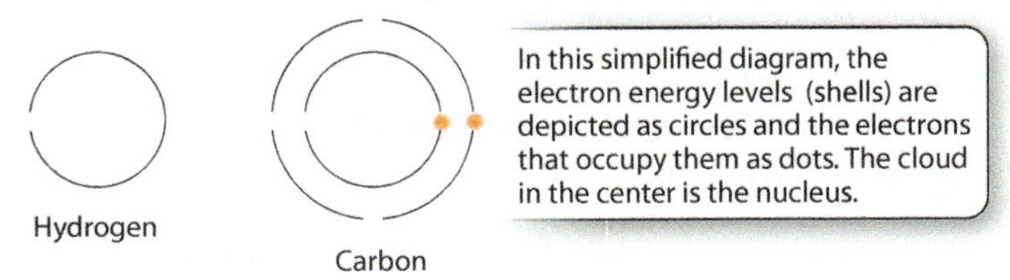

be known, but it is possible to identify a region in space, called an **orbital**, where an electron is present most of the time. For example, **Fig. 2.2a** shows the orbital for hydrogen, which is simply a sphere occupied by a single electron. Most of the time, the electron is found within the space defined by the sphere, although its exact location at any instant is unpredictable.

The maximum number of electrons in any orbital is two. Most atoms have more than two electrons and so have several orbitals positioned at different distances from the nucleus. These orbitals differ in size and shape. Electrons in orbitals close to the nucleus have less energy than do electrons in orbitals farther away, so electrons fill up orbitals close to the nucleus before occupying those farther away. Several orbitals can exist at a given energy level, or **shell.** The first shell consists of the spherical orbital shown in Fig. 2.2a.

Fig. 2.2b shows electron orbitals for carbon. Of carbon's six electrons, two occupy the small spherical orbital representing the lowest energy level. The remaining four are distributed among four possible orbitals at the next highest energy level: One of these four orbitals is a sphere (larger in diameter than the orbital at the lowest energy level) and three are dumbbell-shaped. In carbon, the outermost spherical orbital has two electrons, two of the dumbbell-shaped orbitals have one electron each, and one of the dumbbell-shaped orbitals is empty. Because a full orbital contains two electrons, it would take a total of four additional electrons to completely fill all of the orbitals at this energy level. Therefore, after the first shell, the maximum number of electrons per energy level is eight. **Fig. 2.2c** shows simplified diagrams of atoms in which the highest energy level, or shell, of hydrogen, represented by the outermost circle, contains one electron, and that of carbon contains four electrons.

→ **Quick Check 1** In the early 1900s, Ernest Rutherford produced a beam of very small positive particles and directed it at a thin piece of gold foil just a few atoms thick. Most of the particles passed through the foil without changing their path; very rarely, a particle was deflected. What conclusions can you draw from this experiment about the structure of an atom?

Elements have recurring, or periodic, chemical properties.

The chemical elements are often arranged in a tabular form known as the **periodic table of the elements,** shown in **Fig. 2.3** and generally credited to the nineteenth-century Russian chemist Dmitri Mendeleev. The table provides a way to organize all the chemical elements in terms of their chemical properties.

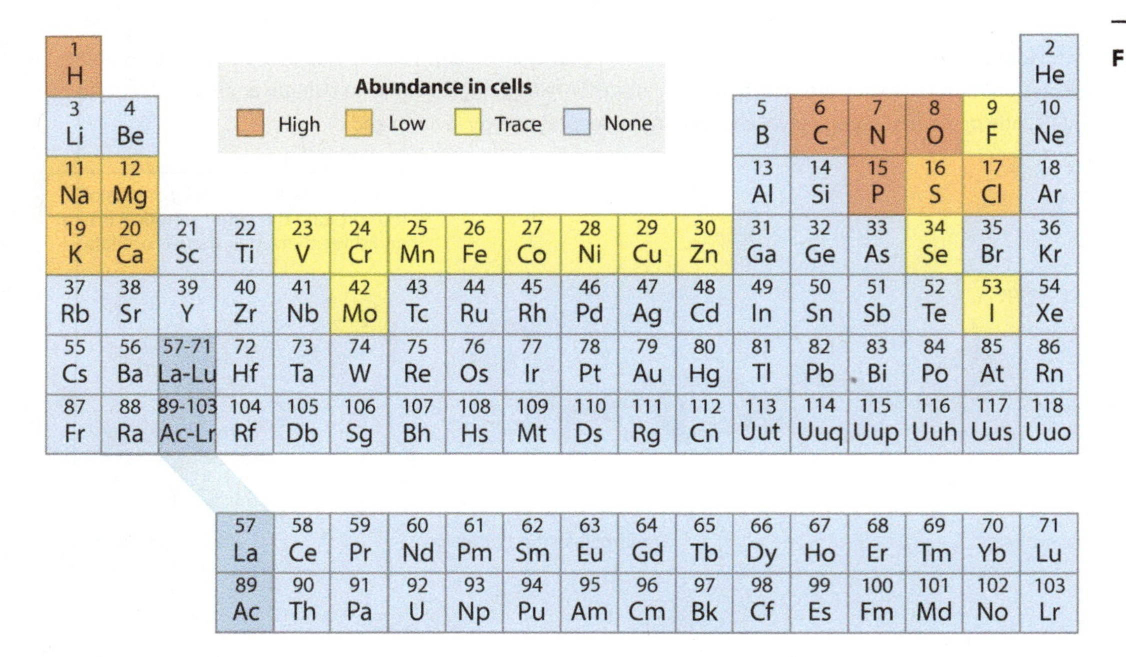

FIG. 2.3 The periodic table of the elements. Elements are arranged by increasing number of protons, the atomic number. The elements in a column share similar chemical properties.

In the periodic table, the elements are indicated by their chemical symbols and arranged in order of increasing atomic number. For example, the second row of the periodic table begins with lithium (Li) with 3 protons and ends with neon (Ne) with 10 protons.

For the first three horizontal rows in the periodic table, elements in the same row have the same number of shells, and so also have the same number and types of orbitals available to be filled by electrons. Across a row, therefore, electrons fill the shell until a full complement of electrons is reached on the right-hand side of the table. **Fig. 2.4** shows the filling of the shells for elements in the second row of the periodic table.

The elements in a vertical column are called a group or family. Members of a group all have the same number of electrons in their outermost shell. For example, carbon (C) and lead (Pb) both have four electrons in their outermost shell. The number of electrons in the outermost shell determines in large part how elements interact with other elements to form a diversity of molecules, as we will see in the next section.

2.2 MOLECULES AND CHEMICAL BONDS

Atoms can combine with other atoms to form **molecules,** which are groups of two or more atoms attached together. that act as a single unit. When two atoms form a molecule, the individual atoms interact through what is called a **chemical bond,** a form of attraction between atoms that holds them together. The ability of atoms to form bonds with other atoms explains in part why just a few types of element can come together in many different ways to make a variety of molecules that can carry out diverse functions in a cell. There are several ways in which atoms can interact with one another, and therefore many different types of chemical bond.

FIG. 2.4 Energy levels (shells) of row 2 of the periodic table. The complete complement of electrons in the outer shell of this row of elements is eight.

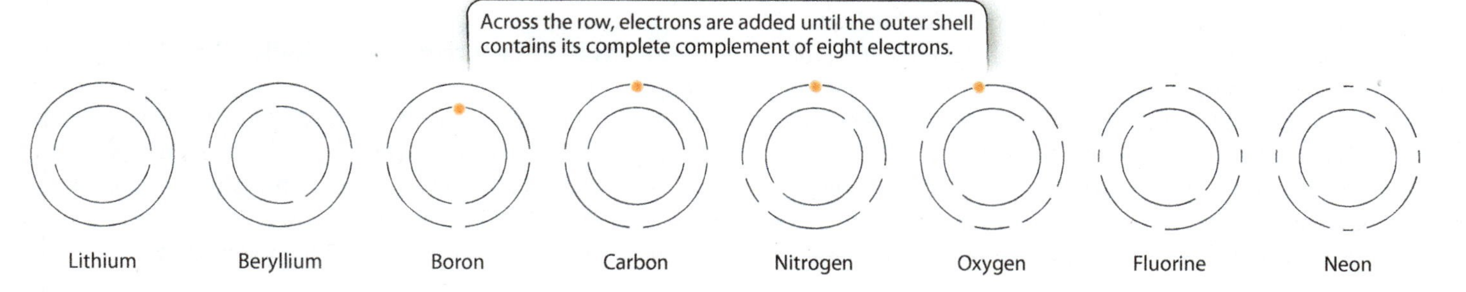

A covalent bond results when two atoms share electrons.

The ability of atoms to combine with other atoms is determined in large part by the electrons farthest from the nucleus—those in the outermost orbitals of an atom. These electrons are called **valence electrons,** and they are at the highest energy level. When atoms combine with other atoms to form a molecule, the atoms share valence electrons with each other. Specifically, when the outermost orbitals of two atoms come into proximity, two atomic orbitals each containing one electron merge into a single orbital containing a full complement of two electrons. The merged orbital is called a **molecular orbital,** and each shared pair of electrons constitutes a **covalent bond** that holds the atoms together.

We can represent a specific molecule by its chemical formula, which is written as the letter abbreviation for each element followed by a subscript giving the number of that type of atom in the molecule. Among the simplest molecules is hydrogen gas, illustrated in **Fig. 2.5,** which consists of two covalently bound hydrogen atoms indicated by the chemical formula H_2. Each hydrogen atom has a single electron in a spherical orbital. When the atoms join into a molecule, the two orbitals merge into a single molecular orbital containing two electrons that are shared by the hydrogen atoms. A covalent bond between atoms is denoted by a single line connecting the two chemical symbols for the atoms, as shown in the structural formula in Fig. 2.5.

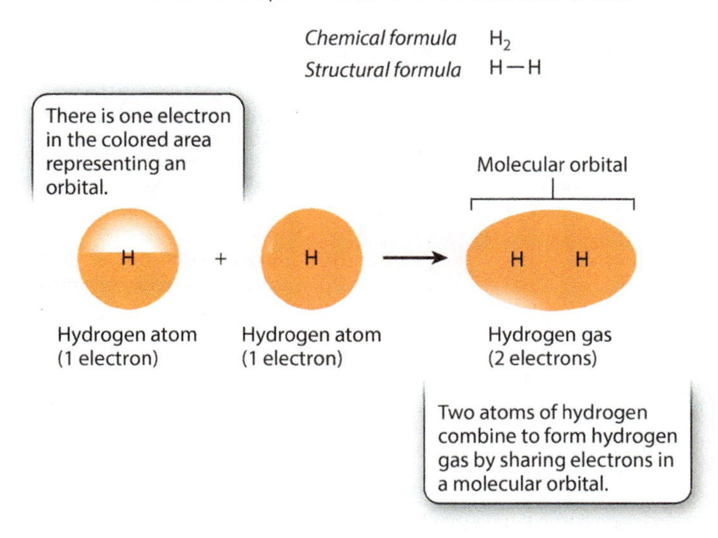

FIG. 2.5 A covalent bond. A covalent bond is formed when two atoms share a pair of electrons in a molecular orbital.

→ **Quick Check 2** From their positions in the periodic table (see Fig. 2.3), can you predict how many lithium atoms and hydrogen atoms can combine to form a molecule?

Two adjacent atoms can sometimes share two pairs of electrons, forming a **double bond** denoted by a double line connecting the two chemical symbols for the atoms. In this case, four orbitals occupied by a single electron merge to form two molecular orbitals.

Molecules tend to be most stable when the two atoms forming a bond share enough electrons to completely occupy the outermost energy level or shell. The outermost shell of a hydrogen atom can hold two electrons, whereas for carbon, nitrogen, and oxygen atoms the number is eight. This simple rule for forming stable molecules is known as the octet rule and applies to many, but not all, elements. For example, as shown in **Fig. 2.6,** one carbon atom (C, with four valence electrons) combines with four hydrogen atoms (H, with one valence electron each) to form CH_4 (methane); nitrogen (N, with five valence electrons) combines with three H atoms to form NH_3 (ammonia); and oxygen (O, with six valence electrons) combines with two H atoms to form H_2O (water). Interestingly, the elements of the same column in the next row behave similarly. This is just one example of the recurring, or periodic, behavior of the elements.

A polar covalent bond is characterized by unequal sharing of electrons.

In hydrogen gas (H_2), the electrons are shared equally by the two hydrogen atoms. In many bonds, however, the electrons are not shared equally by the two atoms. A notable example is provided by the bonds in a water molecule (H_2O), which consists of two

FIG. 2.6 Four molecules. Atoms tend to combine in such a way as to complete the complement of electrons in the outer shell.

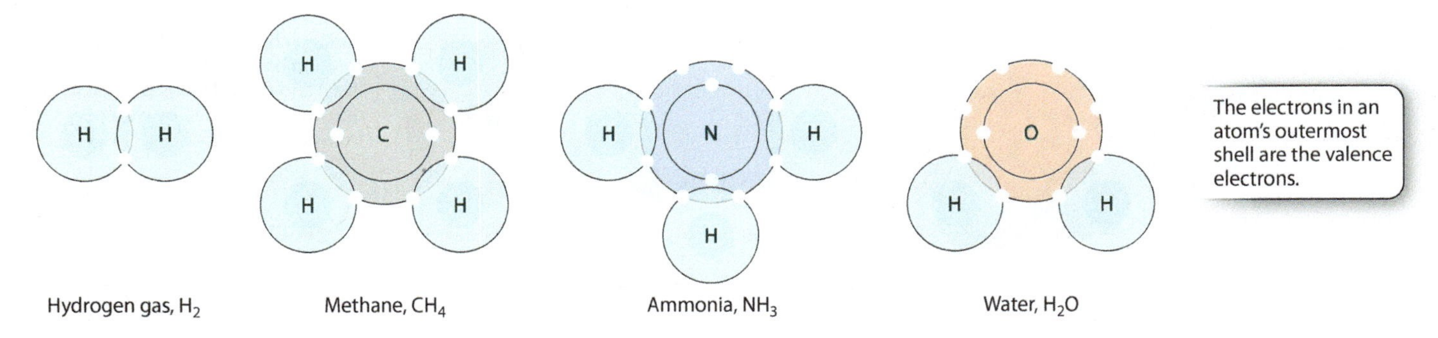

FIG. 2.7 A polar covalent bond. In a polar covalent bond, the two atoms do not share the electrons equally.

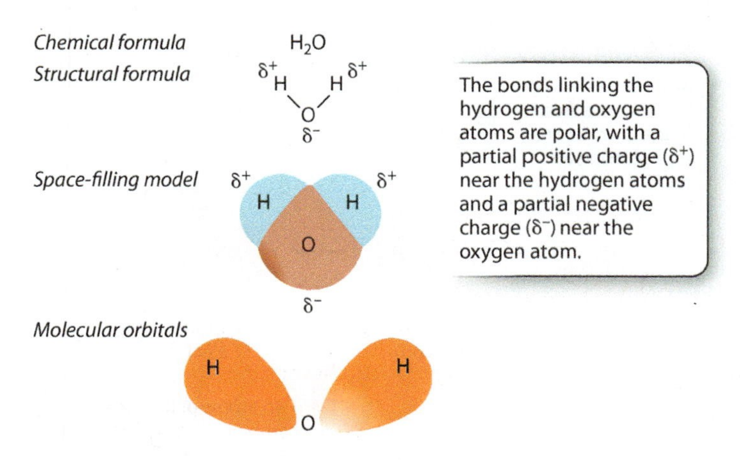

hydrogen atoms each covalently bound to a single oxygen atom (**Fig. 2.7**).

In a molecule of water, the electrons are more likely to be located near the oxygen atom. The unequal sharing of electrons results from a difference in the ability of the atoms to attract electrons, a property known as **electronegativity.** Electronegativity tends to increase across a row in the periodic table; as the number of protons across a row increases, electrons are held more tightly to the nucleus. Therefore, oxygen is more electronegative than hydrogen and attracts electrons more readily than does hydrogen. In a molecule of water, oxygen has a slight negative charge, while the two hydrogen atoms have a slight positive charge (Fig. 2.7). When electrons are shared unequally between the two atoms, the resulting interaction is described as a **polar covalent bond.**

A covalent bond between atoms that have the same, or nearly the same, electronegativity is a **nonpolar covalent bond,** which means that the atoms share the bonding electron pair almost equally. Nonpolar covalent bonds include those in gaseous hydrogen (H_2) and oxygen (O_2), as well as carbon–carbon (C–C) and carbon–hydrogen (C–H) bonds. Molecules held together by nonpolar covalent bonds are important in cells because they do not mix well with water.

An ionic bond forms between oppositely charged ions.

In water, the difference in electronegativity between the oxygen and hydrogen atoms leads to unequal sharing of electrons. In more extreme cases, when an atom of very high electronegativity is

FIG. 2.8 An ionic bond. (a) Sodium chloride (salt) is formed by the attraction of two ions. (b) In solution, the ions are surrounded by water molecules.

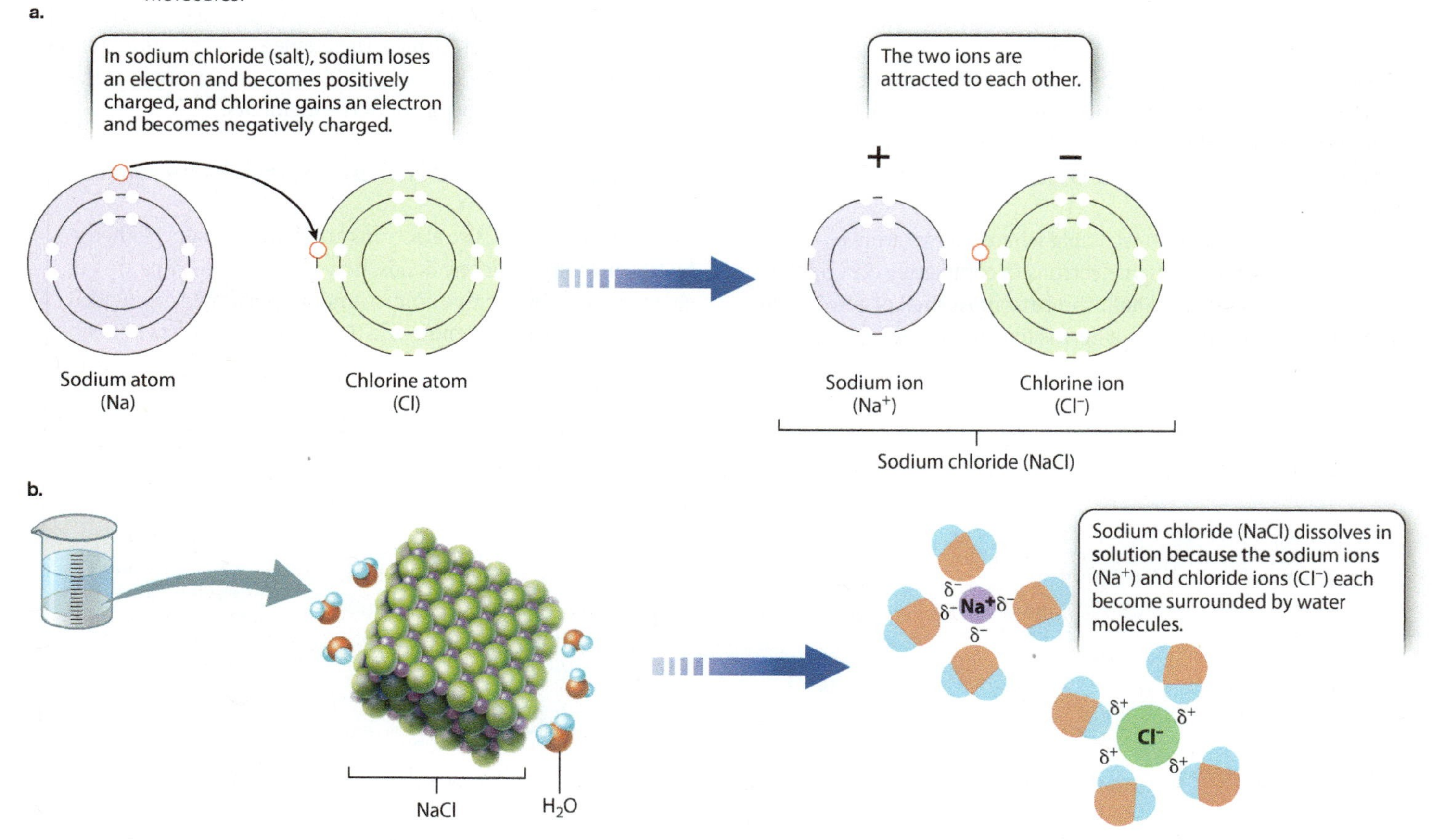

paired with an atom of very low electronegativity, the difference in electronegativity is so great that the electronegative atom "steals" the electron from its less electronegative partner. In this case, the atom with the extra electron has a negative charge and is a negative ion. The atom that has lost an electron has a positive charge and is a positive ion. The two ions are not covalently bound, but because opposite charges attract they associate with each other in what is called an **ionic bond.** An example of a compound formed by the attraction of a positive ion and a negative ion is table salt, or sodium chloride (NaCl) (**Fig. 2.8a**).

When sodium chloride is placed in water, the salt dissolves to form sodium ions (written as "Na^+") that have lost an electron and so are positively charged, and chloride ions (Cl^-) that have gained an electron and so are negatively charged. In solution, the two ions are pulled apart and become surrounded by water molecules: The negatively charged ends of water molecules are attracted to the positively charged sodium ion, and the positively charged ends of other water molecules are attracted to the negatively charged chloride ion (**Fig. 2.8b**). Only as the water evaporates do the concentrations of Na^+ and Cl^- increase to the point where the ions join and precipitate as salt crystals.

A chemical reaction involves breaking and forming chemical bonds.

The chemical bonds that link atoms in molecules can change in a **chemical reaction,** a process by which atoms or molecules, called **reactants,** are transformed into different molecules, called **products.** During a chemical reaction, atoms keep their identity but change which atoms they are bonded to.

For example, two molecules of hydrogen gas ($2H_2$) and one molecule of oxygen gas (O_2) can react to form two molecules of water ($2H_2O$), as shown in **Fig. 2.9.** In this reaction, the numbers of each type of atom are conserved, but their arrangement is different in the reactants and the products. Specifically, the H–H bond in hydrogen gas and the O=O bond in oxygen are broken. At the same time, each oxygen atom forms new covalent bonds with two hydrogen atoms, forming two molecules of water. In fact, this reaction is the origin of the name "hydrogen," which literally means "water former." The reaction releases a good deal of energy; it was used in the main engine of the space shuttle.

FIG. 2.9 A chemical reaction. During a chemical reaction, atoms retain their identity, but their connections change as bonds are broken and new bonds are formed.

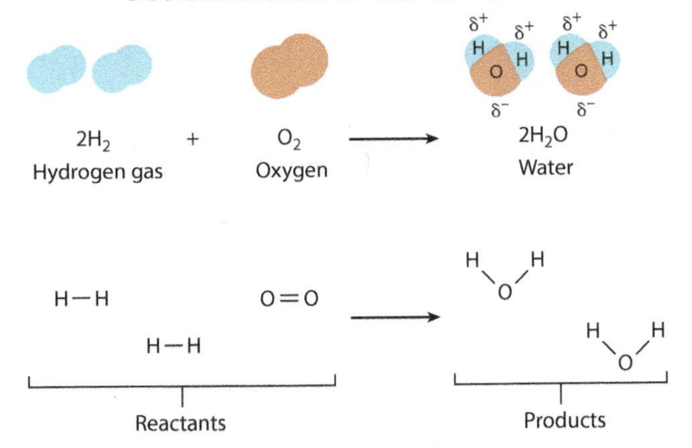

In biological systems, chemical reactions provide a way to build and break down molecules for use by the cell, as well as to harness energy, which can be held in chemical bonds (Chapter 6).

2.3 WATER: THE MEDIUM OF LIFE

On Earth, all life depends on water. Indeed, life originated in water, and the availability of water strongly influences the environmental distributions of different species. Furthermore, water is the single most abundant molecule in cells, so water is the medium in which the molecules of life interact. So important to life is water that, in the late 1990s, the National Aeronautic and Space Administration (NASA) announced that the search for extraterrestrial life would be based on a simple strategy: Follow the water. NASA's logic makes sense because, within our own solar system, Earth stands out both for its abundance of water and the life it supports. What makes water so special as the medium of life?

Water is a polar molecule.

As we saw earlier, water molecules have polar covalent bonds, characterized by an uneven distribution of electrons. A molecule like water that has regions of positive and negative charge is called a **polar** molecule. Molecules, or even different regions of the same molecule, fall into two general classes, depending on how they interact with water: **hydrophilic** ("water loving") and **hydrophobic** ("water fearing").

Hydrophilic compounds are polar; they dissolve readily in water. That is, water is a good **solvent,** capable of dissolving many substances. Think of what happens when you stir a teaspoon of sugar into water: The sugar seems to disappear as it dissolves. What is happening is that the sugar molecules are dispersing through the water and becoming separated from one another, forming a solution in the watery, or **aqueous,** environment.

By contrast, hydrophobic compounds are **nonpolar.** Nonpolar compounds do not have regions of positive and negative charge. As a result, they arrange themselves to minimize their contact with water. For example, oil molecules are hydrophobic, and when oil and water are mixed, the oil molecules organize themselves into droplets that limit the oil–water interface. This **hydrophobic effect,** in which polar molecules like water exclude nonpolar ones, drives such biological processes as the folding of proteins (Chapter 4) and the formation of cell membranes (Chapter 5).

A hydrogen bond is an interaction between a hydrogen atom and an electronegative atom.

Because the oxygen and hydrogen atoms have slight charges, water molecules orient themselves to minimize the repulsion of

FIG. 2.10 Hydrogen bonds in liquid water. Because of thermal motion, hydrogen bonds in water are continually breaking and forming between different pairs of molecules.

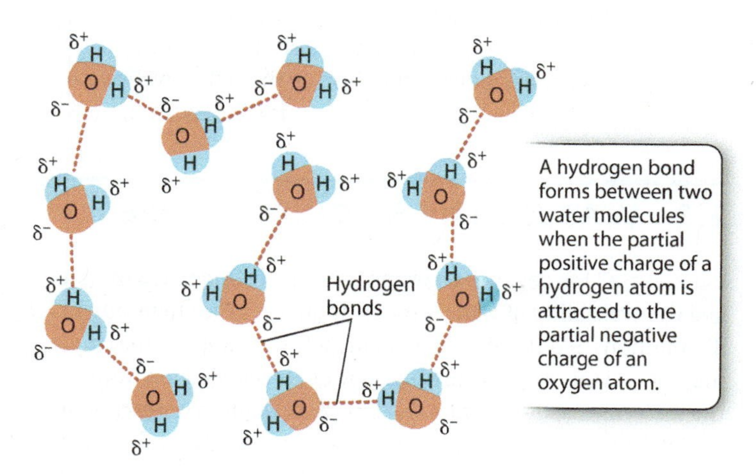

like charges so that positive charges are near negative charges. For example, because of its slight positive charge, a hydrogen atom in a water molecule tends to orient toward the slightly negatively charged oxygen atom in another molecule.

A **hydrogen bond** is the name given to this interaction between a hydrogen atom with a slight positive charge and an electronegative atom of another molecule. In fact, any hydrogen atom covalently bound to an electronegative atom (such as oxygen or nitrogen) will have a slight positive charge and can form a hydrogen bond. In the case of water, the result of many such interactions is a molecular network stabilized by hydrogen bonds. Typically, a hydrogen bond is depicted by a dotted line, as in **Fig. 2.10.**

Hydrogen bonds are much weaker than covalent bonds, but it is hydrogen bonding that gives water many interesting properties, described next. In addition, the presence of many weak hydrogen bonds can help stabilize biological molecules, as in the case of nucleic acids and proteins.

Hydrogen bonds give water many unusual properties.

Hydrogen bonds influence the structure of both liquid water (**Fig. 2.11a**) and ice (**Fig. 2.11b**). When water freezes, most water molecules become hydrogen bonded to four other water molecules, forming an open crystalline structure we call ice. As the temperature increases and the ice melts, some of the hydrogen bonds are destabilized and break, allowing the water molecules to pack more closely and explaining why liquid water is denser than ice. As a result, ice floats on water, and ponds and lakes freeze from the top down, and therefore do not freeze completely. This property allows fish and aquatic plants to survive winter in the cold water under the layer of ice.

→ **Quick Check 3** Why do containers of water, milk, soda, or other liquids sometimes burst when frozen?

Hydrogen bonds also give water molecules the property of **cohesion,** meaning that they tend to stick to one another. A consequence of cohesion is high surface tension, a measure of the difficulty of breaking the surface of a liquid. Cohesion between molecules contributes to water movement in plants. As water

FIG. 2.11 Liquid water and ice. Hydrogen bonds create (a) a dense structure in water, and (b) a highly ordered, less dense, crystalline structure in ice.

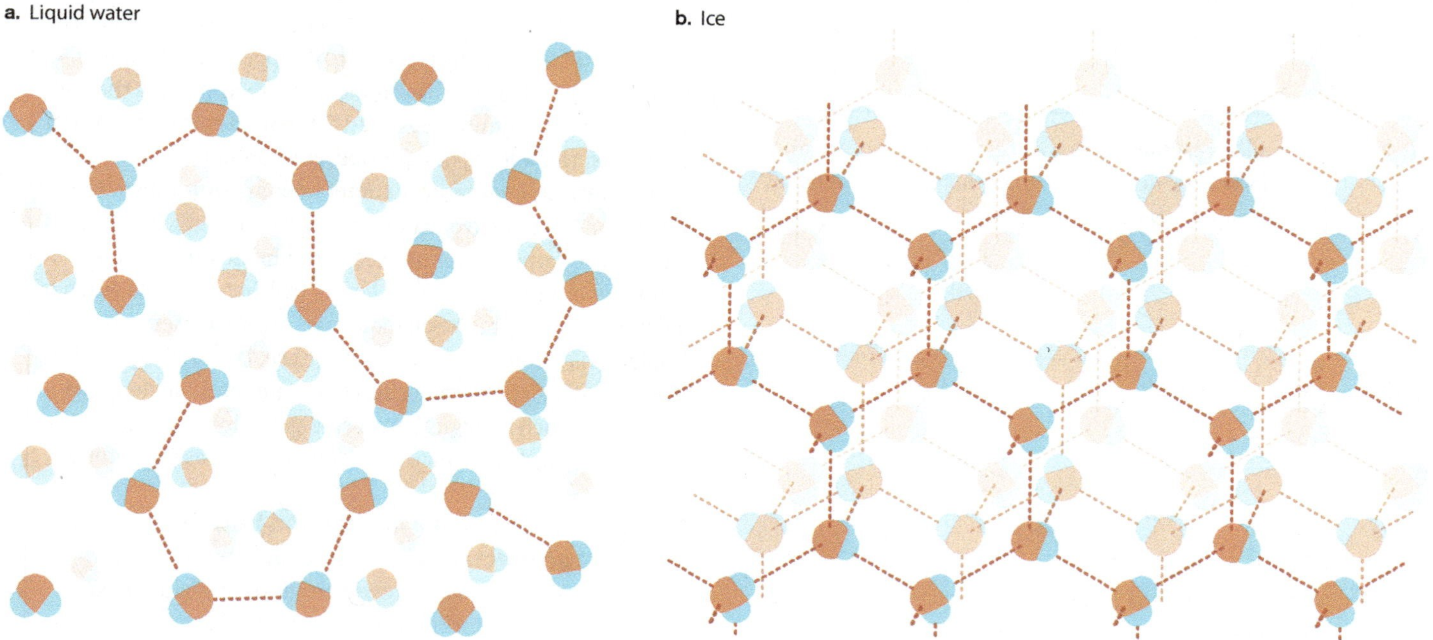

evaporates from leaves, water is pulled upward, sometimes as high as 100 meters above the ground in giant sequoia and coast redwood trees, which are among the tallest trees on Earth.

The hydrogen bonds of water also influence how water responds to heating. Molecules are in constant motion, and this motion increases as the temperature increases. When water is heated, some of the energy added by heating is used to break hydrogen bonds instead of causing more motion among the molecules, so the temperature increases less than if there were no hydrogen bonding. The abundant hydrogen bonds make water more resistant to temperature changes than other substances, a property that is important for living organisms on a variety of scales. In the cell, water resists temperature variations that would otherwise result from numerous biochemical reactions. On a global scale, the oceans minimize temperature fluctuations, stabilizing the temperature on Earth in a range compatible with life.

In short, water is clearly the medium of life on Earth, but is this because water is uniquely suited for life, or is it because life on Earth has adapted through time to a watery environment? We don't know the answer, but probably both explanations are partly true. Chemists have proposed that under conditions of high pressure and temperature, other small molecules, among them ammonia (NH_3) and some simple carbon-containing molecules, might display similar characteristics friendly to life. However, under the conditions that exist on Earth, water is the only molecule uniquely suited to life. Water is a truly remarkable substance, and life on Earth would not be possible without it.

pH is a measure of the concentration of protons in solution.

In any solution of water, a small proportion of the water molecules exist as protons (H^+) and hydroxide ions (OH^-). The pH of a solution measures the proton concentration ($[H^+]$), which is important as the pH influences many chemical reactions and biological processes. It is calculated by the following formula:

$$pH = -\log [H^+]$$

The pH of a solution can range from 0 to 14. Since the pH scale is logarithmic, a difference of one pH unit corresponds to a tenfold difference in hydrogen ion concentration. A solution is neutral (pH = 7) when the concentrations of protons (H^+) and hydroxide ions (OH^-) are equal. When the concentration of protons is higher than that of hydroxide ions, the pH is lower than 7 and the solution is **acidic.** When the concentration of protons is lower than that of hydroxide ions, the pH is higher than 7 and the solution is **basic.** An acid can therefore be described as a molecule that releases a proton (H^+), and a base is a molecule that accepts a proton in aqueous solution.

Pure water has a pH of 7—that is, it is neutral, with an equal concentration of protons and hydroxide ions. The pH of most cells is approximately 7 and is tightly regulated, as most chemical reactions can be carried out only in a narrow pH range. Certain cellular compartments, however, have a much lower pH. The pH of blood is slightly basic, with a pH around 7.4. This value is sometimes referred to in medicine as physiological pH. Freshwater lakes, ponds, and rivers tend to be slightly acidic because carbon dioxide from the air dissolves in the water and forms carbonic acid (Chapter 6).

2.4 CARBON: LIFE'S CHEMICAL BACKBONE

Hydrogen and helium are far and away the most abundant elements in the universe. In contrast, the solid Earth is dominated by silicon, oxygen, aluminum, iron, and calcium (Chapter 1). In other words, Earth is not a typical sample of the universe. Nor is the cell a typical sample of the solid Earth. **Fig. 2.12** shows the relative abundance by mass of chemical elements present in human cells after all the water has been removed. Note that just four elements—carbon (C), oxygen (O), hydrogen (H), and nitrogen (N)—constitute 94% of the total dry mass, and that the most abundant element is carbon. The elemental composition of human cells is typical of all cells. Human life, and all life as we know it, is based on carbon. Carbon molecules play such an important role in living organisms that carbon-containing molecules have a special name—they are called **organic molecules.** Their central role in life implies that there must be something very special about carbon, and there is. Carbon has the ability to combine with many other elements to form a wide

FIG. 2.12 Approximate proportions by dry mass of chemical elements found in human cells.

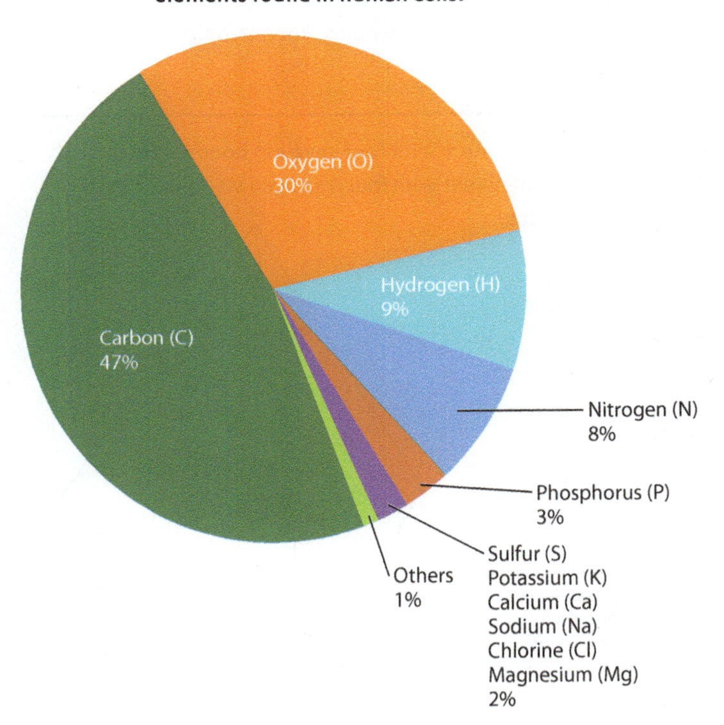

variety of molecules, each specialized for the functions it carries out in the cell.

Carbon atoms form four covalent bonds.

One of the special properties of carbon is that, in forming molecular orbitals, a carbon atom behaves as if it had four unpaired electrons. This behavior occurs because one of the electrons in the outermost spherical orbital moves into the empty dumbbell-shaped orbital (see Fig. 2.2). In this process, the single large spherical orbital and three dumbbell-shaped orbitals change shape, becoming four equivalent hybrid orbitals, each with one electron.

Fig. 2.13 shows the molecular orbitals that result when one atom of carbon combines with four atoms of hydrogen to form the gas methane (CH_4). Each of the four valence electrons of carbon shares a new molecular orbital with the electron of one of the hydrogen atoms. These bonds can rotate freely about their axis. Furthermore, because of the shape of the orbitals, the carbon atom lies at the center of a three-dimensional structure called a tetrahedron, and the four molecular orbitals point toward the four corners of this structure. The ability of carbon to form four covalent bonds, the spatial orientation of these bonds in the form of a tetrahedron, and the ability of each bond to rotate freely all contribute importantly to the structural diversity of carbon-based molecules.

Carbon-based molecules are structurally and functionally diverse.

Carbon has other properties that contribute to its ability to form a diversity of molecules. For example, carbon atoms can link with each other by covalent bonds to form long chains. These chains can be branched, or two carbons at the ends of the chain or within the chain can link to form a ring.

Among the simplest chains is ethane, shown in **Fig. 2.14a.** Ethane is formed when two carbon atoms become connected by a covalent bond. In this case, the orbitals of unpaired electrons in two carbon atoms form the covalent bond. Each carbon atom is also bound to three hydrogen atoms. Some more complex examples of carbon-containing molecules are shown in **Fig. 2.14b,** as both structural formulas and in simplified form.

Two adjacent carbon atoms can also share two pairs of electrons, forming a double bond, as shown in **Fig. 2.15.** Note that each carbon atom has exactly four covalent bonds, but in this case two are shared between adjacent carbon atoms. The double bond is shorter than a single bond and is not free to rotate, so all of the covalent bonds formed by the carbon atoms connected by

FIG. 2.14 Diverse carbon-containing molecules. (a) The molecule ethane contains one C–C bond. (b) A linear chain of carbon atoms and a ring structure that also contains oxygen contain multiple C–C bonds.

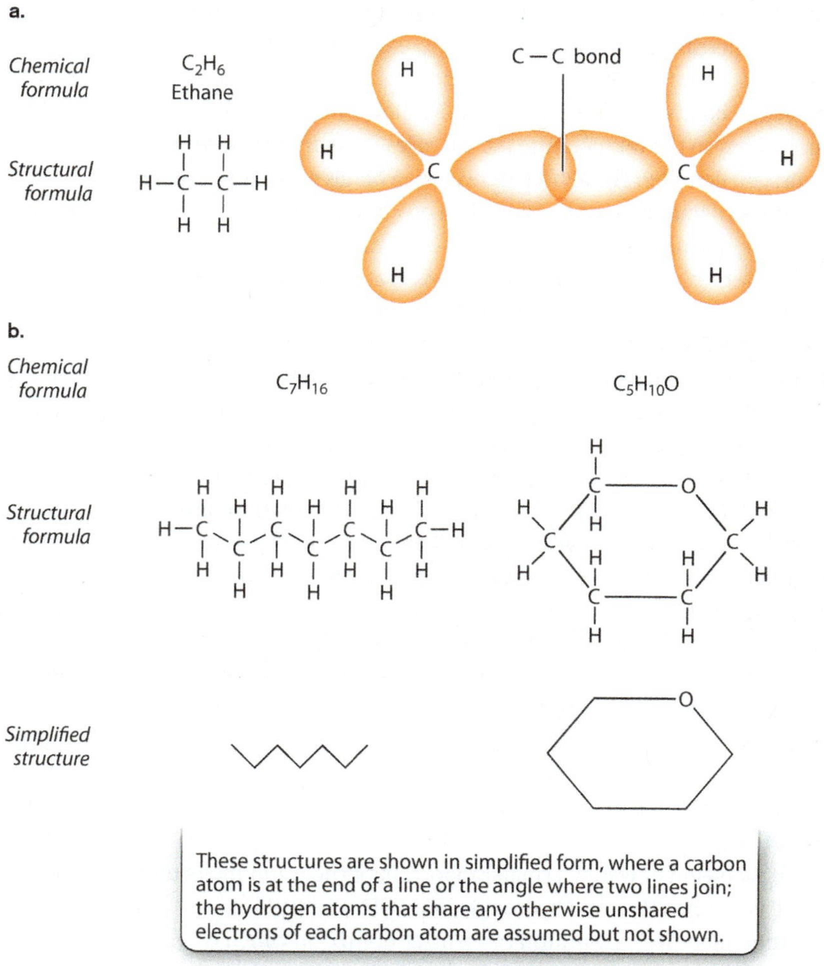

a.

Chemical formula C_2H_6 Ethane

Structural formula

C—C bond

b.

Chemical formula C_7H_{16} $C_5H_{10}O$

Structural formula

Simplified structure

These structures are shown in simplified form, where a carbon atom is at the end of a line or the angle where two lines join; the hydrogen atoms that share any otherwise unshared electrons of each carbon atom are assumed but not shown.

FIG. 2.13 A carbon atom with four covalent bonds. One carbon atom combines with four hydrogen atoms to form methane.

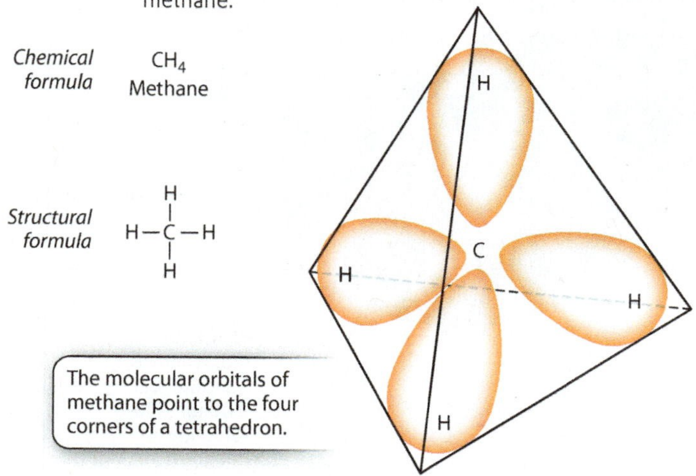

FIG. 2.15 **Molecules containing double bonds between carbon atoms.**

Chemical formula	C_2H_4	C_7H_{14}	C_6H_6
Structural formula			

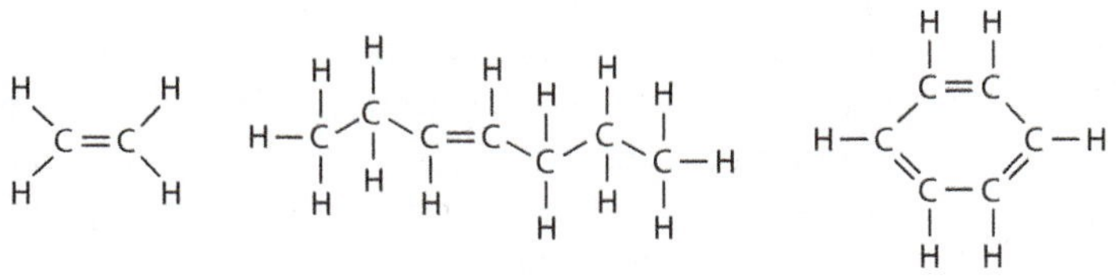

a double bond are in the same geometrical plane. As with single bonds, double bonds can be found in chains of atoms or ring structures.

While the types of atoms making up a molecule help characterize the molecule, the spatial arrangement of atoms is also important. For example, 6 carbon atoms, 13 hydrogen atoms, 2 oxygen atoms, and 1 nitrogen atom can join covalently in many different arrangements to produce molecules with different structures. Two of these many arrangements are shown in **Fig. 2.16.** Note that some of the connections between atoms are identical in the two molecules (black) and some are different (green), even though the chemical formulas are the same ($C_6H_{13}O_2N_1$). Molecules that have the same chemical formula but different structures are known as **isomers.**

We have seen that carbon-containing molecules can adopt a wide range of arrangements, a versatility that helps us to understand how a limited number of elements can create an astonishing variety of molecules. We might ask whether carbon is uniquely versatile. Put another way, if we ever discover life on a distant planet, will it be carbon based, like us? Silicon, just below carbon in the periodic table (see Fig. 2.3), is the one other element that is both reasonably abundant and characterized by four atomic orbitals with one electron each. Some scientists have speculated that silicon might provide an alternative to carbon as a chemical basis for life. However, silicon readily binds oxygen. On Earth, nearly all of the silicon atoms found in molecules are covalently bound to oxygen. Studies of Mars and meteorites show that silicon is tightly bound to oxygen throughout our solar system, and that is likely to be true everywhere we might explore. There are about 1000 different silicate minerals on Earth, but this diversity pales before the millions of known carbon-based molecules. If we ever discover life beyond Earth, very likely its chemistry will be based on carbon.

FIG. 2.16 **The isomers isoleucine and leucine.** Isoleucine and leucine are isomers: Their chemical formulas are the same, but their structures differ.

Chemical formula	$C_6H_{13}O_2N_1$ Isoleucine	$C_6H_{13}O_2N_1$ Leucine
Structural formula		

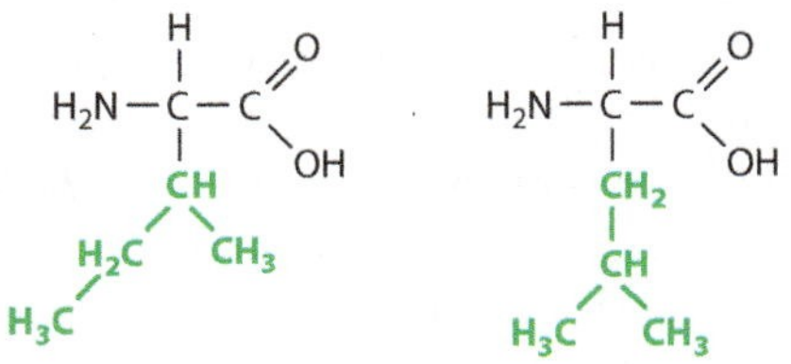

2.5 ORGANIC MOLECULES

Chemical processes in the cell depend on just a few classes of carbon-based molecules. **Proteins** provide structural support and act as catalysts that facilitate chemical reactions. **Nucleic acids** encode and transmit genetic information. **Carbohydrates** provide a source of energy and make up the cell wall in bacteria, plants, and algae. **Lipids** make up cell membranes, store energy, and act as signaling molecules.

These molecules are all large, consisting of hundreds or thousands of atoms, and many are **polymers,** complex molecules made up of repeated simpler units connected by covalent bonds. Proteins are polymers of **amino acids,** nucleic acids are polymers of **nucleotides,** and carbohydrates such as starch are polymers of simple **sugars.** Lipids are a bit different, as we will see, in that they are defined by a property rather than by their chemical structure. The lipid membranes that define cell boundaries consist of **fatty acids** bonded to other organic molecules.

Building macromolecules from simple, repeating units provides a means of generating virtually limitless chemical diversity. Indeed, in macromolecules, the building blocks of polymers play a role much like that of the letters in words. In written language, a change in the content or order of letters changes the meaning of the word (or renders it meaningless). For example, by reordering the letters of the word SILENT you can write LISTEN, a word with a different meaning. Similarly, rearranging the building blocks that make up macromolecules provides an important way to make a large number of diverse macromolecules whose functions differ from one to the next.

In the following sections, we focus on the building blocks of these four key molecules of life, reserving a discussion of the structure and function of the macromolecules for later chapters.

Functional groups add chemical character to carbon chains.

The simple repeating units of polymers are often based on a nonpolar core of carbon atoms. But attached to these carbon atoms are **functional groups.** Functional groups are groups of one or more atoms that have particular chemical properties on their own, regardless of what they are attached to. Among the functional groups frequently encountered in biological molecules are amine (=NH), amino ($-NH_2$), carboxyl (–COOH), hydroxyl (–OH), ketone (=O), phosphate ($-O-PO_3H_2$), sulfhydryl (–SH), and methyl ($-CH_3$). The nitrogen, oxygen, phosphorus, and sulfur atoms in these functional groups are more electronegative than the carbon atoms, and functional groups containing these atoms are polar. The methyl group ($-CH_3$), on the other hand, is nonpolar.

Because many functional groups are polar, otherwise nonpolar molecules containing these groups become polar and so become soluble in the cell's aqueous environment. In other words, they disperse in solution throughout the cell. Moreover, because many functional groups are polar, they are also reactive. Notice in the following sections that the reactions joining simpler molecules into polymers usually take place between functional groups.

Proteins are composed of amino acids.

Proteins do much of the cell's work. Some function as catalysts that accelerate the rates of chemical reactions (in which case they are called **enzymes**), and some act as structural components necessary for cell shape and movement. Your body contains many thousands of distinct types of protein that perform a wide range of functions. Since proteins consist of amino acids linked covalently to form a chain, we need to examine the chemical features of amino acids to understand the diversity and versatility of proteins.

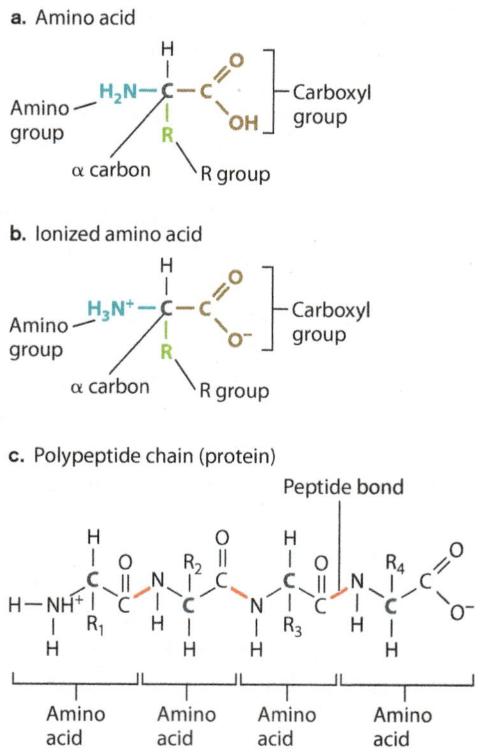

The general structure of an amino acid is shown in **Fig. 2.17a.** Each amino acid contains a central carbon atom, called the **α (alpha) carbon,** covalently linked to four groups: an **amino group** ($-NH_2$; blue), a **carboxyl group** (–COOH; brown), a hydrogen atom (H), and an **R group,** or **side chain,** (green) that differs from one amino acid to the next. The identity of each amino acid is determined by the structure and composition of the side chain. The side chain of the amino acid glycine is simply H, for example, and that of alanine is CH_3. In most amino acids, the α carbon is covalently linked to four different groups. Glycine is the exception, since its R group is a hydrogen atom.

At the pH commonly found in a cell (pH 7.4), the amino and carboxyl groups are ionized (charged), with the amino group gaining a proton ($-NH_3^+$; blue) and the carboxyl group losing a proton ($-COO^-$; brown), as shown in **Fig. 2.17b.**

Amino acids are linked in a chain to form a protein (**Fig. 2.17c**). The carbon atom in the carboxyl group of one amino acid is joined to the nitrogen atom in the amino group of the next by a covalent linkage called a **peptide bond.** In Fig. 2.17c, the chain of amino acids includes four amino acids, and the peptide bonds are indicated in red. The formation of a peptide bond involves the loss of a water molecule since in order to form a C–N bond, the carbon atom of the carboxyl group must release an oxygen atom and the nitrogen atom of the amino group must release two hydrogen atoms. These can then combine to form a water molecule (H_2O). The loss of a water molecule also occurs in the linking of subunits to form polymers such as nucleic acids and complex carbohydrates.

Cellular proteins are composed of combinations of 20 different amino acids, each of which can be classified according to the chemical properties of its R group. The particular sequence, or order, in which amino acids are present in a protein determines how it folds into its three-dimensional structure. The three-dimensional structure, in turn, determines the protein's function. In Chapter 4, we examine how the sequence of amino acids in a particular protein is specified and discuss how proteins fold into their three-dimensional shapes.

FIG. 2.17 **Amino acids and peptide bonds.** (a) An amino acid contains four groups attached to a central carbon atom. (b) In the environment of a cell, the amino group gains a proton and the carboxyl group loses a proton. (c) Peptide bonds link amino acids to form a protein.

Nucleic acids encode genetic information in their nucleotide sequence.

Nucleic acids are examples of informational molecules—that is, they are large molecules that carry information in the sequence of nucleotides that make them up. This molecular information is much like the information carried by the letters in an alphabet, but in the case of nucleic acids, the information is in chemical form.

The nucleic acid **deoxyribonucleic acid (DNA)** is the genetic material in all organisms. It is transmitted from parents to offspring, and it contains the information needed to specify the amino acid sequence of all the proteins synthesized in an organism. The nucleic acid **ribonucleic acid (RNA)** has multiple functions; it is a key player in protein synthesis and the regulation of gene expression.

FIG. 2.18 A ribonucleotide and a deoxyribonucleotide, the units of RNA and DNA.

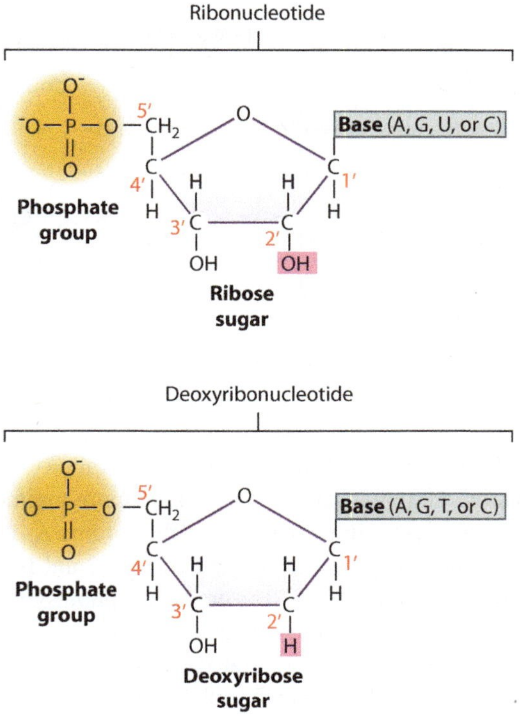

Ribonucleotide

Base (A, G, U, or C)

Phosphate group

Ribose sugar

Deoxyribonucleotide

Base (A, G, T, or C)

Phosphate group

Deoxyribose sugar

DNA and RNA are long molecules consisting of nucleotides bonded covalently one to the next. Nucleotides, in turn, are composed of three components: a 5-carbon sugar, a nitrogen-containing compound called a **base,** and one or more phosphate groups (**Fig. 2.18**). The sugar in RNA is ribose, and the sugar in DNA is deoxyribose. The sugars differ in that ribose has a hydroxyl (OH) group on the second carbon (designated the 2′ carbon), whereas deoxyribose has a hydrogen atom at this position (hence, *deoxy*ribose). (By convention, the carbons in the sugar are numbered with primes—1′, 2′, etc.—to distinguish them from carbons in the base—1, 2, etc.)

The bases are built from nitrogen-containing rings and are of two types. The **pyrimidine** bases (**Fig. 2.19a**) have a single ring and include **cytosine (C), thymine (T),** and **uracil (U).** The **purine** bases (**Fig. 2.19b**) have a double-ring structure and include **guanine (G)** and **adenine (A).** DNA contains the bases A, T, G, and C, and RNA contains the bases A, U, G, and C. Just as the order of amino acids provides the information carried in proteins, so, too, does the sequence of nucleotides determine the information in DNA and RNA molecules.

In DNA and RNA, each adjacent pair of nucleotides is connected by a **phosphodiester bond,** which forms when a phosphate group in one nucleotide is covalently joined to the sugar unit in another nucleotide (**Fig. 2.20**). As in the formation of a peptide bond, the formation of a phosphodiester bond involves the loss of a water molecule.

DNA in cells usually consists of two strands of nucleotides twisted around each other in the form of a **double helix**

FIG. 2.19 Pyrimidine bases and purine bases. (a) Pyrimidines have a single-ring structure, and (b) purines have a double-ring structure.

a. Pyrimidine bases

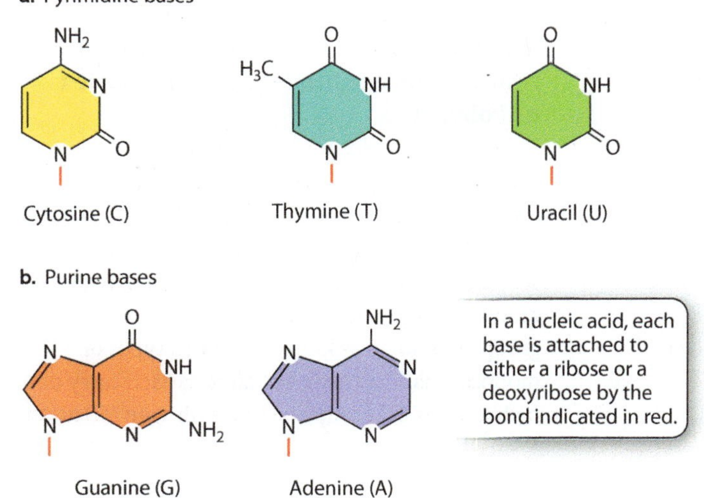

FIG. 2.20 The phosphodiester bond. Phosphodiester bonds link successive deoxyribonucleotides, forming the backbone of the DNA strand.

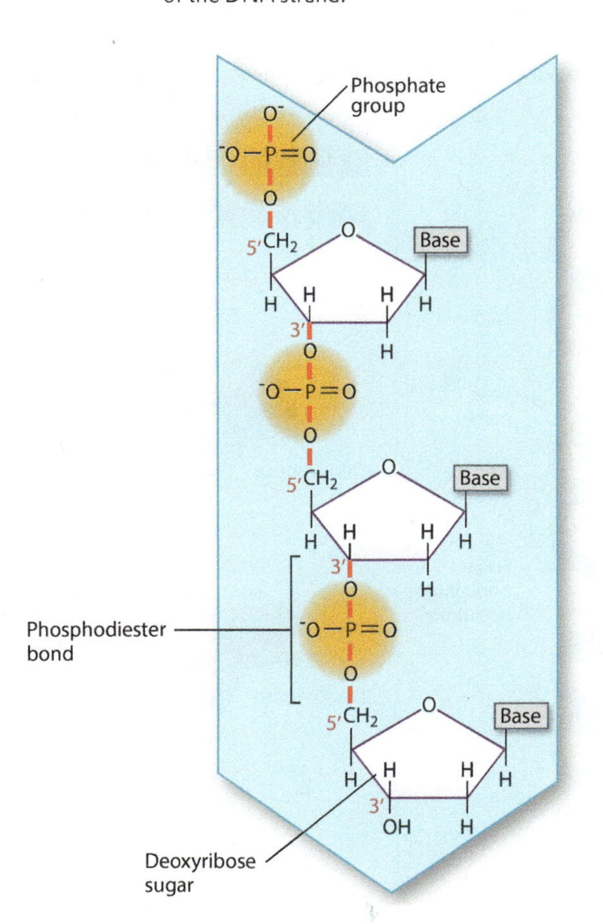

(**Fig. 2.21a**). The sugar-phosphate backbones of the strands wrap like a ribbon around the outside of the double helix, and the bases point inward. The bases form specific purine–pyrimidine pairs that are **complementary:** Where one strand carries an A, the other carries a T; and where one strand carries a G, the other carries a C (**Fig. 2.21b**). Base pairing results from hydrogen bonding between the bases (**Fig. 2.21c**).

Genetic information in DNA is contained in the sequence, or order, in which successive nucleotides occur along the molecule. Successive nucleotides along a DNA strand can occur in *any* order, and hence a long molecule could contain any of an immense number of possible nucleotide sequences. This is one reason why DNA is an efficient carrier of genetic information. In Chapter 3, we consider the structure and function of DNA and RNA in greater detail.

Complex carbohydrates are made up of simple sugars.

Many of us, when we feel tired, reach for a candy bar for a quick energy boost. The energy in a candy bar comes from sugars, which are quickly broken down to release energy. Sugars belong to a class of molecules called **carbohydrates,** distinctive molecules composed of C, H, and O atoms, usually in the ratio 1:2:1. Carbohydrates provide a principal source of energy for metabolism.

FIG. 2.21 The structure of DNA. (a) DNA is most commonly in the form of a double helix, with the sugar and phosphate groups forming the backbone and the bases oriented inward. (b) The bases are complementary, with A paired with T and G paired with C. (c) Base pairing results from hydrogen bonds.

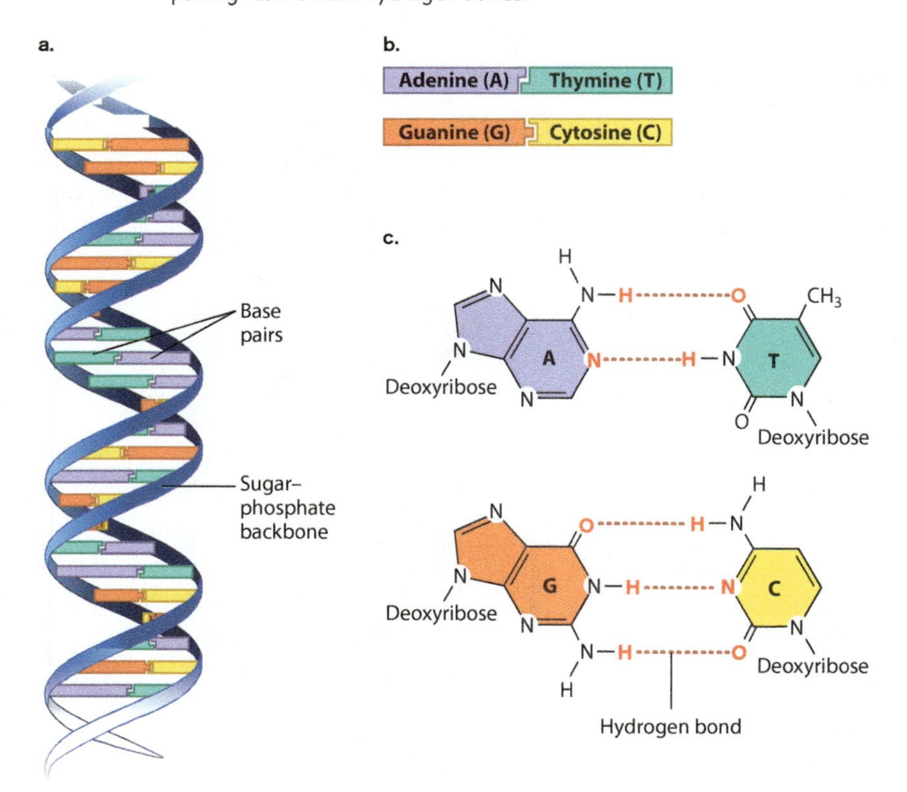

FIG. 2.22 Structural formulas for some 6-carbon aldoses and ketoses.

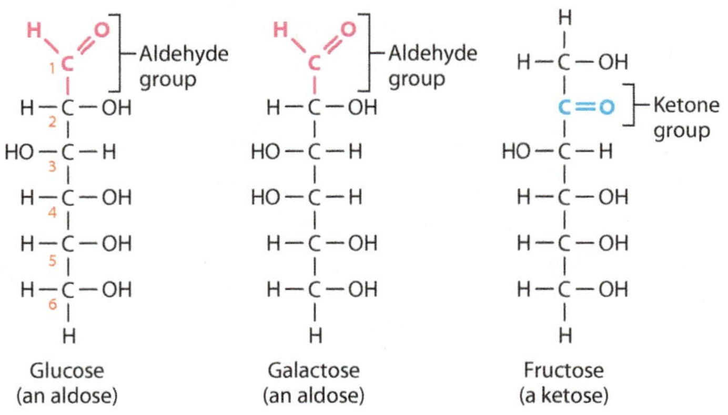

Aldehyde group | Aldehyde group | Ketone group

Glucose (an aldose) | Galactose (an aldose) | Fructose (a ketose)

The simplest carbohydrates are sugars (also called **saccharides**). Simple sugars are linear or, far more commonly, cyclic molecules containing five or six carbon atoms. All 6-carbon sugars have the same chemical formula ($C_6H_{12}O_6$) and differ only in configuration. Glucose (the product of photosynthesis), galactose (found in dairy products), and fructose (a commercial sweetener) are examples; they share the same formula ($C_6H_{12}O_6$) but differ in the arrangement of their atoms (**Fig. 2.22**).

→ **Quick Check 4** Take a close look at Fig. 2.22. How is glucose different from galactose?

A simple sugar is also called a **monosaccharide** (*mono* means "one"), and two simple sugars linked together by a covalent bond is called a disaccharide (*di* means "two"). Sucrose ($C_{12}H_{22}O_{11}$), or table sugar, is a disaccharide that combines one molecule each of glucose and fructose. Simple sugars combine in many ways to form polymers called **polysaccharides** (*poly* means "many") that provide long-term energy storage (starch and glycogen) or structural support (cellulose in plant cell walls). Long, branched chains of monosaccharides are called **complex carbohydrates.**

Let's take a closer look at monosaccharides, the simplest sugars. Monosaccharides are unbranched carbon chains with either an aldehyde (HC=O) or a ketone (C=O) group (Fig. 2.22). Monosaccharides with an aldehyde group are called aldoses and those with a ketone group are known as ketoses. In both types of monosaccharide, the other carbons each carry one hydroxyl (–OH) group and one hydrogen (H) atom.

FIG. 2.23 Formation of the cyclic form of glucose.

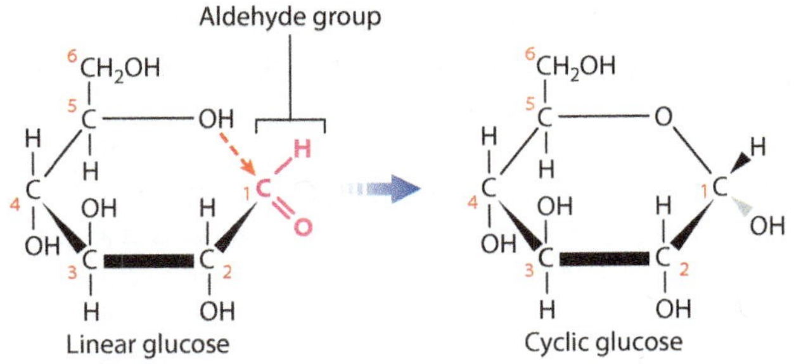

When the linear structure of a monosaccharide is written with the aldehyde or ketone group at the top, the carbons are numbered from top to bottom.

Virtually all of the monosaccharides in cells are in ring form (**Fig. 2.23**), not linear structures. To form a ring, the carbon in the aldehyde or ketone group forms a covalent bond with the oxygen of a hydroxyl group carried by another carbon in the same molecule. For example, cyclic glucose is formed when the oxygen atom of the hydroxyl group on carbon 5 forms a covalent bond with carbon 1, which is part of an aldehyde group. The cyclic structure is approximately flat, and you can visualize it perpendicular to the plane of the paper with the covalent bonds indicated by the thick lines in the foreground. The groups attached to any carbon therefore project either above or below the ring. When the ring is formed, the aldehyde oxygen becomes a hydroxyl group. The presence of the polar hydroxyl groups through the sugar ring makes these molecules highly soluble in water.

Monosaccharides, especially 6-carbon sugars, are the building blocks of complex carbohydrates. Monosaccharides are attached to each other by covalent bonds called **glycosidic bonds** (**Fig. 2.24**). As with peptide bonds, the formation of glycosidic bonds involves the loss of a water molecule. A glycosidic bond is formed between carbon 1 of one monosaccharide and a hydroxyl group carried by a carbon atom in a different monosaccharide molecule.

Carbohydrate diversity stems in part from the monosaccharides that make up carbohydrates, similar to the way that protein and nucleic acid diversity stems from the sequence of their subunits. Some complex carbohydrates are composed of a single type of monosaccharide, while others are a mix of different kinds of monosaccharide. Starch, for example, is a sugar storage molecule in plants composed completely of glucose molecules, whereas pectin, a component of the cell wall, contains up to five different monosaccharides.

Lipids are hydrophobic molecules.

Proteins, nucleic acids, and carbohydrates all are polymers made up of smaller, repeating units with a defined structure. Lipids are different. Instead of being defined by a chemical structure, they share a particular property: Lipids are all hydrophobic. Because they share a property and not a structure, lipids are a chemically diverse group of molecules. They include familiar fats that make up part of our diet, components of cell membranes, and signaling molecules. Let's briefly consider each in turn.

Triacylglycerol is an example of a lipid that is used for energy storage. It is the major component of animal fat and vegetable oil. A triacylglycerol molecule is made up of three fatty acids joined to glycerol (**Fig. 2.25**). A fatty acid is a long chain of carbon atoms attached to a carboxyl group (–COOH) at one end (Figs. 2.25a and 2.25b). **Glycerol** is a 3-carbon molecule with OH groups attached to each carbon (Fig. 2.25c). The carboxyl end of each fatty acid chain attaches to glycerol at one of the OH groups (2.25d), releasing a molecule of water.

Fatty acids differ in the length of their hydrocarbon chain—that is, they differ in the number of carbon atoms in the chain. (A hydrocarbon is a molecule composed entirely of carbon and hydrogen atoms.) Most fatty acids in cells contain an even number of carbon atoms because they are synthesized by the stepwise addition of 2-carbon units.

Some fatty acids have one or more carbon–carbon double bonds; these double bonds can differ in number and location. Fatty acids that do not contain double bonds are described as **saturated.** Because there are no double bonds, the maximum number of

FIG. 2.24 **Glycosidic bonds.** Glycosidic bonds link carbohydrate molecules together; in this example they link glucose monomers together to form the polysaccharide starch.

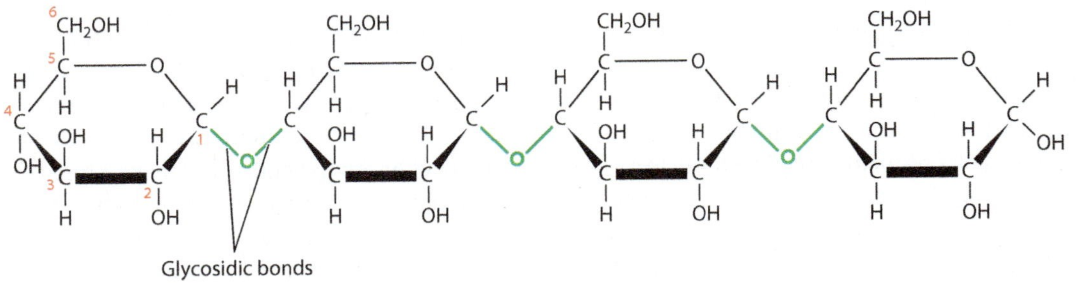

FIG. 2.25 Triacylglycerol and its components.

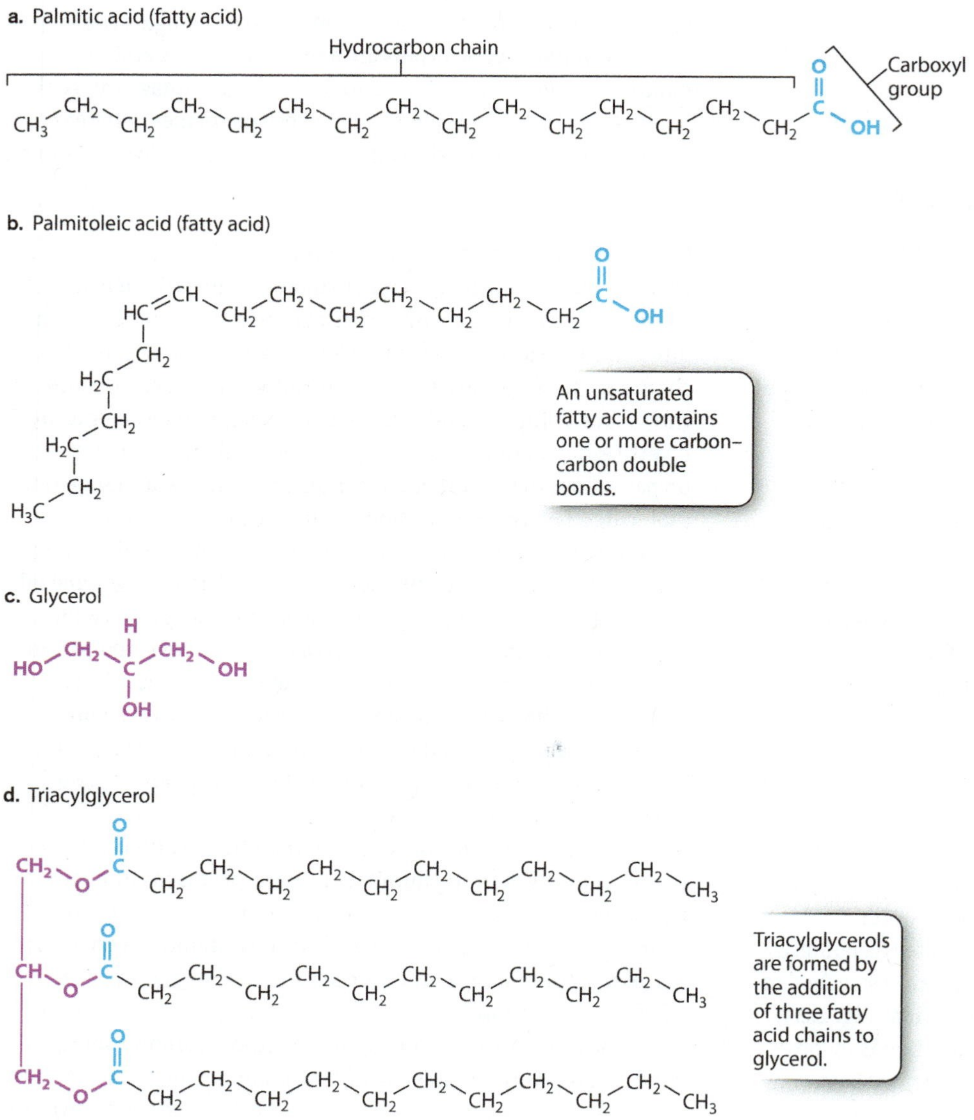

a. Palmitic acid (fatty acid)

Hydrocarbon chain

Carboxyl group

b. Palmitoleic acid (fatty acid)

An unsaturated fatty acid contains one or more carbon–carbon double bonds.

c. Glycerol

d. Triacylglycerol

Triacylglycerols are formed by the addition of three fatty acid chains to glycerol.

hydrogen atoms is attached to each carbon atom, so all of the carbon atoms are said to be "saturated" with hydrogen atoms (Fig. 2.25a). Fatty acids that contain carbon–carbon double bonds are **unsaturated** (Fig. 2.25b). The chains of saturated fatty acids are straight, while the chains of unsaturated fatty acids have a kink at each double bond.

Triacylglycerols can contain different types of fatty acids attached to the glycerol backbone. The hydrocarbon chains of fatty acids do not contain polar covalent bonds like those in a water molecule. Their electrons are distributed uniformly over the whole molecule and thus these molecules are uncharged. As a consequence, triacylglycerols are all extremely hydrophobic and, therefore, form oil droplets inside the cell. Triacylglycerols are an efficient form of energy storage because, by excluding water molecules, a large number can be packed into a small volume.

Although fatty acid molecules are uncharged, the constant motion of electrons leads to regions of slight positive and slight negative charges (**Fig. 2.26**). These charges in turn attract or repel electrons in neighboring molecules, setting up areas of positive and negative charge in those molecules as well. The molecules are temporarily polarized molecules. Such molecules weakly bind to one another because of the attraction of opposite charges. These interactions are known as **van der Waals forces.** The van der Waals forces come into play only when atoms are sufficiently close to one another, and they are weaker than hydrogen bonds, but many van der Waals forces acting together help to stabilize molecules.

Because of van der Waals forces, the melting points of fatty acids depend on their length and level of saturation. As the length of the hydrocarbon chains increases, the number of van der Waals interactions between the chains also increases. The melting temperature increases because more energy is needed to break the greater number of van der Waals interactions. Kinks introduced by double bonds reduce the tightness of the molecular packing and thus the number of intermolecular interactions. As a result, the melting temperature is lower. Therefore, an unsaturated fatty acid has a lower melting point than a saturated fatty acid of the same length. Animal fats such as butter are composed of triacylglycerols with saturated fatty acids and are solid at room temperature, whereas plant fats and fish oils are composed of triacylglycerols with unsaturated fatty acids and are liquid at room temperature.

Steroids such as cholesterol are a second type of lipid (**Fig. 2.27**). Like other steroids, cholesterol has a core composed of 20 carbon atoms bonded to form four fused rings, and it is hydrophobic. Cholesterol is a component of animal cell membranes (Chapter 5) and serves as a precursor for the synthesis of steroid hormones such as estrogen and testosterone (Chapter 38).

FIG. 2.26 **Van der Waals forces.** Transient asymmetry in the distribution of electrons along fatty acid chains leads to asymmetry in neighboring molecules, resulting in weak electrostatic attractions.

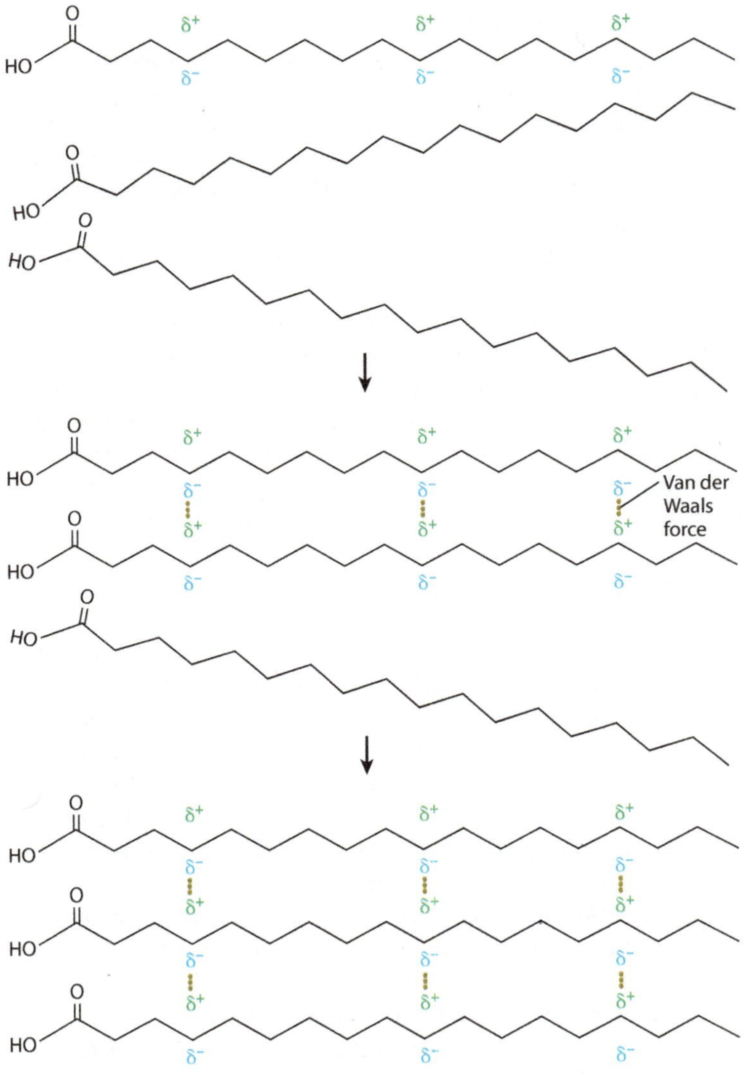

FIG. 2.27 The chemical structure of cholesterol.

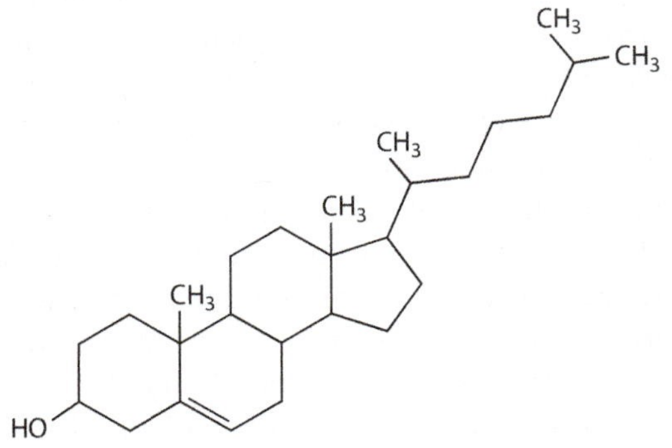

Phospholipids are a third type of lipid. They are a major component of the cell membrane and are described in Chapter 5.

? CASE 1 THE FIRST CELL: LIFE'S ORIGINS

2.6 HOW DID THE MOLECULES OF LIFE FORM?

In Chapter 1, we considered the similarities and differences between living and nonliving things. Four billion years ago, however, the differences may not have been so pronounced. Scientists believe that life originated early in our planet's history, formed by a set of chemical processes that, through time, produced organisms that could be distinguished from their nonliving surroundings. How can we think scientifically about one of biology's deepest and most difficult problems? One idea is to approach life's origins experimentally, asking whether chemical reactions likely to have taken place on the early Earth can generate the molecules of life. It is important to note that even the simplest organisms living today are far more complicated than our earliest ancestors. No one suggests that cells as we know them emerged directly from primordial chemical reactions. Rather, the quest is to discover simple molecular systems able to replicate themselves and subject to natural selection.

A key starting point is the observation, introduced earlier in this chapter, that the principal macromolecules found in organisms are themselves made of simpler molecules joined together. Thus, if we want to understand how proteins might have emerged on the early Earth, we should begin with the synthesis of amino acids, and if we are interested in nucleic acids, we should focus on nucleotides.

The building blocks of life can be generated in the laboratory.

Research into the origins of life was catapulted into the experimental age in 1953 with an elegant experiment carried out by Stanley Miller, then a graduate student in the laboratory of Nobel laureate Harold Urey. Miller started with gases such as water vapor, methane, and hydrogen gas, all thought to have been present in the early atmosphere. He put these gases into a sealed flask and then passed a spark through the mixture (**Fig. 2.28**). On the primitive Earth, lightning might have supplied the energy needed to drive chemical reactions, and the spark was meant to simulate its effects. Analysis of the contents of the flask showed that a number of amino acids were generated.

HOW DO WE KNOW?

FIG. 2.28

Could the building blocks of organic molecules have been generated on the early Earth?

BACKGROUND In the 1950s, Earth's early atmosphere was widely believed to have been rich in water vapor, methane, ammonia, and hydrogen gas, with no free oxygen.

EXPERIMENT Stanley Miller built an apparatus, shown below, designed to simulate Earth's early atmosphere. Then he passed a spark through the mixture to simulate lightning.

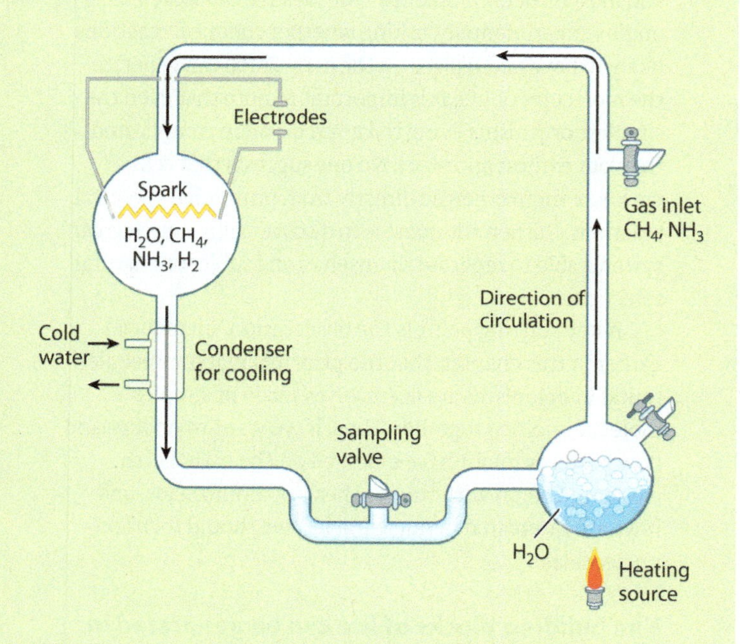

RESULTS As the experiment proceeded, reddish material accumulated on the walls of the flask. Analysis showed that the brown matter included a number of amino acids.

CONCLUSION Amino acids can be generated in conditions that mimic those of the early Earth.

FOLLOW-UP WORK Recent analysis of the original extracts, saved by Miller, shows that the experiment actually produced about 20 different amino acids, not all of them found in organisms.

SOURCES Miller, S. L. 1953. "Production of Amino Acids Under Possible Primitive Earth Conditions." *Science* 117:528–529; Johnson, A. P., et al. 2008. "The Miller Volcanic Spark Discharge Experiment." *Science* 322:404.

Miller and others conducted many variations on his original experiment, all with similar results. Today, many scientists doubt that the early atmosphere had the composition found in Miller's experimental apparatus, but amino acids and other biologically important molecules can form in a variety of simulated atmospheric compositions. If oxygen gas (O_2) is absent and hydrogen is more common in the mixture than carbon, the addition of energy generates diverse amino acids. The absence of oxygen gas is critical since these types of reactions cannot run to completion in modern air or seawater. Here, however, geology supports the experiments: Chemical analyses of Earth's oldest sedimentary rocks indicate that, for its first 2 billion years, Earth's surface contained little or no oxygen.

Later experiments have shown that other chemical reactions can generate simple sugars, the bases found in nucleotides, and the lipids needed to form primitive membranes. Independent evidence that simple chemistry can form the building blocks of life comes from certain meteorites, which provide samples of the early solar system and contain diverse amino acids, lipids, and other organic molecules.

Experiments show how life's building blocks can form macromolecules.

From the preceding discussion, we have seen that life's simple building blocks can be generated under conditions likely to have been present on the early Earth, but can these simple units be linked together to form polymers? Once again, careful experiments have shown how polymers could have formed in the conditions of the early Earth. Clay minerals that form from volcanic rocks can bind nucleotides on their surfaces (**Fig. 2.29a**). The clays provide a surface that places the nucleotides near one another, making it possible for them to join to form chains or simple strands of nucleic acid.

In a classic experiment, biochemist Leslie Orgel placed a short nucleic acid sequence into a reaction vessel and then added individual chemically modified nucleotides. The nucleotides spontaneously joined into a polymer, forming the sequence complementary to the nucleic acid already present (**Fig. 2.29b**).

Such experiments show that nucleic acids can be synthesized experimentally from nucleotides, but until recently the synthesis of nucleotides themselves presented a formidable problem for research on the origins of life. Many tried to generate nucleotides from their sugar, base, and phosphate components, but no one succeeded until 2009. That year, John Sutherland and his colleagues showed that nucleotides can be synthesized under conditions thought to be like those on the young Earth. These chemists showed how simple organic molecules likely to have formed in abundance on the early Earth react in the presence of phosphate molecules, yielding the long-sought nucleotides.

Such humble beginnings eventually gave rise to the abundant diversity of life we see around us, described memorably by Charles Darwin in the final paragraph of *On the Origin of Species:*

> It is interesting to contemplate an entangled bank, clothed with many plants of many kinds, with birds singing on the bushes, with various insects flitting about, and with worms crawling through the damp earth, and to reflect that these elaborately constructed forms, so different from each other, and dependent on each other in so complex a manner, have all been produced by laws acting around us. . . . There is grandeur in this view of life . . . from so simple a beginning endless forms most beautiful and most wonderful have been, and are being, evolved. ■

FIG. 2.29 Spontaneous polymerization of nucleotides. (a) Clays may have played an important role in the origin of life by providing surfaces for nucleotides to form nucleic acids. (b) Addition of a short RNA molecule to a flask containing modified nucleotides results in the formation of a complementary RNA strand. *Source: a. Ray L Frost, Professor of Physical Chemistry, Queensland University of Technology. Australia.*

a.

b.

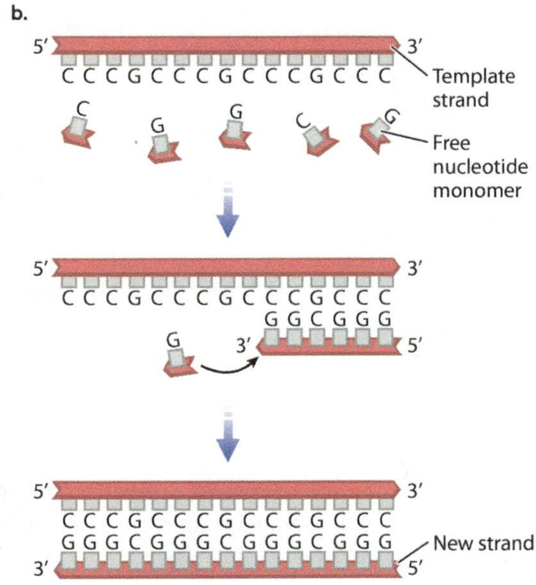

Core Concepts Summary

2.1 The atom is the fundamental unit of matter.

Atoms consist of positively charged protons and electrically neutral neutrons in the nucleus, as well as negatively charged electrons moving around the nucleus. page 30

The number of protons determines the identity of an atom. page 30

The number of protons and neutrons together determines the mass of an atom. page 30

The number of protons versus the number of electrons determines the charge of an atom. page 30

Negatively charged electrons travel around the nucleus in regions called orbitals. page 30

The periodic table of the elements reflects a regular and repeating pattern in the chemical behavior of elements. page 31

2.2 Atoms can combine to form molecules linked by chemical bonds.

Valence electrons occupy the outermost energy level (shell) of an atom and determine the ability of an atom to combine with other atoms to form molecules. page 33

A covalent bond results from the sharing of electrons between atoms to form a molecular orbital. page 33

A polar covalent bond results when two atoms do not share electrons equally as a result of a difference in the ability of the atoms to attract electrons, a property called electronegativity. page 33

An ionic bond results from the attraction of oppositely charged ions. page 34

2.3 Water is abundant and essential for life.

Water is a polar molecule because shared electrons are distributed asymmetrically between the oxygen and hydrogen atoms. page 35

Hydrophilic molecules dissolve readily in water, whereas hydrophobic molecules in water tend to associate with one another, minimizing their contact with water. page 35

A hydrogen bond results when a hydrogen atom covalently bonded to an electronegative atom interacts with an electronegative atom of another molecule. page 35

Water forms hydrogen bonds, which help explain its high cohesion, surface tension, and resistance to rapid temperature change. page 36

The pH of an aqueous solution is a measure of the acidity of the solution. page 37

2.4 Carbon is the backbone of organic molecules.

A carbon atom can form up to four covalent bonds with other atoms. page 38

The geometry of these covalent bonds helps explain the structural and functional diversity of organic molecules. page 38

2.5 Organic molecules include proteins, nucleic acids, carbohydrates, and lipids, each of which is built from simpler units.

Amino acids are linked by covalent bonds to form proteins. page 40

An amino acid consists of a carbon atom (the α carbon) attached to a carboxyl group, an amino group, a hydrogen atom, and a side chain. page 40

The side chain determines the properties of an amino acid. page 40

Nucleotides assemble to form nucleic acids, which store and transmit genetic information. page 40

Nucleotides are composed of a 5-carbon sugar, a nitrogen-containing base, and a phosphate group. page 41

Nucleotides in DNA incorporate the sugar deoxyribose, and nucleotides in RNA incorporate the sugar ribose. page 41

The bases are pyrimidines (cytosine, thymine, and uracil) and purines (guanine and adenine). page 41

Sugars are carbohydrates, molecules composed of C, H, and O atoms, usually in the ratio 1:2:1, and are a source of energy. page 42

Monosaccharides assemble to form disaccharides or longer polymers called complex carbohydrates. page 42

Lipids are hydrophobic. page 43

Triacylglycerols store energy and are made up of glycerol and fatty acids. page 43

Fatty acids consist of a linear hydrocarbon chain of variable length with a carboxyl group at one end. page 43

Fatty acids are either saturated (no carbon–carbon double bonds) or unsaturated (one or more carbon–carbon double bonds). page 43

The tight packing of fatty acids in lipids is the result of van der Waals forces, a type of weak, noncovalent bond. page 44

2.6 Life likely originated on Earth by a set of chemical reactions that gave rise to the molecules of life.

In 1953, Stanley Miller and Harold Urey demonstrated that amino acids can be generated in the laboratory in conditions that mimic those of the early Earth. page 45

Other experiments have shown that sugars, bases, and lipids can be generated in the laboratory. page 46

Once the building blocks were synthesized, they could join together in the presence of clay minerals to form polymers. page 46

Self-Assessment

1. Name and describe the components of an atom.
2. Explain how the periodic table of the elements is organized.
3. Differentiate between covalent bonds and generally weaker interactions such as polar covalent, hydrogen, and ionic bonds.
4. List three unusual properties of water and explain why these properties make water conducive to life.
5. List the four most common elements in organic molecules and state which common macromolecules always contain all four of these elements.
6. List features of carbon that allow it to form diverse structures.
7. List essential functions of proteins, nucleic acids, carbohydrates, and lipids.
8. Describe how diversity is achieved in polymers, using proteins as an example.
9. Sketch the basic structures of amino acids, nucleotides, monosaccharides, and fatty acids.
10. What evidence is there for the hypothesis that life originated on Earth by the creation and polymerization of small organic molecules by natural processes?

Log in to LaunchPad to check your answers to the Self-Assessment questions, and to access additional learning tools.

CHAPTER 3

Nucleic Acids and Transcription

Core Concepts

3.1 Deoxyribonucleic acid (DNA) stores and transmits genetic information.

3.2 DNA is a polymer of nucleotides and forms a double helix.

3.3 Transcription is the process by which RNA is synthesized from a DNA template.

3.4 The primary transcript is processed to become messenger RNA (mRNA).

So much in biology depends on shape. Take your hand, for example. You can pick up a pin, text on a smartphone, or touch your pinky to your thumb. These activities are made possible by the coordinated movement of dozens of bones, muscles, nerves, and blood vessels that give shape to your hand. The functional abilities of your hand emerge from its structure. A causal connection between structure and function exists in many molecules, too. Proteins are a good example. Composed of long, linear strings of 20 different kinds of amino acids in various combinations, each protein folds into a specific three-dimensional shape due to chemical interactions between the amino acids along the chain. The three-dimensional structure of the protein determines its functional properties, such as what other molecules it can bind with, and enables the protein to carry out its job in the cell.

Another notable example is the macromolecule **deoxyribonucleic acid (DNA),** a linear polymer of four different subunits. DNA molecules from all cells and organisms have a very similar three-dimensional structure, reflecting their shared ancestry. This structure, called a **double helix,** is composed of two strands coiled around each other to form a sort of spiral staircase. The banisters of the spiral staircase are formed by the linear backbone of the paired strands, and the steps are formed by the pairing of the subunits at the same level in each strand.

The spiral-staircase structure is common to all cellular DNA molecules, and its structure gave immediate clues to its function. First, DNA *stores* information. Some of the information in DNA encodes for proteins that provide structure and do much of the work of the cell. Information in DNA is called **genetic information,** and it is organized in the form of **genes,** as textual information is organized in the form of words. Genes can exist in different forms in different individuals, even within a single species. Differences in genes can affect the shape of the hand, for example, yielding long or short fingers or extra or missing fingers. As we will see, it is the order of individual subunits (bases) of DNA that accounts for differences in genes. Genes usually have no effect on the organism unless they are "turned on" and their product is made. The turning on of a gene is called **gene expression.** The molecular processes that control whether gene expression occurs at a given time or in a given cell constitute **gene regulation.**

Second, DNA *transmits* genetic information from one generation to the next. The transmission of genetic information from parents to their offspring enables species of organisms to maintain their identity through time. The genetic information in DNA guides the development of the offspring, ensuring that parental apple trees give rise to apple seedlings and parental geese give rise to goslings. As we will see, determining the double-helical structure of DNA provided one of the first hints of how genetic information could be faithfully copied from cell to cell, and from one generation to the next.

In this chapter, we examine the structure of DNA in more detail and show how its structure is well suited to its biological function as the carrier and transmitter of genetic information.

3.1 MAJOR BIOLOGICAL FUNCTIONS OF DNA

DNA is the molecule by which hereditary information is transmitted from generation to generation. Today the role of DNA is well known, but at one time hardly any biologist would have bet on it. Any poll of biologists before about 1950 would have shown overwhelming support for the idea of proteins as life's information molecule. Compared with the seemingly monotonous, featureless structure of DNA, the three-dimensional structures of proteins are highly diverse. Proteins carry out most of the essential activities in the life of a cell, and so it seemed logical to assume that they would play a key role in heredity, too. But while proteins do play a role in heredity, they play a supporting role in looking after the DNA—rather like the way worker bees are essential in maintaining the queen bee, who alone is able to reproduce.

DNA can transfer biological characteristics from one organism to another.

The first experiments to demonstrate that molecules can transfer genetic information from one organism to another were carried out in 1928 by Frederick Griffith, working with the bacterium *Streptococcus pneumoniae.* This organism causes a variety of infections in humans and a deadly form of pneumonia in mice. In the original experiments, illustrated in **Fig. 3.1,** Griffith studied two strains of the bacterium, one a virulent (harmful) strain that caused pneumonia and death when injected into mice, and the other a mutant, nonvirulent strain that allowed injected mice to survive. When the debris of dead virulent cells was mixed with live nonvirulent cells, some of the nonvirulent cells became virulent. Griffith concluded that some type of molecule in the debris carried the genetic information for virulence, but he did not identify the molecule.

Experiments carried out in 1944 by Oswald Avery, Colin MacLeod, and Maclyn McCarty showed that the molecule responsible for the conversion, or **transformation,** of nonvirulent cells into virulent cells is DNA (**Fig. 3.2**). These scientists killed virulent cells with heat, and then purified the remains to make a solution, or preparation. They found that the preparation could carry out transformation. When the preparation was treated with enzymes that destroy any trace of protein or RNA, the transforming ability of the preparation remained. But when the preparation was treated with an enzyme that destroys DNA, the transforming ability was lost.

These experiments, along with others, established that DNA is the genetic material. Today, transformation is widely used in

HOW DO WE KNOW?

FIG.3.1

What is the nature of the genetic material?

BACKGROUND In the 1920s, it was not clear what biological molecule carries genetic information. Fred Neufeld, a German microbiologist, identified several strains of the bacterium *Streptococcus pneumoniae,* one of which was virulent and caused death when injected into mice (Fig. 3.1a), and another which was nonvirulent and did not cause illness when injected into mice (Fig. 3.1b).

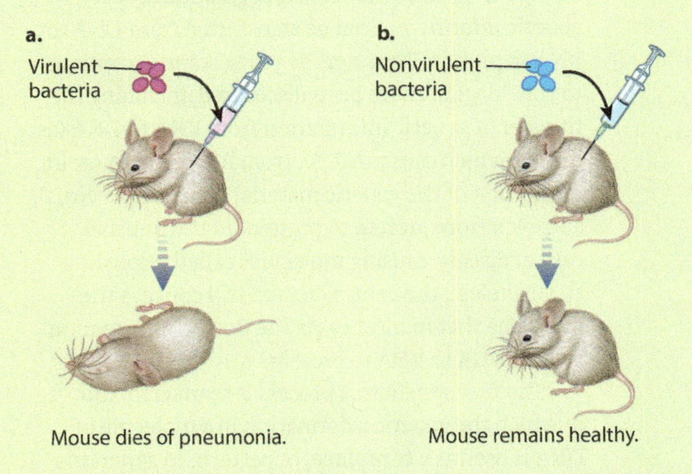

EXPERIMENT Frederick Griffith was also a microbiologist interested in bacterial virulence. He made a puzzling observation. He noted that nonvirulent bacteria do not cause mice to get sick (Fig. 3.1b) and virulent bacteria that had been killed do not cause mice to get sick (Fig. 3.1c), but when the two were mixed, the injected mice got sick and died (Fig. 3.1d). Furthermore, when he isolated bacteria from the dead mice, they had the appearance of the virulent strain, even though he had injected nonvirulent bacteria.

RESULTS

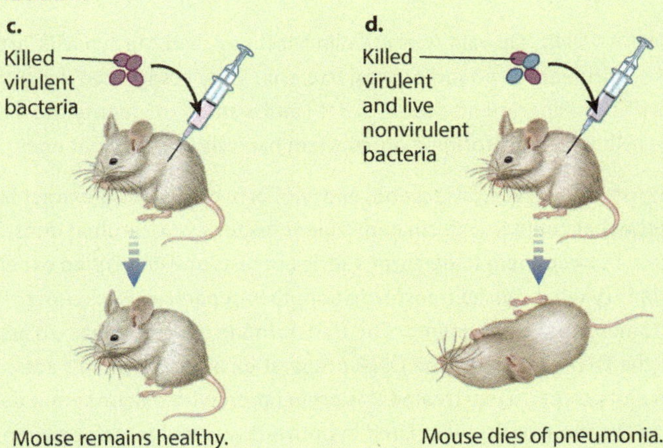

CONCLUSION One strain of bacteria (nonvirulent) can be transformed into another (virulent) by an unknown molecule from the virulent cells. In other words, the unknown molecule carries information that causes virulence.

FOLLOW-UP WORK Griffith's experiments were followed up by many researchers, most notably Oswald Avery, Colin MacLeod, and Maclyn McCarty, who identified DNA as the molecule responsible for transforming bacteria from one strain to the other (see Fig. 3.2). In addition, the process in which DNA is taken up by cells, called transformation, is now a common technique used in molecular biology.

SOURCE Griffith, F. 1928. "The Significance of Pneumococcal Types." *Journal of Hygiene* 27:113–159.

biological research and in the genetic modification of agricultural plants and animals (Chapter 12).

DNA molecules are copied in the process of replication.

DNA can serve as the genetic material because it is unique among cellular molecules in being able to specify exact copies of itself. This copying process, known as **replication,** discussed in detail in Chapter 12, allows the genetic information from one DNA molecule to be copied into that of another DNA molecule. Faithful replication is critical in that it allows DNA to pass genetic information from cell to cell and from parent to offspring. The copying must reproduce the sequence of subunits almost exactly because, as we will see in Chapter 14, mistakes in DNA replication that go unrepaired may be harmful to the cell or organism.

An unrepaired error in DNA replication results in a **mutation,** which is a change in the genetic information in DNA. A mutation in DNA causes the genetic difference between virulent and nonvirulent *Streptococcus pneumoniae.* While most mutations in genes are harmful, rare favorable mutations are essential in the process of evolution because they allow populations of organisms to change through time and adapt to their environment.

Genetic information flows from DNA to RNA to protein.

Biologists often say that genetic information in DNA directs the activities in a cell or guides the development of an organism, but these effects of DNA are indirect. Most of the active molecules in cells and development are proteins, including the enzymes that convert energy into usable forms and the proteins that provide

HOW DO WE KNOW?

FIG.3.2

What is the nature of the genetic material?

BACKGROUND Oswald Avery, Colin MacLeod, and Maclyn McCarty also studied virulence in pneumococcal bacteria. They recognized the significance of Griffith's experiments (see Fig. 3.1) and wanted to identify the molecule responsible for transforming nonvirulent bacteria into virulent ones.

EXPERIMENT Avery, MacLeod, and McCarty prepared an extract from virulent bacteria that could transform nonvirulent bacteria into virulent ones. This extract allowed them to perform a series of tests and controlled experiments. To identify what caused transformation, they separated the extract into its macromolecular components. The transforming activity remained associated with the DNA; however, the DNA preparation also contained trace amounts of RNA and protein. They treated this preparation with enzymes that destroyed one of the three molecules. Their hypothesis was that transformation would not occur if they destroyed the molecule responsible for it.

RESULTS

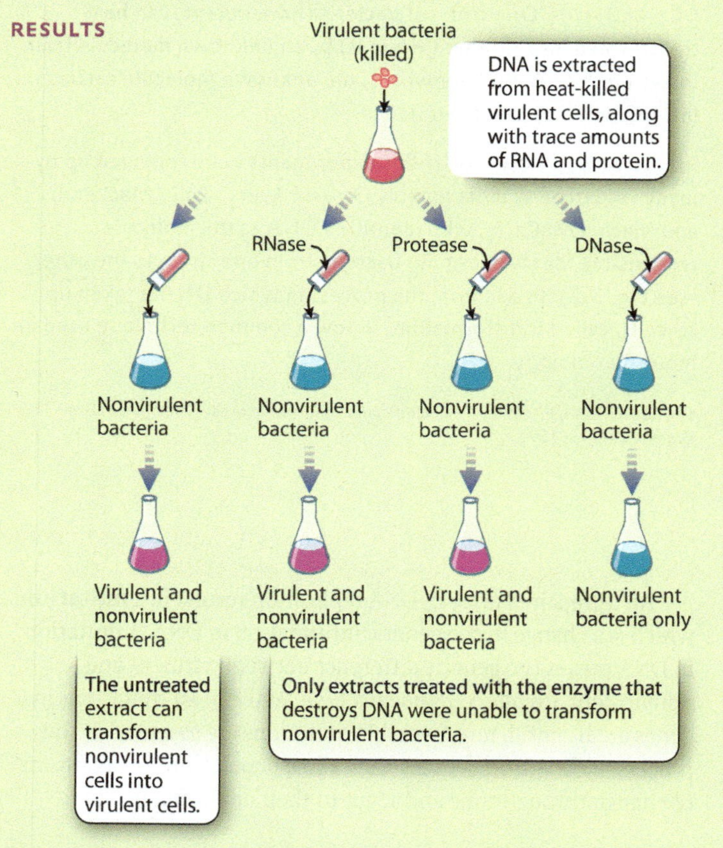

CONCLUSION DNA is the molecule responsible for transforming nonvirulent bacteria into virulent bacteria. This experiment provided a key piece of evidence that DNA is the genetic material.

FOLLOW-UP WORK These experiments were followed up by Alfred Hershey and Martha Chase, who used a different system to confirm that DNA is the genetic material.

SOURCE Avery, O., C. MacLeod, and M. McCarty. 1944. "Studies on the Chemical Nature of the Substance Inducing Transformation of Pneumococcal Types." *Journal of Experimental Medicine* 79:137–158.

structural support for the cell. DNA acts indirectly by specifying the sequence of amino acid subunits of which each protein is composed, and this sequence in turn determines the three-dimensional structure of the protein, its chemical properties, and its biological activities.

In specifying the amino acid sequence of proteins, DNA acts through an intermediary molecule known as **ribonucleic acid (RNA),** another type of linear polymer. As we saw in Chapter 1, the flow of information from DNA to RNA to protein has come to be known as the **central dogma** of molecular biology (**Fig. 3.3**). The central dogma states that genetic information can be transferred from DNA to RNA to protein. Through the years, some exceptions to this "dogma" have been discovered, including the transfer of genetic information from RNA to DNA (as in HIV, which causes AIDS), from RNA to RNA (as in replication of the genetic material of influenza virus), and even from protein to protein (in the unusual case of disease-causing molecules called prions). Nevertheless, the central dogma still conveys the basic idea that in most cases the flow of information is from DNA to RNA to protein.

The first step in this process is **transcription,** in which the genetic information in a molecule of DNA is used as a **template,** or pattern, to generate a molecule of RNA. The term "transcription" is used because it emphasizes that both molecules use the same language of nucleic acids. Transcription is the first step in gene expression, which is the production of a functional gene product. The second step in the readout of genetic information is **translation,** in which a molecule of RNA is used as a code for the sequence of amino acids in a protein. The term "translation" is used to indicate a change of languages, from nucleotides that make up nucleic acids to amino acids that make up proteins.

The processes of transcription and translation are regulated, meaning that they do not occur at all times in all cells, even though all cells in an individual contain the same DNA. Genes are expressed, or "turned on," only at certain times and places, and not expressed, or "turned off," at other times and places. In multicellular organisms, for instance, cells are specialized for certain functions, and these different functions depend on which genes are on and which genes are off in specific cells. Muscle cells express genes that encode for proteins involved in muscle contraction, but these genes are not expressed in skin cells or liver cells, for example. Similarly, during development of a multicellular organism, genes may

FIG. 3.3 The central dogma of molecular biology, which defines the usual flow of information in a cell from DNA to RNA to protein.

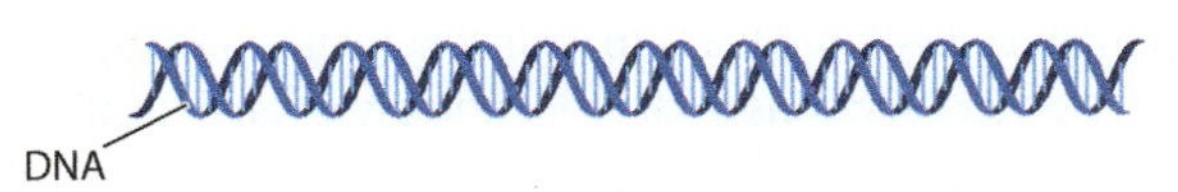

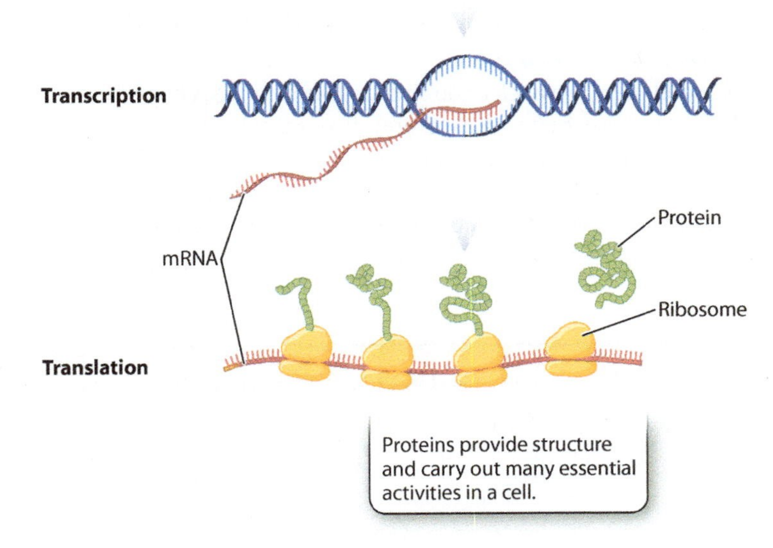

be required at certain times but not at others. In this case, the timing of expression is carefully controlled.

In prokaryotes, transcription and translation occur in the cytoplasm, but in eukaryotes, the two processes are separated from each other, with transcription occurring in the nucleus and translation in the cytoplasm. The separation of transcription and translation in time and space in eukaryotic cells allows for additional levels of gene regulation that are not possible in prokaryotic cells. In spite of this and other differences in the details of transcription and translation between prokaryotes and eukaryotes, the processes are sufficiently similar that they must have evolved early in the history of life.

3.2 CHEMICAL COMPOSITION AND STRUCTURE OF DNA

By about 1950, some biologists were convinced by the work of Griffith, Avery, and others that DNA is the genetic material. To serve as the genetic material, DNA would have to be able to replicate itself, undergo rare mutations, replicate mutant forms as faithfully as the original forms, and direct the synthesis of other macromolecules in the cell. How could one molecule do all this? Part of the answer emerged in 1953, when James D. Watson and Francis H. C. Crick announced a description of the three-dimensional structure of DNA. This discovery marked a turning point in modern biology. The discovery of the structure of DNA opened the door to understanding how genetic information is stored, faithfully replicated, and altered by rare mutations.

A DNA strand consists of subunits called nucleotides.

In the sixty years since the publication of Watson and Crick's paper, we have all become familiar with the iconic double helix of DNA. The elegant shape of the twisting strands relies on the structure of DNA's subunits, called **nucleotides.** As we saw in Chapter 2, nucleotides consist of three components: a 5-carbon **sugar,** a **base,** and one or more **phosphate groups** (**Fig. 3.4**). Each component plays an important role in DNA structure. The 5-carbon sugars and phosphate groups form the backbone of the molecule, with each sugar linked to the phosphate group of the neighboring nucleotide. The bases sticking out from the sugar give each nucleotide its chemical identity. Each strand of DNA consists of an enormous number of nucleotides linked one to the next.

DNA had been discovered 85 years before its three-dimensional structure was determined, and in the meantime a great deal had been learned about its chemistry, specifically the chemistry of nucleotides. Fig. 3.4 illustrates a nucleotide. In the figure, the 5-carbon sugar is indicated by the pentagon, in which four of the five vertices represent the position of a carbon atom. By convention the carbon atoms of the sugar ring are numbered clockwise with primes (1′, 2′, and so forth, read as "one prime," "two prime," and so forth). Technically, the sugar in DNA is 2′-deoxyribose because the chemical group projecting downward from the 2′ carbon is a hydrogen atom (–H) rather than a hydroxyl group (–OH), but for our purposes the term **deoxyribose** will suffice.

Note in Fig. 3.4 that the phosphate group attached to the 5′ carbon has negative charges on two of its oxygen atoms. These charges are present because at cellular pH (around 7), the free hydroxyl groups attached to the phosphorus atom are ionized by the loss of a proton, and hence are negatively charged. It is these negative charges that make DNA a mild acid, which you will recall from Chapter 2 is a molecule that tends to lose protons to the aqueous environment.

Each base is attached to the 1′ carbon of the sugar and projects above the sugar ring. A nucleotide normally contains one of four

FIG. 3.4 Nucleotide structure.

Ionized hydroxyl groups
5′ 4′ 3′ 2′ 1′
Phosphate group
Deoxyribose sugar
Base (A, G, T, or C)

FIG. 3.5 Bases normally found in DNA.

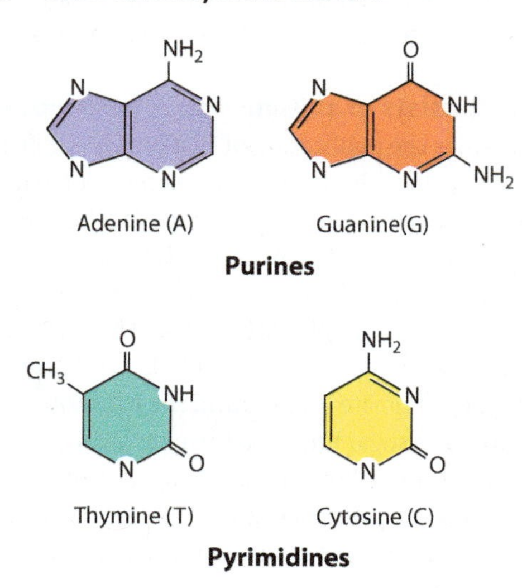

kinds of bases, denoted A, G, T, and C (**Fig. 3.5**). Two of the bases are double-ring structures known as **purines;** these are the bases **adenine (A)** and **guanine (G),** shown across the top of the figure. The other two bases are single-ring structures known as **pyrimidines;** these are the bases **thymine (T)** and **cytosine (C),** shown across the bottom.

The combination of sugar and base is known as a **nucleoside,** which is shown in simplified form in **Fig. 3.6.** A nucleoside with one or more phosphate groups constitutes a nucleotide. More specifically, a nucleotide with one, two, or three phosphate groups is called a nucleoside monophosphate, diphosphate, or triphosphate, respectively (Fig. 3.6). The nucleoside triphosphates are particularly important because, as we will see later in this chapter, they are the molecules that are used to form nucleotide polymers: DNA and RNA. In addition, nucleoside triphosphates have other functions in the cell, notably as carriers of chemical energy in the form of ATP and GTP.

FIG. 3.6 **Nucleosides and nucleotides.** A nucleoside is a sugar attached to a base, and nucleotides are nucleosides with one, two, or three phosphate groups attached. A nucleotide is therefore a nucleoside phosphate.

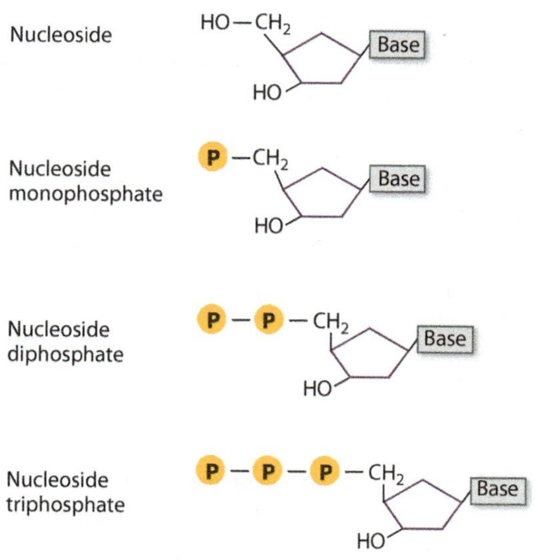

Nucleoside — HO—CH_2 — Base — HO

Nucleoside monophosphate — P—CH_2 — Base — HO

Nucleoside diphosphate — P—P—CH_2 — Base — HO

Nucleoside triphosphate — P—P—P—CH_2 — Base — HO

DNA is a linear polymer of nucleotides linked by phosphodiester bonds.

Not only were the nucleotide building blocks of DNA known before the structure was discovered, it was also known how they were linked into a polymer. The chemical linkages between nucleotides in DNA are shown in **Fig. 3.7.** The characteristic covalent bond that connects one nucleotide to the next is

FIG. 3.7 Nucleotides linked by phosphodiester bonds to form a DNA strand.

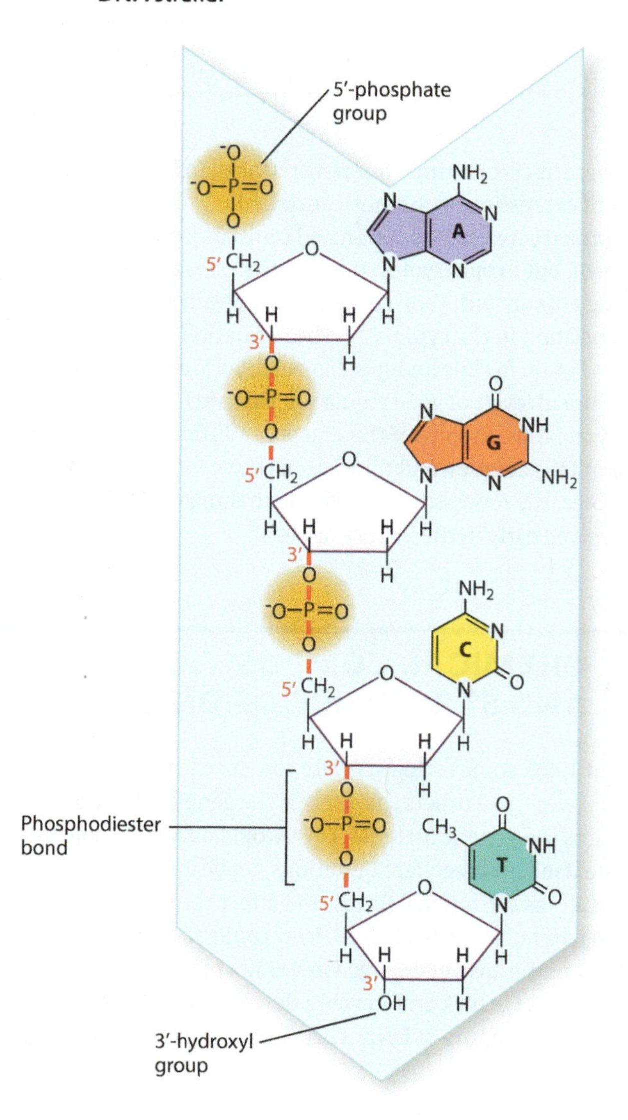

indicated by the vertical red lines that connect the 3′ carbon of one nucleotide to the 5′ carbon of the next nucleotide in line through the 5′-phosphate group. This C–O–P–O–C linkage is known as a **phosphodiester bond,** which in DNA is a relatively stable bond that can withstand stress like heat and substantial changes in pH that would break weaker bonds. The succession of phosphodiester bonds traces the backbone of the DNA strand.

The phosphodiester linkages in a DNA strand give it **polarity,** which means that one end differs from the other. In Fig. 3.7, the nucleotide at the top has a free 5′ phosphate, and is known as the **5′ end** of the molecule. The nucleotide at the bottom has a free 3′ hydroxyl and is known as the **3′ end.** The DNA strand in Fig. 3.7 has the sequence of bases AGCT from top to bottom, but because of strand polarity we need to specify which end is which. For this strand of DNA, we could say that the base sequence is 5′-AGCT-3′ or equivalently 3′-TCGA-5′. When a base sequence is stated without specifying the 5′ end, by convention the end at the left is the 5′ end. Hence, we could say the sequence in Fig. 3.7 is AGCT, which means 5′-AGCT-3′.

Cellular DNA molecules take the form of a double helix.

To the knowledge of the chemical makeup of the nucleotides and their linkages in a DNA strand, Watson and Crick added results from earlier physical studies indicating that DNA is a long molecule. They also relied on important information from X-ray diffraction studies by Rosalind Franklin implying that DNA molecules form a helix with a simple repeating structure. Analysis of the pattern of X-rays diffracted from a crystal of a molecule can indicate the arrangement of atoms in the molecules.

With these critical pieces of information in hand, Watson and Crick set out to build a model of DNA that could account for the results of all previous chemical and physical experiments, using sheet metal cutouts of the bases and wire ties for the sugar–phosphate backbone. After many false starts and much disappointment, they finally found a structure that worked. They realized immediately that they had made one of the most important discoveries in all of biology, and that day, February 28, 1953, they lunched at the Eagle, a pub across the street from their laboratory, where Crick loudly pronounced, "We have discovered the secret of life." The Eagle is still there in Cambridge, England, and sports on its wall a commemorative plaque marking the table where the two ate.

Why all the fuss (and why the Nobel Prize nine years later)? First, let's look at the structure, and you will see that the structure itself tells you how DNA carries and transmits genetic information. The Watson–Crick structure, now often called the double helix, is shown in **Fig. 3.8.** Fig. 3.8a is a space-filling model, in which each atom is represented as a color-coded sphere. The big surprise of the structure is that it consists of two DNA strands like those in Fig. 3.7, each wrapped around the other in the form of a helix coiling to the right, with the sugar–phosphate backbones winding around the outside of the molecule and the bases pointing inward. In the double helix, there are 10 base pairs per complete turn, and the diameter of the molecule is 2 nm, a measurement that is hard to relate to everyday objects, but it might help to know that the cross section of a bundle of 100,000 DNA molecules would be about the size of the period at the end of this sentence. The outside contours of the twisted strands form an uneven pair of grooves, called the **major groove** and the

FIG. 3.8 Structure of DNA. The DNA double helix can be shown with (a) the atoms as solid spheres or (b) the backbones as ribbons.

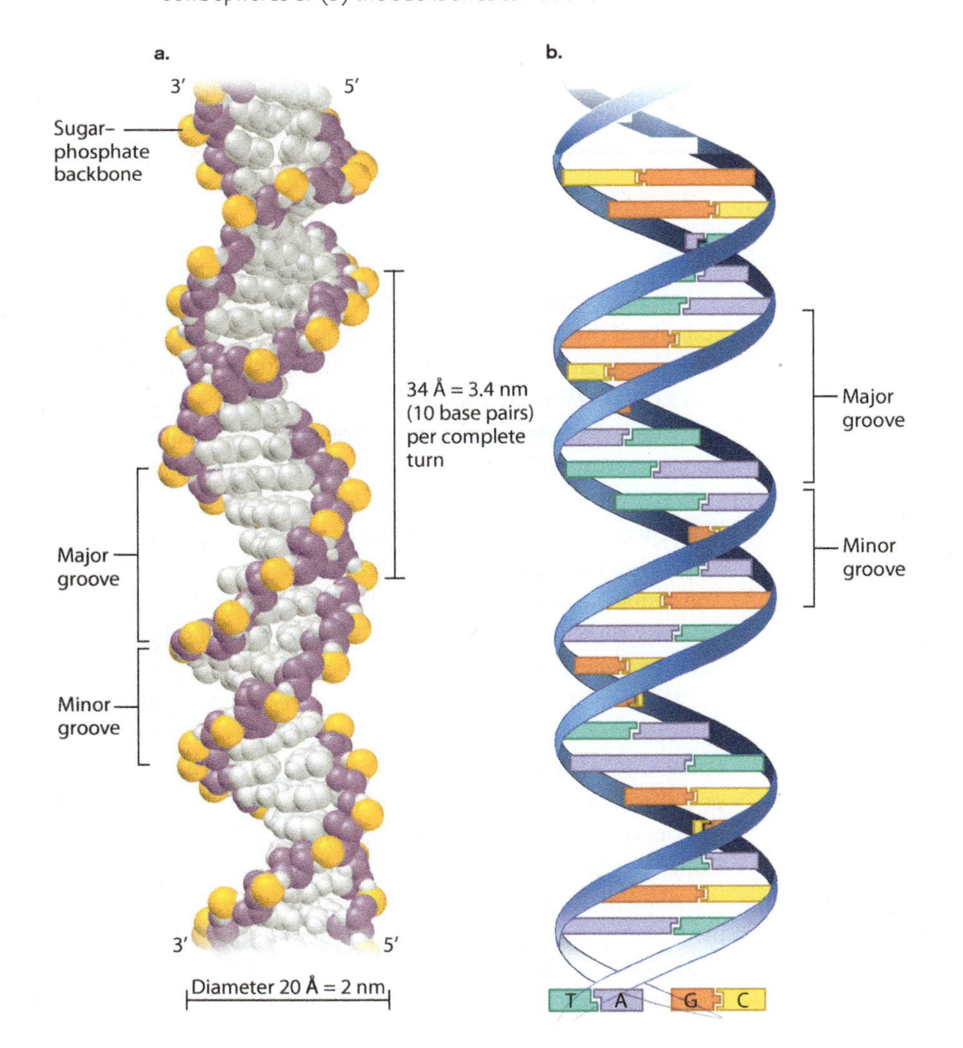

minor groove. These grooves are important because proteins that interact with DNA often recognize a particular sequence of bases by making contact with the bases via the major or minor groove or both.

Importantly, the individual DNA strands in the double helix are **antiparallel,** which means that they run in opposite directions. That is, the 3′ end of one strand is opposite the 5′ end of the other. In Fig. 3.8a, the strand that starts at the bottom left and coils upward begins with the 3′ end and terminates at the top with the 5′ end, whereas its partner strand begins with its 5′ end at the bottom and terminates with the 3′ end at the top.

Fig. 3.8b shows a different depiction of double-stranded DNA, called a ribbon model, which clearly shows the sugar-phosphate backbones winding around the outside with the bases paired between the strands. The ribbon model of the structure closely resembles a spiral staircase, with the backbones forming the banisters and the base pairs the steps. If the amount of DNA in a human egg or sperm (3 billion base pairs) were scaled to the size of a real spiral staircase, it would reach from Earth to the moon.

Note that, as shown in Fig. 3.8b, an A in one strand pairs only with a T in the other, and G pairs only with C. Each base pair contains a purine and a pyrimidine. This precise pairing maintains the structure of the double helix since pairing two purines would cause the backbones to bulge and pairing two pyrimidines would cause them to narrow. The pairing of one purine with one pyrimidine preserves the distance between the backbones along the length of the entire molecule.

Because they form specific pairs, the bases A and T are said to be **complementary,** as are the bases G and C. The formation of only A–T and G–C base pairs means that the paired strands in a double-stranded DNA molecule have different base sequences. The strands are paired like this:

5′-ATGC-3′
3′-TACG-5′

Where one strand has the base A, the other strand across the way has the base T, and where one strand has a G, the other has a C. In other words, the paired strands are not identical but complementary. Because of the A–T and G–C base pairing, knowing the base sequence in one strand immediately tells you the base sequence in its partner strand.

Why is it that A pairs only with T, and G only with C? **Fig. 3.9** illustrates the answer. The specificity of base pairing is brought about by hydrogen bonds that form between A and T (two hydrogen bonds) and between G and C (three hydrogen bonds). A hydrogen bond in DNA is formed when an electronegative atom (O or N) in one base shares a hydrogen atom (H) with another electronegative atom in the base across the way. Hydrogen bonds are relatively weak bonds, typically 5% to 10% of the strength of covalent bonds, and can be disrupted by high pH or heat. However, in total they contribute to the stability of the DNA double helix.

FIG. 3.9 Base pairing. Adenine pairs with thymine, and guanine pairs with cytosine. These base pairs differ in the number of hydrogen bonds.

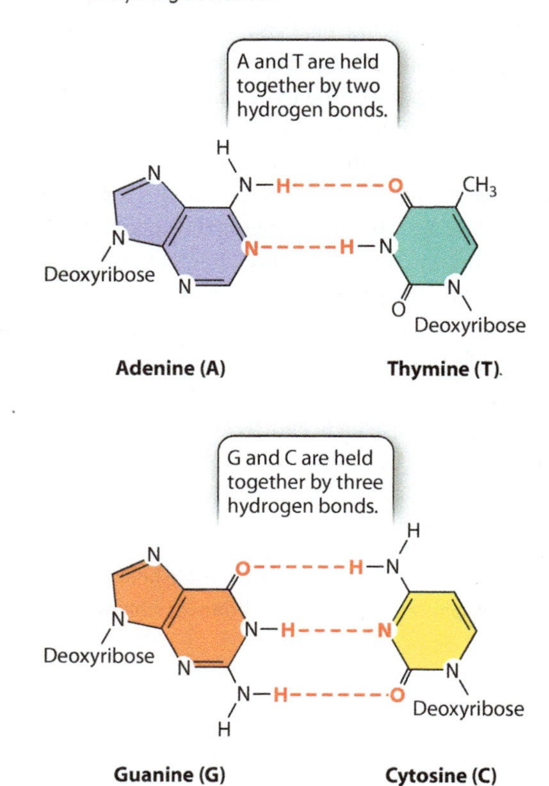

An almost equally important factor contributing to the stability of the double helix is the interactions between bases in the same strand (**Fig. 3.10**). This stabilizing force is known as **base stacking,** and it occurs because the nonpolar, flat surfaces of the bases tend to group together away from water molecules, and hence stack on top of one another as tightly as possible.

The three-dimensional structure of DNA gave important clues about its functions.

The double helix is a good example of how chemical structure and biological function come together. The structure of the molecule itself immediately suggested how genetic information is stored in DNA. One of the most important features of DNA structure is that there is no restriction on the sequence of bases along a DNA strand. Any A, for example, can be followed by another A or a T, C or G. The lack of sequence constraint suggested that the genetic information in DNA could be encoded in the sequence of bases along the DNA, much as textual information in a book is stored in a sequence of letters of the alphabet. With any of four possible bases at each nucleotide site, the information-carrying capacity of a DNA molecule is unimaginable. The number of possible base sequences of a DNA molecule only 133 nucleotides in length is

FIG. 3.10 Interactions stabilizing the double helix. Hydrogen bonds between the bases in opposite strands and base stacking of bases within a strand contribute to the stability of the DNA double helix.

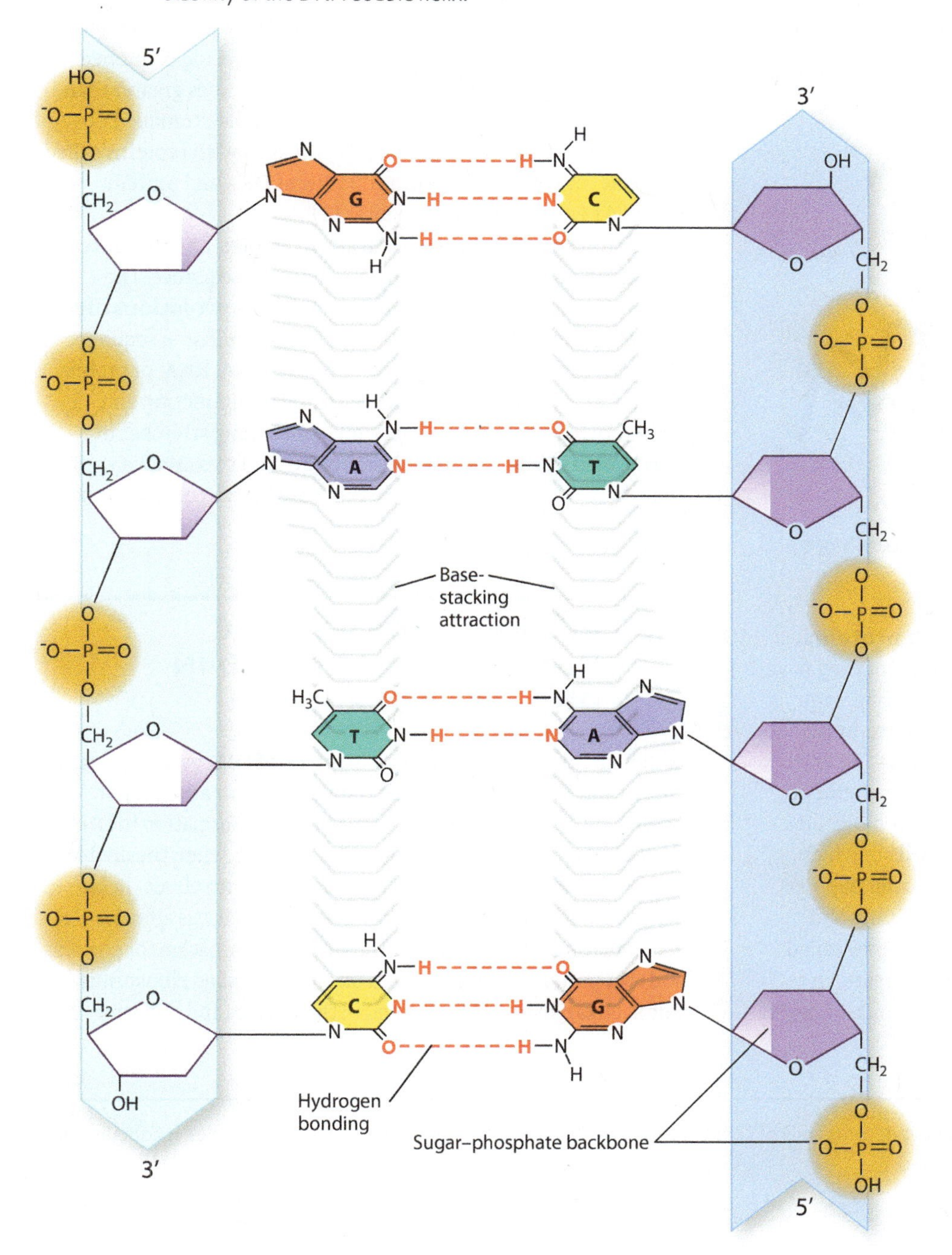

equal to the estimated number of electrons, protons, and neutrons in the entire universe! This is the secret of how DNA can carry the genetic information for so many different types of organisms, and how variation in DNA sequence even within a single species can underlie genetic differences among individuals.

Watson and Crick's model still left many questions unanswered in regard to how the information is read out and what it does, but in the years following it became clear that the major processes in the readout of genetic information were transcription and translation (see Fig. 3.3).

The sequence of bases along either strand completely determines that of the other because wherever one strand carries an A, the other must carry a T, and wherever one carries a G, the other must carry a C. The complementary base sequences of the strands means that, in any double-stranded DNA molecule, the total number of A bases must equal that of T, and the total number of G bases must equal that of C. These equalities are often written in terms of the percent (%) of each base in double-stranded DNA. In these terms, the base-pairing rules imply that %A = %T and %G = %C. Interestingly, these equalities were known and described by the American biochemist Erwin Chargaff before the double helix was discovered, and they provided important clues to the structure of DNA. It was the double helix that finally showed why they are observed.

→ **Quick Check 1** The letter R is conventionally used to represent any purine base (A or G) and Y to represent any pyrimidine base (T or C). In double-stranded DNA, what is the relation between %R and %Y?

The complementary, double-stranded nature of the double helix also suggested a mechanism by which DNA replication could take place. In fact, in their 1953 paper describing the structure of DNA, Watson and Crick wrote, "It has not escaped our notice that the specific pairing we have postulated immediately suggests a possible copying mechanism for the genetic material."

A simplified outline of DNA replication is shown in **Fig. 3.11.** In brief, the two strands of a parental double helix unwind, and as they do each of the parental strands serves as a template for the synthesis of a complementary daughter strand. When the process is complete, there are two molecules, each of which is identical in sequence to the original molecule, except possibly for rare errors (mutations) that cause one base pair to be replaced with another. Although the process of replication as depicted in Fig. 3.11 looks exceedingly simple and straightforward,

FIG. 3.11 DNA replication. The structure of the double helix gave an important clue to how it replicates.

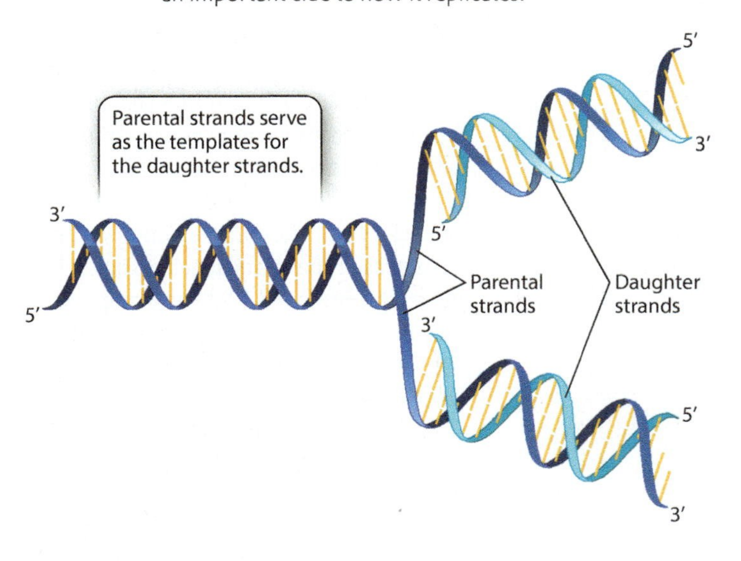

there are technical details that make the actual process more complex. These are discussed in Chapter 12.

Cellular DNA is coiled and packaged with proteins.

The DNA molecules inside cells are highly convoluted. They have to be because DNA molecules in cells have a length far greater than the diameter of the cell itself. The DNA of a bacterium known as *Mycoplasma,* for example, if stretched to its full linear extent would be about 1000 times longer than the diameter of the bacterial cell. Many of the double-stranded DNA molecules in prokaryotic cells are circular and form **supercoils** in which the circular molecule coils upon itself, much like what happens to a rubber band when you twist it between your thumb and forefinger (**Fig. 3.12**). Supercoiling is caused by enzymes called **topoisomerases** that cleave, partially unwind, and reattach a DNA strand, which puts strain on the DNA double helix. Supercoils then relieve the strain and help to preserve the 10 base pairs per turn in the double helix.

In eukaryotic cells, most DNA molecules in the nucleus are linear, and each individual molecule forms one **chromosome.** There is a packaging problem here, too, which you can appreciate by considering that the length of the DNA molecule contained in a single human chromosome is roughly 6000 times greater than the average diameter of the cell nucleus. Double-stranded DNA molecules in eukaryotes are usually packaged with proteins called histones, and others, to form a complex of DNA and proteins referred to as **chromatin** (Chapter 13).

Histone proteins are found in all eukaryotes, and they interact with double-stranded DNA without regard to sequence. The reason for this ability is that these proteins are **evolutionarily conserved,** which means that they are very similar in sequence from one organism to the next. Conserved DNA, RNA, or protein sequences indicate that they serve an essential function and therefore have not changed very much over long stretches of evolutionary time. The more distantly related two organisms are that share conserved sequences, the more highly conserved the sequence is.

3.3 RETRIEVAL OF GENETIC INFORMATION STORED IN DNA: TRANSCRIPTION

Although the three-dimensional structure of DNA gave important clues about how DNA stores and transmits information, it left open many questions about how the genetic information in DNA is read out to control cellular processes. In 1953, when the double helix was discovered, virtually nothing was known about these processes. Within a few years, however, evidence was already accumulating that DNA carries the genetic information for proteins, and that proteins are synthesized on particles called **ribosomes.** But in eukaryotes, DNA is located in the nucleus and ribosomes are

FIG. 3.12 Supercoils. A highly twisted rubber band forms coils of coils (supercoils), much as a circular DNA molecule does when it contains too many base pairs per helical turn.

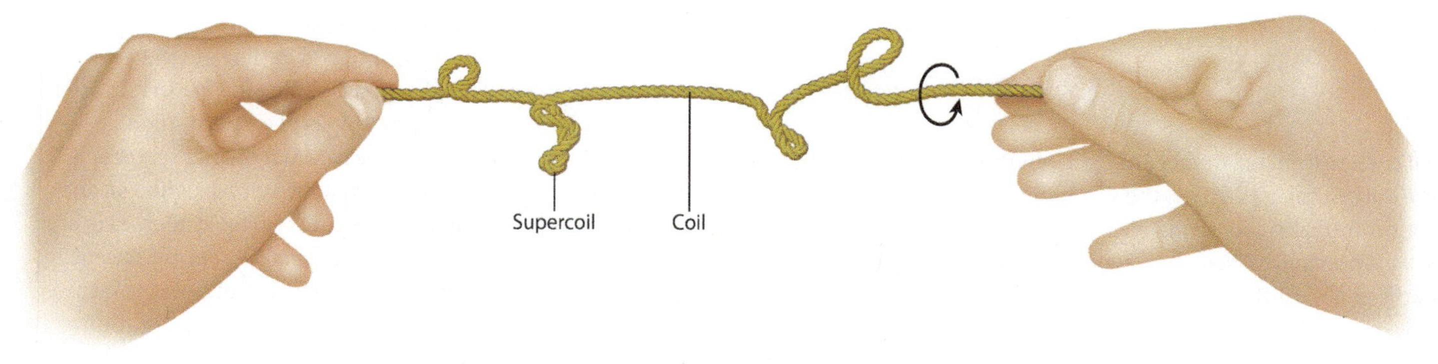

located in the cytoplasm. There must therefore be an intermediary molecule by which the genetic information is transferred from the DNA in the nucleus to ribosomes in the cytoplasm, and some researchers began to suspect that this intermediary was another type of nucleic acid called ribonucleic acid (RNA).

The hypothesis of an RNA intermediary that carries genetic information from DNA to the ribosomes was supported by a clever experiment carried out in 1961 by Sydney Brenner, François Jacob, and Matthew Meselson. They used the virus T2, which infects cells of the bacterium *Escherichia coli* and hijacks the cellular machinery to produce viral proteins. The researchers found that while T2 DNA never associates with bacterial ribosomes, the infected cells produce a burst of RNA molecules shortly after infection and before viral proteins are made. This finding and others suggested that RNA retrieves the genetic information stored in DNA for use in protein synthesis. The transfer of genetic information from DNA to RNA constitutes the key step of transcription in the central dogma of molecular biology (see Fig. 3.2). In this section, we examine RNA and the process of transcription.

? CASE 1 THE FIRST CELL: LIFE'S ORIGINS

What was the first nucleic acid molecule, and how did it arise?

RNA is a remarkable molecule. Like DNA, it can store information in its sequence of nucleotides. In addition, some RNA molecules can actually act as enzymes that facilitate chemical reactions. Because RNA has properties of both DNA (information storage) and proteins (enzymes), many scientists think that RNA, not DNA, was the original information-storage molecule in the earliest forms of life on Earth. This idea, sometimes called the **RNA world hypothesis,** is supported by other evidence as well. Notably, as we will see, RNA is involved in key cellular processes, including DNA replication, transcription, and translation. Many scientists believe that this involvement is a remnant of a time when RNA played a more central role in life's fundamental processes.

Ingenious experiments carried out by Jack W. Szostak and collaborators show how RNA could have evolved the ability to catalyze a simple reaction. A strand of RNA was synthesized in the laboratory and then replicated many times to produce a large population of identical RNA molecules. Next, the RNA was exposed to a chemical that induced random changes in the identity of some of the nucleotides in these molecules. These random changes were mutations that created a population of diverse RNA molecules, much in the way that mutation builds genetic variation in cells.

Next, all of these RNA molecules were placed into a container, and those RNA variants that successfully catalyzed a simple reaction—cleaving a strand of RNA, for example, or joining two strands together—were isolated, and the cycle was repeated. In each round of the experiment, the RNA molecules that functioned best were retained, replicated, subjected to treatments that induced additional mutations, and then tested for the ability to catalyze the same reaction. With each generation, the RNA catalyzed the reaction more efficiently, and after only a few dozen rounds of the procedure, very efficient RNA catalysts had evolved. Experiments such as this one suggest that RNA molecules can evolve over time and act as catalysts. Therefore, many scientists believe that RNA, with its dual functions of information storage and catalysis, was a key molecule in the very first forms of life.

If RNA played a key role in the origin of life, why do cells now use DNA for information storage and proteins to carry out other cellular processes? RNA is much less stable than DNA, and proteins are more versatile, so a plausible explanation is that life evolved from an RNA-based world to one in which DNA, RNA, and proteins are specialized for different functions.

RNA is a polymer of nucleotides in which the 5-carbon sugar is ribose

RNA is a polymer of nucleotides linked by phosphodiester bonds similar to those in DNA (see Fig. 3.7). Each RNA strand therefore has a polarity determined by which end of the chain carries the 3′ hydroxyl (–OH) and which end carries the 5′ phosphate. There are a number of important differences that distinguish RNA from DNA, however (**Fig. 3.13**). First, the sugar in RNA is **ribose,** which carries a hydroxyl group on the 2′ carbon (highlighted in pink in Fig. 3.13a). Hydroxyls are reactive functional groups, so the additional hydroxyl group on ribose in part explains why RNA is a less stable molecule than DNA. Second, the base **uracil** in RNA

FIG. 3.13 **RNA.** RNA differs from DNA in that (a) RNA contains the sugar ribose rather than deoxyribose and (b) the base uracil rather than thymine.

a.

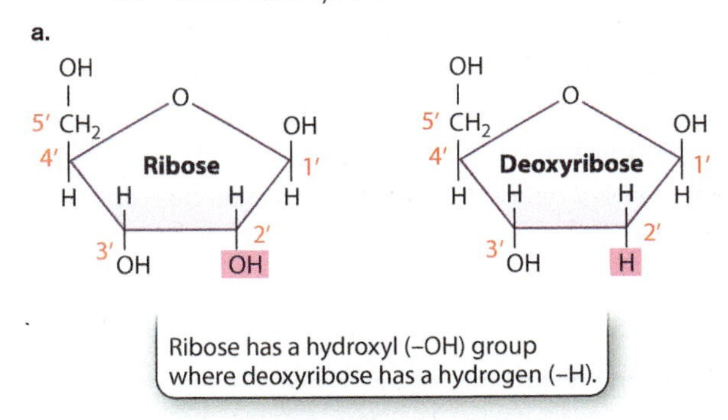

b.

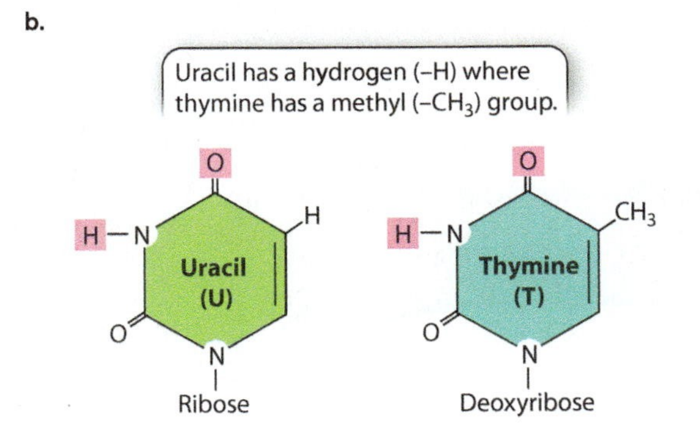

replaces thymine in DNA (Fig. 3.13b). The groups that participate in hydrogen bonding (highlighted in pink in Fig. 3.13b) are identical so that uracil pairs with adenine (U–A) just as thymine pairs with adenine (T–A). Third, while the 5′ end of a DNA strand is typically a monophosphate, the 5′ end of an RNA molecule is typically a triphosphate.

Two other features that distinguish RNA from DNA are physical rather than chemical. One is that RNA molecules are usually much shorter than DNA molecules. A typical RNA molecule used in protein synthesis consists of a few thousand nucleotides, whereas a typical DNA molecule consists of millions or tens of millions of nucleotides. The other major distinction is that most RNA molecules in the cell are single stranded, whereas DNA molecules, as we saw, are double stranded. Single-stranded RNA molecules often form complex three-dimensional structures by folding back upon themselves, which enhances their stability.

In transcription, DNA is used as a template to make complementary RNA.

Conceptually, the process of transcription is straightforward. As a region of the DNA duplex unwinds, one strand is used as a template for the synthesis of an **RNA transcript** that is complementary in sequence to the template according to the base-pairing rules, except that the transcript contains U (uracil) where the template has an A (**Fig. 3.14**). The transcript is produced by polymerization of ribonucleoside triphosphates. The enzyme that carries out the polymerization is known as **RNA polymerase,** which acts by adding successive nucleotides to the 3′ end of the growing transcript (Fig. 3.14). Only the template strand of DNA is transcribed. Its partner, called the **nontemplate strand,** is not transcribed.

It is important to keep in mind the direction of growth of the RNA transcript and the direction that the DNA template is read. All nucleic acids are synthesized by addition of nucleotides to the 3′ end. That is, they grow in a 5′-to-3′ direction, also described simply as the 3′ direction. Just like two strands of DNA in a double helix, the DNA template and the RNA strand transcribed from it are antiparallel, meaning the DNA template runs in the opposite direction from the RNA (Fig. 3.14).

FIG. 3.14 Transcription. The DNA double helix unwinds for transcription, and usually only one strand, the template strand, is transcribed.

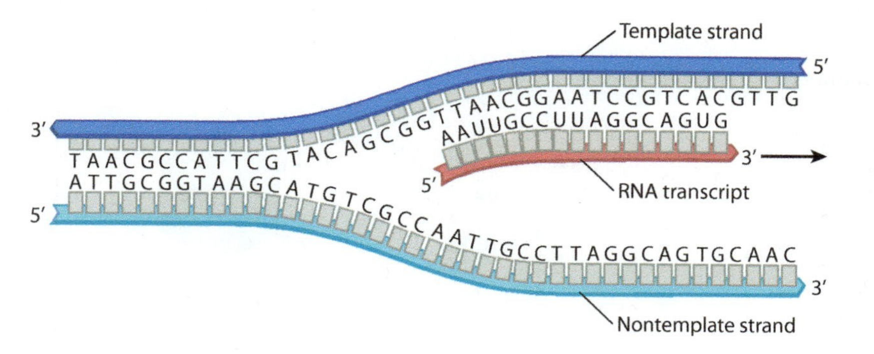

→ **Quick Check 2** A segment of one strand of a double-stranded DNA molecule has the sequence 5′-ACTTTCAGCGAT-3′. What is the sequence of an RNA molecule synthesized from this DNA template?

Transcription starts at a promoter and ends at a terminator.

A long DNA molecule typically contains thousands of genes, most of them coding for proteins or RNA molecules with specialized functions, and hence thousands of different transcripts are produced. For example, the DNA molecule in the bacterium *E. coli* has about 4 million base pairs and produces about 4000 RNA transcripts, most of which code for proteins. A typical map of a small part of a long DNA molecule is shown in **Fig. 3.15.** Each green segment indicates the position where a transcription is initiated, and each purple segment indicates the position where it ends.

The green segments are **promoters,** regions of typically a few hundred base pairs where RNA polymerase and associated proteins bind to the DNA duplex. Many eukaryotic and archaeal promoters contain a sequence similar to 5′-TATAAA-3′, which is known as a **TATA box** because the TATA sequence is usually present. The first nucleotide to be transcribed is usually positioned about 25 base pairs from the TATA box, and transcription takes place as the RNA polymerase moves along the template strand in the 3′-to-5′ direction.

Transcription continues until the RNA polymerase encounters a sequence known as a **terminator** (shown in purple in Fig. 3.15). Transcription stops at the terminator, and the transcript is released. A long DNA molecule contains the genetic information for hundreds or thousands of genes. For any one gene, usually only one DNA strand is transcribed; however, different genes in the same double-stranded DNA molecule can be transcribed from opposite strands. Which strand is transcribed depends on the position of the promoter. As shown in Fig. 3.15, promoters in opposite strands result in transcription occurring in opposite directions. The DNA strand that contains the promoter sequence matters because, as noted earlier, transcription can proceed only by successive addition of nucleotides to the 3′ end of the transcript.

Transcription does not take place indiscriminately from promoters but is a regulated process. For genes called **housekeeping genes,** whose products are needed at all times in all cells, transcription takes place continually. But most genes are transcribed only at certain times, under certain conditions, or in certain cell types. In *E. coli,* for example, the genes that encode proteins needed to utilize the sugar lactose (milk sugar) are transcribed only when lactose is present in the environment. For such genes, regulation of transcription often depends on whether the RNA polymerase and associated proteins are able to bind with the promoter (Chapter 19).

In bacteria, promoter recognition is mediated by a protein called **sigma factor,** which associates with

FIG. 3.15 Transcription along a stretch of DNA. A DNA molecule usually contains many genes that are transcribed individually and at different times, often from opposite strands.

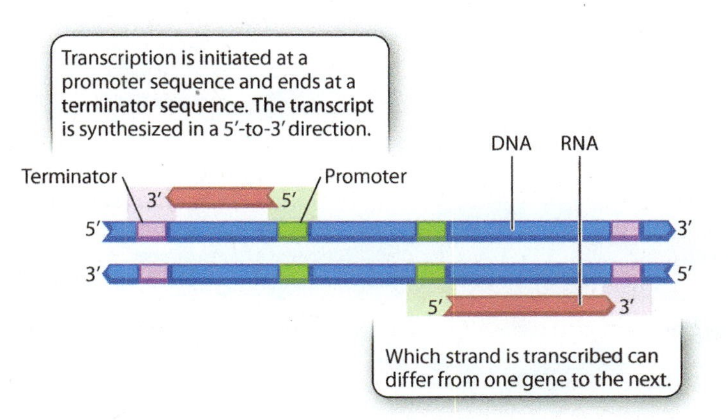

RNA polymerase and facilitates its binding to specific promoters. One type of sigma factor is used for transcription of housekeeping genes and many others, but there are other sigma factors for genes whose expression is needed under special environmental conditions such as lack of nutrients or excess heat.

Promoter recognition in eukaryotes is considerably more complicated. Transcription requires the combined action of at least six proteins known as **general transcription factors** that assemble at the promoter of a gene. Assembly of the general transcription factors is necessary for transcription to occur, but not sufficient. Also needed is the presence of one or more types of **transcriptional activator protein,** each of which binds to a specific DNA sequence known as an **enhancer** (**Fig. 3.16**). Transcriptional activator proteins help control when and in which cells transcription of a gene will occur. They are able to bind with enhancer DNA sequences in or near the gene, and also bind with proteins that allow transcription to begin. The presence of the transcriptional activator proteins that bind with enhancers

FIG. 3.16 The eukaryotic transcription complex, composed of many different proteins.

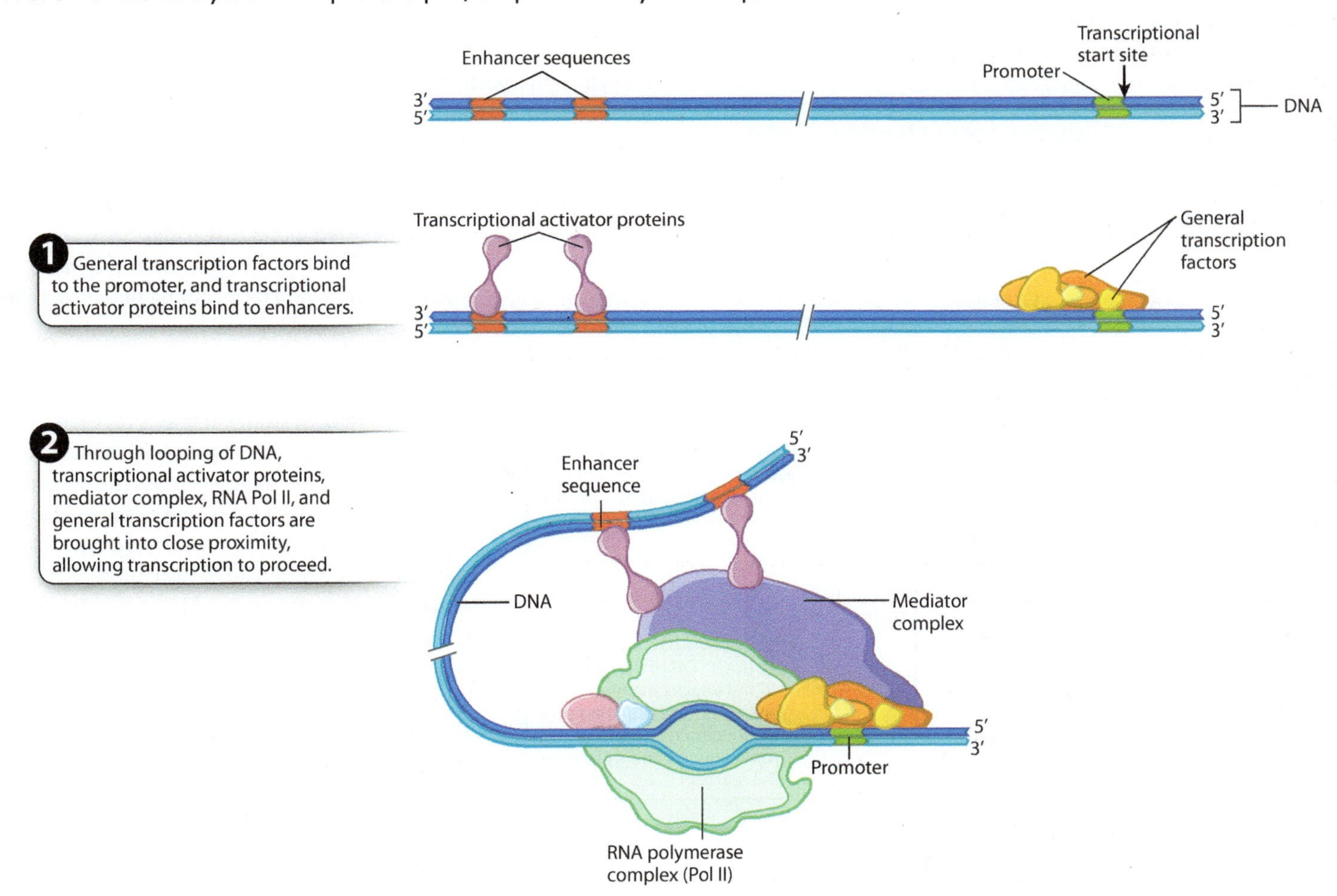

controlling the expression of the gene is therefore required for transcription of any eukaryotic gene to begin (Fig. 3.16).

Once transcriptional activator proteins have bound to enhancer DNA sequences, they can attract, or recruit, a **mediator complex** of proteins, which in turn recruits the RNA polymerase complex to the promoter. Because enhancers can be located almost anywhere in or near a gene, the recruitment of the mediator complex and the RNA polymerase complex may require the DNA to loop around as shown in Fig. 3.16. Cells have several different types of RNA polymerase enzymes, but in both prokaryotes and eukaryotes all protein-coding genes are transcribed by just one of them. In eukaryotes, the RNA polymerase complex responsible for transcription of protein-coding genes is called **Pol II.** Once the mediator complex and Pol II complex are in place, transcription begins, and this process is called transcriptional **initiation.**

FIG. 3.17 Transcription bubble. Within the polymerase, the two strands of DNA separate and the growing RNA strand forms a duplex with the DNA template.

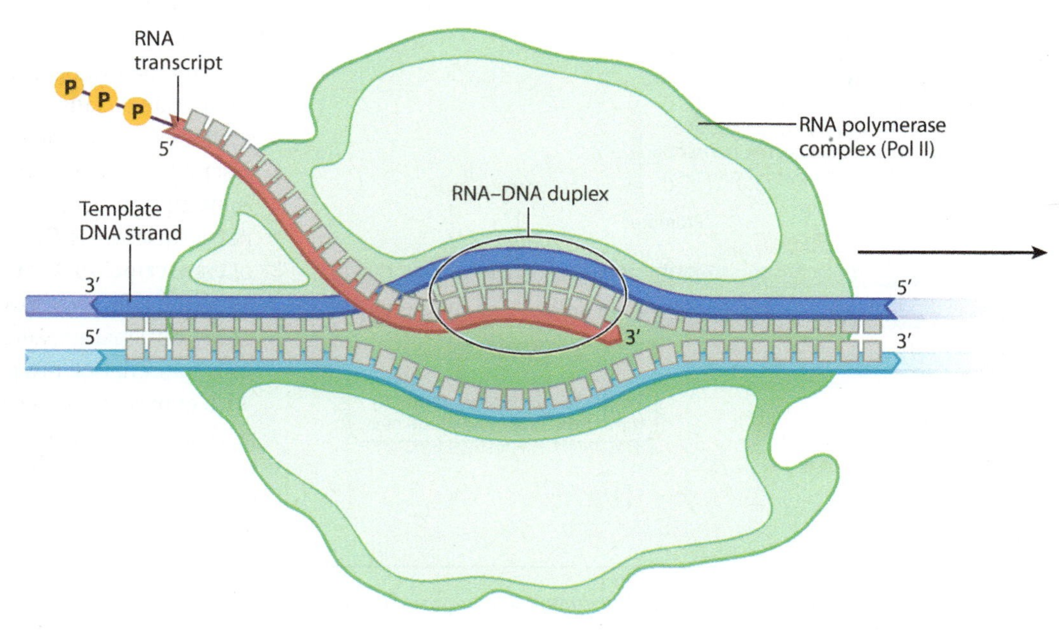

RNA polymerase adds successive nucleotides to the 3′ end of the transcript.

Once transcriptional initiation takes place, successive ribonucleotides are added to grow the transcript. This step is known as **elongation.** Transcription takes place in a sort of bubble in which the strands of the DNA duplex are separated and the growing end of the RNA transcript is paired with the template strand, creating an RNA–DNA duplex (**Fig. 3.17**). In bacteria, the total length of the transcription bubble is about 14 base pairs, and the length of the RNA–DNA duplex in the bubble is about 8 base pairs.

FIG. 3.18 The polymerization reaction that allows the RNA transcript to be elongated.

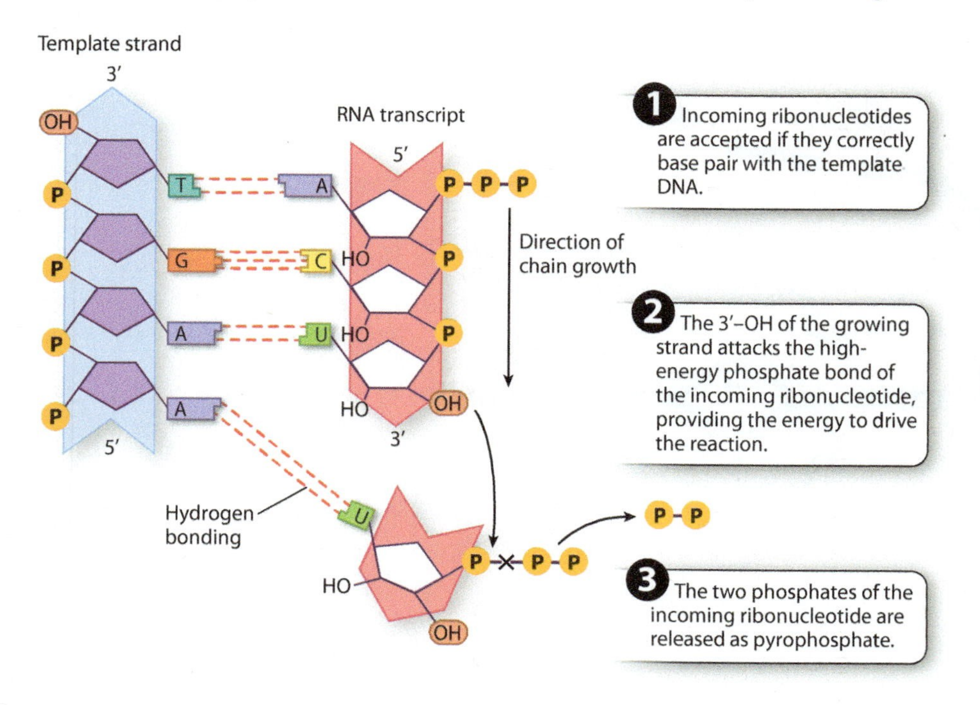

Details of the polymerization reaction are shown in **Fig. 3.18.** The incoming ribonucleoside triphosphate, shown at the bottom right, is accepted by the RNA polymerase only if it undergoes proper base pairing with the base in the template DNA strand. In Fig. 3.18, there is a proper match because U pairs with A. At this point, the RNA polymerase orients the oxygen in the hydroxyl group at the 3′ end of the growing strand into a position from which it can attack the innermost phosphate of the triphosphate of the incoming ribonucleoside, competing for the covalent bond. The bond connecting the innermost phosphate to the next is a high-energy phosphate bond, which when cleaved provides the energy to drive the reaction that creates the phosphodiester bond attaching the incoming nucleotide to the 3′ end of the growing chain. The term "high-energy" here refers to the amount of energy released when the phosphate bond is broken that can be used to drive other chemical reactions.

The polymerization reaction releases a phosphate–phosphate group (pyrophosphate),

FIG. 3.19 The RNA polymerase complex in prokaryotes. This molecular machine has channels for DNA input and output, nucleotide input, and RNA output, and features that disrupt the DNA double helix, stabilize the RNA–DNA duplex, and allow the DNA double helix to re-form.

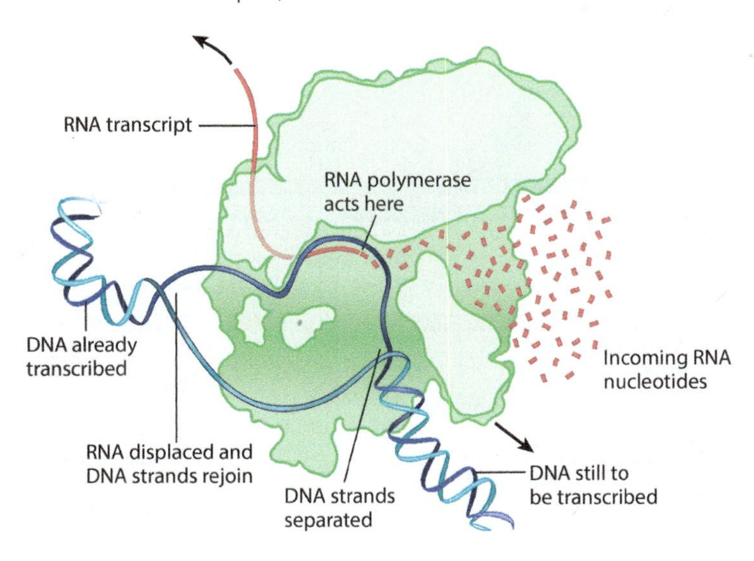

shown at the lower right in Fig. 3.18, which also has a high-energy phosphate bond that is cleaved by another enzyme. Cleavage of the pyrophosphate molecule makes the polymerization reaction irreversible, and the next ribonucleoside triphosphate that complements the template is brought into line.

→ **Quick Check 3** What is the consequence for a growing RNA transcript if an abnormal nucleotide with a 3′ H is incorporated rather than a 3′ OH? How about a 2′ H rather than a 2′ OH?

The RNA polymerase complex is a molecular machine that opens, transcribes, and closes duplex DNA.

Transcription does not take place spontaneously. It requires template DNA, a supply of ribonucleoside triphosphates, and RNA polymerase, a large multiprotein complex in which transcription occurs. To illustrate RNA polymerase in action, we consider here bacterial RNA polymerase because of its relative simplicity.

The transcription bubble forms and transcription takes place inside the polymerase (**Fig. 3.19**). RNA polymerase contains structural features that separate the DNA strands, allow an RNA–DNA duplex to form, elongate the transcript nucleotide by nucleotide, release the finished transcript, and restore the original DNA double helix. Fig. 3.19 shows how structure and function come together in the bacterial RNA polymerase.

RNA polymerase is a remarkable molecular machine capable of adding thousands of nucleotides to a transcript before dissociating from the template. It is also very accurate, with only about 1 incorrect nucleotide incorporated per 10,000 nucleotides.

3.4 FATE OF THE RNA PRIMARY TRANSCRIPT

The RNA transcript that comes off the template DNA strand is known as the **primary transcript,** and it contains the genetic information of the gene that was transcribed. For protein-coding genes, this means that the primary transcript includes the information needed to direct the ribosome to produce the protein corresponding to the gene (Chapter 4). The RNA molecule that combines with the ribosome to direct protein synthesis is known as the **messenger RNA (mRNA)** because it serves to carry the genetic "message" (information) from the DNA to the ribosome. As we will see in this section, there is a major difference between prokaryotes and eukaryotes in the manner in which the primary transcript relates to the mRNA. We will also see that some genes do not code for proteins, but for RNA molecules that have functions of their own.

Messenger RNA carries information for the synthesis of a specific protein.

In prokaryotes, the relation between the primary transcript and the mRNA is as simple as can be: The primary transcript *is* the mRNA. Even as the 3′ end of the primary transcript is still being synthesized, ribosomes bind with special sequences near its 5′ end and begin the process of protein synthesis (**Fig. 3.20**). This intimate connection between transcription and translation can take place because prokaryotes have no nuclear envelope to spatially separate transcription from translation; the two processes are coupled, which means that they are connected in space and time.

FIG. 3.20 Fate of the primary transcript for protein-coding genes in prokaryotes.

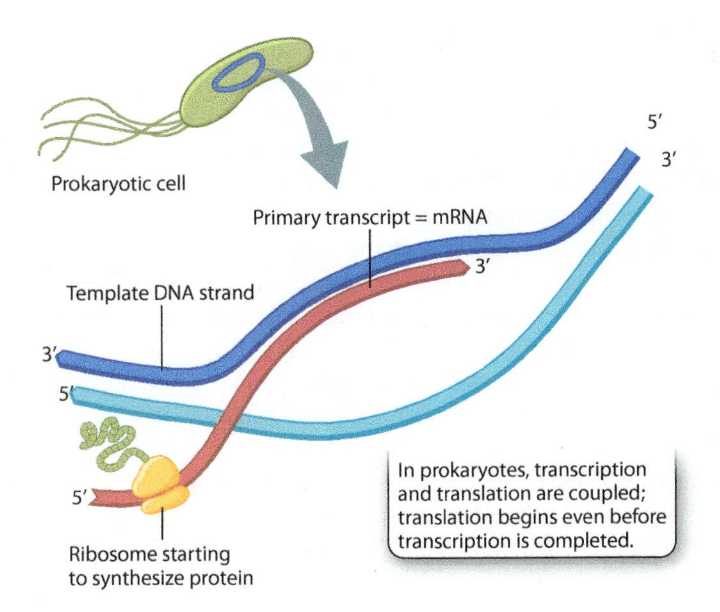

Primary transcripts for protein-coding genes in prokaryotes have another feature not shared with those in eukaryotes: They often contain the genetic information for the synthesis of two or more different proteins, usually proteins that code for successive steps in the biochemical reactions that produce small molecules needed for growth, or successive steps needed to break down a small molecule used for nutrients or energy. Molecules of mRNA that code for multiple proteins are known as **polycistronic mRNA** because the term "cistron" was once widely used to refer to a protein-coding sequence in a gene.

Primary transcripts in eukaryotes undergo several types of chemical modification.

In eukaryotes, the nuclear envelope is a barrier between the processes of transcription and translation. Transcription takes place in the nucleus, and translation in the cytoplasm. The separation allows for a complex chemical modification of the primary transcript, known as **RNA processing,** which converts the primary transcript into the finished mRNA, which can then be translated by the ribosome.

RNA processing consists of three principal types of chemical modification, illustrated in **Fig. 3.21.** First, the 5′ end of the primary transcript is modified by the addition of a special nucleotide attached in an unusual linkage. This addition is called the **5′ cap,** and it consists of a modified nucleotide called 7-methylguanosine. An enzyme attaches the modified nucleotide to the 5′ end of the primary transcript essentially backward: in a normal linkage between two nucleotides, the phosphodiester bond forms between the 5′ carbon of one and the 3′-OH group of the next, but here the cap is linked to the RNA transcript by a triphosphate bridge between the 5′ carbons of both ribose sugars (**Fig. 3.22**). The 5′ cap is essential for translation because in eukaryotes the ribosome recognizes an mRNA by its 5′ cap. Without the cap, the ribosome would not attach the mRNA and translation would not occur.

The second major modification of eukaryotic primary transcripts is **polyadenylation,** the addition of a string of about 250 consecutive A-bearing ribonucleotides to the 3′ end, forming a **poly(A) tail** (see Fig. 3.21). Polyadenylation plays an important role in export of the mRNA into the cytoplasm. In addition, both the 5′ cap and poly(A) tail help to stabilize the RNA transcript. Single-stranded nucleic acids can be unstable and are even susceptible to enzymes that break them down. In eukaryotes, the 5′ cap and poly(A) tail protect the two ends of the transcript and increase the stability of the RNA transcript until it is translated in the cytoplasm.

Not every stretch of the RNA transcript ends up being translated into protein. Transcripts in

FIG. 3.21 Fate of the primary transcript for protein-coding genes in eukaryotes.

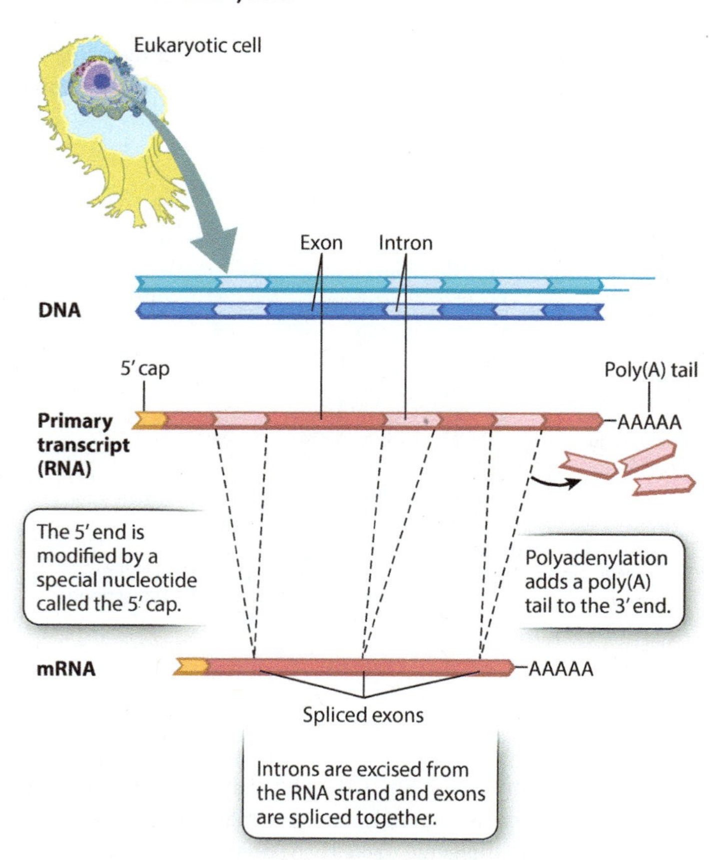

FIG. 3.22 Structure of the 5′ cap on eukaryotic messenger RNA.

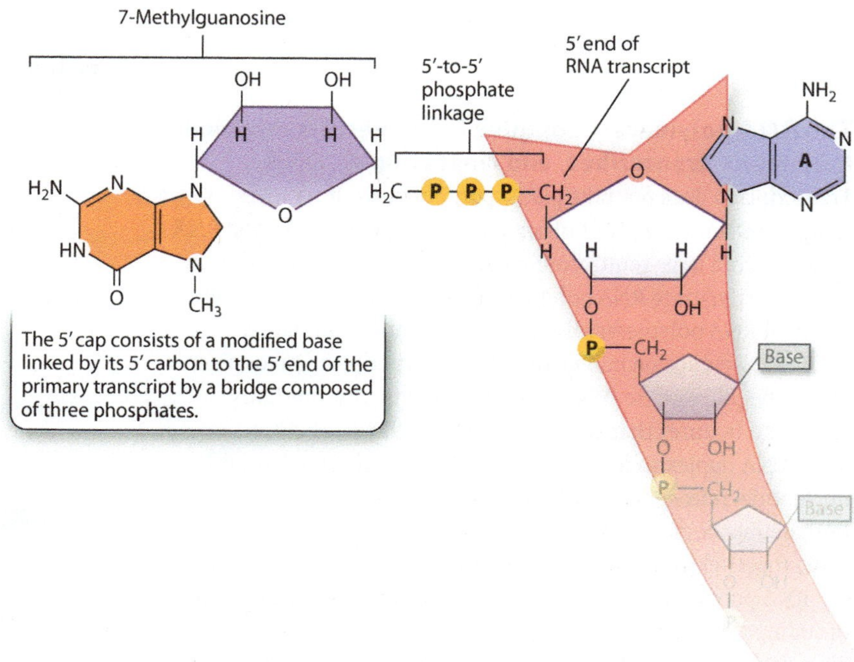

FIG. 3.23 RNA splicing.

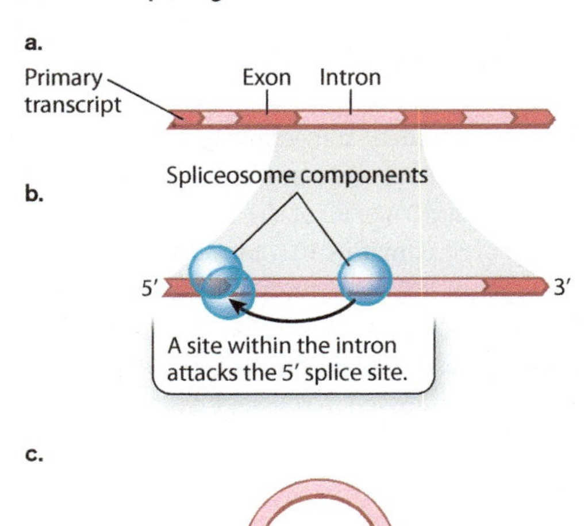

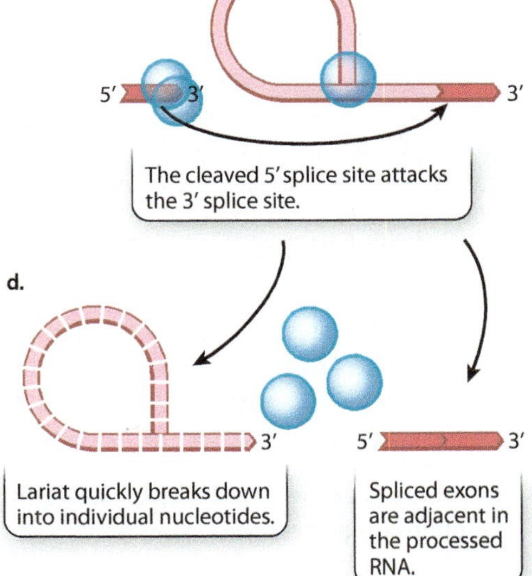

eukaryotes often contain regions of protein-coding sequence, called **exons,** interspersed with noncoding regions called **introns.** The third type of modification of the primary transcript is the removal of the noncoding introns (see Fig. 3.21). The process of intron removal is known as **RNA splicing,** which is catalyzed by a complex of RNA and protein known as the **spliceosome.**

The mechanism of splicing is outlined in **Fig. 3.23.** In the first step, the spliceosome brings a specific sequence within the intron into proximity with the 5′ end of the intron, at a site known as the 5′ splice site (Fig. 3.23a). The proximity enables a reaction that cuts the RNA at the 5′ splice site, and the cleaved end of the intron connects back on itself forming a loop and tail called a **lariat** (Fig. 3.23b). In the next step, the spliceosome brings the 5′ splice site close to the splice site at the 3′ end of the intron. The 5′ splice site attacks the 3′ splice site (Fig. 3.23c), cleaving the bond that holds the lariat on the transcript and attaching the ends of the exons to each other. The result is that the exons are connected and the lariat is released (Fig. 3.23d). The lariat making up the intron is quickly broken down into its constituent nucleotides.

About 90% of all human genes contain at least one intron. Although most genes contain 6 to 9 introns, the largest number is 147, found in a muscle gene. Most introns are just a few thousand nucleotides in length, but about 10% are longer than 10,000 nucleotides. The presence of multiple introns in most genes allows for a process known as **alternative splicing,** in which primary transcripts from the same gene can be spliced in different ways to yield different mRNAs and therefore different protein products (**Fig. 3.24**). More than 80% of human genes are alternatively spliced. In most cases, the alternatively spliced forms differ in whether a particular exon is or is not removed from the primary transcript along with its flanking introns. Alternative splicing allows the same transcript to be processed in diverse ways to produce mRNA molecules with different combinations of exons coding for different proteins.

→ **Quick Check 4** When a region of DNA that contains the genetic information for a protein is isolated from a bacterial cell and inserted into a eukaryotic cell in a proper position between a promoter and a terminator, the resulting cell usually produces the correct protein. But when the experiment is done in the reverse direction (eukaryotic DNA into a bacterial cell), the correct protein is often not produced. Can you suggest an explanation?

Some RNA transcripts are processed differently from protein-coding transcripts and have functions of their own.

Not all primary transcripts are processed into mRNA. Some RNA transcripts have functions of their own, and many of these transcripts are produced by RNA polymerases other than

FIG. 3.24 **Alternative splicing.** A single primary transcript can be spliced in different ways.

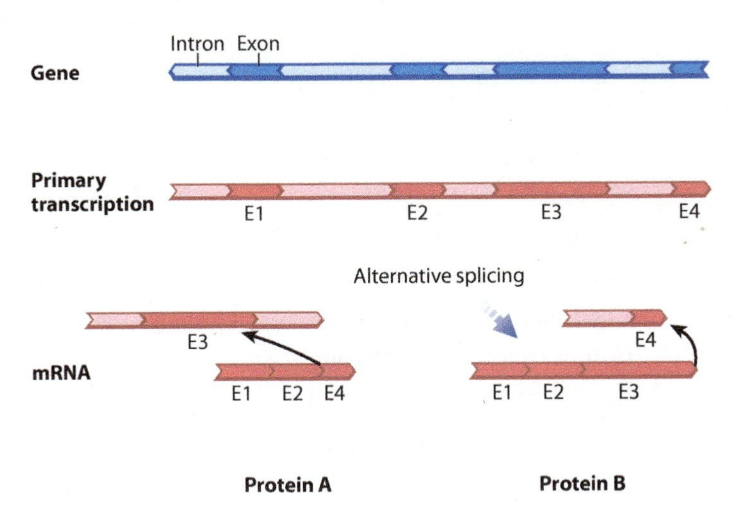

Pol II. These primary transcripts undergo different types of RNA processing, and their processed forms include such important noncoding RNA types as:

* **Ribosomal RNA (rRNA),** found in all ribosomes that aid in translation. In eukaryotic cells, the genes and transcripts for ribosomal RNA are concentrated in the **nucleolus,** a distinct, dense, non–membrane-bound spherical structure observed within the nucleus.
* **Transfer RNA (tRNA)** that carries individual amino acids for use in translation.
* **Small nuclear RNA (snRNA),** found in eukaryotes and involved in splicing, polyadenylation, and other processes in the nucleus.
* Small, regulatory RNA molecules that can inhibit translation or cause destruction of an RNA transcript. Two major types of small regulatory RNA are known as **microRNA (miRNA)** and **small interfering RNA (siRNA)**.

By far, the most abundant transcripts in mammalian cells are those for ribosomal RNA and transfer RNA. In a typical mammalian cell, about 80% of all of the RNA consists of ribosomal RNA, and another approximately 10% consists of transfer RNA. Why are these types of RNA so abundant? The answer is that they are needed in large amounts to synthesize the proteins encoded in the messenger RNA. The roles of mRNA, tRNA, and rRNA in protein synthesis are discussed in the next chapter. ■

Core Concepts Summary

3.1 Deoxyribonucleic acid (DNA) stores and transmits genetic information.

Experiments carried out by Griffith in 1928 demonstrated that bacteria can transmit genetic information from one strain to another. page50

Experiments performed by Avery, MacLeod, and McCarty in 1944 showed that DNA is the molecule that transmits genetic information. page 50

DNA is copied in the process of replication. page 51

Ribonucleic acid (RNA) is synthesized from a DNA template. page 51

The central dogma of molecular biology states that the usual flow of genetic information is from DNA to RNA to protein. DNA is transcribed to RNA, and RNA is translated to protein. page 51

3.2 DNA is a polymer of nucleotides and forms a double helix.

A nucleotide consists of a 5-carbon sugar, a phosphate group, and a base. page 53

The four bases of DNA are adenine, guanine, cytosine, and thymine. page 53

Successive nucleotides are linked by phosphodiester bonds to form a linear DNA molecule. page 54

DNA strands have polarity, with a 5′-phosphate group at one end and a 3′ hydroxyl group at the other end. page 54

Cellular DNA molecules consist of a helical spiral of two paired, antiparallel strands called a double helix. page 55

In a DNA double helix, A pairs with T, and G pairs with C. page 56

The structure of DNA relates to its function. Information is coded in the sequence of bases, and the structure suggests a mechanism for replication, in which each parental strand serves as a template for a daughter strand. page 56

DNA in eukaryotic cells is packaged with evolutionarily conserved proteins called histones. page 58

3.3 Transcription is the process by which RNA is synthesized from a DNA template.

RNA, like DNA, is a polymer of nucleotides linked by phosphodiester bonds. page 59

Some types of RNA can store genetic information and other types can catalyze chemical reactions. These characteristics have led to the RNA world hypothesis, the idea that RNA played a critical role in the early evolution of life on Earth. page 59

Unlike DNA, RNA incorporates the sugar ribose instead of deoxyribose and the base uracil instead of thymine. page 59

RNA is synthesized from one of the two strands of DNA, called the template strand. page 60

RNA is synthesized by RNA polymerase in a 5′-to-3′ direction, starting at a promoter and ending at a terminator in the DNA template. page 60

RNA polymerase separates the two DNA strands, allows an RNA–DNA duplex to form, elongates the transcript, releases the transcript, and restores the DNA duplex. page 62

3.4 The primary transcript is processed to become messenger RNA (mRNA).

In prokaryotes, the primary transcript is immediately translated into protein. page 63

In eukaryotes, transcription and translation are separated in time and space, with transcription occurring in the nucleus and then translation in the cytoplasm. page 64

In eukaryotes, there are three major types of modification to the primary transcript—the addition of a 5′ cap, polyadenylation, and splicing. page 64

The 5′ cap is a 7-methylguanosine added to the 5′ end of the transcript. page 64

The poly(A) tail is a stretch of adenines added to the 3′ end of the transcript. page 64

Splicing is the excision of introns from the transcript, bringing exons together. page 65

Alternative splicing is a process in which primary transcripts from the same gene are spliced in different ways to yield different protein products. page 65

Some RNAs, called noncoding RNAs, do not code for proteins, but instead have functions of their own. page 65

Self-Assessment

1. Explain how the functions of DNA emerge from the structure of its monomers and its antiparallel, double-helical, three-dimensional structure.
2. Describe how DNA molecules are replicated.
3. Explain how the sequence of a molecule of DNA, made up of many monomers of only four possible nucleotides, can encode the enormous amount of genetic information stored in the chromosomes of living organisms.
4. Describe the usual flow of genetic information in a cell.
5. Name four differences between the structure of DNA and RNA.
6. Describe how a molecule of RNA is synthesized using a DNA molecule as a template.
7. Explain the relationship between RNA structure and function.
8. Name and describe three mechanisms of RNA processing in eukaryotes, and explain their importance to the cell.
9. List three types of noncoding RNA and describe their functions.

Log in to LaunchPad to check your answers to the Self-Assessment questions, and to access additional learning tools.

CHAPTER 4

Translation and Protein Structure

Core Concepts

4.1 Proteins are linear polymers of amino acids that form three-dimensional structures with specific functions.

4.2 Translation is the process in which the sequence of bases in messenger RNA specifies the order of successive amino acids in a newly synthesized protein.

4.3 Proteins evolve through mutation and selection and by combining functional units.

Hardly anything happens in the life of a cell that does not require proteins. They are the most versatile of macromolecules, each with its own built-in ability to carry out a cellular function. Some proteins aggregate to form relatively stiff filaments that help define the cell's shape and hold organelles in position. Others span the cell membrane and form channels or pores through which ions and small molecules can move. Many others are enzymes that catalyze the thousands of chemical reactions needed to maintain life. Still others are signaling proteins that enable cells to coordinate their internal activities or to communicate with other cells.

Want to see some proteins? Look at the white of an egg. Apart from the 90% or so that is water, most of what you see is protein. The predominant type of protein is ovalbumin. Easy to obtain in large quantities, ovalbumin was one of the first proteins studied by the scientific method (Chapter 1). In the 1830s, ovalbumin was shown to consist largely of carbon, hydrogen, nitrogen, and oxygen. Each molecule of ovalbumin was estimated to contain at least 400 carbon atoms. Leading chemists of the time scoffed at this number, believing that no organic molecule could possibly be so large. Little did they know: The number of carbon atoms in ovalbumin is actually closer to 2000 than to 400!

The reason that proteins can be such large organic molecules began to become clear only about a hundred years ago, when scientists hypothesized that proteins are polymers (large molecules made up of repeated subunits). Now we know that proteins are linear polymers of any combination of 20 amino acids, each of which differs from the others in its chemical characteristics. In size, ovalbumin is actually an average protein, consisting of a chain of 385 amino acids.

In Chapter 3, we discussed how genetic information flows from the sequence of bases of DNA into the sequence of bases in a transcript of RNA. For protein-coding genes, the transcript is processed into messenger RNA (mRNA). In this chapter, we examine how amino acid polymers are assembled by ribosomes by means of a template of messenger RNA and transfer RNAs. We also discuss how the amino acid sequences of proteins help determine their three-dimensional structures and diverse chemical activities, as well as how proteins change through evolutionary time.

4.1 MOLECULAR STRUCTURE OF PROTEINS

If you think of a protein as analogous to a word in the English language, then the amino acids are like letters. The comparison is not altogether fanciful, as there are about as many amino acids in proteins as letters in the alphabet, and the order of both amino acids and letters is important. For example, the word PROTEIN has the same letters as POINTER, but the two words have completely different meanings. Similarly, the exact order of amino acids in a protein determines the protein's shape and function.

Amino acids differ in their side chains.

The general structure of an amino acid was discussed in Chapter 2 and is shown again in **Fig. 4.1.** It consists of a central carbon atom, called the α **(alpha) carbon,** connected by covalent bonds to four different chemical groups (Fig. 4.1a): an **amino group** ($-NH_2$, shown in dark blue), a **carboxyl group** (-COOH, shown in brown), a hydrogen atom (-H, shown in light blue), and a variable **side chain** or **R group** (shown in green). In the environment of a cell, where the pH is in the range 7.35–7.45 (called physiological pH), the amino group gains a proton to become $-NH_3^+$ and the carboxyl group loses a proton to become $-COO^-$ (Fig. 4.1b). The four covalent bonds from the α carbon are at equal angles. As a result, an amino acid forms a tetrahedron, a pyramid with four triangular faces (Fig. 4.1c).

The R groups of the amino acids differ from one amino acid to the next. They are what make the "letters" of the amino acid "alphabet" distinct from one another. Just as letters differ in their shapes and sounds—vowels like E, I, and O and hard consonants like B, P, and T—amino acids differ in their chemical and physical properties.

The chemical structures of the 20 amino acids commonly found in proteins are shown in **Fig. 4.2.** The R groups (shown in green) are chemically diverse and are grouped according to their properties, with a particular emphasis on whether they are hydrophobic or hydrophilic, or have special characteristics that might affect a protein's structure. These properties strongly

FIG. 4.1 **Structure of an amino acid.** The central carbon atom is attached to an amino group, a carboxyl group, an R group or side chain, and a hydrogen atom.

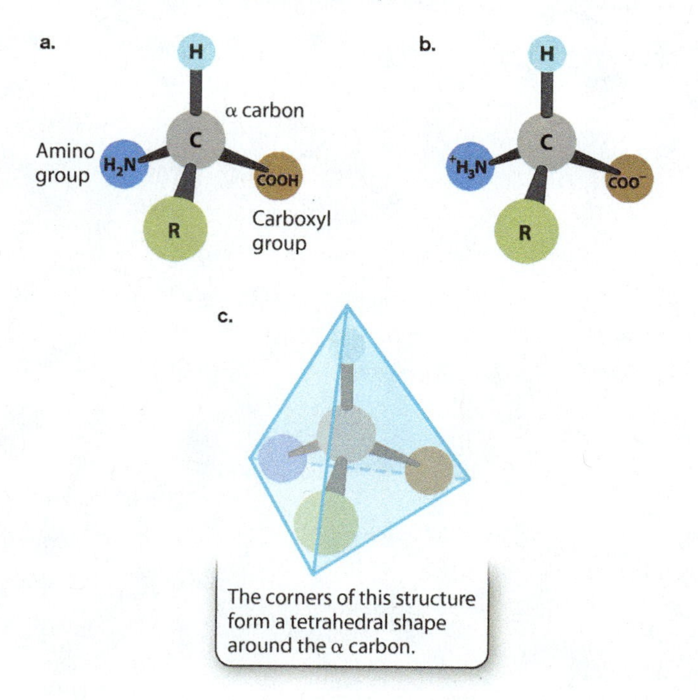

FIG. 4.2 Structures of the 20 amino acids commonly found in proteins.

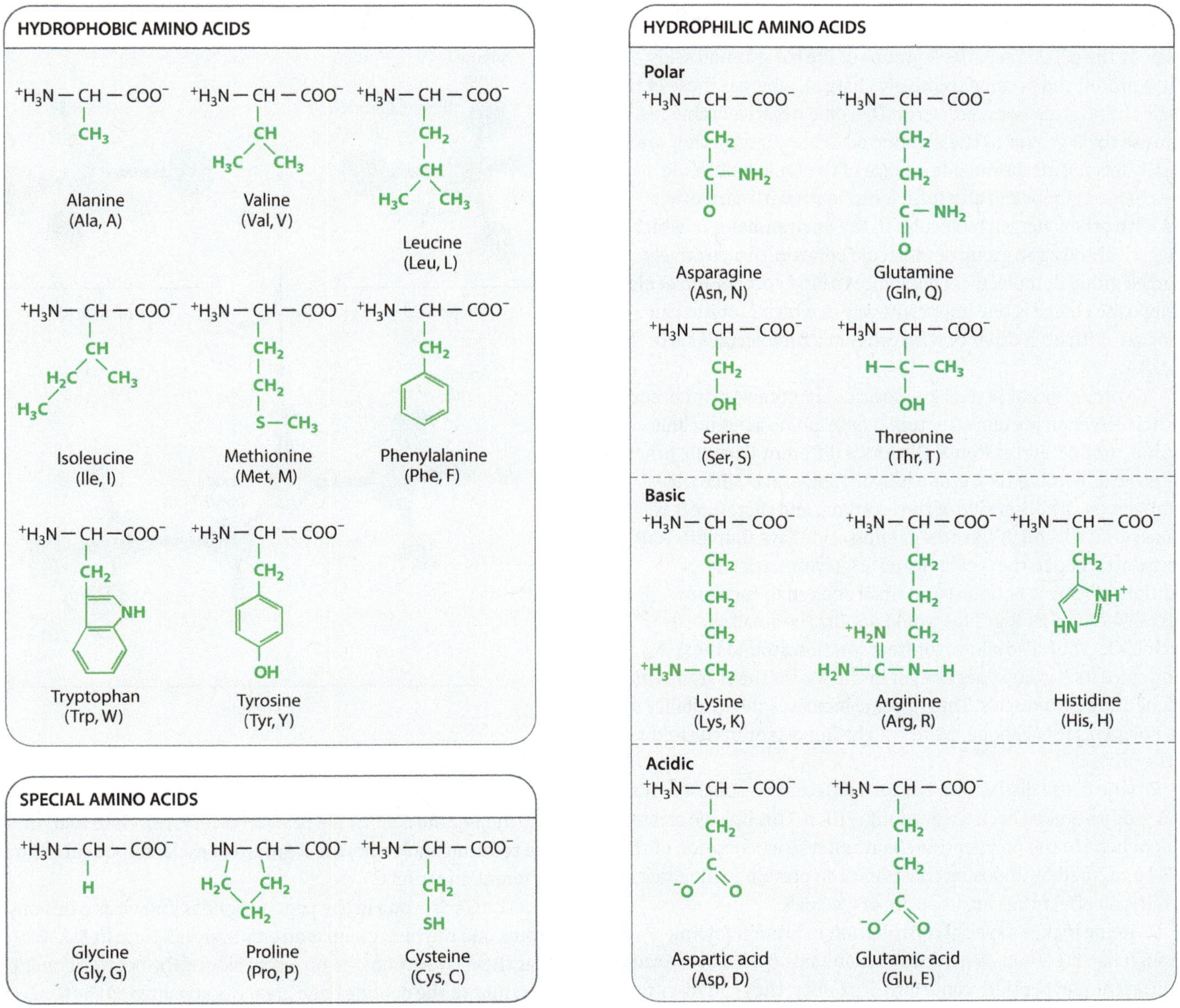

HYDROPHOBIC AMINO ACIDS

Alanine (Ala, A)
Valine (Val, V)
Leucine (Leu, L)
Isoleucine (Ile, I)
Methionine (Met, M)
Phenylalanine (Phe, F)
Tryptophan (Trp, W)
Tyrosine (Tyr, Y)

SPECIAL AMINO ACIDS

Glycine (Gly, G)
Proline (Pro, P)
Cysteine (Cys, C)

HYDROPHILIC AMINO ACIDS

Polar

Asparagine (Asn, N)
Glutamine (Gln, Q)
Serine (Ser, S)
Threonine (Thr, T)

Basic

Lysine (Lys, K)
Arginine (Arg, R)
Histidine (His, H)

Acidic

Aspartic acid (Asp, D)
Glutamic acid (Glu, E)

influence how a polypeptide folds, and hence the three-dimensional shape of the protein.

Hydrophobic amino acids are those that do not readily interact with water or form hydrogen bonds. Most hydrophobic amino acids have nonpolar R groups composed of hydrocarbon chains or uncharged carbon rings. Because water molecules in the cell form hydrogen bonds with each other instead of with the hydrophobic R groups, the hydrophobic R groups tend to aggregate with each other. Their aggregation is also stabilized by weak van der Waals forces (Chapter 2), in which asymmetries in electron distribution create temporary charges in the interacting molecules, which are then attracted to each other. The tendency for hydrophilic water molecules to interact with each other and for hydrophobic molecules to interact with each other is the very same tendency that leads to the formation of oil droplets in water. This is also the reason why most hydrophobic amino acids tend to be buried in the interior of folded proteins, where they are kept away from water.

Amino acids with polar R groups have a permanent charge separation, in which one end of the R group is slightly more negatively charged than the other. As we saw in Chapter 2, polar molecules are hydrophilic, and they tend to form hydrogen bonds with each other or with water molecules.

The R groups of the basic and acidic amino acids are strongly polar. At the pH of a cell, the R groups of the basic amino acids gain a proton and become positively charged, whereas those of the acidic amino acids lose a proton and become negatively charged. Because the R groups of these amino acids are charged, they are usually located on the outside surface of the folded molecule. The charged groups can also form ionic bonds with each other and with other charged molecules in the environment, in which a negatively charged group or molecule bonds with a positively charged group or molecule. This ability to bind another molecule of opposite charge is one important way in which proteins can associate with each other or with other macromolecules such as DNA.

The properties of several amino acids are noteworthy because of their effect on protein structure. These amino acids include glycine, proline, and cysteine. Glycine is different from the other amino acids because its R group is hydrogen, exactly like the hydrogen on the other side of the α carbon, and therefore it is not asymmetric. All of the other amino acids have four different groups attached to the α carbon and are asymmetric. In addition, glycine is nonpolar and small enough to tuck into spaces where other R groups would not fit. The small size of glycine's R group also allows for freer rotation around the C–N bond since its R group does not get in the way of the R groups of neighboring amino acids. Thus, glycine increases the flexibility of the polypeptide backbone, which can be important in the folding of the protein.

Proline is also distinctive, but for a different reason. Note how its R group is linked back to the amino group. This linkage creates a kink or bend in the polypeptide chain and restricts rotation of the C–N bond, thereby imposing constraints on protein folding in its vicinity, an effect the very opposite of glycine's.

Cysteine makes a special contribution to protein folding through its –SH group. When two cysteine side chains in the same or different polypeptides come into proximity, they can react to form an S–S disulfide bond, which covalently joins the side chains. Such disulfide bonds are stronger than the ionic interactions of other pairs of amino acid, and form cross-bridges that can connect different parts of the same protein or even different proteins. This property contributes to the overall structure of single proteins or combinations of proteins.

Successive amino acids in proteins are connected by peptide bonds.

Amino acids are linked together to form proteins. **Fig. 4.3** shows how amino acids in a protein are bonded together. The bond formed between the two amino acids is a **peptide bond,** shown in red in Fig. 4.3. In forming the peptide bond, the carboxyl group of one amino acid reacts with the amino group of the next amino acid in line, and a molecule of water is released. Note that, in the resulting molecule, the R groups of each amino acid point in different directions.

FIG. 4.3 Formation of a peptide bond. A peptide bond forms between the carboxyl group of one amino acid and the amino group of another.

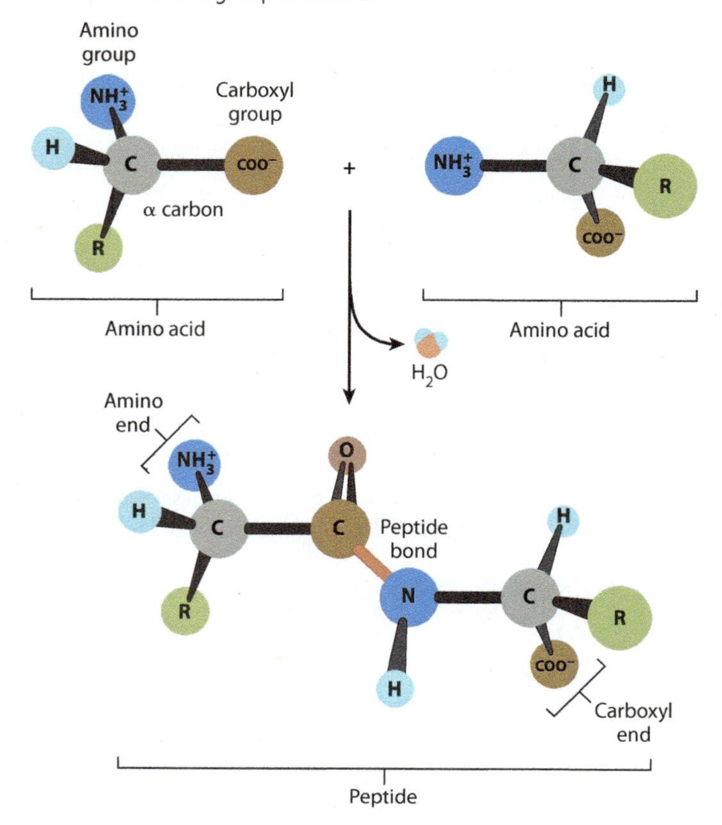

The C=O group in the peptide bond is known as a carbonyl group, and the N–H group is an amide group. Note in Fig. 4.3 that these two groups are on either side of the peptide bond. The electrons of the peptide bond are more attracted to the C=O group than to the NH group because of the greater electronegativity of the oxygen atom. The result is that the peptide bond has some of the characteristics of a double bond. The peptide bond is shorter than a single bond, for example, and it is not free to rotate like a single bond. The other bonds are free to rotate around their central axes.

Polymers of amino acids ranging from as few as two to many hundreds share a chemical feature common to individual amino acids: namely, that the ends are chemically distinct from each other. One end, shown at the left in Fig. 4.3, has a free amino group; this is the **amino end** of the molecule. The other end has a free carboxyl group, which is the **carboxyl end** of the molecule. More generally, a polymer of amino acids connected by peptide bonds is known as a **polypeptide.** Typical polypeptides produced in cells consist of a few hundred amino acids. In human cells, the shortest polypeptides are about 100 amino acids in length; the longest is the muscle protein titin, with 34,350 amino acids. The term **protein** is often used as a synonym for polypeptide, especially when the polypeptide chain has folded into a stable, three-dimensional conformation. Amino acids that are incorporated into a protein are often referred to as amino acid **residues.** In a polypeptide chain at physiological pH, the amino and carboxyl ends are in their charged states of NH_3^+ and COO^-, respectively. However, for simplicity, we denote the ends as NH_2 and COOH.

The sequence of amino acids dictates protein folding, which determines function.

Up to this point, we have considered the sequence of amino acids that make up a protein. This is the first of several levels of protein structure, illustrated in **Fig. 4.4.** The sequence of amino acids in a protein is its **primary structure.** The sequence of amino acids ultimately determines how a protein folds. Interactions between stretches of amino acids in a protein form local **secondary structures.** Longer-range interactions between these secondary structures in turn support the overall three-dimensional shape of the polypeptide, which is its **tertiary structure.** Finally, some proteins are made up of several individual polypeptides that interact with each other, and the resulting ensemble is the **quaternary structure.**

Proteins have a remarkably wide range of functions in the cell, from serving as structural elements to communicating with the external environment to accelerating the rate of chemical reactions. No matter what the function of a protein is, the ability to carry out this function depends on the three-dimensional shape of the protein. When fully folded, some proteins contain pockets with positively or negatively charged side chains at just the right positions to trap small molecules; others have surfaces that can bind another protein or a sequence of nucleotides in DNA or RNA; some form rigid rods for structural support; and still others keep their hydrophobic side chains away from water molecules by inserting into the cell membrane.

The sequence of amino acids in a protein (its primary structure) is usually represented by a series of three-letter or one-letter abbreviations for the amino acids (abbreviations for the 20 common amino acids are given in Fig. 4.2). By convention, the amino acids in a protein are listed in order from left to right, starting at the amino end and proceeding to the carboxyl end. The amino and the carboxyl ends are different, so the order matters. Just as TIPS is not the same word as SPIT, the sequence Thr–Ile–Pro–Ser is not the same polypeptide as Ser–Pro–Ile–Thr.

Secondary structures result from hydrogen bonding in the polypeptide backbone.

Hydrogen bonds can form between the carbonyl group in one peptide bond and the amide group in another, thus allowing

FIG. 4.4 Levels of protein structure. The manner in which a polypeptide folds determines its shape and function.

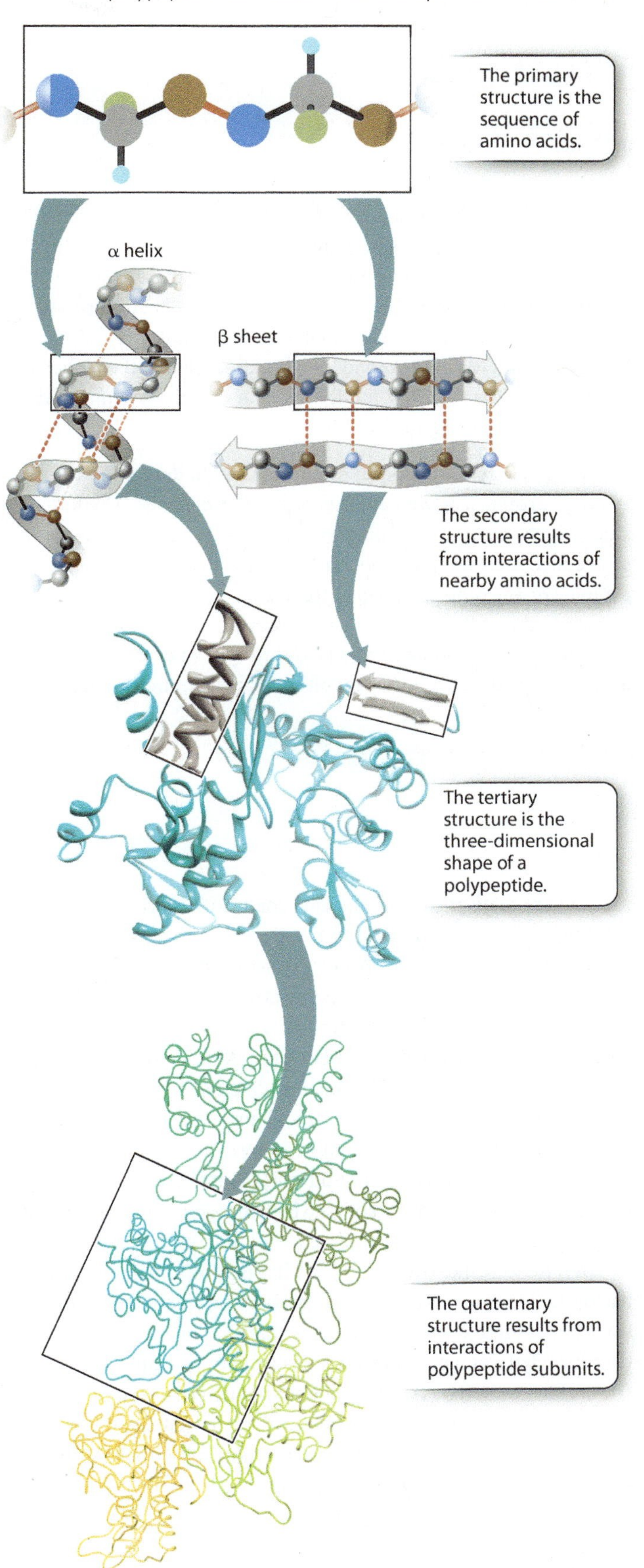

HOW DO WE KNOW?

FIG. 4.5

What are the shapes of proteins?

BACKGROUND The three-dimensional shapes of proteins can be determined by X-ray crystallography. One of the pioneers in this field was Dorothy Crowfoot Hodgkin, who used this technique to define the structures of cholesterol, vitamin B_{12}, penicillin, and insulin. She was awarded the Nobel Prize in Chemistry in 1964 for her early work. Max Perutz and John Kendrew shared the Nobel Prize in Chemistry in 1962 for defining the structures of myoglobin and hemoglobin using this method.

METHOD X-ray crystallography can be used to determine the shape of proteins, as well as other types of molecules. The first step, which can be challenging, is to make a crystal of the protein molecules. A crystal is a solid structure in which the atoms of a protein (or any other molecule) are in an ordered and repeating pattern in three dimensions. Then X-rays are aimed at the crystal while it is rotated. Some X-rays pass through the crystal, while others are scattered in different directions when they hit atoms of the proteins. A film or other detector records the pattern as a series of spots, which is known as a diffraction pattern. The locations and intensities of these spots can be used to infer the position and arrangement of the atoms in the molecule.

RESULTS The X-ray diffraction pattern for hemoglobin looks like this:

Source: Courtesy of William E. Royer, University of Massachusetts Medical School and Vukica Srajer, BioCARS, Center for Advanced Radiation Sources, The University of Chicago.

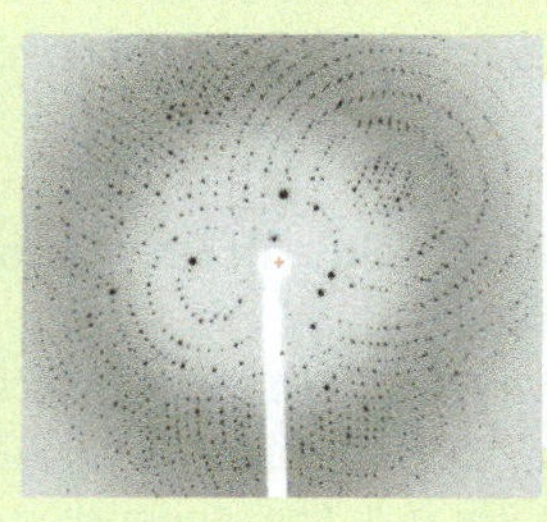

From this two-dimensional pattern, researchers can use mathematical methods to determine the three-dimensional shape of the protein.

FOLLOW-UP WORK Linus Pauling and Robert Corey used X-ray crystallography to determine two types of secondary structures commonly found in proteins—the α helix and the β sheet. Today, this technique is a common method for determining the shape of proteins.

SOURCES Crowfoot, D. 1935. "X-Ray Single Crystal Photographs of Insulin." Nature 135:591–592; Kendrew, J. C., et al. 1958. "A Three-Dimensional Model of the Myoglobin Molecule Obtained by X-Ray Analysis." *Nature* 181:662–666.

localized regions of the polypeptide chain to fold. This localized folding is a major contributor to the secondary structure of the protein. In the early 1950s, American structural biologists Linus Pauling and Robert Corey used X-ray crystallography to study the structure of proteins. This technique was pioneered by British biochemists Dorothy Crowfoot Hodgkin, Max Perutz, and John Kendrew, among others (**Fig. 4.5**). Pauling and Corey studied crystals of highly purified proteins and discovered that two types of secondary structure are found in many different proteins. These are the **α (alpha) helix** and the **β (beta) sheet.** Both these secondary structures are stabilized by hydrogen bonding along the polypeptide backbone.

In α helices, like the one shown in **Fig. 4.6,** the polypeptide backbone is twisted tightly in a right-handed coil with 3.6 amino acids per complete turn. The helix is stabilized by hydrogen bonds that form between each amino acid's carbonyl group (C=O) and the amide group (N–H) four residues ahead in the sequence, as indicated by the dashed lines in Fig. 4.6. Note that the R groups project outward from the α helix. The chemical properties of the projecting R groups largely determine where the α helix is positioned in the folded protein, and how it might interact with other molecules.

The other secondary structure that Pauling and Corey found is the β sheet, depicted in **Fig. 4.7.** In a β sheet, the polypeptide folds back and forth on itself, forming a pleated sheet that is stabilized by hydrogen bonds between carbonyl groups in one chain and amide groups in the other chain across the way (dashed lines). The R groups project alternately above and below the plane of the β sheet. β sheets typically consist of 4 to 10 polypeptide chains aligned side by side, with the amides in each chain hydrogen-bonded to the carbonyls on either side (except for those at the ends of each strand).

β sheets are typically denoted by broad arrows, where the direction of the arrow runs from the amino end of the polypeptide segment to the carboxyl end. In Fig. 4.7, the arrows run in opposite directions, and the polypeptide chains are said to be antiparallel. β sheets can also be formed by hydrogen bonding between polypeptide chains that are parallel (pointing in the same direction). However, the antiparallel configuration is more stable because the carbonyl and amide groups are more favorably aligned for hydrogen bonding.

FIG. 4.6 **An α helix.** Hydrogen bonding between carbonyl and amide groups in the backbone stabilize the helix.

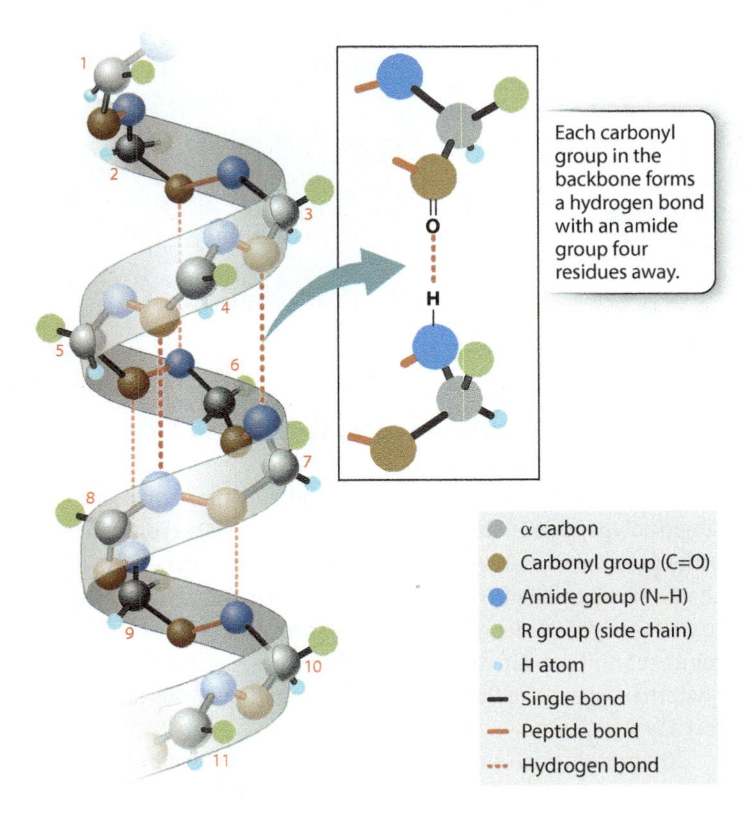

Tertiary structures result from interactions between amino acid side chains.

The tertiary structure of a protein is the three-dimensional conformation of a single polypeptide chain, usually made up of several secondary structure elements. The shape of a protein is defined largely by interactions between the amino acid R groups. By contrast, the formation of secondary structures relies on interactions in the polypeptide backbone and is relatively independent of the R groups. Tertiary structure is determined by the spatial distribution of hydrophilic and hydrophobic R groups along the molecule, as well as by different types of chemical bonds and interactions (ionic, hydrogen, and van der Waals) that form between various R groups. The amino acids whose R groups form bonds with each other may be far apart in the polypeptide chain, but can end up near each other in the folded protein. Hence, the tertiary structure usually includes loops or turns in the backbone that allow these R groups to sit near each other in space and for bonds to form.

The three-dimensional shapes of proteins can be illustrated in different ways, as shown in **Fig. 4.8.** A ball-and-stick model (Fig. 4.8a) draws attention to the atoms in the amino acid chain. A ribbon model (Fig. 4.8b) emphasizes secondary structures, with α helices depicted as twisted ribbons and β sheets as broad arrows. Finally, a space-filling model (Fig. 4.8c) shows the overall shape and contour of the folded protein.

Remember that the folding of a polypeptide chain is determined by the sequence of amino acids. The primary structure determines the secondary and tertiary structures. Furthermore, tertiary structure determines function because it is the three-dimensional shape of the molecule—the contours and distribution of charges on the outside of the molecule and the presence of pockets that might bind with smaller molecules on

FIG. 4.7 **A β sheet.** Hydrogen bonds between neighboring strands stabilize the structure.

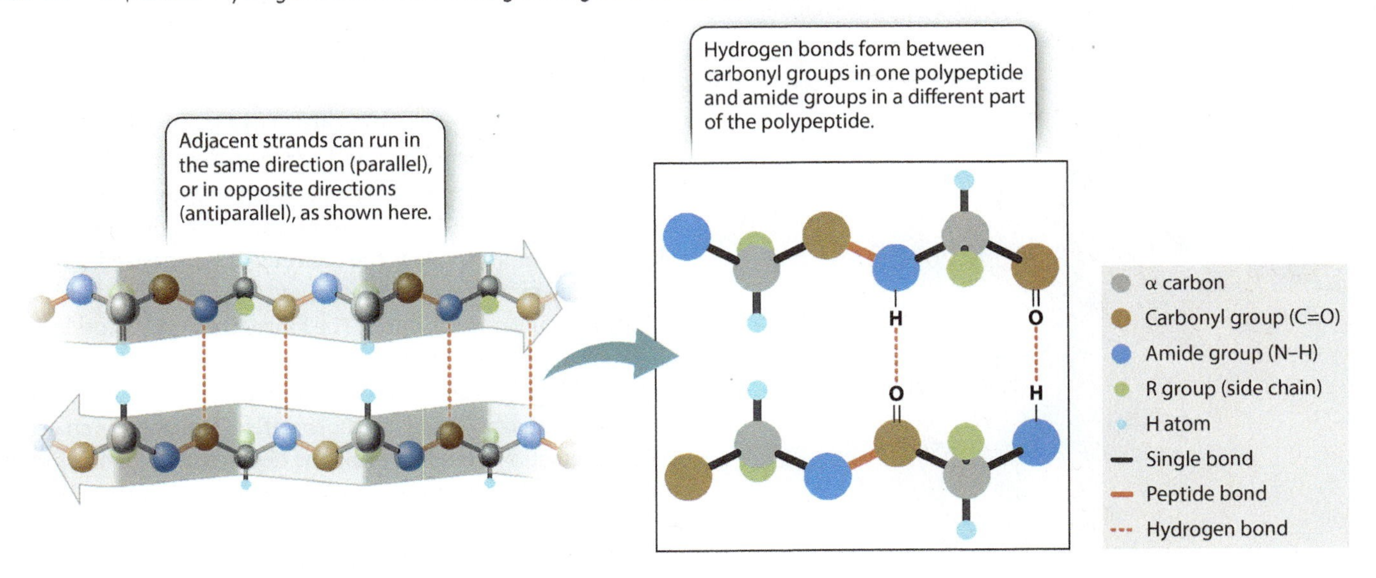

FIG. 4.8 Three ways of showing the structure of the protein tubulin: (a) ball-and-stick model; (b) ribbon model; (c) space-filling model.

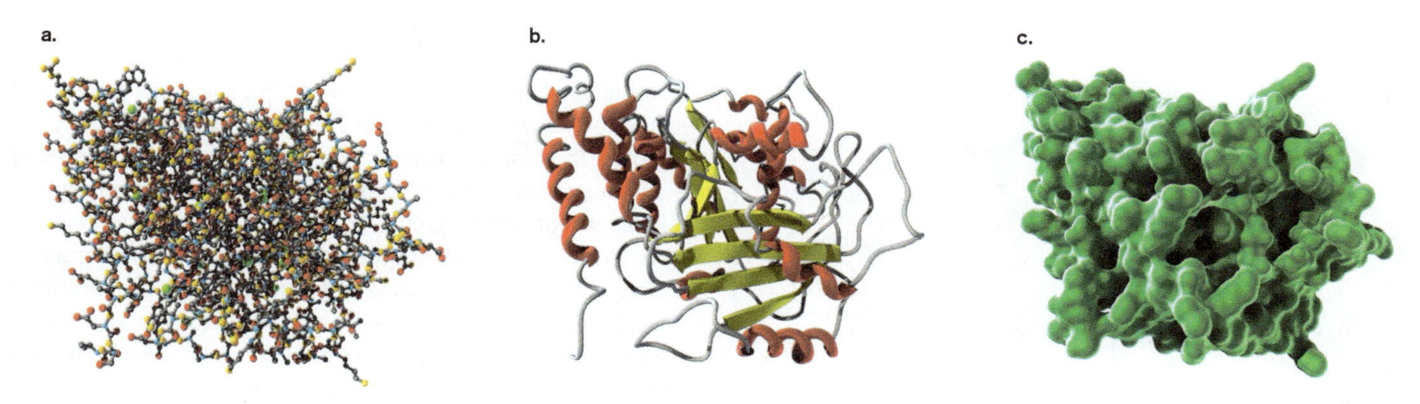

the inside—that enables the protein to serve as structural support, membrane channel, enzyme, or signaling molecule. **Fig. 4.9** shows the tertiary structure of a bacterial protein that contains a pocket in the center in which certain R groups can form hydrogen bonds with a specific small molecule and hold it in place.

The principle that structure determines function can be demonstrated by many observations. For example, most proteins can be unfolded, or **denatured,** by chemical treatment or high temperature that disrupts the hydrogen and ionic bonds holding the tertiary structure together. Under these conditions, the proteins lose their functional activity. Similarly, mutant proteins containing an amino acid that prevents proper folding are often inactive or don't function properly.

→ **Quick Check 1** A mutation leads to a change in one amino acid in a protein. The result is that the protein no longer functions properly. How is this possible?

FIG. 4.9 Tertiary structure determines function. This bacterial protein has a cavity that can bind with a small molecule (shown as a ball-and-stick model in the center).

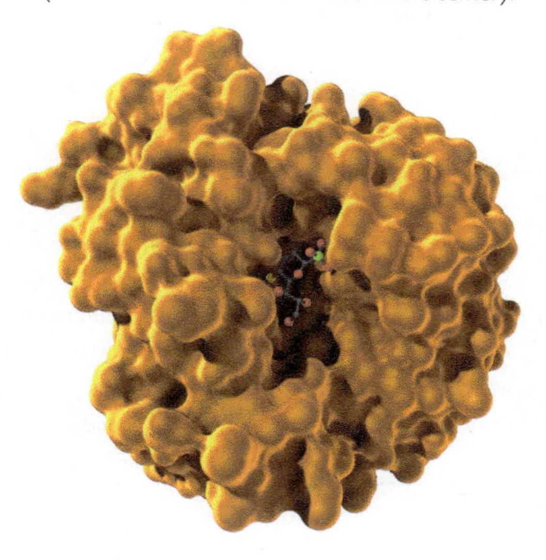

Polypeptide subunits can come together to form quaternary structures.

Although many proteins are complete and fully functional as a single polypeptide chain with a tertiary structure, there are many other proteins that are composed of two or more polypeptide chains or subunits with a tertiary structure that come together to form a higher-order quaternary structure. In the case of a multi-subunit protein, the activity of the complex depends on the quaternary structure formed by the combination of the various tertiary structures.

In a protein with quaternary structure, the polypeptide subunits may be identical or different (**Fig. 4.10**). Fig. 4.10a shows an example of a protein produced by HIV that consists of two identical polypeptide subunits. By contrast, many proteins, such as hemoglobin (shown in Fig. 410b), are composed of different subunits. In either case, the subunits can influence each other in subtle ways and influence their function. For example, the hemoglobin in red blood cells that carries oxygen has four subunits. When one of these binds oxygen, a slight change in its structure is transmitted to the other subunits, making it easier for them to take up oxygen. In this way, oxygen transport from the lungs to the tissues is improved.

Chaperones help some proteins fold properly.

The amino acid sequence (primary structure) of a protein determines how it forms its secondary, tertiary, and quaternary structures. For about 75% of proteins, the folding process takes place within milliseconds as the molecule is synthesized. Some proteins fold more slowly, however, and for these molecules folding is a dangerous business. The longer these polypeptides remain in a denatured (unfolded) state, the longer their hydrophobic groups are exposed to other macromolecules in the crowded cytoplasm. The hydrophobic effect, along with van der Waals interactions, tends to bring the exposed hydrophobic groups together, and their inappropriate aggregation may prevent proper folding. Correctly folded proteins can sometimes unfold because of elevated temperature, for example, and in the denatured state they are subject to the same risks of aggregation.

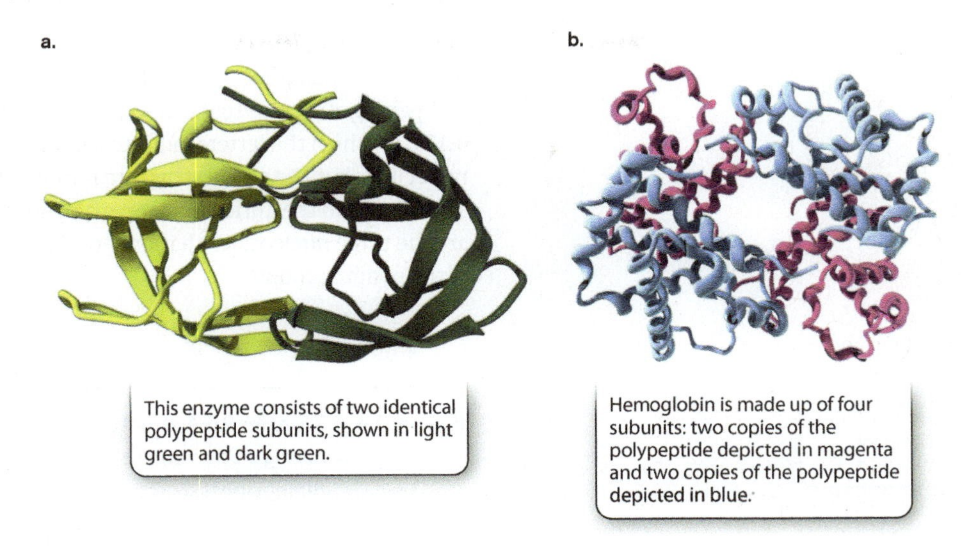

FIG. 4.10 Quaternary structure. Polypeptide units of proteins may be identical, as in (a) an enzyme from HIV, or different, as in (b) hemoglobin.

Cells have evolved proteins called **chaperones** that help protect slow-folding or denatured proteins until they can attain their proper three-dimensional structure. Chaperones bind with hydrophobic groups and nonpolar R groups to shield them from inappropriate aggregation, and in repeated cycles of binding and release they give the polypeptide time to find its correct shape.

4.2 TRANSLATION: HOW PROTEINS ARE SYNTHESIZED

The three-dimensional structure of a protein determines what it can do and how it works, and the immense diversity in the tertiary and quaternary structures among proteins explains their wide range of functions in cellular processes. Yet it is the sequence of amino acids along a polypeptide chain—its primary structure—that governs how the molecule folds into a stable three-dimensional configuration. How is the sequence of amino acids specified? It is specified by the sequence of nucleotides in the DNA, in coded form. The decoding of the information takes place according to the central dogma of molecular biology, which defines information flow in a cell from DNA to RNA to protein (**Fig. 4.11**). In transcription, the sequence of bases along part of a DNA strand is used as a template in the synthesis of a complementary sequence of bases in a molecule of RNA, as described in Chapter 3. In **translation,** the sequence of bases in an RNA molecule known as **messenger RNA** (mRNA) is used to specify the order in which successive amino acids are added to a newly synthesized polypeptide chain.

Translation uses many molecules found in all cells.

Translation requires many components. Well over 100 genes encode components needed for translation, some of which are shown in Fig. 4.11. What are these needed components? First, the cell needs **ribosomes,** which are complex structures of RNA and protein that bind with mRNA and are the site of translation. In prokaryotes, translation occurs as soon as the mRNA comes off the DNA template. In eukaryotes, the processes of transcription and translation are physically separated: Transcription takes place in the nucleus, and translation takes place in the cytoplasm.

In both eukaryotes and prokaryotes, the ribosome consists of a small subunit and a large subunit, each composed of 1 to 3 types of ribosomal RNA and 20 to 50 types of ribosomal protein.

FIG. 4.11 The central dogma, showing how information flows from DNA to RNA to protein. Note the large number of cellular components required for translation.

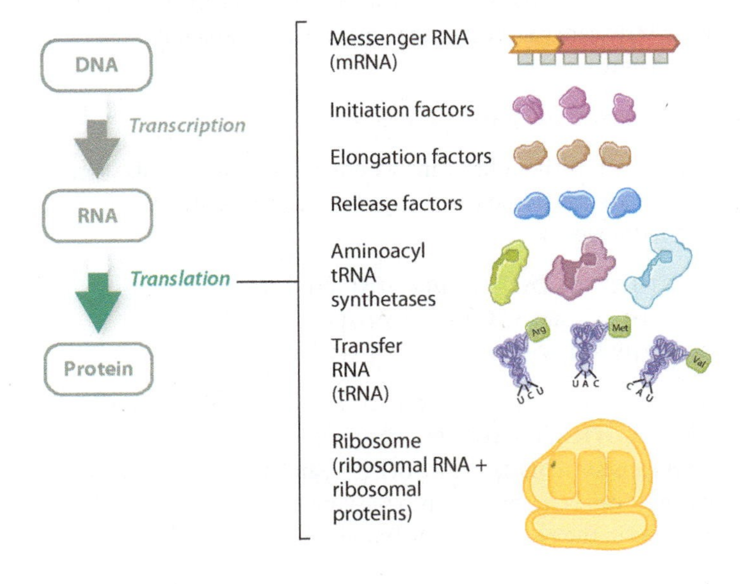

FIG. 4.12 **Simplified structure of ribosomes in prokaryotes and eukaryotes.**

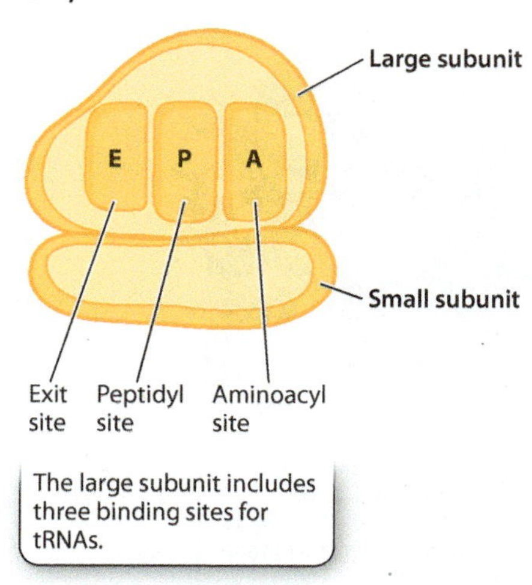

Eukaryotic ribosomes are larger than prokaryotic ribosomes. As indicated in **Fig. 4.12,** the large subunit of the ribosome includes three binding sites for molecules of transfer RNA, which are called the **A (aminoacyl) site,** the **P (peptidyl) site,** and the **E (exit) site.**

A major role of the ribosome is to ensure that, when the mRNA is in place on the ribosome, the sequence in the mRNA coding for amino acids is read in successive, non-overlapping groups of three nucleotides, much as you would read the sentence

THEBIGBOYSAWTHEBADMANRUN

Each non-overlapping group of three adjacent nucleotides (like THE or BIG or BOY in our sentence analogy) constitutes a **codon,** and each codon in the mRNA codes for a single amino acid in the polypeptide chain.

In the example above, it is clear that the sentence begins with THE. However, in a long linear mRNA molecule, the ribosome could begin at any nucleotide. As an analogy, if we knew that the letters THE were the start of the phrase, then we would know immediately how to read

ZWTHEBIGBOYSAWTHEBADMANRUN

However, without knowing where to start reading this string of letters, we could find three ways to break the sentence into three-letter words:

ZWT HEB IGB OYS AWT HEB ADM ANR UN
Z WTH EBI GBO YSA WTH EBA DMA NRU N
ZW THE BIG BOY SAW THE BAD MAN RUN

The different ways of parsing the string into three-letter words are known as **reading frames.** The protein-coding sequence in an mRNA consists of its sequence of bases, and just as in our sentence analogy, it can be translated into the correct protein only if it is translated in the proper reading frame.

While the ribosome establishes the correct reading frame for the codons, the actual translation of each codon in the mRNA into one amino acid in the polypeptide is carried out by means of transfer RNA (tRNA). Transfer RNAs are small RNA molecules of 70 to 90 nucleotides (**Fig. 4.13**). Each has a characteristic self-pairing structure that can be drawn as a cloverleaf, as in Fig. 4.13a, in which the letters indicate the bases common to all tRNA molecules, but the actual structure is more like that in Fig. 4.13b. Three bases in the anticodon loop make up the **anticodon;** these are the three nucleotides that undergo base pairing with the corresponding codon.

FIG. 4.13 **Transfer RNA structure depicted in (a) a cloverleaf configuration and (b) a more realistic three-dimensional structure.**

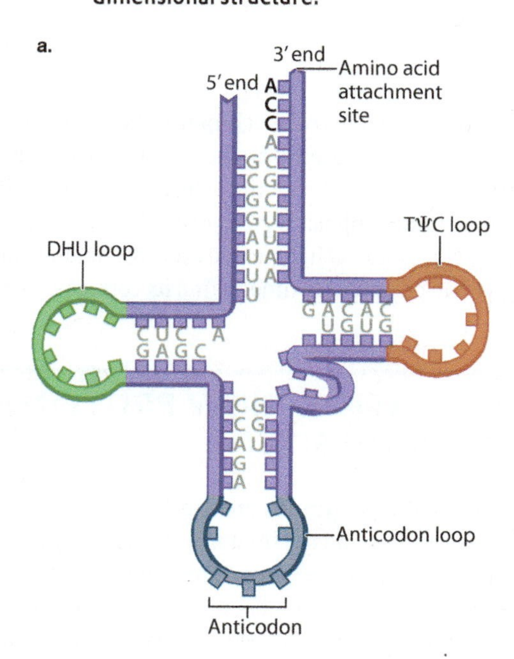

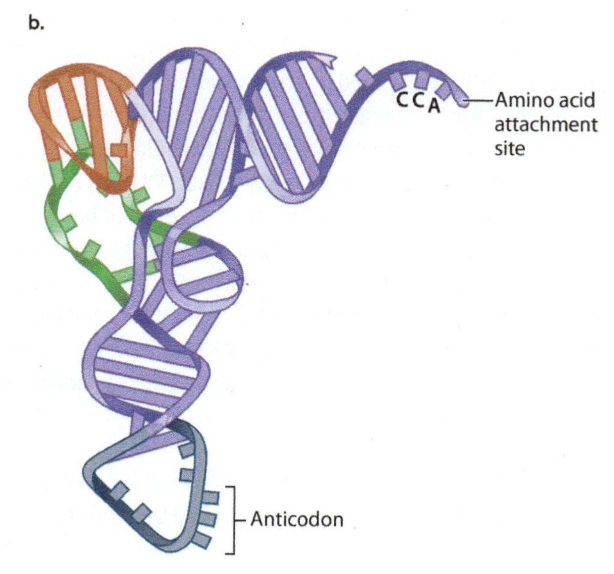

Each tRNA has the nucleotide sequence CCA at its 3′ end (Fig. 4.13), and the 3′ hydroxyl of the A is the attachment site for the amino acid corresponding to the anticodon. Enzymes called **aminoacyl tRNA synthetases** connect specific amino acids to specific tRNA molecules (**Fig. 4.14**). Therefore, these enzymes are directly responsible for actually translating the codon sequence in a nucleic acid to a specific amino acid in a polypeptide chain. Most organisms have one aminoacyl tRNA synthetase for each amino acid. The enzyme binds to multiple sites on any tRNA that has an anticodon corresponding to the amino acid, and it catalyzes formation of the covalent bond between the amino acid and tRNA. A tRNA that has no amino acid attached is said to be uncharged, and one with its amino acid attached is said to be charged. Amino acid tRNA synthetases are very accurate and attach the wrong amino acid far less often than 1 time in 10,000.

Although the specificity for attaching an amino acid to the correct tRNA is a property of aminoacyl tRNA synthetase, the specificity of DNA–RNA and codon–anticodon interactions result from base pairing. **Fig. 4.15** shows the relationships for one codon in double-stranded DNA, in the corresponding mRNA, and in the codon–anticodon pairing between the mRNA and the tRNA. Note that the first (5′) base in the codon in mRNA pairs with the last (3′) base in the anticodon because, as noted in Chapter 3, nucleic acid strands that undergo base pairing must be antiparallel.

FIG. 4.15 Role of base pairing. Base pairing determines the relationship between a three-base triplet in double-stranded DNA to a codon in mRNA, and between a codon in mRNA to an anticodon in tRNA.

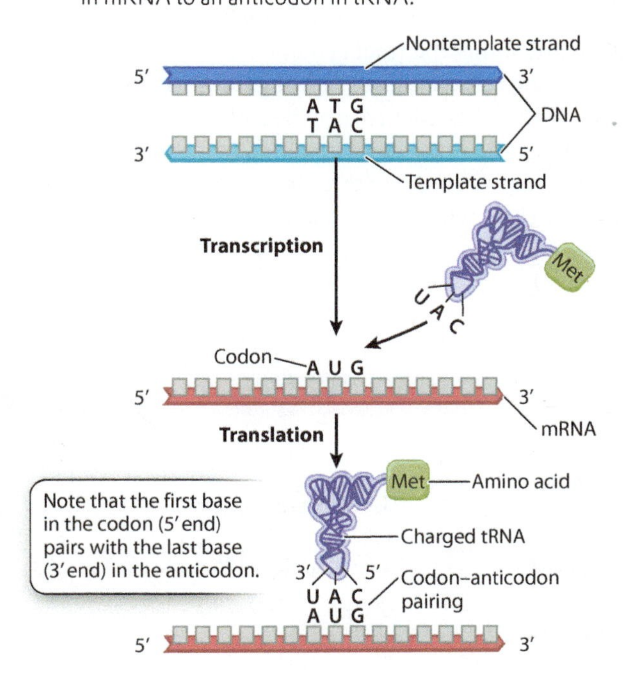

FIG. 4.14 Function of aminoacyl tRNA synthetase enzymes. Aminoacyl tRNA synthetases attach specific amino acids to tRNAs and are therefore responsible for translating the codon sequence in an mRNA into an amino acid sequence in a protein.

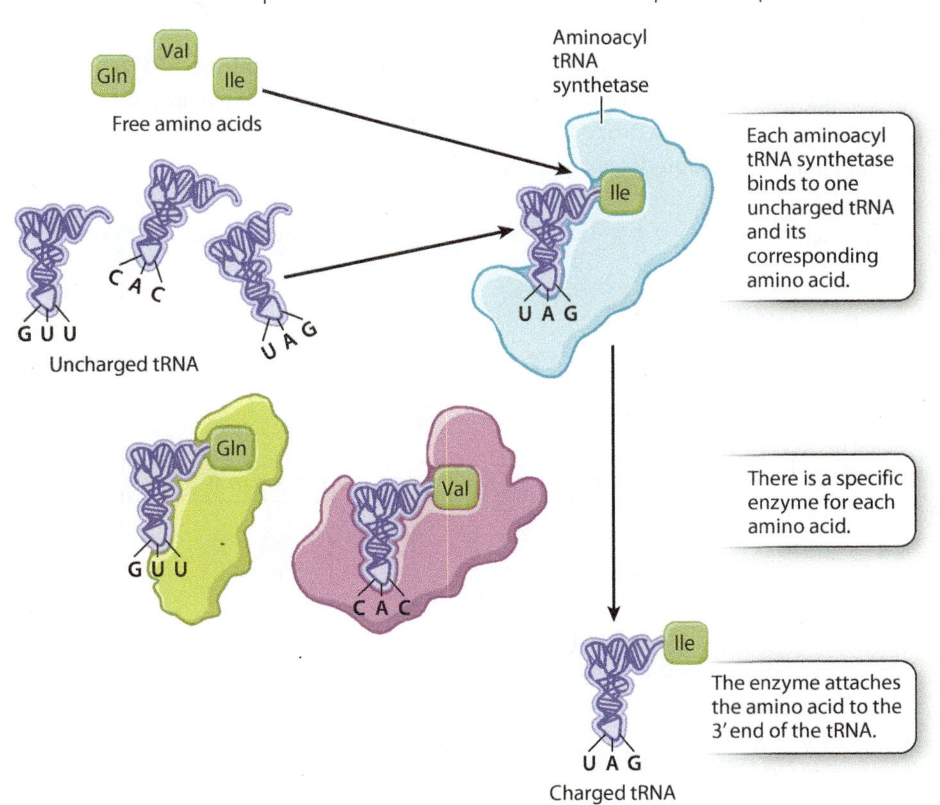

The genetic code shows the correspondence between codons and amino acids.

Fig. 4.15 shows how the codon AUG specifies the amino acid methionine (Met) by base pairing with the anticodon of a charged tRNA, denoted $tRNA^{Met}$. Most codons specify an amino acid according to a **genetic code.** This code is sometimes called the "standard" genetic code because, while it is used by almost all cells, some minor differences are found in a few organisms as well as in mitochondria.

The codon at which translation begins is called the initiation codon, and it is coded by AUG, which specifies Met. The polypeptide is synthesized from the amino end to the carboxyl end, and so Met forms the amino end of any polypeptide being synthesized; however, in many cases the Met is cleaved off by an enzyme after synthesis is complete. The AUG codon is also used to specify the incorporation of Met at internal sites within the polypeptide chain.

As is apparent in Fig. 4.15, the AUG codon that initiated translation is preceded by a region in the mRNA that is not translated. The position of the initiator AUG codon in the mRNA establishes the reading frame that determines how the downstream codons (those following the AUG) are to be read.

Once the initial Met creates the amino end of a new polypeptide chain, the downstream codons are read one by one in non-overlapping groups of three bases. At each step, the ribosome binds to a tRNA with an anticodon that can base pair with the codon, and the amino acid on that tRNA is attached to the growing chain to become the new carboxyl end of the polypeptide chain. This process continues until one of three "stop" codons is encountered: UAA, UAG, or UGA. (The stop codons are also called termination codons or sometimes nonsense codons.) At this point, the polypeptide is finished and released into the cytosol.

The standard genetic code was deciphered in the 1960s by a combination of techniques, but among the most ingenious were chemical methods for making synthetic RNAs of known sequence by American biochemist Har Gobind Khorana and his colleagues. This experiment is illustrated in **Fig. 4.16.**

HOW DO WE KNOW?

FIG. 4.16

How was the genetic code deciphered?

BACKGROUND The genetic code is the correspondence between three-letter nucleotide codons in RNA and amino acids in a protein. American biochemist Har Gobind Khorana performed key experiments that helped to crack the code. For this work he shared the Nobel Prize in Physiology or Medicine in 1968 with Robert W. Holley and Marshall E. Nirenberg.

METHOD Khorana and his group made RNAs of known sequence. They then added these synthetic RNAs to a solution containing all of the other components needed for translation. By adjusting the concentration of magnesium and other factors, the researchers could get the ribosome to initiate synthesis with any codon, even if not AUG.

EXPERIMENT 1 AND RESULTS When a synthetic poly(U) was used as the mRNA, the resulting polypeptide was polyphenylalanine (Phe–Phe–Phe…):

a.

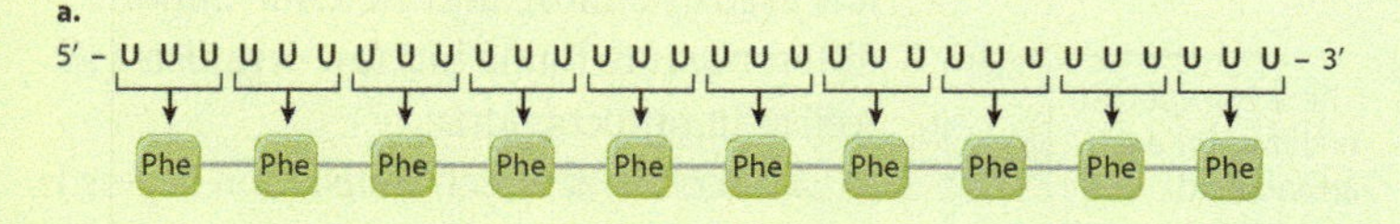

CONCLUSION The codon UUU corresponds to Phe. The poly(U) mRNA can be translated in three possible reading frames, depending on which U is the 5′ end of the start codon, but in each of them, all the codons are UUU.

EXPERIMENT 2 AND RESULTS When a synthetic mRNA with alternating U and C was used, the resulting polypeptide had alternating serine (Ser) and leucine (Leu):

b.

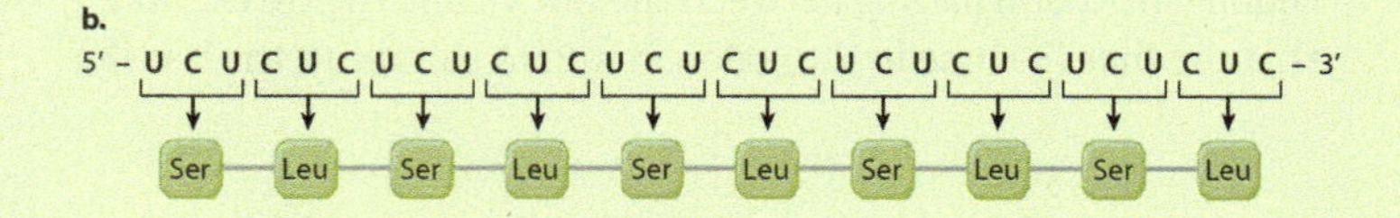

CONCLUSION Here again there are three reading frames, but each of them has alternating UCU and CUC codons. The researchers could not deduce from this result whether UCU corresponds to Ser and CUC to Leu or the other way around; the correct assignment came from experiments using other synthetic mRNA molecules.

EXPERIMENT 3 AND RESULTS When a synthetic mRNA with repeating UCA was used, three different polypeptides were produced—polyserine (Ser), polyhistidine (His), and polyisoleucine (Ile).

c.

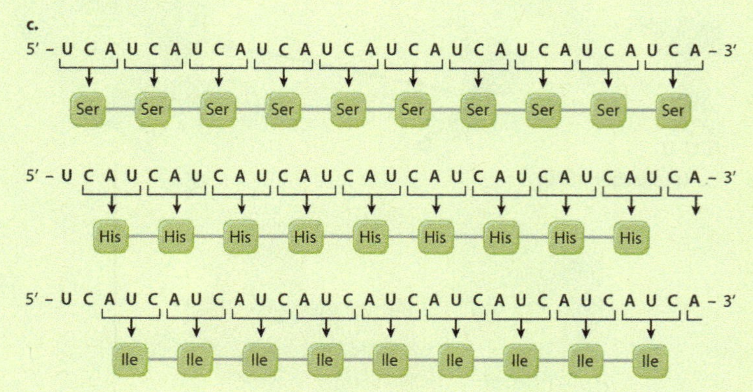

CONCLUSION The results do not reveal which of the three reading frames corresponds to which amino acid, but this was sorted out by studies of other synthetic polymers.

SOURCE Khorana, H. G. 1972. "Nucleic Acid Synthesis in the Study of the Genetic Code." In *Nobel Lectures, Physiology or Medicine* 1963–1970. Amsterdam: Elsevier.

TABLE 4.1 The standard genetic code.

First position (5′ end)	Second position: U	Second position: C	Second position: A	Second position: G	Third position (3′ end)
U	UUU Phe	UCU Ser	UAU Tyr	UGU Cys	U
	UUC Phe — F	UCC Ser	UAC Tyr — Y	UGC Cys — C	C
	UUA Leu	UCA Ser — S	UAA Stop	UGA Stop	A
	UUG Leu — L	UCG Ser	UAG Stop	UGG Trp W	G
C	CUU Leu	CCU Pro	CAU His	CGU Arg	U
	CUC Leu	CCC Pro	CAC His — H	CGC Arg	C
	CUA Leu — L	CCA Pro — P	CAA Gln	CGA Arg — R	A
	CUG Leu	CCG Pro	CAG Gln — Q	CGG Arg	G
A	AUU Ile	ACU Thr	AAU Asn	AGU Ser	U
	AUC Ile — I	ACC Thr	AAC Asn — N	AGC Ser — S	C
	AUA Ile	ACA Thr — T	AAA Lys	AGA Arg	A
	AUG Met M	ACG Thr	AAG Lys — K	AGG Arg — R	G
G	GUU Val	GCU Ala	GAU Asp	GGU Gly	U
	GUC Val	GCC Ala	GAC Asp — D	GGC Gly	C
	GUA Val — V	GCA Ala — A	GAA Glu	GGA Gly — G	A
	GUG Val	GCG Ala	GAG Glu — E	GGG Gly	G

Nonpolar Polar Basic Acidic Stop codon

The standard genetic code shown in **Table 4.1** has 20 amino acids specified by 64 codons. Many amino acids are therefore specified by more than one codon, and hence the genetic code is redundant, or degenerate. The redundancy has strong patterns, however:

- The redundancy results almost exclusively from the third codon position.
- When an amino acid is specified by two codons, they differ either in whether the third position is a U or a C (both pyrimidine bases), or in whether the third position is an A or a G (both purine bases).
- When an amino acid is specified by four codons, the identity of the third codon position does not matter; it could be U, C, A, or G.

The chemical basis of these patterns results from two features of translation. First, in many tRNA anticodons the 5′ base that pairs with the 3′ (third) base in the codon is chemically modified into a form that can pair with two or more bases at the third position in the codon. Second, in the ribosome, there is less than perfect alignment between the third position of the codon and the base that pairs with it in the anticodon, so the requirements for base pairing are somewhat relaxed; this feature of the codon–anticodon interaction is referred to as wobble.

→ **Quick Check 2** What polypeptide sequences would you expect to result from a synthetic mRNA with the repeating sequence 5′-UUUGGGUUUGGGUUUGGG-3′.

Translation consists of initiation, elongation, and termination.

Translation is usually divided into three separate processes. The first is **initiation,** in which the initiator AUG codon is recognized and Met is established as the first amino acid in the new polypeptide chain. The second process is **elongation,** in which successive amino acids are added one by one to the growing chain. And the third process is **termination,** in which the addition of amino acids stops and the completed polypeptide chain is released from the ribosome.

Initiation of translation (**Fig. 4.17**) requires a number of protein **initiation factors** that bind to the mRNA. In eukaryotes, one group of initiation factors binds to the 5′ cap that is added to the mRNA during processing. These recruit a small subunit of the ribosome, and other initiation factors bring up a transfer RNA charged with Met (Fig. 4.17a). The initiation complex then moves along the mRNA until it encounters the first AUG triplet. The position of this AUG establishes the translational reading frame.

When the first AUG codon is encountered, a large ribosomal subunit joins the complex, the initiation factors are released, and the next tRNA is ready to join the ribosome (Fig. 4.17b). Note

FIG. 4.17 Translational initiation and elongation. In eukaryotes, the codon used for initiation is the AUG codon nearest the 5′ end of the mRNA. Note that the growing polypeptide chain remains attached to a tRNA throughout translation.

a.

Met
UAC
AUG

Initiation factors recruit the small ribosomal subunit and tRNA^{Met} and scan the mRNA for an AUG codon.

b.

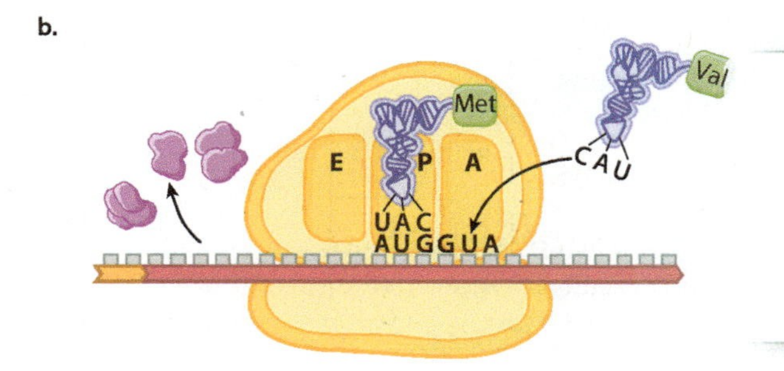

When the complex reaches an AUG, the large ribosomal subunit joins, the initiation factors are released, and a tRNA complementary to the next codon binds to the A site.

c.

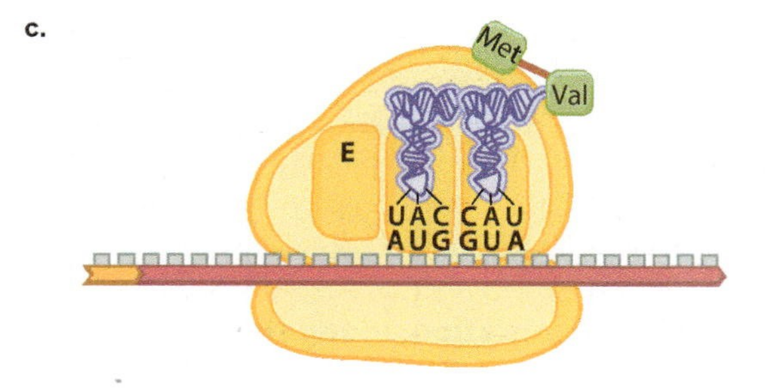

A reaction transfers the Met to the amino acid on the tRNA in the A site, forming a peptide bond.

d.

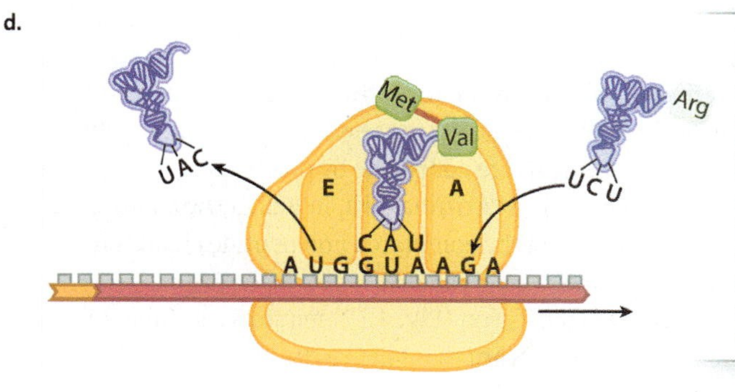

The ribosome moves down one codon, which puts the amino acid carrying the polypeptide into the P site and the now-uncharged tRNA into the E site, where it is ejected. A new tRNA complementary to the next codon binds to the A site.

e.

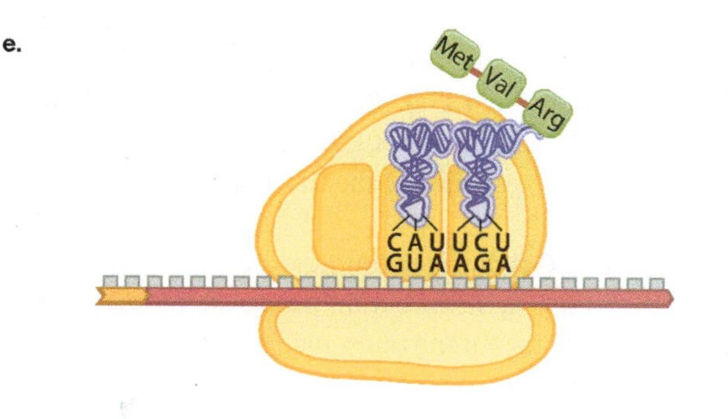

The polypeptide transfers to the amino acid on the tRNA in the A site. The polypeptide is elongated by repeating steps (d) and (e).

in Fig. 4.17b that the tRNA^{Met} binds with the P (peptidyl) site in the ribosome and that the next tRNA in line comes in at the A (aminoacyl) site. Once the new tRNA is in place, a reaction takes place in which the bond connecting the Met to its tRNA is transferred to the amino group of the next amino acid in line, forming a peptide bond between the two amino acids (Fig. 4.17c). An RNA in the large subunit is the actual catalyst for this reaction. The new polypeptide is now attached to the tRNA in the A site. The ribosome then shifts one codon to the right (Fig. 4.17d). With this move, the uncharged tRNA^{Met} shifts to the E site and is released into the cytoplasm, and the peptide-bearing tRNA shifts to the P site; the movement of the ribosome also empties the A site, making it available for the next charged tRNA in line to come into place.

At this point, steps (d) and (e) in Fig. 4.17 are repeated over and over again in the process of elongation, during which the polypeptide chain grows in length by the addition of successive amino acids. Ribosome movement along the mRNA and formation of the peptide bonds require energy, which is obtained with the help of proteins called **elongation factors.** Elongation factors are bound to GTP molecules, and break their high-energy bonds to provide energy for the elongation of the polypeptide.

The elongation process shown in Fig. 4.17 continues until one of the stop codons (UAA, UAG, or UGA) is encountered; these codons signal termination of polypeptide synthesis. Termination takes place because the stop codons do not have corresponding tRNA molecules. Rather, when a stop codon is encountered, a protein **release factor** binds to the A site of the ribosome. The release factor causes the bond connecting the polypeptide to the tRNA to break, creating the carboxyl terminus of the polypeptide and completing the chain. Once the finished polypeptide is released, the small and large ribosomal subunits disassociate from the mRNA and from each other.

Although elongation and termination are very similar in prokaryotes and in eukaryotes, translation initiation differs between the two (**Fig. 4.18**). In eukaryotes, the initiation complex forms at the 5′ cap and scans along the mRNA until the first AUG is encountered

FIG. 4.18 Initiation in eukaryotes and in prokaryotes. (a) In eukaryotes, translation is initiated only at the 5′ cap. (b) In prokaryotes, initiation takes place at any Shine–Dalgarno sequence; this mechanism allows a single mRNA to include coding sequences for multiple polypeptides.

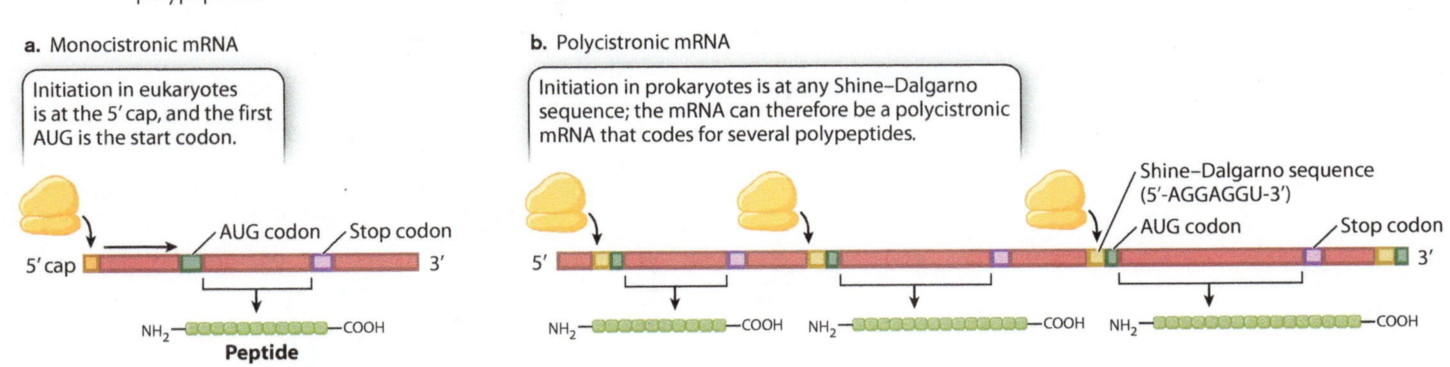

(Fig. 4.18a). In prokaryotes, the mRNA molecules have no 5′ cap. Instead, the initiation complex is formed at one or more internal sequences present in the mRNA known as a Shine–Dalgarno sequence (Figure 4.18b). In *E. coli*, the Shine–Dalgarno sequence is 5′-AGGAGGU-3′, and it is followed by an AUG codon eight nucleotides farther downstream that serves as an initiation codon for translation. The ability to initiate translation internally allows prokaryotic mRNAs to code for more than one protein. Such an mRNA is known as a **polycistronic mRNA.** In Fig. 4.18b, the polycistronic mRNA codes for three different polypeptide chains, each with its own AUG initiation codon preceded eight nucleotides upstream by its own Shine–Dalgarno sequence. Each Shine–Dalgarno sequence can serve as an initiation site for translation, and so all three polypeptides can be translated.

A polycistronic mRNA results from transcription of a group of functionally related genes located in tandem along the DNA and transcribed as a single unit from one promoter. This type of gene organization is known as an **operon.** Prokaryotes have many of their genes organized into operons because the production of a polycistronic mRNA allows all the protein products to be expressed together whenever they are needed. Typically, the genes organized into operons are those whose products are needed either for successive steps in the synthesis of an essential small molecule, such as an amino acid, or else for successive steps in the breakdown of a source of energy, such as a complex carbohydrate.

→ **Quick Check 3** Bacterial DNA containing an operon encoding three enzymes is introduced into chromosomal DNA in yeast (a eukaryote) in such a way that it is properly flanked by a promoter and a transcriptional terminator. The bacterial DNA is transcribed and the RNA correctly processed, but only the protein nearest the promoter is produced. Can you suggest why?

? **CASE 1 THE FIRST CELL: LIFE'S ORIGINS**

How did the genetic code originate?

During transcription and translation, proteins and nucleic acids work together to convert the information stored in DNA into proteins. If we think about how such a system might have originated, however, we immediately confront a chicken-and-egg problem: Cells need nucleic acids to make proteins, but proteins are required to make nucleic acids. Which came first? In Chapter 3, we discussed the special features that make RNA an attractive candidate for both information storage and catalysis in early life. Early in evolutionary history, then, proteins had to be added to the mix. No one fully understands how they were incorporated, but researchers are looking closely at tRNA, the molecule involved in the "translating" step of translation.

In modern cells, tRNA shuttles amino acids to the ribosome, but an innovative hypothesis suggests that in early life tRNA-like molecules might have served a different function. This proposal holds that the early precursors of the ribosome were RNA molecules that facilitated the replication of other RNAs, not proteins. In this version of an RNA world, precursors to tRNA would have shuttled nucleotides to growing RNA strands. Researchers hypothesize that tRNAs bound to amino acids may have acted as simple catalysts, facilitating more accurate RNA synthesis. Through time, amino acids brought into close proximity in the process of building RNA molecules might have polymerized to form polypeptide chains. From there, natural selection would favor the formation of polypeptides that enhanced replication of RNA molecules, bringing proteins into the chemistry of life.

All of the steps in gene expression, including transcription and translation, are summarized in **Fig. 4.19** on the next pages.

VISUAL SYNTHESIS Gene Expression

FIG. 4.19 **Integrating concepts from Chapters 3 and 4**

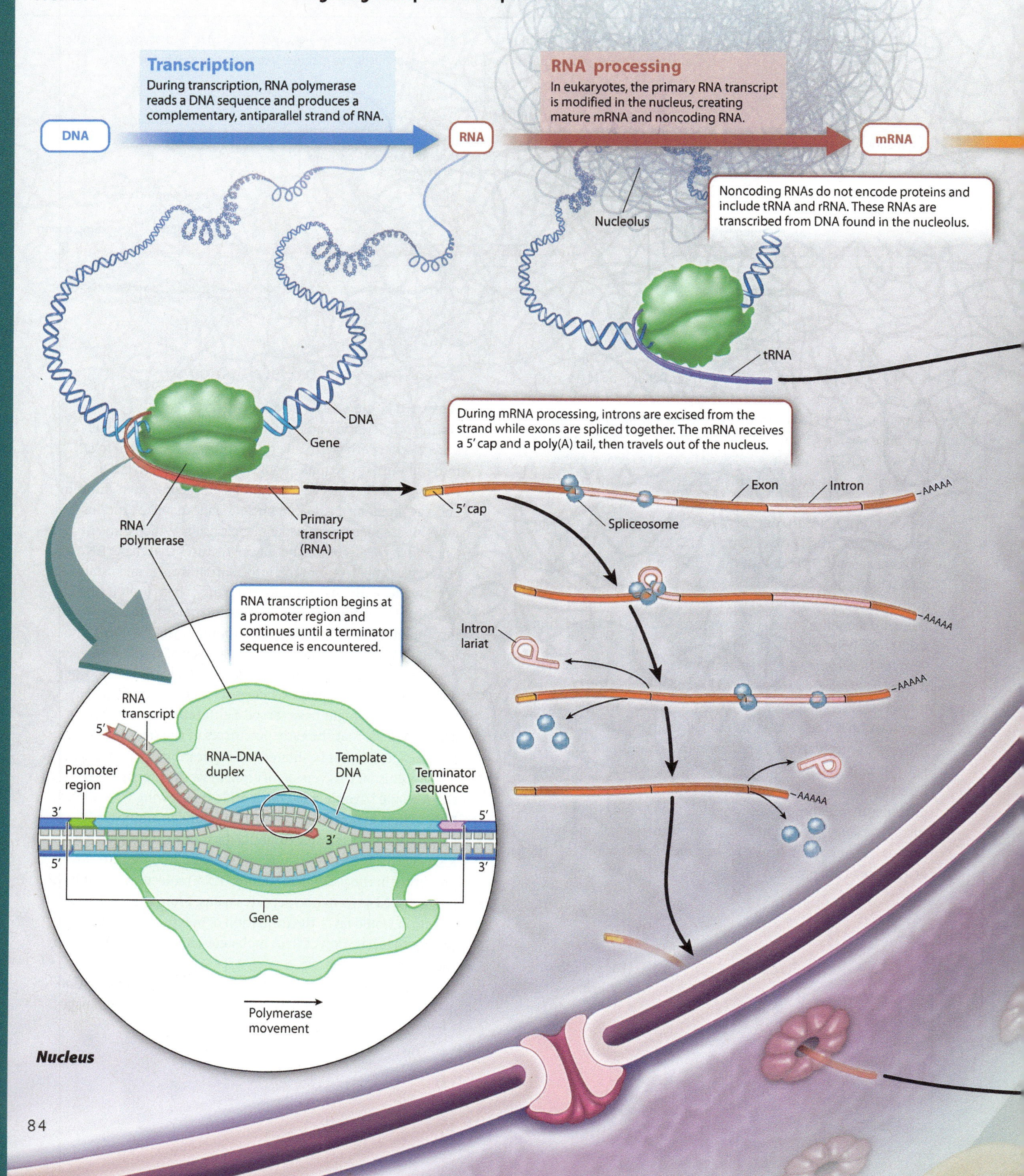

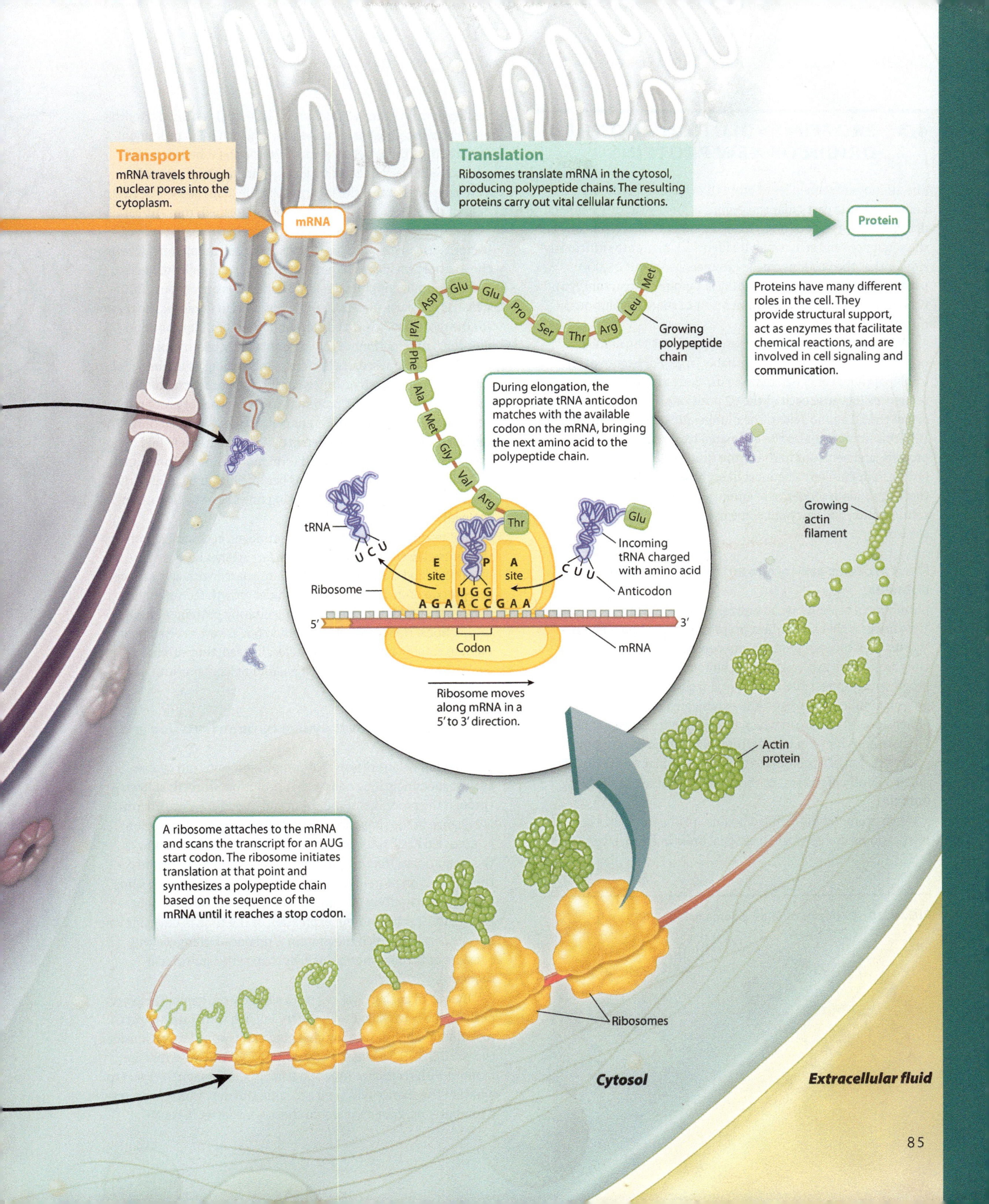
Transport
mRNA travels through nuclear pores into the cytoplasm.
mRNA
Translation
Ribosomes translate mRNA in the cytosol, producing polypeptide chains. The resulting proteins carry out vital cellular functions.
Protein
Proteins have many different roles in the cell. They provide structural support, act as enzymes that facilitate chemical reactions, and are involved in cell signaling and communication.
Met
Leu
Arg
Thr
Ser
Pro
Glu
Glu
Asp
Val
Phe
Ala
Met
Gly
Val
Arg
Thr
Growing polypeptide chain
During elongation, the appropriate tRNA anticodon matches with the available codon on the mRNA, bringing the next amino acid to the polypeptide chain.
tRNA
U C U
E site
P
A site
Glu
Incoming tRNA charged with amino acid
C U U
Anticodon
Ribosome
U G G
A G A A C C G A A
5′
3′
Codon
mRNA
Ribosome moves along mRNA in a 5′ to 3′ direction.
Growing actin filament
Actin protein
A ribosome attaches to the mRNA and scans the transcript for an AUG start codon. The ribosome initiates translation at that point and synthesizes a polypeptide chain based on the sequence of the mRNA until it reaches a stop codon.
Ribosomes
Cytosol
Extracellular fluid

4.3 PROTEIN EVOLUTION AND THE ORIGIN OF NEW PROTEINS

The amino acid sequences of more than a million proteins are known, and the particular three-dimensional structure has been determined for each of more than 10,000 proteins. While few of the sequences and structures are identical, many are sufficiently similar that the proteins can be grouped into about 25,000 **protein families.** A protein family is a group of structurally and functionally related proteins as a result of shared evolutionary history.

Why are there not more types of proteins? The number of possible sequences is unimaginably large. For example, for a polypeptide of only 62 amino acids, there are 20^{62} possible sequences (because each of the 62 positions could be occupied by any of the 20 amino acids). The number 20^{62} equals approximately 10^{80}; this number is also the estimated total number of electrons, protons, and neutrons in the entire universe! So why are there so few protein families? The most likely answer is that the chance that any random sequence of amino acids would fold into a stable configuration and carry out some useful function in the cell is very close to zero.

Most proteins are composed of modular folding domains.

If functional proteins are so unlikely, how could life have evolved? The answer is that the earliest proteins were probably much shorter than modern proteins and needed only a trace of function. Only as proteins evolved through billions of years did they become progressively longer and more specialized in their functions. Many protein families that exist today exhibit small regions of three-dimensional structure in which the protein folding is similar. These regions range in length from 25 to 100 or more amino acids. A region of a protein that folds in a similar way relatively independently of the rest of the protein is known as a **folding domain.**

Two examples of folding domains are illustrated in **Fig. 4.20.** Many folding domains are functional units in themselves. The folding domain called a TIM barrel (Fig. 4.20a) is named after the enzyme triose phosphate isomerase, in which it is a prominent feature. The TIM barrel consists of alternating α helices and parallel β sheets connected by loops. In many enzymes with a TIM barrel, the active site is formed by the loops at the carboxyl ends of the sheets. Fig. 4.20b is a β barrel formed from antiparallel β sheets. β barrel structures occur in proteins in some types of bacteria, usually in proteins that span the cell membrane, where the β barrel provides a channel that binds hydrophilic molecules.

The number of known folding domains is only about 2500, which is far fewer than the number of protein families. The reason for the discrepancy is that different protein families contain different combinations of folding domains. Modern protein families are composed of different combinations of a number of folding domains, each of which contributes some structural or functional feature of the protein. Different types of protein folds occur again and again in different contexts and combinations. The earliest proteins may have been little more than single folding domains that could aggregate to form more complex functional units. As life evolved, the proteins became longer by joining the DNA coding for the individual folding units together into a single molecule.

For example, human tissue plasminogen activator, a protein that is used in treating strokes and heart attacks because it dissolves blood clots, contains domains shared with cell-surface receptors, a domain shared with cellular growth factors, and a domain that folds into large loops facilitating protein–protein interactions. Hence, novel proteins do not always evolve from random combinations of amino acids; instead, they often evolve by joining already functional folding domains into novel combinations.

FIG. 4.20 **Examples of folding domains:** (a) TIM barrel; (b) β barrel.

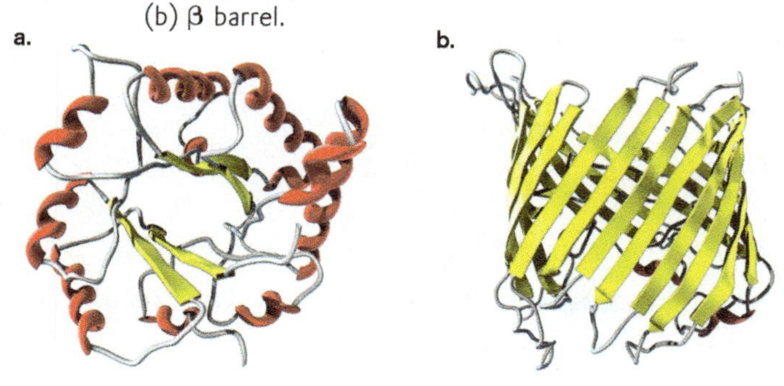

Amino acid sequences evolve through mutation and selection.

Another important reason that complex proteins can evolve against seemingly long odds is that evolution proceeds stepwise through the processes of mutation and selection. A **mutation** is a change in the sequence of a gene. The process of mutation is discussed in Chapter 14, but for now all you need to know is that mutations affecting proteins occur at random in regard to their effects on protein function. In protein-coding genes, some mutations may affect the amino acid sequence; others might change the level of protein expression or the time in development or type of cell in which the protein is produced. Here, we will consider only those mutations that change the amino acid sequence.

By way of analogy, we can use a simple word game. The object of the game is to change an ordinary English word into another meaningful English word by changing exactly one letter. Consider the word GONE. To illustrate "mutations" of the word that are random with respect to function (that is, random with respect to whether the change will yield a meaningful new word), we wrote a computer program that would choose one letter in GONE at random and replace it with a different random letter. The first 24 "mutants" of GONE are:

UONE	GNNE	GONJ	GOZE
GONH	GOLE	GFNE	XONE
NONE	GKNE	GJNE	DONE
GCNE	GONB	GOIE	GGNE
GONI	GFNE	GPNE	GENE
BONE	GOWE	OONE	GYNE

Most of the mutant words are gibberish, corresponding to the biological reality that most random amino acid replacements impair protein function to some extent. On the other hand, some mutant proteins function just as well as the original, and a precious few change function. In the word-game analogy, the mutants that can persist correspond to meaningful words, those words shown in red.

In a population of organisms, random mutations are retained or eliminated through the process of **selection** among individuals on the basis of their ability to survive and reproduce. This process was introduced in Chapter 1 and is considered in greater detail in Chapter 21, but the principle is straightforward. Most mutations that impair protein function will be eliminated because, if the function of the nonmutant protein contributes to survival and reproduction, the individuals carrying these mutations will leave fewer offspring than others. Mutations that do not impair function may remain in the population for long periods because their carriers survive and reproduce in normal numbers; a mutation of this type has no tendency to either increase or decrease in frequency over time. In contrast, individuals that carry the occasional mutation that improves protein function will reproduce more successfully than others. Because of the enhanced reproduction, the mutant gene encoding the improved protein will gradually increase in frequency and spread throughout the entire population.

In the word game, any of the mutants in red may persist in the population, but suppose that one of them, GENE for example, is actually superior to GONE (considered more euphonious, perhaps). Then GENE will gradually displace GONE, and eventually GONE will be gone. In a similar way that one meaningful word may replace another, one amino acid sequence may be replaced with a different one in the course of evolution.

A real-world example that mirrors the word game is found in the evolution of resistance of the malaria parasite to the drug pyrimethamine. This drug inhibits an enzyme known as dihydrofolate reductase, which the parasite needs to survive and reproduce inside red blood cells. Resistance to pyrimethamine is known to have evolved through a stepwise sequence of four amino acid replacements. In the first replacement, serine (S) at the 108th amino acid in the polypeptide sequence (position 108) was replaced with asparagine (N); then cysteine (C) at position 59 was replaced with arginine (R); asparagine (N) at position 51 was then replaced with isoleucine (I); finally, isoleucine (I) at position 164 was replaced with leucine (L). If we list the amino acids according to their single-letter abbreviation in the order of their occurrence in the protein, the evolution of resistance followed this pathway:

NCSI → NCNI → NRNI → IRNI → IRNL

where the mutant amino acids are shown in red. Each successive amino acid replacement increased the level of resistance so that a greater concentration of drug was needed to treat the disease. The quadruple mutant IRNL is resistant to such high levels that the drug is no longer useful.

Depicted according to stepwise amino acid replacements, the analogy between the evolution of pyrimethamine resistance and the word game is clear. It should also be clear from our earlier discussion that hundreds of other mutations causing amino acid replacements in the enzyme must have occurred in the parasite during the course of evolution, but only these amino acid changes occurring in this order persisted and increased in frequency because they conferred greater survival and reproduction of the parasite under treatment with the drug.

→ **Quick Check 4** What do you think happened to the mutations that decreased survival or reproduction of the parasites? ■

Core Concepts Summary

4.1 Proteins are linear polymers of amino acids that form three-dimensional structures with specific functions.

An amino acid consists of an α carbon connected by covalent bonds to an amino group, a carboxyl group, a hydrogen atom, and a side chain or R group. page 70

There are 20 common amino acids that differ in their R groups. Amino acids are categorized by the chemical properties of their R groups—hydrophobic, basic, acidic, polar—and by special structures. page 70

Amino acids are connected by peptide bonds to form proteins. page 72

The primary structure of a protein is its amino acid sequence. The primary structure determines how a protein folds, which in turn determines how it functions. page 73

The secondary structure of a protein results from the interactions of nearby amino acids. Examples include the α helix and β sheet. page 73

The tertiary structure of a protein is its three-dimensional shape, which results from long-range interactions of amino acid R groups. page 75

Some proteins are made up of several polypeptide subunits; this group of subunits is the protein's quaternary structure. page 76

Chaperones help some proteins fold properly. page 76

4.2 Translation is the process by which the sequence of bases in messenger RNA specifies the order of successive amino acids in a newly synthesized protein.

Translation requires many cellular components, including ribosomes, tRNAs, and proteins. page 77

Ribosomes are composed of a small and a large subunit, each consisting of RNA and protein; the large subunit contains three tRNA-binding sites that play different roles in translation. page 77

An mRNA transcript of a gene has three possible reading frames composed of three-nucleotide codons. page 78

tRNAs have an anticodon that base pairs with the codon in the mRNA and carries a specific amino acid. page 78

Aminoacyl tRNA synthetases attach specific amino acids to tRNAs. page 79

The genetic code defines the relationship between the three-letter codons of nucleic acids and their corresponding amino acids. It was deciphered using synthetic RNA molecules. page 79

The genetic code is redundant in that many amino acids are specified by more than one codon. page 81

Translation consists of three steps: initiation, elongation, and termination. page 81

4.3 Proteins evolve through mutation and selection and by combining functional units.

A protein family is a group of proteins that are structurally and functionally related. page 86

There are far fewer protein families than the total number of possible proteins because the probability that a random sequence of amino acids will fold properly to carry out a specific function is very small. page 86

A region of a protein that folds in a particular way and that carries out a specific function is called a folding domain. page 86

Proteins evolve by combining different folding domains. page 86

Proteins also evolve by changes in amino acid sequence, which occurs by mutation and selection. page 87

Self-Assessment

1. Draw one of the 20 amino acids and label the amino group, the carboxyl group, the R group (side chain), and the α carbon.
2. Name four major groups of amino acids, categorized by the properties of their R groups. Explain how the chemical properties of each group affect protein shape.
3. Describe how peptide bonds, hydrogen bonds, ionic bonds, disulfide bridges, and noncovalent interactions (van der Waals forces and the hydrophobic effect) define a protein's four levels of structure.
4. Explain how the order of amino acids determines the way in which a protein folds.
5. Explain the relationship between protein folding and protein function.
6. Describe the relationship between the template strand of DNA, the codons in mRNA, anticodons in tRNA, and amino acids.
7. Describe the steps of translation initiation, elongation, and termination.
8. Name and describe two ways that proteins can acquire new functions in the course of evolution.

Log in to LaunchPad to check your answers to the Self-Assessment questions, and to access additional learning tools.

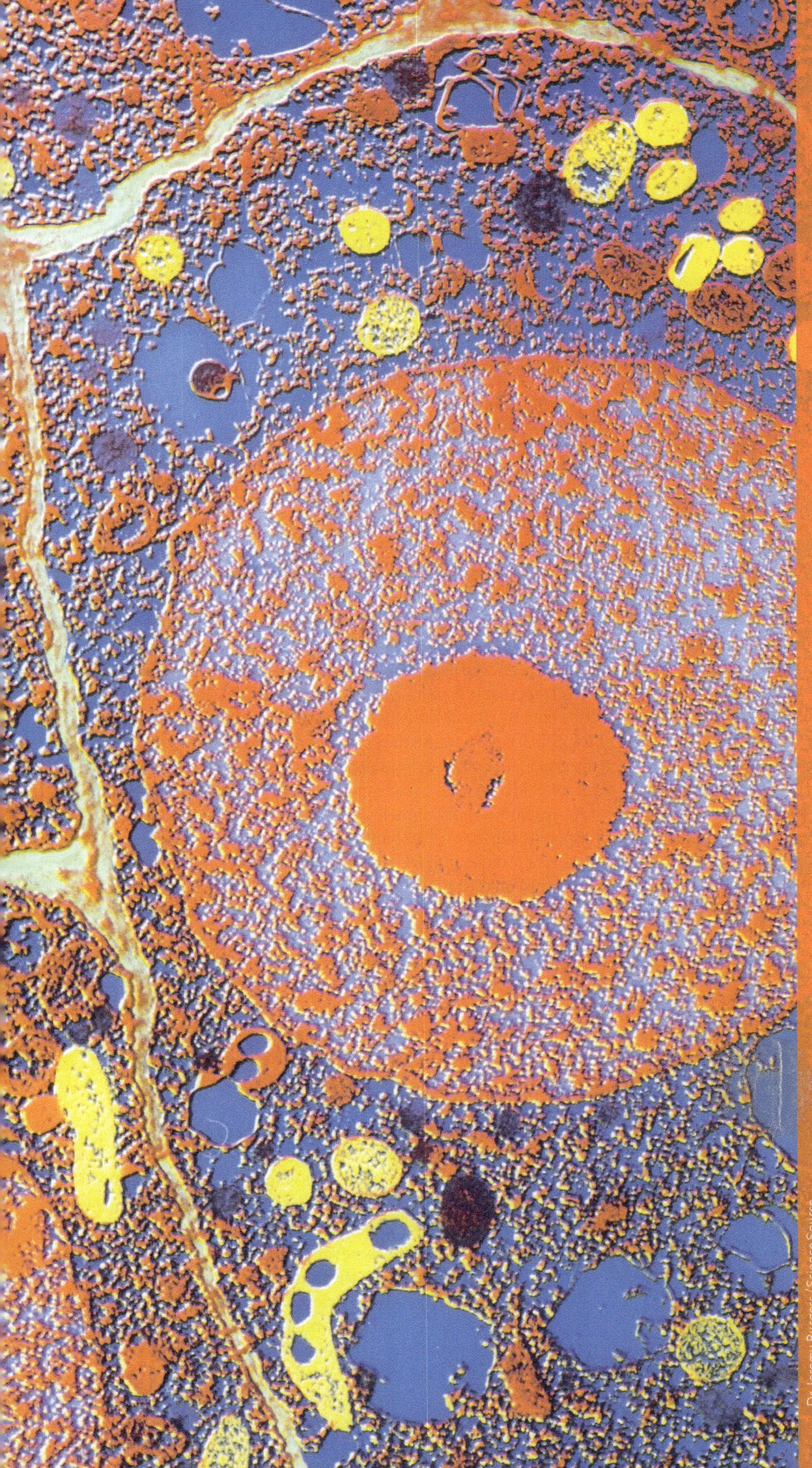

Dr. Jeremy Burgess/Science Source.

CHAPTER 5

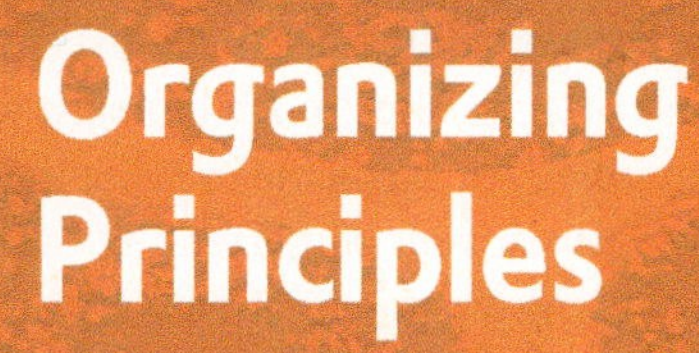

Organizing Principles

Lipids, Membranes, and Cell Compartments

Core Concepts

5.1 Cell membranes are composed of lipids, proteins, and carbohydrates.

5.2 The plasma membrane is a selective barrier that controls the movement of molecules between the inside and the outside of the cell.

5.3 Cells can be classified as prokaryotes or eukaryotes, which differ in the degree of internal compartmentalization.

5.4 The endomembrane system is an interconnected system of membranes that includes the nuclear envelope, endoplasmic reticulum, Golgi apparatus, lysosomes, vesicles, and plasma membrane.

5.5 Mitochondria and chloroplasts are organelles involved in harnessing energy, and likely evolved from free-living prokaryotes.

"With the discovery of the cell, biologists found their atom." So stated François Jacob, the French biologist who shared the Nobel Prize in Physiology or Medicine in 1965. Just as the atom is the smallest, most basic unit of matter, the cell is the smallest, most basic unit of living organisms. All organisms, from single-celled algae to complex multicellular organisms like humans, are made up of cells. Therefore, essential properties of life, including growth, reproduction, and metabolism, must be understood in terms of cell structure and function.

Cells were first seen sometime around 1665, when the English scientist Robert Hooke built a microscope that he used to observe thin sections of dried cork tissue derived from plants. In these sections, Hooke observed arrays of small cavities and named them "cells" (**Fig. 5.1**). Although Hooke was probably looking at the cell walls of empty (rather than living) cells, his observations nevertheless led to the concept that cells are the fundamental unit of life. Later in the seventeenth century, the Dutch microbiologist Anton van Leeuwenhoek greatly improved the magnifying power of microscope lenses, enabling him to see and describe unicellular organisms, including bacteria, protists, and algae. Today, modern microscopy provides unprecedented detail and a deeper understanding of the inner architecture of cells.

Cells differ in size and shape, but they share many features. This similarity in the microscopic organization of all living organisms led to the development in the middle of the nineteenth century of one of the pillars of modern biology: the **cell theory.** Based on the work and ideas of Matthias Schlieden, Theodor Schwann, Rudolf Virchow, and others, the cell theory states that all organisms are made up of cells, that the cell is the fundamental unit of life, and that cells come from preexisting cells. There is no life without cells, and the cell is the smallest unit of life.

This chapter focuses on cells and their internal organization. We pay particular attention to the key role that membranes play in separating a cell from the external environment and defining structural and functional spaces within cells.

FIG. 5.1 The first observation of cells. Robert Hooke used a simple microscope to observe small chambers in a sample of cork tissue that he described as "cells." *Sources: (left) Science Museum/SSPL/The Image Works; (right) Ted Kinsman/Science Source.*

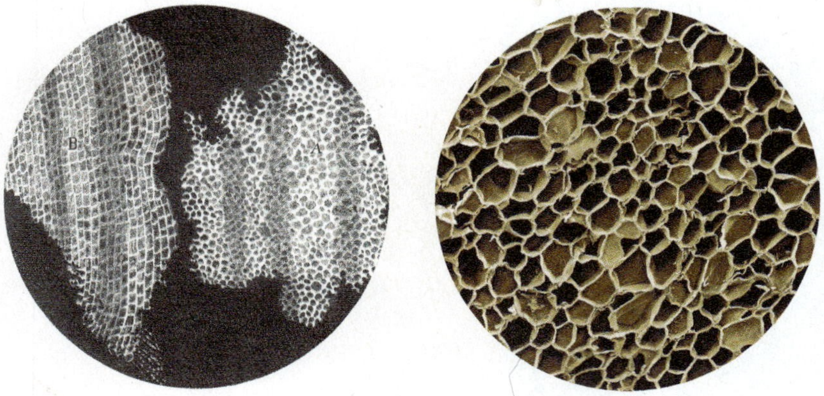

Drawing by Hooke

Cork tissue

5.1 STRUCTURE OF CELL MEMBRANES

Cells are defined by membranes. After all, membranes physically separate cells from their external environment. In addition, membranes define spaces within many cells that allow them to carry out their diverse functions.

Lipids are the main component of cell membranes. They have properties that allow them to form a barrier in an aqueous (watery) environment. Proteins are often embedded in or associated with the membrane, where they perform important functions such as transporting molecules. Carbohydrates can also be found in cell membranes, usually attached to lipids (glycolipids) and proteins (glycoproteins).

Cell membranes are composed of two layers of lipids.

The major types of lipid found in cell membranes are phospholipids, introduced in Chapter 2. Most phospholipids are made up of a glycerol backbone attached to a phosphate group and two fatty acids (**Fig. 5.2**). The phosphate head group is hydrophilic ("water-loving") because it is polar, enabling it to form hydrogen bonds with water. By contrast, the two fatty acid tails are hydrophobic ("water-fearing") because they are nonpolar and do not form hydrogen bonds with water. Molecules with both hydrophilic and hydrophobic regions in a single molecule are termed **amphipathic.**

In an aqueous environment, amphipathic molecules such as phospholipids behave in an interesting way. They spontaneously arrange themselves into various structures in which the polar head groups on the outside interact with water and the nonpolar tail groups come together on the inside away from water. This arrangement results from the tendency of polar molecules like water to exclude nonpolar molecules or nonpolar groups of molecules.

The shape of the structure is determined by the bulkiness of the head group relative to the hydrophobic tails. For example, lipids with bulky heads and a single hydrophobic fatty acid tail are wedge-shaped and pack into spherical structures called **micelles** (**Fig. 5.3a**). By contrast, lipids with less bulky head groups and two hydrophobic tails form a **bilayer** (**Fig. 5.3b**). A lipid bilayer is a structure formed of two layers of lipids in which the hydrophilic heads are the outside surfaces of the bilayer and the hydrophobic tails are sandwiched in between, isolated from contact with the aqueous environment.

The bilayers form closed structures with an inner space since free edges would expose the hydrophobic chains to the aqueous environment. This organization in part explains why bilayers are effective cell membranes. It also explains why membranes are self-healing. Small tears in a membrane are rapidly

FIG. 5.2 Phospholipid structure. Phospholipids, the major component of cell membranes, are made up of glycerol attached to a phosphate-containing head group and two fatty acid tails. They are amphipathic because they have both hydrophilic and hydrophobic domains.

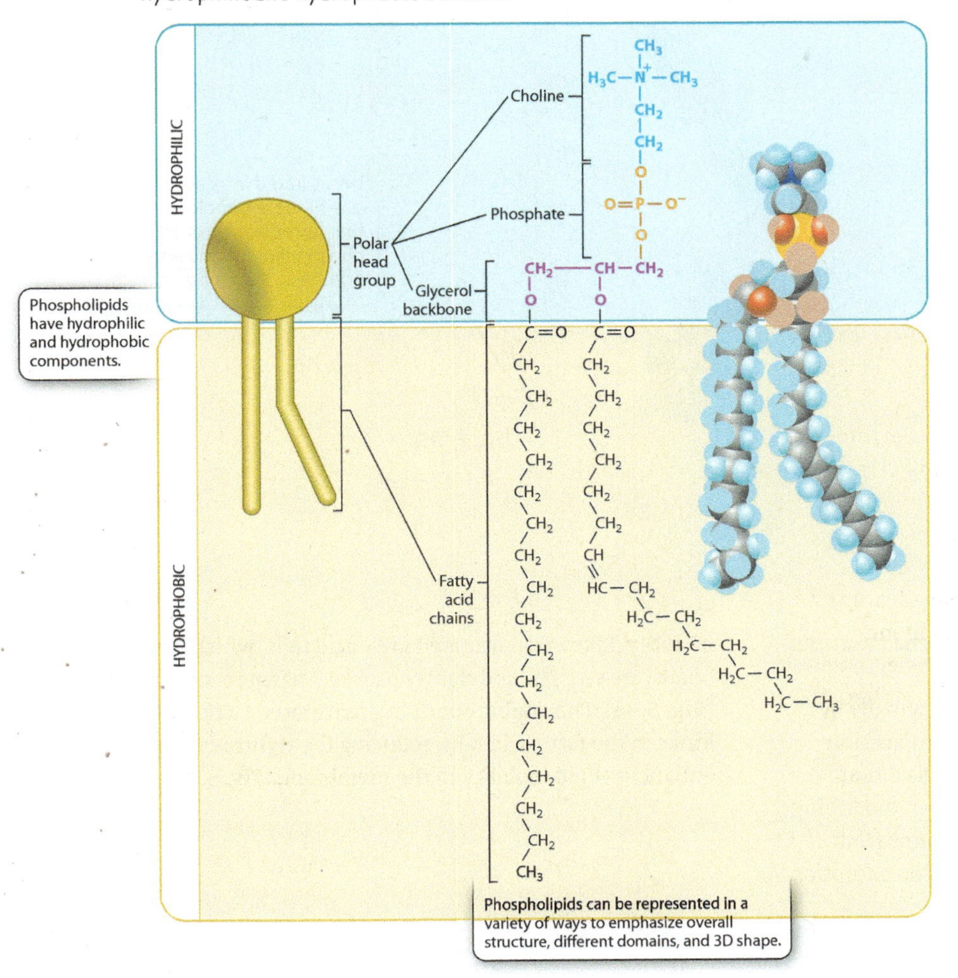

sealed by the spontaneous rearrangement of the lipids surrounding the damaged region because of the tendency of water to exclude nonpolar molecules.

? CASE 1 THE FIRST CELL: LIFE'S ORIGINS

How did the first cell membranes form?

What are the consequences of the ability of phospholipids to form a bilayer when placed in water? The bilayer structure forms spontaneously, dependent solely on the properties of the phospholipid and without the action of an enzyme, as long as the concentration of free phospholipids is high enough and the pH of the solution is similar to that of a cell. The pH is important because it ensures that the head groups are in their ionized (charged) form and thus suitably hydrophilic. Thus, if phospholipids are added to a test tube of water at neutral pH, they spontaneously form spherical bilayer structures called **liposomes** that surround a central space (**Fig. 5.3c**). As the liposomes form, they may capture macromolecules present in solution.

Such a process may have been at work in the early evolution of life on Earth. Experiments show that liposomes can form, break, and re-form in environments like tidal flats that are repeatedly dried and flooded with water. The liposomes can even grow, incorporating more and more lipids from the environment, and capture nucleic acids and other molecules in their interiors. Depending on their chemical composition, early membranes might have been either leaky or almost impervious to the molecules of life. Over time, they evolved in such a way as to allow at least limited molecular traffic between the environment and cell interior. At some point, new lipids no longer had to be incorporated from the environment. Instead, proteins guided lipid synthesis within the cell, although

FIG. 5.3 Phospholipid structures. Phospholipids can form (a) micelles, (b) bilayers, or (c) liposomes when placed in water.

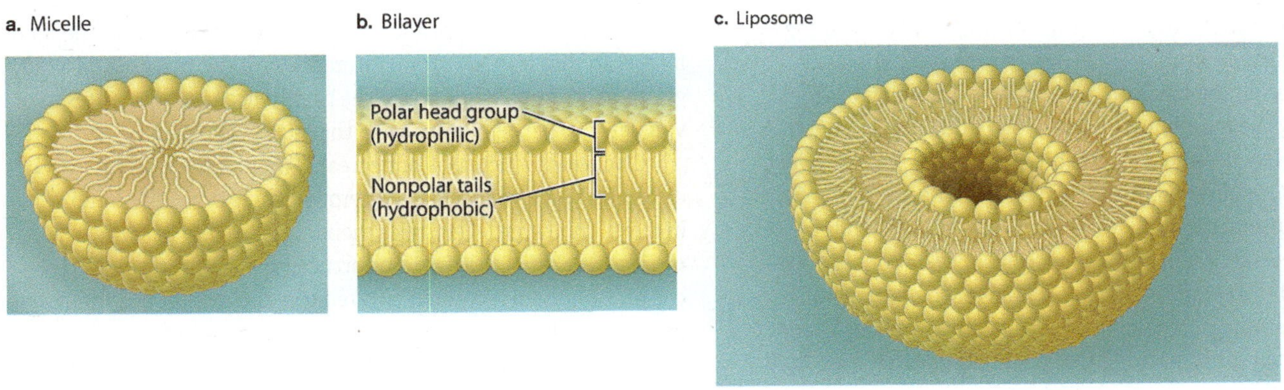

a. Micelle

b. Bilayer

Polar head group (hydrophilic)

Nonpolar tails (hydrophobic)

c. Liposome

FIG. 5.4 Saturated and unsaturated fatty acids in phospholipids. The composition of cell membranes affects the tightness of packing.

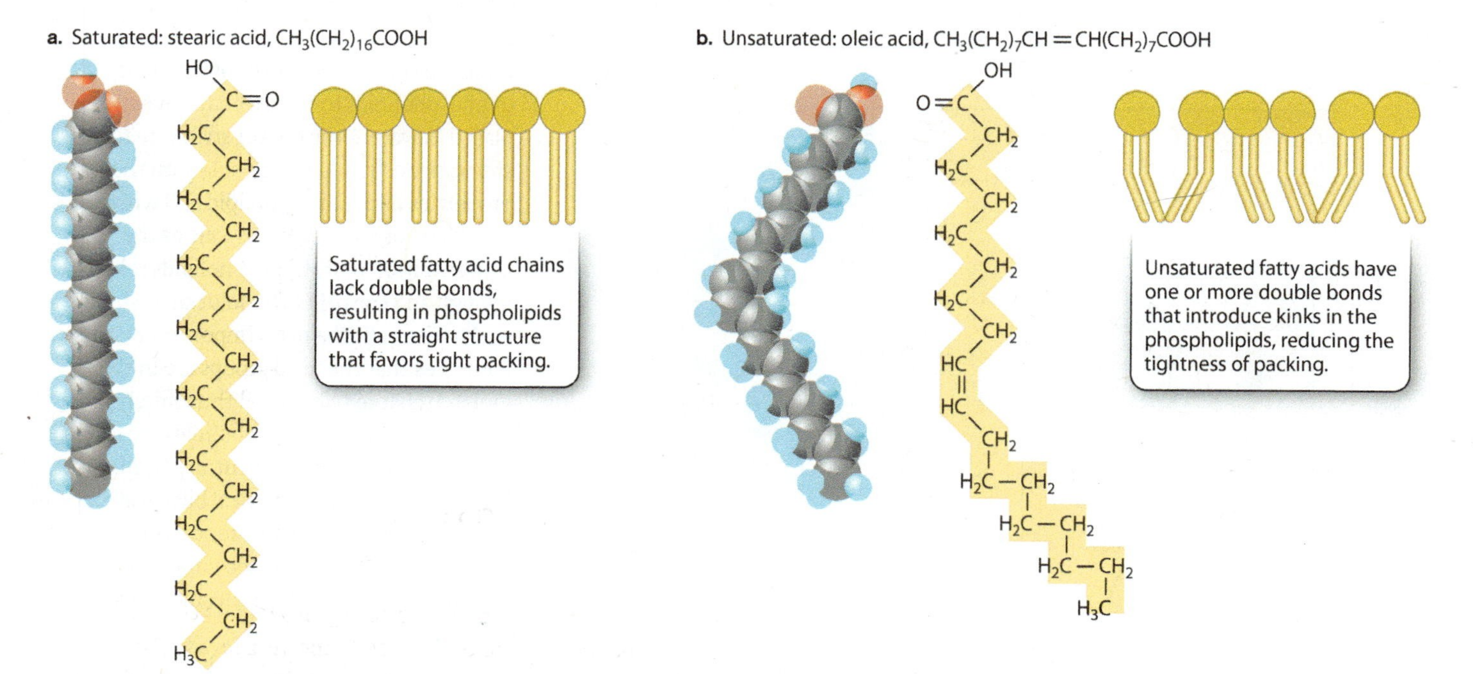

how this switch to protein-mediated synthesis happened remains uncertain.

All evidence suggests that membranes formed originally by straightforward physical processes, but that their composition and function evolved over time. François Jacob once said that evolution works more like a tinkerer than an engineer, modifying already existing materials rather than designing systems from scratch. It seems that the evolution of membranes is no exception to this pattern.

Cell membranes are dynamic.

Lipids freely associate with one another because of extensive van der Waals forces between their fatty acid tails (Chapter 2). These weak interactions are easily broken and re-formed, so lipid molecules are able to move within the plane of the membrane, sometimes very rapidly: A single phospholipid can move across the entire length of a bacterial cell in less than a second. Lipids can also rapidly rotate around their vertical axis, and individual fatty acid chains are able to flex, or bend. As a result, membranes are dynamic: they are continually moving, forming, and re-forming during the lifetime of a cell.

Because membrane lipids are able to move in the plane of the membrane, the membrane is said to be **fluid.** The degree of membrane fluidity depends on which types of lipid make up the membrane. In a single layer of the lipid bilayer, the strength of the van der Waals interactions between the lipids' tails depends on the length of the fatty acid tails and the presence of double bonds between neighboring carbon atoms. The longer the fatty acid tails, the more surface is available to participate in van der Waals interactions. The tighter packing that results tends to reduce lipid mobility. Likewise, saturated fatty acid tails, which have no double bonds, are straight and tightly packed—again reducing mobility (**Fig. 5.4a**). The double bonds in unsaturated fatty acids introduce kinks in the fatty acid tails, reducing the tightness of packing and enhancing lipid mobility in the membrane (**Fig. 5.4b**).

→ **Quick Check 1** Most animal fats are solid at room temperature, whereas plant and fish oils tend to be liquid. Both contain fatty acids. Can you predict which type of fat contains saturated fatty acids, and which type contains unsaturated fatty acids?

In addition to phospholipids, cell membranes often contain other types of lipid, and these can also influence membrane fluidity. For example, **cholesterol** is a major component of animal cell membranes, representing about 30% by mass of the membrane lipids. Like phospholipids, cholesterol is amphipathic, with both hydrophilic and hydrophobic groups in the same molecule. In cholesterol, the hydrophilic region is simply a hydroxyl group (–OH) and the hydrophobic region consists of four interconnected carbon rings with an attached hydrocarbon chain (**Fig. 5.5**). This structure allows cholesterol to insert into the lipid bilayer so that its head group interacts with the hydrophilic head group of phospholipids, while the ring structure participates in van der Waals interactions with the fatty acid chains.

Cholesterol increases or decreases membrane fluidity depending on temperature. At temperatures typically found in a cell, cholesterol decreases membrane fluidity because the interaction of the rigid ring structure of cholesterol with the phospholipid fatty acid tails reduces the mobility of the phospholipids. However, at low temperatures, cholesterol increases membrane fluidity because it prevents phospholipids

FIG. 5.5 Cholesterol in the lipid bilayer. Cholesterol molecules embedded in the lipid bilayer affect the fluidity of the membrane.

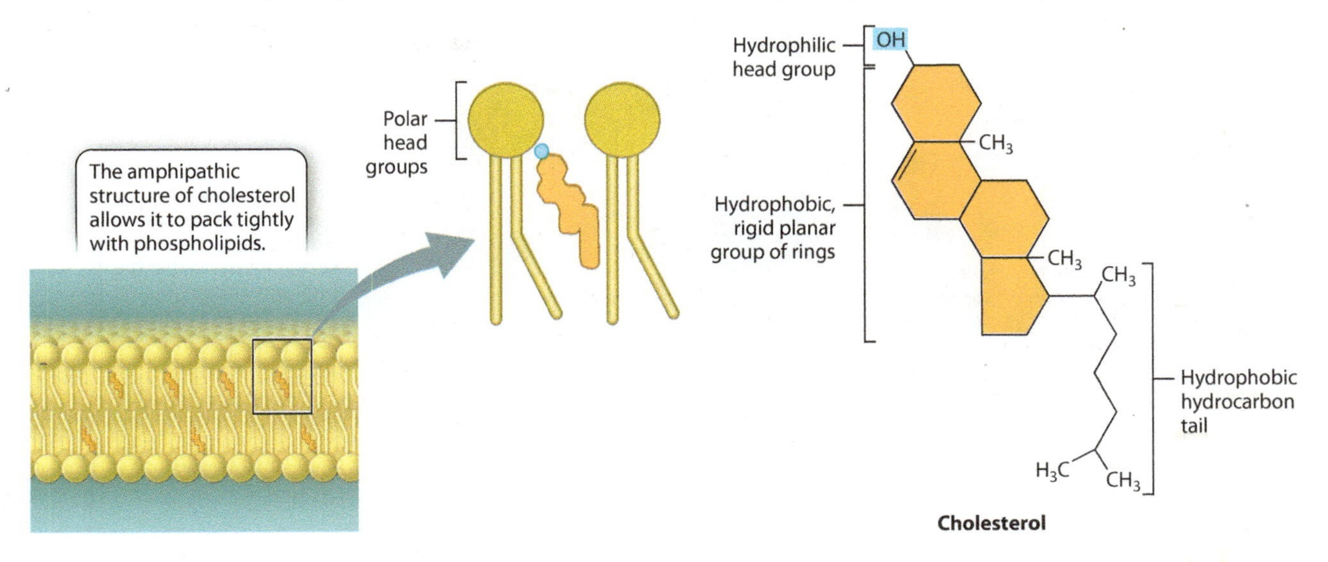

from packing tightly with other phospholipids. Thus, cholesterol helps maintain a consistent state of membrane fluidity by preventing dramatic transitions from a fluid to solid state.

For many decades, it was thought that the various types of lipid found in the membrane were randomly distributed throughout the bilayer. However, more recent studies show that specific types of lipid, such as sphingolipids, sometimes assemble into defined patches called **lipid rafts.** Cholesterol and other membrane components such as proteins also appear to accumulate in some of these regions. Thus, membranes are not always a uniform fluid bilayer, but instead can contain regions with discrete components.

Although lipids are free to move in the plane of the membrane, the spontaneous transfer of a lipid between layers of the bilayer, known as lipid flip-flop, is very rare. This is not surprising since flip-flop requires the hydrophilic head group to pass through the hydrophobic interior of the membrane. As a result, there is little exchange of components between the two layers of the membrane, which in turn allows the two layers to differ in composition. In fact, in many membranes, different types of lipid are present primarily in one layer or the other.

Proteins associate with cell membranes in different ways.

Most membranes contain proteins as well as lipids. For example, proteins represent as much as 50% by mass of the membrane of a red blood cell. Membrane proteins serve different functions (**Fig. 5.6**). Some act as **transporters,** moving ions or other molecules across the membrane. Other membrane proteins

FIG. 5.6 Functions of membrane proteins.

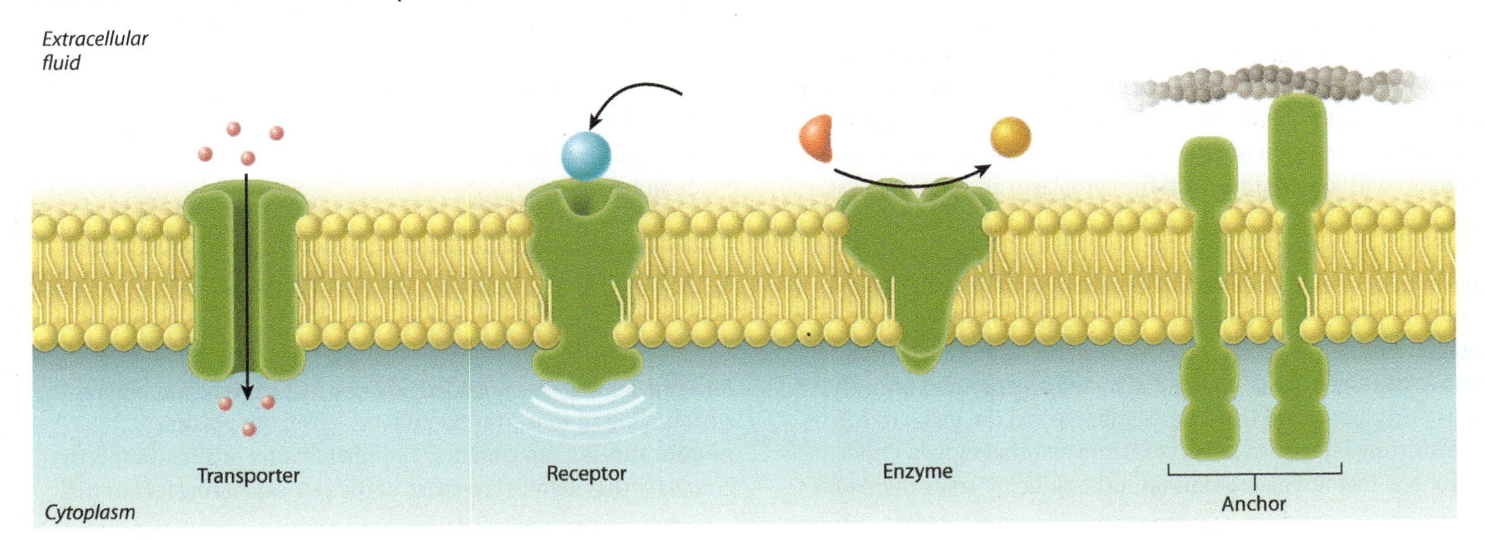

FIG. 5.7 Integral and peripheral membrane proteins. Integral membrane proteins are permanently associated with the membrane. Peripheral membrane proteins are temporarily associated with one or other of the two lipid bilayers or with an integral membrane protein.

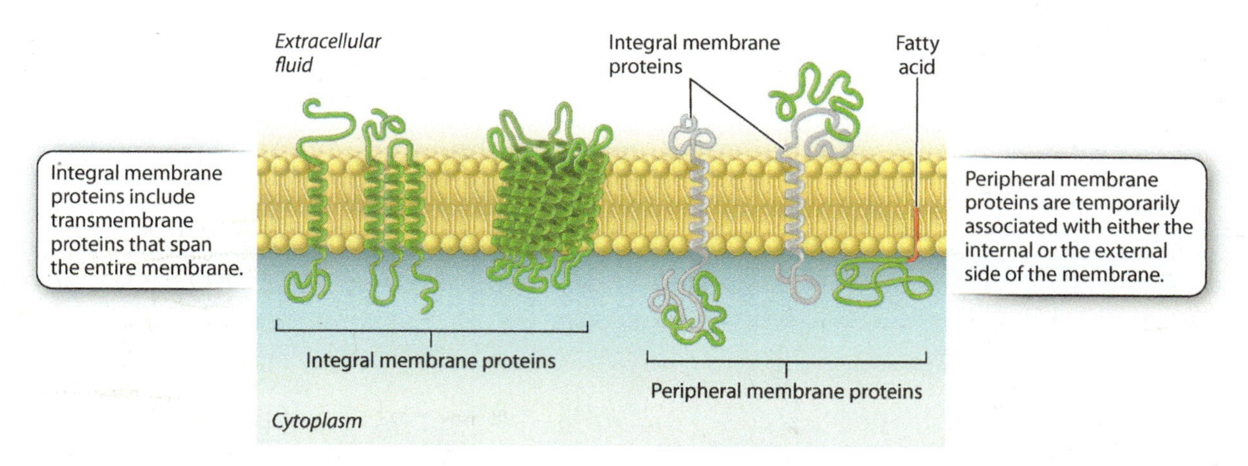

act as **receptors** that allow the cell to receive signals from the environment. Still others are **enzymes** that catalyze chemical reactions or **anchors** that attach to other proteins and help to maintain cell structure and shape.

These various membrane proteins can be classified into two groups depending on how they associate with the membrane (**Fig. 5.7**). **Integral membrane proteins** are permanently associated with cell membranes and cannot be separated from the membrane experimentally without destroying the membrane itself. **Peripheral membrane proteins** are temporarily associated with the lipid bilayer or with integral membrane proteins through weak noncovalent interactions. They are easily separated from the membrane by simple experimental procedures that leave the structure of the membrane intact.

Most integral membrane proteins are **transmembrane proteins** that span the entire lipid bilayer, as shown in Fig. 5.7. These proteins are composed of three regions: two hydrophilic regions, one protruding from each face of the membrane, and a connecting hydrophobic region that spans the membrane. This structure allows for separate functions and capabilities of each end of the protein. For example, the hydrophilic region on the external side of a receptor can interact with signaling molecules, whereas the hydrophilic region on the internal side of the membrane often interacts with other proteins in the cytoplasm of the cell to pass along the message.

Peripheral membrane proteins may be associated with either the internal or external side of the membrane (Fig. 5.7). These proteins interact either with the polar heads of lipids or with integral membrane proteins by weak noncovalent interactions such as hydrogen bonds. Peripheral membrane proteins are only transiently associated with the membrane and can play a role in transmitting information received from external signals. Other peripheral membrane proteins limit the ability of transmembrane proteins to move within the membrane and assist proteins in clustering in lipid rafts.

Proteins, like lipids, are free to move in the membrane. How do we know this? The mobility of proteins in the cell membrane can be demonstrated using an elegant experimental technique called fluorescence recovery after photobleaching, or FRAP (**Fig. 5.8**). In this technique, proteins embedded in the cell membrane are labeled with fluorescent dye molecules. Labeling all the proteins in a membrane creates a fluorescent cell that can be visualized with a fluorescence microscope. A laser is then used to bleach the fluorescent dye molecules in a small area of the cell membrane, leaving a nonfluorescent spot on the surface of the cell. If proteins in the cell membrane were not capable of movement, the bleached area would remain nonfluorescent. However, over time, fluorescence appears in the bleached area, telling us that fluorescent proteins that were not bleached moved into the bleached area.

The idea that lipids and proteins coexist in the membrane, and that both are able to move in the plane of the membrane, led American biologists S. Jonathan Singer and Garth Nicolson to propose the **fluid mosaic model** in 1972. According to this model, the lipid bilayer is a fluid structure within which molecules move laterally, and is a mosaic (a mixture) of two types of molecules, lipids and proteins.

5.2 THE PLASMA MEMBRANE AND CELL WALL

Phospholipids with embedded proteins make up the membrane surrounding all cells. This membrane, called the **plasma membrane,** is a fundamental, defining feature of all cells. It is the boundary that defines the space of the cell, separating its internal

HOW DO WE KNOW?

FIG. 5.8

Do proteins move in the plane of the membrane?

BACKGROUND Fluorescent recovery after photobleaching (FRAP) is a technique used to measure mobility of molecules in the plane of the membrane. A fluorescent dye is attached to proteins embedded in the cell membrane in a process called labeling. A laser is then used to bleach a small area of the membrane.

HYPOTHESIS If membrane components such as proteins move in the plane of the membrane, the bleached spot should become fluorescent over time as unbleached fluorescent molecules move into the bleached area. If membrane components do not move, the bleached spot should remain intact.

Photo source: FRAP of cytoplasmic EGFP in living HeLa cells, performed using an UltraVIEW® spinning disk confocal system (PerkinElmer Inc.). HeLa cells were transfected with pEGFP-C1 (Clontech Laboratories, Inc.) using GeneJuice® transfection reagent (Novagen®). A region of interest in the cytoplasm was photobleached using the UltraVIEW® photokinesis unit, and the recovery of fluorescence in this region was observed. (Fluorescence Recovery After Photobleaching (FRAP) using the UltraVIEW PhotoKinesis accessory, PerkinElmer Technical Note).

EXPERIMENT AND RESULTS

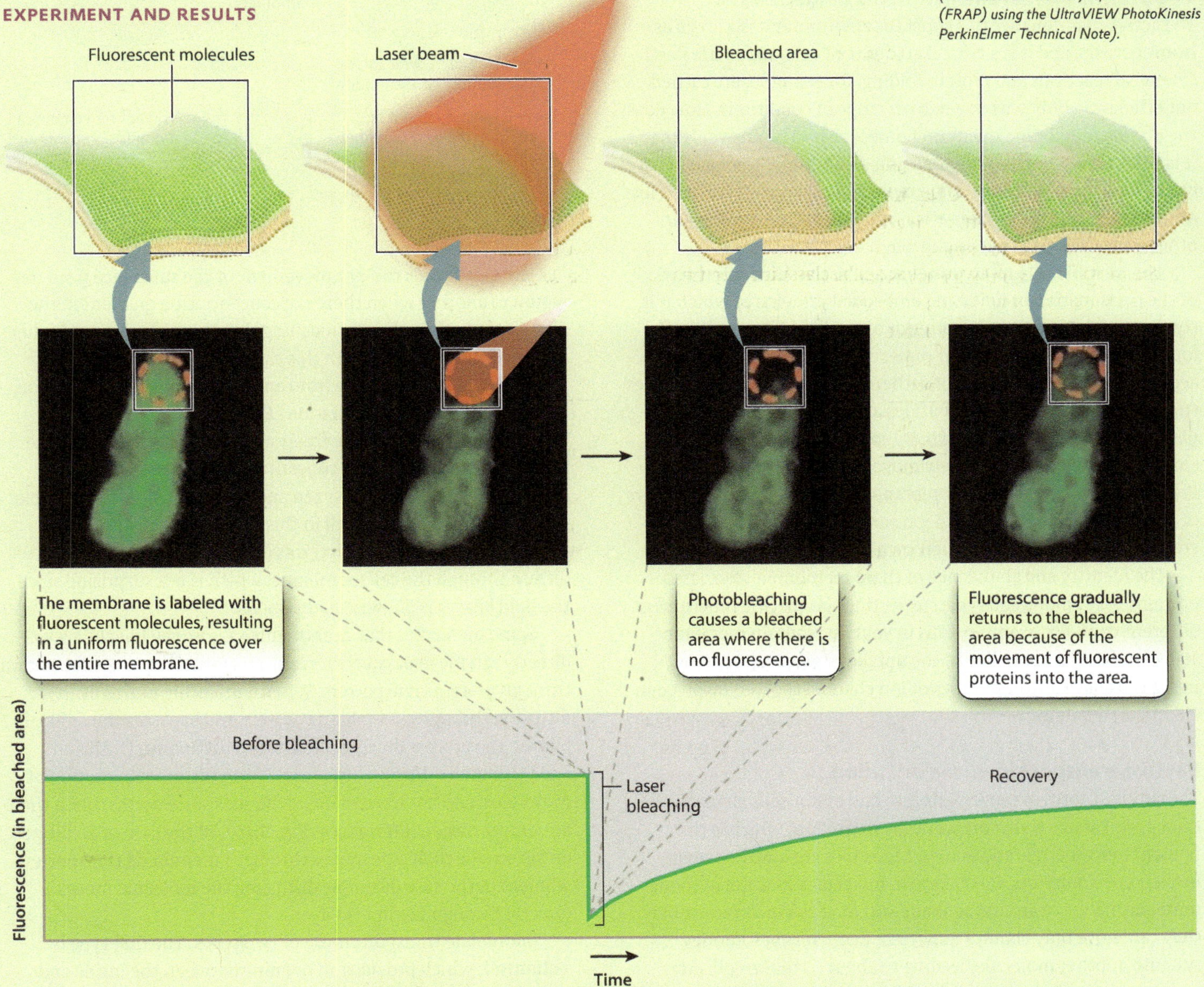

CONCLUSION The gradual recovery of fluorescence in the bleached area indicates that proteins move in the plane of the membrane.

SOURCE Peters, R., et al. 1974. "A Microfluorimetric Study of Translational Diffusion in Erythrocyte Membranes." *Biochim Biophys Acta* 367:282–294.

contents from the surrounding environment. But the plasma membrane is not simply a passive boundary or wall. Instead, it serves an active and important function. The environment outside the cell is changing all the time. In contrast, the internal environment of a cell operates within a narrow window of conditions, such as a particular pH range or salt concentration. It is the plasma membrane that actively maintains intracellular conditions compatible with life.

In addition, the cells of many organisms have a **cell wall** external to the plasma membrane. The cell wall plays an important role in maintaining the shape of these cells. In this section, we consider the functions performed by these two key structures.

The plasma membrane maintains homeostasis.

The active maintenance of a constant environment is known as **homeostasis,** and it is a critical attribute of cells and of life itself. Chemical reactions and protein folding, for example, are carried out efficiently only within a narrow range of conditions. How does the plasma membrane maintain homeostasis? The answer is that it is **selectively permeable** (or semipermeable). This means that the plasma membrane lets some molecules in and out freely; it lets others in and out only under certain conditions; and it prevents other molecules from passing through at all.

The membrane's ability to act as a selective barrier is the result of the combination of lipids and embedded proteins of which it is composed. The hydrophobic interior of the lipid bilayer prevents ions as well as charged or polar molecules from diffusing freely across the plasma membrane. Furthermore, many macromolecules such as proteins and polysaccharides are too large to cross the plasma membrane on their own. By contrast, gases, lipids, and small polar molecules can freely move across the lipid bilayer. Protein transporters in the membrane allow the export and import of molecules, including certain ions and nutrients, that cannot cross the cell membrane on their own.

The identity and abundance of these membrane-associated proteins vary among cell types, reflecting the specific functions of different cells. For example, cells in your gut contain membrane transporters that specialize in the uptake of glucose, whereas nerve cells have different types of ion channel that are involved in electrical signaling.

Passive transport involves diffusion.

The simplest form of movement into and out of cells is passive transport. Passive transport works by **diffusion,** which is the random movement of molecules. Molecules are always moving in their environments. For example, molecules in water at room temperature move around at about 500 m/sec, which means that they can move only about 3 molecular diameters before they run into another molecule, leading to about 5 trillion collisions per second. The frequency with which molecules collide have important consequences for chemical reactions, which depend on the interaction of molecules (Chapter 6).

FIG. 5.9 Diffusion, the movement of molecules due to random motion. Net movement of molecules results only when there are concentration differences.

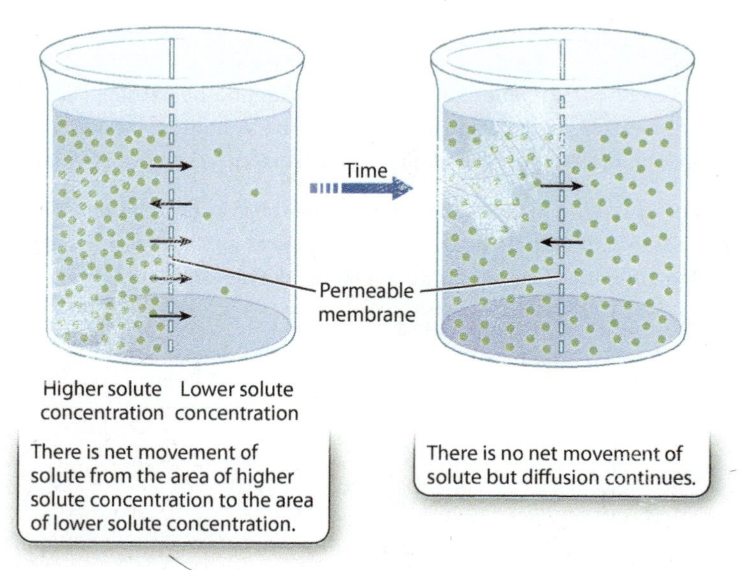

Diffusion leads to a net movement of the substance from one region to another when there is a concentration gradient in the distribution of a molecule, meaning that there are areas of higher and lower concentrations. In this case, diffusion results in net movement of the molecule from an area of higher concentration to an area of lower concentration (**Fig. 5.9**).

Some molecules diffuse freely across the plasma membrane as a result of differences in concentrations between the inside and the outside of a cell. Oxygen and carbon dioxide, for example, move into and out of the cell in this way. Certain hydrophobic molecules, such as triacylglycerols (Chapter 2), are also able to diffuse through the cell membrane, which is not surprising since the lipid bilayer is likewise hydrophobic.

Some molecules that cannot move across the lipid bilayer directly can move passively toward a region of lower concentration through protein transporters. When a molecule moves by diffusion through a membrane protein and bypasses the lipid bilayer, the process is called **facilitated diffusion.** Diffusion and facilitated diffusion both result from the random motion of molecules, and net movement of the substance occurs when there are concentration differences (**Fig. 5.10**). In the case of facilitated diffusion, the molecule moves through a membrane transporter, whereas in the case of simple diffusion, the molecule moves directly through the lipid bilayer.

Membrane transporters are of two types. The first type is a **channel,** which provides an opening between the inside and outside of the cell within which certain molecules can pass, depending on their shape and charge. Some membrane channels are gated, which means that they open in response to some sort

FIG. 5.10 Simple diffusion and facilitated diffusion through the cell membrane.

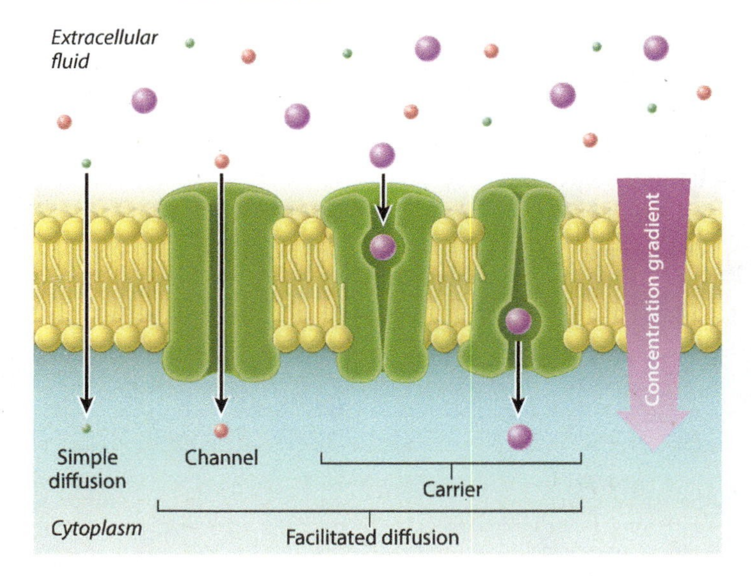

of signal, which may be chemical or electrical (Chapter 9). The second type of transporter is a **carrier,** which binds to and then transports specific molecules. Membrane carriers exist in two conformations, one that is open to one side of the cell, and another that is open to the other side of the cell. Binding of the transported molecule induces a conformational change in the membrane protein, allowing the molecule to be transported across the lipid bilayer, as shown on the right in Fig. 5.10.

Up to this point, we have focused our attention on the movement of molecules (the solutes) in water (the solvent). We can take a different perspective and focus instead on water movement. Water itself also moves into and out of cells by passive transport. Although the plasma membrane is hydrophobic, water molecules are small enough to move passively through the membrane to a limited extent by simple diffusion. In addition, many cells have specific protein channels, known as **aquaporins,** for transporting water molecules. These channels allow water to move more readily across the plasma membrane by facilitated diffusion than is possible by simple diffusion.

The net movement of a solvent such as water across a selectively permeable membrane such as the plasma membrane is known as **osmosis.** As in any form of diffusion, water moves from regions of higher *water* concentration to regions of lower *water* concentration (**Fig. 5.11**). Because water is a solvent within which nutrients such as glucose or ions such as sodium or potassium are dissolved, water concentration drops as solute concentration rises. Therefore, it is sometimes easier to think about water moving from regions of lower *solute* concentration toward regions of higher *solute* concentration. Either way, the direction of water movement is the same. During osmosis, the net movement of water toward the side of the membrane with higher solute concentration continues until it is opposed by another force. This force could be pressure due to gravity (in the case of Fig. 5.11) or the cell wall (in the case of plants, fungi, and bacteria, as described below).

→ **Quick Check 2** A container is divided into two compartments by a membrane that is fully permeable to water and small ions. Water is added to one side of the membrane (side A), and a 5% solution of sodium chloride (NaCl) is added to the other (side B). In which direction will water molecules move? In which direction will sodium and chloride ions move? When the concentration is equal on both sides, will diffusion stop?

Primary active transport uses the energy of ATP.

Passive transport works to the cell's advantage only if the concentration gradient is in the right direction, from higher on the outside to lower on the inside for nutrients that the cell

FIG. 5.11 Osmosis. Osmosis is the net movement of a solvent such as water across a selectively permeable membrane from an area of lower solute concentration to an area of higher solute concentration.

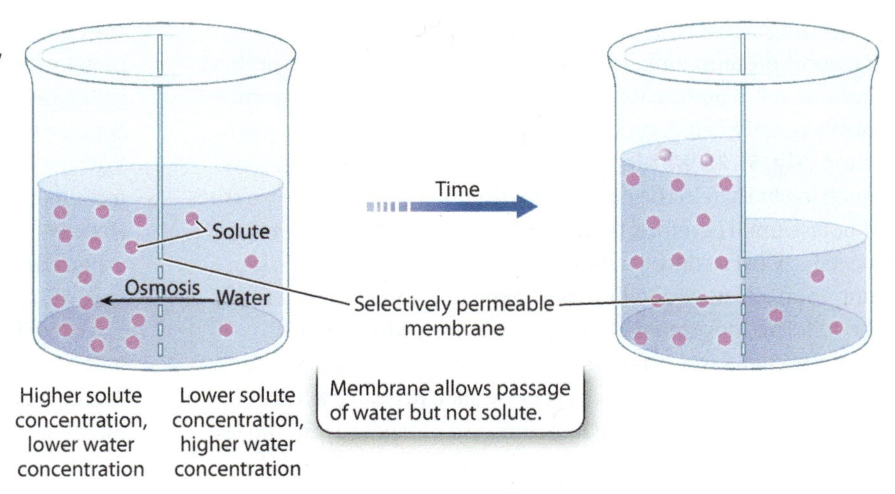

FIG. 5.12 Primary active transport. The sodium-potassium pump is a membrane protein that uses the energy stored in ATP to move sodium and potassium ions against their concentration gradients.

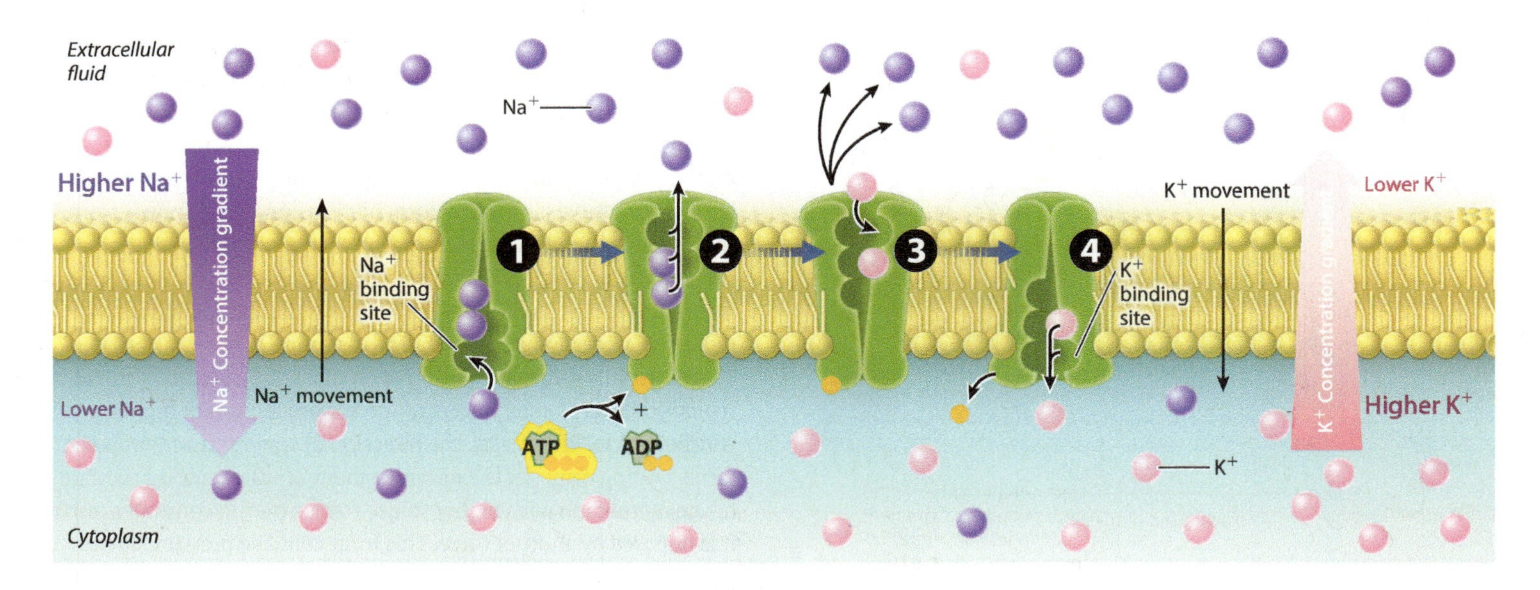

needs to take in, and from higher on the inside and lower on the outside for wastes that the cell needs to export. However, many of the molecules that cells require are not highly concentrated in the environment. Although some of these molecules can be synthesized by the cell, others must be taken up from the environment. In other words, cells have to move these substances from areas of lower concentration to areas of higher concentration. The "uphill" movement of substances against a concentration gradient, called **active transport,** requires energy. The transport of many kinds of molecules across membranes requires energy, either directly or indirectly. In fact, most of the energy used by a cell goes into keeping the inside of the cell different from the outside, a function carried out by proteins in the plasma membrane.

During active transport, cells move substances through transport proteins embedded in the cell membrane. Some of these proteins act as pumps, using energy directly to move a substance into or out of a cell. A good example is the sodium-potassium pump (**Fig. 5.12**). Within cells, sodium is kept at concentrations much lower than in the exterior environment; the opposite is true of potassium. Therefore, both sodium and potassium have to be moved against a concentration gradient. The sodium-potassium pump actively moves sodium out of the cell and potassium into the cell. This movement of ions takes energy, which comes from the chemical energy stored in ATP. Active transport that uses energy directly in this manner is called **primary active transport.** Note that the sodium ions and potassium ions move in opposite directions. Protein transporters that work in this way are referred to as antiporters. Other transporters move two molecules in the same direction, and are referred to as symporters or cotransporters.

Secondary active transport is driven by an electrochemical gradient.

Active transport can also work in another way. Because small ions cannot cross the lipid bilayer, many cells use a transport protein to build up the concentration of a small ion on one side of the membrane. The resulting concentration gradient stores potential energy that can be harnessed to drive the movement of other substances across the membrane against their concentration gradient.

For example, some cells actively pump protons (H^+) across the cell membrane using ATP (**Fig. 5.13a**). As a result, in these cells the concentration of protons is higher on one side of the membrane and lower on the other side. In other words, the pump generates a concentration gradient, also called a chemical gradient because the entity forming the gradient is a chemical (**Fig. 5.13b**). We have already seen that concentration differences favor the movement of protons back to the other side of the membrane. However, the lipid bilayer blocks the movement of protons to the other side and therefore stores potential energy, just like a dam or battery.

In addition to the chemical gradient, another force favors the movement of protons back across the membrane: a difference in charge. Because protons carry a positive charge, the side of the membrane with more protons is more positive than the other side. This difference in charge is called an electrical gradient. Protons (and other ions) move from areas of like charge to areas of unlike

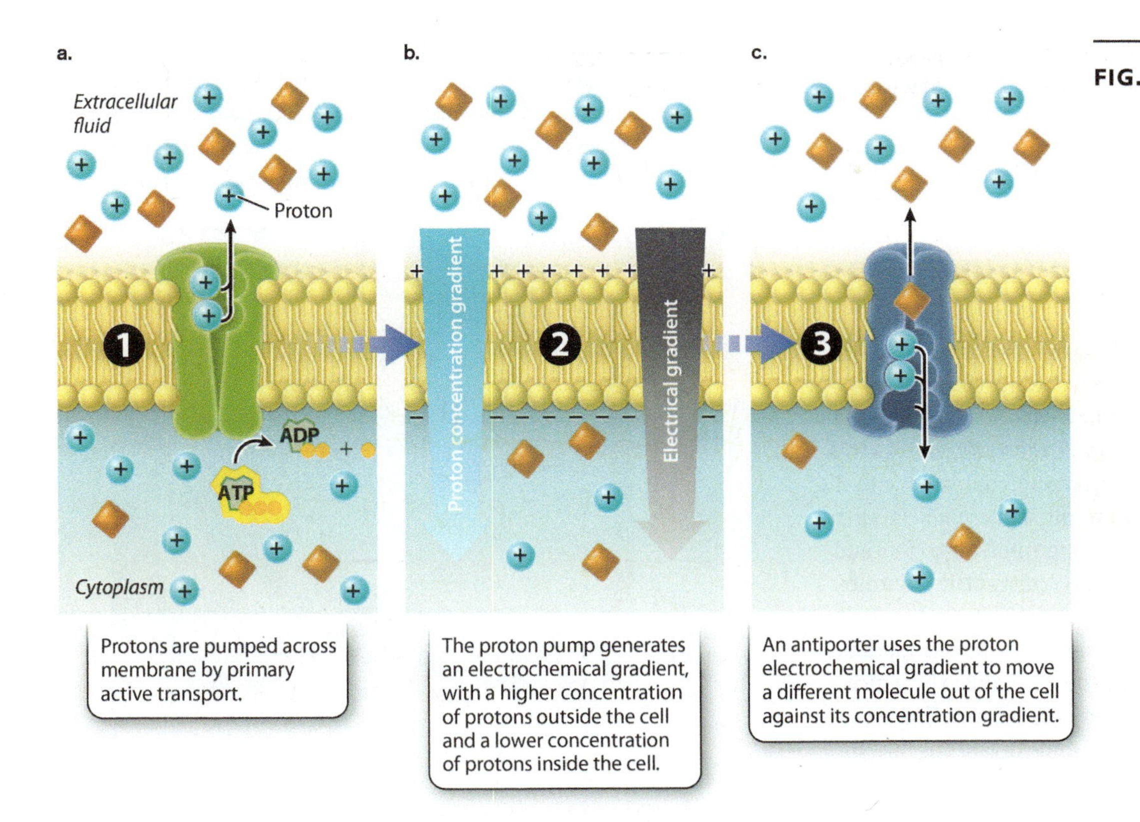

FIG. 5.13 Secondary active transport. Protons are pumped across a membrane by (a) primary active transport, resulting in (b) an electrochemical gradient, which drives (c) the movement of another molecule against its concentration gradient.

charge, driven by an electrical gradient. A gradient that has both charge and chemical components is known as an **electrochemical gradient** (Fig. 5.13b).

If protons are then allowed to pass through the cell membrane by a transport protein, they will move down their electrochemical gradient toward the region of lower proton concentration. These transport proteins can use the movement of protons to drive the movement of other molecules against their concentration gradient (**Fig. 5.13c**). The movement of protons is always from regions of higher to lower concentration, whereas the movement of the coupled molecule is from regions of lower to higher concentration. Because the movement of the coupled molecule is driven by the movement of protons and not by ATP directly, this form of transport is called **secondary active transport.** Secondary active transport uses the potential energy of an electrochemical gradient to drive the movement of molecules; by contrast, primary active transport uses the chemical energy of ATP directly.

The use of an electrochemical gradient as a temporary energy source is a common cellular strategy. For example, cells use a sodium electrochemical gradient generated by the sodium-potassium pump to transport glucose and amino acids into cells. In addition, cells use a proton electrochemical gradient to move other molecules and, as we discuss below and in Chapter 7, to synthesize ATP.

Many cells maintain size and composition using active transport.

Many cells use active transport to maintain their size. Consider human red blood cells placed in a variety of different solutions (**Fig. 5.14**). If a red blood cell is placed in a hypertonic solution (one with a higher solute concentration than that inside the cell), water leaves the cell by osmosis and the cell shrinks. By contrast, if a red blood cell is placed in a hypotonic solution (one with a lower solute concentration than that inside the cell), water moves into the cell by osmosis and the cell lyses, or bursts. Animal cells solve the problem of water movement in part by keeping the intracellular fluid isotonic (that is, at the same solute concentration) as the extracellular fluid. Cells use the active transport of ions to maintain equal concentrations

FIG. 5.14 Changes in red blood cell shape due to osmosis. Red blood cells shrink, swell, or burst because of net water movement driven by differences in solute concentration between the inside and the outside of the cell.

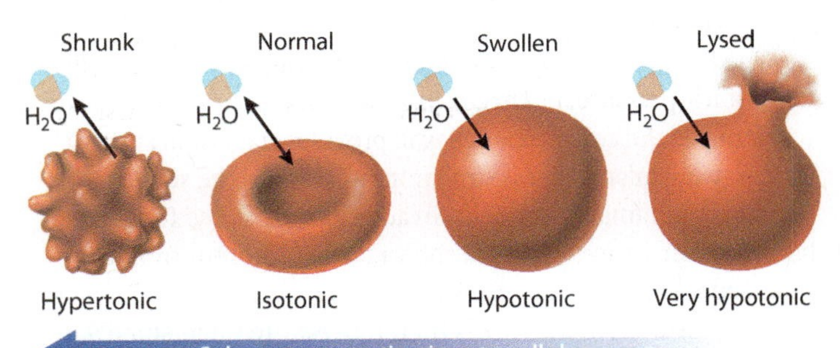

inside and out, and the sodium-potassium pump plays an important role in keeping the inside of the cell isotonic with the extracellular fluid.

→ **Quick Check 3** In the absence of the sodium-potassium pump, the extracellular solution becomes hypotonic relative to the inside of the cell. Poisons such as the snake venom ouabain can interfere with the action of the sodium-potassium pump. What are the consequences for the cell?

Human red blood cells avoid shrinking or bursting by maintaining an intracellular environment isotonic with the extracellular environment, the blood. But what about a single-celled organism, like *Paramecium*, swimming in a freshwater lake? In this case, the extracellular environment is hypotonic compared with the concentration in the cell's interior. As a result, *Paracemium* faces the risk of bursting from water moving in by osmosis. *Paramecium* and some other single-celled organisms contain **contractile vacuoles** that solve this problem. Contractile vacuoles are compartments that take up excess water from inside the cell and then, by contraction, expel it into the external environment. The mechanism by which water moves into the contractile vacuoles differs depending on the organism. The contractile vacuoles of some organisms take in water through aquaporins, while the contractile vacuoles of other organisms first take in protons through proton pumps, with water following by osmosis.

The cell wall provides another means of maintaining cell shape.

As we have seen, animals often maintain cell size by using active transport to keep the inside and outside of the cell isotonic. How do organisms such as plants, fungi, and bacteria maintain cell size and shape? A key feature of these organisms is the cell wall surrounding the plasma membrane (**Fig. 5.15**). The cell wall plays a critical role in the maintenance of cell size and shape. When Hooke looked at cork through his microscope, what he saw was not living cells but the remains of cell walls.

The cell wall provides structural support and protection to the cell. Because the cell wall resists expansion, it allows pressure to build up in a cell when it absorbs water. Recall that when an animal cell, like a red blood cell, is placed in pure water, it swells until it bursts. By contrast, when a plant cell is placed in pure water, water enters the cell by osmosis until the pressure created by the cell wall's resisting further expansion opposes the driving force for the water to enter.

The force exerted by water pressing against an object is called hydrostatic pressure, or **turgor pressure.** The pressure exerted by water inside the cell on the cell wall provides structural support for many organisms that is similar in function to the support provided by animals' skeletons. In addition, plant and fungal cells have another structure, called the **vacuole** (different from the contractile vacuole), that can absorb water and also contribute to turgor pressure (Fig. 5.15). It is therefore easy to understand why plants wilt when dehydrated—the loss of water from the vacuoles reduces turgor pressure and the cells can no longer maintain their shape within the cell wall. Plant vacuoles have many other functions and are often the most conspicuous feature of plant cells. They also explain in part why plant cells are typically larger than animal cells and can store water, as well as nutrients, ions, and wastes.

FIG. 5.15 The plant cell wall and vacuoles. The plant cell wall is a rigid structure that maintains the shape of the cell. The pressure exerted by water absorbed by vacuoles also provides support. *Source: Biophoto Associates/Getty Images.*

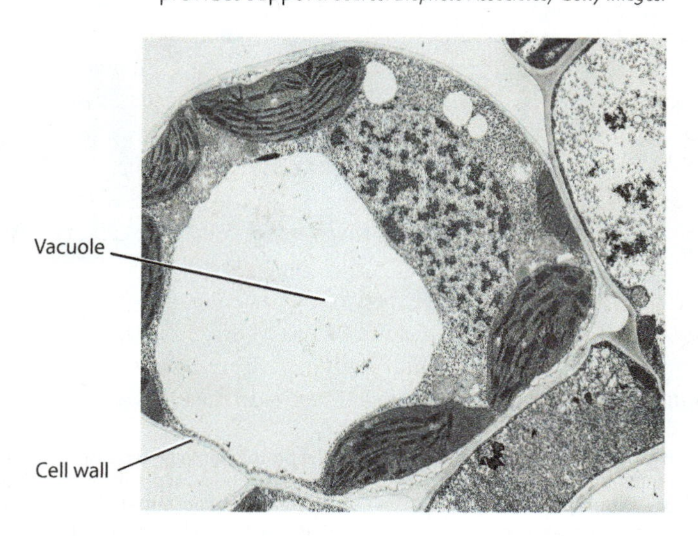

The cell wall is made up of many different components, including carbohydrates and proteins. The specific components differ depending on the organism. The plant cell wall is composed of polysaccharides, one of the best known of which is cellulose, a polymer of the sugar glucose. Cellulose is the most abundant biological material in nature. Many types of algae have cell walls made up of cellulose, as in plants, but others have cell walls made of silicon or calcium carbonate. Fungi have cell walls made of chitin, another polymer based on sugars. In bacteria, the cell wall is made up primarily of peptidoglycan, a mixture of amino acids and sugars.

5.3 THE INTERNAL ORGANIZATION OF CELLS

In addition to the plasma membrane, many cells contain membrane-bound regions within which specific functions are carried out. Such a cell can be compared to a large factory with many rooms and different departments. Each department has a specific function and internal organization that contribute to the overall "life" of the factory. In this section, we give an overview of two broad classes of cells that can be distinguished by the presence or absence of these membrane-enclosed compartments.

Eukaryotes and prokaryotes differ in internal organization.

All cells have a plasma membrane and contain genetic material. In some cells, the genetic material is housed in a membrane-bound space called the **nucleus.** Cells can be divided into two classes based on the absence or presence of a nucleus (Chapter 1). **Prokaryotes,** including bacteria and archaeons, lack a nucleus; **eukaryotes,** including animals, plants, fungi, and protists, have a nucleus (**Fig. 5.16**). The presence of a nucleus in eukaryotes allows for the processes of transcription and translation to be separated in time and space. This separation in turn allows for more complex ways to regulate gene expression than are possible in prokaryotes (Chapter 19).

There are other differences between prokaryotes and eukaryotes. For example, we saw in Chapter 3 that promoter recognition during transcription is different in prokaryotes and eukaryotes. In addition, there are differences in the types of lipid that make up their cell membranes. In mammals, as we have seen, cholesterol is present in cell membranes. Cholesterol belongs to a group of chemical compounds known as sterols, which are molecules containing a hydroxyl group attached to a four-ringed structure. In eukaryotes other than mammals, diverse sterols are synthesized and present in cell membranes. Most prokaryotes do not synthesize sterols, but some synthesize compounds called hopanoids. These five-ringed structures are thought to serve a function similar to that of cholesterol in mammalian cell membranes.

In spite of all of these and other differences between prokaryotes and eukaryotes, it is the absence or presence of a nucleus that defines the two groups. However, from an evolutionary perspective, archaeons and eukaryotes are more closely related to each other than either are to bacteria. For example, archaeons share with eukaryotes many genes involved in transcription and translation, and DNA is packaged with histones in both groups.

Prokaryotic cells lack a nucleus and extensive internal compartmentalization.

Prokaryotes do not have a nucleus—that is, there is no physical barrier separating the genetic material from the rest of the cell. Instead, the DNA is concentrated in a discrete region of the cell interior known as the **nucleoid.** Bacteria often contain additional small circular molecules of DNA known as **plasmids** that carry a few genes. Plasmids are commonly transferred between bacteria through the action of threadlike structures known as **pili** (singular, **pilus**), which extend from one cell to another. Genes for antibiotic resistance are commonly transferred in this way, which accounts for the quick spread of antibiotic resistance among bacterial populations.

Although the absence of a nucleus is a defining feature of prokaryotes, other features also stand out. For example, prokaryotes are small, typically just 1–2 microns (a micron is 1/1,000,000 of a meter) in diameter or smaller. By contrast, eukaryotic cells are commonly much larger, on the order of 10 times larger in diameter and 1000 times larger in volume. The small size of prokaryotic cells means that they have a relatively high ratio of surface area to volume, which makes sense for an organism that absorbs nutrients from the environment. In other words, there is a large amount of membrane surface area available for absorption relative to the volume of the cell that it serves. In addition, most prokaryotes lack the extensive internal organization characteristic of eukaryotes.

FIG. 5.16 Prokaryotic and eukaryotic cells. Prokaryotic cells lack a nucleus and extensive internal compartmentalization. Eukaryotic cells have a nucleus and extensive internal compartmentalization.

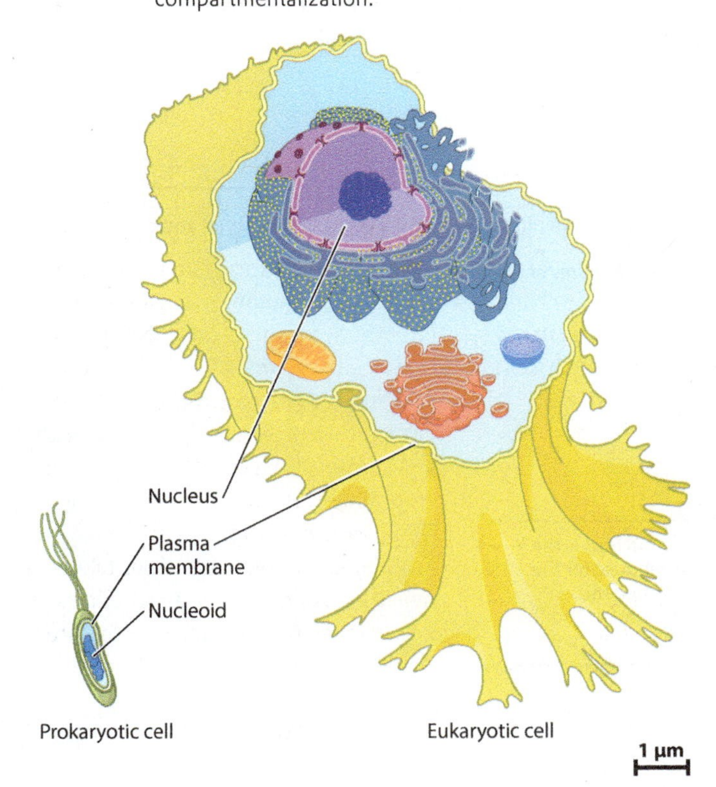

Eukaryotic cells have a nucleus and specialized internal structures.

Eukaryotes are defined by the presence of a nucleus, which houses the vast majority of the cell's DNA. The nuclear membrane allows for more complex regulation of gene expression than is possible in prokaryotic cells (Chapters 3 and 19). In eukaryotes, DNA is transcribed to RNA inside the nucleus, but the RNA molecules carrying the genetic message travel from inside to outside the nucleus, where they instruct the synthesis of proteins.

Eukaryotes have a remarkable internal array of membranes. These membranes define compartments, called **organelles,** that divide the cell contents into smaller spaces specialized for different functions. **Fig. 5.17a** shows a macrophage, a type of animal cell, with various organelles. The **endoplasmic reticulum (ER)** is the organelle in which proteins and lipids are synthesized. The **Golgi apparatus** modifies proteins and lipids produced by the ER and acts as a sorting station as they move to their final destinations. **Lysosomes** contain enzymes that break down macromolecules such as proteins, nucleic acids, lipids, and complex carbohydrates. **Peroxisomes** also contain many different enzymes and are involved in important metabolic reactions, including the breakdown of fatty acids and the synthesis of certain types of phospholipid. **Mitochondria** are specialized organelles that harness energy for the cell. Many cell membranes that define these organelles are

FIG. 5.17 An animal cell and a plant cell. Animal and plant cells have many cell components in common.

a. Typical features of an animal cell

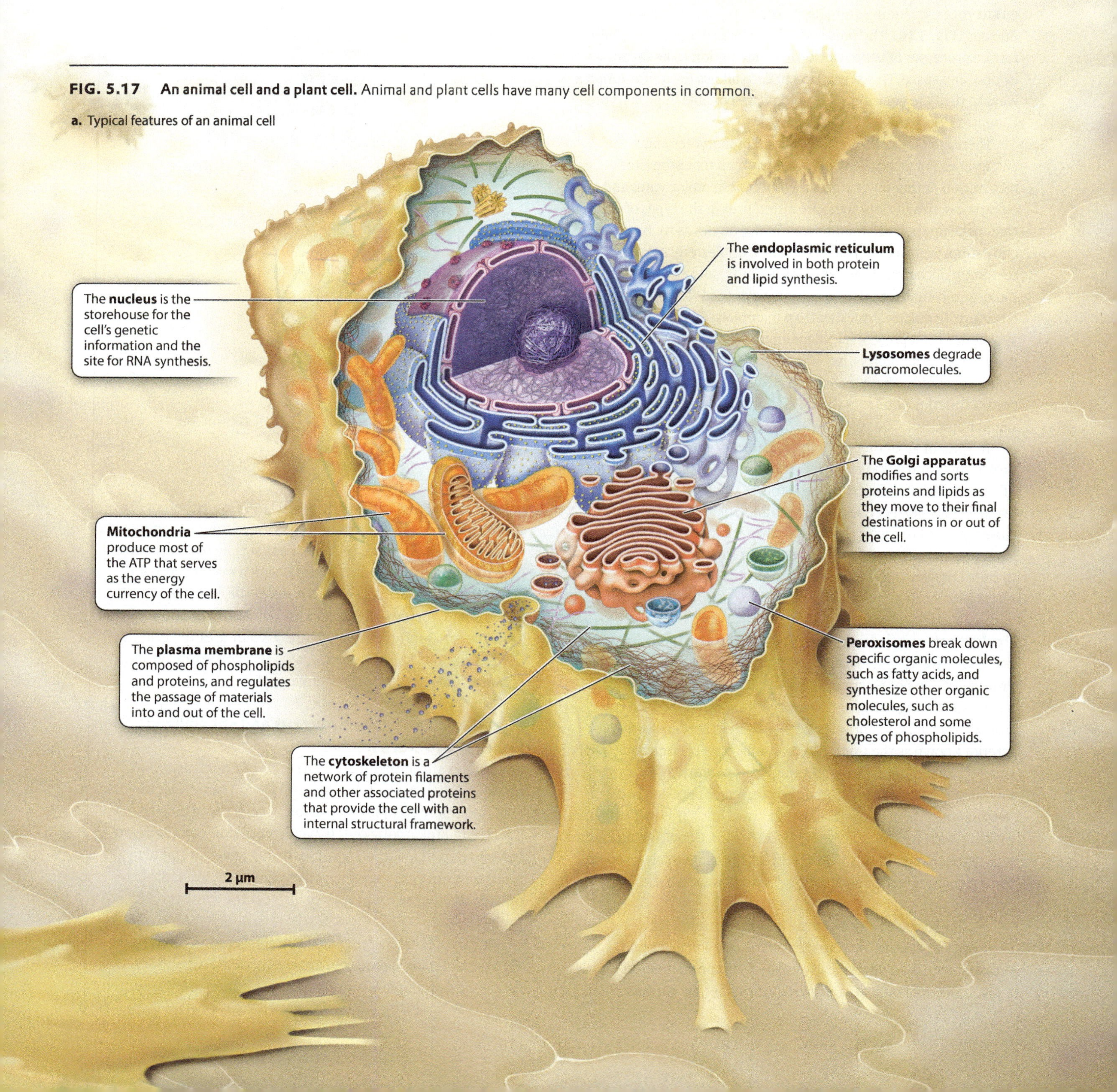

associated with a protein scaffold called the **cytoskeleton** that helps cells to maintain their shape and serves as a network of tracks for the movement of substances within cells (Chapter 10).

Fig. 5.17b shows a typical plant cell. In addition to the organelles described above, plant cells have a cell wall outside the plasma membrane, vacuoles specialized for water uptake, and **chloroplasts** that convert energy of sunlight into chemical energy.

The entire contents of a cell other than the nucleus make up the **cytoplasm.** The jelly-like internal environment of the cell that surrounds the organelles inside the plasma membrane is referred to as the **cytosol.**

b. Typical features of a plant cell

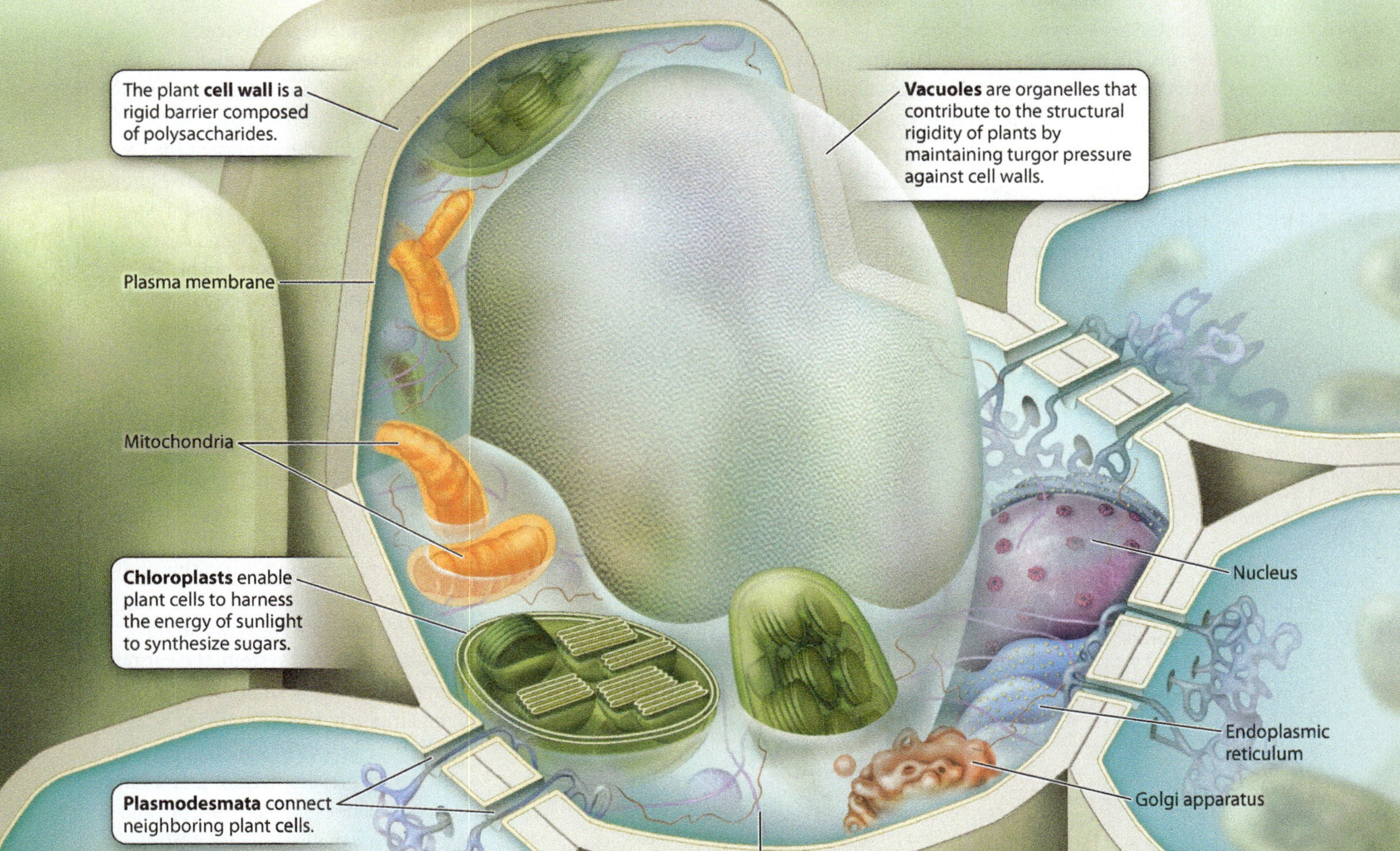

In the next two sections, we consider these organelles in more detail, focusing on the role of membranes in forming distinct compartments within the cell.

5.4 THE ENDOMEMBRANE SYSTEM

In eukaryotes, the total surface area of intracellular membranes is about tenfold greater than that of the plasma membrane. This high ratio of internal membrane area to plasma membrane area underscores the significant degree to which a eukaryotic cell is divided into internal compartments. Many of the organelles inside cells are not isolated entities, but instead communicate with one another. In fact, the membranes of these organelles are either physically connected by membrane "bridges" or they are transiently connected by **vesicles,** small membrane-enclosed sacs that transport substances within a cell or from the interior to the exterior of the cell. These vesicles form by budding off an organelle, taking with them a piece of the membrane and internal contents of the organelle from which they derive. They then fuse with another organelle or the plasma membrane, re-forming a continuous membrane and unloading their contents.

In total, these interconnected membranes make up the **endomembrane system.** The endomembrane system includes the nuclear envelope, endoplasmic reticulum, Golgi apparatus, lysosomes, the plasma membrane, and the vesicles that move between them (**Fig. 5.18**). In plants, the endomembrane system is actually continuous between cells through connecting pores called plasmodesmata (see Fig. 5.17; Chapters 10 and 28).

Extensive internal membranes are not common in prokaryotic cells. However, photosynthetic bacteria have internal membranes that are specialized for harnessing light energy (Chapters 8 and 26).

The endomembrane system compartmentalizes the cell.

Because many types of molecules are unable to cross cell membranes on their own, the endomembrane system divides the interior of a cell into two distinct "worlds," one inside the spaces defined by these membranes and one outside these spaces. A molecule within the interior of the ER can stay in the ER or end up in the interior of the Golgi apparatus or even outside the cell by the budding off and fusing of a vesicle between these organelles. Similarly, a molecule associated with the ER membrane can move to the Golgi membrane or the plasma membrane by vesicle transport. Molecules in the cytosol are in a different physical space, separated by membranes of the endomembrane system. This physical separation allows specific functions to take place within the spaces defined by the membranes and also within the membrane itself.

In spite of forming a continuous and interconnected system, the various compartments have unique properties and maintain distinct identities determined in part by which lipids and proteins are present in their membranes.

Vesicles not only bud off from and fuse with organelles but also with the plasma membrane. When a vesicle fuses with the plasma membrane, the process is called **exocytosis.** It provides a way for a vesicle to empty its contents to the extracellular space or to deliver proteins embedded in the vesicle membrane to the plasma membrane (Fig. 5.18). The process also works in reverse: A vesicle

FIG. 5.18 The endomembrane system. The endomembrane system is a series of interconnected membrane-bound compartments in eukaryotic cells.

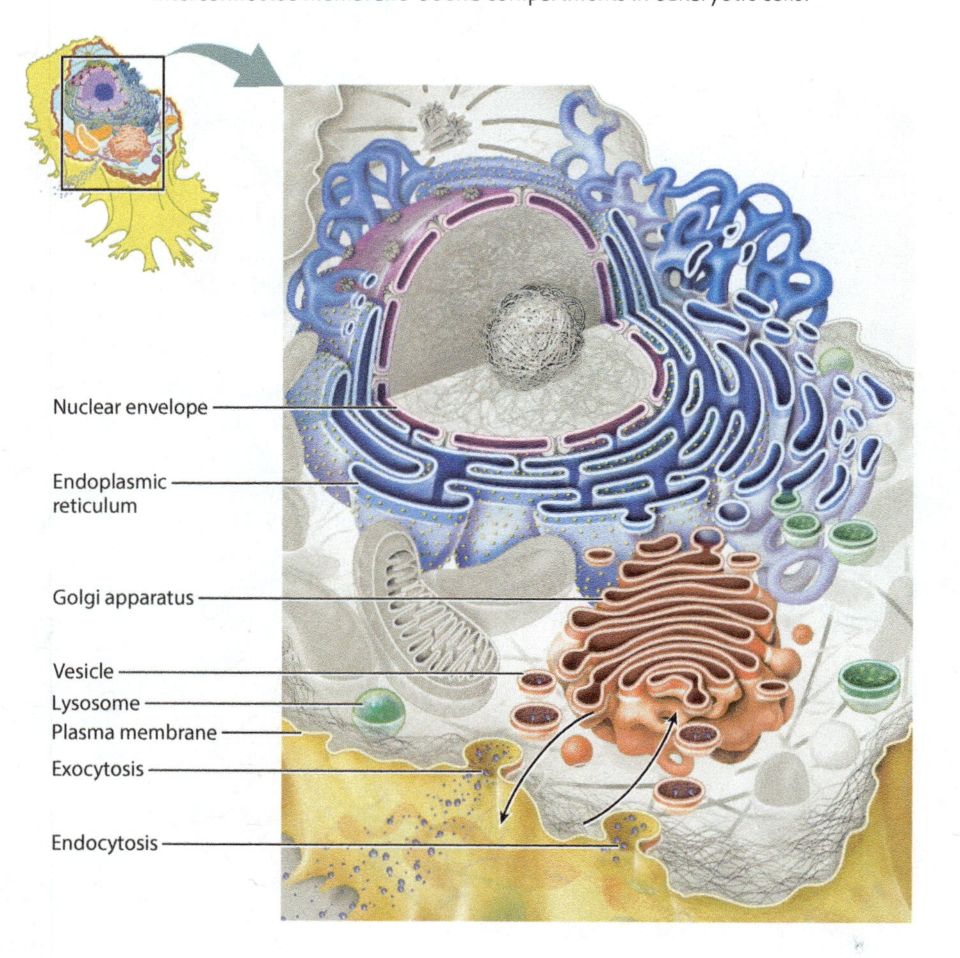

can bud off from the plasma membrane, enclosing material from outside the cell and bringing it into the cell interior. This process is called **endocytosis.** Together, exocytosis and endocytosis provide a way to move material into and out of cells without passing through the cell membrane.

The nucleus houses the genome and is the site of RNA synthesis.

The innermost organelle of the endomembrane system is the **nucleus,** which stores DNA, the genetic material that encodes the information for all the activities and structures of the cell. The **nuclear envelope** defines the boundary of the nucleus (**Fig. 5.19**). It actually consists of two membranes, the inner and outer membranes, and each is a lipid bilayer with associated proteins.

These two membranes are continuous with each other at openings called **nuclear pores.** These pores are large protein complexes that allow molecules to move into and out of the nucleus, and thus are essential for the nucleus to communicate with the rest of the cell. For example, some proteins that are synthesized in the cytosol, such as transcription factors, move through nuclear pores to enter the nucleus, where they control how and when genetic information is expressed.

In addition, the transfer of information encoded by DNA depends on the movement of mRNA (messenger RNA) molecules out of the nucleus through these pores. After exiting the nucleus, mRNA binds to free ribosomes in the cytosol or ribosomes associated with the endoplasmic reticulum (ER). **Ribosomes** are the sites of protein synthesis, in which amino acids are assembled into polypeptides guided by the information stored in mRNA (Chapter 4). In this way, the nuclear envelope and its associated protein pores regulate which molecules move into and out of the nucleus.

FIG. 5.19 A surface view of the nuclear envelope. The nucleus is surrounded by a double membrane and houses the cell's DNA. *Source: Don W. Fawcett/Science Source.*

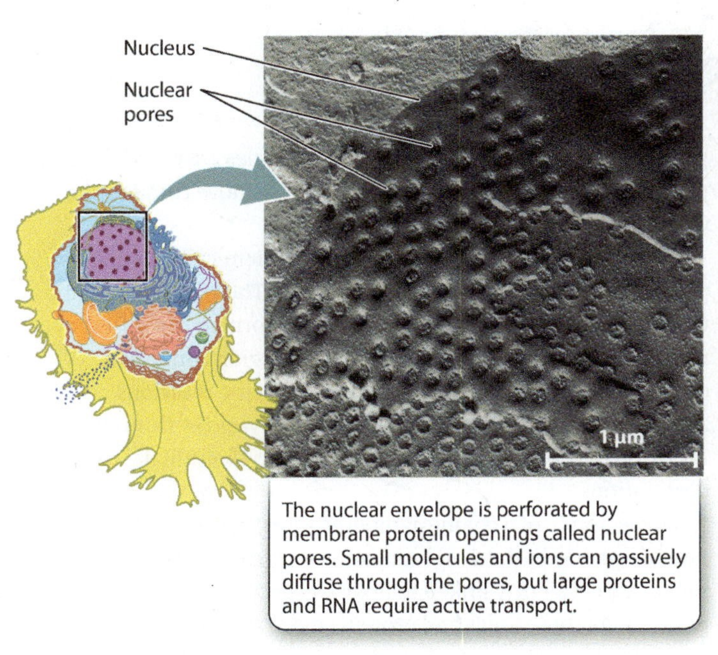

The endoplasmic reticulum is involved in protein and lipid synthesis.

The outer membrane of the nuclear envelope is physically continuous with the endoplasmic reticulum (ER), an organelle bounded by a single membrane (**Fig. 5.20**). The ER is a conspicuous feature of many eukaryotic cells, accounting in some cases for as much as half of the total amount of membrane. The ER produces and transports many of the lipids and proteins used inside and outside the cell, including all transmembrane proteins, as well as proteins destined for the Golgi apparatus, lysosomes, or export out of the cell. The ER is also the site of production of most of the lipids that make up the various internal and external cell membranes.

Unlike the nucleus, which is a single spherical structure in the cell, the ER consists of a complex network of interconnected tubules and flattened sacs. The interior of the ER is continuous throughout and is called the **lumen.** As shown in Fig. 5.20, the ER has an almost mazelike appearance when sliced and viewed in cross section. Its membrane is extensively convoluted, allowing a large amount of membrane surface area to fit within the cell.

When viewed through an electron microscope, ER membranes have two different appearances (Fig. 5.20). Some look rough because they are studded with ribosomes. This portion of the ER is referred to as **rough endoplasmic reticulum (RER).** The rough ER synthesizes transmembrane proteins, proteins that end up in the interior of organelles, and proteins destined for secretion. As a result, cells that secrete large quantities of protein have extensive rough ER, including cells of the gut that secrete digestive enzymes and cells of the pancreas that produce insulin. All cells have at least some rough ER for the production of transmembrane and organelle proteins.

There is a small amount of ER membrane in most cells that appears smooth because it lacks ribosomes. This portion of the ER is therefore called **smooth endoplasmic reticulum (SER)** (Fig. 5.20). Smooth ER is the site of fatty acid and phospholipid biosynthesis. Thus, this type of ER predominates in cells specialized for the production of lipids. For example, cells that synthesize steroid hormones have a well-developed SER that produces large quantities of cholesterol. Enzymes within the SER convert cholesterol into steroid hormones.

The Golgi apparatus modifies and sorts proteins and lipids.

Although it is not physically continuous with the ER, the Golgi apparatus is often the next stop for vesicles that bud off the ER.

FIG. 5.20 **The endoplasmic reticulum (ER).** The ER is a major site for lipid and protein synthesis. (Proteins are also synthesized in the cytoplasm.) *Sources: (top left) Biophoto Associates/Science Source; (bottom left) David M. Phillips/Science Source.*

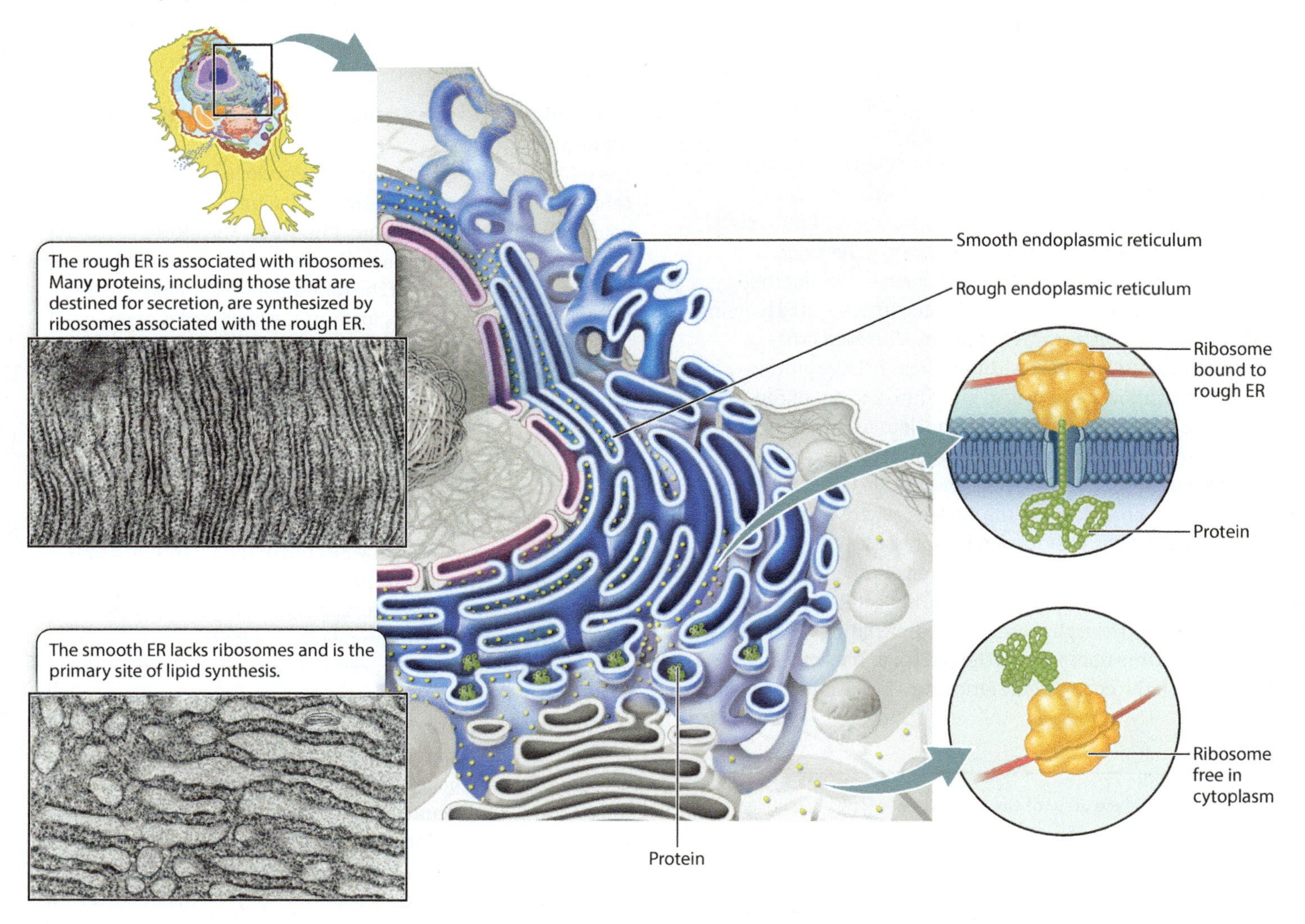

These vesicles carry lipids and proteins, either within the vesicle interior or embedded in their membranes. The movement of these vesicles from the ER to the Golgi apparatus and then to the rest of the cell is part of a biosynthetic pathway in which lipids and proteins are sequentially modified and delivered to their final destinations. The Golgi apparatus has three primary roles: (1) It further modifies proteins and lipids produced by the ER; (2) it acts as a sorting station as these proteins and lipids move to their final destinations; and (3) it is the site of synthesis of most of the cell's carbohydrates.

Under the microscope, the Golgi apparatus looks like stacks of flattened membrane sacs, called **cisternae,** surrounded by many small vesicles (**Fig. 5.21**). These vesicles transport proteins and lipids from the ER to the Golgi apparatus, and then between the various cisternae, and finally from the Golgi apparatus to the plasma membrane or other organelles. Vesicles are therefore the primary means by which proteins and lipids move through the Golgi apparatus to their final destinations.

Enzymes within the Golgi apparatus chemically modify proteins and lipids as they pass through it. These modifications take place in a sequence of steps, each performed in a different region of the Golgi apparatus, since each region contains a different set of enzymes that catalyzes specific reactions. As a result, there is a general movement of vesicles from the ER through the Golgi apparatus and then to their final destinations.

An example of a chemical modification that occurs predominantly in the Golgi apparatus is glycosylation, in which sugars are covalently linked to lipids or specific amino acids of proteins. As these lipids and proteins move through the Golgi

FIG. 5.21 The Golgi apparatus. The Golgi apparatus sorts proteins and lipids to other organelles, the plasma membrane, or the cell exterior.
Source: Biophoto Associates/Science Source.

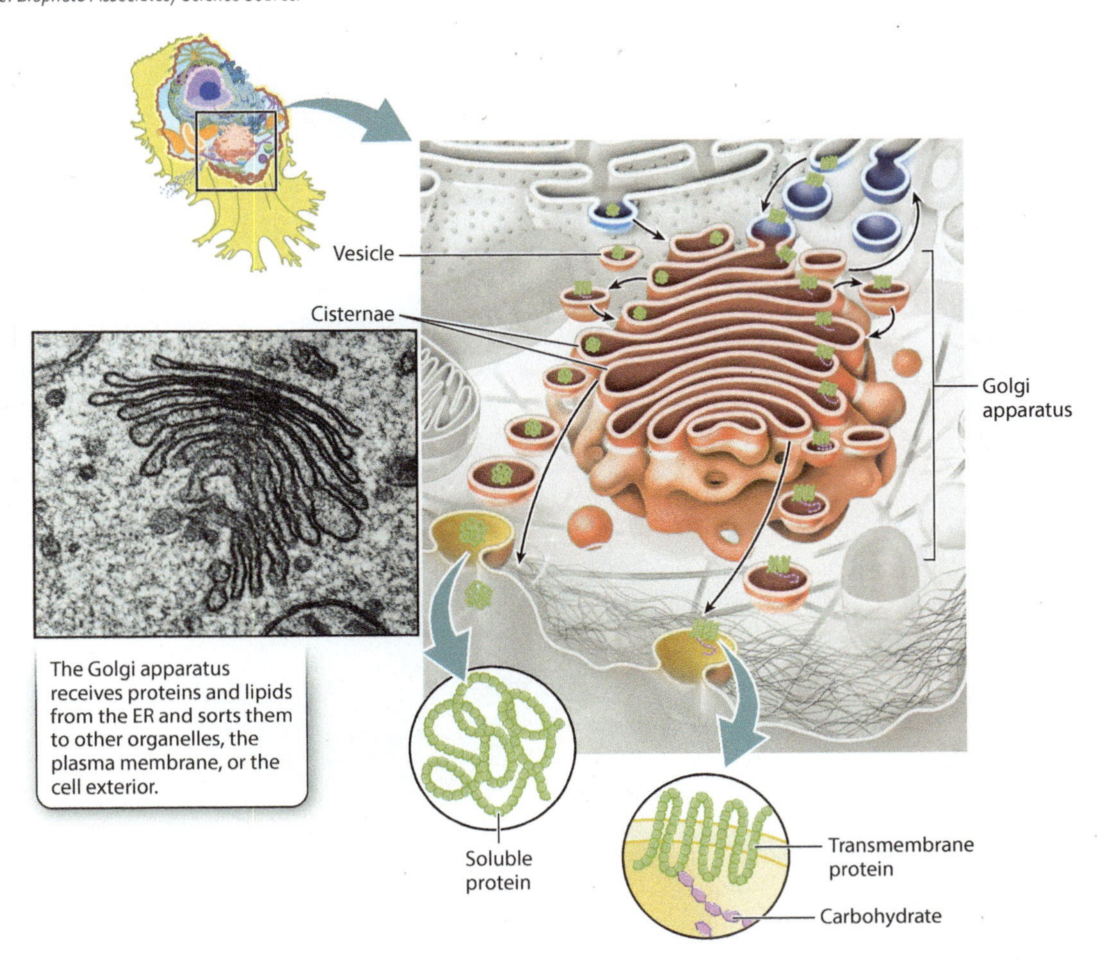

apparatus, they encounter different enzymes in each region that add or trim sugars. Glycoproteins are important components of the eukaryotic cell surface. The sugars attached to the protein can protect the protein from enzyme digestion by blocking access to the peptide chain. As a result, glycoproteins form a relatively flexible and protective coating over the plasma membrane. The distinctive shapes that sugars contribute to glycoproteins and glycolipids also allow them to be recognized specifically by other cells and molecules in the external environment. For example, human blood types (A, B, AB, and O) are defined by the particular sugars that are linked to proteins and lipids on the surface of red blood cells.

While traffic usually travels from the ER to the Golgi apparatus, a small amount of traffic moves in the reverse direction, from the Golgi apparatus to the ER. This reverse pathway is important to retrieve proteins in the ER or Golgi that were accidentally moved forward and to recycle membrane components.

Lysosomes degrade macromolecules.

The ability of the Golgi apparatus to sort and dispatch proteins to particular destinations is dramatically illustrated by lysosomes. **Lysosomes** are specialized vesicles derived from the Golgi apparatus that degrade damaged or unneeded macromolecules (**Fig. 5.22**). They contain a variety of enzymes that break down macromolecules such as proteins, nucleic acids, lipids, and complex carbohydrates. Macromolecules destined for degradation are packaged by the Golgi apparatus into vesicles. The vesicles then fuse with lysosomes, delivering their contents to the lysosome interior.

The formation of lysosomes also illustrates the ability of the Golgi apparatus to sort key proteins. The enzymes inside the lysosomes are synthesized in the RER, sorted in the Golgi apparatus, and then packaged into lysosomes. In addition, the Golgi apparatus sorts and delivers specialized proteins that become embedded in lysosomal membranes. These include proton pumps that keep the internal environment at an acidic pH of about 5,

FIG. 5.22 Lysosomes. Lysosomes are vesicles that degrade macromolecules. *Source: Don W. Fawcett/Getty Images.*

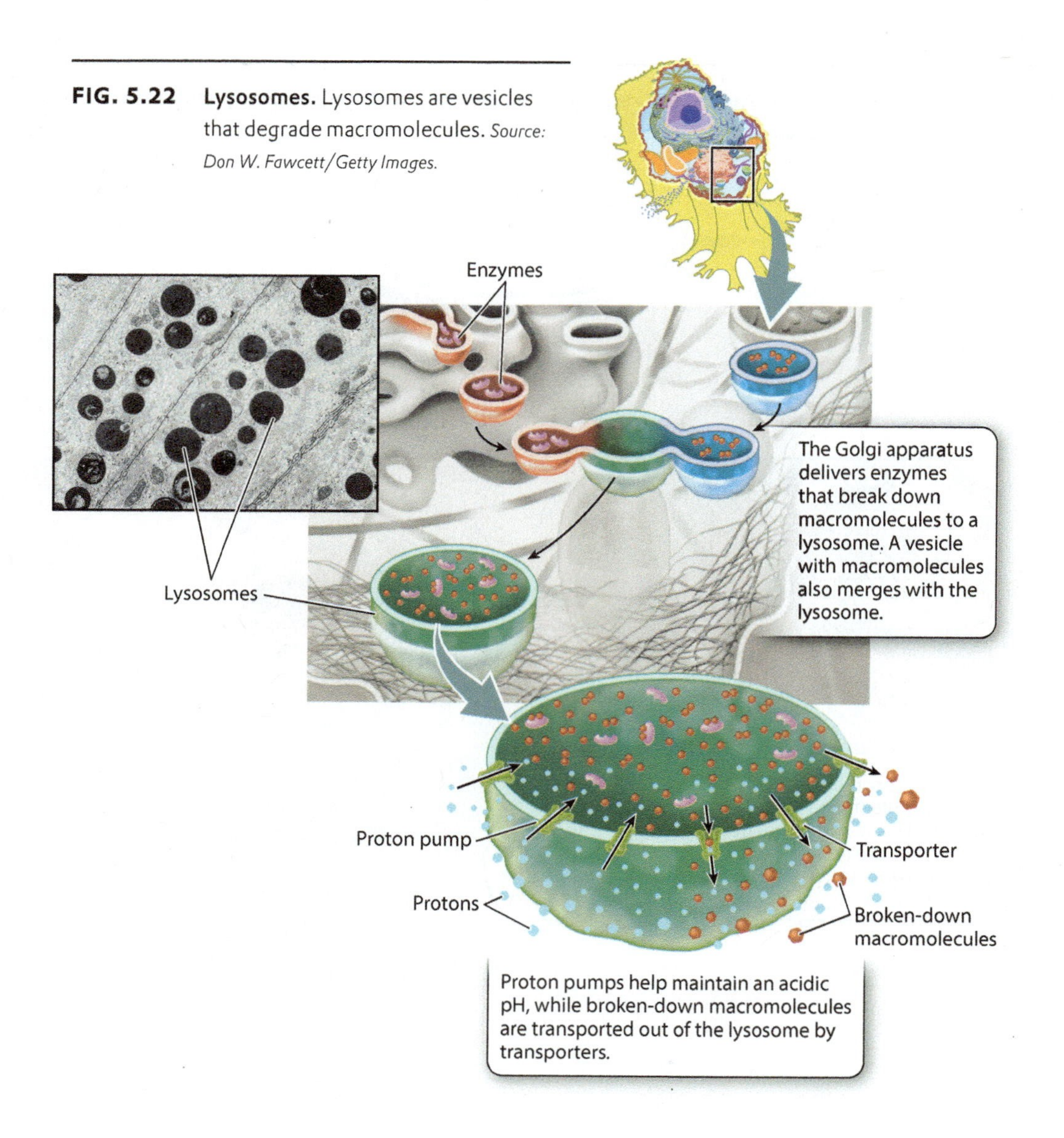

the optimum pH for the activity of the enzymes inside. Other proteins in the lysosomal membranes transport the breakdown products of macromolecules, such as amino acids and simple sugars, across the membrane to the cytosol for use by the cell.

The function of lysosomes underscores the importance of having separate compartments within the cell bounded by selectively permeable membranes. Lysosomal enzymes cannot function in the normal cellular environment, which has a pH of about 7, and many of a cell's enzymes and proteins would unfold and degrade if the entire cell were at the pH of the inside of a lysosome. By restricting the activity of these enzymes to the lysosome, the cell protects proteins and organelles in the cytosol from degradation.

Protein sorting directs proteins to their proper location in or out of the cell.

As we have seen, eukaryotic cells have many compartments, and different proteins function in different places, such as enzymes in lysosomes or transmembrane proteins embedded in the plasma membrane. **Protein sorting** is the process by which proteins end up where they need to be to perform their function. Protein sorting directs proteins to the cytosol, the lumen of organelles, the membranes of the endomembrane system, or even out of the cell entirely.

Recall that proteins are produced in two places: free ribosomes in the cytosol and membrane-bound ribosomes on the rough ER. Proteins produced on free ribosomes are sorted after they are translated. These proteins often contain amino acid sequences, called **signal sequences,** that allow them to be recognized and sorted. As shown in **Fig. 5.23,** there are several types of signal sequences that direct proteins synthesized on free ribosomes to different cellular compartments. Proteins with no signal sequence remain in the cytosol (Fig. 5.23a). Proteins destined for mitochondria or chloroplasts often have a signal sequence at their amino ends (Fig. 5.23b). Proteins targeted to the nucleus usually have signal sequences located internally (Fig. 5.23c). These nuclear signal sequences, called **nuclear localization signals,** enable proteins to move through pores in the nuclear envelope.

FIG. 5.23 Signal sequences on proteins synthesized by free ribosomes. (a) Most proteins with no signal peptide remain in the cytosol. Other signal sequences direct proteins to (b) mitochondria and chloroplasts or to (c) the nucleus.

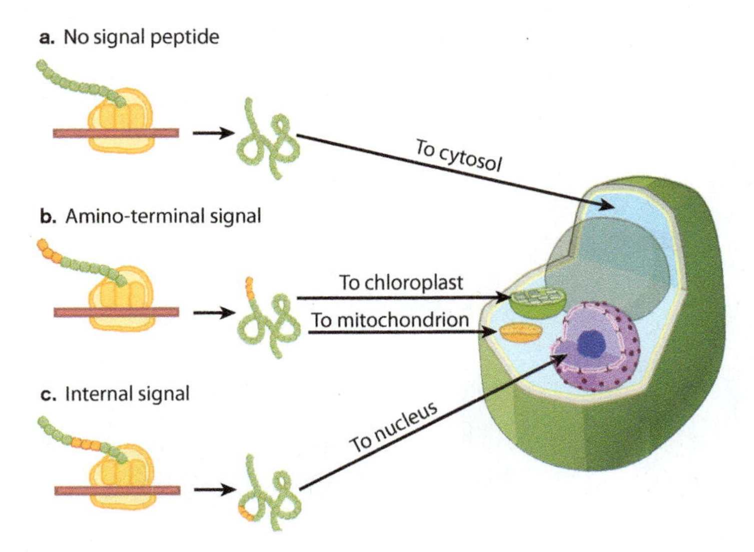

Proteins produced on the rough ER, as we have seen, end up in the lumen of the endomembrane system, secreted out of the cell, or as transmembrane proteins (**Fig. 5.24**). These proteins are sorted as they are translated. They begin translation on free ribosomes, but a specific signal sequence at their amino terminal end directs the ribosome to the rough ER and into a membrane channel that leads into the ER lumen. If the polypeptide contains no other signal sequence, it continues into the lumen. If it contains a second sequence, called a **signal-anchor sequence,** it does not continue all the way into the lumen and ends up in the membrane. Let's consider in more detail how sorting of these proteins happens.

Proteins destined for the ER lumen or secretion have an amino-terminal signal sequence, shown in **Fig. 5.25.** As a free ribosome translates the protein, this sequence is recognized by an RNA–protein complex known as a **signal-recognition particle (SRP).** The SRP binds to both the signal sequence and the free ribosome, and brings about a pause in translation (Fig. 5.25a). The SRP then binds with a receptor on the RER so that the ribosome is now associated with the RER (Fig. 5.25b). The SRP receptor brings the ribosome to a channel in the membrane of the RER. The SRP then dissociates and translation continues, allowing the growing polypeptide chain to be threaded through the channel (Fig. 5.25c). A specific protease cleaves the signal sequence as it emerges in the lumen of the ER (Fig. 5.25d). Some proteins are retained in the interior of the ER and others are transported in vesicles to the interior of the Golgi apparatus. Some of these proteins are secreted by exocytosis.

FIG. 5.24 Pathways for proteins destined (left) to be secreted or (right) transported to the plasma membrane.

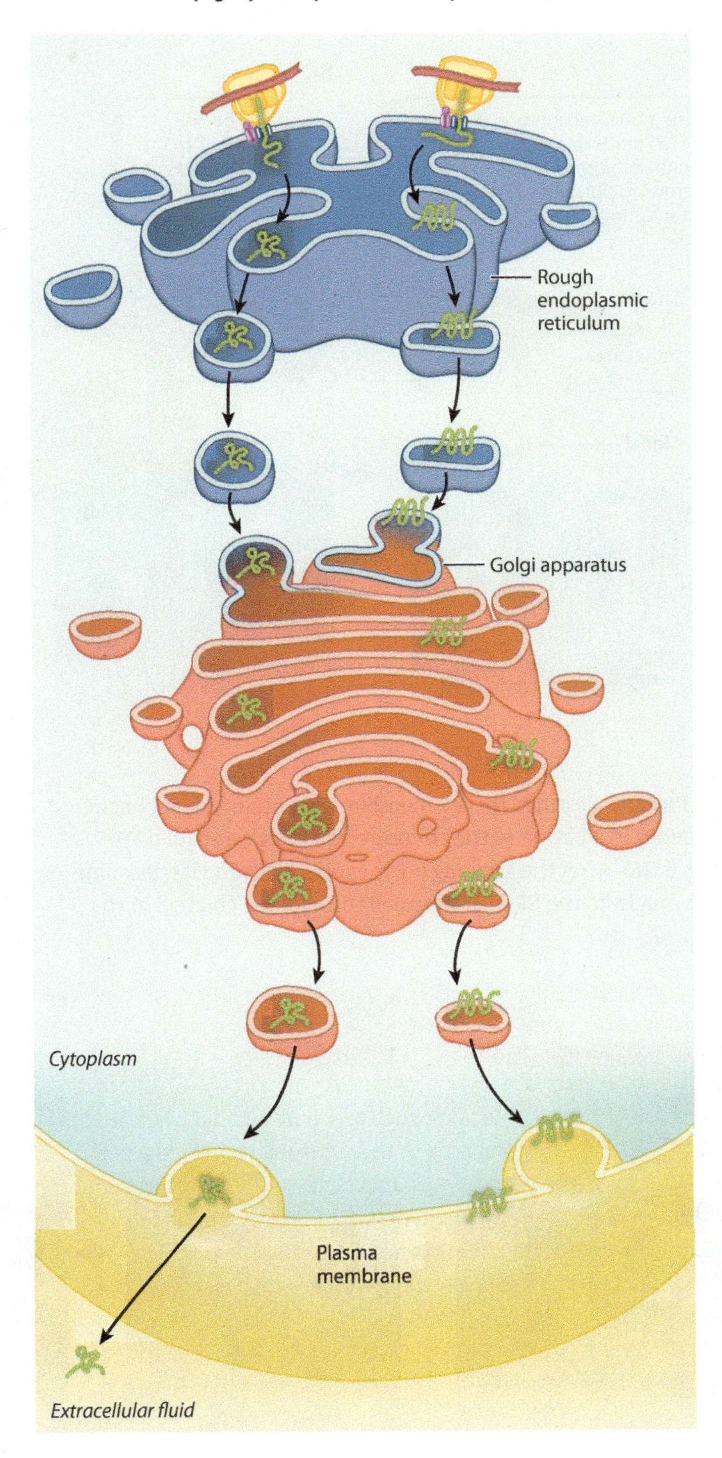

FIG. 5.25 Interaction of a signal sequence, signal-recognition particle (SRP), and SRP receptor. Binding of a signal sequence with a signal-recognition particle (SRP) halts translation, followed by docking of the ribosome on the ER membrane, release of the SRP, and continuation of translation.

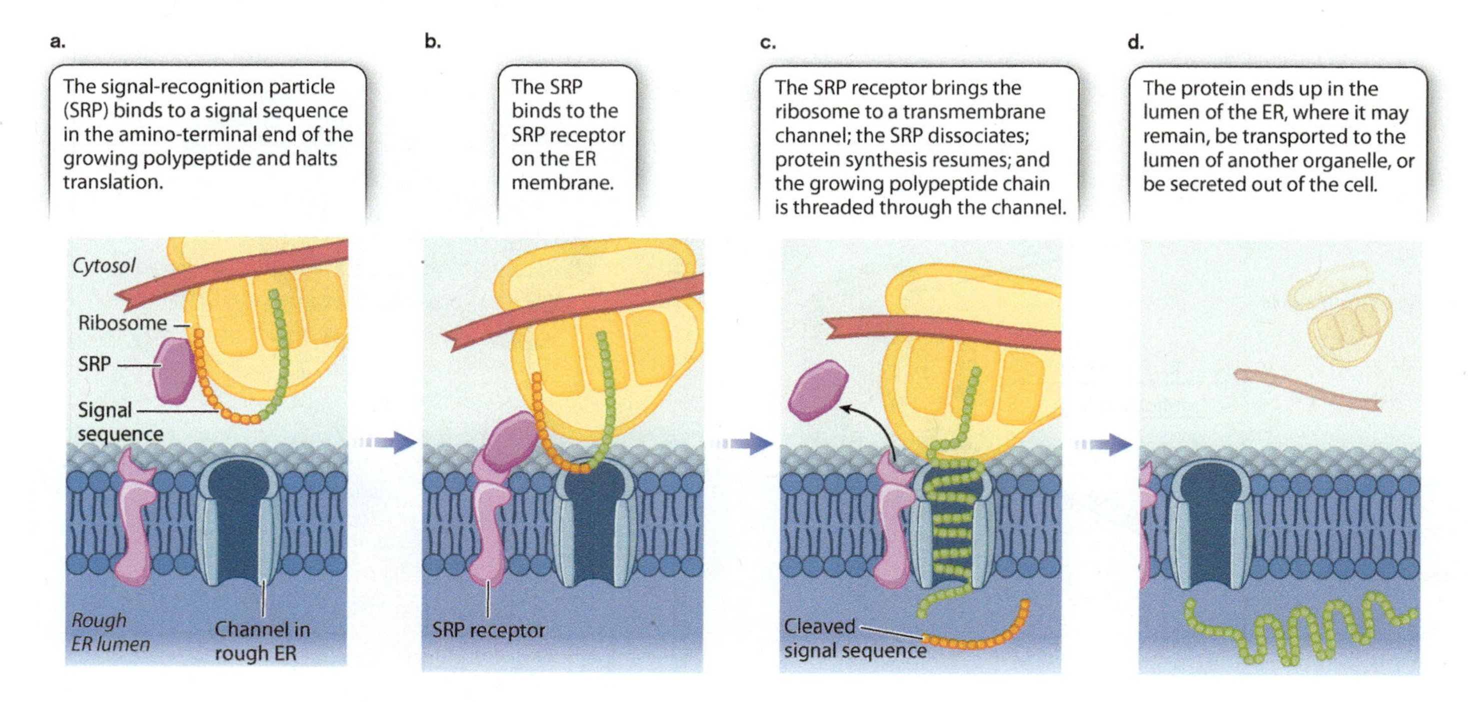

Proteins destined for cell membranes contain a signal-anchor sequence in addition to the amino-terminal signal sequence (**Fig. 5.26**). After the growing polypeptide chain and its ribosome are brought to the ER, it is threaded through the channel in the ER membrane until the signal-anchor sequence is encountered (Fig. 5.26a). The signal-anchor sequence is hydrophobic and is therefore able to diffuse laterally in the lipid bilayer (Fig. 5.26b). At this point, the ribosome dissociates from the channel while

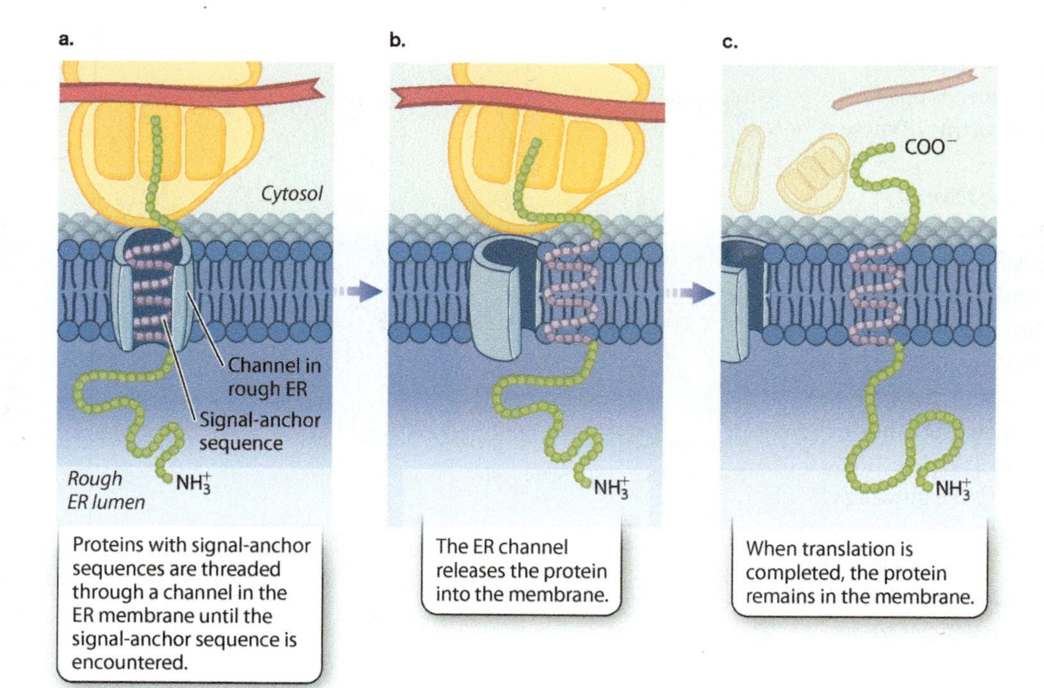

FIG. 5.26 Targeting of a transmembrane protein by means of a hydrophobic signal-anchor sequence. Proteins with a signal-anchor sequence end up embedded in the membrane.

translation continues. When translation is completed, the carboxyl end of the chain remains on the cytosolic side of the ER membrane, the amino end is in the ER lumen, and the region between them resides in the membrane (Fig. 5.26c). Transmembrane proteins such as these may stay in the membrane of the ER or end up in other internal membranes or the plasma membrane, where they serve as transporters, pumps, receptors, or enzymes.

5.5 MITOCHONDRIA AND CHLOROPLASTS

The membranes of two organelles, **mitochondria** and **chloroplasts,** are not part of the endomembrane system. Both of these organelles are specialized to harness energy for the cell. Interestingly, they are both semi-autonomous organelles that grow and multiply independently of the other membrane-bound compartments, and they contain their own genomes. As we discuss in Chapter 27, the similarities between the DNA of these organelles and the DNA of certain bacteria have led scientists to conclude that these organelles originated as bacteria that were captured by a eukaryotic cell and, over time, evolved to their current function.

Mitochondria provide the eukaryotic cell with most of its usable energy.

Mitochondria are organelles that harness energy from chemical compounds like sugars and convert it into ATP, which serves as the universal energy currency of the cell. ATP is able to drive the many chemical reactions in the cell. Mitochondria are present in nearly all eukaryotic cells.

Mitochondria are rod-shaped organelles with an outer membrane and a highly convoluted inner membrane whose folds project into the interior (**Fig. 5.27**). A proton electrochemical gradient is generated across the inner mitochondrial membrane, and the energy stored in the gradient is used to synthesize ATP for use by the cell. In the process of breaking down sugar and synthesizing ATP, oxygen is consumed and carbon dioxide is released. Does this process sound familiar? It also describes your own breathing, or respiration. Mitochondria are the site of cellular respiration, and the oxygen that you take in with each breath is used by mitochondria to produce ATP. Cellular respiration is discussed in greater detail in Chapter 7.

Chloroplasts capture energy from sunlight.

Both animal and plant cells have mitochondria to provide them with life-sustaining ATP. In addition, plant cells and

FIG. 5.27 Mitochondria. Mitochondria synthesize most of the ATP required to meet the cell's energy needs. *Source: Keith R. Porter/Science Source.*

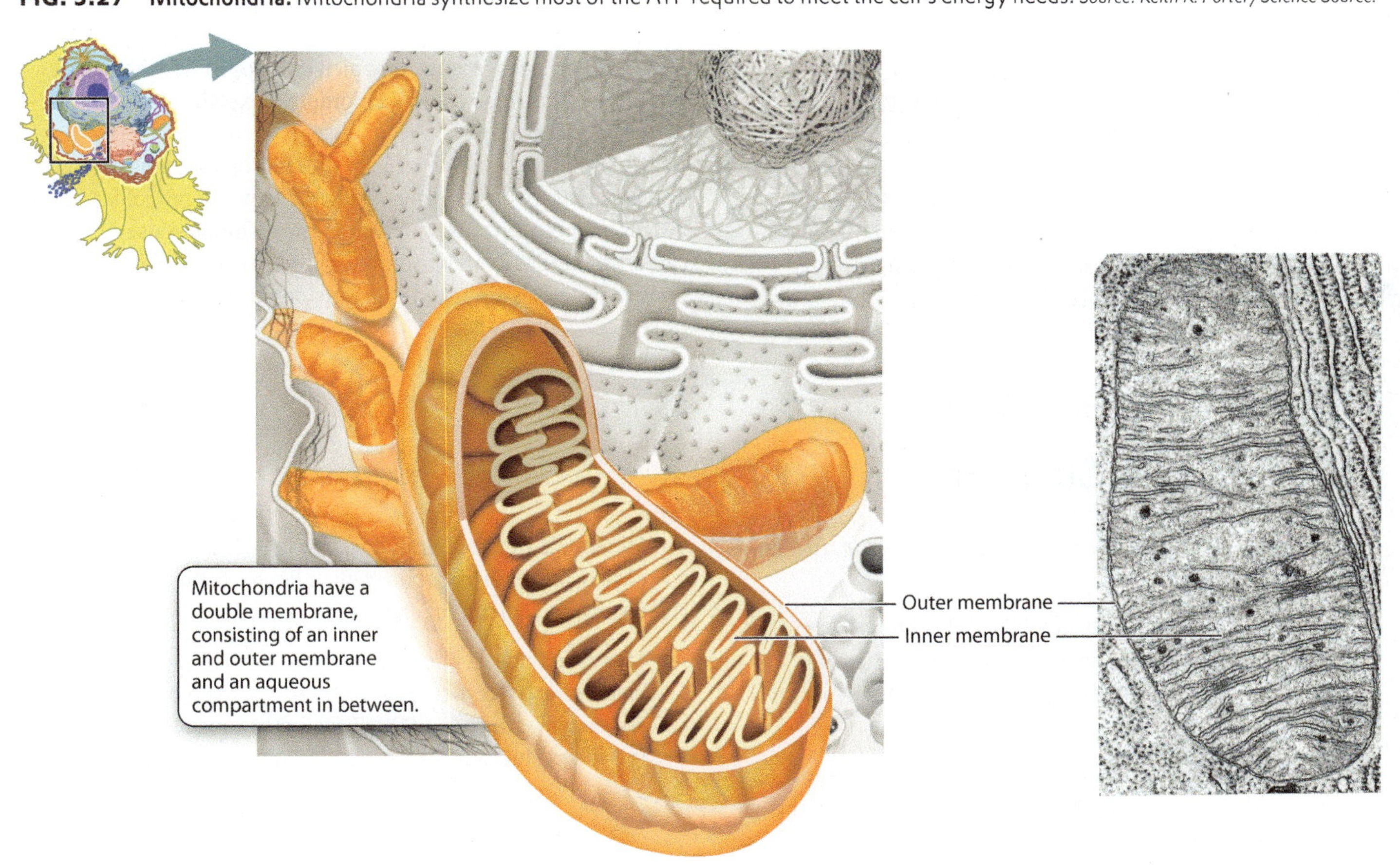

FIG. 5.28 Chloroplasts. Chloroplasts capture energy from sunlight and use it to synthesize sugars. *Source: Dr. Jeremy Burgess/Science Source.*

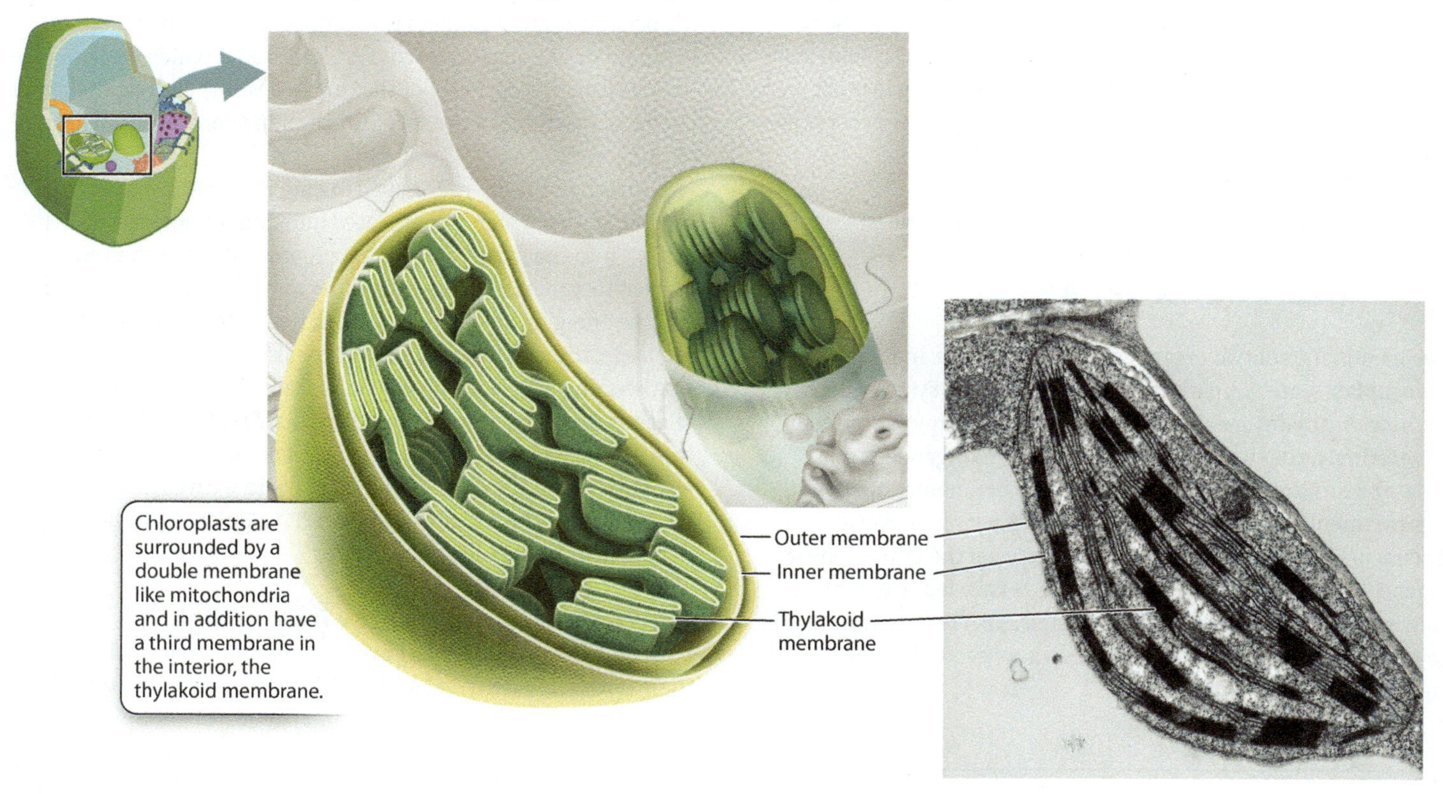

green algae have organelles called chloroplasts that capture the energy of sunlight to synthesize simple sugars (**Fig. 5.28**). This process, called **photosynthesis,** results in the release of oxygen as a waste product. Like the nucleus and mitochondria, chloroplasts are surrounded by a double membrane. They also have a third, internal membrane. This membrane defines a separate internal compartment called the **thylakoid.** The thylakoid membrane contains specialized light-collecting molecules called pigments, of which **chlorophyll** is the most important. The green color of chlorophyll explains why so many plants have green leaves.

Chlorophyll plays a key role in the chloroplast's ability to capture energy from sunlight. Using the light energy collected by this pigment, enzymes present in the chloroplast use carbon dioxide as a carbon source to produce carbohydrates. Photosynthesis is discussed in greater detail in Chapter 8. ■

Core Concepts Summary

5.1 Cell membranes are composed of lipids, proteins, and carbohydrates.

Phospholipids have both hydrophilic and hydrophobic regions. As a result, they spontaneously form structures such as micelles and bilayers when placed in an aqueous environment. page 90

Membranes are fluid, meaning that membrane components are able to move laterally in the plane of the membrane. page 92

Membrane fluidity is influenced by length of fatty acid chains, presence of carbon–carbon double bonds in fatty acid chains, and amount of cholesterol. page 92

Many membranes also contain proteins that span the membrane (transmembrane proteins) or are temporarily associated with one or other layer of the lipid bilayer (peripheral proteins). page 93

5.2 The plasma membrane is a selective barrier that controls the movement of molecules between the inside and the outside of the cell.

Selective permeability results from the combination of lipids and proteins that makes up cell membranes. page 96

Passive transport is the movement of molecules by diffusion, the random movement of molecules. There is a net movement of molecules from regions of higher concentration to regions of lower concentration. page 96

Passive transport can occur by the diffusion of molecules directly through the plasma membrane (simple diffusion) or be aided by protein transporters (facilitated diffusion). page 96

Active transport moves molecules from regions of lower concentration to regions of higher concentration and requires energy. page 97

Primary active transport uses energy stored in ATP; secondary active transport uses the energy stored in an electrochemical gradient. page 98

Animal cells often maintain size and shape by protein pumps that actively move ions in and out of the cell. page 99

Plants, fungi, and bacteria have a cell wall outside the plasma membrane that maintains cell size and shape. page 100

5.3 Cells can be classified as prokaryotes or eukaryotes, which differ in the degree of internal compartmentalization.

Prokaryotic cells lack a nucleus and other internal membrane-enclosed compartments. page 101

Prokaryotic cells include bacteria and archaeons and are much smaller than eukaryotes. page 101

Eukaryotic cells have a nucleus and other internal compartments called organelles. page 101

Eukaryotes include animals, plants, fungi, and protists. page 102

5.4 The endomembrane system is an interconnected system of membranes that includes the nuclear envelope, endoplasmic reticulum, Golgi apparatus, lysosomes, vesicles, and plasma membrane.

The nucleus, which is enclosed by a double membrane called the nuclear envelope, houses the genome. page 105

The endoplasmic reticulum is continuous with the outer nuclear envelope and manufactures proteins and lipids for use by the cell or for export out of the cell. page 105

The Golgi apparatus communicates with the endoplasmic reticulum by transport vesicles. It receives proteins and lipids from the endoplasmic reticulum and directs them to their final destinations. page 105

Lysosomes break down macromolecules like proteins to simpler compounds that can be used by the cell. page 107

Protein sorting directs proteins to their final destinations in or out of the cell. page 108

Proteins synthesized on free ribosomes are sorted after translation and proteins synthesized on ribosomes associated with the rough endoplasmic reticulum are sorted during translation. page 108

Proteins synthesized on free ribososomes are often sorted by means of a signal sequence and are destined for the cytosol, mitochondria, chloroplasts, or nucleus. page 108

Proteins synthesized on ribosomes on the rough endoplasmic reticulum have a signal sequence that is recognized by a signal-recognition particle. These proteins end up as transmembrane proteins, in the interior of various organelles, or secreted. page 109

5.5 Mitochondria and chloroplasts are organelles involved in harnessing energy, and likely evolved from free-living prokaryotes.

Mitochondria harness energy from chemical compounds for use by both animal and plant cells. page 111

Chloroplasts harness the energy of sunlight to build sugars. page 111

Self-Assessment

1. Describe how lipids with hydrophilic and hydrophobic regions behave in an aqueous environment.
2. Describe two types of association between proteins and membranes.
3. Describe an experiment that demonstrates that proteins move in membranes.
4. Name three parameters that need to be stably maintained inside a cell.
5. Explain the role of lipids and proteins in maintaining the selective permeability of membranes.

6. Distinguish between passive and active transport mechanisms across cell membranes.
7. Describe three different ways in which cells maintain size and shape.
8. Compare the organization, degree of compartmentalization, and size of prokaryotic and eukaryotic cells.
9. Name the major organelles in eukaryotic cells and describe their functions.
10. Explain how a protein ends up free in the cytosol, embedded in the plasma membrane, or secreted from the cell.

Log in to LaunchPad to check your answers to the Self-Assessment questions, and to access additional learning tools.

CHAPTER 6

Making Life Work

Capturing and Using Energy

Core Concepts

6.1 Metabolism is the set of biochemical reactions that transforms biomolecules and transfers energy.

6.2 Kinetic energy is energy of motion, and potential energy is stored energy.

6.3 The laws of thermodynamics govern energy flow in biological systems.

6.4 Chemical reactions involve the breaking and forming of bonds.

6.5 The rate of biochemical reactions is increased by protein catalysts called enzymes.

Purestock/Getty Images.

We have seen that cells require a way to encode and transmit information and a membrane to separate inside from out. The third requirement of a cell is energy, which cells need to do work. Cells grow and divide, move, change shape, pump ions in and out, transport vesicles, and synthesize macromolecules such as DNA, RNA, proteins, and complex carbohydrates. All of these activities are considered work, and they therefore require energy.

We are all familiar with different forms of energy—the sun and wind provide sources of energy, as do fossil fuels such as oil and natural gas. We have learned to harness the energy from these sources and convert it to other forms, such as electricity, to provide needed power to our homes and cities.

Cells are faced with similar challenges. They must harness energy from the environment and convert it to a form that allows them to do the work necessary to sustain life. Cells harness energy from the sun and from chemical compounds, including carbohydrates, lipids, and proteins. Although the source of energy may differ among cells, all cells convert energy to a form that can be easily used to drive cellular processes. All cells use energy in the form of a molecule called **adenosine triphosphate (ATP).**

ATP is often called the universal "currency" of cellular energy to indicate that ATP provides energy in a form that all cells can readily use to perform the work of the cell. Although "currency" provides a useful analogy for the role that ATP plays, keep in mind an important distinction between actual currency, such as a dollar bill, and ATP. A dollar bill *represents* a certain value but does not in fact have any value in itself. By contrast, ATP does not represent energy; it actually *contains* energy in its chemical bonds. Nevertheless, the analogy is useful because ATP, like a dollar bill, engages in a broad range of energy "transactions" in the cell, as we discuss below.

In this chapter, we consider energy in the context of cells. What exactly is energy? What principles govern its flow in biological systems? And how do cells make use of it?

6.1 AN OVERVIEW OF METABOLISM

When considering a cell's use of energy, it is helpful to also consider the cell's sources of carbon because carbon is the backbone of the organic molecules that make up cells and because cells often use carbon-based compounds as a stable form of energy storage. How organisms obtain the energy and carbon needed for growth and other vital functions is so fundamental that it is sometimes used to provide a metabolic classification of life (**Fig. 6.1**). Simply put, organisms have two ways of harvesting energy from their environment and two sources of carbon. Together, this means that there are four principal ways in which organisms acquire the energy and materials needed to grow, function, and reproduce.

Organisms can be classified according to their energy and carbon sources.

Organisms have two ways of harvesting energy from their environment: They can obtain energy either from the sun or from chemical compounds (Fig. 6.1). Organisms that capture energy from sunlight are called **phototrophs.** Plants are the most familiar example. Plants use the energy of sunlight to convert carbon dioxide and water into sugar and oxgen (Chapter 8). Sugars, such as glucose, contain energy in their chemical bonds that can be used to synthesize ATP, which in turn can power the work of the cell.

Other organisms derive their energy directly from chemical compounds. These organisms are called **chemotrophs,** and animals are familiar examples. Animals ingest other organisms, obtaining organic molecules such as glucose that they break down in the presence of oxygen to produce carbon dioxide and water. In this process, the energy in the chemical bonds of the organic molecule is converted to energy carried in the bonds of ATP (Chapter 7).

In drawing this distinction between plants and animals, we have to be careful. Although sunlight provides the energy that plants use to synthesize glucose, plants still power most cellular processes by breaking down the sugar they make, just as animals do. And although chemotrophs harness energy from organic molecules, the energy in these organic molecules is generally derived from the sun. Bearing these two points in mind, we use the terms "phototroph" and "chemotroph" because they call our attention to the flow of energy from the sun to organisms and then from one organism to the next (discussed more fully in Chapter 25).

Organisms can also be classified in terms of where they get their carbon (Fig. 6.1). Some organisms are able to convert carbon dioxide (an inorganic form of carbon) into glucose (an organic form of carbon). These organisms are **autotrophs,** or "self feeders," because they make their own organic carbon using inorganic carbon as the starting material. Plants again are an example, so plants are both phototrophs and autotrophs, or photoautotrophs.

Other organisms do not have the ability to convert carbon dioxide into organic forms of carbon. Instead, they obtain their carbon from organic molecules synthesized by other organisms, called preformed organic molecules. In other words, these organisms eat other organisms or molecules derived from other organisms. Such organisms are **heterotrophs,** or "other feeders," as they rely on other organisms for their organic forms of carbon. Animals obtain carbon in this way, and so animals are both chemotrophs and heterotrophs, or chemoheterotrophs. In fact, animals get their energy and carbon from the same molecule. That is, a molecule of glucose supplies both energy and carbon.

As we move away from the familiar examples of plants and animals and begin in Part 2 to explore the diversity of life, we will see that not all organisms fit into the two categories of photoautotrophs and chemoheterotrophs (Fig. 6.1). Some microorganisms gain energy from sunlight but obtain their carbon from preformed organic molecules; such organisms are called photoheterotrophs. Other microorganisms extract energy from inorganic sources but build their own organic molecules; these organisms are called chemoautotrophs. They are often found in extreme environments, such as deep-sea vents, where sunlight is absent and inorganic compounds such as hydrogen sulfide are plentiful.

FIG. 6.1 A metabolic classification of organisms. Organisms can be classified according to their energy and carbon sources. *Photo sources: (clockwise from top-left) Dr. Tony Brain/Science Source; ScienceFoto.DE/Dr. Andre Kemp/Getty Images; Copyright 1997 Microbial Diversity, Rolf Schauder; Martin Oeggerli/Science Source; DNY59/iStockphoto; David J. Patterson; EM image by Manfred Rohde, Helmholtz Centre for Infection Research, Braunschweig, Germany; Anna Omelchenko/Dreamstime.com.*

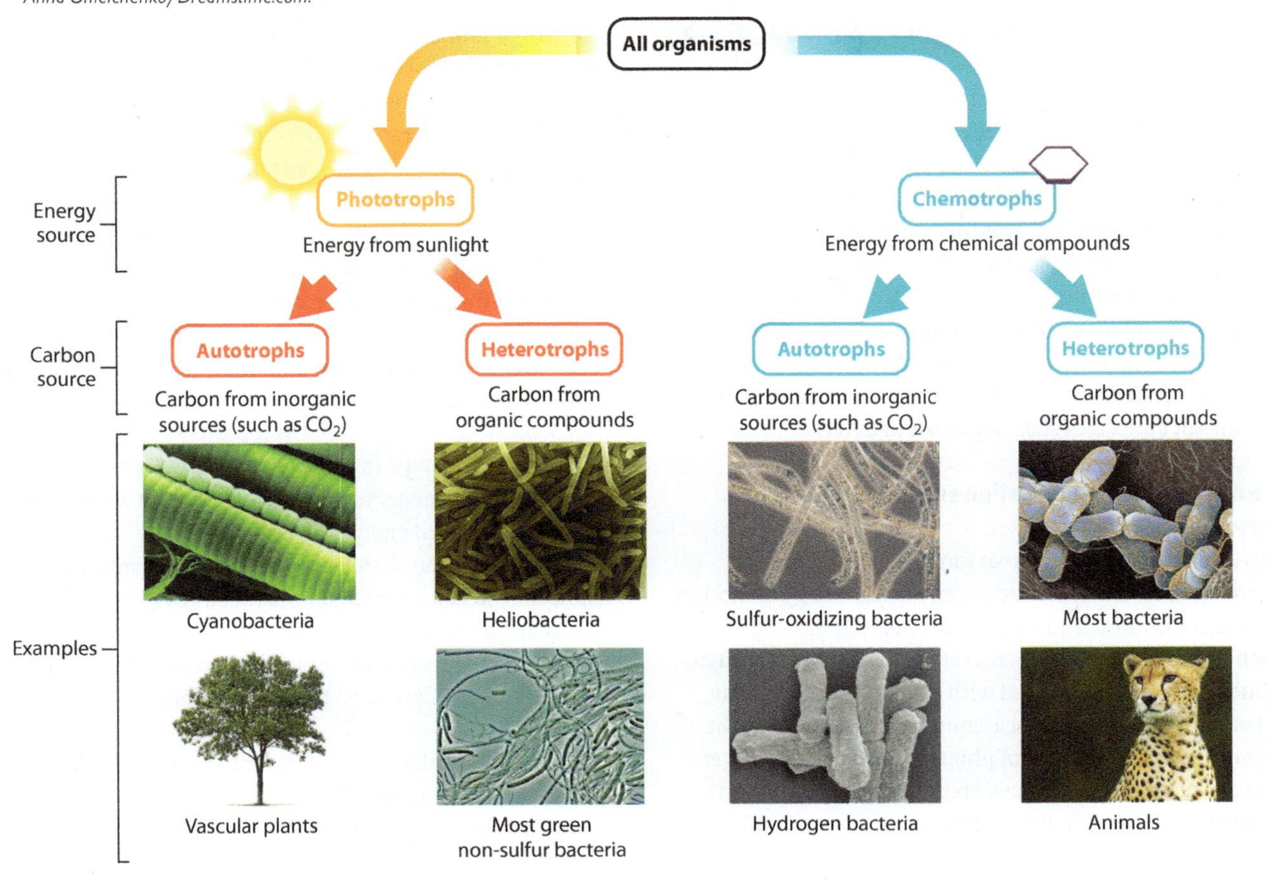

Metabolism is the set of chemical reactions that sustain life.

The building and breaking down of sugars such as glucose andthe harnessing and release of energy in the process are driven by chemical reactions in the cell. The term **metabolism** encompasses the entire set of these chemical reactions that convert molecules into other molecules and transfer energy in living organisms. These chemical reactions are occurring all the time in your cells. Many of these reactions are linked, in that the products of one are the reactants of the next, forming long pathways and intersecting networks.

Metabolism is divided into two branches: **Catabolism** is the set of chemical reactions that break down molecules into smaller units and, in the process, produce ATP, and **anabolism** is the set of chemical reactions that build molecules from smaller units and require an input of energy, usually in the form of ATP (**Fig. 6.2**). For example, carbohydrates can be broken down, or catabolized, into sugars, fats into fatty acids and glycerol, and proteins into amino acids. These initial products can be broken down further to

FIG. 6.2 Catabolism and anabolism. The energy harvested as ATP during the breakdown of molecules in catabolism can be used to synthesize molecules in anabolism.

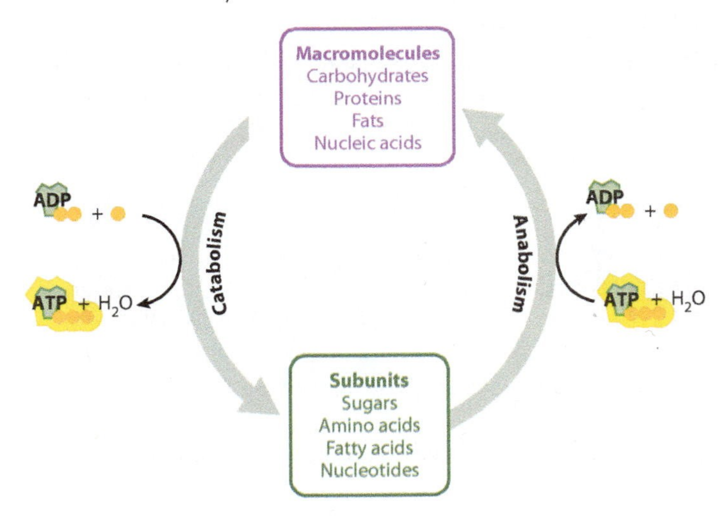

release energy stored in their chemical bonds. The synthesis of macromolecules such as carbohydrates and proteins, by contrast, is anabolic.

6.2 KINETIC AND POTENTIAL ENERGY

There are many different sources of energy, including sunlight, wind, electricity, and fossil fuels. The food we eat also contains energy. **Energy** can be defined as the capacity to do work. For a cell, work involves processes we discussed in earlier chapters, such as synthesizing DNA, RNA, and proteins, pumping substances across the plasma membrane, and moving vesicles between various compartments of a cell.

Although there are many different sources of energy, energy comes in just two major forms. In this section, we consider these two forms of energy. In addition, we focus on how energy is held in chemical bonds of molecules such as glucose and ATP.

Kinetic energy and potential energy are two forms of energy.

Energy can be classified as one of two forms: kinetic energy or potential energy (**Fig. 6.3**). **Kinetic energy** is the energy of motion, and it is perhaps the most familiar form of energy. A moving object, such as a ball bouncing down a set of stairs, possesses kinetic energy. Kinetic energy is associated with any kind of movement, such as a person running or a muscle contracting. Similarly, light is associated with the movement of photons, electricity with the movement of electrons, and thermal energy (perceived as heat) with the movement of molecules, so these, too, are forms of kinetic energy.

Energy is not always associated with motion. An immobile object can still possess a form of energy called **potential energy,** or stored energy. Potential energy depends on the structure of the object or its position relative to its surroundings, and it is released by a change in the object's structure or position. For example, the potential energy of a ball is higher at the top of a flight of stairs than at the bottom (Fig. 6.3). If it were not blocked by the floor, it would move from the position of higher potential energy (the top of the stairs) to the position of lower potential energy (the bottom of the stairs) because of the force of gravity. Similarly, an electrochemical gradient of molecules across a cell membrane is a form of potential energy. Given a pathway through the membrane, the molecules move down their concentration and electrical gradients from higher to lower potential energy (Chapter 5).

Energy can be converted from one form to another. The ball at the top of the stairs has a certain amount of potential energy because of its position. As it rolls down the stairs, this potential energy is converted to kinetic energy associated with movement of the ball and the surrounding air. When the ball reaches the bottom of the stairs, the remaining energy is stored as potential energy. Conversely, it takes an input of energy to move the ball back to the top of the stairs, and this input of energy is stored as potential energy.

FIG. 6.3 Potential and kinetic energy. Some of the high potential energy of a ball at the top of a set of stairs is transformed to kinetic energy as the ball rolls down the stairs.

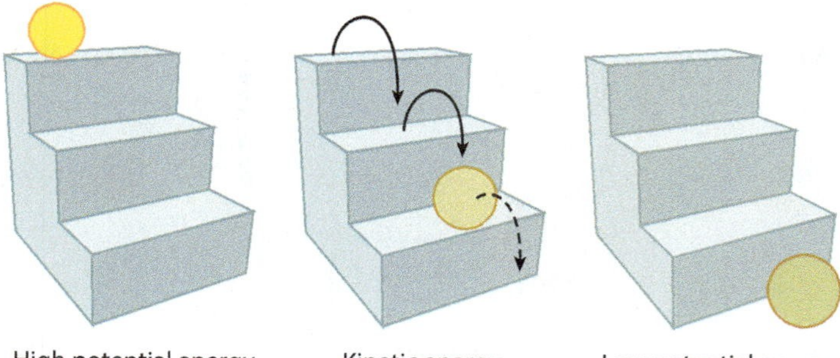

Chemical energy is a form of potential energy.

We obtain the energy we need from the food we eat, which contains chemical energy. **Chemical energy** is a form of potential energy held in the chemical bonds between pairs of atoms in a molecule. Recall from Chapter 2 that a covalent bond results from the sharing of electrons between two atoms. Covalent bonds form when the sharing of electrons between two atoms results in a more stable configuration than if the orbitals of the two atoms did not overlap.

The more stable configuration will always be the one with lower potential energy. As a result, energy is required to break a covalent bond because going from a lower energy state to a higher one requires an input of energy. Conversely, energy is released when a covalent bond forms.

Some bonds are stronger than other ones. A strong bond is hard to break because the arrangement of orbitals in these molecules is much more stable than the two atoms would be on their own. As a result, strong bonds do not contain very much chemical energy, similar to the potential energy of a ball at the bottom of a flight of stairs (Fig. 6.3). This may at first seem counterintuitive, but makes sense when you consider that strong bonds have a very stable arrangement of orbitals and therefore do not require a lot of energy to remain intact. Examples of molecules with strong covalent bonds that contain relatively little chemical energy are carbon dioxide (CO_2) and water (H_2O).

Conversely, some covalent bonds are relatively weak. These bonds are easily broken because the arrangement of orbitals in these molecules is only somewhat more stable than if the two atoms did not share any electrons. As a result, weak covalent bonds require a lot of energy to stay intact and contain a lot of chemical energy, similar to the potential energy of a ball at the top of a flight of stairs (Fig. 6.3). Organic molecules such as carbohydrates, lipids, and proteins contain relatively weak covalent bonds, including many carbon–carbon (C–C) bonds and

FIG. 6.4 The structure of ATP.

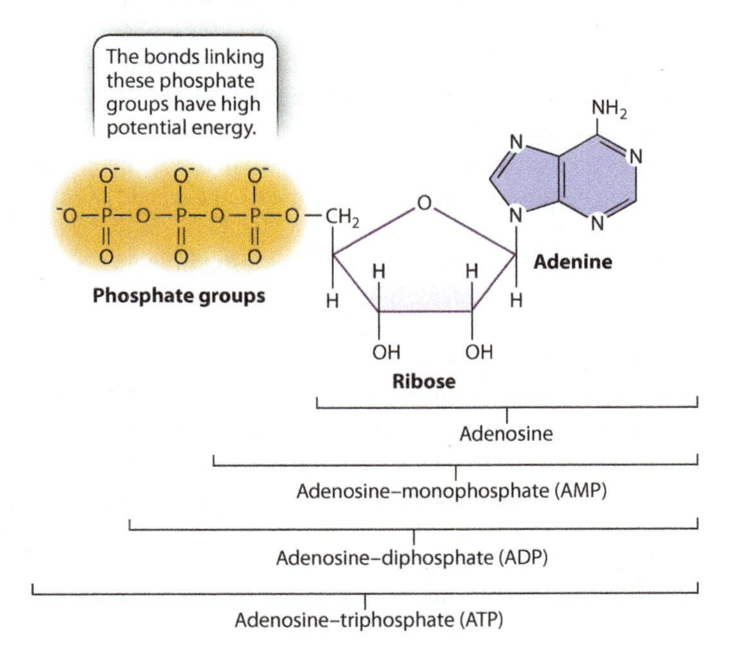

carbon–hydrogen (C–H) bonds. Therefore, organic molecules are rich sources of chemical energy and are called fuel molecules.

ATP is a readily accessible form of cellular energy.

The chemical energy carried in carbohydrates, lipids, and proteins is harnessed by cells to do work. But cells do not use this energy all at once. Instead, through the series of chemical reactions described in Chapter 7, they package this energy into a chemical form that is readily accessible to the cell. One form of chemical energy is adenosine triphosphate, or ATP, shown in **Fig. 6.4.** The chemical energy in the bonds of ATP is used in turn to drive many cellular processes, such as muscle contraction, cell movement, and membrane pumps. In this way, ATP serves as a go-between, acting as an intermediary between fuel molecules that store a large amount of potential energy in their bonds and the activities of the cell that require an input of energy.

ATP is composed in part of adenosine, which is made up of the base adenine and the five-carbon sugar ribose. The ribose is attached to triphosphate, or three phosphate groups (Fig. 6.4). Its chemical relatives are ADP (adenosine diphosphate) and AMP (adenosine monophosphate), with two phosphate groups and one phosphate group, respectively. The use of ATP as an energy source in nearly all cells reflects its use early in the evolution of life on Earth.

The chemical energy of ATP is held in the bonds connecting the phosphate groups. At physiological pH, these phosphate groups are negatively charged and have a tendency to repel each other. The chemical bonds connecting the phosphate groups therefore store chemical energy. This energy is released when new, more stable bonds are formed that contain less chemical energy (section 6.4). The released energy in turn can be harnessed to power the work of the cell.

6.3 THE LAWS OF THERMODYNAMICS

Energy from the sun can be converted by plants and other phototrophs into chemical potential energy held in the bonds of molecules. Chemical reactions can transfer this chemical energy between different molecules. And energy from these molecules can be used by the cell to do work. In all these instances, energy is subject to the laws of thermodynamics. In fact, all processes in a cell, from diffusion to osmosis to the pumping of ions to cell movement, are subject to these laws. There are four laws of thermodynamics, but two of them are particularly relevant to biological processes and are discussed here.

The first law of thermodynamics: Energy is conserved.

The **first law of thermodynamics** is the law of conservation of energy, which states that the universe contains a constant amount of energy. Therefore, energy is neither created nor destroyed. New energy is never formed, and energy is never lost. Instead, energy changes from one form to another. For example, kinetic energy can change to potential energy and vice versa, but the total amount of energy always remains the same (**Fig. 6.5**).

In the simple example of a ball rolling down a set of stairs (see Fig. 6.3), the potential energy stored at the top of the stairs is converted to kinetic energy associated with the movement of the ball and the surrounding air. The total amount of energy in this transformation is constant—that is, the difference in potential energy at the top and the bottom of the stairs is equal to the total amount of kinetic energy associated with the movement of the ball and air molecules.

The second law of thermodynamics: Energy transformations always result in an increase in disorder in the universe.

When energy changes forms, the *total amount of energy* remains constant. However, in going from one form of energy to another,

FIG. 6.5 **The first law of thermodynamics.** When energy is transformed from one form to another, the total amount of energy remains the same.

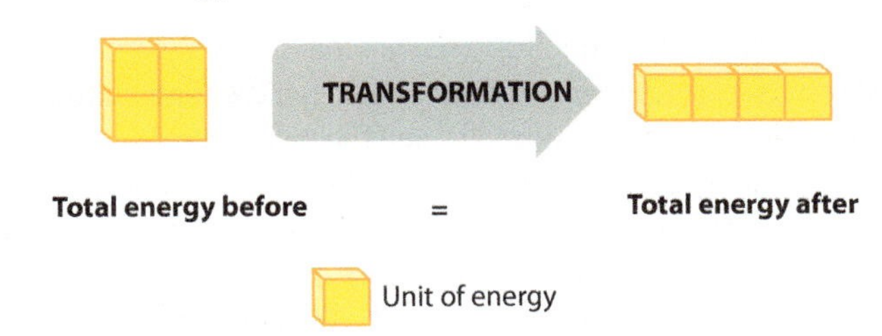

FIG. 6.6 The second law of thermodynamics. Because the amount of disorder increases when energy is transformed from one form to another, some of the total energy is not available to do work.

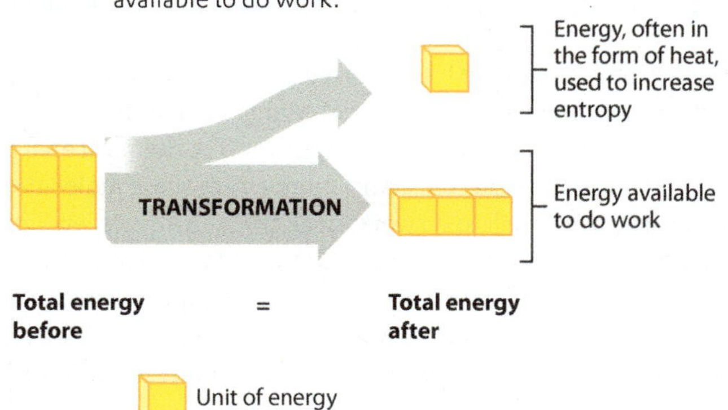

the *energy available to do work* decreases (**Fig. 6.6**). How can the total energy remain the same, but the energy available to do work decrease? The answer is that some of the energy is available to do work, and some is not available to do work. So, energy transformations are never 100% efficient since the amount of energy available to do work decreases every time energy changes forms.

The energy that is not available to do work takes the form of an increase in disorder. Thus, there is a universal price to pay in transforming energy from one form to another. For example, when kinetic energy is changed into potential energy, the amount of disorder always increases. This principle is summarized by the **second law of thermodynamics,** which states that the transformation of energy is associated with an increase in disorder of the universe. The degree of disorder is called **entropy.**

The degree of disorder is just one way to describe entropy. Another way to think about entropy is to consider the number of possible positions and motions (collectively called microstates) a molecule can take on in a given system. As entropy increases, so does the number of positions and motions available to the molecule. Consider the expansion of a gas after the lid is removed from its container. Given more space, the molecules of the gas are less constrained and able to move about more freely. They have more positions available to them and move about at a larger range of speeds, so they have more entropy.

In chemical reactions, most of the entropy increase occurs through the transformation of various forms of energy into thermal energy, which we experience as heat. Thermal energy is a type of kinetic energy corresponding to the random motion of molecules and results in a given temperature. The higher the temperature, the more rapidly the molecules move, and the higher the disorder.

Consider the contraction of a muscle, a form of kinetic energy associated with the shortening of muscle cells. The contraction of muscle is powered by chemical potential energy in fuel molecules. Some of this potential energy is transformed into kinetic energy (movement), and the rest is dissipated as thermal energy (which is why your muscles warm as you exercise). The amount of chemical potential energy expended is equal to the amount of kinetic energy plus the amount of thermal energy, consistent with the first law of thermodynamics. In addition, not all of the chemical energy stored in molecules is converted to kinetic energy, as thermal energy flowing as heat is a necessary by-product of the energy transformation, consistent with the second law of thermodynamics.

In living organisms, catabolic reactions result in an increase of entropy as a single ordered biomolecule is broken down into several smaller ones with more freedom to move around. Anabolic reactions, by contrast, might seem to decrease entropy because they use individual building blocks to synthesize more ordered biomolecules such as proteins or nucleic acids. Do these anabolic reactions violate the second law of thermodynamics? No, because the second law of thermodynamics always applies to the universe as a whole, not to a chemical reaction in isolation. A local decrease in entropy is always accompanied by an even higher increase in the entropy of the surroundings. The production of heat in a chemical reaction increases entropy because heat is associated with the motion of molecules. The combination of the decrease in entropy associated with the building of a macromolecule and the increase in entropy associated with heat always results in a net increase in entropy.

The key point here is that the maintenance of the high degree of function and organization of a single cell or a multicellular organism requires a constant input of energy. This input, as we have seen, comes either from the sun or from the energy stored in chemical compounds. This energy allows molecules to be built and other work to be carried out, but also leads to greater entropy in the surroundings.

→ **Quick Check 1** Cold air has less entropy than hot air. The second law of thermodynamics states that entropy always increases. Do air conditioners violate this law?

6.4 CHEMICAL REACTIONS

Living organisms build and break down molecules, pump ions across membranes, move, and in general function largely through chemical reactions. As we have seen, organisms break down food molecules, such as glucose, storing energy in the bonds of ATP, which then powers chemical reactions that sustain life. Chemical reactions are therefore central to life processes. In this section, we describe what a chemical reaction is and how the laws of thermodynamics apply.

A chemical reaction occurs when molecules interact.

As we saw in Chapter 2, a chemical reaction is the process by which molecules, called reactants, are transformed into other

FIG. 6.7 **A chemical reaction.** Atoms retain their identity during a chemical reaction as bonds are broken and new bonds are formed to yield new molecules.

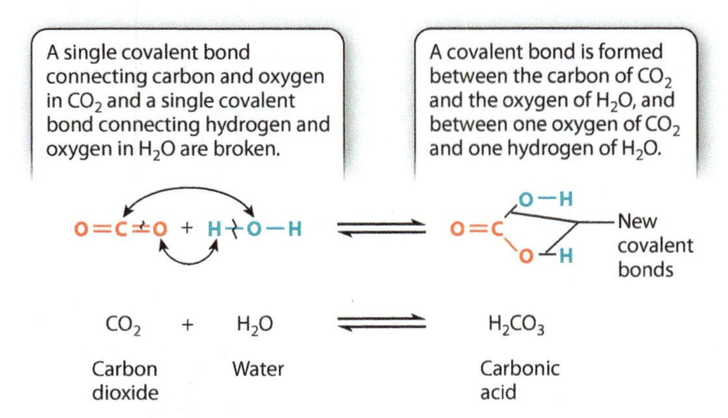

molecules, called products. During a chemical reaction, atoms keep their identity, but the bonds linking the atoms change. For example, carbon dioxide (CO_2) and water (H_2O) can react to produce carbonic acid (H_2CO_3), as shown in **Fig. 6.7.**

This reaction is common in nature. For example, it occurs when carbon dioxide in the air enters ocean waters, and explains why the oceans are becoming more acidic with increasing carbon dioxide levels in the atmosphere. This reaction also takes place in the bloodstream. When cells break down glucose, they produce carbon dioxide. In animals, this carbon dioxide diffuses out of cells and into the blood. Rather than remaining dissolved in solution, most of the carbon dioxide is converted into carbonic acid by the preceding reaction. In an aqueous (watery) environment like the ocean or the blood, carbonic acid exists as bicarbonate ions (HCO_3^-) and protons (H^+).

Most chemical reactions in cells are readily reversible: The products can react to form the reactants. For example, carbon dioxide and water react to form carbonic acid, and carbonic acid can dissociate to produce carbon dioxide and water. This reverse reaction also occurs in the blood, allowing carbon dioxide to be removed from the lungs or gills (Chapter 39). The reversibility of the reaction is indicated by a double arrow (Fig. 6.7).

The way the reaction is written defines forward and reverse reactions: A forward reaction proceeds from left to right and the reactants are located on the left side of the arrow; a reverse reaction proceeds from right to left and the reactants are located on the right side of the arrow.

The direction of a reaction can be influenced by the concentrations of reactants and products. For example, increasing the concentration of the reactants or decreasing the concentration of the products favors the forward reaction. This effect explains how many reactions in metabolic pathways proceed: The products of many reactions are quickly consumed by the next reaction, helping to drive the first reaction forward.

The laws of thermodynamics determine whether a chemical reaction requires or releases energy available to do work.

We have seen that cells use chemical reactions to perform much of the work of the cell. The amount of energy available to do work is called **Gibbs free energy (*G*).** In a chemical reaction, we can compare the free energy of the reactants and products to determine whether the reaction releases energy that is available to do work. The difference between two values is denoted by the Greek letter delta (Δ). In this case, ΔG is the free energy of the products minus the free energy of the reactants (**Fig. 6.8**). If the products of a reaction have more free energy than the reactants, then ΔG is positive and a net input of energy is required to drive the reaction forward (Fig. 6.8a). By contrast, if the products of a reaction have less free energy than the reactants, ΔG is negative and energy is released and available to do work (Fig. 6.8b).

Reactions with a negative ΔG that release energy and proceed spontaneously are called **exergonic,** and reactions with a positive ΔG that require an input of energy and are not spontaneous are called **endergonic.** Note that the term "spontaneous" does not imply instantaneous or even rapid. "Spontaneous" in this context means that a reaction releases energy; "non-spontaneous" means that a reaction requires a sustained input of energy.

Recall that the total amount of energy is equal to the energy available to do work plus the energy that is not available to do work because of the increase in entropy. We can write this relationship as an equation in which the total amount of energy is **enthalpy (*H*),** the energy available to do work is Gibbs free energy (*G*), and the degree of disorder is **entropy (*S*)** multiplied by the **absolute temperature (*T*)** (measured in degrees Kelvin), since temperature influences the movement of molecules (and hence the degree of disorder). Therefore,

Total amount of energy (H) =
energy available to do work (G) + energy lost to entropy (TS)

FIG. 6.8 **Endergonic and exergonic reactions.** An endergonic reaction requires an input of energy; an exergonic reaction releases energy.

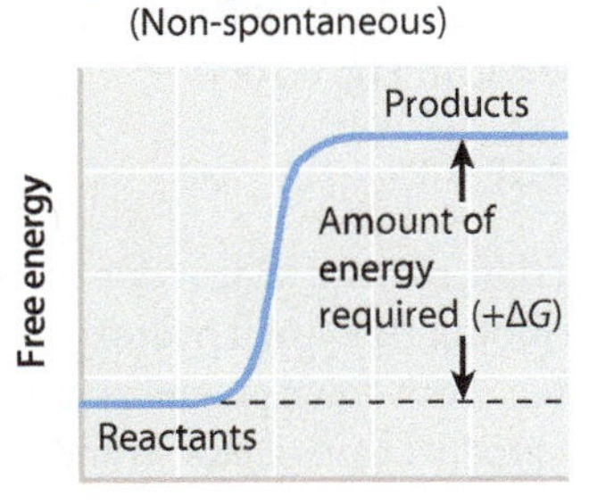

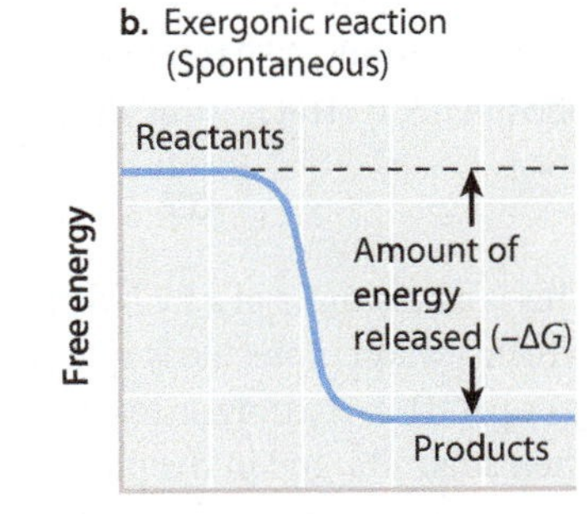

We are interested in the energy available for a cell to do work, or G. Therefore, we express G in terms of the other two parameters, H and TS, as follows:

$$G = H - TS$$

In a chemical reaction, we can compare the total energy and entropy of the reactants with the total energy and entropy of the products to see if there is energy available to do work. As a result, we get

$$\Delta G = \Delta H - T\Delta S$$

This equation is a useful way to see if a chemical reaction takes place spontaneously, what direction the reaction proceeds in, and whether net energy is required or released. Let's take a step back and make sure it makes sense intuitively. The value of ΔG depends on *both* the change in enthalpy and the change in disorder. Catabolic reactions are those in which the products have less chemical energy (lower enthalpy) in their bonds than the reactants have, and the products are more disordered (higher entropy) than the reactants are. In other words, such reactions have a negative value of ΔH and a positive value of ΔS. Notice that, in the equation $\Delta G = \Delta H - T\Delta S$, there is a negative sign in front of ΔS, so both the negative ΔH and positive ΔS contribute to a negative ΔG. Therefore, these reactions proceed spontaneously (**Fig. 6.9**).

The reverse is also true: Increasing chemical energy (positive ΔH) and decreasing disorder (negative ΔS), as in the synthesis of proteins from individual amino acids and other anabolic reactions, results in a positive value of ΔG and requires a net input of energy (Fig. 6.9).

There are also cases in which the change in enthalpy and the change in entropy are both positive or both negative. In these cases, the absolute value of these parameters determines whether ΔG is positive or negative and therefore whether a reaction is spontaneous or not.

→ **Quick Check 2** How does increasing the temperature affect the change in free energy (ΔG) of a chemical reaction?

The hydrolysis of ATP is an exergonic reaction.

Let's apply these concepts to a specific chemical reaction. Earlier, we introduced ATP, the molecule that drives many cellular processes using the chemical potential energy in its chemical bonds. ATP reacts with water to form ADP and inorganic phosphate, P_i (HPO_4^{2-}), as shown here and in **Fig. 6.10**:

$$ATP + H_2O \rightarrow ADP + P_i$$

This is an example of a hydrolysis reaction, a chemical reaction in which a water molecule is split into a proton (H^+) and a hydroxyl group (OH^-). Hydrolysis reactions often break down polymers into their subunits, and in the process one product gains a proton and the other gains a hydroxyl group.

FIG. 6.9 Energy in catabolism and anabolism. Catabolic reactions have a negative ΔG and release energy, often in the form of ATP. Anabolic reactions have a positive ΔG and require an input of energy, often in the form of ATP.

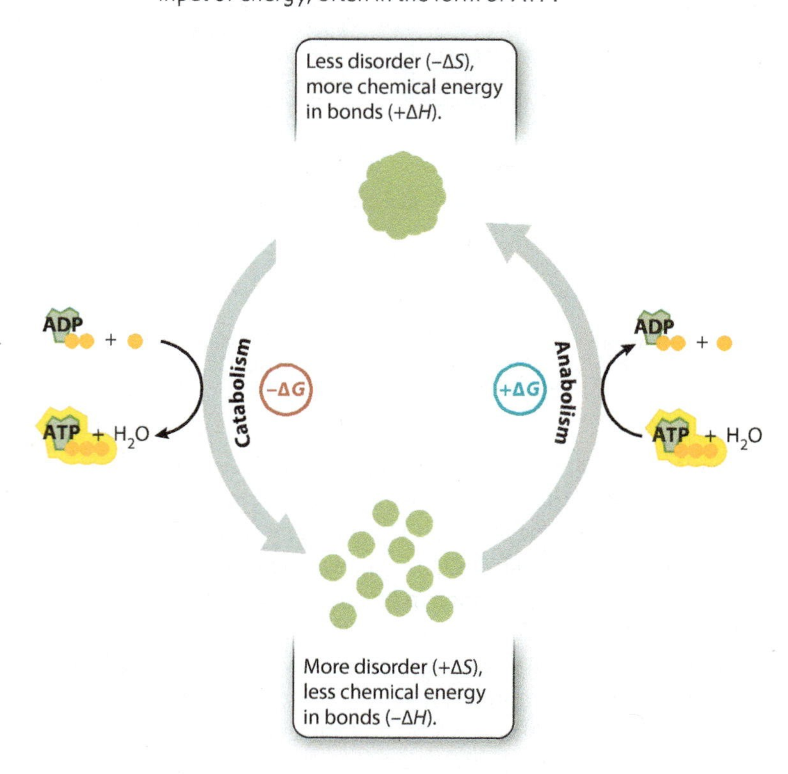

The reaction of ATP with water is an exergonic reaction because there is less free energy in the products compared to the reactants. The free energy difference can be explained by referring to the formula we derived in the last section. Recall that the phosphate groups of ATP are negatively charged at physiological pH and repel each other. ATP has three phosphate groups, and ADP has two. Therefore, ADP is more stable (contains less chemical energy in its bonds) than ATP, resulting in a negative value of ΔH. In addition, a single molecule of ATP is broken down into two molecules, ADP and P_i. Therefore, the reaction is also associated with an increase in entropy, or a positive value of ΔS. Since $\Delta G = \Delta H - T\Delta S$, ΔG is negative and the reaction is a spontaneous one that releases energy available to do work.

The free energy difference for ATP hydrolysis is approximately −7.3 kcal per mole (kcal/mol) of ATP. This value is influenced by several factors, including the concentration of reactants and products, the pH of the solution in which the reaction occurs, and the temperature and pressure. The value −7.3 kcal/mol is the value under standard laboratory conditions in which the concentrations of reactants and products are equal and pressure is held constant. In a cell, it is likely higher, on the order of −12 kcal/mol.

FIG. 6.10 ATP hydrolysis. ATP hydrolysis is an exergonic reaction that releases free energy.

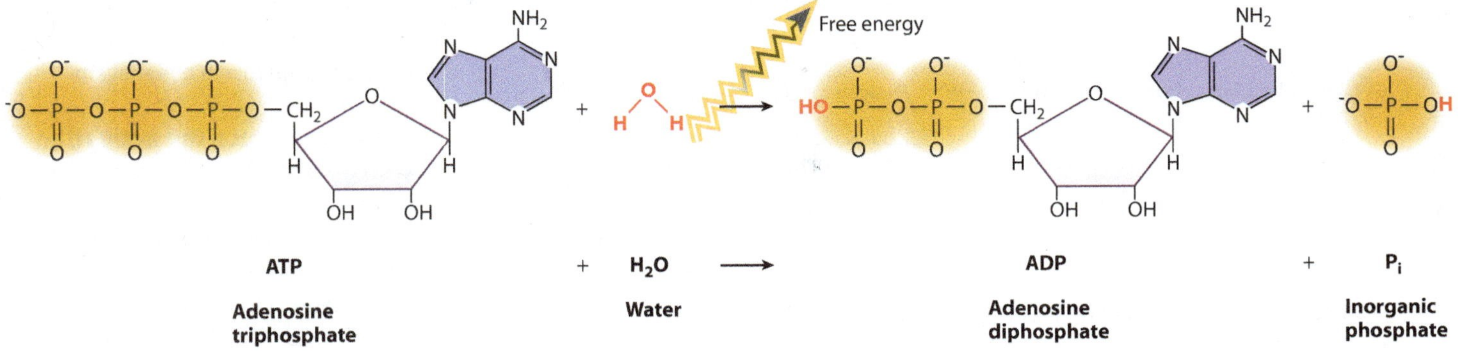

Free energy

ATP (**Adenosine triphosphate**) + **H_2O** (**Water**) ⟶ **ADP** (**Adenosine diphosphate**) + **P_i** (**Inorganic phosphate**)

Keep in mind that the release of free energy during ATP hydrolysis comes from breaking weaker bonds (with more chemical energy) in the reactants and forming more stable bonds (with less chemical energy) in the products. The release of free energy then drives chemical reactions and other processes that require a net input of energy, as we discuss next.

Non-spontaneous reactions are often coupled to spontaneous reactions.

If the conversion of reactant A into product B is spontaneous, the reverse reaction converting reactant B into product A is not. The ΔG's for the forward and reverse reactions have the same absolute value but opposite signs. You might expect that the direction of the reaction would always be from A to B. However, in living organisms, not all chemical reactions are spontaneous. Anabolic reactions are a good example; they require an input of energy to drive them in the right direction. This raises the question: What drives non-spontaneous reactions?

Energetic coupling is a process in which a spontaneous reaction (negative ΔG) drives a non-spontaneous reaction (positive ΔG). It requires that the net ΔG of the two reactions be negative. In addition, the two reactions must occur together. In some cases, this coupling can be achieved if the two reactions share an intermediate.

For example, ATP hydrolysis can be used to drive a non-spontaneous reaction, as shown in **Fig. 6.11a.** In this case, the phosphate group released during ATP hydrolysis is transferred to glucose to produce glucose 6-phosphate. In addition, the net ΔG for the two reactions is negative. So ATP hydrolysis provides the thermodynamic driving force for the non-spontaneous reaction, and the shared phosphate group couples the two reactions together.

Following ATP hydrolysis, the cell needs to replenish its ATP so that it can carry out additional chemical reactions. The synthesis of ATP from ADP and P_i is an endergonic reaction with a positive ΔG, requiring an input of energy. In some cases, exergonic reactions can drive the synthesis of ATP by energetic coupling (**Fig. 6.11b**). The sum of the ΔG's of the two reactions is negative and the reactions share a phosphate group, allowing the two reactions to proceed.

Like ATP, other phosphorylated molecules can be hydrolyzed, releasing free energy. Hydrolysis reactions can be ranked by their

FIG. 6.11 Energetic coupling. A spontaneous (exergonic) reaction drives a non-spontaneous (endergonic) reaction. (a) The hydrolysis of ATP drives the formation of glucose 6-phosphate from glucose. (b) The hydrolysis of phosphoenolpyruvate drives the synthesis of ATP.

The coupled reaction proceeds because ΔG is negative and P_i is shared between the two reactions.

a.

ATP + H_2O ⟶ ADP + P_i	$\Delta G_1 = -7.3$ kcal/mol	**Exergonic reaction**
Glucose + P_i ⟶ Glucose 6-phosphate + H_2O	$\Delta G_2 = +3.3$ kcal/mol	**Endergonic reaction**
Glucose + ATP ⟶ Glucose 6-phosphate + ADP	$\Delta G = -4$ kcal/mol	**Coupled reaction**

b.

Phosphoenolpyruvate + H_2O ⟶ Pyruvate + P_i	$\Delta G_1 = -14.8$ kcal/mol	**Exergonic reaction**
ADP + P_i ⟶ ATP + H_2O	$\Delta G_2 = +7.3$ kcal/mol	**Endergonic reaction**
Phosphoenolpyruvate + ADP ⟶ Pyruvate + ATP	$\Delta G = -7.5$ kcal/mol	**Coupled reaction**

FIG. 6.12 **ΔG of common hydrolysis reactions in a cell.** ATP hydrolysis has an intermediate value of ΔG compared to other common hydrolysis reactions.

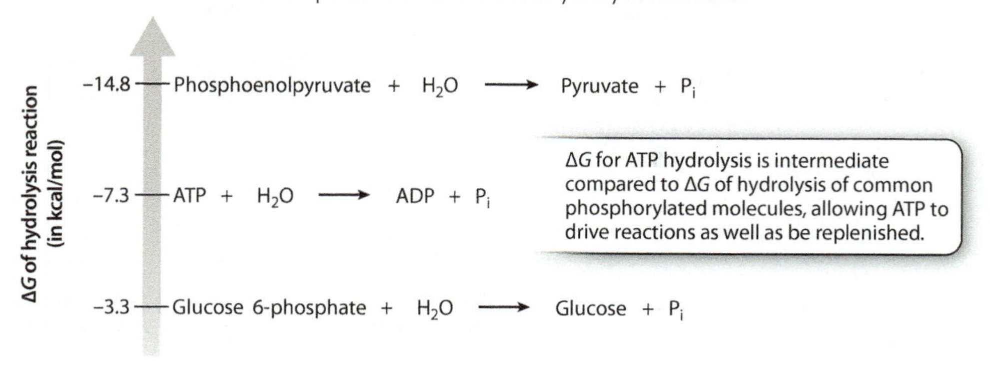

free energy differences (ΔG). **Fig. 6.12** shows that ATP hydrolysis has an intermediate free energy difference compared with the free energy difference for the hydrolysis of other common phosphorylated molecules. Those reactions that have a ΔG more negative than that of ATP hydrolysis transfer a phosphate group to ADP by energetic coupling, and those reactions that have a ΔG less negative than that of ATP hydrolysis receive a phosphate group from ATP by energetic coupling. Therefore, ADP is an energy acceptor and ATP an energy donor. The ATP–ADP system is at the core of energetic coupling between catabolic and anabolic reactions.

6.5 ENZYMES AND THE RATE OF CHEMICAL REACTIONS

Up to this point, we have focused on the spontaneity and direction of chemical reactions. Now we address their rate. As mentioned earlier, a spontaneous reaction is not necessarily a fast one. For example, the breakdown of glucose into carbon dioxide and water is spontaneous with a negative ΔG, but the rate of the reaction is close to zero and the breakdown of glucose is imperceptible. However, glucose is readily broken down inside cells all the time. How is this possible? The answer is that chemical reactions in a cell are accelerated by chemical catalysts.

The rate of a chemical reaction is defined as the amount of product formed (or reactant consumed) per unit of time. Catalysts are substances that increase the rate of chemical reactions without themselves being consumed. In biological systems, the catalysts are usually proteins called **enzymes,** although, as we saw in Chapter 2, some RNA molecules have catalytic activity as well. Enzymes can increase the rate of chemical reactions dramatically. Moreover, because they are highly specific, acting only on certain reactants and catalyzing only some reactions, enzymes play a critical role in determining which chemical reactions take place from all the possible reactions that could occur in a cell. In this section, we discuss how enzymes increase the rate of chemical reactions and how this ability gives them a central role in metabolism.

Enzymes reduce the activation energy of a chemical reaction.

Earlier, we saw that exergonic reactions release free energy and endergonic reactions require free energy. Nevertheless, all chemical reactions require an input of energy to proceed, even exergonic reactions that release energy. For an exergonic reaction, the energy released is more than the initial input of energy, so there is a net release of energy.

Why is an input of energy needed for all chemical reactions? As a chemical reaction proceeds, existing chemical bonds break and new ones form. For an extremely brief period of time, a compound is formed in which the old bonds are breaking and the new ones are forming. This intermediate stage between reactants and products is called the **transition state.** It is highly unstable and therefore has a large amount of free energy.

In all chemical reactions, reactants adopt at least one transition state before their conversion into products. The graph in **Fig. 6.13** represents the free energy levels of the reactant, transition state, and product. This reaction is spontaneous since the free energy of the reactant is higher than the free energy of the product and ΔG is negative. However, the highest free energy value corresponds to the transition state.

To reach the transition state, the reactant must absorb energy from its surroundings (the uphill portion of the curve in Fig. 6.13). As a result, all chemical reactions, even spontaneous ones that release energy, require an input of energy that we can think of as an "energy barrier." The energy input necessary to reach the transition state is called the **activation energy (E_A).** Once the transition state is reached, the reaction proceeds, products are formed, and energy is released into the surroundings (the downhill portion of the curve in Fig. 6.13). There is an inverse correlation between the rate of a reaction and the height of the energy barrier: the lower the energy barrier, the faster the reaction; the higher the barrier, the slower the reaction.

In chemical reactions that take place in the laboratory, heat is a common source of energy used to overcome the energy barrier. In living organisms, reactions are accelerated by the action of enzymes. However, enzymes do not act by supplying heat. Instead, they reduce the activation energy by stabilizing the transition state and decreasing its free energy (the red curve in Fig. 6.13). As the activation energy decreases, the speed of the reaction increases.

FIG. 6.13 An enzyme-catalyzed reaction. An enzyme accelerates a reaction by lowering the activation energy, E_A.

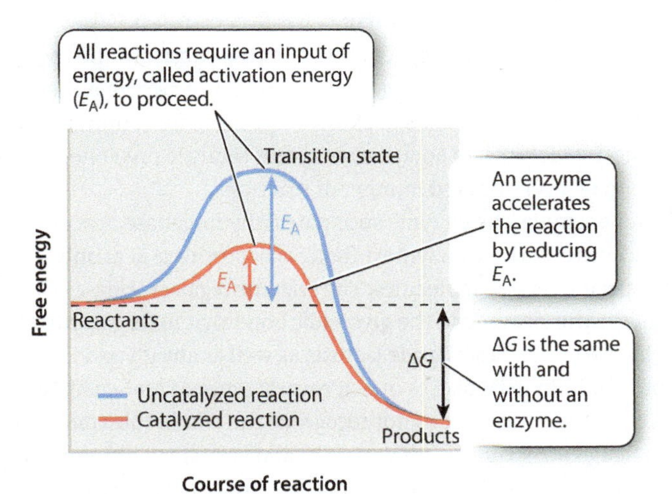

Although an enzyme accelerates a reaction by reducing the activation energy, the difference in free energy between reactants and products (ΔG) does not change. In other words, an enzyme changes the path of the reaction between reactants and products, but not the starting or end point (Fig. 6.13). Consider the breakdown of glucose into carbon dioxide and water. The ΔG of the reaction is the same whether it proceeds by combustion or by the action of multiple enzymes in a metabolic pathway in a cell.

→ **Quick Check 3** Which of the following do enzymes change? ΔG; reaction rate; types of product generated; activation energy; the laws of thermodynamics.

Enzymes form a complex with reactants and products.

An important characteristic of enzymes is that, as catalysts, they participate in a chemical reaction but are not consumed in the process. In other words, they emerge unchanged from a chemical reaction, ready to catalyze the same reaction again. How do enzymes increase the reaction rate without being consumed? The answer is that enzymes form a complex with the reactants and products.

In a chemical reaction catalyzed by an enzyme, the reactant is often referred to as the **substrate.** In such a reaction, the substrate (S) is converted to a product (P):

$$S \leftrightharpoons P$$

In the presence of an enzyme (E), the substrate first forms a complex with the enzyme (enzyme–substrate, or ES). While still part of the complex, the substrate is converted to product (enzyme–product, or EP). Finally, the complex dissociates, releasing the enzyme and product. Therefore, a reaction catalyzed by an enzyme can be described as

$$S + E \leftrightharpoons ES \leftrightharpoons EP \leftrightharpoons E + P$$

The formation of this complex is critical for accelerating the rate of a chemical reaction. Recall from Chapter 4 that proteins adopt three-dimensional shapes and that the shape of a protein is linked to its function. Enzymes are folded into three-dimensional shapes that bring particular amino acids into close proximity to form an **active site.** The active site of the enzyme is the portion of the enzyme that binds substrate and catalyzes its conversion to the product (**Fig. 6.14**). In the active site, the enzyme and substrate form transient covalent bonds and/or weak noncovalent interactions. Together, these interactions stabilize the transition state and decrease the activation energy.

Enzymes also reduce the energy of activation by positioning two substrates to react. The formation of the enzyme–substrate complex promotes the reaction between two substrates by aligning their reactive chemical groups and limiting their motion relative to each other.

The size of the active site is extremely small compared with the size of the enzyme. If only a small fraction of the enzyme is necessary for the catalysis of a reaction, why are enzymes so large? Of the many amino acids that form the active site, only a few actively contribute to catalysis. Each of these amino acids has to occupy a very specific spatial position to align with the correct reactive group on the substrate. If the few essential amino acids were part of a short peptide, the alignment of chemical groups between the peptide and the substrate would be difficult or even impossible because the length of the bonds and the bond angles in the peptide would constrain its three-dimensional structure. In fact, in many cases the catalytic amino acids are spaced far apart in the primary structure of the enzyme, but brought close

FIG. 6.14 The active site of the enzyme catalase, shown schematically and as a surface model. Substrate binds and is converted to product at the active site.

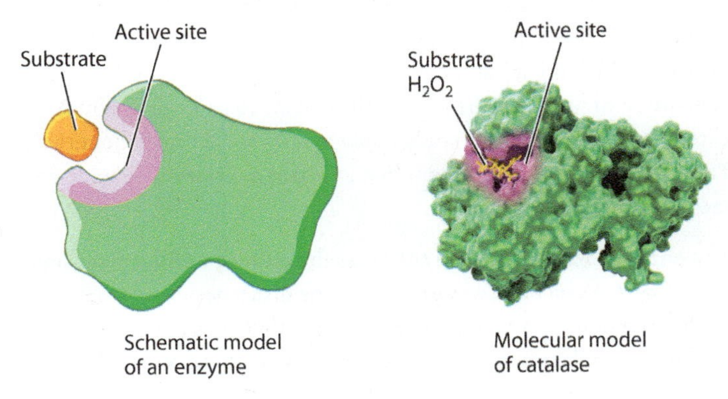

FIG. 6.15 Formation of the active site of an enzyme by protein folding.

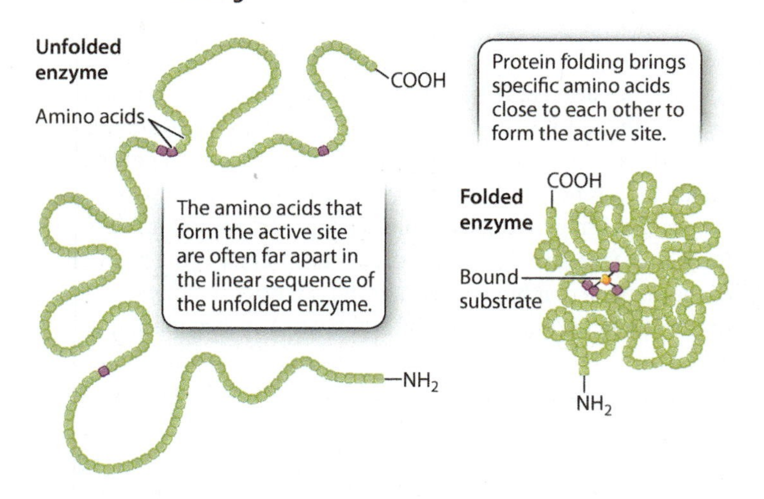

together in the formation of the active site by protein folding (**Fig. 6.15**). In other words, the large size of many enzymes is required at least in part to bring the catalytic amino acids into very specific positions in the active site of the folded enzyme.

The formation of a complex between enzymes and substrates can be demonstrated experimentally, as illustrated in **Fig. 6.16.**

Enzymes are highly specific.

Enzymes are remarkably specific both for the substrate and the reaction that is catalyzed. In general, enzymes recognize either a unique substrate or a class of substrates that share common chemical structures. In addition, enzymes catalyze only one reaction or a very limited number of reactions.

For example, the enzyme succinate dehydrogenase acts only on succinate, and the enzyme β-(beta-)galactosidase acts only on a specific class of molecules. The enzyme β-galactosidase catalyzes the cleavage of the glycosidic bond that links galactose to glucose in the disaccharide lactose, as well as any glycosidic bond that links galactose to one of several types of molecule. In this case, the enzyme does not recognize the whole substrate, but a particular structural motif within it, and very small differences in the structure of this motif affect the activity of the enzyme. For example, β-galactosidase cleaves β-galactoside but does not cleave α-(alpha-)galactoside, which differs from β-galactoside only in the orientation of the glycosidic bond. Hence, the enzyme is able to discriminate between two identical bonds with different

HOW DO WE KNOW?

FIG. 6.16

Do enzymes form complexes with substrates?

BACKGROUND The idea that enzymes form complexes with substrates to catalyze a chemical reaction was first proposed in 1888 by the Swedish chemist Svante Arrhenius. One of the earliest experiments that supported this idea was performed by American chemist Kurt Stern in the 1930s. He studied an enzyme called catalase, which is very abundant in animal and plant tissues. In the conclusion of his paper, he wrote, "It remains to be seen to which extent the findings of this study apply to enzyme action in general." Therefore, another experiment that analyzes a different enzyme and technique are described here.

The enzyme β-galactosidase catalyzes the cleavage of the glycosidic bond that links galactose to glucose in the disaccharide lactose. Lactose belongs to a family of molecules called β-galactosides since it contains a galactose unit attached to the rest of the molecule by a glycosidic bond. In a related compound called β-thiogalactoside, the oxygen atom in the glycosidic bond is replaced by sulfur. The enzyme β-galactosidase binds β-thiogalactoside but cannot cleave or release it.

CH_2OH, HO, O, H, OH, H, H, H, H, OH, O—R
β-galactoside

CH_2OH, HO, O, H, OH, H, H, H, H, OH, S—R
β-thiogalactoside

METHOD A container is separated into two compartments by a selectively permeable membrane. The membrane is permeable to β-galactoside and β-thiogalactoside, but not permeable to the enzyme.

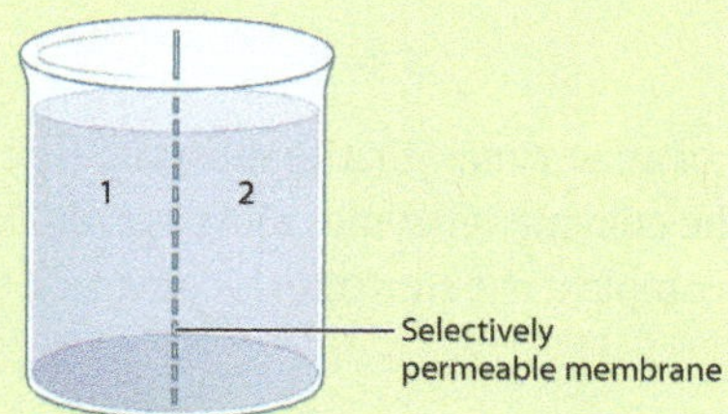

orientations within a molecule. The specificity of enzymes can be attributed to the structure of their active sites. The enzyme active site interacts only with substrates having a precise three-dimensional structure.

Enzyme activity can be influenced by inhibitors and activators.

The activity of enzymes can be influenced by **inhibitors** and **activators.** Inhibitors decrease the activity of enzymes, whereas activators increase the activity of enzymes. Enzyme inhibitors are quite common. They are synthesized naturally by many plants and animals as a defense against predators. Similarly, pesticides and herbicides often target enzymes to inactivate them. Many drugs used in medicine are enzyme inhibitors, including drugs used to treat infections as well as drugs used to treat cancer and other diseases. Given the importance of chemical reactions and the role of enzymes in metabolism, it is not surprising that enzyme inhibitors have such widespread applications.

There are two classes of inhibitors. Irreversible inhibitors usually form covalent bonds with enzymes and irreversibly inactivate them. Reversible inhibitors form weak bonds with enzymes and therefore easily dissociate from them.

Inhibitors can act in many different ways, two of which are shown in **Fig. 6.17.** In some cases, an inhibitor is similar in structure to the substrate and therefore is able to bind to the active site of the enzyme (Fig. 6.17a). Binding of the inhibitor prevents the binding of the substrate. In other words, the inhibitor competes with the substrate for the active site of the enzyme. These types of inhibitors can often be overcome by increasing the concentration of substrate. Other inhibitors bind to a site other than the active site of the enzyme, but still inhibit the activity of the enzyme (Fig. 6.17b). In this case, binding of the inhibitor changes the shape and activity of the enzyme. This type of inhibitor usually has a structure very different from that of the substrate.

Enzymes that are regulated by molecules that bind at sites other than their active sites are called **allosteric enzymes.** The activity of allosteric enzymes can be influenced by both inhibitors and activators. They play a key role in metabolic pathways, as we discuss next.

Allosteric enzymes regulate key metabolic pathways.

Enzyme activators and inhibitors are sometimes important in the normal operation of a cell—for example, to regulate a metabolic pathway. Consider the synthesis of isoleucine from threonine,

EXPERIMENT 1 Radioactively labeled β-thiogalactoside (S) is added to compartment 1 and the movement of S is followed by measuring the level of radioactivity in the two compartments.

RESULT Over time, the level of radioactivity is the same in the two compartments.

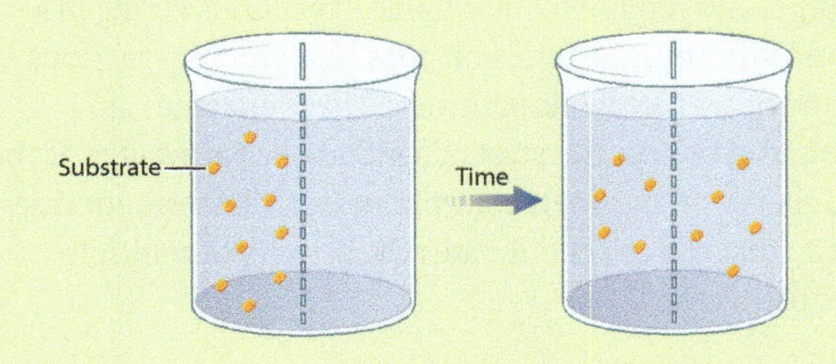

EXPERIMENT 2 Radioactively labeled β-thiogalactoside (S) is added to compartment 1, enzyme (E) is added to compartment 2, and the movement of S is followed by measuring the level of radioactivity in the two compartments.

RESULT Over time, the level of radioactivity is greater in compartment 2 than in compartment 1.

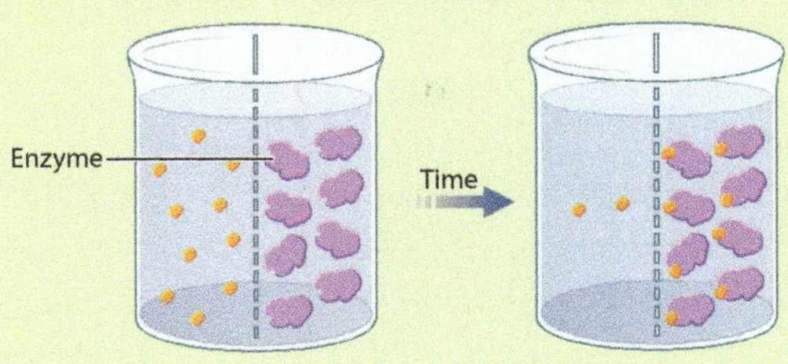

CONCLUSION These results can be explained if the substrate diffuses from compartment 1 to compartment 2, forms a complex with the enzyme, and is not released because the enzyme cannot catalyze the conversion of substrate to product. In other words, E and S form a complex.

SOURCES Adapted from Doherty, D. G., and F. Vaslow. 1952. "Thermodynamic Study of an Enzyme–Substrate Complex of Chymotrypsin." *J. Am. Chem. Soc.* 74: 931–936. Stern, K. G. 1936. "On the Mechanism of Enzyme Action: A Study of the Decomposition of Monoethyl Hydrogen Peroxide by Catalase and of an Intermediate Enzyme–Substrate Compound." *J. Biol. Chem.* 114:473–494.

FIG. 6.17 Two mechanisms of inhibitor function. (a) Some inhibitors bind to the active site of the enzyme and (b) other inhibitors bind to a site that is different from the active site. Both types of inhibitor reduce the activity of an enzyme and therefore decrease the rate of the reaction.

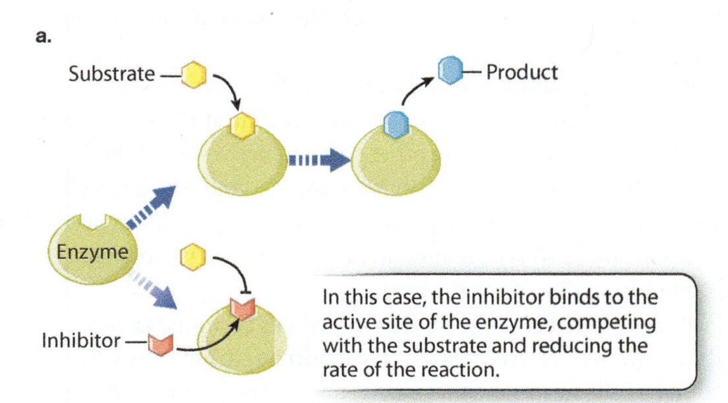

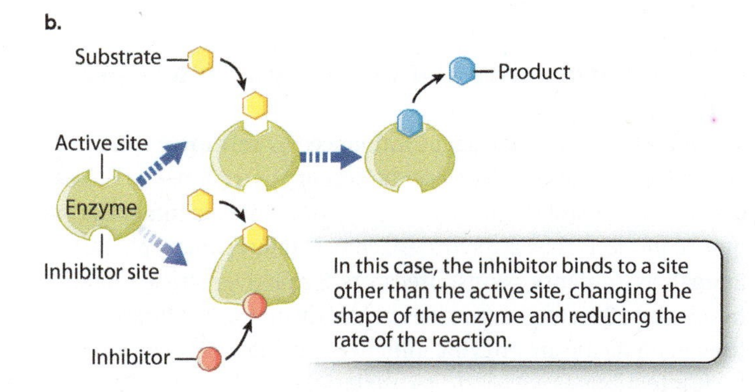

a pathway found in some bacteria. This conversion requires five reactions, each catalyzed by a different enzyme (**Fig. 6.18**).

Once the bacterium has enough isoleucine for its needs, it would be a waste of energy to continue synthesizing the amino acid. To shut down the pathway once it is no longer needed, the cell relies on an enzyme inhibitor. The inhibitor is isoleucine, the final product of the five reactions. Isoleucine binds to the first enzyme in the pathway, threonine dehydratase, at a site distinct from the active site. As a result, threonine dehydratase is an example of an allosteric enzyme. The binding of isoleucine changes the shape of the enzyme and in this way inhibits its function.

FIG. 6.18 Regulation of threonine dehdydratase, an allosteric enzyme. The enzyme is inhibited by the final product, isoleucine, which binds to a site distinct from the active site.

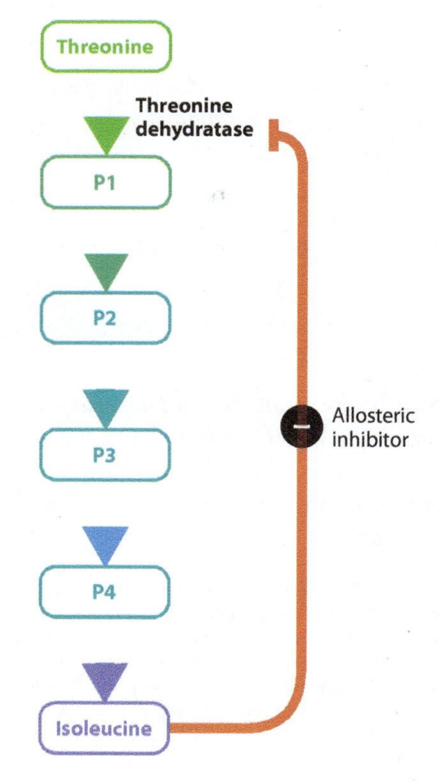

The isoleucine pathway also provides an example of **negative feedback,** in which the final product inhibits the first step of the reaction. This is a common mechanism used widely in organisms to maintain homeostasis, that is, the active maintenance of stable conditions or steady levels of a substance (Chapter 5).

Threonine dehydratase can also adjust the rate of the reaction, depending on the concentration of substrate. At a low concentration of threonine, the rate of the reaction is very slow. As the concentration of threonine increases, the activity of the enzyme increases. At a particular threshold, a small increase in threonine concentration results in a large increase in reaction rate. Finally, when there is excess substrate, the reaction rate slows down.

In the next two chapters, we examine more closely two key metabolic processes, cellular respiration and photosynthesis. Both of these processes require many chemical reactions acting in a coordinated fashion. Allosteric enzymes catalyze key reactions in these and other metabolic pathways. These enzymes are usually found at or near the start of a metabolic pathway or at the crossroads between two metabolic pathways. Allosteric enzymes are one way that the cell coordinates the activity of multiple metabolic pathways.

? CASE 1 THE FIRST CELL: LIFE'S ORIGINS

What naturally occurring elements might have spurred the first reactions that led to life?

Many enzymes are not just made up of protein, but also contain metal ions. These ions are one type of **cofactor,** a substance that associates with an enzyme and plays a key role in its function.

Metallic cofactors, especially iron, magnesium, manganese, cobalt, copper, zinc, and molybdenum, bind to diverse proteins, including enzymes used in DNA synthesis and nitrogen metabolism. With this in mind, scientists have asked whether metal ions might, by themselves, catalyze chemical reactions thought to have played a role in the origin of life. They do. For example, magnesium and zinc ions added to solutions can accelerate the linking of nucleotides to form RNA and DNA molecules.

Metallic cofactors also bind to enzymes used in the transport of electrons for cellular respiration and photosynthesis, processes that are discussed in the next two chapters. Enzymes that contain iron and sulfur clustered together are particularly important in the transport of electrons within cells. Iron–sulfur minerals, especially pyrite (or fool's gold, FeS_2), form commonly in mid-ocean hydrothermal vent systems and other environments where oxygen is absent. It has been proposed that reactions now carried out in cells by iron–sulfur proteins are the evolutionary descendants of chemical reactions that took place spontaneously on the early Earth.

The idea that your cells preserve an evolutionary memory of ancient hydrothermal environments may seem like science fiction, but it finds support in laboratory experiments. For example, the reaction of H_2S and FeS to form pyrite has been shown to catalyze a number of plausibly pre-biotic chemical reactions, including the formation of pyruvate (a key intermediate in energy metabolism discussed in Chapter 7). Thus, the metals in enzymes help connect the chemistry of life to the chemistry of Earth. ■

Core Concepts Summary

6.1 Metabolism is the set of biochemical reactions that transforms biomolecules and transfers energy.

Organisms can be grouped according to their source of energy: Phototrophs obtain energy from sunlight and chemotrophs obtain energy from chemical compounds. page 116

Organisms can also be grouped according to the source of carbon they use to build organic molecules: Heterotrophs obtain carbon from organic molecules, and autotrophs obtain carbon from inorganic sources, such as carbon dioxide. page 116

Catabolism is the set of reactions that break down molecules and release energy, and anabolism is the set of reactions that build molecules and require energy. page 117

6.2 Kinetic energy is energy of motion and potential energy is stored energy.

Kinetic energy is due to motion. page 118

Potential energy depends on the structure of an object or its position relative to its surroundings. page 118

Chemical energy is a form of potential energy held in the bonds of molecules. page 118

6.3 The laws of thermodynamics govern energy flow in biological systems.

The first law of thermodynamics states that energy cannot be created or destroyed. page 119

The second law of thermodynamics states that there is an increase in entropy in the universe over time. page 120

6.4 Chemical reactions involve the breaking and forming of bonds.

In a chemical reaction, atoms themselves do not change, but which atoms are linked to each other changes, forming new molecules. page 121

The direction of a chemical reaction is influenced by the concentration of reactants and products. page 121

Gibbs free energy (G) is the amount of energy available to do work. page 121

Three thermodynamic parameters define a chemical reaction: Gibbs free energy (G), enthalpy (H), and entropy (S). page 121

Exergonic reactions are spontaneous ($\Delta G < 0$) and release energy. page 121

Endergonic reactions are non-spontaneous ($\Delta G > 0$) and require energy. page 121

The change of free energy in a chemical reaction is described by $\Delta G = \Delta H - T\Delta S$. page 122

The hydrolysis of ATP is an exergonic reaction that drives many endergonic reactions in a cell. page 122

In living systems, non-spontaneous reactions are often coupled to spontaneous ones. page 123

6.5 The rate of biochemical reactions is increased by protein catalysts called enzymes.

Enzymes reduce the free energy level of the transition state between reactants and products, thereby reducing the energy input, or activation energy, required for a chemical reaction to proceed. page 124

During catalysis, the substrate and product form a complex with the enzyme. Transient covalent bonds and/or weak noncovalent interactions stabilize the complex. page 125

The size of the active site of an enzyme is small compared to the size of the enzyme as a whole and the active site amino acids occupy a very specific spatial arrangement. page 125

An enzyme is highly specific for its substrate and for the types of reaction it catalyzes. page 126

Inhibitors reduce the activity of enzymes and can act irreversibly or reversibly. page 127

Activators increase the activity of enzymes. page 127

Allosteric enzymes bind activators and inhibitors at sites other than the active site, resulting in a change in their shape and activity. page 127

Allosteric enzymes are often found at or near the start of a metabolic pathway or at the crossroads of multiple pathways. page 128

Self-Assessment

1. Name and describe ways that organisms obtain energy and carbon from the environment.
2. Distinguish between catabolism and anabolism.
3. Name and describe the two forms of energy and provide an example of each.
4. Explain the relationship between strength of a covalent bond and the amount of chemical energy it contains.
5. Draw the structure of ATP, indicating the bonds that are broken during hydrolysis.
6. Describe the first and second laws of thermodynamics and how they relate to chemical reactions.
7. If the difference between the enthalpy of the products and that of the reactants is positive and the difference between the entropy of the products and reactants is negative, predict whether the reaction is spontaneous or not.
8. Describe how the hydrolysis of ATP can drive non-spontaneous reactions in a cell.
9. Give three characteristics of enzymes and describe how they permit chemical reactions to occur in cells.
10. Explain how protein folding allows for enzyme specificity.

Log in to LaunchPad to check your answers to the Self-Assessment questions, and to access additional learning tools.

Image Source/Getty Images.

CHAPTER 7

Cellular Respiration

Harvesting Energy from Carbohydrates and Other Fuel Molecules

Core Concepts

7.1 Cellular respiration is a series of catabolic reactions that convert the energy in fuel molecules into ATP.

7.2 Glycolysis is the partial oxidation of glucose and results in the production of pyruvate, as well as ATP and reduced electron carriers.

7.3 Pyruvate is oxidized to acetyl-CoA, connecting glycolysis to the citric acid cycle.

7.4 The citric acid cycle results in the complete oxidation of fuel molecules and the generation of ATP and reduced electron carriers.

7.5 The electron transport chain transfers electrons from electron carriers to oxygen, using the energy released to pump protons and synthesize ATP by oxidative phosphorylation.

7.6 Glucose can be broken down in the absence of oxygen by fermentation, producing a modest amount of ATP.

7.7 Metabolic pathways are integrated, allowing control of the energy level of cells.

The ability to harness energy from the environment is a key attribute of life. We have seen that energy is needed for all kinds of tasks—among them cell movement and division, muscle contraction, growth and development, and the synthesis of macromolecules. Organic molecules such as carbohydrates, lipids, and proteins are good sources of chemical energy. Some organisms, like humans and other heterotrophs, obtain organic molecules by consuming them in their diet. Others, like plants and other autotrophs, synthesize these molecules from inorganic molecules like carbon dioxide, as we saw in Chapter 6. Regardless of how organic molecules are obtained, nearly all organisms—animals, plants, fungi, and microbes—break them down in the process of **cellular respiration,** releasing energy that can be used to do the work of the cell. Cellular respiration is a series of chemical reactions that convert the chemical energy in fuel molecules into the chemical energy of **adenosine triphosphate (ATP)**, which can be readily used by cells.

It is tempting to think that organic molecules are converted into energy in this process, but this is not the case. Recall from Chapter 6 that the first law of thermodynamics (the law of conservation of energy) states that energy cannot be created or destroyed. Biological processes, like all processes, are subject to the laws of thermodynamics. As a result, the process of cellular respiration converts the chemical potential energy stored in organic molecules to chemical potential energy that is useful to cells: the chemical potential energy in ATP. ATP is the universal energy currency for all cells.

It is also easy to forget that organisms other than animals, such as plants, use cellular respiration. If plants use sunlight as a source of energy, why would they need cellular respiration? As we will see in the next chapter, plants use the energy of sunlight to make carbohydrates. Plants then break down these carbohydrates in the process of cellular respiration to produce ATP.

In this chapter, we discuss the metabolic pathways that supply the energy needs of a cell: the breakdown, storage, and mobilization of sugars such as glucose, the synthesis of ATP, and the coordination and regulation of these metabolic pathways.

7.1 AN OVERVIEW OF CELLULAR RESPIRATION

In the last chapter, we saw that catabolism describes the set of chemical reactions that break down molecules into smaller units. In the process, these reactions release chemical energy that can be stored in molecules of ATP. Anabolism, by contrast, is the set of chemical reactions that build molecules from smaller units. Anabolic reactions require an input of energy, usually in the form of ATP.

Cellular respiration is one of the major sets of catabolic reactions in a cell. During cellular respiration, fuel molecules such as glucose, fatty acids, and proteins are catabolized into smaller units, releasing the energy stored in their chemical bonds to power the work of the cell.

Cellular respiration uses chemical energy stored in molecules such as carbohydrates and lipids to produce ATP.

Cellular respiration is a series of catabolic reactions that converts the energy stored in food molecules, such as glucose, into the energy stored in ATP, and produces carbon dioxide as a waste or by-product. It can occur in the presence of oxygen (termed aerobic respiration) or in the absence of oxygen (termed anaerobic respiration). Most organisms that you are familiar with are capable of aerobic respiration; some bacteria respire anaerobically (Chapter 26). Here we focus on aerobic respiration. Oxygen is consumed in aerobic respiration, and carbon dioxide and water are produced, as follows:

$$\underset{\text{Glucose}}{C_6H_{12}O_6} + \underset{\text{Oxygen}}{6O_2} \rightarrow \underset{\text{Carbon dioxide}}{6CO_2} + \underset{\text{Water}}{6H_2O} + \text{energy}$$

In Chapter 6, we saw that molecules such as carbohydrates and lipids have a large amount of potential energy in their chemical bonds. In contrast, molecules like carbon dioxide and water have less potential energy in their bonds. Cellular respiration releases a large amount of energy because the sum of the potential energy in all of the chemical bonds of the reactants (glucose and oxygen) is higher than that of the products (carbon dioxide and water). The maximum amount of free energy—energy available to do work—released during cellular respiration is –686 kcal per mole of glucose. Recall from Chapter 6 that when $\Delta G < 0$, energy is released.

The overall reaction for cellular respiration helps us focus on the starting reactants, final products, and release of energy. However, it misses the many intermediate steps that take place as the cell catabolizes glucose. Tossing a match into the gas tank of a car releases a tremendous amount of energy in the form of an explosion, but this energy is not used to do work. Instead, it is released as light and heat. Similarly, if all the energy stored in glucose were released at once, most of it would be released as heat and the cell would not be able to harness it to do work.

In cellular respiration, energy is released gradually in a series of chemical reactions (**Fig. 7.1**). This allows some of this energy to be used to form ATP. On average, 32 molecules of ATP are produced from the aerobic respiration of a single molecule of glucose. The energy needed (ΔG of the reaction) to form one mole of ATP from ADP and P_i is at least 7.3 kcal. Thus, cellular respiration harnesses at least $32 \times 7.3 = 233.6$ kcal of energy in ATP for every mole of glucose that is broken down in the presence of oxygen.

About 34% of the total energy released by aerobic respiration is harnessed in the form of ATP ($233.6/686 = 34\%$), with the remainder of the energy given off as heat. This degree of efficiency

FIG. 7.1 An overview of cellular respiration. In most organisms, cellular respiration consumes oxygen and produces ATP by substrate-level phosphorylation and oxidative phosphorylation, as well as carbon dioxide and water.

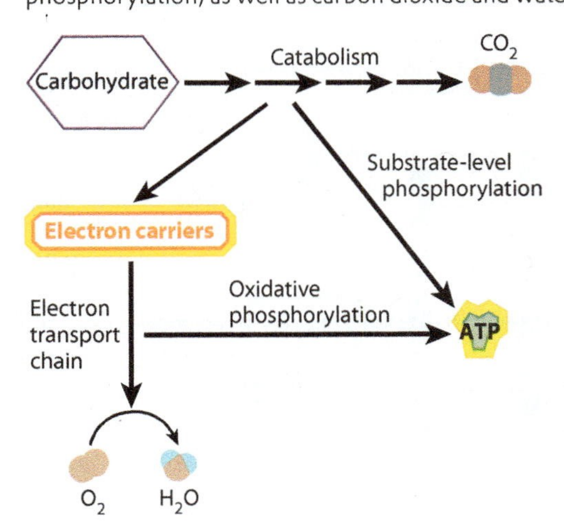

compares favorably with that of a gasoline engine, which operates with an efficiency of about 25%.

ATP is generated by substrate-level phosphorylation and oxidative phosphorylation.

In cellular respiration, the chemical energy stored in a molecule of glucose is used to produce ATP in two different ways (Fig. 7.1). In the first, a phosphorylated organic molecule directly transfers a phosphate group to ADP, as we saw in Chapter 6. In this case, there are two coupled reactions carried out by a single enzyme: the hydrolysis of a phosphorylated organic molecule and the addition of a phosphate group to ADP. The hydrolysis reaction releases enough free energy to drive the synthesis of ATP. This way of generating ATP is called **substrate-level phosphorylation** because a phosphate group is transferred to ADP from an enzyme substrate, in this case an organic molecule.

Substrate-level phosphorylation produces only a small amount of the total ATP generated in cellular respiration, about 12% if the fuel molecule is glucose. Most of the ATP produced during cellular respiration (the remaining 88%) is produced in a different way (Fig. 7.1). In this case, the chemical energy of organic molecules is transferred first to **electron carriers.** The role of these electron carriers is exactly what their name suggests—they carry electrons (and energy) from one set of reactions to another. In cellular respiration, electron carriers transport electrons released during the catabolism of organic molecules to the respiratory **electron transport chain.** Electron transport chains in turn transfer electrons along a series of membrane-associated proteins to a final electron acceptor and in the process harness the energy released to produce ATP. In aerobic respiration, oxygen is the final electron acceptor, resulting in the formation of water. This way of generating ATP is called **oxidative phosphorylation.**

How electron movement through the electron transport chain is coupled to ATP synthesis is central to the energy economy of most cells and is discussed later in the chapter. Electron transport chains are used in respiration to harness energy from fuel molecules such as glucose and in photosynthesis to harness energy from sunlight (Chapter 8).

Redox reactions play a central role in cellular respiration.

Chemical reactions in which electrons are transferred from one atom or molecule to another are referred to as **oxidation–reduction reactions** ("redox reactions" for short). Oxidation is the loss of electrons, and reduction is the gain of electrons. The loss and gain of electrons always occur together in a coupled oxidation–reduction reaction: Electrons are transferred from one molecule to another so that one molecule loses electrons and one molecule gains those electrons. The molecule that loses electrons is oxidized and the molecule that gains electrons is reduced.

We can illustrate oxidation–reduction reactions by looking at the role played by electron carriers in cellular respiration. Two important electron carriers are nicotinamide adenine dinucleotide and flavin adenine dinucleotide. These electron carriers exist in two forms—an oxidized form (NAD^+ and FAD) and a reduced form (NADH and $FADH_2$). When fuel molecules such as glucose are catabolized, some of the steps are oxidation reactions. These oxidation reactions are coupled with the reduction of electron carrier molecules. The reduction reactions in which electrons are transferred to electron carriers can be written as:

$$NAD^+ + 2e^- + H^+ \rightarrow NADH$$
$$FAD + 2e^- + 2H^+ \rightarrow FADH_2$$

Here, NAD^+ and FAD accept electrons, resulting in the production of their reduced forms, NADH and $FADH_2$. Note that in redox reactions involving organic molecules such as NAD^+ or FAD, the gain (or loss) of electrons is often accompanied by the gain (or loss) of protons (H^+). As a result, reduced molecules can easily be recognized by an increase in C–H bonds, and the corresponding oxidized molecules by a decrease in C–H bonds.

In their reduced forms, NADH and $FADH_2$ can donate electrons. The oxidation of NADH and $FADH_2$ allows electrons (and energy) to be transferred to the electron transport chain:

$$NADH \rightarrow NAD^+ + 2e^- + H^+$$
$$FADH_2 \rightarrow FAD + 2e^- + 2H^+$$

In addition, these reactions produce NAD^+ and FAD, which can then accept electrons from the breakdown of fuel molecules. In this way, electron carriers act as shuttles, transferring electrons derived from the oxidation of fuel molecules such as glucose to the electron transport chain.

Cellular respiration can itself be understood as a redox reaction, even though it consists of many steps. In aerobic

FIG. 7.2 Oxidation and reduction in the overall reaction for cellular respiration. Electrons are partially lost or gained in the formation of polar covalent bonds.

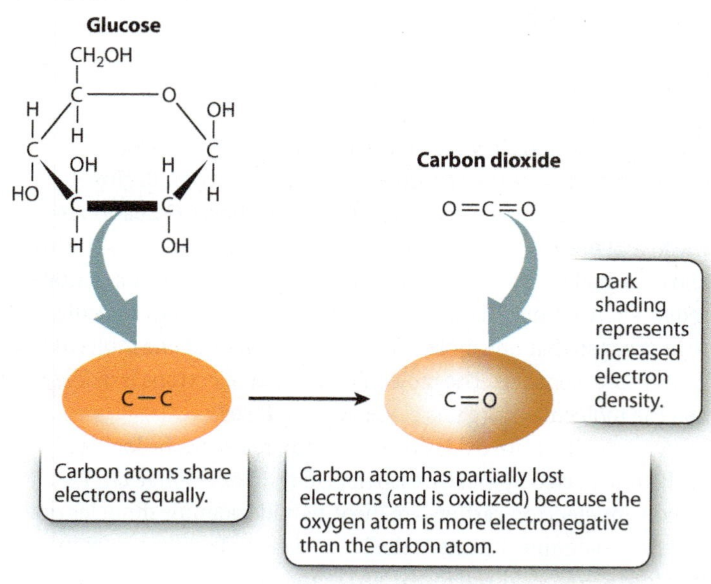

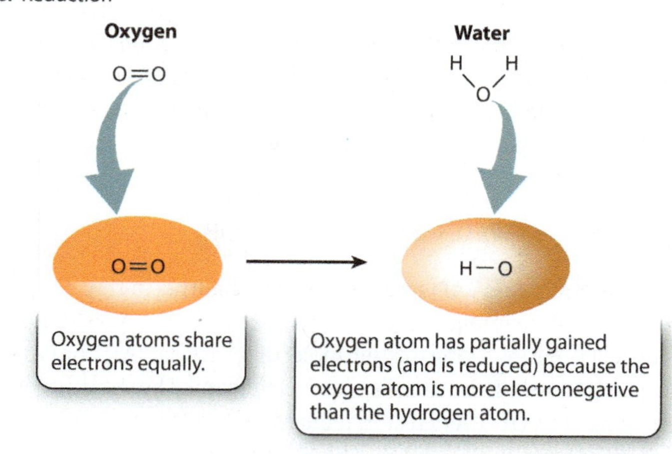

FIG. 7.3 The four stages of cellular respiration. Cellular respiration consists of glycolysis, pyruvate oxidation, the citric acid cycle, and oxidative phosphorylation.

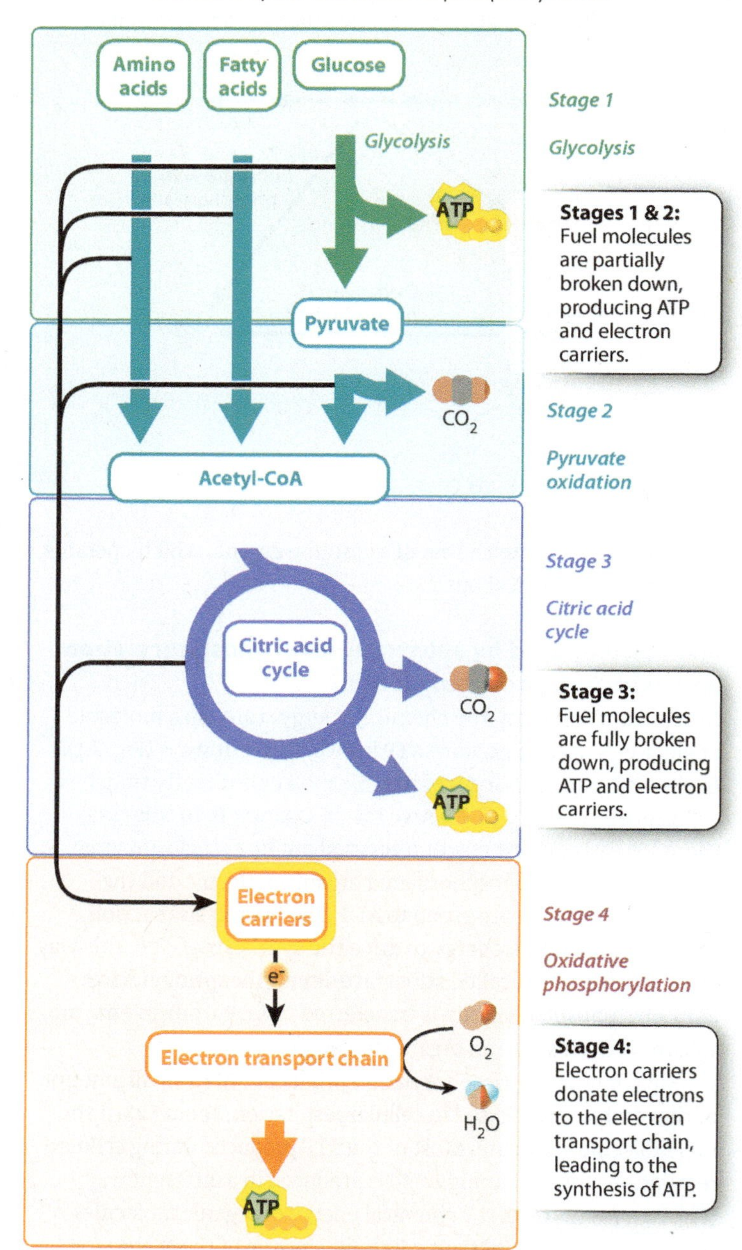

respiration, glucose is oxidized, releasing carbon dioxide, and at the same time oxygen is reduced, forming water:

$$C_6H_{12}O_6 + 6O_2 \rightarrow 6CO_2 + 6H_2O + \text{energy}$$

(Oxidation: $C_6H_{12}O_6 \rightarrow 6CO_2$; Reduction: $6O_2 \rightarrow 6H_2O$)

Let's first consider the oxidation reaction (**Fig. 7.2a**). In glucose, there are many C–C and C–H covalent bonds, in which electrons are shared about equally between the two atoms. By contrast, in carbon dioxide, electrons are not shared equally. The oxygen atom is more electronegative than the carbon atom, so the electrons are more likely to be found near the oxygen atom. As a result, carbon has partially lost electrons to oxygen and is oxidized.

Let's now consider the reduction reaction (**Fig. 7.2b**). In oxygen gas, electrons are shared equally between two oxygen atoms. In water, the electrons that are shared between hydrogen and oxygen are more likely to be found near oxygen because oxygen is more electronegative than hydrogen. As a result, oxygen has partially gained electrons and is reduced.

Glucose is a good electron donor because its oxidation to carbon dioxide releases a lot of energy, whereas oxygen is a good electron acceptor because it has a high affinity for electrons. In cellular respiration, glucose is not oxidized all at once to carbon dioxide. Instead, it is oxidized slowly in a series of reactions, some of which are redox reactions. In this way, energy is released in a controlled manner. Some of this energy can be used to synthesize ATP directly and some of it is stored temporarily in reduced electron carriers and then used to generate ATP by oxidative phosphorylation.

→ **Quick Check 1** For each of the following pairs of molecules, indicate which member of the pair is reduced and which is oxidized, and which has more chemical energy and which has less chemical energy: NAD^+/NADH; FAD/$FADH_2$; CO_2/$C_6H_{12}O_6$.

Cellular respiration occurs in four stages.

Up to this point, we have focused on the overall reaction of cellular respiration and the chemistry of redox reactions. Let's now look at the entire process, starting with glucose. Cellular respiration occurs in four stages (**Fig. 7.3**).

In stage 1, glucose is partially broken down to make pyruvate and energy is transferred to ATP and reduced electron carriers, a process known as **glycolysis.**

In stage 2, pyruvate is oxidized to another molecule called acetyl-coenzyme A (acetyl-CoA), producing reduced electron carriers and releasing carbon dioxide.

Acetyl-CoA enters stage 3, the **citric acid cycle,** also called the tricarboxylic (TCA) cycle or the Krebs cycle. In this series of chemical reactions, the acetyl group is completely oxidized to carbon dioxide and energy is transferred to ATP and reduced electron carriers. The amount of energy transferred to ATP and reduced electron carriers in this stage is nearly twice that of stages 1 and 2 combined.

Stage 4 is **oxidative phosphorylation.** In this series of reactions, reduced electron carriers generated in stages 1–3 donate electrons to the electron transport chain and a large amount of ATP is produced.

In eukaryotes, glycolysis takes place in the cytoplasm, and pyruvate oxidation, the citric acid cycle, and oxidative phosphorylation all take place in mitochondria. The electron transport chain is made up of proteins and small molecules associated with the inner mitochondrial membrane (Chapter 5).

FIG. 7.4 Change in free energy in cellular respiration. Glucose is oxidized through a series of chemical reactions, releasing energy in the form of ATP and reduced electron carriers.

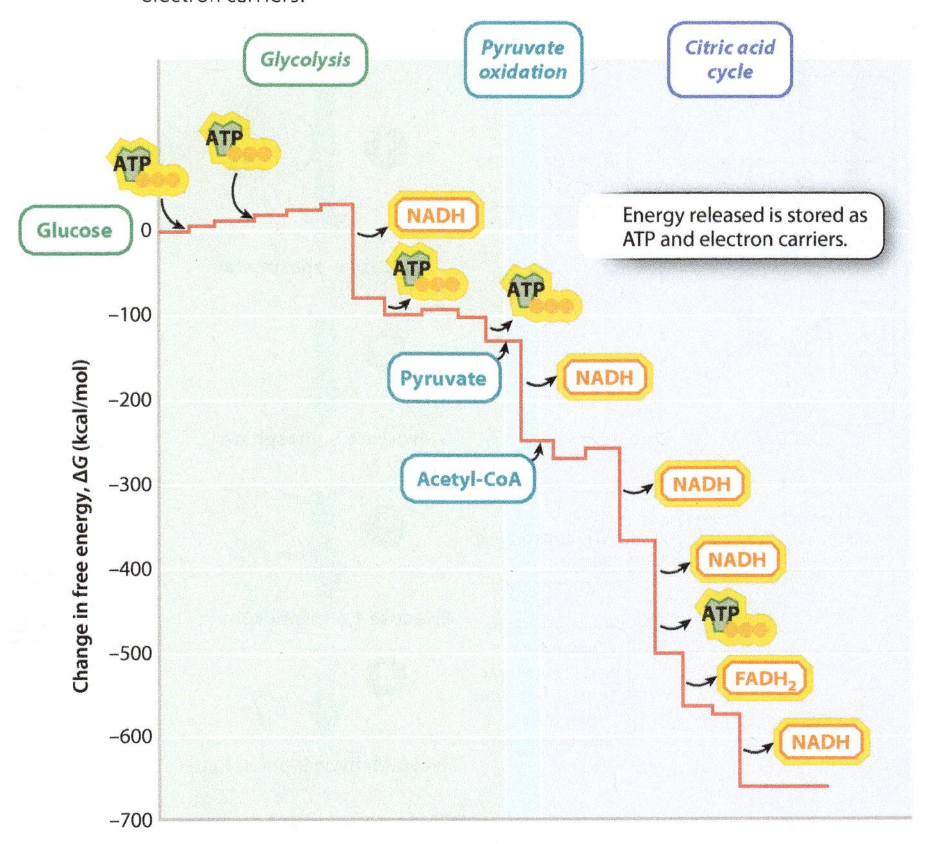

In some bacteria, these reactions take place in the cytoplasm, and the electron transport chain is located in the plasma membrane.

Fig. 7.4 shows the change of free energy at each step in catabolism of glucose. As you can see, the individual reactions allow the initial chemical energy present in a molecule of glucose to be "packaged" into molecules of ATP and reduced electron carriers. Note that the change in free energy is much greater for the steps that generate reduced electron carriers compared to those that produce ATP directly.

7.2 GLYCOLYSIS: THE SPLITTING OF SUGAR

Glucose is the most common fuel molecule in animals, plants, and microbes. It is the starting molecule for glycolysis, which results in the partial oxidation of glucose and the synthesis of a relatively small amount of both ATP and reduced electron carriers. Glycolysis literally means "splitting sugar," an apt name because in glycolysis a 6-carbon sugar (glucose) is split in two, yielding two 3-carbon molecules. The process is anaerobic because oxygen is

FIG. 7.5 Glycolysis. Glucose is partially oxidized to pyruvate, with the net production of 2 ATP and 2 NADH.

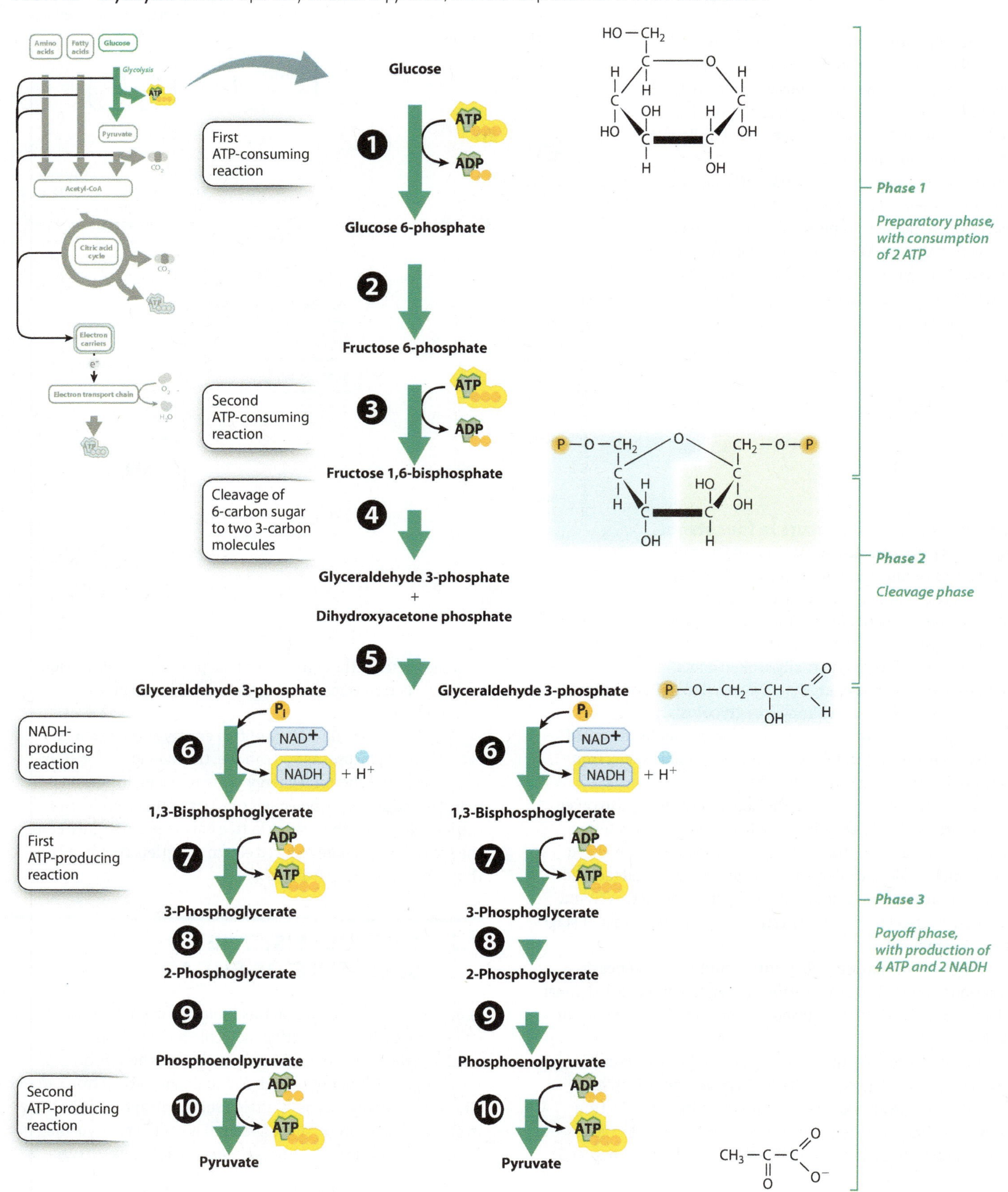

not consumed. Glycolysis evolved very early in the evolution of life, when oxygen was not present in Earth's atmosphere. It occurs in nearly all living organisms and is therefore probably the most widespread metabolic pathway among organisms.

Glycolysis is the partial breakdown of glucose.

Glycolysis begins with a molecule of glucose and produces two 3-carbon molecules of pyruvate and a net total of two molecules of ATP and two molecules of the electron carrier NADH. ATP is produced directly by substrate-level phosphorylation.

Glycolysis is a series of 10 chemical reactions (**Fig. 7.5**). These reactions can be divided into three phases. The first phase prepares glucose for the next two phases by the addition of two phosphate groups to glucose. This phase requires an input of energy. To supply that energy and provide the phosphate groups, two molecules of ATP are hydrolyzed per molecule of glucose. In other words, the first phase of glycolysis is an endergonic process. The phosphorylation of glucose has two important consequences. Whereas glucose enters and exits cells through specific membrane transporters, phosphorylated glucose is trapped inside the cell. In addition, the presence of two negatively charged phosphate groups in proximity destabilizes the molecule so that it can be broken apart in the second phase of glycolysis.

The second phase is the cleavage phase, in which the 6-carbon molecule is split into two 3-carbon molecules. For each molecule of glucose entering glycolysis, two 3-carbon molecules enter the third phase of glycolysis.

The third and final phase of glycolysis is sometimes called the payoff phase because ATP and the electron carrier NADH are produced. Later, NADH will contribute to the synthesis of ATP during oxidative phosphorylation. This phase ends with the production of two molecules of pyruvate.

In summary, glycolysis begins with a single molecule of glucose (six carbons) and produces two molecules of pyruvate (three carbons each). These reactions yield four molecules of ATP and two molecules of NADH. However, two ATP molecules are consumed during the initial phase of glycolysis, resulting in a net gain of two ATP molecules and two molecules of NADH (Fig. 7.5).

→ **Quick Check 2** At the end of glycolysis, but before the subsequent steps in cellular respiration, which molecules contain some of the energy held in the original glucose molecule?

7.3 PYRUVATE OXIDATION

Glycolysis occurs in almost all living organisms, but it does not generate very much energy in the form of ATP. The end product, pyruvate, still contains a good deal of chemical potential energy in its bonds. In the presence of oxygen, pyruvate can be further oxidized to release more energy, first to acetyl-CoA and then even further in the series of reactions in the citric acid cycle. Pyruvate oxidation is a key step that links glycolysis to the citric acid cycle. In eukaryotes, this is the first step that takes place inside the mitochondria.

FIG. 7.6 Mitochondrial membranes and compartments.

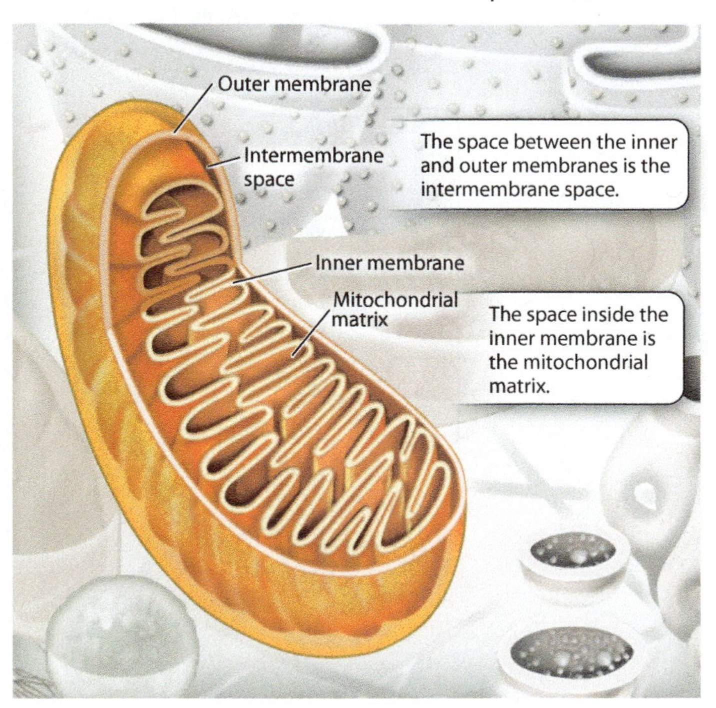

The oxidation of pyruvate connects glycolysis to the citric acid cycle.

The end product of glycolysis is pyruvate, which can be transported into mitochondria. Mitochondria are rod-shaped organelles surrounded by a double membrane (**Fig. 7.6**; Chapter 5). The inner and outer mitochondrial membranes are not close to each other in all areas because the inner membrane has folds that project inward. These membranes define two spaces. The space between the inner and outer membranes is called the **intermembrane space,** and the space enclosed by the inner membrane is called the **mitochondrial matrix.**

Pyruvate is transported into the mitochondrial matrix, where it is converted into acetyl-CoA (**Fig. 7.7**). First, part of the pyruvate molecule is oxidized and splits off to form carbon dioxide, the most oxidized (and therefore the least energetic) form of carbon. The electrons lost in this process are donated to NAD^+, which is reduced to NADH. The remaining part of the pyruvate molecule—an acetyl group ($COCH_3$)—still contains a large amount of potential energy that can be harnessed. It is transferred to coenzyme A (CoA), a molecule that carries the acetyl group to the next set of reactions.

Overall, the synthesis of one molecule of acetyl-CoA from pyruvate results in the formation of one molecule of carbon

FIG. 7.7 Pyruvate oxidation. Pyruvate is oxidized in the mitochondrial matrix, forming acetyl-CoA, the first substrate in the citric acid cycle.

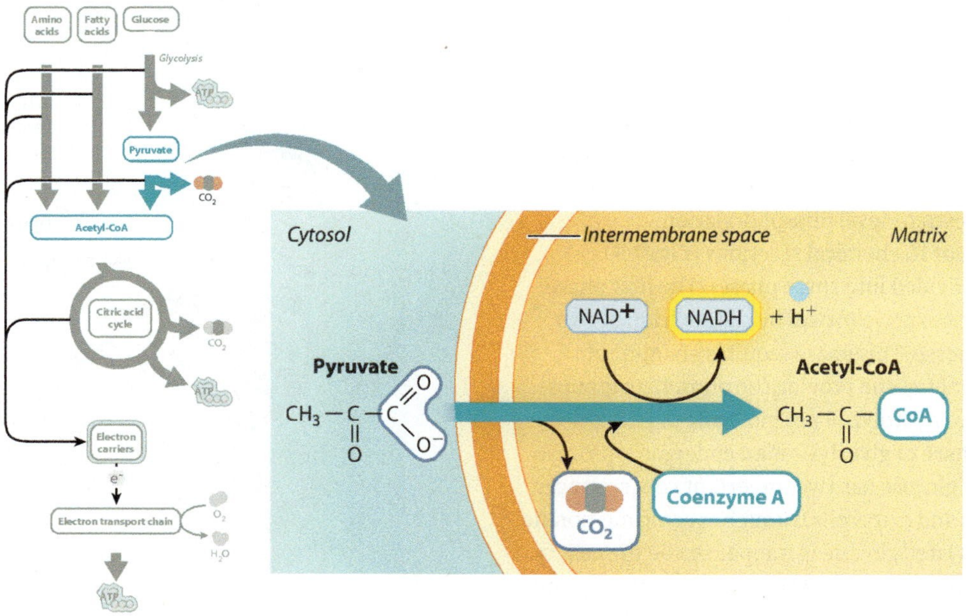

dioxide and one molecule of NADH. Recall, however, that a single molecule of glucose forms two molecules of pyruvate during glycolysis. Therefore, two molecules of carbon dioxide, two molecules of NADH, and two molecules of acetyl-CoA are produced from a single starting glucose molecule in this stage of cellular respiration. Acetyl-CoA is the substrate of the first step in the citric acid cycle.

7.4 THE CITRIC ACID CYCLE

The citric acid cycle is the stage in cellular respiration in which fuel molecules are completely oxidized. Specifically, the acetyl group of acetyl-CoA is completely oxidized to carbon dioxide and the chemical energy is transferred to ATP by substrate-level phosphorylation and to the reduced electron carriers NADH and $FADH_2$. In this way, the citric acid cycle supplies electrons to the electron transport chain, leading to the production of much more energy in the form of ATP than is obtained by glycolysis alone.

The citric acid cycle produces ATP and reduced electron carriers.

Like the synthesis of acetyl-CoA, the citric acid cycle takes place in the mitochondrial matrix. It is composed of eight reactions and is called a cycle because the starting molecule, oxaloacetate, is regenerated at the end (**Fig. 7.8**).

In the first reaction, the 2-carbon acetyl group of acetyl-CoA is transferred to a 4-carbon molecule of oxaloacetate to form the 6-carbon molecule citric acid or tricarboxylic acid (hence the variant names "citric acid cycle" and "tricarboxylic acid cycle"). The molecule of citric acid is then oxidized in a series of reactions. The last reaction of the cycle regenerates a molecule of oxaloacetate, joining to a new acetyl group and allowing the cycle to continue.

The citric acid cycle results in the complete oxidation of the acetyl group of acetyl-CoA. Since the first reaction creates a molecule with six carbons and the last reaction regenerates a 4-carbon molecule, two carbons are eliminated during the cycle. These carbons are released as carbon dioxide. Along with the release of carbon dioxide from pyruvate during pyruvate oxidation, these reactions are the sources of carbon dioxide released during cellular respiration and therefore the sources of the carbon dioxide that we exhale when we breathe.

The oxidation reactions that produce carbon dioxide are coupled with the reduction of the electron carrier NAD^+ to NADH (Fig. 7.8). In this way, energy released in the oxidation reactions is transferred to NADH. More reduced electron carriers (NADH and $FADH_2$) are produced in two additional redox reactions. In fact, the citric acid cycle produces a large quantity of reduced electron carriers: three molecules of NADH and one molecule of $FADH_2$ per turn of the cycle. These electron carriers donate electrons to the electron transport chain, which leads to the production of ATP by oxidative phosphorylation.

One of the reactions of the citric acid cycle is a substrate-level phosphorylation reaction that generates a molecule of GTP (Fig. 7.8). GTP can transfer its terminal phosphate to a molecule of ADP to form ATP. This is the only substrate-level phosphorylation in the citric acid cycle.

FIG. 7.8 The citric acid cycle. The acetyl group of acetyl-CoA is completely oxidized, with the net production of one ATP, three NADH, and one $FADH_2$.

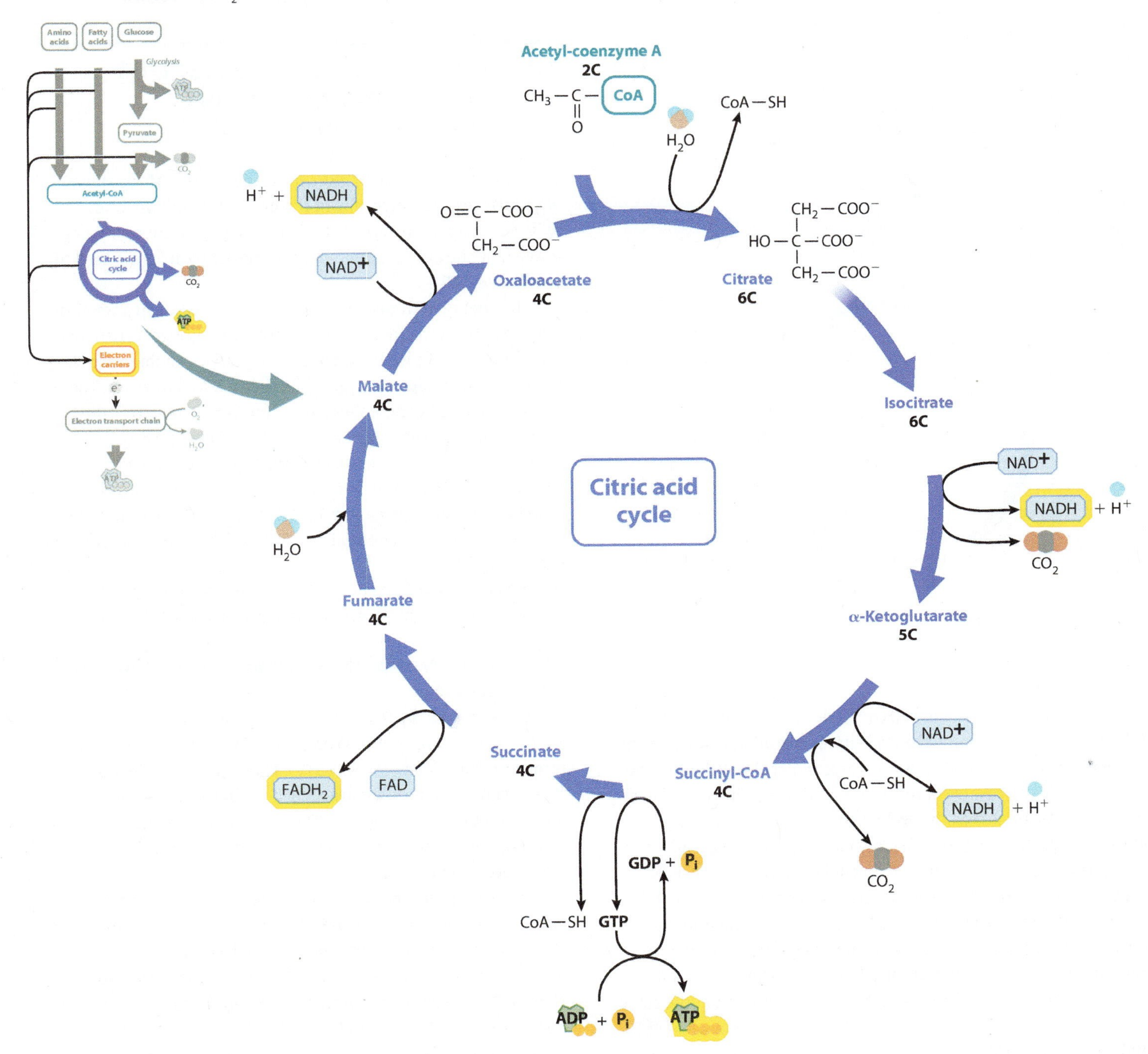

Overall, two molecules of acetyl-CoA produced from a single molecule of glucose yield two molecules of ATP, six molecules of NADH, and two molecules of $FADH_2$ in the citric acid cycle.

→ **Quick Check 3** At the end of the citric acid cycle, but before the subsequent steps of cellular respiration, which molecules contain the energy held in the original glucose molecule?

? CASE 1 THE FIRST CELL: LIFE'S ORIGINS

What were the earliest energy-harnessing reactions?

Some bacteria run the citric acid cycle in reverse, incorporating carbon dioxide into organic molecules instead of liberating it. Running the citric acid cycle in reverse requires energy, which is supplied by sunlight (Chapter 8) or chemical reactions (Chapter 26).

FIG. 7.9 Molecules produced by the citric acid cycle. Many organic molecules can be synthesized from citric acid cycle intermediates.

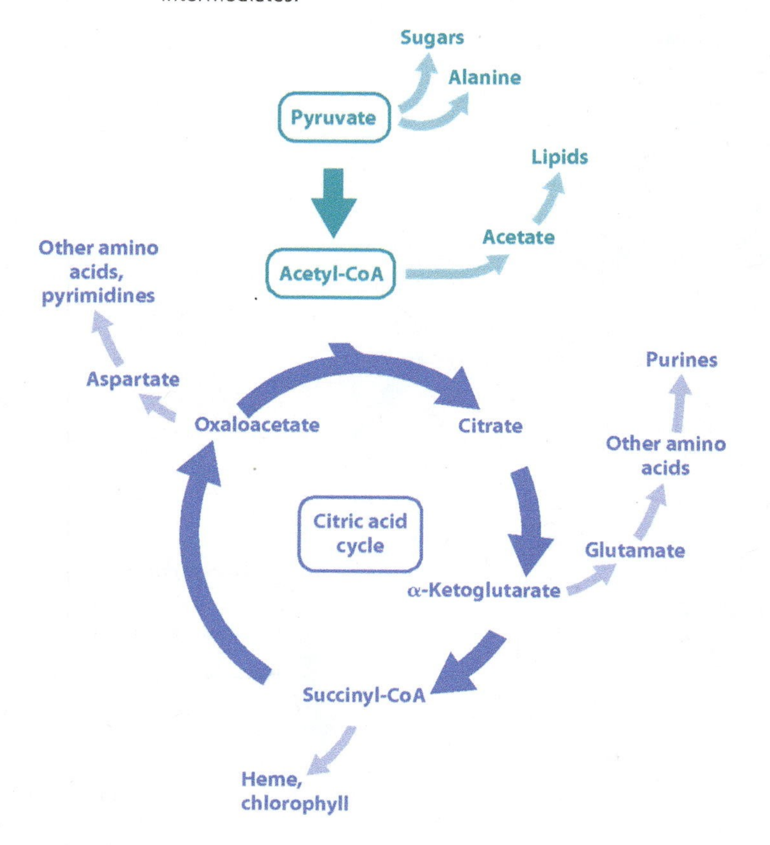

Why would an organism run the citric acid cycle backward? The answer is that running the cycle in reverse allows an organism to build, rather than break down, organic molecules. Whether the cycle is run in the reverse or forward direction, the intermediates generated step by step as the cycle turns provide the building blocks for synthesizing the cell's key biomolecules (**Fig. 7.9**). Pyruvate, for example, is the starting point for the synthesis of sugars and the amino acid alanine; acetate is the starting point for the synthesis of the cell's lipids; oxaloacetate is modified to form different amino acids and pyrimidine bases; and α-(alpha-) ketoglutarate is modified to form other amino acids. As a result, for organisms that run the cycle in the forward direction, the citric acid cycle is used to generate both energy-storing molecules (ATP and reduced electron carriers) and intermediates in the synthesis of other molecules. For organisms that run the cycle in reverse, it is used to generate intermediates in the synthesis of other molecules and also to incorporate carbon into organic molecules.

The centrality of the citric acid cycle to both the synthesis of biomolecules and to meeting the energy requirements of cells suggests that this cycle evolved early, appearing in some of the first cells to feature metabolism. This early appearance of the citric acid cycle, in turn, implies that the great variety of biosynthetic and energy-yielding pathways found in modern cells evolved through the extension and modification of this deeply rooted cycle. As typical of evolution, new cellular capabilities arose by the modification of simpler, more general sets of reactions.

7.5 THE ELECTRON TRANSPORT CHAIN AND OXIDATIVE PHOSPHORYLATION

The complete oxidation of glucose during the first three stages of cellular respiration results in the production of two kinds of reduced electron carriers: NADH and $FADH_2$. We are now going to see how the energy stored in these electron carriers is used to synthesize ATP.

The energy in these electron carriers is released in a series of redox reactions that occur as electrons pass through a chain of protein complexes in the inner mitochondrial membrane to the final electron acceptor, oxygen, which is reduced to water. The energy released by these redox reactions is not converted directly into the chemical energy of ATP, however. Instead, the passage of electrons is coupled to the transfer of protons (H^+) across the inner mitochondrial membrane, creating a concentration and charge gradient (Chapter 5). This electrochemical gradient provides a source of potential energy that is then used to drive the synthesis of ATP.

We next explore the properties of the electron transport chain, the proton gradient, and the synthesis of ATP.

The electron transport chain transfers electrons and pumps protons.

Electrons donated by NADH and $FADH_2$ are transported along a series of four large protein complexes that form the electron transport chain (complexes I to IV). These are shown in **Fig. 7.10.** These membrane proteins are embedded in the mitochondrial inner membrane (see Fig. 7.6). The inner mitochondrial membrane contains one of the highest concentrations of proteins found in eukaryotic membranes.

Electrons enter the electron transport chain at either complex I or II. Electrons donated by NADH enter through complex I, and electrons donated by $FADH_2$ enter through complex II. (Complex II is the same enzyme that catalyzes step 6 in the citric acid cycle.) These electrons are transported through either complex I or II to complex III and then through complex IV.

Within each protein complex of the electron transport chain, electrons are passed from electron donors to electron acceptors. Each donor and acceptor is a redox couple, consisting of an oxidized and a reduced form of a molecule. The electron transport chain contains many of these redox couples. When oxygen accepts electrons at the end of the electron transport chain, it is reduced to form water:

$$O_2 + 4e^- + 4H^+ \rightarrow 2H_2O$$

This reaction is catalyzed by complex IV.

FIG. 7.10 The electron transport chain. (a) The electron transport chain consists of four complexes (I to IV) in the inner mitochondrial membrane. (b) Electrons flow from electron carriers to oxygen, the final electron acceptor. (c) The proton gradient formed from the electron transport chain has potential energy that is used to synthesize ATP.

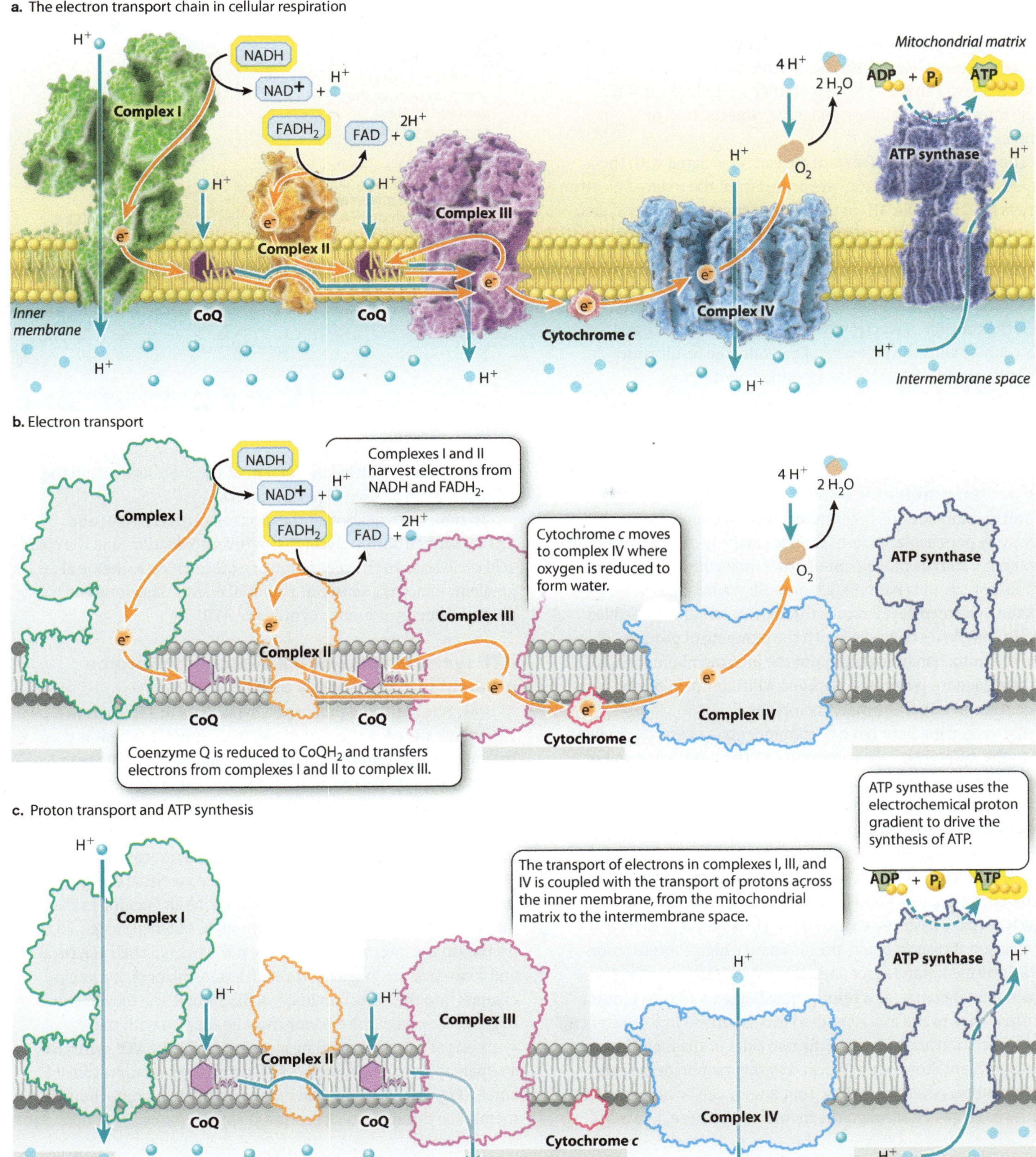

Electrons also must be transported *between* the four complexes (Fig. 7.10). **Coenzyme Q (CoQ),** also called ubiquinone, accepts electrons from both complexes I and II. In this reaction, two electrons and two protons are transferred to CoQ from the mitochondrial matrix, forming $CoQH_2$. Once $CoQH_2$ is formed, it diffuses in the inner membrane to complex III. In complex III, electrons are transferred from $CoQH_2$ to **cytochrome *c*** and protons are released into the intermembrane space. When it accepts an electron, cytochrome *c* is reduced, diffuses in the intermembrane space, and passes the electron to complex IV.

These electron transfer steps are each associated with the release of energy as electrons are passed from the reduced electron carriers NADH and $FADH_2$ to the final electron acceptor, oxygen. Some of this energy is used to reduce the next carrier in the chain, but in complexes I, III, and IV, some of it is used to pump protons (H^+) across the inner mitochondrial membrane, from the mitochondrial matrix to the intermembrane space (Fig. 7.10). Thus, the transfer of electrons through complexes I, III, and IV is coupled with the pumping of protons. The result is an accumulation of protons in the intermembrane space.

→ **Quick Check 4** Animals breathe in air that contains more oxygen than the air they breathe out. Where is oxygen consumed?

FIG. 7.11 ATP synthase. ATP synthase drives the synthesis of ATP by means of an electrochemical proton gradient.

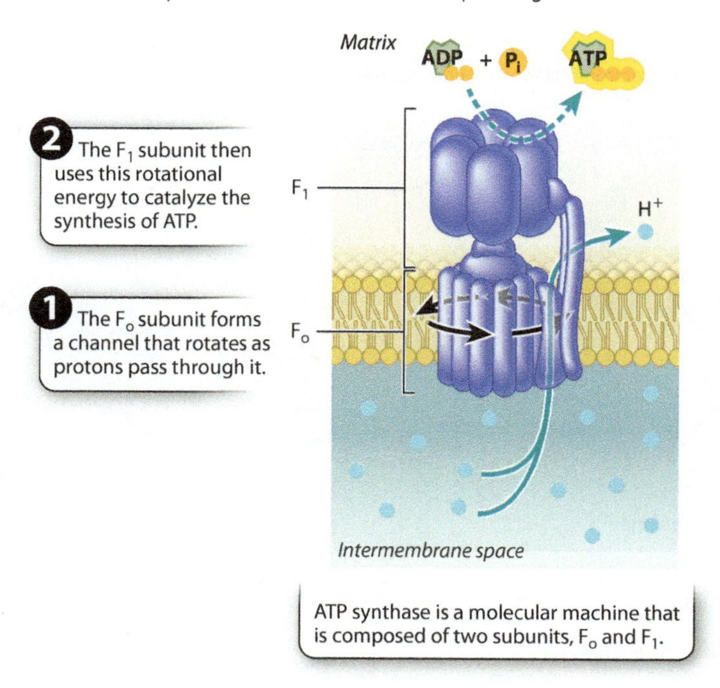

The proton gradient is a source of potential energy.

Like all membranes, the inner mitochondrial membrane is selectively permeable: Protons cannot passively diffuse across this membrane, and the movement of other molecules is controlled by transporters and channels (Chapter 5). We have just seen that the movement of electrons through membrane-embedded protein complexes is coupled with the pumping of protons from the mitochondrial matrix into the intermembrane space. The consequence is a proton gradient, a difference in proton concentration across the inner membrane.

The proton gradient has two components: a chemical gradient due to the difference in concentration and an electrical gradient due to the difference in charge between the two sides of the membrane. To reflect the dual contribution of the concentration gradient and the electrical gradient, the proton gradient is also called an electrochemical gradient.

The proton gradient is a source of potential energy, as discussed in Chapters 5 and 6. It stores energy much in the same way that a battery or a dam does. Through the actions of the electron transport chain, protons have a high concentration in the intermembrane space and a low concentration in the mitochondrial matrix. As a result, there is a tendency for protons to diffuse back to the mitochondrial matrix, driven by a difference in concentration and charge on the two sides of the membrane. This movement, however, is blocked by the membrane, so the gradient stores potential energy. That energy can be harnessed if a pathway is opened through the membrane because, as we will see shortly, the resulting movement of the protons through the membrane can be used to perform work.

In sum, the oxidation of the electron carriers NADH and $FADH_2$ formed during glycolysis, pyruvate oxidation, and the citric acid cycle leads to the generation of a proton electrochemical gradient, which is a source of potential energy. This source of potential energy is used to synthesize ATP.

ATP synthase converts the energy of the proton gradient into the energy of ATP.

In 1961, Peter Mitchell proposed a hypothesis to explain how the energy stored in the proton electrochemical gradient is used to synthesize ATP. In 1978, he was awarded the Nobel Prize in Chemistry for work that fundamentally changed the way we understand how energy is harnessed by a cell.

According to Mitchell's hypothesis, the gradient of protons provides a source of potential energy that is converted into chemical energy stored in ATP. First, for the potential energy of the proton gradient to be released, there must be an opening in the membrane for the protons to flow through. Mitchell suggested that protons in the intermembrane space diffuse down their electrical and concentration gradients through a transmembrane protein channel into the mitochondrial matrix. Second, the movement of protons through the enzyme must be coupled with the synthesis of ATP. This coupling is made possible by **ATP synthase,** a remarkable enzyme composed of two distinct subunits called F_o and F_1 (**Fig. 7.11**). F_o forms the channel in the inner mitochondrial membrane through which protons flow; F_1 is the catalytic unit that

HOW DO WE KNOW?

FIG. 7.12

Can a proton gradient drive the synthesis of ATP?

BACKGROUND Peter Mitchell's hypothesis that a proton gradient can drive the synthesis of ATP was met with skepticism because it was proposed before experimental evidence supported it. In the 1970s, biochemist Efraim Racker and his collaborator Walther Stoeckenius tested the hypothesis.

EXPERIMENT Racker and Stoeckenius built an artificial system consisting of a membrane, a bacterial proton pump activated by light, and ATP synthase.

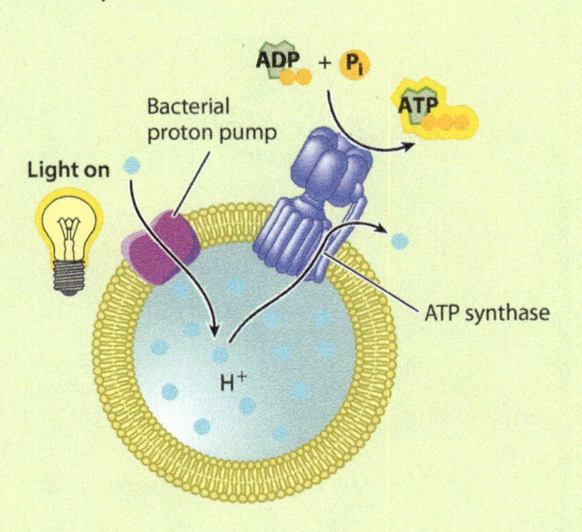

They measured the concentration of protons in the external medium and the amount of ATP produced in the presence and absence of light.

RESULTS In the presence of light, the concentration of protons increased inside the vesicles, suggesting that protons were taken up by the vesicles.

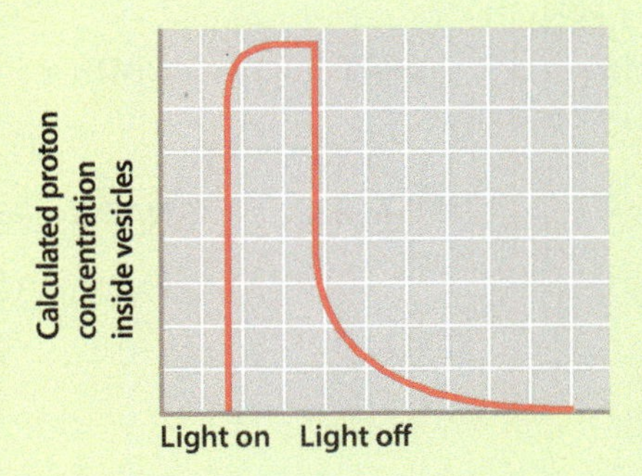

In the dark, the concentration of protons returned to the starting level. ATP was generated in the light, but not in the dark.

Condition	Relative level of ATP
Light	594
Dark	23

INTERPRETATION In the presence of light, the proton pump was activated and protons were pumped to one side of the membrane, leading to the formation of a proton gradient. The proton gradient, in turn, powered synthesis of ATP by ATP synthase.

CONCLUSION A membrane, proton gradient, and ATP synthase are sufficient to synthesize ATP. This result provided experimental evidence for Mitchell's hypothesis.

SOURCES Mitchell, P. 1961. "Coupling of Phosphorylation to Electron and Hydrogen Transfer by a Chemiosmotic Type of Mechanism." *Nature* 191:144–148; Racker, E., and W. Stoeckenius. 1974. "Reconstitution of Purple Membrane Vesicles Catalyzing Light-Driven Proton Uptake and Adenosine Triphosphate Formation." *J. Biol. Chem.* 249:662–663.

synthesizes ATP. Proton flow through the channel (F_0) makes it possible for the enzyme (F_1) to synthesize ATP.

Proton flow through the F_0 channel causes it to rotate, converting the energy of the proton gradient into mechanical rotational energy, a form of kinetic energy. The rotation of the F_0 subunit leads to rotation of the F_1 subunit in the mitochondrial matrix (Fig. 7.11). The rotation of the F_1 subunit in turn causes conformational changes that allow it to catalyze the synthesis of ATP from ADP and P_i. In this way, mechanical rotational energy is converted into the chemical energy of ATP.

Direct experimental evidence for Mitchell's idea, called the **chemiosmotic hypothesis,** did not come for over a decade. One of the key experiments that provided support for his idea is illustrated in **Fig. 7.12.**

→ **Quick Check 5** Uncoupling agents are proteins spanning the inner mitochondrial membrane that allow protons to pass through the membrane and bypass the channel of ATP synthase. Describe the consequences to the proton gradient and ATP production.

TABLE 7.1 Approximate Total ATP Yield in Cellular Respiration.

PATHWAY	SUBSTRATE-LEVEL PHOSPHORYLATION	OXIDATIVE PHOSPHORYLATION	TOTAL ATP
Glycolysis (glucose → 2 pyruvate)	2 ATP	2 NADH = 5 ATP	7
Pyruvate oxidation (2 pyruvate → 2 acetyl-CoA)	0 ATP	2 NADH = 5 ATP	5
Citric acid cycle (2 turns, 1 for each acetyl-CoA)	2 ATP	6 NADH = 15 ATP 2 $FADH_2$ = 3 ATP	20
Total	4 ATP	28 ATP	32

Approximately 2.5 molecules of ATP are produced for each NADH that donates electrons to the chain and 1.5 molecules of ATP for each $FADH_2$. Therefore, overall, the complete oxidation of glucose yields about 32 molecules of ATP from glycolysis, pyruvate oxidation, the citric acid cycle, and oxidative phosphorylation (**Table 7.1**). Some of the energy held in the bonds of glucose is now available in a molecule that can be readily used by cells.

It is worth taking a moment to follow the flow of energy in cellular respiration, illustrated in its full form in **Fig. 7.13.** We began with glucose and noted that it holds chemical potential energy in its covalent bonds. This energy is released in a series of reactions and captured in chemical form. Some of these reactions generate ATP directly by substrate-level phosphorylation. Others are redox reactions that transfer energy to the electron carriers NADH and $FADH_2$. These electron carriers donate electrons to the electron transport chain, which uses the energy stored in the electron carriers to pump protons across the inner membrane of the mitochondria. In other words, the energy of the reduced electron carriers is transformed into energy stored in a proton electrochemical gradient. ATP synthase then converts the energy of the proton gradient to rotational energy, which drives the synthesis of ATP. The cell now has a form of energy that it can use in many ways to perform work.

FIG. 7.13 **The flow of energy in cellular respiration.** A single glucose molecule yields 32 ATP molecules.

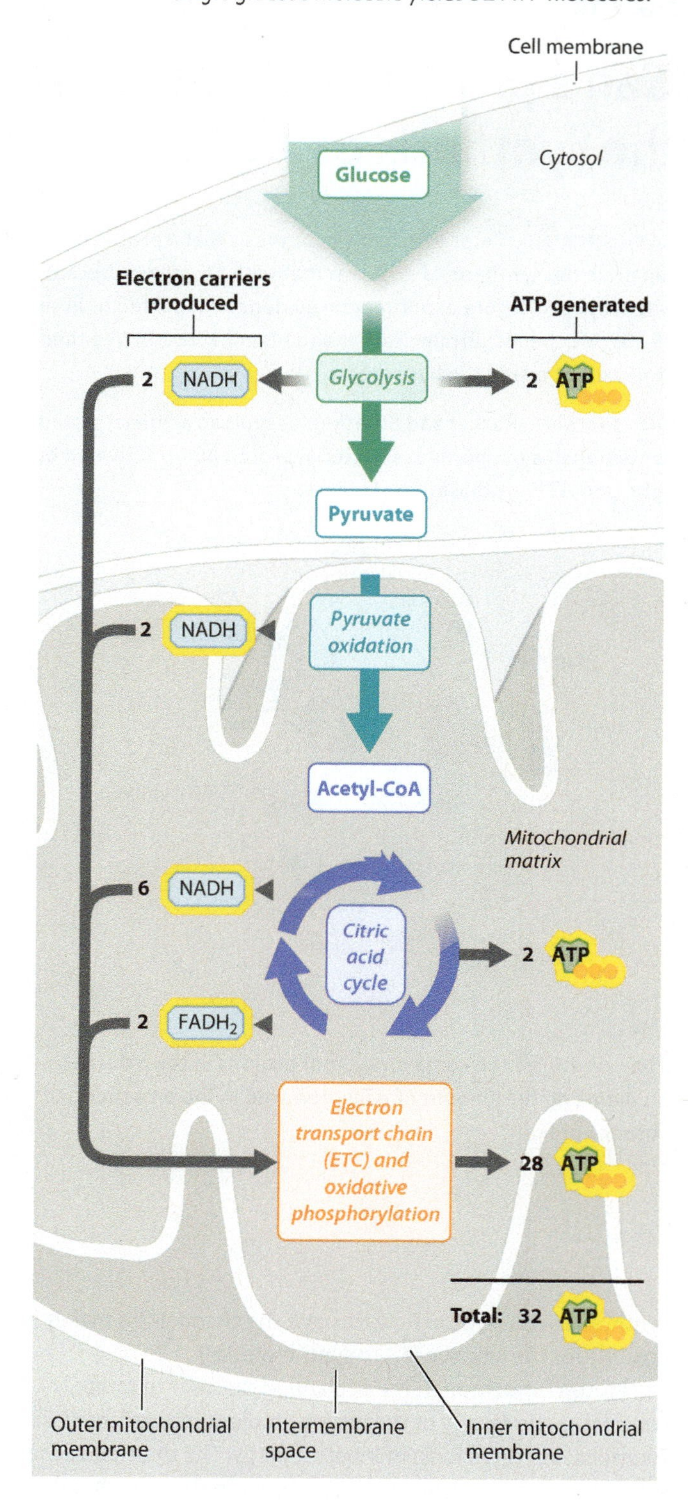

7.6 ANAEROBIC METABOLISM AND THE EVOLUTION OF CELLULAR RESPIRATION

Up to this point, we have followed a single metabolic path: the breakdown of glucose in the presence of oxygen to produce carbon dioxide and water. However, metabolic pathways more often resemble intersecting roads rather than a single, linear path. We saw this earlier in the discussion of the citric acid cycle, where intermediates in the cycle often feed into other metabolic pathways.

One of the major forks in the metabolic road occurs at pyruvate, the end product of glycolysis (section 10.2). When oxygen is present, it is converted to acetyl-CoA, which then enters the citric acid cycle, resulting in the production of ATP and reduced electron carriers to fuel the electron transport chain, as we saw. When oxygen is not present, however, pyruvate is metabolized along a number of different pathways. These pathways occur in many living organisms today and played an important role in the early evolution of life on Earth.

Fermentation extracts energy from glucose in the absence of oxygen.

Pyruvate has many possible fates in the cell. In the absence of oxygen, it can be broken down by **fermentation,** which does not rely on oxygen or any other electron acceptor. Fermentation is accomplished through a wide variety of metabolic pathways that extract energy from fuel molecules such as glucose. Fermentation pathways are important for anaerobic organisms that live without oxygen, as well as some organisms such as yeast that favor fermentation over oxidative phosphorylation, even in the presence of oxygen. It is also sometimes used in aerobic organisms when oxygen cannot be delivered fast enough to meet the cell's metabolic needs, as in exercising muscle.

Recall that during glycolysis, glucose is oxidized to form pyruvate, and NAD^+ is reduced to form NADH. For glycolysis to continue, NADH must be oxidized to NAD^+. If that did not happen, glycolysis would grind to a halt. In the presence of oxygen, NAD^+ is regenerated when NADH donates its electrons to the electron transport chain. In the absence of oxygen during fermentation, NADH is oxidized to NAD^+ when pyruvate or a derivative of pyruvate is reduced.

There are many fermentation pathways, especially in bacteria. Two of the major pathways are **lactic acid fermentation** and **ethanol fermentation** (**Fig. 7.14**). Lactic acid fermentation occurs in animals and bacteria. During lactic acid fermentation, electrons from NADH are transferred to pyruvate to produce lactic acid and NAD^+ (Fig. 7.14a). The overall chemical reaction is written as follows:

$$\text{Glucose} + 2\,\text{ADP} + 2\,\text{P}_i \rightarrow 2\,\text{lactic acid} + 2\,\text{ATP} + 2\text{H}_2\text{O}$$

Ethanol fermentation occurs in plants and fungi. During ethanol fermentation, pyruvate releases carbon dioxide to form acetaldehyde, and electrons from NADH are transferred to acetaldehyde to produce ethanol and NAD^+ (Fig. 7.14b). The overall chemical reaction is written as follows:

$$\text{Glucose} + 2\,\text{ADP} + 2\,\text{P}_i \rightarrow 2\,\text{ethanol} + 2\text{CO}_2 + 2\,\text{ATP} + 2\text{H}_2\text{O}$$

FIG. 7.14 Lactic acid and ethanol fermentation pathways.

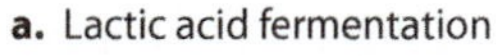

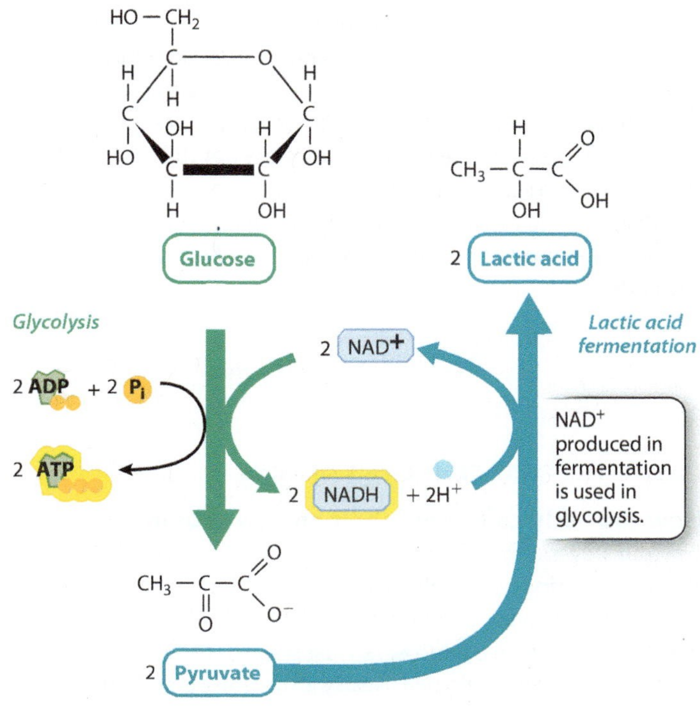

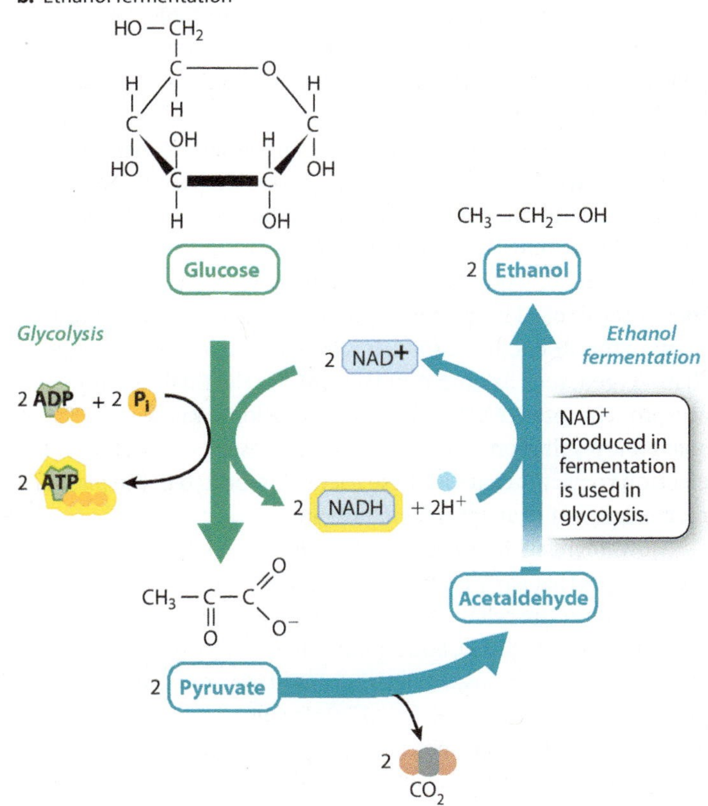

In both fermentation pathways, NADH is oxidized to NAD^+. However, NADH and NAD^+ do not appear in the overall chemical equations because there is no net production or loss of either molecule. NAD^+ molecules that are reduced during glycolysis are oxidized when lactic acid or ethanol is formed.

The breakdown of a molecule of glucose by fermentation yields only two molecules of ATP. The energetic gain is relatively small compared with the total yield of aerobic respiration because the end products, lactic acid and ethanol, are not fully oxidized and still contain a large amount of chemical energy in their bonds. The modest yield explains why organisms that produce ATP by fermentation must consume a large quantity of fuel molecules to power the cell.

→ **Quick Check 6** Bread making involves ethanol fermentation and typically uses yeast, sugar, flour, and water. Why are yeast and sugar used?

? CASE 1 THE FIRST CELL: LIFE'S ORIGINS

How did early cells meet their energy requirements?

The four stages of cellular respiration lead to the full oxidation of glucose, resulting in the release of a large amount of energy stored in its chemical bonds. The first stage, glycolysis, results in only the partial oxidation of glucose, so just some of the energy held in its chemical bonds is released. Nearly all organisms are capable of partially breaking down glucose, suggesting that glycolysis evolved very early in the history of life.

Life first evolved about 4 billion years ago in the absence of atmospheric oxygen. The earliest organisms probably used one of the fermentation pathways to generate the ATP necessary to power cellular processes because fermentation does not require atmospheric oxygen. Fermentation occurs in the cytoplasm and does not require proteins embedded in specialized membranes.

As we have seen, cellular respiration involves an electron transport chain, composed of proteins embedded in a membrane and capable of transferring electrons from one protein to the next and pumping protons. The resulting proton gradient powers the synthesis of ATP. Like fermentation, cellular respiration can occur in the absence of oxygen, but in that case molecules other than oxygen, such as sulfate and nitrate, are the final electron acceptor (Chapter 26). This form of respiration is known as anaerobic respiration and occurs in some present-day bacteria. The electron transport chain in these bacteria is located in the plasma membrane, not in an internal membrane.

How might such a system have evolved? An intriguing possibility is that early prokaryotes evolved pumps to drive protons out of the cell in response to an increasingly acidic environment (**Fig. 7.15**). Some pumps might have used the energy of ATP to pump protons, while others used electron transport proteins to pump protons (Fig. 7.15a). At some point, proton pumps powered by electron transport might have generated a large enough electrochemical gradient that the protons could pass back through the ATP-driven pumps, running them in reverse to synthesize ATP (Fig. 7.15b).

FIG. 7.15 The possible evolution of the electron transport chain and oxidative phosphorylation.

a.

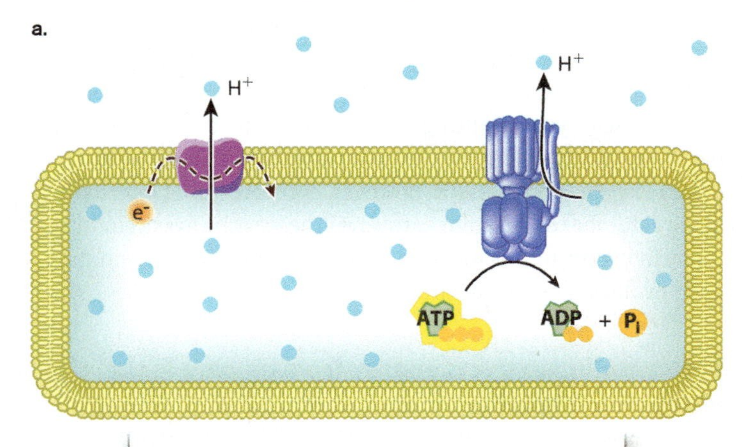

Early cells evolved mechanisms to pump protons out of the cell, powered by ATP and electron transport.

b.

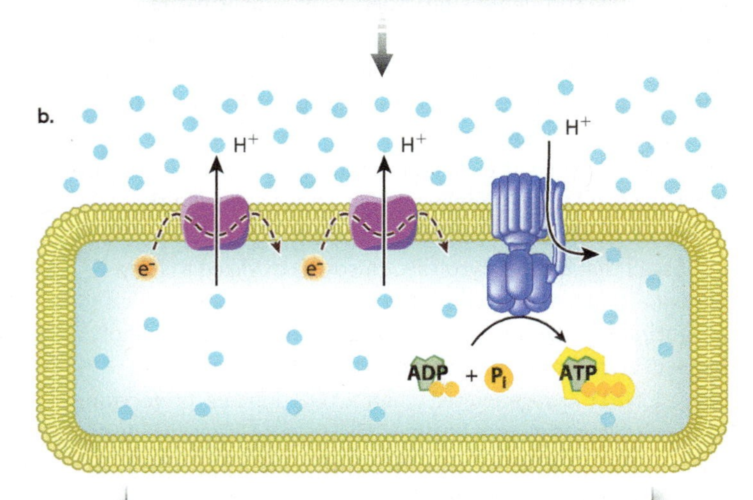

Eventually, electron-transport-powered pumps became efficient enough to run the ATP-driven pump in reverse.

Organisms capable of producing oxygen, the cyanobacteria, did not evolve until about 2.5 billion years ago, maybe earlier. The evolution of this new form of life introduced oxygen into Earth's atmosphere. This dramatic change led to the evolution of new life-forms with new possibilities for extracting energy from fuel molecules such as glucose. Aerobic respiration, in which oxygen serves as the final electron acceptor in the electron transport chain, generates much more energy than does anaerobic respiration or fermentation.

The evolution of cellular respiration illustrates that evolution often works in a stepwise fashion, building on what is already present. In this case, aerobic respiration picked up where anaerobic respiration left off, making it possible to harness more energy from organic molecules to power the work of the cell.

7.7 METABOLIC INTEGRATION

In this chapter, we have focused on the breakdown of glucose. What happens if there is more glucose than is needed by the cell? As well as glucose, you probably consume diverse carbohydrates, lipids, and proteins. How are these broken down? And how are these various metabolic pathways coordinated so that the intracellular level of ATP is maintained in a narrow range? In this final section, we consider how the cell responds to these challenges.

Excess glucose is stored as glycogen in animals and starch in plants.

Glucose is a readily available form of energy in organisms, but it is not always broken down immediately. Excess glucose can be stored in cells and then mobilized—that is, broken down—when necessary. Glucose can be stored in two major forms: as **glycogen** in animals and as **starch** in plants (**Fig. 7.16**). Both these molecules are large branched polymers of glucose.

Carbohydrates that are consumed by animals are broken down into simple sugars and circulate in the blood. The level of glucose in the blood is tightly regulated. When the blood glucose level is high, as it is after a meal, glucose molecules that are not consumed by glycolysis are linked together to form glycogen in liver and muscle. Glycogen stored in muscle is used to provide ATP for muscle contraction. By contrast, the liver does not store glycogen primarily for its own use, but is a central glycogen storehouse for the whole body, able to release glucose into the bloodstream when it is needed elsewhere. Glycogen provides a source of glucose 6-phosphate to feed glycolysis when the level of blood glucose is low. Glucose molecules located at the end of glycogen chains can be cleaved one by one, and they are released in the form of glucose 1-phosphate. Glucose 1-phosphate is then converted into glucose 6-phosphate, an intermediate in glycolysis. One glucose molecule cleaved off a glycogen chain produces three and not two molecules of ATP by glycolysis because the ATP-consuming step 1 of glycolysis is bypassed.

Sugars other than glucose contribute to glycolysis.

The carbohydrates in your diet are digested to produce a variety of sugars (**Fig. 7.17**). Some of these are disaccharides (maltose, lactose, and sucrose) with two sugar units; others are monosaccharides (fructose, mannose, and galactose) with a single sugar unit. The disaccharides are hydrolyzed into monosaccharides, which are transported into cells.

The hydrolysis of some disaccharides produces glucose molecules that directly enter glycolysis. What happens to other monosaccharides? They, too, enter glycolysis, although not as glucose. Instead, they are converted into intermediates of glycolysis that come later in the pathway. For example, fructose is produced by the hydrolysis of sucrose (table sugar) and receives a phosphate group to form either fructose 6-phosphate or fructose 1-phosphate. In the liver, fructose 1-phosphate is cleaved and converted into glyceraldehyde 3-phosphate, which enters glycolysis at reaction 6 (see Fig. 7.5).

FIG. 7.16 Storage forms of glucose. (a) Glycogen is a storage form of glucose in animal cells, and (b) starch is a storage form of glucose in plant cells.

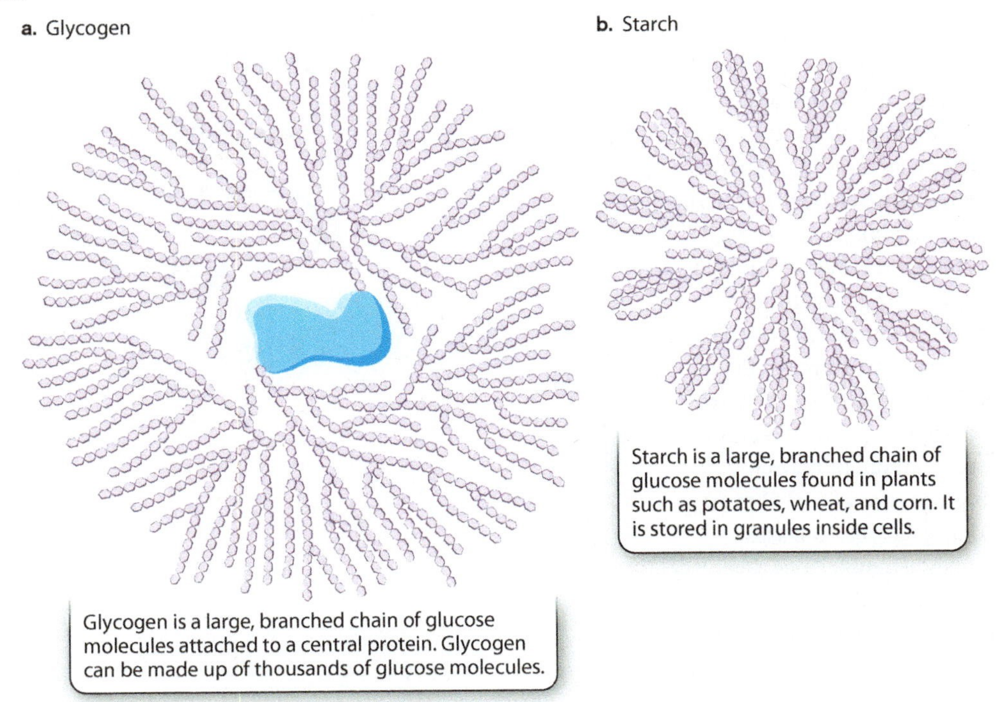

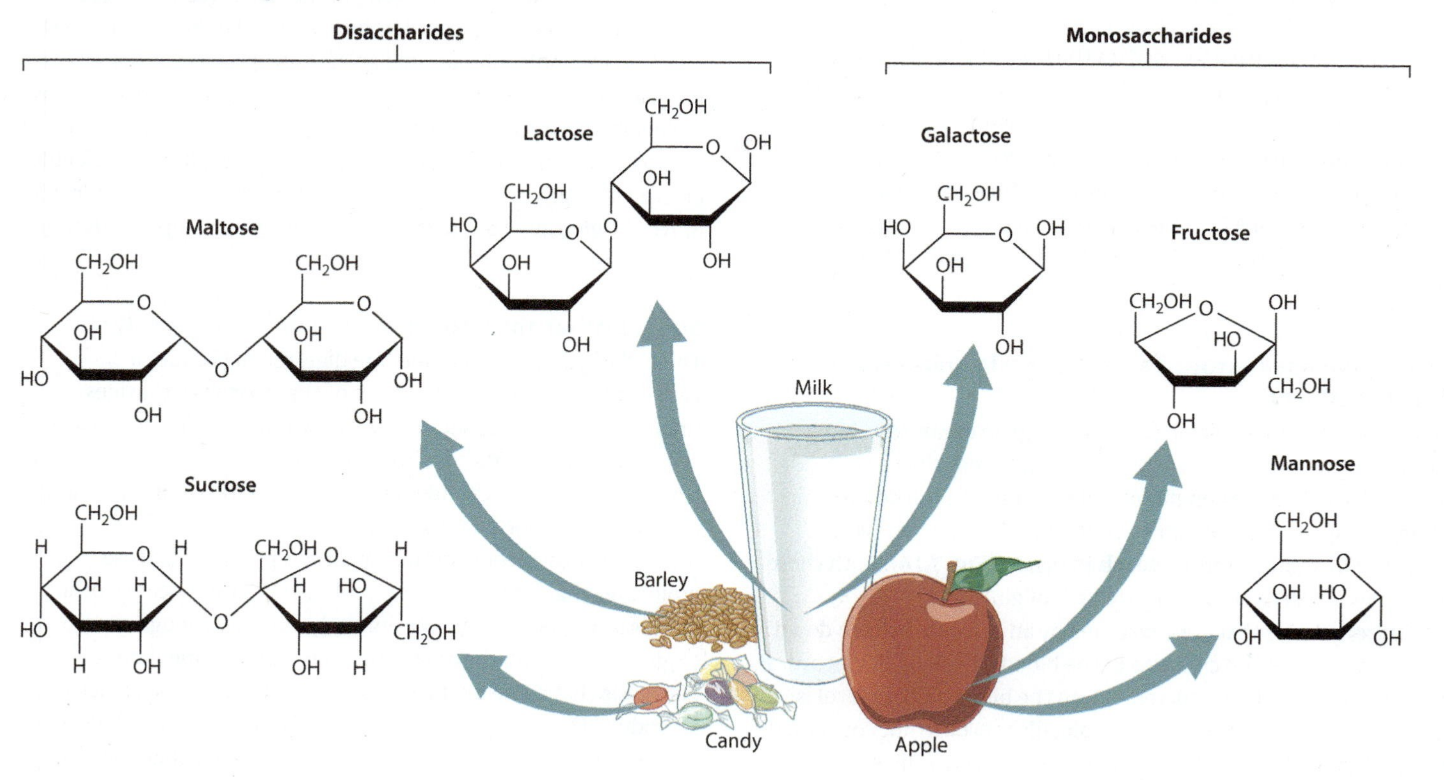

FIG. 7.17 Common sugars in your diet.

Fatty acids and proteins are useful sources of energy.

In addition to carbohydrates, lipids are also a good source of energy. We know this from common experience. Butter, oils, ice cream, and the like all contain lipids and are high in calories, which are units of energy. We can also infer that lipids are a good source of energy from their chemical structure. Recall from Chapter 2 that a type of fat called triacylglycerol is composed of three fatty acid molecules bound to a glycerol backbone. These fatty acid molecules are rich in carbon–carbon and carbon–hydrogen bonds, which, as we saw earlier, carry chemical potential energy.

Following a meal, the small intestine very quickly absorbs triacylglycerols, which are then transported by the bloodstream and either consumed or stored in fat (adipose) tissue. Triacylglycerols are broken down inside cells to glycerol and fatty acids. Then, the fatty acids themselves are shortened by a series of reactions that sequentially remove two carbon units from their ends (**Fig. 7.18**). This process is called β-**(beta-) oxidation.** It does not produce ATP, but releases a large number of NADH and $FADH_2$ molecules that provide electrons for the synthesis of ATP by oxidative phosphorylation. In addition, the end product of the reaction is acetyl-CoA, which feeds the citric acid cycle

FIG. 7.18 **β-oxidation of fatty acids.** The fatty acids from lipids are broken down to produce NADH and $FADH_2$, as well as acetyl-CoA.

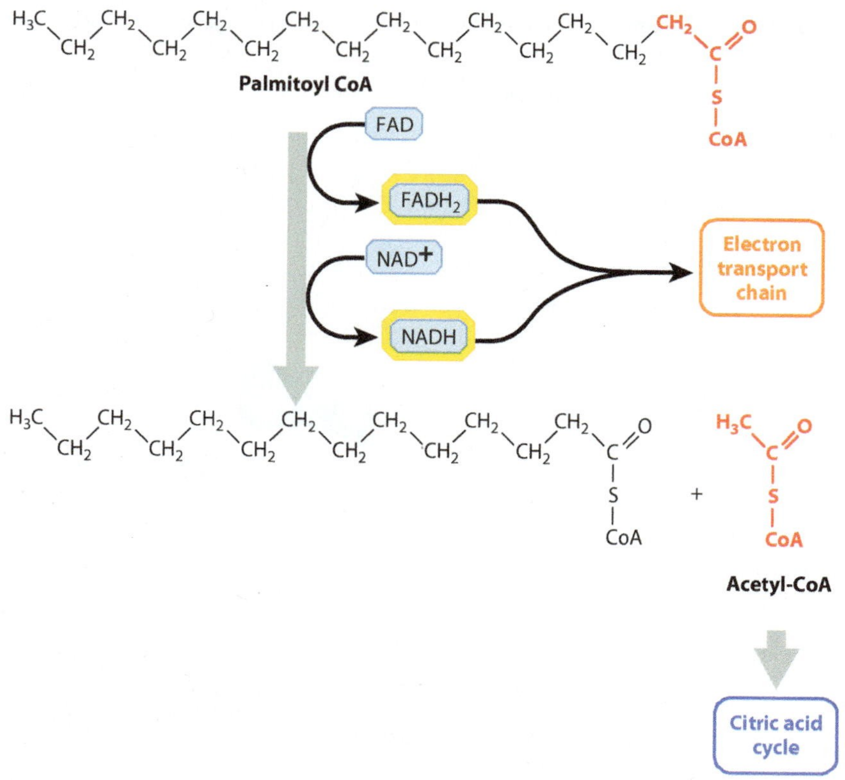

Palmitoyl CoA
FAD
$FADH_2$
NAD^+
NADH
Electron transport chain
Acetyl-CoA
Citric acid cycle

and leads to the production of an even larger quantity of reduced electron carriers.

The oxidation of fatty acids produces a large amount of ATP. For example, the complete oxidation of a molecule of palmitic acid, a fatty acid containing 16 carbons, yields about 106 molecules of ATP. By contrast, glycolysis yields just 2 molecules of ATP, and the complete oxidation of a glucose molecule produces about 32 molecules of ATP (Table 7.1). Fatty acids therefore are a useful and efficient source of energy, but they cannot be used by all tissues of the body. Notably, the brain and red blood cells depend primarily on glucose for energy.

Proteins, like fatty acids, are a source of chemical energy that can be broken down, if necessary, to power the cell. Proteins are typically first broken down to amino acids, some of which can then enter glycolysis and others the citric acid cycle.

The intracellular level of ATP is a key regulator of cellular respiration.

ATP is the key end product of cellular respiration, holding in its bonds energy that can be used for all kinds of cellular processes. ATP is constantly being turned over in a cell, broken down to ADP and P_i to supply the cell's energy needs, and re-synthesized by fermentation and cellular respiration. The level of ATP inside a cell can therefore be an indicator of how much energy a cell has available. When ATP levels are high, the cell has a high amount of free energy and is poised to carry out cellular processes. In this case, pathways that generate ATP are slowed, or down-regulated. By contrast, when ATP levels are low, the cell activates, or up-regulates, pathways that lead to ATP synthesis. Other intermediates of cellular respiration, such as NADH, have a similar effect in that high NAD^+ levels stimulate cellular respiration, whereas high NADH levels inhibit it (**Fig. 7.19**).

How is this kind of coordinated response of the cell possible? The cell uses several mechanisms, one of which is the regulation of enzymes that control key steps of the pathway. One of these key reactions is reaction 3 of glycolysis. In this reaction, fructose 6-phosphate is converted to fructose 1,6-bisphosphate, and a molecule of ATP is consumed. This is a key step in glycolysis because it is highly endergonic and irreversible. As a result, it is considered a "committed" step and is subject to tight control. This reaction is catalyzed by the enzyme phosphofructokinase-1 (PFK-1), which can be thought of as a metabolic valve that regulates the rate of glycolysis.

PFK-1 is an allosteric enzyme with many activators and inhibitors (**Fig. 7.20**). Recall from Chapter 6 that an allosteric enzyme changes its shape and activity in response to the binding of molecules at a site other than the active site. ADP and AMP are allosteric activators of PFK-1. When ADP and AMP are abundant, one or the other binds to the enzyme and causes the enzyme's shape to change. The shape change activates the enzyme, increasing the rate of glycolysis and the synthesis of ATP. When ATP is in abundance, it binds to the same site on the enzyme as ADP and AMP, but in this case binding inhibits the enzyme's

FIG. 7.19 Regulation of cellular respiration. Cellular respiration is inhibited by its products, including ATP and NADH, and activated by its substrates, including ADP and NAD^+.

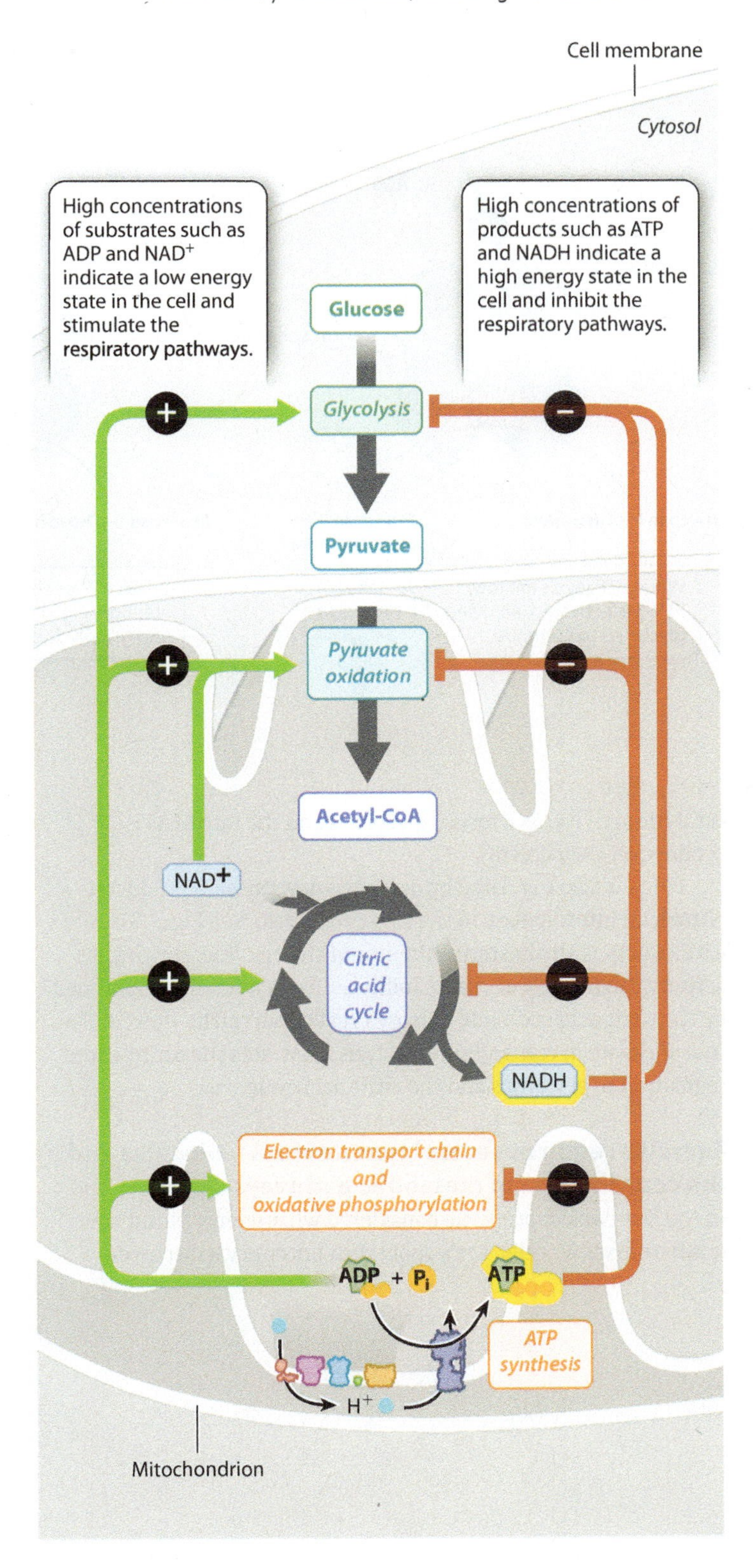

FIG. 7.20 Regulation of PFK-1. The regulation of the glycolytic enzyme phosphofructokinase-1 (PFK-1) is an example of integrated metabolic control.

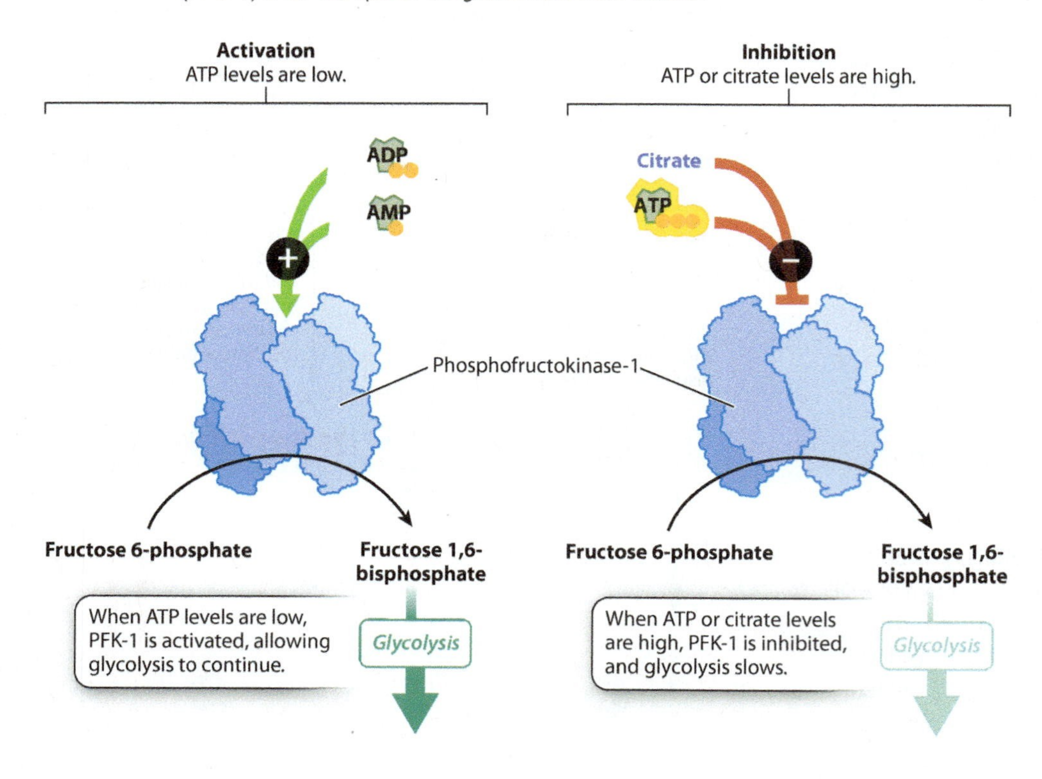

catalytic activity. As a result, glycolysis and the rate of ATP production slow down.

PFK-1 is also regulated by one of its downstream products, citrate, an intermediate in the citric acid cycle (see Fig. 7.8). Citrate acts as an allosteric inhibitor of the enzyme, slowing its activity. High levels of citrate indicate that is not being consumed by the citric acid cycle and glucose breakdown can be slowed. The role of citrate in controlling glycolysis illustrates the coordinated regulation of glycolysis and the citric acid cycle.

Exercise requires several types of fuel molecules and the coordination of metabolic pathways.

In the last two chapters, we considered what energy is and how it is harnessed by cells. Let's apply the concepts we discussed to a familiar example: exercise. Exercise such as running, walking, and swimming is a form of kinetic energy, powered by ATP in muscle cells. Where does this ATP come from?

Muscle cells, like all cells, do not contain a lot of ATP, and stored ATP is depleted by exercise in a matter of seconds. As a result, muscle cells rely on fuel molecules to generate ATP. For a short sprint or a burst of activity, muscle can convert stored glycogen to glucose, and then break down glucose anaerobically to pyruvate and lactic acid by lactic acid fermentation. This pathway is rapid, but it does not generate a lot of ATP. In addition, it is limited by the production of lactic acid, which lowers the pH of the blood.

For longer, more sustained exercise, other metabolic pathways come into play. Muscle cells contain many mitochondria, which produce ATP by aerobic respiration. The energy yield of aerobic respiration is much greater than that of fermentation, but the process is slower. This slower production of ATP by aerobic respiration in part explains why runners cannot maintain the pace of a sprint for longer runs.

For even longer exercise, liver glycogen supplements muscle glycogen: The liver releases glucose into the blood that is taken up by muscle cells and oxidized to produce ATP. In addition, fatty acids are released from adipose tissue and taken up by muscle cells, where they are broken down by β-oxidation. β-oxidation yields even more ATP than does the complete oxidation of glucose, but the process is again slower. Storage forms of energy molecules, such as fatty acids and glycogen, contain large reservoirs of energy, but are slow to mobilize. Thus, exercise takes coordination between different cells, tissues, and metabolic pathways to ensure adequate ATP to meet the needs of working muscle. ■

Core Concepts Summary

7.1 Cellular respiration is a series of catabolic reactions that convert the energy in fuel molecules into ATP.

During cellular respiration, sugar molecules like glucose are broken down in the presence of oxygen to produce carbon dioxide and water. page 132

Cellular respiration releases energy because the potential energy of the reactants is greater than that of the products. page 132

ATP is generated in two ways during cellular respiration: substrate-level phosphorylation and oxidative phosphorylation. page 133

Cellular respiration is an oxidation–reduction reaction. page 133

In oxidation–reduction reactions, electrons are transferred from one molecule to another. Oxidation is the loss of electrons, and reduction is the gain of electrons. page 133

Electron carriers transfer electrons to an electron transport chain, which harnesses the energy of these electrons to generate ATP. page 133

Cellular respiration is a four-stage process that includes (1) glycolysis; (2) pyruvate oxidation; (3) the citric acid cycle; and (4) oxidative phosphorylation. page 135

7.2 Glycolysis is the partial oxidation of glucose and results in the production of pyruvate, as well as ATP and reduced electron carriers.

Glycolysis takes place in the cytoplasm. page 135

Glycolysis is a series of 10 reactions in which glucose is oxidized to pyruvate. page 137

Glycolysis consists of preparatory, cleavage, and payoff phases. page 137

For each molecule of glucose broken down during glycolysis, a net gain of two molecules of ATP and two molecules of NADH is produced. page 137

The synthesis of ATP in glycolysis results from the direct transfer of a phosphate group from a substrate to ADP, a process called substrate-level phosphorylation. page 137

7.3 Pyruvate is oxidized to acetyl-CoA, connecting glycolysis to the citric acid cycle.

The conversion of pyruvate to acetyl-CoA results in the production of one molecule of NADH and one molecule of carbon dioxide. page 137

Pyruvate oxidation occurs in the mitochondrial matrix. page 137

7.4 The citric acid cycle results in the complete oxidation of fuel molecules and the generation of ATP and reduced electron carriers.

The citric acid cycle takes place in the mitochondrial matrix. page 138

The acetyl group of acetyl-CoA is completely oxidized in the citric acid cycle. page 138

The citric acid cycle is a cycle because the acetyl group of acetyl-CoA combines with oxaloacetate, and then a series of reactions regenerates oxaloacetate. page 138

A complete turn of the citric acid cycle results in the production of one molecule of GTP (which is converted to ATP), three molecules of NADH, and one molecule of $FADH_2$. page 138

Citric acid cycle intermediates are starting points for the synthesis of many different organic molecules. page 140

7.5 The electron transport chain transfers electrons from electron carriers to oxygen, using the energy to pump protons and synthesize ATP by oxidative phosphorylation.

NADH and $FADH_2$ donate electrons to the electron transport chain. page 140

In the electron transport chain, electrons move from one redox couple to the next. page 140

The electron transport chain is made up of four complexes. Complexes I and II accept electrons from NADH and $FADH_2$, respectively. The electrons are transferred from these two complexes to coenzyme Q. page 140

Reduced coenzyme Q transfers electrons to complex III and cytochrome c transfers electrons to complex IV. Complex IV reduces oxygen to water. page 142

The transfer of electrons through the electron transport chain is coupled with the movement of protons across the inner mitochondrial membrane into the intermembrane space. page 142

The buildup of protons in the intermembrane space results in a proton electrochemical gradient, which stores potential energy. page 142

The movement of protons back into the mitochondrial matrix through the F_o subunit of ATP synthase is coupled with the formation of ATP, a reaction catalyzed by the F_1 subunit of ATP synthase. page 142

7.6 Glucose can be broken down in the absence of oxygen by fermentation, producing a modest amount of ATP.

Pyruvate, the end product of glycolysis, is processed differently in the presence and the absence of oxygen. page 145

In the absence of oxygen, pyruvate enters one of several fermentation pathways. page145

In lactic acid fermentation, pyruvate is reduced to lactic acid. page 145

In ethanol fermentation, pyruvate is converted to acetaldehyde, which is reduced to ethanol. page 145

During fermentation, NADH is oxidized to NAD^+, allowing glycolysis to proceed. page 145

Glycolysis and fermentation are ancient biochemical pathways and were likely used in the common ancestor of all organisms living today. page 146

7.7 Metabolic pathways are integrated, allowing control of the energy level of cells.

Excess glucose molecules are linked together and stored in polymers called glycogen (in animals) and starch (in plants). page 147

Other monosaccharides derived from the digestion of dietary carbohydrates are converted into intermediates of glycolysis. page 147

Fatty acids contained in triacylglycerols are an important form of energy storage in cells. The breakdown of fatty acids is called β-oxidation. page 148

Phosphofructokinase-1 controls a key step in glycolysis. It has many allosteric activators, including ADP and AMP, and allosteric inhibitors, including ATP and citrate. page 149

The ATP in muscle cells used to power exercise is generated by lactic acid fermentation, aerobic respiration, and β-oxidation. page 150

Self-Assessment

1. Name and describe the four major stages of cellular respiration.
2. Explain what an oxidation–reduction reaction is and why the breakdown of glucose in the presence of oxygen to produce carbon dioxide and water is an example of an oxidation–reduction reaction.
3. Describe two different ways in which ATP is generated in cellular respiration.
4. Write the overall chemical equation for glycolysis, noting the starting and ending products and highlighting the energy-storing molecules that are produced.
5. Describe two different metabolic pathways that pyruvate can enter.
6. Name the products of the citric acid cycle.
7. Describe how the movement of electrons along the electron transport chain leads to the generation of a proton gradient.
8. Describe how a proton gradient is used to generate ATP.
9. Explain how muscle tissue generates ATP during short-term and long-term exercise.

CHAPTER 8

Photosynthesis

Using Sunlight to Build Carbohydrates

Core Concepts

8.1 Photosynthesis is the major pathway by which energy and carbon are incorporated into carbohydrates.

8.2 The Calvin cycle is a three-step process that uses carbon dioxide to synthesize carbohydrates.

8.3 The light-harvesting reactions use sunlight to produce the ATP and NADPH required by the Calvin cycle.

8.4 Challenges to the efficiency of photosynthesis include excess light energy and the oxygenase activity of rubisco.

8.5 The evolution of photosynthesis had a profound impact on life on Earth.

Walk through a forest and you will be struck, literally if you aren't careful, by the substantial nature of trees. Where does the material to construct these massive organisms come from? Because trees grow upward from a firm base in the ground, a reasonable first guess is the soil. In the first recorded experiment on this question, the Flemish chemist and physiologist Jan Baptist van Helmont (1580–1644) found that the 200 pounds of dry soil into which he had planted a small willow tree decreased by only 2 ounces over a 5-year period. During this same period, the tree gained 164 pounds. Van Helmont concluded that water must be responsible for the tree's growth. He was, in fact, half right: A tree is roughly half liquid water. But what he missed completely is that the other half of his tree had been created almost entirely out of thin air.

The process that allowed Van Helmont's tree to increase in mass using material pulled from the air is called **photosynthesis.** Photosynthesis is a biochemical process for building carbohydrates using energy from sunlight and carbon dioxide (CO_2) taken from the air. These carbohydrates are used both as starting points for the synthesis of other molecules and as a means of storing energy that can be converted into ATP through cellular respiration.

8.1 PHOTOSYNTHESIS: AN OVERVIEW

Photosynthesis is the major entry point for energy into biological systems. It is the source of all of the food we eat, both through the direct consumption of plant material and its indirect consumption as meat. It is also the source of all the oxygen that we breathe, as well as fuels for heating and transportation. Fossil fuels are the legacy of ancient photosynthesis: Oil has its origin in the bodies of marine algae and the organisms that graze on them, while coal represents the geologic remains of terrestrial (land) plants. Thus, one motivation to understand photosynthesis is the sheer magnitude and importance of this process for life on Earth. Before exploring the details of how photosynthesis actually occurs, let's look at what types of organism carry out photosynthesis and where they live, as well as what structural components are needed to allow cells to capture energy in this remarkable way.

Photosynthesis is widely distributed.

The photosynthesis that is most evident to us is carried out by plants on land. Trees, grasses, and shrubs are all examples of photosynthetic organisms. However, photosynthesis occurs among prokaryotic as well as eukaryotic organisms, on land as well as in the sea. Approximately 60% of global photosynthesis is carried out by terrestrial organisms, with the remaining 40% taking place in the ocean. The majority of photosynthetic organisms in marine environments are unicellular. About half of oceanic photosynthesis is carried out by single-celled marine eukaryotes, while the other half is carried out by photosynthetic bacteria.

Photosynthesis takes place almost everywhere sunlight is available to serve as a source of energy. In the ocean,

FIG. 8.1 Photosynthesis in extreme environments. (a) Desert crust in the Colorado Plateau formed by photosynthetic bacteria and algae. (b) A hot spring in Yellowstone National Park. The yellow color is due to photosynthetic bacteria. (c) The surface of a permanent snow pack. The red color is due to photosynthetic algae. *Sources: a. Jayne Belnap/USGS; b. f11photo/Shutterstock; c. Shattil & Rozinski/Naturepl.com.*

photosynthesis occurs in the surface layer extending to about 100 m deep, called the **photic zone,** through which enough sunlight penetrates to enable photosynthesis. On land, photosynthesis occurs most readily in environments that are both moist and warm. Tropical rain forests have high photosynthetic productivity, as do grasslands and forests in the temperate zone. However, photosynthetic organisms have evolved adaptations that allow them to tolerate a wide range of environmental conditions, like those illustrated in **Fig. 8.1.** In very dry regions, a combination of photosynthetic bacteria and unicellular algae forms an easily disturbed layer on the surface of the soil known as desert crust. Photosynthetic bacteria are also found in the hot springs of Yellowstone National Park at temperatures up to 75°C. At the other extreme, unicellular algae can grow on the surfaces of glaciers, causing the surface of the snow to appear red. In section 8.4, we discuss why photosynthetic organisms growing in such extreme environments often appear red or yellow rather than green.

Photosynthesis is a redox reaction.

Carbohydrates are synthesized from CO_2 molecules during photosynthesis, yet they have more energy stored in their chemical bonds than is contained in the bonds of CO_2 molecules. Therefore, to build carbohydrates using CO_2 requires an input of energy. This energy comes from sunlight.

How does energy from sunlight become incorporated into chemical bonds? The answer to this question arises from the fact that the synthesis of carbohydrates from CO_2 and water is a reduction–oxidation, or redox, reaction. In Chapter 7, we saw that **reduction** reactions are reactions in which a molecule acquires electrons and gains energy, whereas **oxidation** reactions are reactions in which a molecule loses electrons and releases energy. During photosynthesis, CO_2 molecules are reduced to form higher-energy carbohydrate molecules. This requires both an input of energy from ATP and the transfer of electrons from an **electron donor.** In photosynthesis, energy from sunlight is used to produce ATP and electron donor molecules capable of reducing CO_2 (**Fig. 8.2**).

Where do the electrons used to reduce CO_2 come from? In photosynthesis carried out by plants and many algae, the ultimate electron donor is water. The oxidation of water results in the production of electrons, protons, and O_2. Thus, oxygen is formed in photosynthesis as a by-product of water's role as a source of electrons. We can demonstrate that water is the source of the oxygen released during photosynthesis using isotopes, molecules that can be distinguished on the basis of their molecular mass (**Fig. 8.3**).

Overall, then, the equation for photosynthesis leading to the synthesis of glucose ($C_6H_{12}O_6$) can be described as follows:

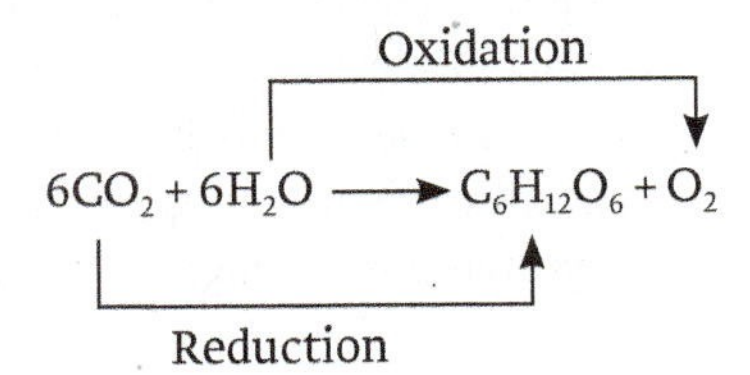

FIG. 8.2 Overview of photosynthesis. In photosynthesis, energy from sunlight is used to synthesize carbohydrates from CO_2.

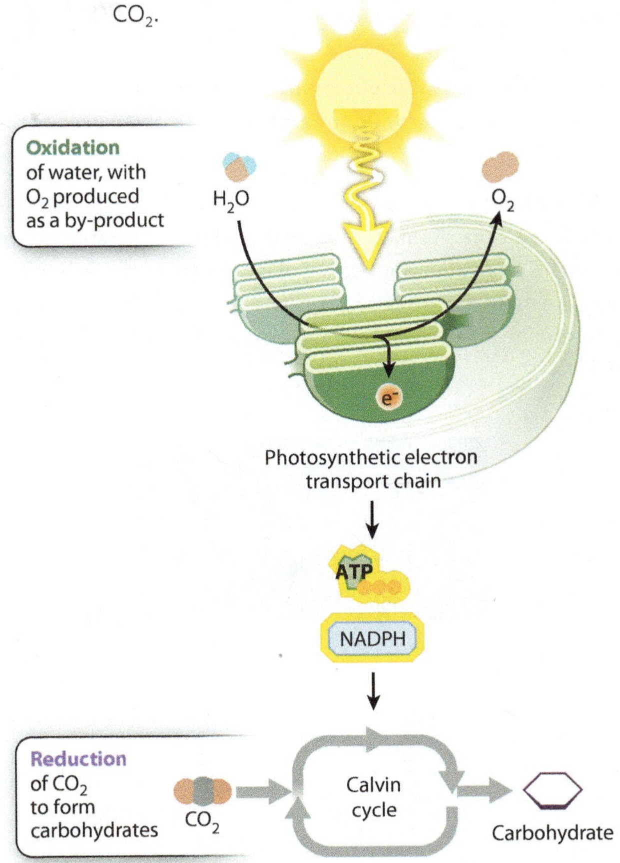

The oxidation of water is linked with the reduction of CO_2 through a series of redox reactions in which electrons are passed from one compound to another. This series of reactions constitutes the **photosynthetic electron transport chain.** The process begins with the absorption of sunlight by protein–pigment complexes. The absorbed sunlight provides the energy that drives electrons through the photosynthetic electron transport chain. In turn, the movement of electrons through this transport chain is used to produce ATP and NADPH. And finally, ATP and NADPH are the energy sources needed to synthesize carbohydrates using CO_2 in a process called the **Calvin cycle** (see Fig. 8.2).

→ **Quick Check 1** If you want to produce carbohydrates containing the heavy oxygen (^{18}O) isotope, should you water your plants with $H_2{}^{18}O$ or inject $C^{18}O_2$ into the air?

The photosynthetic electron transport chain takes place on specialized membranes.

Electron transport chains play a key role in both photosynthesis and respiration (Chapter 7). In both cases, electrons move within and between large protein complexes embedded in specialized membranes. In photosynthetic bacteria, the photosynthetic

HOW DO WE KNOW?

FIG. 8.3

Does the oxygen released by photosynthesis come from H_2O or CO_2?

BACKGROUND The reactants in photosynthesis are water and carbon dioxide. Both contain oxygen, so it is unclear which one is the source of the oxygen that is produced in the reaction.

METHOD Most of the oxygen in the atmosphere is ^{16}O, a stable isotope containing 8 protons and 8 neutrons. A small amount (0.2%) is ^{18}O, a stable isotope with 8 protons and 10 neutrons. The relative abundance of molecules containing ^{16}O versus ^{18}O can be measured using a mass spectrometer. H_2O and CO_2 containing a high percentage of ^{18}O can be used to determine whether the oxygen produced in photosynthesis comes from water or carbon dioxide.

EXPERIMENT

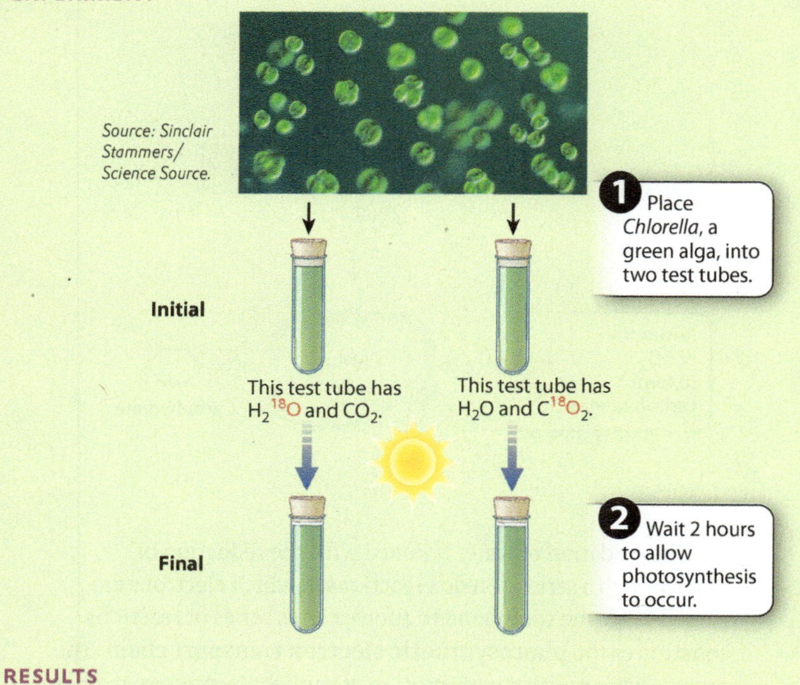

RESULTS

CONCLUSION The percentage of ^{18}O increases only when water contains ^{18}O, but not when carbon dioxide contains ^{18}O. This finding indicates that the oxygen produced in photosynthesis comes from water, not carbon dioxide.

FOLLOW-UP WORK Carbon also has several isotopes, and their measurements have been used to determine the source of increased CO_2 in the atmosphere today (Chapter 25).

SOURCE Adapted from Ruben, S., M. Randall, M. Kamen, and J. L. Hyde. 1941. "Heavy Oxygen (O^{18}) as a Tracer in the Study of Photosynthesis." *Journal of the American Chemical Society* 63:877–879.

electron transport chain is located in membranes within the cytoplasm or, in some cases, directly in the plasma membrane. In eukaryotic cells, photosynthesis takes place in chloroplasts. In the center of the chloroplast is the highly folded **thylakoid membrane** (**Fig. 8.4**). The photosynthetic electron transport chain is located in the thylakoid membrane.

The name "thylakoid" is derived from *thylakois*, the Greek word for "sac." Thylakoid membranes form structures that resemble flattened sacs, and these sacs are grouped into structures called **grana** (singular, granum) that look like stacks of interlinked pancakes. Grana are connected to one another by membrane bridges in such a way that the thylakoid membrane encloses a single interconnected compartment called the **lumen.** The region surrounding the thylakoid membrane is called the **stroma.** Carbohydrate synthesis takes place in the stroma, whereas sunlight is captured and transformed into chemical energy by the photosynthetic electron transport chain in the thylakoid membrane.

Although photosynthetic organisms are correctly described as autotrophs because they can form carbohydrates from CO_2, they also require a constant supply of ATP to meet each cell's energy requirements. Although ATP is produced within chloroplasts, only carbohydrates (and not ATP) are exported from chloroplasts to the cytosol. This explains why cells that have chloroplasts also contain mitochondria. In mitochondria, carbohydrates are broken down to generate ATP (Chapter 7). Cellular respiration is therefore one of several features that heterotrophic organisms like ourselves share with photosynthetic organisms.

We now consider the underlying biochemistry of photosynthesis. Though in this chapter we focus on photosynthesis as carried out by eukaryotic cells, there is a remarkable diversity in the way photosynthesis is carried out in bacteria. We discuss this diversity in Chapter 26. Because carbohydrates are the major product of photosynthesis, we first examine the Calvin cycle, the biochemical pathway used in photosynthesis to synthesize carbohydrates from CO_2. Once we understand what energy forms are needed to drive this autotrophic pathway, we turn our attention to how energy is captured from sunlight. We then examine some

FIG. 8.4 Chloroplast structure. Chloroplasts contain highly folded thylakoid membranes. *Photo source: Biophoto Associates/Science Source.*

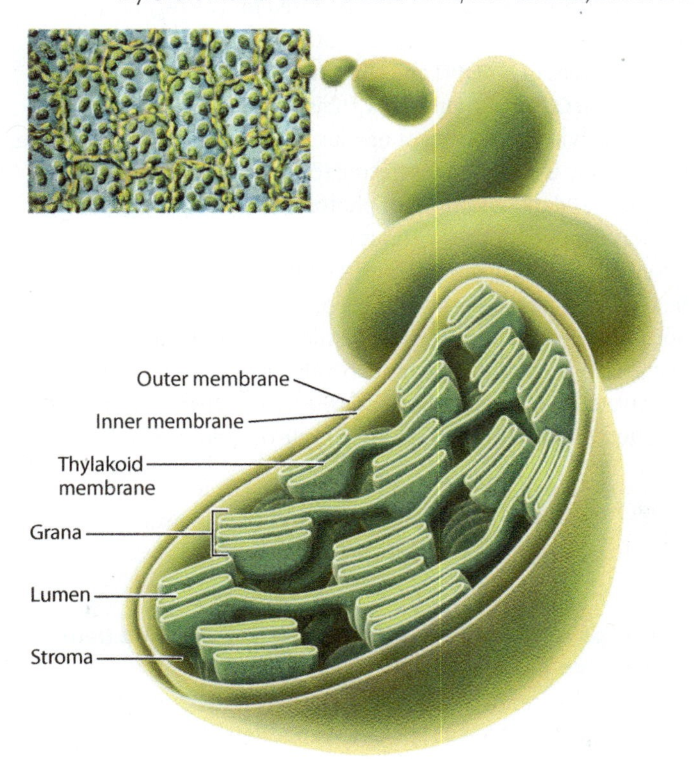

of the challenges of coordinating these two metabolic stages and consider ways in which photosynthetic organisms cope with the inherent biochemical challenges of photosynthesis. Finally, we provide a short overview of the evolutionary history of photosynthesis.

8.2 THE CALVIN CYCLE

The Calvin cycle consists of 15 chemical reactions that synthesize carbohydrates from CO_2. These reactions can be grouped into three main steps: (1) **carboxylation,** in which CO_2 is added to a 5-carbon molecule; (2) **reduction,** in which energy and electrons are transferred to the compounds formed in step 1; and (3) **regeneration** of the 5-carbon molecule needed for carboxylation (**Fig. 8.5**).

The incorporation of CO_2 is catalyzed by the enzyme rubisco.

In the first step of the Calvin cycle (Fig. 8.5), CO_2 is added to a 5-carbon sugar called **ribulose 1,5-bisphosphate (RuBP).** This step is catalyzed by the enzyme **ribulose bisphosphate carboxylase oxygenase,** or **rubisco** for short. An enzyme that

FIG. 8.5 The Calvin cycle. CO_2 is the input and triose phosphate is the output of the Calvin Cycle.

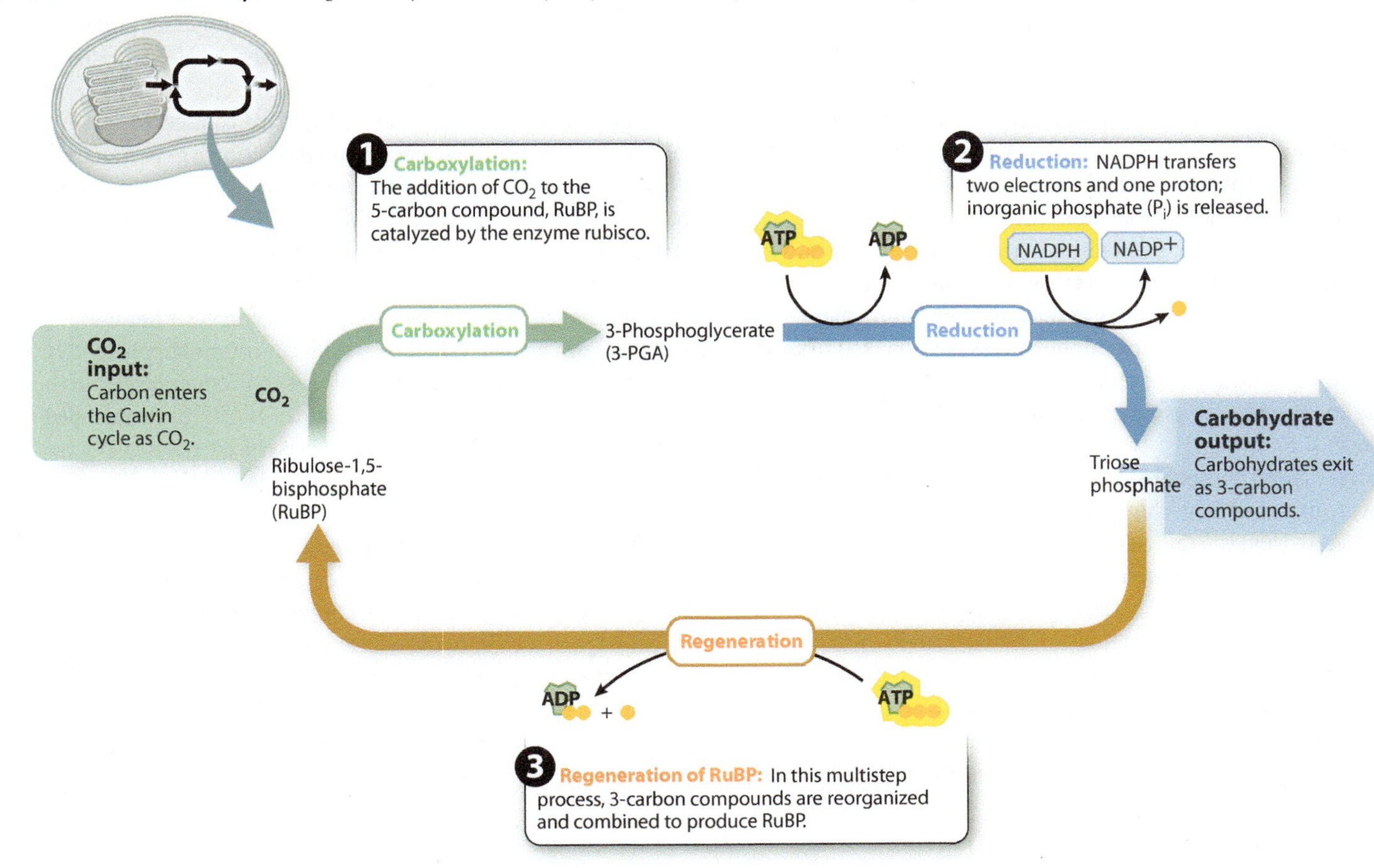

adds CO_2 to another molecule is called a carboxylase, explaining part of rubisco's long name.

Before rubisco can act as a carboxylase, RuBP and CO_2 must diffuse into its active site. Once the active site is occupied, the addition of CO_2 to RuBP proceeds spontaneously in the sense that no addition of energy is required. The product is a 6-carbon compound that immediately breaks into two molecules of **3-phosphoglycerate (3-PGA).** These 3-carbon molecules are the first stable products of the Calvin cycle.

NADPH is the reducing agent of the Calvin cycle.

Rubisco is responsible for the addition of the carbon atoms needed for the formation of carbohydrates, but by itself rubisco does not increase the amount of energy stored within the newly formed bonds. For this energy increase to take place, the carbon compounds formed by rubisco must be reduced. **Nicotinamide adenine dinucleotide phosphate (NADPH)** is the reducing agent used in the Calvin cycle. NADPH transfers the electrons that allow carbohydrates to be synthesized from CO_2 (Fig. 8.5).

Like all components of the Calvin cycle, NADPH can move freely within the stroma of the chloroplast. Although NADPH is a powerful reducing agent, energy and electrons are transferred from NADPH only under the catalysis of a specific enzyme, thus providing a high degree of control over the fate of these electrons. In the Calvin cycle, the reduction of 3-PGA involves two steps: (1) ATP donates a phosphate group to 3-PGA, and (2) NADPH transfers two electrons plus one proton (H^+) to the phosphorylated compound, which releases one phosphate group (P_i). Because two molecules of 3-PGA are formed each time rubisco catalyzes the incorporation of one molecule of CO_2, two ATP and two NADPH are required for each molecule of CO_2 incorporated by rubisco. NADPH provides most of the energy incorporated in the bonds of the carbohydrate molecules produced by the Calvin cycle. Nevertheless, ATP plays an essential role in preparing 3-PGA for the addition of energy and electrons from NADPH.

These energy transfer steps result in the formation of 3-carbon carbohydrate molecules known as **triose phosphates.** Triose phosphates are the true products of the Calvin cycle and they are the principal form in which carbohydrates are exported from the chloroplast during photosynthesis. Larger sugars, such as glucose and sucrose, are assembled from triose phosphates in the cytoplasm.

If every triose phosphate molecule produced by the Calvin cycle were exported from the chloroplast, RuBP could not be regenerated and the Calvin cycle would grind to a halt. In fact, most of the triose phosphate molecules must be used to regenerate RuBP. For every six triose phosphate molecules that are produced, only one can be withdrawn from the Calvin cycle.

The regeneration of RuBP requires ATP.

Of the 15 chemical reactions that make up the Calvin cycle, 12 occur in the last step, the regeneration of RuBP (Fig. 8.5). A large number of reactions is needed to rearrange the carbon atoms from five 3-carbon triose phosphate molecules into three 5-carbon RuBP molecules. ATP is required for the regeneration of RuBP, raising the Calvin cycle's total energy requirements to two molecules of NADPH and three molecules of ATP for each molecule of CO_2 incorporated by rubisco.

The Calvin cycle does not use sunlight directly. For this reason, this pathway is sometimes referred to as the light-independent or even the "dark" reactions of photosynthesis. However, this pathway cannot operate without the energy input provided by a steady supply of NADPH and ATP. Both are supplied by the photosynthetic electron transport chain, in which light is captured and transformed into chemical energy. In addition, several Calvin cycle enzymes are regulated by cofactors that must be activated by the photosynthetic electron transport chain. Thus, in a photosynthetic cell, the Calvin cycle occurs only in the light.

→ **Quick Check 2** The Calvin cycle requires both ATP and NADPH. Which of these molecules provides the major input of energy needed to synthesize carbohydrates?

The steps of the Calvin cycle were determined using radioactive CO_2.

In a series of experiments conducted between 1948 and 1954, the American chemist Melvin Calvin and colleagues identified the carbon compounds produced during photosynthesis (**Fig. 8.6**). They supplied radioactively labeled CO_2 ($^{14}CO_2$) to the unicellular green alga *Chlorella* and then plunged the cells into boiling alcohol, thereby halting all enzymatic reactions. The carbon compounds produced during photosynthesis were thus radioactively labeled and could be identified by their radioactivity (Experiment 1 in Fig. 8.6).

Figuring out the chemical reactions that connected these labeled compounds, however, required additional experiments. For example, by using a very short exposure to $^{14}CO_2$, Calvin and colleagues determined that 3-PGA was the first stable product of the Calvin cycle (Experiment 2 in Fig. 8.6).

To determine how 3-PGA is formed, they first supplied $^{14}CO_2$ so that all of the molecules in the Calvin cycle became radioactively labeled. When they then cut off the supply of CO_2, the amount of RuBP increased relative to the amount seen in the first experiment. Based on this buildup of RuBP, they concluded that the first step in the Calvin cycle was the addition of CO_2 to RuBP (Experiment 3 in Fig. 8.6).

Carbohydrates are stored in the form of starch.

The Calvin cycle is capable of producing more carbohydrates than the cell needs or, in a multicellular organism, more than the cell is able to export. If carbohydrates accumulated in the cell, they would cause water to enter the cell by osmosis, perhaps damaging the cell. Instead, excess carbohydrates are converted to starch, a storage form of carbohydrates discussed in Chapter 2. Because starch molecules are not soluble, they provide a means of carbohydrate

HOW DO WE KNOW?

FIG. 8.6

How is CO_2 used to synthesize carbohydrates?

BACKGROUND In the 1940s, radioactive $^{14}CO_2$ became available in quantities that allowed experiments. Melvin Calvin and Andrew Benson used $^{14}CO_2$ to follow the incorporation of CO_2 into carbohydrates.

EXPERIMENTS AND RESULTS

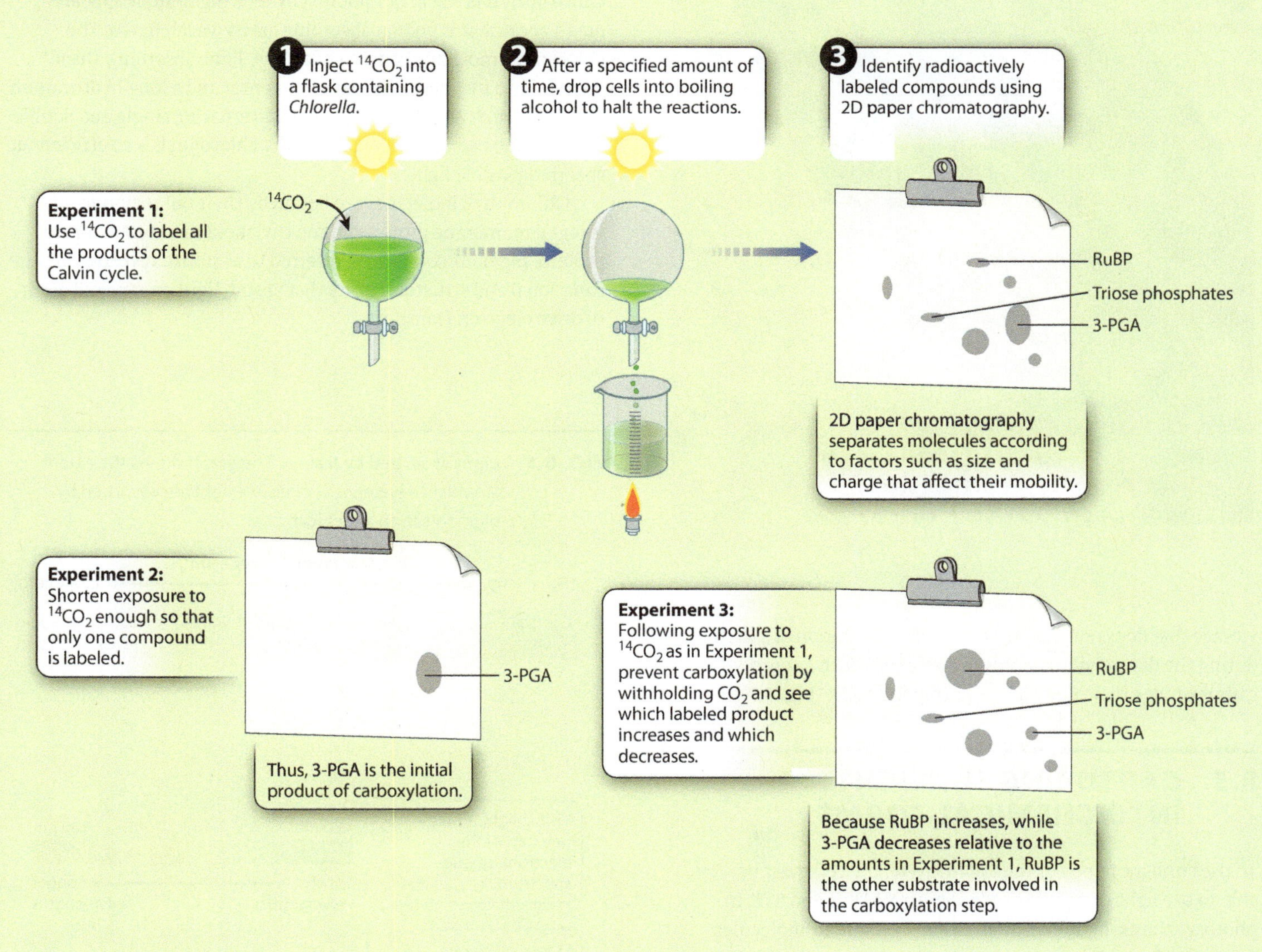

CONCLUSION The initial step in the Calvin cycle unites the 5-carbon RuBP with CO_2, resulting in the production of two molecules of 3-PGA.

FOLLOW-UP WORK In the 1950s, Marshall Hatch and colleagues showed that some plants, including corn and sugarcane, accumulate a 4-carbon compound as the first product in photosynthesis. In Chapter 29, we explore how C_4 photosynthesis allows plants to avoid the oxygenase reaction of rubisco.

SOURCE Calvin, M., and H. Benson. 1949. "The Path of Carbon in Photosynthesis IV: The Identity and Sequence of the Intermediates in Sucrose Synthesis." *Science* 109:140–142.

FIG. 8.7 A chloroplast containing starch granules, shown here in yellow. *Source: Biophoto Associates/Science Source.*

storage that does not lead to osmosis. The formation of starch during the day provides photosynthetic cells with a source of carbohydrates that they can use during the night (**Fig. 8.7**).

8.3 CAPTURING SUNLIGHT INTO CHEMICAL FORMS

To use sunlight to power the Calvin cycle, the cell must be able to use light energy to produce both NADPH and ATP. In photosynthesis, light energy absorbed by pigment molecules drives the flow of electrons through the photosynthetic electron transport chain. The movement of electrons through the photosynthetic electron chain leads to the formation of both NADPH and ATP.

Chlorophyll is the major entry point for light energy in photosynthesis.

To understand how light energy is captured and stored by photosynthesis, we need to know a little about light. The sun, like all stars, produces a broad spectrum of electromagnetic radiation ranging from gamma rays to radio waves. Each point along the electromagnetic spectrum has a different energy level and a corresponding wavelength. **Visible light** is the portion of the electromagnetic spectrum apparent to our eyes, and it includes the range of wavelengths used in photosynthesis. The wavelengths of visible light range from 400 nm to 700 nm. Approximately 40% of the sun's energy that reaches Earth's surface is in this range.

Pigments are molecules that absorb some wavelengths of visible light (**Fig. 8.8**). Pigments look colored because they reflect light enriched in the wavelengths that they do not absorb. **Chlorophyll** is the major photosynthetic pigment; it appears green because it is poor at absorbing green wavelengths. The chlorophyll molecule consists of a large, light-absorbing "head" containing a magnesium atom at its center and a long hydrocarbon "tail" (**Fig. 8.9**). The large number of alternating single and double bonds in the head region explains why chlorophyll is so efficient at absorbing visible light.

Chlorophyll molecules are bound by their tail region to integral membrane proteins in the thylakoid membrane. These protein–pigment complexes, referred to as **photosystems,** are the functional and structural units that absorb light energy and use it to drive electron transport.

FIG. 8.8 Light absorbed by leaves. The graph shows the extent to which wavelengths of visible light are absorbed by pigments in an intact leaf.

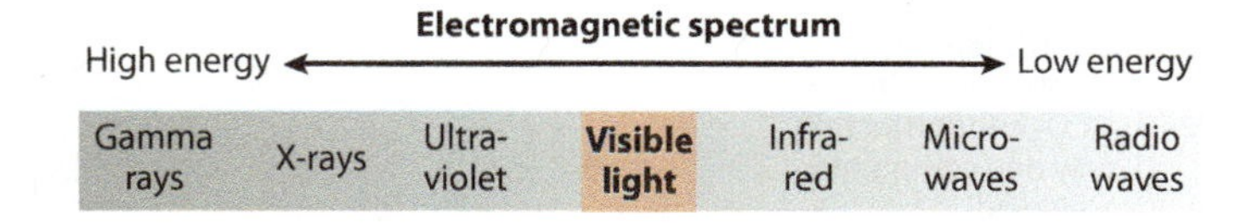

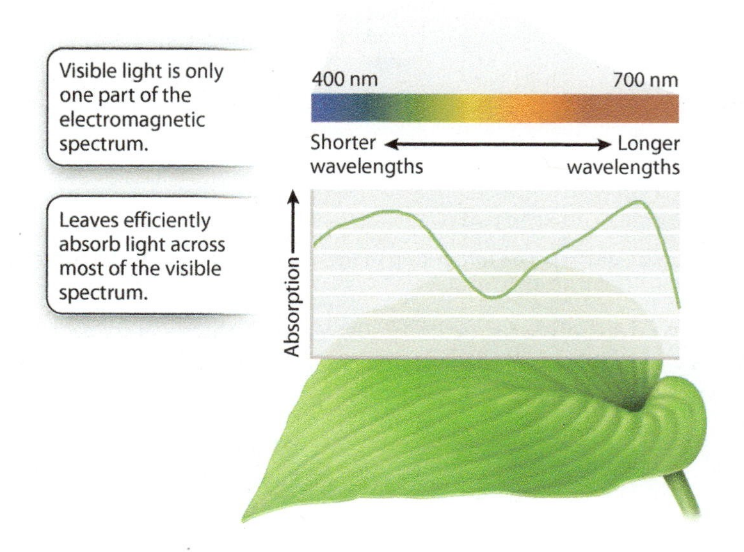

Photosystems contain pigments other than chlorophyll, called **accessory pigments.** The most notable are the orange-yellow carotenoids, which can absorb light from regions of the visible spectrum that are poorly absorbed by chlorophyll. Thus, the presence of these accessory pigments allows photosynthetic cells to absorb a broader range of visible light than would be possible with just chlorophyll alone. As we will see in section 8.4, carotenoids play an important role in protecting the photosynthetic electron transport chain from damage.

Photosystems use light energy to drive the photosynthetic electron transport chain.

When visible light is absorbed by a chlorophyll molecule, one of its electrons is elevated to a higher energy state (**Fig. 8.10**). For

FIG. 8.9 Chemical structure of chlorophyll. Shown is chlorophyll *a*, found in all photosynthetic eukaryotes and cyanobacteria.

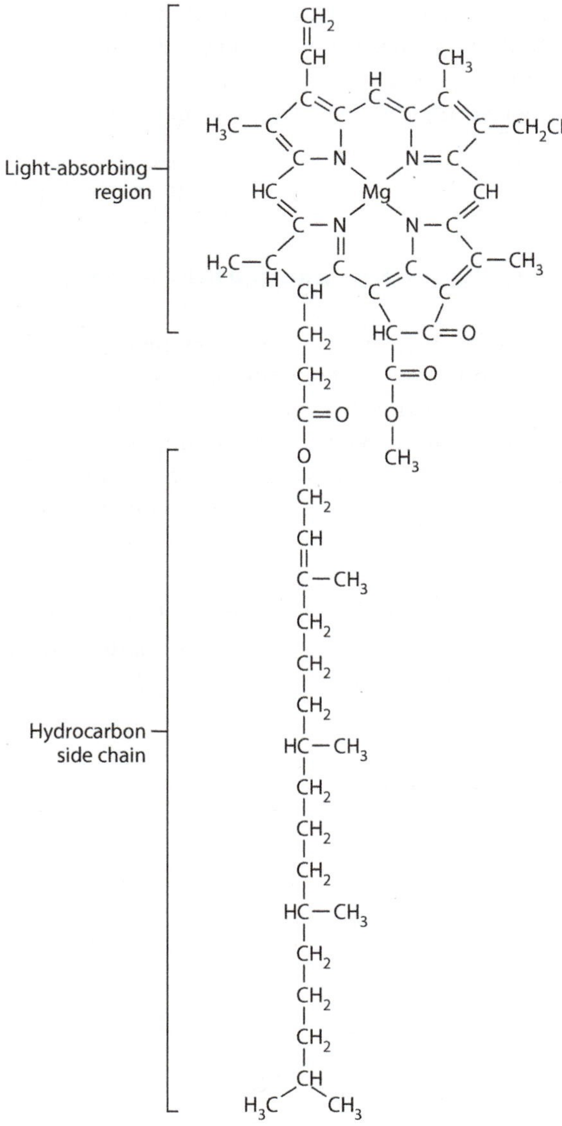

Light-absorbing region

Hydrocarbon side chain

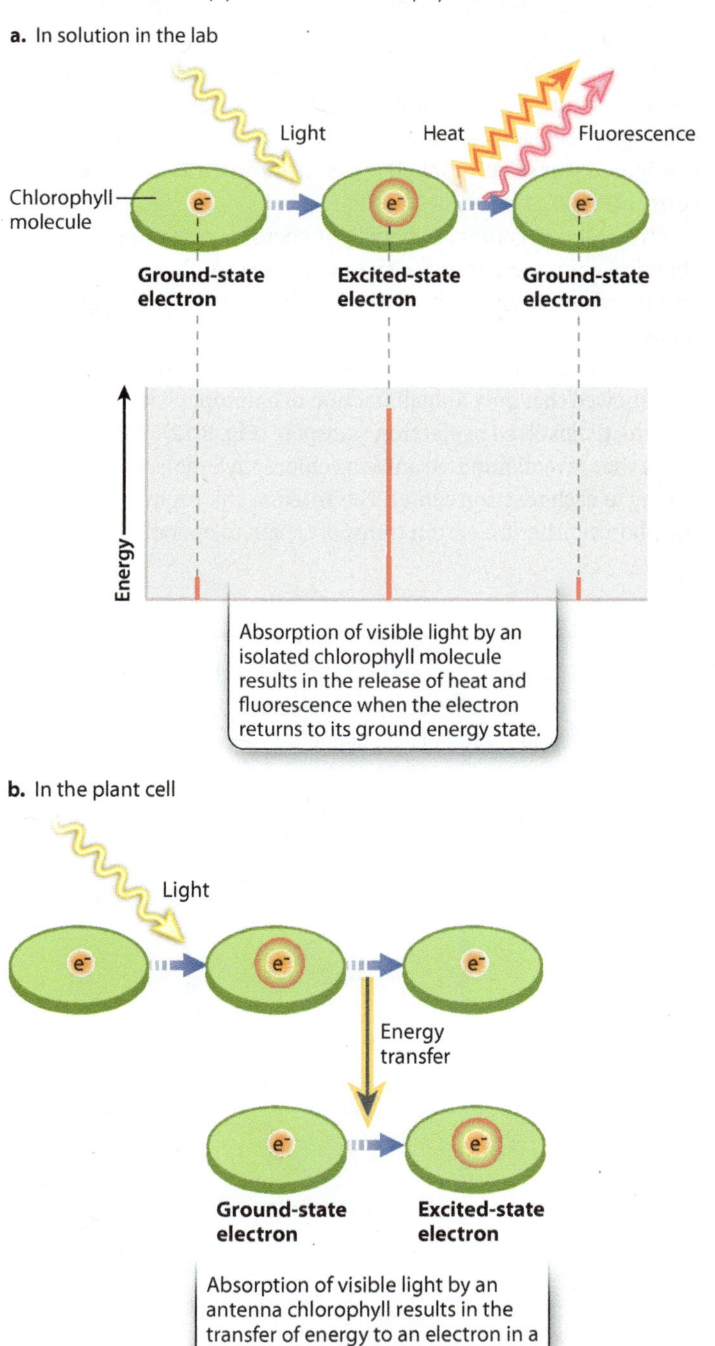

FIG. 8.10 Absorption of light energy by chlorophyll. Absorption of light energy by (a) an isolated chlorophyll molecule in the lab and (b) an antenna chlorophyll molecule.

chlorophyll molecules that have been extracted from chloroplasts in the laboratory, this absorbed light energy is rapidly released, allowing the electron to return to its initial "ground" energy state (Fig. 8.10a). Most of the energy (>95%) is converted into heat; a small amount is reemitted as light (fluorescence).

By contrast, for chlorophyll molecules within an intact chloroplast, energy can be transferred to an adjacent chlorophyll molecule instead of being lost as heat (Fig. 8.10b). When this happens, the energy released as an excited electron returns to its ground state raises the energy level of an electron in an adjacent chlorophyll molecule. This mode of energy transfer is extremely efficient (that is, very little energy is lost as heat), allowing energy initially absorbed from sunlight to be transferred from one chlorophyll molecule to another and then on to another.

Most of the chlorophyll molecules in the thylakoid membrane function as an antenna: Energy is transferred between chlorophyll molecules until it is finally transferred to a specially configured pair of chlorophyll molecules known as the **reaction center** (**Fig. 8.11**).

The reaction center is where light energy is converted into chemical energy as a result of the excited electron's transfer to an adjacent molecule. This division of labor among chlorophyll molecules was discovered in the 1940s in a series of experiments by the American biophysicists Robert Emerson and William Arnold, who showed that only a small fraction of chlorophyll molecules are directly involved in electron transport (**Fig. 8.12**). We now know that several hundred antenna chlorophyll molecules transfer energy to each reaction center. The antenna chlorophylls allow the photosynthetic electron transport chain to operate efficiently. Without the antennae to gather light energy, reaction centers would sit idle much of the time, even in bright sunlight.

The reaction center chlorophylls have a configuration distinct from that of the antenna chlorophylls. As a result, when excited, the reaction center transfers an electron to an adjacent molecule that acts as an electron acceptor (Fig. 8.11a). When the transfer takes place, the reaction center becomes oxidized and the adjacent electron-acceptor molecule is reduced. The result is the conversion of light energy into a chemical form. This electron transfer initiates a light-driven chain of redox reactions that leads ultimately to the formation of NADPH.

Once the reaction center has lost an electron, it can no longer absorb light or contribute additional electrons. Thus, for the photosynthetic electron transport chain to continue, another electron must be delivered to take the place of the one that has entered the transport chain (Fig. 8.11b). As we will see below, these replacement electrons ultimately come from water.

→ **Quick Check 3** How do antenna chlorophylls differ from reaction center chlorophylls?

The photosynthetic electron transport chain connects two photosystems.

In many ways, water is an ideal source of electrons for photosynthesis. Water is so abundant within cells that it is always available to serve as an electron donor in photosynthesis. In addition, O_2, the by-product of pulling electrons from water, diffuses readily away rather than accumulates. However, from an energy perspective, water is a challenging electron donor: It takes

FIG. 8.11 The reaction center. (a) Antenna chlorophylls deliver absorbed light energy to the reaction center, allowing electrons to be transferred to an electron acceptor molecule. (b) After the reaction center has lost an electron, it is reduced by gaining an electron, so it is ready to absorb additional light energy.

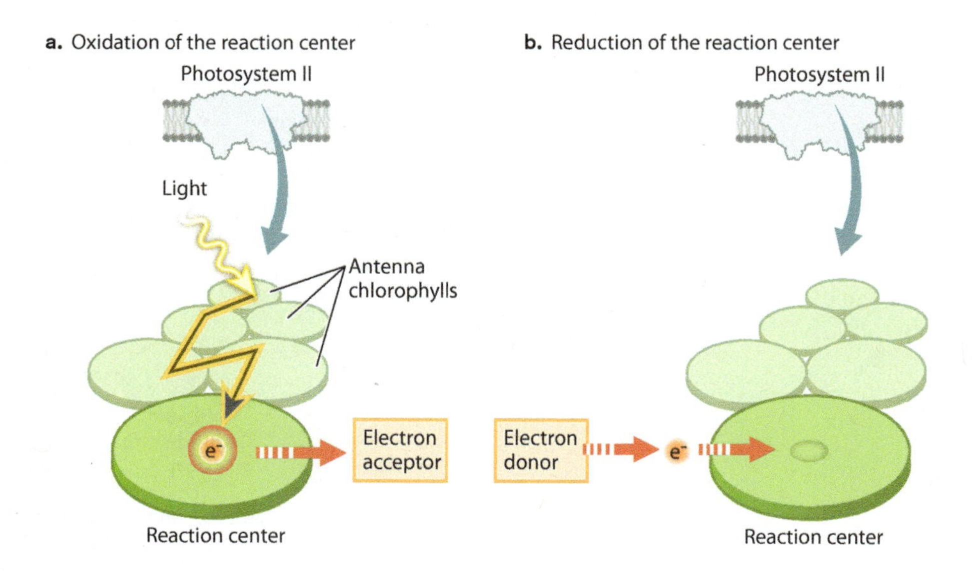

HOW DO WE KNOW?

FIG. 8.12

Do chlorophyll molecules operate on their own or in groups?

BACKGROUND By about 1915, scientists knew that chlorophyll was the pigment responsible for absorbing light energy in photosynthesis. However, it was unclear how these pigments contributed to the reduction of CO_2. The American physiologists Robert Emerson and William Arnold set out to determine the nature of the "photochemical unit" by quantifying how many chlorophyll molecules were needed to produce one molecule of O_2.

EXPERIMENT Emerson and Arnold exposed flasks of the green alga *Chlorella* to flashes of light of such short duration (10^{-5} s) and high intensity (several times that of full sunlight) that each chlorophyll molecule would be "excited" only once. They also made sure that the time between flashes was long enough to allow the reactions resulting from each flash to run to completion. They then measured O_2 production per flash of high intensity light, and determined the concentration of chlorophyll present in their solution of cells. Finally, they divided the number of chlorophyll molecules per mm^3 of solution by O_2 production per flash per mm^3 of cells in solution.

RESULTS

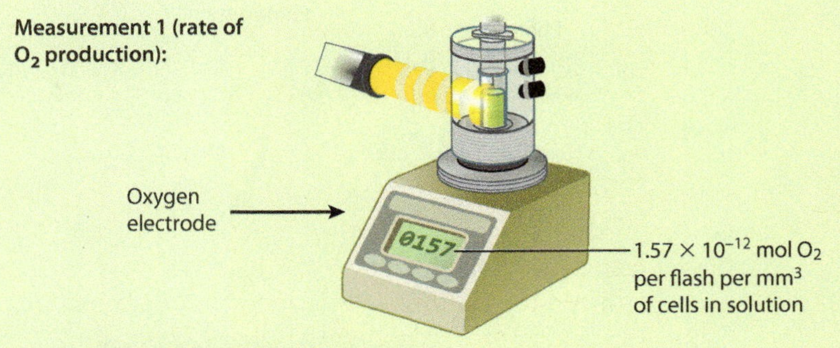

CONCLUSION Because the amount of chlorophyll in their flask was much greater than O_2 production per flash, Emerson and Arnold concluded that each photochemical unit contains many chlorophyll molecules.

FOLLOW-UP WORK Emerson and Arnold's work was followed by studies that demonstrated that the photosynthetic electron transport chain contains two photosystems arranged in series.

SOURCE Emerson, R., and W. Arnold. 1932. "The Photochemical Reaction in Photosynthesis." *Journal of General Physiology* 16:191–205.

a great deal of energy to pull electrons from water. The amount of energy that a single photosystem can capture from sunlight is not enough both to pull an electron from water and produce an electron donor capable of reducing $NADP^+$. The solution is to use two photosystems arranged in series. The energy supplied by the first photosystem allows electrons to be pulled from water, and the energy supplied by the second photosystem step allows electrons to be transferred to $NADP^+$.

If you follow the flow of electrons from water through both photosystems and on to $NADP^+$, as shown in **Fig. 8.13,** you can see a large increase in energy as the electrons pass through each of the two photosystems. You can also see that at every other step along the photosynthetic electron transport chain there is a small decrease in energy. This decrease in energy indicates that these are exergonic reactions (Chapter 6) and thus explains why electrons move in one "direction" through the series of redox reactions that make up the photosynthetic electron transport chain. To run these reactions in the opposite direction would require an input of energy. Because the overall energy trajectory has an up-down-up configuration resembling a "Z," the photosynthetic electron transport chain is sometimes referred to as the **Z scheme.**

For the two photosystems to work together to move electrons from water to NADPH, they must have distinct chemical properties. **Photosystem II** supplies electrons to the beginning of the electron transport chain. When photosystem II loses an electron (that is, when it is itself oxidized), it is able to pull electrons from water. In contrast, **photosystem I** energizes electrons with a second input of light energy so they can be used to reduce $NADP^+$. The key point here is that photosystem I when oxidized is not a sufficiently strong oxidant to split water, whereas photosystem II is not a strong enough reductant to form NADPH.

The major protein complexes of the photosynthetic electron transport chain include the two photosystems as well as the

FIG. 8.13 **The Z scheme.** The use of water as an electron donor requires input of light energy at two places in the photosynthetic electron transport chain.

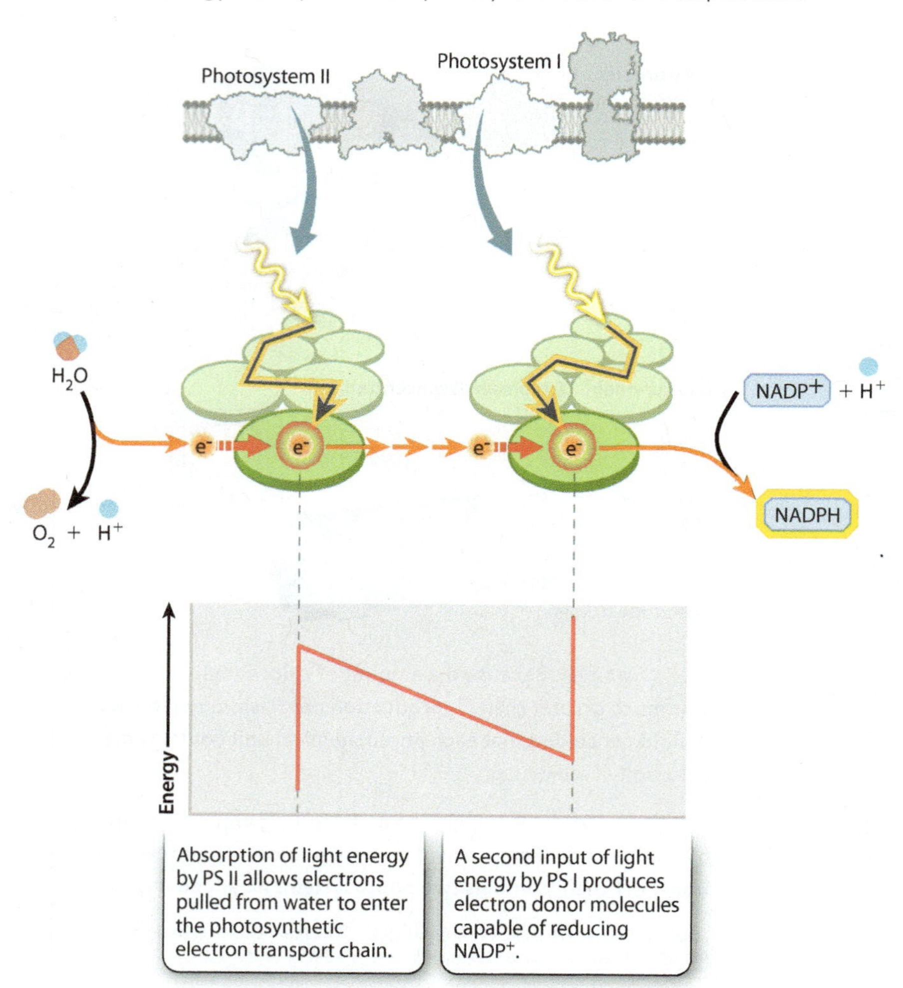

cytochrome-b_6f complex (cyt), through which electrons pass between photosystem II and photosystem I (**Fig. 8.14**). Small, relatively mobile compounds convey electrons between these protein complexes. Plastoquinone (Pq), a lipid-soluble compound similar in structure to coenzyme Q (Chapter 7), carries electrons from photosystem II to the cytochrome-b_6f complex by diffusing through the membrane, while plastocyanin (Pc), a water-soluble protein, carries electrons from the cytochrome-b_6f complex to photosystem I by diffusing through the thylakoid lumen.

Water donates electrons to one end of the photosynthetic electron transport chain, whereas $NADP^+$ accepts electrons at the other end. The enzyme that pulls electrons from water, releasing both H^+ and O_2, is located on the lumen side of photosystem II. The mechanism by which water splitting occurs is not known, despite the considerable industrial value of developing a way to use sunlight to generate hydrogen gas (H_2). NADPH is formed when electrons are passed from photosystem I to a membrane-associated protein called ferredoxin (Fd) (Fig. 8.14b). The enzyme ferredoxin–$NADP^+$ reductase then catalyzes the formation of NADPH by transferring two electrons from two molecules of reduced ferredoxin to $NADP^+$ as well as a proton from the surrounding solution:

$$NADP^+ + 2e^- + H^+ \rightarrow NADPH$$

→ **Quick Check 4** Why are two photosystems needed if H_2O is used as an electron donor?

The accumulation of protons in the thylakoid lumen drives the synthesis of ATP.

So far, we have considered only how the photosynthetic electron transport chain leads to the formation of NADPH. However, we know that the Calvin cycle also requires ATP. In chloroplasts, as in mitochondria, ATP is synthesized by ATP synthase, a transmembrane protein powered by a proton gradient (Chapter 7). In chloroplasts, the ATP synthase is oriented such that the synthesis of ATP is the result of the movement of protons from the thylakoid lumen to the stroma, as shown in Fig. 8.14c.

How do protons accumulate in the thylakoid lumen? Two features of the photosynthetic electron transport chain are responsible for the buildup of protons in the thylakoid lumen (Fig. 8.14c). First, the oxidation of water releases protons and O_2 into the lumen. Second, the cytochrome-b_6f complex, the protein complex situated between photosystem II and photosystem I, and plastoquinone together function as a proton pump that is functionally and evolutionarily related to proton pumping in the electron transport chain of cellular respiration (Chapter 7).

In photosynthesis, the proton pump involves: (1) the transport of two electrons and two protons, by the diffusion of plastoquinone, from the stroma side of photosystem II to the lumen side of the cytochrome-b_6f complex and (2) the transfer of electrons within the cytochrome-b_6f complex to a different molecule of plastoquinone, which results in additional protons being picked up from the stroma and subsequently released into the lumen.

Together, these mechanisms are quite powerful. When the photosynthetic electron transport chain is operating at full capacity, the concentration of protons in the lumen can be more than 1000 times greater than their concentration in the stroma (equivalent to a difference of 3 pH units). This accumulation of

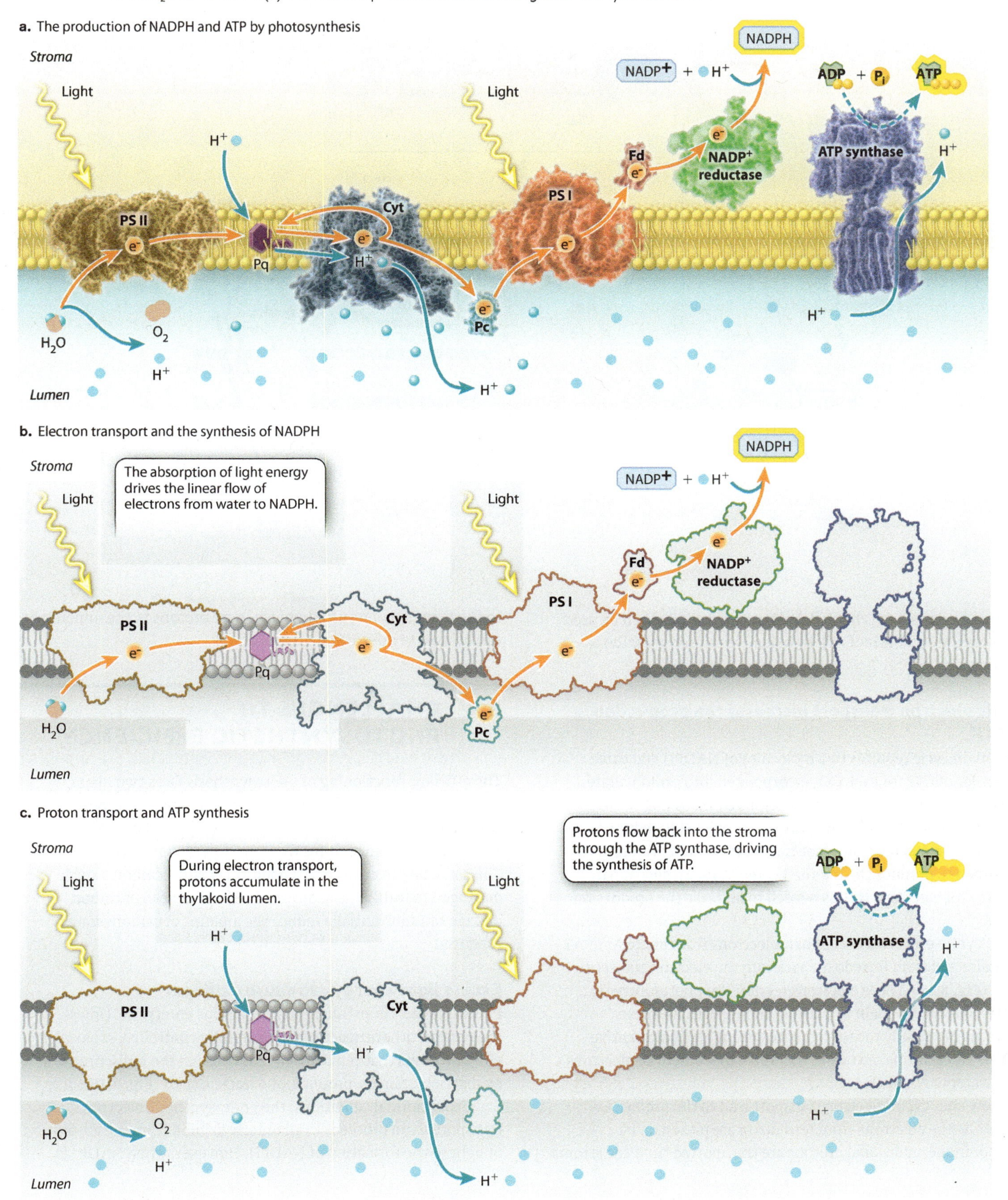

FIG. 8.14 The photosynthetic electron transport chain. (a) An overview of the production of NADPH and ATP. (b) The linear flow of electrons from H_2O to NADPH. (c) The use of a proton electrochemical gradient to synthesize ATP.

FIG. 8.15 Cyclic electron transport. To increase ATP production, some electrons from photosystem I cycle back into the electron transport chain.

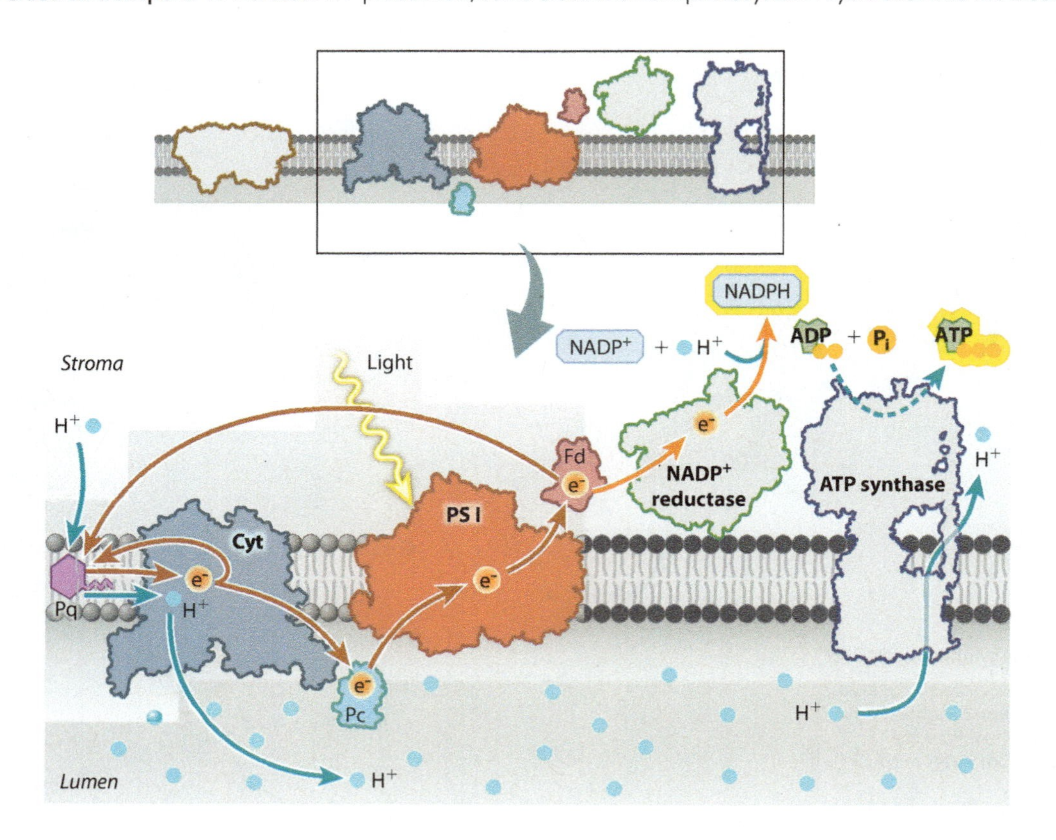

protons on one side of the thylakoid membrane can then be used to power the synthesis of ATP by oxidative phosphorylation as described in Chapter 7.

Cyclic electron transport increases the production of ATP.

The Calvin cycle requires two molecules of NADPH and three molecules of ATP for each CO_2 incorporated into carbohydrates. However, the transport of four electrons through the photosynthetic electron transport chain, needed to reduce two $NADP^+$ molecules, does not transport enough protons into the lumen to produce the required three ATPs. An additional pathway for electrons is thus needed to increase the production of ATP.

In **cyclic electron transport,** electrons from photosystem I are redirected from ferredoxin back into the electron transport chain (**Fig. 8.15**). These electrons reenter the photosynthetic electron transport chain by plastoquinone. Because these electrons eventually return to photosystem I, this alternative pathway is cyclic in contrast to the linear movement of electrons from water to NADPH.

How does cyclic electron transport lead to the production of ATP? As the electrons from ferredoxin are picked up by plastoquinone, additional protons are transported from the stroma to the lumen. As a result, there are more protons in the lumen that can be used to drive the synthesis of ATP.

8.4 CHALLENGES TO PHOTOSYNTHETIC EFFICIENCY

The efficient functioning of photosynthesis faces two major challenges. The first is that if more light energy is absorbed than the Calvin cycle can use, excess energy can damage the cell. The second challenge stems from a property of rubisco: This enzyme can catalyze the addition of *either* carbon dioxide *or* oxygen to RuBP. The addition of oxygen instead of carbon dioxide can substantially reduce the amount of carbohydrate produced.

Excess light energy can cause damage.

Photosynthesis is an inherently dangerous enterprise. Unless the photosynthetic reactions are carefully controlled, molecules will be formed that can damage cells through the indiscriminate oxidization of lipids, proteins, and nucleic acids (**Fig. 8.16**).

Under normal conditions, the photosynthetic electron transport chain proceeds in an orderly fashion from the absorption of light to the formation of NADPH. However, when $NADP^+$ is

FIG. 8.16 Defenses against reactive oxygen species. (a) Reactive oxygen species are generated when light energy or electrons are transferred to oxygen. (b) Defenses against the reactive oxygen species include antioxidants that neutralize reactive oxygen species and xanthophylls that convert excess light energy into heat.

a. The problem: Harmful reactive oxygen species are generated.

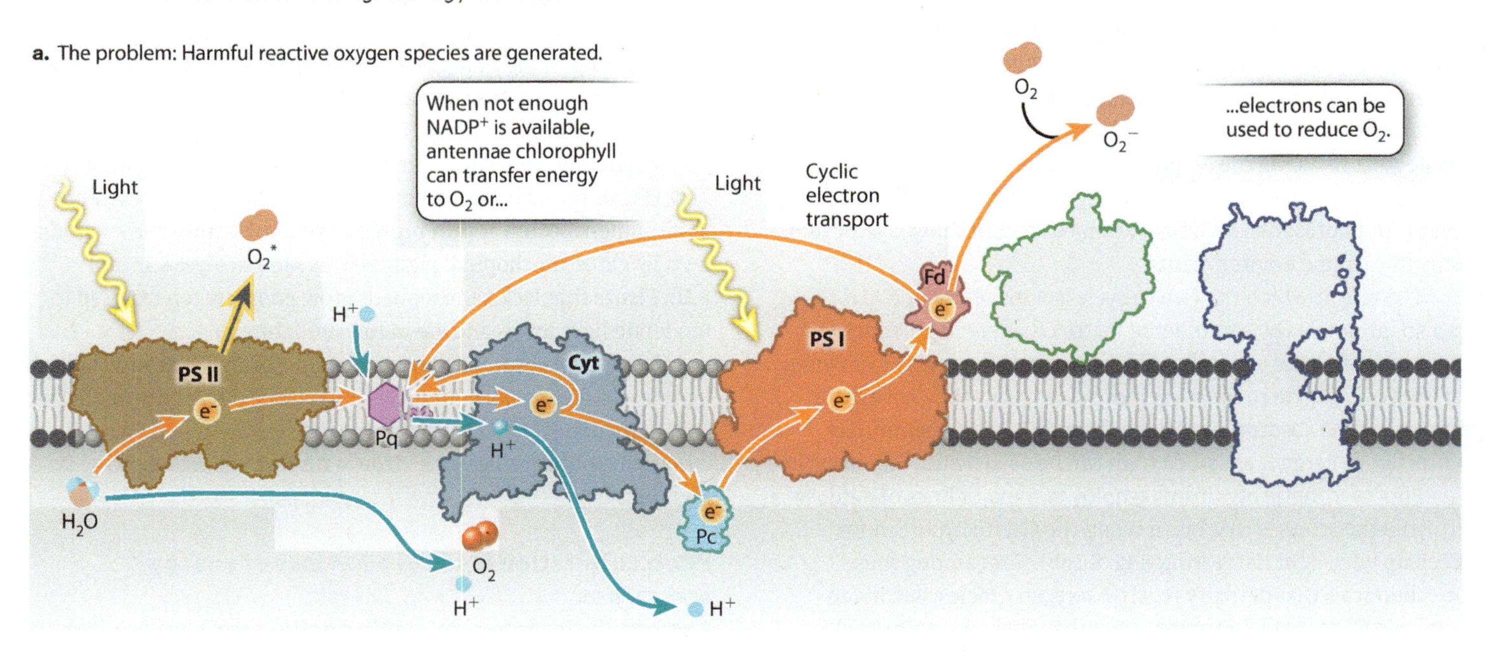

b. The solution: Xanthophylls and antioxidants reduce the amount of reactive oxygen species.

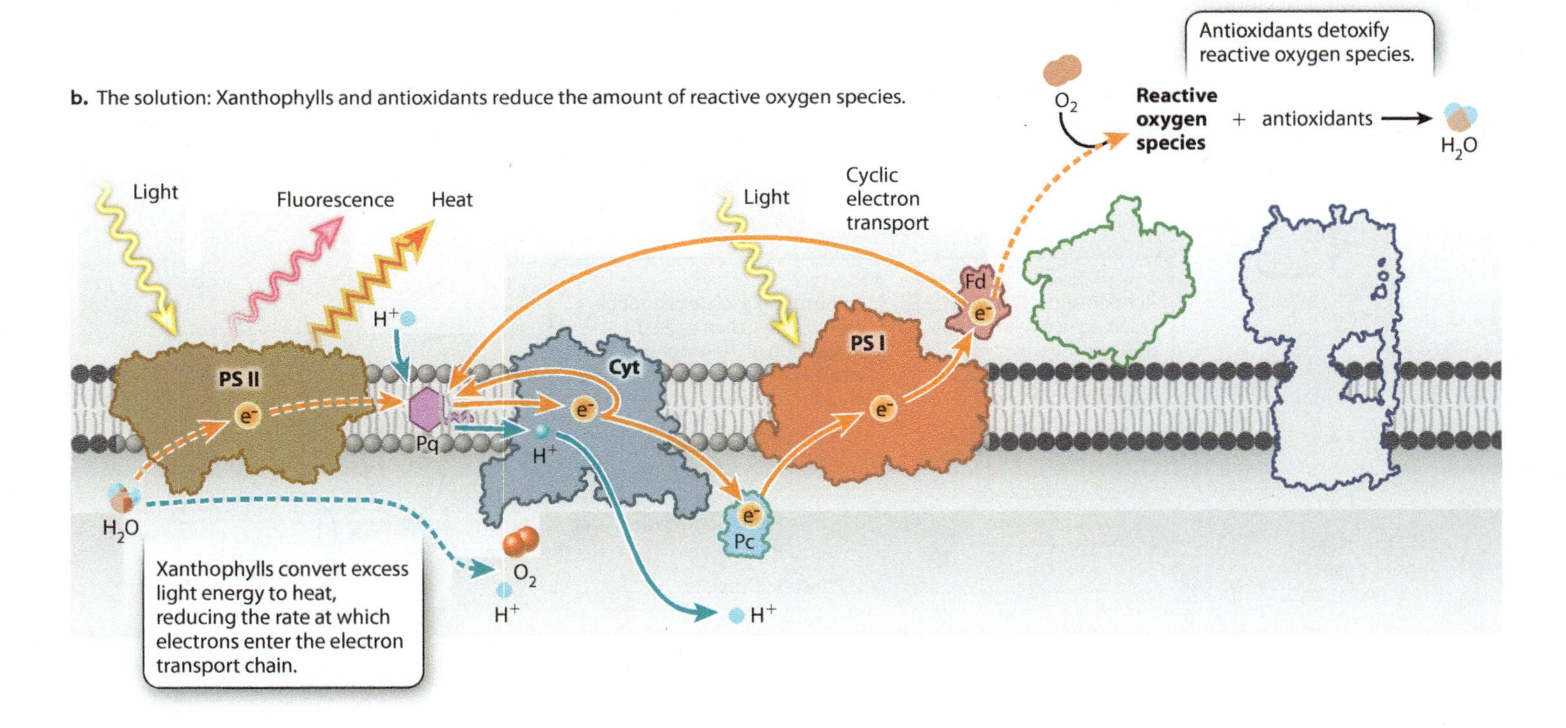

in short supply, the electron transport chain "backs up," greatly increasing the probability of creating highly reactive forms of oxygen known collectively as **reactive oxygen species** (Fig. 8.16a). These highly reactive molecules can be formed either by the transfer of absorbed light energy from antenna chlorophyll directly to O_2 or by the transfer of an electron, forming O_2^-. Both forms of O_2 can cause substantial damage to the cell.

$NADP^+$ is returned to the photosynthetic electron transport chain by the Calvin cycle's use of NADPH. Thus, any factor that causes the rate of NADPH use to fall behind the rate of light-driven electron transport can potentially lead to damage. Such an imbalance is likely to occur, for example, in the middle of the day when light intensity is highest. Could the cell right the balance by supplying $NADP^+$ more quickly? Photosynthetic cells could speed up the resupply of $NADP^+$ by synthesizing more Calvin cycle enzymes. This strategy, however, would be energetically expensive. When light levels are low, such as in the morning and late afternoon, Calvin cycle enzymes would sit idle. An alternative strategy of reducing the amount of chlorophyll in the leaf runs into similar problems. Thus, excess light energy is an everyday event for photosynthetic cells, rather than something that occurs only in extreme environments.

The rate at which the Calvin cycle can make use of NADPH is also influenced by a number of factors that are independent of light intensity. For example, cold temperatures cause the enzymes of the Calvin cycle to function more slowly, but they have little impact on the absorption of light energy. On a cold, sunny day, more light energy is absorbed than can be used by the Calvin cycle.

Photosynthetic organisms employ two major lines of defense to avoid the stresses that occur when the Calvin cycle cannot keep up with light harvesting (Fig. 8.16b). First among these are chemicals that detoxify reactive oxygen species. Ascorbate (vitamin C), β-(beta-)carotene, and other antioxidants are able to neutralize reactive oxygen species. These compounds exist in high concentration in chloroplasts. Some of these antioxidant molecules are brightly colored, like the red pigments found in algae that live on snow shown in Fig. 8.1c. The presence of antioxidant compounds is one of the many reasons that eating green, leafy vegetables is good for your health.

A second line of defense is to prevent reactive oxygen species from forming in the first place. **Xanthophylls** are yellow-orange pigments that slow the formation of reactive oxygen species by reducing excess light energy. These pigments accept absorbed light energy directly from chlorophyll and then convert this energy to heat (Fig. 8.16b). Photosynthetic organisms that live in extreme environments often appear brown or yellow because they contain high levels of xanthophyll pigments, as seen in Figs. 8.1a and 8.1b. Plants that lack xanthophylls grow poorly when exposed to moderate light levels and die in full sunlight.

Converting absorbed light energy into heat is beneficial at high light levels, but at low light levels it would decrease the production of carbohydrates. Therefore, this capability is switched on only when the photosynthetic electron transport chain is working at high capacity.

Photorespiration leads to a net loss of energy and carbon.

A second challenge to photosynthetic efficiency is the fact that rubisco can use both CO_2 and O_2 as substrates. If O_2 instead of CO_2 diffuses into the active site of rubisco, the reaction can

FIG. 8.17 Photorespiration. Carbon and energy are lost when rubisco acts as an oxygenase in photorespiration.

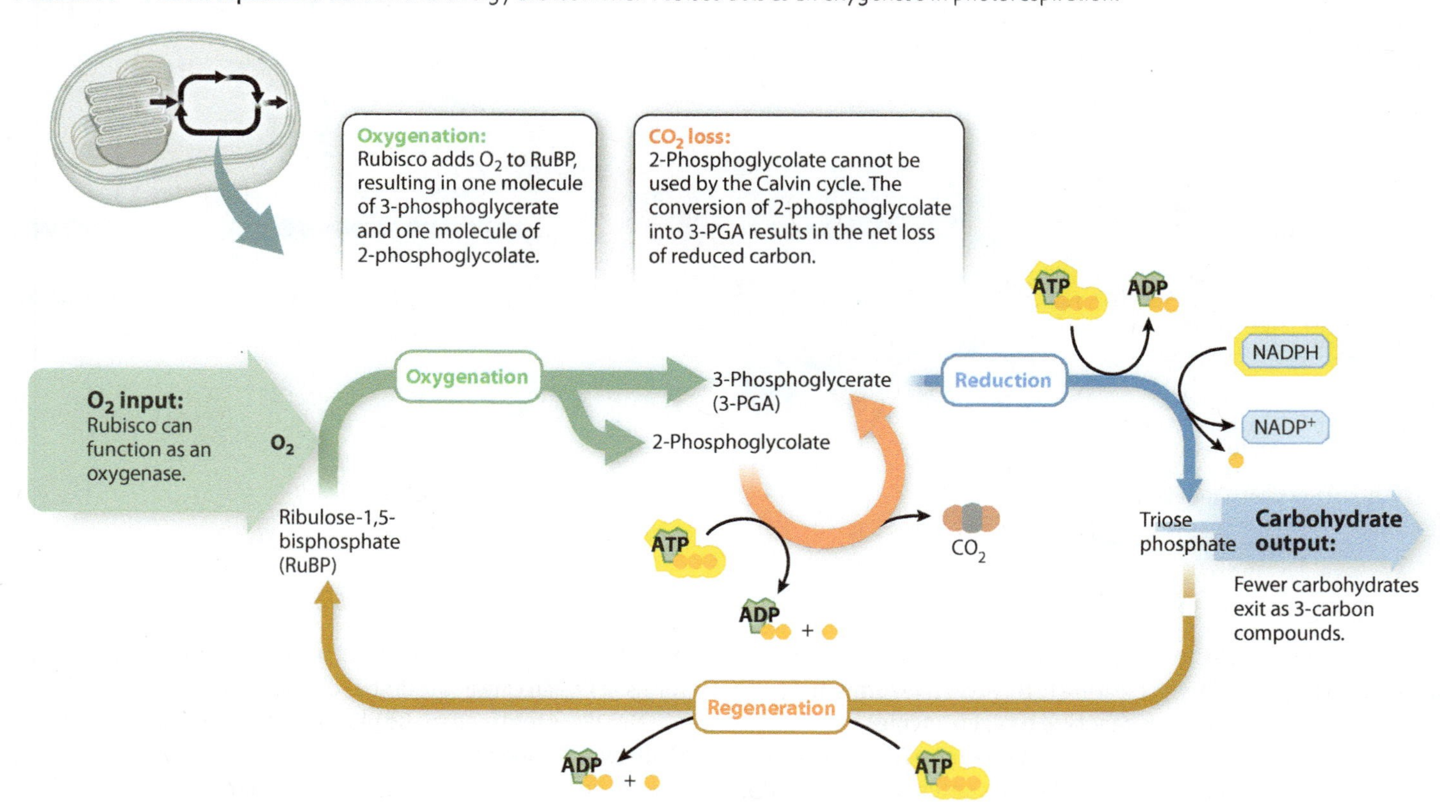

still proceed, although O_2 is added to RuBP in place of CO_2. An enzyme that adds O_2 to another molecule is called an oxygenase. Recall that rubisco is shorthand for RuBP carboxylase oxygenase, reflecting rubisco's ability to catalyze two different reactions.

When rubisco adds O_2 instead of CO_2 to RuBP, the result is one molecule with three carbon atoms (3-PGA) and one molecule with only two carbon atoms (2-phosphoglycolate). The production of 2-phosphoglycolate creates a serious problem because this molecule cannot be used by the Calvin cycle either to produce triose phosphate or to regenerate RuBP.

A metabolic pathway to recycle 2-phosphoglycolate is present in photosynthetic cells. A portion of the carbon atoms in 2-phosphoglycolate are converted into 3-PGA, which can reenter the Calvin cycle. However, this pathway is not able to return all of the carbon atoms in 2-phosphoglycolate to the Calvin cycle; some are released as CO_2. Because the overall effect is the consumption of O_2 and release of CO_2 in the presence of light, this process is referred to as **photorespiration** (**Fig. 8.17**). However, whereas respiration produces ATP, photorespiration consumes ATP. In photorespiration, ATP drives the reactions that recycle 2-phosphoglycolate into 3-PGA. Thus, photorespiration represents a net energy drain on two accounts: First, it results in the oxidation and loss, in the form of CO_2, of carbon atoms that had previously been incorporated and reduced by the Calvin cycle, and second, it consumes ATP.

→ **Quick Check 5** In what ways is photorespiration similar to cellular respiration (Chapter 7) and in what ways does it differ?

The Calvin cycle originated long before the accumulation of oxygen in Earth's atmosphere, providing an explanation why an enzyme with these properties might have initially evolved. Still, why would photorespiration persist in the face of what must be strong evolutionary pressure to reduce or eliminate the unwanted reaction with oxygen?

The difficulty is that for rubisco to favor the addition of CO_2 over O_2 requires that the enzyme be highly selective, and the price of high selectivity is speed. CO_2 and O_2 are similar in size and chemical structure, and for this reason selectivity can only be achieved by rubisco binding more tightly with the transition state (Chapter 6) of the carboxylation reaction. As a result, the better rubisco is at discriminating between CO_2 and O_2, the slower its catalytic rate.

Nowhere is this trade-off more evident than in land plants, whose photosynthetic cells acquire CO_2 from an O_2-rich and CO_2-poor atmosphere. The rubiscos of land plants are highly selective: If they are exposed to equal concentrations of CO_2 and O_2, the ratio of CO_2 addition to O_2 addition is approximately 80:1. As a result, rubisco is a very slow enzyme, with catalytic rates on the order of three reactions per second. To put this rate in perspective, it is not uncommon for metabolic enzymes to achieve a catalytic rate of tens of thousands of reactions per second.

This trade-off between selectivity and speed is a key constraint for photosynthetic organisms. For land plants, rubisco's low catalytic rate means that photosynthetic cells must produce huge amounts of this enzyme; as much as 50% of the total protein within a leaf is rubisco, and it is estimated to be the most abundant protein on Earth. At the same time, because O_2 is approximately 500 times more abundant in the atmosphere than CO_2, as much as one-quarter of the reduced carbon formed in photosynthesis can be lost through photorespiration.

→ **Quick Check 6** Why does rubisco have such a low catalytic rate (that is, why is it so slow)?

Photosynthesis captures just a small percentage of incoming solar energy.

Typically, only 1% to 2% of the sun's energy that lands on a leaf ends up in carbohydrates. Does this mean that photosynthesis is incredibly wasteful? Or is this process, the product of billions of years of evolution, surprisingly efficient? This is not an idle question. Photosynthesis is relevant to solving several pressing global issues: the effects of rising CO_2 concentrations on Earth's climate, the search for a renewable, carbon-neutral fuel to power our transportation sector, and the agricultural demands of our skyrocketing human population.

Photosynthetic efficiency is typically calculated relative to the total energy output of the sun (**Fig. 8.18**). However, only visible light has the appropriate energy levels to raise the energy state of electrons in chlorophyll. Most of the sun's output

FIG. 8.18 Photosynthetic efficiency. Maximum photosynthetic efficiency is theoretically about 4% of incoming solar energy, but actual yields are closer to 1% to 2%.

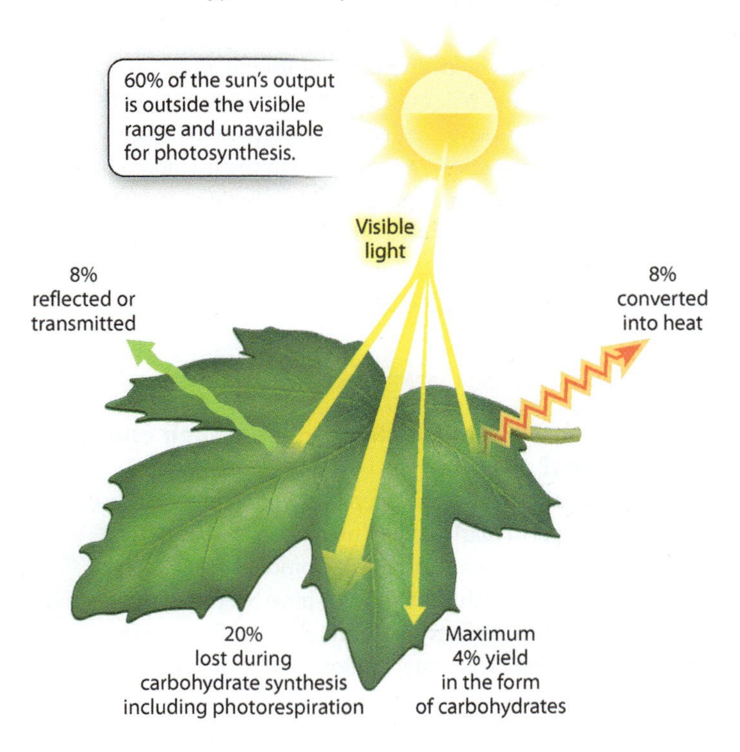

(~60%) is not absorbed by chlorophyll and thus cannot be used in photosynthesis. In addition, leaves are not perfect at absorbing visible light—about 8% is either reflected or passes through the leaf. Finally, even under optimal conditions, not all of the light energy absorbed by chlorophyll can be transferred to the reaction center and instead is given off as heat (also ~8%). As we have seen, when light levels are high, excess light is actively converted into heat by xanthophyll pigments.

The photosynthetic electron transport chain therefore captures at most about 24% of the sun's usable energy arriving at the surface of a leaf (100% − 60% − 8% − 8% = 24%). While this number may appear low, it is on a par with the efficiency of high-performance photovoltaic cells in solar panels, which convert sunlight into electricity. This comparison is even more impressive when you consider that photosynthetic organisms must build and maintain all their biochemical machinery. However, energy is lost at a later step as well. The incorporation of CO_2 into carbohydrates results in considerable loss in free energy, equivalent to ~20% of the total incoming solar radiation. Some of this loss in free energy is due to photorespiration.

In total, therefore, the maximum energy conversion efficiency of photosynthesis is calculated to be around 4% (24% − 20%). Efficiencies achieved by real plants growing in nature, however, are typically much lower, on the order of 1% to 2%. In Chapter 29, we explore the many factors that can constrain the photosynthetic output of land plants, and see how some plants have evolved ways to minimize losses in productivity due to drought and photorespiration.

8.5 THE EVOLUTION OF PHOTOSYNTHESIS

The evolution of photosynthesis had a profound impact on the history of life on Earth. Not only did photosynthesis provide organisms with a new source of energy, but it also released oxygen into the atmosphere. As discussed in the previous chapter, evolution often works in a stepwise fashion, building on what is already present. Here we consider hypotheses for how the photosynthetic pathways that are the dominant entry point for energy into the biosphere today may have evolved.

? CASE 1 THE FIRST CELL: LIFE'S ORIGINS

How did early cells use sunlight to meet their energy requirements?

Sunlight is valuable as a source of energy, but it can also cause damage. This is particularly true of ultraviolet wavelengths, which can damage DNA and other macromolecules. Thus, the earliest interactions with sunlight may have been the evolution of UV-absorbing compounds that could shield cells from the sun's damaging rays. Over time, random mutations could have produced chemical variants of these UV-absorbing molecules. One or more of these variant compounds might have been capable of using sunlight to meet the energy needs of the cell—perhaps by transferring electrons to another molecule as a present-day reaction center does.

The earliest reaction centers may have used light energy to drive the movement of electrons from an electron donor outside the cell in the surrounding medium to an electron-acceptor molecule within the cell. In this way, energy from sunlight could have been used to synthesize carbohydrates. The first electron donor could have been a soluble inorganic ion like reduced iron, Fe^{2+}, which is thought to have been abundant in the early ocean. Alternatively, the first forms of light-driven electron transport may have been cyclic and thus not required an electron donor. In either configuration, light-driven electron transport could have been coupled to the net movement of protons across the membrane, allowing for the synthesis of ATP.

Similarly, it is unlikely that these first photosynthetic organisms employed chlorophyll as a means of absorbing sunlight for the simple reason that the biosynthetic pathway for chlorophyll is complex, consisting of at least 17 enzymatic steps. Yet some of the intermediate compounds leading to chlorophyll are themselves capable of absorbing light. Perhaps each of these now-intermediate compounds was, at one time, a functional end product used as a pigment by an early photosynthetic organism. The biosynthetic pathway may have gained steps as chemical variants, produced by random mutations, were selected because they were more efficient or able to absorb new portions of the visible spectrum. Selection would have eventually resulted in the chlorophyll pigments that are used by photosynthetic organisms today.

The ability to use water as an electron donor in photosynthesis evolved in cyanobacteria.

The most ancient forms of photosynthesis have only a single photosystem in their photosynthetic electron transport chains. However, as we have seen, a single photosystem cannot capture enough energy from sunlight both to pull electrons from water and also raise their energy level enough that they can be used to reduce CO_2. Thus, photosynthetic organisms with a single photosystem must use more easily oxidized compounds, such as H_2S, as electron donors. These organisms can exist only in environments where the electron-donor molecules are abundant. Because these organisms do not use water as an electron donor, they do not produce O_2 during photosynthesis.

A major event in the history of life was the evolution of photosynthetic electron transport chains that use water as an electron donor. The first organisms to accomplish this feat were the cyanobacteria. These photosynthetic bacteria incorporated two different photosystems into a single photosynthetic electron transport chain, one to pull electrons from water molecules and one to raise the energy level of the electrons so that they can be used to reduce CO_2.

How did cyanobacteria end up with two photosystems? We cannot say for sure, but one relevant piece of information is that each of the two photosystems present in cyanobacteria is similar in structure to photosystems found in groups of photosynthetic bacteria that contain only a single photosystem. Thus, it is highly unlikely that the photosystems in cyanobacteria evolved independently. One hypothesis is that the genetic material associated with one photosystem was transferred to a bacterium that already had the other photosystem, resulting in a single bacterium with the genetic material to produce both types of photosystems (shown on the left in **Fig. 8.19**). The mechanisms by which genetic material is transferred between bacteria are discussed more fully in Chapter 26. Another hypothesis is that the genetic material associated with one photosystem underwent duplication. Over time, one of the two photosystems diverged slightly in sequence and function through mutation and selection, giving rise to two distinct but related photosystems (shown on the right in Fig. 8.19). This mechanism, called duplication and divergence, is discussed more fully in Chapter 14.

The ability to use water as an electron donor in photosynthesis had two major impacts on life on Earth. First, it meant that photosynthesis could occur anywhere there was both sunlight and sufficient water for cells to survive. Second, using water as an electron donor results in the release of oxygen. Before the evolution of "oxygenic" photosynthesis, there was little or no free oxygen in Earth's atmosphere. All the oxygen in Earth's atmosphere results from photosynthesis by organisms containing two photosystems.

Eukaryotic organisms are believed to have gained photosynthesis by endosymbiosis.

Photosynthesis is hypothesized to have gained a foothold among eukaryotic organisms when a free-living cyanobacterium took up residence inside a eukaryotic cell (Fig. 8.19). Over time, the cyanobacterium lost its ability to survive outside its host cell and evolved into the chloroplast. The outer membrane of the chloroplast is thought to have originated from the plasma membrane of the ancestral eukaryotic cell, which surrounded the ancestral cyanobacterium as it became incorporated into the cytoplasm of the eukaryotic cell. The inner chloroplast membrane is thought to correspond to the plasma membrane of the ancestral free-living cyanobacterium. The thylakoid membrane then corresponds to the internal photosynthetic membrane found in cyanobacteria. Finally, the stroma corresponds to the cytoplasm of the ancestral cyanobacterium.

The process in which one cell takes up residence inside of another cell is called endosymbiosis. Therefore, the idea that chloroplasts and mitochondria (Chapter 7) arose in this way is called the endosymbiotic hypothesis. It is discussed in Chapter 27.

FIG. 8.19 The evolutionary history of photosynthesis. (a) Hypotheses for the origin of the two photosystems in cyanobacteria. (b) Later, by endosymbiosis, free-living cyanobacteria became the chloroplasts of eukaryotic cells.

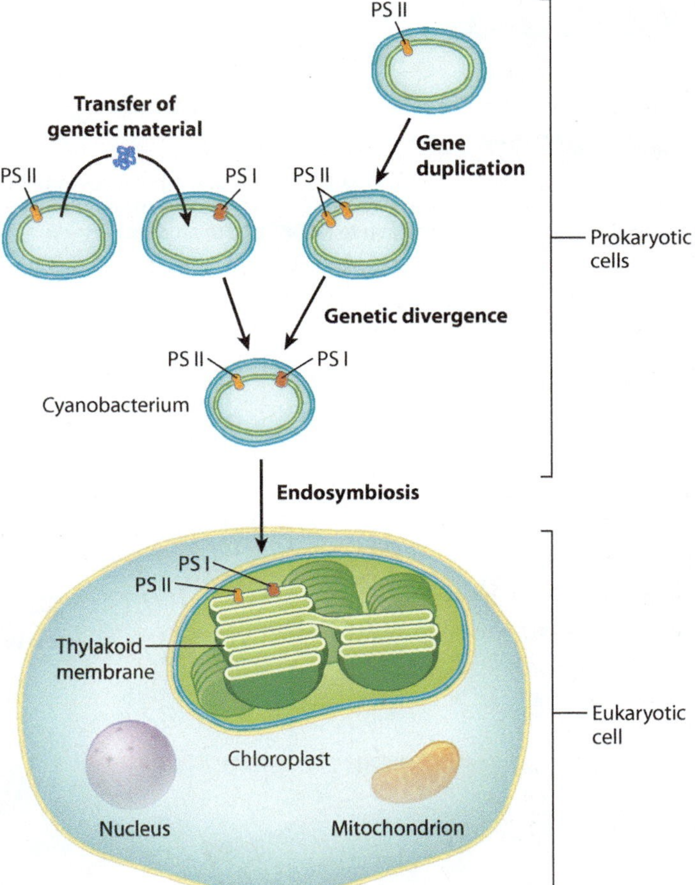

Cellular respiration and photosynthesis are complementary metabolic processes. Cellular respiration breaks down carbohydrates in the presence of oxygen to supply the energy needs of the cell, producing carbon dioxide and water as byproducts, while photosynthesis uses carbon dioxide and water in the presence of sunlight to build carbohydrates, releasing oxygen as a byproduct. We summarize the two processes in **Fig. 8.20**. ■

VISUAL SYNTHESIS

FIG. 8.20

Harnessing Energy: Photosynthesis and Cellular Respiration

Integrating concepts from Chapters 6–8

Photosynthesis

Energy captured from sunlight is stored in carbohydrates.

Photosynthesis requires light energy, CO_2, and H_2O, and produces carbohydrates and O_2. The ***photosynthetic electron transport chain*** uses energy from sunlight to drive the movement of electrons from water to $NADP^+$ and to produce ATP.

Carbon dioxide is reduced by NADPH and, using the energy provided by ATP, is incorporated into a carbohydrate molecule in the ***Calvin cycle***.

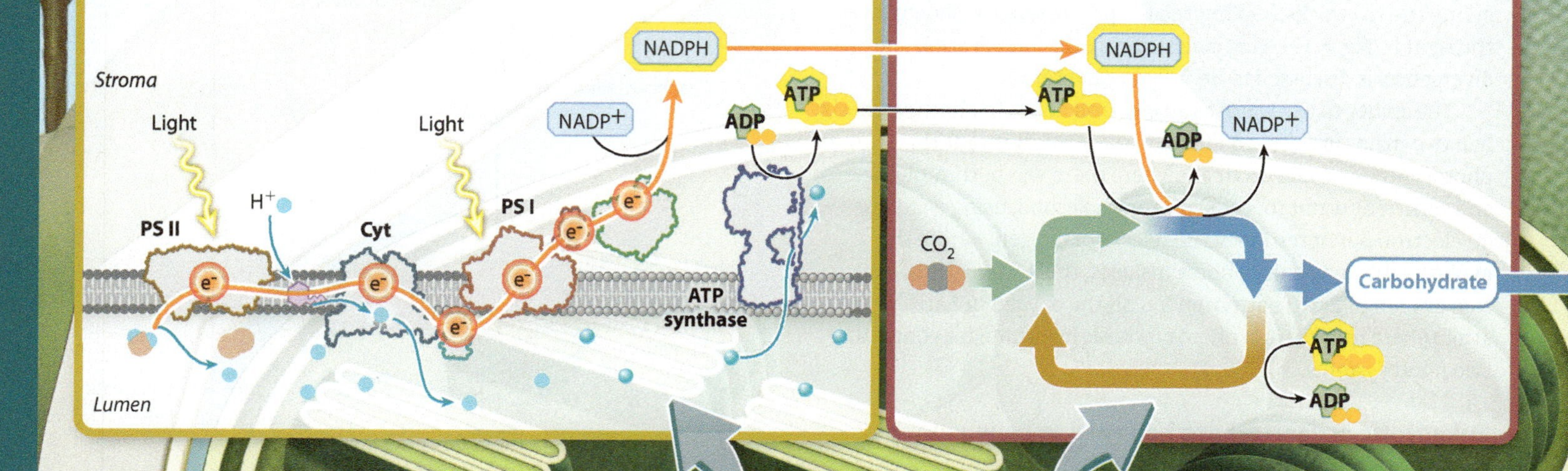

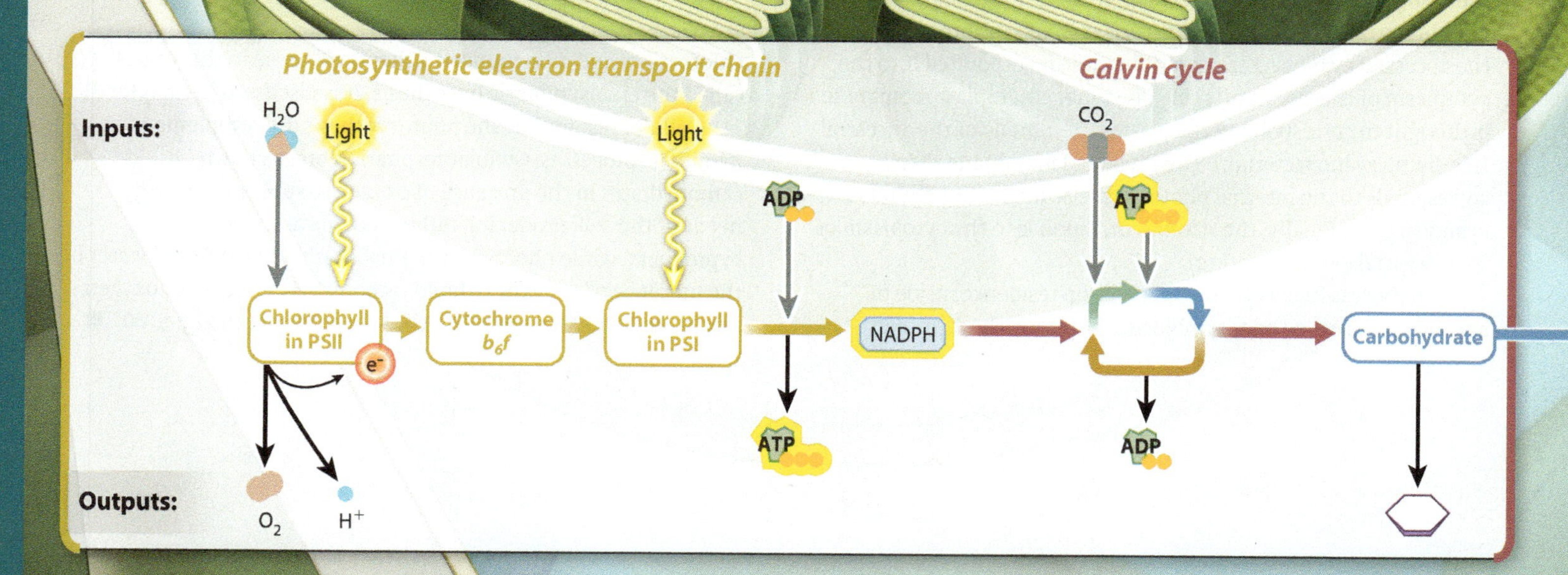

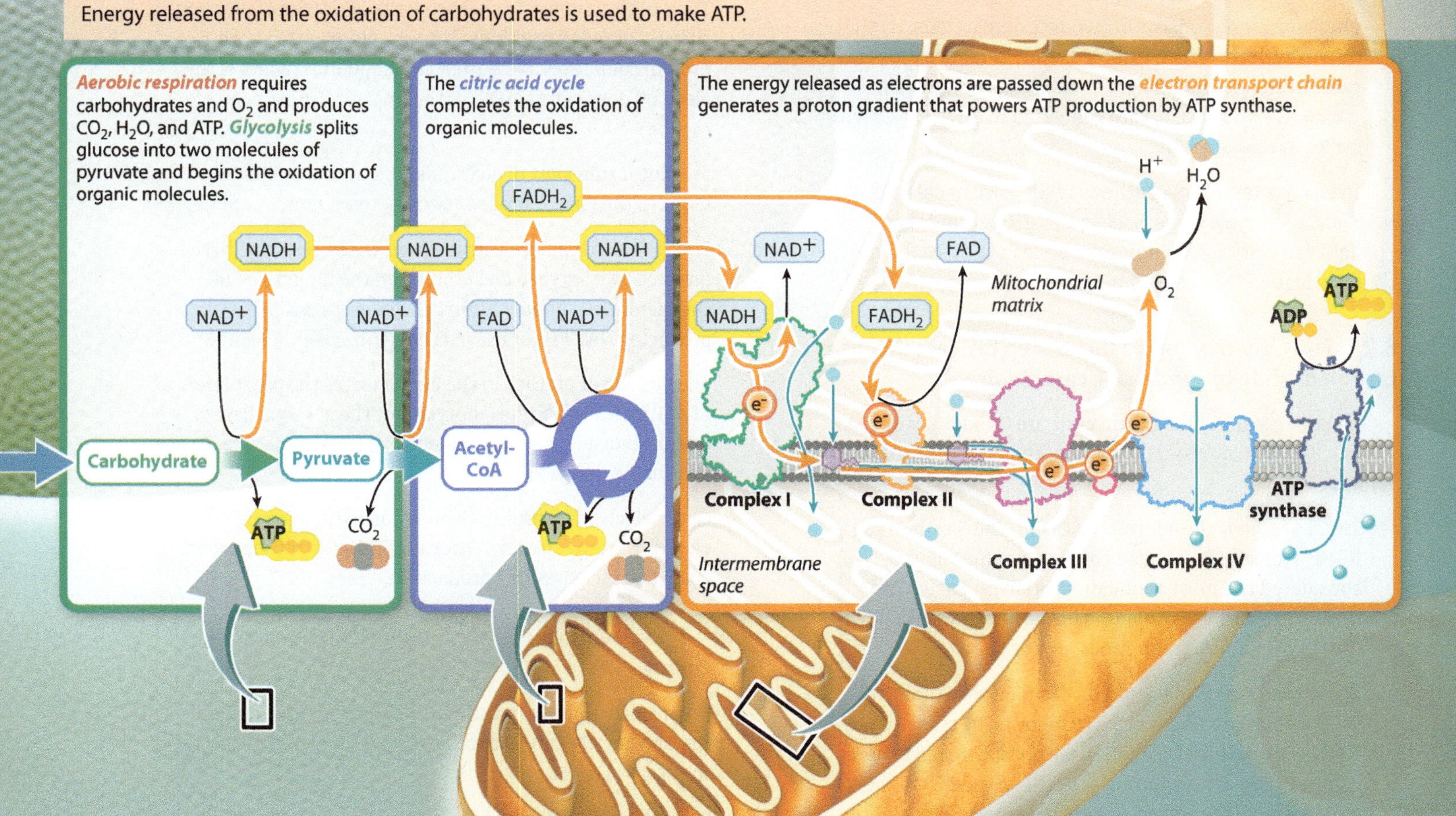

Aerobic respiration
Energy released from the oxidation of carbohydrates is used to make ATP.
Aerobic respiration requires carbohydrates and O_2 and produces CO_2, H_2O, and ATP. Glycolysis splits glucose into two molecules of pyruvate and begins the oxidation of organic molecules.
The citric acid cycle completes the oxidation of organic molecules.
The energy released as electrons are passed down the electron transport chain generates a proton gradient that powers ATP production by ATP synthase.
NADH
NAD+
FADH$_2$
FAD
Carbohydrate
Pyruvate
Acetyl-CoA
ATP
ADP
CO_2
H^+
H_2O
O_2
e⁻
Mitochondrial matrix
Intermembrane space
Complex I
Complex II
Complex III
Complex IV
ATP synthase

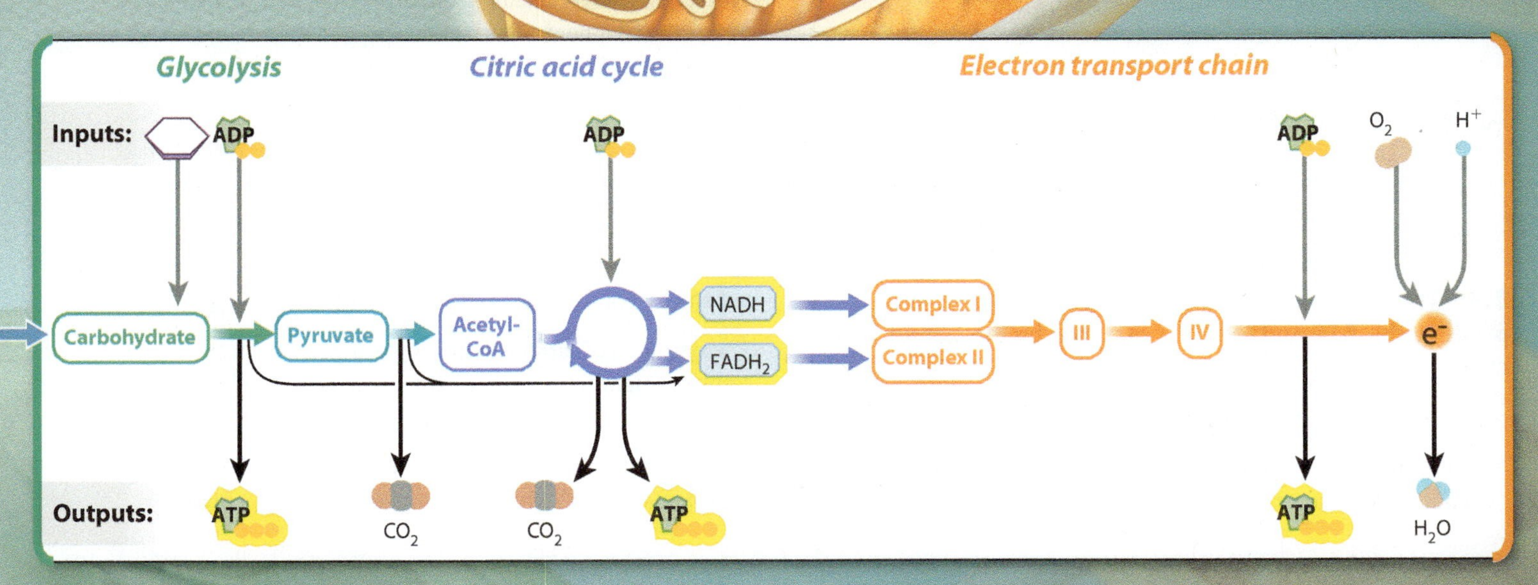

Glycolysis
Citric acid cycle
Electron transport chain
Inputs:
ADP
O_2
H^+
Carbohydrate
Pyruvate
Acetyl-CoA
NADH
FADH$_2$
Complex I
Complex II
III
IV
e⁻
Outputs:
ATP
CO_2
H_2O

Core Concepts Summary

8.1 Photosynthesis is the major pathway by which energy and carbon are incorporated into carbohydrates.

In photosynthesis, water is oxidized, releasing oxygen, and carbon dioxide is reduced, forming carbohydrates. page 155

Photosynthesis consists of two sets of reactions: (1) the Calvin cycle, in which carbon dioxide is reduced to form carbohydrates, and (2) light-harvesting reactions, in which ATP and NADPH are generated to drive the Calvin cycle. page 155

In eukaryotes, photosynthesis takes place in chloroplasts: The Calvin cycle takes place in the stroma, and the light-harvesting reactions take place in the thylakoid membrane. page 156

8.2 The Calvin cycle is a three-step process that uses carbon dioxide to synthesize carbohydrates.

The three steps of the Calvin cycle are (1) addition of CO_2 (carboxylation); (2) reduction; and (3) regeneration. page 157

The first step is the addition of CO_2 to the 5 carbon sugar RuBP. This step is catalyzed by the enzyme rubisco, considered the most abundant protein on Earth. The resulting 6-carbon compound immediately breaks down into two 3-carbon compounds. page 157

The second step is the donation of a phosphate group to the 3 carbon compounds by ATP followed by reduction by NADPH to produce 3 carbon triose phosphate molecules. Some of these triose phosphates are exported from the chloroplast to the cytosol, where they are used to build larger sugars. page 158

The third step is the regeneration of RuBP from five 3-carbon triose phosphates. page 158

Starch formation provides chloroplasts with a way of storing carbohydrates that will not cause water to enter the cell by osmosis. page 158

8.3 The light-harvesting reactions use sunlight to produce the ATP and NADPH required by the Calvin cycle.

Visible light is absorbed by chlorophyll. page 160

Antenna chlorophyll molecules transfer absorbed light energy to the reaction center. page 162

Reaction centers are located within pigment-protein complexes known as photosystems. Special chlorophyll molecules in the reaction center transfer excited-state electrons to an electron-acceptor molecule, thus initiating the photosynthetic electron transport chain. page 162

The electron transport chain consists of a series of electron transfer or redox reactions that take place within both protein complexes and diffusible compounds. Water is the electron donor and $NADP^+$ is the final electron acceptor. page 163

The linear transport of electrons from water to NADPH requires the energy input of two photosystems. page 163

Photosystem II pulls electrons from water, resulting in the production of oxygen and protons on the lumen side of the membrane. Photosystem I passes electrons to $NADP^+$, producing NADPH for use in the Calvin cycle. page 163

The buildup of protons in the lumen drives the production of ATP by oxidative phosphorylation. The ATP synthase is oriented such that ATP is produced on the stroma side of the membrane. page 164

Cyclic electron transport involves the redirection of electrons from ferredoxin back into the electron transport chain and increases ATP production. page 166

8.4 Photosynthesis faces several challenges to its efficiency.

An imbalance between the light-harvesting reactions and the Calvin cycle can lead to the formation of reactive oxygen species. page 166

Protection from excess light energy includes antioxidant molecules that neutralize reactive oxygen species and xanthophyll pigments that dissipate excess light energy as heat. page 168

Rubisco can act catalytically on oxygen as well as on carbon dioxide. When it acts on oxygen, there is a loss of energy and of reduced carbon from the Calvin cycle. page 168

Rubisco has evolved to favor carbon dioxide over oxygen, but the cost of this selectivity is reduced speed. page 169

The synthesis of carbohydrates through the Calvin cycle results in significant energy losses, which are due in part to photorespiration. page 169

The maximum theoretical efficiency of photosynthesis is approximately 4% of total incident solar energy. page 170

8.5 The evolution of photosynthesis had a profound impact on life on Earth.

The ability to use water as an electron donor in photosynthesis evolved in cyanobacteria. page 170

Cyanobacteria evolved two photosystems either by the transfer of genetic material, or by gene duplication and divergence. page 171

Photosynthesis in eukaryotes likely evolved by endosymbiosis. page 171

All of the oxygen in Earth's atmosphere results from photosynthesis by organisms containing two photosystems. page 171

Self-Assessment

1. Write the overall photosynthetic reaction and identify which molecules are oxidized and which molecules are reduced.
2. Compare the overall reactions of photosynthesis and cellular respiration.
3. Name the major inputs and outputs of the Calvin cycle.
4. Describe the three major steps in the Calvin cycle and the role of the key enzyme rubisco.
5. Contrast what happens when antenna chlorophylls absorb light energy with what happens when the reaction center absorbs light energy. Why are antenna chlorophylls so important in photosynthesis?
6. Explain why using water as an electron donor requires a photosynthetic electron transport chain with two photosystems.
7. Show in a diagram how energy from sunlight is used to produce ATP.
8. List the products of linear electron transport and cyclic electron transport, and describe the role of cyclic electron transport.
9. Describe two strategies that plants use to limit the formation and effects of reactive oxygen species.
10. Explain the trade-off that rubisco faces in terms of selectivity and enzymatic speed.
11. Estimate the overall efficiency of photosynthesis and describe where in the pathway energy is dissipated.
12. List three major steps that are hypothesized to have occurred in the evolutionary history of photosynthesis.

Log in to LaunchPad to check your answers to the Self-Assessment questions, and to access additional learning tools.

CASE 2

Cancer

When Good Cells Go Bad

Imagine a simple vaccine that could prevent about 500,000 cases of cancer worldwide every year. You'd think such a discovery would be hailed as a miracle. Not quite. A vaccine to prevent cervical cancer, which affects about half a million women and kills as many as 275,000 annually, has been available since 2006. However, in the United States, this vaccine (and others) has stirred controversy.

Cancer is uncontrolled cell division. Cell division is a normal process that occurs during development of a multicellular organism and subsequently as part of the maintenance and repair of adult tissues. It is carefully regulated so that it occurs only at the right time and place. But sometimes, this careful regulation can be disrupted. When the normal checks on cell division become derailed, cancer can result.

When the normal checks on cell division become derailed, cancer can result.

Most cancers are caused by inherited or acquired mutations, but some are caused by a virus. In fact, nearly all cases of cervical cancer are caused by a virus called human papillomavirus (HPV). There are hundreds of strains of HPV. Some of these strains cause minor problems such as common warts and plantar warts. Other strains are sexually transmitted. Some can cause genital warts but aren't associated with cancer. However, a handful of "high-risk" HPV strains are strongly tied to cancer of the cervix. More than 99% of cervical cancer cases are believed to arise from HPV infections.

HPV infects epithelial cells, a type of cell that lines the body cavities and covers the outer surface of the body. Once inside a cell, the virus hijacks the cellular machinery to produce new viruses. The high-risk strains of HPV that infect the cervix go one step further: Those viruses that aren't fought off by the immune system can permanently integrate their own eight-gene DNA sequence into the host cell's DNA.

Once integrated into the DNA of human epithelial cells, two viral genes, E6 and E7, are expressed and produce two proteins. These viral proteins inhibit the products of key tumor suppressor genes. Tumor suppressor genes code for proteins called tumor suppressors that keep cell division in check by slowing down cell division, repairing DNA replication errors, or instructing defective cells to die. When tumor suppressors are prevented from doing these jobs, cancer can result.

HPV strikes in two ways. The viral E6 protein inhibits a protein called p53, an important tumor suppressor that, among other functions, prevents cell division in healthy cells when there is DNA damage. However, when bound by E6, p53 becomes essentially inactive. Meanwhile, the viral E7 protein inhibits a protein called Rb, which normally blocks transcription factors that promote cell division. Without Rb and p53 to put the brakes on cell division, cervical cells divide uncontrollably.

As they grow and multiply, the abnormal cells push through the basal lamina, a thin layer that separates the epithelial cells lining the cervix from the connective tissue beneath. Untreated, invasive cancer can spread to other organs. Once cancer travels beyond its primary location, the situation is grim. Most cancers cannot be cured after they have spread.

The U.S. Food and Drug Administration (FDA) has approved two vaccines for HPV, both of which protect against the high-risk strains responsible for cervical cancer. Medical groups such as the American Academy of Pediatrics and government organizations such as the Centers for Disease Control and Prevention (CDC) recommend vaccinating girls at age 11 or 12, before

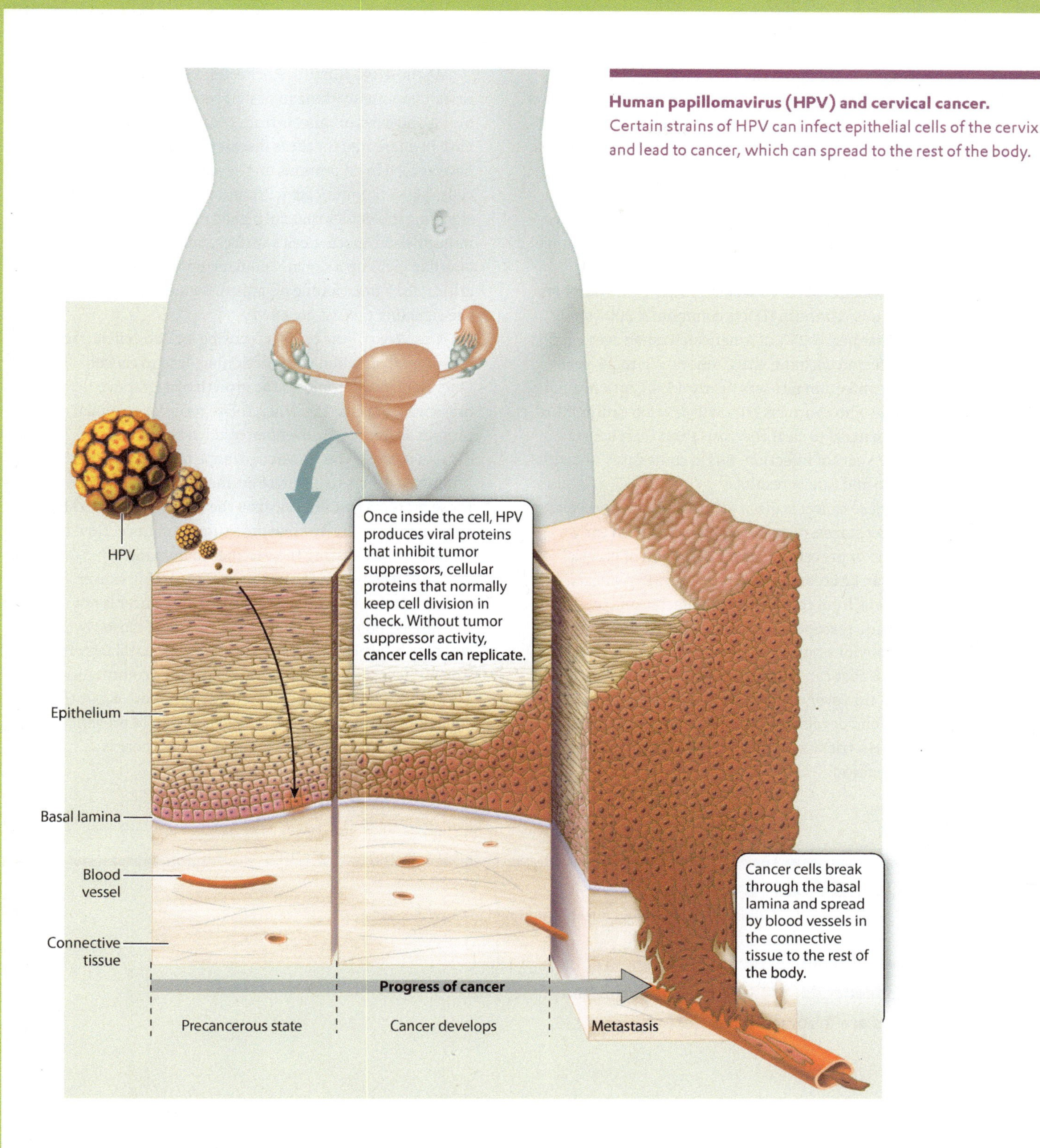

Human papillomavirus (HPV) and cervical cancer. Certain strains of HPV can infect epithelial cells of the cervix and lead to cancer, which can spread to the rest of the body.

they become sexually active. In addition, recent CDC guidelines also recommend vaccinating boys because they can transmit the virus to women when they become sexually active. On the basis of these recommendations, many states are considering legislation that would require middle-school girls to receive the vaccine.

However, there are critics of these recommendations and mandates. Some say the government shouldn't be making medical decisions. Others argue that the use of the vaccine condones sexual activity among young girls and boys. Still others have expressed concern over the vaccine's safety.

Studies have found the HPV vaccine to be safe, a conclusion backed by a 2010 report from the Institute of Medicine, an independent nonprofit organization that advises the U.S government on issues of health. In spite of these findings, the HPV vaccine has not been widely accepted. The CDC found that by 2010, a full 5 years after the vaccine was introduced, only 32% of teenage girls had been fully vaccinated. In 2013, the percentage increased to 38%, but remained low.

Proponents of HPV vaccination say those numbers should cause concern. HPV is common: A 2007 study found that nearly 27% of American women ages 14–59 were infected with the virus. Among 20- to 24-year-olds, the infection rate was nearly 45%. Certainly, not everyone who contracts HPV will develop cancer. Most people manage to clear the virus from their bodies within 2 years of infection. But in some cases, the virus hangs on and cancer results.

Because cancer is often difficult to treat, especially in its later stages, many researchers focus on early detection or prevention. Early detection of cervical cancer can be accomplished by routine Pap smears, in which cells from the cervix are collected and observed under a microscope. As we have seen, cervical cancer is also an obvious target for prevention since it is caused by a virus that can be averted with a conventional vaccine. However, most cancers are not caused by viruses, and developing vaccines to prevent those cancers is a trickier proposition—but researchers are pushing ahead.

At the Mayo Clinic in Rochester, Minnesota, researchers are working to design vaccines to prevent breast and ovarian cancers from recurring in women who have been treated for these diseases. The researchers have zeroed in on proteins on the surface of cancerous cells. Cells communicate with one another by releasing signaling molecules that are picked up by receptor molecules on another cell's surface, much the way a radio antenna picks up a signal. Cellular communication is critical for a functioning organism. Sometimes, though, the signaling process goes awry.

The Mayo Clinic team is focusing on two cell-surface receptors that, when malfunctioning, lead to cancer. One, Her2/neu, promotes the growth of aggressive breast cancer cells. The other, folate receptor α (alpha) protein, is frequently overexpressed in breast and ovarian tumors. The researchers hope to train patients' immune systems to generate antibodies that recognize these proteins and then destroy the cancerous cells. The approach successfully prevented tumors in mice. Now the researchers are testing the vaccines in humans.

Even if these new therapies are successful, cancer researchers have much more work to do. Cancer is not one disease but many, and most of them are caused by a complex interplay of genetic and environmental factors. Any number of things can go wrong as cells communicate, grow, and divide. But as researchers learn more about the cellular processes involved, they can step in to prevent or treat cancer. As the HPV vaccine shows, there is significant progress to be made.

? CASE 2 QUESTIONS

Special sections in Chapters 9–11 discuss the following questions related to Case 2.

1. **How do cell signaling errors lead to cancer?** See page 192.
2. **How do cancer cells spread throughout the body?** See page 213.
3. **What genes are involved in cancer?** See page 236.

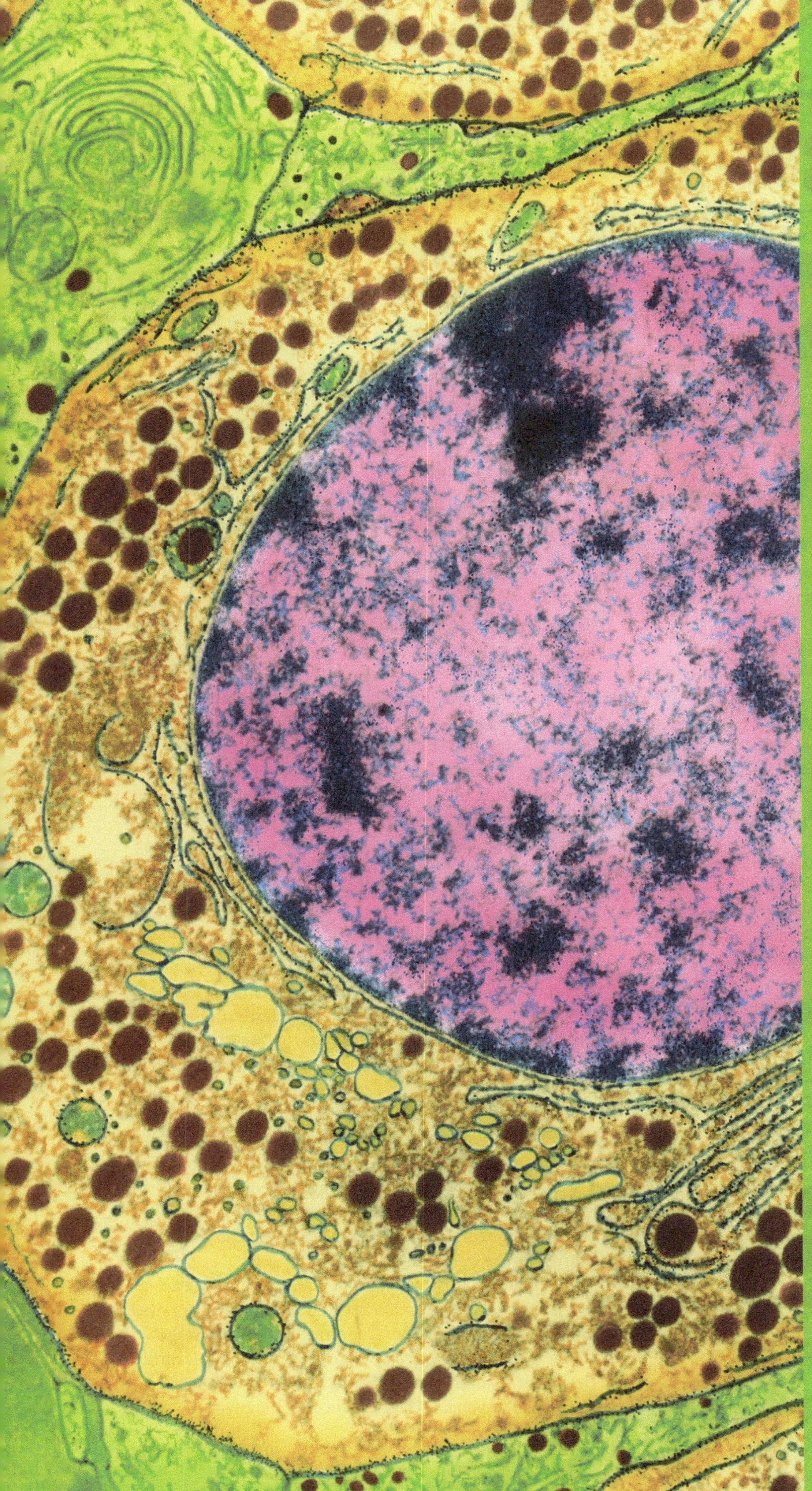

Quest/Science Source.

CHAPTER 9

Cell Signaling

Core Concepts

9.1 Cells communicate primarily by sending and receiving chemical signals.

9.2 Cells can communicate over long and short distances.

9.3 Signaling molecules bind to and activate cell-surface and intracellular receptors.

9.4 G protein-coupled receptors are a large, conserved family of receptors that often lead to short-term responses.

9.5 Receptor kinases are widespread and often lead to long-term responses.

Up to this point, we have considered how life works by looking mainly at what happens inside individual cells. We have seen how a cell uses the information coded in genes to synthesize the proteins necessary to carry out diverse functions. We have also seen how the plasma membrane actively keeps the environment inside the cell different from the environment outside it. Finally, we have explored how a cell harvests and uses energy from the environment.

In the next three chapters, we zoom out from individual cells and consider cells in context. Cells always exist in a particular environment and their behavior is often strongly influenced by this environment. For example, the environment of unicellular organisms consists of their physical surroundings and other cells nearby. The ability of cells to sense and respond to this environment is critical for survival. A unicellular organism senses information that signals when to feed, when to move, and when to divide.

In multicellular organisms, cells do not exist by themselves but as part of cellular communities. These cells may physically adhere to one another, forming tissues and organs. Often multicellular organisms are highly complex, made up of cells of many types with specialized functions. These cells respond to signals, which may come from the outside environment or from other cells of the organism. Cell communication helps to coordinate the activities of the thousands, millions, and trillions of cells that make up the organism. These cells all come about through the process of cell division. Cells divide as part of growth and development, as well as tissue maintenance and repair. Cell division is an example of a cellular activity that is often regulated by signals that tell a cell when to start dividing and when to stop dividing.

In Chapter 9, we look at how cells send, receive, and respond to signals. In Chapter 10, we focus on how cells maintain their shape and, in some cases, physically adhere to one another. As we saw in the case of molecules, form is inextricably linked to function. Finally, in Chapter 11, we consider how one cell becomes two through the process of cell division.

9.1 PRINCIPLES OF CELL COMMUNICATION

The activities of virtually all cells are influenced by their surroundings. Cells receive large amounts of information from numerous sources in their surroundings. One source of information is the physical environment. Another key source of information is other cells—neighboring cells, nearby cells, and even very distant cells. Both unicellular and multicellular organisms sense information from their surroundings and respond by changing their activity or even dividing.

In this section, we take a look at general principles of cell communication, including how cells send and receive signals and how a cell responds after it receives a signal. These mechanisms first evolved in unicellular organisms as a way to sense and respond to their environment, and also as a way to influence the behavior and activity of other cells. As result, the basic principles apply to all cells, both prokaryotic and eukaryotic, and to both unicellular and multicellular organisms.

FIG. 9.1 Four elements required for cellular communication: a signaling cell, signaling molecule, receptor protein, and responding cell.

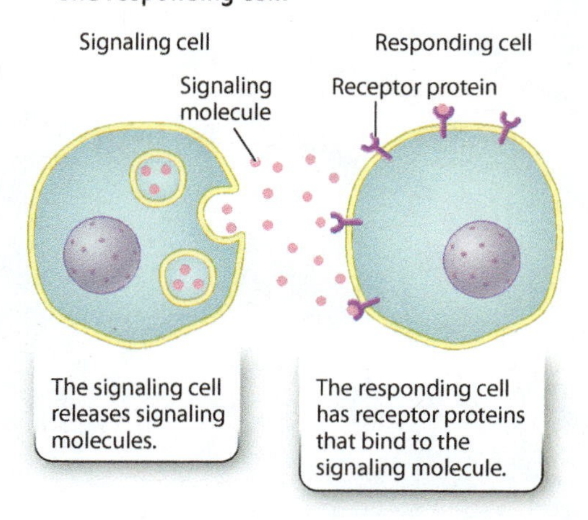

Cells communicate using chemical signals that bind to specific receptors.

Cellular communication consists of four essential elements: a **signaling cell,** a **signaling molecule,** a **receptor protein,** and a **responding cell** (**Fig. 9.1**). The signaling cell is the source of a signaling molecule. Signaling molecules vary immensely and include peptides, lipids, and gases. In all cases, the signaling molecule carries information from one cell to the next. The signaling molecule binds to a receptor protein on or in the responding cell. As a result, there is a change in activity or behavior of the responding cell.

Let's consider a simple example that we are all familiar with. You're startled or scared, and you experience a strange feeling in the pit of your stomach and your heart beats faster. Collectively, these physiological changes are known as the "fight or flight" response. For these changes to occur, a signal goes from signaling cells to responding cells in the stomach, heart, and other organs. This signal is the hormone adrenaline (also called epinephrine). In response to stress, adrenaline is released from the adrenal glands, located above the kidneys. Adrenaline circulates through the body and acts on many types of cell, including the cells of your heart, causing it to beat more strongly and quickly. When the heart beats more strongly and quickly, it is able to deliver oxygen more effectively to the rest of the body.

This example illustrates the four basic elements involved in cellular communication. In this case, the signaling cells are specific cells within the adrenal glands. The signaling molecule is the hormone adrenaline released by these cells. Adrenaline is carried in the bloodstream and interacts with receptor proteins located on the surface of responding cells, like those of the heart.

FIG. 9.2 Communication among bacterial cells. The binding of receptors by the signaling molecule stimulates uptake of DNA.

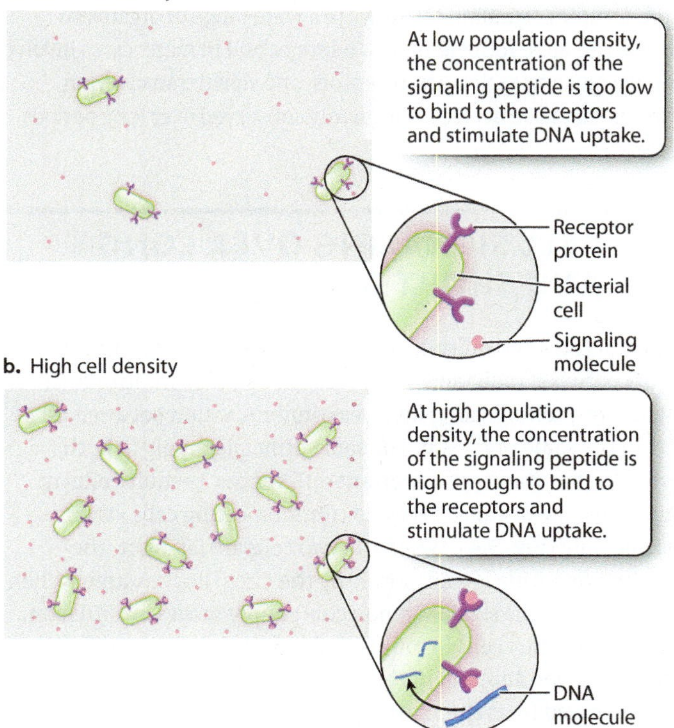

→ **Quick Check 1** If a hormone is released into the bloodstream and therefore comes into contact with many cells, what determines which cells in the body respond to the hormone?

Let's look at a second example, this time involving communication among bacteria (**Fig. 9.2**). *Streptococcus pneumoniae* (also known as pneumococcus) is a disease-causing bacterium that is associated with pneumonia, meningitis, and some kinds of arthritis. Many bacteria, including pneumococcus, are able to take up DNA from the environment and incorporate it into their own genome (Chapter 26). By this means, individual bacterial cells can acquire genes with advantageous properties, including antibiotic resistance.

In the 1960s, it was observed that the rate of DNA uptake by pneumococcal cells increased sharply once the bacterial population reached a certain density. Scientists concluded that the bacteria were able to coordinate DNA uptake across the population so that uptake occurred only at high bacterial population density. How is it possible that these bacteria "know" how many other bacteria are present? It turns out that these bacteria are able to communicate this information to one another through the release of a small peptide.

In the 1990s, scientists discovered a short peptide consisting of 17 amino acids that is continuously synthesized and released by pneumococcal cells. Not long after, a receptor for this peptide was discovered on the surface of the pneumococcal cells. The binding of this peptide to its receptor causes a bacterium to express the genes required for DNA uptake. When the bacteria are at low density, the peptide is at too low a concentration to bind to the receptor (Fig. 9.2a). As the population density increases, so does the concentration of the peptide, until it reaches a level high enough to bind to enough receptors to cause the cells to turn on genes necessary for DNA uptake (Fig. 9.2b).

This is an example of quorum sensing, a process by which bacteria are able to determine whether they are at low or high population density and then turn on specific genes across the entire community. It is used to control and coordinate many different types of bacterial behaviors, such as bioluminescence, antibiotic production, and biofilm formation.

This example also illustrates the four essential elements involved in communication between cells. In this case, however, each bacterial cell is both a signaling cell and a responding cell because each cell makes the signaling molecule and its receptor. Nevertheless, the general idea that cells are able to communicate by sending a signaling molecule that binds to a receptor on a responding cell is universal among both prokaryotes and eukaryotes. These four elements act in very much the same way in diverse types of cellular communication.

Signaling involves receptor activation, signal transduction, response, and termination.

What happens when a signaling molecule binds to a receptor on a responding cell? The first step is **receptor activation.** On binding the signal, the receptor is turned on, or activated (**Fig. 9.3**). The signal usually activates the receptor by causing a conformational

FIG. 9.3 Steps in cell signaling: receptor activation, signal transduction, response, and termination.

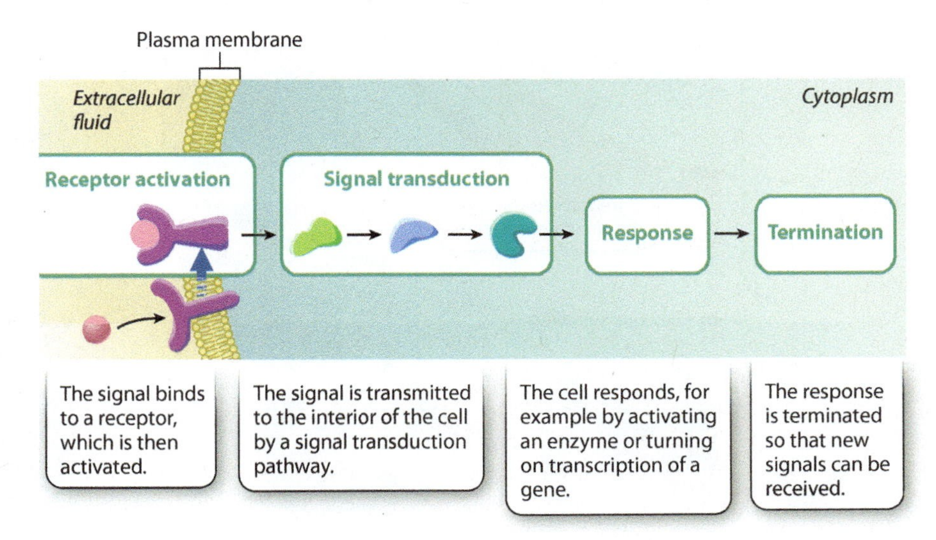

change in the receptor. As a result, some receptors bind to and activate other proteins located inside the cell. Other receptors are themselves enzymes, and binding of the signal changes the shape and activity of the enzyme. Still other receptors are channels that open or close in response to binding a signaling molecule.

Once activated, the receptor often triggers a series of downstream events in a process called **signal transduction.** During signal transduction, one molecule activates the next molecule, which activates the next, and so on. In this way, signal transduction can be thought of as a chain reaction or cascade of biochemical events set off by the binding and activation of the receptor. An important aspect of signal transduction is that the signal is often amplified at each step in the pathway. As a result, a low signal concentration can have a large effect on the responding cell.

Next, there is a cellular **response,** which can take different forms depending on the nature of the signal and the type of responding cell. For example, signaling pathways can activate enzymes involved in metabolic pathways, or turn on genes that cause the cell to divide, change shape, or signal other cells.

The last step is **termination,** in which the cellular response is stopped. The response can be terminated at any point along the signaling pathway. Termination protects the cell from overreacting to existing signals and therefore helps the cell to have an appropriate level of response. It also allows the cell to respond to new signals.

In the case of pneumococcal cells, the peptide binds to and activates a receptor on the cell surface. When enough receptors are bound by the signaling molecule, the message is relayed by signal transduction pathways to the nucleoid. There, genes are turned on that express proteins involved in DNA uptake from the environment. Eventually, when the density of bacteria is low, the initiating signal falls below a critical threshold and gene expression is turned back off.

This example is relatively simple. Remarkably, however, other more complex signaling pathways in a wide range of organisms involve the same four elements and steps, and in many cases involve similar signaling molecules, receptors, and signal transduction systems that have been evolutionarily conserved over long periods of time.

9.2 CELL SIGNALING OVER LONG AND SHORT DISTANCES

In prokaryotes and unicellular eukaryotes, cell communication occurs between individual organisms. In complex multicellular eukaryotes, cell signaling involves communication between cells within the same organism. The same principles apply in both instances, but there are important differences. In multicellular organisms, the distance between communicating cells varies considerably (**Fig. 9.4**). When the two cells are far apart, the signaling molecule is transported by the circulatory system. When they are close, the signaling molecule simply moves by diffusion. In addition, many cells in multicellular organisms are physically attached to one another; in this case, the signaling molecule is not released from the signaling cell at all. In this section, we explore communication over long and short distances, as well as communication between cells that are physically associated with one another.

FIG. 9.4 Signaling distances. Cell communication can be classified according to the distance between the signaling and responding cells.

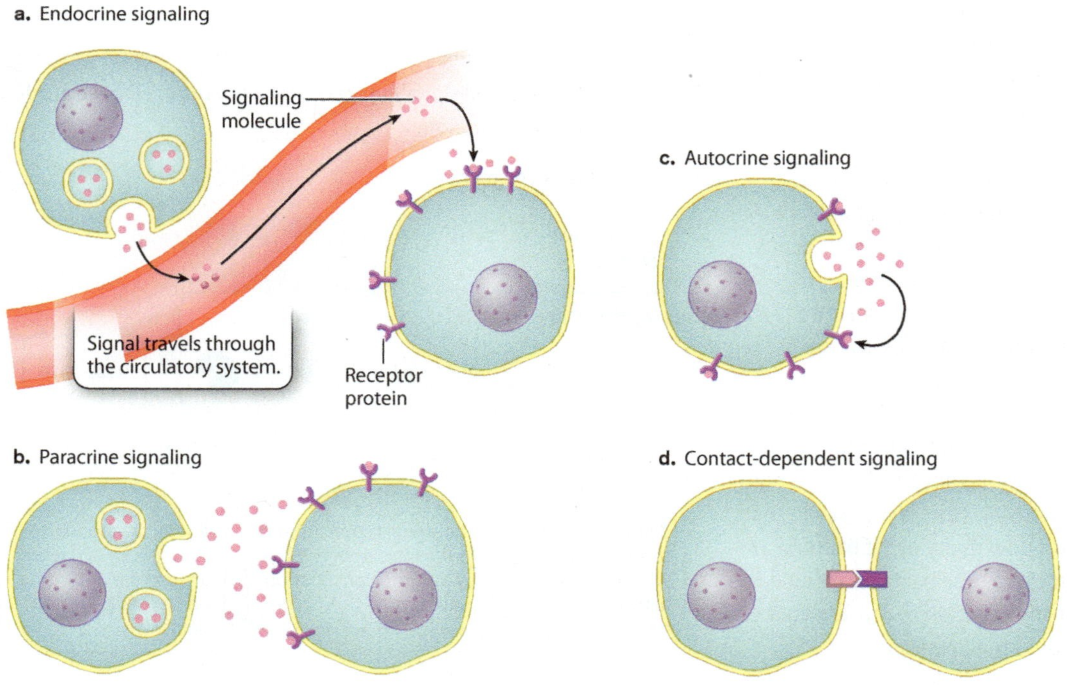

Endocrine signaling acts over long distances.

Signaling molecules released by a cell may have to travel great distances to reach receptor cells in the body. In this case, they are often carried in the circulatory system. Signaling by means of molecules that travel through the bloodstream is called **endocrine signaling** (Fig. 9.4a; Chapter 38).

Adrenaline (section 9.1) provides a good example of endocrine signaling. Adrenaline, which is produced in the adrenal glands, is carried by the bloodstream to target cells that are far from the signaling cells. Other examples of endocrine signaling involve the mammalian steroid hormones estradiol (an estrogen) and testosterone (an androgen). These hormones travel from the ovaries and the testes, respectively (although there are other minor sources of these hormones), through the bloodstream, to target cells in various tissues throughout the body. The increased amount of estrogen in girls during puberty causes the development of breast tissue and the beginning of menstrual cycles. The increased amount of testosterone in boys during puberty causes the growth of muscle cells, deepening of the voice, and growth of facial hair (Chapter 42).

Signaling can occur over short distances.

Signaling can also occur between two cells that are close to each other (Fig. 9.4b). When cells are close to each other, they do not require a circulatory system to deliver the signaling molecule. Instead, the signaling molecule can simply move by diffusion between the two cells. This form of signaling is called **paracrine signaling.** In this case, signaling molecules travel distances of about 20 cell diameters, or a few hundred micrometers.

In paracrine signaling, the signal is usually a small, water-soluble molecule such as a **growth factor.** A growth factor is a type of signaling molecule that causes the responding cell to grow, divide, or differentiate. One of the first growth factors discovered was found by scientists who were attempting to understand how to maintain cultures of cells in the laboratory. For decades, medical research has worked with cells maintained in culture. Initially, these cultured cells had limited use because they failed to divide outside the body unless they were supplied with unidentified factors from mammalian blood serum. In 1974, American scientists Nancy Kohler and Allan Lipton discovered that one of these factors is secreted by platelets (**Fig. 9.5**). Consequently, the growth factor was named "platelet-derived growth factor," or PDGF. We now know of scores of molecules secreted by cells that function as growth factors, and in most cases their effects are confined to neighboring cells.

Growth factors secreted by cells in an embryo work over short distances to influence the kind of cells their neighbors will become. In this way, they help shape the structure of the adult's tissues, organs, and limbs. For example, in developing vertebrates, paracrine signaling by the growth factor Sonic Hedgehog (yes, it's named after a video game character) ensures that the motor neurons in your spinal cord are located properly, that the bones of your vertebral column form correctly, and that your thumb and pinky fingers are on the correct sides of your hands.

A specialized form of short-range signaling is the communication between neurons (nerve cells), or between neurons and muscle cells (Chapters 35 and 37). Neurotransmitters are a type of signaling molecule released from a neuron. After release, they diffuse across a small space, called a synapse, between the signaling cell and the responding cell. If the adjacent cell is a neuron, it often responds by transmitting a nerve impulse further

HOW DO WE KNOW?

FIG. 9.5

Where do growth factors come from?

BACKGROUND Cells can be grown outside the body in culture. However, they survive and grow well only under certain conditions. Researchers hypothesized that there are substances that are required for cells to grow in culture, but the identity and source of these substances were for a long time unknown. A key insight came from the observation that chicken cells grow much better if they are cultured in the presence of blood serum rather than blood plasma. Blood serum is the liquid component of blood that is collected after blood has been allowed to clot. Blood plasma is also the liquid component of blood, but it is collected from blood that has not clotted. American biologists Nancy Kohler and Allan Lipton were interested in identifying the source of the factor in blood serum that allows cells to survive in culture.

HYPOTHESIS Since Kohler and Lipton knew that clotting depends on the release of substances from platelets, they hypothesized that a growth-promoting factor was introduced into the blood by platelets during clotting.

EXPERIMENT 1 The investigators first confirmed earlier observations using cells that are easily grown in culture called fibroblasts, which they obtained from mice. They cultured two sets of fibroblasts in small plastic dishes. To one of the cultures they added serum; to the other culture they added plasma. Then they monitored the rate of cell division in both culture dishes over time.

(continued on following page)

HOW DO WE KNOW?

(continued from previous page)

RESULTS 1 They observed that the rate of cell division in the fibroblasts cultured in serum was far greater than that of the cells cultured in plasma, as expected based on earlier experiments (Fig. 9.5a).

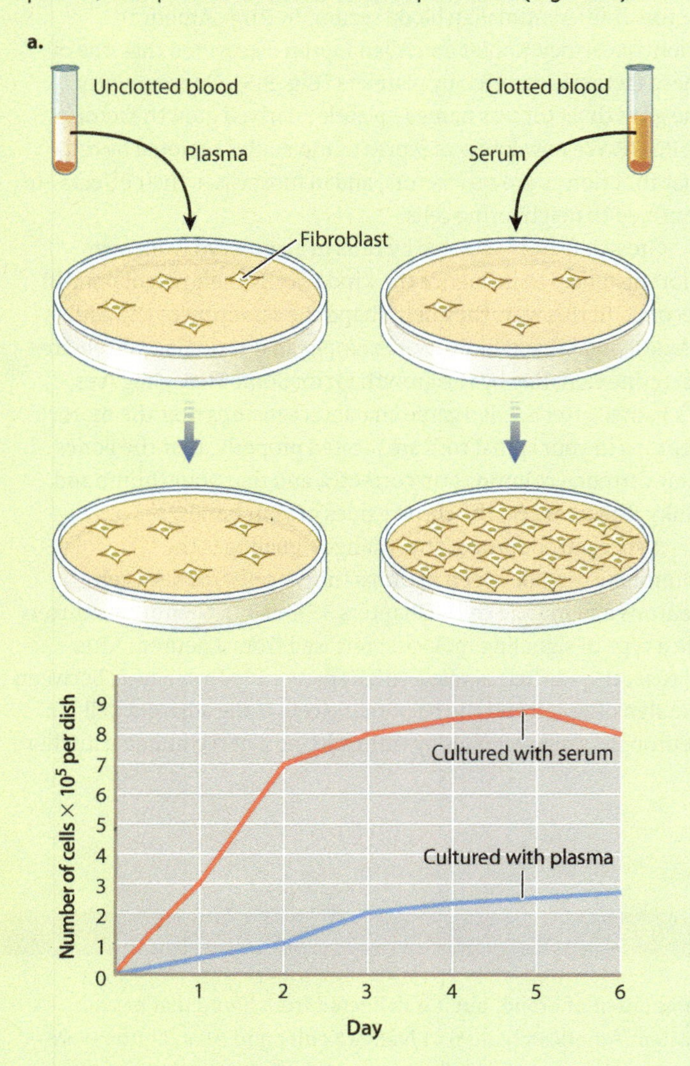

EXPERIMENT 2 To see if the factor is released directly from platelets, they prepared a solution of proteins made from purified platelets, added these proteins to cultured fibroblasts, and measured cell growth.

RESULTS 2 They found that the solution of platelet proteins also caused the growth of fibroblasts to increase compared to the growth of fibroblasts in plasma (Fig. 9.5b).

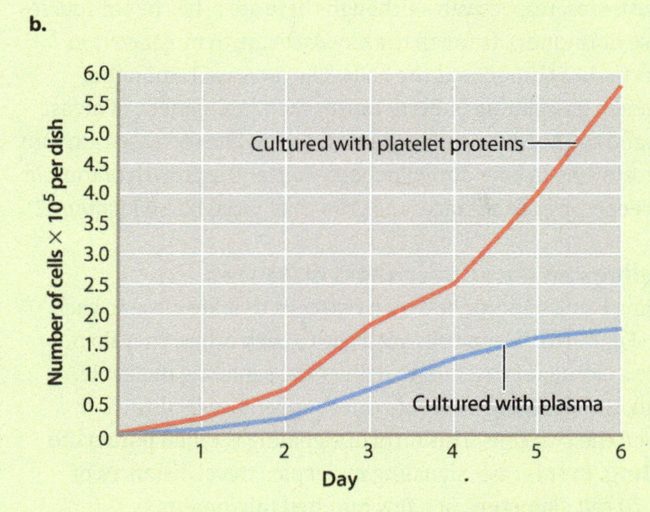

CONCLUSION Kohler and Lipton concluded that the growth-promoting factor is a protein that is released by platelets on clot formation, and therefore normally present in serum.

FOLLOW-UP WORK Over the next few years, these and other scientists purified and characterized the growth factor we know today as platelet-derived growth factor, or PDGF.

SOURCE Kohler, N., and A. Lipton. 1974. "Platelets as a Source of Fibroblast Growth-Promoting Activity." *Experimental Cell Research* 87:297–301.

and then releasing additional neurotransmitters. If the responding cell is a muscle cell, it often responds by contracting.

In some cases, signaling molecules may be released by a cell and then bind to receptors on the very same individual cell. Such cases, where signaling cell and responding cell are one and the same, are examples of **autocrine signaling** (Fig. 9.4c). Autocrine signaling is especially important to multicellular organisms during the development of the embryo (Chapters 20 and 42). For example, once a cell differentiates into a specialized cell type, autocrine signaling is sometimes used to maintain this developmental decision. In addition, autocrine signaling can be used by cancer cells to promote cell division.

Signaling can occur by direct cell–cell contact.

In some cases, a cell communicates with another cell through direct contact, without diffusion or circulation of the signaling molecule. This form of signaling requires that the two communicating cells be in physical contact with each other. A transmembrane protein on the surface of one cell acts as the signaling molecule, and a transmembrane protein on the surface of an adjacent cell acts as the receptor (Fig. 9.4d). In this case, the signaling molecule is not released from the cell, but instead remains associated with the plasma membrane of the signaling cell.

This form of signaling is important during embryonic development. As an example, let's look at the development of the central nervous system of vertebrate animals. In the brain and spinal cord, neurons transmit information in the form of electrical signals that travel from one part of the body to another. The neurons in the central nervous system are greatly outnumbered by supporting cells, called glial cells, which nourish and insulate the neurons. Both the neurons and the glial cells start out as similar cells in the embryo, but some of these undifferentiated cells become neurons and many more become glial cells.

During brain development, the amount of a transmembrane protein called Delta dramatically increases on the surface of some of these undifferentiated cells. These cells will become neurons. Delta proteins on each new neuron bind to transmembrane proteins called Notch on the surface of adjacent, undifferentiated cells. In this case, the signaling cell is the cell with elevated levels of Delta protein. The Delta protein in turn is the signaling molecule, and Notch is its receptor. Cells with activated Notch receptors become glial cells and not neurons. Because one signaling cell sends this same message to all the cells it contacts, it is easy to understand how there can be so many more glial cells than neurons in the central nervous system.

As you can see from these examples, the same fundamental principles are at work when signaling guides a developing embryo, allows neurons to communicate with other neurons or muscles, triggers DNA uptake by pneumococcal cells, or allows your body to respond to stress. All these forms of communication are based on signals that are sent from a signaling cell to a responding cell. These signaling molecules are the language of cellular communication.

9.3 CELL-SURFACE AND INTRACELLULAR RECEPTORS

Receptors are proteins that receive and interpret information carried by signaling molecules. Regardless of the distance between communicating cells, a message is received by a responding cell when the signaling molecule binds to a receptor protein on or in the responding cell. For this reason, the signaling molecule is often referred to as a **ligand** (from the Latin *ligare,* which means "to bind"). The signaling molecule binds to a specific part of the receptor protein called the **ligand-binding site.** The bond is noncovalent and highly specific: The signaling molecule binds only to a receptor with a ligand-binding site that recognizes the molecule.

Almost without exception, the binding of a signaling molecule to the ligand-binding site of a receptor causes a conformational change in the receptor. We say that the conformational change "activates" the receptor because it is through this change that the receptor passes the message from the signaling molecule to the interior of the cell. In many ways, this change in receptor shape is similar to the change that occurs when a substrate binds to the active site of an enzyme (Chapter 6). The conformational change in the receptor ultimately triggers chemical reactions or other changes in the cytosol, and is therefore a crucial step in the reception and interpretation of communications from other cells.

Receptors for polar signaling molecules are on the cell surface.

The location of a particular receptor in a cell depends largely on whether the signaling molecule is polar or nonpolar (**Fig. 9.6**). Many signaling molecules, such as the growth factors we just discussed, are small, polar proteins that cannot pass through the

FIG. 9.6 Cell-surface and intracellular receptors. (a) Cell-surface receptors interact with polar signaling molecules that cannot cross the plasma membrane. (b) Intracellular receptors interact with nonpolar signaling molecules that can cross the plasma membrane.

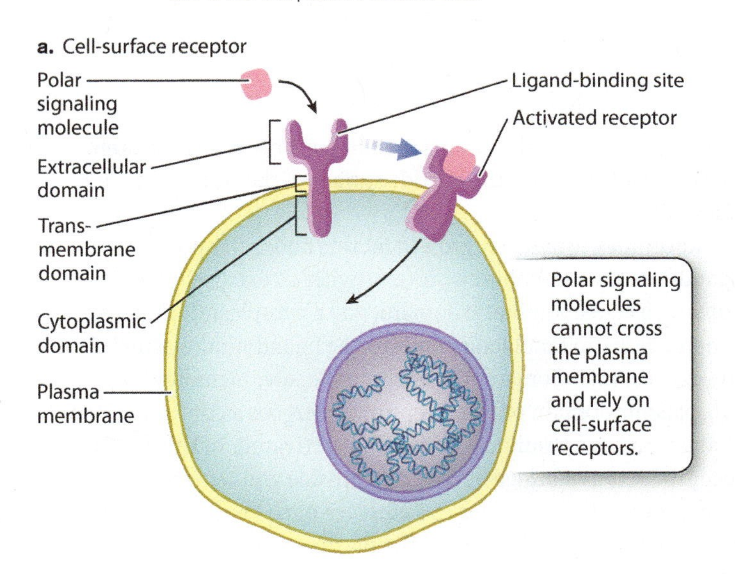

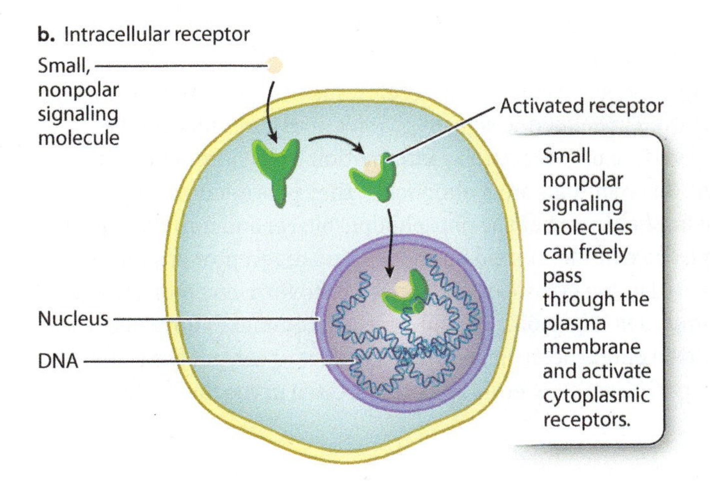

FIG. 9.7 Three types of cell-surface receptor. Receptors act like (a) a light switch. (b) G protein-coupled receptors, (c) receptor kinases, and (d) ligand-gated ion channels can be either "off" (inactive) or "on" (active).

a. A light switch

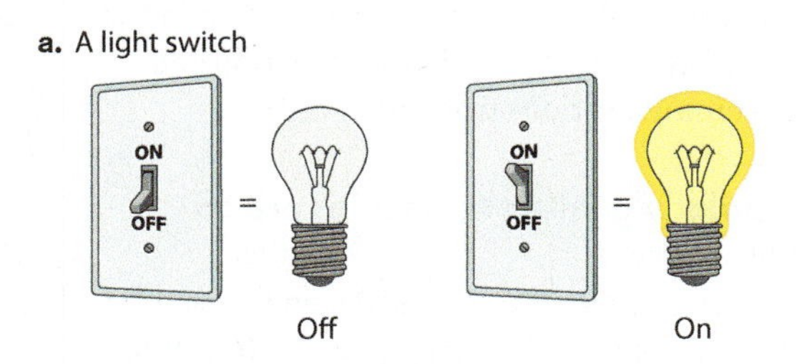

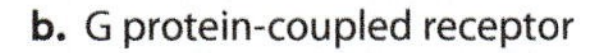

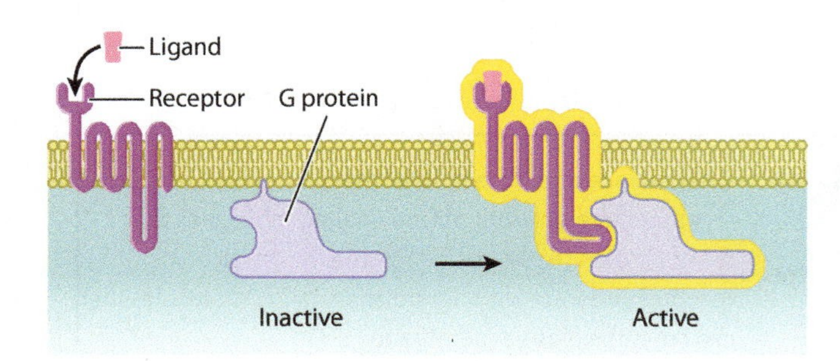

c. Receptor kinase

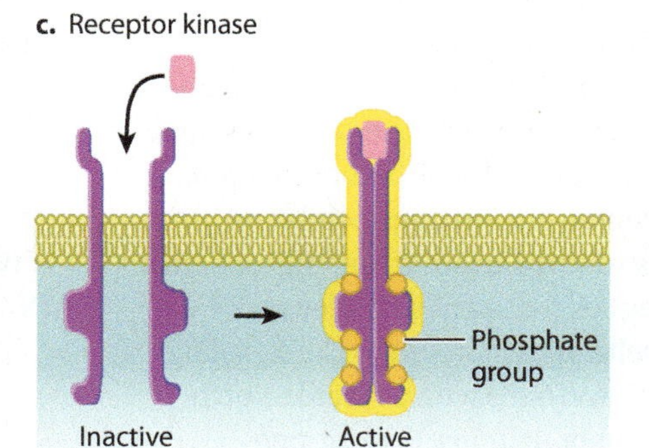

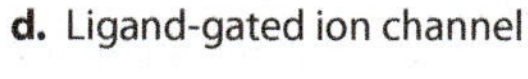

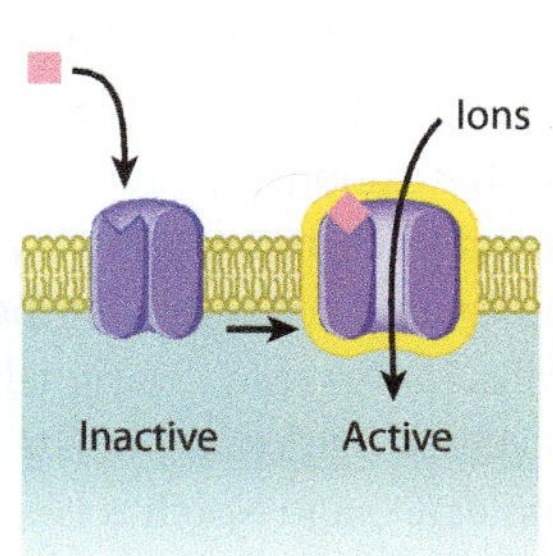

hydrophobic core of the plasma membrane. The receptor proteins for these signals are on the outside surface of the responding cell (Fig. 9.6a).

Receptor proteins for growth factors and other polar ligands are transmembrane proteins with an extracellular domain, a transmembrane domain, and a cytoplasmic domain. When a signaling molecule binds to the ligand-binding site in the extracellular domain, the entire molecule, including the cytoplasmic domain of the receptor, undergoes a conformational change, and as a result the molecule is activated. In this way, the receptor acts as a bridge between the inside and outside of the responding cell that carries the message of the hydrophilic signal across the hydrophobic core of the plasma membrane.

Receptors for nonpolar signaling molecules are in the interior of the cell.

Some nonpolar signaling molecules, such as the steroid hormones involved in endocrine signaling, don't need a receptor on the cell surface in order to relay information to the interior of the cell. Since steroids are hydrophobic, they pass easily through the hydrophobic core of the phospholipid bilayer and into the target cell. Once inside, steroid hormones bind to receptor proteins located in the cytosol or in the nucleus to form receptor–steroid complexes (Fig. 9.6b). Steroid–receptor complexes formed in the cytosol enter the nucleus, where they act to control the expression of specific genes. Steroid receptors located in the nucleus are often already bound to DNA and need only to bind their steroid counterpart to turn on gene expression.

There are many examples of steroid hormones, including sex hormones, glucocorticoids (which raise blood glucose levels), and ecdysone (involved in insect molting). However, since much of the information received by cells is transmitted across the plasma membrane through transmembrane receptors, we focus our attention here on the sequence of events that takes place when receptors on the surface of cells bind their ligands.

Cell-surface receptors act like molecular switches.

As we saw earlier, a receptor is activated after a signaling molecule binds to its ligand-binding site. Many receptors act as binary molecular switches, existing in two alternative states, either "on" or "off" (**Fig. 9.7**). In this way, receptors behave similarly to a light switch (Fig. 9.7a). When bound to their signaling molecule, the molecular switch is turned on. When the signaling molecule is no longer bound, the switch is turned off.

There are thousands of different receptor proteins on the surface of any given cell. Most of them can be placed into one of three groups on the basis of their structures and what occurs immediately after the receptor binds its ligand. One type of cell-surface receptor is called a **G protein-coupled receptor** (Fig. 9.7b; section 9.4). When a ligand binds to a G protein-coupled receptor, the receptor couples to, or associates with, a **G protein,** as its name suggests. G protein-coupled receptors are evolutionary conserved and all have a similar molecular structure (section 9.4)

A second group of cell-surface receptors are themselves enzymes, which are activated when the receptor binds its ligand. Most of these are **receptor kinases** (Fig. 9.7c and section 9.5). A kinase is an enzyme that catalyzes the transfer of a phosphate group from ATP to a substrate. To catalyze this reaction, it binds both ATP and the substrate. This process is called phosphorylation (**Fig. 9.8**). Phosphorylation is important because it affects the activity of the substrate: When a protein is phosphorylated by a kinase, it typically becomes active and is switched on. The addition of a phosphate group to a protein can activate it by altering its shape or providing a new site for other proteins to bind. **Phosphatases** remove a phosphate group, a process called dephosphorylation (Fig. 9.8). When a protein is dephosphorylated by a phosphatase, it typically becomes inactive and is switched off.

Receptors in the third group, **ion channels,** alter the flow of ions across the plasma membrane. These channels can be opened in different ways. Some open in response to changes in voltage across the membrane; these are called voltage-gated ion channels and are discussed in Chapter 35. Other ion channels open when bound by their ligand; these are called ligand-gated ion channels (Fig. 9.7d) and are discussed in Chapter 37. Recall from Chapter 5 that channel proteins help ions and other molecules diffuse into and out of the cell by providing a hydrophilic pathway through the hydrophobic core of the plasma membrane. Most of the time, the channels are closed. However, when a signaling molecule binds to the extracellular portion of a ligand-gated ion channel, the channel undergoes a conformational change that opens it and allows ions to flow in and out. This type of signaling is especially important for nerve and muscle cells since their functions depend on a rapid change in ion flow across the plasma membrane.

FIG. 9.8 **Phosphorylation and dephosphorylation.** A kinase transfers a phosphate group from ATP to a protein, typically activating the protein. A phosphatase removes a phosphate group from a protein, typically deactivating the protein.

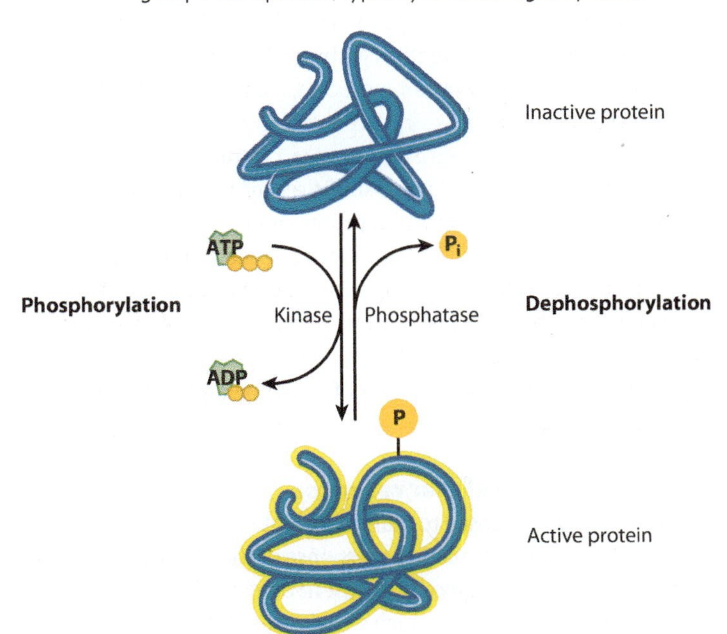

What happens after a signaling molecule binds to its receptor and flips a molecular switch? Following receptor activation, signaling pathways transmit the signal to targets in the interior of the cell, the cell responds, and eventually the signal is terminated. In the next two sections, we examine the signaling pathways activated by G protein-coupled receptors and receptor kinases.

9.4 G PROTEIN-COUPLED RECEPTORS AND SHORT-TERM RESPONSES

G protein-coupled receptors, introduced in section 9.3, are a very large family of cell-surface molecules. They are found in virtually every eukaryotic organism. In humans, for example, approximately 800 different G protein-coupled receptors have been found. They all have two characteristics in common. First, they have a similar structure, consisting of a single polypeptide chain that has seven transmembrane spanning regions, with the ligand-binding site on the outside of the cell and the portion that binds to the G protein on the inside of the cell. Second, when activated, they associate with a G protein. In this way, they are able to transmit the signal from the outside to the inside of the cell. These characteristics result from their shared evolutionary history.

In spite of their similarity, different G protein-coupled receptors are able to respond to a diverse set of different signaling molecules, including hormones, neurotransmitters, and small molecules. In addition, their effects are quite diverse. For example, signaling through these receptors is responsible for our senses of sight, smell, and taste (Chapter 36). In fact, it is thought that G protein-coupled receptors used for cell communication in multicellular organisms evolved from sensory receptors in unicellular eukaryotes.

The first step in cell signaling is receptor activation.

As we saw in section 9.3, G protein-coupled receptors are transmembrane proteins that bind signaling molecules. When a ligand binds to a G protein-coupled receptor, it is on, or active. In its active state, it couples to (that is, it binds to) a G protein, which is located on the cytoplasmic side of the plasma membrane.

The G protein also has two states, on or off. A G protein is able to bind to the guanine nucleotides GTP and GDP. When it is bound to GTP, it is on, or active, and when it is bound to GDP, it is off, or inactive.

When a ligand binds to a G protein-coupled receptor, the receptor in turn binds to and activates a G protein by causing it

FIG. 9.9 Activation of a G protein by a G protein-coupled receptor. A G protein-coupled receptor is activated when it binds a signaling molecule, which leads to exchange of GDP for GTP on the G protein's α subunit, separation of the α subunit, and downstream effects.

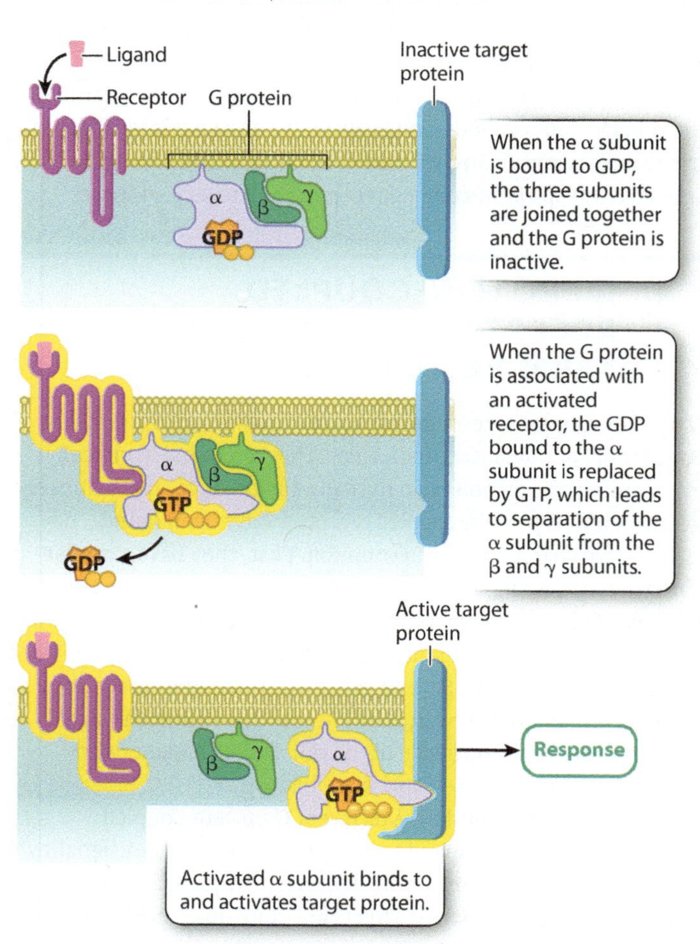

FIG. 9.10 Adrenaline signaling in heart muscle. Adrenaline binds to a G protein-coupled receptor, leading to production of the second messenger cAMP and activation of protein kinase A. The cell's response is increased heart rate.

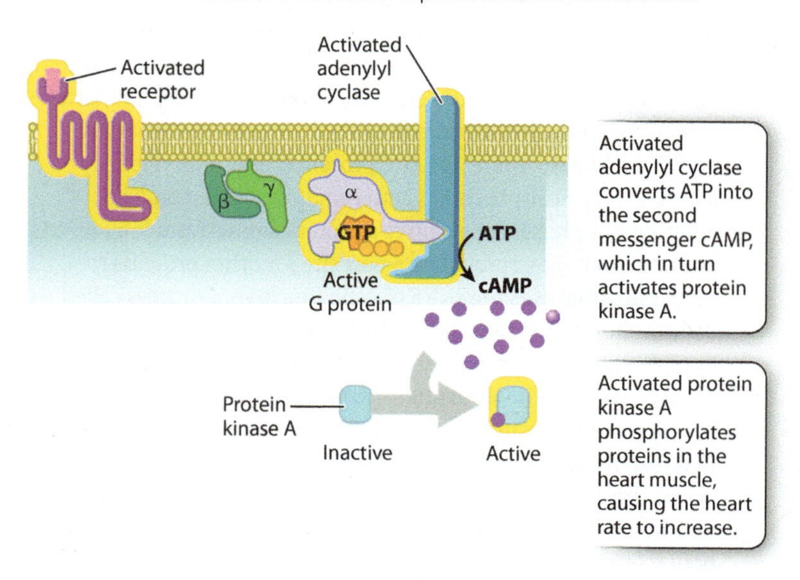

to release GDP and bind GTP. As long as the G protein is bound to GTP, it is in the on position. The activated G protein goes on to activate additional proteins in the signaling pathway.

Some G proteins are composed of three subunits, called the α (alpha), β (beta), and γ (gamma) subunits (**Fig. 9.9**). The α subunit is the part of the G protein that binds to either GDP or GTP. When GDP bound to the α subunit is replaced by GTP, the α subunit separates from the β and γ subunits. In most cases, it is the isolated GTP-bound α subunit that is now active and able to bind to target proteins in the cell.

→ **Quick Check 2** Is the term "G protein" just a shorter name for a G protein-coupled receptor?

Signals are often amplified in the cytosol.

Because different G protein-coupled receptors have different effects in different cells, we will follow the steps in cell signaling by following a specific example. Adrenaline, discussed earlier, binds to a G protein-coupled receptor. When adrenaline binds to its G protein-coupled receptor on cardiac muscle cells, GDP in the G protein is replaced by GTP and the G protein is activated.

The GTP-bound α subunit then binds to and activates an enzyme in the cell membrane called adenylyl cyclase (**Fig. 9.10**). Adenylyl cyclase converts the nucleotide ATP into cyclic AMP (cAMP). Cyclic AMP is known as a **second messenger.** The term "second messenger" was coined to differentiate it from signaling molecules like adrenaline and growth factors, which are considered first messengers. Second messengers are signaling molecules found inside cells that relay information to the next target in the signal transduction pathway. In the case of cAMP, it binds to and activates another molecule, a kinase called protein kinase A (PKA).

A little adrenaline goes a long way as a result of signal amplification (**Fig. 9.11**). First, a single receptor bound to adrenaline can activate several G protein molecules. Second, each molecule of adenylyl cyclase catalyzes the production of large amounts of the second messenger cAMP. And third, each PKA molecule activates multiple protein targets by phosphorylation. These sequential molecular changes in the cytosol amplify the signal so that a very small amount of signaling molecule has a large effect on a responding cell.

Signals lead to a cellular response.

Cells are typically exposed to many different types of signaling molecules. What determines the response of the cell? In part, that depends on the types of receptor present on the surface of the cell. These receptors determine which signals the cell is able to respond to. The response of the cell also depends on the set of proteins that is found in it, as different cell types have different sets of intracellular

FIG. 9.11 Amplification of G protein-coupled signaling. Signaling through G protein-coupled receptors is amplified at several places, so a small amount of signal can produce a large response in the cell.

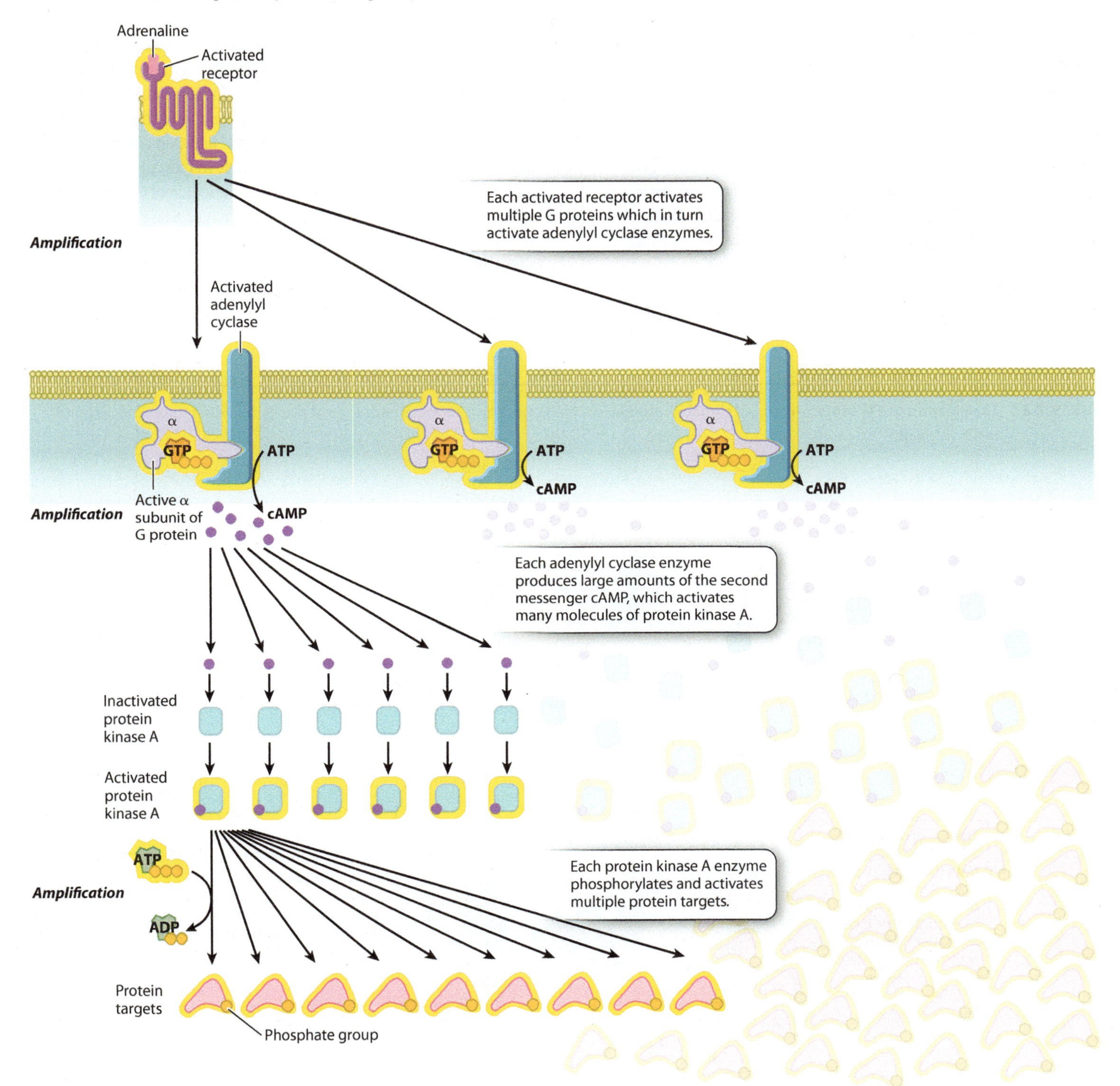

proteins and signaling pathways. As a result, the same signaling molecule can have different effects in different types of cells.

In the case of the heart, activated PKA leads to the opening of calcium channels that are present in heart muscle cells. The resulting influx of calcium ions results in shorter intervals between muscle contractions and thus a faster heart rate. As long as adrenaline is bound to its receptor, the heart rate remains rapid. This increase in the heart rate in turn results in increased blood flow to the brain and skeletal muscles to deal with the stress that set off the signal in the first place.

This example is typical of signaling through G protein-coupled receptors. These receptors tend to activate downstream enzymes or, in some cases, open ion channels. Because they often modify proteins that are already synthesized in the cell, their effects tend to be rapid, short-lived, and easily reversible, as we will see next.

Signaling pathways are eventually terminated.

After a good scare, we eventually calm down and our heartbeat returns to normal. This change means that the signaling pathway initiated by adrenaline has been terminated. How does this happen? First, most ligands, including adrenaline, do not bind to their receptors permanently. The length of time a signaling molecule remains bound to its receptor depends on how tightly the receptor holds on to it, a property called **binding affinity.** Once adrenaline leaves the receptor, the receptor reverts to its inactive conformation and no longer activates G proteins (**Fig. 9.12**).

Even when a receptor is turned off, a signal will continue to be transmitted unless the other components of the signaling pathway are also inactivated. A second place where the signal is terminated is at the G protein itself (Fig. 9.12). G proteins can catalyze the hydrolysis of GTP to GDP and inorganic phosphate. This means that an active, GTP-bound α subunit in the "on" position automatically turns itself "off" by converting GTP to GDP. In fact, the α subunit converts GTP to GDP almost as soon as a molecule of GTP binds to it. Thus, a G protein is able to activate adenylyl cyclase, and adenylyl cyclase is able to make cAMP, only during the very short time it takes the α subunit to convert GTP to GDP. Without an active receptor to generate more active G protein α subunits, transmission of the signal quickly comes to a halt.

FIG. 9.12 Termination of a G protein-coupled signal. G protein-coupled signaling is terminated at several places, allowing the cell to respond to new signals.

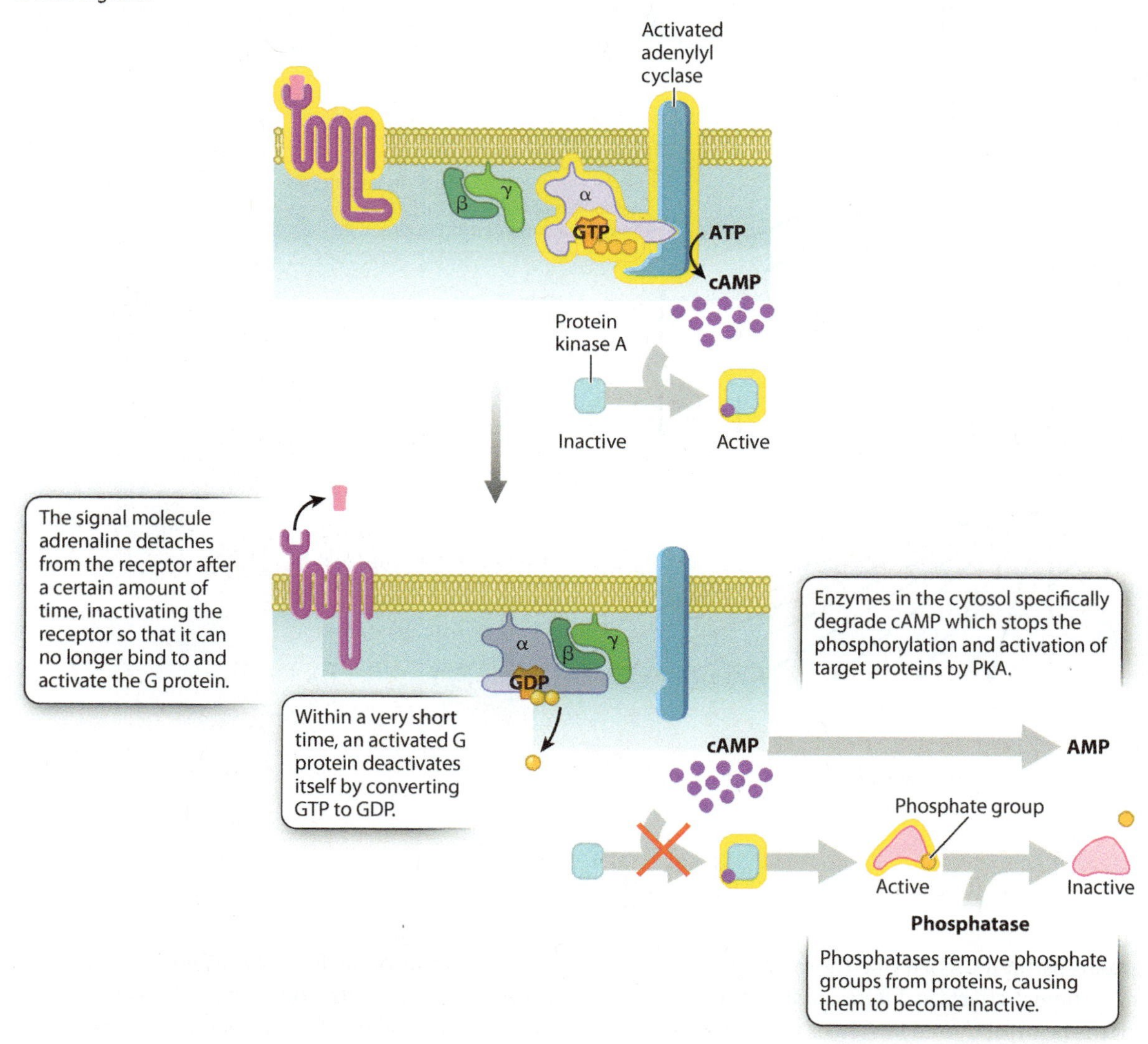

Farther down the pathway, an enzyme converts the second messenger cAMP to AMP, which no longer activates protein kinase A. Phosphatases remove the phosphate groups added by PKA, inactivating PKA's target proteins (Fig. 9.12). In fact, most signaling pathways are counteracted at one or more points as a means of decreasing or terminating the response of the cell to the signal.

→ **Quick Check 3** Name four ways in which the adrenaline signal to the heart is terminated.

9.5 RECEPTOR KINASES AND LONG-TERM RESPONSES

Like the communication that takes place through G protein-coupled receptors, signaling through receptor kinases causes cells to respond in many ways. During embryonic development, receptor kinase signaling is responsible for the formation and elongation of limb buds that eventually become our arms and legs. When we cut a finger, platelet-derived growth factor (PDGF) is released from platelets in the blood and binds to its receptor kinase on the surface of cells at the site of the wound, where it triggers cell division necessary to repair the wound.

The cellular responses that result from receptor kinase activation tend to involve changes in gene expression, which in turn allow cells to grow, divide, differentiate, or change shape. In contrast, the activation of G protein-coupled receptors typically leads to shorter-term changes in the cell, like activating enzymes or opening ion channels.

Signaling through receptor kinases takes place in most eukaryotic organisms, and the structure and function of these receptors have been conserved as organisms have evolved over hundreds of millions of years. A well-studied receptor kinase called Kit provides an example. In vertebrates, signaling through the Kit receptor kinase is important for the production of pigment in skin, feathers, scales, and hair. The conserved function of this receptor can be seen in individuals with mutations in the *kit* gene, as shown in **Fig. 9.13.** (By convention, the name of a protein, like Kit, is capitalized and in roman type. The name of the gene that encodes the protein, like *kit,* is lower case and italicized.) As you can see from Fig. 9.13, mammals, reptiles, birds, and fish with a mutation in the *kit* gene have a similar appearance. This observation indicates that the function of this gene has remained

FIG. 9.13 **Mutations in the Kit receptor kinase.** Similar patterns of incomplete pigmentation are present in mammals, reptiles, birds, and fish that have this mutated receptor. *Sources: (clockwise) Mark Boulton/Science Source; Mark Smith/Science Source; © Phillip Colla/Oceanlight.com; Werner Bollmann/Oxford Scientific/Getty Images.*

FIG. 9.14 Receptor kinase activation and signaling. Receptor kinases bind signaling molecules, dimerize, phosphorylate each other, and activate intracellular signal molecules.

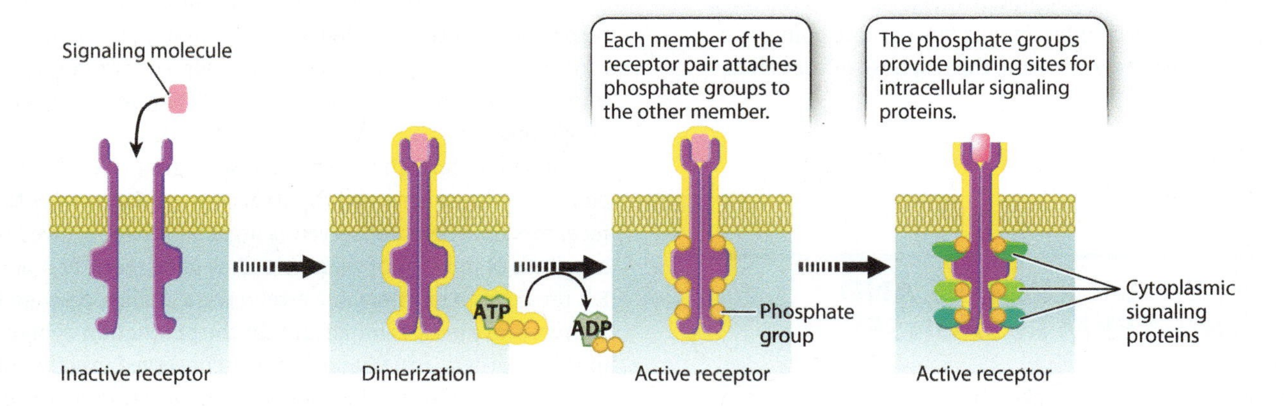

fairly constant since the appearance of the last common ancestor of these groups, more than 500 million years ago.

Receptor kinases phosphorylate each other, activate intracellular signaling pathways, lead to a response, and are terminated.

Signaling through receptor kinases follows the same basic sequence of events that we saw in signaling though G protein-coupled receptors, including receptor activation, signal transduction, cellular response, and termination. Let's consider an example. Think about the last time you got a cut. The cut likely bled for a minute or two, and then the bleeding stopped. The signaling molecule platelet-derived growth factor (PDGF) helps the healing process get started. When platelets in the blood encounter damaged tissue, they release a number of proteins, including PDGF. PDGF is the signaling molecule that binds to PDGF-specific receptor kinases on the surface of cells at the site of a wound.

Receptor kinases have an extracellular portion that binds the signaling molecule and an intracellular portion that is a kinase, an enzyme that transfers a phosphate group from ATP to another molecule. A single molecule of PDGF binds to the extracellular portion of two receptors, causing the receptors to partner with each other. This partnering of two similar or identical molecules is called dimerization. Dimerization activates the cytoplasmic kinase domains of the paired receptors, causing them to phosphorylate each other at multiple sites on their cytoplasmic tails (**Fig. 9.14**). The addition of these phosphate groups provides places on the receptor where other proteins bind and become active.

One of the downstream targets of an activated receptor kinase is Ras. Ras is a G protein, but in contrast to the three-subunit G proteins discussed earlier, Ras consists of a single subunit, similar to the α subunit of the three-subunit G proteins. In the absence of a signal, Ras is bound to GDP and is inactive. However, when Ras is activated by a receptor kinase, it exchanges GDP for GTP. Activated GTP-bound Ras triggers the activation of a protein kinase that is the first in a series of kinases that are activated in turn, as each kinase phosphorylates the next in the series. The series of kinases collectively are called the mitogen-activated protein kinase pathway, or MAP kinase pathway (**Fig. 9.15**). The final activated kinase in the series enters the nucleus, where it phosphorylates target proteins. Some of these proteins include transcription factors that turn on genes needed for cell division so that your cut can heal.

The signals received by receptor kinases are amplified as the signal is passed from kinase to kinase. Each phosphorylated kinase in the series activates multiple molecules of the downstream kinase, and the downstream kinase in turn activates many molecules of another kinase still farther downstream. In this way, a very small amount of signaling molecule (PDGF in our example) can cause a large-scale response in the cell.

Receptor kinase signaling is terminated by the same basic mechanisms that are at work in G protein-coupled receptor pathways. For example, protein phosphatases inactivate receptor kinases and other enzymes of the MAP kinase pathway. Furthermore, Ras hydrolyzes GTP to GDP, just like the G protein α subunit. Shortly after Ras binds to GTP and becomes active, Ras converts GTP to GDP and becomes inactive. Without an active receptor kinase to generate more active Ras, activation of the MAP kinase pathway stops.

? CASE 2 CANCER: WHEN GOOD CELLS GO BAD

How do cell signaling errors lead to cancer?

Many cancers arise when something goes wrong with the way a cell responds to a signal that leads to cell division or, in some cases, when a cell behaves as if it has received a signal for cell division when in fact it hasn't. Defects in cell signaling that lead to cancer can take place at just about every step in the cell signaling process.

FIG. 9.15 The MAP kinase pathway. Some receptor kinases signal through Ras, which in turn activates the MAP kinase pathway, leading to changes in gene expression.

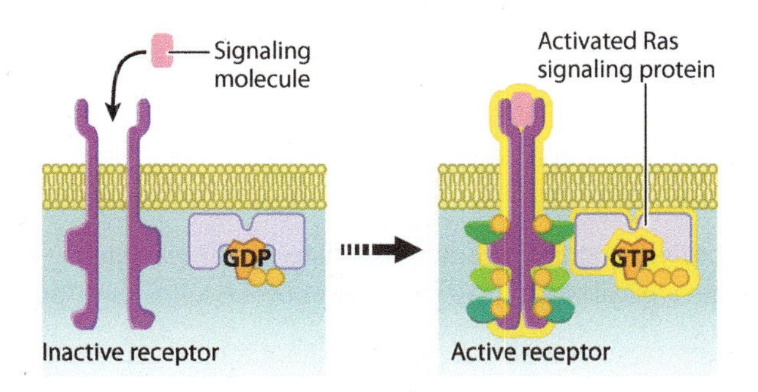

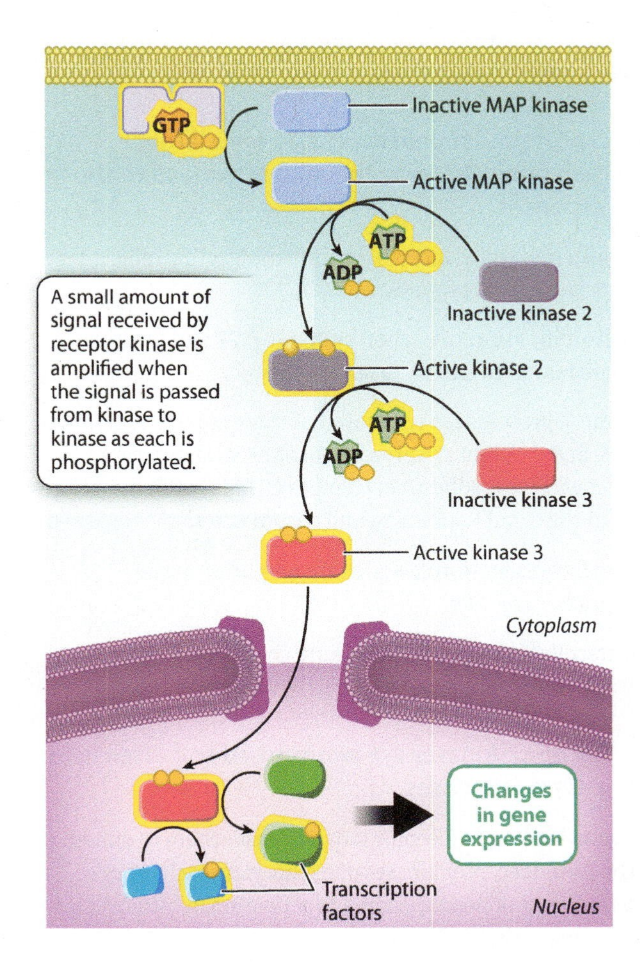

In some cases, a tumor may form as the result of the overproduction of a signaling molecule or the production of an altered form of a signaling molecule. In other cases, the source of the problem is the receptor. For example, individuals with some forms of cancer have from 10 to 100 times the normal number of receptors for a signaling molecule called epidermal growth factor (EGF). As a result, the cell is more responsive to the normally low levels of EGF in the body. The EGF receptor is a receptor kinase, like the PDGF receptor. Under normal conditions, binding of EGF to its receptor leads to the controlled division of cells. However, in cancer, the presence of excess receptors heightens the response of the signaling pathway, leading to abnormally high gene expression and excess cell division. This is the case, for example, in certain breast cancers, in which an EGF receptor called HER2/neu is overexpressed.

Farther down the pathway, mutant forms of the Ras protein are often present in cancers. One especially harmful mutation prevents Ras from converting its bound GTP to GDP. The protein remains locked in the active GTP-bound state, causing the sustained activation of the MAP kinase pathway. More than 30% of all human cancers involve abnormal Ras activity as a result of one or more mutations in the *ras* gene.

Signaling pathways are integrated to produce a response in a cell.

In this chapter, we have focused on individual signaling pathways to illustrate the general principles of communication between cells. In each case, we saw that a cell releases a signaling molecule or in some cases retains the signaling molecule attached to its surface. The signaling molecule binds to a cell-surface or intracellular receptor. Binding of the signaling molecule induces a conformational change in the receptor, causing it to become active. The activated receptor in turn causes changes in the interior of the cell, frequently turning on a signal transduction cascade that is amplified and leads to a cellular response. Eventually, the response terminates, and the cell is poised to receive new signals.

Focusing on each pathway one at a time allowed us to understand how each pathway operates. But in the context of an entire organism or even a single cell, cell signaling can be quite complex, with multiple pathways acting at once and interacting with one another.

Multiple types of signaling molecules can bind to a single cell and activate several signaling pathways simultaneously. In this case, a cell's final response depends on how the pathways intersect with one another. The integration of different signals gives cells a wide range of possible responses to their environment. Receiving two different signals may enhance a particular response, such as cell growth, or one signal may inhibit the signaling pathway triggered by the other signal, weakening the response.

For example, studies have shown that enzymes in the MAP kinase pathway can be inhibited by active PKA. Recall that PKA is activated by elevated cAMP levels associated with the G protein-coupled receptor pathway. The regulation of the MAP kinase pathway by PKA illustrates how one signaling pathway can inhibit another.

Researchers have started to make use of this molecular crosstalk. Many patients with breast cancer have elevated MAP kinase activity in their tumor cells. When human breast cancer cells are genetically modified to express an activated G protein α subunit, their growth after transplantation into mice is significantly inhibited. In addition, in cell culture, elevated cAMP levels block cells from responding to growth factors that signal through Ras to the MAP kinase pathway. As we better understand how different signaling pathways integrate in specific cell types, we have a chance to alter the activity of particular pathways and ultimately the response of the cell. ■

Core Concepts Summary

9.1 Cells communicate primarily by sending and receiving chemical signals.

There are four essential players in communication between two cells: a signaling cell, a signaling molecule, a receptor protein, and a responding cell. page 180

The signaling molecule binds to its receptor on the responding cell, leading to receptor activation, signal transduction and amplification, a cellular response, and eventually termination of the response. page 181

9.2 Cells can communicate over long and short distances.

Endocrine signaling takes place over long distances and often relies on the circulatory system for transport of signaling molecules. page 183

Paracrine signaling takes place over short distances between neighboring cells and relies on diffusion. page 183

Autocrine signaling occurs when a cell signals itself. page 184

Some forms of cell communication depend on direct contact between two cells. page 184

9.3 Signaling molecules bind to and activate cell-surface and intracellular receptors.

A signaling molecule, or ligand, binds specifically to the ligand-binding site of the receptor. Binding causes the receptor to undergo a conformational change that activates the receptor. page 185

Receptors for polar signaling molecules, including growth factors, are located on the plasma membrane. page 186

Receptors for nonpolar signaling molecules, such as steroid hormones, are located in the cytosol or in the nucleus. page 186

There are three major types of cell-surface receptor: G protein-coupled receptors, receptor kinases, and ion channels. All act as molecular switches. page 186

G protein-coupled receptors associate with G proteins, which relay the signal to the interior of the cell. page 186

Receptor kinases phosphorylate each other and activate target proteins. page 187

Ligand-gated ion channels open in response to a signal, allowing the movement of ions across the plasma membrane. page 187

9.4 G protein-coupled receptors are a large, conserved family of receptors that often lead to short-term responses.

G protein-coupled receptors bound to signaling molecules associate with G proteins. page 187

G proteins are active when bound to GTP and inactive when bound to GDP. page 187

Some G proteins are composed of three subunits, denoted α, β, and γ. When a G protein encounters an activated receptor, the α subunit exchanges GDP for GTP, dissociates from the β and γ subunits, and becomes active. page 188

Signal transduction cascades are amplified in the cytosol. page 188

Intracellular, cytosolic signals are short-lived before they are terminated. page 189

9.5 Receptor kinases are widespread and often lead to long-term responses.

Ligand binding to receptor kinases causes them to dimerize. The paired receptor kinases phosphorylate each other's cytoplasmic domains, leading to activation of intracellular signaling pathways. page 192

The phosphorylated receptors bind other proteins, which in turn activate other cytosolic signaling molecules, such as Ras. page 192

When Ras is activated, the GDP to which it is bound is released and is replaced by GTP. The active, GTP-bound Ras binds to and activates the first in a series of kinases. page 192

Cancer can be caused by errors in any step of the signaling pathways involved in cell division. page 192

Mutations in receptor kinases and Ras are frequently associated with human cancers. page 193

Several signaling pathways can take place simultaneously in a single cell, and the cellular response depends on the integration of these pathways. page 193

Self-Assessment

1. Name the four essential elements in cell communication.
2. Name the steps that occur when a signal binds to a receptor on a responding cell.
3. Describe one way in which endocrine, paracrine, autocrine, and contact-dependent signaling pathways are similar to one another and one way in which they are different from one another.
4. Explain how cells respond to external signals, even when those signals cannot enter the cell.
5. Explain how signals can specifically target only some cells, even if they are released into the bloodstream and come into contact with many cells.
6. Describe three different responses of a cell-surface receptor on binding a signaling molecule and undergoing a conformational change.
7. List several responses a cell might have to a signaling molecule.
8. Compare and contrast receptors associated with polar and nonpolar signaling molecules.
9. List three ways in which a signal is amplified in a G protein-coupled receptor signaling pathway.
10. Describe three ways in which the response of a cell to a signal can be terminated.

Log in to LaunchPad to check your answers to the Self-Assessment questions, and to access additional learning tools.

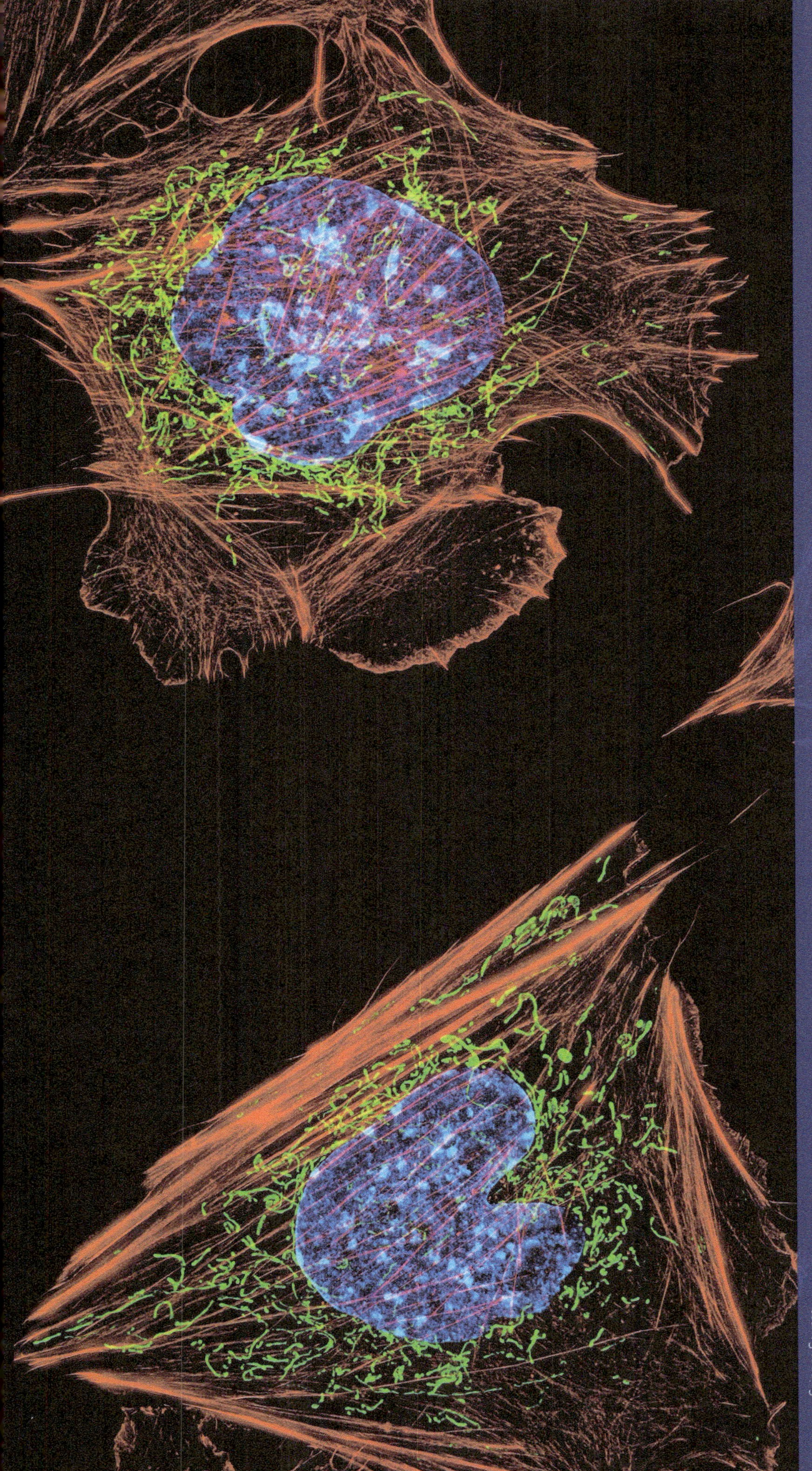

Science Source.

CHAPTER 10

Cell and Tissue Architecture

Cytoskeleton, Cell Junctions, and Extracellular Matrix

Core Concepts

10.1 Tissues and organs are communities of cells that perform specific functions.

10.2 The cytoskeleton is composed of microtubules, microfilaments, and intermediate filaments that help maintain cell shape.

10.3 Cell junctions connect cells to one another to form tissues.

10.4 The extracellular matrix provides structural support and informational cues.

A recurring theme in this book, and in all of biology, is that form and function are inseparably linked. This is true at every level of structural organization, from molecules to organelles to cells and to organisms themselves. We saw in Chapter 4 that the function of proteins depends on their shape. In Chapters 7 and 8, we saw that the function of organelles like mitochondria and chloroplasts depends in part on the large internal surface areas of the inner membrane of mitochondria and the thylakoid membrane of chloroplasts.

The functions of different cell types are also reflected in their shape and internal structural features (**Fig. 10.1**). Consider a red blood cell. It is shaped like a disk that is slightly indented in the middle, and it lacks a nucleus and other organelles (Fig. 10.1a). This unusual shape and internal organization allow it to deform readily, so it can pass through blood vessels with diameters smaller than that of the red blood cell itself. Spherical cells in the liver (Fig. 10.1b) that synthesize proteins and glycogen look and function very differently from long, slender muscle cells (Fig. 10.1c) that contract to exert force. A neuron (Fig. 10.1d), with its long and extensively branched extensions that communicate with other cells, is structurally and functionally quite distinct from a cell lining the intestine (Fig. 10.1e) that absorbs nutrients.

In multicellular organisms, cells often exist in communities, forming tissues and organs. Again, form and function are intimately linked. Think of the branching structure and large surface area of the mammalian lung, which is well adapted for gas exchange, or the muscular heart, adapted for pumping blood through large circulatory systems.

In this chapter, we look at what determines cell shape. We also examine the different ways cells connect to one another to build tissues and organs in multicellular organisms. Finally, we discuss the physical environment outside cells that is synthesized by cells and influences their behavior.

FIG. 10.1 **Diverse cell types.** Cells differ in shape and are well adapted for their various functions. *Sources: a. Cheryl Power/Science Source; b. PROFESSORS P. MOTTA & T. NAGURO/Science Source; c. Innerspace Imaging/Science Source; d. Biophoto Associates/Science Source; e. Don W. Fawcett/Science Source.*

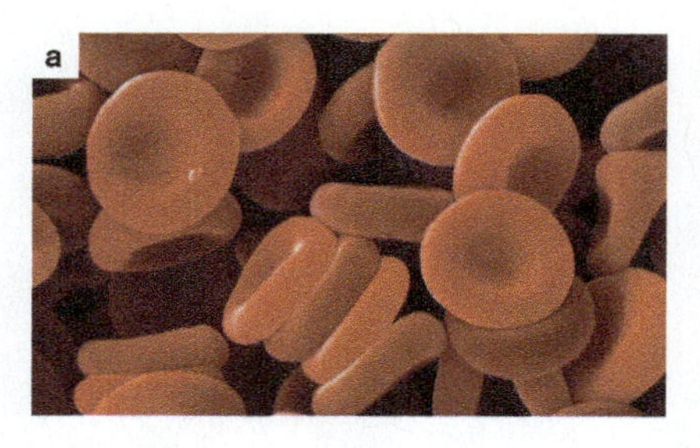

The biconcave shape of a red blood cell maximizes its surface area for gas exchange and allows it to deform as it passes through the circulatory system.

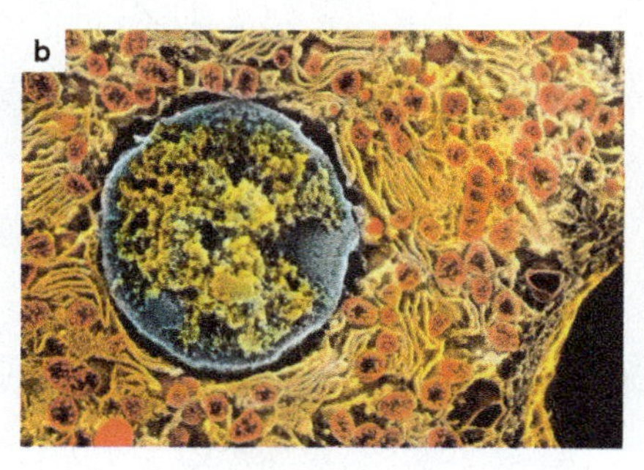

Hepatocytes (liver cells) contain large amounts of rough ER, shown here surrounding the central nucleus, needed for protein synthesis.

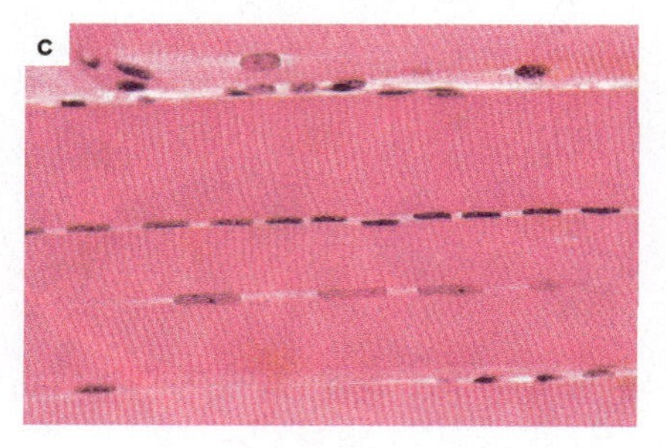

Long, multinucleated muscle cells have a striped appearance and are specialized for contraction.

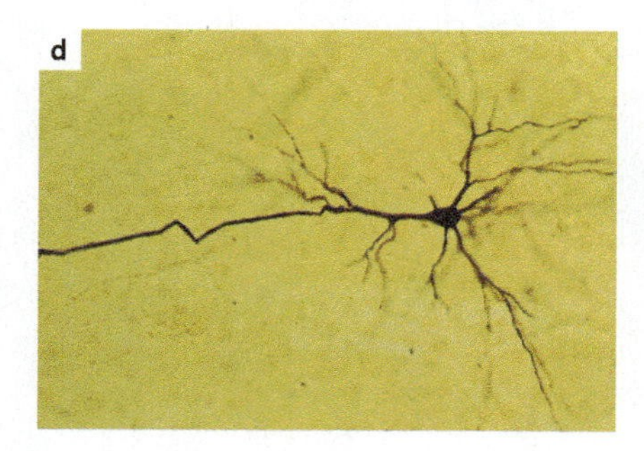

Long, slender extensions of the plasma membrane allow neurons to communicate with other cells.

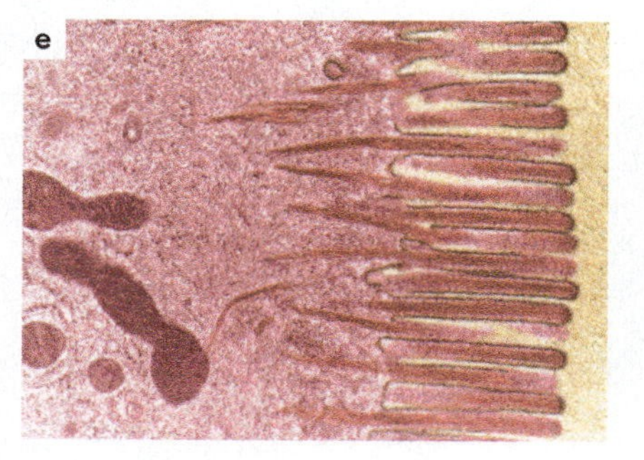

Thin, comb-like projections, called microvilli, on intestinal cells increase their absorptive surface area.

10.1 TISSUES AND ORGANS

Multicellular organisms consist of communities of cells with several notable features (see below and Chapter 28). First, and most obviously, cells adhere to one another. Most cells in multicellular organisms are physically attached to other cells or to substances in their surroundings. Second, cells communicate with one another. As we saw in Chapter 9, cells respond to signals from neighboring cells and the physical environment. In addition, as we will see, cells in tissues have pathways for the movement of molecules from one cell to another. Third, cells are specialized to carry out different functions as a result of development and differentiation. The different shapes of cells often reflect these specialized functions.

Tissues and organs are communities of cells.

A biological **tissue** is a collection of cells that work together to perform a specific function. Animals and plants have tissues that

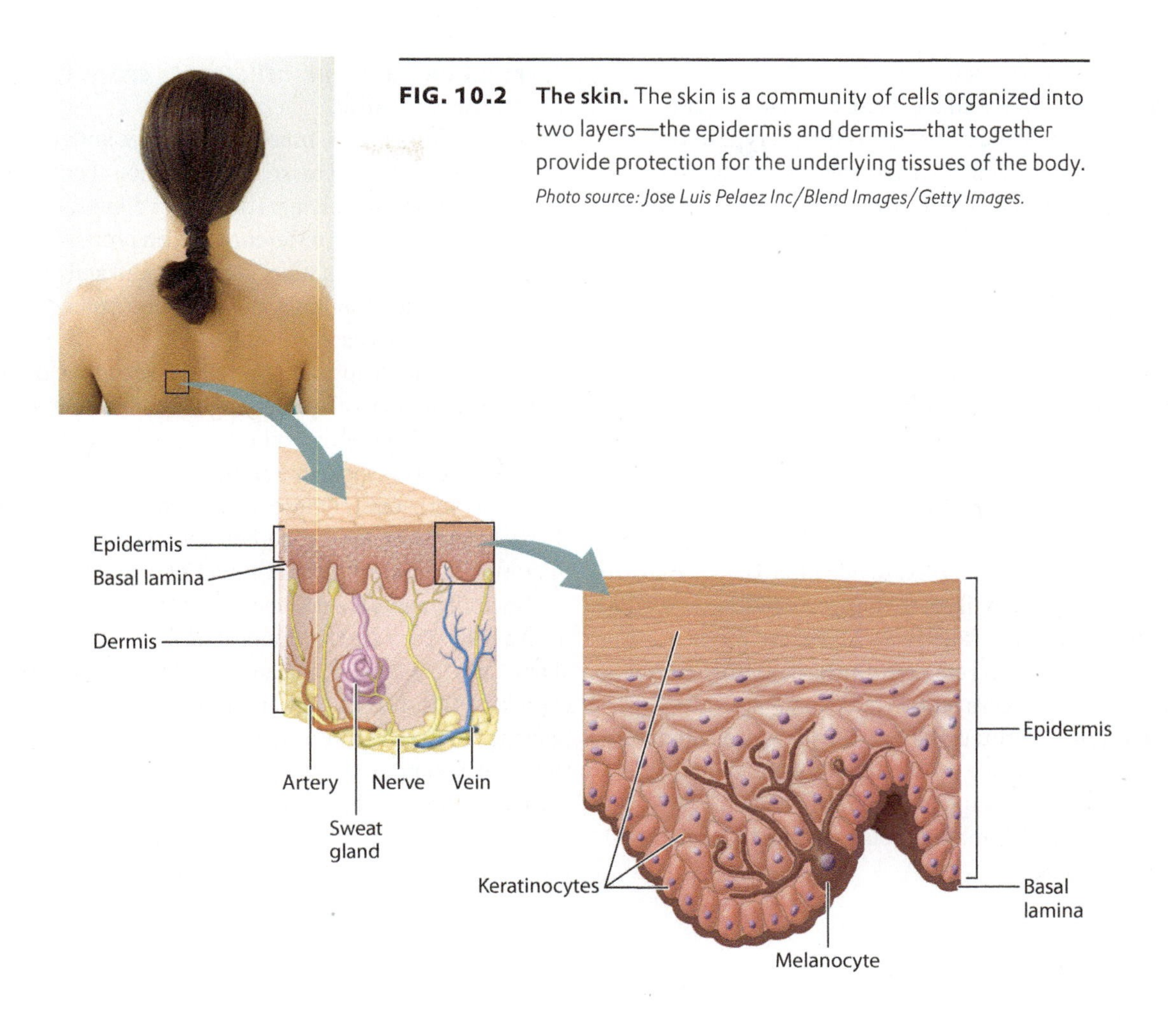

FIG. 10.2 The skin. The skin is a community of cells organized into two layers—the epidermis and dermis—that together provide protection for the underlying tissues of the body. *Photo source: Jose Luis Pelaez Inc/Blend Images/Getty Images.*

allow them to carry out the various processes necessary to sustain them. In animals, for example, four types of tissue—epithelial, connective, nervous, and muscle—combine to make up all the organs of the body. Two or more tissues often combine and function together as an **organ,** such as a heart or lung.

Tissues and organs have distinctive shapes that reflect how they work and what they do. In the same way, the different cell types that make up these organs have distinctive shapes based on what they do in the organ. In animals, the shape of cells is determined and maintained by structural protein networks in the cytoplasm called the **cytoskeleton** (section 10.2). The shape and structural integrity of tissues and organs depend on the ability of cells to connect to one another. In turn, the connection of cells to one another depends on structures called **cell junctions** (section 10.3). Equally important to a strong, properly shaped tissue or organ is the ability of cells to adhere to a meshwork of proteins and polysaccharides outside the cell called the **extracellular matrix** (section 10.4).

The structure of skin relates to its function.

To start our investigation of the cytoskeleton, cell junctions, and the extracellular matrix, let's consider a community of cells very familiar to all of us—our own skin (**Fig. 10.2**). The structure of mammalian skin is tied to its function. Skin has two main layers. The outer layer of skin, the **epidermis,** serves as a water-resistant, protective barrier. The layer beneath the epidermis is the **dermis.** This layer of the skin supports the epidermis, both physically and by supplying it with nutrients. It also provides a cushion surrounding the body.

As you can see in Fig. 10.2, the epidermis is several cell layers thick. Cells arranged in one or more layers are called epithelial cells and together make up a type of animal tissue called epithelial tissue. Epithelial tissue covers the outside of the body and lines many internal structures, such as the digestive tract and vertebrate blood vessels. The epidermal layer of skin is primarily composed of epithelial cells called keratinocytes. The epidermis also contains melanocytes that produce pigment that gives skin its coloration.

Keratinocytes in the epidermis are specialized to protect underlying tissues and organs. They are able to perform this function in part because of their elaborate system of cytoskeletal filaments. These filaments are often connected to the cell junctions that hold adjacent keratinocytes together. Cell junctions also connect the bottom layer of keratinocytes to a specialized form of extracellular matrix called the **basal lamina** (also called the basement membrane, although it is not in fact a membrane), which underlies and supports all epithelial tissues.

The second layer of skin, the dermis, is made up mostly of connective tissue, a type of tissue characterized by few cells and substantial amounts of extracellular matrix. The main type of cell in the dermis is the fibroblast, which synthesizes the extracellular matrix. The dermis is strong and flexible because it is composed of tough protein fibers of the extracellular matrix. The dermis also has many blood vessels and nerve endings.

We will come back to the skin several times in this chapter to show more precisely how these multiple connections among cytoskeleton, cell junctions, and the extracellular matrix make the skin a watertight and strong protective barrier.

10.2 THE CYTOSKELETON

Just as the bones of vertebrate skeletons provide internal support for the body, the protein fibers of the cytoskeleton provide internal support for cells (**Fig. 10.3**). All eukaryotic cells have at least two cytoskeletal elements, **microtubules** and **microfilaments.** Animal cells have a third element, **intermediate filaments.** All three of these cytoskeletal elements are long chains, or polymers, made up of protein subunits. In addition to providing structural support, microtubules and microfilaments enable the movement of substances within cells as well as changes in cell shape.

Microtubules and microfilaments are polymers of protein subunits.

Microtubules are hollow tubelike structures with the largest diameter of the three cytoskeletal elements, about 25 nm (Fig. 10.3a). They are polymers of protein dimers. Each dimer is made up of two slightly different **tubulin** proteins, called α (alpha) and β (beta) tubulin. One α tubulin and one β tubulin combine to make a tubulin dimer and the tubulin dimers are assembled to form the microtubule.

Microtubules help maintain cell shape and the cell's internal structure. In animal cells, microtubules radiate outward to the cell periphery from a microtubule organizing center called the **centrosome.** This spokelike arrangement of microtubules helps cells withstand compression and thereby maintain their shape. Many organelles are tethered to microtubules, and thus microtubules guide the arrangement of organelles in the cell.

Microfilaments are polymers of **actin** monomers, arranged to form a helix. They are the thinnest of the three cytoskeletal fibers, about 7 nm in diameter, and are present in various locations in the cytoplasm (Fig. 10.3b). They are relatively short and extensively branched in the cell cortex, the area of the cytoplasm just beneath the plasma membrane. At the cortex, microfilaments reinforce the plasma membrane and organize proteins associated with it.

These cortical microfilaments are also important in part for maintaining the shape of a cell, such as the biconcave shape

FIG. 10.3 Three types of cytoskeletal element. (a) Microtubule; (b) microfilament; and (c) intermediate filament.

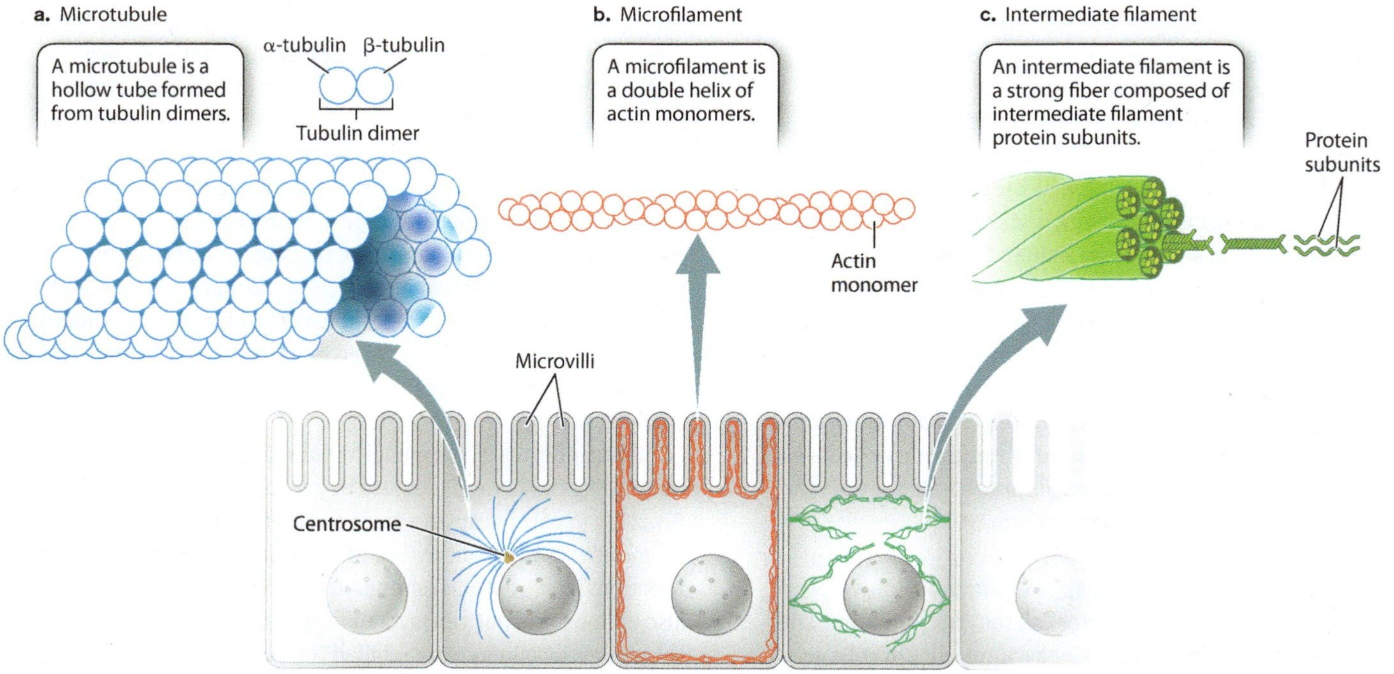

FIG. 10.4 Microfilaments in intestinal microvilli. Actin microfilaments help to maintain the structure of microvilli in the intestine. *Photo source: ©1982, Rockefeller University Press. Originally published in The Journal of Cell Biology. 94:425–443.*

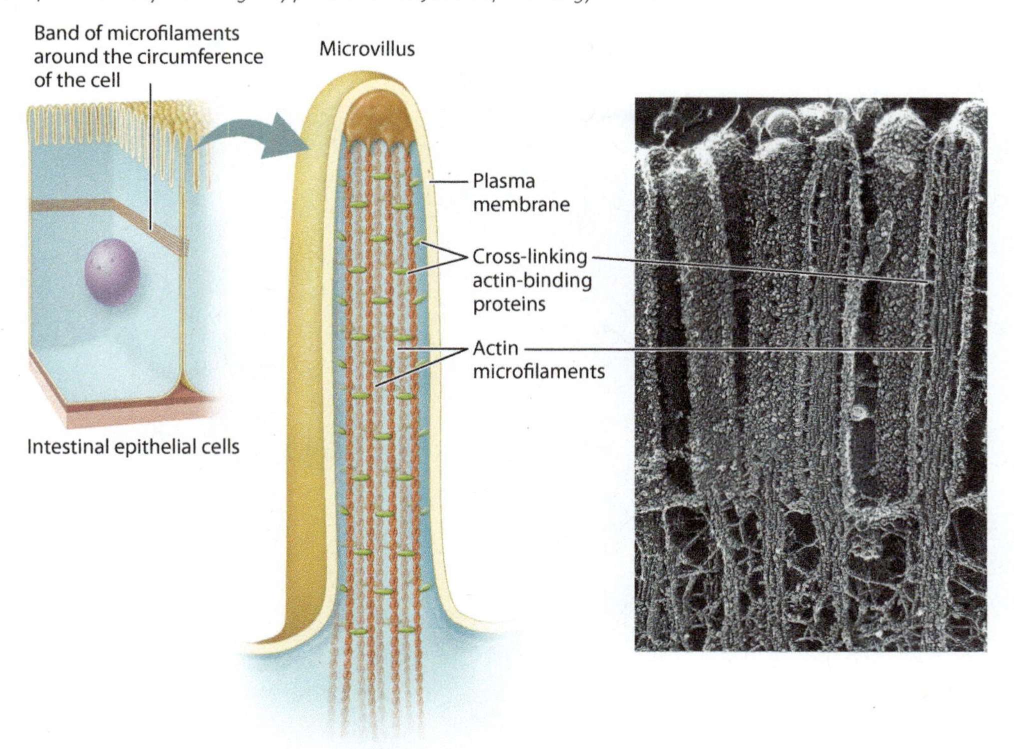

of red blood cells discussed earlier (see Fig. 10.1a). The shape of absorptive epithelial cells such as those in the small intestine is also maintained with the help of microfilaments. In these cells, bundles of microfilaments are found in microvilli, hairlike projections that extend from the surface of the cell, and longer bundles of microfilaments form a band that extends around the circumference of epithelial cells (**Fig. 10.4**).

Microtubules and microfilaments are dynamic structures.

When we think of the parts of a skeleton, we are inclined to think of our bones. Unlike our bones, which undergo remodeling but do not change rapidly, microtubules and microfilaments are dynamic. They become longer by the addition of subunits to their ends, and become shorter by the loss of subunits.

The rate at which protein subunits are added depends on the concentrations of tubulin and actin in that region of the cell. At high concentrations of subunits, microtubules and microfilaments can become longer at both ends, although the subunits are assembled more quickly on one end than the other. The faster-assembling end is called the plus end and the slower-assembling end is called the minus end (**Fig. 10.5**).

The dynamic nature of microfilaments and microtubules is important for some of their functions. For example, some forms

FIG. 10.5 Plus and minus ends of microtubules and microfilaments. The two ends assemble at different rates.

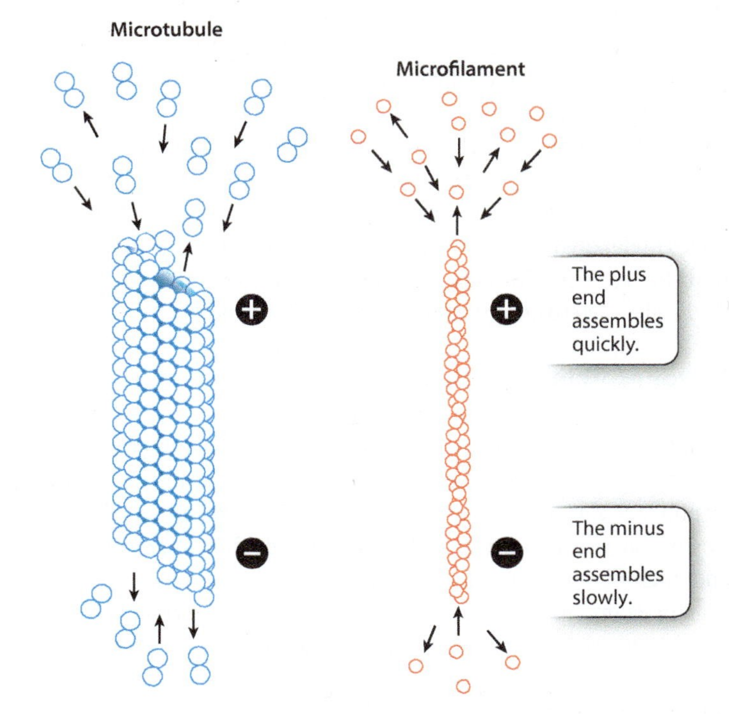

FIG. 10.6 Dynamic instability. The plus ends of microtubules undergo cycles of rapid disassembly followed by slow assembly.

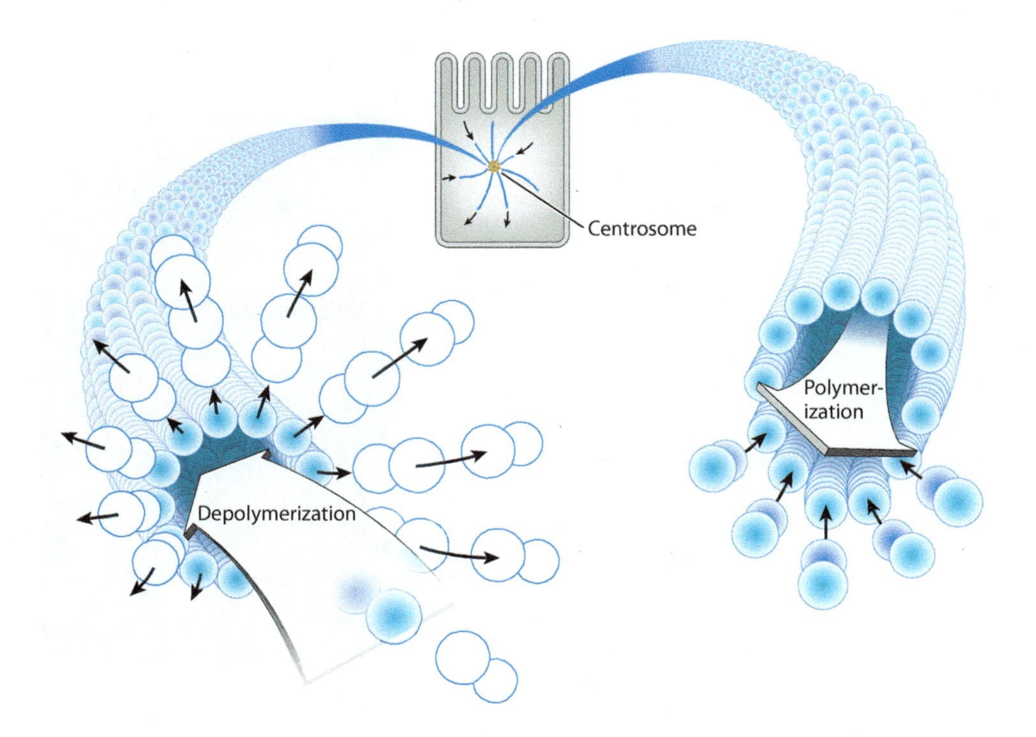

of cell movement, such as a single-celled amoeba foraging for food or a mammalian white blood cell chasing down foreign bacteria, depend on actin polymerization and depolymerization. It is also required when a single cell divides in two during cytokinesis (Chapter 11).

Microtubules make up the spindles that attach to chromosomes during cell division. In this case, the ability of spindle microtubules to "explore" the space of the cell and encounter chromosomes is driven by a unique property of microtubules: Their plus ends undergo seemingly random cycles of rapid depolymerization followed by slower polymerization. These cycles of depolymerization and polymerization are called **dynamic instability** (**Fig. 10.6**). They allow spindle microtubules to quickly find and attach to chromosomes during cell division.

Motor proteins associate with microtubules and microfilaments to cause movement.

A motor is a device that imparts motion. We saw that microtubules and microfilaments have some capacity to lengthen and shorten by polymerization and depolymerization. However, when joined by small accessory proteins called **motor proteins,** microtubules and microfilaments are capable of causing amazing movements.

For example, microtubules function as tracks for transport within the cell. Two motor proteins that associate with these microtubule tracks are **kinesin** and **dynein.** Kinesin transports cargo toward the plus end of microtubules, located at the periphery of the cell (**Fig. 10.7**). By contrast, dynein carries its load away from the plasma membrane toward the minus end, located at the centrosome in the interior of the cell. Movement along microtubules by kinesin and dynein is driven by conformational changes in the motor proteins and is powered by energy harvested from ATP.

Let's look at an especially striking example of this system at work in the specialized skin cells called melanophores present in some vertebrates. Melanophores are similar to the melanocytes in our own skin that produce the pigment melanin. However, rather than hand off their melanin to other cells as in humans, melanophores keep their pigment granules and move them around the cell in response to hormones or neuronal signals. This redistribution of melanin within the cell allows animals such as fish or amphibians to change color. For example, at night the melanin granules in the skin of a zebrafish embryo are dispersed throughout the melanophores, making it darkly colored. As morning comes and the day brightens, the pigment granules aggregate at the center of the cell around the centrosome, causing the embryo's color to lighten (**Fig. 10.8**).

The melanin granules in the melanophores move back and forth along microtubules, transported by kinesin and dynein. Kinesin moves the granules out toward the plus end of the microtubule during dispersal, and dynein moves them back toward the minus end during aggregation. The daytime and nighttime camouflage provided by this color change helps prevent young, developing organisms from being spotted by hungry predators lurking below.

FIG. 10.7 Intracellular transport. The motor protein kinesin interacts with microtubules to move vesicles in the cell.

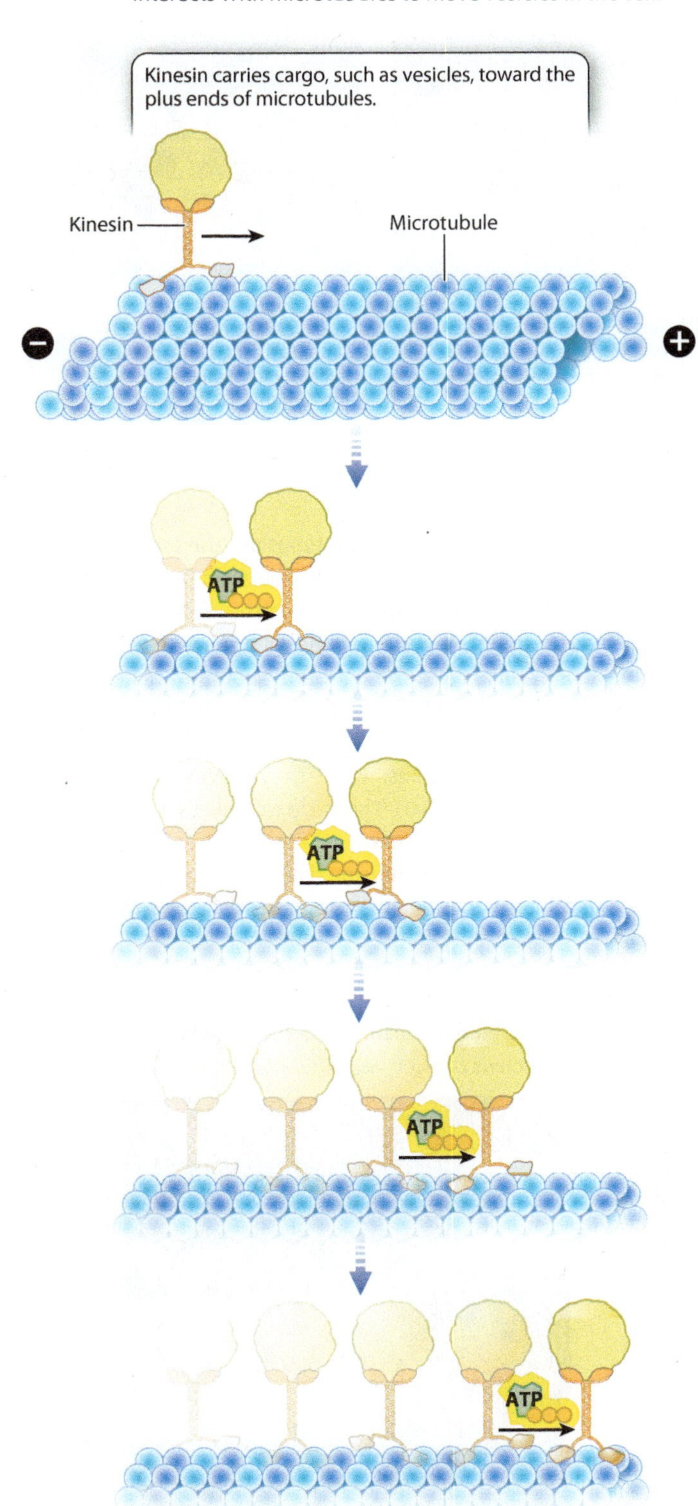

FIG. 10.8 Color change in zebrafish embryos, driven by motor proteins kinesin and dynein. Melanin granules are redistributed along microtubules in the melanophores of the skin. *Photo sources: (left and right) Jane Bradbury, "Small Fish, Big Science", PLOS Biology, 2, (5), May 2004, pg. 0569, Image courtesy of Adam Amsterdam, Massachusetts Institute of Technology, Boston, MA; (bottom) Courtesy Darren Logan, Wellcome Trust Sanger Institute.*

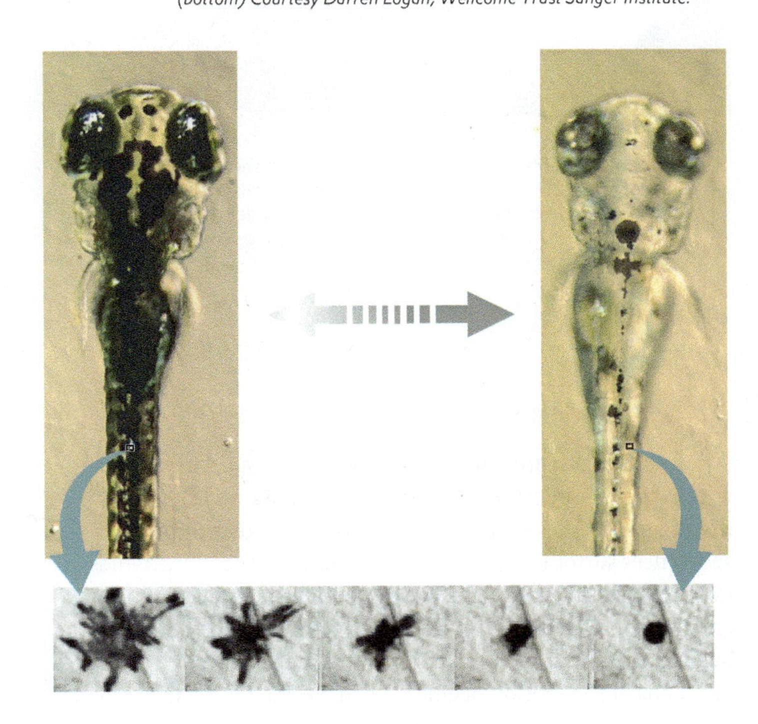

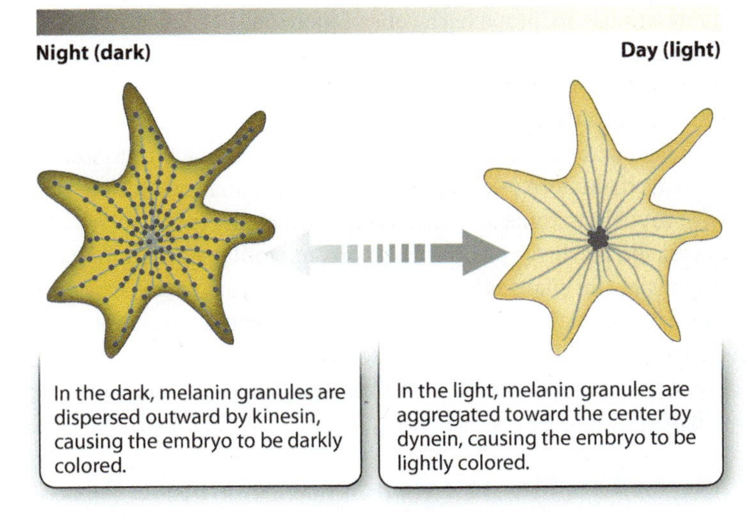

FIG. 10.9 Cilia and flagella in diverse cell types. Cilia and flagella, composed of microtubules, move cells or allow cells to propel substances. *Sources: (from left to right) SPL/Science Source; Eye of Science/Science Source; Andrew Syred/Science Source; Juergen Berger/Science Source.*

Coordinated beating of the cilia that cover the paramecium moves the cell through its environment.

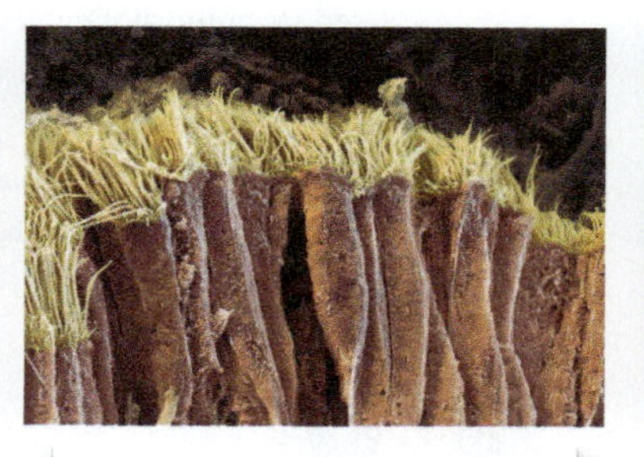

The cilia in these human airway epithelial cells propel mucus containing debris out of the lungs.

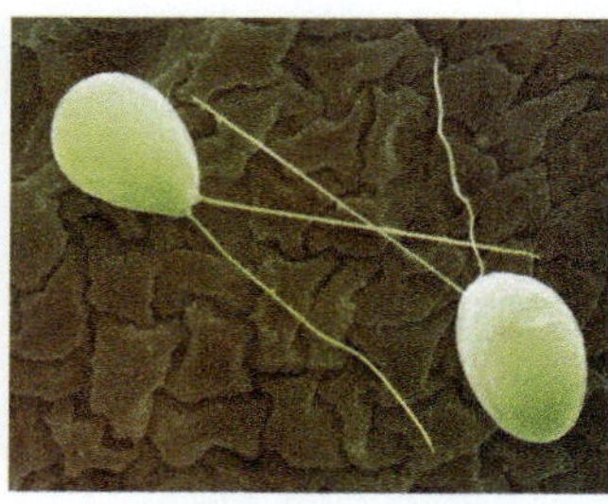

These unicellular algae are propelled by two flagella.

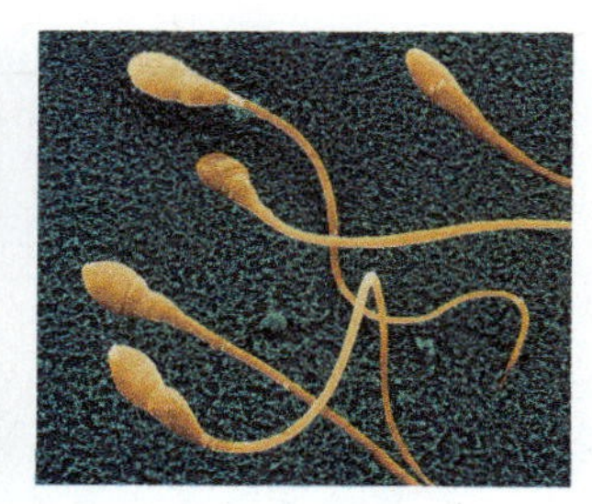

Some cells of multicellular organisms, including these sperm cells, swim by movement of a flagellum.

→ **Quick Check 1** Would a defect in dynein or in kinesin cause a zebrafish embryo to remain darkly colored after daybreak?

In addition to providing tracks for the transport of material within the cell, microtubules are found in **cilia** and **flagella,** fiberlike organelles that propel the movement of cells or substances surrounding the cell (**Fig. 10.9**). In these organelles, microtubules associate with the motor protein dynein, which causes movement. Many single-celled eukaryotic organisms that live in aquatic environments propel themselves through the water by means of the motion of cilia (which are short) or flagella (which are long). Some cells of multicellular organisms also have cilia or flagella. For example, the sperm cells of algae, some plants, and many animals are propelled by one or more flagella. Epithelial cells in a number of animal tissues, such as the lining of the trachea and the upper respiratory tract, have cilia that move substances along the surface of the cell layer.

Like microtubules, microfilaments also associate with motor proteins to produce movement. Actin microfilaments associate with **myosin** to transport various types of cellular cargo, such as vesicles, inside of cells. Furthermore, microfilaments are responsible for changes in the shape of many types of cell. One of the most dramatic examples of cell shape change is the shortening (contraction) of a muscle cell. Muscle contraction depends on the interaction of myosin with microfilaments, and is powered by ATP (Chapter 37).

Intermediate filaments are polymers of proteins that vary according to cell type.

The intermediate filaments of animal cells have a diameter that is intermediate between those of microtubules and microfilaments, about 10 nm (see Fig. 10.3c). They are polymers of intermediate filament proteins that combine to form strong, cable-like structures in the cell. As a result, they provide cells with mechanical strength.

We have seen that different cell types use the same tubulin dimers to form microtubules and the same actin monomers to form microfilaments. By contrast, the proteins making up intermediate filaments differ from one cell type to another. For example, in epithelial cells, these protein subunits are keratins; in fibroblasts, they are vimentins; and in neurons, they are neurofilaments. Some intermediate filaments, called lamins, are even found inside the nucleus, where they provide support for the nuclear envelope (**Fig. 10.10**). There are well over 100 different kinds of intermediate filaments.

Once assembled, many intermediate filaments become attached to cell junctions at their cytoplasmic side, providing

FIG. 10.10 Intermediate filaments in the cytoplasm and the nucleus of a cell. *Source: Courtesy of R. D. Goldman.*

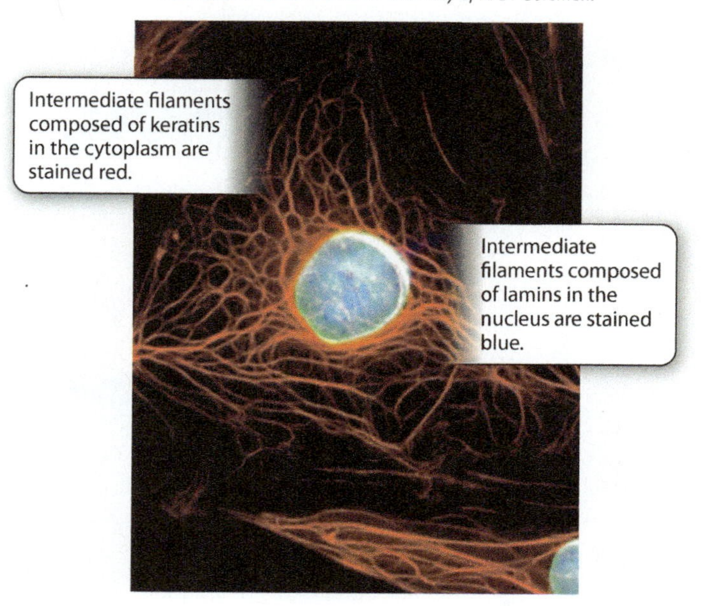

FIG. 10.11 Intermediate filaments in the epidermis of the skin. Intermediate filaments bind to cell junctions called desmosomes, forming a strong, interconnected network. Epidermolysis bullosa is a group of blistering diseases of the skin that can be caused by a defect in intermediate filaments. *Source: Helen Osler/Northscot/Rex USA.*

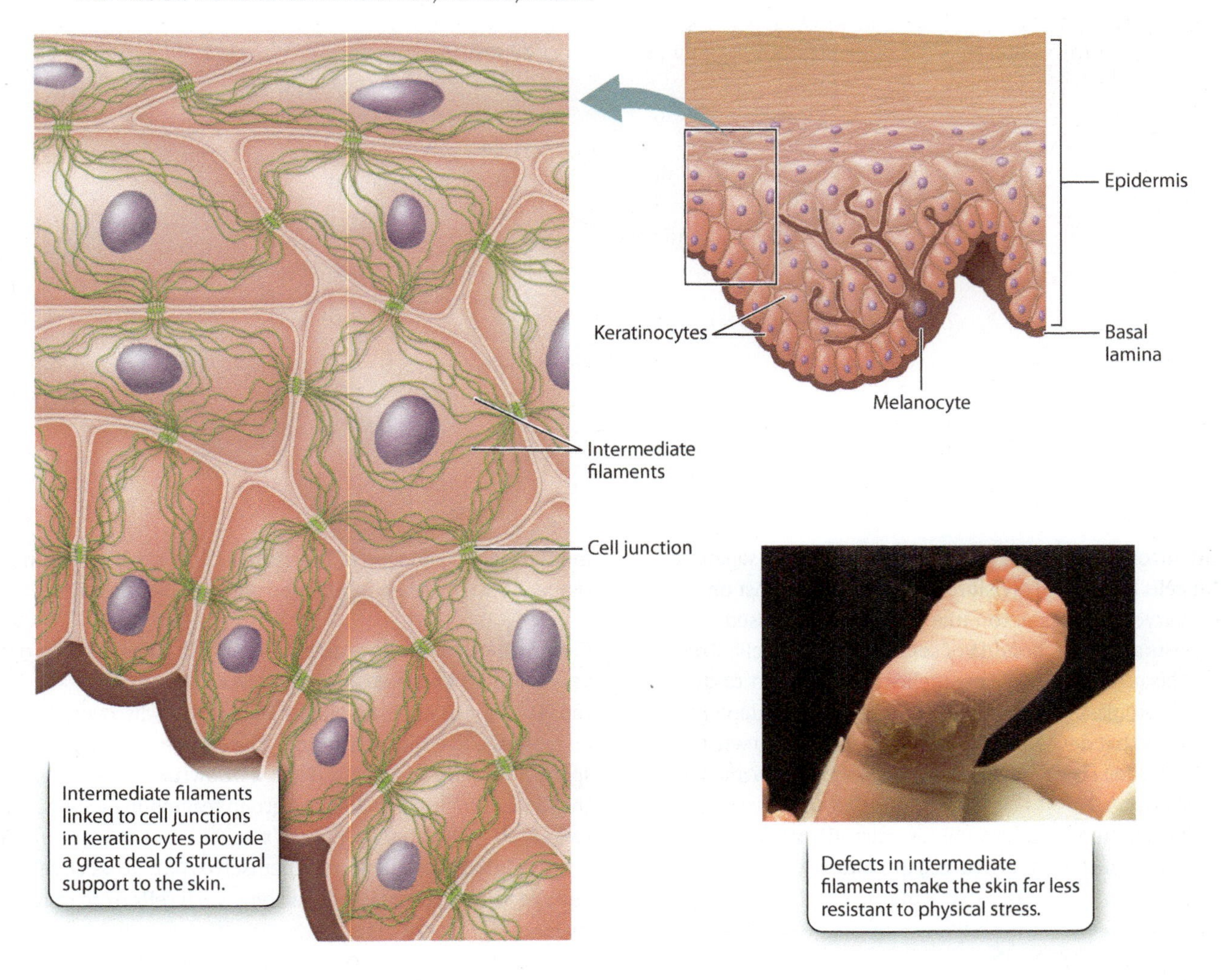

strong support for the cells (**Fig. 10.11**). In the case of epithelial cells, this anchoring results in structural continuity from one cell to another that greatly strengthens the entire epithelial tissue. This is especially important for tissues that are regularly subject to physical stress, such as the skin and the lining of the intestine.

Genetic defects that disrupt the intermediate filament network can have severe consequences. For example, some individuals with epidermolysis bullosa, a group of rare genetic diseases, have defective keratin genes. Intermediate filaments do not polymerize properly in these individuals, thus forming weaker connections between the layers of cells that make up the epidermis. As a consequence, the outer layers can detach, resulting in extremely fragile skin that blisters in response to the slightest trauma (Fig. 10.11). The sensitivity to physical stress is so extreme that infants with epidermolysis bullosa often suffer significant damage to the skin during childbirth. Therefore, Caesarean section is sometimes recommended in cases where the disease is diagnosed during pregnancy.

The cytoskeleton is an ancient feature of cells.

Actin and tubulin are found in all eukaryotic cells, and their structure and function have remained relatively unchanged throughout the course of evolution. The amino acid sequences of yeast tubulin and human tubulin are 75% identical. Similar comparisons of actin from amoebas and animals show that they are 80% identical after close to a billion years of evolution. In fact, a mixture of yeast and human actin monomers forms hybrid microfilaments able to function normally in the cell.

Not long ago, it was believed that cytoskeletal proteins were present only in eukaryotic cells. However, a number of studies have shown that many prokaryotes also have a system of proteins similar in structure to the cytoskeletal elements of eukaryotic

TABLE 10.1 Major Functions of Cytoskelatal Elements.

CYTOSKELETAL ELEMENT	SUBUNITS	MAJOR FUNCTIONS
Microtubules	Tubulin dimers	Cell shape and support Cell movement (by cilia, flagella) Cell division (chromosome segregation) Vesicle transport Organelle arrangement
Microfilaments	Actin monomers	Cell shape and support Cell movement (by crawling) Cell division (cytokinesis) Vesicle transport Muscle contraction
Intermediate filaments	Diverse	Cell shape and support

cells and are involved in similar processes, including the separation of daughter cells during cell division. Interestingly, at least one of these prokaryotic cytoskeleton-like proteins is expressed in the chloroplasts and mitochondria of some eukaryotic cells. The presence of this protein in these organelles lends support to the theory that chloroplasts and mitochondria were once independent prokaryotic cells that developed a symbiotic relationship with another cell. This idea is called the endosymbiotic theory and it is discussed in Chapter 27.

The major functions of microtubules, microfilaments, and intermediate filaments are summarized in **Table 10.1.**

10.3 CELL JUNCTIONS

The cell is the fundamental unit of living organisms (Chapter 1). Some estimates place the number of cells in an adult human being at between 50 and 75 trillion, whereas others place the number at well above 100 trillion. Whichever estimate is more accurate, humans, as well as all complex multicellular organisms, are made up of a lot of cells! What then keeps us (or any other multicellular organism) from slumping into a pile of cells? And what keeps cells organized into tissues, and tissues into organs?

Tissues are held together and function as a unit because of cell junctions. Cell junctions physically connect one cell to the next and anchor cells to the extracellular matrix. Some tissues have cell junctions that perform roles other than adhesion. For example, cell junctions in the outer layer of the skin and the lining of intestine provide a seal so that the epithelial sheet can act as a selective barrier. Other cell junctions allow communication between adjacent cells so that they work together as a unit. In this section, we look more closely at the roles of the various types of cell junction in tissues, as well as their interaction with the cytoskeleton.

Cell adhesion molecules allow cells to attach to other cells and to the extracellular matrix.

In 1907, American embryologist H. V. Wilson discovered that if he pressed a live sponge through fine cloth he could break up the sponge into individual cells. Then if he swirled the cells together, they would coalesce back into a group resembling a sponge. If he swirled the cells from sponges of two different species together, he observed that the cells sorted themselves out—that is, cells from one species of sponge associated only with cells from that same species (**Fig. 10.12a**). Fifty years later, German-born embryologist Johannes Holtfreter observed that if he took neuronal cells and skin cells from an amphibian embryo and treated them the same way that Wilson had treated sponge cells, the embryonic cells would sort themselves according to tissue type (**Fig. 10.12b**).

Cells are able to sort themselves because of the presence of various proteins on their surface called **cell adhesion molecules** that attach cells to one another or to the extracellular matrix. While a number of cell adhesion molecules are now known, the **cadherins** (calcium-dependent adherence proteins) are especially important in the adhesion of cells to other cells. There are many different kinds of cadherin, and a given cadherin may bind only to another cadherin of the same type. This property explains Holtfreter's observations of the cells from amphibian embryos. E-cadherin (for "epidermal cadherin") is present on the surface of embryonic epidermal cells, and N-cadherin (for "neural cadherin") is present on neuronal cells. The epidermal cells adhered to one another through E-cadherin, and the neuronal cells adhered to each other through N-cadherin.

FIG. 10.12 Cell type–specific cell adhesion. Experiments showed that cells from (a) sponges and from (b) amphibian embryos are in each case able to adhere to one another in a specific manner. *Photo sources: a. (top) Borut Furlan/WaterFrame/age fotostock; (bottom) Franco Banfi/WaterFrame/age fotostock.*

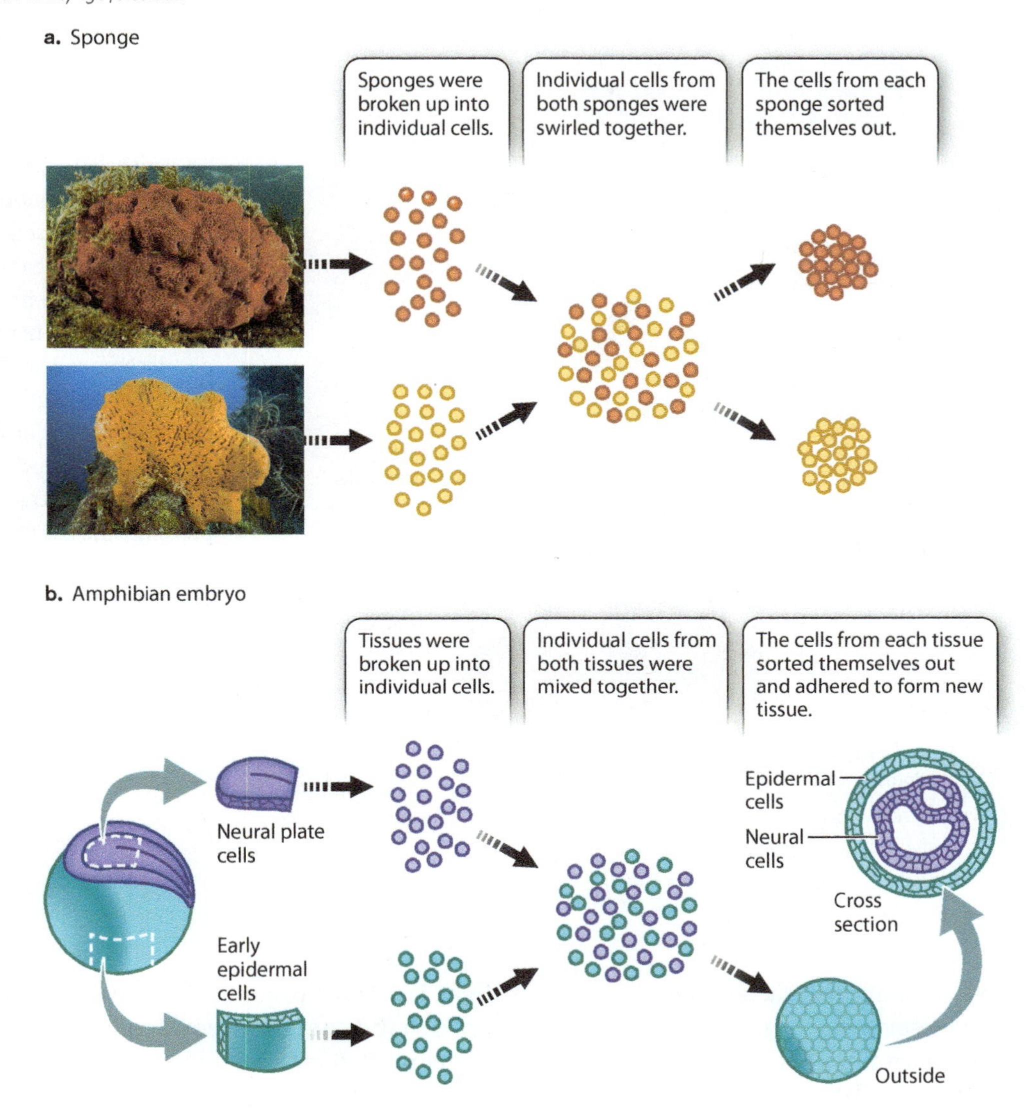

Cadherins are transmembrane proteins (Chapter 5). The extracellular domain of a cadherin molecule binds to the extracellular domain of a cadherin of the same type on an adjacent cell. The cytoplasmic portion of the protein is linked to the cytoskeleton, including microfilaments and intermediate filaments (**Fig. 10.13a**). This arrangement provides structural continuity from the cytoskeleton of one cell to the cytoskeleton of another, increasing the strength of tissues and organs.

As well as being stably connected to other cells, cells also attach to proteins of the extracellular matrix. Cell adhesion molecules that enable cells to adhere to the extracellular matrix are called **integrins.** Like cadherins, integrins are transmembrane proteins, and their cytoplasmic domain is linked to microfilaments or intermediate filaments (**Fig. 10.13b**). Also like the cadherins, integrins are of many different types, each binding to a specific extracellular matrix protein. Integrins are present on the surface of virtually every animal cell. In addition to their role in adhesion, integrins also act as receptors that communicate information about the extracellular matrix to the interior of the cell (section 10.4)

Anchoring junctions connect adjacent cells and are reinforced by the cytoskeleton.

Cadherins and integrins are often organized into cell junctions, complex structures in the plasma membrane that allow cells to

FIG. 10.13 Cell adhesion by cadherins and integrins. (a) Cadherins are transmembrane proteins that connect cells to other cells; (b) integrins are transmembrane proteins that connect cells to the extracellular matrix.

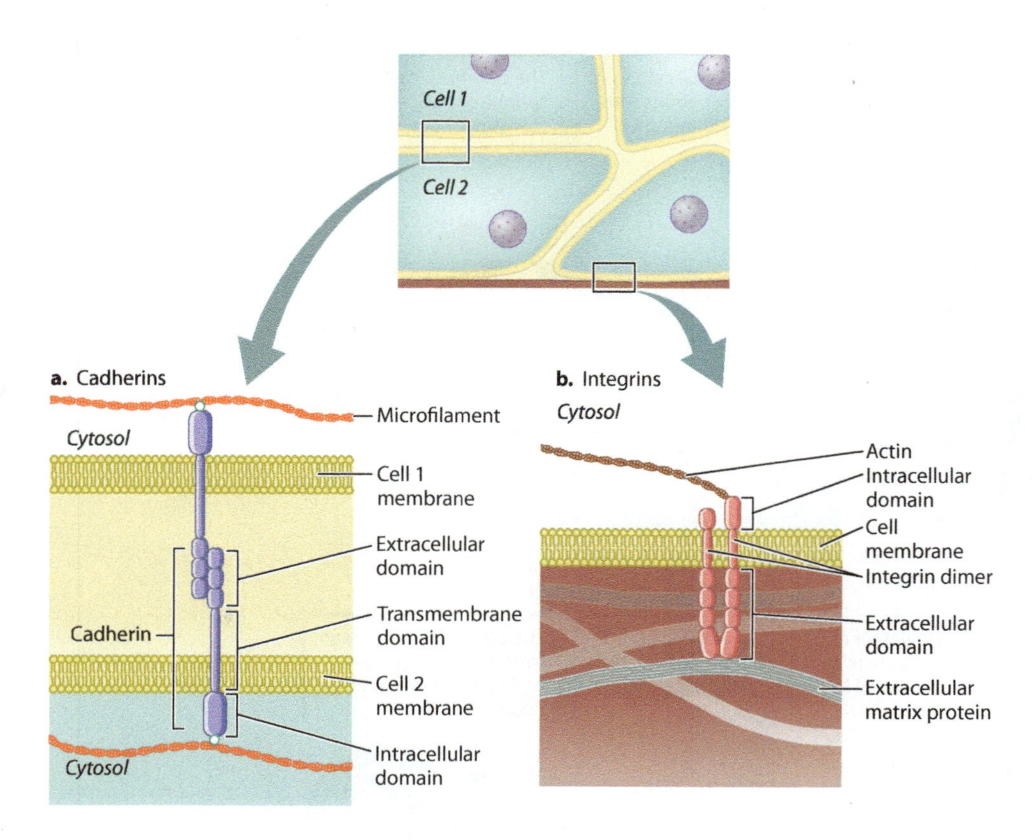

adhere to one another. These anchoring cell junctions are of two types: adherens junctions and desmosomes (**Fig. 10.14**).

In our earlier discussion of microfilaments (section 10.2), we saw that a long bundle of actin microfilaments forms a band that extends around the circumference of epithelial cells, such as the epithelial cells that line the intestine. This band of actin is attached to the plasma membrane by cadherins in a beltlike structure called an **adherens junction** (Fig. 10.14). The cadherins in the adherens junction of one cell attach to the cadherins in the adherens junctions of adjacent cells. This arrangement establishes a physical connection among the actin cytoskeletons of all cells present in an epithelial layer of cells.

Like adherens junctions, **desmosomes** are cell junctions that allow cells to adhere to one another. However, unlike adherens junctions that form a belt around the circumference of cells, desmosomes are buttonlike points of adhesion (Fig. 10.14). Cadherins are at work here, too, strengthening the connection between cells in a manner similar to adherens junctions. Cadherins in the desmosome of one cell bind to cadherins in the desmosomes of adjacent cells. The cytoplasmic domain of these cadherins connects to intermediate filaments in the cytoskeleton. This second type of physical connection among neighboring cells greatly enhances the structural integrity of epithelial cell layers.

Epithelial cells are not only attached to one another, but also to the underlying extracellular matrix (specifically, the basal lamina). In this case, the cells are firmly anchored to the extracellular matrix by a type of desmosome called a **hemidesmosome** (Fig. 10.14). Integrins are the prominent cell adhesion molecules in hemidesmosomes. The extracellular domains bind extracellular matrix proteins, and the cytoplasmic domains connect to intermediate filaments. These intermediate filaments connect to desmosomes in other parts of the plasma membrane. The result is a firmly anchored and reinforced layer of cells.

→ **Quick Check 2** Adherens junctions and desmosomes both attach cells to other cells and are made up of cadherins. How, then, are they different?

Tight junctions prevent the movement of substances through the space between cells.

Epithelial cells form sheets or boundaries that line tissues and organs, including the digestive tract, respiratory tract, and outer layer of the skin. Like any effective boundary, a layer of epithelial cells must limit or control the passage of material across it. Adherens junctions and desmosomes provide strong adhesion between cells, but they do not prevent materials from passing freely through the spaces between the cells. This function is provided by a different type of cell junction. In vertebrates, these are called **tight junctions** (Fig. 10.14). Tight junctions establish a seal between cells so that the only way a substance can travel from one side of a sheet of epithelial cells to the other is by moving *through* the cells by means of one of the cellular transport mechanisms discussed in Chapter 5.

A tight junction is a band of interconnected strands of integral membrane proteins, particularly proteins called claudins and occludins. Like adherens junctions, tight junctions encircle the epithelial cell. The proteins forming the tight junction in one cell

FIG. 10.14 Cell junctions. Cell junctions connect cells to other cells or to the basal lamina and are reinforced by the cytoskeleton.

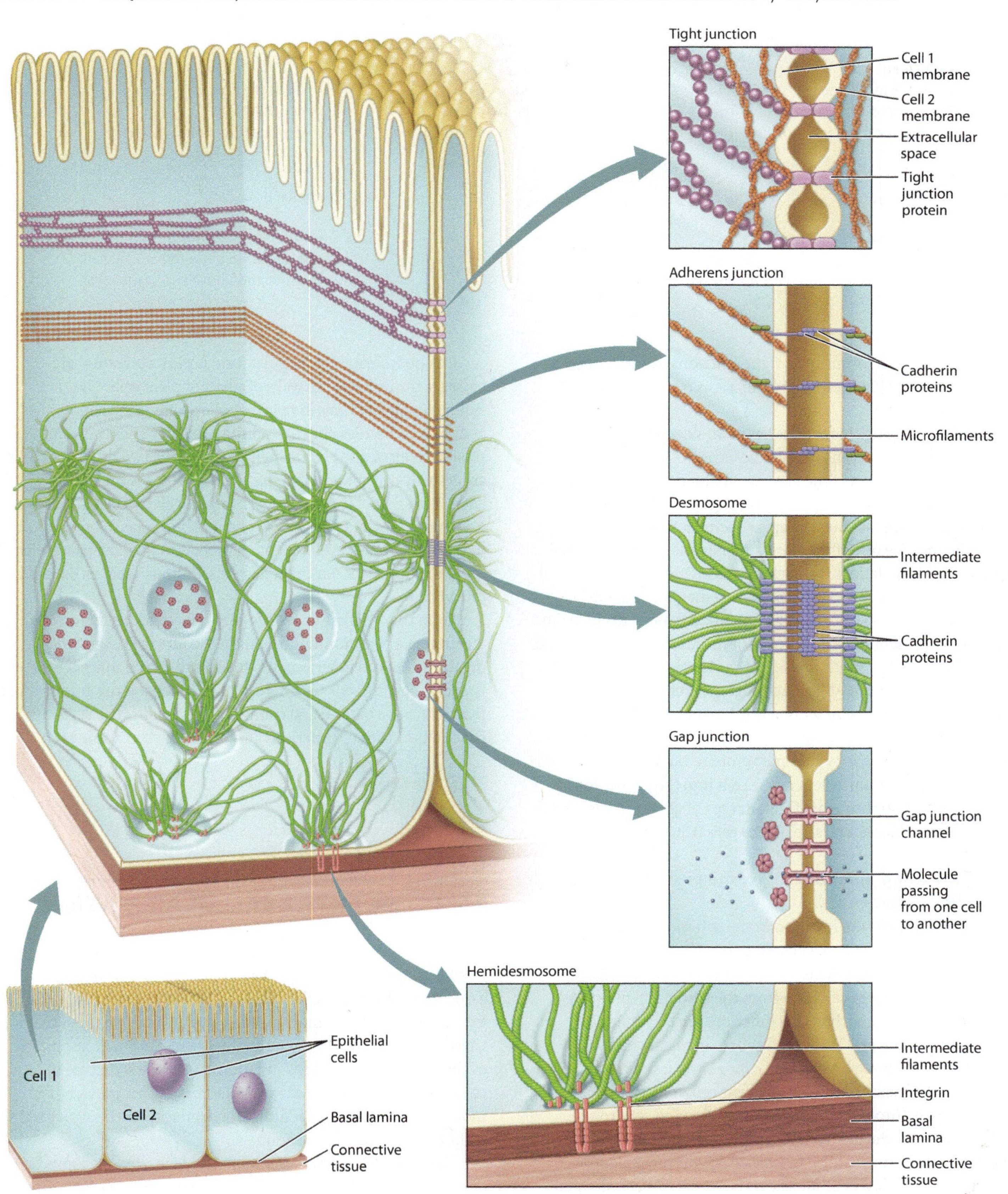

TABLE 10.2 Types and Functions of Cell Junctions.

CELL JUNCTION	MAJOR COMPONENT	CYTOSKELETAL ATTACHMENT	PRIMARY FUNCTION
Anchoring			
Adherens junction	Cadherins	Microfilaments	Cell–cell adhesion
Desmosome	Cadherins	Intermediate filaments	Cell–cell adhesion
Hemidesmosome	Integrins	Intermediate filaments	Cell–extracellular matrix adhesion
Barrier			
Tight junction	Claudins, occludins		Epithelial boundary
Communicating			
Gap junction	Connexins		Communication between animal cells
Plasmodesma	Cell membrane		Communication between plant cells

bind to the proteins forming the tight junctions in adjacent cells. Also like adherens junctions, tight junctions connect to actin microfilaments.

Cells that have tight junctions have two sides because the tight junction divides the plasma membrane into two distinct regions (Fig. 10.14). The portion of the plasma membrane in contact with the lumen, or the inside of any tubelike structure like the gut, is called the apical membrane. The apical membrane defines the "top" side of the cell. The rest of the plasma membrane is the basolateral membrane, which defines the bottom ("baso") and sides ("lateral") of the cell. These two regions of the plasma membrane are of different composition because the tight junction prevents lipids and proteins in the membrane on one side of the junction from diffusing to the other side. As a result, the apical and basolateral membranes of a cell are likely to have different integral membrane proteins, which causes them to be functionally different as well. In the small intestine, for example, glucose is transported from the lumen into intestinal epithelial cells by transport proteins on the apical side of the cells, and is transported out of the cells into the circulation by facilitated diffusion through a different type of glucose transporter restricted to the basolateral sides of the cells (Chapter 40).

Communicating junctions allow the passage of molecules between cells.

Not all cell junctions are involved in the adhesion of cells to each other or in sealing a layer of cells. **Gap junctions** (Fig. 10.14) of animal cells and **plasmodesmata** of plant cells (see Fig. 5.17b) permit materials to pass directly from the cytoplasm of one cell to the cytoplasm of another. Gap junctions are a complex of integral membrane proteins called connexins arranged in a ring. The ring of connexin proteins connects to a similar ring of proteins in the membrane of an adjacent cell. Ions and signaling molecules pass through these junctions, allowing cells to communicate. In the heart, for example, ions pass though gap junctions connecting cardiac muscle cells. This rapid electrical communication allows the muscle cells to beat in a coordinated fashion (Chapter 39).

Plasmodesmata (the singular form is **plasmodesma**) are passages through the cell walls of adjacent plant cells. They are similar to gap junctions in that they allow cells to exchange ions and small molecules directly, but the similarity ends there. In plasmodesmata, the plasma membranes of the two connected cells are actually continuous. The size of the opening is considerably larger than that of gap junctions, large enough for cells to transfer RNA molecules and proteins, an ability that is especially important during embryonic development. Plasmodesmata allow plant cells to send signals to one another despite being enclosed within rigid cell walls.

From this discussion, we see that cell junctions interact to create stable communities of cells in the form of tissues and organs. These cell junctions are important for the functions of tissues, allowing cells to adhere to each other and the extracellular matrix, act as a barrier, and communicate rapidly. The types and functions of the cell junctions are summarized in **Table 10.2**.

→ **Quick Check 3** Which type(s) of cell junction prevent(s) substances from moving through the space between cells? Which type(s) of cell junction attach(es) cells to one another?

10.4 THE EXTRACELLULAR MATRIX

Up to this point, we have looked at how the cytoskeleton maintains the shape of cells. We have also seen how the stable association of animal cells with one another and with the extracellular matrix is made possible by cell junctions, and that these junctions are reinforced by the cytoskeleton. As important as the cytoskeleton and cell junctions are to the structure of cells and tissues, it is the extracellular matrix that provides the molecular framework that ultimately determines the structural architecture of plants and animals.

The extracellular matrix is an insoluble meshwork composed of proteins and polysaccharides. Its components are synthesized, secreted, and modified by many different cell types. There are many different forms of extracellular matrix, which differ in the amount, type, and organization of the proteins and polysaccharides that make them up. In both plants and animals, the extracellular matrix not only contributes structural support but also provides informational cues that determine the activity of the cells that are in contact with it.

The extracellular matrix of plants is the cell wall.

The paper we write on, the cotton fibers in the clothes we wear, the wood in the chairs we sit on are, in fact, the extracellular matrix of plants. In plants, the extracellular matrix forms the cell wall, and the main component of the plant cell wall is the polysaccharide cellulose (Chapters 2 and 5). Its presence in the cell wall of every plant makes cellulose the most widespread organic macromolecule on Earth.

The plant cell wall represents possibly one of the most complex examples of an extracellular matrix. It is certainly one of the most diverse in the functions it performs. Cell walls maintain the shape and turgor pressure of plant cells and act as a barrier that prevents foreign materials and pathogens from reaching the plasma membrane. In many plants, cell walls collectively serve as a skeletal support structure for the entire plant.

The plant cell wall is composed of as many as three layers: the outermost middle lamella, the primary cell wall, and the secondary cell wall, located closest to the plasma membrane (**Fig. 10.15**). The middle lamella is synthesized first, during the late stages of cell division. It is composed of a gluelike complex carbohydrate, and it is the main mechanism by which plant cells adhere to one another. The primary cell wall is formed next and consists mainly of cellulose, but it also contains a number of other molecules, including pectin. The primary cell wall is laid down while the cells are still growing. It is assembled by enzymes on the surface of the cell and remains thin and flexible. Once cell growth has stopped, the secondary cell wall is constructed in many, but not all, plant cells. It also is made largely of cellulose but in addition contains a substance called lignin. Lignin hardens the cell wall and makes it water resistant. In woody plants, the cell wall can be up to 25% lignin. The rigid secondary cell wall permits woody plants to grow to tremendous heights. Giant sequoia trees grow to more than 300 feet and are supported entirely by the lignin-reinforced cellulose fibers of the interconnected cell walls.

As a plant cell grows, additional cell wall components must be synthesized to expand the area of the wall. Unlike the extracellular matrix components that are secreted by animal cells, the cellulose polymer is assembled outside the cell, on the extracellular surface of the plasma membrane. Both the glucose monomers that form the polymer and the enzymes that attach them are delivered to the cell surface by arrays of microtubules. Here is yet another example of how the cytoskeleton plays an indispensable role in regulating the shape of a cell.

FIG. 10.15 The three layers of the plant cell wall: middle lamella, primary cell wall, and secondary cell wall. The major component of the plant cell wall is cellulose, a polymer of glucose. *Photo source: Biophoto Associates/Science Source.*

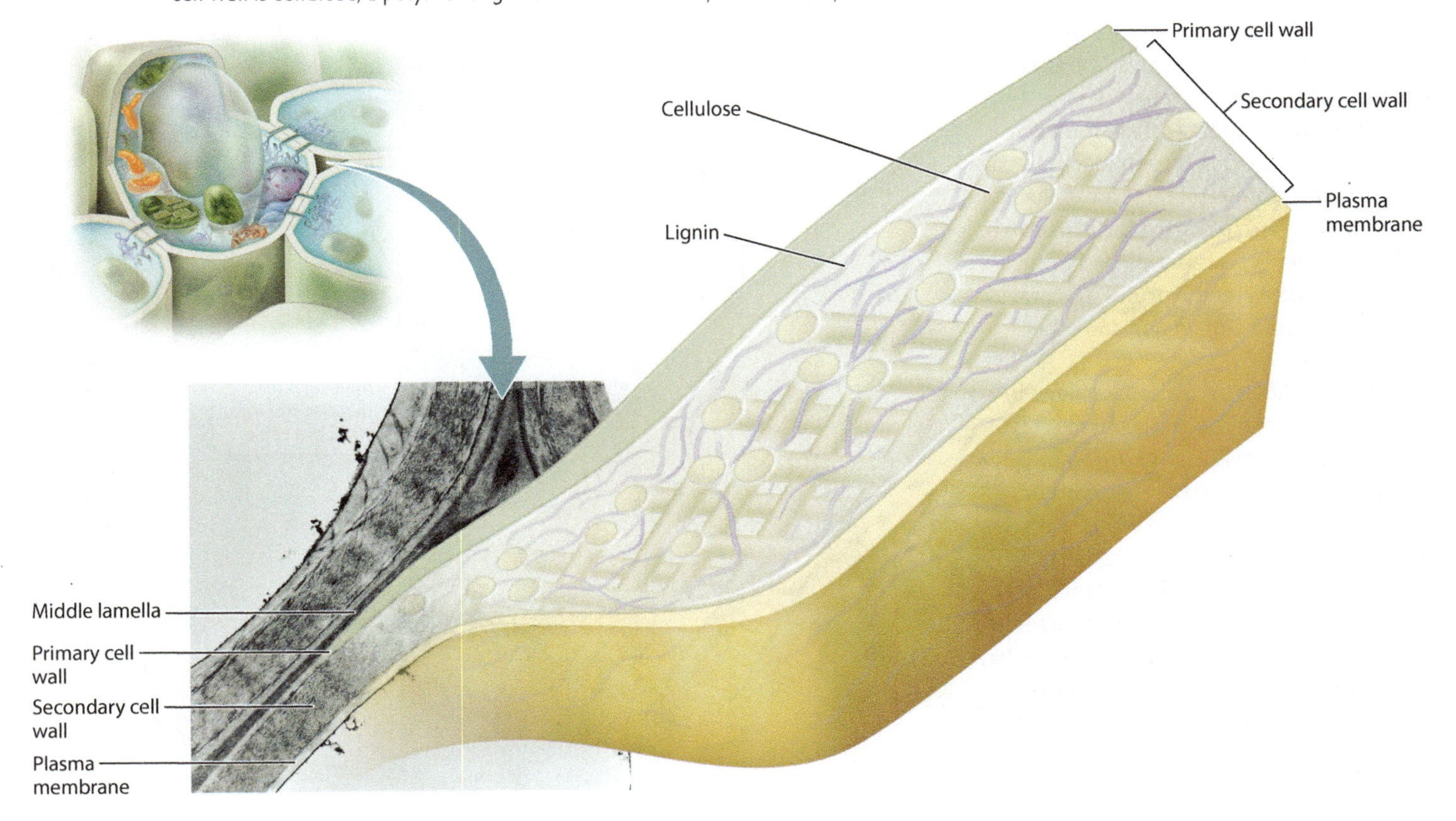

FIG. 10.16 Animal connective tissue. Connective tissue is composed of protein fibers in a gel-like polysaccharide matrix. *Photo source: Biophoto Associates/Science Source.*

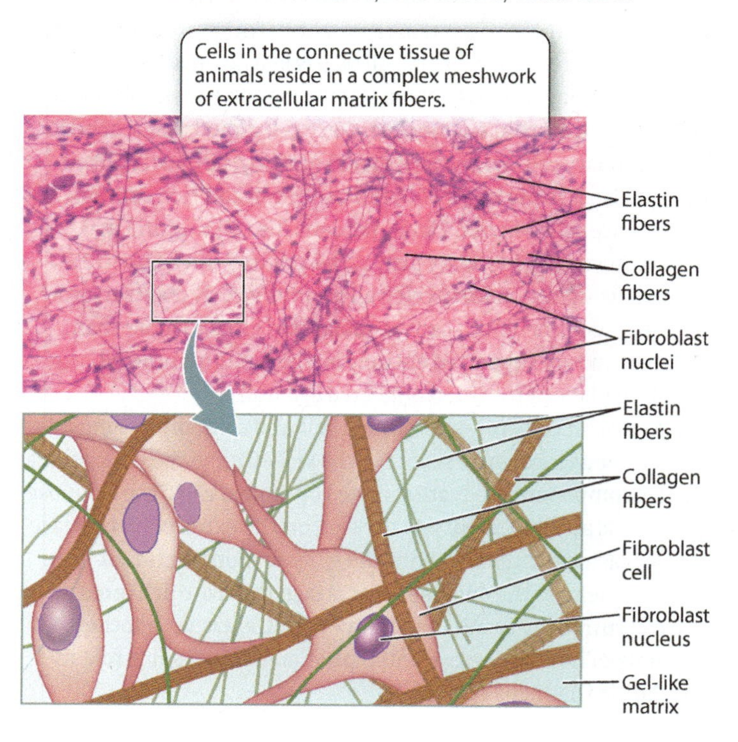

The extracellular matrix is abundant in connective tissues of animals.

The extracellular matrix of animals, like that of plants, is a mixture of proteins and polysaccharides secreted by cells. The animal extracellular matrix is composed of large fibrous proteins, including collagen, elastin, and laminin, which impart tremendous tensile strength. These fibrous proteins are embedded in a gel-like polysaccharide matrix. The matrix is negatively charged, attracting positively charged ions and water molecules that provide protection against compression and other physical stress.

The extracellular matrix can be found in abundance in animal connective tissue (**Fig. 10.16**). Connective tissue structurally integrates and supports various parts of the body and, in this way, is necessary for multicellularity. All animals express similar connective tissue proteins, highlighting their importance and evolutionarily conserved function. Connective tissue underlies all epithelial tissues, as we have seen (section 10.1). For example, the dermis of the skin is connective tissue. It provides support and nutrients to the overlying epidermis.

The main type of cell in the dermis is the fibroblast. Fibroblasts synthesize most of the extracellular matrix proteins.

Connective tissue is unusual compared to other tissue types in that it is dominated by the extracellular matrix and has a low cell density. Consequently, the extracellular matrix determines the properties of different types of connective tissue. Other more specialized types of animal connective tissue include bone, cartilage, and tendon.

Collagen is the most abundant protein in the extracellular matrix of animals. There are more than 20 different forms of collagen, and in humans collagen accounts for almost a quarter of the protein present in the body. Over 90% of this collagen is type I collagen, which is present in the dermis of your skin, where it provides strong, durable support for the overlying epidermis. The tendons that connect your muscles to bones and the ligaments that connect your bones to other bones are able to withstand the physical stress placed on them because they are made up primarily of collagen.

Collagen's strength is related to its structure. Like a rope or a cable, collagen is composed of intertwined fibers that make it

FIG. 10.17 Collagen. Type I collagen molecules are organized in a triple helix and grouped into bundles called fibrils, which in turn are grouped into bundles called fibers. This type of arrangement, seen in fibers and steel cables, imparts tremendous strength. *Photo sources: (top to bottom) Tom Grundy/Alamy; Egon Bömsch/imageBROKER/age fotostock; Eye of Science/Science Source.*

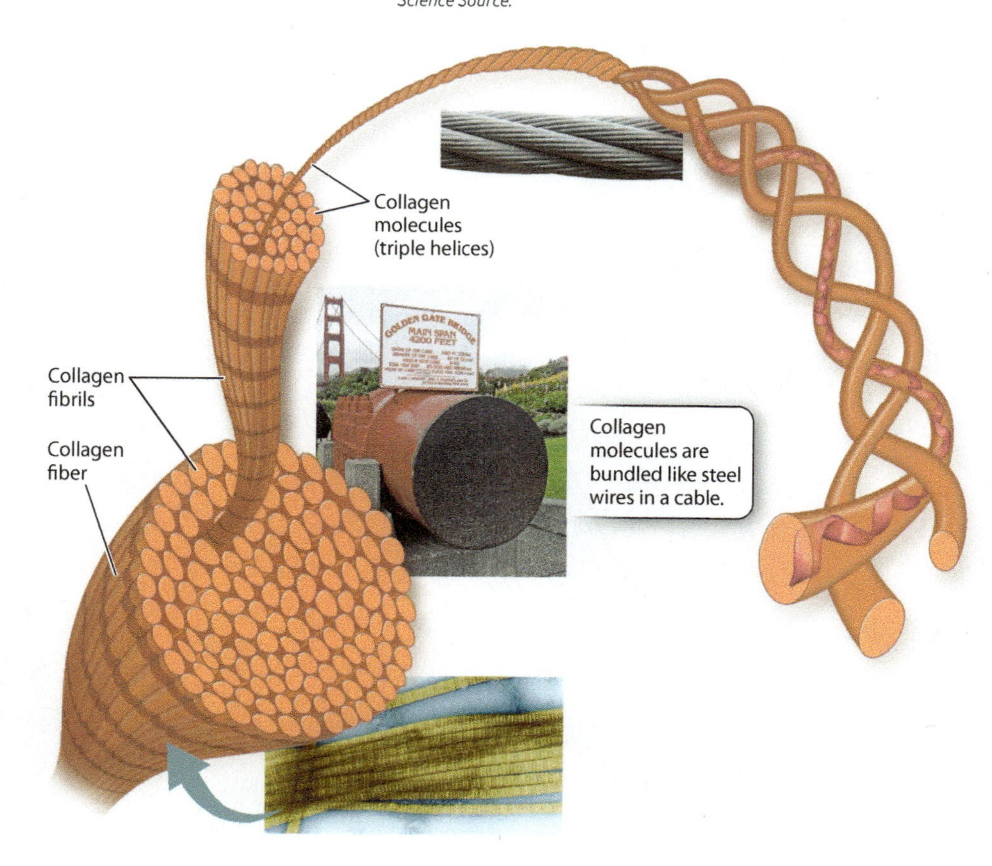

much stronger than if it were a single fiber of the same diameter. A collagen molecule consists of three polypeptides wound around one another in a triple helix. A bundle of collagen molecules forms a fibril, and the fibrils are assembled into fibers (**Fig. 10.17**). Once multiple collagen fibers are assembled into a ligament or tendon, the final structure is incredibly strong.

The basal lamina is a specialized layer of extracellular matrix that is present beneath all epithelial tissues, including the lining of the digestive tract, epidermis of the skin, and endothelial cells that line the blood vessels of vertebrates (**Fig. 10.18**). The role of the basal lamina is to provide a structural foundation for these epithelial tissues. The basal lamina is made of several proteins, including a special type of collagen. The triple-helical structure of collagen provides flexible support to the epithelial sheet.

? CASE 2 CANCER: WHEN GOOD CELLS GO BAD

How do cancer cells spread throughout the body?

Nonmalignant, or benign, tumors are encapsulated masses of cells that divide continuously because regulation of cell division has gone awry (Chapter 11). As the tumor grows, its border pushes outward against adjacent tissues. Benign tumors are rarely life threatening unless the tumor interferes with the function of a vital organ.

Malignant tumors are more dangerous. They contain some cells that can metastasize, that is, break away from the main tumor and colonize distant sites in the body. Metastatic tumor cells have an enhanced ability to adhere to extracellular matrix proteins, especially those in the basal lamina. This is significant because for a cell to metastasize, it must enter and leave the bloodstream through capillaries or other vessels. Since all blood vessels, including capillaries, have a basal lamina, a metastatic tumor cell needs to cross a basal lamina at least twice—once on the way into the bloodstream and again on the way out (**Fig. 10.19**). Since cells attach to basal lamina proteins by means of integrins, many studies have compared the integrins in metastatic and non-metastatic cells in the search for potential targets for treatment.

FIG. 10.18 **The basal lamina.** The basal lamina, found beneath epithelial tissue, is a specialized form of extracellular matrix. *Photo source: ISM/Phototake.*

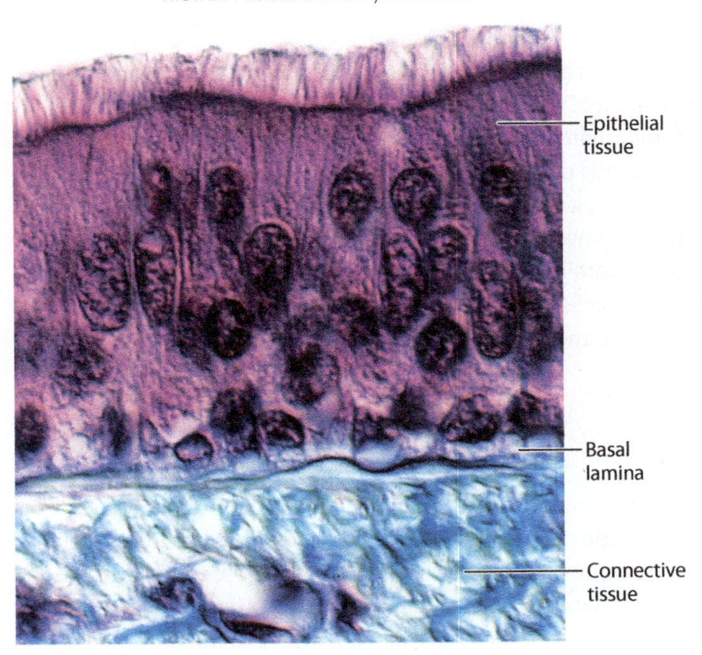

FIG. 10.19 **Metastatic cancer cells.** Some cancer cells spread from the original site of cancer formation to the bloodstream and then to distant organs of the body.

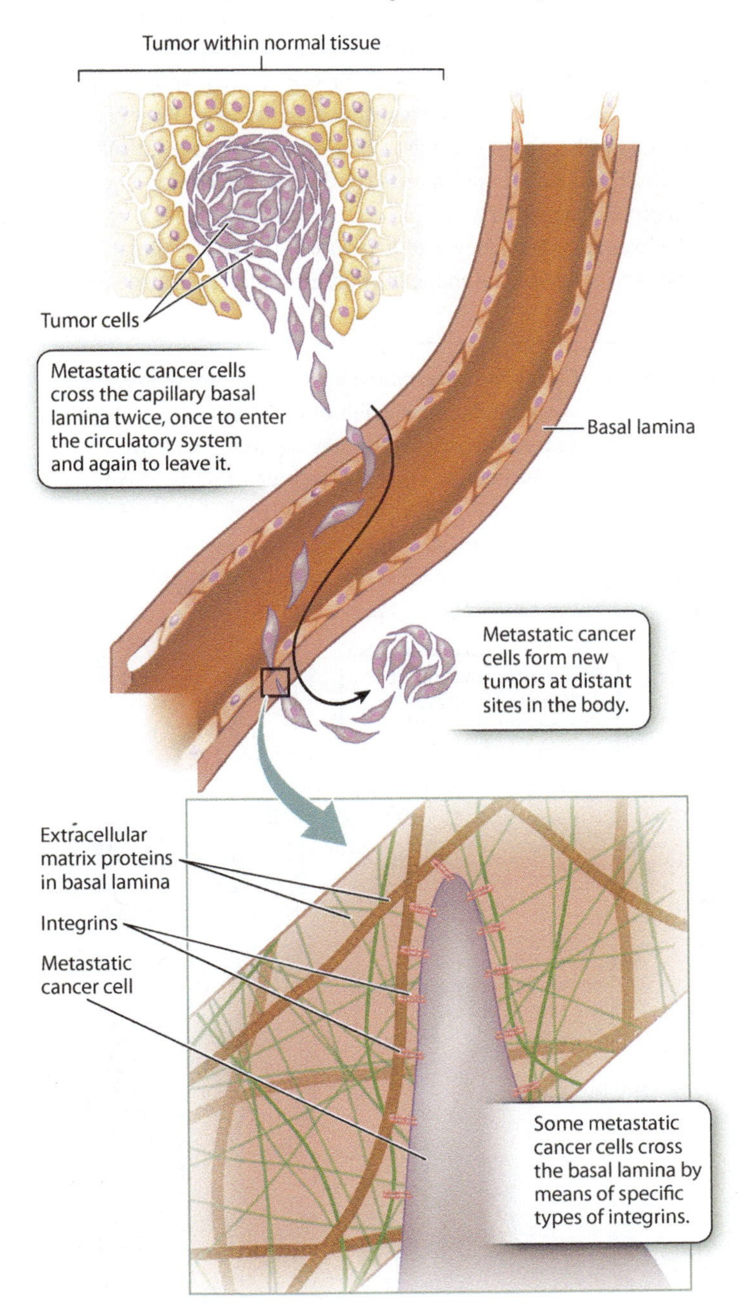

In some types of cancer, the number of specific integrins on the cell surface is an indicator of metastatic potential. Melanoma provides an example. A specific type of integrin is present in high amounts on metastatic melanoma cells but is absent on non-metastatic cells from the same tumor. In laboratory tests, blocking these integrins eliminates the melanoma cell's ability to cross an artificial basal lamina. Drugs targeting this integrin protein are currently in clinical trials.

Extracellular matrix proteins influence cell shape and gene expression.

Cells continue to interact with the extracellular matrix long after they have synthesized it or moved into it, and these interactions can have profound effects on cell shape and gene expression. Some of these cellular responses are the result of interactions between the extracellular matrix and integrins on the cell surface. The integrins act as receptors that relay the signal to the interior of the cell as the first step in a signal transduction pathway, like those discussed in Chapter 9. Biologists have observed the results of these interactions in experiments conducted with cells grown in culture in the laboratory.

An example shows how the *structure* of the extracellular matrix can influence the shape of cells. Fibroblasts cultured on a two-dimensional surface coated with extracellular matrix proteins attach to the matrix and flatten out as they maximize their adhesion to the matrix. By contrast, the same cells cultured in a three-dimensional gel of extracellular matrix look and behave like the spindle-shaped, highly migratory fibroblasts present in living connective tissue (**Fig. 10.20**).

FIG. 10.20 Cell shape determined by the structure of the extracellular matrix. Fibroblasts adopt different shapes depending on whether they are grown on a two-dimensional or a three-dimensional matrix. *Photo source: F. Tao, S. Chaudry, B. Tolloczko, J. G. Martin, and S. M. Kelly Modulation of smooth muscle phenotype in vitro by homologous cell substrate Am J Physiol Cell Physiol June 1, 2003 284:(6) C1531-C1541; published ahead of print March 5, 2003, doi:10.1152/ajp.*

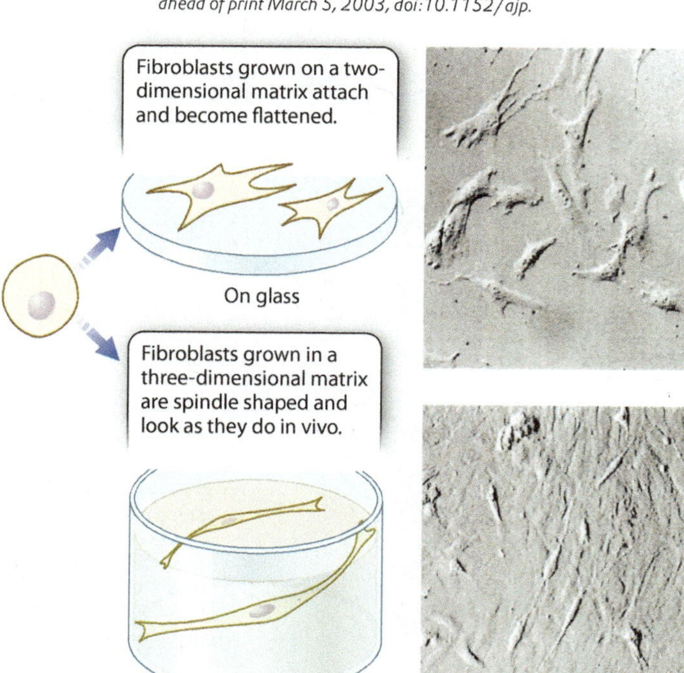

A second example shows how the *composition* of the extracellular matrix can influence cell shape. When nerve cells are grown in culture on a plastic surface, they attach to the surface of the dish but do not take on a neuron-like shape. However, when these cells are grown on the same surface coated with the extracellular matrix protein laminin, they develop long extensions that resemble the axons and dendrites of normal nerve cells (**Fig. 10.21**).

FIG. 10.21 Cell shape determined by composition of the extracellular matrix. Neurons adopt different shapes depending on whether or not they are cultured with the extracellular matrix protein laminin. *Source: Courtesy Motoyoshi Nomizu.*

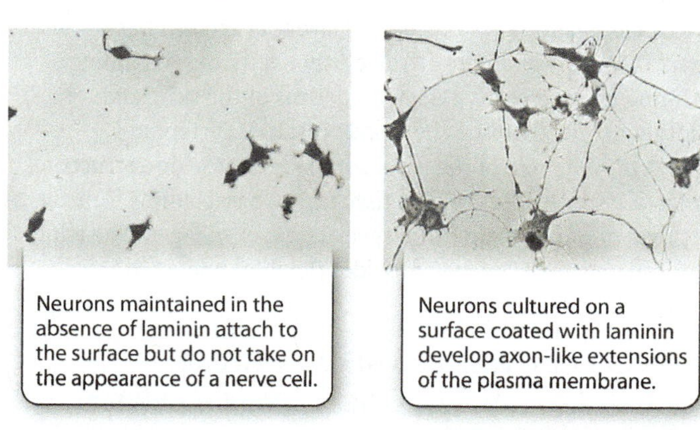

In addition to influencing cell shape, the structure and composition of the extracellular matrix can influence gene expression of the cells that are grown in it. When stimulated by the milk-inducing hormone prolactin, mouse mammary epithelial cells express the genes for milk proteins. These cells grown in culture on plain glass coated with collagen remain alive and apparently healthy, and even make and secrete a small amount of milk proteins, including β-casein. However, if the mammary epithelial cells are grown in three-dimensional collagen gels, they synthesize and secrete up to 10 times more β-casein.

→ **Quick Check 4** Do you think cadherins or integrins are responsible for the dependence of mammary cells on the extracellular matrix for their ability to produce milk proteins? Why?

An experiment that explores the importance of the composition of the extracellular matrix proteins in the regulation of gene expression is described in **Fig. 10.22.** This experiment, and the others described here, demonstrate that there is a dynamic interplay between the extracellular matrix and the cells that synthesize it. ■

HOW DO WE KNOW?

FIG. 10.22

Can extracellular matrix proteins influence gene expression?

BACKGROUND The adhesion of cells to the extracellular matrix is required for cell division, DNA synthesis, and proper cell shape. Research by Iranian-American cell biologist Mina Bissell and colleagues indicated that a cell's interaction with extracellular matrix proteins influences gene expression. Bissell discovered that mammary cells synthesize and secrete high levels of the milk protein β-casein when grown in a three-dimensional collagen matrix but not in a two-dimensional collagen matrix. American cell biologist Joan Caron followed up these studies using liver cells (hepatocytes).

HYPOTHESIS Caron hypothesized that a specific protein in the extracellular matrix is necessary for the expression of the protein albumin from hepatocytes grown in culture. Albumin is a major product of hepatocytes.

EXPERIMENT Caron cultured hepatocytes on a thin layer of type I collagen, which does not induce albumin synthesis. Next, she added a mixture of several different extracellular matrix proteins to the culture and looked for changes in albumin gene expression and protein secretion into the media. She then tested individual extracellular matrix proteins from the mixture to see which one was responsible for the increase in albumin gene expression.

RESULTS Caron found that when she cultured cells on collagen with a combination of three extracellular matrix proteins—laminin, type IV collagen, and heparin sulfate proteoglycan (HSPG)—the cells synthesized albumin mRNA and secreted albumin protein for several weeks, but if she cultured the cells on collagen alone, they did not (top and middle graphs). In addition, when she tested individual extracellular matrix proteins, she found that laminin, but not any of the other proteins, caused an increase in albumin gene expression (bottom graph).

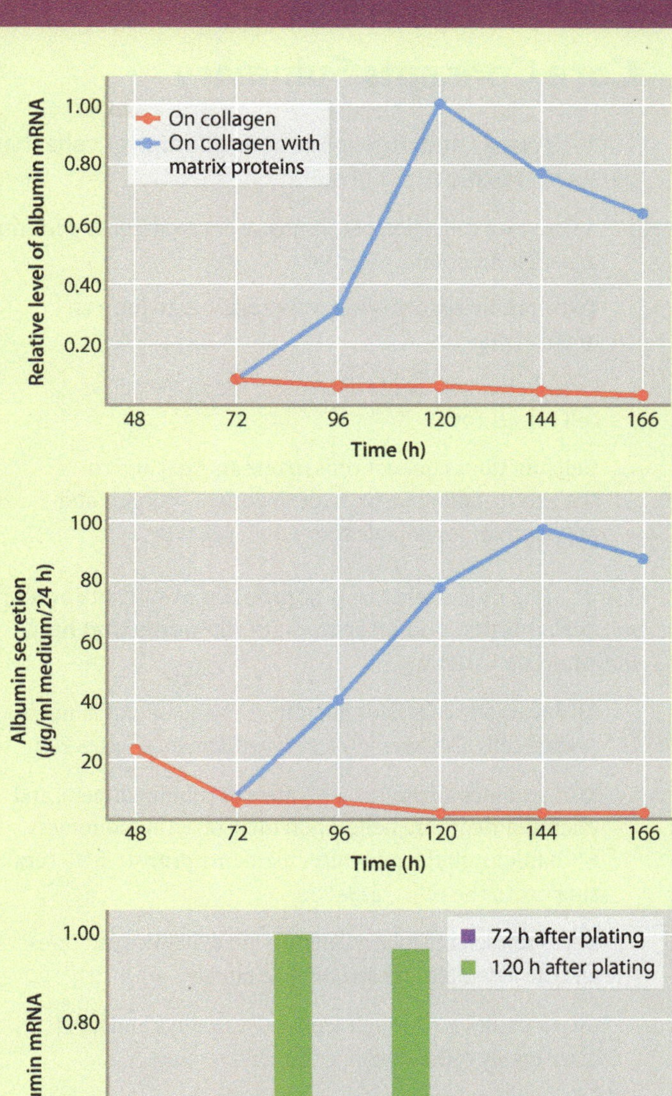

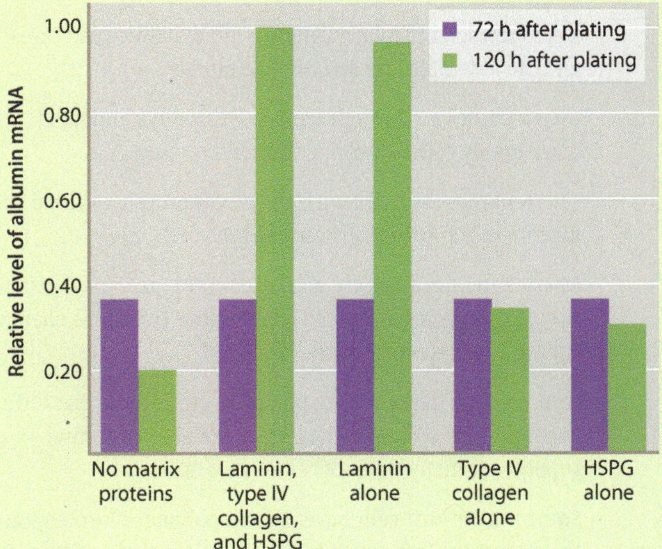

CONCLUSION Caron's hypothesis was supported by the experiments. A specific extracellular matrix protein, laminin, influences the expression of albumin by hepatocytes.

FOLLOW-UP WORK Bissell continued her work with mammary cells and found that the expression of the β-casein gene was also increased by laminin in the same way as the albumin gene in hepatocytes.

SOURCES Lee, E. Y., et al. 1985. "Interaction of Mouse Mammary Epithelial Cells with Collagen Substrata: Regulation of Casein Gene Expression and Secretion." *Proceedings of the National Academy of Sciences, USA* 82:1419–1423; Caron, J. M. 1990. "Induction of Albumin Gene Transcription in Hepatocytes by Extracellular Matrix Proteins." *Molecular and Cellular Biology* 10:1239–1243.

Core Concepts Summary

10.1 Tissues and organs are communities of cells that perform specific functions.

A tissue is a collection of cells that work together to perform a specific function. page 198

Two or more tissues often work together to form an organ. page 199

Cytoskeletal elements determine the shape of the cell. page 199

Cell junctions connect cells to one another and to the extracellular matrix, a meshwork of proteins and polysaccharides outside the cell. page 199

10.2 The cytoskeleton is composed of microtubules, microfilaments, and intermediate filaments that help maintain cell shape.

All eukaryotic cells have microtubules and microfilaments. Animal cells also have intermediate filaments. page 200

Microtubules are hollow polymers of tubulin dimers, and microfilaments are helical polymers of actin monomers. Both microtubules and microfilaments provide structural support to the cell. page 200

Microtubules and microfilaments are dynamic structures and can assemble and disassemble rapidly. page 201

Microtubules go through rounds of assembly and rapid disassembly called dynamic instability. page 202

Microtubules associate with the motor proteins dynein and kinesin to transport substances in the cell. page 202

Microfilaments associate with the motor protein myosin to transport vesicles in the cell and to cause cell shape changes, such as muscle contraction. page 204

Intermediate filaments are polymers of proteins that differ depending on cell type. They provide stable structural support for many types of cells. page 204

Some prokaryotic cells have protein polymer-like elements that function similarly to microtubules and microfilaments. page 205

10.3 Cell junctions connect cells to one another to form tissues.

Cell junctions anchor cells to each other and to the extracellular matrix, allow sheets of cells to act as a barrier, and permit communication between cells in tissues. page 206

Anchoring cell junctions include adherens junctions and desmosomes. page 208

Adherens junctions form a belt around the circumference of a cell. They are composed of cell adhesion molecules called cadherins and connect to microfilaments. page 208

Desmosomes are button-like points of adhesion between cells. They are composed of cadherins and connect to intermediate filaments. page 208

Hemidesmosomes are composed of cell adhesion molecules called integrins and connect cells to the extracellular matrix and intermediate filaments. page 208

Tight junctions prevent the passage of substances through the space between cells and divide the plasma membrane into apical and basolateral regions. page 208

Gap junctions (in animals) and plasmodesmata (in plants) allow cells to communicate rapidly with one another. page 210

10.4 The extracellular matrix provides structural support and informational cues.

The extracellular matrix is an insoluble meshwork of proteins and polysaccharides secreted by the cells it surrounds. It provides structural support to cells, tissues, and organs. page 210

In plants, the extracellular matrix is found in the cell wall, and the main component of the plant cell wall is the polysaccharide cellulose. page 211

In animals, the extracellular matrix is found in abundance in connective tissue. page 212

Collagen is the primary component of connective tissues in animals and is exceptionally strong. page 212

A specialized extracellular matrix called the basal lamina is present under all epithelial cell layers. page 213

In addition to providing structural support for cells, the extracellular matrix can influence cell shape and gene expression. page 214

Self-Assessment

1. Name three types of cytoskeletal element, the subunits they are composed of, their relative sizes, and the major functions of each type.

2. Explain how the dynamic nature of microtubules and microfilaments is important for their functions.
3. Describe the functions of the three major motor proteins and state which cytoskeletal element each interacts with.
4. Describe three major types and functions of cell junctions.
5. Identify the cytoskeletal element that interacts with adherens junctions, desmosomes, and hemidesmosomes.
6. Predict the effects of interfering with the function of cadherins and integrins.
7. Name two places where the extracellular matrix can be found in plants and animals.
8. Describe two effects that the extracellular matrix can have on the cells that synthesize it.

Log in to LaunchPad to check your answers to the Self-Assessment questions, and to access additional learning tools.

CHAPTER 11

Cell Division

Variations, Regulation, and Cancer

Core Concepts

11.1 During cell division, a single parental cell divides into two daughter cells.

11.2 Mitotic cell division is the basis of asexual reproduction in unicellular eukaryotes and the process by which cells divide in multicellular eukaryotes.

11.3 Meiotic cell division is essential for sexual reproduction, the production of offspring that combine genetic material from two parents.

11.4 The cell cycle is regulated so that cell division occurs only at appropriate times and places.

11.5 Cancer is uncontrolled cell division that results from mutations in genes that control cell division.

Dr. Torsten Wittmann/Science Source.

Cells come from preexisting cells. This is one of the fundamental principles of biology and a key component of the cell theory, which was introduced in Chapter 1. **Cell division** is the process by which cells make more cells. Multicellular organisms begin life as a single cell, and then cell division produces the millions, billions, or in the case of humans, trillions of cells that make up the fully developed organism. Even after a multicellular organism has achieved its adult size, cell division continues. In plants, cell division is essential for continued growth. In many animals, cell division replaces worn-out blood cells, skin cells, and cells that line much of the digestive tract. If you fall and scrape your knee, the cells at the site of the wound begin dividing to replace the damaged cells and heal the scrape.

Cell division is also important in reproduction. In bacteria, for example, a new generation is produced when the parent cell divides and forms two daughter cells. The parent cell first makes identical copies of its genetic material so that each of the two daughter cells has the same genetic material as the parent cell. The type of reproduction that occurs when offspring receive genetic material from a single parent is called **asexual reproduction.** Because DNA replication is not completely error free, the daughter cells may carry small genetic differences or mutations compared to the parent cell.

By contrast, **sexual reproduction** results in offspring that receive genetic material from two parents. Half the genetic material is supplied by the female parent and is present in the egg and the other half is supplied by the male parent and is contributed by the male's sperm. Eggs and sperm are specialized cells called **gametes.** A female gamete and a male gamete merge during fertilization to form a new organism (Chapter 42). If the egg and the sperm each contain the complete genetic material from a parent, what prevents the offspring from having twice as many chromosomes as each parent? A unique feature of gametes is that they contain half the number of chromosomes as the other cells in the parent organism. So when fertilization occurs, the combination of genetic material from the egg and the sperm results in a new organism with the same number of chromosomes as the parents. The production of gametes comes about by a form of cell division that results in daughter cells with half the number of chromosomes as the parent cell. As we will see, the products of this cell division are not genetically identical to the parent.

What determines when cells divide and, importantly, when they should not? And what determines which cells divide? To answer these questions, we must understand the process of cell division and how it is controlled. This discussion will lay the groundwork for exploring how cancer results from a loss of control of cell division.

11.1 CELL DIVISION

Cell division is the process by which a single cell becomes two daughter cells. While this process may seem simple, successful cell division must satisfy several important requirements. First, the two daughter cells must each receive the full complement of genetic material (DNA) present in the single parent cell. Second, the parent cell must be large enough to divide in two and still contribute sufficient cytoplasmic components such as proteins, lipids, and other macromolecules to each daughter cell. Satisfying these requirements means that key cellular components must be duplicated before cell division takes place. This duplication of material is achieved in a series of steps that constitutes the life cycle of every cell. When you think of a life cycle, you might think of various stages beginning with birth and ending with death. In the case of a single cell, the life cycle begins and ends with cell division.

In this section, we explore the different mechanisms by which prokaryotic and eukaryotic cells divide. Prokaryotic cells divide by **binary fission.** When eukaryotic cells divide, they first divide the nucleus by **mitosis,** and then divide the cytoplasm into two daughter cells by **cytokinesis.** As we discuss, it is likely that mitosis evolved from binary fission.

Prokaryotic cells reproduce by binary fission.

The majority of prokaryotic cells, namely bacteria and archaeons, divide by binary fission. In this form of cell division, a cell replicates its DNA, increases in size, and divides into two daughter cells. Each daughter cell receives one copy of the replicated parental DNA. The molecular mechanisms that drive binary fission have been studied most extensively in bacteria. The process of binary fission is similar in archaeons, as well as in chloroplasts and mitochondria, organelles within plant, fungal, and animal cells that evolved from free-living prokaryotic cells (Chapters 5 and 27).

Let's consider the process of binary fission in the intestinal bacterium *Escherichia coli* (**Fig. 11.1**). The circular genome of *E. coli* is attached by proteins to the inside of the plasma membrane. DNA replication is initiated at a specific location on the circular DNA molecule, called the origin of replication, and proceeds in opposite directions around the circle. The result is two DNA molecules, each of which is attached to the plasma membrane at a different site. The two attachment sites are initially close together. The cell then elongates and, as it does so, the two DNA attachment sites move apart. When the cell is about twice its original size and the DNA molecules are well separated, a constriction forms at the midpoint of the cell. Eventually, new membrane and cell wall are synthesized at the site of the constriction, dividing the single cell into two. The result is two daughter cells, each having the same genetic material as the parent cell.

Like most cellular processes, binary fission requires the coordination of many components in both time and space. Recent research has identified several genes whose products play a key role in bacterial cell division. One of these genes, called *FtsZ,* has been especially well studied. Many copies of the protein it encodes assemble and form a ring at the site of constriction where the new cell wall forms between the two daughter cells. *FtsZ* is present in the genomes of diverse bacteria and archaeons, suggesting that it plays a fundamental role in prokaryotic cell division. Interestingly, it appears to be evolutionarily related to tubulin, which you will recall from Chapter 10 makes up the dynamic microtubules found

FIG. 11.1 Binary fission. Cell division in bacteria and archaeons occurs by binary fission.

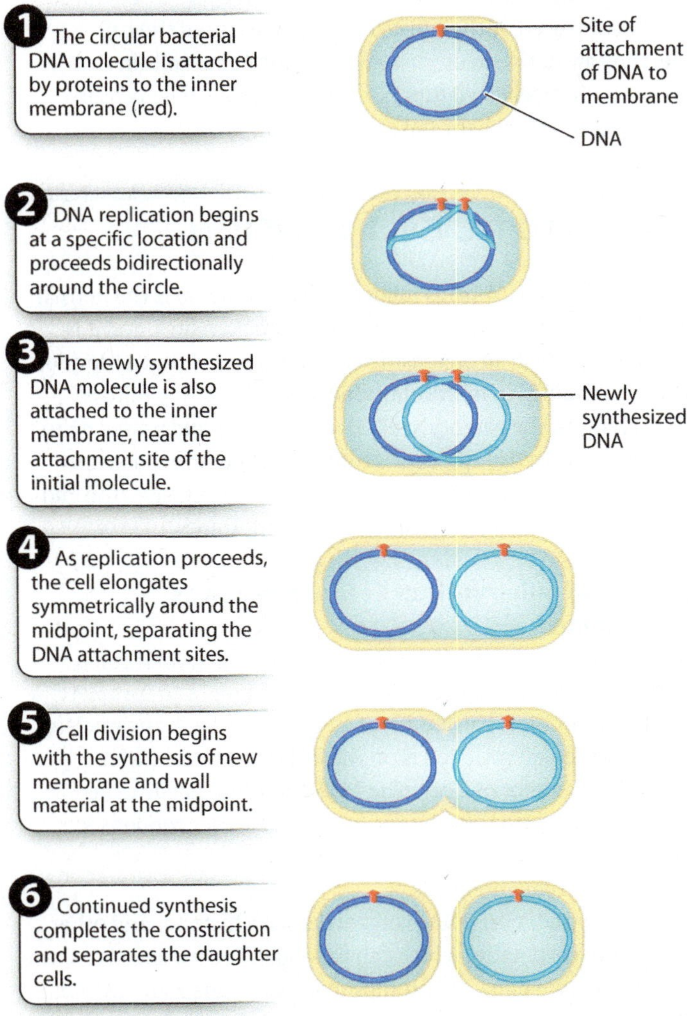

1. The circular bacterial DNA molecule is attached by proteins to the inner membrane (red).

Site of attachment of DNA to membrane

DNA

2. DNA replication begins at a specific location and proceeds bidirectionally around the circle.

3. The newly synthesized DNA molecule is also attached to the inner membrane, near the attachment site of the initial molecule.

Newly synthesized DNA

4. As replication proceeds, the cell elongates symmetrically around the midpoint, separating the DNA attachment sites.

5. Cell division begins with the synthesis of new membrane and wall material at the midpoint.

6. Continued synthesis completes the constriction and separates the daughter cells.

in eukaryotic cells that are important in intracellular transport, cell movement, and cell division.

→ **Quick Check 1** What do you predict would be the consequence of a mutation in *FtsZ* that disrupts the function of the protein it encodes?

Eukaryotic cells reproduce by mitotic cell division.

The basic steps of binary fission that we just saw—replication of DNA, segregation of replicated DNA to daughter cells, and division of one cell into two—occur in all forms of cell division. However, cell division in eukaryotes (by mitosis) is more complicated than cell division in prokaryotes (by binary fission). Compared with the single, relatively small, circular DNA molecule that is the genome of prokaryotic cells, the genome of eukaryotic cells is typically much larger and is organized into one or more linear chromosomes, each of which must be replicated and separated into daughter cells. And whereas the DNA of prokaryotes is attached to the inside of the plasma membrane, allowing replicated DNA to be separated into daughter cells by cell growth, the DNA of eukaryotes is located in the nucleus. As a result, eukaryotic cell division requires first the breakdown and then the re-formation of the nuclear envelope, as well as mechanisms other than cell growth to separate replicated DNA. As we saw in Chapter 10 and discuss in more detail in section 11.2, chromosomes of dividing eukaryotic cells attach to the mitotic spindle, which separates them into daughter cells.

Interestingly, some unicellular eukaryotes exhibit forms of cell division that have characteristics of binary fission and mitosis. For example, dinoflagellates, like all eukaryotes, have a nucleus and linear chromosomes. However, unlike most eukaryotes, the nuclear envelope does not break down but stays intact during cell division. Furthermore, the replicated DNA is attached to the nuclear envelope. The nucleus then grows and divides in a manner reminiscent of binary fission. These and other observations of intermediate forms of cell division in additional organisms strongly suggest that mitosis evolved from binary fission.

The cell cycle describes the life cycle of a eukaryotic cell.

Cell division in eukaryotic cells proceeds through a number of steps that make up the **cell cycle** (**Fig. 11.2**). The cell cycle consists of two distinct stages: **M phase** and **interphase.** During M phase, the parent cell divides into two daughter cells. M phase

FIG. 11.2 The cell cycle. The eukaryotic cell cycle consists of M phase (mitosis and cytokinesis) and interphase.

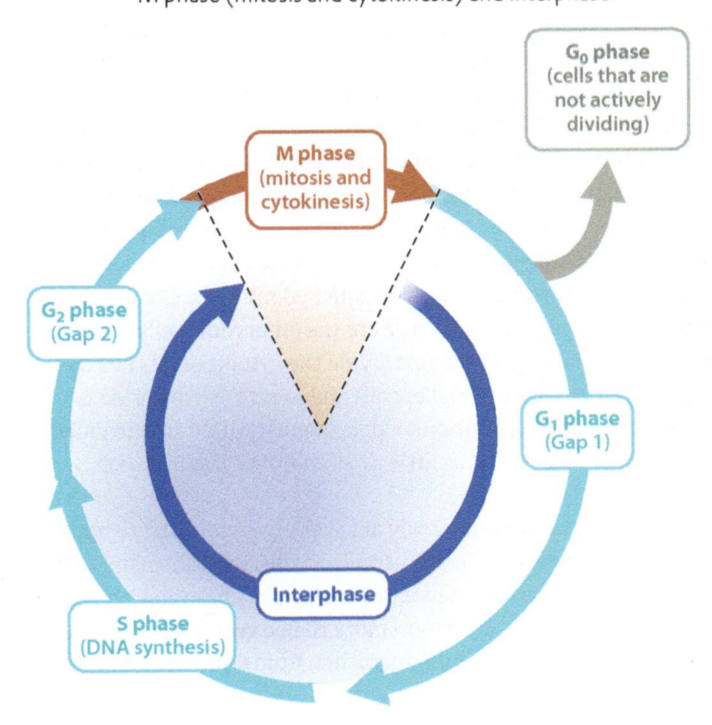

consists of two different events: (1) mitosis, the separation of the chromosomes into two nuclei, and (2) cytokinesis, the division of the cell itself into two separate cells. Usually, these two processes go hand in hand, with cytokinesis typically beginning even before mitosis is complete. In most mammalian cells, M phase lasts about an hour.

The second stage of the cell cycle, called interphase, is the time between two successive M phases (Fig. 11.2). For many years, it was thought that the relatively long period of interphase is uneventful. Today, we know that during this stage the cell makes many preparations for division. These preparations include replication of the DNA in the nucleus so that each daughter cell receives a copy of the genome, and an increase in cell size so that each daughter cell receives sufficient amounts of cytoplasmic and membrane components to allow it to survive on its own.

Interphase can be divided into three phases, as shown in Fig. 11.2. Among the many preparations that the cell must make during interphase, one particularly important task is the replication of the entire DNA content of the nucleus. Since replication involves the synthesis of DNA, this stage is called **S phase** ("S" for "synthesis").

In most cells, S phase does not immediately precede or follow mitosis but is separated from it by two gap phases: **G_1 phase** between the end of M phase and the start of S phase, and **G_2 phase** between the end of S phase and the start of M phase. Many essential processes occur during both "gap" phases, despite the name. For example, during the G_1 phase, specific regulatory proteins are made and activated. Once active, the regulatory proteins, many of which are kinases, then promote the activity of enzymes that synthesize DNA. In the G_2 phase, both the size and protein content of the cell increase in preparation for division. Thus, G_1 is a time of preparation for S-phase DNA synthesis, and G_2 is a time of preparation for M-phase mitosis and cytokinesis.

How long does a cell take to pass through the cell cycle? That depends on the type of cell and the organism's stage of development. Actively dividing cells in some human tissues such as the intestine and skin require frequent replenishing. It usually takes cells in these tissues about 12 hours to complete the cell cycle. Most other actively dividing cells in your body take about 24 hours to complete the cycle. A unicellular eukaryote like yeast can complete an entire cell cycle in just 90 minutes. Champions in the race through the cell cycle are the embryonic cells of some frog species. Early cell divisions divide the cytoplasm of the large frog egg cell into many smaller cells and so no growth period is needed between cell divisions. Consequently, there are virtually no G_1 and G_2 phases, and as little as 30 minutes pass between cell divisions.

Not all the cells in your body are actively dividing since not all tissues require the rapid replenishing of cells. Instead, many cells pause in the cell cycle somewhere between M phase and S phase for periods ranging from days to more than a year. This period is called the **G_0 phase** and is distinguished from G_1 by the absence of preparations for DNA synthesis (Fig. 11.2). Liver cells remain in G_0 for as much as a year. Other cells such as nerve cells and those that form the lens of the eye enter G_0 permanently; these cells are nondividing. Thus, many brain cells lost to disease or damage cannot be replaced. Although cells in G_0 have exited the cell cycle, they are active in other ways—in particular, cells in G_0 still perform their specialized functions. For example, liver cells in G_0 still carry out metabolism and detoxification.

11.2 MITOTIC CELL DIVISION

Mitotic cell division (mitosis followed by cytokinesis) is the normal mode of asexual reproduction in unicellular eukaryotes, and it is the means by which an organism's cells, tissues, and organs develop and are maintained in multicellular eukaryotes. During mitosis and cytokinesis, the parental cell's DNA is divided and passed on to two daughter cells. This process is continuous, but is divided into discrete steps marked by dramatic changes in the cytoskeleton and in the packaging and movement of the chromosomes.

The DNA of eukaryotic cells is organized as chromosomes.

One of the key challenges faced by a dividing eukaryotic cell is ensuring that both daughter cells receive an equal and complete set of chromosomes. The length of DNA contained in the nucleus of an average eukaryotic cell is on the order of 1 to 2 m, well beyond the diameter of a cell. The DNA therefore needs to be condensed to fit into the nucleus, and then further condensed during cell division so that it does not become tangled as it segregates into daughter cells.

In eukaryotic cells, DNA is organized with histones and other proteins into chromatin, which can be looped and packaged to form the structures we know as chromosomes (Chapters 3 and 13). One of the earliest events in mitosis is the condensing of chromosomes from long, thin, threadlike structures typical of interphase to short, dense forms that are identifiable under the microscope during M phase.

Every species is characterized by a specific number of chromosomes, and each chromosome contains a single molecule of DNA carrying a specific set of genes. When chromosomes condense and become visible during mitosis, they adopt characteristic shapes and sizes that allow each chromosome to be identified by its appearance in the microscope. The portrait formed by the number and shapes of chromosomes representative of a species is called its **karyotype.** Most of the cells in the human body, with the exception of the gametes, contain 46 chromosomes (**Fig. 11.3**). In contrast, cells from horses have 64 chromosomes, and cells from corn have 20.

In a normal human karyotype, the 46 chromosomes can be arranged into 23 pairs, 22 pairs of **homologous chromosomes** numbered 1 to 22 from the longest to the shortest chromosome

FIG. 11.3 A human karyotype. This karyotype shows 22 pairs of chromosomes plus 2 sex chromosomes, or 46 chromosomes in total. *Source: ISM/Phototake.*

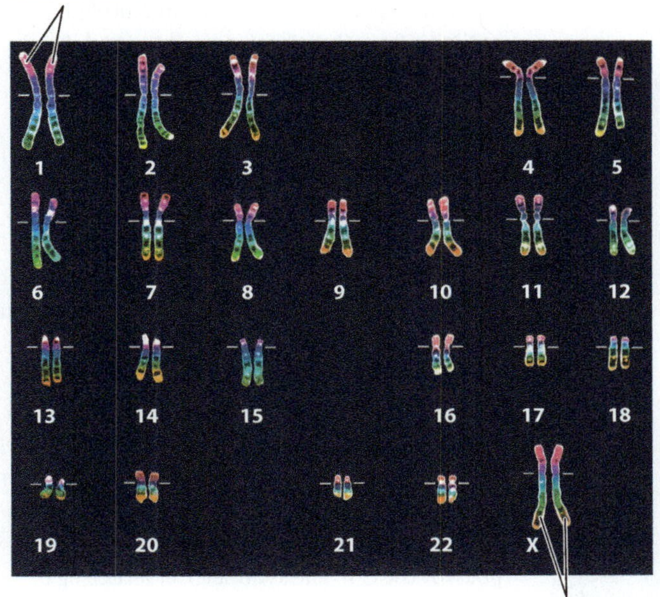

FIG. 11.4 Homologous chromosomes and sister chromatids. Sister chromatids result from the duplication of chromosomes. They are held together at the centromere.

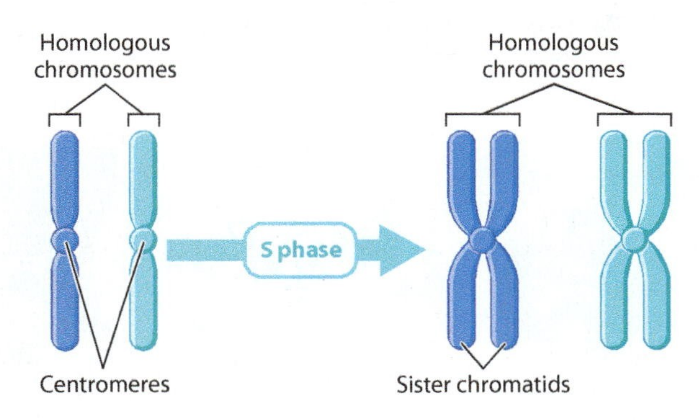

and 1 pair of **sex chromosomes** (Fig. 11.3). Each pair of homologous chromosomes represents two of the same type of chromosome (both carrying the same set of genes), one of which was received from the mother and the other from the father. The sex chromosomes are the X and Y chromosomes. Individuals with two X chromosomes are female, and those with an X and a Y chromosome are male.

The number of complete sets of chromosomes in a cell is known as its ploidy. A cell with one complete set of chromosomes is **haploid,** and a cell with two complete sets of chromosomes is **diploid.** Some organisms, such as plants, can have four or sometimes more complete sets of chromosomes. Such cells are polyploid.

In order for cell division to proceed normally, every chromosome in the parent cell must be duplicated so that each daughter cell receives a full set of chromosomes. This duplication occurs during S phase. Even though the DNA in each chromosome duplicates, the two identical copies, called **sister chromatids,** do not separate. They stay side by side, physically held together at a constriction called the **centromere.** At the beginning of mitosis, the nucleus of a human cell contains 46 chromosomes, each of which is a pair of identical sister chromatids linked together at the centromere (**Fig. 11.4**). Thus, counting chromosomes is simply a matter of counting centromeres.

During mitosis, the sister chromatids separate from each other and go to opposite ends of the cell, so that each daughter cell receives the same number of chromosomes as present in the parent cell, as we describe now.

→ **Quick Check 2** Which DNA sequences are more alike: a pair of sister chromatids or a pair of homologous chromosomes?

Prophase: Chromosomes condense and become visible.

Mitosis takes place in five stages, each of which is easily identified by events that can be observed in the microscope (**Fig. 11.5**). When you look in a microscope at a cell in interphase, specific chromosomes cannot be distinguished because they are long and thin. As the cell moves from G_2 phase to the start of mitosis, the chromosomes condense and become visible in the nucleus. The first stage of mitosis is known as **prophase** and is characterized by the appearance of visible chromosomes.

Outside the nucleus, in the cytosol, the cell begins to assemble the **mitotic spindle,** a structure made up predominantly of microtubules that pull the chromosomes into separate daughter cells. Recall from Chapter 10 that the **centrosome** is a compact structure that is the microtubule organizing center for animal cells. The centrosome is thus the structure from which the spindles radiate. Plant cells also have microtubule-based mitotic spindles, but they lack centrosomes.

As part of the preparation for mitosis during S phase in animal cells, the centrosome duplicates and each one begins to migrate around the nucleus, the two ultimately halting at opposite poles in the cell at the start of prophase. The final locations of the centrosomes define the opposite ends of the cell that will eventually be separated into two daughter cells. As the centrosomes make their way to the poles of the cell, tubulin dimers assemble around them, forming microtubules that radiate from each centrosome. These radiating filaments form the mitotic spindle and later serve as the guide wires for chromosome movement.

FIG. 11.5 Mitosis, or nuclear division. Mitosis can be divided into separate steps, but the process is continuous. *Source: Jennifer Waters/Science Source.*

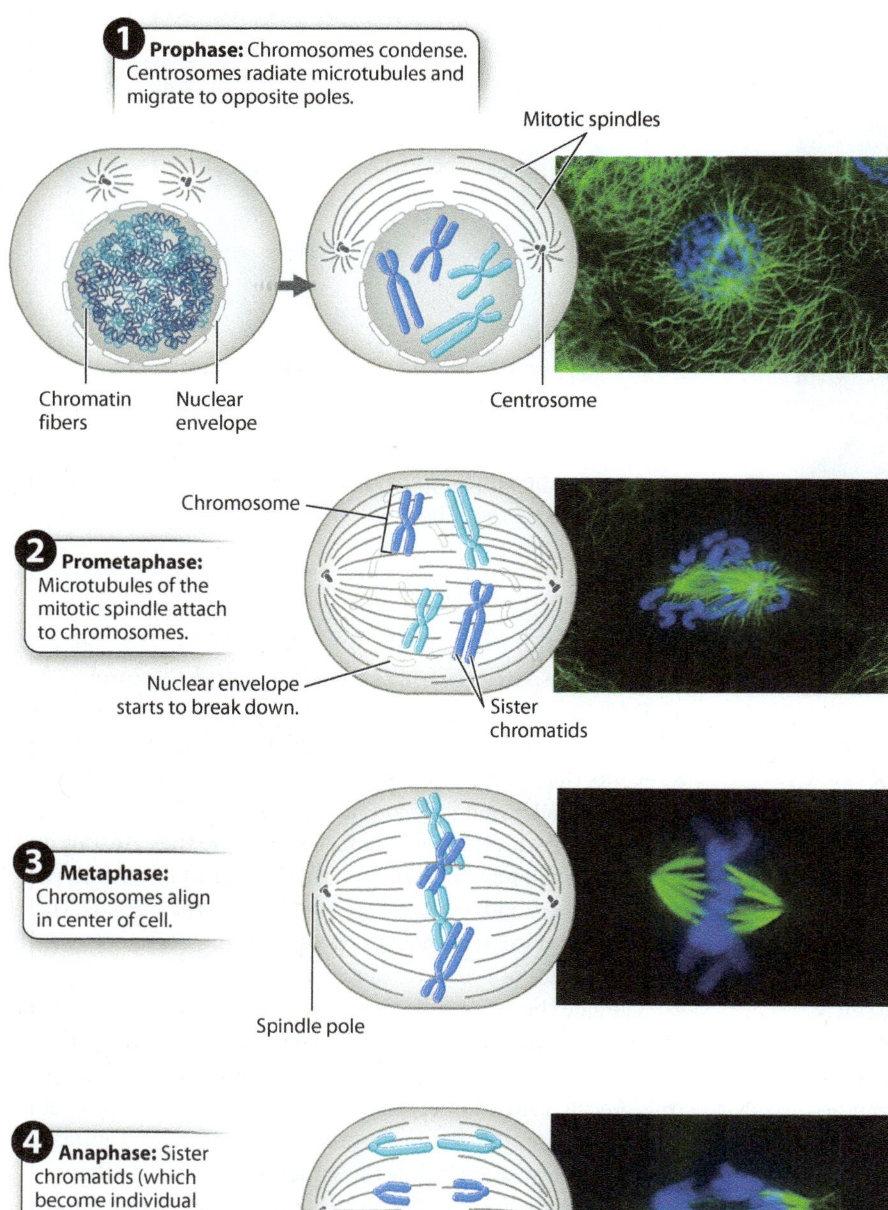

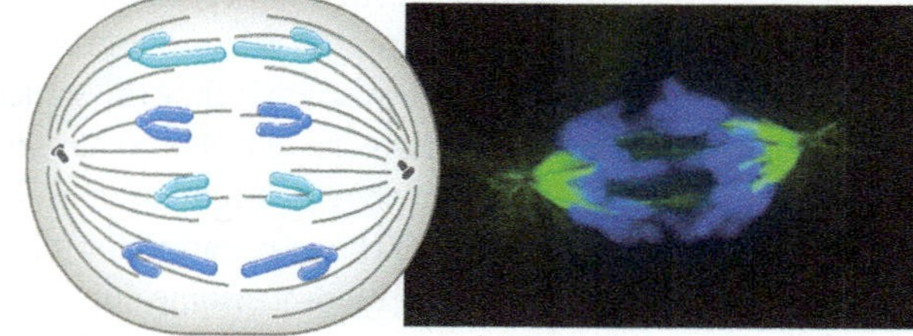

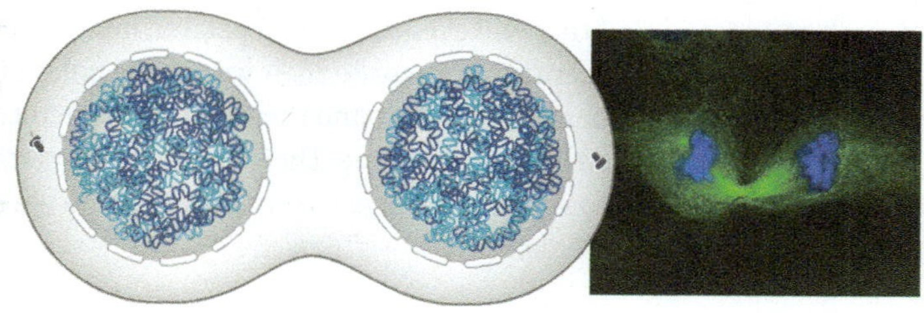

Prometaphase: Chromosomes attach to the mitotic spindle.

In the next stage of mitosis, known as **prometaphase,** the nuclear envelope breaks down and the microtubules of the mitotic spindle attach to chromosomes (Fig. 11.5). The microtubules radiating from the centrosomes grow and shrink as they explore the region of the cell where the nucleus once was. This process of growing and shrinking depends on the dynamic instability of microtubules, discussed in Chapter 10.

As the ends of the microtubules encounter chromosomes, they attach to the chromosomes at their centromeres. Associated with the centromere of each chromosome are two protein complexes called **kinetochores,** one located on each side of the constriction (**Fig. 11.6**). Each kinetochore is associated with one of the two sister chromatids and forms the site of attachment for a single spindle microtubule. This arrangement ensures that each sister chromatid is attached to a spindle microtubule radiating from one of the poles of the cell. The symmetrical tethering of each chromosome to the two poles of the cell is essential for proper chromosome segregation.

Metaphase: Chromosomes align as a result of dynamic changes in the mitotic spindle.

Once each chromosome is attached to the mitotic spindles from both poles of the cell, the microtubules of the mitotic spindle lengthen or shorten to move the chromosomes into position in the middle of the cell. There the chromosomes are lined up in a single plane that is roughly equidistant from both poles of the cell. This stage of mitosis, when the chromosomes are aligned in the middle of the dividing cell, is called **metaphase** (see Fig. 11.5). It is one of the most visually distinctive stages under the microscope.

Anaphase: Sister chromatids fully separate.

In the next stage of mitosis, called **anaphase,** the sister chromatids separate (see Fig. 11.5). The centromere holding a pair of sister chromatids together splits, allowing the two sister chromatids to separate from each other. After separation, each chromatid is considered to be a full-fledged chromosome. The spindle microtubules attached to the kinetochores gradually shorten, pulling the newly separated chromosomes to the opposite poles of the cell.

FIG. 11.6 Kinetochores. Kinetochores are sites of spindle attachment and are located on both sides of the centromere.

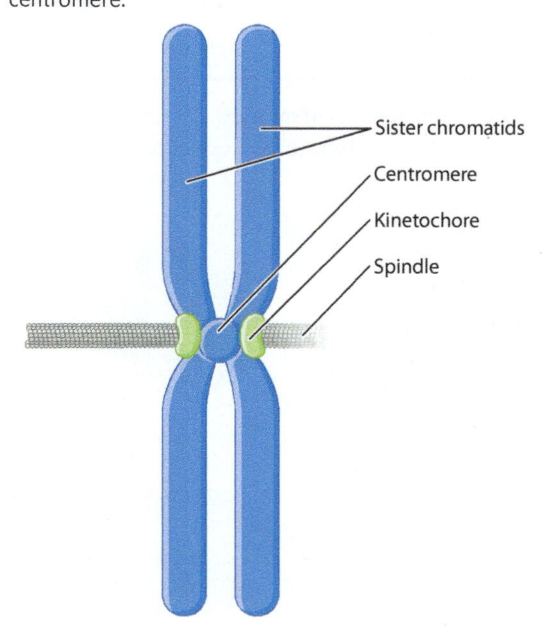

This event is the basis for the equal segregation of chromosomes between the two daughter cells. During S phase in a human cell, each of the 46 chromosomes is duplicated to yield 46 pairs of identical sister chromatids. Thus, when the chromatids are separated at anaphase, an identical set of 46 chromosomes arrives at each spindle pole, the complete genetic material for one of the daughter cells.

Telophase: Nuclear envelopes re-form around newly segregated chromosomes.

Once a complete set of chromosomes arrives at a pole, the chromosomes have entered the area that will form the cytosol of a new daughter cell. This event marks the beginning of **telophase,** during which the cell prepares for its division into two new cells (see Fig. 11.5). The microtubules of the mitotic spindle break down and disappear, while a nuclear envelope reforms around each set of chromosomes, creating two new nuclei. As the nuclei become increasingly distinct in the cell, the chromosomes contained within them decondense, becoming less visible in the microscope. This stage marks the end of mitosis.

The parent cell divides into two daughter cells by cytokinesis.

Usually, as mitosis is nearing its end, cytokinesis begins and the parent cell divides into two daughter cells (**Fig. 11.7**). In animal cells, this stage begins when a ring of actin filaments, called the **contractile ring,** forms against the inner face of the cell membrane at the equator of the cell perpendicular to the axis of what was the spindle (Fig. 11.7a). As if pulled by a drawstring, the ring contracts, pinching the cytoplasm of the cell and dividing it in two. This process is similar to what occurs in binary fission, though in the case of binary fission, the process is driven by FtsZ protein, a homolog of tubulin, not by actin. The constriction of the contractile ring is driven by motor proteins that slide bundles of actin filaments in opposite directions. Successful division results in two daughter cells, each with its own nucleus. The daughter cells are now free to enter G_1 phase and start the process anew.

For the most part, mitosis is similar in animal and in plant cells, but cytokinesis is different (Fig. 11.7b). Since plant cells have a cell wall, the division of the cell is achieved by constructing a new cell wall. During telophase, dividing plant cells form a structure called the phragmoplast in the middle of the cell. The phragmoplast

FIG. 11.7 Cytokinesis, or cytoplasmic division. (a) In animal cells, cytokinesis involves a contractile ring made of actin. (b) In plant cells, it involves the growth of a new cell wall called a cell plate. *Sources: a. Dr. Paul Andrews, University of Dundee/Science Source; b. Carolina Biological Supply Company/Phototake.*

a.

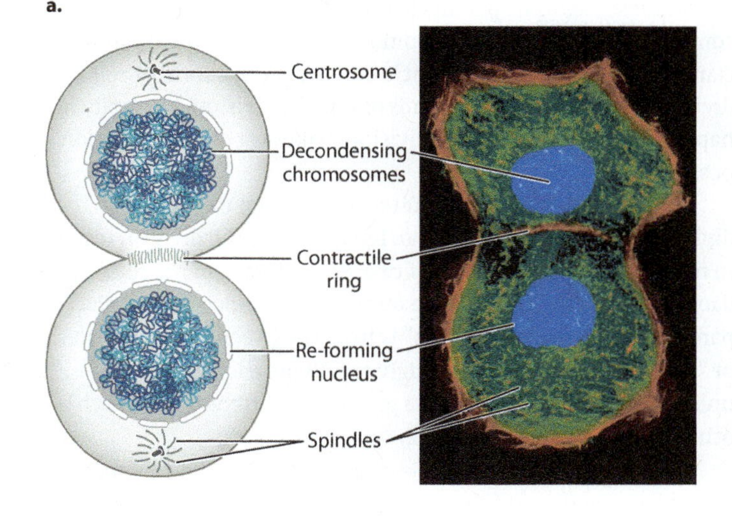

b.

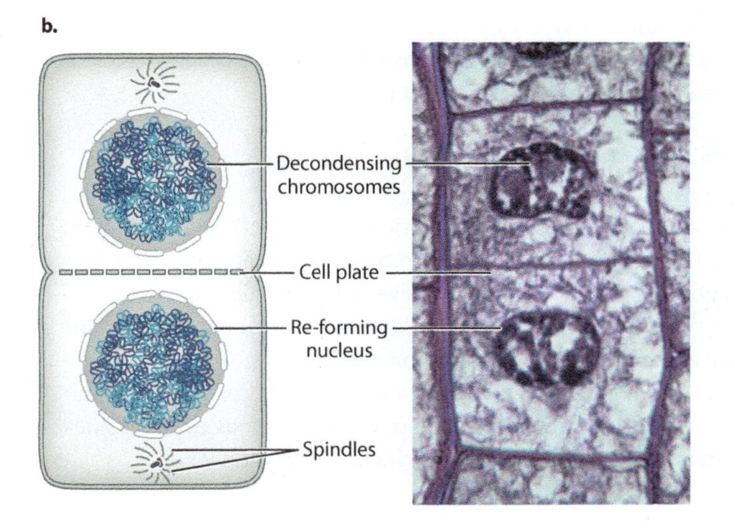

consists of overlapping microtubules that guide vesicles containing cell wall components to the middle of the cell. During late anaphase and telophase, these vesicles fuse to form a new cell wall, called the cell plate, in the middle of the dividing cell. Once this developing cell wall is large enough, it fuses with the original cell wall at the perimeter of the cell. Cytokinesis is then complete and the plant cell has divided into two daughter cells.

→ **Quick Check 3** What would be the consequence if a cell underwent mitosis but not cytokinesis?

11.3 MEIOTIC CELL DIVISION

As we discussed, mitotic cell division is important in the development of a multicellular organism and in the maintenance and repair of tissues and organs. Mitotic cell division is also the basis of asexual reproduction in unicellular eukaryotes. We now turn to the basis of sexual reproduction. In sexual reproduction, gametes fuse during fertilization to form a new organism. A new organism produced by sexual reproduction has the same number of chromosomes as its parents because the egg and sperm each contain half the number of chromosomes as each diploid parent. Gametes are produced by **meiotic cell division,** a form of cell division that includes two rounds of nuclear division. By producing haploid gametes, meiotic cell division makes sexual reproduction possible.

There are several major differences between meiotic cell division and mitotic cell division. First, meiotic cell division results in four daughter cells instead of two. Second, each of the four daughter cells contains half the number of chromosomes as the parent cell. (The word "meiosis" is from the Greek for "diminish" or "lessen.") Third, the four daughter cells are each genetically unique. In other words, they are genetically different from each other and from the parental cell.

In multicellular animals, the cells produced by meiosis are the haploid eggs and sperm that fuse in sexual reproduction. In other organisms, such as fungi, the products are spores, and in some unicellular eukaryotes, the products are new organisms. In this section, we consider the steps by which meiosis occurs, its role in sexual reproduction, and how it likely evolved.

Pairing of homologous chromosomes is unique to meiosis.

Like mitotic cell division, meiotic cell division follows one round of DNA synthesis, but, unlike mitotic cell division, meiotic cell division consists of two successive cell divisions. The two cell divisions are called **meiosis I** and **meiosis II,** and they occur one after the other. Each cell division results in two cells, so that by the end of meiotic cell division a single parent cell has produced four daughter cells. During meiosis I, homologous chromosomes separate from each other, reducing the total number of chromosomes by half. During meiosis II, sister chromatids separate, as in mitosis.

Meiosis I begins with **prophase I,** illustrated in **Fig. 11.8.** The beginning of prophase I marks the earliest visible manifestation

FIG. 11.8 Prophase I of meiosis. Chromosomes condense and homologous chromosomes pair.

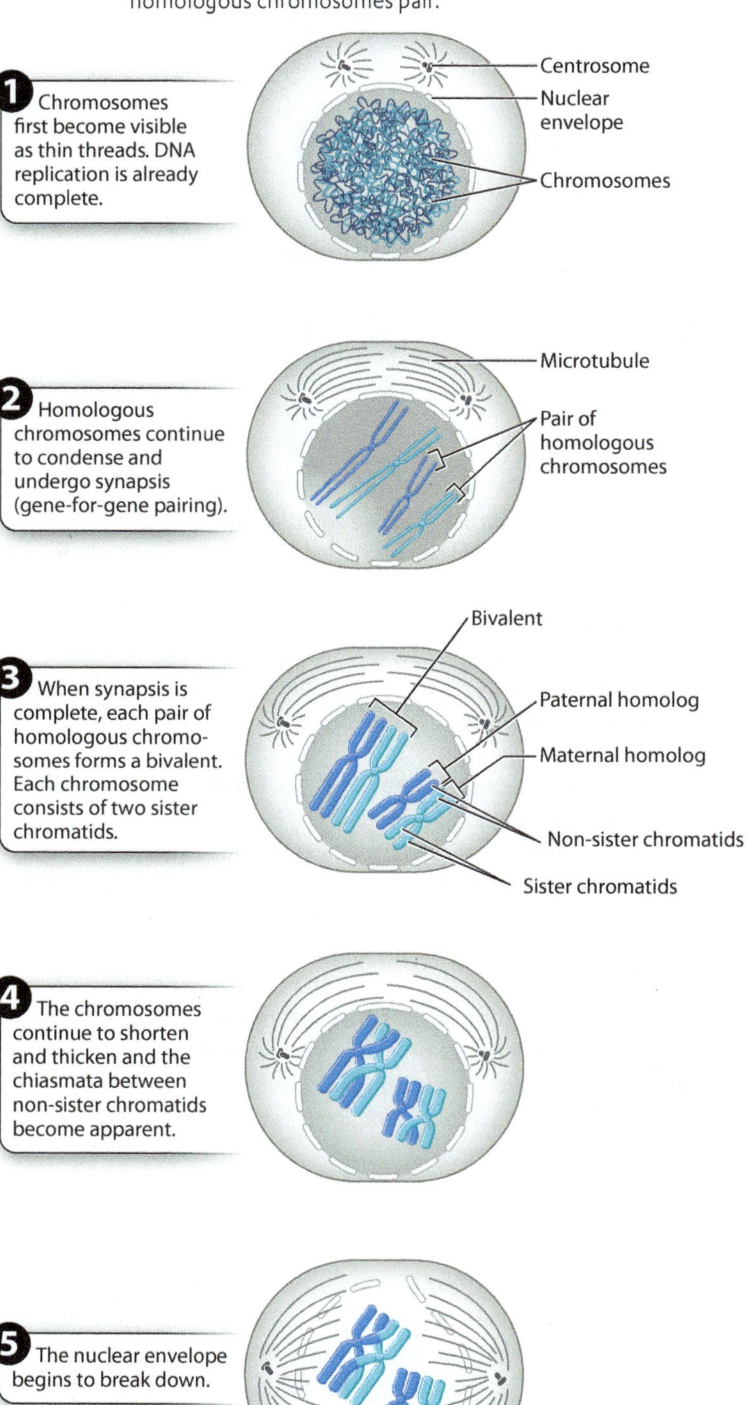

of chromosome condensation. The chromosomes first appear as long, thin threads present throughout the nucleus. By this time, DNA replication has already taken place, so each chromosome has become two sister chromatids held together at the centromere.

What happens next is an event of enormous importance, and it is unique to meiosis. The homologous chromosomes pair with each other, coming together to lie side by side, gene for gene, in a process known as **synapsis.** Even the X and Y chromosomes pair, but only at the tip where their DNA sequences are nearly identical. Because one of each pair of homologs is maternal in origin and the other is paternal in origin, chromosome pairing provides an opportunity for the maternal and paternal chromosomes to exchange genetic information, as described in the next section.

Because each homologous chromosome is a pair of sister chromatids attached to a single centromere, a pair of synapsed chromosomes creates a four-stranded structure: two pairs of sister chromatids aligned along their length. The whole unit is called a **bivalent,** and the chromatids attached to different centromeres are called **non-sister chromatids** (Fig. 11.8). Non-sister chromatids result from the replication of homologous chromosomes (one is maternal and the other is paternal in origin), so they have the same set of genes in the same order, but are not genetically identical. By contrast, sister chromatids result from replication of a single chromosome, so are genetically identical.

→ **Quick Check 4** In a human cell at the end of prophase I, how many chromatids, centromeres, and bivalents are present?

Crossing over between DNA molecules results in exchange of genetic material.

Within the bivalents are cross-like structures, each called a **chiasma** (from the Greek meaning a "cross piece"; the plural is "chiasmata") (**Fig. 11.9**). Each chiasma is a visible manifestation of a **crossover,** the physical breakage and reunion between non-sister chromatids.

Through the process of crossing over, homologous chromosomes of maternal origin and paternal origin exchange DNA segments. The positions of these exchanges along the chromosome are essentially random, and therefore each chromosome that emerges from meiosis is unique, containing some DNA segments from the maternal chromosome and others from the paternal chromosome. The process is very precise: Usually, no nucleotides are gained or lost as homologous chromosomes exchange material. Occasionally, the exchange is imprecise and portions of the chromatids may be gained or lost, resulting in loss or duplication of material. Note the results of crossing over as shown in Fig. 11.9: The recombinant chromatids are those that carry partly paternal and partly maternal segments. In this way, crossing over increases genetic diversity.

The number of chiasmata that are formed during meiosis depends on the species. In humans, the usual range is 50–60 chiasmata per meiosis. Most bivalents have at least one chiasma. Even the X and Y chromosomes are joined by a chiasma in the small region where they are paired. In addition to their role in exchanging genetic material, the chiasmata also play a mechanical role in meiosis by holding the bivalents together while they become properly oriented in the center of the cell during metaphase, the stage we turn to next.

FIG. 11.9 **Chiasmata.** Crossing over at chiasmata between non-sister chromatids results in recombinant chromatids.

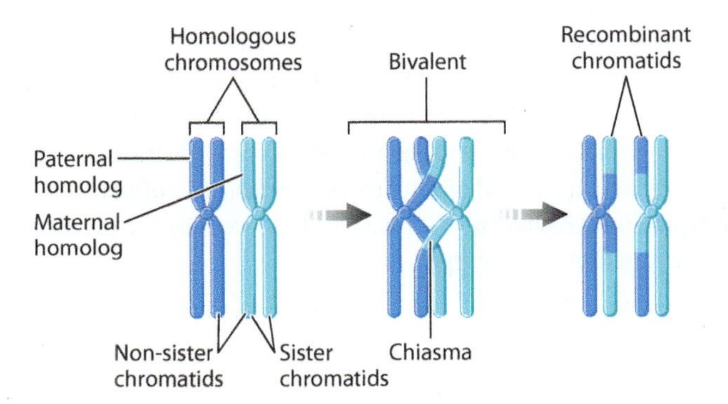

The first meiotic division brings about the reduction in chromosome number.

At the end of prophase I, the chromosomes are fully condensed and have formed chiasmata, the nuclear envelope has begun to disappear, and the meiotic spindle is forming. We are now ready to move through the remaining stages of meiosis I, which are illustrated in **Fig. 11.10.**

In **prometaphase I,** the nuclear envelope breaks down and the meiotic spindles attach to kinetochores on chromosomes. In **metaphase I,** the bivalents move so that they come to lie on an imaginary plane cutting transversely across the spindle. Each bivalent lines up so that its two centromeres lie on opposite sides of this plane, pointing toward opposite poles of the cell. Importantly, the orientation of these bivalents is random with respect to each other. For some, the maternal homolog is attached to the spindle radiating from one pole and the paternal homolog is attached to the spindle originating from the other pole. For others, the orientation is reversed. As a result, when the homologous chromosomes separate from each other in the next step, a complete set of chromosomes moves toward each pole, and that chromosome set is a random mix of maternal and paternal homologs. The random alignment of chromosomes on the spindle in metaphase I further increases genetic diversity in the products of meiosis.

At the beginning of **anaphase I,** the two homologous chromosomes of each bivalent separate as they are pulled in

FIG. 11.10 Meiosis I. Meiosis I is the reductional division: The number of chromosomes is halved.

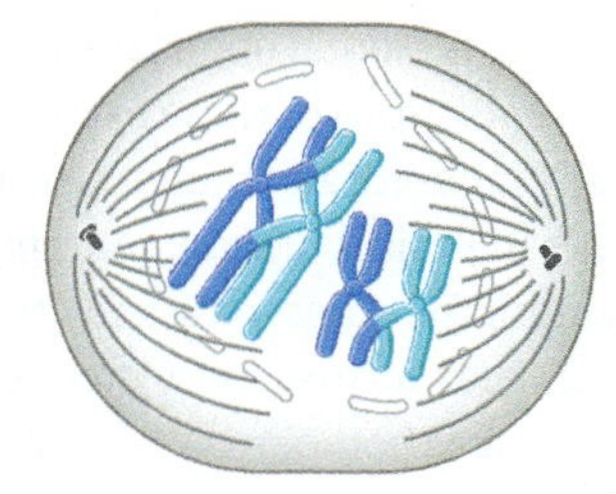

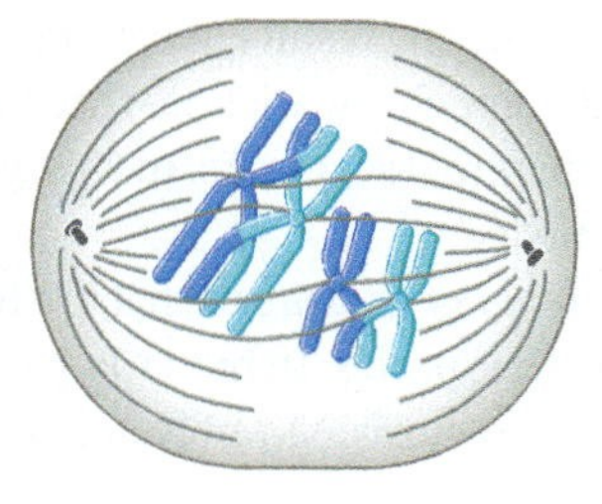

3 Metaphase I: Homologous pairs line up in center of cell, with bivalents oriented randomly with respect to each other.

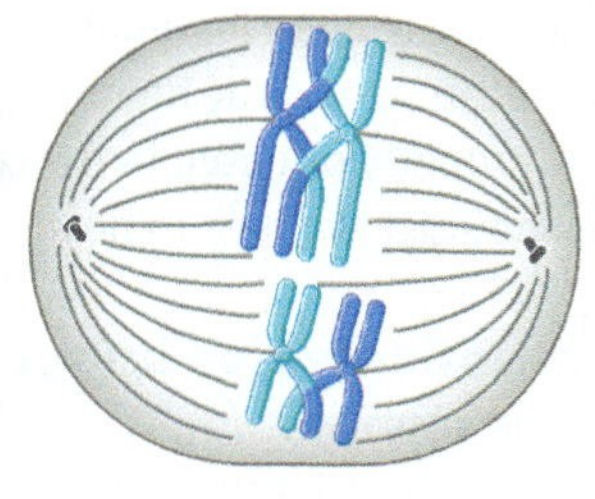

4 Anaphase I: Homologous chromosomes separate, but sister chromatids do not separate.

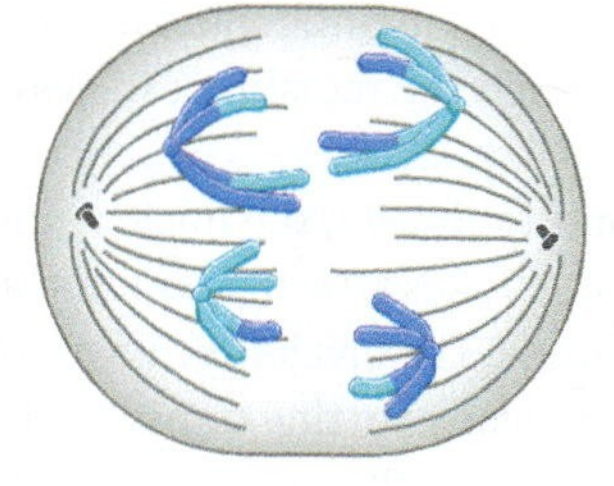

5 Telophase I and cytokinesis: Daughter cells are ready to move into prophase II.

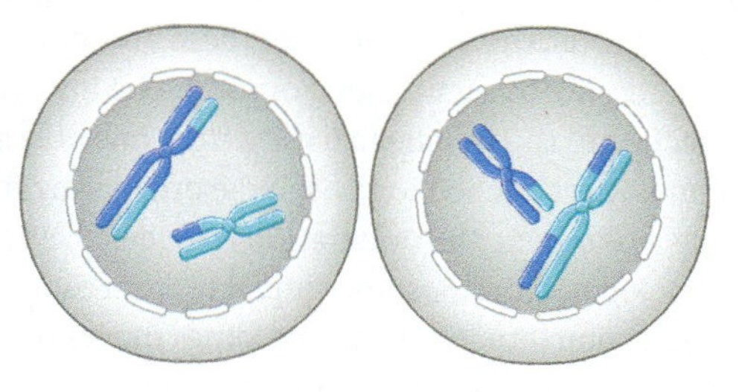

opposite directions. The key feature of anaphase I is that the centromeres do not split and the two chromatids that make up each chromosome remain together. This occurs because spindle microtubules from one pole of the cell attach to both kinetochores of a given chromosome during prometaphase I. Anaphase I is thus very different from anaphase of mitosis, in which the centromeres split and each pair of chromatids separates.

The end of anaphase I coincides with the arrival of the chromosomes at the poles of the cell. Only one of the two homologous chromosomes goes to each pole, so in human cells there are 23 chromosomes at each pole at the end of meiosis I, which is the haploid number of chromosomes. Each of these chromosomes consists of two chromatids attached to a single centromere. Meiosis I is sometimes called the **reductional division** since it reduces the number of chromosomes in daughter cells by half.

In **telophase I,** the chromosomes may uncoil slightly and a nuclear envelope briefly reappears. In many species (including humans), the cytoplasm divides, producing two separate cells. The chromosomes do not completely decondense, however, and so telophase I blends into the first step of the second meiotic division. Importantly, there is no DNA synthesis between the two meiotic divisions.

→ **Quick Check 5** List three ways in which meiosis I differs from mitosis.

The second meiotic division resembles mitosis.

Now let's turn to the second meiotic division, meiosis II, shown in **Fig. 11.11.** In this division, sister chromatids separate, creating haploid daughter cells. Starting with **prophase II,** the second meiotic division is in many respects like a normal mitotic division, except that the nuclei in prophase II have the haploid number of chromosomes, not the diploid number. In prophase II, the chromosomes recondense to their maximum extent. Toward the end of prophase II, the nuclear envelope begins to disappear (in those species in which it has formed), and the spindle begins to be set up.

In **prometaphase II,** spindles attach to kinetochores and, in **metaphase II,** the chromosomes line up so that their centromeres lie on an imaginary plane cutting across the spindle.

In **anaphase II,** the centromere of each chromosome splits. The separated chromatids, now each regarded as a full-fledged chromosome, are pulled toward opposite poles of the spindle. In this sense, anaphase II resembles anaphase of mitosis.

Finally, in **telophase II,** the chromosomes uncoil and become decondensed and a nuclear envelope re-forms around each set of chromosomes. The nucleus of each cell resulting from telophase II has the haploid number of chromosomes. Because cells in meiosis II have the same number of chromosomes at the beginning and at the end of the process, meiosis II is often called the **equational division.** Telophase II is followed by the division of the cytoplasm in many species.

→ **Quick Check 6** The genetic constitution of each cell after telophase II is different from the others. What two processes during meiosis result in these differences?

FIG. 11.11 Meiosis II. Meiosis II is the equational division: The number of chromosomes stays the same.

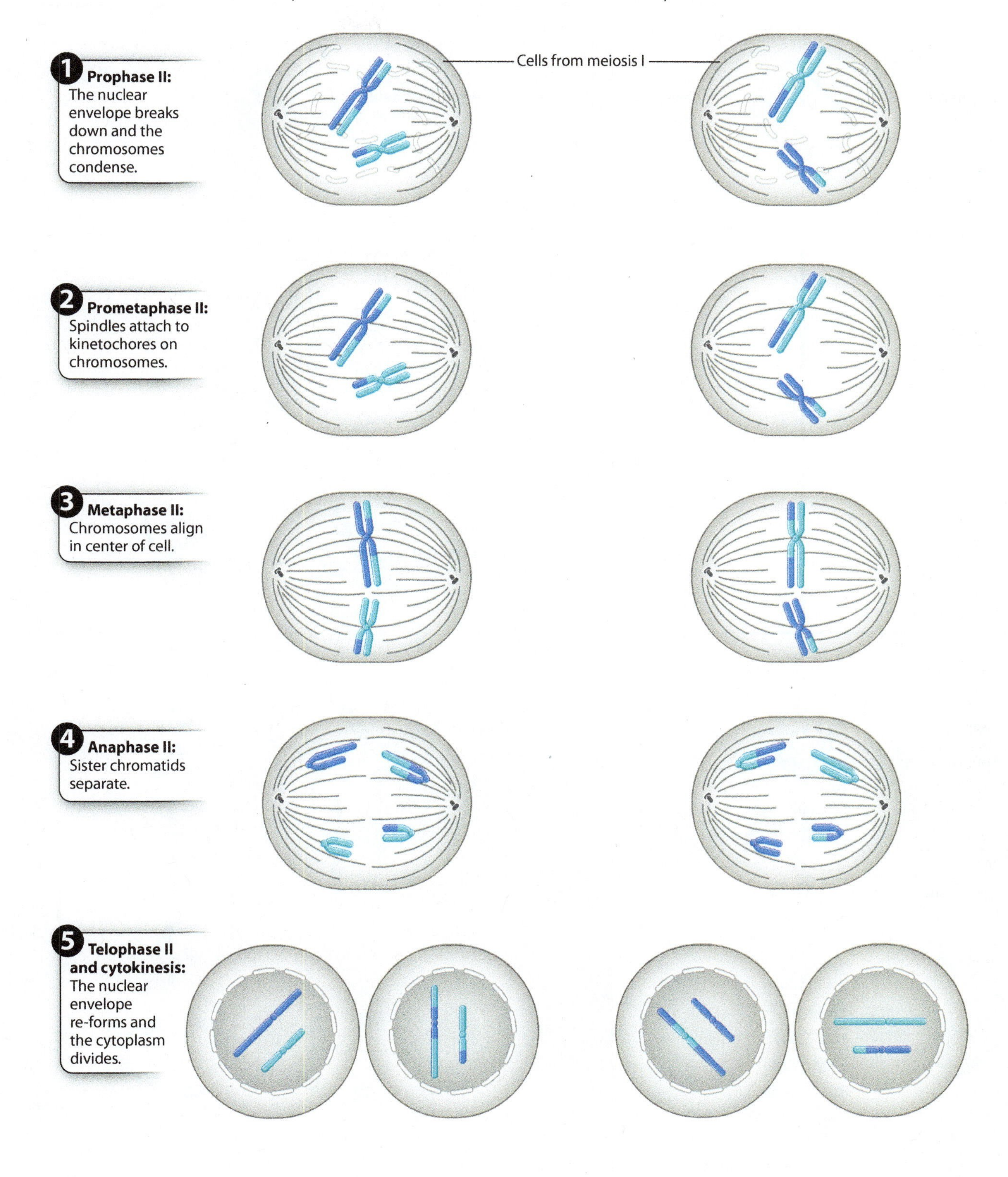

FIG. 11.12 Comparison of mitosis and meiosis.

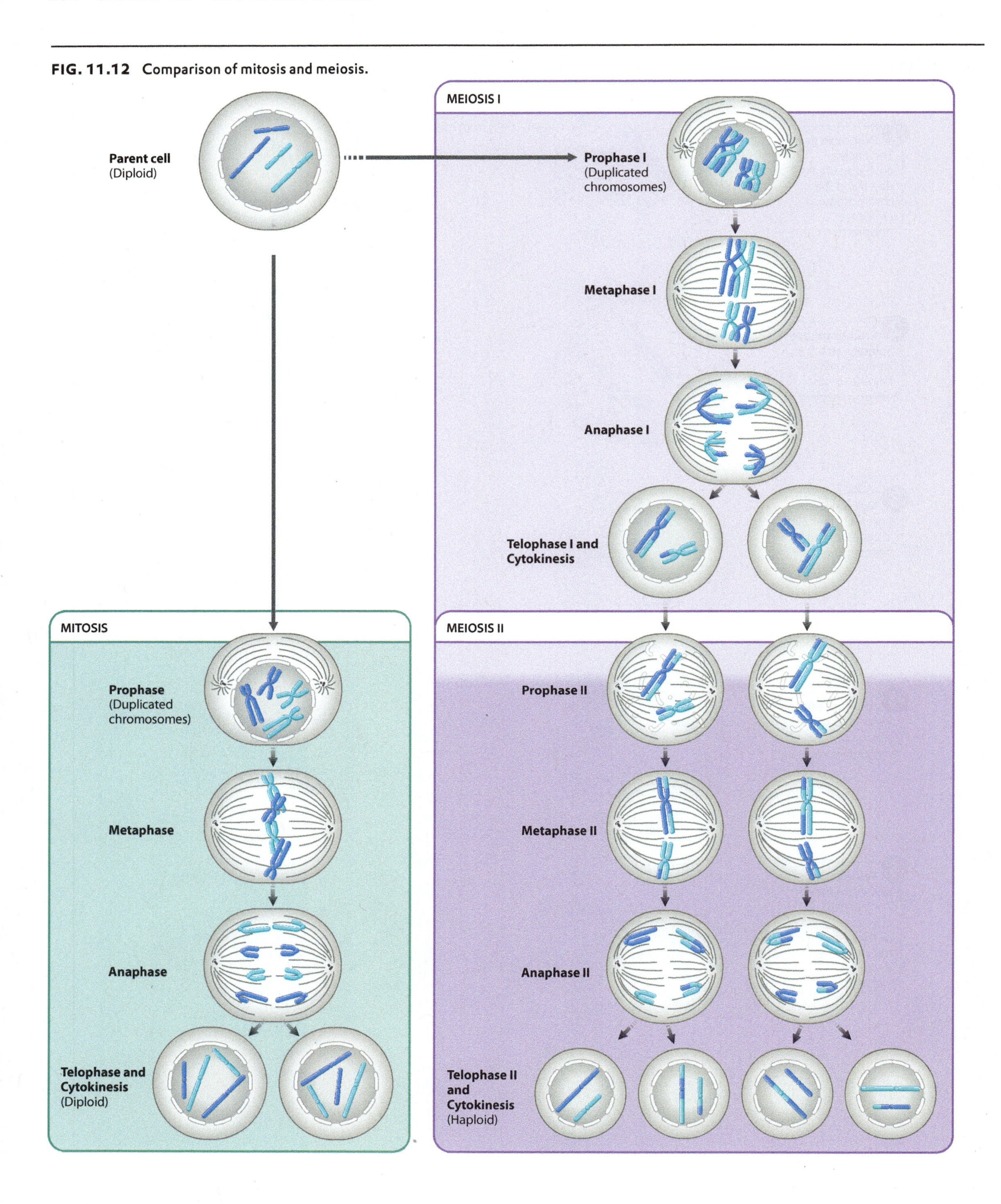

TABLE 11.1 Comparison of Mitosis and Meiosis.

	MITOSIS	MEIOSIS
Function	Asexual reproduction in unicellular eukaryotes Development in multicellular eukaryotes Tissue regeneration and repair in multicellular eukaryotes	Sexual reproduction Production of gametes and spores
Organisms	All eukaryotes	Most eukaryotes
Number of rounds of DNA synthesis	1	1
Number of cell divisions	1	2
Number of daughter cells	2	4
Chromosome complement of daughter cell compared with parent cell	Same	Half
Pairing of homologous chromosomes	No	Meiosis I—Yes Meiosis II—No
Crossing over	No	Meiosis I—Yes Meiosis II—No
Separation of homologous chromosomes	No	Meiosis I—Yes Meiosis II—No
Centromere splitting	Yes	Meiosis I—No Meiosis II—Yes
Separation of sister chromatids	Yes	Meiosis I—No Meiosis II—Yes

A comparison of mitosis and meiosis gives us hints about how meiosis might have evolved (**Fig. 11.12** and **Table 11.1**). During meiosis I, maternal and paternal homologs separate from each other, whereas during meiosis II, sister chromatids separate from each other, similar to mitosis. The similarity of meiosis II and mitosis suggests that meiosis likely evolved from mitosis. Mitosis occurs in all eukaryotes and was certainly present in the common ancestor of all living eukaryotes. Meiosis is present in most, but not all, eukaryotes. Because the steps of meiosis are the same in all eukaryotes, meiosis is thought to have evolved in the common ancestor of all eukaryotes and has been subsequently lost in some groups.

Division of the cytoplasm often differs between the sexes.

In multicellular organisms, division of the cytoplasm in meiotic cell division differs between the sexes. In female mammals (**Fig. 11.13a**), the cytoplasm is divided very unequally in both meiotic divisions. Most of the cytoplasm is retained in one meiotic product, a very large cell called the oocyte, which can develop into the functional egg cell, and the other meiotic products receive only small amounts of cytoplasm. These smaller cells are called **polar bodies.** In male mammals (**Fig. 11.13b**), the cytoplasm divides about equally in both meiotic divisions, and each of the resulting meiotic products goes on to form a functional sperm. During the development of the sperm, most of the cytoplasm is eliminated, and what is left is essentially a nucleus in the sperm head equipped with a long whiplike flagellum to help propel it toward the egg.

Meiosis is the basis of sexual reproduction.

Sexual reproduction involves two processes: meiotic cell division and fertilization. Meiotic cell division, as we just saw, produces cells with half the number of chromosomes present in the parent cell. In multicellular animals, the products of meiotic cell division are gametes: An egg cell is a gamete and a sperm cell is a gamete. Each gamete is haploid, containing a single set of chromosomes. In humans, meiosis takes place in the ovaries of the female and the testes of the male, and each resulting gamete contains 23 chromosomes, including one each of the 22 numbered chromosomes plus either an *X* or a *Y* chromosome.

FIG. 11.13 Cytoplasmic division in females and in males. (a) In females, cytoplasmic division results in one oocyte and three polar bodies; (b) in males, it results in four sperm cells.

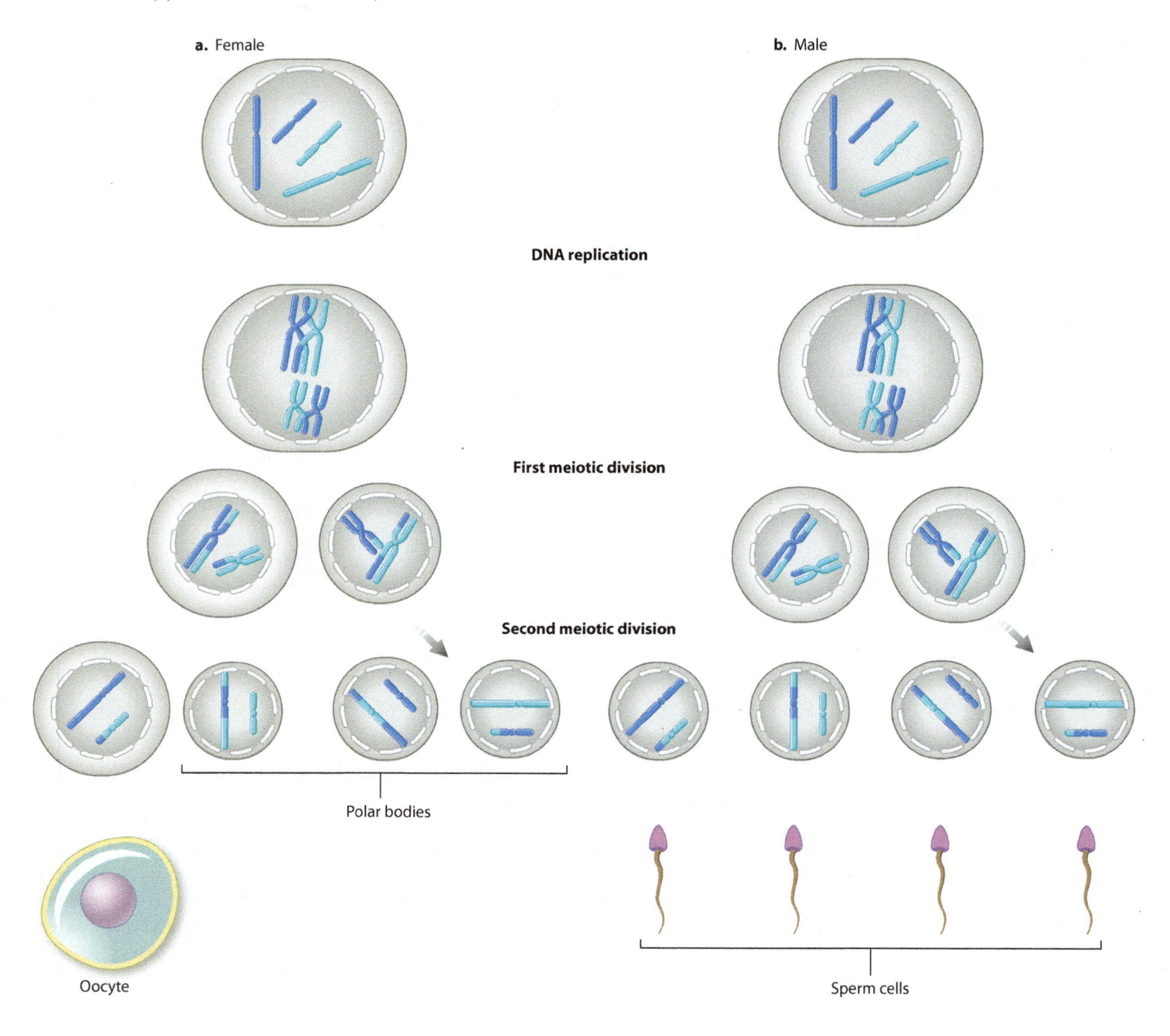

During fertilization, these gametes fuse to form a single cell called a **zygote.** The zygote is diploid, having two complete sets of chromosomes, one from each parent. Therefore, fertilization restores the original chromosome number.

As we discuss further in Chapters 16 and 42, sexual reproduction plays a key role in increasing genetic diversity. Genetic diversity results from meiotic cell division (the cells that are produced are each genetically different from one another as a result of crossing over and the random segregation of homologous chromosomes), and from fertilization (different gametes are combined to produce a new, unique individual). The increase in genetic diversity made possible by sexual reproduction allows organisms to evolve and adapt more quickly to their environment than is possible with asexual reproduction.

In the life cycle of multicellular animals like humans, then, the diploid organism produces single-celled haploid gametes that fuse to make a diploid zygote. In this case, the only haploid cells in the life cycle are the gametes, and the products of meiotic cell division do not undergo mitotic cell division but instead fuse to become a diploid zygote. However, there are a number of life cycles in other organisms that differ in the timing of meiotic cell division and fertilization, discussed more fully in Chapter 27. Some organisms, like most fungi, are haploid. These haploid cells can fuse to produce a diploid zygote, but this cell immediately undergoes meiotic cell division to produce haploid cells, so that the only diploid cell in the life cycle is the zygote (Chapter 34). Other organisms, like plants, have both multicellular haploid and diploid phases (Chapter 30). In this case, meiotic cell division produces haploid spores that divide by mitotic cell division to produce a multicellular haploid phase; subsequently, haploid cells fuse to form a diploid zygote that also divides by mitotic cell division to produce a multicellular diploid phase.

11.4 REGULATION OF THE CELL CYCLE

Both mitotic and meiotic cell division must occur only at certain times and places. Mitotic cell division, for example, occurs during growth of a multicellular organism, wound healing, or in the maintenance of actively dividing tissues such as the skin or lining of the intestine. Similarly, meiotic cell division occurs only at certain times during development. Even for unicellular organisms, cell division must be regulated so that it takes place only when conditions are favorable—for example, when enough nutrients are present in the environment. Thus, a cell may have to receive a signal before it will divide. In Chapter 9, we saw how cells respond to signals. Growth factors, for example, bind to cell-surface receptors and activate intracellular signaling pathways that lead to cell division.

Even when a cell receives a signal to divide, it does not divide until it is ready. Has all of the DNA been replicated during S phase? Has the cell grown to a size sufficient to support division into viable daughter cells? If these and other preparations have not been accomplished, the cell halts its progression through the cell cycle.

So, cells have regulatory mechanisms that initiate cell division, as well as mechanisms for spotting faulty or incomplete preparations and arresting cell division. When these mechanisms fail—for example, dividing in the absence of a signal or when the cell is not ready—the result is uncontrolled cell division, a hallmark of cancer. In this section, we consider how cells control their passage through the cell cycle. Our focus is on control of mitotic cell division. However, many of the same factors also regulate meiotic cell division, a reminder of the close evolutionary connection between these two forms of cell division.

Protein phosphorylation controls passage through the cell cycle.

Early animal embryos, such as those of frogs and sea urchins, are useful models for studying cell cycle control because they are large and undergo many rapid mitotic cell divisions following fertilization. During these rapid cell divisions, mitosis and S phase alternate with virtually no G_1 and G_2 phases in between. Studies of animal embryos revealed two interesting patterns. First, as the cells undergo this rapid series of divisions, several proteins appear and disappear in a cyclical fashion. Researchers interpreted this observation to mean that these proteins might play a role in the control of the progression through the cell cycle. Second, several enzymes become active and inactive in cycles. These enzymes are kinases, proteins that phosphorylate other proteins (Chapter 9). The timing of kinase activity is delayed slightly relative to the appearance of the cyclical proteins.

These and many other observations led to the following view of cell cycle control: Proteins are synthesized that activate the kinases. These regulatory proteins are called **cyclins** because their levels rise and fall with each turn of the cell cycle. Once activated by cyclins, the kinases phosphorylate target proteins involved in promoting cell division (**Fig. 11.14**). These kinases, **cyclin-dependent kinases,** or CDKs, are always present within the cell but are active only when bound to the appropriate cyclin. It is the kinase activity of the cyclin–CDK complexes that triggers the

FIG. 11.14 Cyclins and cyclin-dependent kinases (CDKs). Cyclins and CDKs control progression through the cell cycle.

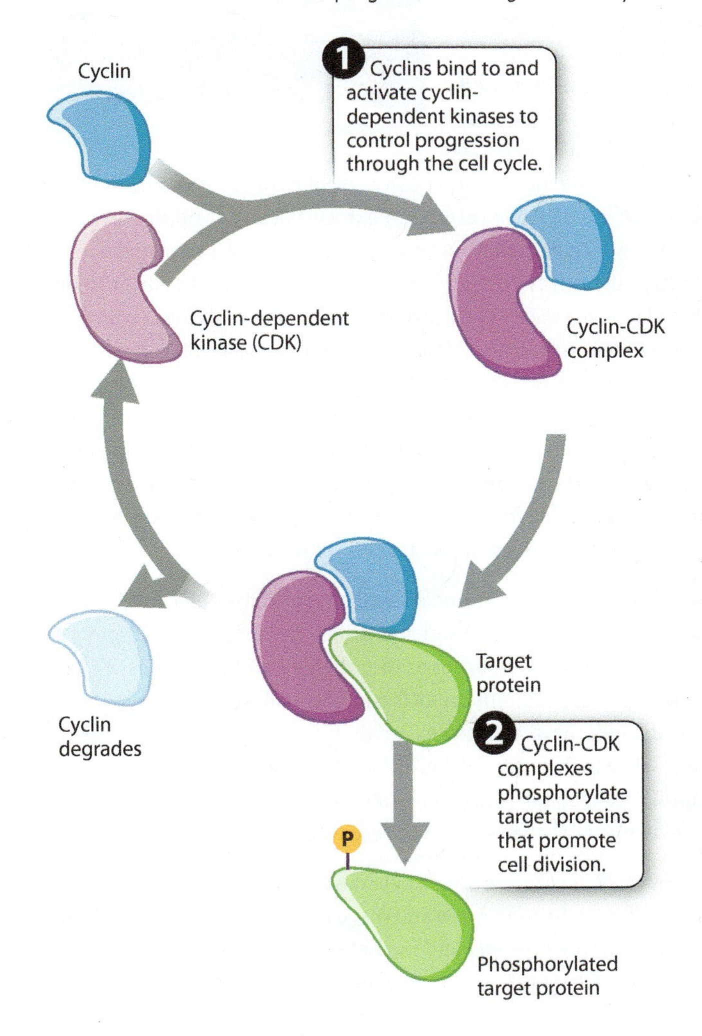

required cell cycle events. Therefore, the cyclical change in cyclin–CDK activity depends on the cyclical levels of the cyclins. Cyclins in turn may be synthesized in response to signaling pathways that promote cell division.

These cell cycle proteins, including cyclins and CDKs, are widely conserved across eukaryotes, reflecting their fundamental role in controlling cell cycle progression, and have been extensively studied in yeast, sea urchins, mice, and humans (**Fig. 11.15**).

HOW DO WE KNOW?

FIG. 11.15

How is progression through the cell cycle controlled?

Source: SeaPics.com.

BACKGROUND The cell cycle is characterized by cyclical changes in many components of the cell: Chromosomes condense and decondense, the mitotic spindle forms and breaks down, the nuclear envelope breaks down and re-forms. How these regular and cyclical changes are controlled is the subject of active research. First, clues came from studies in yeast in which mutations in certain genes blocked progression through the cell cycle, suggesting that these genes encode proteins that play a role in cell cycle progression. Additional clues came from studies in embryos of the sea urchin *Arbacia punctulata*. These embryos are large and divide rapidly by mitosis, making them a good model system for the study of the control of cell division. In the early 1980s, it was known that inhibition of protein synthesis blocks key steps of cell division in sea urchins. It was also known that an enzyme called MPF, or M-phase promotion factor, is important for the transition from G_2 to M phase.

EXPERIMENT To better understand events in the cell cycle, English biochemist Tim Hunt and colleagues measured protein levels in sea urchin embryos as they divide rapidly by mitosis. He added radioactive methionine (an amino acid) to eggs, which became incorporated into any newly synthesized proteins. The eggs were then fertilized and allowed to develop. Samples of the rapidly dividing embryos were taken every 10 minutes and run on a gel to visualize the levels of different proteins.

RESULTS Most protein bands became darker as cell division proceeded, indicating more and more protein synthesis. However, the level of one protein band oscillated, increasing in intensity and then decreasing with each cell cycle, as shown in the graph.

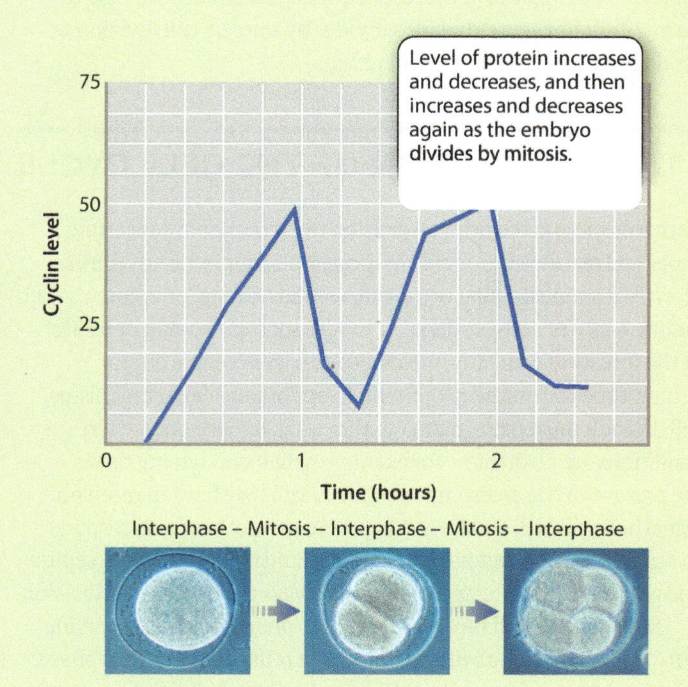

Source: Biology Pics/Science Source.

CONCLUSION Hunt and colleagues called this new protein "cyclin." Although they did not know its function, its fluctuating level suggested it might play a role in the control of the cell cycle. However, more work was needed to figure out if cyclins actually cause progression through the cell cycle or whether their levels oscillate in response to progression through the cell cycle.

FOLLOW-UP WORK Hunt and colleagues found a similar rise and fall in the levels of certain proteins in the sea urchin *Lytechinus pictus* and surf clam *Spisula solidissima*. MPF was found to consist of a cyclin protein and a cyclin-dependent kinase that play a key role in the G_2–M transition. In other words, cyclins do in fact control progression through the cell cycle by their interactions with cyclin-dependent kinases. Hunt shared the Nobel Prize in Physiology or Medicine in 2001 for his work on cyclins.

SOURCE T. Evans et al. 1983. "Cyclin: A Protein Specified by Maternal mRNA in Sea Urchin Eggs That Is Destroyed at Each Cleavage Division." *Cell* 33:389–396.

FIG. 11.16 Three cyclin–CDK complexes. Each one is active at different times in the cell cycle.

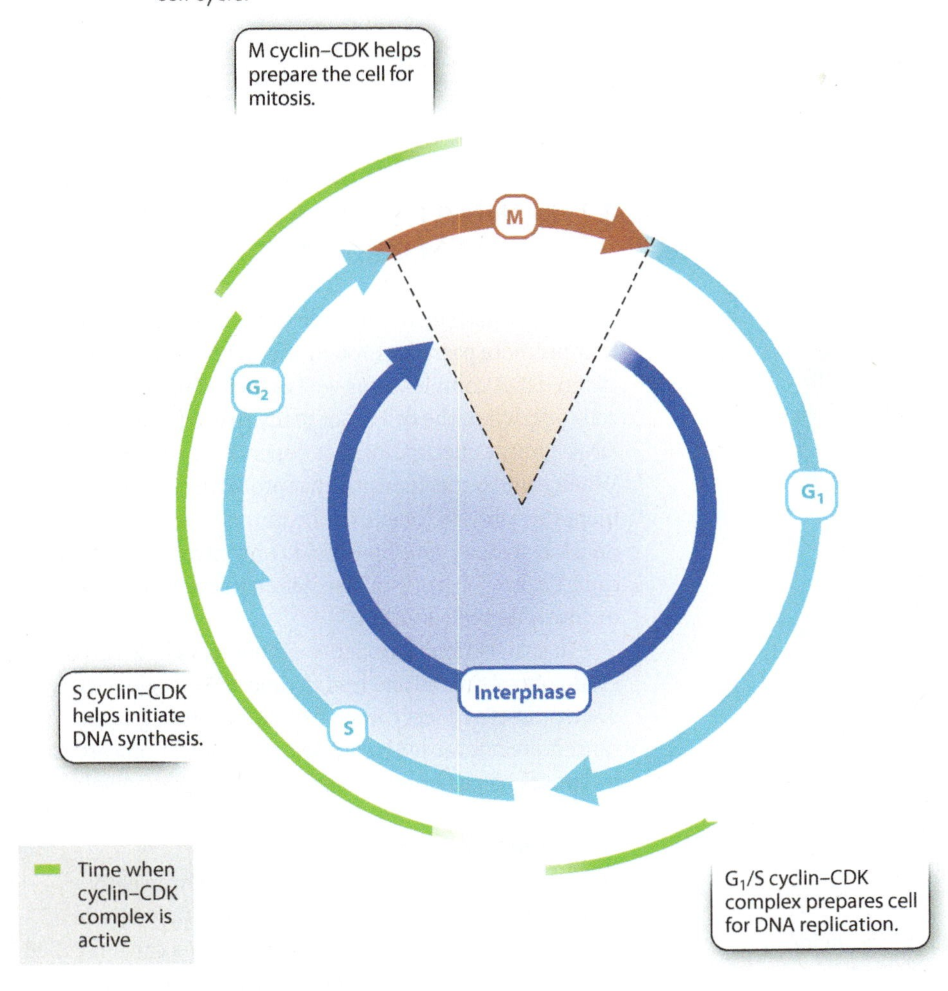

Different cyclin–CDK complexes regulate each stage of the cell cycle.

In mammals, there are several different cyclins and CDKs that act at specific steps of the cell cycle. Three steps in particular are subject to cyclin–CDK regulation in all eukaryotes (**Fig. 11.16**).

The G_1/S cyclin–CDK complex, which is active at the end of the G_1 phase, is necessary for the cell to enter S phase. For example, this cyclin–CDK complex activates a protein that promotes the expression of histone proteins needed for packaging the newly replicated DNA strands.

The S cyclin–CDK complex is necessary for the cell to initiate DNA synthesis. It activates enzymes and other proteins necessary for DNA replication. Once replication has begun at a particular place on the DNA, S cyclin–CDK activity prevents the replication proteins from reassembling at the same place and re-replicating the same DNA sequence. Synthesizing the same sequence repeatedly would be dangerous because cells with too much DNA could die or become cancerous.

The M cyclin–CDK complex initiates multiple events associated with mitosis. For example, M cyclin–CDK phosphorylates structural proteins in the nucleus, triggering the breakdown of the nuclear envelope in prometaphase. M cyclin–CDK also phosphorylates proteins that regulate the assembly of tubulin into microtubules, promoting the formation of the mitotic spindle.

Cell cycle progression requires successful passage through multiple checkpoints.

Cyclins and CDKs not only allow a cell to progress through the cell cycle, but also give the cell opportunities to halt the cell cycle should something go wrong. In other words, if the preparations for the next stage of the cell cycle are incomplete or if there is some kind of damage, there are mechanisms that block the cyclin–CDK activity required for the next step, pausing cell division until preparations are complete or the damage is repaired. Each of these mechanisms is called a **checkpoint.**

Cells have many cell cycle checkpoints. Three major checkpoints that have been well studied are illustrated in **Fig. 11.17.** The presence of damaged DNA arrests the cell at the end of G_1 before DNA synthesis; the presence of unreplicated DNA arrests the cell at the end of G_2 before the cell enters mitosis; and abnormalities in chromosome attachment to the spindle arrest the cell in early mitosis. By way of illustration, we focus on the key checkpoint that occurs at the end of G_1 in response to the presence of damaged DNA.

DNA can be damaged by environmental insults such as ultraviolet radiation or chemical agents. Typically, damage takes the form of double-stranded breaks in the DNA. If the cell progresses through mitosis with DNA damage, the damage might be inherited by the daughter cells, or the chromosomes might not segregate normally. Some checkpoints delay progression through the cell cycle until DNA damage is repaired. An important one occurs in late G_1. This DNA damage checkpoint depends on several regulatory proteins, some of which recognize damaged DNA while others arrest cell cycle progression before S phase (Fig. 11.17).

When DNA is damaged by radiation, a specific protein kinase is activated that phosphorylates a protein called p53. Phosphorylated p53 binds to DNA, where it turns on the expression of several genes. One of these genes codes for a protein that binds to and blocks the

FIG. 11.17 Cell cycle checkpoints. Cell cycle checkpoints monitor key steps in the cell cycle.

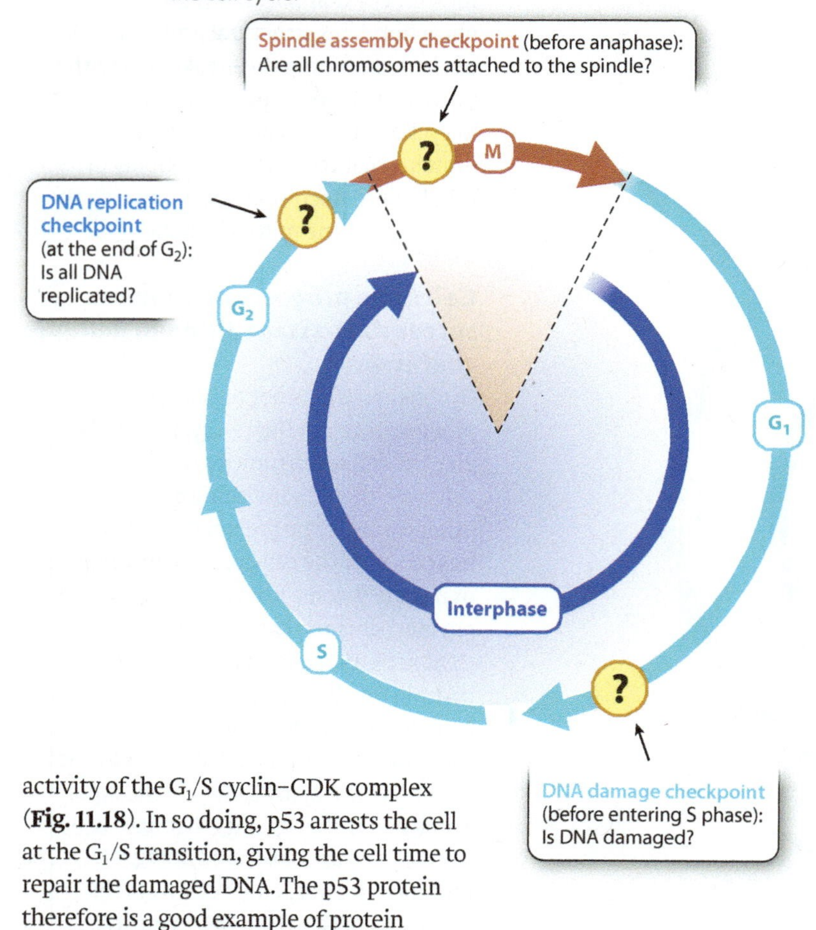

activity of the G_1/S cyclin–CDK complex (**Fig. 11.18**). In so doing, p53 arrests the cell at the G_1/S transition, giving the cell time to repair the damaged DNA. The p53 protein therefore is a good example of protein involved in halting the cell cycle when the cell is not ready to divide. Because of its role in protecting the genome from accumulating DNA damage, p53 is sometimes called the "guardian of the genome." Mutations in p53 are common in cancer, a topic we turn to next.

→ **Quick Check 7** Can you think of two ways in which the function of p53 can be disrupted?

? CASE 2 WHEN GOOD CELLS GO BAD

11.5 WHAT GENES ARE INVOLVED IN CANCER?

As we have just seen, cells have evolved mechanisms that promote passage through the cell cycle, such as the cyclin–CDK complexes, as well as mechanisms that halt the cell cycle when the cell is not ready to divide, such as the DNA damage checkpoint that depends on the p53 protein. When cellular mechanisms that promote cell division are inappropriately activated or the normal checks on cell division are lost, cells may divide uncontrollably. The result can be cancer, a group of diseases characterized by rapid, uncontrolled cell division.

In this section, we explore various ways in which cell cycle control can fail and lead to cancer. We begin with a discussion of a cancer caused by a virus. Although most cancers are not caused by viruses, the study of cancers caused by viruses helped us to understand how cancers develop.

Oncogenes promote cancer.

Our understanding of cancer is based partly on early observations of cancers in animals. In the first decade of the twentieth century, Peyton Rous studied cancers called sarcomas in chickens (**Fig. 11.19**). His work and that of

FIG. 11.18 DNA damage checkpoint controlled by p53.

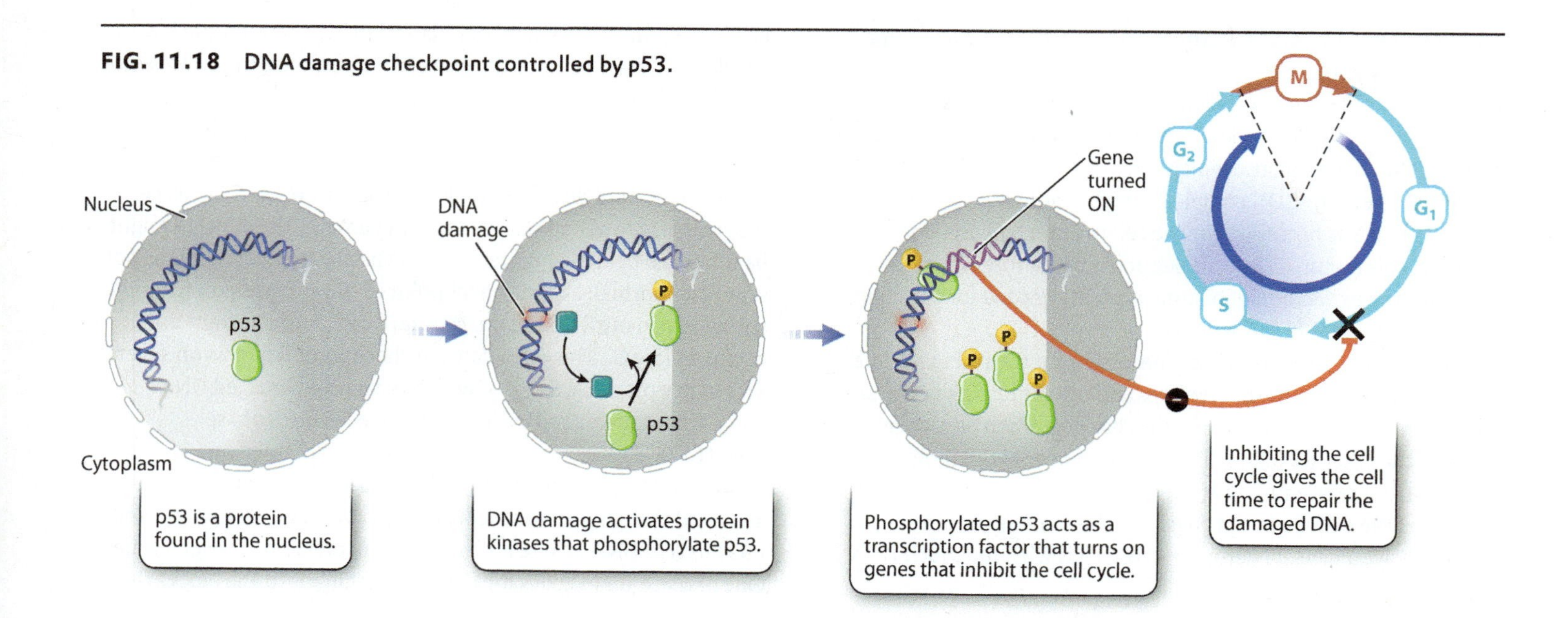

FIG. 11.19

Can a virus cause cancer?

BACKGROUND In the early 1900s, little was known about the cause of cancer or the nature of viruses. Peyton Rous, an American pathologist, studied a form of cancer called a sarcoma in chickens. First, he moved a cancer tumor from a diseased chicken to a healthy chicken, and found that the cancer could be transplanted. Then, he tried to isolate the factor that causes the cancer in chickens. He made an extract of the tumor, filtered it to remove all the cells, and injected the extract into a healthy chicken to see if it could induce cancer.

HYPOTHESIS Experiments in other organisms, such as mice, rats, and dogs, showed that an extract free of cells from a tumor does not cause cancer. Therefore, Rous hypothesized that a cell-free extract of the chicken sarcoma would not induce cancer in healthy chickens.

EXPERIMENT Rous took a sample of the sarcoma from a chicken, ground it up, suspended it in solution, and centrifuged it to remove the debris. In one experiment, he injected this extract into a healthy chicken. In a second experiment, he passed the extract through a filter to remove all cells, including cancer cells and bacterial cells, and then injected this cell-free extract into a healthy chicken.

RESULTS In both cases, healthy chickens injected with the extract from the sarcoma developed cancer at the site of injection. Microscopic examination of the cancer showed it to be the same type of cancer as the original one.

CONCLUSIONS AND FOLLOW-UP WORK Rous concluded, contrary to his hypothesis, that a small agent—a virus or chemical—is capable of causing cancer. Later experiments confirmed that the cause of the cancer is a virus. This result was surprising, controversial, and dismissed at the time. A second cancer-causing virus was not found until the 1930s. Although most cancers are not caused by viruses, work with cancer-causing viruses helped to identify cellular genes that, when mutant, can lead to cancer. In 1966, Rous shared the Nobel Prize in Physiology or Medicine for his discovery.

SOURCES Rous, P. 1910. "A Transmissable Avian Neoplasm (Sarcoma of the Common Fowl)." *J. Exp. Med.* 12:696–705; Rous, P. 1911. "Transmission of a malignant new growth by means of a cell-free filtrate." *JAMA* 56:198.

Source: Biology Pics/Science Source.

others led to the discovery of the first virus known to cause cancer in animals, named the Rous sarcoma virus.

Viruses are assemblages of protein surrounding a core of either RNA or DNA. They multiply by infecting cells and using the biochemical machinery of their host to synthesize proteins encoded in their genome and to make more copies of themselves. Viruses typically carry only a handful of genes, making it relatively easy to identify which of those genes is involved in cancer. The investigation of cancer-causing viruses therefore provided major insights into our understanding of cancer.

As discussed in Chapter 9, growth factors activate several types of proteins inside the cell that promote cell division. The gene from the Rous sarcoma virus that promotes uncontrolled cell division encodes an overactive protein kinase similar to the kinases in the cell that function as signaling proteins. This viral gene is named *v-src,* for *viral-src* (pronounced "sarc" and short for "sarcoma," the type of cancer it causes).

The *v-src* gene is one of several examples of an **oncogene,** or cancer-causing gene, found in viruses. A real surprise was the discovery that the *v-src* oncogene is found not just in the Rous sarcoma virus. It is an altered version of a gene normally found in the host animal cell, known as *c-src* (*cellular-src*). The *c-src* gene plays a role in the normal control of cell division during embryonic development.

Proto-oncogenes are genes that when mutated may cause cancer.

The discovery that the *v-src* oncogene has a normal counterpart in the host cell was an important step toward determining the cellular genes that participate in cell growth and division. These normal cellular genes are called **proto-oncogenes.** They are involved in cell division, but do not themselves cause cancer. Only when they are mutated do they have the potential to cause cancer. Today, we know of scores of proto-oncogenes, most of which were identified through the study of cancer-causing viruses in chickens, mice, and cats.

Oncogenes also play a major role in human cancers. Most human cancers are not caused by viruses. Instead, human proto-oncogenes can be mutated into cancer-causing oncogenes by environmental agents such as chemical pollutants. For example, organic chemicals called aromatic amines present in cigarette smoke can enter cells and damage DNA, resulting in mutations that can convert a proto-oncogene into an oncogene.

What types of functions are performed by the products of proto-oncogenes? Nearly every protein that performs a key step in a signaling cascade that promotes cell division can be the product of a proto-oncogene. These include growth factors, cell-surface receptors, G proteins, and protein kinases. Each of these can be mutated to become oncogenes.

Let's consider an example of a proto-oncogene. In Chapter 9, we discussed platelet-derived growth factor, or PDGF. This protein promotes cell division by binding to and dimerizing a receptor kinase in the membrane of the target cell. Dimerization of the receptor leads to the activation of several signaling pathways and the promotion of cell division. In one type of leukemia (a cancer of blood cells), a mutation in the gene that encodes the PDGF receptor results in an altered receptor that is missing the extracellular portion needed to bind the growth factor. The mutant receptor dimerizes on its own, independent of PDGF binding, and is therefore always turned on. The overactive PDGF receptor activates too many target proteins over too long a time period, leading to uncontrolled proliferation of blood cells.

Tumor suppressors block specific steps in the development of cancer.

Up to this point, we have considered what happens when mechanisms that promote cell division are inappropriately activated. Now let's consider what happens when mechanisms that usually prevent cell cycle progression are removed.

Earlier, we discussed cell cycle checkpoints that halt the cell cycle until the cell is ready to divide. One of these checkpoints depends on the p53 protein, which normally arrests cell division in response to DNA damage. When the p53 protein is mutated or its function is inhibited, the cell can divide before the DNA damage is repaired. The result is that cells continue to divide in the presence of damaged DNA, leading to the accumulation of mutations that promote cell division. The p53 protein is mutated in many types of human cancer, highlighting its critical role in regulating the cell cycle.

The p53 protein is one example of a **tumor suppressor.** Tumor suppressors are proteins whose normal activities inhibit cell division. Some tumor suppressors participate in cell cycle checkpoints, as is the case for p53. Other tumor suppressors repress the expression of genes that promote cell division, while still others trigger cell death.

Tumor suppressors act in opposition to proto-oncogenes. Therefore, whether a cell divides or not depends on the activities of both proto-oncogenes and tumor suppressors: Proto-oncogenes must be turned on and tumor suppressors must be turned off. Given the importance of controlling cell division, it is not surprising that cells have two counterbalancing systems that must be in agreement before cell division takes place.

→ **Quick Check 8** How do oncogenes differ from tumor suppressor genes?

Most cancers require the accumulation of multiple mutations.

Most human cancers require more than the overactivation of one oncogene or the inactivation of a single tumor suppressor. Given the multitude of different tumor suppressor proteins that are produced in the cell, it is likely that one will compensate for

FIG. 11.20 Multiple-mutation model for the development of cancer. Cells that become cancerous typically carry mutations in several different genes.

Inactivation of first tumor suppressor gene

Normal cells

Activation of oncogene

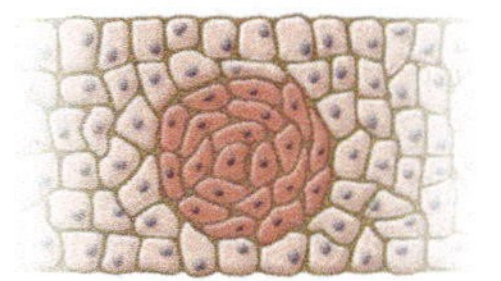

Benign cancer

Inactivation of second tumor suppressor gene

Malignant cancer

Inactivation of third tumor suppressor gene

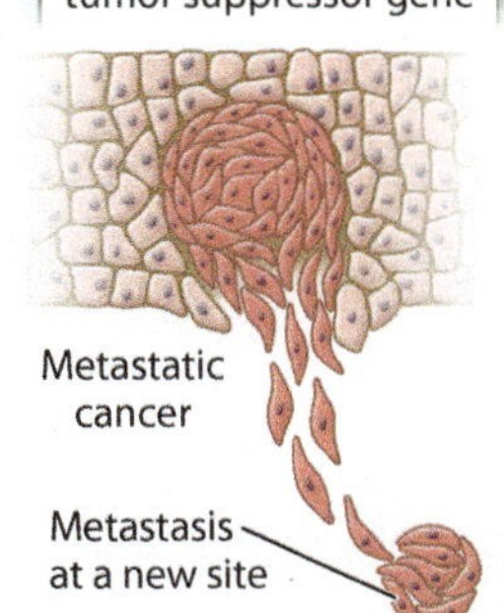

even the complete loss of another. Cells have evolved a redundant arrangement of these control mechanisms to ensure that cell division is properly regulated.

When several different cell cycle regulators fail, leading to both the overactivation of oncogenes and the loss of tumor suppressor activity, cancer will likely develop. The cancer may be benign, which means that it is relatively slow growing and does not invade the surrounding tissue, or it may be malignant, which means that it grows rapidly and invades surrounding tissues. In many cases of malignant colon cancer, for example, tumor cells contain at least one overactive oncogene and several inactive tumor suppressor genes (**Fig. 11.20**). The gradual accumulation of these mutations over a period of years can be correlated with the stepwise progression of the cancer from a benign form to full malignancy.

Taken together, we can now define some of the key characteristics that make a cell cancerous. Uncontrolled cell division is certainly important, but given the communities of cells and extracellular matrix we have discussed over the last few chapters, additional characteristics should also be considered. In 2000, American biologists Douglas Hanahan and Robert Weinberg wrote a paper in the journal *Cell* (*Cell* 100: 57–70) called "Hallmarks of Cancer" in which they highlighted key features of cancer cells. These include the ability to divide on their own in the absence of growth signals; resistance to signals that inhibit cell division or promote cell death; the ability to invade local and distant tissues (metastasis); and the production of signals to promote new blood vessel growth for nutrients to support cell division.

Considered in this light, a cancer cell is one that no longer plays by the rules of a normal cellular community. Cancer therefore serves to remind us of the normal controls and processes that are required to allow cells to exist in a community. These processes, which work together, are summarized in **Fig. 11.21** on the following pages. ■

VISUAL SYNTHESIS

FIG. 11.21

Cellular Communities

Integrating concepts from Chapters 9–11

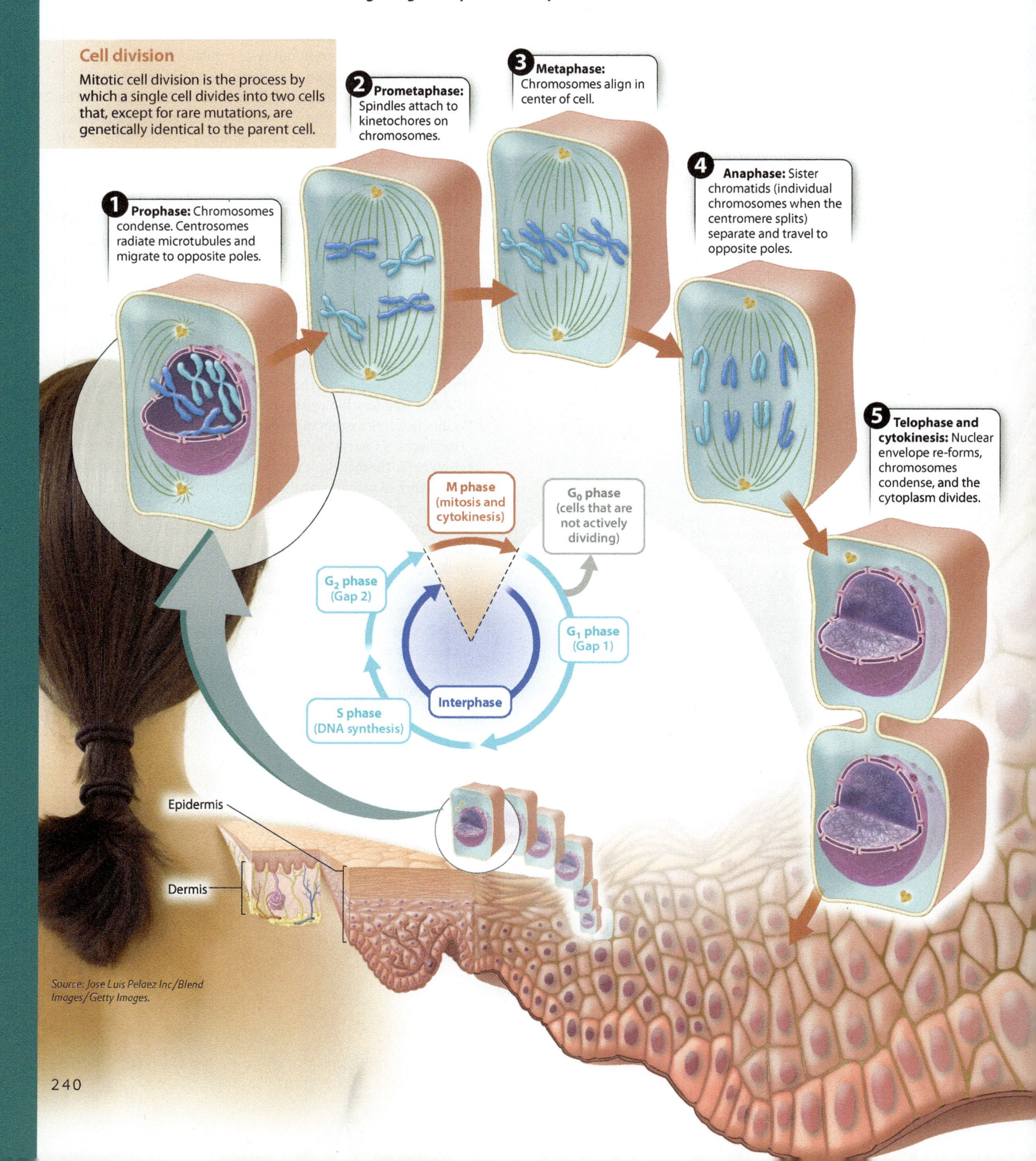

Source: Jose Luis Pelaez Inc/Blend Images/Getty Images.

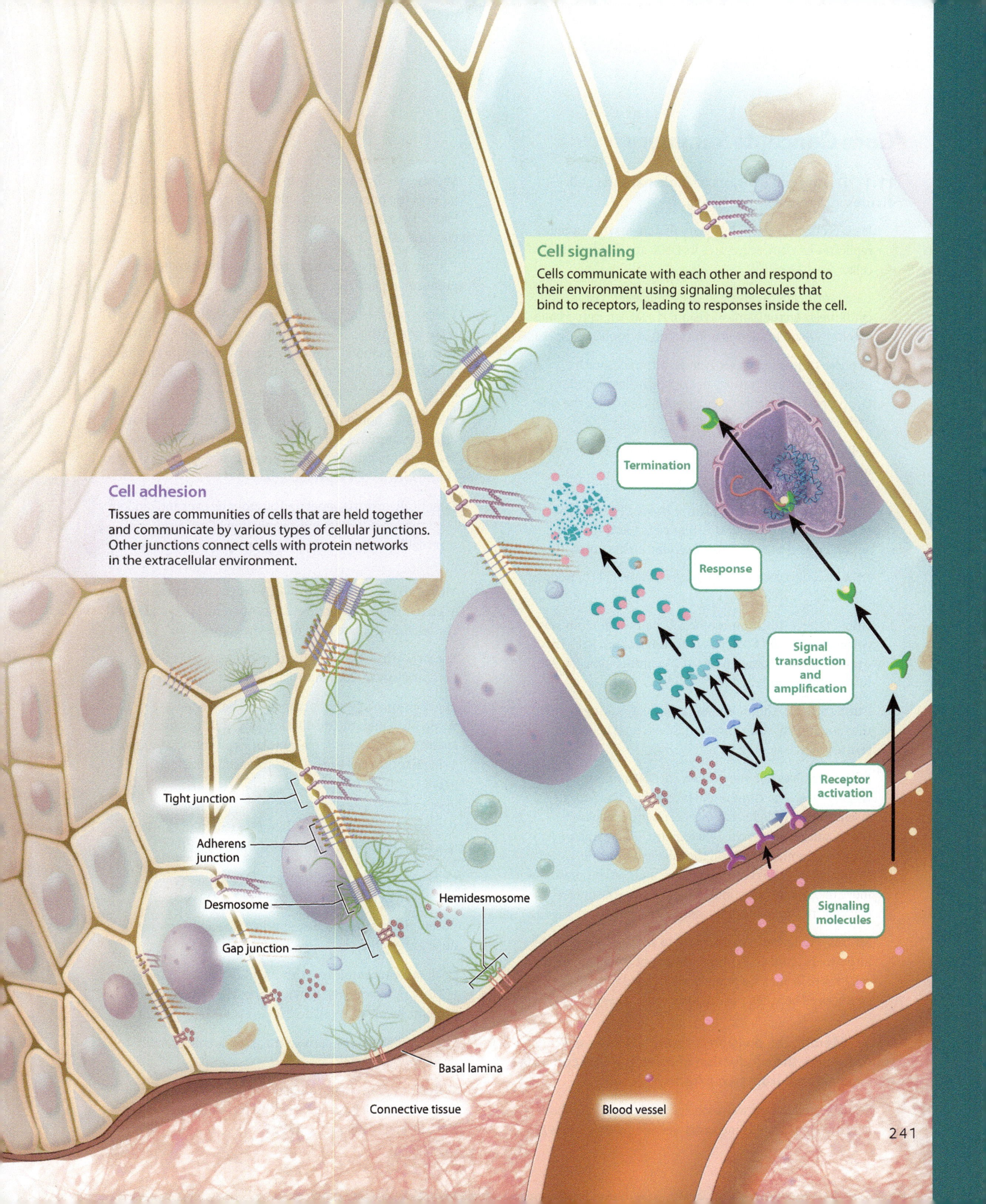

Cell signaling
Cells communicate with each other and respond to their environment using signaling molecules that bind to receptors, leading to responses inside the cell.
Cell adhesion
Tissues are communities of cells that are held together and communicate by various types of cellular junctions. Other junctions connect cells with protein networks in the extracellular environment.
Termination
Response
Signal transduction and amplification
Receptor activation
Signaling molecules
Tight junction
Adherens junction
Desmosome
Gap junction
Hemidesmosome
Basal lamina
Connective tissue
Blood vessel

Core Concepts Summary

11.1 During cell division, a single parental cell divides into two daughter cells.

Prokaryotic cells divide by binary fission, in which a cell replicates its DNA, segregates its DNA, and divides into two cells. page 220

Eukaryotic cells divide by mitosis (nuclear division) and cytokinesis (cytoplasmic division). Together, mitosis and cytokinesis are known as mitotic cell division. page 222

M phase (mitosis and cytokinesis) alternates with interphase, which consists of G_1, S (synthesis), and G_2 phases. These four stages together constitute the cell cycle. page 222

Cells that do not need to divide exit the cell cycle and are in G_0. page 222

11.2 Mitotic cell division is the basis of asexual reproduction in unicellular eukaryotes and the process by which cells divide in multicellular eukaryotes.

DNA in a eukaryotic cell is packaged as linear chromosomes. page 222

Humans have 46 chromosomes: 22 pairs of homologous chromosomes and 1 pair of sex chromosomes. Each parent contributes one complete set of 23 chromosomes at fertilization. page 222

During S phase, chromosomes replicate, resulting in the formation of sister chromatids held together at the centromere. page 223

Mitosis involves five steps following DNA replication: (1) prophase—the chromosomes condense and become visible under the light microscope; (2) prometaphase—the spindles attach to the centromeres; (3) metaphase—the chromosomes line up in the middle of the cell; (4) anaphase—the centromeres split and the chromosomes move to opposite poles; and (5) telophase—the nuclear envelope re-forms and chromosomes decondense. page 223

Mitosis is followed by cytokinesis, in which one cell divides into two. In animals, a contractile ring of actin pinches the cell in two. In plants, a new cell wall, called the cell plate, is synthesized between the daughter cells. page 225

11.3 Meiotic cell division is essential for sexual reproduction, the production of offspring that combine genetic material from two parents.

Sexual reproduction involves meiosis and fertilization, both of which are important in increasing genetic diversity. page 226

Meiotic cell division is a form of cell division that reduces the number of chromosomes by half to produce haploid gametes or spores that have one copy of each chromosome. page 226

Fertilization involves the fusion of haploid gametes to produce a diploid cell. page 226

Meiosis consists of two successive cell divisions: The first is reductional (the chromosome number is halved), and the second is equational (the chromosome number stays the same). Each division consists of prophase, prometaphase, metaphase, anaphase, and telophase. page 226

In meiosis I, homologous chromosomes pair and exchange genetic material at chiasmata, or regions of crossing over. In contrast to mitosis, centromeres do not split and sister chromatids do not separate. page 227

Genetic diversity is generated by crossing over and random alignment and subsequent segregation of maternal and paternal homologs on the metaphase plate in meiosis I. page 227

Meiosis II is similar to mitosis, in which chromosomes align on the metaphase plate, centromeres split, and sister chromatids separate from each other. page 228

The similarity of meiosis II and mitosis suggests that meiosis evolved from mitosis. page 231

The division of the cytoplasm differs between the sexes: Male meiotic cell division produces four functional sperm cells, whereas female meiotic cell division produces a single functional egg cell and three polar bodies. page 231

11.4 The cell cycle is regulated so that cell division occurs only at appropriate times and places.

Cells have regulatory mechanisms for promoting and preventing cell division. page 233

Levels of proteins called cyclins increase and decrease during the cell cycle. page 233

Cyclins form complexes with cyclin-dependent kinases (CDKs), activating the CDKs to phosphorylate target proteins involved in cell division. page 233

These complexes are activated by signals that promote cell division. page 234

Different cyclin–CDK complexes control progression through the cell cycle at key steps, including G_1/S phase, S phase, and M phase. page 235

The cell also has checkpoints that halt progression through the cell cycle if something is not right. page 235

The p53 protein is an example of a checkpoint protein; it prevents cell division in the presence of DNA damage. page 235

11.5 Cancer is uncontrolled cell division that results from mutations in genes that control cell division.

Cancer results when mechanisms that promote cell division are inappropriately activated or the normal checks on cell division are lost. page 236

Cancers can be caused by certain viruses carrying oncogenes that promote uncontrolled cell division. page 238

Viral oncogenes have cellular counterparts called proto-oncogenes that play normal roles in cell growth and division and that, when mutated, can cause cancer. page 238

Oncogenes and proto-oncogenes often encode proteins involved in signaling pathways that promote cell division. page 238

Tumor suppressors encode proteins, such as p53, that block cell division and inhibit cancers. page 238

Cancers usually result from several mutations in proto-oncogenes and tumor suppressor genes that have accumulated over time within the same cell. page 238

Self-Assessment

1. Compare and contrast the ways in which prokaryotic cells and eukaryotic cells divide.
2. Describe three situations in which mitotic cell division occurs.
3. Name the five steps of mitosis, and draw the structure and position of the chromosomes at each step.
4. Describe how chromosomes behave in meiosis. State when chromosomes are duplicated (forming sister chromatids) and when they are not duplicated.
5. Compare and contrast mitotic cell division and meiotic cell division in terms of number of products, number of cell divisions, and the processes unique to each.
6. Name two ways in which meiotic cell division creates genetic diversity, and explain how each occurs.
7. Explain how cytokinesis differs between animal and plant cells.
8. Describe the roles of cyclins and cyclin-dependent kinases in the cell cycle.
9. Give three examples of checkpoints that the cell monitors before proceeding through the cell cycle.
10. Describe what an oncogene, a proto-oncogene, and a tumor suppressor gene do.

Log in to LaunchPad to check your answers to the Self-Assessment questions, and to access additional learning tools.

CASE 3

You, from A to T

Your Personal Genome

Answers in the genetic sequence. Claudia Gilmore found that her genome contains a mutation that can lead to breast cancer. *Source: Photo by Dominic Gutierrez, courtesy Claudia Gilmore.*

As an American Studies major at Georgetown University, Claudia Gilmore had plenty of experience taking exams. But at the age of 21, she faced an altogether different kind of test—and no amount of studying could have prepared her for the result.

Gilmore had decided to be tested for a mutation in a gene known as *BRCA1*. Specific mutations in the *BRCA1* and *BRCA2* genes are associated with an increased risk of breast and ovarian cancers.

Gilmore's grandmother had battled breast cancer and later died from ovarian cancer. Before her death, she had tested positive for the *BRCA1* mutation. Her son, Gilmore's father, was tested and discovered he had inherited the mutation. After talking with a genetic counselor to help her understand the implications of the test, Gilmore gave a sample of her own blood and crossed her fingers.

Mutations that increase the risk of developing a particular disease are called genetic risk factors.

"I had a fifty-fifty chance of inheriting the mutation. I knew there was a great possibility it would be a part of my future," she says. "But I was 21, I was healthy. A part of me also thought this could never really happen to me."

Unfortunately, it could. Two weeks after her blood was drawn, she learned that she, too, carried the mutated gene.

Genetic testing is becoming increasingly common—in some cases, even routine. In 2003, after 13 years of painstaking work, scientists published the first draft of the complete human genome. The human genome consists of the DNA in one complete set of 23 chromosomes. It contains about 3 billion base pairs, the structural units of DNA. In the years that followed, much attention has been placed on understanding the genetic differences between individuals.

In reality, there isn't one single human genome. Everyone on Earth (with the exception of identical twins) has his or her own unique genetic sequence. Your personal genome, consisting of two sets of chromosomes, is the blueprint that codes for your hair color, the length of your nose, and your susceptibility to certain diseases. On average, any two human genomes are 99.9% identical, meaning that they differ at about 3 million base pairs.

Oftentimes, those differences have no impact on health. In some cases, however, a particular genetic signature is associated with disease. Sometimes a gene mutation makes a given illness inevitable. Certain mutations in a gene called *HTT*, for instance, always result in Huntington's disease, a degenerative brain condition that usually appears in middle age.

The link between mutations and disease isn't always so clear-cut, however. Most genetic diseases are complex in origin, and it may take multiple genetic mutations, as well as other non-genetic factors, for the disease to develop. Mutations that increase the risk of developing a particular disease are called genetic risk factors.

Certain mutations in the *BRCA1* and *BRCA2* genes are known genetic risk factors for breast and ovarian cancers, for instance. But not everyone with these mutations develops cancer. According to the National Cancer Institute, about 60% of women with a harmful *BRCA1* mutation will develop breast cancer in her lifetime, compared with about 12% of women in the general population. And 15% to 40% of women with a *BRCA1* mutation will be diagnosed with ovarian cancer, compared with just 1.4% of women without that genetic change.

The *BRCA1* mutations are just some of the thousands of harmful genetic changes that geneticists have identified so far. Other common mutations have been shown to elevate

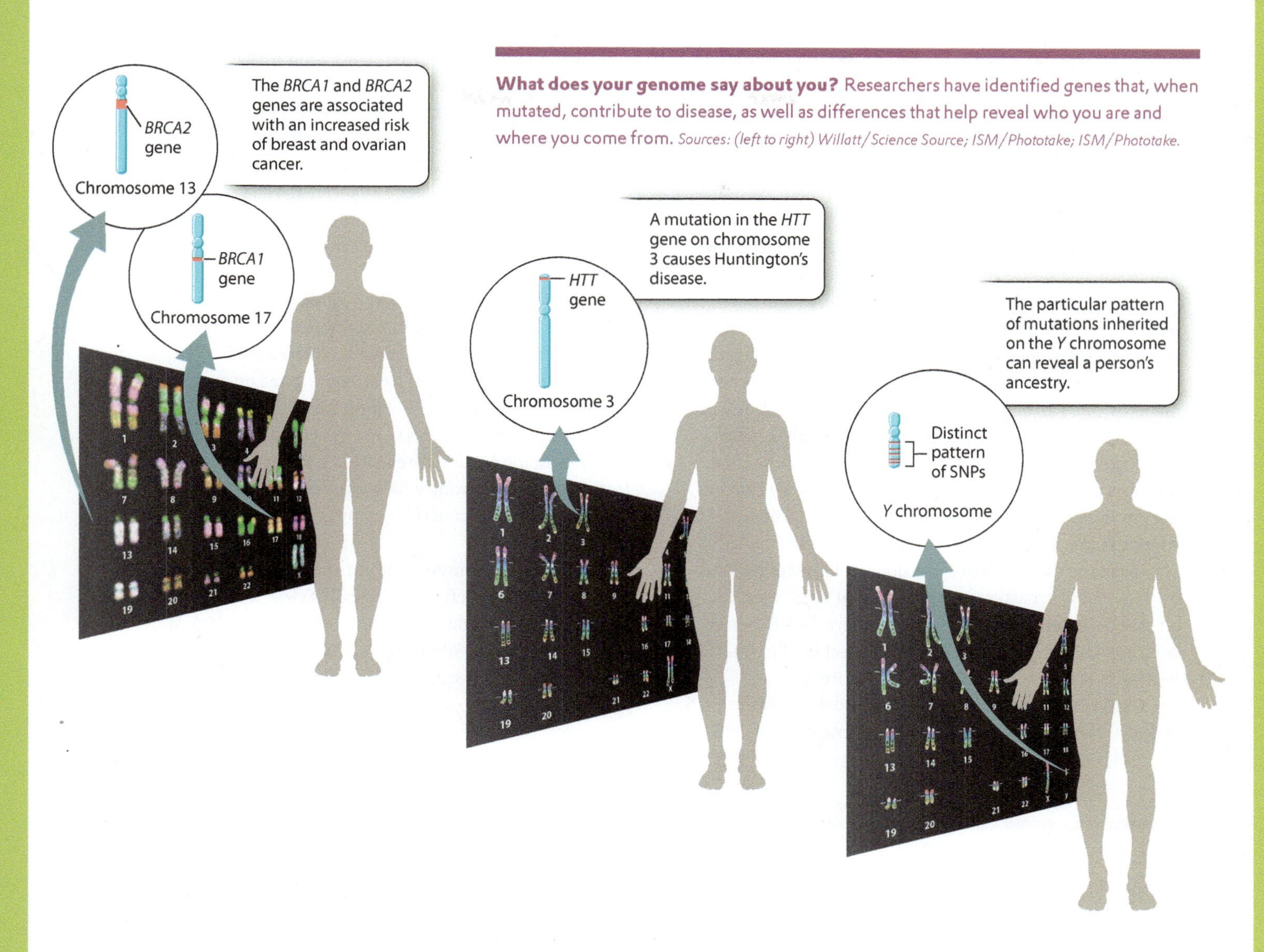

What does your genome say about you? Researchers have identified genes that, when mutated, contribute to disease, as well as differences that help reveal who you are and where you come from. *Sources: (left to right) Willatt/Science Source; ISM/Phototake; ISM/Phototake.*

one's risk of developing heart disease, diabetes, various cancers, and numerous other common illnesses. Often, these mutations involve changes to just a single base pair of DNA. These common single-letter mutations are known as single nucleotide polymorphisms, or SNPs.

Genome sequencing technology has improved significantly over the last decade, making it easier and less expensive to scan an individual's DNA for potentially harmful SNPs. A number of companies now offer genetic tests directly to the public. Unlike tests such as the *BRCA1* blood test that Gilmore was given, these direct-to-consumer (DTC) tests are offered to customers without any involvement of a medical professional, at a cost of a few hundred dollars.

Some of these tests aren't related to health at all. Your personal genome contains many unique features—from the shape of your fingernails to the shade of your skin—that don't affect your health but make you the person you are. Some DTC testing services aim to tell customers about their history by screening genes to identify variations that are more common in certain geographical regions or among members of certain ethnic groups. One such company offers genetic tests to African-Americans to determine in what part of the African continent their ancestors originated.

Other popular DTC tests inspect DNA samples for SNPs associated with certain diseases and physical traits—everything from Parkinson's disease and age-related macular degeneration to earwax type and propensity for baldness.

Advocates of the tests say the technology puts the power of genetic information in the hands of consumers. Critics, on the other hand, argue that the information provided by DTC tests isn't always very meaningful. The presence of some SNPs might raise the risk of a rare disease by just 2% or 3%, for example. In many cases, the precise link between mutation and disease is still being sorted out. Also, without input from a genetic counselor or medical professional, consumers may not know how to interpret the information revealed by the tests. The American Medical Association has recommended that a doctor always be involved when any genetic testing is performed.

Moreover, knowing your genetic risk factors isn't the whole story. When it comes to your health and well-being, the environment also plays a significant role. Someone might override a genetic predisposition for skin cancer by using sunscreen faithfully every day. On the other hand, a person might have a relatively low genetic risk for type 2 diabetes, but still boost the odds of developing the disease by eating a poor diet and getting little physical exercise.

Claudia Gilmore is especially careful to exercise regularly and eat a healthy diet. Still, she can only control her environment to a degree. She knew that, given her genetic status, her risk of breast cancer remained high. She made the extraordinary decision to eliminate that risk by undergoing a mastectomy at the age of 23. It wasn't an easy decision, she says, but she feels privileged to have been able to take proactive steps to protect her health. "I'll be a 'previvor' instead of a survivor," she says.

Two years later the same medical choice was made by the American actress, film director, screenwriter, and author Angelina Jolie, who recounted her experience in the *New York Times*, writing, "Once I knew that this [*BRCA1* mutation] was my reality, I decided to be proactive and to minimize the risk as much I could. I made a decision to have a preventive double mastectomy. I started with the breasts, as my risk of breast cancer is higher than my risk of ovarian cancer, and the surgery is more complex. . . . I am writing about it now because I hope that other women can benefit from my experience."[1]

For now, it's still too costly to sequence every individual's entire genome. But each year, many more genetic tests hit the market. Already, doctors are beginning to design medical treatments based on a patient's personal genome. People with a certain genetic profile, for example, are less likely than others to benefit from statins, medications prescribed to lower cholesterol. Doctors are also choosing which cancer drugs to prescribe based on the unique genetic signatures of patients and their tumors.

We've only just entered the era of personal genomics. While there's much left to decipher, it's clear that each of our individual genomes contains a wealth of biological information. And, as Gilmore says, "I've always been taught that knowledge is power."

[1]Angelina Jolie, "My Medical Choice," *New York Times,* A25, May 14, 2013.

? CASE 3 QUESTIONS

Special sections in Chapters 12–20 discuss the following questions related to Case 3.

1. **What new technologies are being developed to sequence your personal genome?** See page 264.
2. **Why sequence your personal genome?** See page 274.
3. **What can your personal genome tell you about your genetic risk factors?** See page 294.
4. **How can genetic risk factors be detected?** See page 317.
5. **How do genetic tests identify disease risk factors?** See page 342.
6. **How can the *Y* chromosome be used to trace ancestry?** See page 358.
7. **How can mitochondrial DNA be used to trace ancestry?** See page 360.
8. **Can personalized medicine lead to effective treatments of common diseases?** See page 375.
9. **How do lifestyle choices affect expression of your personal genome?** See page 386.
10. **Can cells with your personal genome be reprogrammed for new therapies?** See page 403.

CHAPTER 12

DNA Replication and Manipulation

Core Concepts

12.1 In DNA replication, a single parental molecule of DNA produces two daughter molecules.

12.2 The replication of linear chromosomal DNA requires mechanisms that ensure efficient and complete replication.

12.3 Techniques for manipulating DNA follow from the basics of DNA structure and replication.

12.4 Genetic engineering allows researchers to alter DNA sequences.

Dr. Gopal Murti/Science Source.

One of the overarching themes of biology discussed in Chapter 1 is that the functional unit of life is the cell, a theme that rests in turn on the fundamental concept that all cells come from preexisting cells. In Chapter 11, we considered the mechanics of cell division and how it is regulated. Prokaryotic cells multiply by binary fission, whereas eukaryotic cells multiply by mitosis and cytokinesis. These processes ensure that cellular reproduction results in daughter cells that are like the parental cell. In other words, in cell division, like begets like. The molecular basis for the resemblance between parental cells and daughter cells is that a double-stranded DNA molecule in the parental cell duplicates and gives rise to two double-stranded daughter DNA molecules that are identical to each other except for rare mutations that may have taken place.

The process of duplicating a DNA molecule is called **DNA replication,** and it occurs in virtually the same way in all organisms, reflecting its evolution very early in life's history. The process is conceptually simple—the parental strands separate and new complementary partner strands are made—but the molecular details are more complicated. Once scientists understood the molecular mechanisms of DNA replication, they could devise improved methods for manipulating and studying DNA. An understanding of DNA replication is therefore fundamental not only to understanding how cells and organisms produce offspring like themselves, but also to understanding some of the key experimental techniques in modern biology.

12.1 DNA REPLICATION

You may recall from Chapter 3 that double-stranded DNA consists of a pair of deoxyribonucleotide polymers wound around each other in antiparallel helical coils in such a way that a purine base (A or G) in one strand is paired with a pyrimidine base (T or C, respectively) in the other strand. To say that the strands are antiparallel means that they run in opposite directions: One strand runs in the 5′-to-3′ direction and the other runs in the 3′-to-5′ direction. These key elements of DNA structure are the only essential pieces of information needed to understand the mechanism of DNA replication.

During DNA replication, the parental strands separate and new partners are made.

When Watson and Crick published their paper describing the structure of DNA, they also coyly laid claim to another discovery: "It has not escaped our notice that the specific pairing we have postulated [A with T, and G with C] immediately suggests a copying mechanism for the genetic material." The copying mechanism they had in mind is exquisitely simple. The two strands of the parental duplex molecule separate (**Fig. 12.1**), and each individual parental strand serves as a model, or **template strand,** for the synthesis of a **daughter strand.** As each daughter strand is synthesized, the order of the bases in the template strand determines the order of the complementary bases added to the daughter strand. For example, the sequence 5′-ATGC-3′ in the template strand specifies the sequence 3′-TACG-5′ in the daughter strand because A pairs with T and G pairs with C. (The designations 3′ and 5′ convey the antiparallel orientation of the strands.)

A key prediction of the model shown in Fig. 12.1 is **semiconservative replication.** That is, after replication, each new DNA duplex consists of one strand that was originally part of the parental duplex and one newly synthesized strand. An alternative model is conservative replication, which proposes that the original DNA duplex remains intact and the daughter DNA duplex is completely new. Which model is correct? If there were a way to distinguish newly synthesized daughter DNA strands ("new strands") from previously synthesized parental strands ("old strands"), the products of replication could be observed and the mode of replication determined.

American molecular biologists Matthew S. Meselson and Franklin W. Stahl carried out an experiment to determine how DNA replicates. This experiment, described in **Fig. 12.2,** has been called "the most beautiful experiment in biology" because it so elegantly demonstrated the scientific method (Chapter 1) of hypothesis, prediction, and experimental test. They distinguished "old" from "new" DNA strands by labeling them with nonradioactive isotopes of nitrogen that have different densities: the normal form of nitrogen, ^{14}N, and a heavier form with an extra neutron, denoted ^{15}N.

Meselson and Stahl found that DNA in fact replicates semiconservatively (Fig. 12.2). This finding also predicted the results when cells are allowed to undergo two rounds of replication in a medium containing only light ^{14}N nitrogen. The heavy strand and light strand each serve as templates for a new light daughter strand. The result is that half of the DNA molecules will have one

FIG. 12.1 DNA replication. During DNA replication, each parental strand serves as a template for the synthesis of a complementary daughter strand.

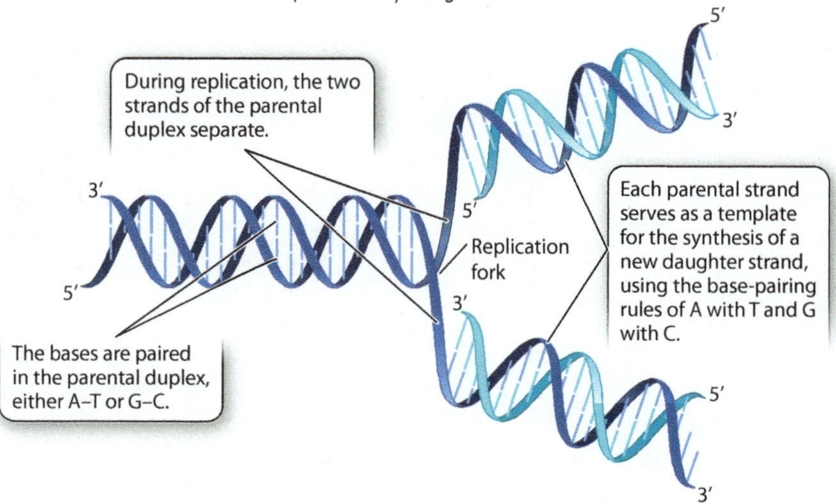

FIG.12.2

How is DNA replicated?

BACKGROUND Watson and Crick's discovery of the structure of DNA in 1953 suggested a mechanism by which DNA is replicated. Experimental evidence came from research by American molecular biologists Matthew Meselson and Franklin Stahl in 1958.

HYPOTHESIS DNA replicates in a semiconservative manner, meaning that each new DNA molecule consists of one parental strand and one newly synthesized strand.

ALTERNATIVE HYPOTHESIS DNA replicates in a conservative manner, meaning that one DNA molecule consists of two parental strands, and the other consists of two newly synthesized strands.

METHOD Meselson and Stahl distinguished parental strands ("old") from newly synthesized strands ("new") using two isotopes of nitrogen atoms. Old strands were labeled with a heavy form of nitrogen with an extra neutron (^{15}N), and new strands were labeled with the normal, lighter form of nitrogen (^{14}N).

EXPERIMENT The researchers first grew bacterial cells on medium containing only the heavy ^{15}N form of nitrogen. As the cells grew, ^{15}N was incorporated into the DNA bases, resulting, after several generations, in DNA containing only ^{15}N. They then transferred the cells into medium containing only light ^{14}N nitrogen. After one round of replication in this medium, cell replication was halted. The researchers could not observe the DNA directly, but instead they measured the density of the DNA by spinning it in a high-speed centrifuge in tubes containing a solution of cesium chloride.

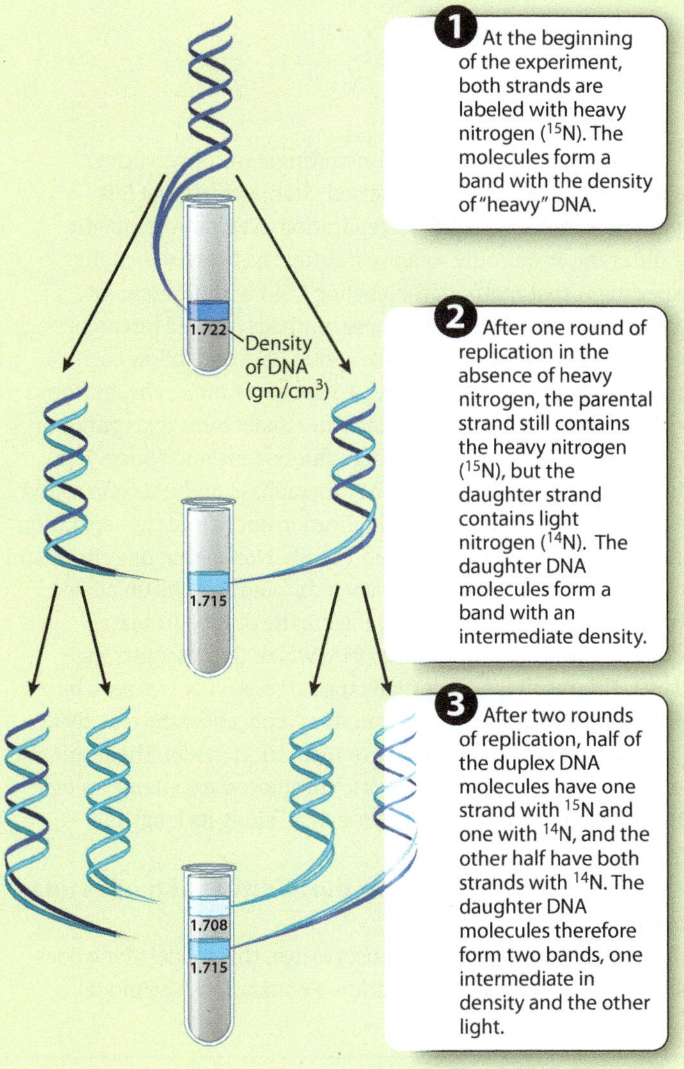

1.722 Density of DNA (gm/cm^3)

1 At the beginning of the experiment, both strands are labeled with heavy nitrogen (^{15}N). The molecules form a band with the density of "heavy" DNA.

1.715

2 After one round of replication in the absence of heavy nitrogen, the parental strand still contains the heavy nitrogen (^{15}N), but the daughter strand contains light nitrogen (^{14}N). The daughter DNA molecules form a band with an intermediate density.

1.708

1.715

3 After two rounds of replication, half of the duplex DNA molecules have one strand with ^{15}N and one with ^{14}N, and the other half have both strands with ^{14}N. The daughter DNA molecules therefore form two bands, one intermediate in density and the other light.

PREDICTION If DNA replicates conservatively, half of the DNA in the cells should be composed of two heavy parental strands containing ^{15}N, and half should be composed of two light daughter strands containing ^{14}N. If DNA replicates semiconservatively, the daughter DNA molecules should each consist of one heavy strand and one light strand.

RESULTS When the fully ^{15}N-labeled parental DNA was spun, it concentrated in a single heavy band with a density of 1.722 gm/cm^3. After one round of replication, the DNA formed a band at a density of 1.715 gm/cm^3, which is the density expected of a duplex molecule containing one heavy (^{15}N-labeled) strand and one light (^{14}N-labeled) strand. DNA composed only of ^{14}N would have a density of 1.708 gm/cm^3. After two rounds of replication, half the molecules exhibited a density of 1.715 gm/cm^3, indicating a duplex molecule with one heavy strand and one light strand, and the other half exhibited a density of 1.708 gm/cm^3, indicating a duplex molecule containing two light strands.

CONCLUSION DNA replicates semiconservatively, supporting the first hypothesis.

SOURCE Meselson, M., and F. W. Stahl. 1958. "The Replication of DNA in *Escherichia coli*." *PNAS* 44:671–82.

heavy old strand and one light new strand and an intermediate density, and half of the DNA molecules will have one light old strand and one light new strand and a low density. This is precisely what they observed (Fig. 12.2).

→ **Quick Check 1** Suppose Meselson and Stahl had done their experiment the other way around, starting with cells fully labeled with ^{14}N light DNA and then transferring them to medium containing only ^{15}N heavy DNA. What density of DNA molecule would you predict after one and two rounds of replication?

Important as it was in demonstrating semiconservative replication in bacteria, the Meselson–Stahl experiment left open the possibility that DNA replication in eukaryotes might be different. It was only some years after the Meselson–Stahl experiment that methods for labeling DNA with fluorescent nucleotides were developed. These methods allowed researchers to visualize entire strands of eukaryotic DNA and follow each strand through replication. **Fig. 12.3** shows a human chromosome with unlabeled DNA that subsequently underwent two rounds of replication in medium containing a fluorescent nucleotide. The chromosome was photographed at metaphase of the second round of mitosis, after chromosome duplication but before the separation of the chromatids into the daughter cells. Notice that one chromatid contains hybrid DNA with one labeled strand and one unlabeled strand, which fluoresces faintly (light); the other chromatid contains two strands of labeled DNA, which fluoresces strongly (dark). This result is conceptually the same as what was seen by Meselson and Stahl after two rounds of replication and exactly as predicted by the semiconservative replication model. The result also demonstrates that each eukaryotic chromosome contains a single DNA molecule that runs continuously all along its length.

New DNA strands grow by the addition of nucleotides to the 3′ end.

Although replication is semiconservative, this model alone does not tell us the details of replication. For example, the model implies that both daughter strands should grow in length by the addition of nucleotides near the site where the parental strands separate, a site called the **replication fork.** As more and more parental DNA is unwound and the replication fork moves forward (to the left in Fig. 12.1), both new strands would also grow in the direction of replication fork movement. But it turns out that this scenario is impossible.

We have seen that the two DNA strands in a double helix run in an antiparallel fashion: One of the template strands (the bottom one in Fig. 12.1) has a left-to-right 5′-to-3′ orientation, whereas the other template strand (the top one in Fig. 12.1) has a left-to-right 3′-to-5′ orientation. Therefore, the new daughter strands also have opposite orientations, so that near the replication fork the daughter strand in the bottom duplex terminates in a 3′ hydroxyl, whereas that in the top duplex terminates in a 5′ phosphate. There's the rub: The strand that terminates in the 5′ phosphate cannot grow in the direction of the replication fork because new DNA strands can grow only by the addition of successive nucleotides to the 3′end. That is, DNA always grows in the 5′-to-3′ direction.

DNA polymerization occurs only in the 5′-to-3′ direction because of the chemistry of nucleic acid synthesis, discussed in Chapter 3. The building blocks of DNA (Chapter 2) are nucleotides, each consisting of a deoxyribose sugar with three phosphate groups attached to the 5′ carbon, a nitrogenous base (A, T, G, or C) attached to the 1′ carbon, and a free hydroxyl (–OH) group attached to the 3′ carbon. DNA polymerization occurs when the

FIG. 12.3 Eukaryotic DNA replication. Further evidence that DNA replication is semiconservative came from observing the uptake of fluorescent nucleotides into chromosomal DNA. *Source: Daniel Hartl.*

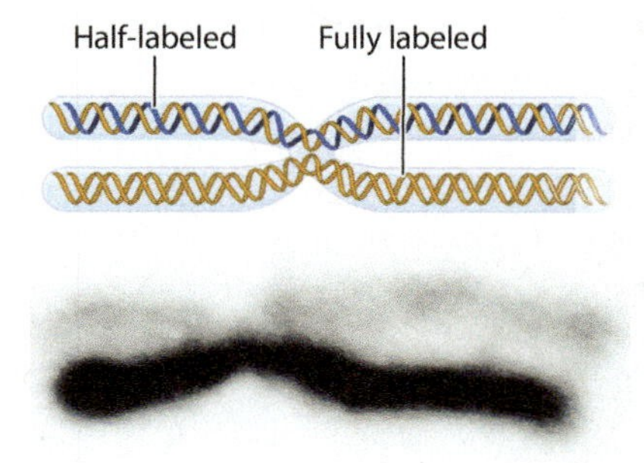

FIG. 12.4 DNA synthesis by nucleotide addition to the 3′ end of a growing DNA strand.

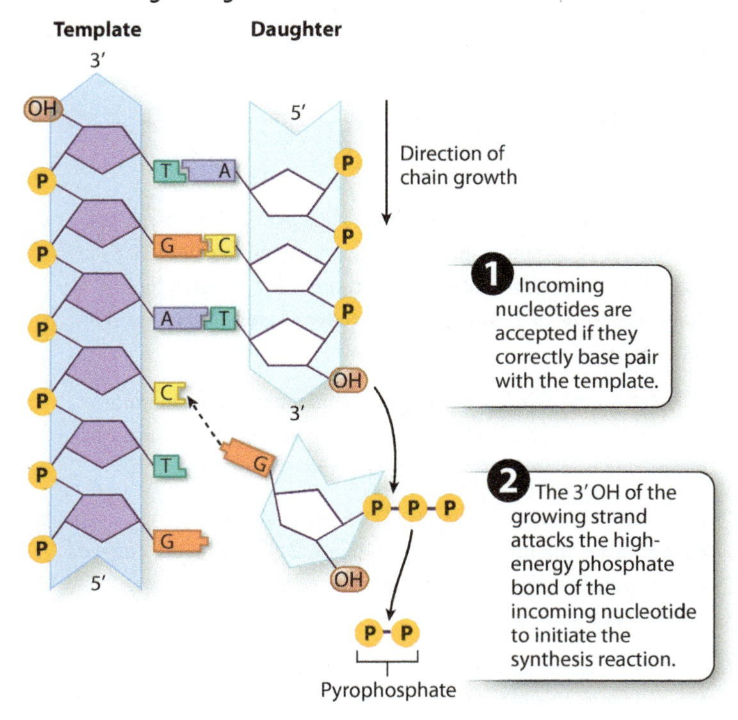

3′ hydroxyl at the growing end of the polynucleotide chain attacks the triphosphate group at the 5′ end of an incoming nucleotide (**Fig. 12.4**). Each of these incoming nucleotides is a nucleotide triphosphate (three phosphate groups attached to the 5′ carbon of the deoxyribose). As the incoming nucleotide triphosphate is added to the growing DNA strand, one of the nucleotide's high-energy phosphate bonds is broken to release the outermost two phosphates (called pyrophosphate), and immediately the high-energy phosphate bond in the pyrophosphate is cleaved to drive the polymerization reaction forward and make it irreversible.

The polymerization reaction is catalyzed by **DNA polymerase,** an enzyme that is a critical component of a large protein complex that carries out DNA replication. DNA polymerases exist in all organisms and are highly conserved, meaning that they vary little from one species to another because they carry out an essential function. A cell typically contains several different DNA polymerase enzymes, each specialized for a particular situation. But all DNA polymerases share the same basic function in that they synthesize a new DNA strand from an existing template. Most, but not all, also correct mistakes in replication, as we will see. DNA polymerases have many practical applications in the laboratory, which we will discuss later in this chapter.

In replicating DNA, one daughter strand is synthesized continuously and the other in a series of short pieces.

Because a new DNA strand can be elongated only at the 3′ end, the two daughter strands are synthesized in quite different ways (**Fig. 12.5**). The daughter strand shown at the bottom of Fig. 12.5 has its 3′ end pointed toward the replication fork, so that as the parental double helix unwinds, nucleotides can be added onto the 3′ end and this daughter strand can be synthesized as one long, continuous polymer. This daughter strand is called the **leading strand.**

FIG. 12.5 Leading and lagging strand synthesis. In DNA replication, because the template strand is made of two antiparallel strands, one daughter strand (the leading strand) is synthesized continuously and the other (the lagging strand) is synthesized in smaller pieces.

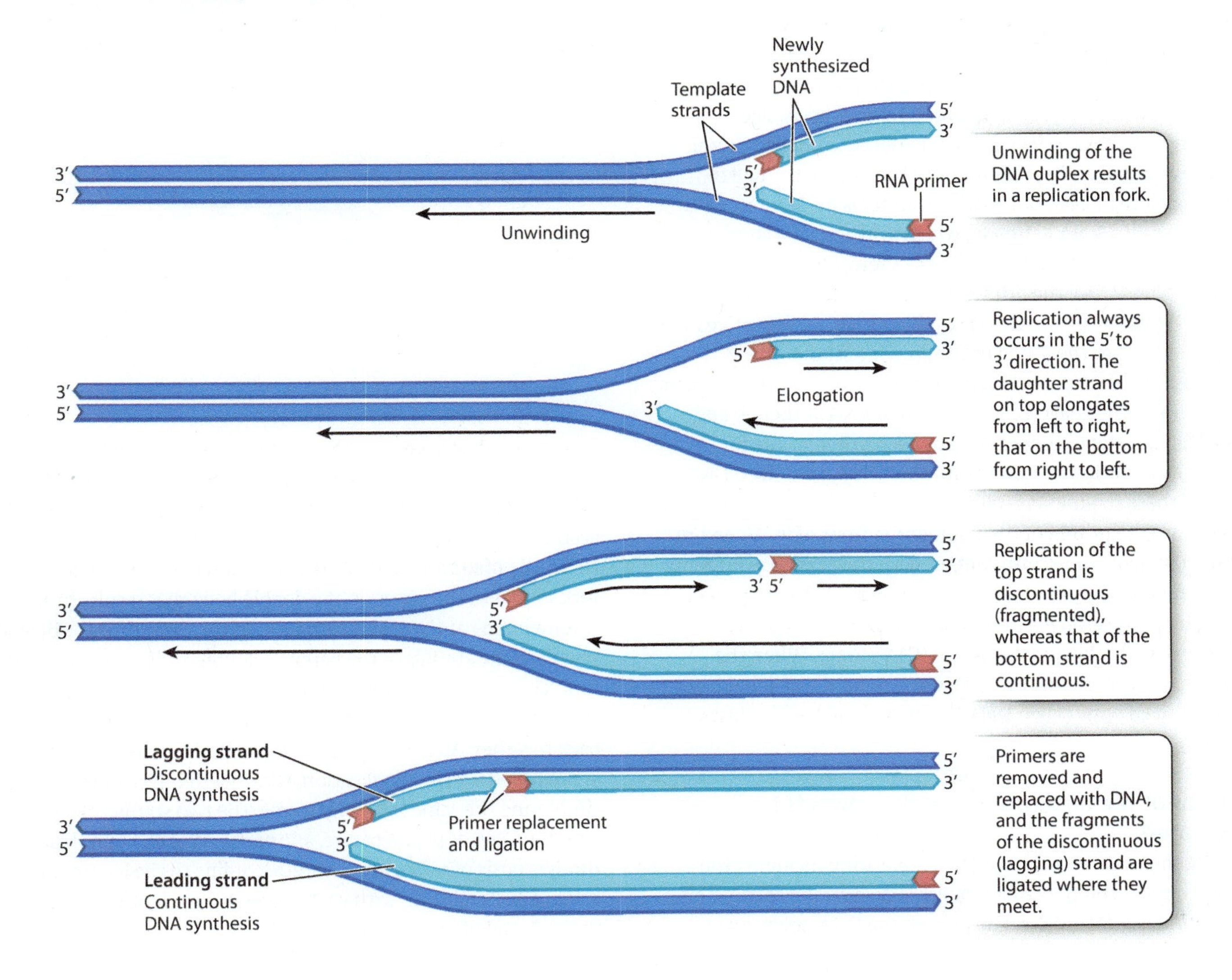

The situation is different for the daughter strand shown at the top in Fig. 12.5. Its 5′ end is pointed toward the replication fork, but the strand cannot grow in that direction. Instead, as the replication fork unwinds, it grows away from the fork and forms a stretch of single-stranded DNA of a few hundred to a few thousand nucleotides, depending on the species. Then, as the parental DNA duplex unwinds further, a new daughter strand is initiated with its 5′ end near the replication fork, and this strand is elongated at the 3′ end as usual. The result is that the daughter strand shown at the top in Fig. 12.5 is actually synthesized in short, discontinuous pieces. As the parental double helix unwinds, a new piece is initiated at intervals, and each new piece is elongated at its 3′ end until it reaches the piece in front of it. This daughter strand is called the **lagging strand.** The short pieces in the lagging strand are sometimes called **Okazaki fragments** after their discoverer, Japanese molecular biologist Reiji Okazaki.

The presence of leading and lagging strands during DNA replication is a consequence of the antiparallel nature of the two strands in a DNA double helix, and the fact that DNA polymerase can synthesize DNA in only one direction.

A small stretch of RNA is needed to begin synthesis of a new DNA strand.

Each new DNA strand must begin with a short stretch of RNA that serves as a **primer,** or starter, for DNA synthesis. The primer is needed because the DNA polymerase complex cannot begin a new strand on its own; it can only elongate the end of an existing piece of DNA or RNA. The primer is made by an RNA polymerase called **RNA primase,** which synthesizes a short piece of RNA complementary to the DNA template and does not require a primer. Once the primer has been synthesized, the DNA polymerase takes over and elongates the primer, adding successive DNA nucleotides to the 3′ end of the growing strand.

Because the DNA polymerase complex extends an RNA primer, all new DNA strands have a short stretch of RNA at their 5′ end. For the lagging strand, there are many such primers, one for each of the discontinuous fragments of newly synthesized DNA. As each of these fragments is elongated by DNA polymerase, it grows toward the primer of the fragment in front of it. When the growing fragment comes into contact with the primer, a different DNA polymerase complex takes over, removing the RNA primer and extending the growing fragment with DNA nucleotides to fill the space left by removal of the RNA primer. When the replacement is completed, the adjacent fragments are joined, or ligated, by an enzyme called **DNA ligase.** This process is illustrated in **Fig. 12.6.**

FIG. 12.6 Synthesis and removal of RNA primers in the lagging strand.

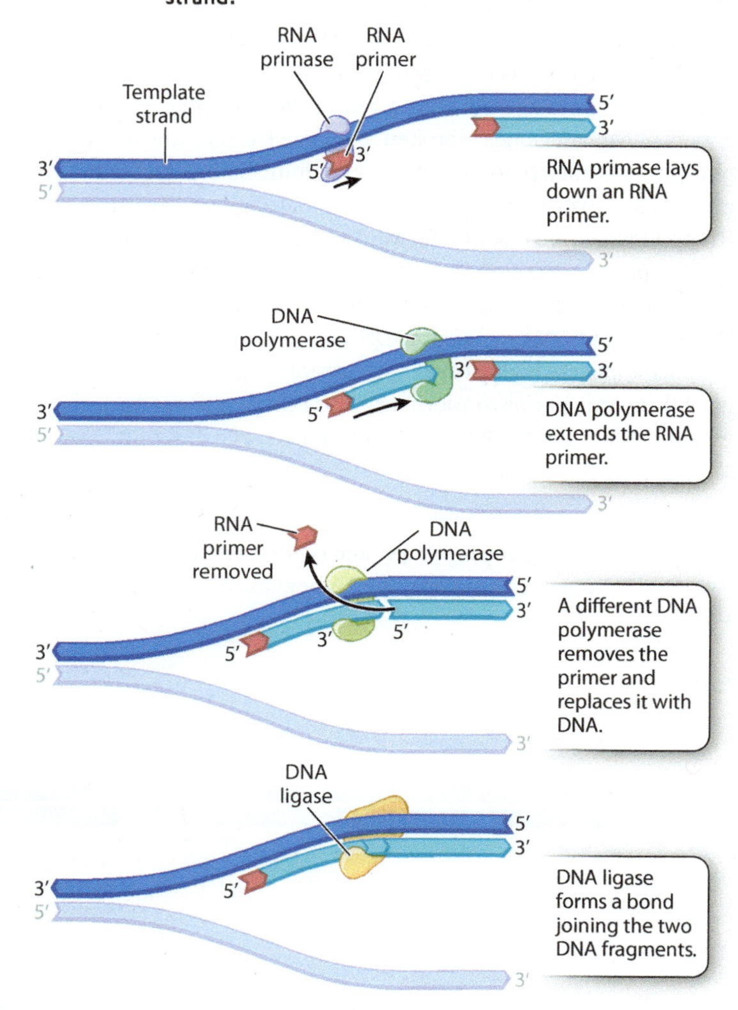

Synthesis of the leading and lagging strands is coordinated.

Fig. 12.7a shows an overview of DNA replication including the DNA polymerase complex that elongates each strand at the 3′ end, the RNA primase that makes the primer, the DNA polymerase complex that replaces the RNA primer with DNA, and the DNA ligase that joins the DNA fragments in the lagging strand. Many other proteins and enzymes work at the same time to ensure accurate and efficient synthesis of the daughter strands. One of these, **topoisomerase II,** works upstream from the replication fork to relieve the stress on the double helix that results from its unwinding at the replication fork. In the meantime, **helicase** separates the strands of the parental double helix at the replication fork. Then, **single-strand binding protein** binds to these single-stranded regions to prevent the template strands from coming back together. Although many evolutionarily conserved proteins are required for DNA replication, the underlying process is quite simple and the same in all organisms. The two strands of the parental DNA duplex separate, and each serves as a template for the synthesis of a daughter strand according to the base-pairing rules of A–T and G–C, with each successive nucleotide being added to the 3′ end of the growing strand.

FIG. 12.7 The replication fork. (a) Many proteins play different roles in replication, including separating and stabilizing the strands of the double helix. (b) In the cell, proteins involved in replication come together to form a complex that replicates both strands at the same pace.

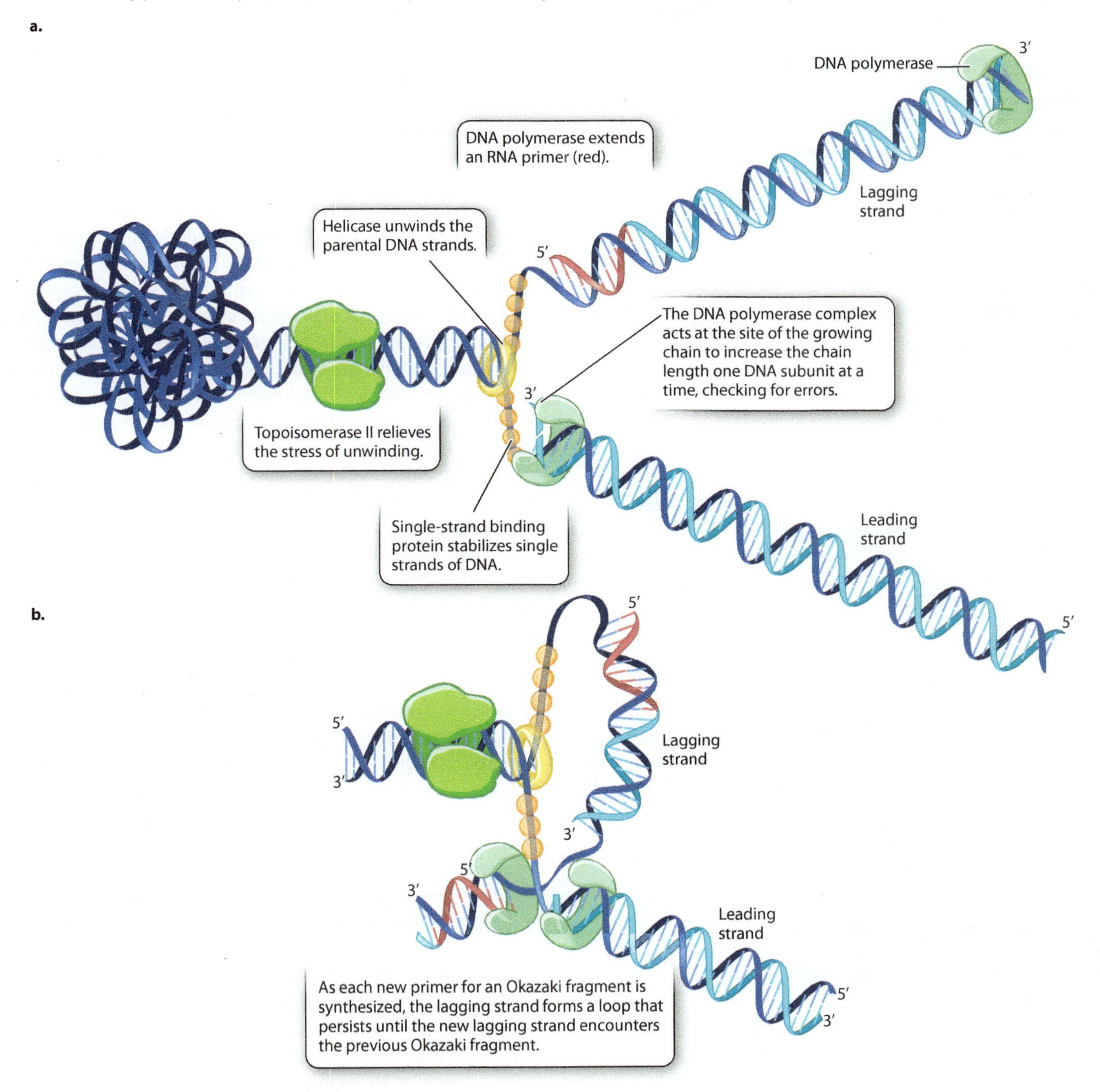

While Figure 12.7a makes it clear how synthesis of the leading strand and the lagging strand take place, it does not show how synthesis of these strands occurs at the same time and rate. To help ensure that both strands of the double helix are replicated at nearly the same rate, synthesis of the leading and lagging strands is coordinated as shown in **Fig. 12.7b.** The DNA polymerase complexes for each strand stay in contact with each other, with the lagging strand's polymerase releasing and retrieving the lagging strand for the synthesis of each new RNA primer. The positioning of the polymerases is such that both the leading strand and the lagging strand pass through in the same direction, which requires that the lagging strand be looped around as shown in Fig. 12.7b. In

FIG. 12.8 Proofreading, the process by which an incorrect nucleotide is removed immediately after it is incorporated.

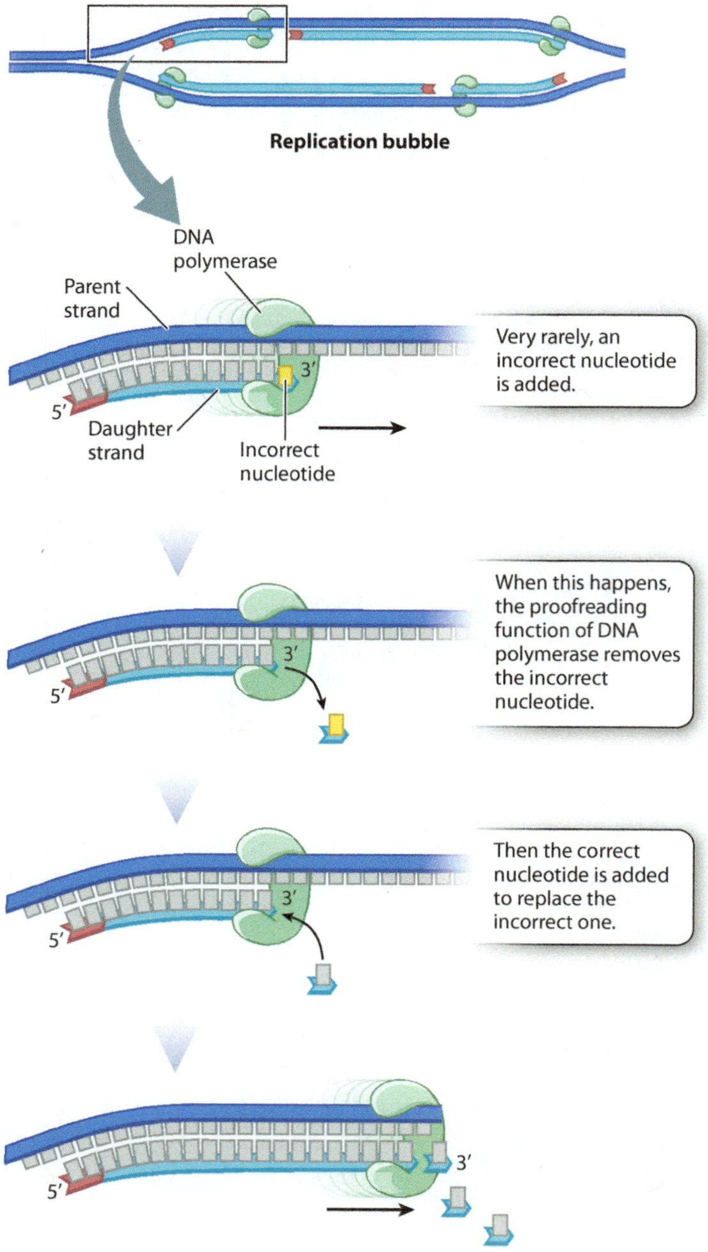

this configuration, the 3′ end of the lagging strand and the 3′ end of the leading strand are elongated together, which ensures that neither strand outpaces the other in its rate of synthesis.

As we will see in Chapter 14, when DNA damage occurs during replication, the rate of synthesis slows down so that the DNA can be repaired. If synthesis of one strand slows down to repair DNA damage, synthesis of the other strand slows down, too. The pairing of the replication complexes is disrupted when the RNA primer of the previous Okazaki fragment is encountered, and then takes place again when a new lagging-strand primer is formed. The lagging-strand loop is sometimes called a "trombone loop."

DNA polymerase is self-correcting because of its proofreading function.

Most DNA polymerases can correct their own errors in a process called **proofreading,** which is a separate enzymatic activity from strand elongation (synthesis) (**Fig. 12.8**).

When each new nucleotide comes into line in preparation for attachment to the growing DNA strand, the nucleotide is temporarily held in place by hydrogen bonds that form between the base in the new nucleotide and the base across the way in the template strand. The strand being synthesized and the template strand therefore have complementary bases—A paired with T, or G paired with C. However, on rare occasions, improper hydrogen bonds form, with the result that an incorrect nucleotide is attached to the new DNA strand. DNA polymerase can correct errors because it detects mispairing between the template and the most recently added nucleotide. Mispairing between a base in the parental strand and a newly added base in the daughter strand activates a DNA-cleavage function of DNA polymerase that removes the incorrect nucleotide and inserts the correct one in its place.

Mutations resulting from errors in nucleotide incorporation still occur, but proofreading reduces their number. In the bacterium *E. coli,* for example, about 99% of the incorrect nucleotides that are incorporated during replication are removed and repaired by the proofreading function of DNA polymerase. Those that slip past proofreading and other repair systems (Chapter 14) lead to mutations, which are then faithfully copied and passed on to daughter cells. Some of these mutations may be harmful, but others are neutral and a rare few may be beneficial. These mutations are the ultimate source of genetic variation that we see among individuals of the same species and among species, as we explore further in Chapter 15.

12.2 REPLICATION OF CHROMOSOMES

The steps involved in DNA replication are universal, suggesting that they evolved in the common ancestor of all living organisms. In addition, they are the same whether they occur in a test tube or in a cell, or whether the segment of DNA being replicated is short or long. However, the replication of an entire linear chromosome poses particular challenges. Here, we consider two such challenges encountered by cells in replicating their chromosomes: how replication starts and how it ends.

Replication of DNA in chromosomes starts at many places almost simultaneously.

DNA replication is relatively slow. In eukaryotes, it occurs at a rate of about 50 nucleotides per second. At this rate, replication from end to end of the DNA molecule in the largest human chromosome would take almost two months. In fact, it takes only a few hours. This fast pace is possible because in a long DNA molecule, replication begins almost simultaneously at many places. Each point at which DNA synthesis is initiated is called an **origin of replication.** The opening of the double helix at each origin of replication forms a **replication bubble** with a replication fork on each side, each with a leading strand and a lagging strand with topoisomerase II, helicase, and single-strand binding protein playing their respective roles (**Fig. 12.9**). DNA synthesis takes place at each replication fork, and as the replication forks move in opposite directions the replication bubble increases in size. When two replication bubbles meet, they fuse to form one larger replication bubble.

Note in Fig. 12.9 that within a single replication bubble, the same daughter strand is the leading strand at one replication fork and the lagging strand at the other replication fork. This situation results from the fact that the replication forks in each replication bubble move away from each other. When two replication bubbles fuse and the leading strand from one meets the lagging strand from the other, the ends of the strands that meet are joined by DNA ligase, just as happens when the discontinuous fragments within the lagging strand meet (see Fig. 12.6).

FIG. 12.9 Origins of replication. The replication of eukaryotic chromosomal DNA is initiated at many different places on the chromosome.

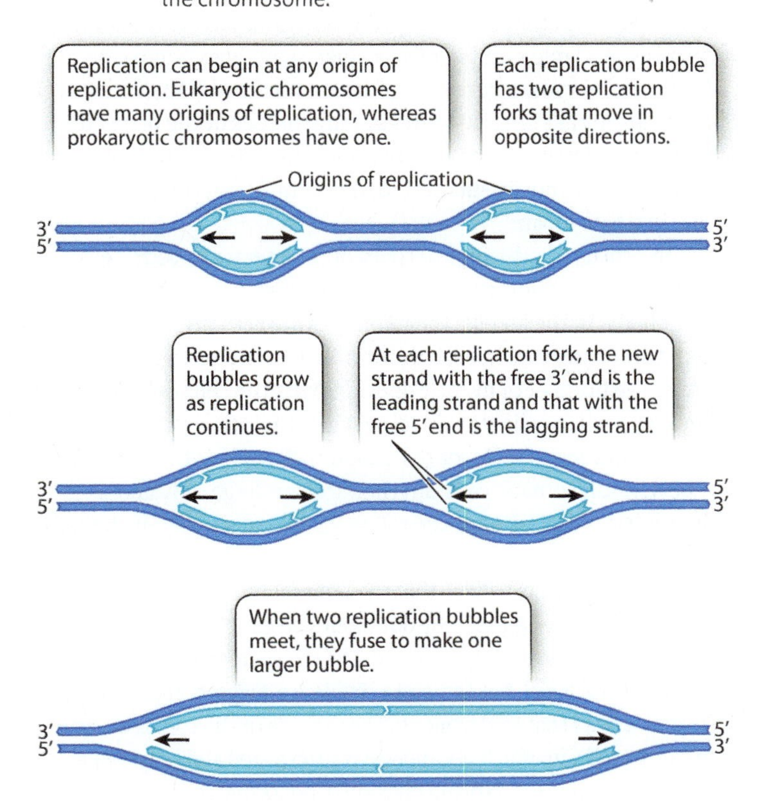

FIG. 12.10 Replication of a circular bacterial chromosome.

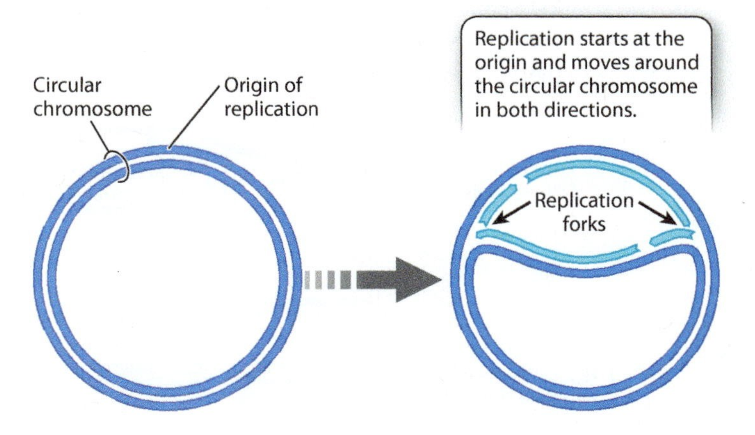

Some DNA molecules, including most of the DNA molecules in bacterial cells and the DNA in mitochondria and chloroplasts (Chapter 3) are small circles, not long linear molecules. Such circular DNA molecules typically have only one origin of replication (**Fig. 12.10**). Replication takes place at both replication forks, and the replication forks proceed in opposite directions around the circle until they meet and fuse on the opposite side, completing one round of replication.

Telomerase restores tips of linear chromosomes shortened during DNA replication.

A circular DNA molecule can be replicated completely because it has no ends, and the replication forks can move completely around the circle (Fig. 12.10). Linear DNA molecules have ends, however, and at each round of DNA replication the ends become slightly shorter. The reason for the shortening is illustrated in **Fig. 12.11.**

Recall that each fragment of newly synthesized DNA starts with an RNA primer. On the leading strand, the only primer required is at the origin of replication when synthesis begins. The leading strand is elongated in the same direction as the moving replication fork and is able to replicate the template strand all the way to the end. But on the lagging strand, which grows away from the replication fork, many primers are required, and the final RNA primer is synthesized about 100 nucleotides from the 3′ end of the template. Because this primer initiates synthesis of the final Okazaki fragment of the lagging strand, there is no other Okazaki fragment to synthesize the missing 100 base pairs and remove the primer. Therefore, when DNA replication is complete, the new daughter DNA strand (light blue in Fig. 12.11) will be missing about 100 base pairs from the tip (plus a few more owing to the length of the primer). When this daughter strand is itself replicated, the newly synthesized strand must terminate at the shortened end of the template strand, and so the new duplex molecule is shortened by about 100 base pairs from the original parental molecule. The

FIG. 12.11 Shortening of linear DNA at the ends. Because an RNA primer cannot be synthesized precisely at the 3′ end of a DNA strand, its partner strand becomes a little shorter in each round of replication.

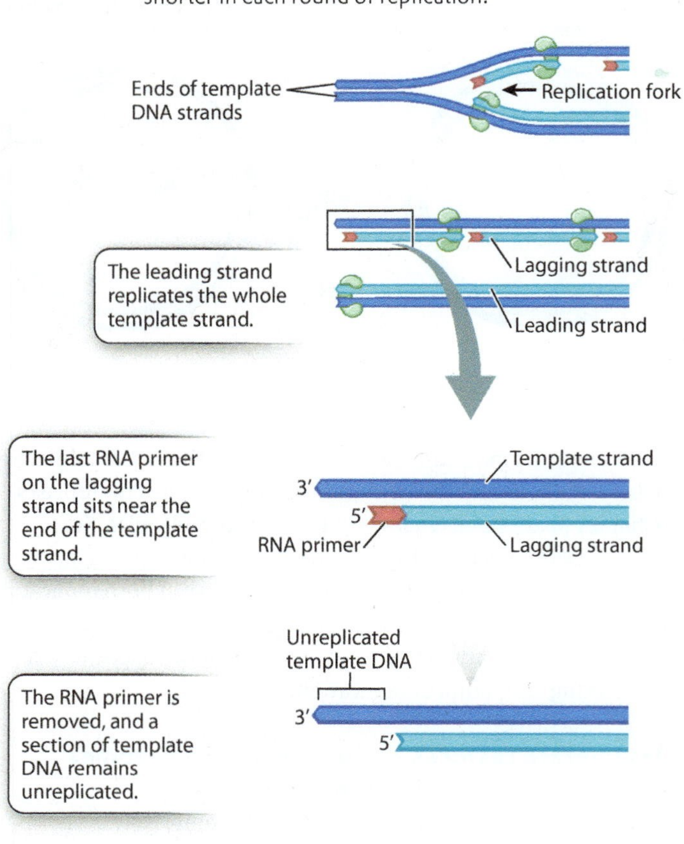

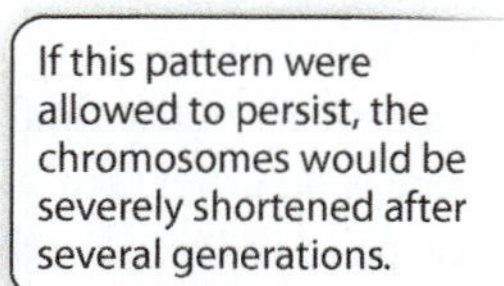

strand shortening in each round of DNA replication is a problem because without some mechanism to restore the tips, the DNA in the chromosome would eventually be nibbled away to nothing.

Fig. 12.12 illustrates the mechanism that eukaryotic organisms have evolved to solve the problem of shortened ends. Each end of a eukaryotic chromosome is capped by a repeating sequence called the **telomere.** The repeating sequence that constitutes the telomere differs from one group of organisms to the next, but in the chromosomes of humans and other vertebrate it consists of the sequence 3′-GGGATT-5′ repeated over and over again in about 1500–3000 copies. For simplicity, just four copies of this telomere sequence are shown in Fig. 12.12. The telomere is slightly shortened in each round of DNA replication, as shown in Fig. 12.11, but the shortened end is quickly restored by an enzyme known as **telomerase,** which contains an RNA molecule complementary to the telomere sequence.

As shown in Fig. 12.12, the telomerase replaces the lost telomere repeats using its RNA molecule as a guide to add successive DNA nucleotides to the 3′ end of the template strand. Once the 3′ end of the template strand has been elongated, an RNA primer and the complementary DNA strand are synthesized. In this way the original telomere is completely restored. Because there are no genes in the telomere, the slight shortening and subsequent restoration that take place have no harmful consequences.

Telomerase activity differs from one cell type to the next. It is fully active in **germ cells,** which produce sperm or eggs, and also in **stem cells,** which are undifferentiated cells that can undergo an unlimited number of mitotic divisions and can differentiate into any of a large number of specialized cell types. Stem cells are found in embryos, where they differentiate into all the various cell types. Stem cells are also found in some tissues of the body after embryonic development, where they replenish cells that have a high rate of turnover, such as blood and intestinal cells, and play a role in tissue repair.

In contrast to the high activity of telomerase in germ cells and stem cells, telomerase is almost inactive in most cells in the adult body. In these cells, the telomeres are actually shortened by about 100 base pairs in each mitotic division. Telomere shortening limits the number of mitotic divisions that the cells can undergo because human cells stop dividing when their chromosomes have telomeres with fewer than about 100 copies of the telomere repeat. Adult somatic cells can therefore undergo only about 50 mitotic divisions until the telomeres are so short that the cells stop dividing.

Many biologists believe that the limit on the number of cell divisions explains in part why our tissues become less youthful and wounds heal more slowly with age. The telomere hypothesis of aging is still controversial, but increasing evidence suggests that it is one of several factors that lead to aging. The flip side of the coin is observed in cancer cells, in which telomerase is reactivated and helps support the uncontrolled growth and division of abnormal cells.

FIG. 12.12 **Telomerase.** Telomerase prevents successive shortening of linear chromosomes.

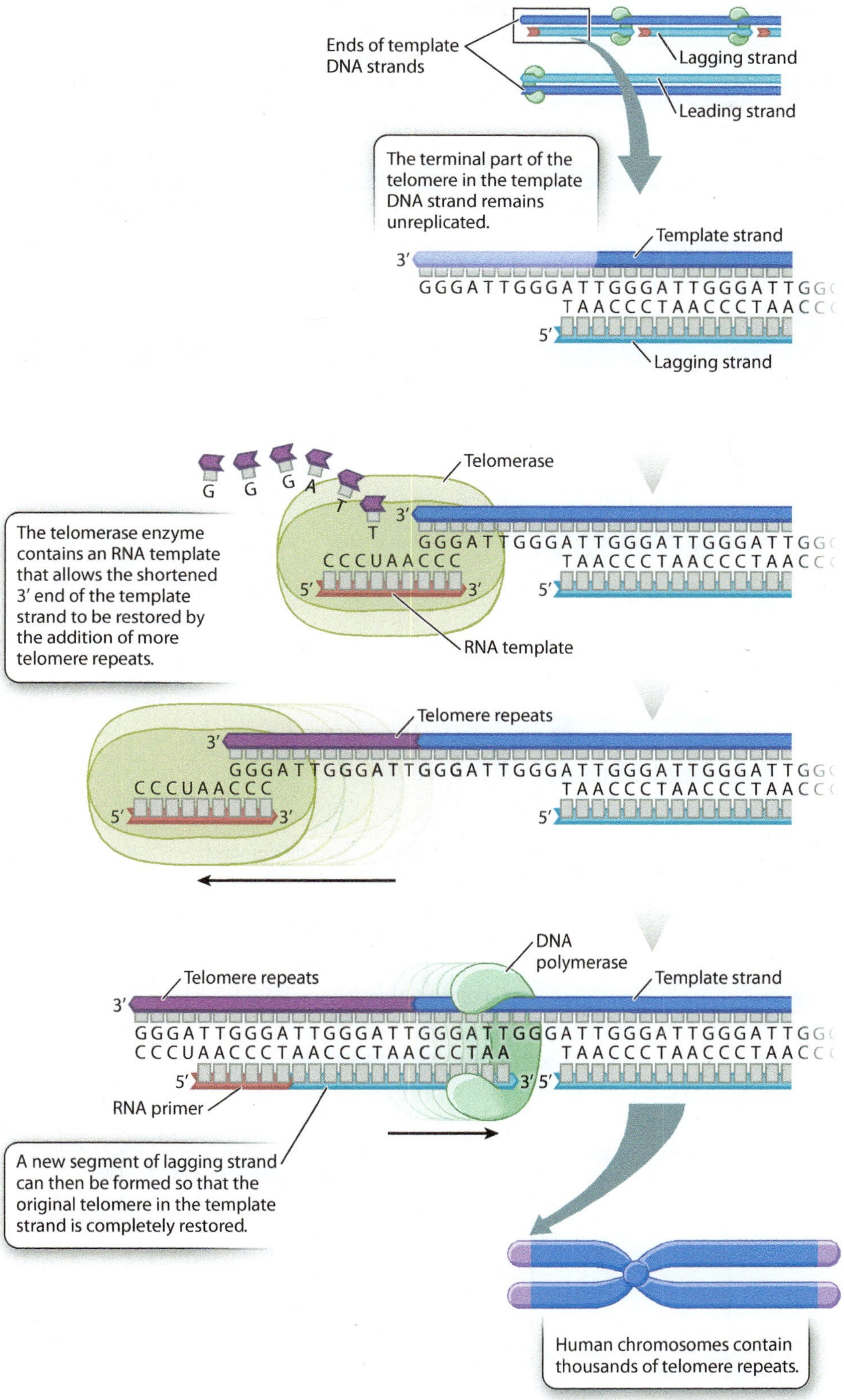

12.3 ISOLATION, IDENTIFICATION, AND SEQUENCING OF DNA FRAGMENTS

Watson and Crick's discovery of the structure of DNA and knowledge of the mechanism of replication allowed biologists not only to understand some of life's central processes, but also to create tools to study how life works. Biologists often need to isolate, identify, and determine the nucleotide sequence of particular DNA fragments. Such procedures can determine whether a genetic risk factor for diabetes has been inherited, blood at a crime scene matches that of a suspect, a variety of rice or wheat carries a genetic factor for insect resistance, or two species of organisms are closely related. Many of the experimental procedures for the isolation, identification, and sequencing of DNA are based on knowledge of DNA structure and physical properties of DNA. Others make use of the principles of DNA replication.

In this section, we discuss how particular fragments of double-stranded DNA can be produced, how DNA fragments of different sizes can be physically separated, and how the nucleotide sequence of a piece of DNA can be determined.

The polymerase chain reaction selectively amplifies regions of DNA.

DNA that exists in the nuclei of your cells is present in two copies per cell, except for DNA in the X and Y chromosomes in males, which are each present in only one copy per cell in males. In the laboratory, it is very difficult to manipulate or visualize a sample containing just one or two copies of a DNA molecule. Instead, researchers typically work with many identical copies of the DNA molecule they are interested in. A common method for making copies of a piece of DNA is the **polymerase chain reaction (PCR)**, which allows a targeted region of a DNA molecule to be replicated (or **amplified**) into as many copies as desired. PCR is both selective and highly sensitive, so it is used to

FIG. 12.13 The polymerase chain reaction (PCR). PCR results in amplification of the DNA sequence flanked by the two primers.

a. Each cycle of amplification includes three steps.

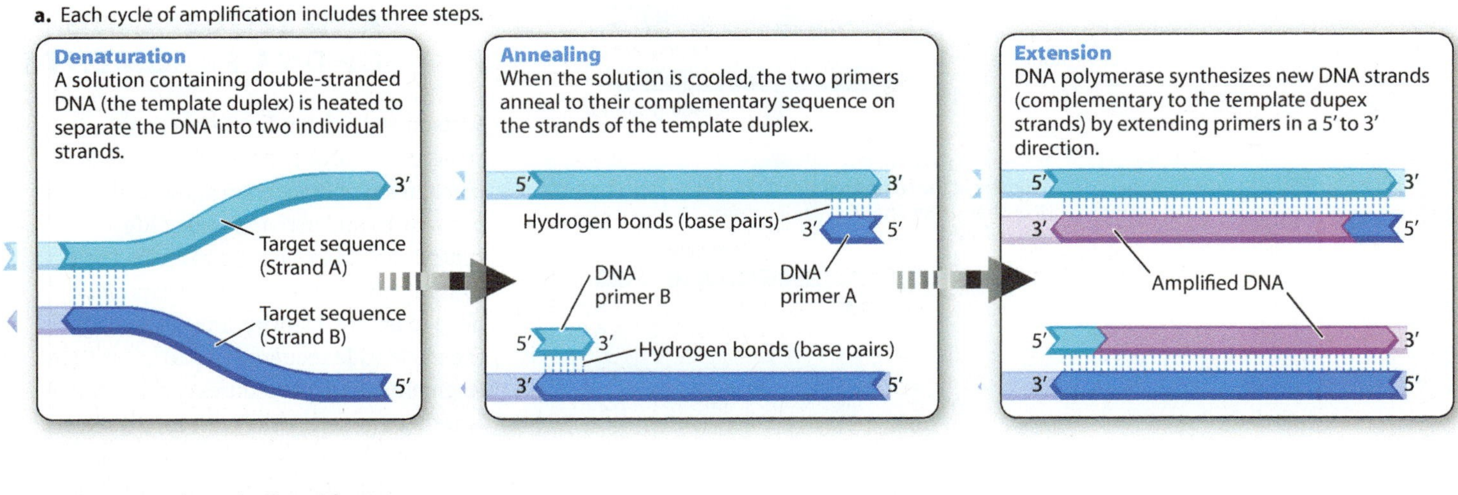

b. PCR repeats the cycle of amplification.

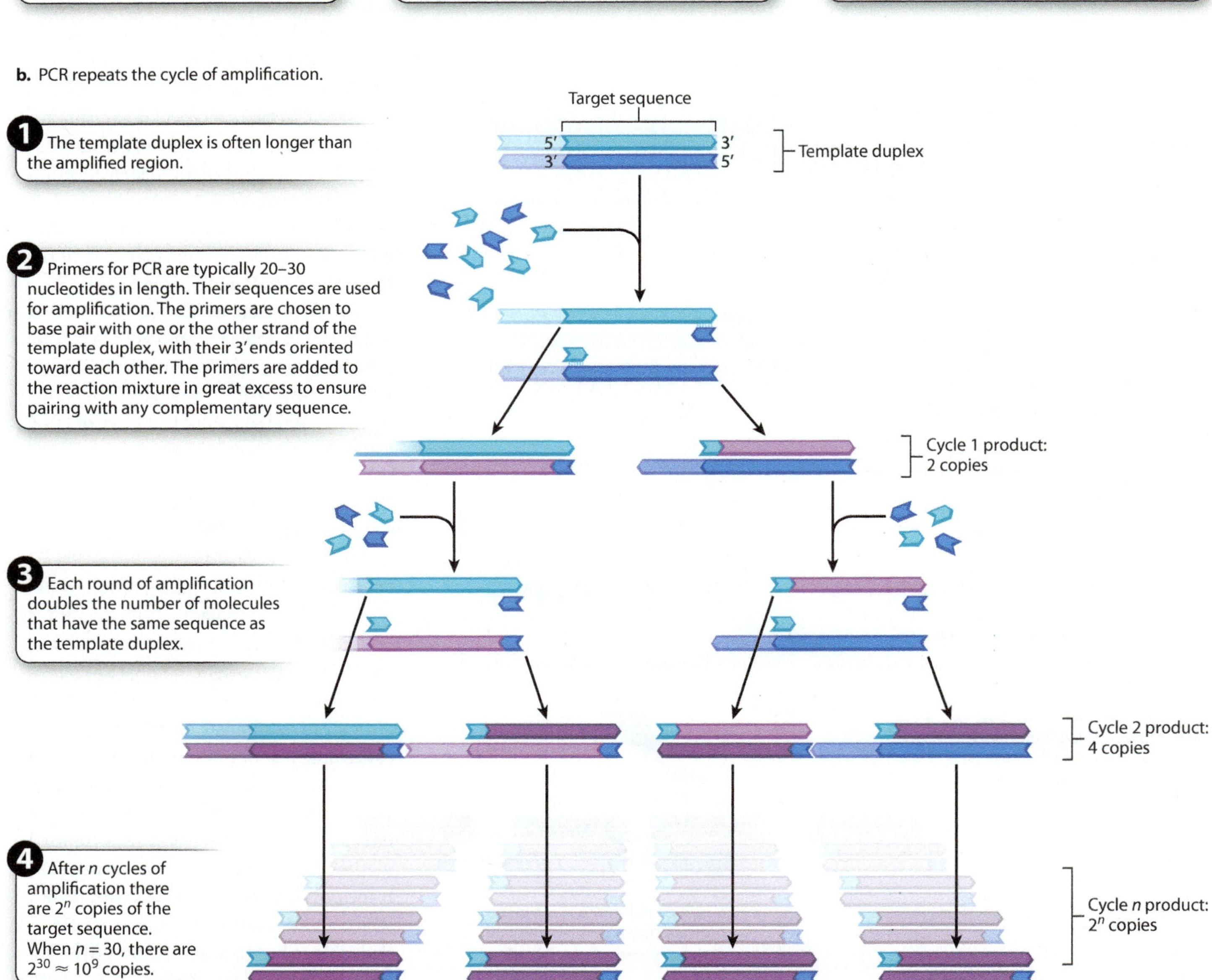

amplify and detect small quantities of nucleic acids, such as HIV in blood-bank supplies, or to study DNA samples as minuscule as those left by a smoker's lips on a cigarette butt dropped at the scene of a crime. The starting sample can be as small as a single molecule of DNA.

The principles of PCR are illustrated in **Fig. 12.13.** Because the PCR reaction is essentially a DNA synthesis reaction, it requires the same basic components used by the cell to replicate its DNA. In this case, the procedure takes place in a small plastic tube containing a solution that includes four essential components:

1. Template DNA. At least one molecule of double-stranded DNA containing the region to be amplified serves as the template for amplification.
2. DNA polymerase. The enzyme DNA polymerase is used to replicate the DNA.
3. All four deoxynucleoside triphosphates. Deoxynucleoside triphosphates with the bases A, T, G, or C are needed as building blocks for the synthesis of new DNA strands.
4. Two primers. Two short sequences of single-stranded DNA are required for the DNA polymerase to start synthesis. Enough primer is added so that the number of primer DNA molecules is much greater than the number of template DNA molecules.

The primer sequences are **oligonucleotides** (*oligos* is the Greek word for "few") produced by chemical synthesis and are typically 20–30 nucleotides long. Their base sequences are chosen to be complementary to the ends of the region of DNA to be amplified. In other words, the primers flank the specific region of DNA to be amplified. The 3′ end of each primer must be oriented toward the region to be amplified so that, when DNA polymerase extends the primer, it creates a new DNA strand complementary to the targeted region. Because the 3′ ends of the primers both point toward the targeted region, one of the primers pairs with one strand of the template DNA and the other pairs with the other strand of the template DNA.

PCR creates new DNA fragments in a cycle of three steps, as shown in Fig. 12.13a. The first step, **denaturation,** involves heating the solution in a plastic tube to a temperature just short of boiling so that the individual DNA strands of the template separate (or "denature") as a result of the breaking of hydrogen bonds between the complementary bases. The second step, **annealing,** begins as the solution is cooled. Because of the great excess of primer molecules, the two primers bind (or "anneal") to their complementary sequence on the DNA (rather than two strands of the template duplex coming back together). In the final step, **extension,** the solution is heated to the optimal temperature for DNA polymerase and the polymerase elongates (or "extends") each primer with the deoxynucleoside triphosphates.

After sufficient time to allow new DNA synthesis, the solution is heated again, and the cycle of denaturation, annealing, and extension is repeated over and over, as indicated in Fig. 12.13b, usually for 25–35 cycles. In each cycle, the number of copies of the targeted fragment is doubled. The first round of PCR amplifies the targeted region into 2 copies, the next into 4, the next into 8, then 16, 32, 64, 128, 256, 512, 1024, and so forth. By the third round of amplification, the process begins to produce molecules that are only as long as the region of the template duplex flanked by the sequences complementary to the primers. After several more rounds of replication the majority of molecules are of this type. The doubling in the number of amplified fragments in each cycle justifies the term "chain reaction."

Although PCR is elegant in its simplicity, the DNA polymerase enzymes from many species (including humans) irreversibly lose both structure and function at the high temperature required to separate the DNA strands. At each cycle, you would have to open the tube and add fresh DNA polymerase. This is possible, and in fact it was how PCR was done when the technique was first developed, but the procedure is time consuming and tedious. To solve this problem, we now use DNA polymerase enzymes that are heat-stable, such as those from the bacterial species *Thermus aquaticus,* which lives at the near-boiling point of water in natural hot springs, including those at Yellowstone National Park. This polymerase, called *Taq* polymerase, remains active at high temperatures. Once the reaction mixtures are set up, the entire procedure is carried out in a fully automated machine. The time of each cycle, temperatures, number of cycles, and other variables can all be programmed. The fact that DNA polymerase from a bacterium that lives in hot springs can be used to amplify DNA from any organism reminds us again of the conserved function and evolution early in the history of life of this enzyme.

Electrophoresis separates DNA fragments by size.

PCR amplification does not always work as you might expect. Sometimes the primers have the wrong sequence and so fail to anneal properly; sometimes they anneal to multiple sites and several different fragments are amplified. To determine whether or not PCR has yielded the expected product, a researcher must determine the size of the amplified DNA molecules. Usually, the researcher knows what the size of the correctly amplified fragment should be, making it possible to compare the expected size to the actual size.

One way to determine the actual size of a DNA fragment is by **gel electrophoresis (Fig. 12.14)**, a procedure in which DNA samples are inserted into slots or wells near the edge of a rectangular slab of porous material resembling solidified agar (the "gel"), which is composed of a tangle of polymers that make it difficult for large molecules to pass through. The gel is then inserted into an apparatus and immersed in a solution that allows an electric current to be passed through it (Fig. 12.14a). Since fragments of double-stranded DNA are negatively charged

FIG. 12.14 **Gel electrophoresis.** (a) A typical electrophoresis apparatus consists of a plastic tray, gel, and solution. (b) DNA bands can be visualized after staining with a dye that fluoresces under ultraviolet light. *Photo source: Guy Tear/Wellcome Images.*

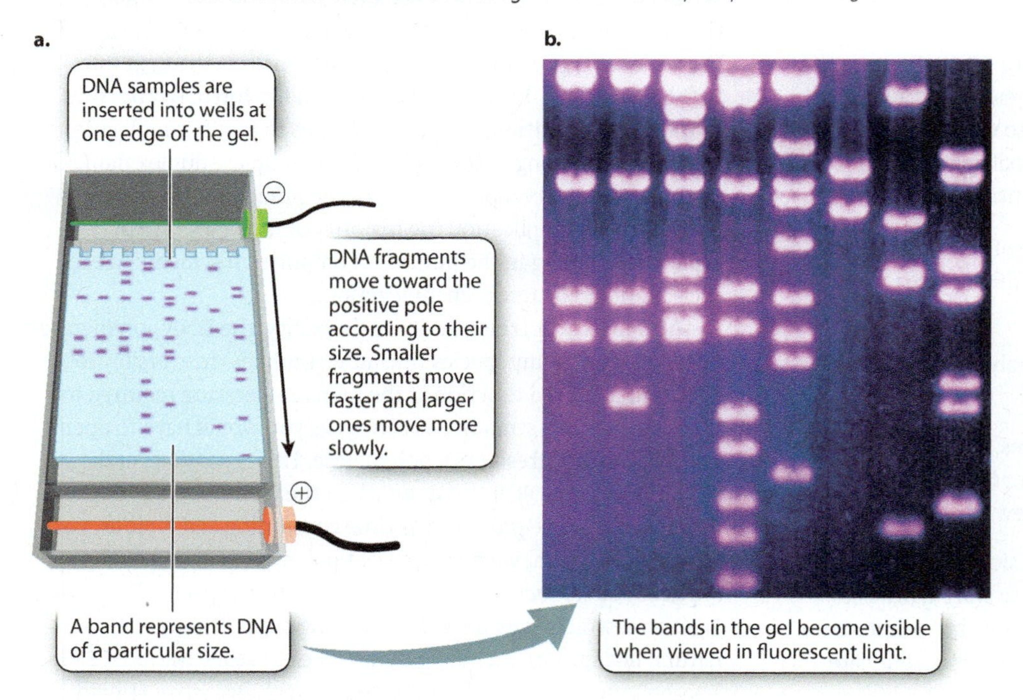

Quick Check 2 You do a PCR reaction, run a sample of the product on a gel, and use a dye to visualize the bands. You expect a single product of 300 bp, which you observe, but you also see a band of about 550 bp. Can you suggest a reason why? Does the size of the unexpected band indicate anything about your target sequence?

because of the ionized phosphate groups along the backbone, the molecules move toward the positive pole of the electric field.

The DNA molecules move according to their size. Short fragments pass through the pores of the gel more readily than large fragments, and so in a given interval of time short fragments move a greater distance in the gel than large fragments. The rate of migration is dependent only on size, not on sequence, and so all fragments of a given size move together at the same rate in a discrete band, which can be made visible by dyes that bind to DNA and fluoresce under ultraviolet light (Fig. 12.14b). A solution of DNA fragments of known sizes is usually placed in one of the wells, resulting in a series of bands, called a ladder, which can be used for size comparison.

Consider a PCR experiment with 25 base-pair (bp) primers flanking a 250-bp length of DNA. PCR amplifies this region, generating many millions of copies of a 300-bp DNA fragment consisting of the 250-bp target sequence and the primer sequences on both ends. To determine if the experiment worked as expected, a sample of the product is checked by gel electrophoresis. This sample is loaded into the well of the gel, a current is applied for a short period of time, and DNA fragments migrate to positions in the gel corresponding to their size, creating bands of DNA that are visualized with a dye. If a single band of 300 bp is seen on the gel, then the experiment worked as expected. Sometimes, however, the reaction might yield no bands, a single band of the incorrect size, or multiple bands. The researcher would then need to go back and investigate why the experiment did not work as expected.

Gel electrophoresis can separate DNA fragments produced by any means, not only PCR. Genomic DNA can be cut with certain enzymes and the resulting fragments separated by gel electrophoresis, as described in the next section. Similar procedures can also be used to separate protein molecules.

Restriction enzymes cleave DNA at particular short sequences.

In addition to amplifying segments of DNA, researchers often also make use of techniques that cut DNA at specific sites. Cutting DNA molecules allows pieces from the same or different organisms to be brought together in recombinant DNA technology, which is discussed in the next section. It also is a way to determine whether or not specific sequences are present in a segment of DNA, as techniques for cutting DNA depend on specific DNA sequences. Finally, cutting DNA allows whole genomes to be broken up into smaller pieces for further analysis, such as DNA sequencing.

The method for cutting DNA makes use of a class of enzyme that recognizes specific, short nucleotide sequences in double-stranded DNA and cleaves the DNA at these sites. The enzymes are known as **restriction enzymes,** of which about 1000 different kinds have been isolated from bacteria and other microorganisms. The recognition sequences the enzymes cleave, often called **restriction sites,** are typically four or six base pairs long. Most restriction enzymes cleave double-stranded DNA at or near the restriction site. For example, the enzyme *Eco*RI has the following restriction site:

$$\begin{array}{l} \;\downarrow \\ 5'\text{-GAATTC-}3' \\ 3'\text{-CTTAAG-}5' \\ \qquad\quad\;\uparrow \end{array}$$

Wherever the enzyme finds this site in a DNA molecule, it cleaves each strand exactly at the position indicated by the vertical arrows. Note that the *Eco*RI restriction site is symmetrical: Reading from the 5′ end to the 3′ end, the sequence of the top strand is exactly the same as the sequence of the bottom strand. This kind of symmetry is **palindromic** (it reads the same in both

TABLE 12.1 Examples of Restriction Enzymes with Six-Base Cleavage Sites

Enzyme	Site	Enzyme	Site
*Bam*HI	↓ 5′-GGATCC-3′ 3′-CCTAGG-5′ ↑	*Hind*III	↓ 5′-AAGCTT-3′ 3′-TTCGAA-5′ ↑
*Kpn*I	↓ 5′-GGTACC-3′ 3′-CCATGG-5′ ↑	*Sst*I	↓ 5′-GAGCTC-3′ 3′-CTCGAG-5′ ↑
*Hpa*I	↓ 5′-GTTAAC-3′ 3′-CAATTG-5′ ↑	*Sma*I	↓ 5′-CCCGGG-3′ 3′-GGGCCC-5′ ↑

directions) and is typical of restriction sites. Note also that the site of cleavage is not in the center of the recognition sequence. The cleaved double-stranded molecules therefore each terminate in a short single-stranded overhang. In this case the overhang is at the 5′ end, as shown below:

```
  ↓                        ↓
5′-G        -3′     +    5′- AATTC-3′
3′-CTTAA  -5′            3′-         G-5′
            ↑                        ↑
```

Table 12.1 shows more examples of restriction enzymes. Some cleave their restriction site to produce a 5′ overhang, others produce a 3′ overhang, and still others cleave in the middle of their restriction site and leave blunt ends with no overhang. The standard symbols for restriction enzymes include both italic and roman letters. The italic letters stand for the species from which the enzyme is derived (*E. coli* in the case of *Eco*RI, *Bacillus amyloliquefaciens* in the case of *Bam*HI), and the Roman letters designate the particular restriction enzyme isolated from that species.

When a particular restriction enzyme is used to break up a whole genome into smaller fragments, the specificity of restriction enzymes ensures that the DNA from each cell in an individual organism yields the same set of fragments and any particular DNA sequence present in the cells is contained in a fragment of the same size. If the genome is small enough that the number and sizes of the fragments are limited, the individual fragments can be visualized directly in a gel (**Fig. 12.15**). These fragments can then be extracted from the gel for further analysis or manipulation. For instance, the extracted fragments can be sequenced or ligated to other fragments.

But most genomes are too large to give such a simple picture. For example, the human genome is cleaved into more than a million different fragments by *Eco*RI, and these fragments appear as one big smear in a gel rather than as a series of discrete bands. In the next section, we discuss how individual bands containing a sequence of interest can be detected in such a gel.

DNA strands can be separated and brought back together again.

A researcher may wish to know whether a gene in one species is present in the DNA of a related species, but the nucleotide sequence of the gene is unknown. In such a case, techniques like PCR that require knowledge of the target DNA sequence cannot be used. Instead, the researcher can determine whether a DNA strand containing the gene in one species can base pair with the complement of a similar gene in a DNA strand from the other species. The base pairing of complementary single-stranded nucleic acids is known as **renaturation** or hybridization, and the process is the opposite of denaturation. In denaturation, the DNA strands in a duplex molecule are separated, and in renaturation, complementary strands come together again. Denaturation and renaturation are also opposites in regard to the experimental conditions in which they occur. When a solution containing DNA molecules is gradually heated to a sufficiently high temperature, the strands denature. When the solution is allowed to cool, the complementary strands renature.

Denatured DNA strands from one source can renature with DNA strands from a different source if their sequences are precisely or mostly complementary. Two very closely related sequences will have more perfectly matched bases and thus more hydrogen bonds holding them together than two sequences that are less closely related. The degree of base pairing between two

FIG. 12.15 *Eco*RI restriction fragments (a) on a circular DNA plasmid and (b) separated on a gel.

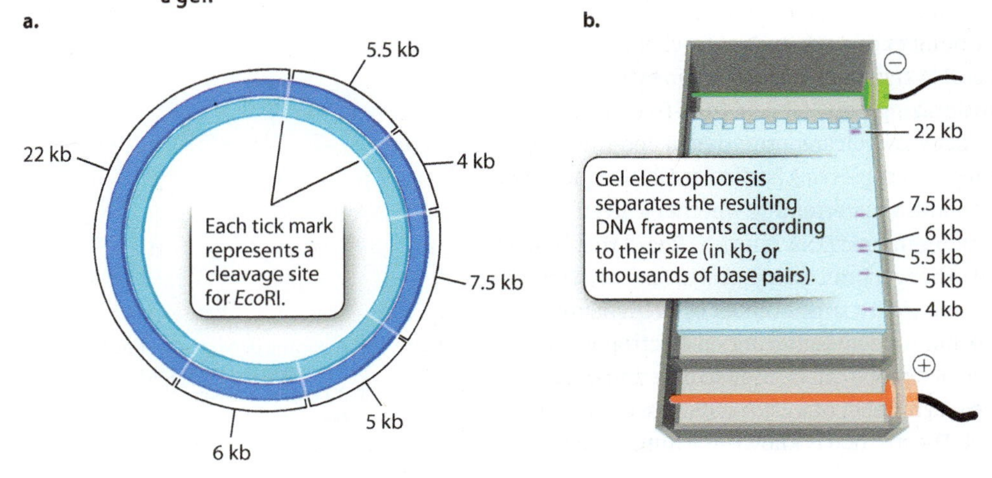

FIG. 12.16 A Southern blot. *Photo source: Reproduced with permission from Da'dara, A.A., Walter, R. D. 1998 Biochem J., 336(Pt 3): 545–550. © The Biochemical Society.*

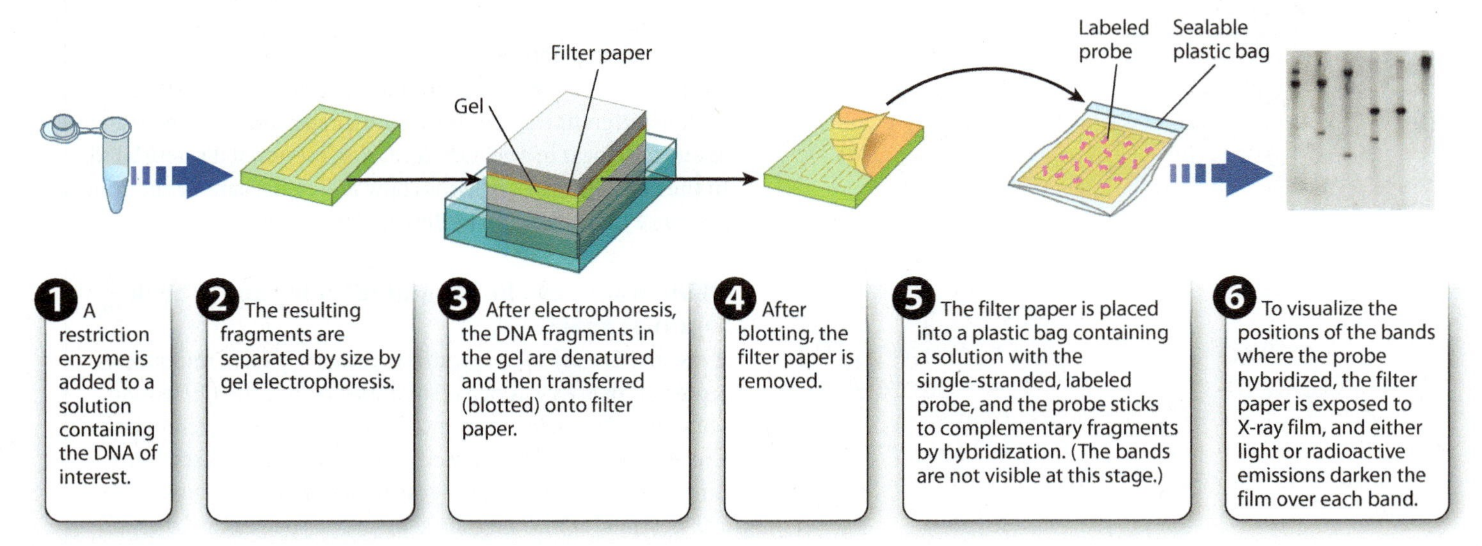

sequences affects the temperature at which they renature: Very closely related sequences renature at a higher temperature than less closely related sequences. Evolutionary biologists have used this principle to estimate the proportion of perfectly matched bases present in the DNA of different species, which is used as a measure of how closely those species are related. In general, the more closely two species are related, the more similar their DNA sequences. Knowledge of species relatedness is important in many applications, including conservation, identifying endangered species, and tracing the evolutionary history of organisms.

Renaturation makes it possible to use a small DNA fragment as a **probe.** This fragment is usually attached to a light-emitting or radioactive chemical that serves as a label. The probe can be used to determine whether or not a sample of double-stranded DNA molecules contains sequences that are complementary to it. Any DNA fragment can be used as a probe, and a probe can be obtained in any number of ways, such as by chemical isolation of a DNA fragment, amplification by PCR, or synthesizing a nucleic acid from free nucleotides.

Let's say you are interested in determining the number of copies of a particular DNA sequence in a genome or determining whether a given gene in human genomic DNA is intact. Recall that human and other genomic DNA cut with a restriction enzyme yield a large number of DNA fragments that look like a smear following gel electrophoresis. A labeled probe can be used to determine the size and number of a DNA sequence of interest in a gel. The method is known as a **Southern blot** (**Fig. 12.16**) after its inventor, the British molecular biologist Edwin M. Southern.

In a Southern blot, single-stranded DNA molecules that have resolved into bands on a gel are transferred to a filter paper, such that they are on the same spot on the paper as they were on the gel. The filter paper is washed with a solution containing

FIG. 12.17 (a) A deoxynucleotide and (b) a dideoxynucleotide. Incorporation of a dideoxynucleotide prevents strand elongation.

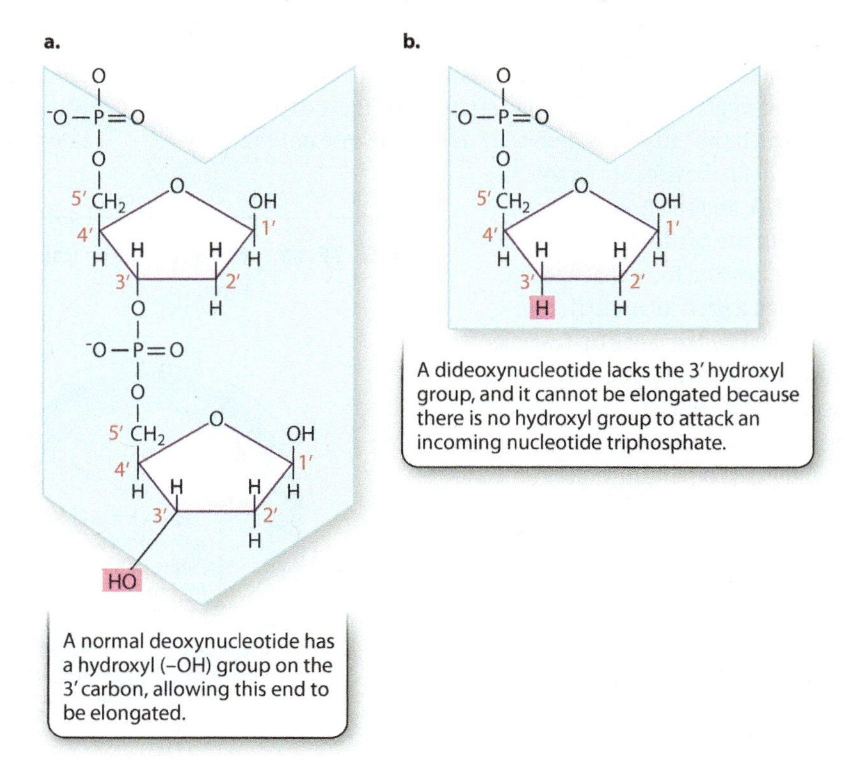

single-stranded probes that hybridize to whichever pieces of DNA on the filter paper are complementary. The paper is then exposed to X-ray film (in the case of radioactively labeled probes) so the bands can be visualized.

The presence of a band or bands on the film therefore indicates the number and size of DNA fragments that are complementary to the probe. This information in turn tells you the sizes and the number of copies of a particular DNA sequence present in the starting sample.

DNA sequencing makes use of the principles of DNA replication.

The ability to determine the nucleotide sequence of DNA molecules has given a tremendous boost to biological research. Techniques used to sequence DNA follow from our understanding of DNA replication. Consider a solution containing identical single-stranded molecules of DNA, each being used as the template for the synthesis of a complementary daughter strand originating at a short primer sequence. The problem is to determine the nucleotide sequence of the template strand. A brilliant answer to this problem was developed by the English geneticist Frederick Sanger, an achievement rewarded with a share in the Nobel Prize in Chemistry in 1980. (It was his second Nobel; he had also been honored with the award in 1958 for his discovery of a method for determining the sequence of amino acids in a polypeptide chain.)

Recall that a free 3′ hydroxyl group is essential for each step in elongation because that is where the incoming nucleotide is attached (**Fig. 12.17a**). Making use of this fact, Sanger synthesized **dideoxynucleotides,** in which the 3′ hydroxyl group on the sugar ring is absent (**Fig. 12.17b**). Whenever a dideoxynucleotide is incorporated into a growing daughter strand, there is no hydroxyl group to attack the incoming nucleotide, and strand growth is stopped dead in its tracks (**Fig. 12.18a**). For this reason, a dideoxynucleotide is known as a **chain terminator.** By including a small amount of each of the chain terminators in a reaction tube along with larger quantities of all four normal nucleotides, a DNA primer, a DNA template, and DNA polymerase, Sanger was able to produce a series of interrupted daughter strands, each terminating at the site at which a dideoxynucleotide was incorporated.

Fig. 12.18b shows how the interrupted daughter strands help us to determine the DNA sequence by the procedure now called **Sanger sequencing.** In a tube containing dideoxy-A and all the other elements required for many rounds of DNA replication, a strand of DNA is synthesized complementary to the template until, when it reaches a T in the template strand, it incorporates an A. Only a small fraction of the A nucleotides in the sequencing reaction are in the dideoxy form, so only a fraction of the daughter strands incorporate a dideoxy-A at that point, resulting in termination. The rest of the strands incorporate a normal deoxy-A and continue synthesis, although most of these will be stopped at some point farther along the line when a T is reached again. Similarly, in a reaction containing dideoxy-C, DNA fragments will

FIG. 12.18 Sanger sequencing. (a) The incorporation of an incoming dideoxynucleotide stops the elongation of a new strand. (b) Dideoxynucleotides terminate strands at different points in the template sequence. (c) Separation of interrupted daughter strands by size shows where each terminator was incorporated and hence the identity of the corresponding nucleotide in the template strand.

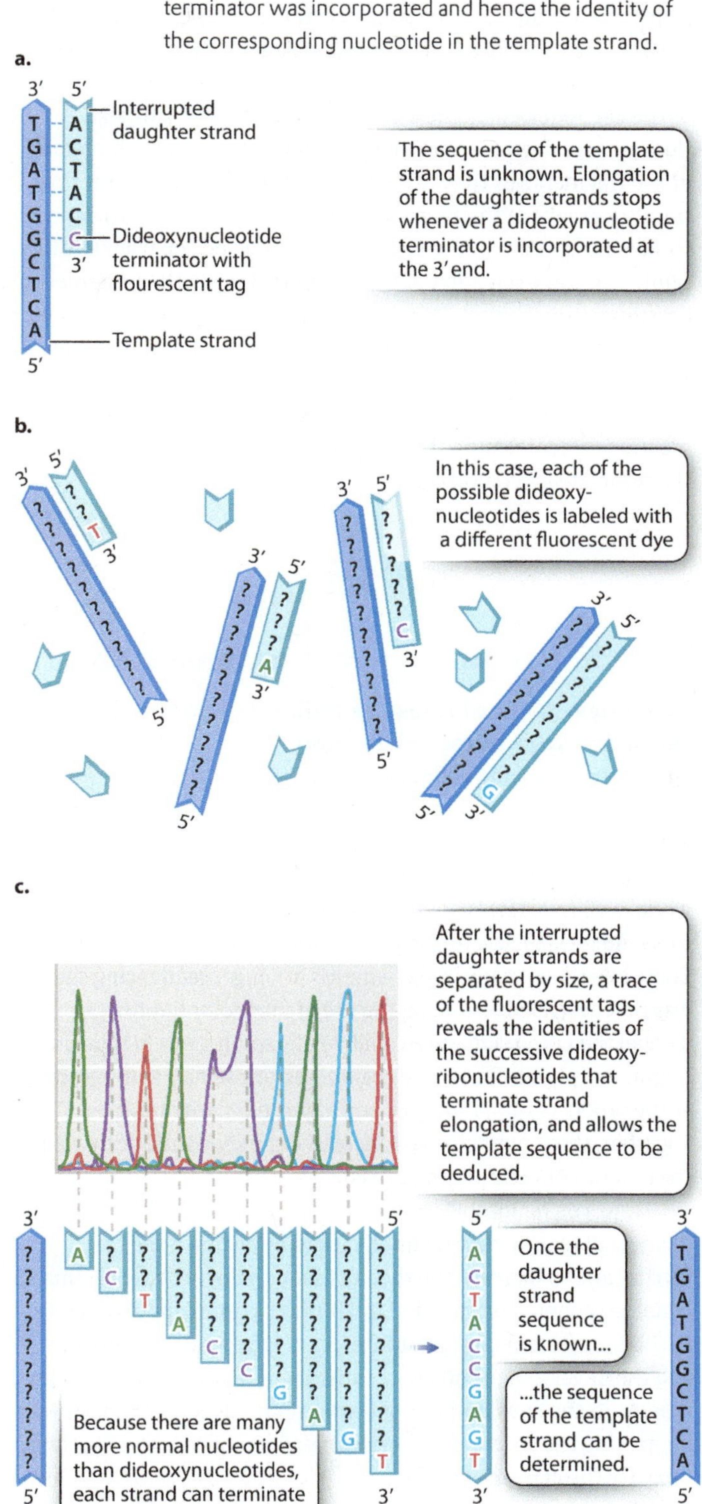

be produced whose sizes correspond to the positions of the Cs, and likewise for dideoxy-T and dideoxy-G.

Each of the four dideoxynucleotides is chemically labeled with a different fluorescent dye, as indicated by the different colors of A, C, T, and G in Fig. 12.18b, and so all four terminators can be present in a single reaction and still be distinguished. After DNA synthesis is complete, the daughter strands are separated by size with gel electrophoresis. The smallest daughter molecules migrate most quickly and therefore are the first to reach the bottom of the gel, followed by the others in order of increasing size. A fluorescence detector at the bottom of the gel "reads" the colors of the fragments as they exit the gel. What the scientist sees is a trace (or graph) of the fluorescence intensities, such as the one shown in **Fig. 12.18c.** The differently colored peaks, from left to right, represent the order of fluorescently tagged DNA fragments emerging from the gel. Thus, a trace showing peaks colored green-purple-red-green-purple-purple-blue-green-blue-red corresponds to a daughter strand having the sequence 5′-ACTACCGAGT-3′ (Fig. 12.18c). In the Sanger sequencing method, each sequencing reaction can determine the sequence of about 1000 nucleotides in the template DNA molecule.

→ **Quick Check 3** You have determined that the newly synthesized strand of DNA in your sequencing reaction has the sequence 5′-ACTACCGAGT-3′. What is the sequence of the template strand?

? CASE 3 YOU, FROM A TO T: YOUR PERSONAL GENOME

What new technologies are being developed to sequence your personal genome?

One of the high points of modern biology has been the determination of the complete nucleotide sequence of the DNA in a large number of species, including ours. The human genome and many others were sequenced by Sanger sequencing. This technique works well and is still the gold standard for accuracy, but it takes time and is expensive for large genomes like the human genome. There have been many improvements in Sanger sequencing since it was first described, including the use of four-color fluorescent dyes to label the DNA fragments, capillary electrophoresis to separate the fragments, photocells to read the fluorescent signals automatically as the products run off the gel, and highly efficient enzymes. Together, these methods have increased the speed and decreased the cost of DNA sequencing considerably.

However, being able to sequence everyone's genome, including yours, will require new technologies to bring down the cost further and to increase the speed of sequencing. The first human genome sequence, completed in 2003 at a cost of approximately $2.7 billion, stimulated great interest in large-scale sequencing and the development of devices that increased scale and decreased cost. As in the development of computer hardware, emphasis was on making the sequencing devices smaller while increasing their capacity through automation.

Modern methods of sequencing include the use of fluorescent nucleotides or light-detection devices that reveal the identity of each base as it is added to the growing end of a DNA strand. Another approach passes individual DNA molecules through a pore in a charged membrane; as each nucleotide passes through the pore it is identified by means of the tiny difference in charge that occurs. These miniaturized and automated devices carry out what is often called massively parallel sequencing, which can determine the sequence of hundreds of millions of base pairs in a few hours.

The goal of technology development was captured in the catchphrase "the $1000 genome," a largely symbolic target indicating a reduction in sequencing cost from about $1 million per megabase to less than $1 per megabase. For all practical purposes, the $1000-genome target has been achieved, but $1000 is still too costly to make genome sequencing a routine diagnostic procedure. The technologies are still evolving and the costs decreasing, and so it is a fair bet that in the coming years your own personal genome can be sequenced quickly and cheaply.

12.4 GENETIC ENGINEERING

Along with methods to manipulate DNA fragments came the capability of isolating genes from one species and introducing them into another. This type of genetic engineering is called **recombinant DNA** technology because it literally recombines DNA molecules from two (or more) different sources into a single molecule. Recombinant DNA technology involves cutting DNA by restriction enzymes, isolating them by gel electrophoresis, and ligating them with enzymes used in DNA replication. This technology is possible because the DNA of all organisms is the same, differing only in sequence but not in chemical or physical structure. When DNA fragments from different sources are combined into a single molecule and incorporated into a cell, they are replicated and transcribed just like any other DNA molecule.

Recombinant DNA technology can combine DNA from any two sources, including different species. DNA from one species of bacteria can be combined with another, or a human gene can be combined with bacterial DNA, or the DNA from a plant and a fungus can be combined into a single molecule. These new sequences may be unlike any found in nature, raising questions about their possible effects on human health and the environment.

This section discusses one of the basic methods for producing recombinant DNA, some of the important applications of recombinant DNA technology such as genetically modified organisms, and a new method that can be used to change the DNA sequence of any organism into any other desired sequence.

Recombinant DNA combines DNA molecules from two or more sources.

The first application of recombinant DNA technology was the introduction of foreign DNA fragments into the cells of bacteria in the early 1970s. The method is simple and straightforward, and remains one of the mainstays of modern molecular research. It can be used to generate a large quantity of a protein for study or therapeutic use.

The method requires a fragment of double-stranded DNA that serves as the donor. The donor fragment may be a protein-coding gene, a regulatory part of a gene, or any DNA segment of interest. If you are interested in generating bacteria that could produce human insulin, you might use the coding region of the human insulin gene as your donor DNA molecule. Also required is a **vector** sequence into which the donor fragment is to be inserted. The vector is the carrier of the donor fragment, and it must have the ability to be maintained in bacterial cells. A frequently used vector is a bacterial **plasmid,** a small circular molecule of DNA found naturally in certain bacteria that can replicate when the bacterial genomic DNA replicates and be transmitted to the daughter bacterial cells when the parental cell divides. Many naturally occurring plasmids have been modified by genetic engineering to make them suitable for use as vectors in recombinant DNA technology.

A common method for producing recombinant DNA is shown in **Fig. 12.19.** In order to make sure donor DNA can be fused with vector DNA, both pieces are cut with the same restriction enzyme so they both have the same overhangs. In the example shown in Fig. 12.19, the donor DNA is a fragment produced by digestion of genomic DNA with the restriction enzyme *Eco*RI, resulting in four-nucleotide 5′ overhangs at the fragment ends. The vector is a circular plasmid that contains a single *Eco*RI cleavage site, so digestion of the vector with *Eco*RI opens the circle with a single cut that also has four-nucleotide 5′ overhangs. Note that the overhangs on the donor fragment and the vector are complementary, allowing the ends of the donor fragment to renature with the ends of the opened vector when the two molecules are mixed. Once this renaturation has taken place, the ends of the donor fragment and the vector are covalently joined by DNA ligase. The joining of the donor DNA to the vector creates the recombinant DNA molecule.

The next step in the procedure is **transformation,** in which the recombinant DNA is mixed with bacteria that have been chemically coaxed into a physiological state in which they take up DNA from outside the cell. Having taken up the recombinant DNA, the bacterial cells are transferred into growth medium, where they multiply. Since the vector part of the recombinant DNA molecule contains all the DNA sequences needed for its replication and partition into the daughter cells, the recombinant DNA multiplies as the bacterial cell multiplies. If the recombinant DNA functions inside the bacterial cell, then new genetic characteristics may be expressed by the bacteria. For example, the recombinant DNA may allow the bacterial cells to produce a human protein, such as insulin or growth hormone.

→ **Quick Check 4** In making recombinant DNA molecules that combine restriction fragments from different organisms, researchers usually prefer restriction enzymes like *Bam*HI or *Hind*III that generate fragments with "sticky ends" (ends with overhangs) rather than enzymes like *Hpa*I or *Sma*I (Table 12.1) that generate fragments with "blunt ends" (ends without overhangs). Can you think of a reason for this preference?

FIG. 12.19 **Recombinant DNA.** Donor DNA fragments are joined with a vector molecule that can replicate in bacterial cells.

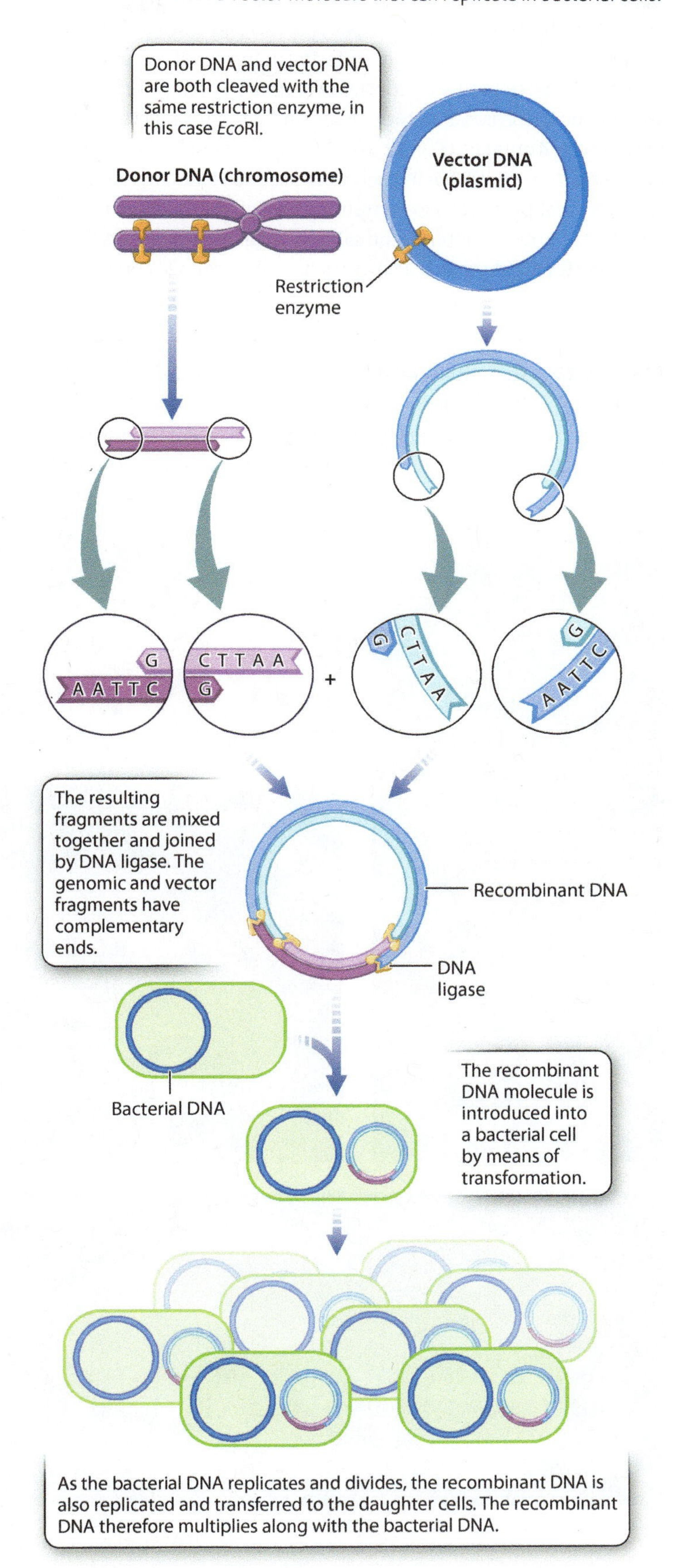

Recombinant DNA is the basis of genetically modified organisms.

Applications of recombinant DNA have gone far beyond genetically engineered bacteria. Using methods that are conceptually similar to those described for bacteria but differing in many details, scientists have been able to produce varieties of genetically engineered viruses and bacteria, laboratory organisms, agricultural crops, and domesticated animals (**Fig. 12.20**). Examples include sheep that produce a human protein in their milk used to treat emphysema, chickens that produce eggs containing human antibodies to help fight harmful bacteria, and salmon with increased growth hormone for rapid growth. Plants such as corn, canola, cotton, and many others have been engineered to resist insect pests, and other engineered products are rice with a high content of vitamin A, tomatoes with delayed fruit softening, potatoes with waxy starch, and sugarcane with increased sugar content. To model disease, researchers have used recombinant DNA to produce organisms such as laboratory mice that have been engineered to develop heart disease and diabetes. By studying these organisms, researchers can better understand human diseases and begin to find new treatments for them.

Genetically engineered organisms are known as **transgenic organisms** or **genetically modified organisms (GMOs).** Transgenic laboratory organisms are indispensable in the study

FIG. 12.20 Genetically modified organisms (GMOs). Organisms can be genetically modified so that (a) wheat resists weed-killing herbicides, (b) soybeans resist insects and have improved oil quality, (c) sheep produce more healthy fats, (d) chickens are unable to spread bird flu, (e) salmon have faster growth, and (f) pigs digest plant phosphorus more efficiently. *Sources: a. Adam Hart-Davis/Science Source; b. Steve Percival/Science Source; c. meirion matthias/Shutterstock; d. John Daniels/Ardea; e. AquaBounty Technologies; f. Carla Gottgens/Bloomberg via Getty Images.*

FIG. 12.21 **DNA editing.** In this example, CRISPR RNA and its associated protein are used to cleave target DNA, and double-stranded template DNA is used in DNA repair to alter its sequence.

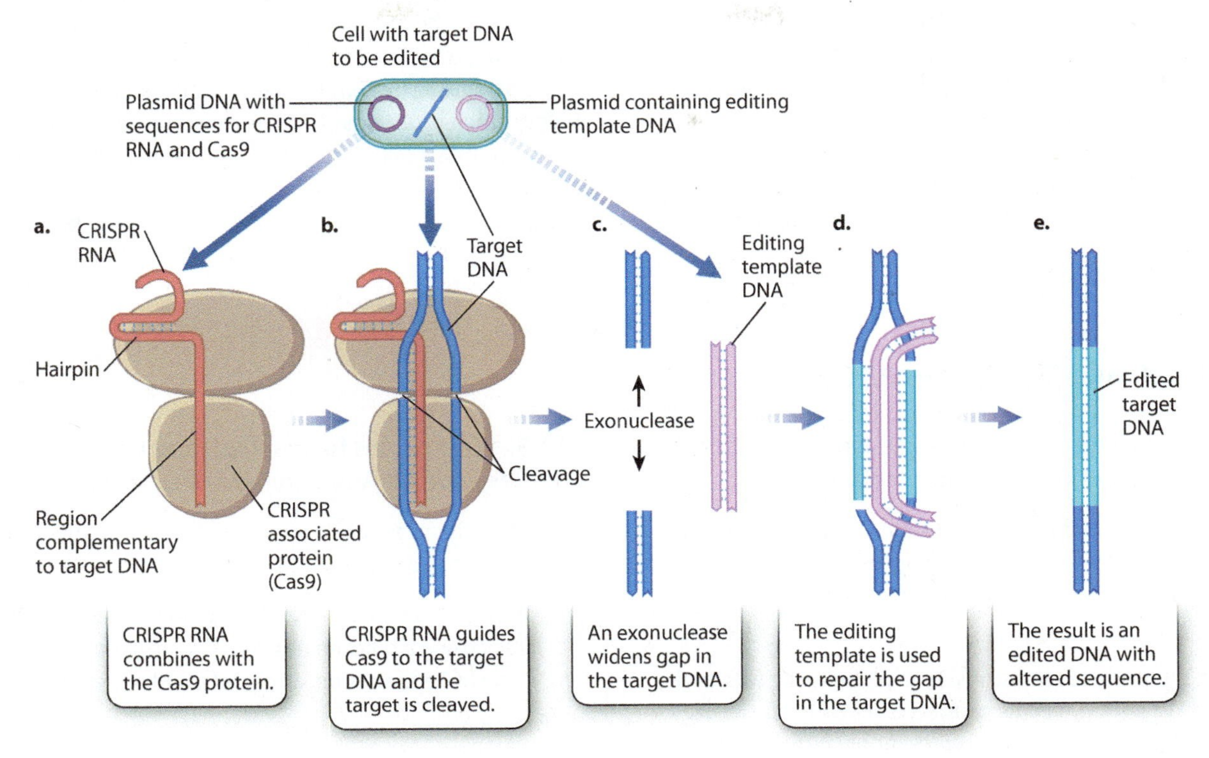

of gene function and regulation and to identify genetic risk factors for disease. In crop plants and domesticated animals, GMOs promise enhanced resistance to disease, faster growth and higher yields, more efficient utilization of fertilizer or nutrients, and improved taste and quality. But there are concerns about unexpected effects on human health or the environment, the increasing power and influence of agribusiness conglomerates, and ethical objections to tampering with the genetic makeup of animals and plants. Nevertheless, more than 250 million acres of GMO crops are grown annually in more than 20 countries. The majority of this acreage is in the United States and South America. Resistance to the use of GMOs in Europe remains strong and vocal.

DNA editing can be used to alter gene sequences almost at will.

Recombinant DNA technology combines existing DNA from two or more different sources. Its usefulness in research, medicine, and agriculture, however, is limited by the fact that it can make use only of existing DNA sequences. Therefore, scientists have also developed many different techniques to alter the nucleotide sequence of almost any gene in a deliberate, targeted fashion. In essence, these techniques allow researchers to "rewrite" the nucleotide sequence so that specific mutations can be introduced into genes to better understand their function, or mutant versions of genes can be corrected to restore normal function. Collectively, these techniques are known as **DNA editing.**

One of the newest and most exciting ways to edit DNA goes by the acronym **CRISPR** (clustered regularly interspaced short palindromic repeats), and it was discovered in an unexpected way. Researchers noted that about half of all species of bacteria and most species of Archaea contain small segments of DNA of about 20–50 base pairs derived from plasmids or viruses, but their function was at first a mystery. Later, it was discovered that they play a role in bacterial defense. When a bacterium is infected by a virus for the first time, it makes a copy of part of the viral genome and incorporates it into its genome. On subsequent infection by the same virus, the DNA copy of the viral genome is transcribed to RNA that combines with a protein that has a DNA-cleaving function. The RNA serves as a guide to identify target DNA in the virus by complementary base pairing, and the protein cleaves the target DNA. In this way, bacteria "remember" and defend themselves from past infections. The phrase "clustered regularly interspaced short palindromic repeats" describes the organization of the viral DNA segments in the bacterial genome.

In modern genetic engineering, the CRISPR mechanism is put to practical use to alter the nucleotide sequence of almost any gene in any kind of cell. One method is outlined in **Fig. 12.21.** The first step is to transform a cell with a plasmid containing sequences that code for a CRISPR RNA as well as the CRISPR-associated protein Cas9. The RNA contains a region that can form a hairpin-shaped structure, as well as a region engineered to have bases complementary to any DNA molecule in the cell to be altered, known as the target DNA (Fig. 12.21a). When the RNA undergoes base pairing with the target DNA, Cas9 cleaves the target DNA (Fig. 12.21b). Exonucleases in the cell then expand the gap (Fig. 12.21c). The gap can be repaired using another DNA molecule that serves as a template for editing the target DNA (Fig. 12.21d). This editing template DNA is introduced to the cell by a plasmid and contains a sequence of interest to replace the degraded sequence of the target DNA, flanked by sequences complementary to the target. The strands of the gapped target DNA undergo base pairing with the complementary ends of the editing template, and DNA synthesis elongates the target DNA strands and closes the gap (Fig. 12.21d). The result is that the target DNA is restored, but its sequence has been altered according to the sequence present in the editing template (Fig. 12.21e).

DNA editing by CRISPR is technically straightforward and highly efficient. The method has generated great interest because of its potential to correct genetic disorders of the blood, immune system, or other tissues and organs in which only a subset of cells with restored function can alleviate symptoms. ■

Core Concepts Summary

12.1 In DNA replication, a single parental molecule of DNA produces two daughter molecules.

DNA replication involves the separation of the two strands of the double helix at a replication fork and the use of these strands as templates to direct the synthesis of new strands. page 248

DNA replication is semiconservative, meaning that each daughter DNA molecule consists of a newly synthesized strand and a strand that was present in the parental DNA molecule. page 248

Nucleotides are added to the 3′ end of the growing strand. Therefore, synthesis proceeds in a 5′-to-3′ direction. page 250

At the replication fork, one new strand is synthesized continuously (the leading strand) and the other is synthesized in small pieces that are ligated together (the lagging strand). page 251

DNA polymerase requires RNA primers for DNA synthesis. page 252

DNA polymerase can correct its own mistakes by detecting a pairing mismatch between a template base and an incorrect new base. page 254

12.2 The replication of linear chromosomal DNA requires mechanisms that ensure efficient and complete replication.

Chromosomal DNA has many origins of replication, and replication proceeds from all of these almost simultaneously. page 254

Telomerase prevents chromosomes from shortening after each round of replication by adding a short stretch of DNA to the ends of chromosomes. page 255

12.3 Techniques for manipulating DNA follow from the basics of DNA structure and replication.

The polymerase chain reaction (PCR) is a technique for amplifying a segment of DNA. page 257

PCR requires a DNA template, DNA polymerase, the four nucleoside triphosphates, and two primers. It is a repeated cycle of denaturation, annealing, and extension. page 259

Gel electrophoresis allows DNA fragments to be separated according to size, with small fragments migrating farther than big fragments in a gel. page 259

Restriction enzymes cut DNA at specific recognition sequences called restriction sites, which cut DNA leaving a single-stranded overhang or a blunt end. page 260

In DNA denaturation, the two strands of a single DNA molecule separate from each other. In DNA renaturation or hybridization, two complementary strands come back together again. page 261

A Southern blot involves using a filter paper as a substrate for DNA fragments that have been cut up by restriction enzymes and hybridized to a probe. page 262

In Sanger sequencing of DNA, dideoxynucleotide chain terminators are used to stop the DNA synthesis reaction and produce a series of short DNA fragments from which the DNA sequence can be determined. page 263

New DNA sequencing technologies are being developed to increase the speed and decrease the cost of sequencing, perhaps making it possible to sequence everyone's personal genomes. page 264

12.4 Genetic engineering allows researchers to alter DNA Sequences.

A recombinant DNA molecule can be made by cutting DNA from two organisms with the same restriction enzyme and then using DNA ligase to join them. page 264

Recombinant DNA is the basis for genetically modified organisms (GMOs), which offer both potential benefits and risks. page 266

Almost any DNA sequence in an organism can be altered by means of a form of DNA editing called CRISPR, which uses modified forms of molecules found in bacteria and archaeons that can cleave double-stranded DNA at a specific site. page 267

Self-Assessment

1. Explain how DNA structure itself suggests a mechanism of DNA replication.
2. Explain how the chemical structure of deoxynucleotides determines the orientation of the DNA strands and how this affects the direction of DNA synthesis.
3. List the differences and similarities in the way the two daughter strands of DNA are synthesized at a replication fork.
4. Explain why replicating the tips of linear chromosomes is problematic and how the cell overcomes this challenge.
5. Describe what PCR does. Name and explain its three steps and give at least two uses for the PCR technique.
6. Explain how the properties of DNA determine how it moves through a gel, is cut by restriction enzymes, and hybridizes to other DNA strands.
7. Describe how DNA molecules are sequenced.
8. Describe how recombinant DNA techniques can be used to express a mammalian gene in bacteria.

Log in to LaunchPad to check your answers to the Self-Assessment questions, and to access additional learning tools.

CHAPTER 13

Genomes

Core Concepts

13.1 A genome is the genetic material of a cell, organism, organelle, or virus, and its sequence is the order of bases along the DNA or (in some viruses) RNA.

13.2 Researchers annotate genome sequences to identify genes and other functional elements.

13.3 The number of genes in a genome and the size of a genome do not correlate well with the complexity of an organism.

13.4 The orderly packaging of DNA allows it to carry out its functions and fit inside the cell.

13.5 Viruses have diverse genomes, but all require a host cell to replicate.

SSPL/Getty Images.

In Chapter 12, we saw how small pieces of DNA are isolated, identified, and sequenced. This technology has advanced to the point where the complete genome sequences for thousands of species have been determined, including those of humans and our closest primate relatives, as well as dozens of other mammals. The term **genome** refers to the genetic material of a specified organism, cell, nucleus, organelle, or virus. Some genomes, like that of HIV, are small, whereas others, like the human genome, are large. Technically, the **human genome** refers to the DNA in the chromosomes present in a sperm or egg. However, the term is often used informally to mean all of the genetic material in an organism. So your personal genome consists of the DNA in two sets of chromosomes, one inherited from each parent, plus a much smaller mitochondrial genome inherited from your mother. The human genome sequence is so long that printing it in the size of the type used in this book would require 1.5 million pages. As we will see, however, the human genome is far from the largest among organisms.

The sequence of a genome is merely a long string of A's, T's, G's, and C's, which represent the order of bases present in successive nucleotides along the DNA molecules in the genome. But a genome sequence, on its own, is not very useful to scientists. Additional research is required to understand what proteins and other molecules are encoded in the genome sequence, and to learn when these molecules are produced during an organism's lifetime and what they do.

In this chapter, we discuss how the sequence of a genome is determined and analyzed to reveal its key biological features, such as the protein-coding genes. We also examine what other kinds of DNA sequences are present in genomes and how these sequences are organized, with special emphasis on the human genome. Finally, we explore the diversity of genome types present in viruses.

13.1 GENOME SEQUENCING

What exactly is a genome? Originally, the term referred to the complete set of chromosomes present in a reproductive cell, like a sperm or an egg, which in the human genome is 23 chromosomes. The word "genome" is almost as old as the word "gene," and it was coined at a time when chromosomes were thought to consist of densely packed genes lined up one after another.

We know now, however, that chromosomes consist primarily of DNA and associated proteins, and that the genetic information in the chromosomes resides in the DNA. One might therefore define the genome as the DNA molecules that are transmitted from parents to offspring. This definition has the advantage of including the DNA in organisms that lack true chromosomes, such as bacteria and archaeons, as well as eukaryotic organelles that contain their own DNA, such as mitochondria and chloroplasts. But a definition restricted to DNA is too narrow because it excludes viruses like HIV, whose genetic material consists of RNA. Defining a genome as the genetic material transmitted from parent to offspring therefore embraces all known cellular forms of life, all known organelles, and all viruses.

HOW DO WE KNOW?

FIG.13.1

How are whole genomes sequenced?

BACKGROUND DNA sequencing technologies can only determine the sequence of DNA fragments far smaller than the genome itself. How can the sequences of these small fragments be used to determine the sequence of an entire genome? In the early years of genome sequencing, many researchers thought that it would be necessary to know first where in the genome each fragment originated before sequencing it. A group at Celera Genomics reasoned that if so many fragments were sequenced that the ends of one would almost always overlap with those of others, then a computer program with sufficient power might be able to assemble the short sequences to reveal the sequence of the entire genome.

HYPOTHESIS A genome sequence can be determined by sequencing small, randomly generated DNA fragments and assembling them into a complete sequence by matching regions of overlap between the fragments.

EXPERIMENT Hundreds of millions of short sequences from the genome of the fruit fly, *Drosophila melanogaster,* were sequenced. Fig. 13.1a shows examples of overlapping fragments, using a sentence from Watson and Crick's original paper on the chemical structure of DNA as an analogy.

In this first section, we focus on how genomes are sequenced, building on the DNA sequencing technology introduced in Chapter 12.

Complete genome sequences are assembled from smaller pieces.

In Chapter 12, we discussed a method of DNA sequencing known as Sanger sequencing. Sequencing methods are now automated to the point where specialized machines can determine the sequence of billions of DNA nucleotides in a single day. However, even with recent advances, the data are obtained in the form of short sequences, typically less than a few hundred nucleotides long. If you are interested in sequencing a short DNA fragment, these technologies work well. However, let's say you are interested in the sequence of human chromosome 1, which is a DNA molecule approximately 250 million nucleotides long. How can you sequence a DNA molecule as long as that?

In one approach, the single long DNA molecule is first broken up into small fragments, each of which is short enough to be sequenced by existing technologies. Even though the sequence

RESULTS The computer program the group had written to assemble the fragments worked. The researchers were able to sequence the entire *Drosophila* genome by piecing together the fragments according to their overlaps. In the sentence analogy, the fragments (Fig. 13.1a) can be assembled into the complete sentence (Fig. 13.1b) by matching the overlaps between the fragments.

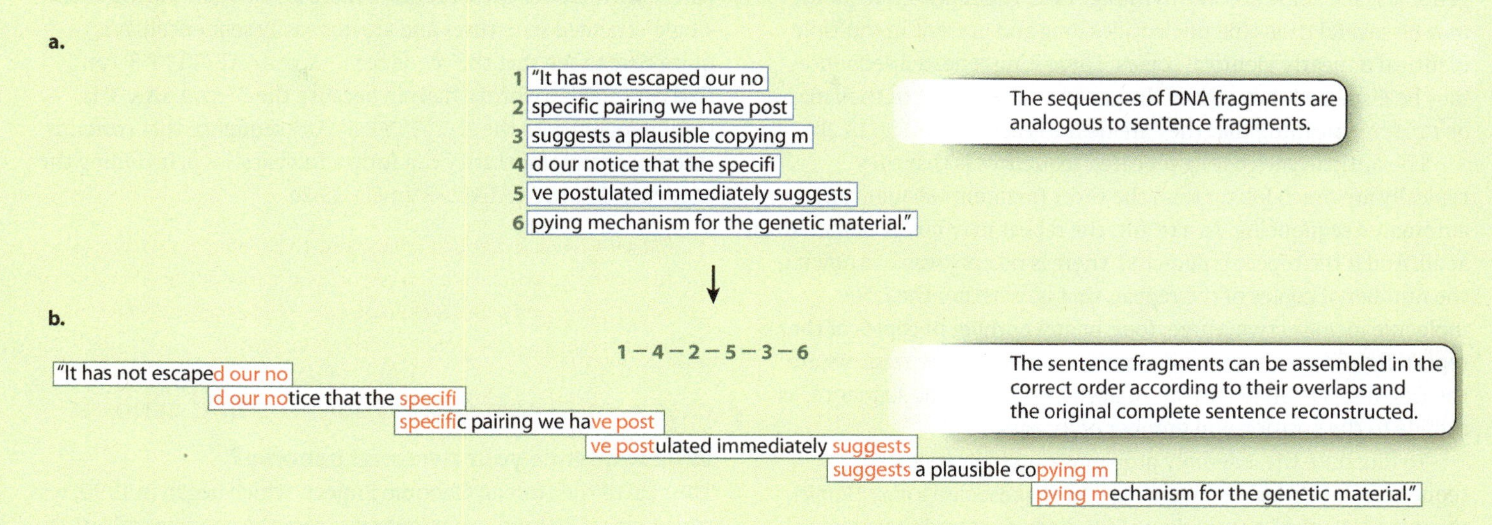

CONCLUSION The hypothesis was supported: Celera Genomics could determine the entire genomic sequence of an organism by sequencing small, random fragments and piecing them together at their overlapping ends.

FOLLOW-UP WORK Today, the computer assembly method is routinely used to determine genome sequences. This method is also used to infer the genome sequences of hundreds of bacterial species simultaneously—for example, in bacterial communities sampled from seawater or from the human gut.

SOURCE Adams, M. D., et al. 2000. "The Genome Sequence of *Drosophila* melanogaster." *Science* 287:2185–2195.

data obtained from a small DNA fragment is only a minuscule fraction of the length of the DNA molecules in most genomes, each run of an automated sequencing machine yields hundreds of millions of these short sequences from random locations throughout the genome. To sequence a whole genome, researchers typically sequence such a large number of random DNA fragments that, on average, any particular small region of the genome is sequenced 10–50 times. This redundancy is necessary to minimize both the number of errors present in the final genome sequence and the number and size of gaps where the genome sequence is incomplete.

When the sequences of a sufficient number of short stretches of the genome have been obtained, the next step is **sequence assembly:** The short sequences are put together in the correct order to generate the long, continuous sequence of nucleotides in the DNA molecule present in each chromosome.

Assembly is accomplished by complex computer programs, but the principle is simple. The short sequences are assembled according to their overlaps, as illustrated in **Fig. 13.1,** which uses a sentence to represent the nucleotide sequence. This approach is called **shotgun sequencing** because the sequenced fragments do not originate from a particular gene or region but from sites scattered randomly across the chromosome.

→ **Quick Check 1** DNA sequencing technology has been around since the late 1970s. Why did sequencing whole genomes present a challenge?

Sequences that are repeated complicate sequence assembly.

Sequence assembly is not quite as straightforward as Fig. 13.1 suggests. Real sequences are composed of the nucleotides A, T, G, and C, and any given short sequence could come from either strand of the double-stranded DNA molecule. Therefore, the overlaps between fragments must be long enough both to ensure that the assembly is correct and to determine from which strand of DNA the short sequence originated.

Some features of genomes present additional challenges to sequence assembly, and the limitations of the computer programs for handling such features require hands-on assembly. Chief among these complicating features is the problem of repeated sequences known collectively as **repetitive DNA.**

There are a variety of types of repeated sequence in eukaryotic genomes, and some are shown in **Fig. 13.2.** The repeated sequence may be several thousand nucleotides long and present in multiple identical or nearly identical copies. These long repeated sequences may be dispersed throughout the genome (Fig. 13.2a), or they may be tandem, meaning that they are next to each other (Fig. 13.2b).

The difficulty with long repeated sequences is that they typically are much longer than the short fragments sequenced by automated sequencing. As a result, the repeat may not be detected at all. And if the repeat is detected, there is no easy way of knowing the number of copies of the repeat, that is, whether the DNA molecule includes two, three, four, or any number of copies of the repeat. Sometimes, researchers can use the ends of repeats, where the fragments overlap with an adjacent, nonrepeating sequence, as a guide to the position and number of repeats.

To illustrate the assembly problems caused by repeated sequences, let's consider an analogy. In Shakespeare's play *Hamlet,* the word "Hamlet" occurs about 500 times scattered throughout the text, like a dispersed repeat (Fig. 13.2a). If you chose short sequences of letters from the play at random and found the sequence "Hamlet" or "amlet" or "Haml," you would have no way of knowing which of the 500 "Hamlets" these letters came from. Only if you found sequences that contained "Hamlet" overlapping with adjacent unique text could you identify their origin. A similar problem occurs in the case of tandem repeats.

In another type of repeat the repeating sequence is short, even as short as two nucleotides, such as AT, repeated over and over again in a stretch of DNA (Fig. 13.2c). Short repeating sequences of this kind are troublesome for sequencing machines because any single-stranded fragment consisting of alternating AT can fold back upon itself to form a double-stranded structure in which A is paired with T. Such structures are more stable than the unfolded single-stranded structures and are not easily sequenced. It is quite easy to see that the sequence 5′-AAAAAATTTTTT-3′ can fold back itself to form a hairpin because the 5′-AAAAAA-3′ is complementary to the 3′-TTTTTT-5′. Any sequence that contains internal complementarity can form a foldback loop, including the sequence 5′-...ATATATAT...-3′ in Fig. 13.2c.

FIG. 13.2 Principal types of sequence repeats found in eukaryotic genomes. Repeats often pose problems in DNA sequencing. (Repeats are not drawn to scale.)

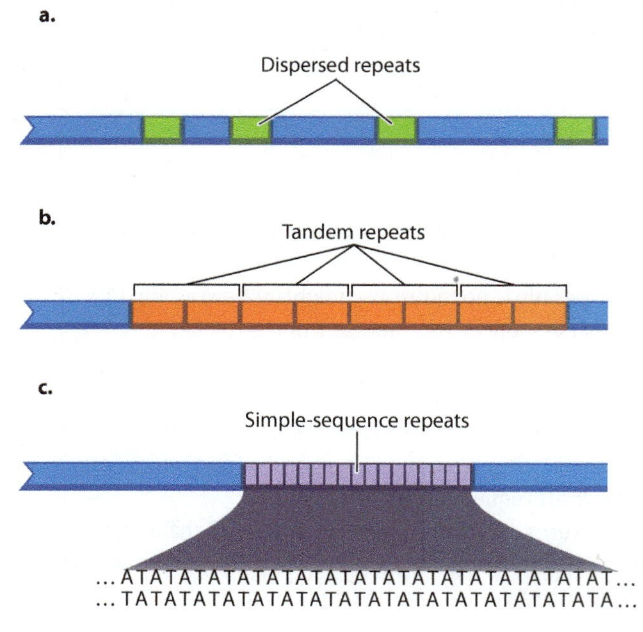

→ **Quick Check 2** Let's say that a stretch of repeated AT is successfully sequenced. From what you know of the difficulties of sequencing long repeated sequences, what other problems might you encounter in assembling these fragments?

? CASE 3 YOU, FROM A TO T: YOUR PERSONAL GENOME

Why sequence your personal genome?

The goal of the Human Genome Project, which began in 1990, was to sequence the human genome as well as the genomes of certain key organisms used as models in genetic research. The model organisms chosen are the mainstays of laboratory biology—a species each of bacteria, yeast, nematode worm, fruit fly, and mouse. By 2003, the genome sequences of these model organisms had been completed, as well as that of the human genome. By then, the cost of sequencing had become so low and the sequence output so high that many more genomes were sequenced than originally planned. Genome sequencing is even cheaper today, and soon you can choose to have your personal genome sequenced.

Why sequence more genomes? And if the human genome is sequenced, why sequence yours? As we saw in Case 3: You, from A to T, there is really no such thing as *the* human genome, any more than there is *the* fruit fly genome or *the* mouse genome. With the exception of identical twins, every person's genome is unique, the product of a fusion of a unique egg with a unique sperm. The sequence that is called "the human genome" is actually a composite of sequences from different individuals. This sequence is nevertheless useful because most of us share the same genes and regulatory regions, organized the same way on chromosomes. Detailed knowledge of your own personal genome can be valuable. Our individual DNA sequences differ at millions of nucleotide sites from one person to the next. Some of these differences account in part for the physical differences we see among us; others have the potential to predict susceptibility to disease and response to medication. For Claudia Gilmore, knowledge of the sequence of her *BRCA1* gene had a significant impact on her life.

Determining these differences is a step toward **personalized medicine,** in which an individual's genome sequence, by revealing his or her disease susceptibilities and drug sensitivities, allows treatments to be tailored to that individual. There may come a time, perhaps within your lifetime, when personal genome sequencing becomes part of routine medical testing. Information about a patient's genome will bring benefits but also raises ethical concerns and poses risks to confidentiality and insurability.

13.2 GENOME ANNOTATION

A goal of biology is to identify all the macromolecules in biological systems and understand their individual functions and the ways in which they interact. This research has practical applications: Increased understanding of the molecular and cellular basis of disease, for example, can lead to improved diagnosis and treatment.

The value of genome sequencing in identifying macromolecules is that the genome sequence contains, in coded form, the nucleotide sequence of all RNA molecules transcribed from the DNA as well as the amino acid sequence of all proteins. There is a catch, however. A genome sequence is merely an extremely long list of A's, T's, G's, and C's that represent the order in which nucleotides occur along the DNA in one strand of the double helix. (Because of complementarity, knowing the sequence of one strand specifies the other.) The catch is that in multicellular organisms, not all the DNA is transcribed into RNA, and not all the RNA that is transcribed is translated into protein. Therefore, genome sequencing is just the first step in understanding the function of any particular DNA sequence. Following genome sequencing, the next step is to identify the locations and functions of the various types of sequence present in the genome.

Genome annotation identifies various types of sequence.

Genomes contain many different types of sequence, among them protein-coding genes. Protein-coding genes are themselves composed of different regions, including regulatory elements that specify when and where an RNA transcript is produced, noncoding introns that are removed from the RNA transcript during RNA processing (Chapter 3), and protein-coding exons that contain the codons that specify the amino acid sequence of a polypeptide chain (Chapter 4). Genomes also contain coding sequences for RNAs that are not translated into protein (noncoding RNAs), such as ribosomal RNA, transfer RNA, and other types of small RNA molecule (Chapter 3). Finally, while much of the DNA in the genomes of multicellular organisms is transcribed at least in some cell types, the functions of a large portion of these transcripts in metabolism, physiology, development, or behavior are unknown.

Genome annotation is the process by which researchers identify the various types of sequence present in genomes. Genome annotation is essentially an exercise in adding commentary to a genome sequence that identifies which types of sequence are present and where they are located. It can be thought of as a form of pattern recognition, where the patterns are regularities in sequence that are characteristic of protein-coding genes or other types of sequence.

An example of genome annotation is shown in **Fig. 13.3.** Genes present in one copy per genome are indicated in orange. Most single-copy genes are protein-coding genes. In some computer programs that analyze genomes, the annotation of a single-copy gene also specifies any nearby regulatory regions

FIG. 13.3 **Genome annotation.** Given the DNA sequence of a genome, researchers can pinpoint locations of various types of sequence.

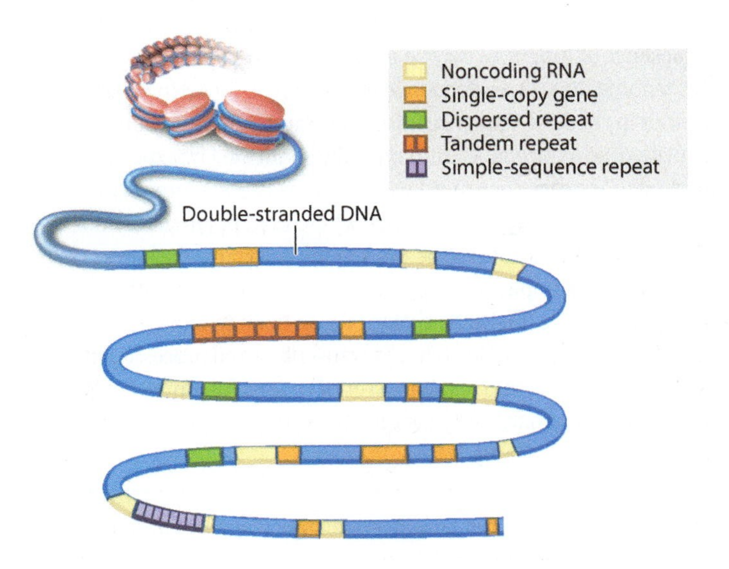

that control transcription, the intron–exon boundaries in the gene, and any known or predicted alternative forms in which the introns and exons are spliced (Chapter 3). Each single-copy gene is given a unique name and its protein product identified. Note in Fig. 13.3 that single-copy genes can differ in size from one gene to the next. The annotations in Fig. 13.3 also specify the locations of sequences that encode RNAs that are not translated into proteins, as well as various types of repeated sequence. Most genomes also contain some genes that are present in multiple copies that originate from gene duplication; however these are usually identifiable in the same way as single-copy genes because the copies often differ enough in sequence that they can be distinguished.

Small genomes such as that of HIV and other viruses can be annotated by hand, but for large genomes like the human genome, computers are essential. In the human genome, some protein-coding genes extend for more than a million nucleotides. Roughly speaking, if the sequence of the approximately 3 billion nucleotides in a human egg or sperm was printed in normal-sized type like this, the length of the ribbon would stretch 4000 miles (6440 kilometers), about the distance from Fairbanks, Alaska, to Miami, Florida. By contrast, for the approximately 10,000 nucleotides in the HIV genome, the ribbon would extend a mere 70 feet (21 meters).

Genome annotation is an ongoing process because, as macromolecules and their functions and interactions become better understood, the annotations to the genome must be updated. A sequence that is annotated as nonfunctional today may be found to have a function tomorrow. For this reason, the annotation of certain genomes—including the human genome—will certainly continue to change.

Genome annotation includes searching for sequence motifs.

Because genome annotation is essentially pattern recognition, it begins with the identification of patterns called **sequence motifs,** telltale sequences of nucleotides that indicate what types of function (or absence of function) may be encoded in a particular region of the genome (**Fig. 13.4**). Sequence motifs can be found in the DNA itself or in the RNA sequence inferred from the DNA sequence. Once identified, sequence motifs are typically confirmed by experimental methods. One sequence motif we have already encountered in Chapter 3 is a promoter, a sequence where RNA polymerase and associated proteins bind to the DNA to initiate transcription.

Another example of a sequence motif is an **open reading frame (ORF)** (Fig. 13.4a). The motif for an open reading frame is a long string of nucleotides that, if transcribed and processed into messenger RNA, would result in a set of codons for amino acids that does not contain a stop codon. The presence of an ORF motif by itself is enough to annotate the DNA segment as potentially protein coding. The qualifier "potentially" is necessary because ORFs identified in a DNA sequence do not necessarily code for protein. For this reason they are often called putative ORFs. A region containing a putative ORF may exist merely by chance (even a random sequence of nucleotides will contain ORFs averaging 21 codons in length); or a putative ORF may not be transcribed; or if a putative ORF is transcribed, it might be in a noncoding RNA or an intron of a protein-coding RNA. In the next section, we discuss how the analysis of messenger RNA sequences can determine whether a putative ORF is an actual ORF in DNA coding for protein.

FIG. 13.4 Some common sequence motifs useful in genome annotation. (a) An open reading frame; (b) a noncoding RNA molecule; (c) transcription factor binding sites.

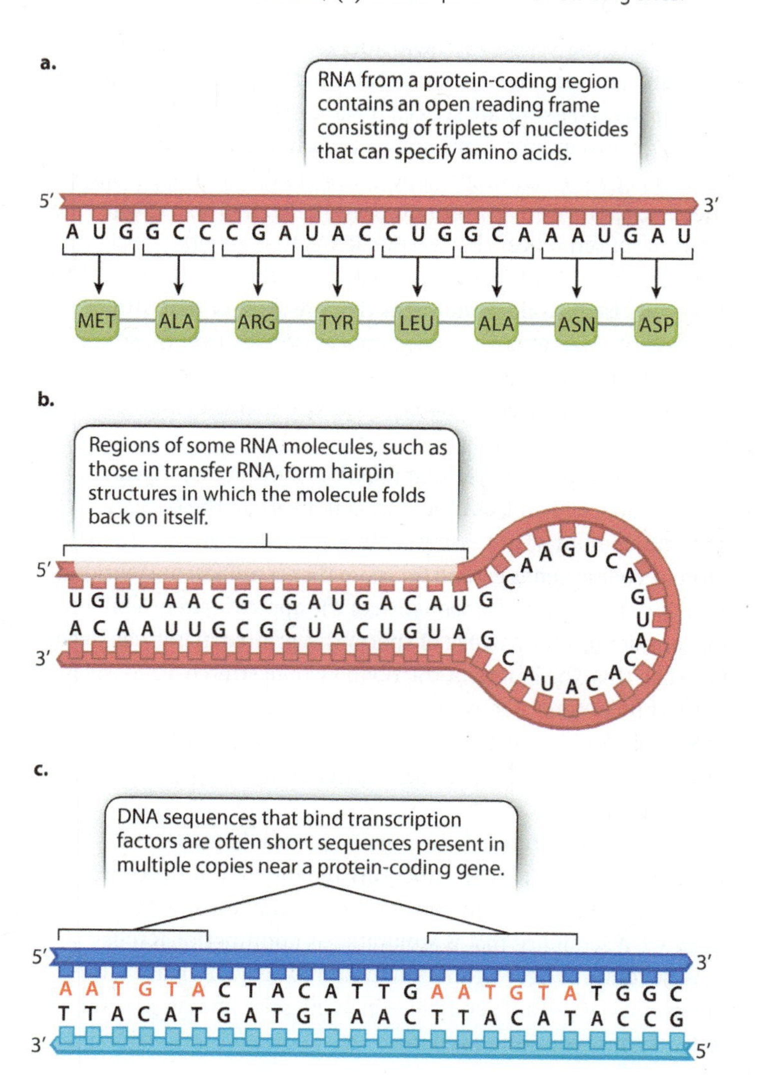

Fig. 13.4b shows another type of sequence motif, this one also present in a hypothetical RNA transcript inferred from the DNA sequence. The nucleotide sequence at one end of the RNA is complementary to that at the other end, so the single-stranded molecule is able to fold back on itself and undergo base pairing to form a hairpin-shaped structure. Such hairpin structures are characteristic of certain types of RNA that function in gene regulation (Chapter 19). The DNA from which this RNA is transcribed has complementary sequences on either end as well.

Some sequence motifs are detected directly in the double-stranded DNA. Fig. 13.4c shows two copies of a short sequence that is a known binding site for DNA-binding proteins called transcription factors (Chapter 3), which initiate transcription. Transcription factor binding sites are often present in multiple copies and in either strand of the DNA. Sometimes they are located near the region of a gene where transcription is initiated because the transcription factor helps determine when the gene will be transcribed. However, they can also be located far upstream of the gene, downstream of the gene, or in introns, and so their identification is difficult.

Comparison of genomic DNA with messenger RNA reveals the intron–exon structure of genes.

In annotating an entire genome, researchers typically make use of information outside the genome sequence itself. This information may include sequences of messenger RNA molecules that are isolated from various tissues or various stages of development of the organism. Recall from Chapter 3 that messenger RNA (mRNA) molecules undergo processing and are therefore usually simpler than the DNA sequences from which they are transcribed—for example, introns are removed and exons are spliced together. The resulting mature mRNA therefore contains a long sequence of codons uninterrupted by a stop codon—in other words, an ORF. The ORF in an mRNA is the region that is actually translated into protein.

One aspect of genome annotation is the determination of which portions of the genome sequence correspond to sequences in mRNA transcripts. An example is shown in **Fig. 13.5,** which compares the DNA and mRNA for the beta (β) chain of hemoglobin, the oxygen-carrying protein in red blood cells. Note that the genomic DNA contains some sequences present in the mRNA, which correspond to exons, and some sequences that are not present in the mRNA, which correspond to introns. Comparison of mRNA with genomic DNA therefore reveals the intron–exon

FIG. 13.5 Identification of exons and introns by comparison of genomic DNA with mRNA sequence.

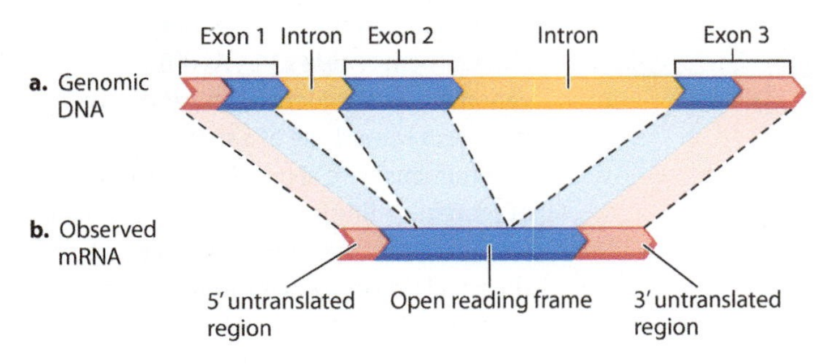

structure of protein-coding genes. In fact, introns were first discovered by comparing β-globin mRNA with genomic DNA.

An annotated genome summarizes knowledge, guides research, and reveals evolutionary relationships among organisms.

Genome annotation, which aims to identify all the functional and repeat sequences present in the genome, is an imperfect science. Even in a well-annotated genome, some protein-coding sequences or other important features may be overlooked, and occasionally the annotation of a sequence motif is incorrect.

Because researchers often have to rely on sequence motifs alone and not experimental data, their descriptions may be vague. For example, a common annotation in large genomes is "hypothetical protein." In some cases, such as the genome of the malaria parasite, this type of annotation accounts for about 50% of the possible protein-coding genes. There is no hint of what a hypothetical protein may do or even whether it is actually produced, since it is determined solely by the presence of a putative ORF in the genomic sequence and not by the presence of actual mRNA or protein. Other annotations might be "DNA-binding protein," "possible hairpin RNA," or "tyrosine kinase"—with no additional detail about these motifs' functions or relationships to other sequences. In short, although some genome annotations summarize experimentally verified facts, many others are hypotheses and guides to future research.

Genome sequences contain information about ancestry and evolution, and so comparisons among genomes can reveal how different species are related. For example, the sequence of the human genome is significantly more similar to that of the chimpanzee than to that of the gorilla, indicating a more recent common ancestry of humans and chimpanzees (Chapter 1).

Analysis of the similarities and differences in protein-coding genes and other types of sequence in the genomes of different species is an area of study called **comparative genomics.** Such studies help us understand how genes and genomes evolve. They can also guide genome annotation because the sequences of important functional elements are often very similar among genomes of different organisms. Sequences that are similar in different organisms are said to be **conserved.** A sequence motif that is conserved is likely to be important, even if its function is unknown, since it has changed very little over evolutionary time.

The HIV genome illustrates the utility of genome annotation and comparison.

HIV and related viruses provide an example of how genomes are annotated and how comparisons among genome sequences can reveal evolutionary relationships. A **virus** is a small infectious agent that contains a nucleic acid genome packaged inside a protein coat called a **capsid.** Viruses can bind to receptor proteins on cells of the host organism, enabling the viral genome to enter the cell. Infection of a host cell is essential to viral reproduction because viruses use cellular ATP and hijack cellular machinery to replicate, transcribe, and translate their genome in order to make more viruses. Later in this chapter we will see that viruses can differ in the types of nucleic acid that make up their genome and how their messenger RNA is produced, but here we consider only the genome of HIV.

Whereas the genome of all cells consists of double-stranded DNA, the genome of HIV is single-stranded RNA. The sequence of the HIV genome identifies it as a retrovirus that replicates by a DNA intermediate that can be incorporated into the host genome. More narrowly, the sequence of the HIV genome groups it among the mammalian lentiviruses, so named because of the long lag between the initial time of infection and the appearance of symptoms (*lenti-* means "slow").

Fig. 13.6 shows the evolutionary relationships among a sample of lentiviruses, grouped according to the similarity of their genome sequences. The evolutionary tree shows that closely related viruses have closely related hosts. For example, simian

FIG. 13.6 Evolutionary relationships among viruses related to HIV. *Data from A. Katzourakis, M. Tristem, O. G. Pybus, and R. J. Gifford, 2007, "Discovery and Analysis of the First Endogenous Lentivirus," Proceedings of the National Academy of Sciences USA 104(15): 6261–6265.*

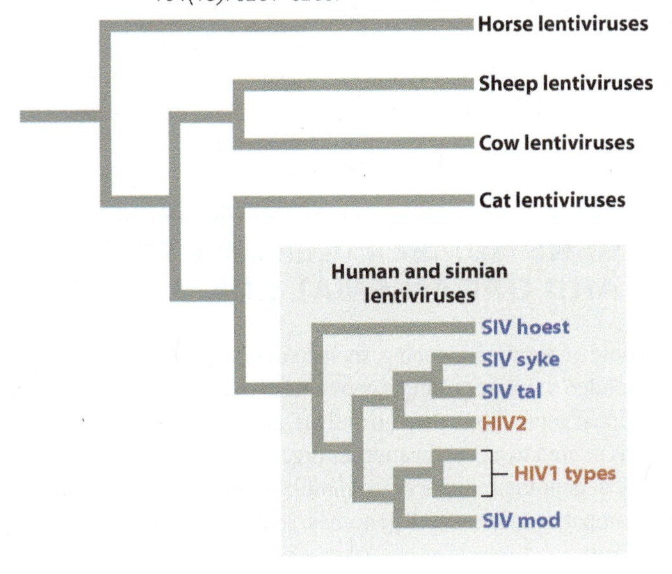

FIG. 13.7 The annotated genome of HIV.

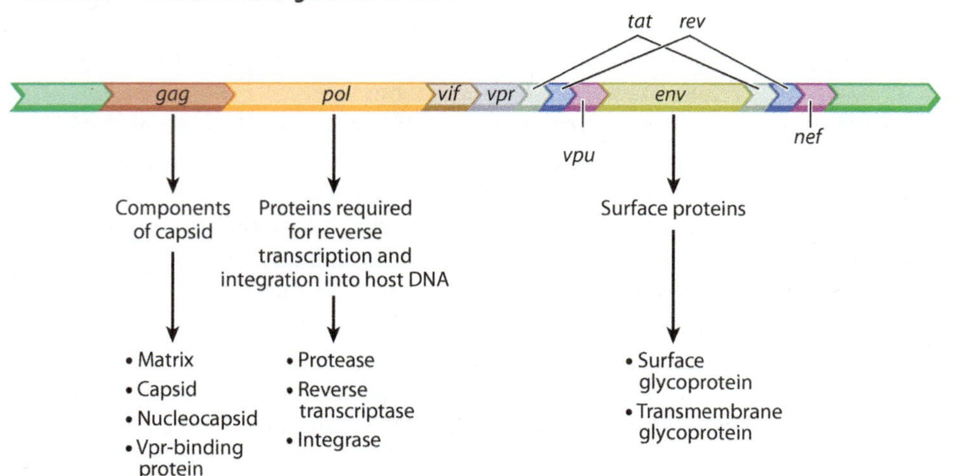

lentiviruses are more closely related to one another than they are to cat lentiviruses. This observation implies that the genomes of the viruses evolve along with the genomes of their hosts. A second feature shown by the evolutionary tree is that human HIV originated from at least two separate simian viruses that switched hosts from simians (most likely chimpanzees) to humans.

The annotated sequence of the HIV genome tells us a lot about the biology of this virus (**Fig. 13.7**). Many details are left out here, but the main point is to show the functional elements of HIV in the form of an annotated genome. The open reading frame denoted *gag* encodes protein components of the capsid, *pol* encodes proteins needed for reverse transcription of the viral RNA into DNA and incorporation into the host genome, and *env* encodes proteins that are embedded in the lipid envelope. The annotation in Fig. 13.7 also includes the genes *tat* and *rev*, encoding proteins essential for the HIV life cycle, as well as the genes *vif*, *vpr*, *vpu*, and *nef*, which encode proteins that enhance virulence in organisms. Identification of the genes necessary to complete the HIV cycle is the first step to finding drugs that can interfere with the cycle and prevent infection.

→ **Quick Check 3** Fig. 13.6 shows that closely related lentiviruses have closely related host organisms. What does this observation suggest about the ability of lentiviruses to infect a wide variety of hosts?

13.3 GENE NUMBER, GENOME SIZE, AND ORGANISMAL COMPLEXITY

Before whole-genome sequencing, molecular biology and evolution research tended to focus on single genes. Researchers studied how individual genes are turned on and off, and how a gene in one organism is related to a gene in another organism. Now, with the availability of genome sequences from multiple species, it is possible to make comparisons across full genomes. Some of the results have been striking.

Gene number is not a good predictor of biological complexity.

The complete genome sequences of many organisms allow us to make comparisons among them (**Table 13.1**). As different genomes were sequenced and annotated, it came as something of a surprise to find that humans have about the same number of protein-coding genes as many organisms with much smaller genomes. For example, the small flowering plant *Arabidopsis thaliana* has a genome only about 5% as large as that of humans, but the plant has at least as many protein-coding genes. And the nematode worm *Caenorhabditis elegans* contains only 959 cells, of which 302 are nerve cells that form the worm's brain. Humans have 100 trillion cells, of which 100 billion are nerve cells. Despite having 100 million times as many cells as the worm, we have roughly the same number of protein-coding genes. Another example of the disconnect between level of complexity and genome size is the observation that the genome size of the mountain grasshopper *Podisma pedestris* is more than 100 times that of the fruit fly *Drosophila melanogaster.* While there is no agreed-upon definition of organismal complexity, it would be hard to make a case that grasshoppers are any more or less complex than fruit flies.

On the other hand, some organisms are able to do more things with the genes they have than other organisms. For example, the expression of protein-coding genes can be regulated in many subtle ways, causing different gene products to be made in different amounts in different cells at different times (Chapters 19 and 20). Differential gene expression allows the same protein-coding genes to be deployed in different combinations to yield a variety of distinct cell types. In addition, proteins can interact

TABLE 13.1 Gene Numbers of Several Organisms. Humans have more genes than fruit flies and nematode worms, but not as many as might be expected on the basis of complexity. *Data from "Sequence Composition of the Human Genome" in International Genome Sequencing Consortium, 2001, "Initial Sequencing and Analysis of the Human Genome" (PDF). Nature 409(6822): 860–921, doi:10.1038/35057062, PMID 11237011.*

COMMON NAME	SPECIES NAME	APPROXIMATE NUMBER OF PROTEIN-CODING GENES
Mustard plant	*Arabidopsis thaliana*	27,000
Human	*Homo sapiens*	25,000
Nematode worm	*Caenorhabditis elegans*	22,000
Fruit fly	*Drosophila melanogaster*	17,000
Baker's yeast	*Saccharomyces cerevisiae*	6000
Gut bacterium	*Escherichia coli*	4000

with one another so that, even though there are relatively few types of protein, they are capable of combining in many different ways to perform different functions. We also have seen that a single gene may yield multiple proteins, either because of alternative splicing (different exons are spliced together to make different proteins) or posttranslational modification (proteins undergo biochemical changes after they have been translated).

Overall, the data in Table 13.1 pose many tantalizing evolutionary questions and suggest that major evolutionary changes can be accomplished not only by the acquisition of whole new genes, but also by modifying existing genes and their regulation in subtle ways.

Viruses, bacteria, and archaeons have small, compact genomes.

As well as comparing numbers of genes, we can also compare sizes of genomes in different organisms. Before making such comparisons, we need to understand how genome size is measured. Genomes are measured in numbers of base pairs, and the yardsticks of genome size are a thousand base pairs (a kilobase, kb), a million base pairs (a megabase, Mb), and a billion base pairs (a gigabase, Gb).

Most viral genomes range in size from 3 kb to 300 kb, but a few are very large. The largest viral genome, found in a virus that infects the amoeba *Acanthamoeba polyphaga,* is 1.2 Mb. This viral genome contains almost 1000 protein-coding genes, including some for sugar, lipid, and amino acid metabolism not found in any other viruses.

The largest viral genome is twice as large as that of the bacterium *Mycoplasma genitalium.* At 580 kb and encoding only 471 genes, the genome of *M. genitalium* is the smallest known among free-living bacteria, those capable of living entirely on their own. The complete sequence of small bacterial genomes has allowed researchers to define the smallest, or minimal, genome (and therefore the minimal set of proteins) necessary to sustain life. Current findings suggest that the small *M. genitalium* is about two times larger than the minimal genome size thought to be necessary to encode all the functions essential to life.

The genomes of bacteria and archaeons are information dense, meaning that most of the genome has a defined function. Roughly speaking, 90% or more of their genomes consist of protein-coding genes (although in many cases the protein has an unknown function). Bacterial genomes range in size from 0.5 to 10 Mb. The bigger genomes have more genes, allowing these bacteria to synthesize small molecules that other bacteria have to scrounge for, or to use chemical energy in the covalent bonds of substances that other bacteria cannot. Archaeons, whose genomes range in size from 0.5 to 5.7 Mb, have similar capabilities.

Among eukaryotes, there is no relationship between genome size and organismal complexity.

In eukaryotes, just as the number of genes does not correlate well with organismal complexity, the size of the genome is unrelated to the metabolic, developmental, and behavioral complexity of the organism (**Table 13.2**). The range of genome sizes is huge, even among similar organisms (**Fig. 13.8**). For comparison, Fig. 13.8 also shows the range of genome sizes among bacteria and archaeons. The largest eukaryotic genome exceeds the size of the smallest by a factor of more than 500,000—and both the smallest and the largest are found among protozoa. The range among flowering plants (angiosperms) is about three orders of magnitude, and the range among animals is about seven orders of magnitude. One species of lungfish has a genome size more than 45 times larger than the human genome (Table 13.2). Clearly, there is no relationship between the size of the genome and the complexity of the organism.

TABLE 13.2 Genome Sizes of Several Organisms. Genome size varies tremendously among eukaryotic organisms, and there is no correlation between genome size and the complexity of an organism. *Data from "Sequence Composition of the Human Genome" in International Human Genome Sequencing Consortium, 2001, "Initial Sequencing and Analysis of the Human Genome" (PDF). Nature 409(6822): 860–921, doi:10.1038/35057062, PMID 11237011.*

COMMON NAME	SPECIES NAME	APPROXIMATE GENOME SIZE (Mb)
Fruit fly	*Drosophila melanogaster*	180
Fugu fish	*Fugu rubripes*	400
Boa constrictor	*Boa constrictor*	2100
Human	*Homo sapiens*	3100
Locust	*Schistocerca gregaria*	9300
Onion	*Allium cepa*	18,000
Newt	*Amphiuma means*	84,000
Lungfish	*Protopterus aethiopicus*	140,000
Fern	*Ophioglossum petiolatum*	160,000
Amoeba	*Amoeba dubia*	670,000

The disconnect between genome size and organismal complexity is called the **C-value paradox.** The C-value is the amount of DNA in a reproductive cell, and the paradox is the apparent contradiction between genome size and organismal complexity, leading to the difficulty of predicting one from the other.

→ **Quick Check 4** Given our knowledge of genome sizes in different organisms, would you predict that *Homo sapiens* or the two-toed salamander (*Amphiuma means*) has the larger genome?

Why are some eukaryotic genomes so large? One reason is **polyploidy,** or having more than two sets of chromosomes in the genome. Polyploidy is especially prominent in many groups of plants. Humans have two sets of 23 chromosomes, giving us 46 chromosomes in total. But the polyploid bread wheat *Triticum aestivum,* for example, has six sets of seven chromosomes.

Polyploidy has played an important role in plant evolution. Many agricultural crops are polyploid, including wheat, potatoes, olives, bananas, sugarcane, and coffee. Among flowering plants, it is estimated that 30% to 80% of existing species have polyploidy

FIG. 13.8 **Ranges of genome size.** Even within a single group, genome sizes of different species can vary by several factors of 10. *After Fig. 1, p. 186 in T. R. Gregory, 2004, "Macroevolution, Hierarchy Theory, and the C-Value Enigma," Paleobiology 30:179–202.*

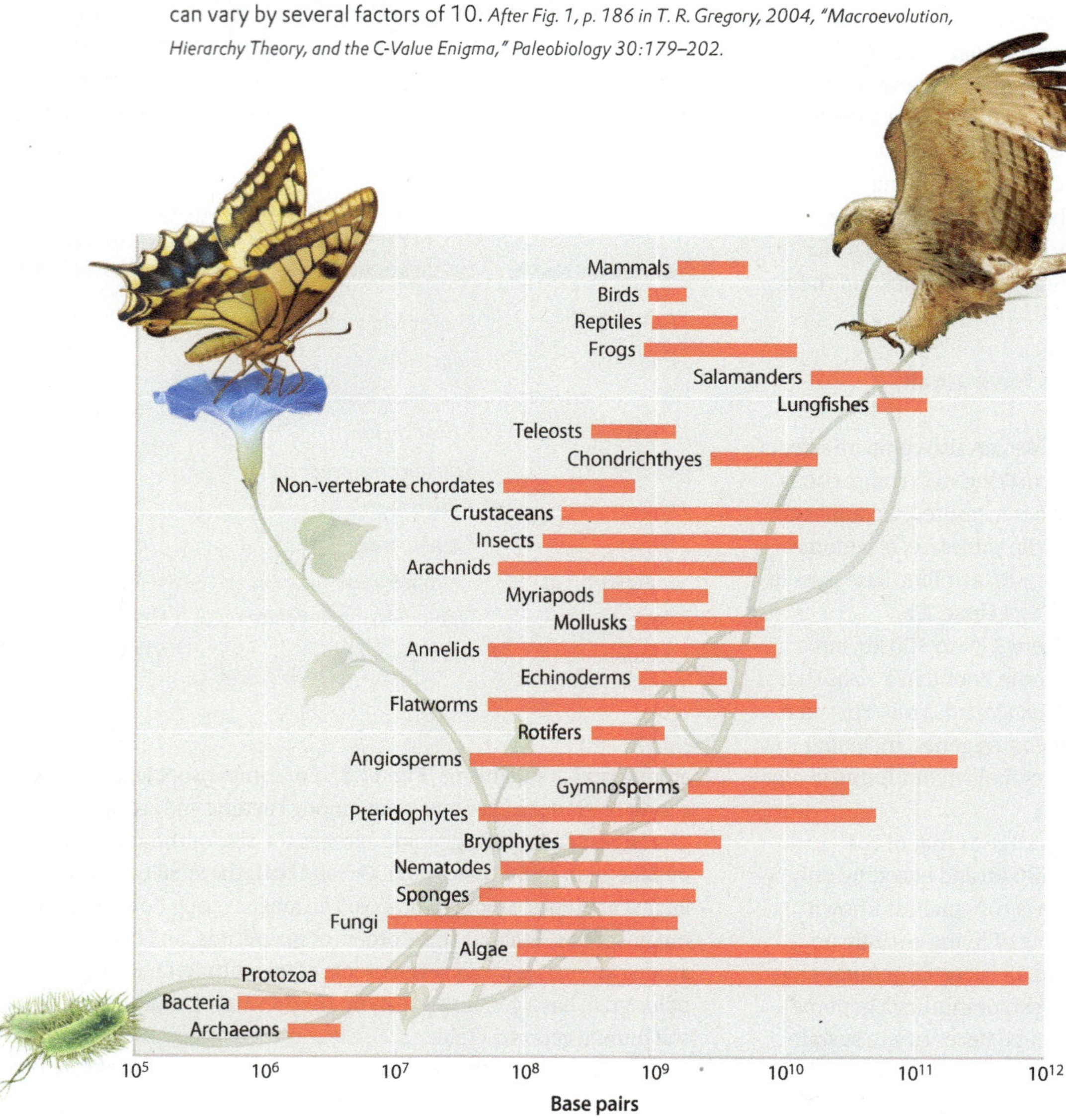

in their evolutionary history, either because of the duplication of the complete set of chromosomes in a single species, or because of hybridization, or crossing, between related species followed by duplication of the chromosome sets in the hybrid (**Fig. 13.9**). Some ferns take polyploidy to an extreme: One species has 84 copies of a set of 15 chromosomes—1260 chromosomes altogether.

However, the principal reason for large genomes among some eukaryotes is that their genomes contain large amounts of DNA that do not code for proteins, such as introns and DNA sequences that are present in many copies. These repeated sequences are discussed next.

About half of the human genome consists of transposable elements and other types of repetitive DNA.

Complete genome sequencing has allowed the different types of noncoding DNA to be specified more precisely in a variety of organisms. It came as a great surprise to learn that in the human genome only about 2.5% of the genome actually codes for proteins. The other 97.5% includes sequences we have encountered earlier, including introns, noncoding DNA, and various other types of repetitive sequences.

In **Fig. 13.10,** we see the composition of the human genome, including the principal types of repeated sequences. Earlier we saw how repeated sequences could be classified as dispersed or tandem according to their organization in the genome. In Fig. 13.10 the repeated sequences are classified according to their function. Among the repeated sequences is alpha (α) satellite DNA, which consists of tandem copies of a 171-bp sequence repeated near each centromere an average of 18,000 times. The α satellite DNA is essential for attachment of spindle fibers to the centromeres during cell division (Chapter 11).

Fig. 13.10 also shows the proportions of several types of repeated sequence collectively known as **transposable elements** (abbreviated *TE;* they are also called **transposons**), which are DNA sequences that can replicate and insert themselves into new positions in the genome. As a result, they have the potential to increase their copy number in the genome over time. Transposable elements are sometimes referred to as "selfish" DNA because it seems that their only function is to duplicate themselves and proliferate in the genome, making them the ultimate parasite.

Transposable elements make up about 45% of the DNA in the human genome. They can be grouped into two major classes based on the way they replicate. One class consists of **DNA transposons,** which replicate and transpose by DNA replication and repair. The other class consists of elements that transpose by means of an RNA intermediate. These are sometimes called **retrotransposons** because their RNA is used as a template to synthesize complementary strands of DNA, a process that reverses

FIG. 13.9 Polyploidy in plants. A type of crocus known as Golden Yellow was formed by hybridization between two different species, and it contains a full set of chromosomes from each parent. The parental chromosomes in the hybrid are indicated in orange and blue in the diagram on the right. *Photo source: McPHOTO/age fotostock..*

the usual flow of genetic information from DNA into RNA (*retro-* means "backward"). More than 40% of the human genome consists of various types of retrotransposons, whereas only about 3% of human DNA consists of DNA transposons.

Over the course of evolutionary time, the amount of repetitive DNA in a genome can change drastically, in large part because of the accumulation of transposable elements (Fig. 13.10). In some genomes repetitive DNA is maintained over long periods of time even as new species evolve and diversify, whereas in other genomes the amount of repetitive DNA is held in check because of deletion and other processes. The result is that the genomes of different species can contain vastly differing amounts of repetitive DNA, and this accounts in large part for the C-value paradox.

FIG. 13.10 Sequence composition of the human genome.

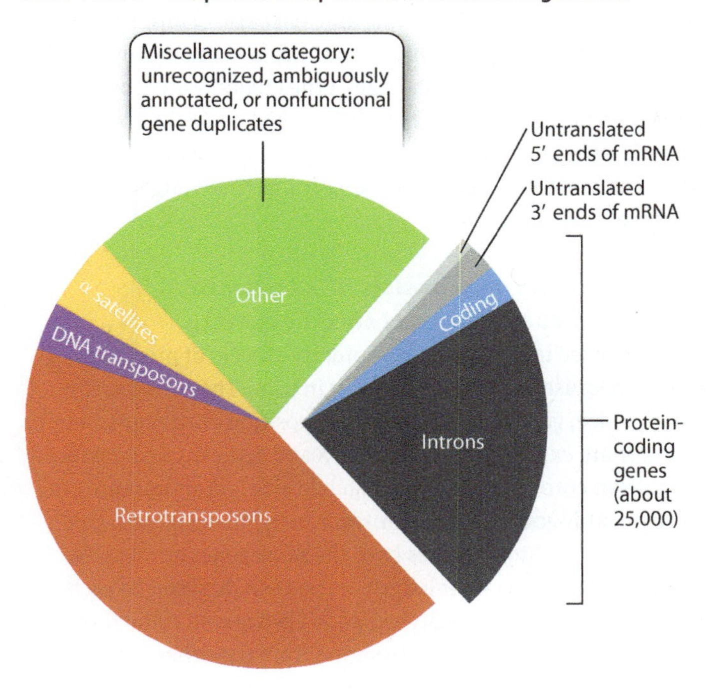

13.4 ORGANIZATION OF GENOMES

Regardless of whether an organism has a large or a small genome, the genomes of all organisms are large relative to the size of the cell. For example, if the circular genome of the intestinal bacterium *Escherichia coli* were fully extended, its length would be 200 times greater than the diameter of the cell itself. The fully extended length of DNA in human chromosome 1, our longest chromosome, would be 10,000 times greater than the diameter of the average human cell. There is consequently a need to package an enormous length of DNA into a form that will fit inside the cell while still allowing the DNA to replicate and carry out its coding functions. The mechanism of packaging differs substantially in bacteria, archaeons, and eukaryotes. We focus here on bacteria and eukaryotes, primarily because less is known about how DNA is packaged in archaeons.

Bacterial cells package their DNA as a nucleoid composed of many loops.

Bacterial genomes are circular, and the DNA double helix is underwound, which means that it makes fewer turns in going around the circle than would allow every base in one strand to pair with its partner base in the other strand. Underwinding is caused by an enzyme, **topoisomerase II,** that breaks the double helix, rotates the ends to unwind the helix, and then seals the break. Underwinding creates strain on the DNA molecule, which is relieved by the formation of **supercoils,** in which the DNA molecule coils on itself. Supercoiling allows all the base pairs to form, even though the molecule is underwound. (You can make your own supercoil by stretching and twirling the ends of a rubber band, then relaxing the stress slightly to allow the twisted part to form coils around itself. See Fig. 3.12.) Supercoils that result from underwinding are called negative supercoils, and those that result from overwinding are positive supercoils. In most organisms, DNA is negatively supercoiled.

In bacteria, the supercoils of DNA form a structure with multiple loops called a **nucleoid** (**Fig. 13.11**). The supercoil loops are bound together by proteins. In *E. coli*, the nucleoid has about 100 loops, each containing about 50 kb of DNA. The protein binding that forms the loops as well as the negative supercoiling of the DNA compress the molecule into a compact volume.

Eukaryotic cells package their DNA as one molecule per chromosome.

In eukaryotic cells, DNA in the nucleus is packaged differently from DNA in bacteria. As discussed in Chapter 3, eukaryotic DNA is linear and each DNA molecule forms a single chromosome. In a chromosome, DNA is packaged with proteins to form a DNA-protein complex called **chromatin.** There are several levels of packaging. First, eukaryotic DNA is wrapped twice around a group of histone proteins called a **nucleosome.** A nucleosome is made up of eight histone proteins: two each of histones H2A, H2B, H3, and H4. The histone proteins are rich in the amino acids lysine and arginine, whose positive charges are attracted to the negative charges of the phosphates along the backbone of each DNA strand.

This first level of packaging of the DNA is sometimes referred to as "beads on a string," with the nucleosomes the beads and the DNA the string. It is also called a **10-nm fiber** in reference to its diameter, which is about five times the diameter of the DNA double helix (**Fig. 13.12**). In general, these are areas of the genome that are transcriptionally active.

The next level of packaging occurs when the chromatin is more tightly coiled, forming a **30-nm fiber** (Fig. 13.12). As the chromosomes in the nucleus condense in preparation for cell division, each chromosome becomes progressively shorter and thicker as the 30-nm fiber coils onto itself to form a 300-nm coil, a 700-nm coiled coil, and finally a 1400-nm condensed chromosome in a manner that is still not fully understood. The progressive packaging constitutes **chromosome condensation,** an active, energy-consuming process requiring the participation of several types of proteins.

Greater detail of the structure of a fully condensed chromosome is revealed when the histones are chemically removed (**Fig. 13.13**). Without histones, the DNA spreads out in loops around a supporting protein structure called the chromosome **scaffold.** Each loop of relaxed DNA is 30 to 90 kb long and anchored to the scaffold at its base. Before removal of the histones, the loops are compact and supercoiled. Each human chromosome contains 2000–8000 such loops, depending on its size.

Despite intriguing similarities between the nucleoid model in Fig. 13.11 and the chromosome scaffold model in Fig. 13.13, the structures evolved independently and make use of different types of protein to bind the DNA and form the folded structure of DNA and protein. Furthermore, the size of the eukaryotic chromosome is vastly greater than the size of the bacterial nucleoid. To appreciate the difference in scale, keep in mind that the volume of a fully condensed human *chromosome* is five times larger than the volume of a bacterial *cell.*

FIG. 13.11 **Bacterial nucleoid.** The circular bacterial chromosome twists on itself to form supercoils, which are anchored by proteins. *Source. Dr. Klaus Boller/Science Source.*

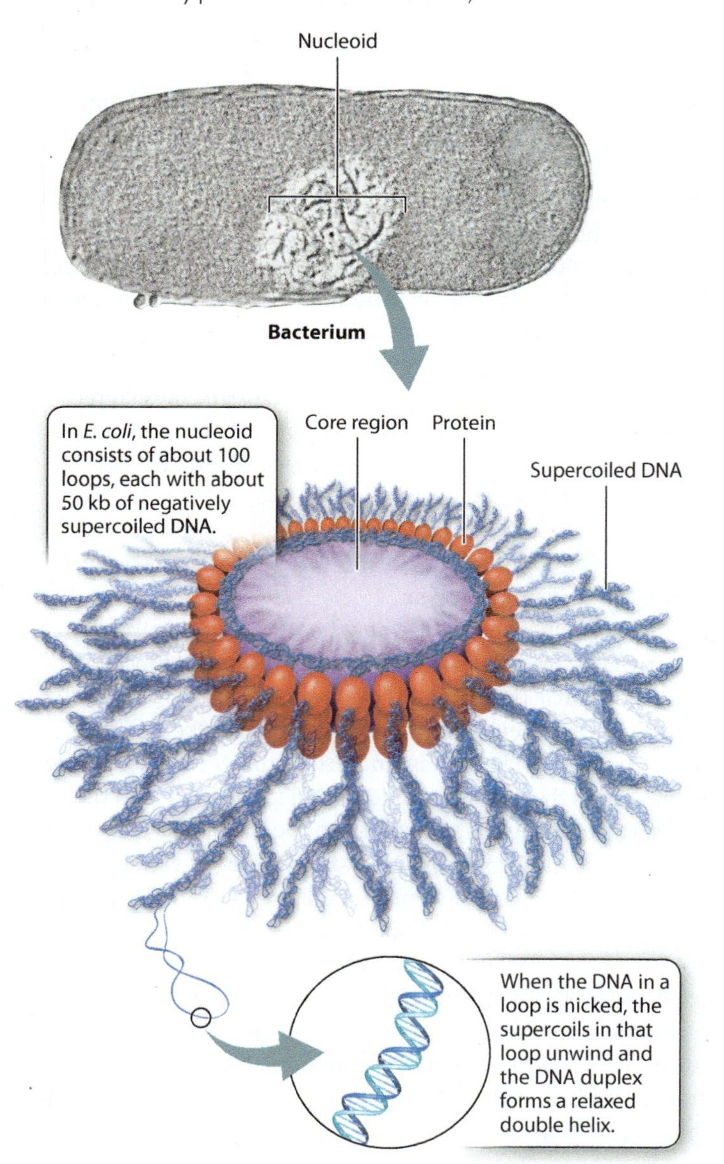

The human genome consists of 22 pairs of chromosomes and two sex chromosomes.

As emphasized in Chapter 11, the orderly process of meiosis is possible because chromosomes occur in pairs. The pairs usually match in size, general appearance, and position of the centromere, but there are exceptions, such as the *X* and *Y* sex chromosomes. The pairs of chromosomes that match in size and appearance are called **homologous chromosomes.** The members of each pair of homologous chromosomes have the same genes arranged in the same order along their length. If the DNA duplexes in each pair of homologs were denatured, each DNA strand could form a duplex with its complementary strand from the other homolog. There would be some differences in DNA sequence due to

FIG. 13.12 Levels of chromosome condensation. DNA is wrapped around nucleosomes to form a 10-nm fiber, which can become coiled into a 30-nm fiber and higher-order structures.

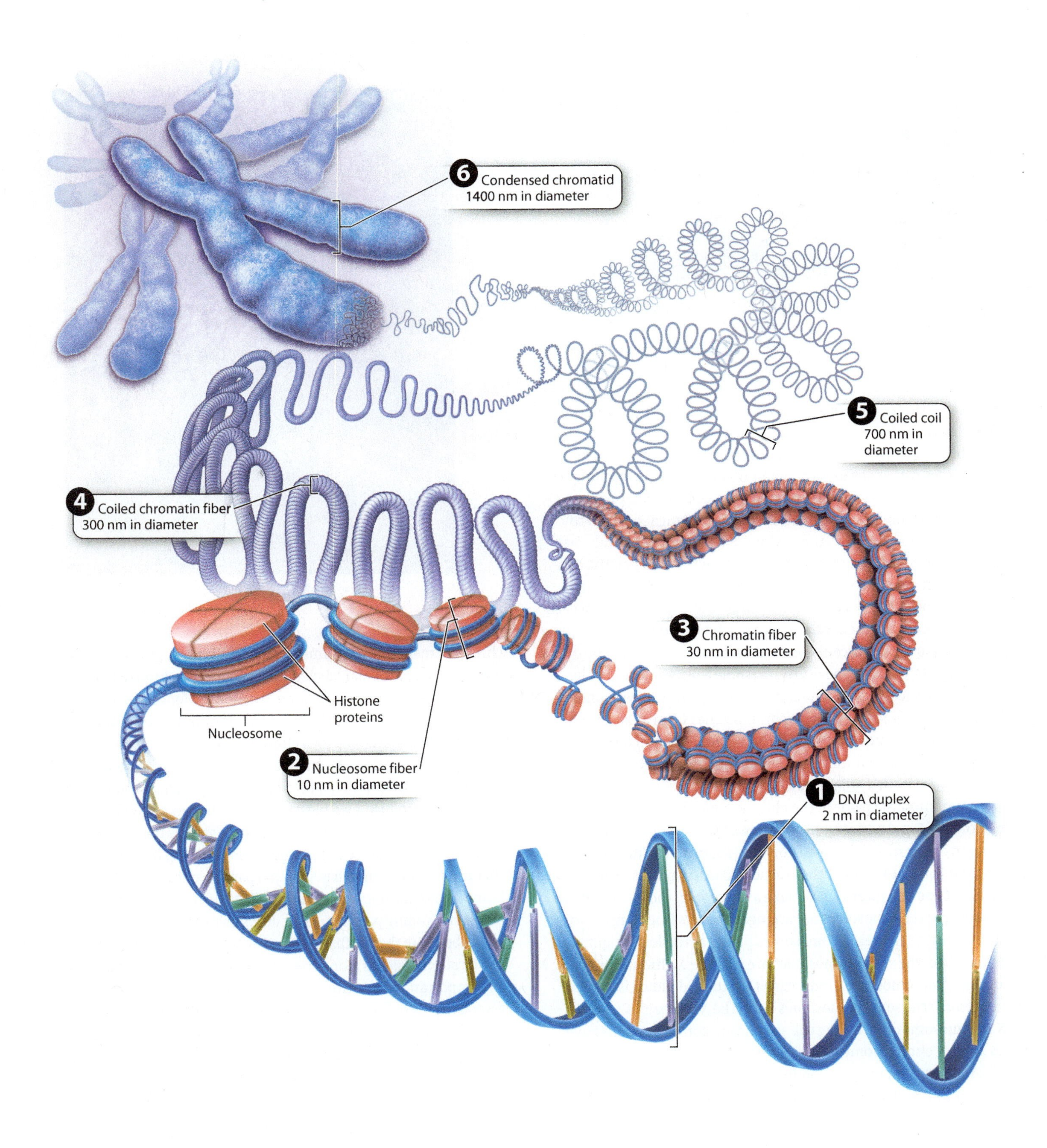

FIG. 13.13 (a) A chromosome with histones and (b) a chromosome depleted of histones, showing the underlying scaffold. *Source: Courtesy of Ulrich Laemmli.*

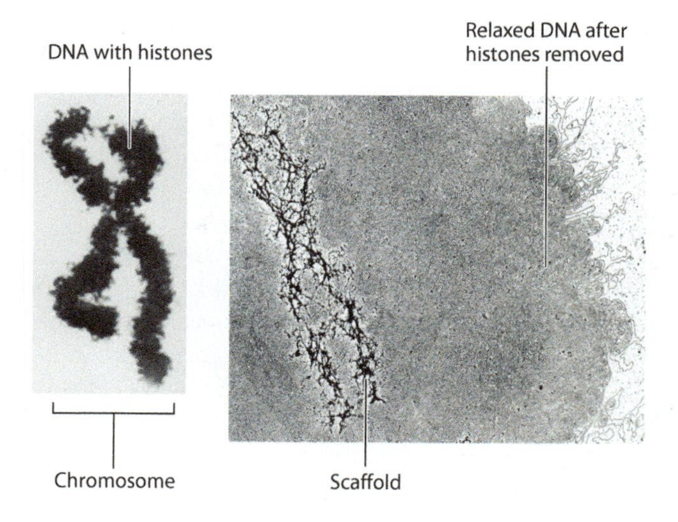

FIG. 13.14 A chromosome paint of chromosomes from a human male. (a) Condensed chromosomes are "painted" with fluorescent dyes. (b) Chromosomes are arranged in the standard form of a karyotype. *Source: NHGRI, www.genome.gov.*

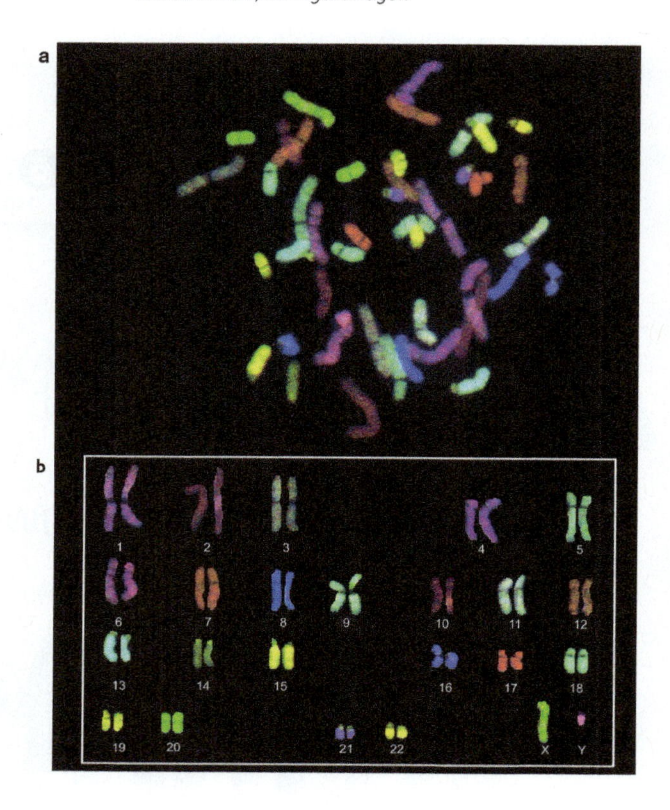

genetic variation, but not so many differences as to prevent DNA hybridization.

Chromosome painting illustrates the nearly identical nature of the DNA molecules in each pair of homologs. In this technique, individual chromosomes are isolated from cells in metaphase of mitosis. Metaphase of mitosis is the easiest stage in which to isolate chromosomes because of the availability of chemicals that prevent the spindle from forming. These chemicals block the cell cycle at metaphase, so cells progress to metaphase and then stop.

Once the chromosomes have been isolated, the DNA from each chromosome is fragmented, denatured, and labeled with a unique combination of fluorescent dyes. Under fluorescent light, the dyes give the DNA in each type of chromosome a different color. The fluorescently labeled DNA fragments are then mixed with and hybridized to intact metaphase chromosomes from another cell. Each labeled fragment hybridizes to its complementary sequence in the metaphase chromosomes, "painting" each metaphase chromosome with dye-labeled DNA fragments.

A chromosome paint of the chromosomes in a human male is shown in **Fig. 13.14.** Fig. 13.14a shows the chromosomes in the random orientation in which they were found in the metaphase cell, and Fig. 13.14b shows them arranged in a standard form called a **karyotype** (Chapter 11). To make a karyotype, the images of the homologous chromosomes are arranged in pairs from longest to shortest, with the sex chromosomes placed at the lower right. In this case, the sex chromosomes are *XY* (one *X* chromosome and one *Y* chromosome), indicating that the individual is male; in a female the sex chromosomes would be *XX* (a pair of *X* chromosomes). Including the sex chromosomes, humans have 23 pairs of chromosomes.

An important observation from the chromosome paint shown in Fig. 13.14 is that the two members of each pair of homologous chromosomes show the same pattern of fluorescent color. This means that a particular labeled DNA fragment hybridized only with the two homologs of one chromosome. Hence, the DNA in each pair of homologous chromosomes is different from that in any other pair of homologous chromosomes.

Higher resolution of human chromosomes can be obtained by the use of stains that bind preferentially to certain chromosomal regions and produce a visible pattern of bands, or crosswise striations, in the chromosomes. One such stain is the Giemsa stain; a Giemsa-stained karyotype is shown in **Fig. 13.15.** Note that each chromosome has a unique pattern of bands and that homologous chromosomes can readily be identified by their identical banding patterns. Procedures using the Giemsa stain yield about 300 bands that are used as landmarks for describing the location of genes along the chromosome. Such landmarks are also important in identifying specific chromosomal abnormalities associated with diseases, including some childhood leukemias.

Every species of eukaryote has its characteristic number of chromosomes. The number differs from one species to the next, with little relation between chromosome number and genome size. Despite variation in number, the rule that chromosomes

FIG. 13.15 **Giemsa bands in a karyotype of a human male.** Each dashed line marks the position of the centromere.
Source: Darryl Leja, NHGRI. www.genome.gov.

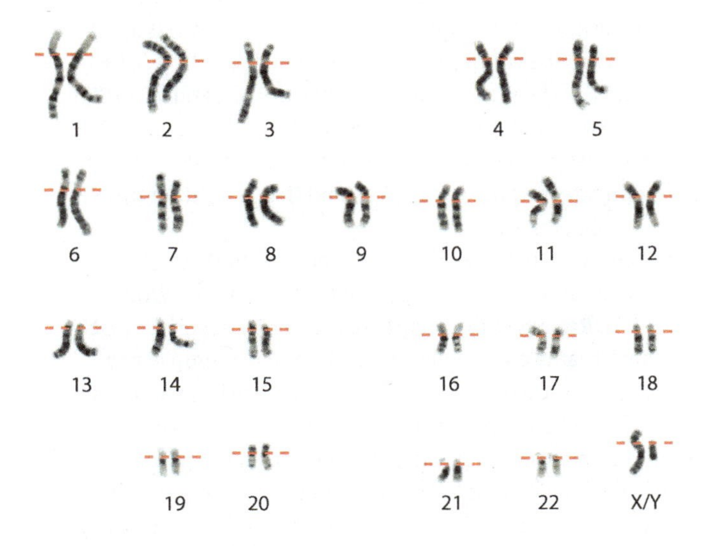

come in pairs holds up pretty well, with the exception of the sex chromosomes. Polyploids, too, are an exception, but even in naturally occurring polyploids the number of copies of each homologous chromosome is usually an even number, so there are pairs of homologs after all.

The occurrence of chromosomes in pairs allows eukaryotes to reproduce sexually. When reproductive cells are formed during meiotic cell division, each cell receives one and only one copy of each of the pairs of homologous chromosomes (Chapter 11). When reproductive cells from two individuals fuse to form an offspring cell, the chromosome number characteristic of the species is reconstituted.

Organelle DNA forms nucleoids that differ from those in bacteria.

Most eukaryotic cells contain mitochondria, and many contain chloroplasts. Each type of organelle has its own DNA, meaning that eukaryotic cells have multiple genomes. Each eukaryotic cell has a **nuclear genome** consisting of the DNA in the chromosomes. Cells with mitochondria also have a **mitochondrial genome,** and those with chloroplasts also have a **chloroplast genome.**

Because the genome organization and mechanisms of protein synthesis in these organelles resemble those of bacteria, most biologists subscribe to the theory that the organelles originated as free-living bacterial cells that were engulfed by primitive eukaryotic cells billions of years ago (Chapter 27). In Chapters 7 and 8, we saw that the likely ancestor of mitochondria resembled a group of today's non-photosynthetic bacteria (a group that includes *E. coli*), and the likely ancestor of chloroplasts resembled today's photosynthetic cyanobacteria. In both cases, the DNA of the organelles became smaller during the course of evolution because most of the genes were transferred to the DNA in the nucleus. If the products of these transferred genes are needed in the organelles, they are synthesized in the cytoplasm and targeted for entry into the organelles by signal sequences (Chapter 5).

DNA in mitochondria and chloroplasts is usually circular and exists in multiple copies per organelle. Among animals, the size of a mitochondrial DNA molecule ranges from 14 to 18 kb. Mitochondrial DNA in plants is generally much larger, up to 100 kb. Plant chloroplast DNA is more uniform in size than mitochondrial DNA, with a range of 130 to 200 kb.

On the basis of size alone, some kind of packaging of organelle DNA is necessary. For example, the mitochondrial DNA in human cells is a circular molecule of 16 kb. Fully extended, it would have a circumference about as large as that of the mitochodrion itself. From their bacterial origin, you might expect organelle DNA to be packaged as a nucleoid rather than as a chromosome, and indeed it is. But the structures of the nucleoid in mitochondria and chloroplasts differ from each other, and also differ from those in free-living bacteria.

13.5 VIRUSES AND VIRAL GENOMES

Now we come to viral genomes—but not because they are particularly complex. Viral genomes are actually rather small and compact with little or no repetitive DNA, and their sequences are relatively easy to assemble, as we have seen in Fig. 13.7 for HIV. What sets viral genomes apart from those of cellular organisms are the types of nucleic acid that make up their genomes and the manner in which these genomes are replicated in the infected cells. In their nucleic-acid composition and manner of replication, viral genomes are far more diverse than the genomes of cellular organisms: All cellular organisms have genomes of double-stranded DNA that replicate by means of the processes described in Chapter 12. Not so with viruses. Furthermore, whereas biologists classify cellular organisms according to their degree of evolutionary relatedness, they classify viruses based on their type of genome and mode of replication.

We mostly think of viruses as causing disease, and indeed they do. Familiar examples include influenza ("flu"), polio, and HIV. In some cases, they cause cancer (Case 2: Cancer). In fact, the term "virus" comes from the Latin for "poison." But viruses play other roles as well. Some viruses transfer genetic material from one cell to another. This process is called horizontal gene transfer to distinguish it from parent-to-offspring (vertical) gene transfer. Horizontal gene transfer has played a major role in the evolution of bacterial and archaeal genomes, as well as in the origin and spread of antibiotic-resistance genes (Chapter 26). Molecular biologists have learned to make use of this ability of viruses to deliver genes into cells.

In this section, we focus on viral diversity with an emphasis on viral genomes. Thousands of viruses have been described in detail, and probably millions more have yet to be discovered. It has been estimated that life on Earth is host to 10^{31} virus particles—ten hundred thousand million more virus particles than grains of sand! Most of these viruses infect bacteria, archaeons, and unicellular eukaryotes, and they are they are especially abundant in the ocean (10^{11} viruses per liter). Viruses are small, consisting of little more than a genome in a package, and they can reproduce only by hijacking host-cell functions, but as a group they have had amazing evolutionary success.

Viruses can be classified by their genomes.

The genomes of viruses are diverse. Some genomes are composed of RNA and others of DNA. Some are single stranded, others are double stranded, and still others have both single- and double-stranded regions. Some are circular and others are made up of a single piece or multiple linear pieces of DNA (called linear and segmented genomes, respectively).

Unlike forms of cellular life, there is no evidence that all viruses share a single common ancestor. Different types of virus may have evolved independently more than once. Since classification of viruses based on evolutionary relatedness is not possible, other criteria are necessary. One of the most useful classifications is the type of nucleic acid the virus contains and how the messenger RNA, which produces viral proteins, is synthesized. The classification is called the Baltimore system after David Baltimore, who devised it.

According to the Baltimore system, there are seven major groups of viruses, designated I–VII, as shown in **Fig. 13.16.** These groups are largely based on whether their nucleic acid is double-stranded DNA, double-stranded RNA, partially double-stranded and partially single-stranded, or single-stranded RNA or DNA with a positive (+) or negative (−) sense. The sense of a nucleic acid molecule is positive if its sequence is the same as the sequence of the mRNA that is used for protein synthesis, and negative if it is the complementary sequence. For example, a (+)RNA strand has the same nucleotide sequence as the mRNA, whereas a (−) RNA strand has the complementary sequence. Similarly, a (+) DNA strand has the same sequence as the mRNA, and a (−)DNA strand has the complementary sequence (except that U in RNA is replaced with T in DNA). Because mRNA is synthesized from a DNA template, it is the (−)DNA strand that is used for mRNA synthesis (Chapter 3).

Two groups synthesize mRNA by the enzyme reverse transcriptase, and therefore are placed into their own groups (VI and VII). **Reverse transcriptase** is an RNA-dependent DNA polymerase that uses a single-stranded RNA as a template to synthesize a DNA strand that is complementary in sequence to the RNA (Fig. 13.16). The reverse transcriptase then displaces the RNA template and replicates the DNA strand to produce a double-stranded DNA molecule that can be incorporated into the host genome. In synthesizing DNA from an RNA template, the enzyme reverses the usual flow of genetic information from DNA to RNA. This capability is so unusual and was so unexpected that many molecular biologists at first doubted whether such an enzyme could exist. Finally the enzyme was purified and its properties verified, for which its discoverers, Howard Temin and David Baltimore, were awarded the Nobel Prize in Physiology or Medicine in 1975, shared with Renato Dulbecco.

As we saw earlier, genome size varies greatly among different viruses. RNA viral genomes tend to be smaller than DNA viral

FIG. 13.16 The Baltimore system of virus classification. This system classifies viruses by type of nucleic acid and the way mRNA is produced.

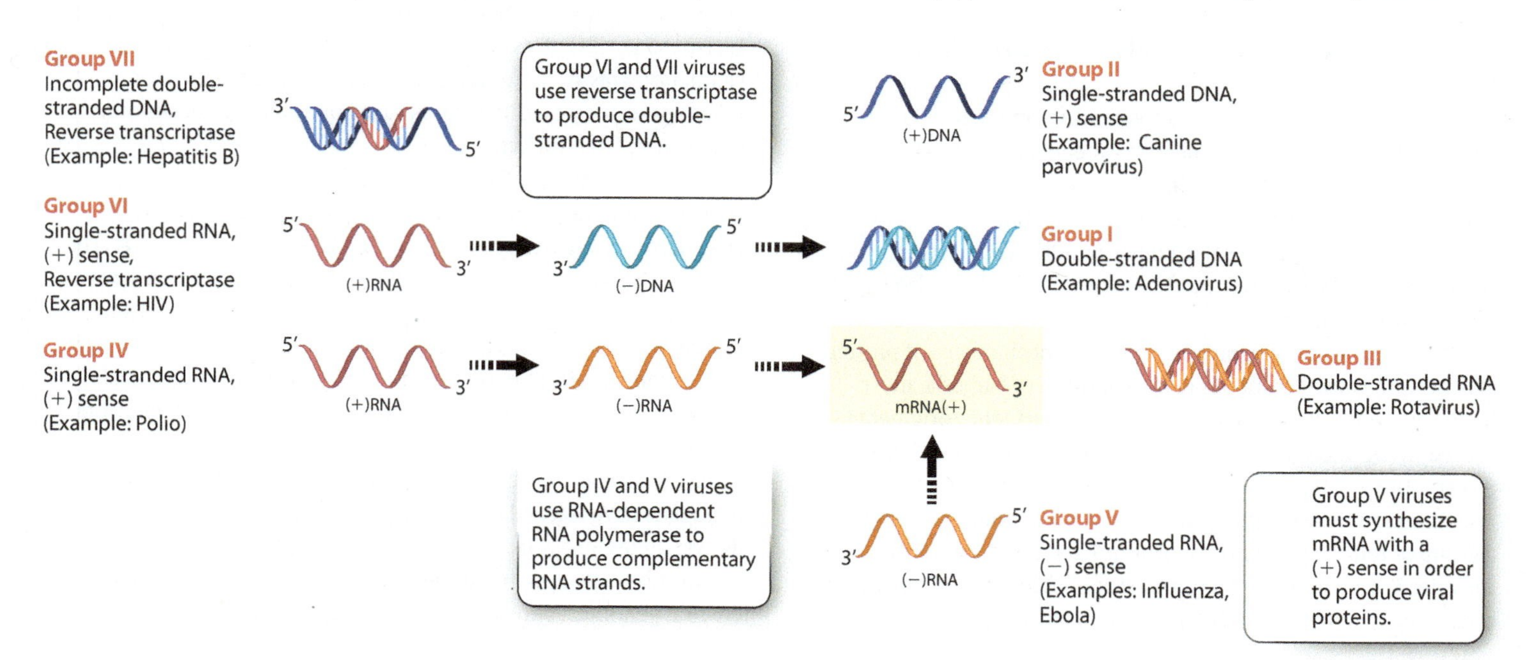

genomes. Most eukaryotic viruses that have RNA genomes and most plant viruses, including tobacco mosaic virus, are in group IV. Among bacterial and archaeal viruses, most genomes consist of double-stranded DNA.

The host range of a virus is determined by viral and host surface proteins.

Although viruses show great diversity in their genomic nucleic-acid composition and manner of replication, they all consist of some type of nucleic acid, a capsid, and sometimes a lipid **envelope** (Chapter 1). Viruses can reproduce only by infecting living cells and exploiting the cell's metabolism and protein synthesis to produce more viruses. Most viruses are tiny, some hardly larger than a ribosome, 25–30 nm in diameter. Roughly speaking, the average size of a virus, relative to that of the host cell it infects, may be compared to the size of an average person relative to that of a commercial airliner.

All known cells and organisms are susceptible to viral infection, including bacteria, archaeons, and eukaryotes. A cell in which viral reproduction occurs is called a **host cell.** Some viruses kill the host cell; others do not. Although viruses can infect all types of living organism, a given virus can infect only certain species or types of cell. At one extreme, a virus can infect just a single species. Smallpox, which infects only humans, is a good example. In cases like this, we say that the virus has a narrow host range. For other viruses, the host range can be broad. For example, rabies infects many different types of mammal, including squirrels, dogs, and humans. Similarly, tobacco mosaic virus, a plant virus, infects more than 100 different species of plants (**Fig. 13.17**). No matter how broad the host range, however, plant viruses cannot infect bacteria or animals, and bacterial viruses cannot infect plants or animals.

Host specificity results from the way that viruses gain entry into cells. Proteins on the surface of the capsid or envelope (if present) bind to proteins on the surface of host cells. These proteins interact in a specific manner, so the presence of the host protein on the cell surface determines which cells a virus can infect. For example, a protein on the surface of the HIV envelope called gp120 (encoded by the *env* gene; see Fig. 13.7) binds to a protein called CD4 on the surface of certain immune cells (Chapter 43), so HIV infects these cells but not other cells.

Following attachment, the virus gains entry into the cell, where it can replicate by using the host cell machinery to make new viruses. In some cases, the viral genome integrates into the host cell genome (Chapter 19).

The host range of a virus can change because of mutation and other mechanisms (Chapter 43). For example, avian, or bird, flu is a type of influenza virus that was once restricted primarily to birds but now can infect humans. The first reported case in humans was in 1996, and the disease has since spread widely. Similarly, HIV once infected only nonhuman primates, but its host range expanded in the twentieth century to include humans. Canine distemper virus, which infects dogs, expanded its host range in the early 1990s, leading to the infection and death of lions in Tanzania.

FIG. 13.17 A plant infected by tobacco mosaic virus. *Courtesy H.D. Shew; Reproduced, by permission, from Shew, H.D., and Lucas, G.B., eds. 1991. Compendium of Tobacco Diseases. American Phytopathological Society, St. Paul, MN.*

Viruses have diverse sizes and shapes.

Viruses show a wide variety of shapes and sizes, which are determined in some cases by the type of genome they carry. Three examples are shown in **Fig. 13.18.** The T4 virus (Fig. 13.18a) infects cells of the bacterium *Escherichia coli.* Viruses that infect bacterial cells are called **bacteriophages,** which literally means "bacteria eaters." The T4 bacteriophage has a complex structure that includes a head composed of protein surrounding a molecule of double-stranded DNA, a tail, and tail fibers. In infecting a host cell, the T4 tail fibers attach to the surface, and the DNA and some proteins are injected into the cell through the tail.

Most viruses are not structurally so complex. Consider the tobacco mosaic virus, which infects plants (see Fig. 13.17). It has a helical shape formed by the arrangement of protein subunits entwined with a molecule of single-stranded RNA (Fig. 13.18b). Tobacco mosaic virus was the first virus discovered, revealed in experiments showing that the infectious agent causing brown spots and discoloration of tobacco leaves was so small that it could pass through the pores of filters that could trap even the smallest bacterial cells.

Many viruses have an approximately spherical shape formed from polygons of protein subunits that come together at their edges to form a polyhedral capsid. Among the most common polyhedral shapes is an icosahedron, which has 20 identical triangular faces. The example in Fig. 13.18c is adenovirus, a common cause of upper respiratory infections in humans. Many viruses that infect eukaryotic cells, such as adenovirus, are surrounded by a glycoprotein envelope composed of a lipid bilayer with embedded proteins and glycoproteins that recognize and attach to host cell receptors.

FIG. 13.18 Virus structures. Viruses come in a variety of shapes, including (a) the head-and-tail structure of a bacteriophage, (b) the helical shape of tobacco mosaic virus, and (c) the icosahedral shape of adenovirus. *Sources: a. Dept. of Microbiology, Biozentrum/Getty Images; b. Biology Pics/Science Source; c. BSIP/Science Source.*

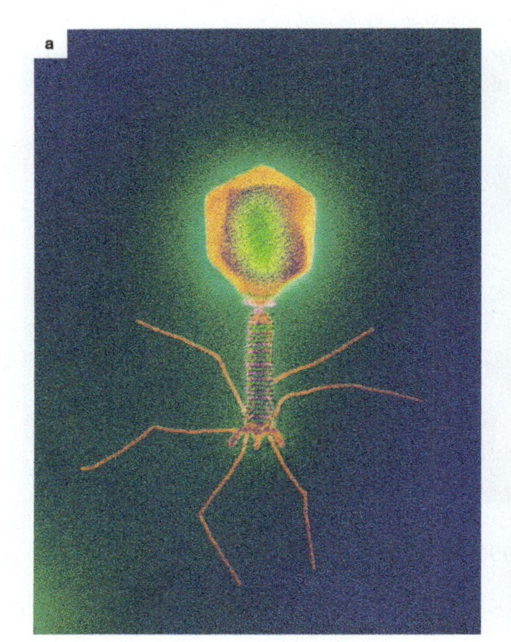

Viruses are capable of self-assembly.

The genome of a virus contains the genetic information needed to specify all the structural components of the virus. Progeny virus particles are formed according to **molecular self-assembly:** When the viral components are present in the proper relative amounts and under the right conditions, the components interact spontaneously to assemble themselves into the mature virus particle.

Tobacco mosaic virus illustrates the process of self-assembly. In the earliest stages, the coat-protein monomers assemble into two circular layers forming a cylindrical disk. This disk binds with the RNA genome, and the combined structure forms the substrate for polymerization of all the other protein monomers into a helical filament that incorporates the rest of the RNA as the filament grows (**Fig. 13.19**). The mature virus particle consists of 2130 protein monomers and 1 single-stranded RNA molecule of 6400 ribonucleotides. ■

FIG. 13.19 Tobacco mosaic virus. The protein and RNA components assemble themselves inside a cell.

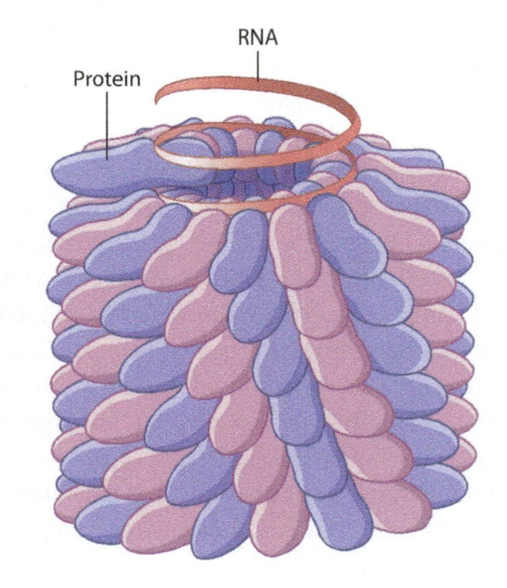

Core Concepts Summary

13.1 A genome is the genetic material of a cell, organism, organelle, or virus, and its sequence is the order of bases along the DNA or (in some viruses) RNA.

The sequence of an organism's genome can be determined by breaking up the genome into small fragments, sequencing these fragments, and then putting the sequences together at their overlaps. page 272

Sequences that are repeated in the genome can make sequence assembly difficult. page 273

13.2 Researchers annotate genome sequences to identify genes and other functional elements.

Genome annotation is the process by which the types and locations of the different kinds of sequences, such as protein-coding genes, are identified. page 275

Genome annotation sometimes involves scanning the DNA sequence for characteristic sequence motifs. page 276

Comparison of DNA sequences with messenger RNA sequences reveals the intron–exon structure of protein-coding genes. page 276

By comparing annotated genomes of different organisms, we can gain insight into their ancestry and evolution. page 277

The annotated HIV genome shows it is a retrovirus, and it contains the genes *gag*, *pol*, and *env*. page 277

13.3 The number of genes in a genome and the size of a genome do not correlate well with the complexity of an organism.

The C-value paradox describes the observation that the size of a genome (measured by its C-value) does not correlate with an organism's complexity. page 279

In eukaryotes, the C-value paradox can be explained by differences in the amount of noncoding DNA, including repetitive sequences and transposons. Some genomes have a lot of noncoding DNA; others do not. page 280

13.4 The orderly packaging of DNA allows it to carry out its functions and fit inside the cell.

Bacteria package their circular DNA in a structure called a nucleoid. page 281

Eukaryotic cells package their DNA into linear chromosomes. page 282

DNA in eukaryotes is wound around groups of histone proteins called nucleosomes to form a 10-nm fiber, which in turn coils to form higher-order structures, such as the 30-nm fiber. page 282

Diploid organisms have two copies of each chromosome, called homologous chromosomes. page 284

Humans have 23 pairs of chromosomes, including the *X* and *Y* sex-chromosome pair. Females are *XX* and males are *XY*. page 284

The genomes of mitochondria and chloroplasts are organized into nucleoids that resemble, but are distinct from, those of bacteria. page 285

13.5 Viruses have diverse genomes, but all require a host cell to replicate.

Viruses can be classified by the Baltimore system, which defines seven groups on the basis of type of nucleic acid and the way the mRNA is synthesized. page 286

Viruses can infect all types of organism, but a given virus can infect only some types of cell. page 287

The host range of a virus is determined by proteins on viral and host cells. page 287

Viruses have diverse shapes, including head-and-tail, helical, and icosahedral. page 287

Viruses are capable of molecular self-assembly under the appropriate conditions. page 288

Self-Assessment

1. Describe the shotgun method for determining the complete genome sequence of an organism.
2. Repeated sequences can be classified according to their organization in the genome as well as according to their function. Give at least two examples of each.
3. Explain the purpose of genome annotation.
4. Describe how the comparison of genomic DNA to messenger RNA can identify the exons and introns in a gene.
5. Explain how comparing the sequences of two genomes can help to infer evolutionary relationships.

6. What are some reasons why, in multicellular eukaryotes, genome size is not necessarily related to number of protein-coding genes or organismal complexity?
7. Compare and contrast the mechanisms by which bacterial cells and eukaryotic cells package their DNA.
8. Draw a nucleosome, indicating the positions of DNA and proteins.
9. Define "homologous chromosomes" and describe a technique that you could use to show their similarity.
10. Describe the steps necessary to synthesize mRNA from each of the following: double-stranded DNA, single-stranded (+)DNA, single-stranded (−)DNA, single-stranded (+)RNA, and single-stranded (−)RNA.

Log in to LaunchPad to check your answers to the Self-Assessment questions, and to access additional learning tools.

CHAPTER 14

Mutation and DNA Repair

Core Concepts

14.1 Mutations are very rare for any given nucleotide and occur randomly without regard to the needs of an organism.

14.2 Small-scale mutations include point mutations, insertions and deletions, and movement of transposable elements.

14.3 Chromosomal mutations involve large regions of one or more chromosomes.

14.4 DNA can be damaged by mutagens, but most DNA damage is repaired.

The sequencing of the human genome was a great step forward. However, as we saw in the previous chapter, it is a simplification to speak of *the* human genome. Virtually all species have abundant amounts of genetic variation, that is, different nucleotides at the same site in the genome. The reference human genome sequence is actually a composite of several genomes, displaying the most common nucleotide at most sites.

Variation among different individuals' genomes arise from **mutations.** Any heritable change in the genetic material is a mutation, and by "heritable" we mean that the mutation is stable and therefore passed on through cell division (meiotic cell division, mitotic cell division, or binary fission). The process by which mutations occur is fundamental in biology because mutation is the ultimate source of genetic variation. Genetic variation accounts in part for the physical differences we see among individuals, such as differences in hair color, eye color, and height. And, on a much larger scale, genetic variation results in the diversity of organisms on this planet, from bacteria to blue whales.

There are many different types of mutation, from small changes affecting a single base to larger alterations, such as the duplication or deletion of a segment of a chromosome. In this chapter, we examine some of the basic principles of mutation: the different types of mutation, how and when they occur, and how they are repaired. In Chapter 15, we look at common types of mutation, or genetic variation, present in populations, and, in Chapters 16 and 17, at how this variation is inherited from one generation to the next.

14.1 THE RATE AND NATURE OF MUTATIONS

Mutations result from mistakes in DNA replication or from unrepaired damage to DNA. The damage may be caused by reactive molecules produced in the normal course of metabolism, by chemicals in the environment, or by radiation of various types, including X-rays and ultraviolet light. Most genomes also contain DNA sequences that can "jump" from one position to another in the genome, and their insertion into or near genes is a source of mutation. Yet another source of mutation is incorrectly repaired chromosome breaks caused by reactive chemicals or radiation.

Most mutations are **spontaneous,** occurring by chance in the absence of any assignable cause. They occur randomly, unconnected to an organism's needs—it makes no difference whether or not a given mutation would benefit the organism. Whether a favorable mutation does or does not occur is purely a matter of chance. This key principle, that mutations are spontaneous and random, is the focus of this section.

For individual nucleotides, mutation is a rare event.

We'll see later in this chapter that there are several different types of mutation, but the most common mutation is the substitution

FIG. 14.1 Rate of mutation per nucleotide per replication. *Data from: J. W. Drake, B. Charlesworth, D.Charlesworth, J. F. Crow, 1998, "Rates of Spontaneous Mutation," Genetics 148:1667–1686.*

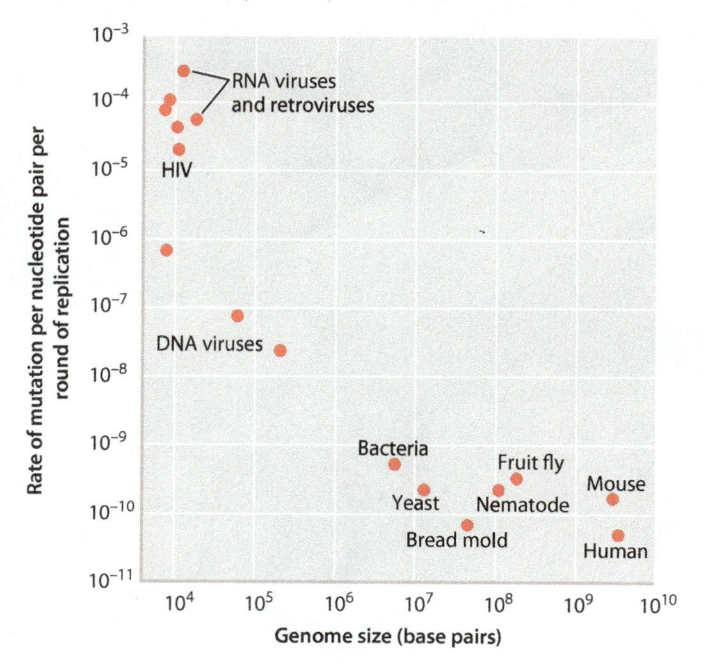

of one nucleotide pair for a different nucleotide pair. Nucleotide-substitution mutations are nevertheless relatively rare, and their rate of occurrence differs among organisms. **Fig. 14.1** compares the rates of newly arising mutations in a given base pair in a single round of replication in viruses and several types of organisms. Most of these mutations are due to errors in replication. As seen in the graph, the mutation rates for different organisms range across almost eight orders of magnitude.

The highest rates of mutation per nucleotide per replication are found among RNA viruses and retroviruses, including HIV. Lower rates occur in DNA viruses, and even lower rates in unicellular organisms such as bacteria and yeast. The rates of mutation per nucleotide per DNA replication are nearly the same for all multicellular animals, including mice and humans.

Mutation rates vary over a large range for a number of reasons. RNA viruses and retroviruses have a relatively high rate of mutation because RNA is a less stable molecule than DNA, but more importantly because the replication of these genomes lacks a proofreading function. For the other genomes, the rate of mutation per nucleotide per DNA replication reflects differences in fidelity of replication and the efficiency of proofreading and other mechanisms of DNA repair.

For the cellular organisms plotted in Fig. 14.1, the average mutation rate is about 10^{-10}, which means that, on average, only 1 nucleotide in every 10 billion is mistakenly substituted for another.

But averaging conceals many details. First, certain nucleotides are especially prone to mutation and can exhibit rates of mutation that are greater than the average by a factor of 10 or more. Sites in the genome that are especially mutable are called **hotspots.** Second, in some multicellular organisms, the rates of mutation differ between the sexes. In humans, for example, the rate of mutation is substantially greater in males than in females. Finally, the rates for the multicellular animals plotted in Fig. 14.1 depend on the type of cell: A distinction must be made between mutations that occur in **germ cells** (haploid gametes and the diploid cells that give rise to them) and mutations that occur in **somatic cells** (the other cells of the body). In mammals, the rate of mutation per nucleotide per replication is greater in somatic cells than in germ cells.

Across the genome as a whole, mutation is common.

We can also look at mutation rate at another level of scale. Instead of considering the rate per nucleotide per replication, we can examine the rate of mutation across an entire genome in one generation. While it is clear from Fig. 14.1 that the rate of mutation per nucleotide per replication in most multicellular organisms is low, the rate of mutation per genome per generation depends on the size of the genome and the number of cell divisions per generation. Taking into account genome size and cell divisions per generation, the rate of new mutations across the whole genome per generation is shown in **Fig. 14.2.** Note that, while humans have the smallest rate of mutation per nucleotide per replication (Fig. 14.1), humans also have the largest rate of mutation per genome per generation (Fig. 14.2). This seeming paradox arises because humans have a large genome and undergo many cell divisions per generation.

In humans, the average number of newly arising nucleotide-substitution mutations per genome in one generation is about 30, or about 60 per diploid zygote. However, about 80% of the newly arising mutations in a zygote come from the father. This is because, in a human male at age 30, the diploid germ cells have gone through about 400 cycles of DNA replication and cell division before meiosis takes place, as compared with about 30 cycles in females. Moreover, while the number of newly arising mutations from the mother remains approximately constant with mother's age, the number of newly arising mutations from the father increases with age. Sperm from men of age 40 contain about twice as many newly arising mutations as sperm from men of age 20.

Such a large number of new mutations as occurs in the human genome in one generation would be intolerable in organisms with a high density of protein-coding genes, such as bacteria or fungi. It is tolerable in humans and other mammals only because more than 90% of the nucleotides in the genome seem free to vary without deleterious consequences for the organism, and a mere 2.5% of the genome code for protein (Chapter 13). The vast majority of newly arising mutations therefore occur in noncoding DNA and are likely neutral or very nearly neutral in their effects. It is these mutations, accumulated through many generations and shuffled by recombination and independent assortment (Chapter 11), that account for the genetic diversity of most populations and the genetic uniqueness of each individual.

FIG. 14.2 **Rate of mutation per genome per generation.** *Data from: J. W. Drake, B. Charlesworth, D.Charlesworth, J. F. Crow, 1998, "Rates of Spontaneous Mutation," Genetics 148:1667–1686.*

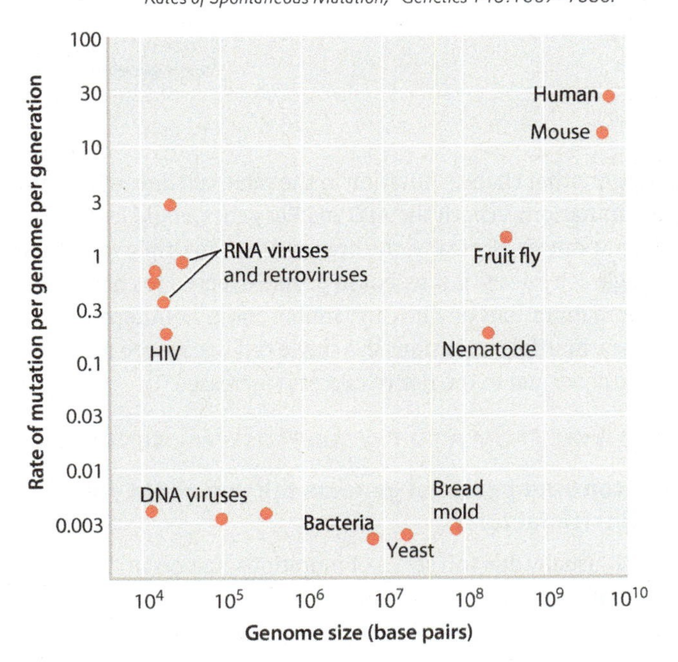

Only germ-line mutations are transmitted to progeny.

Which is more important—the rate of mutation per nucleotide per replication or the rate per genome per generation? That depends on context. Mutations can take place in any type of cell. Those that occur in eggs and sperm and the cells that give rise to them are called **germ-line mutations,** and those in nonreproductive cells are called **somatic mutations.** This distinction is important because somatic mutations affect only the individual in which they occur—they are not transmitted to future generations. In contrast, germ-line mutations are transmitted to future generations because they occur in reproductive cells.

For germ-line mutations, it is the rate of mutation per genome per generation that matters more. Germ-line mutations are important to the evolutionary process because, as they are passed from one generation of organism to the next, they may eventually come to be present in many individuals descended from the original carrier.

For somatic mutations, the mutation rate that matters is the rate of mutation per nucleotide per replication. Although somatic mutations are not transmitted to future generations, they are transmitted to daughter cells in mitotic cell divisions (Chapter 11). Hence, a somatic mutation affects not only the cell in which it occurs, but also all the cells that descend from it. The areas of different color or pattern that appear in "sectored" flowers, valued as ornamental plants (**Fig. 14.3**), are usually due to somatic mutations in flower-color genes. A mutation in a flower-color gene occurs in one cell, and as the cell replicates during development of the flower, all its descendants in the cell lineage—the generations of cells that originate from a single ancestral cell—carry that mutation, producing a sector with altered coloration.

Most cancers result from mutations in somatic cells (Chapter 11). In some cases, the mutation increases the activity of a gene that promotes cell growth and division, while in other cases, it decreases the activity of a gene that restrains cell growth and division. In either event, the mutant cell and its descendants escape from one of the normal control processes. Fortunately, a single somatic mutation is usually not sufficient to cause cancer—usually two or three or more mutations in different genes are required to derail control of normal cell division so extensively that cancer results.

To cause cancer, the mutations must occur sequentially in a single cell line. **Fig. 14.4** shows three key mutations that have been implicated in the origin of invasive colon cancer: *p53*, *Ras*, and *APC*. Each mutation occurs randomly, but if, by chance, a mutation in the *Ras* gene occurs in a cell that is derived from one in which the *APC* gene has been mutated, that cell's progeny forms

FIG. 14.3 **Somatic mutation.** Somatic mutations in the Japanese morning glory (*Ipomoea nil*) in cell lineages that differ in their ability to make purple pigment cause sectors of different pigmentation in the flower. *Source: Courtesy Atsushi Hoshino, National Institute for Basic Biology, Japan.*

FIG. 14.4 **Three somatic mutations implicated in the origin of invasive colon cancer.** These mutations must occur in the same cell lineage for cancer to develop. *Source: Kathleen R. Cho, University of Michigan Medical School.*

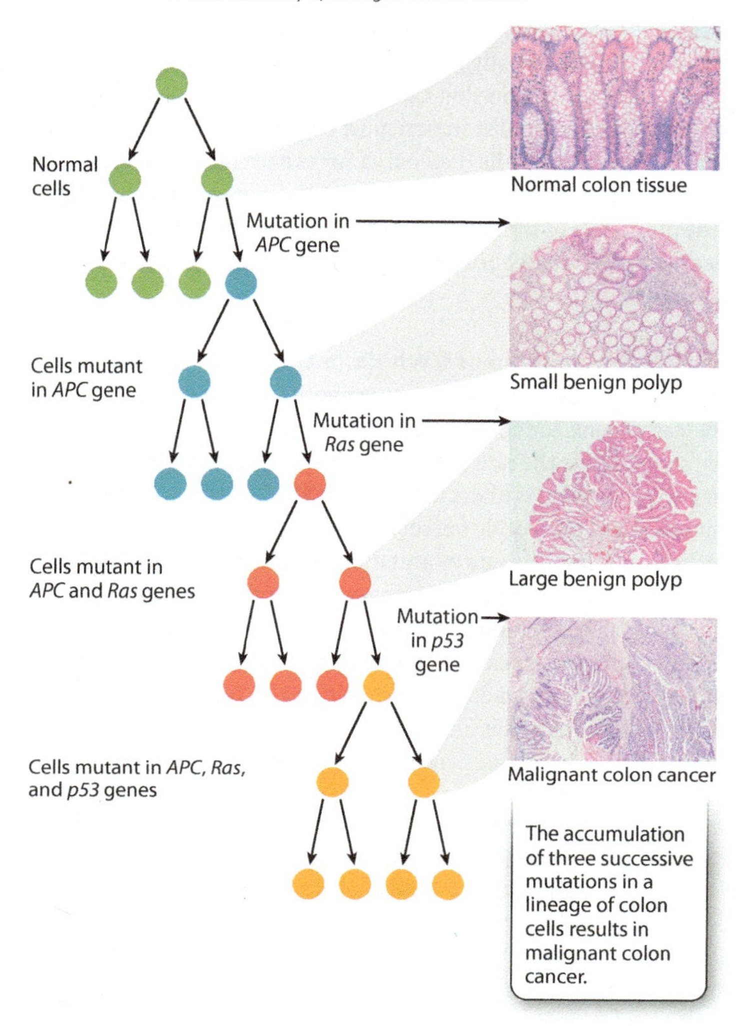

a polyp. Another chance mutation in the same cell line, which now carries mutations in both the *APC* and *Ras* genes, could lead to malignant cancer. Normally, the occurrence of multiple mutations in a single cell lineage is rare, but in people exposed to chemicals that cause mutations or who carry mutations in DNA repair processes, multiple mutations in a single cell lineage are more likely to occur and so the risk of cancer is increased.

? CASE 3 YOU, FROM A TO T: YOUR PERSONAL GENOME

What can your personal genome tell you about your genetic risk factors?

Cancer is usually due to a series of mutations that occur sequentially in a single lineage of somatic cells, as illustrated for colon cancer in Fig. 14.4. Each type of cancer is caused by its own particular sequence of mutations, although mutations in some genes are implicated in several different types of cancer. An example is the *p53* gene, the nonmutant product of which detects DNA damage and slows the cell cycle to allow time for DNA repair (Chapter 11). Mutations in *p53* are one step in the mutational progression of many different types of cancer, including colon cancer and breast cancer.

In most individuals with cancer, all the sequential mutations that cause the cancer are spontaneous mutations that take place in somatic cells. They are not transmitted through the germ line, and so there is little or no increased risk of cancer in the offspring. In some families, however, there is a germ-line mutation in one of the genes implicated in cancer that is transmitted from parents to their children. In any child who inherits the mutation, all cells in the body contain the defective gene, and hence the cells already have taken one of the mutational steps that lead to cancer. The effect of such a germ-line mutation is therefore to reduce the number of additional mutations that would otherwise be necessary to produce cancer cells.

Any mutation that increases the risk of disease in an individual is known as a **genetic risk factor** for that disease. For colon cancer, the major genetic risk factors are mutations in *APC, Ras,* and *p53*. For breast cancer, the major genetic risk factors are mutations in the *BRCA1* and *BRCA2* genes (Case 3: You, from A to T). A risk factor does not cause the disease, but it makes the disease more likely to occur. For the genes implicated in colon cancer, each is a risk factor because when it is mutated, fewer additional mutations are needed to bring about tumor growth.

The DNA sequence of each of our personal genomes can reveal the genetic risk factors that each of us carries, not only for cancer but for many other diseases as well. Not all genetic risk factors are known for all diseases, and a great deal of current research aims to identify new ones. But many genetic risk factors are already known for a large number of common diseases, including high blood pressure, diabetes, inflammatory bowel disease, age-related macular degeneration, Alzheimer's disease, and many forms of cancer. Therefore, your personal genome can be of great value in identifying diseases for which you carry risk factors, as was the case for Claudia Gilmore (Case 3: You, from A to T).

Our personal genomes can identify only *genetic* risk factors, however. In many cases, disease risk is substantially increased by environmental risk factors as well, especially lifestyle choices. While there are genetic risk factors for lung cancer, for example, the single biggest environmental risk factor is smoking tobacco. For skin cancer, the greatest environmental risk factor is exposure to the damaging ultraviolet rays in sunlight or in the sunlamps used in tanning beds. For heart disease, it is smoking, lack of physical activity, and obesity. For diabetes, it is an unhealthy diet.

For breast cancer, the environmental risk factors include certain forms of hormone therapy, lack of physical activity, and alcohol. While we may not be able to do much about the genetic risk factors for any of these conditions, knowing that we have them may make us more careful about the lifestyle choices that we make.

Mutations are random with regard to an organism's needs.

How do mutations arise? Consider the following example: If an antibiotic is added to a liquid culture of bacterial cells that are growing and dividing, most of the cells are killed, but a few survivors continue to grow and divide. These survivors are found to contain mutations that confer resistance to the antibiotic. This simple observation raises a profound question. Does this experiment reveal the presence of individual bacteria with mutations that had arisen spontaneously and were already present? Or do the antibiotic-resistant mutants arise in response to the presence of the antibiotic?

These alternative hypotheses have deep implications for all of biology because they suggest two very different ways in which mutations might arise. The first suggests that mutations occur without regard to the needs of an organism. According to this hypothesis, the presence of the antibiotic in the experiment with bacterial cells does not direct or induce antibiotic resistance in the cells, but instead allows the small number of preexisting antibiotic-resistant mutants to flourish. The second hypothesis suggests that there is some sort of feedback between the needs of an organism and the process of mutation, and the environment directs specific mutations that are beneficial to the organism.

To distinguish between these two hypotheses, Joshua and Esther Lederberg in 1952 carried out a now-famous experiment, described in **Fig. 14.5.** Bacterial cells were grown and formed colonies on agar plates in the absence of antibiotic. Then, using replica plating, a technique they invented, the Lederbergs transferred these colonies to new plates containing antibiotic. Only bacteria that were resistant to the antibiotic grew on the new plates. Because replica plating preserved the arrangement of the colonies, the Lederbergs were able to go back to the original plate and identify the colony that produced the antibiotic-resistant colony on the replica plate. From that original colony, they were then able to isolate a pure culture of antibiotic-resistant bacteria.

Replica plating allowed the Lederbergs to isolate a pure culture of antibiotic-resistant bacteria, even though the original bacteria never were exposed to antibiotic. This result supported the first hypothesis: Mutations occur randomly, and without regard to the needs of the organisms. The role of the environment is not to create specific mutations, but instead to select for them. The principle the Lederbergs demonstrated is true of all organisms so far examined.

→ **Quick Check 1** If mutations occur at random with respect to an organism's needs, how does a species become more adapted to its environment over time?

HOW DO WE KNOW?

FIG. 14.5

Do mutations occur randomly, or are they directed by the environment?

BACKGROUND Researchers have long observed that beneficial mutations tend to persist in environments where they are useful—in the presence of antibiotic, bacterial populations become antibiotic resistant; in the presence of insecticides, insect populations become insecticide resistant.

HYPOTHESIS These observations lead to two hypotheses about how a mutation, such as one that confers antibiotic resistance on bacteria, might arise. The first suggests that mutations occur randomly in bacterial populations and over time become more common in the population in the presence of antibiotic (which destroys those bacteria without the mutation). In other words, they occur randomly with respect to the needs of an organism. The second hypothesis suggests that the environment, in this case the application of antibiotic, induces or directs antibiotic resistance.

a.

Agar plate 1

Agar plate 2

Incubation to allow growth of colonies

b.

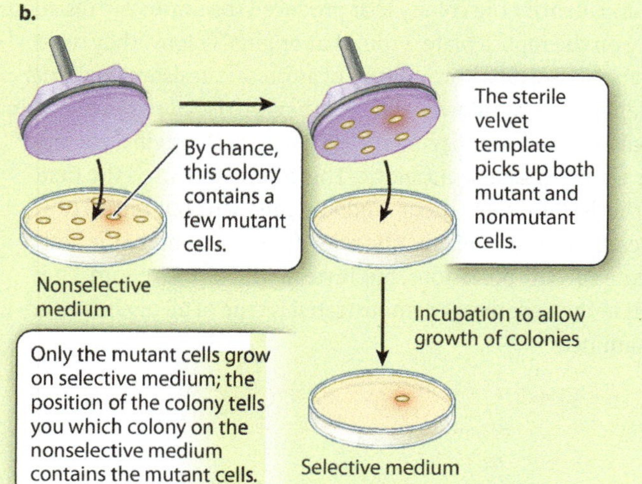

c.

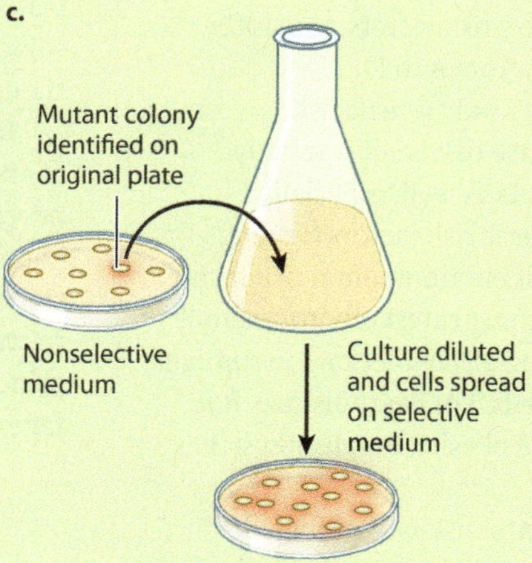

METHOD To distinguish between these two hypotheses, Joshua and Esther Lederberg developed replica plating. In this technique, bacteria are grown on agar plates, where they form colonies (Fig. 14.5a). The cells in any one colony result from the division of a single original cell, and thus they constitute a group of cells that are genetically identical except for rare mutations that occur in the course of growth and division. Then a disk of sterilized velvet is pressed onto the plate. Cells from each colony stick to the velvet disk (in mirror image, but the relative positions of the colonies are preserved). The disk is then pressed onto the surface of a fresh plate, transferring to the new plate a few cells that originate from each colony on the first agar plate, in their initial positions.

EXPERIMENT First, the Lederbergs grew bacterial colonies on medium without antibiotic, called a nonselective medium because all cells are able to grow and form colonies on it. Then, by replica plating, they transferred some cells from each colony to a plate containing antibiotic, so only antibiotic-resistant cells could multiply and form colonies. (Medium containing antibiotic is a selective medium because it "selects" for a particular attribute or element, in this case antibiotic-resistant cells.) Because replica plating preserves the arrangement of the colonies, the location of an antibiotic-resistant colony on the selective medium reveals the location of its parental colony on the nonselective plate (Fig. 14.5b). Finally, the Lederbergs were able to go back to the parental colony and demonstrate that it was a pure culture of antibiotic-resistant bacteria by plating cultures of this colony on selective medium (Fig. 14.5c).

CONCLUSION The Lederbergs' replica-plating experiments demonstrated that antibiotic-resistant mutants can arise in the absence of antibiotic because at no time in the experiments did the cells on nonselective medium come into contact with the antibiotic. Only the successive generations of daughter cells carried over to selective medium by replica plating were exposed to the antibiotic. Nevertheless, by their procedure the Lederbergs were able to isolate pure colonies of antibiotic-resistant cells.

FOLLOW-UP WORK These results have been extended to other types of mutation and other organisms, suggesting that mutations are random and not directed by the environment.

SOURCE Lederberg, J., and E. M. Lederberg. 1952. "Replica Plating and Indirect Selection of Bacterial Mutants." *Journal of Bacteriology* 63:399–406.

14.2 SMALL-SCALE MUTATIONS

At the molecular level, a mutation is a change in the nucleotide sequence of a genome. Such changes can be small, affecting one or a few bases, or large, affecting entire chromosomes. We begin by considering the origin and effects of small-scale changes to the DNA sequence. While mutation provides the raw material that allows evolution to take place, it can play this role only because of an important feature of living systems: Once a mutation has taken place in a gene, the mutant genome is replicated as faithfully as the nonmutant genome.

Point mutations are changes in a single nucleotide.

Most DNA damage or errors in replication are immediately removed or corrected by specialized enzymes in the cell. (Some examples of DNA repair will be discussed in section 14.4.) We have already seen one example of DNA repair, the proofreading function of DNA polymerase during the process of replication, which acts to remove an incorrect nucleotide from the 3′ end of growing DNA strand (see Fig. 12.7). Damage that is corrected is not regarded as a mutation because the DNA sequence is immediately restored to its original state.

If a change in DNA is to become stable and subsequently inherited through mitotic or meiotic cell divisions, it must escape correction by the DNA repair systems. The example in **Fig. 14.6** shows how a mutation incorporated during replication can become a permanent change to the genome. If, during replication, DNA polymerase mistakenly adds a G (instead of a T) to a growing strand across from an A and its proofreading function does not catch the error, the result is a T–G mismatch in the double-stranded DNA. At the next replication the G in the new template strand specifies a C in the daughter strand, with the result that the daughter DNA duplex has a perfectly matched C–G base pair. From this point forward, the DNA molecule containing the mutant C–G base pair will replicate as faithfully as the original molecule bearing the nonmutant T–A base pair. A mutation in which one base pair (in this example, T–A) is replaced by a different base pair (in this example, C–G) is called a **nucleotide substitution** or **point mutation.** This is the most frequent type of mutation.

The effect of a point mutation depends in part on where in the genome it occurs. In many multicellular eukaryotes, including humans, the vast majority of DNA in the genome does not code for protein or RNA (Chapter 13). Most of the sequences in noncoding DNA have no known function, which may explain why many mutations in noncoding DNA have no detectable effects on the organism.

FIG. 14.6 **A point mutation.** A point mutation is a change in a single nucleotide.

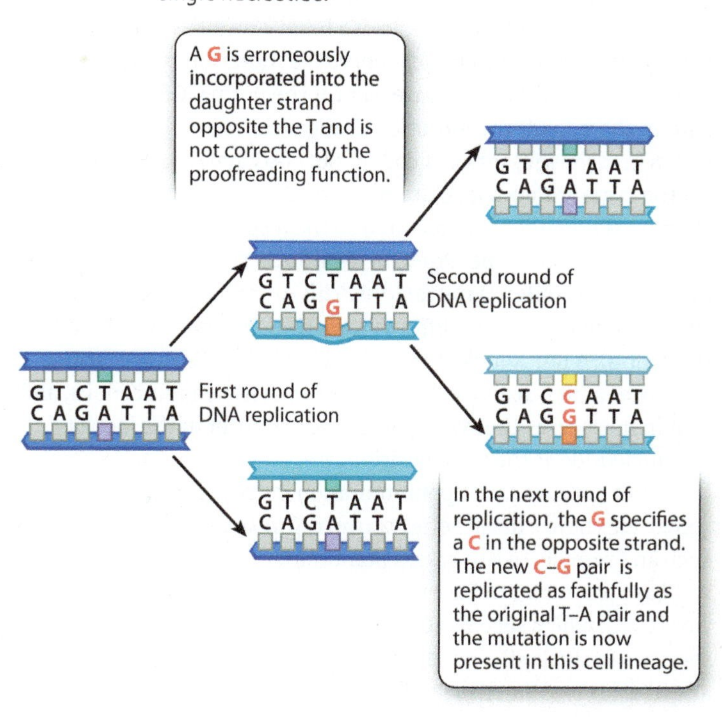

On the other hand, mutations in coding sequences do have predictable consequences in an organism. **Fig. 14.7** shows an example in the human DNA sequence coding for the amino acid chain of β-(beta-)globin, a subunit of the protein hemoglobin, which carries oxygen in red blood cells. Fig. 14.7a shows a

FIG. 14.7 **Synonymous and nonsynonymous mutations.** Synonymous mutations do not change the amino acid sequence of the resulting protein. Nonsynonymous mutations change the amino acid sequence.

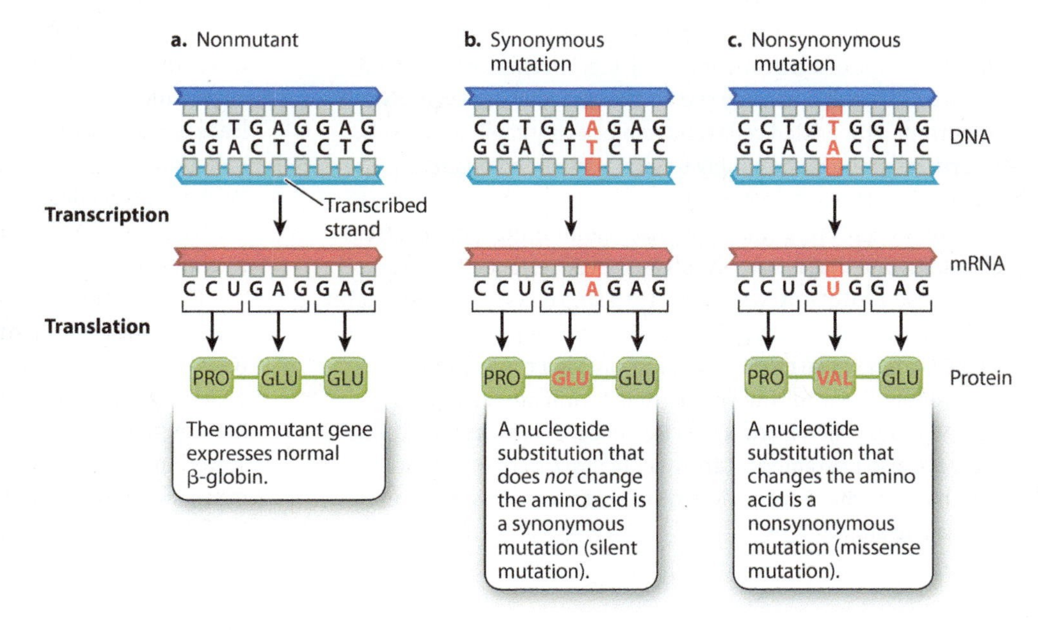

small part of the nonmutant DNA sequence, transcription and translation of which results in incorporation of the amino acids Pro–Glu–Glu in the β-globin polypeptide.

Fig. 14.7b shows an example of a harmless nucleotide substitution in this coding sequence. The mutation consists of the substitution of an A–T base pair for the normal G–C base pair. In the mRNA, the mutation changes the normal GAG codon into the mutant GAA codon. But GAG and GAA both code for the same amino acid, glutamic acid (Glu). In other words, they are synonymous codons, and so the resulting amino acid sequences are the same: Pro–Glu–Glu. Such mutations are called **synonymous (silent) mutations.** This example is typical in that the synonymous codons differ at their third position (the 3′ end of the codon). A quick look at the genetic code (see Table 4.1) shows that most amino acids can be specified by synonymous codons, and that in most cases the synonymous codons differ in the identity of the nucleotide at the third position.

On the other hand, a point mutation in coding sequences can sometimes have a drastic effect on an organism. Fig. 14.7c shows such an example. In this case, a point mutation substitutes the first A–T base pair for a T–A base pair. The result is a change in the mRNA from GAG, which specifies Glu (glutamic acid), into GUG, which specifies Val (valine). The resulting protein therefore contains the amino acids Pro–Val–Glu instead of Pro–Glu–Glu. In other words, the nucleotide substitution results in an **amino acid replacement.** Point mutations that cause amino acid replacements are called **nonsynonymous (missense) mutations.**

Since the complete β-globin chain consists of 146 amino acids, it may seem that a change in only one amino acid would have little effect. But the change in even a single amino acid can affect the three-dimensional structure of a protein, and therefore change its ability to function. Individuals who inherit two copies of the mutant β-globin gene that specifies the Glu-to-Val replacement have a disease known as **sickle-cell anemia.** In this condition, the hemoglobin molecules tend to crystallize when exposed to lower than normal levels of oxygen. The crystallization of hemoglobin causes the cell to collapse from its normal ellipsoidal shape into the shape of a half-moon, or "sickle." In this form, the red blood cell is unable to carry the normal amount of oxygen. More important, the sickled cells tend to block tiny capillary vessels, interrupting the blood supply to vital tissues and organs and resulting in severe pain and fever.

Fig. 14.8 shows a third way that a point mutation can affect a protein, one that nearly always has severe effects. This is a **nonsense mutation,** which creates a stop codon that terminates translation. In the example in Fig. 14.8, the mutation creates a UAG codon in the mRNA. Because UAG is a translational stop codon, the resulting polypeptide terminates after Pro. Polypeptides that are truncated are nearly always nonfunctional. They are also unstable and are quickly destroyed. Eukaryotic cells have mechanisms to destroy mRNA molecules that contain premature stop codons.

FIG. 14.8 A nonsense mutation. A nonsense mutation can change an amino acid to a stop codon, resulting in a shortened and unstable protein.

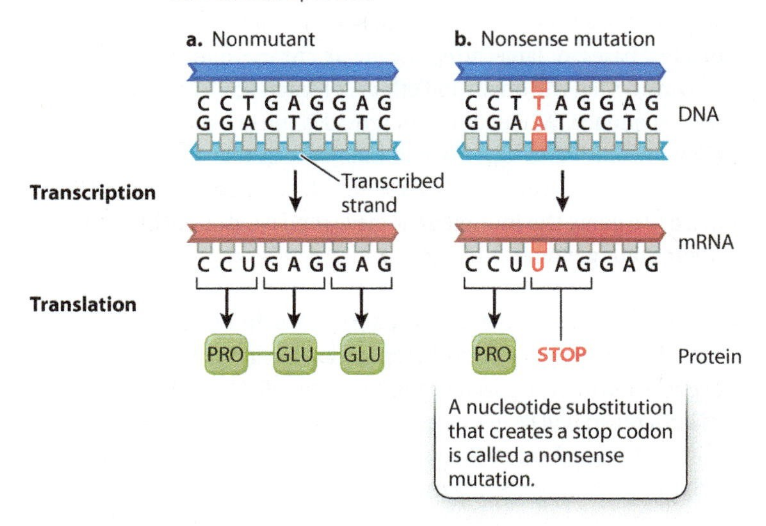

Small insertions and deletions involve several nucleotides.

Another relatively common type of mutation is the deletion or insertion of a small number of nucleotides. In noncoding DNA, such mutations have little or no effect. In protein-coding regions, their effects depend on their size. A small deletion or insertion that is an exact multiple of three nucleotides results in a polypeptide with as many fewer (in the case of a deletion) or more (in the case of an insertion) amino acids as there are codons deleted or inserted. Thus, a deletion of three nucleotides eliminates one amino acid, and an insertion of six adds two amino acids.

The effects of a deletion of three nucleotides can be seen in cystic fibrosis. This disease is characterized by the production of abnormal secretions in the lungs, liver, and pancreas and other glands. Its chief symptoms are recurrent respiratory infections, malnutrition resulting from incomplete digestion and absorption of fats and proteins, and liver disease. Patients with cystic fibrosis have an accumulation of thick, sticky mucus in their lungs, which often leads to respiratory complications, including recurrent bacterial infections. Untreated, 95% of affected children die before age 5. With proper medical care, including regular physical therapy to clear the lungs, antibiotics, pancreatic enzyme supplements, and good nutrition, the average life expectancy is currently 30 to 40 years.

The mutations responsible for cystic fibrosis are in the gene encoding the cystic fibrosis transmembrane conductance regulator (CFTR) (**Fig. 14.9**). The CFTR protein is a chloride channel, which acts as a transporter to pump chloride ions out of the cell. Malfunction of CFTR causes ion imbalances that result in abnormal secretions from the many cell types in which the *CFTR* gene is expressed. About 70% of the mutations associated with cystic fibrosis have a specific mutation known as *Δ508* (*delta 508*), which is a deletion of three nucleotides that eliminates a phenylalanine normally present at position 508 in the protein. The missing amino acid results in a CFTR protein that does not fold properly and is degraded before reaching the membrane. Some researchers hope that future therapy will include drugs that stabilize the mutant protein or even treatments that repair the defective gene in affected cells.

Small deletions or insertions that are not exact multiples of 3 can cause major changes in amino acid sequence because they do not insert or delete entire codons. The effect of such a mutation can be appreciated by seeing how deletion of a single letter turns a perfectly sensible sentence of three-letter words into gibberish. Consider the sentence:

THE BIG BOY SAW THE CAT EAT THE BUG

If the red E is deleted, the new reading frame for three-letter words is as follows:

THB IGB OYS AWT HEC ATE ATT HEB UG

FIG. 14.9 Effect of a single amino acid deletion in CFTR, the cystic fibrosis transmembrane conductance regulator.

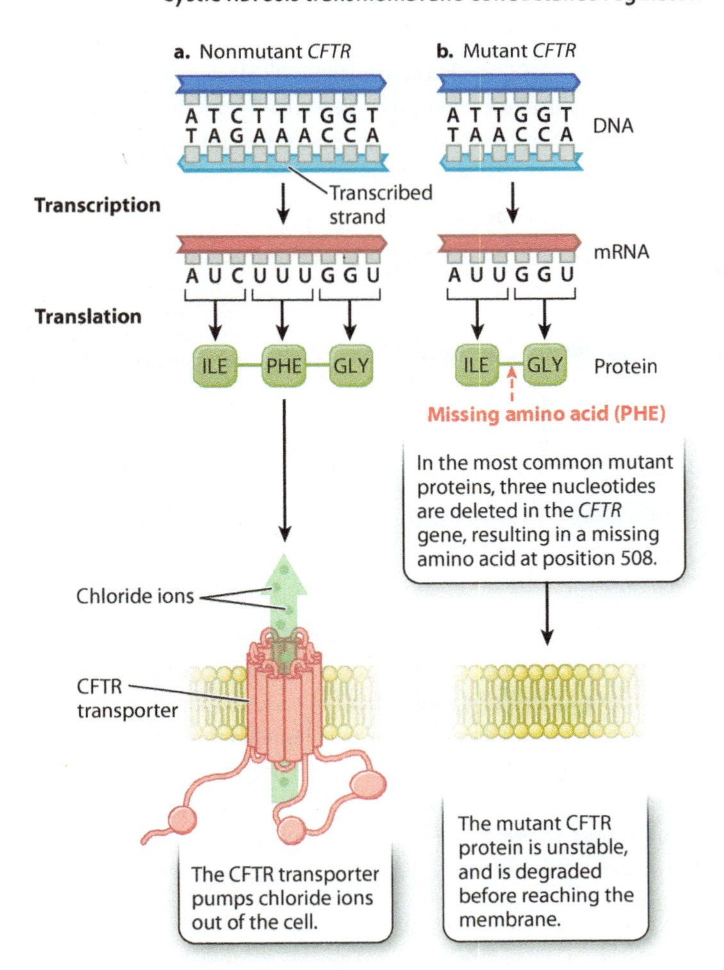

FIG. 14.10 A frameshift mutation. A frameshift mutation changes the translational reading frame.

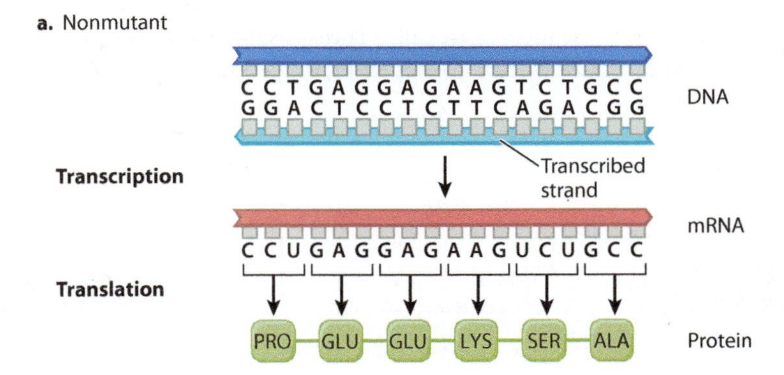

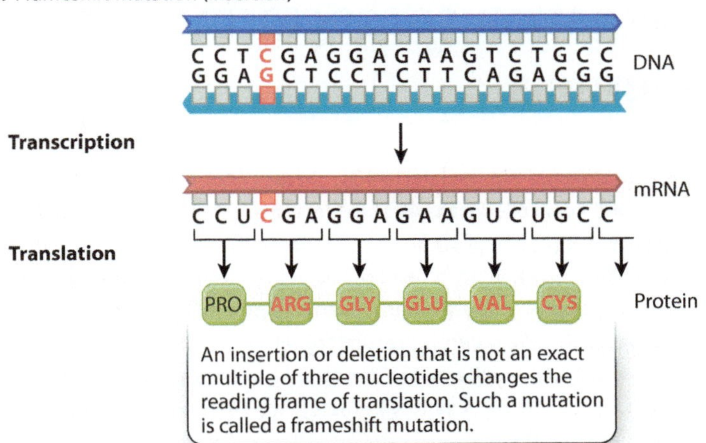

The result is unintelligible. Similarly, an insertion of a single nucleotide causes a one-nucleotide shift in the reading frame of the mRNA, and it changes all codons following the site of insertion. For this reason, such mutations are called **frameshift mutations.**

Fig. 14.10 shows the consequences of a frameshift mutation in the β-globin gene. The normal sequence in Fig. 14.10a corresponds to amino acids 5–10. The frameshift mutation in Fig. 14.10b is caused by the insertion of a C–G base pair. The mRNA transcript of the DNA therefore also has a single-base insertion. When this mRNA is translated, the one-nucleotide shift in the reading frame results in an amino acid sequence that has no resemblance to the original protein. All amino acids downstream of the site of insertion are changed, resulting in loss of protein function.

→ **Quick Check 2** The coding sequence of genes that specify proteins with the same function in related species sometimes differ from each other by an insertion or deletion of contiguous nucleotides. The number of nucleotides that are inserted or deleted is almost always an exact multiple of 3. Why is this expected?

Some mutations are due to the insertion of a transposable element.

An important source of new mutations in many organisms is the insertion of movable DNA sequences into or near a gene. Such movable DNA sequences are called **transposable elements** or **transposons.** As we saw in Chapter 13, the genomes of virtually all organisms contain several types of transposable element, each present in multiple copies per genome.

Transposable elements were discovered by American geneticist Barbara McClintock in the 1940s (**Fig. 14.11**). She studied corn (maize) because genetic changes that affect pigment formation can be observed directly in the kernels. The normal color of

HOW DO WE KNOW?

FIG. 14.11

What causes sectoring in corn kernels?

BACKGROUND In the late 1940s, Barbara McClintock discovered what are now called transposable elements, DNA sequences that can move from one position to any other in the genome. She studied corn (*Zea mays*). Wild-type corn has purple kernels, resulting from expression of purple anthocyanin pigment (Fig. 14.11a). A mutant with yellow kernels results from lack of purple anthocyanin pigment. McClintock noticed that streaks of purple pigmentation could be seen in many yellow kernels (Fig. 14.11b). This observation indicated that the mutation causing yellow color was unstable and that the gene could revert to the normal purple color.

Photo sources: a. photo_journey/Shutterstock; b. Rob Martienssen, Cold Spring Harbor Laboratory.

HYPOTHESIS McClintock hypothesized that the yellow mutant color resulted from a transposable element, which she called *Dissociator* (*Ds*), jumping into a site near or in the anthocyanin gene and disrupting its function. She attributed the purple streaks to cell lineages in which the transposable element had jumped out again, restoring the anthocyanin gene.

EXPERIMENT AND RESULTS By a series of genetic crosses, McClintock showed that the genetic instability of *Ds* was due to something on another chromosome that she called *Activator* (*Ac*). She set up crosses in which she could track the *Ac*-bearing chromosome. She observed that in the presence of *Ac*, cells in mutant yellow kernels reverted to normal purple, resulting in purple sectors in an otherwise yellow kernel. From this observation, she inferred that the *Ds* element had jumped out of the anthocyanin gene, restoring its function. She also demonstrated that restoration of the original purple color was associated with mutations elsewhere in the genome. From this observation, she inferred that the *Ds* element had integrated elsewhere in the genome, where it disrupted the function of another gene.

CONCLUSION McClintock's conclusion is illustrated in Fig. 14.11c: Transposable elements can be excised from their original position in the genome and inserted into another position.

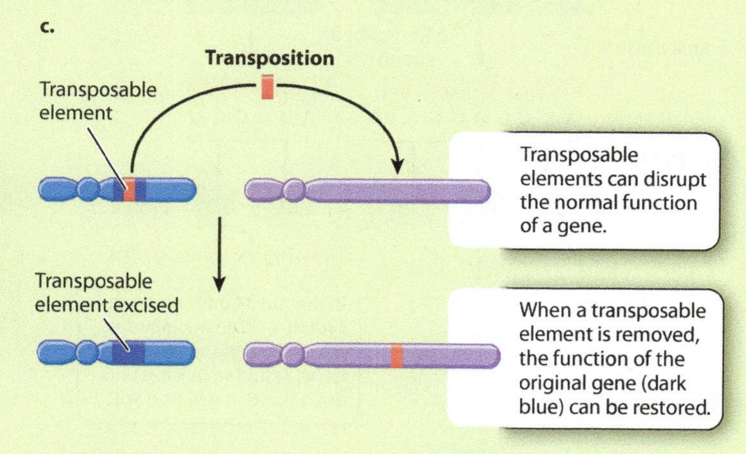

FOLLOW-UP WORK McClintock won the Nobel Prize in Physiology or Medicine in 1983. Later experiments showed that *Ds* is a transposable element that lacks a functional gene for transposase, the protein needed for the element to move, and *Ac* is a transposable element that encodes transposase. Presence of *Ac* produces active transposase that allows *Ds* to move. Much additional work showed that there are many different types of transposable element and that they are ubiquitous among organisms.

SOURCE McClintock, B. 1950. "The Origin and Behavior of Mutable Loci in Maize." *Proceedings of the National Academy of Sciences of the USA*. 36:344–355.

maize kernels is purple. Each kernel consists of many cells, and if no mutations affecting pigmentation occur in a kernel, it will be uniformly purple. The kernel pigments are synthesized by several different enzymes in a metabolic pathway, and any of these enzymes can be rendered nonfunctional by mutations in their genes, including the insertion of transposable elements. Since McClintock's work, we have learned that most transposable elements are segments of DNA a thousand or more base pairs long. When such a large piece of DNA inserts into a gene, it can interfere with transcription, cause errors in RNA processing, or disrupt the open reading frame. The result in the case of maize is that the cell is unable to produce pigment, and so the kernels will be yellow.

Among mutants affecting kernel pigmentation, McClintock observed certain mutants that resulted in kernels that were mostly yellow but speckled with purple. She suspected that these particular mutants might result from a transposable element jumping into and out of a gene. Each colored sector consists of a lineage of daughter cells from a single ancestral cell in which pigment synthesis had been restored. McClintock realized that, just as the inability of the kernel cells to produce pigment was caused by a transposable element jumping into a gene, restoration of that ability could be caused by the transposable element jumping out again. Her hypothesis that transposable elements are responsible for the pigment mutations was confirmed when she found that in cells where pigment had been restored, mutations affecting other genes had also occurred. She deduced that these other mutations were due to a transposable element jumping out of a pigment gene and into a different gene in the same cell.

The movement, or **transposition,** of transposable elements occurs by different mechanisms according to the type of transposon (Chapter 13). McClintock's original discovery was of a DNA transposon that transposes by a cut-and-paste mechanism in which the transposon is cleaved from its original location in the genome by a specific enzyme (**transposase**) and inserted into a different position. Removal of the transposon from its original position and repair of the cleavage leads to restoration of gene function, which in McClintock's experiment was the ability to produce purple pigment. Unstable mutations due to DNA transposons can occur in almost any organism, including the Japanese morning glory, whose sectored flowers are shown in Fig. 14.3.

Not all transposons undergo transposition by a cut-and-paste mechanism. As discussed in Chapter 13, retrotransposons undergo transposition by an RNA intermediate, and when these types of transposable elements move, the retrotransposon used as a template for transcription stays behind in its original location. This mode of transposition might be called a copy-and-paste mechanism. It is mediated by two enzymes. One is reverse transcriptase (Chapter 13), which produces a double-stranded DNA copy of the retrotransposon from its RNA transcript, much in the same way that reverse transcriptase produces double-stranded DNA from viral RNA genomes. The other enzyme is an integrase, which cuts genomic DNA and inserts the retrotransposon at the cut site. By this mechanism, a retrotransposon can be copied and pasted at various sites in the genome, potentially disrupting the function of many different genes.

14.3 CHROMOSOMAL MUTATIONS

Whereas most mutations involve only one or a few nucleotides, some affect larger regions extending over hundreds of thousands or millions of nucleotides and have effects on chromosome structure that are often large enough to be visible through an ordinary optical microscope. Double-stranded breaks in DNA that are incorrectly repaired can lead to chromosomal mutations. The breaks may result from interactions between DNA and reactive molecules produced in metabolism or from reactive chemicals in the environment or by radiation (especially X-rays). Chromosomal mutations can also arise from errors in DNA replication, particularly in sequences that are tandemly repeated along the DNA (Chapter 13).

Chromosomal mutations can delete or duplicate regions of a chromosome containing several or many genes, and the resulting change in gene copy number also changes the amount of the products of these genes in the cell. Chromosomal mutations can also alter the linear order of genes along a chromosome or interchange the arms of nonhomologous chromosomes. While these types of chromosomal mutations do not change gene copy number, they do affect chromosome pairing and segregation in meiosis. These effects distinguish chromosome abnormalities from nucleotide substitutions, small-scale deletions and duplications, transpositions, and other submicroscopic mutations. In this section, we briefly consider the major types of chromosome abnormalities in more detail.

Duplications and deletions result in gain or loss of DNA.

Among the most common chromosomal abnormalities are those in which a segment of the chromosome is either present in two copies or is missing altogether (**Fig. 14.12**). A chromosome in which a region is present twice instead of once is said to contain a **duplication** (Fig. 14.12a). Although large duplications that include hundreds or thousands of genes are usually harmful and quickly eliminated from the population, small duplications including only

FIG. 14.12 **Duplication and deletion.** (a) In a duplication, a segment of chromosome is repeated. (b) In a deletion, a segment of chromosome is missing.

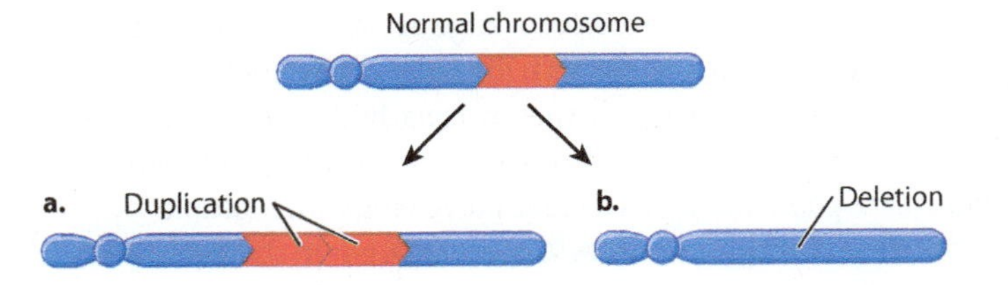

FIG. 14.13 Origin of the β-globin gene family. The β-globin gene family evolved through several episodes of duplication and divergence.
Data from: Y. A. Trusov and P. H. Dear, 1996, "A Molecular Clock Based on the Expansion of Gene Families," Nucleic Acids Research 24(6):995–999.

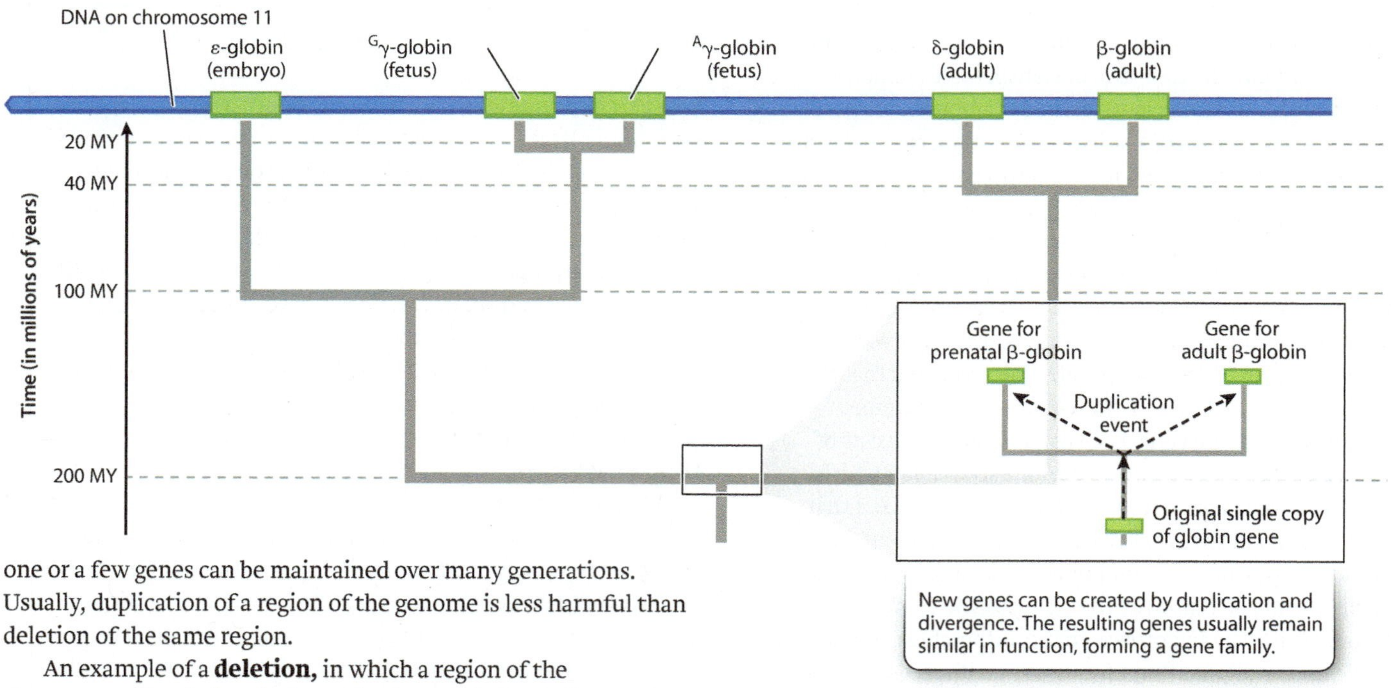

one or a few genes can be maintained over many generations. Usually, duplication of a region of the genome is less harmful than deletion of the same region.

An example of a **deletion,** in which a region of the chromosome is missing, is shown in Fig. 14.12b. A deletion can result from an error in replication or from the joining of breaks in a chromosome that occur on either side of the deleted region. Even though a deletion may eliminate a gene that is essential for survival, the deletion can persist in the population because chromosomes usually occur in homologous pairs. If one member of a homologous pair has a deletion of an essential gene but the gene is present in the other member of the pair, that one copy of the gene is often sufficient for survival and reproduction. In these cases, the deletion can be transmitted from generation to generation and persist harmlessly, as long as the chromosome is present along with a normal chromosome.

But some deletions decrease the chance of survival or reproduction of an organism even when the homologous chromosome is normal. In general, the larger the deletion, the smaller the chance of survival. In the fruit fly *Drosophila,* individuals with deletions of more than 100–150 genes rarely survive even when the homologous chromosome is normal. The interpretation of the reduced survival is that organisms are sensitive to the **dosage,** or number of copies, of each gene. Normal embryonic development requires that genes be present in a particular dosage. Although small deviations from normal gene dosage can be tolerated, as indicated by the survival of individuals containing small duplications or deletions, the cumulative effects of large deviations from normal gene dosage are incompatible with life. It is usually not the total number of copies of each gene that matters, but rather the number of copies of each gene relative to other genes. This explains why some plant populations can include polyploids with multiple complete sets of chromosomes (Chapter 13), but in the same species large duplications and deletions are lethal.

It is worth emphasizing that one rarely observes deletions or duplications that include the **centromere,** the site associated with attachment of the spindle fibers that move the chromosome during cell division (Chapter 11). The reason is that an abnormal chromosome without a centromere, or one with two centromeres, is usually lost within a few cell divisions because it cannot be directed properly into the daughter cells during cell division.

Gene families arise from gene duplication and evolutionary divergence.

Small duplications play an important role in the origin of new genes in the course of evolution. In most cases, when a gene is duplicated, one of the copies is free to change without causing harm to the organism because the other copy continues to carry out the normal function of the gene. Occasionally, a mutation in the "extra" copy of the gene may result in a beneficial effect on survival or reproduction, and gradually a new gene is formed from the duplicate. These new genes usually have a function similar to that of the original gene.

This process of creating new genes from duplicates of old ones is known as **duplication and divergence.** The term **divergence** refers to the slow accumulation of differences between duplicate copies of a gene that occurs on an evolutionary time scale. Multiple rounds of duplication and divergence can give rise to a group of genes with related functions known as a **gene family.** The largest gene family in the human genome has about 400 genes and encodes for proteins that detect odors. These proteins are structurally very similar, but differ in the region that binds small odor molecules. It is the diversity of the odorant binding sites that allows us to identify so many different smells.

Most gene families are not as large and diverse as that for odor detection. **Fig. 14.13** shows the evolutionary origin of the family of globin genes, which in humans are spread across about 50 kb of chromosome 11. The globin gene family consists of five different genes that are expressed at various times during development (embryo, fetus, or adult). The two γ-(gamma-)globin polypeptides are nearly identical in amino acid sequence but expressed at different levels in the fetus, whereas the adult δ-(delta) and β-polypeptides differ somewhat more and are expressed at very different levels. The sequences are all similar enough to imply that the genes arose through duplication and divergence.

Globin sequences have changed through evolutionary time at a relatively constant rate, and so the number of differences between any two sequences is proportional to the time since they were created by duplication. The relative constancy of rates of evolutionary change in a DNA nucleotide sequence or a protein amino acid sequence is known as a **molecular clock,** and it allows molecular differences among genomes to be correlated with the fossil record and dated accordingly (Chapter 21). For example, the earliest duplication event in the tree in Fig. 14.13, which produced distinct genes for fetal and adult hemoglobin, took place at about the same time (200 million years ago) that the common ancestor of today's placental mammals became a distinct species from the common ancestor of today's marsupial mammals.

An inversion has a chromosomal region reversed in orientation.

Chromosomes in which the normal order of a block of genes is reversed contain an **inversion (Fig. 14.14)**. An inversion is typically produced when the region between two breaks in a chromosome is flipped in orientation before the breaks are repaired. Especially in large genomes, the breaks are likely to occur in noncoding DNA rather than within a gene. Whereas large inversions can cause problems in meiosis, small inversions are common in many populations and play an important role in chromosome evolution. A small inversion can have almost no effect on the organism because it still contains all of the genes present in the original chromosome, merely flipped in their order, and is too small to cause problems in meiosis. The accumulation of inversions over evolutionary time explains in part why the order of genes along a chromosome can differ even among closely related species.

FIG. 14.14 Inversion. A chromosome with an inversion has a segment of chromosome present in reverse orientation.

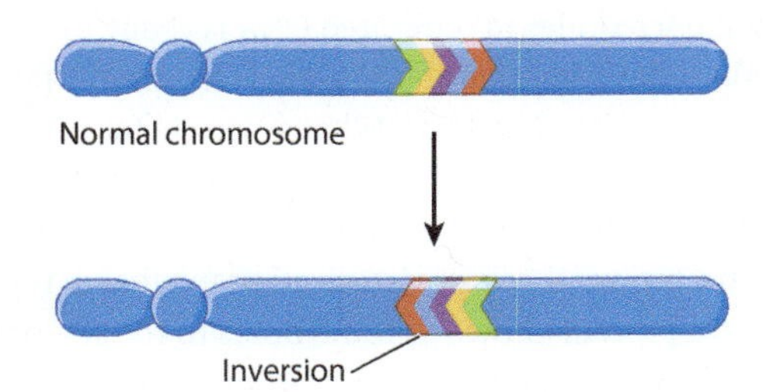

FIG. 14.15 Reciprocal translocation. A reciprocal translocation results from an interchange of parts between nonhomologous chromosomes.

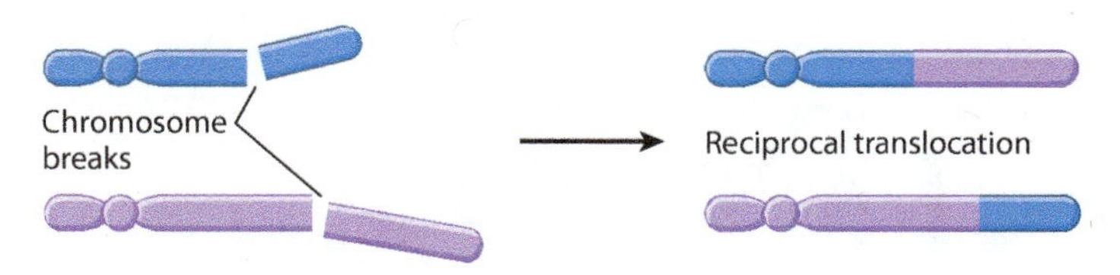

A reciprocal translocation joins segments from nonhomologous chromosomes.

A **reciprocal translocation (Fig. 14.15)** occurs when two different (nonhomologous) chromosomes undergo an exchange of parts. In the formation of a reciprocal translocation, both chromosomes are broken and the terminal segments are exchanged before the breaks are repaired. In large genomes, the breaks are likely to occur in noncoding DNA, so the breaks themselves do not usually disrupt gene function.

Since reciprocal translocations change only the arrangement of genes and not their number, most reciprocal translocations do not affect the survival of organisms. Proper gene dosage requires the presence of both parts of the reciprocal translocation, however, as well as one copy of each of the normal homologous chromosomes. Problems can arise in meiosis because both chromosomes involved in the reciprocal translocation may not move together into the same daughter cells, resulting in gametes with only one part of the reciprocal translocation. This inequality does upset gene dosage, because these gametes have extra copies of genes in one of the chromosomes and are missing copies of genes in the other. In the next chapter, we will see that such abnormalities are observed in a significant number of human embryos.

14.4 DNA DAMAGE AND REPAIR

DNA is a fragile molecule, prone to damage of many different kinds. Every day, in each cell in our bodies, the DNA is damaged in some way at tens of thousands of places along the molecule. Fortunately, cells have evolved mechanisms for repairing various types of damage and restoring the DNA to its original condition. In this section, we examine some of the major types of DNA damage and some of the ways in which it is repaired.

DNA damage can affect both DNA backbone and bases.

Most mutations are spontaneous and occur naturally. However, mutations can also be induced by radiation or chemicals. **Mutagens** are agents that increase the probability of mutation.

FIG. 14.16 Major types of DNA damage. Most types of DNA damage can be repaired by specialized enzymes.

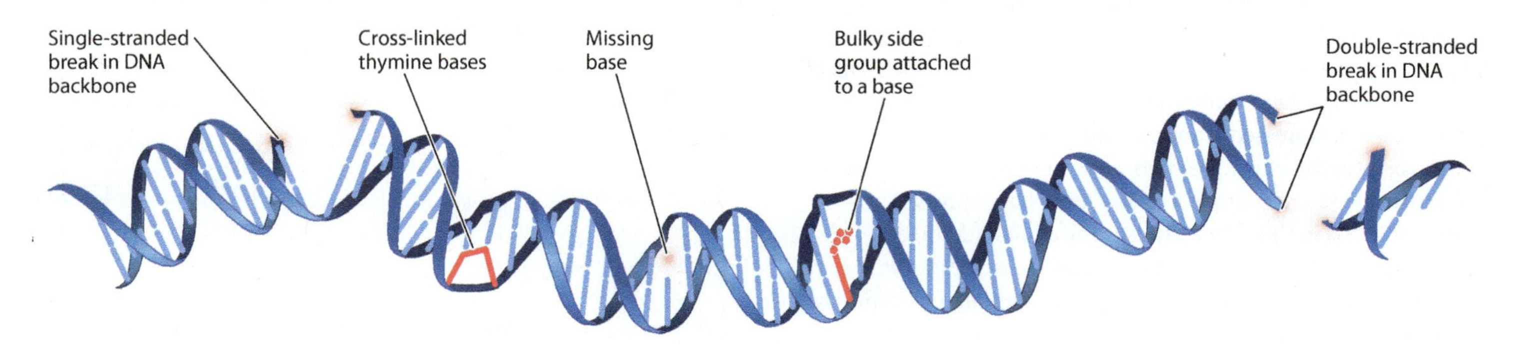

The presence of a mutagen can increase the probability of mutation by a factor of 100 or more.

Some of the most important types of DNA damage induced by mutagens are illustrated in **Fig. 14.16,** which shows a highly damaged DNA molecule. Some types of damage affect the structure of the DNA double helix. These include breaks in the sugar–phosphate backbone, one of the main mutagenic effects of X-rays. Breaks can occur in just one strand of the DNA or both. Ultraviolet light can cause cross-links between adjacent pyrimidine bases, especially thymine, resulting in the formation of thymine dimers. Covalent bonding between adjacent thymines in a DNA strand causes the double helix to become pinched, both the major groove and the minor groove become wider, and the T–A base pairing is weakened.

Yet another type of structural damage is loss of a base from one of the deoxyribose sugars, resulting in a gap in one strand where no base is present. Spontaneous loss of a purine base is one of the most common types of DNA damage, occurring at the rate of about 13,000 purines lost per human cell per day. Most of these mutations result from the interaction between DNA and normal metabolic by-products. The rate increases with age, and it can also be increased by exposure to oxidizing agents such as household bleach or hydrogen peroxide.

Other types of damage affect the bases themselves. Bases that are chemically damaged tend to mispair. Some bases are damaged spontaneously when reaction with a water molecule replaces an amino group ($-NH_2$) with an atom of oxygen ($=O$), which interferes with the base's ability to form hydrogen bonds with a complement. Some naturally occurring molecules mimic bases and can be incorporated into DNA and cause nucleotide substitutions. Caffeine mimics a purine base, for example (although to be mutagenic, the amount of caffeine required is far more than any normal person could possibly consume).

Chemicals that are highly reactive tend to be mutagenic, often because they add bulky side groups to the bases that hinder proper base pairing. The main environmental source of such chemicals is tobacco smoke. Other chemicals can perturb the DNA replication complex and cause the insertion or deletion of one or occasionally several nucleotides.

→ **Quick Check 3** Earlier, we described the Lederbergs' experiment, which demonstrated that mutations are not directed by the environment. But mutagens, which are environmental, can lead to mutations. What's the difference?

Most DNA damage is corrected by specialized repair enzymes.

Cells contain many specialized DNA-repair enzymes that correct specific kinds of damage, and in this section we examine a few examples. Perhaps the simplest is the repair of breaks in the sugar–phosphate backbone, which are sealed by **DNA ligase,** an enzyme that can repair the break by using the energy in ATP to join the 3′ hydroxyl of one end to the 5′ phosphate of the other end. Most organisms have multiple different types of DNA ligase, some of which participate in DNA replication (Chapter 12) and others in DNA repair. One type of ligase seals single-stranded breaks in DNA, and a different type seals double-stranded breaks. Double-stranded breaks often result in chromosomal rearrangements because they are less likely to be repaired than single-stranded breaks. In addition to their importance in DNA replication and repair, ligases are an important tool in research in molecular biology because they allow DNA molecules from different sources to be joined to produce recombinant DNA (Chapter 12).

As we saw earlier in this chapter, mispairing of bases during DNA replication leads to the incorporation of incorrect nucleotides and potentially to nucleotide substitutions. About 99% of the mispaired bases are corrected immediately by the proofreading function of DNA polymerase, in which the mispaired nucleotide is removed immediately after incorporation and replaced by the correct nucleotide (Chapter 12).

Even after proofreading by DNA polymerase, the proportion of mismatched nucleotides in replicated DNA is about 10^{-6}, which is at least 1000 times greater than the actual frequency of errors in cellular organisms (see Fig. 14.1). What accounts for the difference is a second-chance mechanism for catching mismatches that is known as **postreplication mismatch repair** (**Fig. 14.17**).

In the bacterium *E. coli,* postreplication mismatch repair begins when a protein known as MutS binds to the site of a mismatch (Fig. 14.17a) and brings two other proteins (MutL

FIG. 14.17 **Postreplication mismatch repair.** In postreplication mismatch repair, the segment of a DNA strand containing the mismatch is removed and then resynthesized.

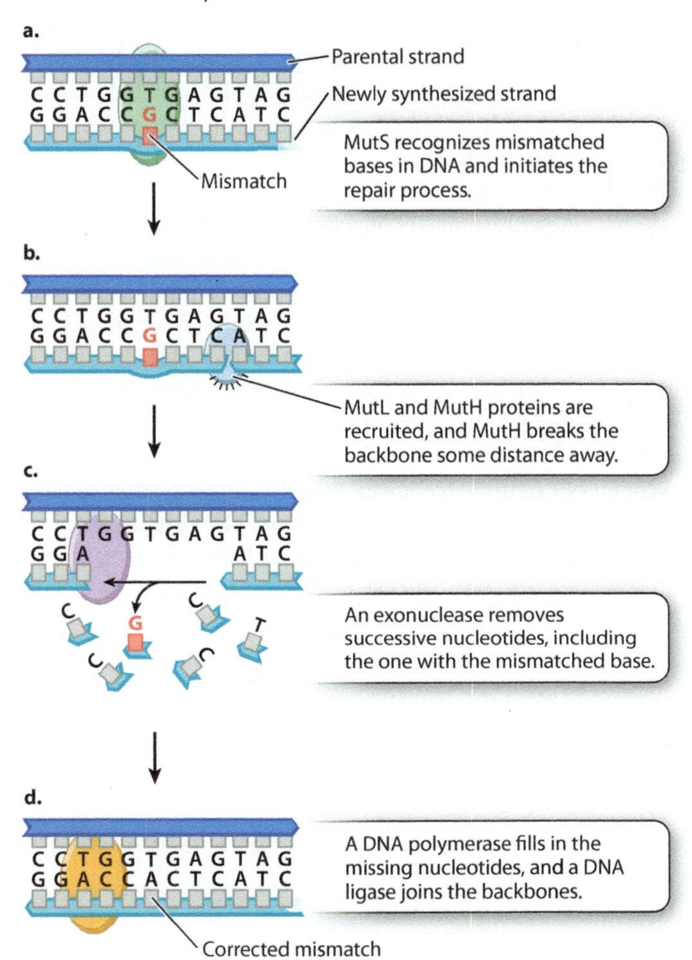

and MutH) to the site. MutL determines which DNA backbone will be cleaved, and MutH cleaves the backbone in the vicinity of the mismatch (Fig. 14.17b). An exonuclease degrades the cleaved strand to a distance beyond the site of the mismatch (Fig. 14.17c). A DNA polymerase then synthesizes new DNA to fill the gap, correcting the mismatch, and a DNA ligase seals the remaining nick in the DNA backbone. Postreplication mismatch repair usually, but not always, cleaves the daughter strand, so the corrected strand matches the parental template strand.

Eukaryotes have proteins corresponding to those in Fig. 14.17, and the importance of postreplication mismatch repair is underlined by the finding that a genetic predisposition to certain types of colon cancer results from mutations in the human counterparts of MutS and MutL.

The result of proofreading by DNA polymerase and postreplication mismatch repair is an increase in the fidelity of DNA replication. Cells have evolved many other systems that repair different types of DNA damage that can occur even in nondividing cells. One example is **base excision repair,** which corrects abnormal or damaged bases. In the first step of base excision repair, an abnormal or damaged base is cleaved from the sugar in the DNA backbone. Then, the baseless sugar is removed from the backbone, leaving a gap of one nucleotide. Finally, a repair polymerase inserts the correct nucleotide into the gap.

An example of base excision repair is shown in **Fig. 14.18,** which depicts a molecule in which uracil instead of cytosine is present in a DNA strand (Fig. 14.18a). Recall that uracil is normally found in mRNA, but not in DNA. Most organisms have an enzyme called DNA uracil glycosylase that cleaves the uracil from its deoxyribose sugar, resulting in a sugar in the DNA backbone lacking a base (Fig. 14.18b). Such baseless sugars are recognized by an enzyme called AP endonuclease (Fig. 14.18c), which cleaves the backbone on both sides of the sugar that lacks a base. This cleavage results in a one-nucleotide gap, which is filled by new synthesis using the intact DNA strand as a template. The result is that uracil is removed and replaced with cytosine. Specialized glycosylases remove other types of abnormal or damaged bases from the DNA, and these enzymes are also coupled with the AP endonuclease.

FIG. 14.18 **Base excision repair.** In base excision repair, an improper base in DNA and its deoxyribose sugar are both removed, and the resulting gap is then repaired.

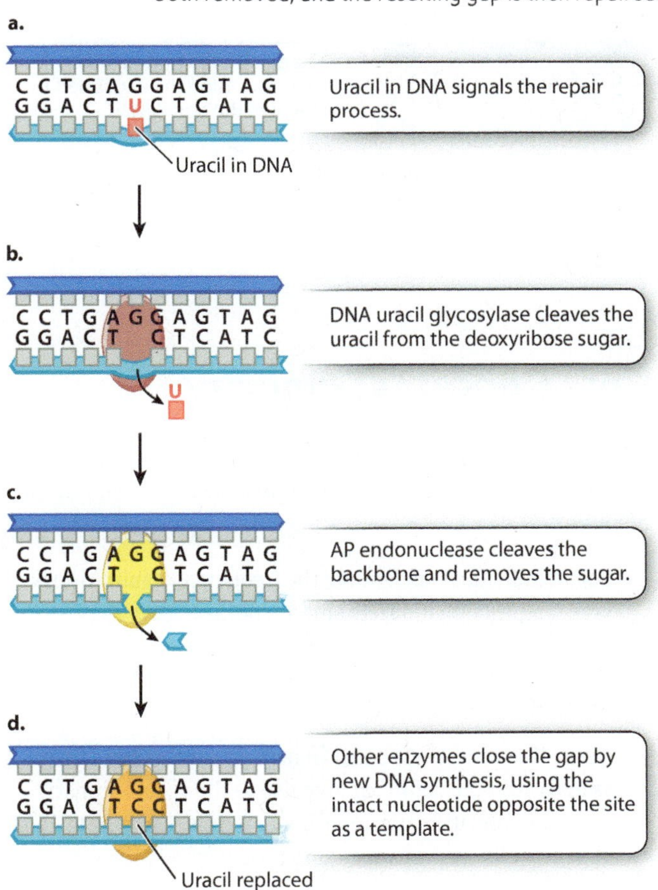

FIG. 14.19 Nucleotide excision repair. In nucleotide excision repair, a damaged segment of a DNA strand is removed and resynthesized using the undamaged partner strand as a template.

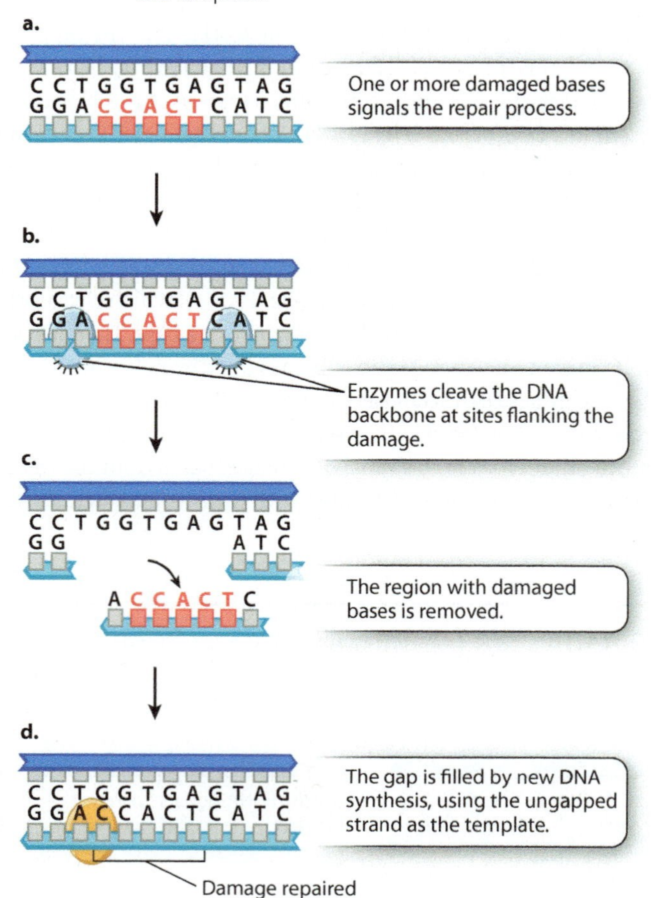

Cells have also evolved mechanisms to repair short stretches of DNA containing mismatched or damaged bases. One of these is known as **nucleotide excision repair,** which has a similar mechanism of action to mismatch repair but uses different enzymes (**Fig. 14.19**). Instead of degrading a DNA strand nucleotide by nucleotide until the mismatch is removed, nucleotide excision repair removes an entire damaged section of a strand at once. Nucleotide excision repair is also used to remove nucleotides with bulky side groups, as well as thymine dimers resulting from ultraviolet light.

The importance of excision repair is illustrated by the disease xeroderma pigmentosum (XP), in which nucleotide excision repair is defective. People with XP are exquisitely sensitive to the UV radiation in sunlight. Because of the defect in nucleotide excision repair, damage to DNA resulting from UV light is not corrected, leading to the accumulation of mutations in skin cells. The result is high rates of skin cancer. People with XP must minimize their exposure to sunlight and in extreme cases stay out of the sun altogether.

When DNA repair mechanisms function properly, they reduce the rate of mutation to a level that is compatible with life. But they are not perfect, and they do not catch all the mistakes. The result is genetic variation, the raw material of evolution. ■

Core Concepts Summary

14.1 Mutations are very rare for any given nucleotide and occur randomly without regard to the needs of an organism.

For an individual nucleotide, the frequency of mutation is very rare, though it differs among species. page 292

Mutations are common across an entire genome and across many cell divisions. page 293

Germ-line mutations occur in haploid gametes as well as in the diploid cells that give rise to them, and somatic mutations occur in nonreproductive cells. Only mutations in germ cells are transmitted to progeny and play a role in evolution. page 293

Mutations are random in that they occur without regard to the needs of an organism. This principle can be demonstrated by replica plating bacterial colonies in the absence and presence of antibiotic. page 295

14.2 Small-scale mutations include point mutations, insertions and deletions, and movement of transposable elements.

A point mutation, or nucleotide substitution, is a change of one base for another. page 297

The effect of a point mutation depends on where it occurs. If it occurs in noncoding DNA, it will likely have no effect on an organism. If it occurs in a protein-coding gene, it can result in a change in the amino acid sequence (nonsynonymous mutation), no change in the amino acid sequence (synonymous mutation), or the introduction of a stop codon (nonsense mutation). page 297

Small insertions or deletions in DNA add or remove one base or a few contiguous bases. Their effect depends on where in the genome they occur and on their size. An insertion or deletion of a single nucleotide in a protein-coding gene results in a frameshift mutation, in which all the codons downstream of the insertion or deletion are changed. page 298

Transposable elements are DNA sequences that can jump from one place in a genome to another. They can affect the expression of a gene if they insert into or near a gene. page 300

14.3 Chromosomal mutations involve large regions of one or more chromosomes.

A duplication is a region of the chromosome that is present two times, and a deletion is the loss of part of a chromosome. page 301

Duplications and deletions both affect gene dosage, which can have important effects on a cell or organism. page 302

Gene duplication followed by evolutionary divergence results in gene families made up of genes with related but not identical functions. page 302

An inversion is a segment of a chromosome in reverse orientation. page 303

A reciprocal translocation involves the exchange of a part of one chromosome with another. Reciprocal translocations do not affect gene dosage, but can lead to problems during meiosis. page 303

14.4 DNA can be damaged by mutagens, but most DNA damage is repaired.

Some types of damage alter the structure of DNA, such as single-stranded or double-stranded breaks, cross-linked thymine dimers, or missing bases. page 304

Other types of DNA damage, such as changes in the side groups that form hydrogen bonds or the addition of side groups that interfere with base pairing, affect the bases themselves. page 304

Many specialized DNA repair enzymes can correct DNA damage. page 304

DNA ligase seals breaks in DNA and is an important tool for molecular biologists. page 304

Postreplication mismatch repair provides a backup mechanism for mistakes not caught by the proofreading function of DNA polymerase. page 304

Base excision repair corrects individual nucleotides and involves several DNA repair enzymes working together. page 305

Nucleotide excision repair functions similarly to postreplication mismatch repair but excises longer stretches of damaged nucleotides. page 306

Self-Assessment

1. Mutations in which types of cell are most likely to contribute to evolutionary change in a population of organisms? Why?
2. Explain the difference between the mutation rate for a given nucleotide and the mutation rate for a given cell.
3. Explain what it means to say that mutations are random.
4. Describe the effects of nonsynonymous, synonymous, and nonsense mutations on a protein and the effects of small insertions or deletions in an open reading frame.
5. Explain how the location of a small-scale mutation in the genome can determine the effect it has on the functions of a cell.
6. Which type of chromosomal abnormality—deletion, duplication, inversion, or translocation—would you expect to have the greatest effect on the organisms that carry it, and why?
7. Explain how a gene family, such as the odorant receptor gene family, is thought to have evolved.
8. What is a mutagen? Name two common mutagens and their effects on DNA.
9. Describe a DNA repair mechanism.

Log in to LaunchPad to check your answers to the Self-Assessment questions, and to access additional learning tools.

Tastyart Ltd Rob White/Getty Images.

CHAPTER 15

Genetic Variation

Core Concepts

15.1 Genetic variation describes common genetic differences (polymorphisms) among the individuals in a population at any given time.

15.2 Human genetic variation can be detected by DNA typing, which can uniquely identify each individual.

15.3 Two common types of genetic variation are single-nucleotide polymorphisms (SNPs) and copy-number variation (CNV).

15.4 Chromosomal variants can also occur but are usually harmful.

In the last chapter, we saw how errors in DNA arise and what common types of error occur. What is the fate of all these mistakes? Most, as we have seen, are corrected by DNA repair mechanisms. In this case, the DNA sequence is not changed. Some errors escape these repair mechanisms and are replicated as faithfully as the original DNA sequence. If such mutations occur in somatic (body) cells, they can be passed on in the individual's cells through mitotic cell divisions, but they will not be passed on to progeny. If they occur in the germ line, they can be passed on to progeny. Over time, through evolution (Chapters 1 and 21), the proportion of individuals in a population carrying these mutations may increase or decrease. Therefore, if we look at any present-day population of organisms, such as the human population, we will find that it harbors lots of genetic differences, all of which result from mutations that occurred sometime in the past. **Genetic variation** refers to genetic differences that exist among individuals in a population at a particular point in time.

In the human population, you can observe the effects of common genetic differences by looking at the people around you. They differ in height, weight, facial features, skin color, eye color, hair color, hair texture, and in many other ways. These traits differ in part because of genetic variation, and in part because of the environment. Weight is affected by diet, for example, and skin color by exposure to sunlight.

In this chapter, we consider examples of major types of DNA variation and chromosomal variation in populations today and examine their consequences for the organism. We also describe key molecular techniques that allow genetic variation to be studied directly in DNA molecules. The emphasis here is primarily, but not exclusively, on human populations.

15.1 GENOTYPE AND PHENOTYPE

We tend to think of mutations as something negative or harmful; the term "mutant" ordinarily connotes something abnormal. But because mutations result in genetic variation among individuals and organisms, we are all mutants, different from one another genetically because of mutations, that is, differences in our DNA. Whereas some mutations are harmful, some have no effect on an organism, and some indeed are beneficial. Without mutations, evolution would not be possible. Mutations generate the occasional favorable variants that allow organisms to evolve and become adapted to their environment over time (Chapter 21).

Genotype is the genetic makeup of a cell or organism; phenotype is its observed characteristics.

The genetic makeup of a cell or organism constitutes its **genotype.** A population with a gene pool that has many variants in many different genes will consist of organisms with numerous different genotypes. For example, any two human genomes are likely to differ at about 3 million nucleotide sites, or about one difference per thousand nucleotides across the genome. In other words, we differ from one person to the next in a very large number but very small fraction of nucleotides.

Mutations, as we have seen, are the ultimate source of differences among genotypes. If we consider any present-day population of organisms, we find that some mutations are very common. Geneticists use the term **polymorphism** to refer to any genetic difference among individuals that is sufficiently common that it would almost certainly be present in a group of 50 randomly chosen individuals. For example, if many people have an A–T pair at a particular site in the genome, but many others have a G–C pair at the same site, this difference is a polymorphism. Of course, the polymorphism is the result of a mutation. All individuals once had the same genotype at this site, but a mutation occurred at this site sometime in the past, and is now commonly found in the population.

Phenotype is an individual's observable characteristics or traits, such as height, weight, eye color, and so forth. The phenotype may be visible, as in these characteristics, or may be seen in the development, physiology, or behavior of a cell or organism. For example, color blindness and lactose intolerance are phenotypes. The phenotype results in part from the genotype: A genotype with a mutated gene for an enzyme that would normally metabolize lactose can lead to the phenotype of lactose intolerance. However, the environment also commonly plays an important role, so it is most accurate to say that a phenotype results from an interaction between the genotype and the environment. These genotype–environment interactions are discussed in Chapter 18.

The effect of a genotype often depends on several factors.

Let's examine how genotype influences phenotype with an example of a genetic polymorphism that we introduced in the previous chapter. **Fig. 15.1** shows genetic variation in the gene for β-(beta-)globin, one of the subunits of hemoglobin that carries oxygen in red blood cells. Three forms of the β-globin gene are shown in the figure: *A*, *S*, and *C*. These three forms of the gene are relatively common in certain African populations. The different forms of any gene are called **alleles,** and they correspond to different DNA sequences (polymorphisms) in the genes. In this case, the most common allele is the *A* allele, which has a GAG codon in the position indicated. This codon translates to glutamic acid (Glu) in the resulting polypeptide.

The allele denoted "*S*" in Fig. 15.1 is associated with sickle-cell anemia (Chapter 14). In this allele, the GAG codon of the *A* allele is instead a GTG codon, with the result that the glutamic acid in the protein is replaced with valine (Val). The third allele in Fig. 15.1 is the *C* allele. It has a variation in the same codon as the *S* allele, but in this case the change is from GAG to AAG, with the result that glutamic acid is replaced by lysine (Lys). Although in each case only one amino acid of the β-globin protein is affected, this change

FIG. 15.1 Three alleles of the gene encoding β-globin, a subunit of hemoglobin. Both the *S* allele and the *C* allele are associated with increased resistance to malaria.

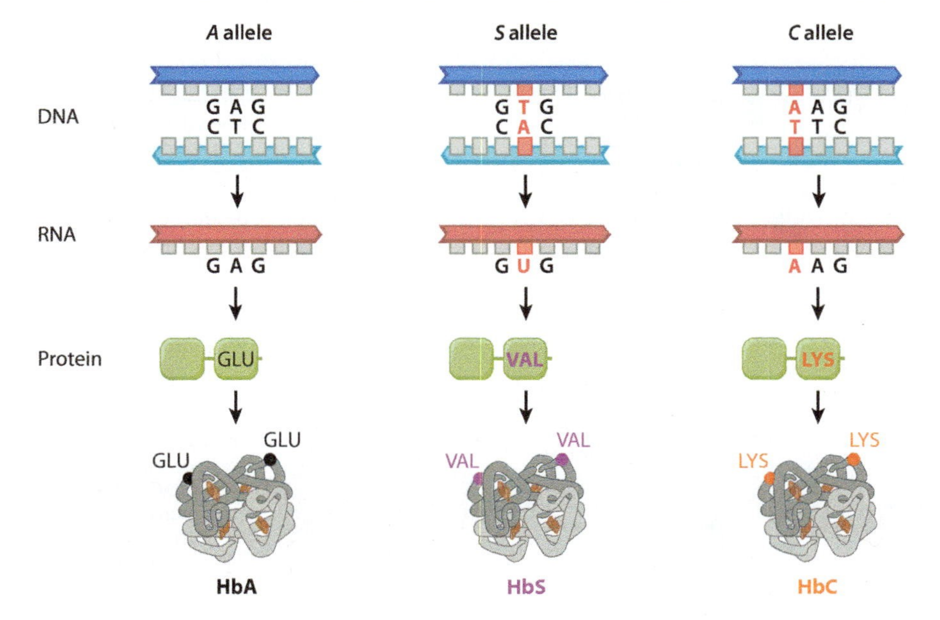

can have a dramatic effect on the function of the protein, since the amino acid sequence determines how a protein folds, and protein folding in turn determines the protein's function (Chapter 4).

An individual who inherits an allele of the same type from each parent is said to be **homozygous.** For the hemoglobin *A, S,* and *C* alleles, there are three possible homozygous genotypes: *AA, SS,* and *CC.* The first letter or symbol in each pair indicates the allele inherited from one parent, and the second indicates the allele inherited from the other parent. By convention, alleles and genotypes are designated by italic letters.

By contrast, individuals who inherit different types of allele from their parents are **heterozygous.** For the hemoglobin *A, S,* and *C* alleles, there are three possible heterozygous genotypes: *AS, AC,* and *SC.* Note that while each individual can have only two alleles of a gene, many more alleles can exist in an entire population. In the hemoglobin example, a person can be homozygous, with two identical alleles, or heterozygous, with two different alleles, but there are three different alleles available in the population can mix and match in an individual. How those alleles are inherited is explored in Chapter 16.

What are the effects of these mutations? Are they beneficial, harmful, or neutral? The short answer is, it depends. Let's consider the *S* allele. When it is inherited from both parents, the individual is homozygous *SS,* which results in the phenotype of sickle-cell anemia. In the absence of proper medical care, patients with sickle-cell anemia usually die before adulthood. However, when the *S* allele is inherited from one parent and the *A* allele is inherited from the other parent, the individual is heterozygous (*AS*) and the phenotype is only a mild form of sickle-cell disease.

Furthermore, in Africa, where malaria is widespread, being a heterozygote is actually beneficial because it affords partial protection against malaria.

Whether the effects of an allele, in this case the *S* allele, are beneficial or harmful illustrates two important principles about the connection between genotype and phenotype. First, the answer often depends on whether the mutation is homozygous or heterozygous. In areas with malaria, the *S* allele is harmful as a homozygous genotype but beneficial as a heterozygous genotype. Second, the effect of a particular genotype may depend on the environment. The *S* allele, as a heterozygous genotype, is beneficial only in malarial-prone regions, where it offers protection from the disease that outweighs its other effects. In areas without malaria, it is harmful.

What about the *C* allele? Individuals who inherit an *A* allele from one parent and a *C* allele from the other parent (genotype *AC*) have the phenotype of partial protection against malaria, and those who inherit a *C* allele from both parents (genotype *CC*) are not only more protected from malaria but also have at worst a mild anemia that usually needs no medical treatment. As with the *S* allele, the phenotype of the *C* allele depends on whether the individual is heterozygous or homozygous for the allele.

→ **Quick Check 1** A mutation arises in a bacterium that confers antibiotic resistance. Is this mutation harmful, beneficial, or neutral?

Some genetic differences are major risk factors for disease.

As we saw in the last chapter, when a mutation occurs in the coding sequence of a gene, it may have no effect on the amino acid sequence, or it may result in a change in the amino acid sequence, introduce a stop codon, or shift the reading frame. Many of the mutations that alter amino acid sequence are harmful. Harmful mutations are often eliminated in one or a few generations because they decrease the survival and therefore the capacity for reproduction of the individuals that carry them.

Sometimes, however, harmful mutations persist in a population. For example, many polymorphisms increase susceptibility to particular diseases. One such polymorphism increases the risk of emphysema, a condition marked by loss of lung tissue. Emphysema patients suffer from shortness of breath even when at rest, a wheezy cough, and increased blood pressure in the arteries of the lungs. Without proper treatment, the disease progresses to respiratory failure or congestive heart failure, and

FIG. 15.2 A harmful mutation. (a) The enzyme alpha-1 antitrypsin (α1AT) normally inhibits the activity of elastase, which breaks down elastin in the lung. (b) Cigarette smoke inhibits α1AT activity and increases the risk of emphysema. (c) A mutation in *α1AT* called *PiZ* is harmful because it reduces α1AT activity and also leads to increased risk of emphysema.

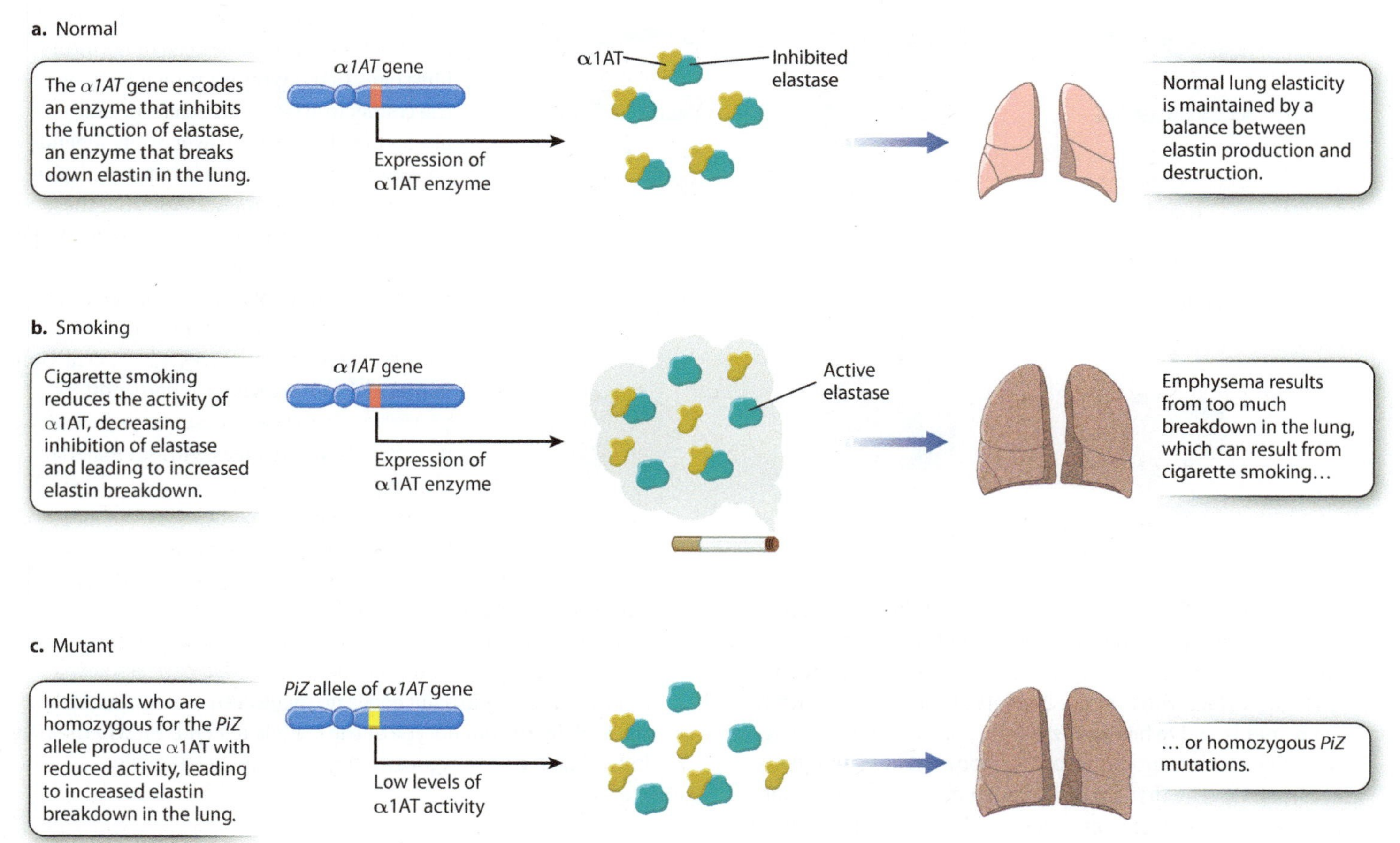

eventual death. Onset of the disease usually occurs in middle age, and the lifespan of affected individuals is shortened by 10 to 30 years depending on the effect of treatment.

About 80% of all cases of emphysema are associated with cigarette smoking, which affects the action of the enzyme alpha-1 antitrypsin (α1AT). The main function of α1AT is to inhibit another enzyme, known as elastase, which breaks down the connective-tissue protein elastin in lungs (**Fig. 15.2**). The elasticity of the lung, which allows normal breathing, requires a balance between the production and the breakdown of elastin. Too-rapid breakdown by elastase is normally prevented by the inhibition of elastase by α1AT (Fig. 15.2a). Cigarette smoke reduces the activity of α1AT, resulting in excessive destruction of elastin, loss of lung elasticity, and emphysema (Fig. 15.2b).

While most cases of emphysema are due to smoking, in some individuals genetics plays a role. An individual's chance of getting emphysema is significantly increased by inheriting a mutation in the gene that encodes α1AT. Among the many different alleles of this gene, a defective allele denoted *PiZ* is particularly common in populations of European descent. Among Caucasians in the United States, about 1 person in 30 is heterozygous for the *PiZ* allele, and about 1 in 3000 is homozygous *PiZ/PiZ*. Individuals with the homozygous genotype produce α1AT with reduced activity and hence have reduced elastase inhibition, leading to severe emphysema and death in more than 70% of affected individuals (Fig. 15.2c). Smoking markedly increases both the severity of the disease and the rapidity of its progression in *PiZ/PiZ* individuals. The life expectancy of *PiZ/PiZ* nonsmokers is 65 years, whereas that of *PiZ/PiZ* smokers is only 40 years.

The combination of homozygous *PiZ* and smoking is an example of a **genotype-by-environment interaction,** in which a phenotype is the result of an interplay between genes and the environment. In this case, a particular combination of genetic and environmental risk factors is much worse than either risk factor acting alone. Chapter 18 includes a more detailed discussion of genotype-by-environment interactions.

Not all genetic differences are harmful.

Some mutations have no effect on the organism, or have effects that are not associated with differences in survival or reproduction. Such mutations are considered neutral. Many mutations have neutral effects on organisms because they occur in noncoding DNA. Neutral mutations are therefore especially likely to occur in organisms with large genomes and abundant noncoding DNA (Chapter 13).

→ **Quick Check 2** Given what you read about the human genome in Chapter 13, would you predict that most mutations in humans are harmful, beneficial, or neutral?

Sometimes common, seemingly harmless genetic variations occur in coding sequences. One example in human populations is the taster phenotype associated with perception of a bitter taste from certain chemicals, including phenylthiocarbamide (PTC). The taster polymorphism was discovered in 1931 when a commercial chemist seeking a new artificial sweetener accidentally released a cloud of fine crystalline PTC and heard his colleague working nearby complain about its bitter taste. The chemist himself tasted nothing. He began testing his own and other families, and set the stage for future genetic studies.

The minimal concentration for PTC tasting varies among individuals, but being able to taste a concentration of 0.5 millimolar or less is often taken as the cutoff between the taster and nontaster phentoypes. In a sample from Utah, the frequency of nontasters is about 30%. This is typical for people of European descent, but the frequency of nontasters differs among populations, from as low as 3% in West Africa to as high as 40% in India.

The ability or inability to taste PTC is due largely, but not exclusively, to alleles of a single gene called *TAS2R38* that encodes a taste receptor in the tongue. Nonhuman primates are homozygous for an allele of this gene known as the *PAV* allele, so called because the protein it encodes has the amino acids proline (P), alanine (A), and valine (V) at specific positions. Humans also have the *PAV* allele, and *PAV/PAV* homozygous genotypes are almost all tasters.

The most common allele associated with the nontaster phenotype is *AVI*, in which the amino acids in the taste receptor are alanine (A), valine (V), and isoleucine (I) instead of proline, alanine, and valine. About 80% of homozygous *AVI/AVI* genotypes are nontasters. One copy of the *PAV* allele is usually sufficient for the taster phenotype; about 98% of *PAV/AVI* heterozygous genotypes are tasters.

The factors contributing to the taster phenotype are more complex than a genotype at a single gene, however. The *AVI/AVI* genotype tips the balance toward the nontaster phenotype, but not completely. Other genes and the environment also play a role.

Why is this strange variation present in the human population? One hypothesis is that an aversion to compounds that contain a thiourea group (N–C=S), as PTC does, may discourage eating certain plants that produce poisonous defense compounds. One class of such compounds is the glycosinolates, which are present in many wild plants and some cultivated vegetables, including broccoli, watercress, turnip, and horseradish. Sure enough, *PAV/PAV* tasters rate such vegetables as significantly more bitter than *AVI/AVI* nontasters, whereas *PAV/AVI* heterozygous genotypes are intermediate in their perception of bitterness.

We hasten to add, however, that the low level of glycosinolates in broccoli and other cultivated vegetables is nontoxic, so you should still eat your vegetables! But if you find broccoli and its relatives somewhat bitter, you may well be a taster.

A few genetic differences are beneficial.

While many mutations are neutral, or nearly so, and many others are harmful, some mutations are beneficial. In human populations, beneficial mutations are often discovered through their effects in protecting against infectious disease. The most widely known example is probably the sickle-cell allele in the gene encoding the β chain of hemoglobin, which when heterozygous protects against malaria (see Fig. 15.1). In this section, we consider another example: This one protects against AIDS, which is caused by the human immunodeficiency virus (HIV).

By means of its surface glycoprotein (a product of the *env* region in the annotated HIV genome shown in Fig. 13.7), HIV combines with a cell-surface receptor called CD4 to gain entry into immune cells called T cells. Interaction with CD4 alone, however, does not enable the virus to infect the T cell. The HIV surface glycoprotein must also interact with another receptor on the T cell, which is denoted CCR5, in the early stages of infection (**Fig. 15.3**). The normal function of CCR5 is to bind certain small secreted proteins that promote tissue inflammation in response to infection. But because CCR5 is also an HIV receptor, cells lacking CCR5 are more difficult to invade.

A beneficial effect of a particular mutation in the *CCR5* gene was discovered in studies focusing on HIV patients whose infection had not progressed to full-blown AIDS after 10 years or more. The protective allele is denoted the *Δ32* allele because the mutation is a 32-base-pair deletion in the coding sequence of the *CCR5* gene (**Fig. 15.4**). Because 32 is not a multiple of 3, the reading frame for translation is shifted at the site of the deletion, and instead of the normal amino acid sequence Ser–Gln–Tyr–Gln–Phe···, the mutant sequence is Ile–Lys–Asp–Ser–His···. Not only is the amino

FIG. 15.3 HIV infection of T cells. HIV gains entry into a T cell by interacting with a CD4 protein and a CCR5 receptor on the surface of T cells.

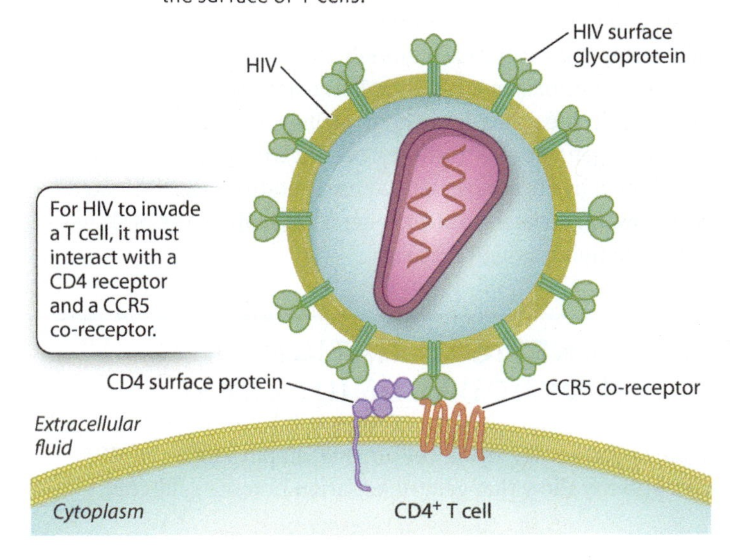

FIG. 15.4 A beneficial mutation in the human population. Mutant *CCR5* has a 32-nucleotide deletion that results in defective CCR5 protein and therefore slows the progression of HIV to AIDS.

a. Nonmutant *CCR5* allele

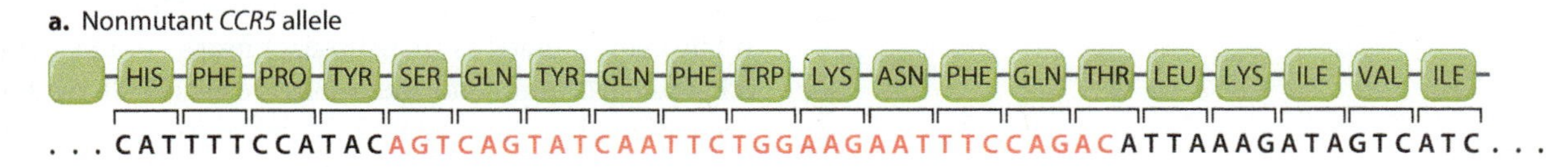

b. *Δ32* mutant *CCR5* allele

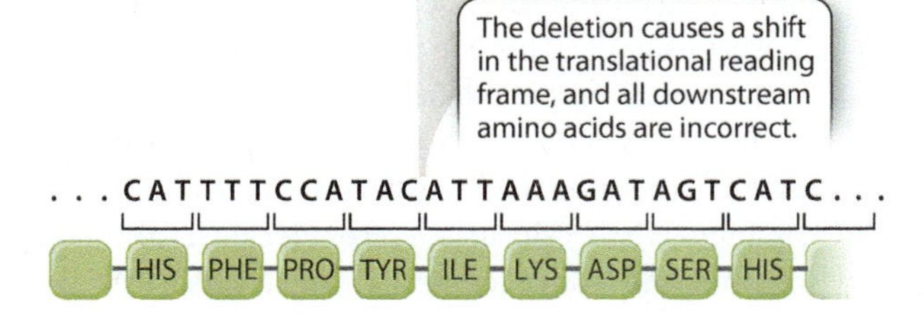

acid sequence different from the nonmutant form, the ribosome encounters a stop codon a mere 26 amino acids farther along and translation terminates. The mutant protein is 215, not 352, amino acids long. The CCR5 protein produced by the *Δ32* allele is completely inactive.

The effect of the *Δ32* allele is pronounced. In individuals with the homozygous *Δ32/Δ32* genotype, HIV progression to AIDS is rarely observed. There is some protection even in individuals with heterozygous *Δ32* genotypes, where progression to AIDS is delayed by an average of about 2 years after infection by HIV.

Much has been written about the evolutionary history of the *Δ32* allele. It is found almost exclusively in European populations, where the frequency of heterozygous genotypes ranges from 10% to 25%. The narrow geographical distribution was originally interpreted to mean that the allele was selected over time because it provided protection against some other infectious agent that also interacted with the CCR5 protein.

Beneficial mutations not only provide protection against disease. Many beneficial mutations permit organisms to become better adapted to their environment. For example, certain birds, such as Rüppell's vulture, commonly fly at altitudes of 20,000 feet and sometimes much higher. This feat is possible because of mutations in the structure of hemoglobin that allow hemoglobin to bind oxygen with high affinity, even at the low pressure of oxygen high in the atmosphere. These mutations were selected and passed on generation after generation, making the birds well adapted to flying at high altitudes.

15.2 GENETIC VARIATION AND INDIVIDUAL UNIQUENESS

While examples like sickle-cell anemia, emphysema, and HIV susceptibility show that genetic variation in some genes has important effects on phenotypes, most of the genetic variation in populations is neutral or has no obvious effects. Much of the neutral variation consists of differences in noncoding DNA. Nonetheless, this variation can be revealed by direct studies of DNA that employ many of the techniques for DNA manipulation described in Chapter 12.

Areas of the genome with variable numbers of tandem repeats are useful in DNA typing.

At many locations in the human genome, short (10–50 base-pair) sequences are repeated in multiple tandem copies. At each location the number of copies of the repeat may differ from chromosome to chromosome. This kind of variation, in which chromosomes have different numbers of a repeated sequence at a particular location, is called a **variable number tandem repeat,** or VNTR. There are about 30,000 different VNTR locations in the human genome. Just as an individual has two copies of each gene, possibly with two different alleles, so each individual has two copies of each of these VNTRs, with copies differing in the number of repeats being analogous to different alleles of a gene. Each VNTR typically has many alleles differing in copy number, so that, except for identical twins, an individual's genotype for 6–8 VNTR locations is usually sufficient to identify the individual uniquely. Differences among individuals, such as VNTRs, are the basis of **DNA typing,** in which the analysis of a small quantity of DNA can be as reliable for identifying individuals as fingerprints. This is why DNA typing is often called **DNA fingerprinting.**

Any of several methods can be used to DNA fingerprint someone on the basis of that individual's VNTRs. Amplification by the polymerase chain reaction (PCR, Chapter 12) is a convenient and reliable technique because each VNTR in the genome is flanked by a unique sequence that can be used to design PCR primers that will amplify one and only one VNTR. Although the number of alleles of a particular VNTR can be very large, an example with only

FIG. 15.5 Variable number tandem repeats (VNTRs). (a) The number of short, repeated sequences at a given site in a human chromosome can vary. (b) These differences can be visualized using PCR followed by gel electrophoresis.

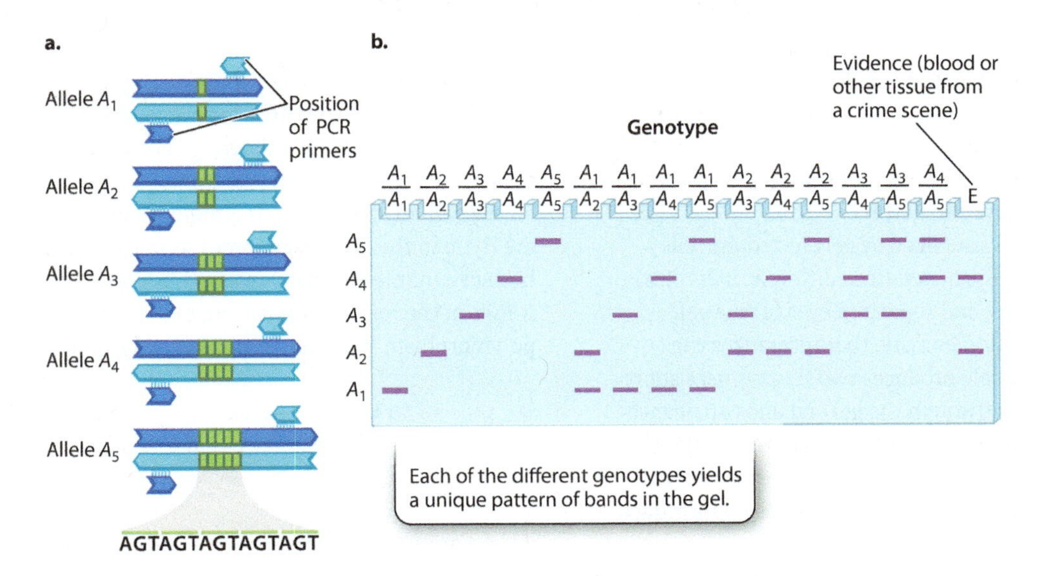

five alleles is shown in **Fig. 15.5**, where the repeated sequence is shown in green between the flanking unique sequences.

No matter how many alleles exist in the population as a whole, any one diploid individual can have only two alleles, which are present at the same positions in the homologous chromosomes inherited from the mother and father. With five alleles there are 5 possible homozygous genotypes and 10 possible heterozygous genotypes. These can be visualized after PCR by separating the resulting molecules by size by gel electrophoresis (Chapter 12). The pattern of bands expected from the DNA in each genotype formed from the alleles in Fig. 15.5a is shown in Fig. 15.5b. Note that each genotype yields a distinct pattern of bands, and so these patterns can be used to identify the genotype of any individual for a particular region of DNA.

With as few as 10 equally frequent alleles, the likelihood that any two individuals would have the same genotype at a particular location by chance alone is about 1 in 50, and with 20 equally frequent alleles the likelihood decreases to about 1 in 200. These numbers suggest why DNA typing can yield a genetic fingerprint that can uniquely match DNA samples.

The lane labeled "E" at the far right in Fig. 15.5b shows the DNA bands for this same polymorphism from biological material collected at the scene of a crime. In this case, the bands in lane E imply that the source of the evidence is an individual of genotype A_2/A_4 because this is the only genotype with bands with vertical positions in the gel that exactly match those of the evidence sample. A single polymorphism of the type shown in Fig. 15.5 is not enough to establish for sure that two pieces of DNA come from the same person, but because the human genome has polymorphic sites scattered throughout, many polymorphisms can be examined. If six or eight of these polymorphisms match between two samples, it is very likely that the samples come from the same individual (or from identical twins). On the other hand, if any of the polymorphisms fails to match band for band, then it is almost certain that the samples come from different individuals (barring such technical mishaps as samples whose DNA is contaminated, degraded, or mislabeled).

One of the advantages of DNA typing is that a large amount of DNA is not needed. Even minuscule amounts of DNA can be typed, so tiny spots of blood, semen, or saliva are sufficient samples. A cotton swab wiped across the inside of your cheek will contain enough DNA for typing; so will a discarded paper cup or a cigarette butt. One enterprising researcher in England used PCR to type the dog feces on his lawn and matched it to the DNA from hair he obtained from neighborhood dogs, much to the chagrin of the guilty dog's owner.

→ **Quick Check 3** In typing DNA from a sample found at a crime scene, how can a DNA mismatch prove that a suspect is not the source of that sample, whereas a DNA match does not necessarily prove that a suspect is the source?

Some polymorphisms add or remove restriction sites in the DNA.

Restriction enzymes, which cleave double-stranded DNA at specific sequences known as restriction sites, typically four or six nucleotides long, were among the first tools used to study genetic variation in DNA. These enzymes are useful in DNA fingerprinting because DNA from different individuals can differ in the distance between adjacent restriction sites or in the presence or absence of a particular restriction site at some location in the genome. This kind of polymorphism, in which differences in restriction sites result in different lengths of restriction fragments, is called a **restriction fragment length polymorphism,** or RFLP.

Fig. 15.6 shows an example. Here, a restriction enzyme cleaves DNA near the sequence GAGGAG. The three sites shown are in and flanking the β-globin gene from the sickle-cell example we explored earlier in the chapter. However, in the *S* allele (see Figure 15.1), the restriction site in the middle includes a GAG that is mutated to GTG, making the sequence unrecognizable to the restriction enzyme. As a result, the DNA of the *S* allele is cleaved at only two sites, whereas the DNA of the *A* allele is cleaved at all three.

These differences can be visualized by gel electrophoresis (Chapter 12), in which DNA is isolated from different individuals, cut with restriction enzymes, and separated by size on a gel. In this case, we use a restriction enzyme that recognizes the GAGGAG sequence. The *A* allele produces two fragments: a short fragment that moves rapidly through the gel and ends up near the bottom, and a longer fragment that ends up near the middle. The fragment produced by the *S* allele is even longer because there is no restriction site in the middle. This fragment ends up forming a band near the top of the gel.

FIG. 15.6 Restriction fragment length polymorphism (RFLP) between the *A* and *S* alleles of the β-globin gene.

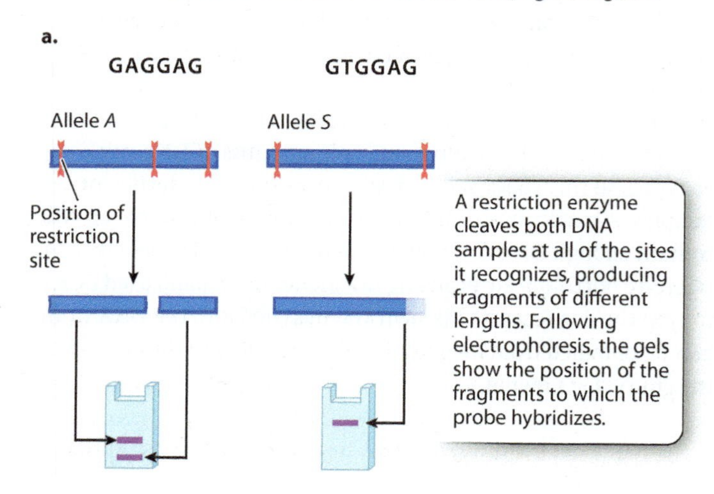

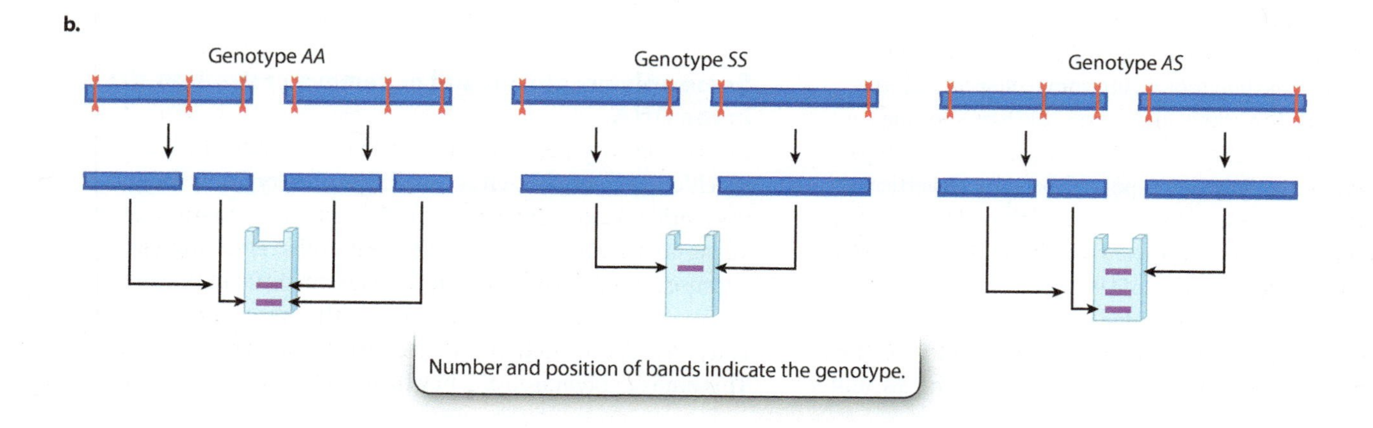

Just as with VNTRs, any individual can carry at most two different alleles of a restriction fragment length polymorphism. As seen in the RFLP in Fig. 15.6b, DNA from homozygous *AA* individuals yields only the short and medium fragments, and that from homozygous *SS* individuals yields only the long fragment. DNA from heterozygous *AS* individuals yields all three fragments. Therefore, each of the three genotypes *AA*, *AS*, and *SS* yields a unique pattern of bands, allowing each of the possible genotypes to be identified.

How many restriction fragment length polymorphisms are there in the human genome? As in the case of VNTRs, the human genome has many restriction sites, so there are numerous RFLPs in the human population, making it another type of polymorphism that can be used in DNA typing.

→ **Quick Check 4** When a segment of DNA containing either a VNTR or an RFLP is analyzed, the result is fragments of DNA of different lengths. How, then, are VNTRs and RFLPs different?

15.3 GENOMEWIDE STUDIES OF GENETIC VARIATION

VNTRs and RFLPs are forms of genetic variation that are still widely used in DNA fingerprinting. But large-scale and relatively inexpensive DNA sequencing methods are now also used to identify and study genetic variation. These methods have revealed two types of genetic variation that are very common in the human genome: variation of individual nucleotides and variation in copy number of regions of DNA that can include one or more genes. In this section, we focus on these two types of genetic variation.

Single-nucleotide polymorphisms (SNPs) are single-base changes in the genome.

One of the most common types of genetic variation is a difference in a nucleotide at a specific site. A **single-nucleotide polymorphism (SNP)** is a site in the genome where either of two different nucleotide pairs can occur and where each nucleotide

FIG. 15.7 Single-nucleotide polymorphisms (SNPs). A SNP in the region neighboring *OCA2* is strongly associated with blue eyes. *Sources: (left) Amos Morgan/Getty Images; (right) Stockbroker xtra/age fotostock.*

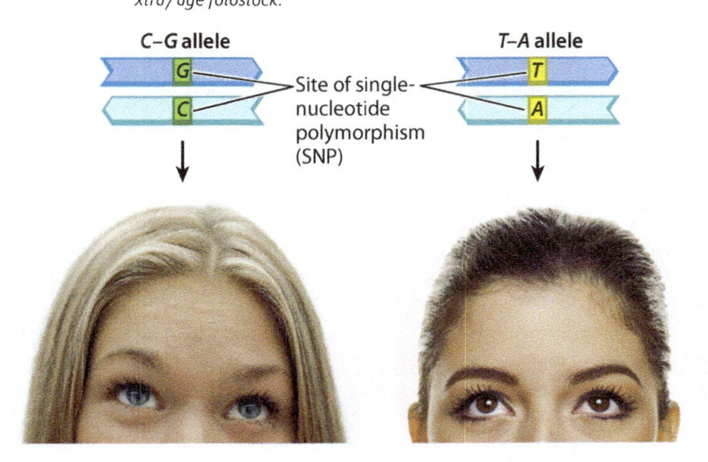

pair is common enough in the population to be present in a random sample of 50 diploid individuals. The differences among the hemoglobin alleles are SNPs. The *A*, *S*, and *C* alleles differ from one another at just one nucleotide site (see Fig. 15.1).

Eye color is also associated with a SNP. The blue eye phenotype results from reduced expression of a gene called *OCA2*, which encodes a membrane protein involved in the transport of small molecules, including the amino acid tyrosine, which is a precursor of the melanin pigment associated with the brown eye phenotype. Although more than a dozen alleles of *OCA2* are known that have an amino acid replacement in the protein, none of these results in blue eyes.

The SNP implicated in blue eyes is a site that is a C–G base pair in one allele. The other common allele has a T–A at this position, which is not associated with blue eyes (**Fig. 15.7**). In one study, the homozygous *C–G* genotype was found in 94% of 183 individuals with blue eyes and in only 2% of 176 individuals with brown eyes. This strong association is quite unexpected because the SNP associated with blue eyes is in an intron of a neighboring gene, and is therefore noncoding. The proposed mechanism for the association is that the *C–G* allele makes the adjacent *OCA2* gene less accessible to transcription factors, thereby reducing the amount of the transporter protein and consequently the production of melanin in the iris. The proposed mechanism may be right or wrong; this example emphasizes that scientists have much to learn about the possible phenotypic effects of noncoding DNA.

The *C–G* versus *T–A* SNP is typical of most SNPs in that only two of the four possible base pairs (*G–C* and *A–T* in addition to *C–G* and *T–A*) are present in the population at any appreciable frequency.

→ **Quick Check 5** What's the difference between a point mutation and a SNP?

? CASE 3 YOU, FROM A TO T: YOUR PERSONAL GENOME

How can genetic risk factors be detected?

SNPs result from point mutations, the most frequent type of mutation, which occurred in the past and then spread through the population. A point mutation is the substitution of one base pair for another in double-stranded DNA (Chapter 14). Each of us is genetically unique partly because of the abundance of SNPs in the human genome. There are approximately 3 million SNPs that distinguish any one human genome from any other. SNPs are abundant in the genomes of most species, and while many have no effect on the organism, some are thought to be the main source of evolutionary innovation and others are major contributors to inherited disease. Practically speaking, SNPs are important because they can be used to detect the presence of a genetic risk factor for a disease before the onset of the disease. For example, the ability to detect the β-globin SNPs shown in Fig. 15.1 allows prenatal identification of the β-globin genotype of a fetus.

While there is great interest in developing ultrafast DNA sequencing machines that can determine anyone's personal genome quickly at relatively low cost, 99.9% of the nucleotides between any two genomes are identical. An alternative is to focus on genotyping just SNPs. As many as one million SNPs at different positions in the genome can be genotyped simultaneously, and the genotyping can be carried out on thousands or tens of thousands of individuals. Such massive genotyping allows any SNP associated with a disease to be identified, which is especially important for complex diseases affected by many different genetic risk factors (Chapter 18).

What does it mean to say that a given SNP is associated with a disease? It means that individuals carrying one of the alleles of that SNP are more likely to develop the disease than those carrying the other allele. The increased risk depends on the disease and can differ from one SNP to the next. Sickle-cell anemia affords an example at one extreme of the spectrum of effects. In this case, the T–A base pair in the *S* allele of the β-globin gene is the SNP that results in the amino acid replacement of glutamic acid with valine in the protein. Because *SS* individuals always have sickle-cell anemia, it would be fair to say that homozygous *S* "causes" sickle-cell anemia.

But except for inherited diseases that result from single mutant genes, which are usually rare, the vast majority of SNPs implicated in disease increase the risk only moderately as compared with individuals lacking the risk factor. We then say that the SNP is "associated" with the disease since the SNP alone does not cause the disease but only increases the risk. For heart disease, diabetes, and some other diseases, many SNPS at different places in the genome, as well as environmental risk factors, can be associated with the disease. Usually, genetic and environmental risk factors act cumulatively: the more you have, the greater the risk.

As emphasized in the case of Claudia Gilmore's genome, certain SNPs in the *BRCA1* and *BRCA2* genes are associated with an increased risk of breast and ovarian cancers. Women who carry a mutation in either of these genes can minimize their risk by frequent mammograms and other tests that enable early treatment. Both genes are large—*BRCA1* codes for a protein of 1863 amino acids and *BRCA2* for one of 3418 amino acids—and many different mutations in these genes can predispose to breast or ovarian cancer. In certain high-risk populations, however, such as Ashkenazi Jews, only a few mutations predominate, and the SNPs associated with these mutations can be detected easily and efficiently.

Copy-number variation constitutes a significant proportion of genetic variation.

In addition to SNPs, another very common form of genetic variation present in the modern human population is **copy-number variation (CNV),** or differences among individuals in the number of copies of a region of the genome. In contrast to the short repeats characteristic of VNTRs, the regions involved in CNVs are large and may include one or more genes. An example is shown in **Fig. 15.8.** In this case, a region of the genome that is normally present in only one copy per chromosome (Fig. 15.8a) may in some chromosomes be duplicated (Fig. 15.8b) or deleted (Fig. 15.8c). The multiple copies of the CNV region are usually adjacent to one another along the chromosome.

One of the surprises that emerged from sequencing the human genome was that CNV is quite common in the human population. Any two individuals' genomes differ in copy number at about five different regions, each with an average length of 200 kb to 300 kb. Across the genome as a whole, about 10% to 15% of the genome is subject to copy-number variation.

Some CNVs occur in noncoding regions, but others consist of genes that are present in multiple tandem copies along the chromosome. An example is the human gene *AMY1* for the salivary gland enzyme amylase, which aids in the digestion of starch. This gene is located in chromosome 1, and the *AMY1* copy number differs from one chromosome 1 to the next. **Fig. 15.9** shows the distribution of *AMY1* copy number along chromosome 1 in two groups of people: societies with a long history of a high-starch diet, and societies with a long history of a low-starch diet. There is a clear tendency for chromosomes from the latter group to have fewer copies of *AMY1* than those from the former group. On average, a chromosome 1 from the low-starch group has a copy number of about 2.5, whereas one from the high-starch group has an average of about 3.5. Because each individual has two copies of chromosome 1, this difference means that the average individual in the low-starch group has about 5 copies of *AMY1*, whereas the average individual in the high-starch group has about 7 copies. A plausible hypothesis is that extra copies were selected in groups with a high-starch diet because of the advantage extra copies conferred in digesting starch.

FIG. 15.9 Copy number variation in a gene that helps break down starch. Copy numbers of *AMY1* vary between societies with high-starch and low-starch diets. *Data from G. H. Perry, N. J. Dominy, K. G. Claw, A. S. Lee, H. Fiegler, R. Redon, J. Werner, F. A. Villanea, J. L. Mountain, R. Misra, N. P. Carter, C. Lee, and A. C. Stone, 2007, "Diet and the Evolution of Human Amylase Copy Number Variation," Nature Genetics 39:1256–1260.*

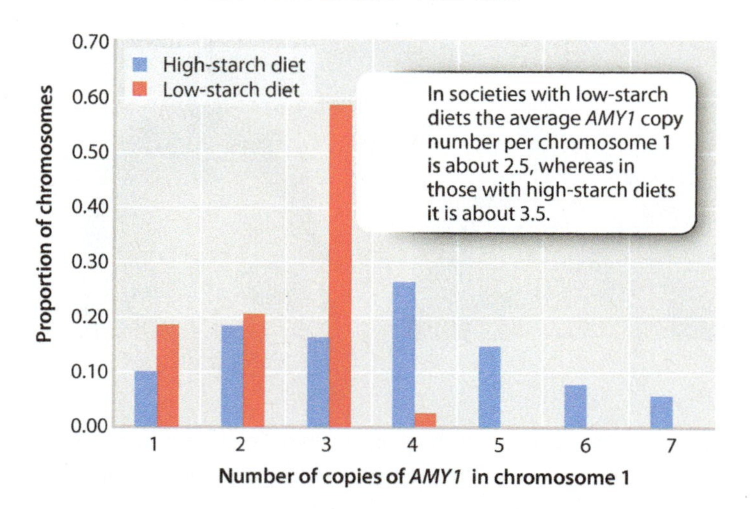

FIG. 15.8 Copy-number variation in a region of a chromosome.

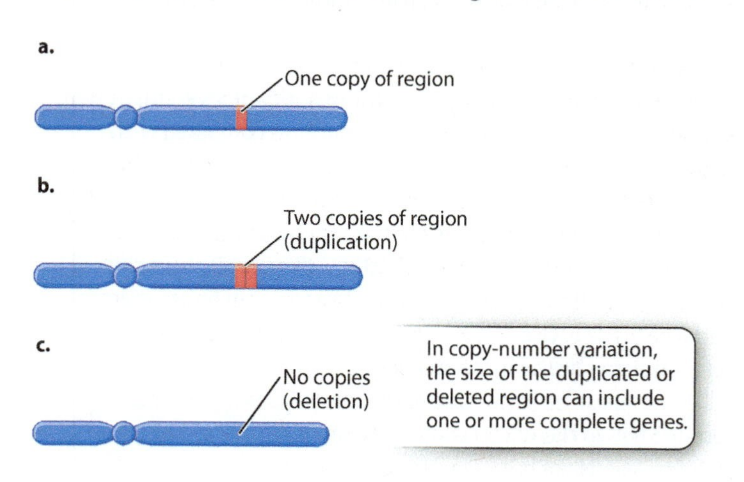

15.4 GENETIC VARIATION IN CHROMOSOMES

Copy-number variation usually involves only one or a small number of genes, and the size of the duplicated or deleted region is physically so small that the differences are undetectable with conventional microscopy. In the human genome, some common variants involve large regions of the chromosome and are big enough to be visible through a microscope. Most of these variations involve regions of the genome with an extremely low density of genes, such as regions around the centromere or on the long arm of the Y chromosome.

FIG. 15.10 **Nondisjunction.** (a) Chromosomes separate evenly into gametes in normal meiotic cell division. (b) Homologous chromosomes fail to separate in first-division nondisjunction. (c) Sister chromatids fail to separate in second-division nondisjunction.

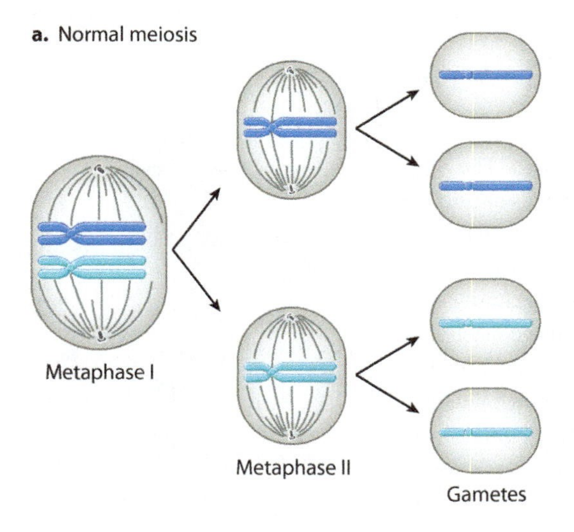

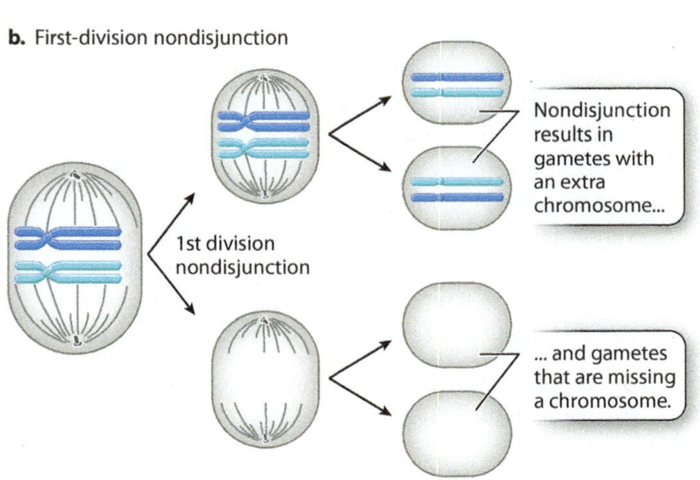

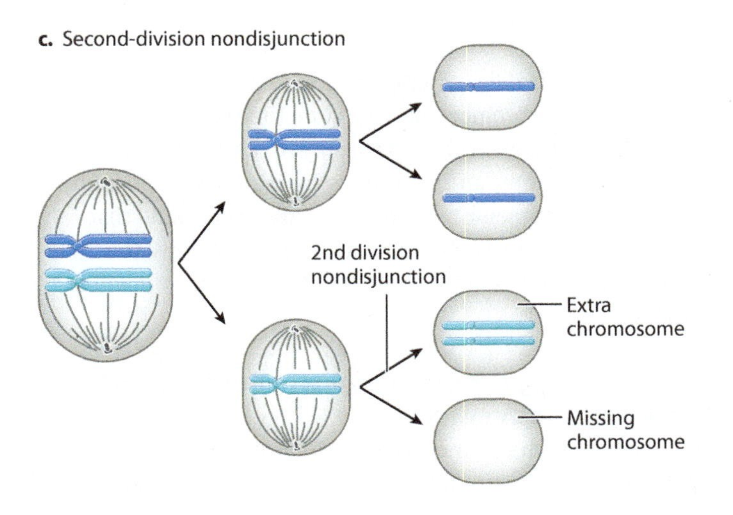

It may come as a surprise to learn that major chromosomal differences occur quite commonly in humans. They are infrequently observed, however, because most of them are lethal. Nevertheless, a few major chromosomal differences are found in the general population. Some are common enough that their phenotypic effects are familiar. Others are less common, and a few have no major phenotypic effects at all. In this section, we focus on the types of difference that are observed, what causes them, and their phenotypic effects.

Nondisjunction in meiosis results in extra or missing chromosomes.

Nondisjunction is the failure of a pair of chromosomes to separate during anaphase of cell division (Chapter 11). The result is that one daughter cell receives an extra copy of one chromosome, and the other daughter cell receives no copy of that chromosome. Nondisjunction can take place in mitosis, and it leads to cell lineages with extra or missing chromosomes—which are often observed in cancer cells.

When nondisjunction takes place in meiosis, the result is reproductive cells (gametes) that have either an extra or a missing chromosome. **Fig. 15.10** illustrates nondisjunction in meiosis. Fig. 15.10a shows key stages in a normal meiosis, in which chromosome separation (disjunction) takes place in both meiotic divisions, with the result that each gamete receives one and only one copy of each chromosome.

Failure of chromosome separation can occur in either of the two meiotic divisions. Both types of nondisjunction result in gametes with an extra chromosome or a missing chromosome, but there is an important difference. In **first-division nondisjunction** (Fig. 15.10b), which is much more common, the homologous chromosomes fail to separate and all of the resulting gametes have an extra or missing chromosome. On the other hand, in **second-division nondisjunction** (Fig. 15.10c), sister chromatids fail to separate, giving rise to products with a normal number of chromosomes, an extra chromosome, or a missing chromosome.

Some human disorders result from nondisjunction.

Approximately 1 out of 155 live-born children (0.6%) has a major chromosome abnormality of some kind. About half of these have a major abnormality in chromosome structure, such as an inversion or translocation (Chapter 14). The others have a chromosome that is present in an extra copy or a chromosome that is missing because of nondisjunction—a parent has produced an egg or a sperm with an extra or a missing chromosome (**Fig. 15.11**).

Perhaps the most familiar of the conditions resulting from nondisjunction is **Down syndrome,** which results from the presence of an extra copy of chromosome 21. Down syndrome is also known as **trisomy 21** because affected individuals have three copies of chromosome 21. The extra chromosome can be seen in the chromosome layout called a **karyotype.** French researchers

FIG. 15.11 Incidence of extra and missing chromosomes in human live births. *After D. L. Hartl and M. Ruvolo, 2012, Genetics: Analysis of Genes and Genomes, 8th ed., Burlington, MA: Jones & Bartlett, Table 8.2, p. 268.*

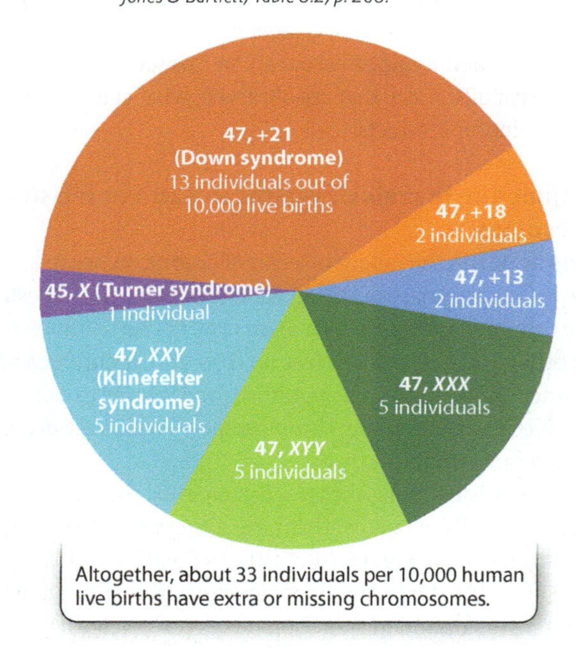

Jérôme Lejeune, Marthe Gautier, and Raymond Turpin first saw the extra chromosome when they applied a technique to visualize a patient's chromosomes (**Fig. 15.12**). The shorthand for the condition is 47,+21, meaning that the individual has 47 chromosomes with an extra copy of chromosome 21.

The presence of an extra chromosome 21 leads to a set of traits that, taken together, constitute a syndrome. These traits presumably result from an increase in dosage of the genes and other elements present on chromosome 21, although not all the specific genes or genetic elements have been identified. Children with Down syndrome are mentally disabled to varying degrees; most are in the mild to moderate range and with special education and support they can acquire basic communication, self-help, and social skills. People with Down syndrome are short because of delayed maturation of the skeletal system; their muscle tone is low; and they have a characteristic facial appearance. About 40% of children with Down syndrome have major heart defects, and life expectancy is currently about 55 years. Among adults with Down syndrome, the risk of dementia is about 15% to 50%, which is about five times greater than the risk in the general population.

Although Down syndrome affects approximately 1 in 750 live-born children overall, there is a pronounced effect of mother's age on the risk of Down syndrome. For mothers of ages 45–50, the risk of having a baby with Down syndrome is approximately 50 times greater than it is for mothers of ages 15–20. The increased risk is the reason why many physicians recommend tests to determine abnormalities in the fetus for pregnant women over the age of 35.

HOW DO WE KNOW?

FIG.15.12

What is the genetic basis of Down syndrome?

Source: Terry Harris/Rex Features/ AP Images.

BACKGROUND The detailed images of cells and chromosomes available today are produced by microscopic and imaging techniques not available during most of the history of genetics and cell biology. Techniques for observing human chromosomes were originally so unreliable that the correct number of human chromosomes was firmly established only in the mid-1950s. Mistakes were common, such as counting two nearby chromosomes as one or including chromosomes in the count of one nucleus when they actually belonged to a nearby nucleus. Despite these limitations, in 1959 French scientists Jérôme Lejeune, Marthe Gautier, and Raymond Turpin decided to investigate whether any common birth abnormalities were associated with chromosomal abnormalities. They chose Down syndrome, one of the most common birth abnormalities, because its genetic basis was an enigma.

HYPOTHESIS Down syndrome is associated with extra or missing chromosomes.

Two other trisomies of autosomes (chromosomes other than the X and Y chromosomes) are sometimes found in live births. These are trisomy 13 and trisomy 18. Both are much rarer than Down syndrome and the effects are more severe. In both cases, the developmental abnormalities are so profound that newborns with these conditions usually do not survive the first year of life. Other trisomies occur, but they are not compatible with life.

Extra or missing sex chromosomes have fewer effects than extra autosomes.

Children born with an abnormal number of autosomes are rare, but extra or missing **sex chromosomes** (*X* or *Y*) are relatively common among live births. In most mammals, females have two

EXPERIMENT The chromosomes of five boys and four girls with Down syndrome were studied. A small patch of skin was taken from each child and placed in growth medium. In this medium, fibroblast cells, which synthesize the extracellular matrix that provides the structural framework of the skin, readily undergo DNA synthesis and divide. After allowing cellular growth and division, the fibroblast cells were prepared in a way that allowed the researchers to count the number of chromosomes under a microscope.

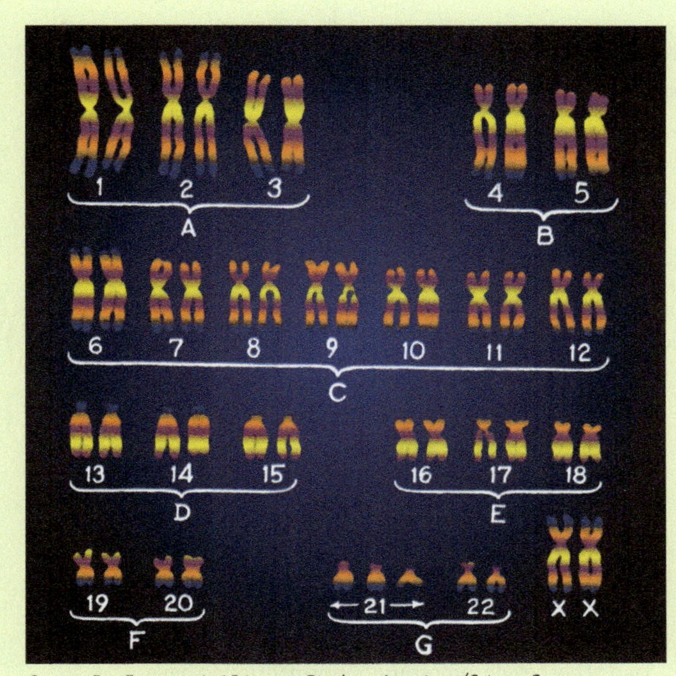

Source: Rex Features via AP Images, Biophoto Associates/Science Source. Colorization by Mary Martin.

RESULTS Chromosomes were counted in 103 cells, yielding counts of 46 chromosomes (10 cells), 47 chromosomes (84 cells), and 48 chromosomes (9 cells). For each cell examined, the researchers noted whether the count was "doubtful" or "absolutely certain." The researchers marked cells as "doubtful" if the chromosomes or nuclei were not well separated from one another. They encountered 57 cells of which they felt "absolutely certain," and in each of these they counted 47 chromosomes. The extra chromosome was always one of the smallest chromosomes in the set.

CONCLUSION The researchers concluded that Down syndrome is the result of a chromosomal abnormality, an extra copy of one of the smallest chromosomes. The discovery helped resolve many of the questions surrounding the genetics of Down syndrome.

FOLLOW-UP WORK Improved techniques including chromosome banding (shown in Fig. 13.15) later indicated that the extra chromosome was chromosome 21. The finding stimulated many other chromosome studies. Today, it is clear that, except for the sex chromosomes, the vast majority of fetuses with extra or missing chromosomes undergo spontaneous abortion. Furthermore, the Human Genome Project has indicated that genes on chromosome 21 have direct roles in many of the signs and symptoms of Down syndrome.

SOURCE J. Lejeune, Gautier, M., and Turpin, R. 1959. "Premier exemple d'aberration autosomique humaine." *Comptes rendus des séances de l'Académie des Sciences.* 248:1721–1722. Translated from the French in *Landmarks in Medical Genetics,* edited by P. S. Harper. Oxford, New York, 2004.

X chromosomes (*XX*), and males have one *X* and one *Y* chromosome (*XY*). The presence of the *Y* chromosome, not the number of *X* chromosomes, leads to male development. The female karyotype 47, *XXX* (47 chromosomes, including three *X* chromosomes) and the male karyotype 47, *XYY* (47 chromosomes, including one *X* and two *Y* chromosomes) are found among healthy females and males. These persons are in the normal range of physical development and mental capability, and usually their extra sex chromosome remains undiscovered until their chromosomes are examined for some other reason.

The reason that 47, *XYY* males show no detectable phenotypic effects is the unusual nature of the *Y* chromosome. The *Y* contains only a few functional genes other than the gene that stimulates the embryo to take the male developmental pathway. The absence of detectable phenotypic effects in 47, *XXX* females has a completely different explanation that has to do with the manner in which the activity of genes in the *X* chromosome is regulated. In the cells of female mammals, all *X* chromosomes except one are inactivated and gene expression is largely repressed. (This process, called *X*-inactivation, is discussed in Chapter 20.) Because of *X*-inactivation, a 47, *XXX* female has one active *X* chromosome per cell, the same number as in a 46, *XX* female.

→ **Quick Check 6** A male baby is born with the sex-chromosome constitution *XYY*. Both parents have normal sex chromosomes (*XY* in the father, *XX* in the mother). In which meiotic division of which parent did the nondisjunction take place that produced the *XYY* baby?

FIG. 15.13 Symptoms associated with (a) Klinefelter syndrome and (b) Turner syndrome.

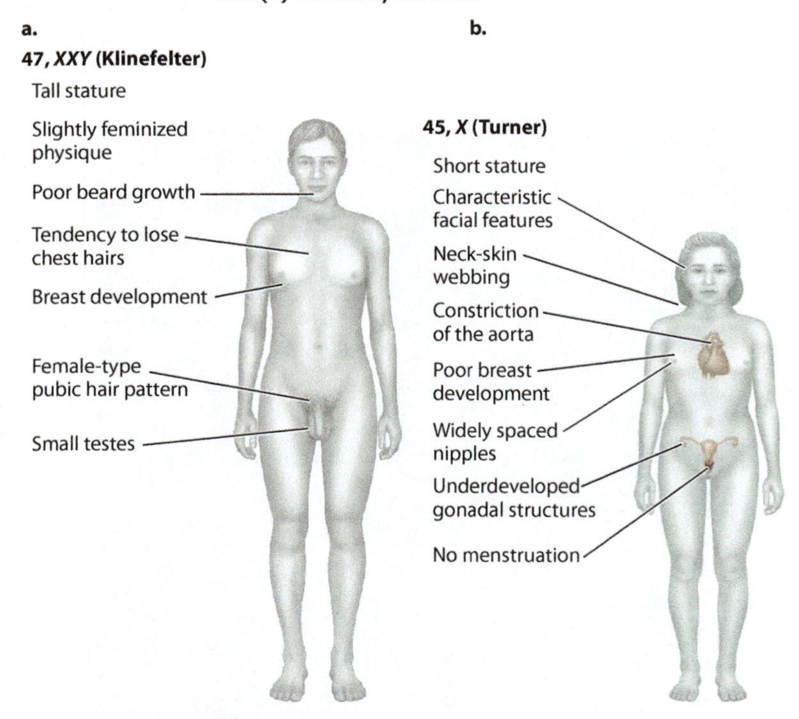

Two other sex chromosome abnormalities do have phenotypic effects, especially on sexual development (**Fig. 15.13**). The more common is 47, *XXY* (47 chromosomes, including two *X* chromosomes and one *Y* chromosome). Individuals with this karyotype are male because of the presence of the *Y* chromosome, but they have a distinctive group of symptoms known as **Klinefelter syndrome** (Fig. 15.13a). Characteristics of the syndrome are very small testes but normal penis and scrotum. Growth and physical development, for the most part, are normal, although affected individuals tend to be tall for their age. About half of the individuals have some degree of mental impairment. When a person with Klinefelter syndrome reaches puberty, the testes fail to enlarge, the voice remains high pitched, pubic and facial hair remain sparse, and there is some enlargement of the breasts. Affected males do not produce sperm and hence are sterile; up to 10% of men who seek aid at infertility clinics turn out to be *XXY*.

Much rarer than all the other sex chromosome abnormalities is 45, *X* (45 chromosomes, with just one *X* chromosome), the karyotype associated with characteristics known as **Turner syndrome** (Fig. 15.13b). Affected females are short and often have a distinctive webbing of the skin between the neck and shoulders. There is no sexual maturation. The external sex organs remain immature, the breasts fail to develop, and pubic hair fails to grow. Internally, the ovaries are small or absent, and menstruation does not occur. Although the mental abilities of affected females are very nearly normal, they have specific defects in spatial abilities and arithmetical skills.

Earlier, we saw that in *XX* females one *X* chromosome undergoes inactivation. Why, then, do individuals with just one *X* chromosome show any symptoms at all? The explanation is that some genes on the inactive *X* chromosome in *XX* females are not completely inactivated. Evidently, for normal development to take place, both copies of some genes that escape complete inactivation must be expressed.

A karyotype with 45, *X* is one of the rarest seen in live-born babies. The reason is that more than 99% of the fetuses that are 45, *X* undergo spontaneous abortion—the subject of the next section.

Nondisjunction is a major cause of spontaneous abortion.

Trisomies and the sex-chromosome abnormalities discussed in the previous section account for most of the simpler chromosomal abnormalities that are found among babies born alive. However, these chromosomal abnormalities represent only a minority of those that actually occur. In fertilized eggs with extra or missing

FIG. 15.14 Chromosomal abnormalities in spontaneous abortion. *After D. L. Hartl and M. Ruvolo, 2012, Genetics: Analysis of Genes and Genomes, 8th ed., Burlington, MA: Jones & Bartlett, Table 8.2, p. 268.*

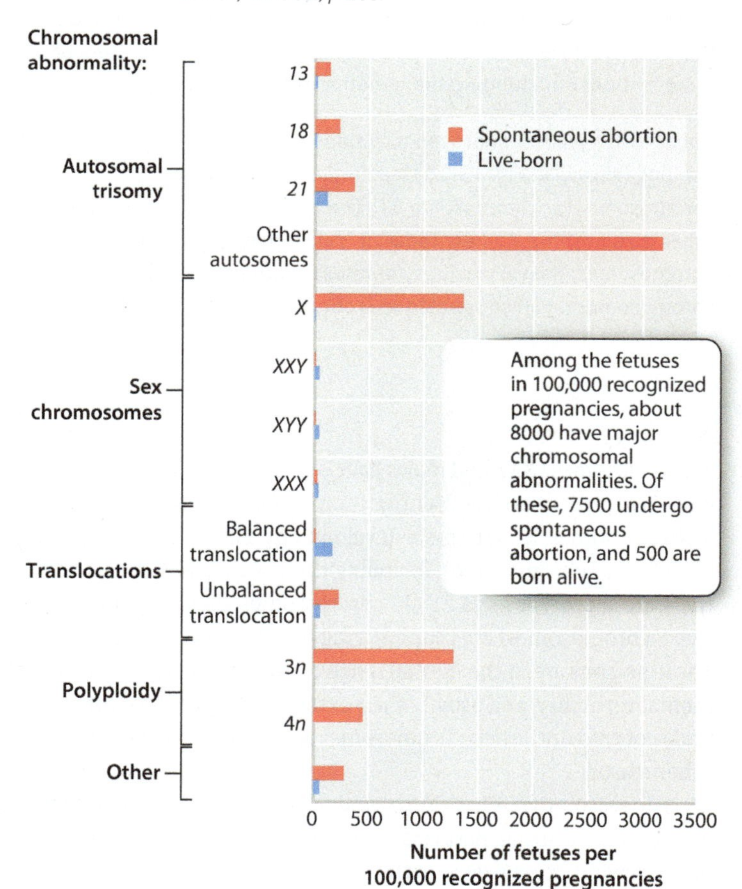

FIG. 15.15 Formation of polyploid organisms. (a) A triploid organism can result from failure of division in meiosis. (b) A tetraploid organism can result from failure of cell division in mitosis.

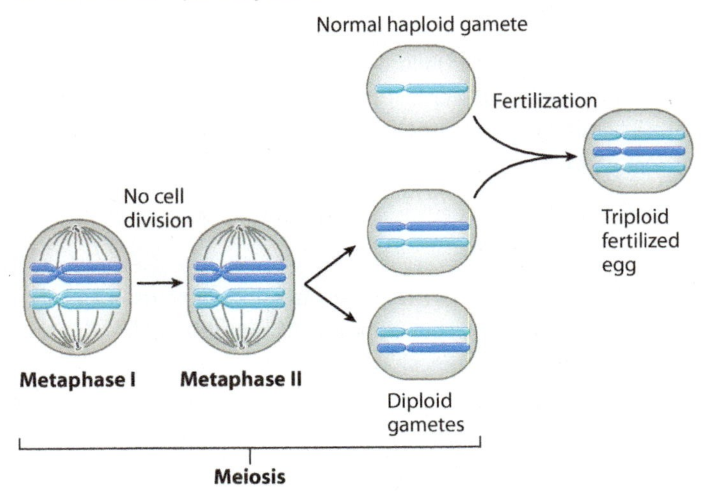

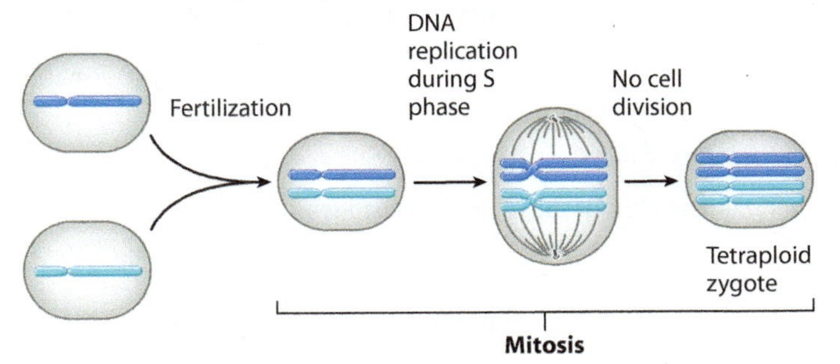

chromosomes, the dosage of genes in these chromosomes is unbalanced relative to the rest of the genome, and the embryos usually fail to complete development. At some time during pregnancy—in some cases very early, in other cases relatively late—the chromosomally abnormal embryo or fetus undergoes spontaneous abortion.

The relative proportions of some of the major chromosomal abnormalities in spontaneous abortion are shown in **Fig. 15.14.** The bars in red represent recognized pregnancies that terminate in spontaneous abortion, and those in blue represent those in which the fetus develops to term and is born alive. Note the large number of autosomal trisomies, none of which (with the exception of trisomies 13, 18, and 21) permits live births. Even among these, about 75% of fetuses with trisomy 21 undergo spontaneous abortion, and the proportions are even greater for trisomies 13 and 18. The 45, *X* karyotype also very infrequently results in live birth.

A surprisingly large number of fetuses that undergo spontaneous abortion are **triploid** (with three complete sets of chromosomes, 69 altogether) or **tetraploid** (four complete sets, 92 altogether). These karyotypes usually result from a defective spindle apparatus and failure of cell division in anaphase. When this occurs in meiosis, the result is a diploid gamete and a triploid fertilized egg (**Fig. 15.15a**). It can also occur after normal fertilization, when the diploid egg goes through mitosis but the cells fail to divide after the DNA is replicated. The result is a tetraploid (**Fig. 15.15b**). Many spontaneous abortions result from an **unbalanced translocation** (Chapter 14), in which only part of a reciprocal translocation (along with one of the nontranslocated chromosomes) is inherited from one of the parents (**Fig. 15.16**).

Altogether, about 15% of all recognized pregnancies terminate with spontaneous abortion of the fetus, and roughly half of these are due to major chromosomal abnormalities. This number tells only part of the story because embryos with a missing autosome are not found among spontaneously aborted fetuses. These must occur at least as frequently as those with an extra autosome because both are created by the same event of nondisjunction (see Fig. 15.12). The explanation seems to be that in fertilized eggs with a missing autosome the abortion occurs shortly after fertilization, and most cases are not recognized. ■

FIG. 15.16 (a) A balanced and (b) an unbalanced translocation.

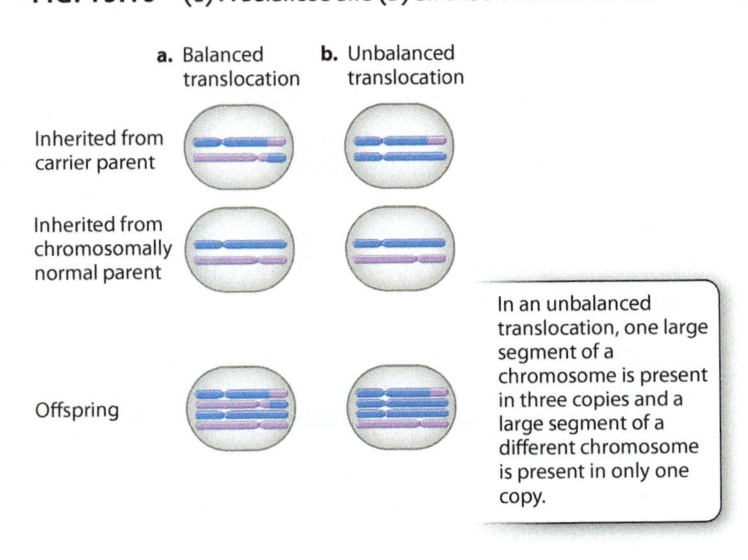

Core Concepts Summary

15.1 Genetic variation describes common genetic differences (polymorphisms) among the individuals in a population at any given time.

The genotype of an organism is its genetic makeup. page 310

The phenotype of an organism is any observable characteristic of that organism, such as its appearance, physiology, or behavior. The phenotype results from a complex interplay of the genotype and the environment. page 310

A person's genotype can be homozygous, in which he or she has two alleles of the same type (one from each parent), or heterozygous, in which he or she has a different type of allele from each parent. page 311

Genetic variation can be neutral (no effect on survival or reproduction), harmful (associated with decreased survival or reproduction), or beneficial (associated with increased survival or reproduction). page 311

15.2 Human genetic variation can be detected by DNA typing, which can uniquely identify each individual.

A variable number tandem repeat (VNTR) results from differences in the number of small-sequence repeats in a given area of the genome. page 314

DNA typing (DNA fingerprinting) analyzes genetic polymorphisms at multiple genes with multiple alleles. Because there are so many possible genotypes in the population, this procedure can uniquely identify an individual with high probability. page 315

A restriction fragment length polymorphism (RFLP) results from small changes in the DNA sequence between chromosomes, such as point mutations, that create or destroy restriction sites. page 316

15.3 Two common types of genetic polymorphism are single-nucleotide polymorphisms (SNPs) and copy-number variation (CNV).

A SNP is a common difference in a single nucleotide. The DNA in any two human genomes differs at about 3 million SNPs. Most SNPs are not associated with any detectable effect on phenotype. page 316

CNV is a difference in the number of copies of a particular DNA sequence among chromosomes. page 318

15.4 Chromosomal variants can also occur but are usually harmful.

Nondisjunction is the failure of a pair of chromosomes to separate during anaphase of cell division. It can occur in mitosis or meiosis and results in daughter cells with extra or missing chromosomes. page 319

Nondisjunction in meiosis leads to gametes with extra or missing chromosomes. page 319

Most human fetuses with extra or missing chromosomes spontaneously abort, but some are viable and exhibit characteristic syndromes. page 319

Trisomy 21 (Down syndrome) is usually characterized by three copies of chromosome 21. page 320

Extra or missing sex chromosomes (X or Y) are common and result in syndromes such as Klinefelter syndrome (47, XXY) and Turner syndrome (45, X). page 322

Self-Assessment

1. Explain the terms "genotype" and "phenotype." What determines a genotype? What determines a phenotype?
2. Describe the relationship between a genotype and a phenotype: Does the same genotype always result in the same phenotype? Must individuals with the same phenotype have the same genotype?
3. With regard to mutations, what is meant by the terms "harmful," "beneficial," and "neutral"? Why it is sometimes an oversimplification to consider a mutation as either harmful, beneficial, or neutral?
4. Describe two types of genetic polymorphism that are useful in DNA typing.
5. Define the term "SNP" and explain why researchers are interested in detecting SNPs.
6. Diagram how nondisjunction in meiosis I or II can result in extra or missing chromosomes in reproductive cells (gametes).
7. Describe the consequences of an extra copy of chromosome 21 (Down syndrome).

Log in to LaunchPad to check your answers to the Self-Assessment questions, and to access additional learning tools.

Koichi Saito/A.collection/Getty Images.

CHAPTER 16

Mendelian Inheritance

Core Concepts

16.1 Early theories of heredity incorrectly assumed the inheritance of acquired characteristics and blending of parental traits in the offspring.

16.2 The study of modern transmission genetics began with Gregor Mendel, who used the garden pea as his experimental organism and studied traits with contrasting characteristics.

16.3 Mendel's first key discovery was the principle of segregation, which states that members of a gene pair separate equally into gametes.

16.4 Mendel's second key finding was the principle of independent assortment, which states that different gene pairs segregate independently of one another.

16.5 The patterns of inheritance that Mendel observed in peas can also be seen in humans.

Except for identical twins, each of us has his or her own personal genome, a unique human genome differing from all that have existed before and from all that will come after. Genetic variation from one person to the next leads in part to our individuality, from differences in appearance to differences in the ways our bodies work. Examples of genetic variation in the human population range from harmless curiosities like the genetic difference in taste receptors that determine whether or not you taste broccoli as being unpleasantly bitter, to mutant forms of genes resulting in serious diseases like sickle-cell anemia or emphysema.

This chapter focuses on how that genetic variation is inherited. **Transmission genetics** deals with the manner in which genetic differences among individuals are passed from generation to generation. We are all aware of the effects of genetics. We know that children resemble their parents and that there are sometimes uncanny similarities among even distant relatives. But some patterns are more difficult to discern. Traits such as eye color, nose shape, or risk for a particular disease may be passed down faithfully generation after generation, but sometimes they are not, and sometimes they appear and disappear in seemingly random ways.

As a modern science, transmission genetics began with the pea-breeding experiments carried out by the monk Gregor Mendel in the 1860s. However, even before then, people understood enough about inheritance that they were able to select crops and livestock with particular characteristics.

16.1 EARLY THEORIES OF INHERITANCE

Thousands of years before Mendel, many societies carried out practical plant and animal breeding. The ancient practices were based on experience rather than on a full understanding of the rules of genetic transmission, but they were nevertheless highly successful. In Mesoamerica, for example, Native Americans chose corn (maize) plants for cultivation that had the biggest ears and softest kernels. Over many generations, their cultivated corn came to have less and less physical resemblance to its wild ancestral species. Similarly, people in the Eurasian steppes selected their horses for a docile temperament suitable for riding or for hitching to carts or sleds. Practical breeding of this kind provided the crops and livestock from which most of our modern domesticated animals and plants derive.

Early theories of heredity predicted the transmission of acquired characteristics.

The first written speculations about mechanisms of heredity were made by the ancient Greeks. Hippocrates (460–377 BCE), considered the founder of Western medicine, proposed that each part of the body in a sexually mature adult produces a substance that collects in the reproductive organs and that determines the inherited characteristics of the offspring. An implication of this theory is that any **trait,** or characteristic, of an individual can be transmitted from parent to offspring. Even traits that are acquired during the lifetime of an individual, such as muscle strength or bodily injury, were thought to be heritable because of the substance supposedly passed from each body part to the reproductive organs.

The theory that acquired characteristics can be inherited was invoked to explain such traits as the webbed feet of ducks, which were thought to result from many successive generations in which adult ducks stretched the skin between their toes while swimming and passed this trait to offspring. A few decades later, however, Aristotle (384–322 BCE) emphasized several observations that the theory of inheritance of acquired characteristics cannot account for:

- Traits such as hair color can be inherited, but it is difficult to see how hair—a nonliving tissue—could send substances to the reproductive organs.
- Traits that are not yet present in an individual can be transmitted to the offspring. For example, a father and his adult son can both be bald, even though the son was born before the father became bald.
- Parts of the body that are lost as a result of surgery or accident are not missing in the offspring.

From these and other observations, Aristotle concluded that the process of heredity transmits only the *potential* for producing traits present in the parents, and not the traits themselves. Nevertheless, Hippocrates' theory influenced biology until well into the 1800s. It was incorporated into an early theory of evolution proposed by the French biologist Jean-Baptiste Lamarck around 1800. Charles Darwin, however, developed an alternative theory—the theory of evolution by natural selection (Chapter 21).

While traits acquired during the lifetime of a parent are not transmitted to the offspring, parental misfortune or misbehavior can nevertheless result in impaired fetal development and in some cases permanent damage. For example, maternal malnutrition, drug addiction, alcoholism, infectious disease, and other conditions can severely affect the fetus, but these effects are due to disruption of fetal development and not to changes in the genome.

Belief in blending inheritance discouraged studies of hereditary transmission.

Darwin subscribed to the now-discredited model of **blending inheritance,** in which traits in the offspring resemble the average of those in the parents. For example, this model predicts that the offspring of plants with blue flowers and those with red flowers will have purple flowers. While traits of offspring are sometimes the average of those of the parents (think of certain cases of human height), the idea of blending inheritance—which implies the blending of the genetic material—as a general rule presents problems. For example, it cannot explain the reappearance of a

FIG. 16.1 Blending inheritance. This model predicts loss of variation over time and blending of genetic material, which is not observed.

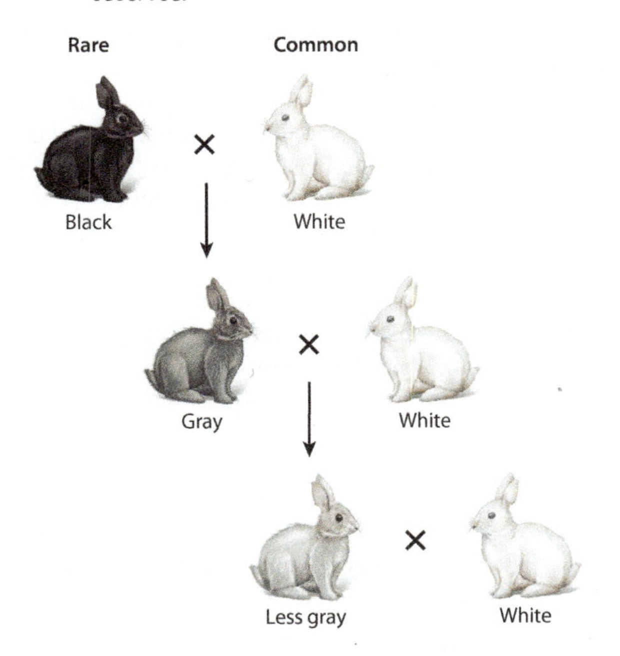

trait several generations after it apparently "disappeared" in a family, such as red hair or blue eyes.

Another difficulty with the concept of blending inheritance is that variation is lost over time. Consider an example where black rabbits are rare and white rabbits are common (**Fig. 16.1**). Black rabbits, being rare, are most likely to mate with the much more common white rabbits. If blending inheritance occurs, the result will be gray rabbits. These will also mate with white rabbits, producing lighter-gray rabbits. Over time, the population will end up being all white, or very close to white, and there will be less variation. The only way that black rabbits will be present is if they are reintroduced into the population through mutation or migration.

Whatever the trait may be, blending inheritance predicts that inheritance will tend to be a homogenizing force, producing in each generation a blend of the original phenotypes. But we know from common experience that variation in most populations is plentiful.

It is ironic that Darwin believed in blending inheritance because this mechanism of inheritance is incompatible with his theory of evolution by means of natural selection. The incompatibility was pointed out by some of Darwin's contemporaries, and Darwin himself recognized it as a serious problem. The problem with blending inheritance is that rare variants, such as the black rabbits in the above example, will have no opportunity to increase in frequency, even if they survive and reproduce more than white rabbits, since they gradually disappear over time.

Although Darwin was convinced that he was right about natural selection, he was never able to reconcile his theory with the concept of blending inheritance. Unknown to him, the solution had already been discovered in experiments carried out by Gregor Mendel (1822–1884), an Augustinian monk in a monastery in the city of Brno in what is now the Czech Republic. Mendel's key discovery was this: It is not *traits* that are transmitted in inheritance—it is *genes* that are transmitted.

16.2 THE FOUNDATIONS OF MODERN TRANSMISSION GENETICS

Mendel's scientific fame rests on the pea-breeding experiments that he carried out in the years 1856 to 1864 (**Fig. 16.2**), in which he demonstrated the basic principles of transmission genetics. As we will see in Chapter 18, most genetic variation in populations is due to multiple genes interacting with one another and with the environment, while traits due to single genes, such as those that Mendel studied, tend to be rare. Nevertheless, Mendel's experiments were so simple and presented with such clarity that they still serve as examples of the scientific method at its best (Chapter 1).

Mendel's experimental organism was the garden pea.

For his experimental material, Mendel used several strains of ordinary garden peas (*Pisum sativum*) that he obtained from local

FIG. 16.2 Gregor Mendel and his experimental organism, the pea plant. *Sources: Juliette Wade/Getty Images; (inset) Photo by Authenticated News/Getty Images.*

seed suppliers. His experimental approach was similar to that of a few botanists of the eighteenth and nineteenth centuries who studied the results of **hybridization,** or interbreeding between two different varieties or species of an organism. Where Mendel differed from his predecessors was in paying close attention to a small number of easily classified traits with contrasting characteristics. For example, where one strain of peas had yellow seeds, another had green seeds; and where one had round seeds, another had wrinkled seeds (**Fig. 16.3**). Altogether Mendel studied seven physical features expressed in contrasting fashion among the strains: seed color, seed shape, pod color, pod shape, flower color, flower position, and plant height.

FIG. 16.3 Contrasting traits. Mendel focused on seven contrasting traits.

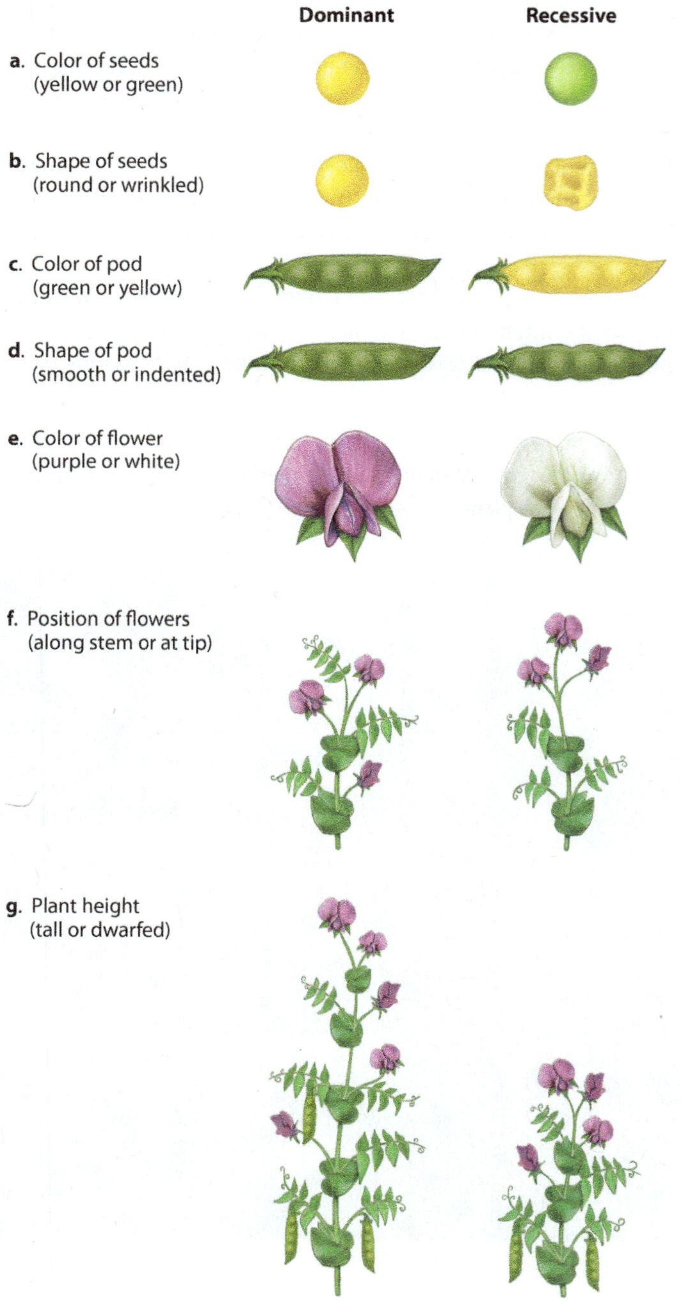

Dominant
Recessive
a. Color of seeds (yellow or green)
b. Shape of seeds (round or wrinkled)
c. Color of pod (green or yellow)
d. Shape of pod (smooth or indented)
e. Color of flower (purple or white)
f. Position of flowers (along stem or at tip)
g. Plant height (tall or dwarfed)

The expression of each trait in each of the original strains that Mendel obtained was **true breeding,** which means that the physical appearance of the offspring in each successive generation is identical to the previous one. For example, plants of the strain with yellow seeds produced only yellow seeds, and those of the strain with green seeds produced only green seeds. Likewise, plants of the strain with round seeds produced only round seeds, and those of the strain with wrinkled seeds produced only wrinkled seeds.

The objective of Mendel's experiments was simple. By means of crosses between the true-breeding strains and crosses among their progeny, Mendel hoped to determine whether there are statistical patterns in the occurrence of the contrasting characteristics, such as yellow seeds or green seeds. If such patterns could be found, he would seek to devise a hypothesis to explain them and then use his hypothesis to predict the outcome of further crosses.

In designing his experiments, Mendel departed from other plant hybridizers of the time in three important ways:

1. Mendel studied true-breeding strains, unlike many other plant hybridizers, who used complex and poorly defined material.

2. Mendel focused on one trait, or a small number of traits, at a time, with characteristics that were easily contrasted among the true-breeding strains. Other plant hybridizers crossed strains differing in many traits and tried to follow all the traits at once. This resulted in amazingly complex inheritance, and no underlying patterns could be discerned.

3. Mendel counted the progeny of his crosses, looking for statistical patterns in the offspring of his crosses. Others typically noted only whether offspring with a particular characteristic were present or absent, but did not keep track of and count all the progeny of a particular cross.

In crosses, one of the traits was dominant in the offspring.

Mendel began his studies with crosses between true-breeding strains that differed in a single contrasting characteristic. Crossing peas is not as easy as it may sound. Pea flowers include both female and male reproductive organs enclosed together within petals (**Fig. 16.4**). Because of this arrangement, pea plants usually fertilize themselves (self-fertilize). Performing crosses between different plants is a tedious and painstaking process. First, the flower of the designated female parent must be opened at an early stage and the immature anthers (male reproductive organs) clipped off and discarded, so the plant cannot self-fertilize. Then mature pollen from the designated male parent must be collected and deposited on the stigma (female reproductive organ) of the female parent, from where it travels to the reproductive

FIG. 16.4 Crossing pea plants. Crossing plants in a way that isolates either the ovules or the pollen controls what genetic material each parent contributes to the offspring.

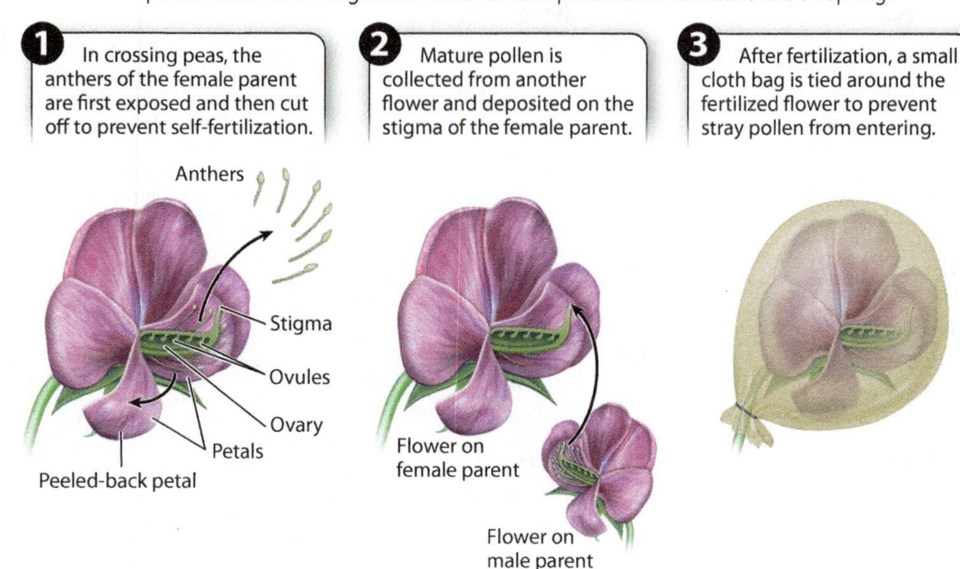

cells in the ovule at the base of the flower. And finally, a cloth bag must be tied around the female flower to prevent stray pollen from entering. No wonder Mendel complained that his eyes hurt!

For each of the seven pairs of contrasting traits, true-breeding strains differing in the trait were crossed. **Fig. 16.5** illustrates a typical result, in this case for a cross between a plant producing yellow seeds and a plant producing green seeds. In these kinds of crosses, the parental generation is referred to as the **P_1 generation,** and the first offspring, or filial, generation is referred to as the **F_1 generation.** In the cross of P_1 yellow × P_1 green, Mendel observed that all the F_1 progeny had yellow seeds. This result was shown to be independent of the seed color, yellow or green, of the pollen donor because both of the crosses shown below yielded progeny plants with yellow seeds:

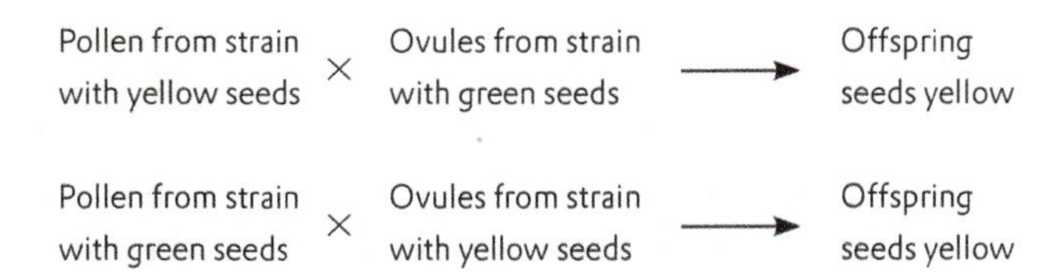

Crosses like these, in which the expressions of the trait in the female and male parents are interchanged, are known as **reciprocal crosses.** Mendel showed that reciprocal crosses yielded the same result for each of his seven pairs of contrasting traits.

Moreover, for each pair of contrasting traits, only one of the traits appeared in the F_1 generation. For simple crosses involving two parents that are true breeding for different traits, the trait that appears in the F_1 generation is said to be **dominant,** and the contrasting trait that does not appear is said to be **recessive.** For each pair of traits illustrated in Fig. 16.3, the dominant trait is shown on the left, and the recessive on the right. Thus, yellow seed is dominant to green seed, round seed is dominant to wrinkled seed, and so forth.

Mendel explained these findings by supposing that there is a hereditary factor for yellow seeds and a different hereditary factor for green seeds, likewise a hereditary factor for round seeds and a different one for wrinkled seeds, and so on. We now know that the hereditary factors that result in contrasting traits are different forms of a gene that affect the trait. The different forms of a gene are called **alleles.** The alleles of a gene are variations of the DNA sequence of a gene that occupies a particular region along a chromosome. The particular combination of alleles present in

FIG. 16.5 The first-generation hybrid (F_1). A cross between two of Mendel's true-breeding plants (the parental, or P_1, generation) yielded first-generation hybrids displaying the dominant trait.

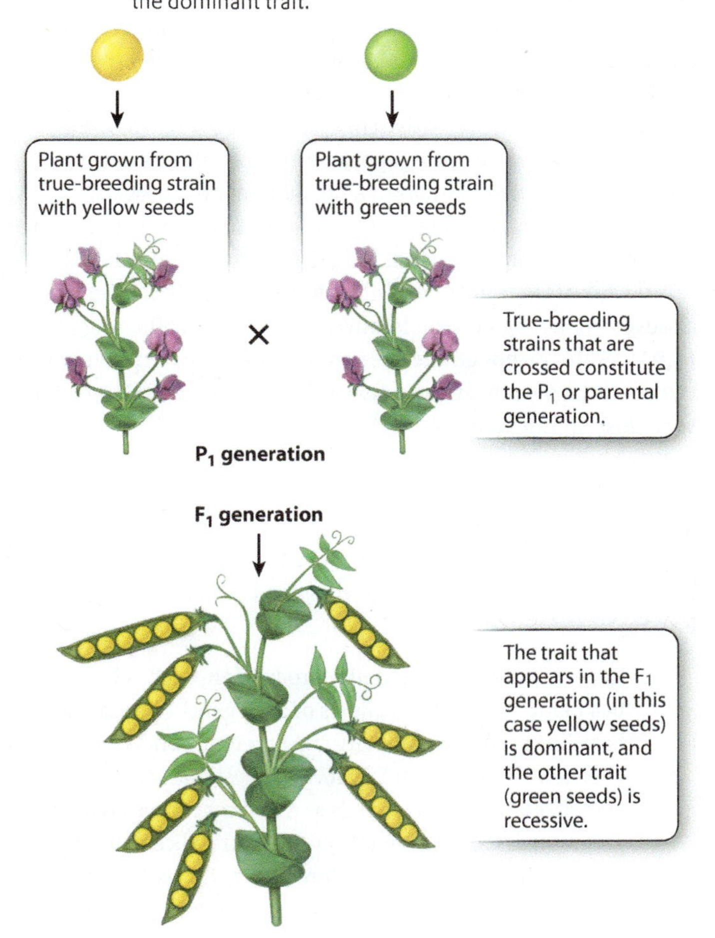

an individual constitutes its **genotype,** and the expression of the trait in the individual constitutes its **phenotype.** As noted, Mendel found that, for each of his traits, one phenotype was dominant and the other phenotype was recessive.

In the cross between true-breeding plants that produce yellow seeds and true-breeding plants that produce green seeds, the genotype of each F_1 seed includes an allele for yellow from the yellow parent and an allele for green from the green parent. The phenotype of each F_1 seed is nevertheless yellow, because yellow is dominant to green.

The molecular basis of dominance in this case is the fact that, in diploid organisms, only one copy of a gene is needed to carry out its normal function. The yellow color in pea seeds results from an enzyme that breaks down green chlorophyll, allowing yellow pigments to show through. Green seeds result from a mutation in this gene that inactivates the enzyme, and so the green chlorophyll is retained. In an F_1 hybrid, such as that in Fig. 16.5, which receives a nonmutant gene from one parent and a mutant gene from the other, the seeds are yellow because one copy of the nonmutant gene produces enough of the enzyme to break down the chlorophyll to yield a yellow seed.

16.3 SEGREGATION: MENDEL'S KEY DISCOVERY

Mendel's most important discovery was that the F_1 progeny of a cross between plants with different traits did *not* breed true. In the **F_2 generation,** produced by allowing the F_1 flowers to undergo self-fertilization, the recessive trait reappeared (**Fig. 16.6**). Not only did the recessive trait reappear, it reappeared in a definite numerical proportion. Among a large number of F_2 progeny, Mendel found that the dominant : recessive ratio was very close to 3 : 1. The results he observed among the F_2 progeny for each of the seven pairs of traits are given in **Table 16.1.** Across experiments for all seven traits, the ratio of dominant : recessive F_2 offspring was 14,949 : 5010. Although there is variation from one experiment to the next, the overall ratio of 14,949 : 5010 equals 2.98 : 1, which is a very close approximation to 3 : 1.

FIG. 16.6 Self-fertilization of the F_1 hybrid, resulting in seeds of the F_2 generation. The recessive trait appeared again in the F_2 generation.

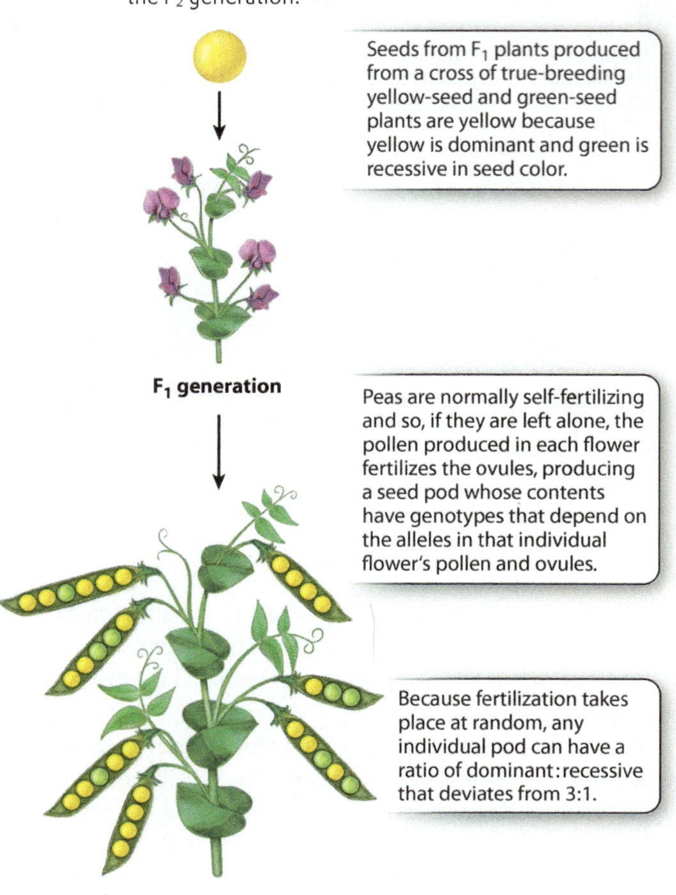

Genes come in pairs that segregate in the formation of reproductive cells.

The explanation for the 3 : 1 ratio in the F_2 and Mendel's observations in general can be summarized with reference to **Fig. 16.7:**

1. Except for cells involved in reproduction, each cell of a pea plant contains two alleles of each gene. In each true-breeding strain constituting the P_1 generation, the two alleles are identical. In Fig. 16.7, we designate the allele associated with yellow seeds as *A* and that associated with green seeds as *a*. The genotype of the true-breeding strain with yellow seeds can therefore be written as *AA*, and that of the true-breeding strain with green seeds as *aa*. The *AA* and *aa* genotypes are said to be homozygous, which means that both alleles inherited from the parents are the same.

→ **Quick Check 1** What are the genotypes and phenotypes for Mendel's true-breeding parent plants?

2. Each reproductive cell, or **gamete,** contains only one allele of each gene. In this case, a gamete can contain the *A* allele or the *a* allele, but not both.
3. In the formation of gametes, the two members of a gene pair **segregate** (or separate) equally into gametes so that half the gametes get one allele and half get the other allele. This separation of alleles into different gametes defines the **principle of segregation.** In the case of homozygous plants (such as *AA* or *aa*), all the gametes from an individual are the same. That is, the homozygous *AA* strain with yellow seeds produces gametes containing the *A* allele, and the homozygous *aa* strain with green seeds produces gametes containing the *a* allele.

TABLE 16.1 Observed F_2 Ratios in Mendel's Experiments

TRAIT	DOMINANT TRAIT	RECESSIVE TRAIT	RATIO
Seed color	6,022	2,001	3.01 : 1
Seed shape	5,474	1,850	2.96 : 1
Pod color	428	152	2.82 : 1
Pod shape	882	299	2.95 : 1
Flower color	705	224	3.15 : 1
Flower position	651	207	3.14 : 1
Plant height	787	277	2.84 : 1

4. The fertilized egg cell, called the **zygote,** is formed from the random union of two gametes, one from each parent. For the cross *AA* × *aa*, the zygote is an F_1 hybrid formed from the union of an *A*-bearing gamete with an *a*-bearing gamete. Each F_1 hybrid therefore has the genotype *Aa* (Fig. 16.7). The *Aa* genotype is **heterozygous,** which means that the two alleles for a given gene inherited from each parent are different. The seeds produced by the F_1 progeny are yellow because yellow is dominant to green. Note that each F_1 progeny contains an *a* allele because its genotype is *Aa*, but the phenotype of the seed is yellow and indistinguishable from that of an *AA* genotype.

→ **Quick Check 2** Is it possible for two individuals to have the same phenotype but different genotypes? The same genotype, but different phenotypes? How?

FIG. 16.7 The principle of segregation.

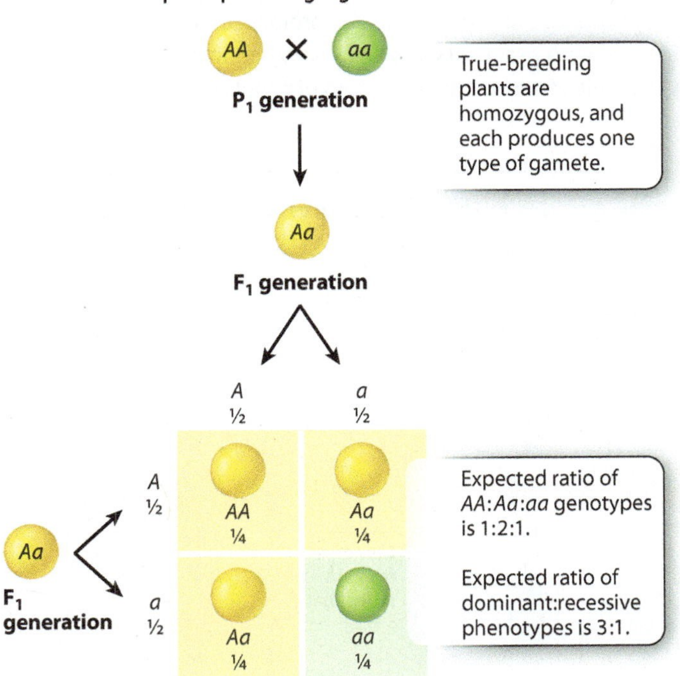

5. When the F_1 progeny (genotype *Aa*) form gametes, by the principle of segregation the *A* and *a* alleles again separate equally so that half the gametes contain only the *A* allele and the other half contain only the *a* allele (Fig. 16.7).

6. In the formation of the F_2 generation, the gametes from the F_1 parents again combine at random. The consequences of random union of gametes can be worked out by means of a checkerboard of the sort shown at the bottom of Fig. 16.7. In this kind of square, known as a **Punnett square** after its inventor, the British geneticist Reginald Punnett, the gametes from each parent, each with its respective frequency, are arranged along the top and sides of a grid. Each box in the grid represents the union of the gametes in the corresponding row and column, showing all the possible genotypes of offspring that can result from random fertilization.

7. The boxes of the Punnett square correspond to all of the possible offspring genotypes of the F_2 generation, with each possible genotype's frequency obtained by multiplication of the gametic frequencies in the corresponding row and column. For the example illustrated in Fig. 16.7, the expected *genotypes* of the offspring are therefore ¼ *AA*, ½ *Aa*, and ¼ *aa* (or 1 : 2 : 1). When there is dominance, however, as there is in this case, the *AA* and *Aa* genotypes have the same phenotype, and so the ratio of dominant : recessive *phenotypes* is 3 : 1. The Punnett square in Fig. 16.7 illustrates the biological basis of the 3 : 1 ratio of phenotypes that Mendel observed in the F_2 generation. Nevertheless, the underlying ratio of *AA* : *Aa* : *aa* genotypes is 1 : 2 : 1.

→ **Quick Check 3** What are the expected progeny (genotypes and phenotypes) from a cross of an *AA* plant with an *Aa* plant?

The principle of segregation was tested by predicting the outcome of crosses.

The model of segregation of gene pairs (alleles) and their random combination in the formation of a zygote depicted in Fig. 16.7 was Mendel's hypothesis, an explanation he put forward to explain an observed result. Further experiments were necessary to support or disprove his hypothesis. The true test of a hypothesis is whether it can predict the results of experiments that have not yet been carried out (Chapter 1). If the predictions are correct, one's confidence in the hypothesis is strengthened. Mendel appreciated this intuitively, even though the scientific method as understood today had not been formalized.

The Punnett square in Fig. 16.7 makes two predictions. The first is that the seeds in the F_2 generation showing the recessive green seed phenotype should be homozygous *aa*. If the green F_2 seeds have the genotype *aa*, then they should breed true. That is, when the seeds are grown into mature plants and self-fertilization is allowed to take place, the self-fertilized *aa* plants should produce only green seeds (*aa*). This prediction was confirmed by examining seeds actually produced by plants grown from the F_2 green seeds.

TABLE 16.2 Genetic Ratios from Self-Fertilization of Plants Showing the Dominant Phenotype

TRAIT	HOMOZYGOUS DOMINANT	HETERO-ZYGOUS	RATIO
Yellow seeds	166	353	0.94:2
Round seeds	193	372	1.04:2
Green pods	40	60	1.33:2
Smooth pods	29	71	0.82:2
Purple flowers	36	64	1.13:2
Flowers along stem	33	67	0.99:2
Tall plants	28	72	0.78:2

The second prediction from the Punnett square in Fig. 16.7 is more complex. It has to do with the seeds in the F_2 generation that show the dominant yellow phenotype. Note that although these seeds have the same phenotype, they have two different genotypes (*AA* and *Aa*). Among just the yellow seeds, ⅓ should have the genotype *AA* and ⅔ should have the genotype *Aa*, for a ratio of 1 *AA*:2 *Aa*. (The proportions are ⅓ : ⅔ because we are considering *only* the seeds that are yellow.) Plants with the *AA* and *Aa* genotypes can be distinguished by the types of seed they produce when self-fertilized. The *AA* plants produce only seeds with the dominant yellow phenotype (that is, they are true breeding), whereas the *Aa* plants yield dominant yellow and recessive green seeds in the ratio 3:1. Mendel did such experiments, and the prediction turned out to be correct. His data confirming the 1:2 ratio of *AA*:*Aa* among F_2 individuals with the dominant phenotype are shown in **Table 16.2.**

A testcross is a mating to an individual with the homozygous recessive genotype.

A more direct test of segregation is to cross the F_1 progeny with the true-breeding recessive strain instead of allowing them to self-fertilize. Any cross of an unknown genotype with a homozygous recessive genotype is known as a **testcross.**

The F_1 progeny show the dominant yellow seed phenotype. A yellow seed phenotype can result from either of two possible genotypes, *Aa* or *AA*. A testcross with plants of the homozygous recessive genotype (*aa*) can distinguish between these two possibilities (**Fig. 16.8**).

Let's first consider what happens in a testcross to an *Aa* individual (Fig. 16.8a). An *Aa* individual produces both *A*-bearing gametes and *a*-bearing gametes. As shown by the Punnett square, the predicted offspring from the testcross are ½ *Aa* zygotes, which yield yellow seeds, and ½ *aa* zygotes, which yield green seeds. By contrast, an *AA* individual produces only *A*-bearing gametes, so all the zygotes have the *Aa* genotype, which yields yellow seeds (Fig. 16.8b). In other words, the testcross gives different results depending on whether the parent is heterozygous (*Aa*) or homozygous (*AA*).

Note that in a testcross the *phenotypes* of the progeny reveal the *alleles* present in the gametes from the tested parent. A testcross with an *Aa* individual yields ½ *Aa* (yellow seeds) and ½ *aa* (green seeds) since the *Aa* parent produces ½ *A*-bearing and ½ *a*-bearing gametes. A testcross with an *AA* individual yields only *Aa* (yellow seeds) since the *AA* parent produces only *A*-bearing gametes. These results are a direct demonstration of the principle of segregation since the ratio of the phenotypes of progeny reflect the equal segregation of alleles into gametes. Some of Mendel's testcross data indicating 1:1 segregation in heterozygous genotypes are shown in **Table 16.3.**

Segregation of alleles reflects the separation of chromosomes in meiosis.

The principles of transmission genetics have a physical basis in the process of meiosis (Chapter 11). During meiosis I, maternal and paternal chromosomes (homologous chromosomes) align on the metaphase plate. Then, during anaphase I, the homologous chromosomes separate, and each chromosome goes to a different pole. Because gene pairs are carried on homologous chromosomes, the segregation of alleles observed by Mendel corresponds to the separation of chromosomes that takes place in anaphase I.

Fig. 16.9 illustrates the separation of a pair of homologous chromosomes in anaphase I. In the configuration shown, the copies of the *A* allele (dark blue) separate from the copies of the *a* allele (light blue) in anaphase I. The separation of chromosomes is the physical basis of the segregation of alleles.

Dominance is not universally observed.

Many traits do not show complete dominance such as Mendel observed with pea plants. Most traits we see are determined by multiple genes or the interaction of genotype and the

FIG. 16.8 **A testcross.** A cross with a homozygous recessive individual reveals the genotype of the other parent.

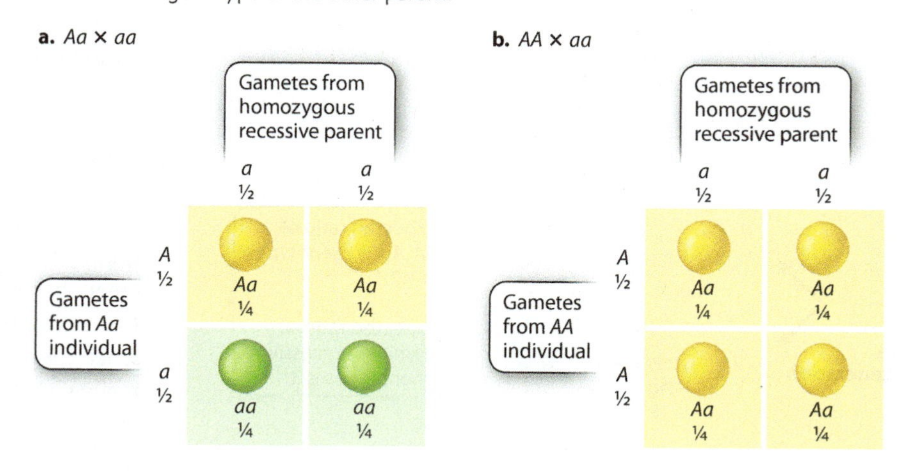

TABLE 16.3 Phenotype of Progeny from Testcrosses of Heterozygotes

TRAIT	DOMINANT TRAIT	RECESSIVE TRAIT	RATIO
Seed color	196	189	1.04:1
Seed shape	193	192	1.01:1
Flower color	85	81	1.05:1
Plant height	87	79	1.10:1

environment and so do not display the expected 3:1 ratio of phenotypes. But even among traits that are determined by a single gene, the 3:1 phenotype ratio is not always observed. In some cases, the trait shows **incomplete dominance,** in which the phenotype of the heterozygous genotype is intermediate between those of the homozygous genotypes. In such cases, the result of segregation can be observed directly because each genotype has a distinct phenotype. An example is flower color in the snapdragon (*Antirrhinum majus*), in which the homozygous genotypes have red (C^RC^R) or white (C^WC^W) flowers, and the heterozygous genotype C^RC^W has pink flowers (**Fig. 16.10**). In notating incomplete dominance, we use superscripts to indicate the alleles, rather than upper-case and lower-case letters, because neither allele is dominant to the other. A cross of homozygous C^RC^R and C^WC^W strains results in hybrid F_1 progeny that are pink (C^RC^W), and when these are crossed the resulting F_2 generation consists of ¼ red (C^RC^R), ½ pink (C^RC^W), and ¼ white (C^WC^W).

Note that in this case the genotype ratio and the phenotype ratio are both 1:2:1, since each genotype has a distinct phenotype. Such a direct demonstration of segregation makes one wonder whether Mendel's work might have been appreciated more readily if his traits had shown incomplete dominance!

FIG. 16.9 Segregation of alleles of a single gene. Homologous chromosomes separate during meiosis, leading to segregation of alleles.

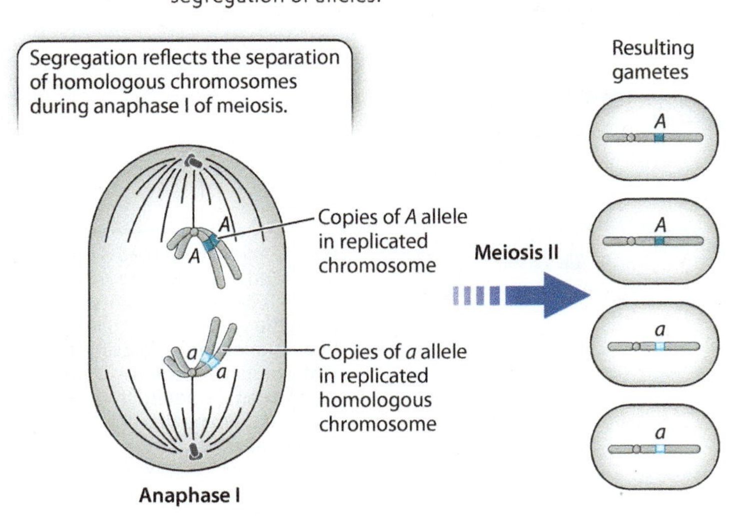

FIG. 16.10 Incomplete dominance for flower color in snapdragons. In incomplete dominance, an intermediate phenotype is seen.

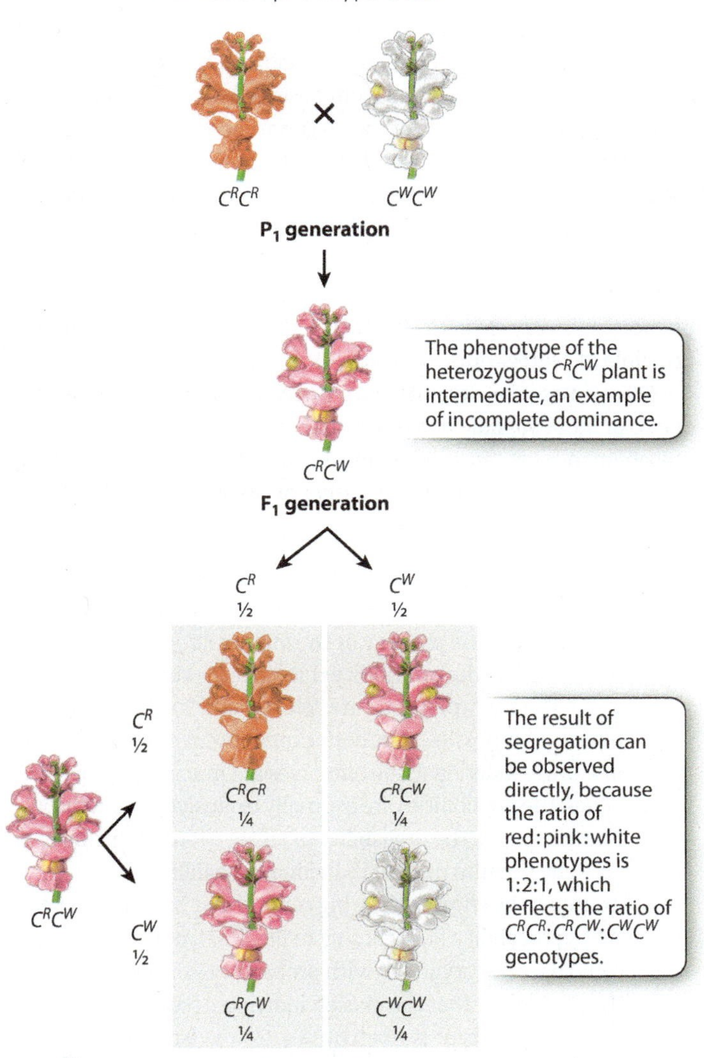

The principles of transmission genetics are statistical and are stated in terms of probabilities.

The element of chance in fertilization implies that the genotype of any particular progeny cannot be determined in advance. However, one can deduce the likelihood, or **probability,** that a specified genotype will occur. The probability of occurrence of a genotype must always lie between 0 and 1; a probability of 0 means that the genotype cannot occur, and a probability of 1 means that the occurrence of the genotype is certain. For example, in the cross *Aa* × *AA*, no offspring can have the genotype *aa*, so in this mating the probability of *aa* is 0. Similarly, in the mating *AA* × *aa*, all offspring must have the genotype *Aa*, so in this mating the probability of *Aa* is 1.

In many cases, the probability of a particular genotype is neither 0 nor 1, but some intermediate value. For one gene, the probabilities for a single individual can be deduced from the parental genotypes in the mating and the principle of segregation. For example, the probability of producing a homozygous recessive individual from the cross *Aa* × *Aa* is ¼ (see Fig. 16.7), and that from the cross *Aa* × *aa* is ½ (see Fig. 16.8a).

The genotype and phenotype probabilities for a single individual can also be inferred from observed data because the overall proportions of two (or more) genotypes among a large number of observations approximates the probability of each of the genotypes for a single observation. For example, in Mendel's F_2 data (see Table 16.1), the overall ratio of dominant : recessive is 2.98 : 1, or very nearly 3 : 1. This result implies that the probability that an individual F_2 plant has the homozygous recessive phenotype is very close to ¼, which is the value inferred from the principle of segregation.

Sometimes it becomes necessary to combine the probabilities of two or more possible outcomes of a cross, as in determining the probability of a genotype occurring based on the probabilities of the gametes occurring. In such cases either of two rules are helpful:

1. **Addition rule.** This principle applies when the possible outcomes being considered cannot occur simultaneously. For example, suppose that a single offspring is chosen at random from the progeny of the mating *Aa* × *Aa*, and we wish to know the probability that the offspring is either *AA* or *Aa*. The key words here are "either" and "or." Each of these outcomes is possible, but both cannot occur simultaneously in a single individual; the outcomes are mutually exclusive. When the possibilities are mutually exclusive, the addition rule states that the probability of either event occurring is given by the sum of their individual probabilities. In this example, the chosen offspring could either have genotype *AA* (with probability ¼, according to Fig. 16.7), or the offspring could have genotype *Aa* (with probability ½). Therefore, the probability that the chosen individual has either the *AA* or the *Aa* genotype is given by ¼ + ½ = ¾ (see Fig. 16.7). Alternatively, the ¾ could be interpreted to mean that, among a large number of offspring from the mating *Aa* × *Aa*, the proportion exhibiting the dominant phenotype will be very close to ¾. This interpretation is verified by the data in Table 16.1.

2. **Multiplication rule.** This principle applies when outcomes can occur simultaneously, and the occurrence of one has no effect upon the likelihood of the other. Events that do not influence one another are independent, and the multiplication rule states that the probability of two independent events occurring together is the product of their respective probabilities. This rule is widely used to determine the probabilities of successive offspring of a cross because each event of fertilization is independent of any other. For example, in the mating *Aa* × *Aa*, one may wish to determine the probability that, among four peas in a pod, the one nearest the stem is green and the others yellow. Here, the word "and" is a simple indicator that the multiplication rule should be used. Because each seed results from an independent fertilization, this probability is given by the product of the probability that the seed nearest the stem is *aa* and the probability that each of the other seeds is either *AA* or *Aa*, and hence the probability is ¼ × ¾ × ¾ × ¾ = 27/256, as shown for the top pod in **Fig. 16.11.**

The addition and multiplication rules are very powerful when used in combination. Consider the following question: In the mating *Aa* × *Aa*, what is the probability that, among four seeds in a pod, exactly one is green? We have already seen in Fig. 16.11 that the multiplication rule gives the probability of the seed nearest the stem being green as 27/256. As illustrated in Fig. 16.11, there are only four possible ways in which exactly one seed can be green, each of which has a probability of 27/256, and these outcomes are mutually exclusive. Therefore, by the addition rule, the probability of there being exactly one green and three yellow seeds in a pod, occurring in any order, is given by 27/256 + 27/256 + 27/256 + 27/256 = 108/256, or approximately 42%.

→ **Quick Check 4** What is the probability that *any* two peas are green and two are yellow in a pea pod with exactly four seeds?

Mendelian segregation preserves genetic variation.

As noted earlier, Darwin was befuddled because blending inheritance would make genetic variation disappear so rapidly that evolution by means of natural selection could not occur. Although

FIG. 16.11 Application of the multiplication and addition rules.

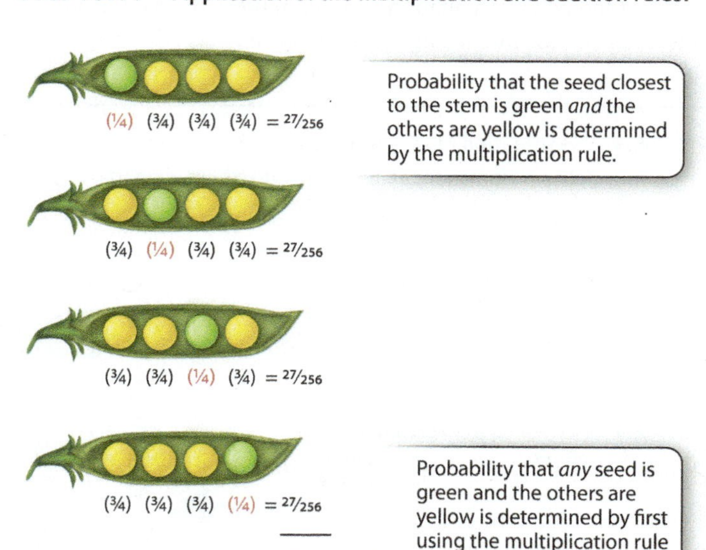

Darwin was completely unaware of Mendel's findings, segregation was the answer to his problem.

An important consequence of segregation is that it demonstrates that the alleles encoding a trait do not alter one another when they are present together in a heterozygous genotype (except in very rare instances). The recessive trait, masked in one generation, can appear in the next, looking exactly as it did in the true-breeding strains. Mendel fully appreciated the significance of this discovery. In one of his letters, he emphasized that "the two parental traits appear, separated and unchanged, and there is nothing to indicate that one of them has either inherited or taken over anything from the other." In other words, no hint of any sort of blending between the parental genetic material takes place and therefore no homogenization of the trait in the population. Because the individual genes maintain their identity down through the generations (except for rare mutations), genetic variation in a population also tends to be maintained through time. The maintenance of genetic variation is discussed further in Chapter 21.

16.4 INDEPENDENT ASSORTMENT

We have seen that segregation of different alleles of a single gene results in a 3:1 ratio of dominant: recessive phenotypes in the F_2 generation. What happens when the parental strains differ in two traits, for example when a strain having yellow and wrinkled seeds is crossed with a strain having green and round seeds? The results of these kinds of experiments constitute Mendel's second key discovery, the **principle of independent assortment.** This principle states that segregation of one set of alleles of a gene pair is independent of the segregation of another set of alleles of a different gene pair. That is, each pair of alleles assorts (segregates) without affecting or being affected by the assortment of any other pair of alleles.

Independent assortment is observed when genes segregate independently of one another.

In the cross between a strain with seeds that are yellow and wrinkled and a strain with seeds that are green and round, the phenotype of the F_1 seeds is easily predicted. Because yellow is dominant to green, and round is dominant to wrinkled, the F_1 seeds are expected to be yellow and round, and in fact they are (**Fig. 16.12**). When these seeds are grown and the F_1 plants are allowed to undergo self-fertilization, the result is as shown in Fig. 16.12. Among 639 seeds from this cross, Mendel observed the following:

yellow round	367
green round	122
yellow wrinkled	113
green wrinkled	37

FIG. 16.12 Mendel's crosses with two traits. Plants heterozygous for two genes affecting different traits produce offspring with a phenotypic ratio of 9 : 3 : 3 : 1.

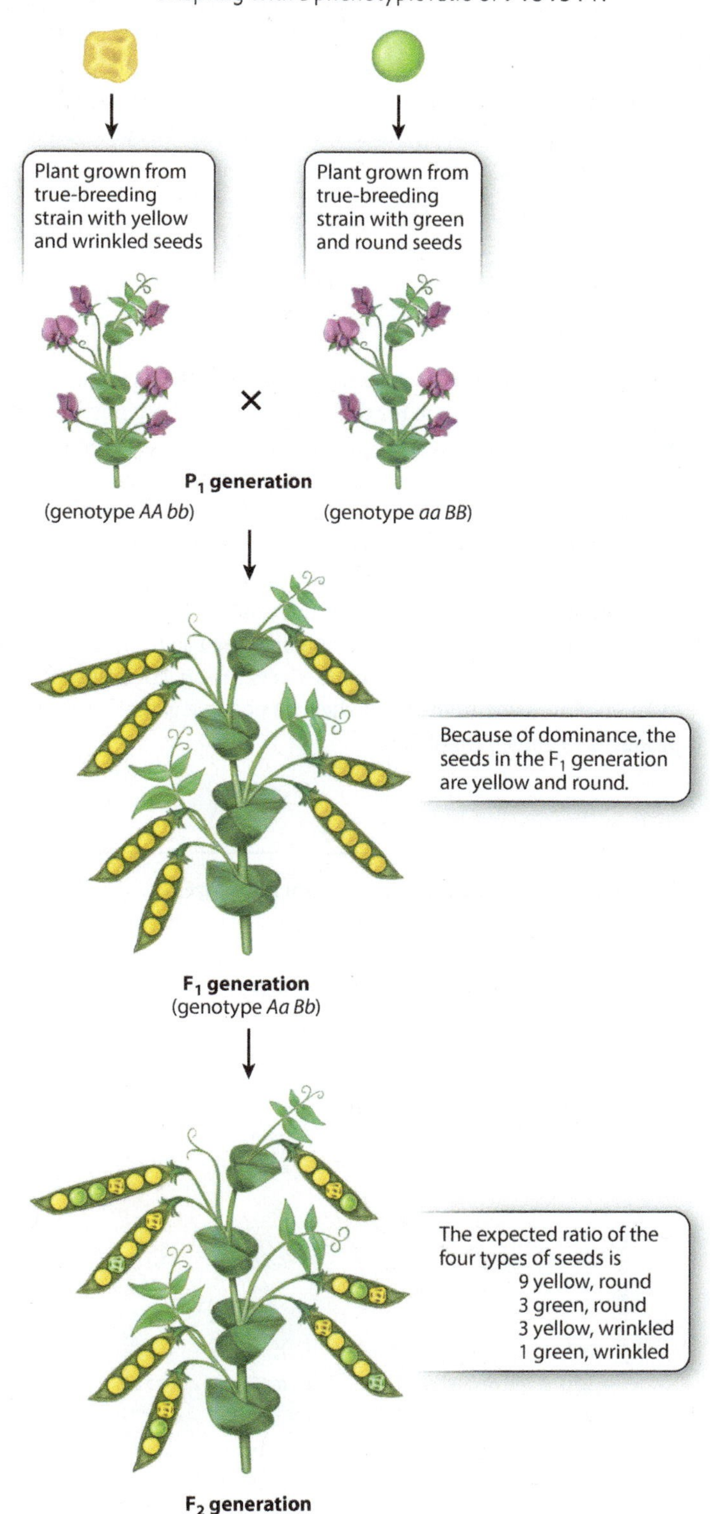

The ratio of these phenotypes is 9.9:3.3:3.1:1.0, which Mendel realized is close to 9:3:3:1. The latter ratio is that expected if the *A* and *a* alleles for seed color undergo segregation and form gametes independently of the *B* and *b* alleles for seed shape. How did Mendel come to expect a 9:3:3:1 ratio of phenotypes? For seed color alone we expect a ratio of ¾ yellow : ¼ green, and for seed shape alone we expect a ratio of ¾ round : ¼ wrinkled. If the traits are independent, then we can use the multiplication rule to predict the outcomes for both traits:

yellow round	$(3/4) \times (3/4) = 9/16$
green round	$(1/4) \times (3/4) = 3/16$
yellow wrinkled	$(3/4) \times (1/4) = 3/16$
green wrinkled	$(1/4) \times (1/4) = 1/16$

Note that 9/16:3/16:3/16:1/16 is equivalent to 9:3:3:1.

The underlying reason for the 9:3:3:1 ratio of phenotypes in the F_2 generation is that the alleles for yellow versus green and those for round versus wrinkled are assorted into gametes independently of each other. In other words, the hereditary transmission of either gene has no effect on the hereditary transmission of the other. A Punnett square depicting independent assortment is shown in **Fig. 16.13.** The *A* and *a* alleles segregate equally into gametes as ½ *A* : ½ *a*, and likewise the *B* and *b* alleles segregate equally into gametes as ½ *B* : ½ *b*. The result of independent assortment is that the four possible gametic types are produced in equal proportions:

AB gametes	$(1/2) \times (1/2) = 1/4$
Ab gametes	$(1/2) \times (1/2) = 1/4$
aB gametes	$(1/2) \times (1/2) = 1/4$
ab gametes	$(1/2) \times (1/2) = 1/4$

As the Punnett square in Fig. 16.13 shows, random union of these gametic types produces the expected ratio of 9 yellow round,

FIG. 16.13 Independent assortment of two genes. The Punnett square reveals the underlying phenotypic ratio of 9:3:3:1.

Pollen gametes (columns); Ovule gametes (rows)

	A B ¼	*A b* ¼	*a B* ¼	*a b* ¼
A B ¼	*AA BB* 1/16	*AA Bb* 1/16	*Aa BB* 1/16	*Aa Bb* 1/16
A b ¼	*AA Bb* 1/16	*AA bb* 1/16	*Aa Bb* 1/16	*Aa bb* 1/16
a B ¼	*Aa BB* 1/16	*Aa Bb* 1/16	*aa BB* 1/16	*aa Bb* 1/16
a b ¼	*Aa Bb* 1/16	*Aa bb* 1/16	*aa Bb* 1/16	*aa bb* 1/16

There are 9 possible genotypes and 4 possible phenotypes. The ratio of phenotypes is 9:3:3:1.

HOW DO WE KNOW?

FIG. 16.14

How are single-gene traits inherited?

BACKGROUND Gregor Mendel's experiments, carried out in the years 1856–1863, are among the most important in all of biology.

EXPERIMENTS Mendel set out to improve upon previous research in heredity. He writes that "among all the numerous experiments made, not one has been carried out to such an extent and in such a way as to make it possible to determine the number of different forms under which the offspring of the hybrids appear, or to arrange these forms with certainty according to their separate generations, or definitely to ascertain their statistical relations." By studying simple traits across several generations of crosses, Mendel observed how these traits were inherited.

RESULTS Mendel concluded that the "statistical relations" were very clear. Crosses between plants that were hybrids of a single trait displayed two phenotypes in a ratio of 3:1. Crosses between plants that were hybrids of two traits displayed four different phenotypes in a ratio of 9:3:3:1.

HYPOTHESIS From observing these ratios among several different traits, Mendel made two key hypotheses about the inheritance of traits, now called Mendel's laws:

1. The principle of segregation states that individuals inherit two copies (alleles) of each gene, one from the mother and one from the father, and when the individual forms reproductive cells, the two copies separate (segregate) equally in the eggs or sperm.

Hypothesis: Segregation

2. The principle of independent assortment states that the two copies of each gene segregate into gametes independently of the two copies of another gene.

Hypothesis: Independent assortment

	A B ¼	*A b* ¼	*a B* ¼	*a b* ¼
A B ¼	*AA BB* 1/16	*AA Bb* 1/16	*Aa BB* 1/16	*Aa Bb* 1/16
A b ¼	*AA Bb* 1/16	*AA bb* 1/16	*Aa Bb* 1/16	*Aa bb* 1/16
a B ¼	*Aa BB* 1/16	*Aa Bb* 1/16	*aa BB* 1/16	*aa Bb* 1/16
a b ¼	*Aa Bb* 1/16	*Aa bb* 1/16	*aa Bb* 1/16	*aa bb* 1/16

ANALYSIS Mendel found support for his two hypotheses in the statistical analysis of the results of his meticulous crosses.

1. The principle of segregation: A prediction of the hypothesis of segregation is that, among seeds with the dominant phenotype, the ratio of homozygous to heterozygous genotypes should be 1:2. Mendel tested this prediction in several ways, one of which was simply to allow plants grown from F_2 seeds to self-fertilize. Any that were true breeding for the dominant phenotype he classified as homozygous, and any that segregated to yield both dominant and recessive phenotypes he classified as heterozygous. In the experimental test, the observed numbers fit the expected values within the margins that would be expected by chance.

Experimental test

	AA	*Aa*
Observed	166	353
Expected	173	346

2. The principle of independent assortment: Although the Punnett square for two pairs of alleles has 16 squares, there are only 9 genotypes, and the hypothesis of independent assortment predicts that these genotypes should appear in the ratio of 1 *AA BB*:2 *AA Bb*:1 *AA bb*:2 *Aa BB*; 4 *Aa Bb*:2 *Aa bb*:1 *aa BB*: 2 *aa Bb*:1 *aa bb*. As before, Mendel tested this hypothesis by self-fertilization of plants grown from the F_2 seeds, classifying any that bred true for either trait as homozygous and any that segregated to yield both dominant and recessive phenotypes as heterozygous. In the experimental test, once again the observed numbers fit the expected values within the margins that would be expected by chance.

Experimental test

	AA BB	*AA Bb*	*AA bb*
Observed	38	60	28
Expected	33	66	33

	Aa BB	*Aa Bb*	*Aa bb*
Observed	65	138	68
Expected	66	132	66

	aa BB	*aa Bb*	*aa bb*
Observed	35	67	30
Expected	33	66	33

FOLLOW-UP WORK Mendel's work was ignored during his lifetime, and its importance was not recognized until 1900, 16 years after his death. The rediscovery marks the beginning of the modern science of genetics.

SOURCE Mendel's paper in English is available at http://www.mendelweb.org/Mendel.html.

3 green round, 3 yellow wrinkled, and 1 green wrinkled. **Fig. 16.14** summarizes how Mendel's experiments led him to formulate his two laws.

Independent assortment reflects the random alignment of chromosomes in meiosis.

Independent assortment of genes on different chromosomes results from the mechanics of meiosis (Chapter 11), in which different pairs of homologous chromosomes align randomly on the metaphase plate in meiosis I. For some pairs of chromosomes, the maternal chromosome goes toward one pole during anaphase I, and the paternal chromosome goes to the other pole, but for other pairs, just the opposite occurs. Because the alignment is random, gene pairs on different chromosomes assort independently of one another.

Fig. 16.15 illustrates two possible alignments that are equally likely. In one alignment, the *B* allele (dark red) goes to the same

FIG. 16.15 Independent assortment of genes in different chromosomes. Chromosomes are sorted into daughter cells randomly during meiosis, resulting in independent assortment of genes.

Independent assortment of genes in different chromosomes reflects the fact that nonhomologous chromosomes can orient in either of two ways that are equally likely.

Anaphase I

Anaphase I

Meiosis II

Meiosis II

Resulting gametes

Resulting gametes

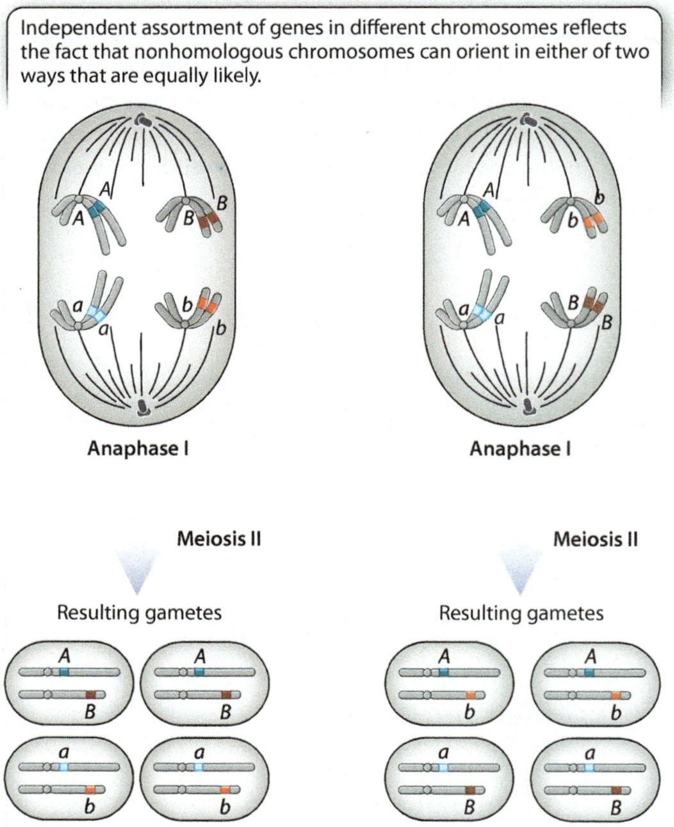

pole as the *A* allele (dark blue), and, in the other alignment, the *b* allele (light red) goes in the same direction as the *A* allele. The first type of alignment results in a 1:1 ratio of *AB* : *ab* gametes, and the second type of alignment results in a 1:1 ratio of *Ab* : *aB* gametes. Because the two orientations are equally likely, the overall ratio of *AB* : *ab* : *Ab* : *aB* from a large number of cells undergoing meiosis is expected to be 1:1:1:1. This is the principle of independent assortment for genes located in different chromosomes.

Not all genes undergo independent assortment. For example, genes that are sufficiently close together in one chromosome do not assort independently of one another. Genes in the same chromosome that fail to show independent assortment are said to be linked and are discussed in Chapter 17.

FIG. 16.16 Epistasis, the interaction of genes affecting the same trait. Epistasis can modify the 9 : 3 : 3 : 1 ratio of phenotypes, in this example to 13 : 3.

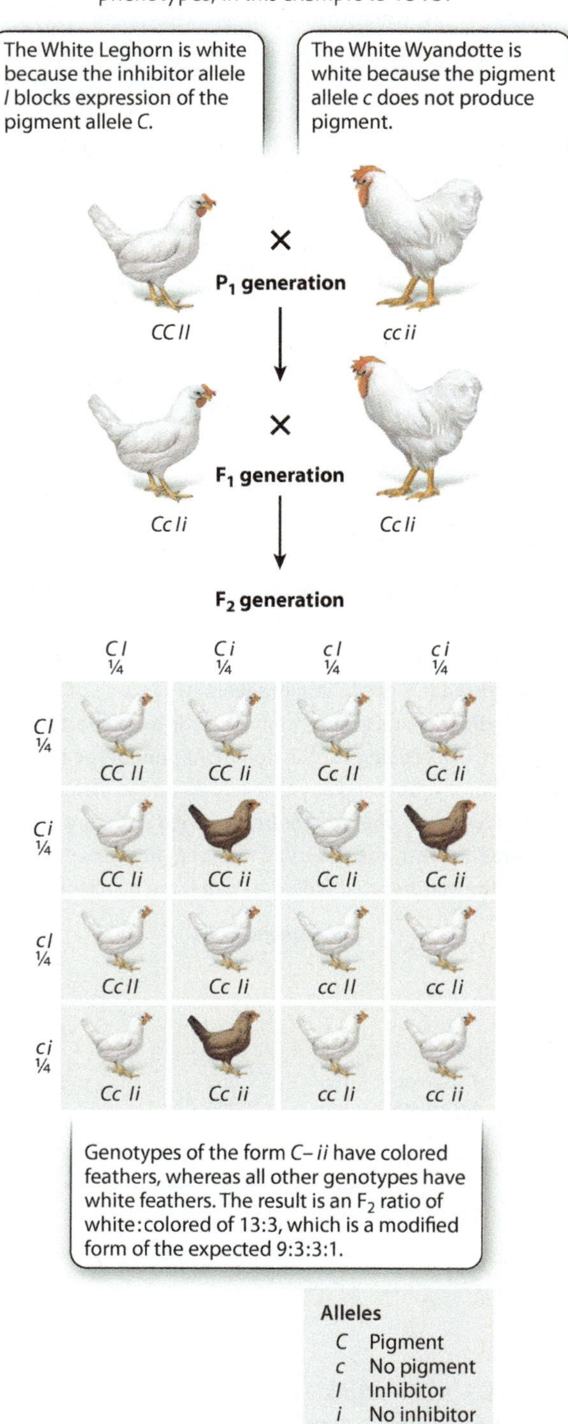

Phenotypic ratios can be modified by interactions between genes.

The 9:3:3:1 ratio results from independent assortment of two genes when one allele of each gene is dominant and when the two genes affect different traits. However, even with complete dominance, the ratio of phenotypes may be different if the two genes affect the same trait. This often happens when the genes code for proteins that act in the same biochemical pathway. In such cases, the gene products can interact to affect the phenotypic expression of the genotypes, resulting in a modification of the expected ratio. Genes that modify the phenotypic expression of other genes are said to show **epistasis.**

There are many types of epistasis leading to different modifications of the 9:3:3:1 ratio. Among the more common modified ratios are 12:3:1, as well as 9:3:4 and 13:3. **Fig. 16.16** shows one example, in which the F_2 generation of a cross between White Leghorn and White Wyandotte chickens displays the modified ratio 13:3. Both breeds are white, but for different genetic reasons. There are two genes involved in pigment production, each with two alleles. The *C* gene encodes a protein that affects coloration in feathers. The dominant allele *C* produces pigment, and the recessive allele *c* does not produce pigment. A different gene, *I*, codes for an inhibitor protein. The product of the dominant allele, *I*, inhibits the expression of *C*, whereas the recessive allele, *i*, does not produce the inhibitor and so does not inhibit the expression of *C*. Therefore, the White Leghorn (genotype *CC II*) is white because the product of the dominant allele *I* inhibits the pigment in the feathers, and the White Wyandotte (genotype *cc ii*) is white because the recessive allele *c* does not produce feather pigment to begin with. The F_1 generation has genotype *Cc Ii* and is also white. With independent assortment, only the three *C – ii* offspring have colored feathers in the F_2 generation (the dash indicates that the second allele could be either *C* or *c*), while the rest have white feathers, and so the ratio of white : colored is 13:3.

→ **Quick Check 5** Why do the F_1 chickens with genotype *Cc Ii* have white feathers?

16.5 PATTERNS OF INHERITANCE OBSERVED IN FAMILY HISTORIES

Segregation of alleles takes place in human meiosis just as it does in peas and most other sexual organisms. The results are not so easily observed as in peas, however, for several reasons. First, humans do not choose their mating partners for the convenience of biologists, and so experimental crosses are not possible. Second, the number of children in human families is relatively small, and so the Mendelian ratios are often obscured by random fluctuations due to chance. Nevertheless, in many cases the occurrence of segregation can be observed even in human families. The patterns are often easiest to spot for inherited traits that are rare because in those cases the mutant alleles occur only in affected individuals and their close relatives (who may not themselves be affected).

In studying human families, the record of the ancestral relationships among individuals is summarized in a diagram of family history called a **pedigree.** Some typical symbols used in pedigrees are shown in **Fig. 16.17.** The same patterns of dominance and recessiveness that Mendel observed in his pea plants can be seen in some pedigrees. Over the next few sections, we'll see how to recognize telltale features of a pedigree that reveal the genotypic nature of a trait.

FIG. 16.17 Some conventions used in drawing human pedigrees.

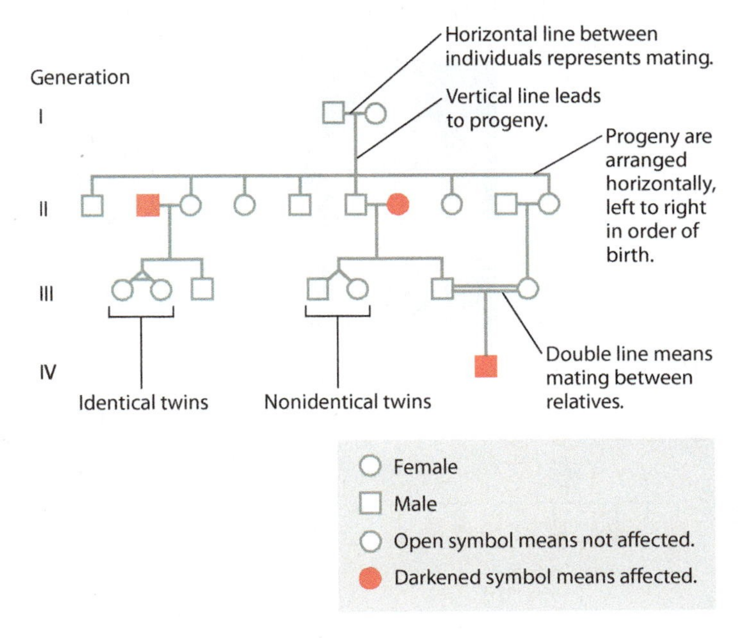

Dominant traits appear in every generation.

The pedigree shown in **Fig. 16.18** is for a rare dominant trait, brachydactyly, in which the middle long bone in the fingers fails to grow and therefore the fingers remain very short. This pedigree, published in 1905, was the first demonstration of dominant Mendelian inheritance in humans. This particular form of brachydactyly results from a mutation in a gene whose normal product is a protein involved in cartilage formation, which is necessary for bone growth.

These are the features of the pedigree that immediately suggest dominant inheritance:

1. Affected individuals are equally likely to be females or males.
2. Most matings that produce affected offspring have only one affected parent. This occurs because the brachydactyly trait is rare, and therefore a mating between two affected individuals is extremely unlikely.
3. Among matings in which one parent is affected, approximately half the offspring are affected.

If a dominant trait is rare, then affected individuals will almost always be heterozygous (*Aa*), not homozygous (*AA*). In a mating in which one parent is heterozygous for the dominant gene (*Aa*) and the other is homozygous recessive (*aa*), half the offspring are expected to be heterozygous (*Aa*) and the other half homozygous recessive (*aa*).

FIG. 16.18 Pedigree of a trait caused by a dominant allele. This pedigree shows the inheritance of shortened fingers associated with a form of brachydactyly (inset). *Photo source: Stefan Mundlos.*

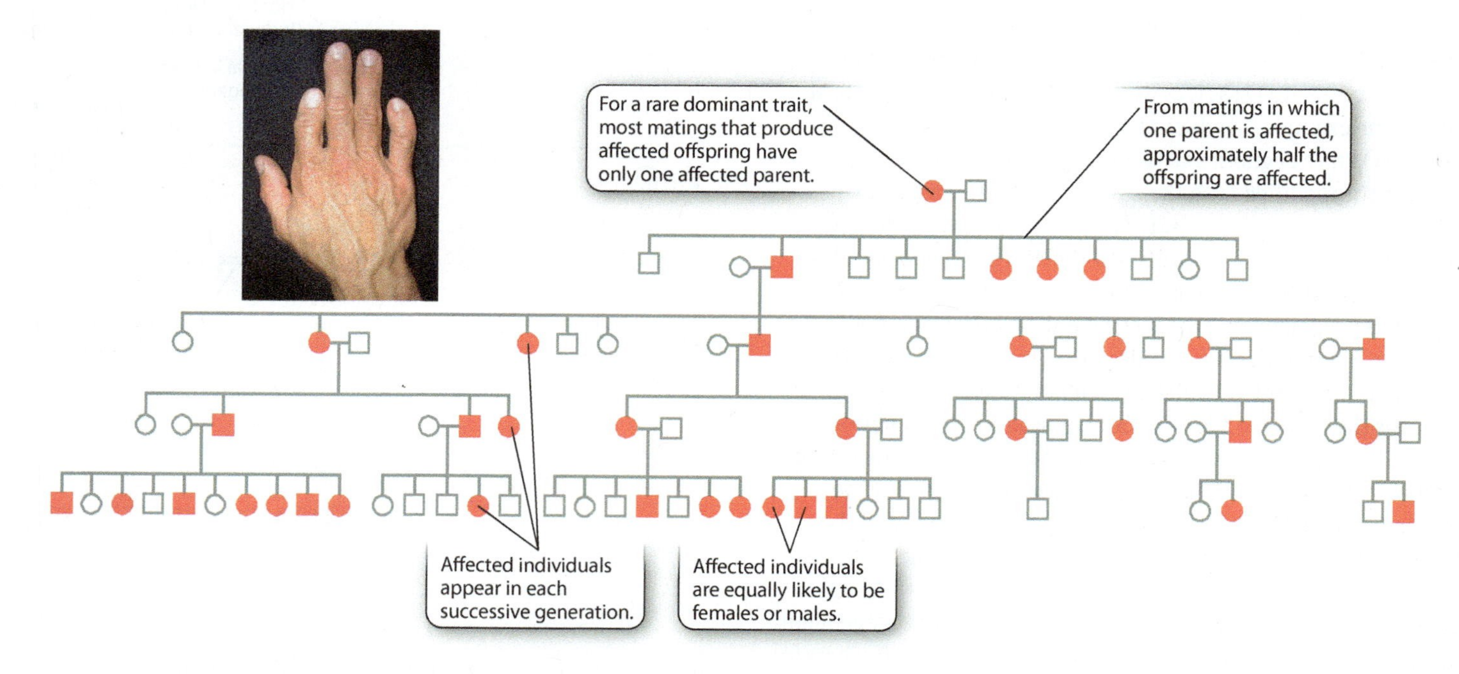

Recessive traits skip generations.

Recessive inheritance shows a pedigree pattern very different from that of dominant inheritance. The pedigree shown in **Fig. 16.19** pertains to albinism, in which the amount of melanin pigment in the skin, hair, and eyes is reduced. In most populations, the frequency of albinism is about 1 in 36,000, but it has a much higher frequency—about 1 in 200—among the Hopi and several other Native American tribes of the Southwest. (It is not unusual for genetic diseases to have elevated frequencies among isolated populations.) This type of albinism is due to a mutation in the gene *OCA2*, which encodes a membrane transporter protein thought to be important in transport of the amino acid tyrosine, which is used in the synthesis of the melanin pigment responsible for skin, hair, and eye color.

As noted in Chapter 15, another type of mutation affecting expression of this same gene is associated with blue eyes. As shown in Fig. 16.19, double lines represent matings between relatives, and in both cases shown here the mating is between first cousins.

These are the principal pedigree characteristics of recessive traits:

1. The trait may skip one or more generations.
2. Affected individuals are equally likely to be females or males.
3. Affected individuals may have unaffected parents, as in the offspring of the second mating in the second generation in Fig. 16.19. For a recessive trait that is sufficiently rare, virtually all affected individuals have unaffected parents.
4. Affected individuals often result from mating between relatives, typically first cousins.

Recessive inheritance has these characteristics because recessive alleles can be transmitted from generation to generation without manifesting the recessive phenotype. In order for an affected individual to occur, the recessive allele must be inherited from both parents. Mating between relatives often allows rare recessive alleles to become homozygous because an ancestor that is shared between the relatives may carry the gene (*Aa*). The recessive allele in the common ancestor can be transmitted to both parents, making them each a carrier of the allele as well. If both parents are unaffected carriers of the allele, they both have the genotype *Aa*, and therefore ¼ of their offspring are expected to be homozygous *aa* and affected.

Many genes have multiple alleles.

The examples discussed so far in this chapter involve genes with only two alleles, such as *A* for yellow seeds and *a* for green seeds. Similarly, as noted in Chapter 15, most single-nucleotide polymorphisms (SNPs) have only two alleles, differing only in which particular base pair is present at a particular position in genomic DNA. On the other hand, because a gene consists of a sequence of nucleotides, any nucleotide or set of nucleotides in

FIG. 16.19 **Pedigree of a trait caused by a recessive allele.** The photograph is of a group of Hopi males, three with albinism, which is caused by a recessive allele. *Source: BAE GN 02458C 06404400, National Anthropological Archives, Smithsonian Institution.*

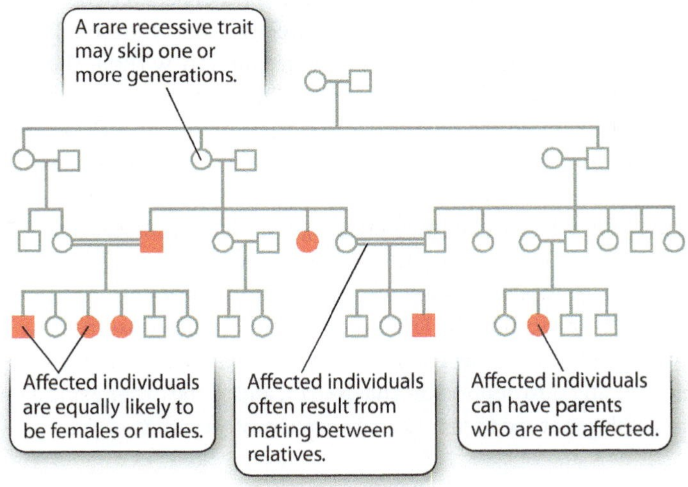

the gene can undergo mutation. Each of the mutant forms that exists in a population constitutes a different allele, and hence a population of organisms may contain many different alleles of the same gene, which are called **multiple alleles.** Some genes have so many alleles that they can be used for individual identification by means of DNA typing, such as the VNTR (variable number tandem repeat) sequences discussed in Chapter 15.

In considering genetic diseases, multiple alleles are often grouped into categories such as "mutant" and "normal." But there are often many different "mutant" and many different "normal" alleles in a population. For the "mutant" alleles, the DNA sequences are different from one another, but each produces a protein product whose function is impaired under the usual environmental conditions. For the "normal" alleles, the DNA sequences are also different, but they all are able to produce functional protein.

For example, more than 400 different recessive alleles that cause phenylketonuria (PKU) have been discovered across the world. PKU is a moderate to severe form of mental retardation caused by mutations in the gene encoding the enzyme phenylalanine hydroxylase. Children affected with PKU are unable to break down the excess phenylalanine present in a normal diet, and the buildup impairs the development of neurons in the brain. About 1 in 10,000 newborns inherits two mutant alleles, which could be two copies of the same mutant allele or two different mutant alleles, and is affected.

The "normal" form of the gene encoding phenylalanine hydroxylase also exists in the form of multiple alleles, each of which differs from the others, but nevertheless encodes a functional form of the enzyme.

→ **Quick Check 6** How is it possible that there are multiple different alleles in a population and yet any individual can have only two alleles?

Incomplete penetrance and variable expression can obscure inheritance patterns.

Many traits with single-gene inheritance demonstrate complications that can obscure the expected patterns in pedigrees. Chief among these are traits with **incomplete penetrance,** which means that individuals with a genotype corresponding to a trait do not actually show the phenotype, either because of environmental effects or because of interactions with other genes. Penetrance is the proportion of individuals with a particular genotype that show the expected trait. If the penetrance is less than 100%, then the trait shows reduced, or incomplete, penetrance. Familial cancers often show incomplete penetrance. For example, retinoblastoma caused by certain mutations in the *Rb* gene and breast cancer caused by certain mutations in the *BRCA1* and *BRCA2* genes are incompletely penetrant. That is, some people who inherit a mutation in these genes that predisposes to cancer do not in fact develop cancer. Similarly, some mutations in the *apoliprotein E* (*APOE*) gene increase the risk of developing Alzheimer's disease. However, not everyone who inherits these forms of the gene develops Alzheimer's disease, so these alleles are also incompletely penetrant.

Another common complication in human pedigrees is **variable expressivity,** which means that a particular phenotype is expressed with a different degree of severity in different individuals. Don't confuse variable expressivity with incomplete penetrance. With variable expressivity, the trait is always expressed, though the severity varies; with incomplete penetrance, the trait is sometimes expressed and sometimes not. Variation among individuals in the expression of a trait can result from the action of other genes, from effects of the environment, or both. An example is provided by deficiency of the enzyme alpha-1 antitrypsin (α1AT) discussed in Chapter 15, which is associated with loss of lung elasticity and emphysema. Among individuals with emphysema due to α1AT deficiency, the severity of the symptoms varies dramatically from one patient to the next.

In this case, tobacco smoking is an environmental factor that increases the severity of the disease.

Incomplete penetrance and variable expressivity both provide examples where a given genotype does not always produce the same phenotype, since the expression of genes is often influenced by other genes, the environment, or a combination of the two.

? CASE 3 YOU, FROM A TO T: YOUR PERSONAL GENOME

How do genetic tests identify disease risk factors?

Your personal genome, as well as that of every human being, contains a unique combination of alleles of thousands of different genes. Most of these have no detectable effects on health or longevity, but many are risk factors for genetic diseases. Molecular studies have discovered particular alleles of genes associated with a large number of such conditions, and the presence of these alleles can be tested. More than a thousand genetic tests have already been deployed, and many more are actively being developed. A **genetic test** is a method of identifying the genotype of an individual. The tests may be carried out on entire populations or restricted to high-risk individuals.

The benefits of genetic testing can be appreciated by an example. Screening of newborns for phenylketonuria identifies babies with high blood levels of phenylalanine. In the absence of treatment, 95% of such newborns will progress to moderate or severe mental retardation, whereas virtually all those placed on a special diet with a controlled amount of phenylalanine will have mental function within the normal range. For recessive conditions like phenylketonuria, tests can be carried out on people with affected relatives to identify the heterozygous genotypes. Testing can also identify genetic risk factors for disease, and carriers can take additional precautions. For example, individuals with α1AT deficiency can prolong and improve the quality of their lives by not smoking tobacco, women with genetic risk factors *BRCA1* and *BRCA2* for breast cancer can have frequent mammograms, and those with the *TCF7L2* risk factor for type 2 diabetes can decrease their risk by lifestyle choices that include weight control and exercise.

While there are many potential benefits to genetic testing, there are also some perils. One major concern is maintaining the privacy of those who choose to be tested. With medical records increasingly going online, who will have access to your test results, and how will this information be used? Could your test results be used to deny you health or life insurance because you have a higher than average risk of some medical condition? Or could an employer who got hold of your genetic test results decide to reassign you to another job, or even eliminate your position because of your genetic predispositions? There are some safeguards designed to protect you from such discrimination. The Genetic Information Nondiscrimination Act (GINA) was signed into law in 2008 and forbids the use of genetic information in decisions concerning employment and health insurance. The protection provided by GINA will, it is hoped, allow for the responsible and productive use of genetic information.

There is also increasing concern about the reliability and accuracy of genetic tests, especially direct-to-consumer (DTC) genetic tests. DTC tests can be purchased directly without the intervention of medical professionals. Since the consumer sends a biological sample and DTC tests are carried out by the provider, the tests are not regarded as medical devices and so are unregulated. One problem is that some DTC tests are based on flimsy and unconfirmed evidence connecting a gene with a disease. Another is that the link between genotype and risk may be exaggerated for marketing purposes. Yet another is lack of information on quality control in the DTC laboratories. Finally, consumer misinterpretation may regard genotype as destiny, at one extreme descending into depression and despair, and at the other using a low-risk genotype to justify an unhealthy lifestyle. ■

Core Concepts Summary

16.1 The earliest theories of heredity incorrectly assumed the inheritance of acquired characteristics and blending of parental traits in the offspring.

The inheritance of acquired characteristics suggests that traits that develop during the lifetime of an individual can be passed on to offspring. With rare exceptions, this mode of inheritance does not occur. page 326

Blending inheritance is the incorrect hypothesis that characteristics in the parents are averaged in the offspring. This model predicts the blending of genetic material, which does not occur. Different forms of a gene maintain their separate identities even when present together in the same individual. page 326

16.2 The study of modern transmission genetics began with Gregor Mendel, who used the garden pea as his experimental organism and studied traits with contrasting characteristics.

Mendel started his experiments with true-breeding plants, ones whose progeny are identical to their parents. He followed just one or two traits at a time, allowing him to discern simple patterns, and he counted all the progeny of his crosses. page 327

In crosses of one true-breeding plant with a particular trait and another true-breeding plant with a contrasting trait, just one of the two characteristics appeared in the offspring. The trait that appeared in this generation is dominant, and the trait that is not seen is recessive. page 328

Mendel explained this result by hypothesizing that there is a hereditary factor for each trait (now called a gene); that each pea plant carries two copies of the gene for each trait; and that one of two different forms of the gene (alleles) is dominant to the other one. page 329

16.3 Mendel's first key discovery was the principle of segregation, which states that members of a gene pair separate equally into gametes.

When Mendel allowed the progeny of the first cross to self-fertilize, he observed a 3 : 1 ratio of the dominant and recessive traits among the progeny. page 330

Mendel reasoned that the parent in this generation must have two different forms of the same gene (*A* and *a*) and that these alleles segregate from each other to form gametes with each gamete getting *A* or *a* but not both. When the gametes combine at random, they produce progeny in the genotypic ratio 1 *AA* : 2 *Aa* : 1 *aa*, which yields a phenotypic ratio of 3 : 1 because *A* is dominant to *a*. This idea became known as the principle of segregation. page 330

The principle of segregation reflects the separation of homologous chromosomes that occurs in anaphase I of meiosis. page 332

Some traits show incomplete dominance, in which the phenotype of the heterozygous genotype is intermediate between those of the two homozygous genotypes. page 333

The expected frequencies of progeny of crosses can be predicted using the addition rule, which states that when two possibilities are mutually exclusive, the probability of either event occurring is the sum of their individual probabilities; and the multiplication rule, which states that when two possibilities occur independently, the probability of both events occurring is the product of the probabilities of each of the two events. page 333

16.4 Mendel's second key finding was the principle of independent assortment, which states that different gene pairs segregate independently of one another.

In a cross with two traits, each with two contrasting characteristics, the two traits behave independently of each other. This idea became known as the principle of independent assortment. For example, in a self-cross of a double heterozygote, the phenotypic ratio of the progeny is 9 : 3 : 3 : 1, reflecting the independent assortment of two 3 : 1 ratios. page 335

Independent assortment results from chromosome behavior during meiosis, specifically from the random orientation of different chromosomes on the meiotic spindle. page 337

In some cases, genes interact with each other, modifying the expected ratios in crosses. Epistasis is a gene interaction in which one gene affects the expression of another. page 338

16.5 The patterns of inheritance that Mendel observed in peas can also be seen in humans.

In a human pedigree, females are represented by circles and males as squares; affected individuals are shown as filled symbols and unaffected individuals as open symbols. Horizontal lines denote matings. page 339

Dominant traits appear in every generation and affect males and females equally. page 339

Recessive traits skip one or more generations and affect males and females equally. Affected individuals often result from matings between close relatives. page 340

Although a given individual has only two alleles of each gene, there can be many alleles of a particular gene in the population as a whole. page 340

Interpreting pedigrees can be difficult because of incomplete penetrance and variable expressivity. Penetrance is the percentage of individuals with a particular genotype who show the expected phenotype, and expressivity is the degree to which a genotype is expressed in the phenotype. page 341

Genetic testing enables the genotype of an individual to be determined for one or more genes. Such tests carry both benefits and risks. page 342

Self-Assessment

1. In his famous paper, Mendel writes that he set out to "determine the number of different forms in which hybrid progeny appear" and to "ascertain their numerical interrelationships." How did his close attention to numbers lead him to discover segregation and independent assortment?

2. Distinguish among gene, allele, genotype, and phenotype.

3. Name and describe Mendel's two laws.

4. Explain how the mechanics of meiosis and the movement of homologous chromosomes underlie Mendel's principles of segregation and independent assortment.

5. Explain how you can predict the genotypes and phenotypes of offspring if you know the genotypes of the parents.

6. Describe an instance in which you would use a testcross, and why.

7. Define the multiplication and addition rules, and explain how these rules can help you predict the outcome of a cross between parents with known genotypes.

8. What are some reasons why a single trait might not show a 3:1 ratio of phenotypes in the F_2 generation of a cross between true-breeding strains, and why a pair of traits might not show a 9:3:3:1 ratio of phenotypes in the F_2 generation of a cross between true-breeding strains?

9. Construct a human pedigree for a dominant and a recessive trait and explain the patterns of inheritance.

10. Discuss the benefits and risks of genetic testing and personal genomics.

Log in to LaunchPad to check your answers to the Self-Assessment questions, and to access additional learning tools.

Inheritance of Sex Chromosomes, Linked Genes, and Organelles

Core Concepts

17.1 Many organisms have a distinctive pair of chromosomes, often called the *X* and *Y* chromosomes, that differ between the sexes and show different patterns of inheritance in pedigrees from other chromosomes.

17.2 *X*-linked genes, which show a crisscross inheritance pattern, provided the first evidence that genes are present in chromosomes.

17.3 In genetic linkage, two genes are sufficiently close together in the same chromosome that the particular combination of alleles present in the chromosome tends to remain together in inheritance.

17.4 Most *Y*-linked genes are passed from father to son.

17.5 Mitochondria and chloroplast DNA follow their own inheritance pattern.

SSPL/ Getty Images.

Mendel's principles of segregation and independent assortment are the foundation of transmission genetics (Chapter 16). For traits such as pea color and seed shape that are encoded by single genes and display simple dominance, these principles predict simple phenotypic ratios in the progeny from self-crosses of heterozygous genotypes. The principle of segregation also defines the inheritance patterns expected in human pedigrees for traits due to recessive or dominant mutations, as we saw in the case of albinism and brachydactyly.

However, we also saw in Chapter 16 that not all crosses are as simple as those for Mendel's pea plants. For example, the 3:1 phenotypic ratio in progeny of self-crosses of heterozygotes is 1:2:1 in the case of alleles that show incomplete dominance. The 9:3:3:1 ratio of phenotypes observed for two genes that show independent assortment can be altered by epistasis, producing ratios such as 12:3:1 or 13:3. None of these is an exception to Mendel's laws since his laws reflect chromosome movement during meiosis (Chapter 11). Instead, they reflect how genes are expressed or how different genes interact to produce a phenotype.

This chapter highlights additional patterns of inheritance that Mendel did not observe owing to his choice of experimental organism and the traits he studied. None of these patterns undermines or invalidates his insight that alternative alleles of a gene can have different effects on the expression of a phenotype. Nor do they contradict later discoveries that genes in the nucleus are present in homologous chromosomes that pair and segregate in meiosis. Since Mendel's time, researchers have observed many inheritance patterns that seem to defy one or both of Mendel's laws. What such examples reveal is that the location of a gene is as important to our predictions about the inheritance of a trait as whether its alleles are dominant or recessive.

In this chapter, we discuss how genes carried in the sex chromosomes are transmitted differently in males and in females and how genes close to each other in the same chromosome do not undergo independent assortment and therefore violate Mendel's second law. Genes located in the genomes of mitochondria and chloroplast appear to defy Mendel's laws altogether since the organelles are inherited differently from the way chromosomes are inherited. Such unique patterns of inheritance expand the types of ratios and predictions we saw in Chapter 16 and draw our attention to the location and organization of genes—in specific chromosomes, relative to other genes, or outside the nucleus altogether.

17.1 THE *X* AND *Y* SEX CHROMOSOMES

Mendel concluded from his experiments that reciprocal crosses yield the same types of progeny in the same proportions. This is in most cases true, but an important exception occurs with the X and Y sex chromosomes. For example, red–green color blindness in humans is due to mutant alleles of a gene in the X chromosome. Reciprocal crosses do not produce the same types and numbers of progeny. In one type of cross, when a color-blind man mates with a woman who is not color blind, all of the sons and daughters have normal color vision. However, in the reciprocal cross, when a color-blind woman mates with a man who is not color blind, all of the daughters have normal color vision but all of the sons are color blind. For the X and Y sex chromosomes, reciprocal crosses are not equivalent.

In many animals, sex is genetically determined and associated with chromosomal differences.

Most chromosomes come in pairs that match in shape and size. The members of each pair are known as homologous chromosomes because they have the same genes along their length (Chapter 11). One member of each pair of homologous chromosomes is inherited from the mother and the other from the father. In many animal species, however, the sex of an individual is determined by a distinctive pair of unmatched chromosomes known as the **sex chromosomes,** which are usually designated as the **X chromosome** and the **Y chromosome** (Chapter 13). Chromosomes other than the sex chromosomes are known as **autosomes.**

In humans, a normal female has two copies of the X chromosome (a sex-chromosome constitution denoted XX), and a normal male has one X chromosome and one Y chromosome (XY). The sizes of the human X and Y chromosomes are very different from each other (**Fig. 17.1**). The X chromosome DNA

FIG. 17.1 Human sex chromosomes. The human *X* and *Y* sex chromosomes differ in size and number of genes. *Source: Science Photo Library/Science Source.*

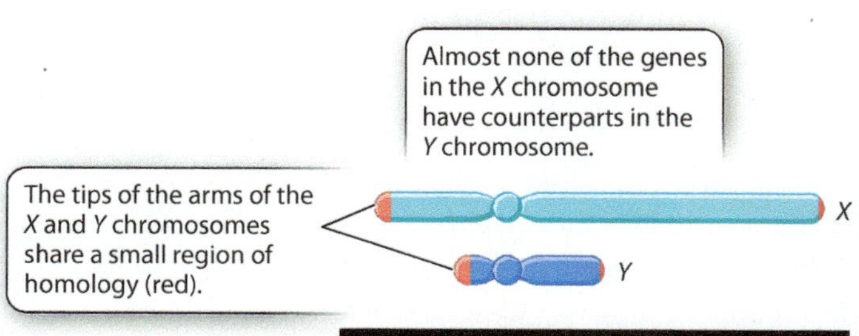

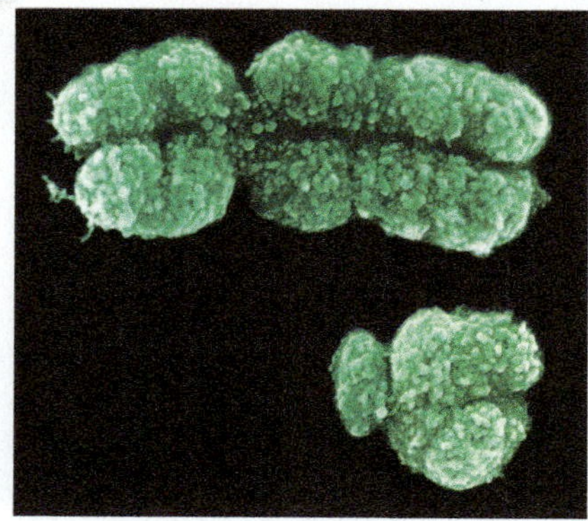

molecule is more than 150 Mb long, while the Y chromosome is only about 50 Mb long. Except for a small region near each tip, the gene contents of the X and Y chromosomes are different from each other. The X chromosome, which includes more than 1000 genes, has a gene density similar to that of most autosomes. The vast majority of these genes have no counterpart in the Y chromosome. In contrast to the X chromosome, the Y chromosome contains only about 50 protein-coding genes.

The regions of homology between the X and Y chromosomes consist of about 2.7 Mb of DNA near the tip of the short arm and about 0.3 Mb of DNA near the tip of the long arm. These regions of homology allow the chromosomes to pair during meiosis (Chapter 11). In most cells undergoing meiosis, a crossover (physical breakage, exchange of parts, and rejoining of the DNA molecules) occurs in the larger of these regions. The crossover allows the chromosomes to move as a unit to align properly at metaphase I so that when their centromeres separate from each other at anaphase I the X and Y chromosomes go to opposite poles.

The relative size and gene content of the X and Y chromosomes differ greatly among species. In some species of mosquitoes, the X and Y chromosomes are virtually identical in size and shape, and the regions of homology include almost the entire chromosome. This situation is unusual, however. In most species, the Y chromosome contains many fewer genes than the X chromosome. Some species, like grasshoppers, have no Y chromosome: Females in these species have two X chromosomes, and males have only one X chromosome. The total number of chromosomes in grasshoppers therefore differs between females and males. In birds, moths, and butterflies, the sex chromosomes are reversed: Females have two different sex chromosomes and males have two of the same sex chromosome.

FIG. 17.2 Inheritance of the *X* and *Y* chromosomes. Segregation of the sex chromosomes in meiosis and random fertilization result in a 1 : 1 ratio of female : male embryos.

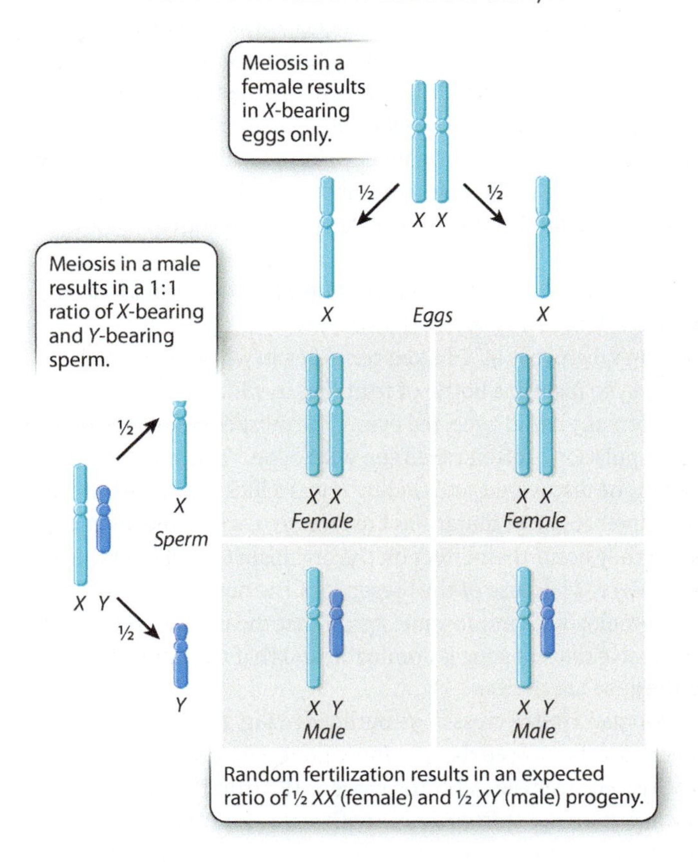

Segregation of the sex chromosomes predicts a 1 : 1 ratio of females to males.

It is ironic that Mendel did not interpret sex as an inherited trait. If he had, he might have realized that sex itself provides one of the most convincing demonstrations of segregation. In human males, segregation of the X chromosome from the Y chromosome during anaphase I of meiosis results in half the sperm bearing an X chromosome and the other half bearing a Y chromosome (**Fig. 17.2**). Meiosis in human females results in eggs that each contains one X chromosome. With random fertilization, as shown in Fig. 17.2, half of the fertilized eggs are expected to be chromosomally XX (and therefore female) and half are expected to be chromosomally XY (and therefore male).

The expected ratio of 1 : 1 of females : males refers to the sex ratio at the time of conception. This is the primary sex ratio, and it is not easily observed because the earliest stages of human fertilization and development are inaccessible to large-scale study. What can be observed is the sex ratio at birth, called the secondary sex ratio, which differs among populations but usually shows a slight excess of males. In the United States, for example, the secondary sex ratio is approximately 100 females : 105 males. The explanation seems to be that, for reasons that are not entirely understood, females are slightly less likely to survive from conception to birth. On the other hand, males are slightly less likely to survive from birth to reproductive maturity, and so, at the age of reproductive maturity, the sex ratio is very nearly 1 : 1. Male mortality continues to be greater than that of females throughout life, so that by age 85 and over, the sex ratio is about 100 females : 50 males.

The sex of each birth appears to be random relative to previous births—that is, there do not seem to be tendencies for some families to have boys or for other ones to have girls. Many people are surprised by this fact, for they know of one or more large families consisting mostly of boys or mostly of girls. But the occasional family with children of predominantly one sex is expected simply by chance. When one occurs, its unusual sex distribution commands attention out of proportion to the actual numbers of such families. In short, families with unusual sex distributions are not more frequent than would be expected by chance.

17.2 INHERITANCE OF GENES IN THE *X* CHROMOSOME

Genes in the *X* chromosome are called **X-linked genes.** These genes have a unique pattern of inheritance first discovered by Thomas Hunt Morgan in 1910. Morgan's pioneering studies of genetics of the fruit fly *Drosophila melanogaster* helped bring Mendelian genetics into the modern era. The discovery of *X*-linked inheritance was not only important in itself, but also provided the first experimental evidence that chromosomes contain genes.

X-linked inheritance was discovered through studies of male fruit flies with white eyes.

Morgan's discovery of *X*-linked genes began when he noticed a white-eyed male in a bottle of fruit flies in which all the others had normal, or wild-type, red eyes. (The most common phenotype in a population is often called the **wild type.**) This was the first mutant he discovered, and finding it was a lucky break. As we saw in Chapter 16, most mutations are recessive, which means that when they occur their effect on the organism (the phenotype) is not observed because of the presence of the nonmutant gene in the homologous chromosome. Recall that the nonmutant form of a recessive mutant gene is dominant, and that the different forms of the gene are alleles.

Morgan's initial crosses are outlined in **Fig. 17.3.** In the first generation, he crossed the mutant white-eyed male with a wild-type red-eyed female. All of the progeny (F_1) fruit flies had red eyes, as you would expect from a cross with any recessive mutation. Morgan then carried out matings between brothers and sisters among the F_1 generation, and he found that the phenotype of white eyes reappeared among the progeny. This result, too, was expected. However, there was a surprise: Morgan observed that the white-eye phenotype was associated with the sex of the fly. In the F_2 generation, all the white-eyed fruit flies were male, and the white-eyed males appeared along with red-eyed males in a ratio of 1:1. No females with white eyes were observed; all the females had red eyes.

FIG. 17.3 Morgan's white-eyed fly. Morgan's discovery of *X*-linked genes derived from his crosses of a white-eyed male in 1910.

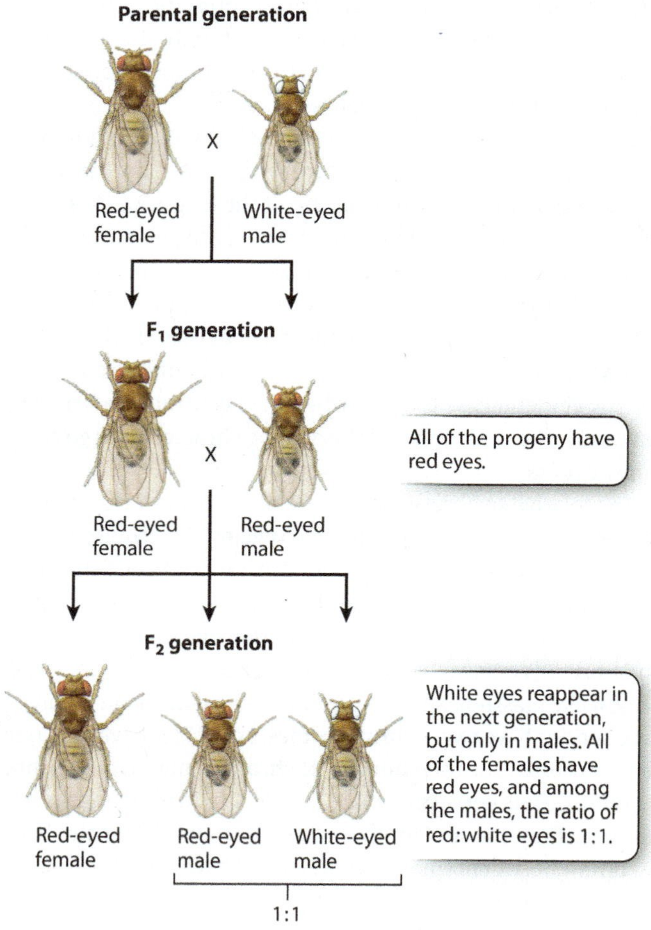

Genes in the *X* chromosome exhibit a "crisscross" inheritance pattern.

When Morgan did his crosses with the white-eyed male, the *X* chromosome had only recently been discovered by microscopic examination of the chromosomes in male and female grasshoppers. Morgan was the first to understand that the pattern of inheritance of the *X* chromosome would be different from that of the autosomes, and he proposed the hypothesis that the white-eyed phenotype was due to a mutation in a gene in the *X* chromosome. This hypothesis could explain the pattern of inheritance shown in Fig. 17.3.

The key features of *X*-linked inheritance are shown in **Fig. 17.4.** In *Drosophila*, as in humans, females are *XX* and males are *XY*. Fig. 17.4a shows an *XY* male in which the *X* chromosome contains a recessive mutation. Because the *Y* chromosome does not carry an allele of this gene, the recessive mutation will be reflected in the male's phenotype—in this case, white eyes.

The Punnett square in Fig. 17.4a explains why all the offspring had red eyes when Morgan crossed the white-eyed male with a red-eyed female in the parental generation. During meiosis in the male, the mutant *X* chromosome segregates from the *Y* chromosome, and each type of sperm, *X* or *Y*, combines with a normal *X*-bearing egg. The result is that the female progeny are heterozygous. They have only one copy of the mutant allele, and because the mutant allele is recessive, the heterozygous females do not express the mutant white-eye trait. The male progeny are also red-eyed because they receive their *X* chromosome from their wild-type red-eyed mother.

The Punnett square in Fig. 17.4a illustrates two important principles governing the inheritance of *X*-linked genes:

1. The phenotypes of the *XX* offspring indicate that a male transmits his *X* chromosome only to his daughters. In this case, the *X* chromosome transmitted by the male carries the white-eye mutation.

FIG. 17.4 ***X*-linkage.** (a) Cross between a homozygous nonmutant female and a male carrying an *X*-linked recessive allele (red). (b) Cross between a heterozygous female and a nonmutant male. *X*-linked recessive alleles are expressed in males because males have only one *X* chromosome.

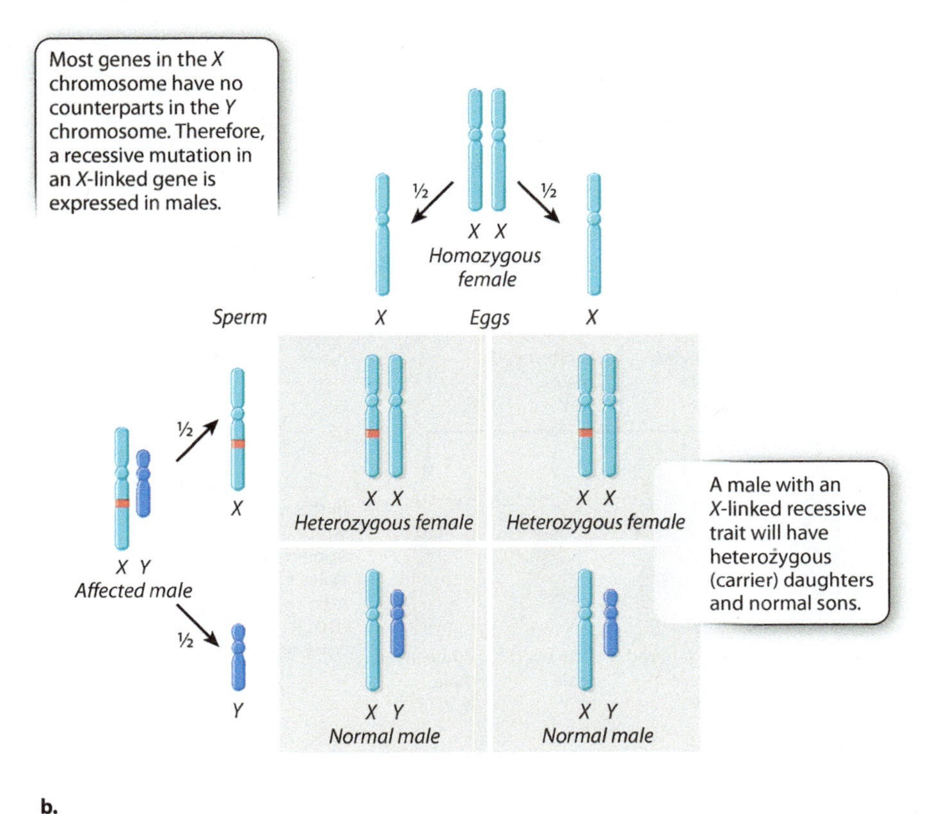

2. The phenotypes of the XY offspring indicate that a male inherits his X chromosome from his mother. In this case, the X chromosome transmitted by the mother carries the nonmutant allele of the gene.

These principles underlie a pattern often referred to as **crisscross inheritance:** An X chromosome present in a male in one generation must be transmitted to a female in the next generation, and in the generation after that can be transmitted back to a male. Therefore, an X chromosome can "crisscross," or alternate, between the sexes in successive generations.

The mating illustrated in Fig. 17.4b shows another important feature of X-linked inheritance, one that explains the results of Morgan's F_1 cross. In this case, the mother is a heterozygous female. During meiosis, the mutant allele and the nonmutant allele undergo segregation. Half of the resulting eggs contain an X chromosome with the mutant allele, and half contain an X chromosome with the nonmutant allele. These combine at random with either X-bearing sperm or Y-bearing sperm. Among the female progeny in the next generation, half are heterozygous for the mutant allele and the other half are homozygous for the normal allele. Thus, none of the females exhibits the white-eye trait because the mutation is recessive. Among the male progeny, half receive the mutant allele and have white eyes, whereas the other half receive the nonmutant allele and have red eyes. The expected progeny from the cross in Fig. 17.4b therefore consist of all red-eyed females, and there is a 1 : 1 ratio of red-eyed to white-eyed males, which is what Morgan observed in the F_2 generation of his crosses, as shown in Fig. 17.3.

Knowing the patterns revealed by the Punnett squares, we can now assign genotypes to Morgan's original crosses (**Fig. 17.5**). In Fig. 17.5, the white-eyed males that Morgan used in his parental generation crosses are given the genotype of w^- Y, and the red-eyed female has a genotype of w^+w^+. The symbol "w^-" stands for the recessive white-eye mutation in one X chromosome, and the symbol "w^+" stands for the dominant nonmutant allele (red eyes) in the other X chromosome. The symbol "Y" stands for the Y chromosome, and it is important to remember that the Y chromosome does not contain an allele of the white-eye gene.

In the cross between a wild-type red-eyed female and a white-eyed male, illustrated in Fig. 17.5a, the male offspring have red eyes because

FIG. 17.5 Genotypes of Morgan's white-eyed fruit fly crosses.

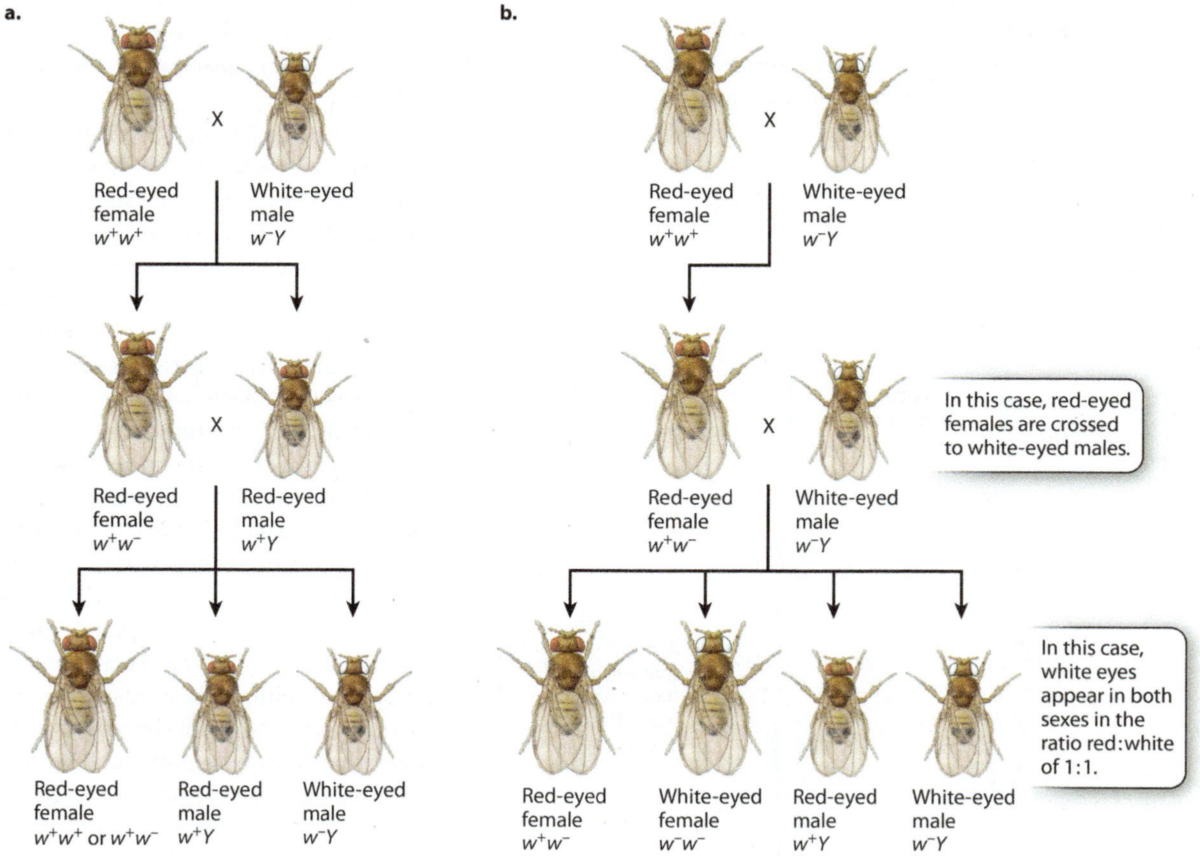

a.

Red-eyed female w^+w^+ X White-eyed male w^-Y

Red-eyed female w^+w^- X Red-eyed male w^+Y

Red-eyed female w^+w^+ or w^+w^- | Red-eyed male w^+Y | White-eyed male w^-Y

b.

Red-eyed female w^+w^+ X White-eyed male w^-Y

Red-eyed female w^+w^- X White-eyed male w^-Y

In this case, red-eyed females are crossed to white-eyed males.

Red-eyed female w^+w^- | White-eyed female w^-w^- | Red-eyed male w^+Y | White-eyed male w^-Y

In this case, white eyes appear in both sexes in the ratio red:white of 1:1.

they receive their X chromosome from their mother. The female offspring from this cross receive one of their X chromosomes from their father, and hence they are heterozygous, w^+w^-. When the male and female progeny are mated together, their offspring consist of all red-eyed females (half of which are heterozygous) and a 1:1 ratio of red-eyed to white-eyed males, exactly as Morgan had observed.

The hypothesis of X-linkage not only explained the original data, but it also predicted the results of other crosses. One important test is outlined in Fig. 17.5b. Here, the parental cross is the same as that in Fig. 17.5a, but instead of mating the F_1 females to their brothers, they are mated to white-eyed males. The prediction is that there should be a 1:1 ratio of red-eyed to white-eyed females as well as a 1:1 ratio of red-eyed to white-eyed males. Again, these were the results observed. By the results of these crosses and others, Morgan demonstrated that the pattern of inheritance of the white-eye mutation parallels the pattern of inheritance of the X chromosome.

X-linkage provided the first experimental evidence that genes are in chromosomes.

Morgan's original experiments indicated that the white-eye mutation showed a pattern of inheritance like that expected of the X chromosome. However, it was one of Morgan's students who showed experimentally that the white-eye mutation was actually a physical part of the X chromosome. Today, it seems obvious that genes are in chromosomes because we know that genes consist of DNA and that DNA in the nucleus is found in chromosomes. But in 1916, when Calvin B. Bridges, who had joined Morgan's laboratory as a freshman, was working on his PhD research under Morgan's direction, neither the chemical nature of the gene nor the chemical composition of chromosomes was known.

In one set of experiments, Bridges crossed mutant white-eyed females with wild-type red-eyed males (**Fig. 17.6**). Usually, the progeny consisted of red-eyed females and white-eyed males (Fig. 17.6a). This is the result expected when the X chromosomes in the mother separate normally at anaphase I in meiosis because all the daughters receive a w^+-bearing X chromosome from their father and all the sons receive a w^--bearing X chromosome from their mother.

But Bridges noted a few rare exceptions among the progeny. He saw that about 1 offspring in 2000 from the cross was "exceptional"—either a female with white eyes or a male with red eyes. The exceptional females were fertile, and the exceptional males were sterile. To explain these exceptional progeny, Bridges proposed the hypothesis diagrammed in Fig. 17.6b: The X chromosomes in a female occasionally fail to separate in anaphase I in

FIG. 17.6 Nondisjunction as evidence that genes are present in chromosomes. (a) Normal chromosome separation yields expected progeny. (b) Nondisjunction yields exceptional progeny.

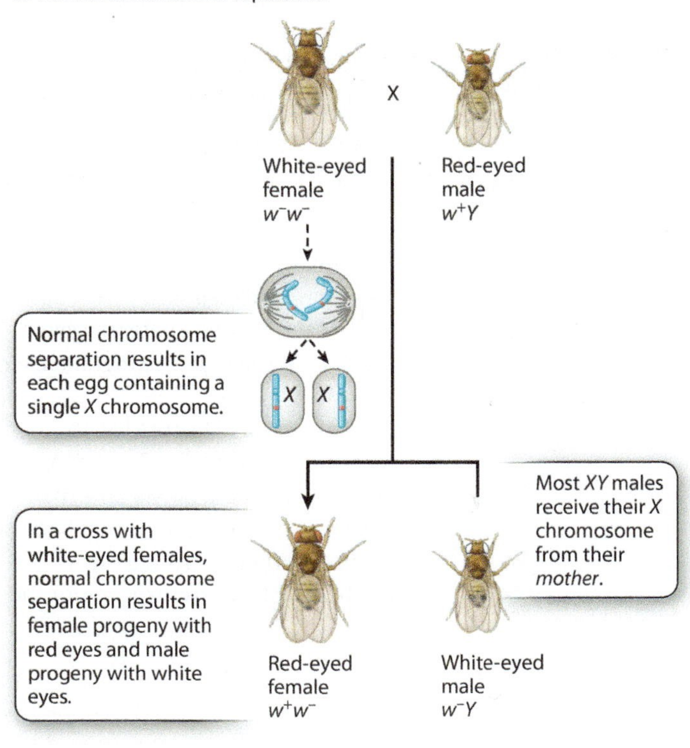

b. Nondisjunction

X

White-eyed female
w^-w^-

Red-eyed male
w^+Y

Nondisjunction results in eggs with either two X chromosomes or no X chromosome.

X X

O

Rare XO males receive their X chromosome from their father.

In a cross with white-eyed females, nondisjunction of the X chromosome results in XXY female progeny with white eyes and XO male progeny with red eyes.

White-eyed female
w^-w^-Y

Red-eyed male
w^+

meiosis, and both X chromosomes go to the same pole. Recall from Chapter 15 that chromosomes sometimes fail to separate normally in meiosis, a process known as **nondisjunction.** Nondisjunction of X chromosomes results in eggs containing either two X chromosomes or no X chromosome. Figure 17.6b shows the implications for eye color in the progeny if the hypothesis is correct. The exceptional white-eyed females would contain two X chromosomes plus a Y chromosome (genotype $w^-/w^-/Y$), and the exceptional red-eyed males would contain a single X chromosome and no Y chromosome (genotype w^+).

Bridges's hypothesis for the exceptional progeny in Fig. 17.6b was bold, as it assumed that *Drosophila* males could develop in the absence of a Y chromosome (*XO* embryos yielding sterile males, where "*O*" indicates absence of a chromosome), and that females could develop in the presence of a Y chromosome (*XXY* embryos yielding fertile females). The hypothesis was accurate as well as bold. Microscopic examination of the chromosomes in the exceptional fruit flies confirmed that the exceptional white-eyed females had *XXY* sex chromosomes and that the exceptional sterile red-eyed males had an X but no Y. Because fruit flies with three X chromosomes (*XXX*) or no X chromosome (*OY*) were never observed, Bridges concluded that embryos with these chromosomal constitutions are unable to survive. Bridges also conducted crosses that showed that nondisjunction can take place in males as well as in females. From the phenotypes of these exceptional fruit flies and their chromosome constitutions, Bridges concluded that the white-eye gene (and by implication any other X-linked gene) is physically present in the X chromosome.

Bridges's demonstration that genes are present in chromosomes was also the first experimental evidence of nondisjunction. *Drosophila* differ from humans in that the Y chromosome is necessary for male fertility but not for male development. As we will see later in this chapter, a gene in the Y chromosome itself is the trigger for male development in humans and other mammals, and so for these organisms, the Y chromosome is needed both for male development and male fertility. Nondisjunction occasionally takes place in meiosis in humans as well as in fruit flies. When nondisjunction takes place in the human sex chromosomes, it results in chromosomal constitutions such as 47, *XXY* and 47, *XYY* males as well as 47, *XXX* and 45, *X* females. Nondisjunction of autosomes can also occur, resulting in fetuses that have extra copies or missing copies of entire chromosomes. The consequences of nondisjunction of human chromosomes were examined in Chapter 15.

Genes in the *X* chromosome show characteristic patterns in human pedigrees.

The features of X-linked inheritance can be seen in human pedigrees for traits due to an X-linked recessive mutation. These are illustrated in **Fig.** 17.7 for red–green color blindness, a condition that affects about 1 in 20 males. An individual with red–green color blindness will have difficulty seeing the number in the colored dots in Fig. 17.7.

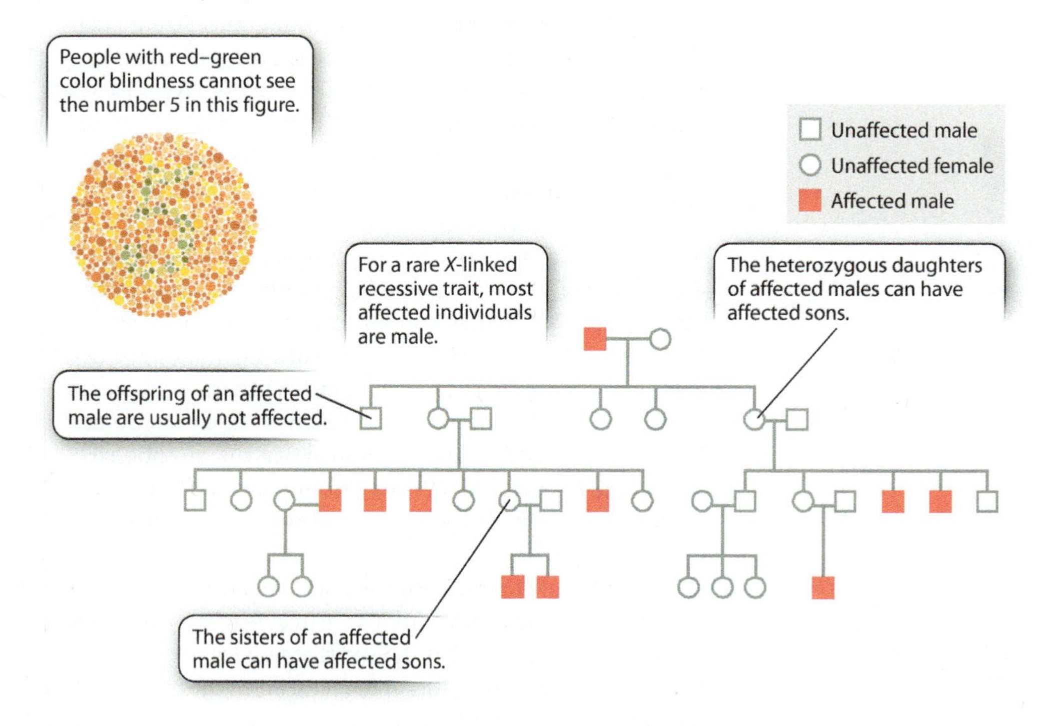

FIG. 17.7 Inheritance of an *X*-linked recessive mutation. This pedigree shows the inheritance of red–green color blindness, an *X*-linked recessive mutation. *Source: Dorling Kindersley/ Getty Images.*

The key features of the inheritance of traits due to rare *X*-linked recessive alleles, which are noted in the pedigree, are listed here:

1. Affected individuals are usually males because males need only one copy of the mutant gene to be affected, whereas females need two copies to be affected.
2. Affected males have unaffected sons because males transmit their *X* chromosome only to their daughters.
3. A female whose father is affected can have affected sons because such a female must be a heterozygous carrier of the recessive mutant allele.

An additional feature worth mentioning is that the sisters of an affected male each have a 50% chance of being a heterozygous carrier because when a brother is affected the mother must be heterozygous for the recessive allele.

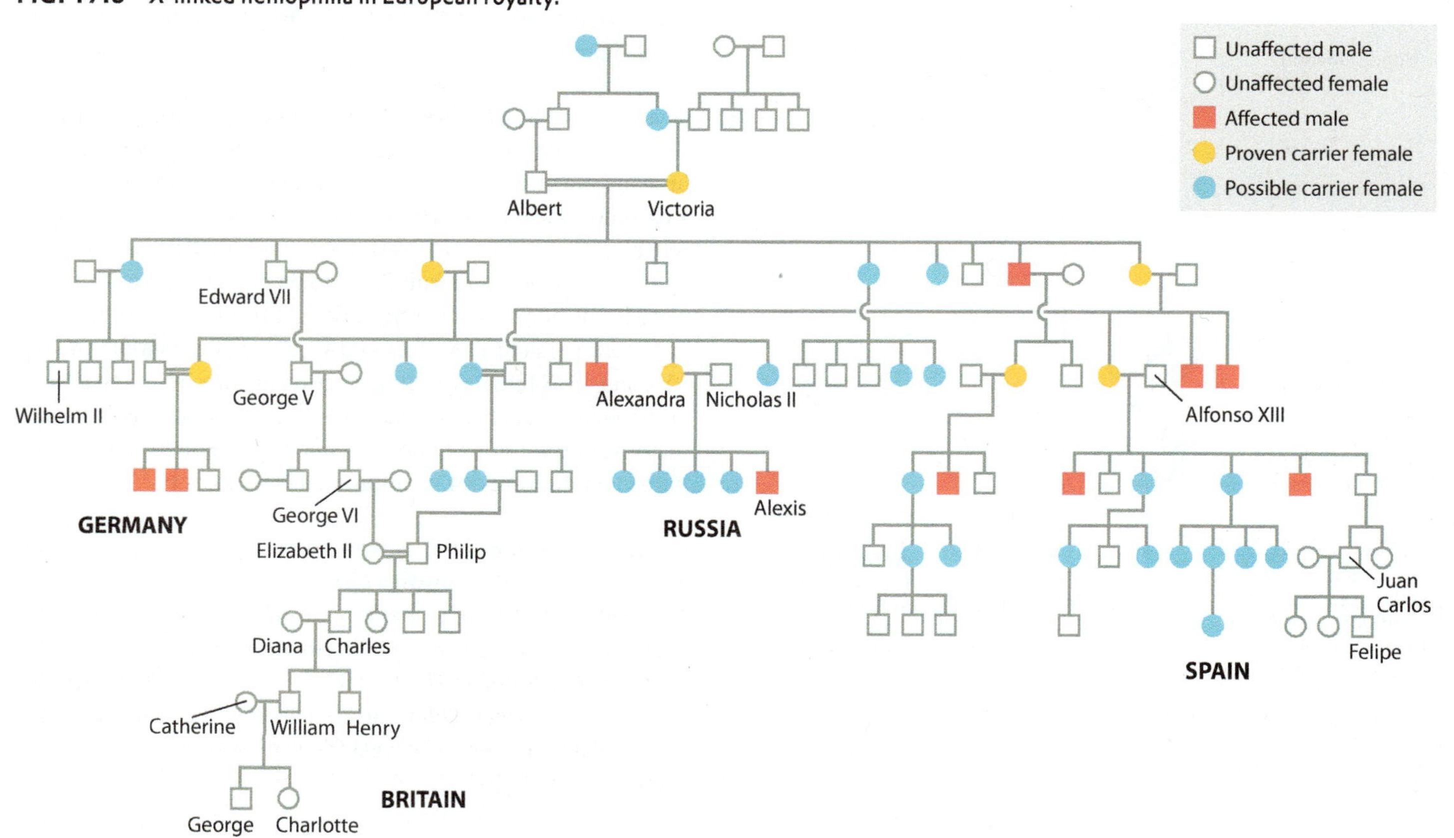

FIG. 17.8 *X*-linked hemophilia in European royalty.

→ **Quick Check 1** Is it possible for an unaffected female to have female offspring with red–green color blindness?

A pedigree for one of the most famous examples of human *X*-linked inheritance is shown in **Fig. 17.8.** The trait is a form of **hemophilia,** which results from a recessive mutation in a gene encoding a protein necessary for blood clotting. Affected individuals bleed excessively from even minor cuts and bruises, and internal bleeding can cause excruciating pain. Affecting about 1 in 7000 males, hemophilia is famous because of its presence in many members of European royalty descended from Queen Victoria of England (1819–1901), who was a heterozygous carrier of the gene. By the marriages of her carrier granddaughters, the gene was introduced into the royal houses of Germany, Russia, and Spain. The mutant allele is not present in the present royal family of Britain, however, because this family descends from King Edward VII, one of Victoria's four sons, who was not himself affected and therefore passed only a normal *X* chromosome to his descendants.

The source of Queen Victoria's hemophilia mutation is not known. None of her ancestors is reported as having a bleeding disorder. Quite possibly the mutation was present for a few generations before Victoria was born but remained hidden because it was passed from heterozygous female to heterozygous female.

17.3 GENETIC LINKAGE AND RECOMBINATION

Mendel was fortunate not only because peas do not possess sex chromosomes, but also because the genes that influenced the traits he studied, such as round/wrinkled and yellow/green seeds, are on separate chromosomes or far apart on the same chromosome. What happens when genes are close to each other in the same chromosome? We explore the answer to this question in this section.

Nearby genes in the same chromosome show linkage.

Genes that are sufficiently close together in the same chromosome are said to be **linked.** That is, they tend to be transmitted together in inheritance and do not assort independently of each other as Mendel observed. Note that *linked genes* refer to two genes that are close together in the same chromosome, which may be an autosome or sex chromosome. This is not to be confused with an *X-linked gene,* which is simply one that is present in the *X* chromosome.

Linkage was discovered in *Drosophila* by Alfred H. Sturtevant, another of Morgan's students. Once again, genes in the *X* chromosome played a key role in the discovery because the phenotype of the males reveals their genotype as in a testcross with an autosomal recessive. An example is shown in **Fig. 17.9.** Sturtevant worked with male fruit flies that have an *X* chromosome carrying two recessive mutations. One is in the *white* gene (*w*) discussed earlier, which when nonmutant results

FIG. 17.9 Linkage of the *white* (*w*) and *crossveinless* (*cv*) genes in the *X* chromosome. Genes are written according to their order along the chromosome, with a horizontal line between homologous chromosomes, in this case the *X* and *Y*. Crossveins are shown in red for clarity.

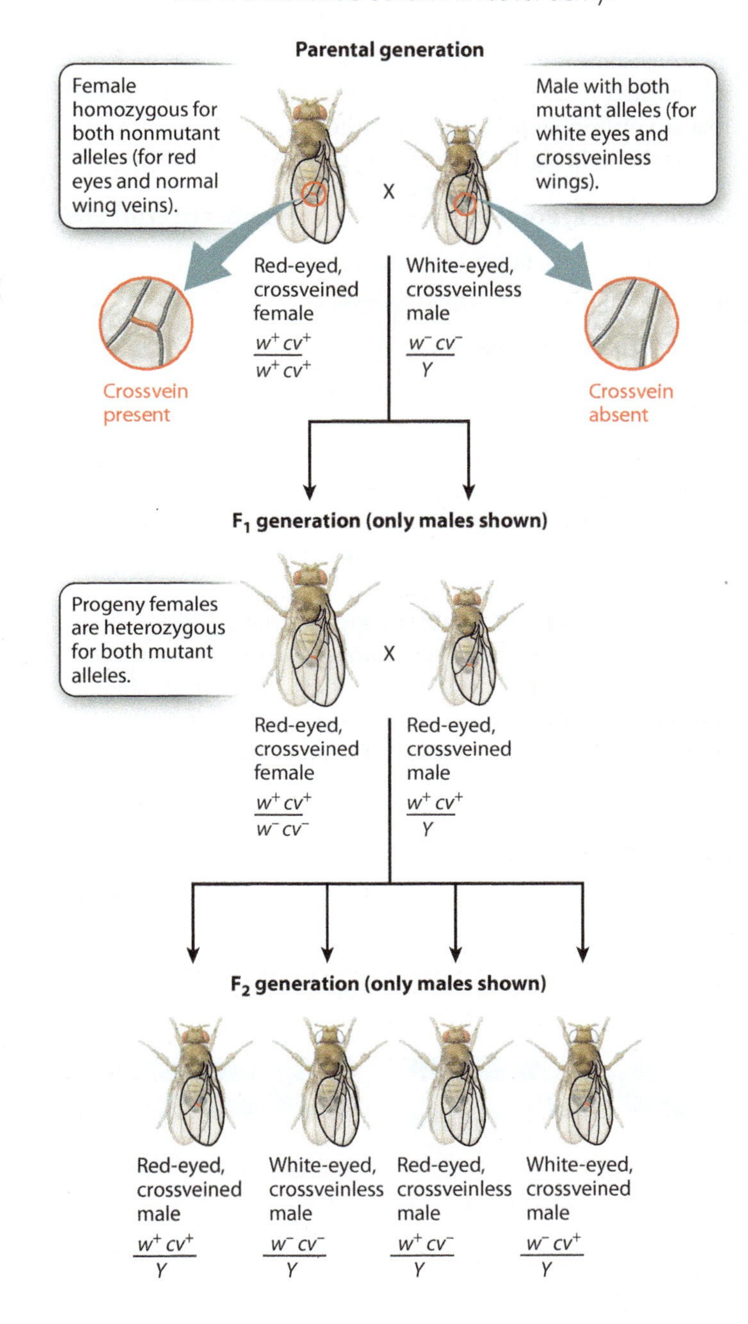

in fruit flies with red eyes and when mutant results in fruit flies with white eyes. The other recessive mutation is in a gene called *crossveinless* (*cv*), which when nonmutant results in fruit flies with tiny crossveins in the wings and when mutant results in the absence of these crossveins.

Sturtevant crossed this doubly mutant male with a female carrying the nonmutant forms of the genes (w^+ and cv^+) in both *X* chromosomes. He saw that the offspring consist of phenotypically wild-type females that are heterozygous for both genes, and phenotypically wild-type males. When these are crossed with each other, the female F_2 progeny do not tell us anything because they are all wild type; each female receives the w^+cv^+ X chromosome from her father and therefore has red eyes and normal crossveins. In the male F_2 progeny, however, the situation is different: Each male progeny receives its X chromosome from the mother and its Y chromosome from the father, and so the phenotype of each male immediately reveals the genetic constitution of the X chromosome that the male inherited from the mother.

As shown in Fig. 17.9, the male F_2 progeny consist of four types:

Genotype of F_2 Progeny	Number of Fruit Flies
w^+cv^+/Y (red eyes, normal crossveins)	357
w^-cv^-/Y (white eyes, missing crossveins)	341
w^+cv^-/Y (red eyes, missing crossveins)	52
w^-cv^+/Y (white eyes, normal crossveins)	45

Although all four possible classes of maternal gametes are observed in the male progeny, they do not appear in the ratio 1:1:1:1 expected when gametes contain two independently assorting genes (Chapter 16). The lack of independent assortment means that the genes show linkage.

The male progeny fall into two groups. One group, represented by larger numbers of progeny, derives from maternal gametes containing either w^+cv^+ or w^-cv^-. These are called **nonrecombinants** because the alleles are present in the same combination as that in the parent. The other group of male progeny consists of w^+cv^- and w^-cv^+ combinations of alleles. These are called **recombinants,** and they result from a **crossover,** the physical exchange of parts of homologous chromosomes, which takes place in prophase I of meiosis (Chapter 11).

Crossing over is a key process in meiosis (Chapter 11). Most chromosomes have one or more crossovers that form between the homologous chromosomes as they pair and undergo meiosis. Human females average about 2.75 crossovers per chromosome pair, and human males average about 2.50 crossovers per chromosome pair. As noted earlier for the X and Y chromosomes, crossovers between homologous chromosomes are important mechanically because they help hold the homologs together so they can align properly at metaphase I and segregate to opposite poles at anaphase I.

Fig. 17.10 shows how recombinant chromosomes arise from crossing over between genes, using the hypothetical genes *A* and *B*. In a cell undergoing meiosis in which a crossover takes place between the genes, the allele combinations are broken up in the chromatids involved in the exchange, and the resulting gametes are *AB*, *Ab*, *aB*, and *ab* (Fig. 17.10a).

Note that crossing over does not result only in recombinant chromosomes. Fig. 17.10a shows that, even when a crossover occurs in the interval between the genes, two of the resulting chromosomes contain the nonrecombinant configuration of alleles. These nonrecombinant configurations occur because crossing over occurs at the four-strand stage of meiosis (when each homologous chromosome is a pair of sister chromatids), but only two of the four strands (one sister chromatid from each homologous chromosome) are included in any crossover.

FIG. 17.10 Linkage and recombination. (a) Crossing over between genes results in recombination between the alleles of the genes. (b) When crossing over occurs outside the interval between genes, there is no recombination between the alleles of the genes.

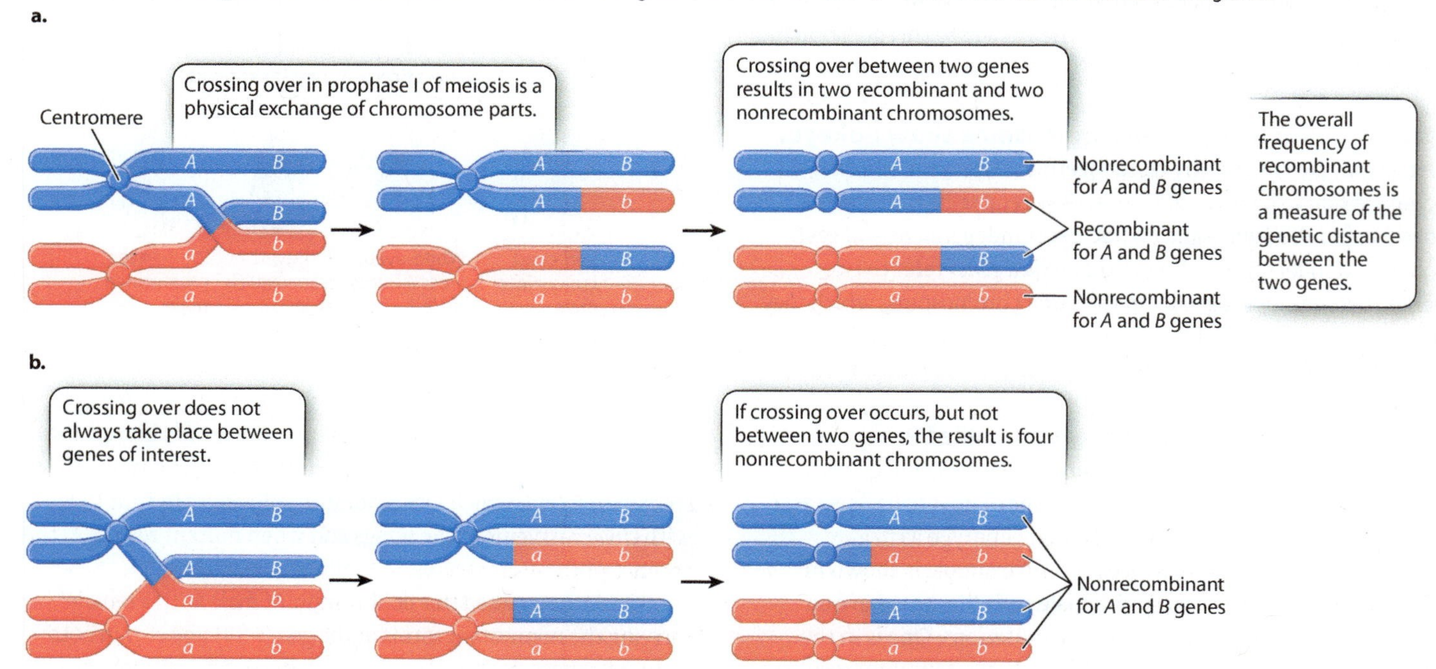

Fig. 17.10b shows a second way in which nonrecombinant chromosomes originate. When two genes are close together in a chromosome, a crossover may not occur in the interval between the genes but at some other location along the chromosome. In this example, the crossover occurs in the region between gene *A* and the centromere, and therefore the combination of alleles *A* and *B* remains intact in one pair of chromatids and the combination of alleles *a* and *b* remains intact in the homologous chromatids. The resulting gametes are equally likely to have either *AB* or *ab*, both of which are nonrecombinant.

Nonrecombinant chromosomes can also be the result of two crossover events occurring between two genes, in which the effects of the first crossover (creating recombinants) is reversed by a second crossover (re-creating nonrecombinants).

The frequency of recombination is a measure of the genetic distance between linked genes.

When two genes are on separate chromosomes, a ratio of 1:1:1:1 is expected for the nonrecombinant (parental) and recombinant (nonparental) gametic types, as described by the principle of independent assortment (Chapter 16). For two genes present in the same chromosome, we can consider two extreme situations. If they are located very far apart from each other, one or more crossovers will almost certainly occur between them, and there will be a 1:1:1:1 ratio of nonrecombinant and recombinant gametes (as shown for the case of a single crossover in Fig. 17.10a). At the other extreme, if two genes are so close together that crossing over never takes place between them, we would expect only nonrecombinant chromosomes (Fig. 17.10b).

What happens in between these extremes? In these cases, in some cells undergoing meiosis, no crossover takes place between the genes, in which case all the resulting chromosomes are nonrecombinant (Fig. 17.10b); and in other cells undergoing meiosis, a crossover occurs between the genes, in which case half the resulting chromosomes are nonrecombinant and half are recombinant (Fig. 17.10a). Since the meioses with no crossover between the genes result only in nonrecombinant chromosomes and those with a crossover result in half nonrecombinant and half recombinant chromosomes, the nonrecombinant chromosomes in the offspring will be more numerous than the recombinant chromosomes.

The actual frequency of recombinants depends on the distance between the genes. The distance between the genes is important because whether or not a crossover occurs between the genes is a matter of chance, and the closer the genes are along the chromosome, the less likely it is that a crossover will take place in the interval between them. Because the formation of recombinant chromosomes requires at least one crossover between the genes, genes that are close together (more tightly linked) show less recombination than genes that are far apart. In fact, the proportion of recombinant chromosomes observed among the total, which is called the **frequency of recombination,** is a measure of genetic distance between the genes along the chromosome.

In the example with the genes *w* and *cv*, the total number of chromosomes observed among the progeny is $357 + 341 + 52 + 45 = 795$, and the number of recombinant chromosomes is $52 + 45 = 97$. The frequency of recombination between *w* and *cv* is therefore $97/795 = 0.122$, or 12.2%, and this serves as a measure of the genetic distance between the genes. In studies of genetic linkage, the distance between genes is not measured directly by physical distance between them, but rather by the frequency of recombination.

The frequency of recombination between any two genes on the same chromosome ranges from 0% (when crossing over between the genes never takes place) to 50% (when the genes are so far apart that a crossover between the genes almost always takes place). Genes that are linked have a recombination frequency somewhere between 0% and 50%. The maximum frequency of recombination is 50% because, when nonsister chromatids involved in crossovers are chosen at random, any new crossover has two equally likely consequences: It can either change a previously recombinant chromatid into a nonrecombinant chromatid, or change a previously nonrecombinant chromatid into a recombinant chromatid. The result is that however many crossovers there may be (as long as there is at least one), the maximum frequency of recombination remains 50%. A frequency of recombination of 50% yields the same ratio of gametic types as observed with independent assortment, which means that genes that are far enough apart in the same chromosome show independent assortment.

→ **Quick Check 2** Why is the upper limit of recombination 50% rather than 100%?

Recombination plays an important role in creating new combinations of alleles in each generation and in ensuring the genetic uniqueness of each individual. If there were no recombination (that is, if all the alleles in each chromosome were completely linked), any individual human would be able to produce only $2^{23} = 8.4$ million types of reproductive cells. While this is a large number, the average number of sperm per ejaculate is much larger—approximately 350 million. Because recombination does occur, and because the crossovers resulting in recombination can occur at any of thousands of different positions in the genome, each of the 350 million sperm is virtually certain to carry a different combination of alleles.

Genetic mapping assigns a location to each gene along a chromosome.

With the exception of a few regions, such as the area near the centromere, the likelihood of a crossover occurring somewhere between two points on a chromosome is approximately proportional to the length of the interval between the points. Therefore, the frequency of recombination can be used as a measure of the physical distance between genes. These distances are used in the construction of a **genetic map,** which is a diagram showing the relative position of genes along a chromosome. The maps are drawn using a scale in which one unit of distance

HOW DO WE KNOW?

FIG. 17.11

Can recombination be used to construct a genetic map of a chromosome?

BACKGROUND In 1910, Thomas Hunt Morgan discovered *X*-linkage by studying the white-eye mutation in *Drosophila*. Soon other *X*-linked mutations were found. Alfred H. Sturtevant, Morgan's student, decided to test whether mutant genes in the same *X* chromosome were inherited together, that is, linked. He found that genes in the *X* chromosome were linked, but not completely, and that different pairs of genes showed great differences in their linkage. Some genes showed almost no recombination, whereas others underwent so much recombination that they showed independent assortment.

HYPHOTHESIS Sturtevant hypothesized that recombination was due to crossing over between the genes, and that genes farther apart in the chromosome would show more recombination.

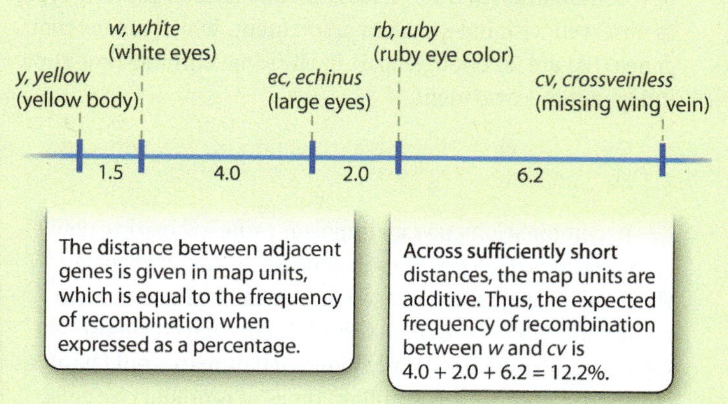

EXPERIMENT Taking this idea a step further, Sturtevant reasoned that if one knew the frequency of recombination between genes *a* and *b*, between *b* and *c*, and between *a* and *c*, then one should be able to deduce the order of genes along the chromosome. He also predicted that, if the order of genes were known to be *a*–*b*–*c*, then if the genes were close enough, the frequency of recombination between *a* and *c* should equal the sum of the frequencies between *a* and *b* and that between *b* and *c*.

RESULTS Sturtevant studied the frequencies of recombination between many pairs of genes along the *X* chromosome, including some of those shown in the illustration shown here.

CONCLUSION The results confirmed Sturtevant's hypothesis and showed that genes could be arranged in the form of a genetic map, depicting their linear order along the chromosome, with the distance between any pair of genes proportional to the frequency of recombination between them. Across sufficiently short regions, the frequencies of recombination are additive.

FOLLOW-UP WORK Genetic mapping remains a cornerstone of genetic analysis, showing which chromosome contains a mutant gene and where along the chromosome the gene is located. The method helped to identify the genes responsible for many single-gene inherited disorders, including Huntington's disease, cystic fibrosis, and muscular dystrophy.

SOURCE Sturtevant, A. H. 1913. "The Linear Arrangement of Six Sex-Linked Factors in *Drosophila*, as Shown by Their Mode of Association." *Journal of Experimental Zoology* 14:43–59.

(called a **map unit**) is the distance between genes resulting in 1% recombination. Thus, in a *Drosophila* genetic map containing the genes *w* and *cv*, the distance between the genes is 12.2 map units.

Genetic maps are built up step by step as new genes are discovered that are genetically linked, as shown in **Fig. 17.11.** Across distances that are less than about 15 map units, the map distances are approximately additive, which means that the distances between adjacent genes can be added to get the distance between the genes at the ends. For example, in Fig. 17.11, there are two genes between *w* and *cv*. The map distance between *w* and the next gene, *ec*, is 4.0, the distance between *ec* and the next gene, *rb*, is 2.0, and the distance between *rb* and *cv* is 6.2. The map distance between *w* and *cv* is therefore 4.0 + 2.0 + 6.2 = 12.2 map units, and hence the expected frequency of recombination between these genes is 12.2%, which is the value observed.

However, for two genes that are farther apart than about 15 map units, the observed recombination frequency is somewhat smaller than the sum of the map distances between the genes. The reason is that, with greater distances, two or more crossovers between the genes may occur in the same chromosome, and thus an exchange produced by one crossover may be reversed by another crossover farther along the way.

→ **Quick Check 3** For two genes that show independent assortment, what is the frequency of recombination?

Genetic risk factors for disease can be localized by genetic mapping.

The discovery of abundant genetic variation in DNA sequences in human populations, such as single-nucleotide polymorphisms

(Chapter 15), made it possible to study genetic linkage in the human genome. At first, the focus was on finding mutations that cause disease, such as the mutation that causes cystic fibrosis (Chapter 14). The method was to study large families extending over three or more generations in which the disease was present and then to identify the genotypes of each of the individuals for thousands of genetic markers (previously discovered DNA polymorphisms) throughout the genome. The goal was to find genetic markers that showed a statistical association with the disease gene, which would indicate genetic linkage and reveal the approximate location of the disease gene along the chromosome.

Fig. 17.12 illustrates the underlying concept. It assumes 100 chromosomes observed among different individuals in a pedigree, of which 50 carry a mutant allele of a gene and 50 carry a nonmutant allele. These chromosomes are tested for a marker of known location, in this case a single-nucleotide polymorphism (SNP) in which one of the nucleotide pairs in the DNA is a G–C base pair in some chromosomes and A–T in others. In Fig. 17.12a, there is clearly an association between the disease gene and the SNP. Almost all of the chromosomes that carry the mutant allele show the G–C nucleotide pair, whereas almost all of the chromosomes that carry the nonmutant allele show the A–T nucleotide pair. The two chromosomes in which the mutant and nonmutant genes are associated with the other SNPs can be attributed to recombination.

FIG. 17.12 Genetic mapping. SNPs associated with a mutant gene show where that gene is located in the genetic map.

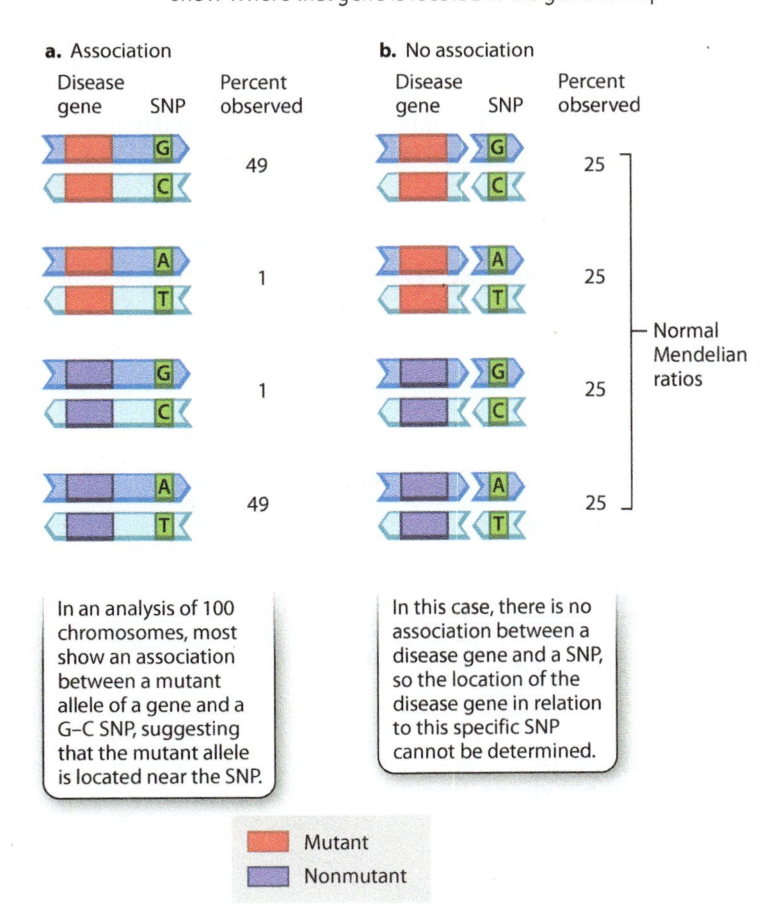

The association in Fig. 17.12a may be contrasted with the pattern in Fig. 17.12b, in which there is no association. In this case, each of the alleles of the disease gene is equally likely to carry either form of the SNP. The failure to find an association means that the SNP is not closely linked to the disease gene, and may in fact be in a different chromosome. In actual studies, an association is almost never observed, but the lucky find of an association helps identify the location of the disease gene in the genetic map. Using such association methods, hundreds of important disease genes have been located by genetic mapping. Once the location of the disease gene is known, the identity and normal function of the gene can be determined.

17.4 INHERITANCE OF GENES IN THE *Y* CHROMOSOME

Like the X chromosome, the Y chromosome exhibits a particular pattern of inheritance because of its association with the male sex. In humans and other mammals, all embryos initially develop immature internal sexual structures of both females and males. *SRY*, a gene in the Y chromosome, encodes a protein that is the trigger for male development. ("*SRY*" stands for "sex-determining region in the Y chromosome.") In the presence of *SRY*, male structures complete their development and female structures degenerate. In the absence of *SRY*, male embryonic structures degenerate and female structures complete their development. The *SRY* gene is therefore the male-determining gene in humans and other mammals. Because they are linked to *SRY*, most genes in the Y chromosome show a distinctive pattern of inheritance in pedigrees, different from the patterns of autosomal genes Mendel observed in his pea plants, in that they are transmitted only from father to son.

Y-linked genes are transmitted from father to son to grandson.

Genes that are present in the unique region of the Y chromosome (the part that cannot cross over with the X) are known as **Y-linked genes,** of which there are not many. As well as the *SRY* male-determining gene, they include a number of genes in which mutations are associated with impaired fertility and low sperm count.

The pedigree characteristics of Y-linked inheritance are striking (**Fig. 17.13**):

1. Only males are affected with the trait.
2. Females never inherit or transmit the trait, regardless of how many affected male relatives they have.
3. All sons of affected males are also affected.

FIG. 17.13 Inheritance of *Y*-linked traits. The pedigree pattern is that of father to son to grandson to great-grandson, and so forth.

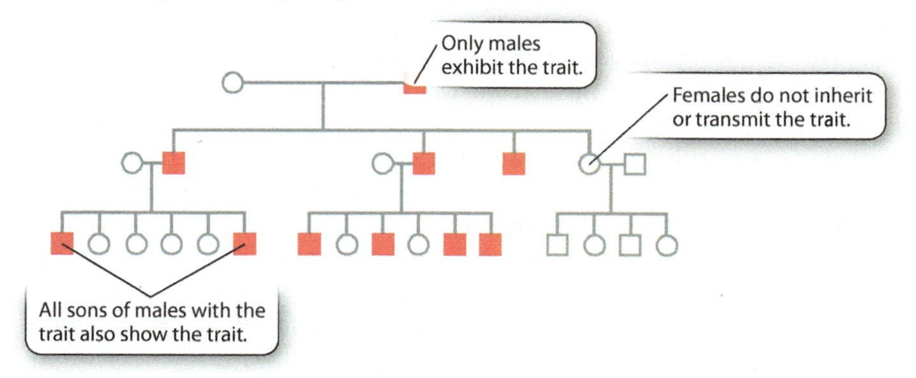

Because the Y chromosome is always transmitted from father to son (and never transmitted to daughters), a trait determined by a Y-linked gene will occur in fathers, sons, grandsons, and so forth. Traits resulting from Y-linked genes cannot be present in females nor can they be transmitted by females. However, other than maleness itself and some types of impaired fertility, no physical traits are known that follow a strict Y-linked pattern of inheritance. This observation emphasizes the extremely low density of functional genes in the Y chromosome.

? CASE 3 YOU, FROM A TO T: YOUR PERSONAL GENOME

How can the *Y* chromosome be used to trace ancestry?

The example of Claudia Gilmore illustrates how, through tests of her personal genome with regard to the *BRCA1* and *BRCA2* mutations, she became aware of her elevated risk of breast cancer. Your personal genome not only can tell you about your genetic risk factors for disease, but it also contains important information about your genetic ancestry. For example, your male ancestors can be traced through your Y chromosome. The regions at the tips in which the X and Y chromosomes share homology is only about 6% of the entire length of the Y chromosome. This means that 94% of the Y chromosome consists of sequences that are completely linked with one another because that portion of the chromosome does not pair with another chromosome and does not undergo crossing over.

Because of this complete linkage, each hereditary lineage of Y chromosomes is separate from every other lineage. As mutations occur along the Y chromosome, they are completely linked to any past mutations that may be present and also completely linked to any future mutations that may take place. The mutations therefore accumulate, and this allows the evolutionary history of a set of sequences to be reconstructed.

Fig. 17.14 shows an evolutionary tree based on the accumulation of mutations at a set of nucleotide sites along the Y chromosome. Each unique combination of nucleotides constitutes a Y-chromosome **haplotype,** or haploid genotype. In the figure, the most ancient Y chromosomes are at the top, and the accumulation of new mutations as the generations proceed results in the successive creation of new haplotypes. Each Y-chromosome lineage may leave some nonmutant descendants as well as some mutant descendants, and hence any or all of the sequences shown may coexist in a present-day population.

In human history, the mutations creating new Y-chromosome haplotypes were occurring at the same time as populations were migrating and founding new settlements across the globe, and so each geographically distinct population came to have a somewhat different set of Y-chromosome haplotypes. The differences among populations are offset to some extent by migration among populations, which mixes the geographical locations of various haplotypes. Nevertheless, the fact that the mutations accumulate through time and are completely linked allows the evolutionary history of the haplotypes to be reconstructed. It also enables Y chromosomes to be traced to their likely ethnic origin.

The worldwide distribution of real Y-chromosome lineages among human populations is shown in **Fig. 17.15.** The different colors in the pie charts represent different haplotypes. Neighboring populations tend to have more closely related Y chromosomes than more distant populations. Four major clusters of Y-chromosome lineages can be recognized in Fig. 17.15. One is concentrated in Africa, another in Southeast Asia and Australia, a third in Europe and central and western Asia, and the fourth in North and South America. These clusters correspond roughly with the spread of human settlements around the globe inferred from archaeological evidence.

FIG. 17.14 *Y*-chromosome haplotypes. Because *Y* chromosomes in a lineage conform to an evolutionary tree, the ancestry of a male's *Y* chromosome can be traced.

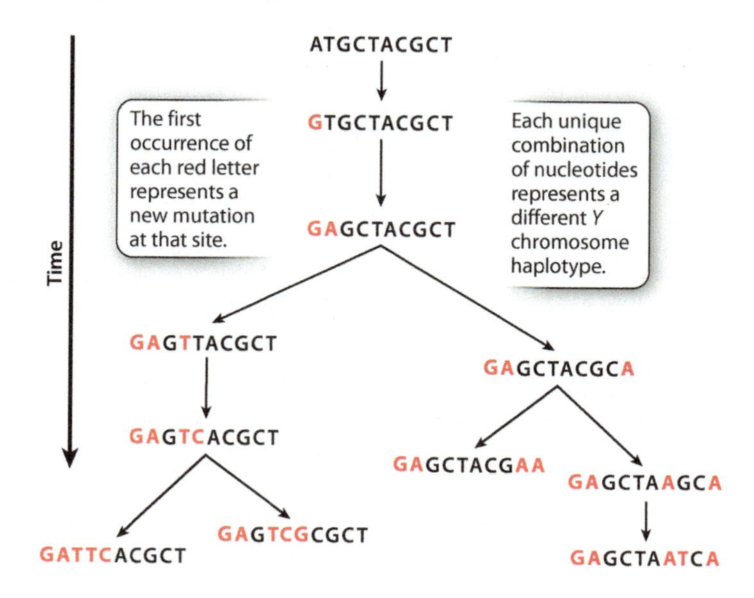

FIG. 17.15 Geographical distribution of *Y*-chromosome haplotypes among native populations. The evolutionary trees of *Y*-chromosome haplotypes reflect the origin and movement of different *Y* chromosomes over time. *Data from M. A. Jobling and C. Tyler-Smith, 2003, "The Human Y Chromosome: An Evolutionary Marker Comes of Age," Nature Reviews Genetics 4:598–612.*

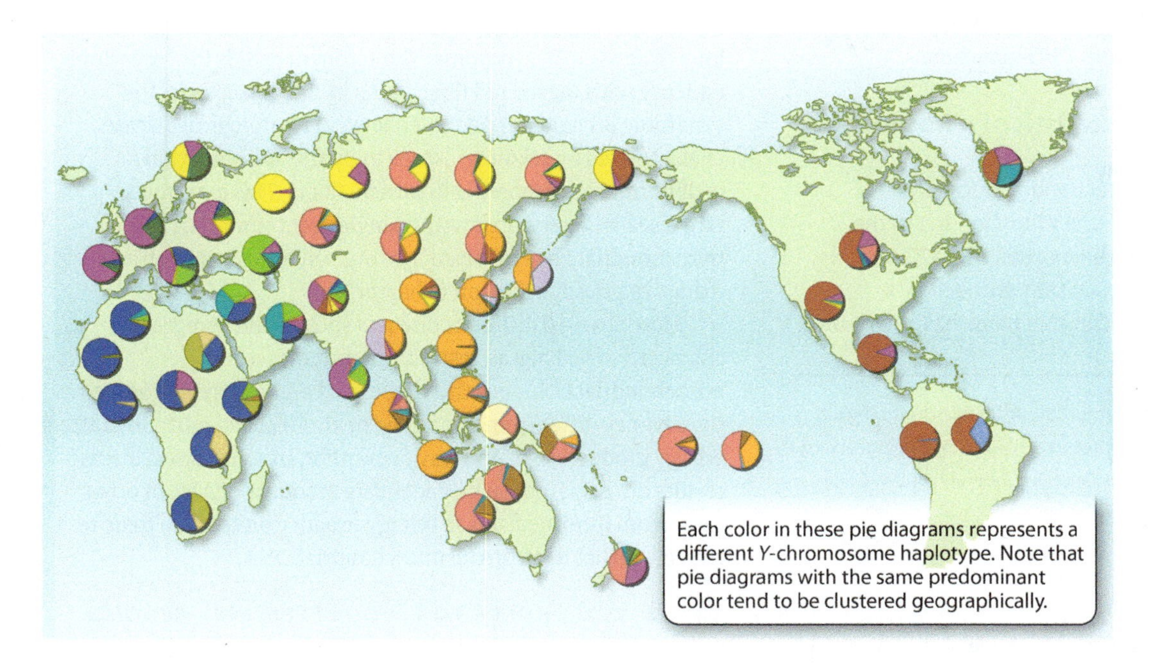

The implication of Fig. 17.15 is that the haplotype of your Y chromosome (if you have a Y chromosome) contains genetic information about its origin. And you can learn what this information is from genetic testing companies that sell direct-to-consumer (DTC) services. Their tests are not regarded as medical devices and so are unregulated, and quality control is sometimes uncertain. Nevertheless, you can send saliva or other biological samples to a DTC provider, which will (for a fee) test your Y chromosome and send you a report that details its possible origin. Of course, because of recombination and independent assortment of genes in other chromosomes, the ethnic origin of your Y chromosome may have little or nothing to do with the ethnic origins of genes in any of your other chromosomes.

17.5 INHERITANCE OF MITOCHONDRIAL AND CHLOROPLAST DNA

Sex chromosomes and linked genes are not the only genes that are inherited in ways that are unlike the patterns of inheritance that Mendel observed in peas. Genes in mitochondria and chloroplasts also show distinct inheritance patterns. Mitochondria and chloroplasts are ancient organelles of eukaryotic cells originally acquired by the engulfing of prokaryotic cells (Chapter 5). Mitochondria generate ATP that cells use for their chemical energy. Chloroplasts are found only in plant cells and eukaryotic algae. These organelles contain chlorophyll, a green pigment that absorbs light and, in the process of photosynthesis (Chapter 8), produces the sugars that are essential for growth of the plant or algal cells and of the organisms that eat them. Mitochondria and chloroplasts have their own genomes that contain genes for many of the enzymes that carry out the organelles' functions. Genes present in these genomes move with the organelle during cell division, independent of the segregation of chromosomes in the nucleus.

Mitochondrial and chloroplast genomes often show uniparental inheritance.

During sexual reproduction, organelles do not show the regular, highly choreographed movements that chromosomes undergo during Mendelian segregation. Organelles are partitioned to the gametes along with other cytoplasmic components, and therefore their mode of inheritance depends on how the gametes are formed, how much cytoplasm is included in the gametes, and the fate of the cytoplasm in each parental gamete after fertilization.

Considering the great diversity in the details of reproduction in different groups of organisms, it is not surprising that there is a diversity of types of inheritance of cytoplasmic organelles. The three most important types are:

- **Maternal inheritance,** in which the organelles in the offspring cells derive from those in the mother.
- **Paternal inheritance,** in which the organelles in the offspring cells derive from those in the father.
- **Biparental inheritance,** in which the organelles in the offspring cells derive from those in both parents.

In most organisms, either maternal inheritance or paternal inheritance predominates, but sometimes there is variation from one offspring to the next. For example, transmission of the chloroplasts ranges from strictly paternal in the giant redwood *Sequoia,* to strictly maternal in the sunflower *Helianthus,* to either maternal or paternal (or less frequently biparental) in the fern *Scolopendrium,* to mostly maternal but sometimes paternal or biparental in the snapdragon *Antirrhinum.*

There is likewise great diversity among organisms in the inheritance of mitochondrial DNA. Most animals show maternal transmission of the mitochondria, as would be expected from their large, cytoplasm-rich eggs and the small, cytoplasm-poor sperm. Among other organisms, there is again much variation, including maternal transmission of mitochondria in flowering plants and paternal transmission in the green alga *Chlamydomonas.*

Maternal inheritance is characteristic of mitochondrial diseases.

In humans and other mammals, mitochondria normally show strictly maternal inheritance—the mitochondria in the offspring cells derive from those in the mother. **Fig. 17.16** shows the characteristic pedigree patterns of a trait encoded by a mitochondrial genome transmitted through maternal inheritance:

FIG. 17.16 **Maternal inheritance.** Human mitochondrial DNA is transmitted from a mother to all of her offspring. *Source: Courtesy of Dr. Kurenai Tanji, Columbia University Medical Center, New York, NY.*

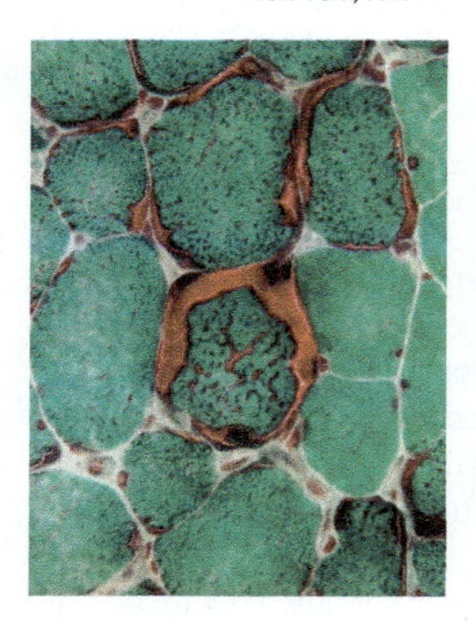

Inherited mitochondrial diseases are often associated with muscle weakness reflecting deficient production of ATP. The red patches in the microscopic image result from clumps of defective mitochondria in muscle fibers observed in one form of epilepsy due to mutation in mitochondrial DNA.

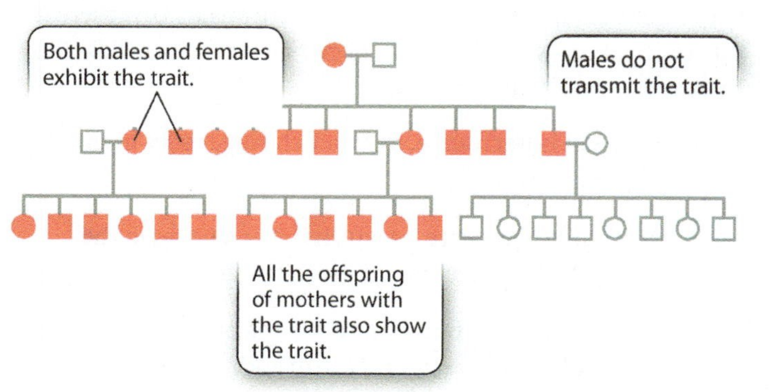

1. Both males and females can show the trait.
2. All offspring from an affected female show the trait.
3. Males do not transmit the trait to their offspring.

The pedigree in Fig. 17.16 follows a mitochondrial disease known as MERRF syndrome (the acronym stands for "myoclonic epilepsy with ragged red fibers"). As its name suggests, the syndrome is characterized by epilepsy, a neurological disease characterized by seizures, as well as by the accumulation of abnormal mitochondria in muscle fibers. This extremely rare disease is associated with a single point mutation in a mitochondrial gene involved in protein synthesis that affects oxidative phosphorylation (Chapter 7).

More than 40 different diseases show these pedigree characteristics. They all result from mutations in the mitochondrial DNA, but the tissues and organs affected as well as the severity differ from one to the next. All of the mutations affect energy production in one way or another. In some cases, disease results directly from lack of adequate amounts of ATP, in other cases from intermediates in energy production that are toxic to the cell or that damage the mitochondrial DNA.

? CASE 3 YOU, FROM A TO T: YOUR PERSONAL GENOME

How can mitochondrial DNA be used to trace ancestry?

Sequencing of mitochondrial DNA reveals mitochondrial haplotypes analogous to those in the Y chromosome. As in the Y chromosome, mutations in mitochondrial DNA accumulate through time, and each mitochondrial haplotype remains intact through successive generations because no recombination between mitochondrial genomes takes place. This means that mitochondrial DNA can be used to trace ancestry and population history, much as described previously for the Y chromosome. The mitochondrial DNA is actually more informative than Y-chromosomal DNA because it shows substantially more genetic variation so its ancestry can be tracked on a finer scale.

The use of mitochondrial DNA to trace human origins and migration is discussed in greater detail in Chapter 24. As with the Y chromosome, your personal mitochondrial genome includes information about its origin. If you want to learn about your mitochondrial DNA, you can send a tissue sample to any of a number of direct-to-consumer genotyping services that will, for a fee, analyze the DNA and provide you with a report. ■

Core Concepts Summary

17.1 Many organisms have a distinctive pair of chromosomes, often called the *X* and *Y* chromosomes, that differ between the sexes and show different patterns of inheritance in pedigrees from other chromosomes.

In humans and other mammals, *XX* individuals are female and *XY* individuals are male. page 346

The human *X* and *Y* chromosomes are different lengths and contain different genes, except for small regions of homology that allow the two chromosomes to pair in meiosis. page 347

Segregation of the *X* and *Y* chromosomes during male meiosis results in half of the sperm receiving an *X* chromosome and half a *Y* chromosome so that random union of gametes predicts a 1:1 female:male sex ratio at the time of fertilization. page 347

17.2 *X*-linked genes, which show a crisscross inheritance pattern, provided the first evidence that genes are present in chromosomes.

Thomas Hunt Morgan studied a mutation in the fruit fly *Drosophila melanogaster* that resulted in fruit flies with white eyes rather than wild-type red eyes. In this species, as in mammals, females are *XX* and males are *XY*. page 348

In a cross of a normal red-eyed female with a mutant white-eyed male, all of the male and female progeny had red eyes. When brothers and sisters of this cross were mated with each other, all of the females had red eyes, but males were red-eyed and white-eyed in a 1:1 ratio. page 348

This pattern of inheritance is observed because the gene Morgan studied is located in the *X* chromosome. The nonmutant w^+ allele is dominant to the mutant w^- allele, and the gene is present only in the *X* chromosome and not in the *Y* chromosome. page 348

X-linked genes show a crisscross inheritance pattern, in which the *X* chromosome with the mutant gene that is present in males in one generation is present in females in the next generation. page 349

Calvin Bridges observed rare fruit flies that did not follow the usual pattern for *X*-linked inheritance and inferred that these exceptional fruit flies resulted from nondisjunction, or failure of homologous chromosomes to segregate, in male or female meiosis. His observations provided evidence that genes are carried in chromosomes. page 350

In humans, *X*-linked inheritance shows a pattern in which affected individuals are almost always males, affected males have unaffected sons, and a female whose father is affected can have affected sons. page 352

17.3 In genetic linkage, two genes are sufficiently close together in the same chromosome that the combination of alleles present in the chromosome tends to remain together in inheritance.

Genes that are close together in the same chromosome are linked and do not undergo independent assortment. page 353

Recombinant chromosomes result from crossing over between genes on the same chromosome and show a nonparental combination of alleles. page 354

Nonrecombinant chromosomes have the same configuration of alleles as one of the parental chromosomes. page 354

In genetic mapping, the observed proportion of recombinant chromosomes is the frequency of recombination and can be used as a measure of distance along a chromosome. A recombination frequency of 1% is 1 map unit. page 355

Gene linkage and mapping are used to identify the locations of disease genes in the human genome. page 357

17.4 Most *Y*-linked genes are passed from father to son.

In humans and other mammals, the *Y* chromosome contains a gene called *SRY* that results in male development. page 357

In Y-linked inheritance, only males are affected and all sons of an affected male are affected. Females are never affected and do not transmit the trait. page 357

Most Y-linked genes show complete linkage, which allows their evolutionary history to be traced. page 358

17.5 Mitochondria and chloroplast DNA follow their own inheritance pattern.

Mitochondria and chloroplasts have their own genomes, which reflect their evolutionary history as free-living prokaryotes. page 359

Mitochondria in humans and other mammals show maternal inheritance, in which individuals inherit their mitochondrial DNA from their mother. page 359

Because mitochondrial DNA does not undergo recombination and is maternally inherited, it can be used to trace human ancestry and migration. page 360

Self-Assessment

1. Explain how the human *X* and *Y* chromosomes can pair during meiosis even though they are of different lengths and most of their genes are different.
2. Describe the biological basis for the 1:1 ratio of males and females at conception in mammals.
3. For a recessive *X*-linked mutation, such as color blindness, explain the pattern of inheritance from an affected male through his daughters into her children.
4. Explain why linked genes do not exhibit independent assortment.
5. Describe how recombination frequency can be used to build a genetic map.
6. Describe the pattern of inheritance expected from a *Y*-linked gene in a human pedigree.
7. Describe the pattern of inheritance expected from a gene present in mitochondrial DNA in a human pedigree.
8. Explain how *Y*-chromosome and mitochondrial DNA data can be used to trace ancestry.

Log in to LaunchPad to check your answers to the Self-Assessment questions, and to access additional learning tools.

CHAPTER 18

The Genetic and Environmental Basis of Complex Traits

Core Concepts

18.1 Complex traits are those influenced both by the action of many genes and by environmental factors.

18.2 Genetic effects on complex traits are reflected in resemblance between relatives.

18.3 Twin studies help separate the effects of genotype and environment on variation in a trait.

18.4 Many common diseases and birth defects are affected by multiple genetic and environmental risk factors.

Ryan McVay/Getty Images.

Biologists initially had a hard time accepting the principles of Mendelian inheritance because they seem so at odds with everyday observations. Common and easily observed traits like height, weight, hair color, and skin color give no evidence of segregation in pedigrees, and simple phenotypic ratios like 3:1 or 9:3:3:1 are not observed for them. The lack of these characteristic ratios raised serious doubt whether Mendel's principles are valid for common traits. Some biologists concluded that they apply only to seemingly trivial traits like round and wrinkled seeds in peas.

At about the time that Mendel was studying inheritance in garden peas, the biologist Francis Galton, a friend and cousin of Charles Darwin, was studying common traits including human height. From studies of height and other common traits in parents and their offspring, Galton discovered general principles in the inheritance of such traits. For example, parents who are tall tend to have offspring who are taller than average but not as tall as themselves. Galton's principles for the inheritance of common traits did not invoke genes, segregation, independent assortment, or other features of Mendelian inheritance, but they did describe Galton's observations. Not only were genes—what Mendel called "hereditary factors"—thought to be unnecessary in Galton's theory of inheritance, but also many biologists thought that Galton's theories and Mendel's were incompatible.

This, it turned out, was not the case. The traits that Mendel studied are now called **single-gene traits** because each one is determined by variation at a single gene and the traits for the most part are not influenced by the environment. By focusing on single-gene traits, Mendel was able to infer underlying mechanisms of inheritance based on physical factors we now call genes. By contrast, **complex traits,** such as human height, are influenced by multiple genes as well as by the environment. As a result, their inheritance patterns are more difficult to follow and simple phenotypic ratios are not observed.

In many ways, complex traits are more important than single-gene Mendelian traits. One reason is their prevalence—complex traits are found in all organisms and include most of the traits we can see around us. By contrast, there are relatively few examples of common single-gene traits. Complex traits are also important in human health and disease. In the most common disorders—among them heart disease, diabetes, and cancer—single-gene Mendelian inheritance is seldom found. It is therefore important to understand the inheritance of complex traits and common disorders, which are the subjects of this chapter. We begin by describing some of the features of complex traits, and then show how these principles are not only compatible with, but are in fact predicted by, Mendelian inheritance. Finally, we describe how modern molecular genetics and genomics have allowed the identification of genes affecting complex traits.

18.1 HEREDITY AND ENVIRONMENT

Complex traits are important not only in humans, but also in agricultural plants and animals. We will examine human height in some detail because it has been widely studied, but equally well known complex traits are number of eggs laid by hens, milk production in dairy cows, and yield per acre of grain (**Fig. 18.1**).

Many common human diseases, including high blood pressure, obesity, diabetes, and depression, are complex traits (**Fig. 18.2**). High blood pressure, for example, affects about one-third of the U.S. population, and obesity another third. Type 2 diabetes affects around 8% of the U.S. population, and an estimated 15% will suffer at least one episode of severe depression in the course of a lifetime. Taken together, about 200 million Americans—two-thirds of the entire population—suffer from one or more of these common disorders. None of these traits shows single-gene Mendelian inheritance.

In many complex traits, the phenotype of an individual is determined by measurement: Human height is measured in inches, milk yield by the gallon, grain yield by the bushel, egg production by

FIG. 18.1 Examples of complex traits. (a) Human height, (b) egg number, (c) milk production, and (d) grain yield. *Sources: a. Randy Faris/Corbis; b. muratart/Shutterstock; c. smereka/Shutterstock; d. Radius Images/Getty Images.*

FIG. 18.2 Examples of human diseases that are complex traits. (a) High blood pressure, (b) obesity, (c) diabetes, and (d) depression. *Sources: a. Burwell and Burwell Photography/iStockPhoto; b. Ocean/Corbis; c. Junophoto/Getty Images; d. Imago/ZUMApress.com.*

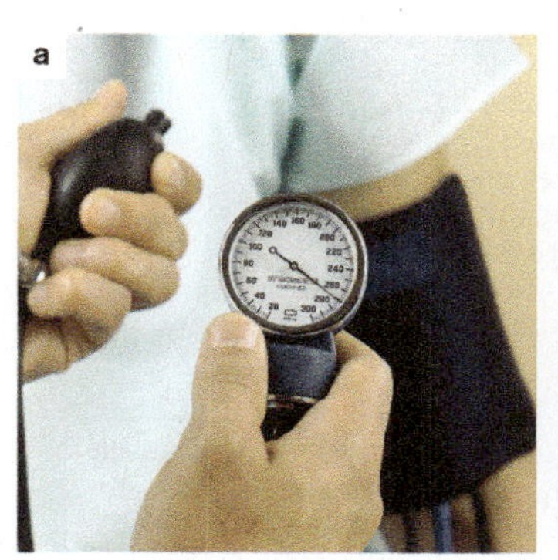

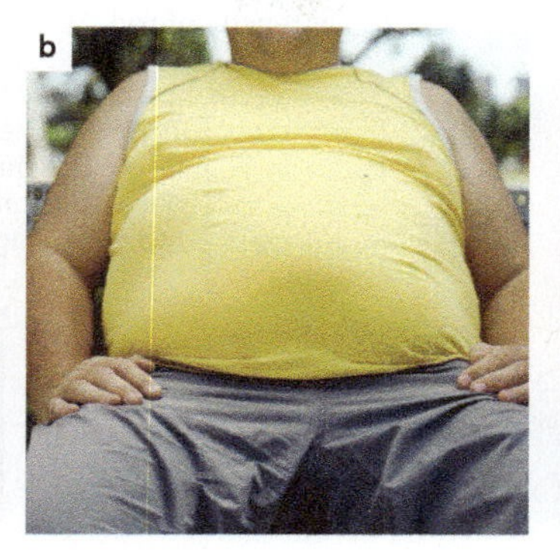

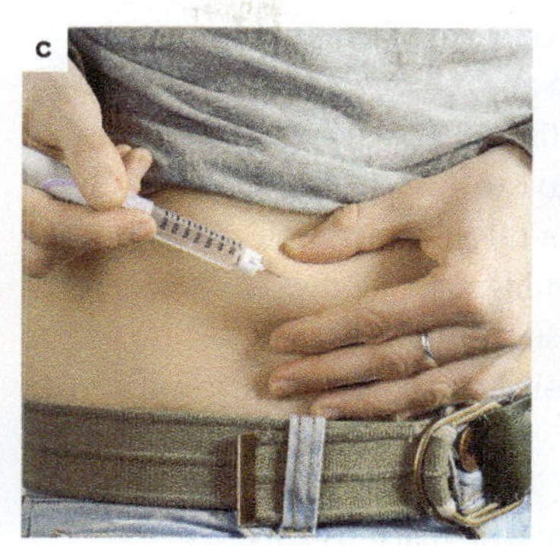

the number of eggs, blood pressure by millimeters of mercury, and blood sugar by millimoles per liter. Because the phenotype of complex traits such as these is measured along a continuum with only small intervals between similar individuals, complex traits like these are often called **quantitative traits.** By contrast, single-gene traits often appear in one of two or more different phenotypes, such as round versus wrinkled seeds, or green versus yellow seeds.

Complex traits are affected by the environment.

Expression of complex traits is notoriously susceptible to lifestyle choices and other environmental factors. Inadequate nutrition is linked to slow growth rate and short stature in adults. Salt intake is associated with an increased likelihood of high blood pressure and is therefore an **environmental risk factor** for this common disorder. An environmental risk factor is a characteristic in a person's surroundings that increases the likelihood of developing a particular disease. For example, a junk-food diet high in fat and carbohydrates is an environmental risk factor for obesity and diabetes.

Environmental effects are also important in agriculture. Farmers are well aware that adequate nutrition is essential to normal growth of chicks, lambs, piglets, and calves, and that continued high-quality feed is necessary for high egg production and milk yield when the animals reach adulthood. In crop plants like grains, adequate soil moisture and nutrients are necessary for sustained high yields.

Environmental factors not only affect the average phenotype for complex traits, but they also affect the variation in phenotype from one individual to the next. For example, most fields of corn you see planted along the roadside come from seeds that are genetically identical to one another, yet there is phenotypic variation in complex traits like plant height (**Fig. 18.3**). Because the plants are genetically identical, the differences in phenotype result from differences in the environment. Some parts of the field may receive more sunlight than others, and some parts may have better water drainage. No matter how uniform an environment may seem, there are always minor differences from one area to the next, and these differences can result in variation in complex traits. In the example in Fig. 18.3, the plants in the foreground are shorter than the others because that corner of the field has poorer drainage, and the plants growing there have to compete for nutrients with the grass growing nearby.

FIG. 18.3 Phenotypic variation due to variation in the environment. Genetically identical corn plants vary in height because of variation in exposure to sun and in soil composition at different locations in the same field. *Source: Bruce Leighty/Getty Images.*

Environmental effects on complex traits in animals are evident in true-breeding, homozygous strains like those used by Mendel in his experiments with pea plants. Such true-breeding strains are called **inbred lines,** and they are often used for research. Even though all animals in any inbred line are genetically identical and are caged in the same facility and fed the same food, there is variation from one animal to the next. In one study of cholesterol levels in an inbred line of mice, for example, average serum cholesterol was 120 mg/dl (milligrams per deciliter), but the range was 60–180 mg/dl. Because the mice are genetically identical, the variation in serum cholesterol resulted entirely from variation in the environment.

Complex traits are affected by multiple genes.

Most complex traits are affected by many genes, in contrast to Mendel's traits, which are primarily affected by only one. Therefore, in complex traits the familiar phenotypic ratios such as 3:1 that Mendel saw when he crossed two true-breeding strains with contrasting characters are not observed. For complex traits, the effects of individual genes are obscured by variation in phenotype that is due to multiple genes affecting the trait and also due to the environment. The number of genes affecting complex traits is usually so large that different genotypes can have very similar phenotypes, which also makes it difficult to see the effects of individual genes on a trait.

In a few traits, however, the effects of the environment are minor and the number of genes is small, and in these cases the genetic basis of the trait can be analyzed. A classic example, studied by Herman Nilsson-Ehle about a century ago, concerns the color of seed casing in wheat (which give the seeds their color), which ranges from nearly white to dark red (**Fig. 18.4**). His experiment demonstrated that complex traits are subject to the same laws that Mendel worked out for single-gene traits, but that the inheritance patterns are more difficult to see because of the number of genes involved.

In studying seed color in true-breeding varieties and their first-generation and second-generation hybrids, Nilsson-Ehle realized that the relative frequencies of different shades of red color could be explained by the effects of three genes that undergo independent assortment, as illustrated by the Punnett square shown in Fig. 18.4. Each of the genes has two alleles, designated by combinations of upper-case and lower-case letters. In the development of seed color, each upper-case allele in a genotype intensifies the red coloration in an additive fashion. Altogether there are seven possible phenotypes, ranging from a phenotype of 0 (nearly colorless, genotype *aa bb cc*) to 6 (dark red, genotype *AA BB CC*).

Nilsson-Ehle first crossed true-breeding dark red plants (*AA BB CC*) with true-breeding colorless plants (*aa bb cc*) to obtain the hybrids with genotype *Aa Bb Cc*. He then crossed the hybrids together, which is the cross shown at the top of Fig. 18.4. Because the three genes show independent assortment (Chapter 16), the distribution of phenotypes expected in progeny from the cross

FIG. 18.4 Multiple genes contributing to a complex trait. Three unlinked genes, each with two alleles, influence the intensity of red coloration of the seed casing in wheat.

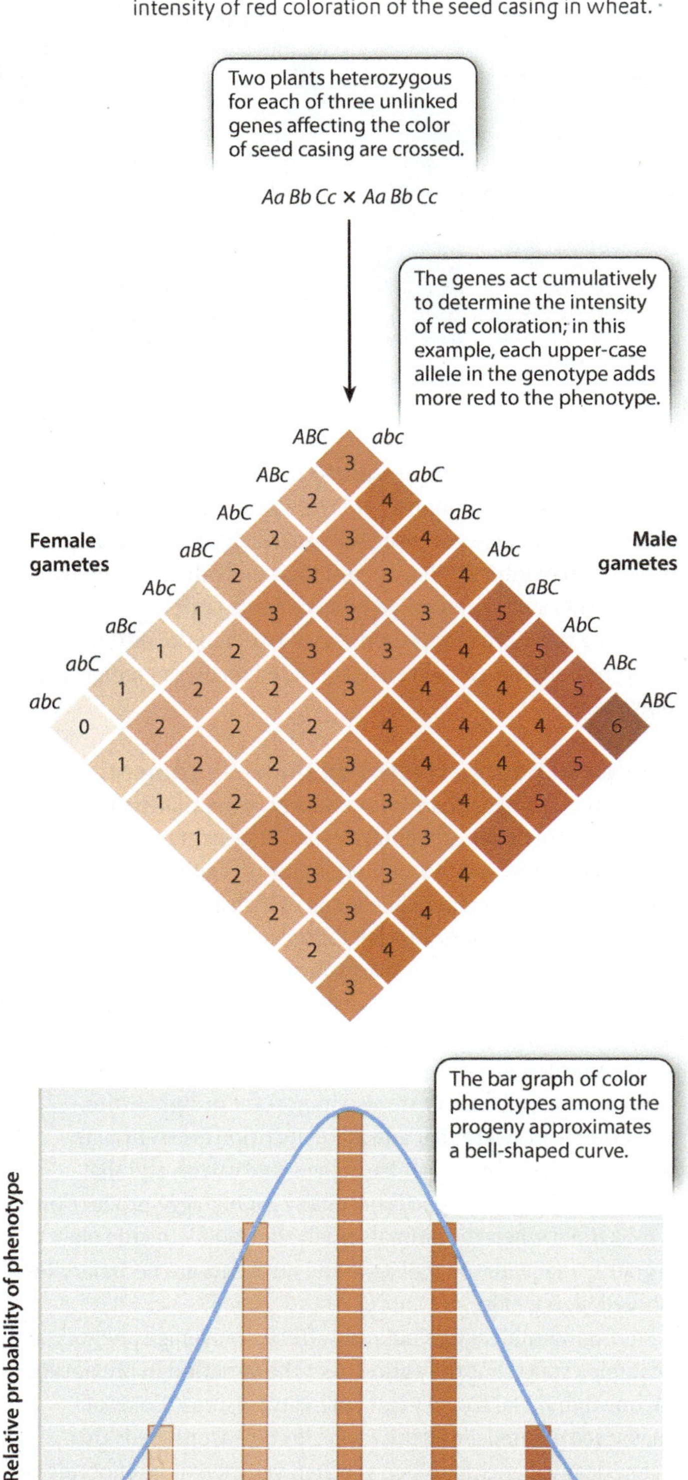

Aa Bb Cc × *Aa Bb Cc* is as shown in the Punnett square. The bar graph below the Punnett square shows the seed-color phenotypes and their relative proportions, from which Nilsson-Ehle inferred three genes with independent assortment. The distribution of seed-color phenotypes is approximated by a bell-shaped curve known as a **normal distribution.** The phenotypes of many complex traits, including human height, conform to the normal distribution. Nilsson-Ehle's seed-color case is exceptional in that virtually all complex traits are affected by many more genes than three (in the case of human height, hundreds, as discussed below).

When differences in phenotype due to the environment can be ignored, the genetic variation affecting complex traits can be detected more easily. And when studying inbred lines, differences in phenotype due to genotype can be ignored because all individuals have the same genotype, and the effects of environment can be observed. In most cases, however, both genetic variation and environmental variation among individuals are present, and it is difficult to quantify how much variation in phenotype is due to genes and how much is due to environment.

It's important to point out that complex traits are not really more "complex" than any other biological trait. The term is used merely to imply that both genetic factors and environmental factors contribute to variation in phenotype among individuals. Just as there are environmental factors that affect complex traits such as height, so there are genetic factors that affect height. Similarly, just as an environmental risk factor increases the likelihood of a common disease, so does a genetic risk factor predispose an individual to the condition. For example, the human gene *ApoE* encodes a protein that helps transport fat and cholesterol. Certain alleles of *ApoE* are associated with high levels of cholesterol, and one particular allele is a genetic risk factor for Alzheimer's disease, the most common serious form of age-related loss of cognitive ability. Lifestyle environmental factors such as diet, physical exercise, and mental stimulation are also thought to play a role in the risk of Alzheimer's disease.

The relative importance of genes and environment can be determined by differences among individuals.

For any one individual, it is impossible to specify the relative roles of genes and environment in the expression of a complex trait. For example, in an individual 66 inches tall it would be meaningless to attribute 33 inches of height to parentage (genes) and 33 inches of height to nutrition (environment). This kind of partitioning makes no sense because genes and environment act together so intimately in each individual that their effects are inseparable. To attempt to separate them for any individual would be like asking to what extent it is breathing or oxygen that keeps us alive. Both are important, and both must take place together if we are to live.

It is nevertheless possible to separate genes and environment in regard to their effects on the *differences,* or variation, among individuals within a particular population. For some traits, the variation seen among individuals is due largely if not exclusively to differences in the environment. For other traits, the variation is due mainly to genetic differences. In the case of human height, roughly 80% of the variation among individuals of the same sex is due to genetic differences, and the remaining 20% to differences in their environment, primarily differences in nutrition during their years of growth.

Genetic and environmental effects can interact in unpredictable ways.

One of the features of complex traits is that genetic and environmental effects may interact, often in unpredictable ways. Consider the example shown in **Fig. 18.5.** In this experiment, two strains of corn were each grown in a series of soils in which the amount of nitrogen had been enriched by the growth of legumes. The yield of strain 1 varies little across these different environments. The yield of strain 2, however, increases dramatically with soil nitrogen. Each of the lines is known as a **norm of reaction,** which for any genotype graphically depicts how the environment (shown on the x-axis) affects phenotype (shown on the y-axis) across a range of environments. Note that in one case in Fig. 18.5, the environment has little to no effect on the phenotype (the norm of reaction is nearly flat), but in the other it has a very noticeable effect. In cases like strain 2, it is impossible to predict the phenotype of a given genotype without knowing what the environmental conditions are and how the phenotype changes in response to variation in the environment.

Such variation in the effects of the environment on different genotypes is known as **genotype-by-environment interaction.** This type of interaction is important because it implies that the effect of a genotype cannot be specified without knowing the environment, and the other way around. A further implication is that there may be no genotype that is the "best" across a broad

FIG. 18.5 Genotype-by-environment interaction for grain yield in corn. The effect of soil nitrogen on grain yield is minimal in strain 1, but dramatic in strain 2. *Data courtesy of J. W. Dudley.*

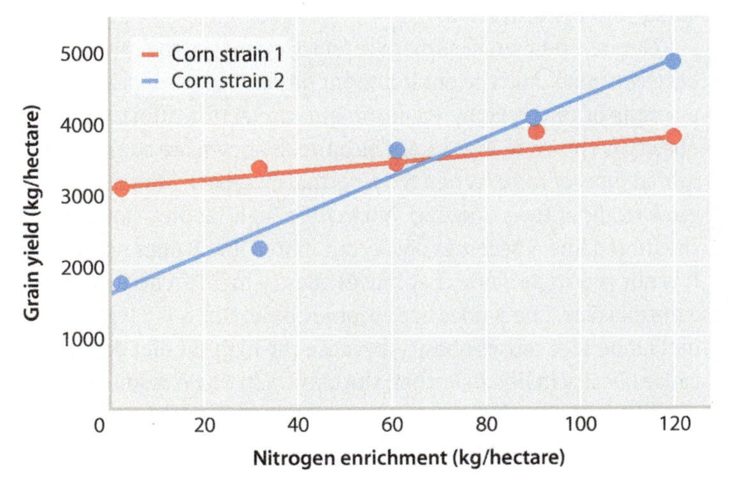

FIG. 18.6 Genotype-by-environment interaction for obesity in mice. When fed a normal diet, both inbred lines grow to about the same size. When fed a high-fat diet, mouse A becomes obese but mouse B does not. *Source: Phil Jones/ Georgia Regents University.*

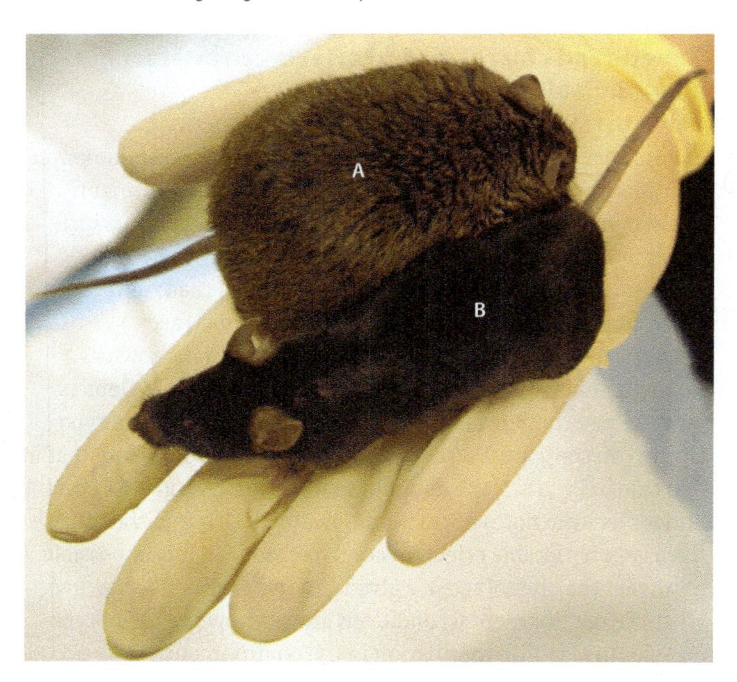

range of environments, and likewise no environment that is "best" for all genotypes. For complex traits, phenotype depends on both genotype *and* environment. For the strains of corn in Fig. 18.5, for example, which strain is "better" depends on the soil nitrogen. With little nitrogen enrichment, strain 1 is better; at intermediate values, both strains yield about the same; and with high nitrogen enrichment, strain 2 is better. In this case, knowing the norms of reaction, and recognizing the magnitude and direction of genotype-by-environment interaction, allow each farmer to use the strains likely to perform best under the available conditions of cultivation.

Genotype-by-environment interaction makes the interplay between genes and the environment difficult to predict. An example of genotype-by-environment interaction affecting obesity is shown in **Fig. 18.6.** The animals shown are adults of two inbred lines of mice. When fed a normal diet, both inbred lines grow to about the same size. When fed a high-fat diet, however, the inbred line A becomes obese, but inbred line B does not. Hence, it is not genotype alone that causes obesity in line A because with a normal diet, line A does not become obese. Nor is it a high-fat diet alone that causes obesity, because the high-fat diet does not cause obesity in line B. Rather, the obesity in line A results from a genotype-by-environment interaction (which in this case is a genotype-by-diet interaction).

18.2 RESEMBLANCE AMONG RELATIVES

Mendel had many advantages over Galton in his studies of inheritance. Mendel's peas were true breeding, produced a new generation each year, and yielded large numbers of progeny. The environment had a negligible effect on the traits, and the genetic effect on each trait was due to alleles of a single gene. Mendel's

FIG. 18.7 Galton's data showing distribution of height of offspring of (a) the tallest parents and (b) the shortest parents. *Data from F. Galton, 1888, "Co-Relations and Their Measurement, Chiefly from Anthropometric Data," Proceedings of the Royal Society, London 45:135–145, reprinted in K. Pearson, 1920, "Notes on the History of Correlation," Biometrika 13:25–45.*

a.

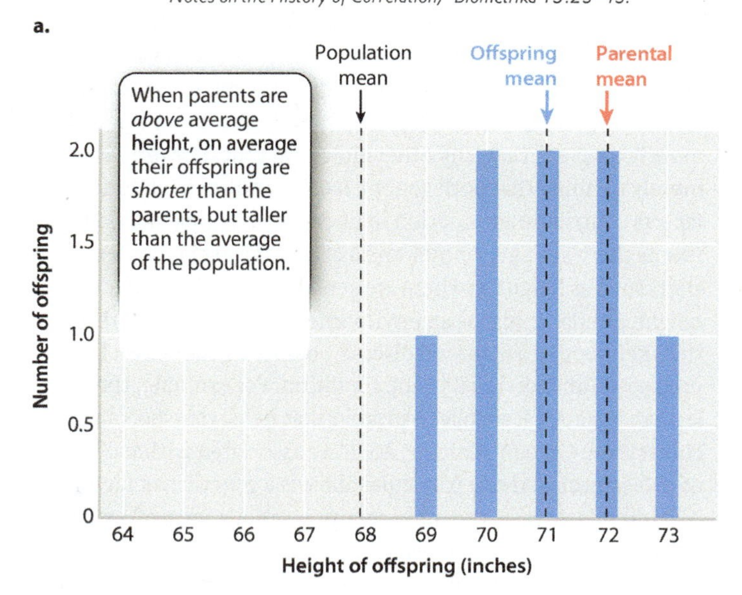

b.

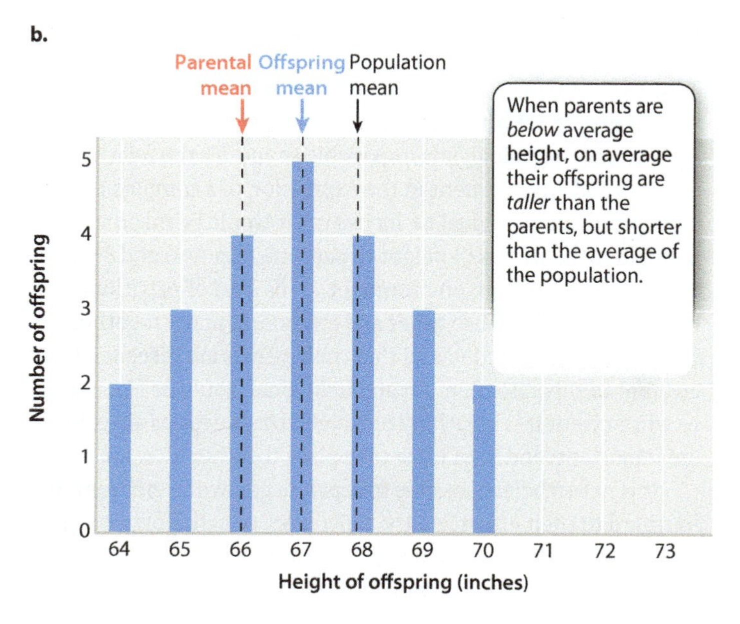

crosses yielded simple ratios such as 3 : 1 or 1 : 1, which could be interpreted in terms of segregation of dominant and recessive alleles.

By contrast, Galton studied variation in such traits as height in humans. Humans are obviously not true breeding and have few offspring. Galton had one big advantage, though: Whereas most simple Mendelian traits are relatively uncommon, Galton's traits are readily observed in everyday life. What did Galton discover?

For complex traits, offspring resemble parents but show regression toward the mean.

Galton's observations are as important in understanding complex traits as Mendel's are in understanding single-gene traits. He studied many complex traits, including human height, strength, and various other physical characteristics, including number of fingerprint ridges. The discovery he regarded as fundamental resulted from his data on adult height of parents and their progeny (**Fig. 18.7**). Galton noted that each category of parent (tall or short) produced a range of progeny forming a distribution with its own mean.

The bar graph in Fig. 18.7a shows the distribution of height among the progeny of the tallest parents, whose average height is 72 inches. The mean height of the offspring is 71 inches, which is greater than the mean height of the whole population of the study (68.25 inches) but less than that of the parents. The bar graph in Fig. 18.7b is the distribution of height of progeny of the shortest parents, who averaged 66 inches. In this case, the mean height of the progeny is 67 inches, which is less than the mean height of the population but greater than that of the parents. Note, however, that *some* of the offspring of tall parents are taller than their parents, and likewise *some* of the offspring of short parents are shorter than their parents. It is only *on average* that the height of the offspring is less extreme than that of the parents.

Galton's results can be plotted on a graph, represented as the blue data points in **Fig. 18.8,** in which the average height of each pair of parents (called the midparent value) is compared with that of their child. For convenience in visualization, midparent heights are grouped into categories, and the mean of each group is plotted along the x-axis. For each group of parents, the height of the offspring is also averaged and plotted along the y-axis.

Galton regarded this observation as his most important discovery, publishing his results in 1886. Today, we call it **regression toward the mean.** The offspring exhibit an average phenotype that is closer to the population mean than the phenotype of the parents. In other words, when the average height of the parents is *smaller* than the population mean, then the average height of the offspring is *greater* than that of the parents (but smaller than the population mean). Likewise, when the average height of the parents is *greater* than the population mean, then the average height of the offspring is *smaller* than that of the parents (but greater than the population mean).

Regression toward the mean is observed for two reasons. The first is that during meiotic cell division (Chapter 11), segregation and recombination break up combinations of genes that result in extreme phenotypes, such as very tall or very short, that are present in the parents. The second reason is that the phenotype of the parents results not only from genes but also from the environment. Environmental effects are not inherited, so any effect of the environment on the parents' phenotypes is not transmitted to the offspring. For example, if the parents are tall or short purely because of an environmental effect like better or worse nutrition, the average height of the offspring will be equal to the population mean.

→ **Quick Check 1** Does regression toward the mean imply that the human population is getting shorter over time?

FIG. 18.8 Regression toward the mean. Galton's data on the average adult height of parents and that of their offspring show that offspring mean height (blue line) falls between the parental mean (dashed red line) and the population mean (dashed black line).

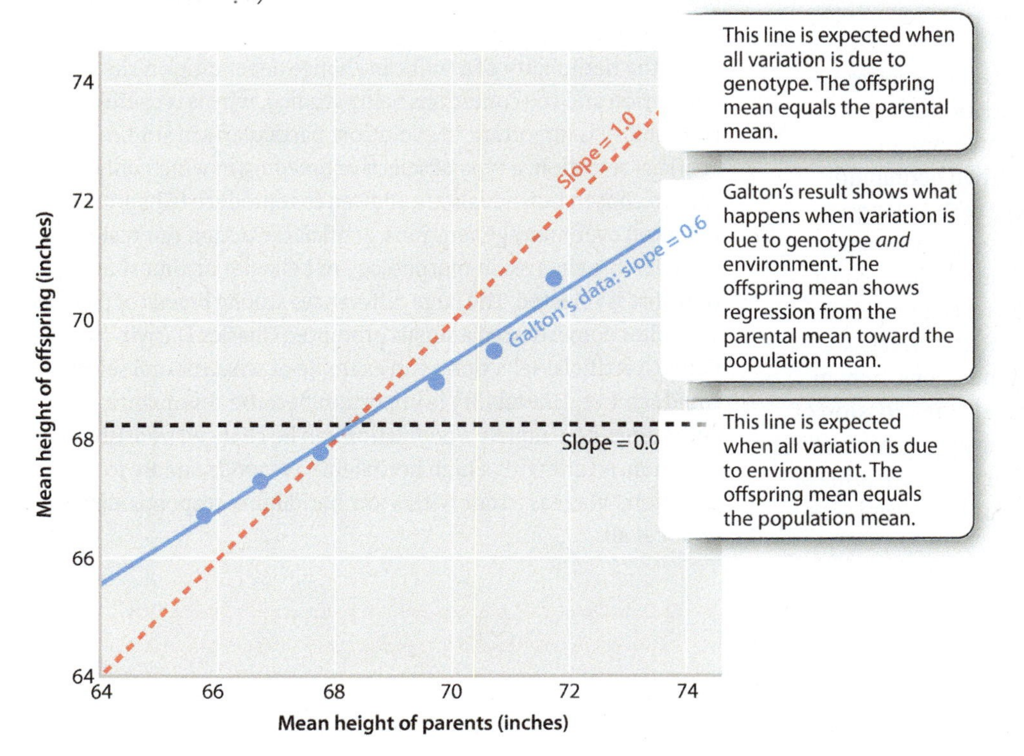

Heritability is the proportion of the total variation due to genetic differences among individuals.

How much of the difference in height among individuals is due to genetic differences, and how much is due to environmental differences? The slope of the line that relates the average phenotype of parents to the average phenotype of their offspring (in the case of Galton's data, the blue line in Fig. 18.8) can answer this question because it provides a measure of the **heritability** of the trait. The heritability of a trait in a population of organisms is the proportion of the total variation in the trait that is due to genetic differences among individuals. For a complex trait, the heritability determines how closely the mean of the progeny resembles that of the parents. In Galton's data, the slope of the line is 0.6 and therefore the heritability of height in this population is 60%.

The red dashed line in Fig. 18.8 presumes a hypothetical ideal trait in which variation is determined completely by genetic differences among individuals and the heritability is 100%. When heritability is 100%, the slope of the line representing the trait is 1. As we do with most complex traits, such as Nilsson-Ehle's wheat seed color, we assume that each of a large number of genes contributing to the trait has two alleles, one that contributes to the trait and one that does not, and the genes undergo independent assortment. The number of alleles in the genotype contributing to the trait determines the phenotype in any individual for that trait. (The genetic model is like that shown in Fig. 18.4 but with many more genes.) In this ideal case in which heritability is 100%, the average phenotype of the offspring from any pair of parents will equal the average phenotype of the parents themselves. In the real world, the ideal is rarely encountered, and deviations from the red line result from complications such as dominance, genetic linkage, and epistasis (non-additive contributions of the alleles of different genes).

The black dashed line in Fig. 18.8 represents the opposite extreme, a hypothetical ideal trait in which variation is determined completely by differences in the environment among individuals and the heritability is 0%. When heritability is 0%, the line representing the trait has a slope of 0. As long as environmental effects are not transmitted from one generation to the next, the average phenotype of the offspring will be equal to the average of the population as a whole, no matter what the phenotypes of the parents. Although environmental effects are usually not transmitted in plant and animal populations, human populations are exceptional in showing **cultural transmission** of some environmental effects. For example, the average wealth of the offspring of rich parents is greater than that of the offspring of poor parents, and this difference is obviously due to transmission of the parents' money, not their genes.

The term "heritability" is often misinterpreted. The problem is that, in non-scientific contexts, the word means "the capability of being inherited or being passed by inheritance." This definition suggests that heritability has something to do with the inheritance of a trait. But "heritability" as used for complex traits means no such thing. It refers only to the *variation* in a trait among individuals, and specifically to the proportion of the variation among individuals in a population due to differences in genotype.

Hence, a heritability of 100% does not imply that the environment cannot affect a trait. As emphasized earlier, the environment is always important, just as oxygen is important to life. What a heritability of 100% means is that *variation* in the environment does not contribute to differences among individuals in a specific population. For example, if genetically different strains of chrysanthemums are grown in a greenhouse and subject to identical environmental conditions, then differences in flowering time have to be due to genetic differences, and the heritability of the trait would be 100%. Similarly, a heritability of 0% does not imply that genotype cannot affect the trait. A heritability of 0% merely means that differences in genotype do not contribute to the *variation* in the trait among individuals in a specific population. If genetically identical strains of chrysanthemums are grown in different environments, then differences in flowering time must be due to the environment, and heritability would be 0%.

Heritability therefore is not an intrinsic property of a trait. For chrysanthemum flowering time, the heritability in one case was 100% and in the second 0%. Heritability applies only to the trait in a particular population across the range of environments that exist at a specific time. Similarly, the heritability depicted by the slope of the blue line in Fig. 18.8 applies only to the population studied (205 pairs of British parents and their 930 adult offspring, in the late nineteenth century) and may be larger or smaller in different populations at different times. In particular, the magnitude of the heritability cannot specify how much of the difference in average phenotype *between* two populations is due to genotype and how much due to environment.

If the heritability of a trait can change depending on the population and the conditions being studied, why is it useful? Heritability is important in evolution, particularly in studies of artificial selection, a type of selective breeding in which only certain chosen individuals are allowed to reproduce (Chapter 21). Practiced over many generations, artificial selection can result in considerable changes in morphology or behavior or almost any trait that is selected. The large differences among breeds of pigeons and other domesticated animals prompted Charles Darwin to point to artificial selection as an example of what natural selection could achieve. Heritability is important because this quantity determines how rapidly a population can be changed by artificial selection. A trait with a high heritability responds rapidly to selection, whereas a trait with a low heritability responds slowly or not at all.

→ **Quick Check 2** Many people are surprised to learn that, while each individual's fingerprints are unique, the total number of fingerprint ridges is highly heritable, about 90% heritability in many populations. What does high heritability of this trait mean?

18.3 TWIN STUDIES

Galton pioneered studies of twins as a way to separate the effects of genotype and environment on phenotypic differences among individuals. Depending on whether they arise from one or from two egg cells, twins can be **identical (monozygotic twins)** or **fraternal (dizygotic twins).** Identical twins arise from a single fertilized egg (the zygote), which, after several rounds of cell division, separates into two distinct but genetically identical embryos. Strikingly similar in overall appearance (**Fig. 18.9a**), identical twins have stimulated the imagination since antiquity, inspiring stories by the Roman playwright Plautus, William Shakespeare (himself the father of twins), Alexander Dumas, Mark Twain, and many others.

FIG. 18.9 **Identical twins.** (a) Identical (monozygotic) twins arise from a single fertilized egg and are genetically identical. (b) Fraternal (dizygotic) twins arise from two different fertilized eggs and are no more closely related than other pairs of siblings. *Sources: a, AP Photo/Macomb Daily, Craig Gaffield; b. Wendy Connett/Getty Images.*

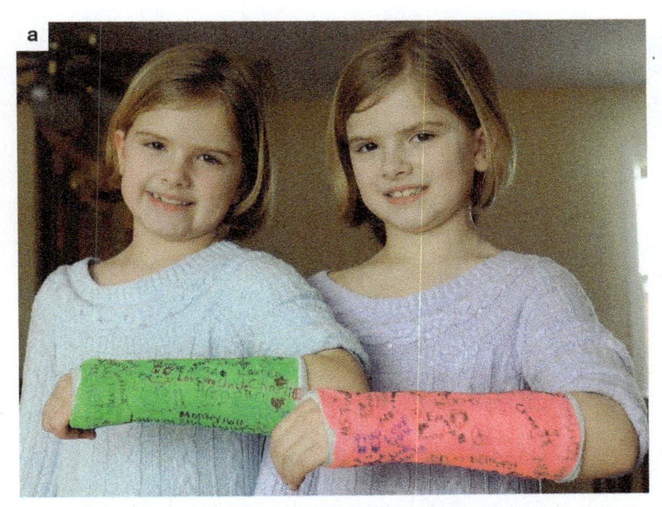

In contrast, fraternal twins result when two separate eggs, produced by a double ovulation, are fertilized by two different sperm. Whereas identical twins are genetically identical, fraternal twins are only as closely related as any other pair of siblings (**Fig. 18.9b**).

Twin studies help separate the effects of genes and environment in differences among individuals.

The utility of twins in the study of complex traits derives from the genetic identity of identical twins. Whereas differences between fraternal twins with regard to any trait may arise because of genetic or environmental factors (or genotype-by-environment interactions), differences between identical twins must be due only to environmental factors because the twins are genetically identical. Consequently, if variation in a trait has an important genetic component, then identical twins will be more similar to each other than fraternal twins are similar to each other. But if the genetic contribution to variation is negligible, then identical twins will not be more similar to each other than fraternal twins are to each other. One caveat of twin studies is that the environment of identical twins is often more similar than that of fraternal twins. This difficulty can be minimized in part by studying identical twins separated from each other shortly after birth and raised in different environments,

For complex traits, such as depression or diabetes, the extent to which twins are alike is measured according to the **concordance** of the trait, which is defined as the percentage of cases in which both members of a pair of twins show the trait when it is known that at least one member shows it. The relative importance of genetic and environmental factors in causing differences in phenotype can be estimated by comparing the concordances of identical and fraternal twins. A significantly higher concordance rate for a given trait for identical twins compared to that for fraternal twins suggests that the trait has a strong genetic component. Similar concordance rates for identical and fraternal twins suggest that the genetic component is less important.

In the study of concordance, twin pairs in which neither member is affected contribute no information, and these twin pairs must be removed from the analysis. The concordance is based solely on twin pairs in which one or both members express the trait. To take a concrete example, let us symbolize each member of a twin pair as a closed square (■) if the member shows the trait and as an open square (□) if the member does not show the trait. Then, in any sample of identical twins, there are three possibilities, illustrated with data for adult-onset diabetes:

■■ 36 identical twin pairs, both members showing adult-onset diabetes

■□ 40 identical twin pairs, only one member showing adult-onset diabetes

□□ 336 identical twin pairs, neither member showing adult-onset diabetes

In this example, the 336 twin pairs who do not show the trait provide no information. The concordance is based only on the first two types of twin pairs, among which in 36 pairs both members show the trait and in 40 pairs only one shows the trait. In this case the concordance among identical twins is 36/(36 + 40) = 47%. Among fraternal twins, by contrast, the concordance for adult-onset diabetes is only 10%. The difference in concordance between identical and fraternal twins (47% for identical twins versus 10% for fraternal twins) implies that differences in the risk of adult-onset diabetes have an important genetic component. Furthermore, the observation that the concordance for identical twins is much less than 100% implies that environmental factors (which we now know to include diet and exercise) are also important in the risk of adult-onset diabetes.

Table 18.1 shows identical and fraternal twin concordance rates for a number of other disorders. Just as with diabetes, the marked difference in concordance rates between identical and fraternal twins for high blood pressure, asthma, rheumatoid arthritis, and epilepsy suggests that all these disorders have an important genetic component. In addition, the fact that the

HOW DO WE KNOW?

FIG. 18.10

What is the relative importance of genes and of the environment for complex traits?

BACKGROUND Twin studies remain important for assessing the relative importance of "nature" (genotype) and "nurture" (environment) in determining variation among individuals for complex traits. The idea of using twins to distinguish nature from nurture is usually attributed to Francis Galton because of an article he wrote about twins in 1875. In fact, the modern twin study does not trace to Galton but to Curtis Merriman in the United States and Hermann Siemens in Germany, who independently hit upon the idea in 1924. In Galton's time, it was not even known that there are genetically two kinds of twins.

EXPERIMENT The rationale of a twin study is to compare identical twins with same-sex fraternal twins. In principle, identical twins differ only because of environment, whereas fraternal twins differ because of genotype as well as environment. The extent to which fraternal twins differ more from each other than identical twins differ from each other measures the effects of genotype. Some twin studies focus on twins reared apart in order to compensate for shared environmental influences that may be stronger for identical twins than for fraternal twins.

RESULTS The bar graph shows the results of typical twin studies for various traits. The concordance between twins is the fraction of twin pairs in which both twins show the trait among all those pairs in which at least one twin shows the trait. Roughly speaking, the difference in the concordance between identical twins and fraternal twins is a measure of the relative importance of genotype. In the data shown here, for example, autism and clinical depression both show strong genetic influences, whereas female alcoholism shows almost no genetic influence.

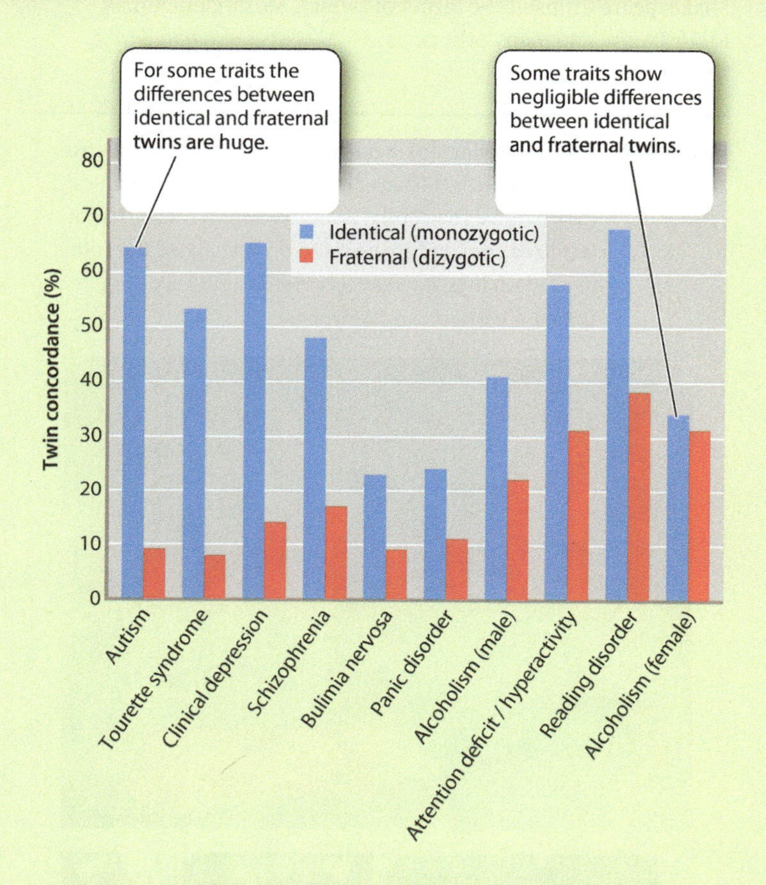

CONCLUSION The examples shown here indicate very large differences in the importance of genotype versus environment among complex traits. These results are typical of most complex traits.

FOLLOW-UP WORK Twin studies must be interpreted in light of other studies comparing complex traits among individuals with various degrees of genetic relatedness. On the whole, twin studies of complex traits have yielded results that are consistent with other available evidence.

SOURCES Rende, R. D., R. Plomin, and S. G. Vandenberg. 1990. "Who Discovered the Twin Method?" *Behavior Genetics* 20:277–285; McGue, M., and T. J. Bouchard, Jr. 1998. "Genetic and Environmental Influences on Human Behavioral Differences." *Annual Review of Neuroscience* 21:1–24.

TABLE 18.1 Twin Concordance Rates for Several Traits.

DISORDER	CONCORDANCE IN IDENTICAL TWINS (%)	CONCORDANCE IN FRATERNAL TWINS (%)
High blood pressure	25	7
Asthma	47	24
Rheumatoid arthritis	34	7
Epilepsy	37	10
Handedness (left or right)	79	77
Measles	95	87
Acute infection leading to death	8	9

Data from M. McGue and T. J. Bouchard, Jr., 1998, "Genetic and Environmental Influences on Human Behavioral Differences," Annual Review of Neuroscience 21:1–24.

concordance is not 100%, even for identical twins, implies that both genes and environment—"nature" *and* "nurture"—play important roles in the differences in risk among people.

Contrast these results with the data in Table 18.1 for handedness, measles, or death from acute infection. Note that in these cases the concordance rates for identical twins are very similar to those for fraternal twins—there is no significant genetic component to variation in the trait, at least insofar as can be determined from twin studies. (Many students are quite surprised to learn that handedness does not have a detectable genetic component.) For infectious diseases, exposure to the disease agents in the environment plays an important role. In the case of measles, for example, the data imply that, whether twins are identical or fraternal, when one twin catches measles, the other will very likely catch it, too.

Fig. 18.10 shows the identical and fraternal concordance rates for a number of common behavioral disorders. For some, such as schizophrenia, depression, and autism, the difference between identical and fraternal twins is very large, again implying an important role for genetic factors. For other traits, such as alcoholism in females, the difference is negligible, suggesting an important role for environmental and social factors. This example also illustrates that complex traits can have quite different risk factors and outcomes in different sexes.

18.4 COMPLEX TRAITS IN HEALTH AND DISEASE

Much evidence beyond twin studies implies an important role for particular alleles as risk factors for diabetes, high blood pressure, asthma, rheumatoid arthritis, epilepsy, schizophrenia, clinical depression, autism, and many other conditions. These are all complex traits, affected by genotype, environment, and genotype-by-environment interactions.

Even the most common birth defects are affected by multiple genetic risk factors. The most common birth anomalies and the numbers of affected babies born in the United States each year—collectively more than 35,000—are indicated in **Fig. 18.11.** About 20% of these anomalies are due to an extra chromosome. Trisomy 21 (Down syndrome) is by far the most common of these chromosome-number abnormalities. Another roughly 10% are due to defects in metabolism, many of which—among them cystic fibrosis, sickle-cell anemia, and phenylketonuria (the inability to break down the amino acid phenylalanine)—are rare simple Mendelian disorders. Birth anomalies due to extra or missing chromosomes and those due to single-gene Mendelian inheritance are not considered complex traits in terms of their inheritance. Together these account for about 30% of the of birth anomalies in Fig. 18.11. The majority (70%) of birth anomalies in Fig. 18.11 are inherited as complex traits that are affected by both genetic and environmental risk factors.

Most common diseases and birth defects are affected by many genes that each have relatively small effects.

Because the most common birth anomalies as well as childhood and adult disorders are complex traits, biologists are keenly interested in identifying the genes that contribute to differences in risk, understanding what these genes do, and translating this knowledge into prevention or treatment. They have therefore

FIG. 18.11 Incidence of the most common birth anomalies.

Data from Centers for Disease Control and Prevention; National Newborn Screening and Genetics Resource Center; March of Dimes.

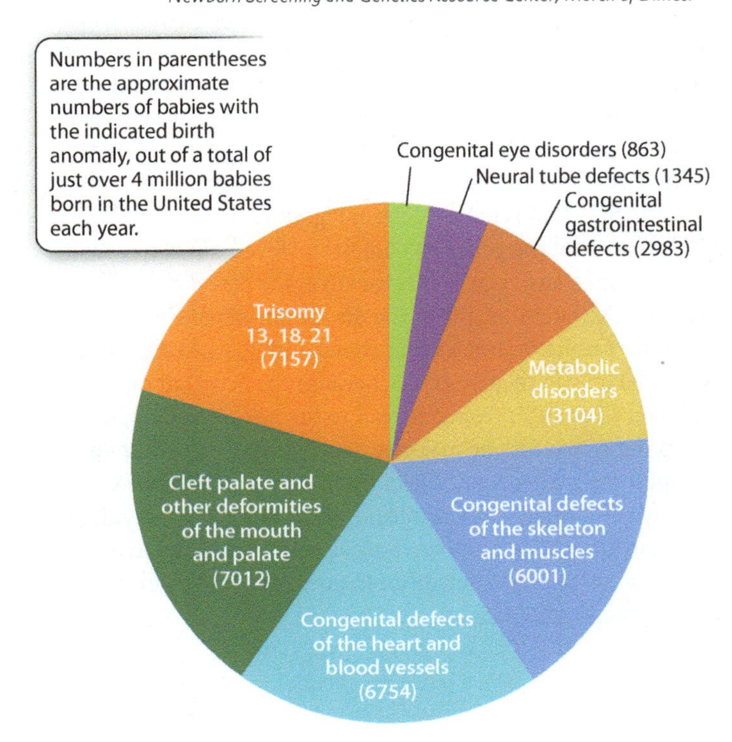

begun to apply modern molecular methods—genome sequencing (Chapter 13), genome annotation (Chapter 13), and genotyping (Chapter 15)—to identifying genes affecting complex traits.

The identification of genes affecting complex traits is not only a major goal of much current research in human genetics, but also in the genetics of domesticated animals and plants. Gene identification is also important in model organisms used in research, especially the laboratory mouse. Much is already known about these model organisms and they are easily manipulated, making it possible to discover the molecular mechanisms by which genes affecting complex traits exert their effects.

The study of complex traits has reached a stage at which patterns are beginning to emerge. Many of these patterns are exemplified by the chromosome map in **Fig. 18.12,** which shows the location of genes in the human genome that affect cholesterol levels. The first observation is that many genes contribute to amounts of different types of cholesterol in humans—specifically, more than 50 genes contribute to the level of HDL (high-density lipoproteins, or "good cholesterol"), LDL (low-density lipoproteins, or "bad cholesterol"), and triglycerides. A second observation is that many of these genes affect two or even all three of the types of molecule. (A single gene that has multiple effects is said to show **pleiotropy.**) Third, many of the genes show epistasis, that is, multiple genes act in the same pathway to affect a trait (Chapter 16). Fourth, many of the genes occur in clusters, being physically close together in the same chromosome. Clustering of genes reflects the fact that many genes arose through the process of duplication of a single gene followed by divergence over time, generating a family of genes near one another with related functions (Chapter 14). Finally, the functions of all of the genes that contribute to a complex trait are not known. Some of the genes in Fig. 18.12 affect serum cholesterol through known metabolic pathways, but many others work through unknown pathways. Many of the human genes shown have counterparts in the mouse genome that have similar effects on serum cholesterol, so these genes are open to direct experimental investigation.

A principle that is not evident in Fig. 18.12 is that the effects of each of the genes on cholesterol levels are very unequal. Some have relatively large effects, whereas others have small ones. The distribution of the magnitude of gene effects for complex traits resembles that shown in **Fig. 18.13.** The axes are labeled only in relative terms, because the actual numbers and effect sizes differ from one trait to the next. However, the main point is that for the majority of genes that contribute to a complex trait, the magnitude of their individual effects is typically quite small. The magnitude of the effects of individual genes also often differs between the sexes, which helps explain why complex traits so often differ in prevalence or severity between males and females.

→ **Quick Check 3** When genes for complex traits have effects that are distributed as shown in Fig. 18.13, which are easier to identify: those that are numerous with small effects, or those that are few with large effects?

FIG. 18.12 Genes affecting the level of serum cholesterol in the human genome. *Data from X. Wang and B. Paigen, 2005, "Genetics of Variation in HDL Cholesterol in Humans and Mice," Circulation Research 96:27–42.*

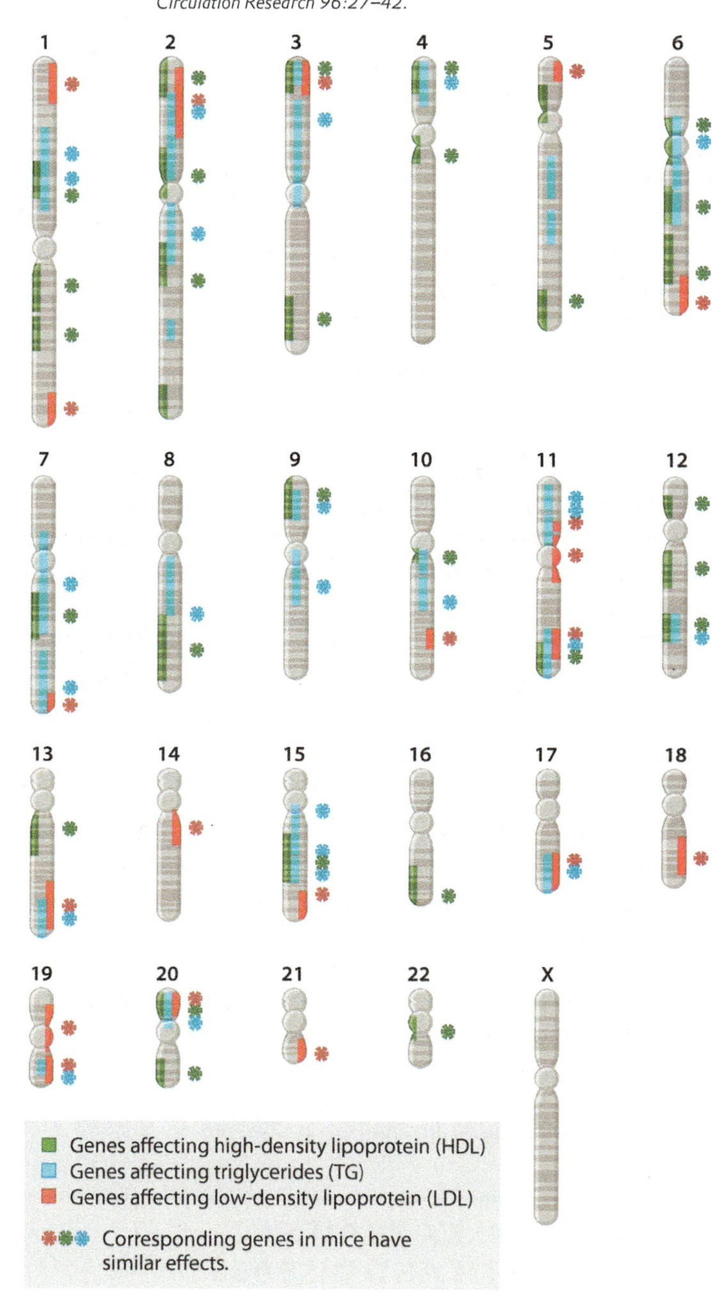

Human height is affected by hundreds of genes.

For most complex traits, the genes that have been identified to date account for only a relatively small fraction of the total variation in the trait. One extreme example is adult height. An

FIG. 18.13 Relationship between number of genes and their effect on complex traits.

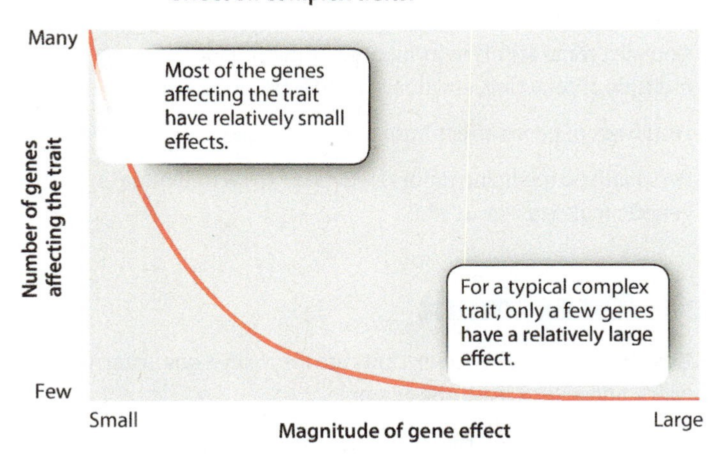

enormous amount of data is available for height, not because height has been studied extensively for its own sake, but because in studies of common diseases the height of each individual is recorded, and hence data on height are available without additional effort or expense.

In one analysis, results from 79 separate studies were combined. These studies included more than 250,000 individuals of European ancestry who were genotyped for common nucleotide variants (SNPs) at about 3 million nucleotide sites. The analysis identified 697 genes affecting height. Some of these genes are known to affect skeletal development, growth hormones, or other growth factors, but most have no obvious connection to the biology of growth. Some of the genes had previously been identified by studies of rare mutations that have pronounced effects on skeletal growth, either in human families or in laboratory mice. An unexpected finding was that a few genes affecting height are known to be associated with bone mineral density, obesity, and rheumatoid arthritis. These and many other of the genes identified might affect height indirectly.

The 697 genes for height identified among the 250,000 individuals account for about 60% of the genetic variation in height, a number that suggests that many more genes with still smaller effect also contribute to variation in height. And this analysis does not address at all the effects of the environment on human height, nor genotype-by-environment interactions, which likely play an important role as well. Complex traits therefore require sophisticated studies to tease apart all the genetic and environmental factors that play a role.

? CASE 3 YOU, FROM A TO T: YOUR PERSONAL GENOME

Can personalized medicine lead to effective treatments of common diseases?

The multiple genetic and environmental factors affecting complex traits imply that different people can have the same disease for different reasons. For instance, one person might develop breast cancer because of a mutation in the *BRCA1* or *BRCA2* genes, while another might develop breast cancer because of other genetic risk factors or even environmental ones. Similarly, one person might develop emphysema because of a mutation in the gene that encodes the enzyme alpha-1 antitrypsin (α1AT) (Chapter 15), and another might have emphysema as a result of cigarette smoking. Other examples where the same disease can be the result of different underlying genetic or environmental factors include elevated cholesterol levels, high blood pressure, and depression. Because the underlying genetic basis for the same disease may be different in different patients, some patients respond well to certain drugs and others do not.

The traditional strategy for treating diseases is to use the same medicine for the same disease. However, because we now know that the same disease may have different causes, another possibility has emerged: identifying each patient's genotype for each of the relevant genes, and then matching the treatment to the genetic risk factors in each patient. The approach is known as **personalized medicine.** Personalized medicine matches the treatment to the patient, not the disease.

Personalized medicine not only aims to identify ahead of time medicine that will work effectively, but also to avoid medicines that may lead to harmful side effects, even death. Advertisements and enclosures with prescription drugs enumerate long lists of side effects that you may get if you take a particular medicine. At present, it is difficult to know in advance who will get one or more of these side effects and who won't—that is, the side effects of taking medicine are themselves complex traits and the result of many underlying genetic and environmental factors. If we could identify ahead of time those patients who will respond negatively to a particular medicine and those who won't, we could minimize potentially harmful effects of medicines on certain individuals.

Someday, it may be possible to determine each patient's genome sequence quickly, reliably, and cheaply. At present, personalized medicine is restricted to studies of a few key genes known to have important effects on treatment outcomes. In the treatment of asthma, for example, some of the differences in response to albuterol inhalation have been traced to genetic variation in the gene *ADRB2,* encoding the β-(beta-)2-adrenergic receptor. Similarly, certain drugs used in the treatment of Alzheimer's disease are less effective in women with a particular apolipoprotein E (*APOE*) genotype than in other classes of patients. Also, more than half of the cases of muscle weakness occurring as an adverse effect of drug treatment used to control high cholesterol can be traced to genetic variation in the gene *SLCO1B1,* which encodes a liver transport protein. Certain variants of this protein have decreased ability to transport the drug, resulting in higher concentrations remaining in the blood and a risk of muscle weakness.

Other factors in addition to genes are involved in these disorders, but genes play an important role. Knowledge of a patient's genotype can help guide treatment. These are only a few of many examples, but they demonstrate the substantial potential benefits of personalized medicine. ■

Core Concepts Summary

18.1 Complex traits are those influenced both by the action of many genes and by environmental factors.

Complex traits that are measured on a continuous scale, like human height, are called quantitative traits. page 364

It is usually difficult to assess the relative roles of genes and the environment ("nature" versus "nurture") in the production of a given trait in an individual, but it is reasonable to consider the relative roles of genetic and environmental variation in accounting for differences among individuals for a given trait. page 365

The relative importance of genes and environment in causing differences in phenotype among individuals differs among traits. For some traits (like height), genetic differences are the more important source of variation, whereas for others (such as cancer), environmental differences can be the more important. page 366

Genetic and environmental factors can interact in unpredictable ways, resulting in genotype-by-environment interactions. page 367

18.2 Genetic effects on complex traits are reflected in resemblance between relatives.

In an analysis of heights of parents and offspring, Galton observed regression toward the mean, in which the offspring exhibit an average phenotype that is less different from the population mean than that of the parents. page 369

"Heritability" refers to the proportion of the total variation in a trait that can be attributed to genetic differences among individuals. page 370

The heritability of a given trait can differ among populations because of differences in genotype or environment. page 370

18.3 Twin studies help separate the effects of genotype and environment on variation in a trait.

Monozygotic, or identical, twins result from the fertilization of a single egg and are genetically identical. page 371

Dizygotic, or fraternal, twins result from the fertilization of two eggs and are genetically related to each other in the same way that other siblings are related to each other. page 371

"Concordance" refers to the percentage of cases in which both members of a pair of twins show the trait when it is known that at least one member shows it. page 371

Comparisons of concordance rates of identical twins and concordance rates of fraternal twins can help to determine to what extent variation in a particular trait has a genetic component. page 371

18.4 Many common diseases and birth defects are affected by multiple genetic and environmental risk factors.

Complex traits are often influenced by many genes with multiple, interacting, and unequal, effects. page 374

Hundreds of genes affect human height. page 374

Personalized medicine tailors treatment to an individual's genetic makeup. page 375

Self-Assessment

1. Explain why some complex traits are also called quantitative traits, and give at least one example.
2. Name several factors that influence variation in complex traits.
3. Explain why it does not make sense to try to separate the effects of genes ("nature") and the environment ("nurture") in a single individual, while it does make sense to separate genetic and environmental effects on differences among individuals.
4. Explain how you would go about determining the relative importance of genes and the environment for variation in risk for a complex trait such as type 2 diabetes.
5. Graph a trait, like human height, with height on the x-axis and number of individuals on the y-axis, and describe the shape of the resulting graph.
6. Explain why the effect of a genotype on a phenotype cannot always be determined without knowing what the environment is, and why the effect of a particular environment on a phenotype cannot always be determined without knowing what the genotype is.
7. Define "regression toward the mean" and explain why it occurs for complex traits.
8. Define the heritability of a trait and explain why the heritability of a given trait depends on the population being studied.
9. Define twin concordance and explain how twin studies can be used to investigate the importance of genetic and environmental factors in the expression of a trait.
10. For a typical complex trait, describe the relationship between the number of genes affecting the trait and the magnitude of their effects on the trait.
11. Explain what personalized medicine is and how it relates to complex traits such as human diseases.

Log in to LaunchPad to check your answers to the Self-Assessment questions, and to access additional learning tools.

CHAPTER 19

Genetic and Epigenetic Regulation

Core Concepts

19.1 The regulation of gene expression in eukaryotes takes place at many levels, including DNA packaging in chromosomes, transcription, and RNA processing.

19.2 After an mRNA is transcribed and exported to the cytoplasm in eukaryotes, gene expression can be regulated at the level of mRNA stability, translation, and posttranslational modification of proteins.

19.3 Transcriptional regulation is illustrated in bacteria by the control of the production of proteins needed for the utilization of lactose, and in viruses by the control of the lytic and lysogenic pathways.

In the discussion of complex traits in Chapter 18, we emphasized the principle that complex traits are not determined by single genes with simple Mendelian inheritance. The traits we usually encounter in ourselves or in others, such as height and weight, diabetes and high blood pressure, are influenced by multiple genes that interact with one another and with environmental factors such as diet and exercise. The number of genes influencing complex traits can be large—for example, hundreds of genes contribute to adult height. The activities of these genes must be coordinated in time (for instance, at a particular point in development) and place (for instance, in a particular type of tissue). For example, growth to normal adult height requires that the production of growth hormone be coordinated in time and place with the production of growth-hormone receptor.

In Chapters 3 and 4, we outlined the basic steps of information flow in a cell, focusing on transcription and translation. In these processes, DNA is a template for the production of messenger RNA (mRNA), which itself is the template for the synthesis of a protein. For these processes to work in a living organism to produce the traits that we see, they must be coordinated so that genes are only **expressed,** or turned on, in the right place and time, and in the right amount. In a multicellular organism, for example, certain genes are expressed in some cells but not in others. Muscle actin and myosin are turned on in muscle cells but not in liver or kidney cells. And even for single-celled organisms like bacteria, certain genes are expressed only in response to environmental signals, such as the availability of nutrients, as we discuss in section 19.3.

Gene regulation encompasses the ways in which cells control gene expression. It can be thought of as the *where? when?* and *how much?* of gene expression. Where (in which cells) are genes turned on? When (during development or in response to changes in the environment) are they turned on? How much gene product is made? This chapter provides an overview of the most common ways in which gene expression is regulated.

One of the important principles we discuss is that gene regulation can occur at almost any step in the path from DNA to mRNA to protein—at the level of the chromosome itself, by controlling transcription or translation, and, perhaps surprisingly, even after the protein is made. Each of these steps or *levels* of gene expression may be subject to regulation. In other words, each successive event that takes place in the expression of a gene is a potential control point for gene expression.

We begin by discussing gene regulation in eukaryotes, and then turn to regulation in prokaryotes and viruses. All life, from the simplest to the most complex, requires gene regulation. Gene regulation relies on similar processes in all organisms because all life shares common ancestry. Nevertheless, certain features of eukaryotes—the packaging of DNA into chromosomes, mRNA processing, and the separation in space of transcription and translation—provide additional levels of gene regulation in eukaryotes that are not possible in prokaryotes.

19.1 CHROMATIN TO MESSENGER RNA IN EUKARYOTES

Gene regulation in multicellular eukaryotes leads to cell specialization: Different types of cell express different genes. The human body contains about 200 major cell types, and although for the most part they share the same genome, they look and function differently from one another because each type of cell expresses different sets of genes. For example, the insulin needed to regulate sugar levels in the blood is produced only by small patches of cells in the pancreas. Every cell in the body contains the genes that encode insulin, but only in these patches of pancreatic cells are they expressed. In this section, we take a look at gene regulation as it occurs in eukaryotic cells, focusing on regulation at the level of DNA, chromatin, and mRNA.

Gene expression can be influenced by chemical modification of DNA or histones.

Fig. 19.1 shows the major places where gene regulation in eukaryotes can take place. The first level of control is at the chromosome, even before transcription takes place. The manner in which DNA is packaged in the nucleus in eukaryotes provides an important opportunity for regulating gene expression. Regulation at this level of gene expression is determined in part by whether the proteins necessary for transcription can gain physical access to the genes they transcribe.

DNA in eukaryotes is packaged as **chromatin,** a complex of DNA, RNA, and proteins that gives chromosomes their structure (Chapters 3 and 13). When chromatin is in its coiled state, the DNA is not accessible to the proteins that carry out transcription. The chromatin must loosen to allow space for transcriptional enzymes and proteins to work. This is accomplished through **chromatin remodeling,** in which the nucleosomes are repositioned to expose different stretches of DNA to the nuclear environment.

One way in which chromatin is remodeled is by chemical modification of the histones around which DNA is wound (**Fig. 19.2**). Modification usually occurs on **histone tails,** strings of amino acids that protrude from the histone proteins in the nucleosome. Individual amino acids in the tails can be modified by the addition (or later removal) of different chemical groups, including methyl groups ($—CH_3$) and acetyl groups ($—COCH_3$). Most often, methylation or acetylation occurs on the lysine residues of the histone tails. Some of these modifications tend to activate transcription and others to repress transcription. The pattern of modifications of the histone tails constitutes a **histone code** that affects chromatin structure and gene transcription. Modification of histones takes place at key times in development to ensure that the proper genes are turned on or off, as well as in response to environmental cues.

In many eukaryotic organisms, gene expression is also affected by chemical modification of certain bases in the DNA, the most

FIG. 19.1 Levels of gene regulation.

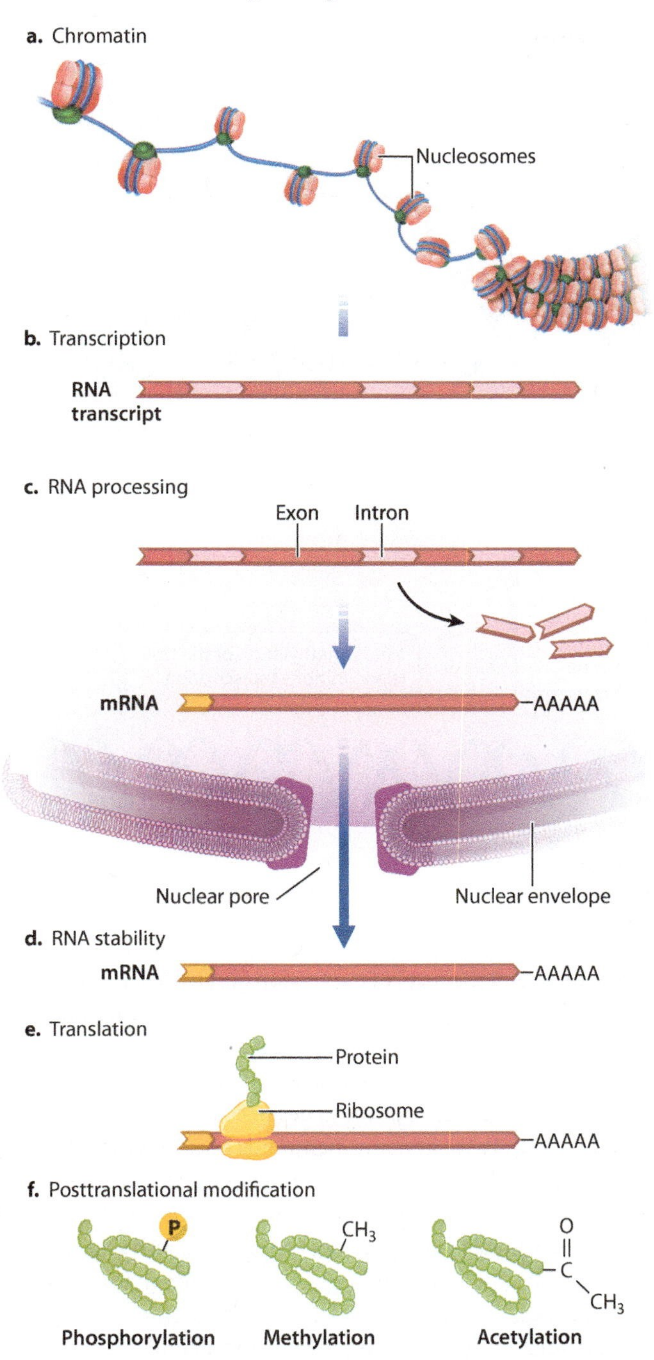

FIG. 19.2 Histone modifications. Typical modifications of the amino acid lysine observed in the histone tails of nucleosomes include addition of a methyl group (Me), addition of three methyl groups (Me_3), and acetylation (Ac).

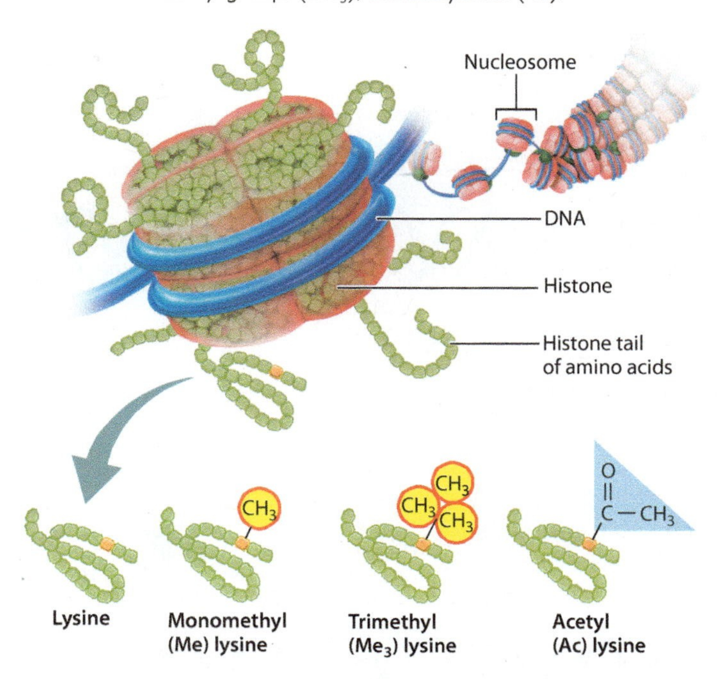

common of which is the addition of a methyl group to the base cytosine (**Fig. 19.3**). DNA methylation recruits proteins that affect changes in chromatin structure, histone modification, and nucleosome positioning that restrict access of transcription factors to gene promoters. Methylation of cytosines often occurs in cytosine bases that are adjacent to guanosine bases on a DNA strand. Cytosine methylation often occurs in **CpG islands,** which are clusters of adjacent CG nucleotides located in or near the promoter of a gene. Heavy cytosine methylation is associated with transcriptional repression of the gene near the CpG island. The methylation state of a CpG island can change over time or in response to environmental signals, providing a way to turn genes on or off. Cells sometimes heavily methylate CpG islands of transposable elements or viral DNA sequences that are integrated into the genome, thus preventing the expression of genes in the foreign DNA (Chapter 14).

Together, the modification of cytosine bases, changes to histones, and alterations in chromatin structure are often termed **epigenetic,** from the Greek *epi-* ("over and above") and "genetic" ("inheritance"). That is, epigenetic mechanisms of gene regulation typically involve changes not to the DNA sequence itself but to the manner in which the DNA is packaged. Epigenetic modifications can in some cases affect gene expression. They can be inherited through mitotic cell divisions, just as genes are, but are often reversible and responsive to changes in the environment. Furthermore, epigenetic modifications present in sperm or eggs are sometimes transmitted from parent to offspring, meaning that these chemical modifications can be inherited.

In humans and other mammals, about 100 genes are repressed by chemical modification such as DNA methylation in the germ line in a sex-specific manner. This sex-specific silencing of gene expression is known as **imprinting.** For some genes, the allele of the gene inherited from the mother is imprinted and therefore

FIG. 19.3 Methylation states of CpG islands in or near the promoter of a gene.

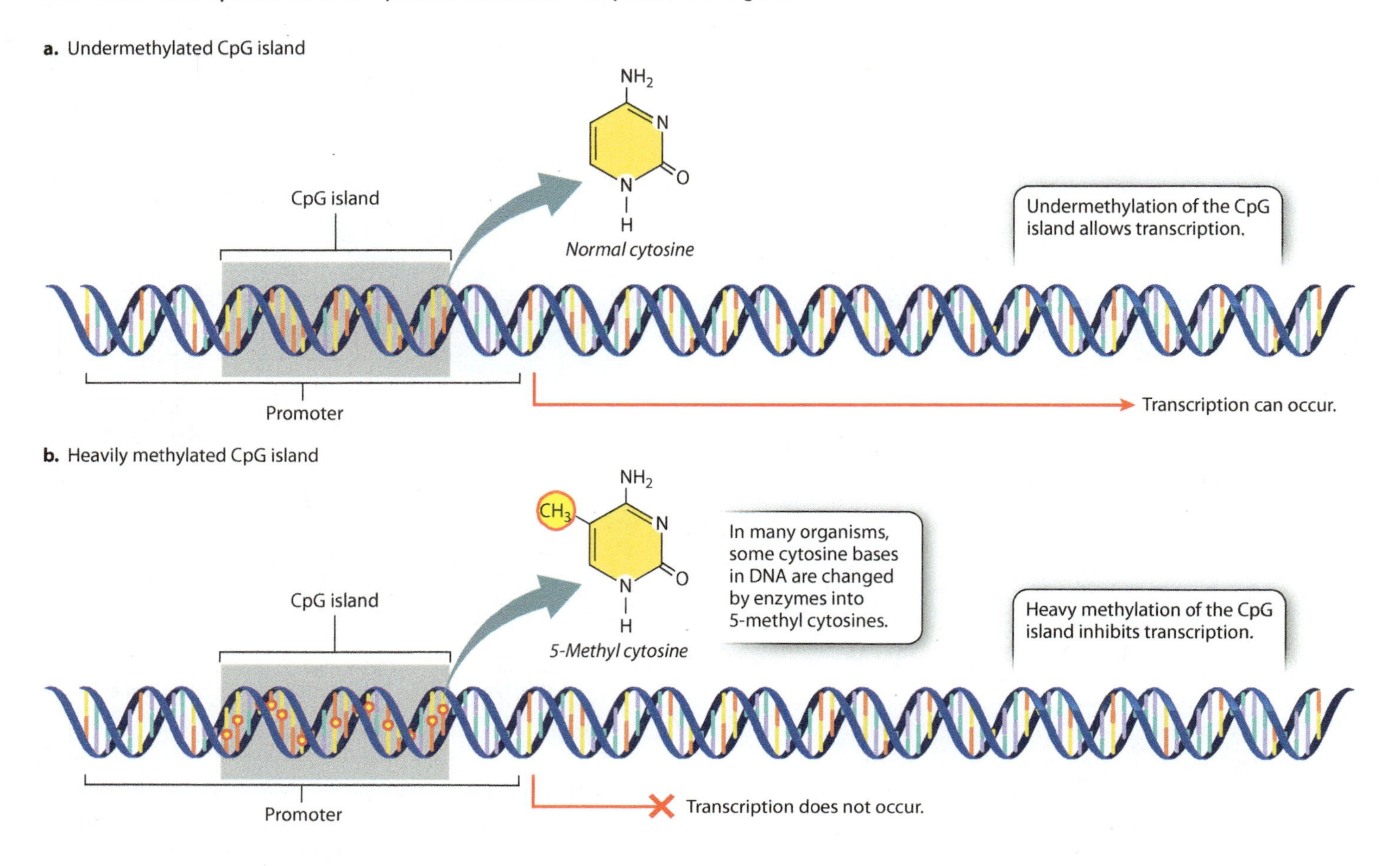

silenced, so only the allele inherited from the father is expressed. For other imprinted genes, the allele of the gene inherited from the father is imprinted and therefore silenced, so only the allele inherited from the mother is expressed. The imprinting persists in somatic cells through the lifetime of the individual but is reset in its germ line according to the individual's sex.

Some of the genes that are imprinted affect the growth rate of the embryo. For example, in matings between lions and tigers that take place in captivity, the offspring of a male tiger with a female lion (a "tigon") is about the same size as its parents, whereas the offspring of a male lion and a female tiger (a "liger") is a giant cat—the largest of any big cat that exists. Much of the size difference between a liger and a tigon is thought to result from whether genes for rapid embryonic growth that are subject to imprinting are inherited from the mother or the father.

Gene expression can be regulated at the level of an entire chromosome.

A striking example of an epigenetic form of gene regulation is the manner in which mammals equalize the expression of X-linked genes in *XX* females and *XY* males. For most genes, there is a direct relation between the number of copies of the gene (the gene dosage) and the level of expression of the gene. An increase in gene dosage increases the level of expression because each copy of the gene is regulated independently of other copies. For example, as we saw in Chapter 15, the presence of an extra copy of most human chromosomes results in spontaneous abortion because of the increase in the expression of the genes in that chromosome.

XX females have twice as many *X* chromosomes as *XY* males. For genes located in the *X* chromosome, the dosage of genes is twice as great in females as it is in males. However, the level of expression of *X*-linked genes is about the same in both sexes. These observations imply that the regulation of genes in the *X* chromosome is different in females and in males. This differential regulation is called **dosage compensation.**

Different species have evolved different mechanisms of dosage compensation. In the fruit fly *Drosophila,* males double the transcription of the single *X* chromosome to achieve equal expression compared to the two *X* chromosomes in females. In the nematode worm *Caenorhabditis,* transcription of both *X* chromosomes in females is decreased to one-half the level of the single *X* chromosome in males. In mammals, including humans, dosage compensation occurs through the inactivation of one *X* chromosome in each cell in females. This process, known as **X-inactivation,** was first proposed by Mary F. Lyon in the early 1960s.

FIG. 19.4 ***X*-inactivation in female mammals.** *X*-inactivation equalizes the expression of most genes in the *X* chromosome between *XX* females and *XY* males. *Source: b. Eyal Nahmias/Alamy.*

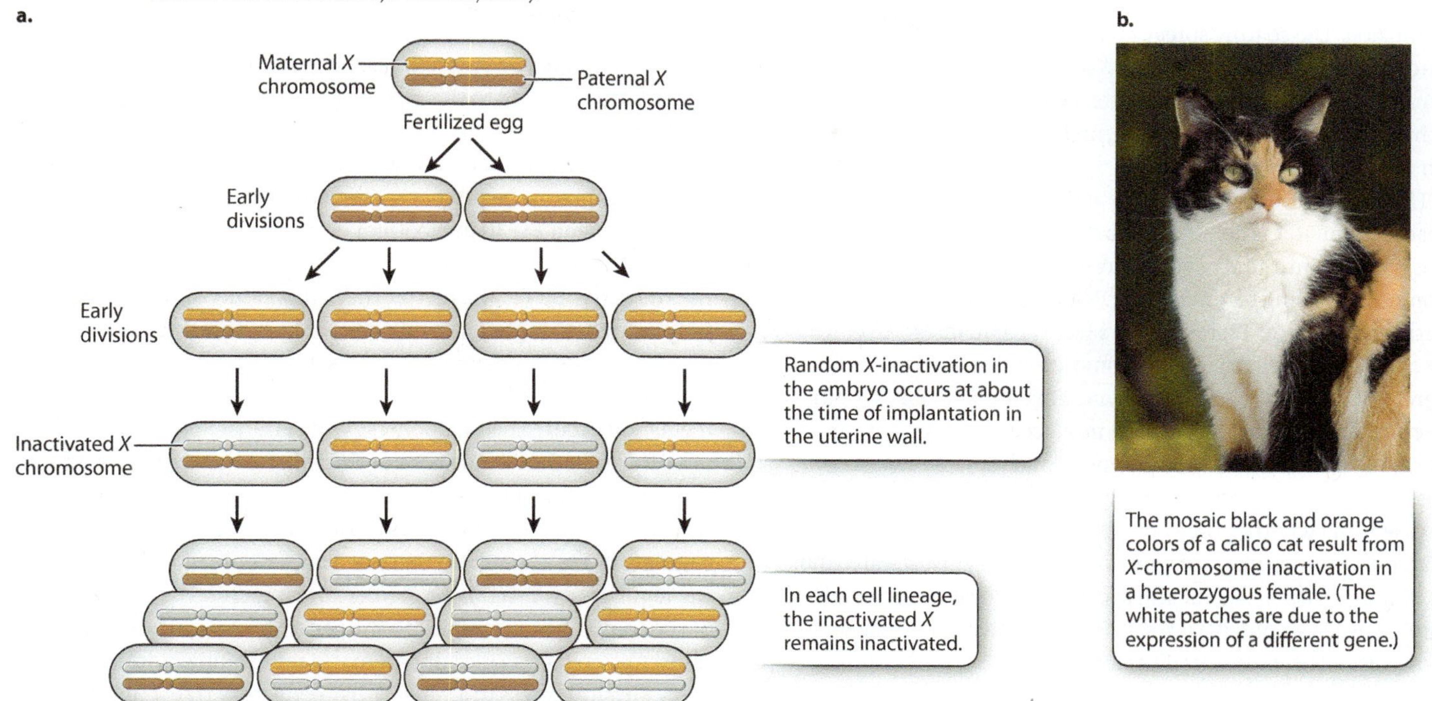

Soon after a fertilized egg with two X chromosomes implants in the mother's uterine wall, one X chromosome at random is inactivated (**Fig. 19.4**). In Fig. 19.4a, the inactive X chromosome is shown in gray. The inactive state persists through cell division, so in each cell lineage, the same X chromosome that was originally inactivated remains inactive. The result is that a normal female is a mosaic, or patchwork, of tissue. In some patches, the genes in the maternal X chromosome are expressed (and the paternal X chromosome is inactivated), whereas in other patches, the genes in the paternal X chromosome are expressed (and the maternal X chromosome is inactivated). The term "inactive X chromosome" is a slight exaggeration since a substantial number of genes are still transcribed, although usually at a low level.

As one argument for her X-inactivation hypothesis, Lyon called attention to calico cats, which are nearly always female (Fig. 19.4b). In calico cats, the orange and black fur colors are due to different alleles of a single gene in the X chromosome. In a heterozygous female, X-inactivation predicts discrete patches of orange and black, and this is exactly what is observed. (The white patches on a calico cat are due to an autosomal gene.)

How does X-inactivation work? Some of the details are still unknown, but the main features of the process are shown in **Fig. 19.5.** A key player is a small region in the X chromosome called the X-chromosome inactivation center (*XIC*), which contains a gene called *Xist* (X-inactivation specific transcript). The *Xist* gene is normally transcribed at a very low level, and the RNA is unstable, but in an X chromosome about to become inactive, *Xist* transcription markedly increases. The transcript undergoes RNA

FIG. 19.5 **The role of *Xist* in *X*-inactivation.** The *Xist* transcript is a noncoding RNA that is expressed from the inactive *X* chromosome and coats the entire chromosome, leading to inactivation of most of the genes in the chromosome.

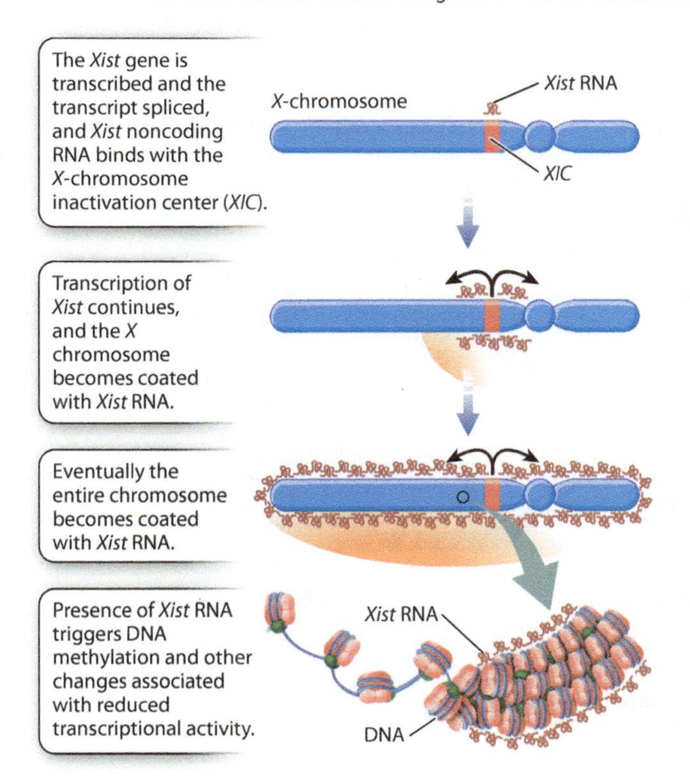

splicing, but it does not encode a protein. *Xist* RNA is therefore an example of a noncoding RNA (Chapter 3). Instead of being translated, the processed *Xist* RNA coats the *XIC* region, and as it accumulates, the coating spreads outward from the *XIC* until the entire chromosome is coated with *Xist* RNA. The presence of *Xist* RNA along the chromosome recruits factors that promote DNA methylation, histone modification, and other changes associated with transcriptional repression.

The *Xist* gene is both necessary and sufficient for X-inactivation: If it is deleted, X-inactivation does not occur; if it is inserted into another chromosome, it inactivates that chromosome. Using recombinant DNA techniques of the sort described in Chapter 12, researchers were able to insert *Xist* into one of the three copies of chromosome 21 present in cells taken from a patient with Down syndrome. Remarkably, *Xist* in its new location in chromosome 21 produced an RNA transcript that coated that copy of chromosome 21 and repressed most transcription of its genes! Whether all genes in chromosome 21 are repressed, and whether they are as fully repressed as those in the inactive X chromosome, are questions still being investigated, but this finding raises clear (but still speculative) ideas about how *Xist* might be used as a potential treatment for Down syndrome and various other chromosomal disorders.

Transcription is a key control point in gene expression.

While access to DNA and appropriate histone modifications are necessary for transcription, they are not sufficient. The molecular machinery that actually carries out transcription is also required once the template DNA is made accessible through chromatin remodeling and histone modification. The mechanisms that regulate whether or not transcription occurs are known collectively as **transcriptional regulation** (see Fig. 19.1b).

Transcriptional regulation in eukaryotic cells requires the coordinated action of many proteins that interact with one another and with DNA sequences near the gene. Let's first review the basic process of transcription in eukaryotes (Chapter 3). First, an important group of proteins called **general transcription factors** are recruited to the gene's promoter, which is the region of a gene where transcription is initiated. The transcription factors are brought there by one of the proteins that bind to a short sequence in the promoter called the TATA box, which is usually situated 25–30 nucleotides upstream of the nucleotide site where transcription begins. Once bound to the promoter, the transcription factors recruit the components of the **RNA polymerase complex,** the enzyme complex that synthesizes the RNA transcript complementary to the template strand of DNA.

Where in the many steps of transcription initiation does regulation occur? The first point at which transcription can be regulated is in the recruitment of the general transcription factors and components of the RNA polymerase complex (**Fig. 19.6**). Recruitment of these elements is controlled by proteins called **regulatory transcription factors.** Transcription does not occur if the regulatory transcription factors do not recruit the components of the transcription complex to the gene. Some regulatory transcription factors recruit chromatin remodeling proteins that allow physical access to a gene. Other regulatory transcription factor have two binding sites, one of which binds with a particular DNA sequence in or near a gene known as an **enhancer** (Fig. 19.6) and the other of which recruits one or more general transcription factors to the promoter region. The general transcription factors then recruit the RNA polymerase complex, and transcription can begin (Chapter 3).

Hundreds of different regulatory transcription factors control the transcription of thousands of genes. Some bind with enhancers and stimulate transcription; others bind with DNA sequences known as **silencers** and repress transcription. Enhancers and silencers are often in or near the genes they regulate, but in some cases they may be many thousands of nucleotides distant from the genes. A typical gene may be regulated by multiple enhancers and silencers of different types, each with one or more regulatory transcription factors that can bind with it. Transcription takes place only when the proper combination of regulatory transcription factors is present in the same cell, as shown in Fig. 19.6. Since transcription of a gene with multiple silencers and enhancers depends on the presence of a particular combination of regulatory transcription factors, this type of regulation is called **combinatorial control.**

FIG. 19.6 **Protein–DNA and protein–protein interactions in the eukaryotic transcription complex.**

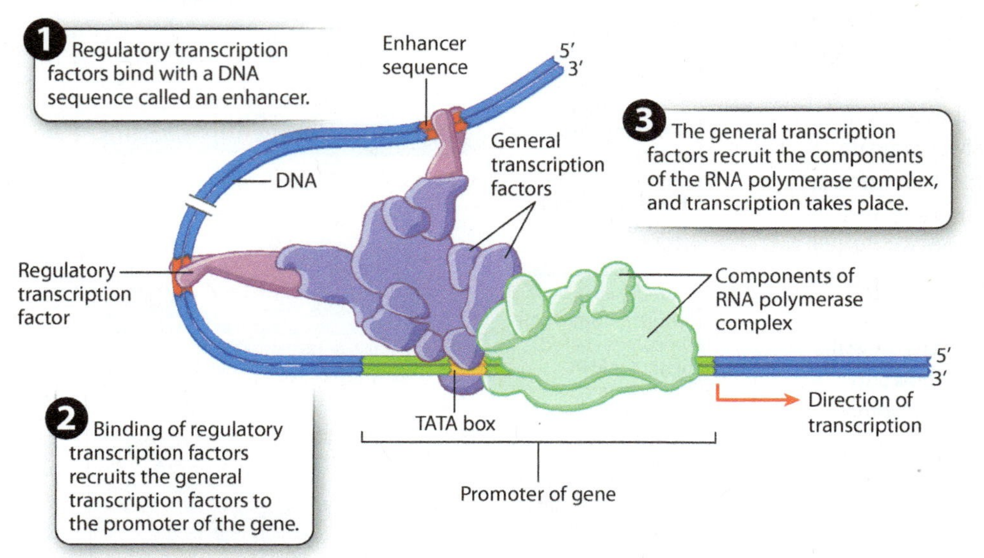

→ **Quick Check 1** The idea that the expression of some genes is controlled by the products of other genes was originally criticized in this way: If *n* genes were to be controlled, then another *n* genes would be needed to control them, and then another *n* genes would be needed to control the controllers, and so on and on. How does combinatorial control help refute this criticism?

RNA processing is also important in gene regulation.

A great deal happens in the nucleus after transcription takes place. The initial transcript, called the **primary transcript,** undergoes several types of modification, collectively called **RNA processing** (Chapter 3), that includes the addition of a nucleotide cap to the 5′ end and a string of tens to hundreds of adenosine nucleotides to the 3′ end to form the poly(A) tail. These modifications are necessary for the RNA molecule to be transported to the cytoplasm and recognized by the translational machinery. The poly(A) tail also helps to determine how long the RNA will persist in the cytoplasm before being degraded. RNA processing is therefore an important point where gene regulation can occur (see Fig. 19.1c).

In eukaryotes, the primary transcript of many protein-coding genes is far longer than the messenger RNA ultimately used in protein synthesis. The long primary transcript consists of regions that are retained in the messenger RNA (the **exons**) interspersed with regions that are excised and degraded (the **introns**). The introns are excised during **RNA splicing** (Chapter 3). The exons are joined together in their original linear order to form the processed messenger RNA.

RNA splicing provides an opportunity for regulating gene expression because the same primary transcript can be spliced in different ways to yield different proteins in a process called **alternative splicing.** This process takes place because what the spliceosome—the splicing machinery—recognizes as an exon in some primary transcripts, it recognizes as part of an intron in other primary transcripts. The alternative-splice forms may be produced in the same cells or in different types of cell. Alternative splicing accounts in part for the observation that we produce many more proteins than our total number of genes. By some estimates, over 90% of human genes undergo alternative splicing.

Fig. 19.7 shows the primary transcript of a gene encoding an insulin receptor found in humans and other mammals. During RNA splicing in liver cells, exon 11 is included in the messenger RNA, and the insulin receptor produced from this messenger RNA has low affinity for insulin. In contrast, in cells of skeletal muscle, the 36 nucleotides of exon 11 are spliced out of the primary transcript along with the flanking introns. The resulting protein is 12 amino acids shorter, and this form of the insulin receptor has high affinity for insulin. The different forms of the protein are important: The higher sensitivity of muscle cells to insulin enables them to absorb enough glucose to fulfill their energy needs.

Some RNA molecules can become a substrate for enzymes that modify particular bases in the RNA, thereby changing its sequence and what it codes for. This process is known as **RNA editing** (**Fig. 19.8**). One type of editing enzyme (Fig. 19.8a) removes the amino group ($-NH_2$) from adenosine and converts it to inosine, a base that in translation functions like guanosine. Another enzyme (Fig. 19.8b) removes the amino group from cytosine and converts it to uracil. In the human genome, hundreds if not thousands of transcripts undergo RNA editing. In many cases, not all copies of the transcript are edited, and some copies may be edited more

FIG. 19.7 Alternative splicing of a mammalian insulin-receptor transcript. Alternative splicing generates different processed mRNAs and different proteins from the same primary transcript.

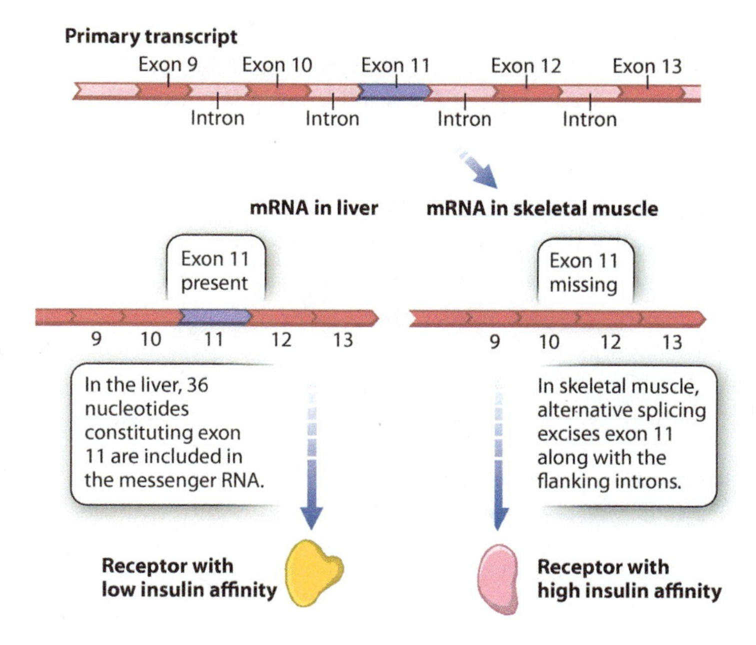

FIG. 19.8 RNA editing. RNA editing results in chemical modifications to the bases in mRNA, which can lead to changes in the amino acid sequence of the protein.

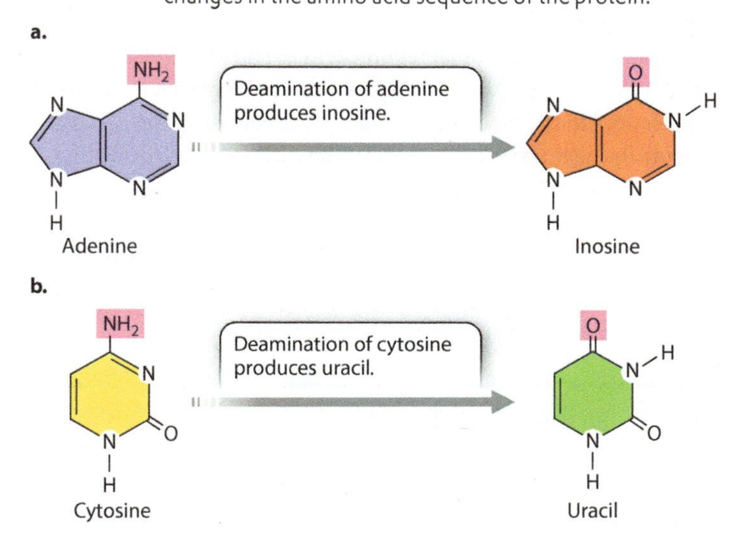

FIG. 19.9 Tissue-specific RNA editing of the human apolipoprotein B transcript. Different RNA editing in (a) the liver and (b) the intestine results in proteins with different functions.

a. Processed apolipoprotein mRNA in liver

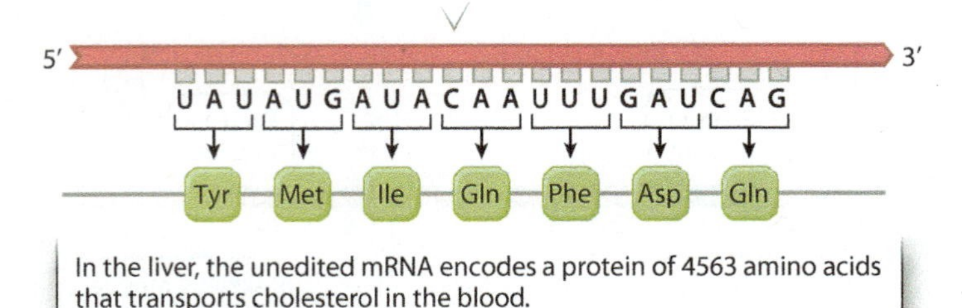

In the liver, the unedited mRNA encodes a protein of 4563 amino acids that transports cholesterol in the blood.

b. Processed apolipoprotein mRNA in intestine

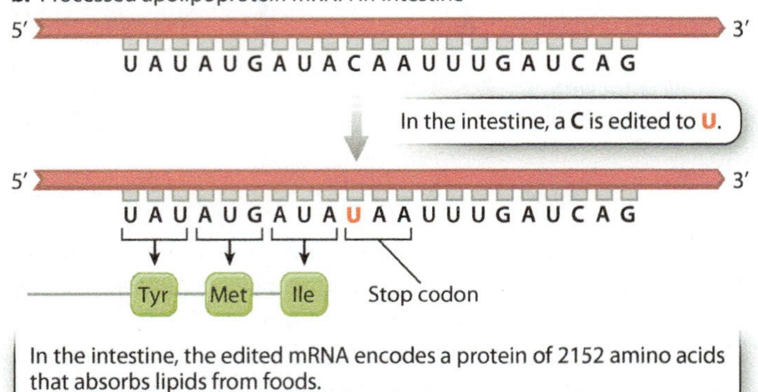

In the intestine, the edited mRNA encodes a protein of 2152 amino acids that absorbs lipids from foods.

extensively than others. The result is that transcripts from the same gene can produce multiple types of proteins even in a single cell.

Transcripts from the same gene may undergo different editing in different cell types. An example of tissue-specific RNA editing is shown in **Fig. 19.9.** The mRNA fragments show part of the coding sequence for apolipoprotein B. The unedited mRNA in the liver (Fig. 19.9a) is translated into a protein that transports cholesterol in the blood. In contrast, RNA editing of the message occurs in the intestine (Fig. 19.9b). The cytosine nucleotide in codon 2153 is edited to uracil. The edited codon is UAA, which is a stop codon. Translation therefore terminates at this point, releasing a protein only about half as long as the liver form. This shorter form of the protein helps the cells of the intestine absorb lipids from the foods we eat.

19.2 MESSENGER RNA TO PHENOTYPE IN EUKARYOTES

In eukaryotes, a processed mRNA must exit the nucleus before the translation step of gene expression can occur. The mRNA migrates to the cytoplasm through one of a few thousand nuclear pores, large protein complexes that span both layers of the nuclear envelope and regulate the flow of macromolecules in and out of the nucleus. Once the mRNA is in the cytoplasm, there are multiple opportunities for gene regulation at the levels of mRNA stability, translation, and protein activity (see Figs. 19.1d–19.1f).

Small regulatory RNAs inhibit translation or promote mRNA degradation.

Regulatory RNA molecules known as **small regulatory RNAs** are among the most exciting recent discoveries in gene regulation. They are of exceptional interest to biologists and drug researchers because their small size allows easy synthesis in the laboratory, and researchers can design their sequences to target transcripts of interest.

Two important types of small regulatory RNAs are known as **siRNA (small interfering RNA)** and **miRNA (microRNA).** While they have somewhat different functions, they share similarities in their production (**Fig. 19.10**). Both types of small RNA are transcribed from DNA and form hairpin structures, or stem-and-loops, stabilized by base pairing in the stem. Enzymes in the cytoplasm specifically recognize these structures and cleave the stem from the hairpin, then further cut the stem into small, double-stranded fragments 20–25 base pairs in length.

One of the two strands from each RNA fragment is incorporated into a protein complex known as **RISC (RNA-induced silencing complex).** The small, single-stranded RNA targets the RISC to specific RNA molecules by base pairing with short regions in the target. Depending on the type of small regulatory RNA, the RNA sequence in the RISC, and the particular type of RISC, the small regulatory RNA may result in chromatin remodeling, degradation of RNA transcripts, or inhibition of mRNA translation (Fig. 19.10). Other functions have also been reported.

Regulation by small RNAs is widespread in eukaryotes, and it is thought to have evolved originally as a defense against viruses and transposable elements. These molecules have also evolved to have important regulatory roles in their own right. Human chromosomes are thought to encode about 1000 microRNAs, each of which can target tens or hundreds of mRNA molecules. Half or more of human proteins may have their synthesis regulated in part by such small regulatory RNAs.

→ **Quick Check 2** How do small regulatory RNAs differ from messenger RNA?

Translational regulation controls the rate, timing, and location of protein synthesis.

Translation of mRNA into protein provides another level of control of gene expression. **Fig. 19.11** shows the structure of a hypothetical mRNA molecule in a eukaryotic cell and highlights some of the features that help regulate its translation (Chapter 4). Not all mRNA molecules have all the features shown, but almost all mRNA molecules have a 5′ cap, a 5′ untranslated region (5′ UTR), an open reading frame (ORF) containing the codons that determine the

FIG. 19.10 **Production of small regulatory RNAs and an overview of their functions.**

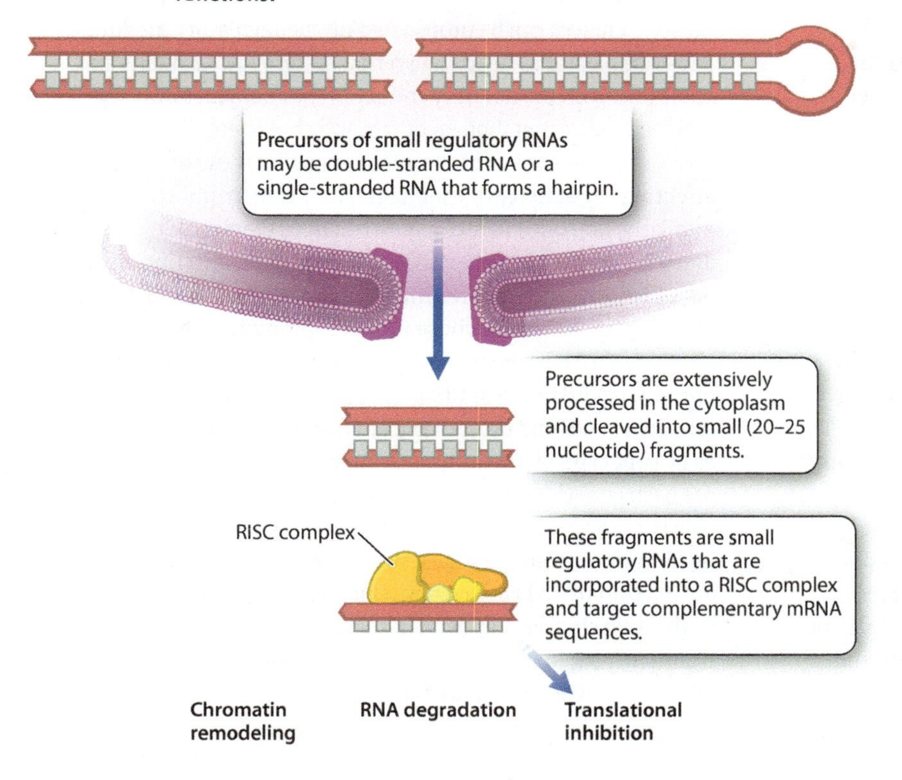

FIG. 19.11 **Some features of mRNA that affect gene expression.** These include the 5′ cap, 5′ untranslated region (5′ UTR), 3′ untranslated region (3′ UTR), and poly(A) tail.

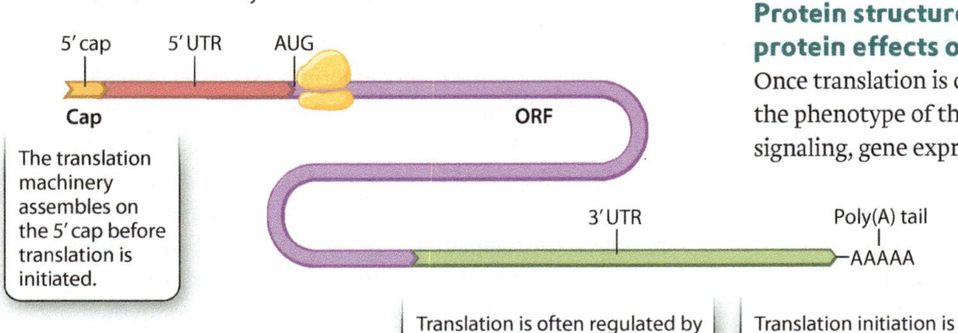

amino acid sequence of the protein, a 3′ untranslated region (3′ UTR), and a poly(A) tail. The 5′ UTR and 3′ UTR may contain regions that bind with proteins. These RNA-binding proteins help control mRNA translation and degradation. The UTRs may also contain binding sites for small regulatory RNAs. During development, as we will see in Chapter 20, some RNA-binding proteins interact with molecular motors that transport the mRNA to particular regions of the cell. In other cases, the proteins are only in particular locations in the cell and repress translation of the mRNAs that are transported there and to which they bind. By either transport or repression, these proteins cause the mRNA to be translated only in certain places in the cell.

The cap structure is one of the main recognition signals for translation initiation, which requires the coordinated action of about 25 proteins. These proteins are present in most cells in limiting amounts, and so at any one time while some mRNAs from a gene are being translated, other mRNAs transcribed from the same gene may not have a translation initiation complex assembled. Upon formation, the initiation complex moves along the 5′ UTR, scanning for an AUG codon (the initiation codon) to allow the complete ribosome to assemble and begin translation.

The 3′ UTR and the poly(A) tail are also important in translation initiation. The efficiency of translation initiation is greatly increased by physical contact between a protein that binds the poly(A) tail and one that binds the 5′ cap. The physical contact creates a loop in the mRNA, bringing the 3′ end of the mRNA into proximity with the start site for translation. In fact, most mRNA sequences that regulate translation are present in the 3′ UTR.

Although translation initiation is the principal mode of translational regulation, not all mRNA molecules are equally accessible to translation. Among the key variables are the secondary (folded) structure of the 5′ UTR, the distance from the 5′ cap to the AUG initiation codon, and the sequences flanking the AUG initiation codon.

Protein structure and chemical modification modulate protein effects on phenotype.

Once translation is completed, the resulting protein can alter the phenotype of the cell or organism by affecting metabolism, signaling, gene expression, or cell structure. After translation, proteins are modified in multiple ways that regulate their structure and function. Collectively, these processes are called **posttranslational modification** (see Fig. 19.1e). Regulation at this level is essential because some proteins are downright dangerous. For example, proteases such as the digestive enzyme trypsin must be kept inactive until secreted out of the cell. If they were not, their activity would kill the cell. These types of protein are often controlled by being translated in inactive forms that are made active by modification after secretion.

Folding and acquiring stability are key control points for some proteins (Chapter 4). While many proteins fold properly as they come off the ribosome, others require help from other proteins, called chaperones, which act as folding facilitators. Correct folding is important because improperly folded proteins may form aggregates that are destructive to cell function. Many diseases are associated with protein aggregates, including Alzheimer's disease, Huntington's disease, and the human counterpart of mad cow disease.

Posttranslational modification also helps regulate protein activity. Many proteins are modified by the addition of one or more sugar molecules to the side chains of some amino acids. This modification can alter the protein's folding and stability, or target the molecule to particular cellular compartments. Reversible addition of a phosphate group to the side groups of amino acids such as serine, threonine, or tyrosine is a key regulator of protein activity (Chapter 9). Introduction of the negatively charged phosphate group alters the conformation of the protein, in some cases switching it from an inactive state to an active state and in other cases the reverse. Because the function of a protein molecule results from its shape and charge (Chapter 4), a change in protein conformation affects protein function.

Marking proteins for enzymatic destruction by the addition of chemical groups after translation is also important in controlling their activity. For example, we have seen how the destruction of successive waves of cyclin proteins helps move the cell through its division cycle (Chapter 11).

? CASE 3 YOU, FROM A TO T: YOUR PERSONAL GENOME

How do lifestyle choices affect expression of your personal genome?

If you examine Fig. 19.1 as a whole and consider the DNA sequence shown as your personal genome, the situation looks pretty grim. You might be led to believe that genes dictate everything, and that biology is destiny. But if you focus on the lower levels of regulation in Fig. 19.1, a different picture emerges. The picture is different because much of the regulation that occurs after transcription (regulation of mRNA stability, regulation of translation, posttranslational modifications) is determined by the physiological state of your cells, which in turn is strongly influenced by your lifestyle choices. For example, your cells can synthesize 12 of the amino acids in proteins, but if any of these is present in sufficient amounts in your diet, it is absorbed during digestion and not synthesized. The amino acid you ingest blocks the synthetic pathway through feedback effects.

The effect of an intervention—genetic or environmental—at any given level can affect regulatory processes at both higher and lower levels. This cascade of regulatory effects in both directions can occur because the expression of any gene is regulated at multiple levels, and because there is much feedback and signaling back and forth between nucleus and cytoplasm. It is because of these feedback and signaling mechanisms that the effects of lifestyle choices can be propagated up the regulatory hierarchy. For example, it has been shown that dietary intake of fats and cholesterol affects not only the activity of enzymes directly involved in the metabolism of fats and cholesterol, but also the levels of transcription of the genes encoding these enzymes by affecting the activity of their regulatory transcription factors. Similarly, lifestyles that combine balanced diets with exercise and stress relief have been shown to increase transcription of genes whose products prevent cellular dysfunction and decrease transcription of genes whose products promote disease.

So far, we have been talking primarily about complex traits of the type discussed in Chapter 18, which are affected by multiple genes and by multiple environmental factors as well as by genotype-by-environment interaction. For example, there are both genetic and environmental risk factors for breast and ovarian cancers, as we have seen. Simple Mendelian traits caused by mutations in single genes, such as cystic fibrosis and alpha-1 antitrypsin (α-1AT) deficiency (Chapter 17), are less responsive to lifestyle choices. But even in these cases, lifestyle matters. For example, people with α-1AT deficiency should not smoke tobacco and should avoid environments with low air quality.

19.3 TRANSCRIPTIONAL REGULATION IN PROKARYOTES

The central message of Fig. 19.1 is that the regulation of gene expression occurs through a hierarchy of regulatory mechanisms acting at different levels (and usually at multiple levels) from DNA to protein. Gene regulation in prokaryotes is simpler than gene regulation in eukaryotes since DNA is not packaged into nucleosomes, mRNA is not processed, and transcription and translation are not separated by a nuclear envelope. In prokaryotes, expression of a protein-coding gene entails transcription of the gene into messenger RNA and translation of the messenger RNA into protein. Each of these levels of gene expression is subject to regulation.

Because gene regulation in prokaryotes is simpler than gene regulation in eukaryotes, prokaryotes have served as model organisms for our understanding of how genes are turned on and off. In this section, we consider in more detail how gene expression is regulated at the level of transcription in bacteria and in viruses that infect bacteria. We focus on two well-studied systems: (1) the regulation of genes in the intestinal bacterium *Escherichia coli* that allows proteins needed to utilize the sugar lactose to be produced only when lactose is present in the environment and only when it is the best nutrient available, and (2) the regulation of genes in a virus that infects *E. coli* that controls whether the virus integrates its DNA into the bacterial host or lyses (breaks open) the cell. In both cases, specific genes are turned on and off in response to environmental conditions.

Transcriptional regulation can be positive or negative.

In both eukaryotes and prokaryotes, transcription can be positively or negatively regulated. In **positive regulation,** a regulatory molecule (usually a protein) must bind to the DNA at a site near the gene in order for transcription to take place. In **negative regulation,** a regulatory molecule (again, usually a protein) must bind to the DNA at a site near the gene in order for transcription to be prevented. Most promoters contain one or more short sequences that function to help recruit the proteins needed for transcription when transcription of the gene is activated. In eukaryotes, many promoters contain the sequence 5′-TATAAA-3′ (or some variant of this sequence) about 25–35 base pairs upstream from the transcription start site, which helps recruit the proteins needed for transcription. Many prokaryotic promoters contain a pair of such regulatory sequences located about 10 and 35 base pairs upstream of the transcription start site. These promoter sequences are called the −10 and −35 sequence motifs.

Fig. 19.12 illustrates positive regulation. The main players are DNA, the RNA polymerase complex, and a regulatory protein called a transcriptional **activator.** As shown in Fig. 19.12a, when the activator protein is present in a state that can interact with its binding site in the DNA, the RNA polymerase complex is recruited to the promoter of the gene and transcription takes place. When the activator is not present, or not able to bind with the DNA (Fig. 19.12b), the RNA polymerase is not recruited to the promoter and transcription does not occur. The binding site for the activator may be upstream of the promoter, as shown in the figure, be downstream of the promoter, or even overlap the promoter.

FIG. 19.12 Role of the activator in positive transcriptional regulation in prokaryotes. When an activator protein binds to DNA, it promotes transcription of a gene.

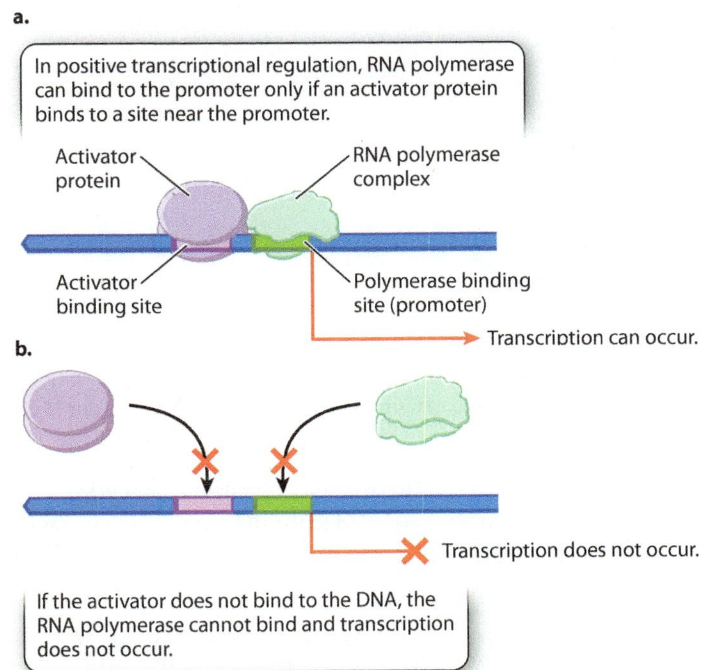

FIG. 19.13 Role of the repressor in negative transcriptional regulation in prokaryotes. When a repressor protein binds to DNA, it prevents transcription of a gene.

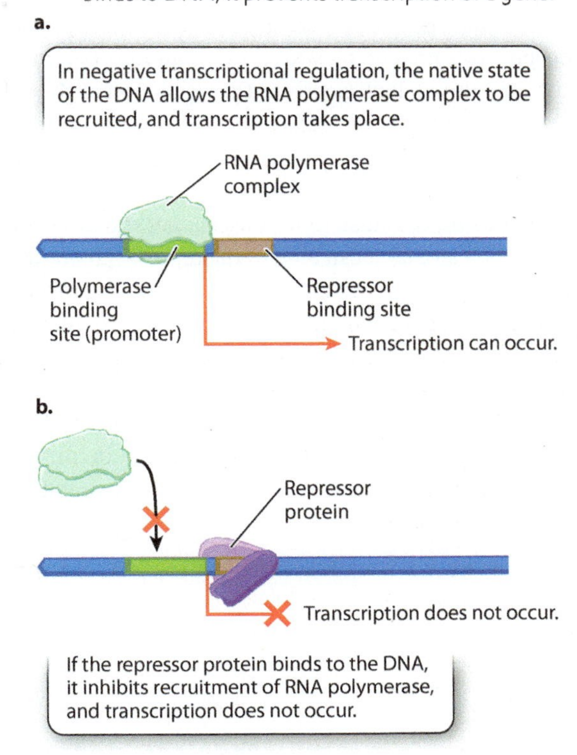

Sometimes, the activator protein combines with a small molecule in the cell and undergoes a change in shape that alters its binding affinity for DNA. The change in shape is an example of an **allosteric effect** (Chapter 6). In some cases, the activator cannot bind with DNA on its own, but combining with the small molecule changes its shape and its ability to bind with particular sequences in the DNA. The presence of the small molecule in the cell therefore results in transcription of the gene. Genes subject to this type of positive control typically encode proteins needed only when the small molecule is present in the cell. For example, in *E. coli*, the genes for the breakdown of the sugar arabinose are normally repressed in the absence of the sugar because a regulatory protein binds the DNA in such a way as to prevent transcription. In the presence of arabinose, however, the shape of the protein is altered so that it binds a different site in the DNA, allowing transcription of the genes. For other genes, activator proteins can bind DNA only when a small molecule is absent from the cell. These genes typically encode proteins needed for synthesis of the small molecule. In *E. coli*, the genes for synthesis of the amino acid cysteine are regulated in this fashion.

Fig. 19.13 illustrates negative regulation. In this case, the DNA in its native state can recruit the RNA polymerase complex, and transcription takes place at a constant rate unless something turns it off (Fig. 19.13a). What turns it off is binding with a protein called a **repressor** (Fig. 19.13b). Again, the binding site for the repressor can be upstream of the promoter, downstream of the promoter, or overlapping with the promoter.

As with positive control, the ability of the repressor to bind DNA is often determined by an allosteric interaction with a small molecule. A small molecule that interacts with the repressor and prevents it from binding DNA is called an **inducer.** In the next section, we will look closely at the regulation of genes for the breakdown of the sugar lactose in *E. coli,* and we will see that a derivative of lactose binds to and inhibits the repressor and is therefore an inducer of these genes.

In other cases, the small molecule changes the conformation of the repressor so that it can bind to the repressor binding site and inhibit transcription. Genes regulated in this way are often needed for synthesis of the small molecule. In *E. coli,* for example, the genes for the synthesis of the amino acid tryptophan are negatively regulated by tryptophan. When tryptophan is present in sufficient amounts, it binds with a regulatory protein to form the functional repressor, and transcription of the genes does not occur. When the level of tryptophan drops too low to form the repressor, transcription of the genes is initiated.

→ **Quick Check 3** Explain how the ability of a large, multisubunit protein molecule to bind a specific DNA sequence can be altered when it binds with a small molecule no larger than a single amino acid.

Lactose utilization in *E. coli* is the pioneering example of transcriptional regulation.

The principle that the product of one gene can regulate transcription of other genes was first discovered in the 1960s by François Jacob and Jacques Monod, who studied how the bacterium *E. coli* regulates production of the proteins needed for utilization of the sugar lactose. Lactose consists of one molecule each of the sugars glucose and galactose covalently linked by a β-(beta-)galactoside bond. An enzyme called β-galactosidase cleaves lactose, releasing glucose and galactose. Both molecules can then be broken down and used as a source of carbon and energy (Chapter 7).

Jacob and Monod began their research with the observation that active β-galactosidase enzyme is observed in cells only in the presence of lactose or certain molecules chemically similar to lactose. Why is this so? **Fig. 19.14** describes and tests two hypotheses. One is that lactose stabilizes an unstable form of β-galactosidase produced by all cells all the time. The other is that lactose leads to the expression of the gene for β-galactosidase. The experiment shown in Fig. 19.14 demonstrated that lactose turns on the β-galactosidase gene and does not stabilize or activate the enzyme encoded by the gene. How lactose activates expression of the β-galactosidase gene was the subject of additional experiments. These Nobel Prize–winning follow-up studies of Jacob and Monod gave the first demonstrated example of transcriptional gene regulation.

To understand how the genes for lactose utilization are regulated, you need to know the main players, and these are shown in **Fig. 19.15.** The overall situation looks quite complicated, but when you take it apart and look at the individual pieces, it is in fact quite

HOW DO WE KNOW?

FIG. 19.14

How does lactose lead to the production of active β-galactosidase enzyme?

BACKGROUND Active β-galactosidase enzyme is observed only in *E. coli* cells that are growing in the presence of lactose.

HYPOTHESIS One hypothesis is that the enzyme is always being produced, but is produced in an unstable form that breaks down rapidly in the absence of lactose. A second hypothesis is that the enzyme is stable, but is produced only in the presence of lactose.

EXPERIMENT François Jacob and Jacques Monod exposed a culture of growing cells to lactose and later removed it. They measured the amount of β-galactosidase present in the culture during the experiment.

RESULTS Almost immediately upon addition of lactose, β-galactosidase began to accumulate, and its amount steadily increased. When lactose was removed, the enzyme did not disappear immediately (as would be the case if it were unstable). Instead, the amount of enzyme remained the same as when lactose was present. This result is expected only if β-galactosidase is a stable enzyme that is synthesized when lactose is added and stops being synthesized when lactose is removed.

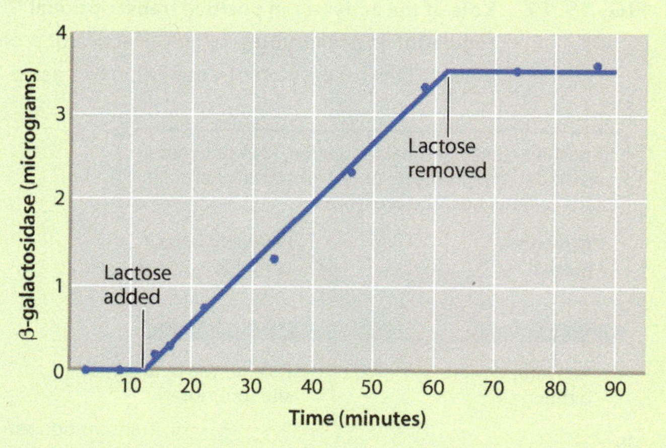

CONCLUSION Synthesis of β-galactosidase is turned on when lactose is added and turned off when lactose is removed, in support of the second hypothesis.

FOLLOW-UP WORK These results stimulated pioneering experiments that ultimately led to the discovery of the lactose operon and the mechanism of transcriptional regulation by a repressor protein.

SOURCE Monod, J. 1965. "From Enzymatic Adaption to Allosteric Transitions." Nobel Prize lecture. http://nobelprize.org/nobel_prizes/medicine/laureates/1965/monod-lecture.htm.

FIG. 19.15 Structural and regulatory elements of the lactose operon. The *lac* operon consists of the promoter, the operator, and all the structural genes that are transcribed into a single mRNA called a polycistronic mRNA.

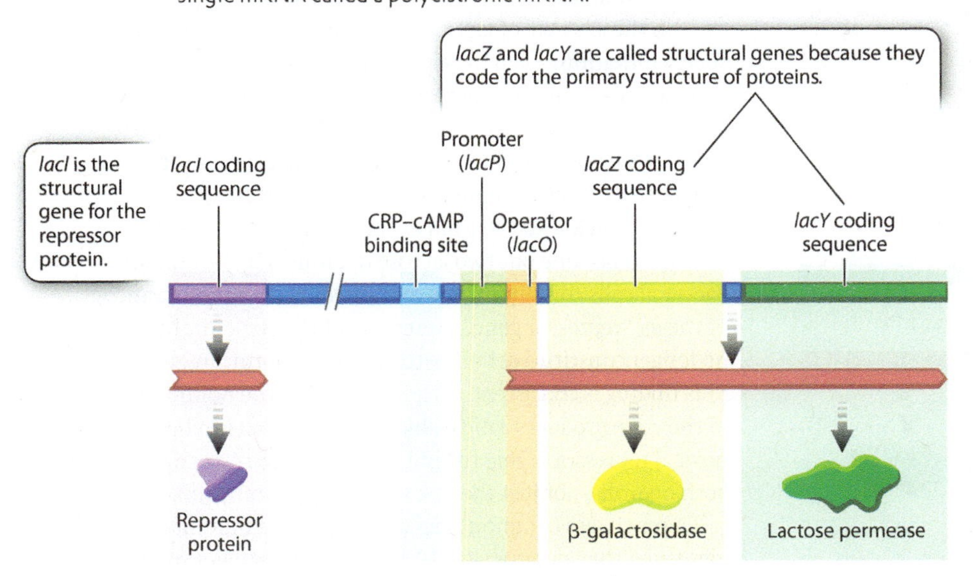

simple. Let us start with the two coding sequences on the right in the figure:

- *lacZ* is the gene (coding sequence) for the enzyme β-galactosidase, which cleaves the lactose molecule into its glucose and galactose constituents. A functional, nonmutant form of the *lacZ* gene is denoted *lacZ*$^+$.
- *lacY* is the gene (coding sequence) for the protein lactose permease, which transports lactose from the external medium into the cell. A functional, nonmutant form of the *lacY* gene is denoted *lacY*$^+$.

These genes are called **structural genes** because they code for the sequence of amino acids making up the primary structure of each protein. Bacteria that contain mutations that eliminate function of the *lacZ* gene (denoted *lacZ*$^-$ mutants), or those that eliminate function of the *lacY* gene (denoted *lacY*$^-$ mutants), cannot utilize lactose as a source of energy. Without a functional product from *lacY*, lactose cannot enter the cell, and without a functional product from *lacZ*, lactose cannot be cleaved into its component sugars. A functional form of β-galactosidase (*lacZ*$^+$) and of permease (*lacY*$^+$) are both essential for the utilization of lactose and for cell growth.

Regulation of the *lacZ* and *lacY* structural genes is controlled by the product of another structural gene, called *lacI*, which encodes a repressor protein. Located between *lacI* and *lacZ* are a series of regulatory sequences for *lacZ* and *lacY* that include a promoter, *lacP*, whose function is to recruit the RNA polymerase complex and initiate transcription, and an **operator,** *lacO*, which is the binding site for the repressor. Another regulatory region is a binding site for a protein called CRP, which is discussed later.

The *lacZ* and *lacY* genes share a single set of regulatory elements and are transcribed together, a gene organization that is common in bacteria. Typically, a group of functionally related genes are located next to one another along the bacterial DNA and share a promoter, and when the coding sequences of the structural genes are transcribed from the promoter, they are transcribed together into a single molecule of messenger RNA. Such an mRNA is called a **polycistronic RNA** ("cistron" is an old term for "coding sequence"). In Fig. 19.15, the polycistronic RNA includes the coding sequences for β-galactosidase and lactose permease. The region of DNA consisting of the promoter, the operator, and the coding sequence for the structural genes is called an **operon.**

Operons are found in bacteria and archaeons, whose cells can translate polycistronic mRNA molecules correctly because their ribosomes can initiate translation anywhere along an mRNA that contains a proper ribosome-binding site (Chapter 4). In a polycistronic mRNA, each of the coding sequences is preceded by a ribosome-binding site, so translation can be initiated there.

The repressor protein binds with the operator and prevents transcription, but not in the presence of lactose.

The lactose operon is negatively regulated by the repressor protein encoded by the *lacI* gene. That is, the structural genes of the lactose operon are always expressed unless the operon is turned off by a regulatory molecule, in this case the repressor. **Fig. 19.16** shows what the operon looks like in the absence of lactose. The *lacI* gene, encoding the repressor protein, is

FIG. 19.16 The lactose operon in the repressed state in the absence of lactose.

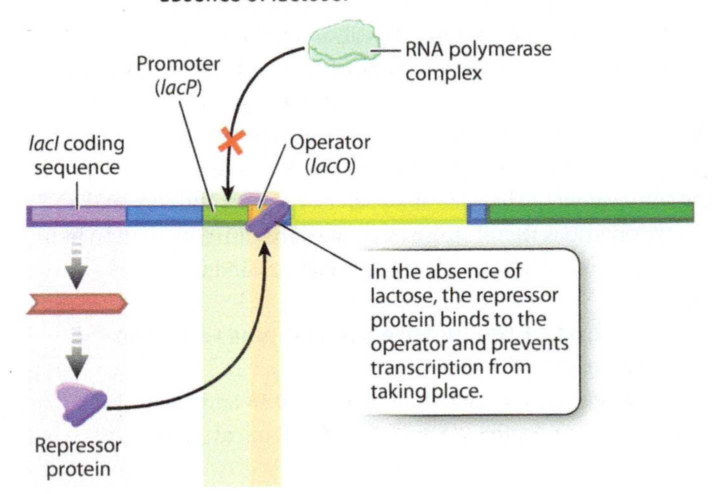

FIG. 19.17 The lactose operon in the induced state in the presence of lactose.

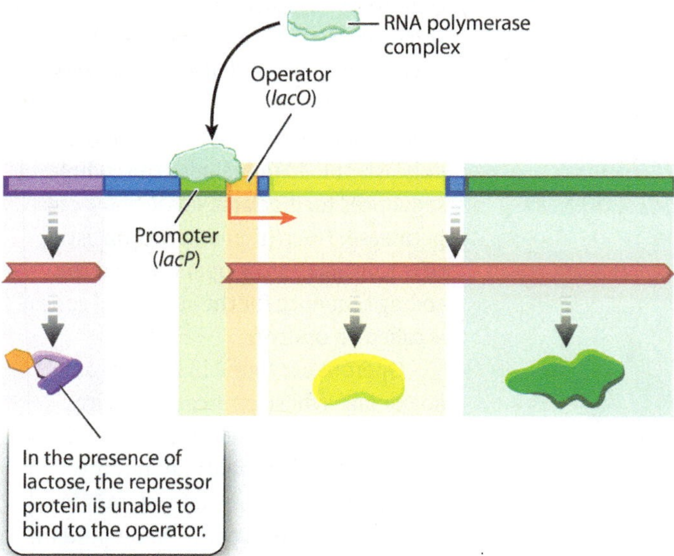

expressed constantly at a low level. The repressor protein binds with the operator (*lacO*), the RNA polymerase complex is not recruited, and transcription does not take place.

The configuration of the lactose operon in the presence of lactose is shown in **Fig. 19.17.** When lactose is present in the cell, the repressor protein is unable to bind to the operator, RNA polymerase is recruited, and transcription occurs. In other words, lactose acts as an inducer of the lactose operon since it prevents binding of the repressor protein. The inducer is not actually lactose itself, but rather an isomer of lactose called allolactose, which differs in the way the sugars are linked. Lactose in the cell is always accompanied by a small amount of allolactose, and so induction of the lactose operon occurs in the presence of lactose.

The binding of the inducer to the repressor results in an allosteric change in repressor structure that inhibits the protein's ability to bind to the operator. The absence of repressor from the operator allows the RNA polymerase complex to be recruited to the promoter, and the polycistronic mRNA is produced. The resulting lactose permease allows lactose to be transported into the cell on a large scale, and β-galactosidase cleaves the molecules to allow the constituents to be used as a source of energy and carbon. The lactose operon is therefore an example of negative regulation by a repressor, whose function is modulated by an inducer.

The function of the lactose operon was revealed by genetic studies.

Although the interactions shown in Fig. 19.16 and Fig. 19.17 have since been confirmed by direct biochemical studies, the inferences about how the repressor and operator work were originally drawn from studies of mutations (**Fig. 19.18**). As part of their investigation, Jacob and Monod identified bacterial mutants that always expressed β-galactosidase and permease, even in the absence of lactose. The phenotype of a cell carrying such a mutation is said to be **constitutive** for production of the proteins. Constitutive expression means that it occurs continuously. The most common constitutive phenotype resulted from a mutation in the *lacI* gene (*lacI*$^-$ mutants) that produced a defective repressor protein (Fig. 19.18a).

Jacob and Monod also did experiments in which *E. coli* contained not one but two lactose operons (Fig. 19.18b). In bacterial cells containing one mutant and one nonmutant copy of the *lacI* repressor gene, expression of the structural genes was no longer constitutive but instead showed normal regulation. This finding is consistent with the idea that the nonmutant form of the gene produces a diffusible protein, which implies that the normal repressor is able to bind to and repress transcription from both operons, not just the one that it is physically linked to.

A much less common class of constitutive mutants identified the operator (Fig. 19.18c). Genetic studies of the mutations in these cells showed that they were not located in the *lacI* gene that encodes the repressor but, instead, closer to the coding sequence of *lacZ*. The genetic element in which the mutations occurred was called the lactose operator (*lacO*), and the constitutive mutations were designated *lacO*^{c} ("*c*" for "constitutive"). When two different lactose operons, one normal and one with *lacO*^{c}, were in the same cell, the operon carrying *lacO*^{c} was transcribed constitutively, even in the presence of normal repressor, because the repressor was unable to bind to the mutant operator site in *lacO*^{c} (Fig. 19.18d). Jacob and Monod wrote that "to explain this effect, it seems necessary to invoke a new type of genetic entity, called an 'operator,' which would be: (a) adjacent to the group of [structural] genes and would control their [transcriptional] activity; and (b) would be sensitive to the repressor produced by a particular regulatory gene."

→ **Quick Check 4** Predict the consequence of a mutation in the *lacI* repressor gene that produces repressor protein that is able to bind to the operator, but not able to bind allolactose.

The lactose operon is also positively regulated by CRP–cAMP.

Fig. 19.16 and Fig. 19.17 show how the ability of the repressor to bind with either the operator (in the absence of lactose) or with the inducer (in the presence of lactose) provides a simple and elegant way for the bacterial cell to transcribe the genes needed for lactose utilization only in the presence of lactose. The elucidation of these interactions was as far as the research tools used by Jacob and Monod could take them. Since the original experiments, the lactose operon has been studied in much greater detail and additional levels of regulation have been discovered.

FIG. 19.18 Lactose operon regulatory mutants. Jacob and Monod found two classes of constitutive mutants, (a) one that affects the repressor and (c) one that affects the operator. In cells containing both a mutant and a nonmutant lactose operon, (b) those with the repressor mutant become regulated, and (d) those with the operator mutant remained constitutive.

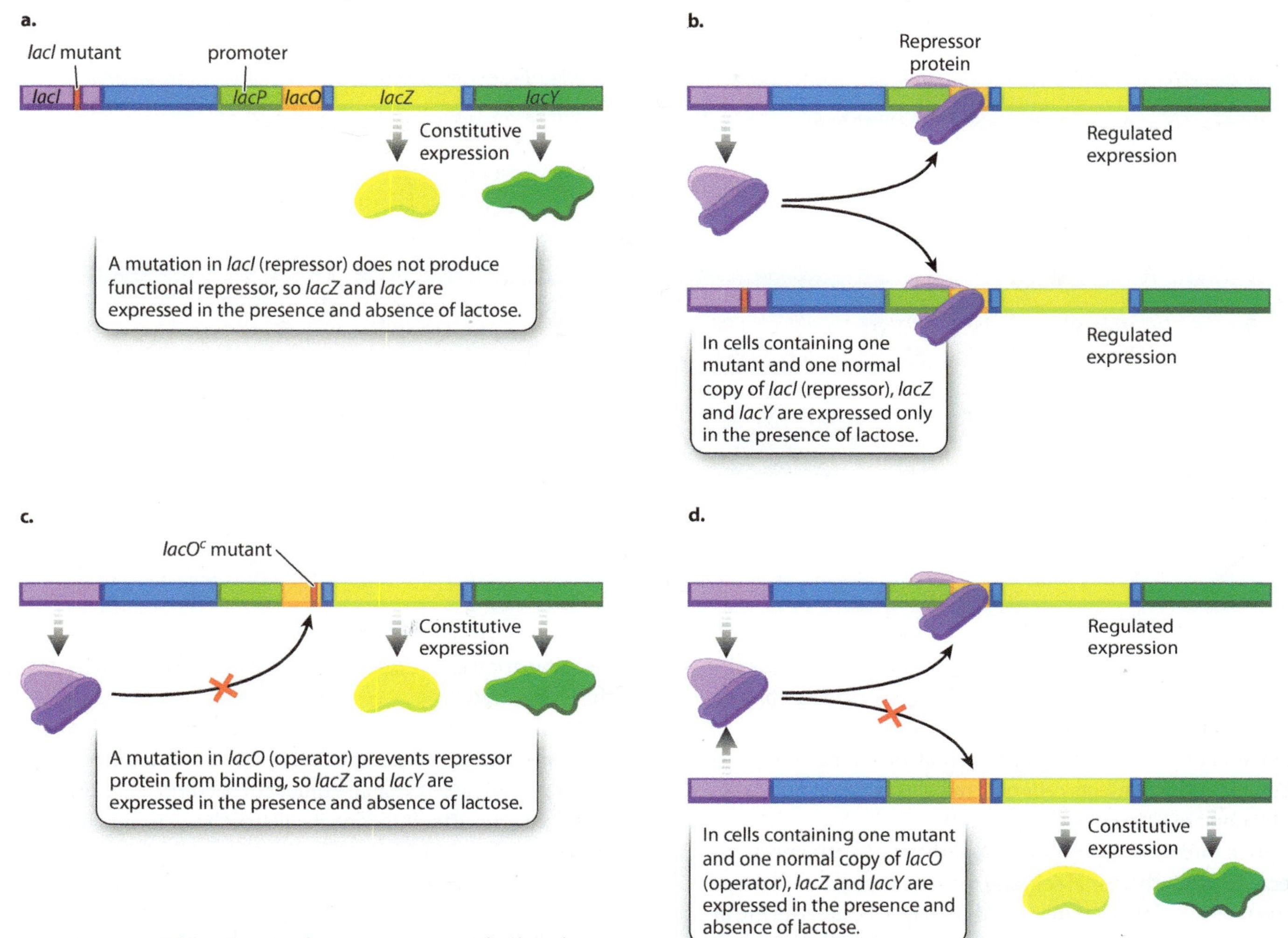

One of these additional levels involves the CRP binding site shown in Fig. 19.15, which in **Fig. 19.19** is occupied by a protein called the CRP–cAMP complex. The CRP–cAMP complex is a positive regulator of the lactose operon, which you will recall is a protein that activates gene expression upon binding DNA. "CRP" stands for "cAMP receptor protein." The role of CRP–cAMP is to provide another level of control of transcription that is more sensitive to the nutritional needs of the cell than the level of control provided by the presence or absence of lactose. *E. coli* can utilize many kinds of molecules as sources of energy. When more than one type of energy source is available in the environment, certain sources are used before others. For example, glucose is preferred to lactose, and lactose is preferred to glycerol. The CRP–cAMP complex helps regulate which compounds are utilized.

The concentration of the small molecule cAMP in the cell is a signal about the nutritional state of the cell. In the absence of glucose, cAMP levels are high, and cAMP binds to CRP, changing the shape of CRP so that it can bind at a site near the operator and stimulate binding of RNA polymerase to transcribe *lacZ* and *lacY* when lactose is present in the cell (Fig. 19.19a). In this way, cAMP is an allosteric activator of CRP binding. However, if lactose is not present in the cell, the lactose repressor binds to the lactose operator and prevents transcription even in the presence of the cAMP–CRP complex (see Fig. 19.16).

In the presence of glucose, cAMP levels are low, and the cAMP–CRP complex does not bind the lactose operon. As a result, even in the presence of lactose, the lactose operon is not transcribed to high levels (Fig. 19.19b). In this way, *E. coli* preferentially utilizes glucose when both glucose and lactose are present and utilizes lactose only when glucose is depleted.

Transcriptional regulation determines the outcome of infection by a bacterial virus.

Transcriptional regulation has also been well studied in viruses. Bacterial cells are susceptible to infection by a variety

FIG. 19.19 The CRP–cAMP complex, a positive regulator of the lactose operon. (a) In the absence of large amounts of glucose, cAMP levels are high and the CRP–cAMP complex binds to a site near the promoter, where it activates transcription. (b) In the presence of large amounts of glucose, cAMP levels are low and the CRP–cAMP complex does not bind, so transcription is not induced to high levels, even in the presence of lactose.

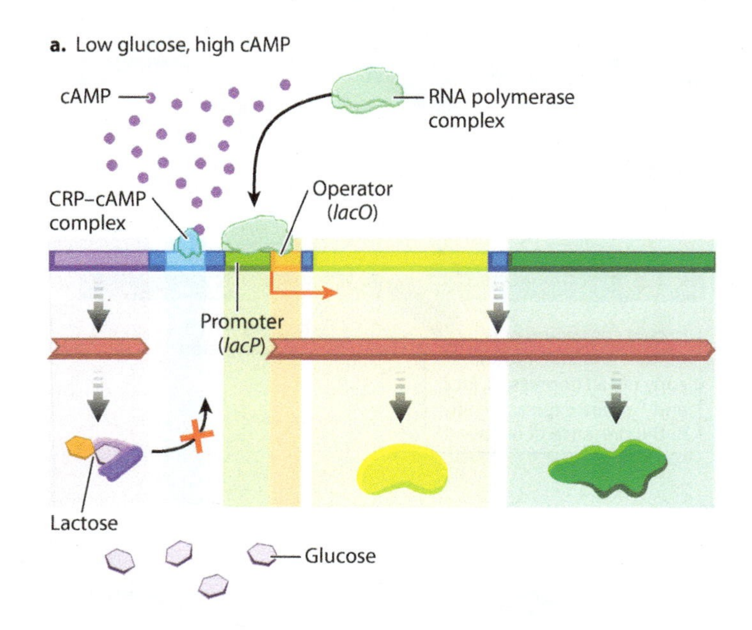

of viruses known as **bacteriophages** ("bacteriophage" literally means "bacteria-eater," and is often shortened to just "phage"). Among bacteriophages is a type that can undergo one of two fates when infecting a cell. The best known example is bacteriophage λ (lambda), which infects cells of *E. coli*. The possible results of λ infection are illustrated in **Fig. 19.20.**

Upon infection, the linear DNA of the phage genome is injected into the bacterial cell, and almost immediately the ends of the molecule join to form a circle. In normal cells growing in nutrient medium, the usual outcome of infection is the **lytic pathway,** shown on the left in Fig. 19.20. In the lytic pathway, the virus hijacks the cellular machinery to replicate the viral genome and produce viral proteins. After about an hour, the infected cell undergoes **lysis** and bursts open to release a hundred or more progeny phage capable of infecting other bacterial cells.

The alternative to the lytic pathway is **lysogeny,** shown on the right in Fig. 19.20. In lysogeny, the bacteriophage DNA and the bacterial DNA undergo a process of recombination at a specific site in both molecules, which results in a bacterial DNA molecule that now includes the bacteriophage DNA. Lysogeny often takes place in cells growing in poor conditions. The relative sizes of the DNA molecules in Fig. 19.20 are not to scale. In reality, the length of the bacteriophage DNA is only about 1% of that of the bacterial DNA. When the bacteriophage DNA is integrated by lysogeny, the only bacteriophage gene transcribed and translated is one that represses the transcription of other phage genes, preventing entry into the lytic pathway. The bacteriophage DNA is replicated along with the bacterial DNA and transmitted to the bacterial progeny when the cell divides. Under stress, such as exposure to ultraviolet light, recombination is reversed, freeing the phage DNA and initiating the lytic pathway.

At the molecular level, the choice between the lytic and lysogenic pathways is determined by the positive and negative regulatory effects of a small number of bacteriophage proteins produced soon after infection. Which pathway results depends on the outcome of a competition between the production of a protein known as cro and that of another protein known as cI. If the production of cro predominates, the lytic pathway results; if cI predominates, the lysogenic pathway takes place.

Fig. 19.21 shows the small region of the bacteriophage DNA in which the key interactions take place. Almost immediately after infection and circularization of the bacteriophage DNA, transcription takes place from the promoters P_L and P_R. Transcription of genes controlled by the P_R promoter results in a transcript encoding the proteins cro and cII. The cro protein represses transcription of a gene controlled by another promoter P_M, which encodes the protein cI. In normal cells growing in nutrient medium, proteases present in the bacterial cell degrade cII and prevent its accumulation. With cro protein preventing *cI* expression and cII protein unable to accumulate, transcription of bacteriophage genes in the lytic pathway takes place, including those genes needed for bacteriophage DNA replication, those encoding proteins in the bacteriophage head and tail, and, finally, those needed for lysis.

Alternatively, in bacterial cells growing in poor conditions, reduced protease activity allows cII protein to accumulate. When cII protein reaches a high enough level, it stimulates transcription from the promoter P_E. The transcript from P_E includes the coding sequence for cI protein, and the cI protein has three functions:

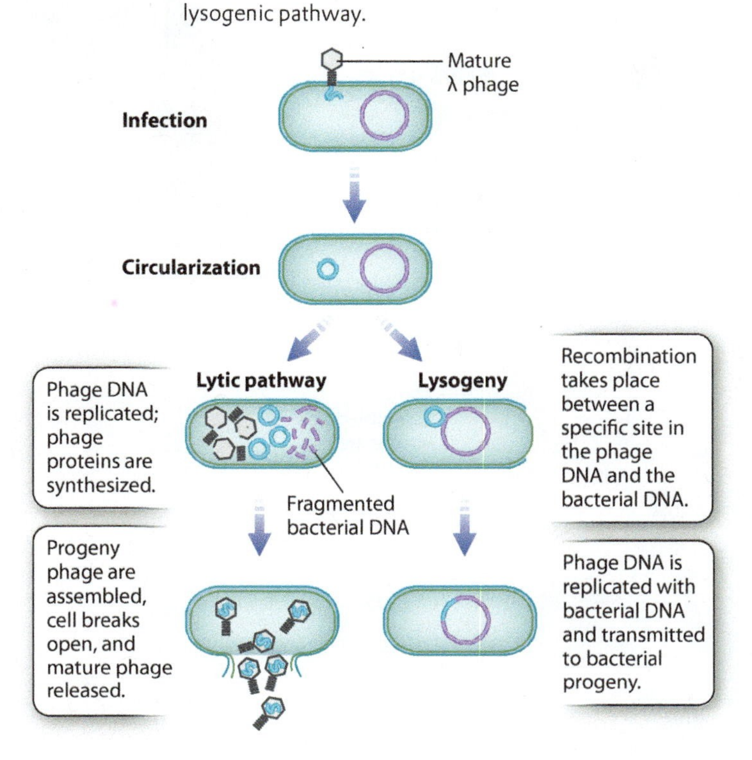

FIG. 19.20 Alternative outcomes of infection by bacteriophage λ. The bacteriophage can enter either the lytic or the lysogenic pathway.

- It binds with the operator O_R and prevents further expression of *cro* and *cII*.
- It stimulates transcription of its own coding sequence from the promoter P_M, establishing a positive feedback loop that keeps the level of cI protein high.
- It binds with the operator O_L and prevents further transcription from P_L.

The result is that cI production shuts down transcription of all bacteriophage genes except its own gene, and this is the regulatory state that produces lysogeny. (The protein needed for recombination between the bacteriophage DNA and the bacterial DNA is produced by transcription from the P_L promoter before it is shut down by cI.) When cells that have undergone lysogeny are exposed to ultraviolet light or certain other stresses, the cI protein is degraded. In this case, cro and cII are produced again, and the lytic pathway follows.

Regulation of the lytic and lysogenic pathways works to the advantage of bacteriophage λ, but the process is not like something an engineer might design. That is because biological systems are not engineered, they evolve. Regulatory mechanisms are built up over time by the selection of successive mutations. Each evolutionary step refines the regulation in such a way as to be better adapted to the environment than it was before. Each successive step occurs only because it increases survival and reproduction.

We summarize these two alternative outcomes of infection by a bacterial virus, and the earlier discussions of viral diversity, replication, host range, and effects on a cell, in **Fig. 19.22** on the following page. ■

FIG. 19.21 Transcriptional regulation of *cI* and *cro* genes, which determine the lytic versus lysogenic pathway.

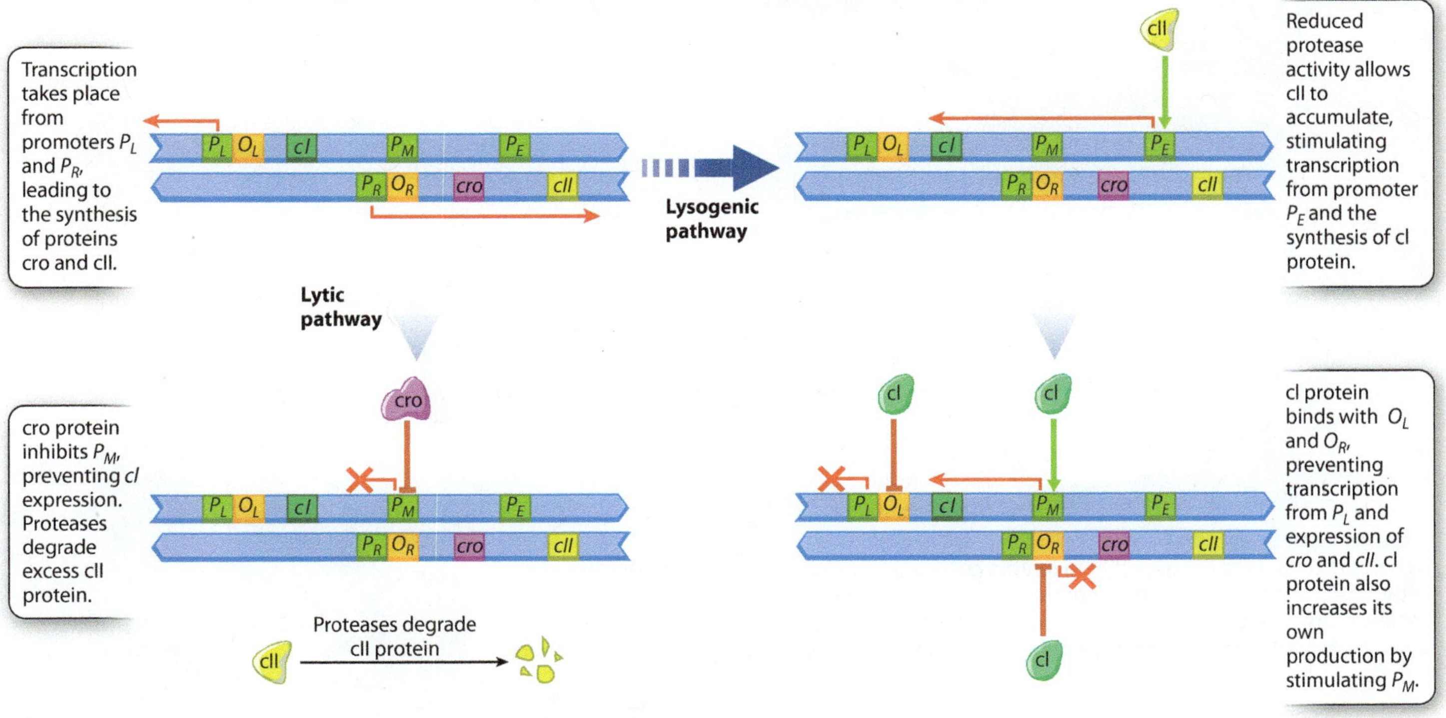

Transcription takes place from promoters P_L and P_R, leading to the synthesis of proteins cro and cII.

P_L O_L *cI* P_M P_E
P_R O_R *cro* *cII*

Lysogenic pathway

cII

Reduced protease activity allows cII to accumulate, stimulating transcription from promoter P_E and the synthesis of cI protein.

Lytic pathway

cro protein inhibits P_M, preventing *cI* expression. Proteases degrade excess cII protein.

cro

Proteases degrade cII protein

cI

cI protein binds with O_L and O_R, preventing transcription from P_L and expression of *cro* and *cII*. cI protein also increases its own production by stimulating P_M.

VISUAL SYNTHESIS Virus: A Genome in Need of a Cell

FIG. 19.22

Integrating concepts from Chapters 11, 13, and 19

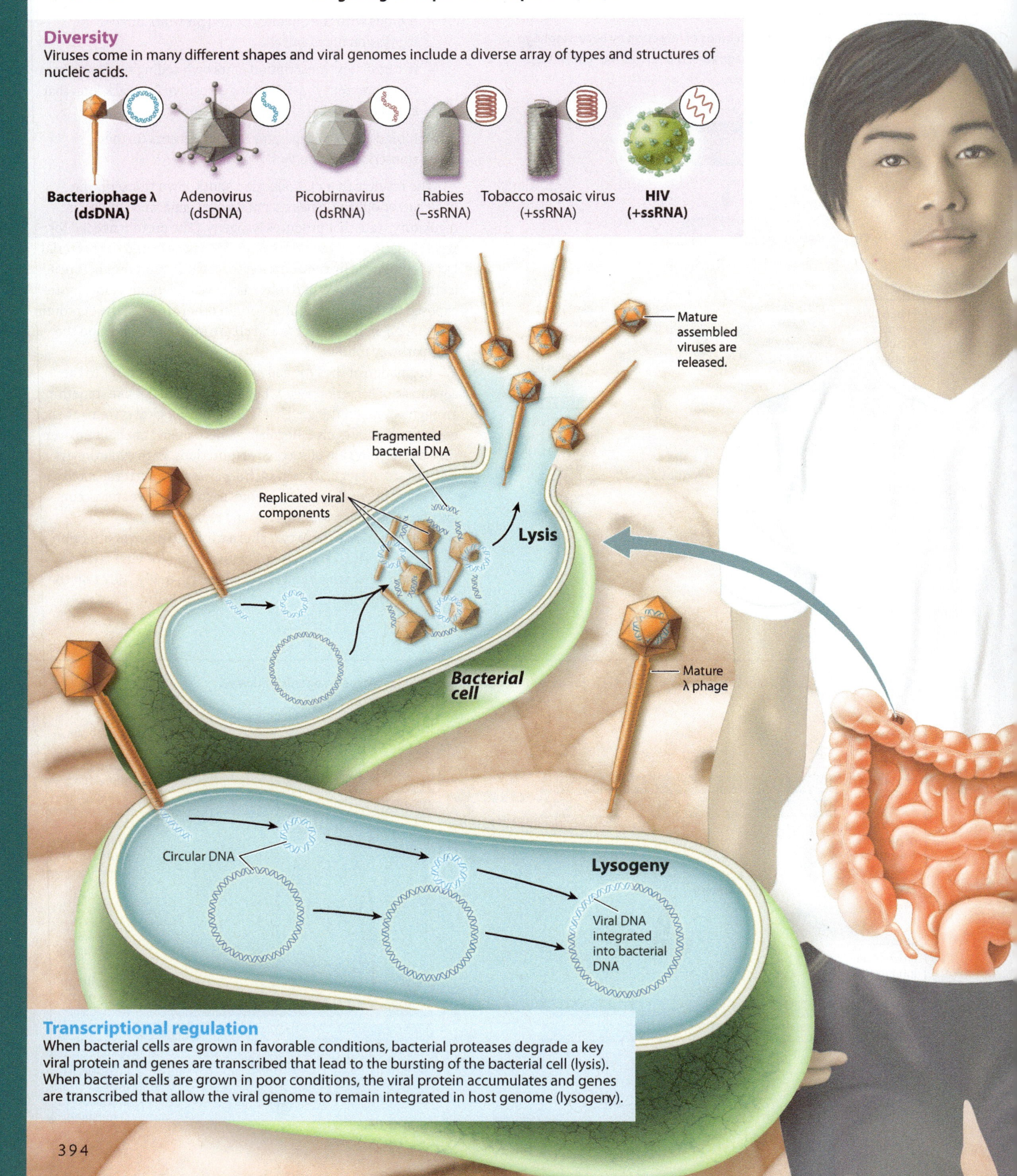

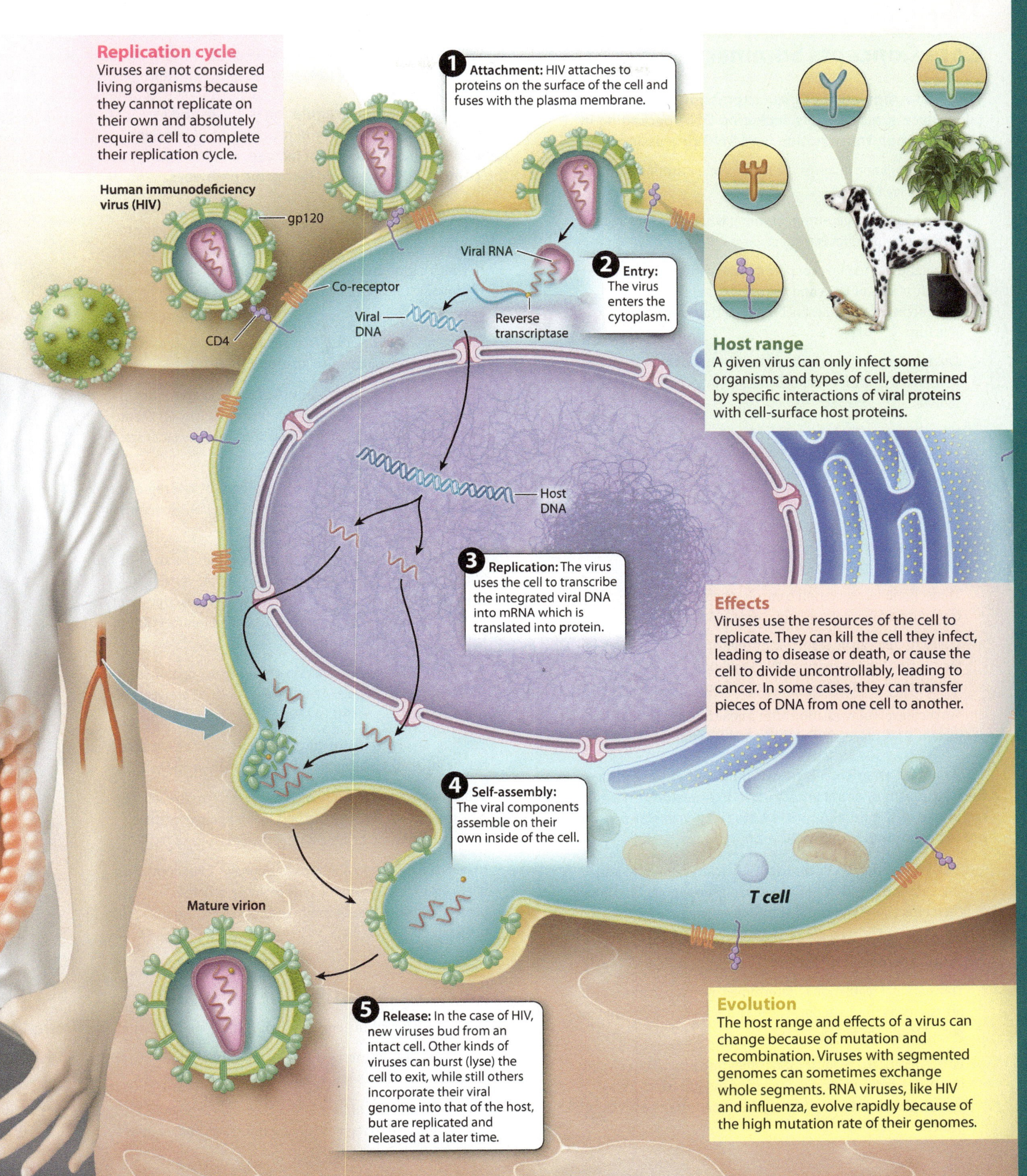
Replication cycle
Viruses are not considered living organisms because they cannot replicate on their own and absolutely require a cell to complete their replication cycle.
Human immunodeficiency virus (HIV)
gp120
Co-receptor
CD4
1 Attachment: HIV attaches to proteins on the surface of the cell and fuses with the plasma membrane.
Viral RNA
2 Entry: The virus enters the cytoplasm.
Viral DNA
Reverse transcriptase
Host DNA
3 Replication: The virus uses the cell to transcribe the integrated viral DNA into mRNA which is translated into protein.
4 Self-assembly: The viral components assemble on their own inside of the cell.
Mature virion
5 Release: In the case of HIV, new viruses bud from an intact cell. Other kinds of viruses can burst (lyse) the cell to exit, while still others incorporate their viral genome into that of the host, but are replicated and released at a later time.
T cell
Host range
A given virus can only infect some organisms and types of cell, determined by specific interactions of viral proteins with cell-surface host proteins.
Effects
Viruses use the resources of the cell to replicate. They can kill the cell they infect, leading to disease or death, or cause the cell to divide uncontrollably, leading to cancer. In some cases, they can transfer pieces of DNA from one cell to another.
Evolution
The host range and effects of a virus can change because of mutation and recombination. Viruses with segmented genomes can sometimes exchange whole segments. RNA viruses, like HIV and influenza, evolve rapidly because of the high mutation rate of their genomes.

Core Concepts Summary

19.1 The regulation of gene expression in eukaryotes takes place at many levels, including DNA packaging in chromosomes, transcription, and RNA processing.

Gene expression involves the turning on or turning off of a gene. page 378

Gene regulation determines where, when, how much, and which gene product is made. page 378

Regulation at the level of chromatin involves chemical modifications of DNA and histones that make a gene accessible or inaccessible to the transcriptional machinery. page 378

Dosage compensation is the process by which the expression of *X*-linked genes is equalized in *XX* individuals and *XY* individuals. page 380

X-inactivation, the mechanism of dosage compensation for sex chromosomes in mammals, involves the inactivation of one of the two *X* chromosomes in females. page 381

X-inactivation occurs by the transcription of a noncoding RNA known as Xist, which coats the entire *X* chromosome, leading to DNA and histone modifications and transcriptional repression. page 381

Transcriptional regulation controls whether or not transcription of a gene occurs. page 382

Transcription can be regulated by regulatory transcription factors that bind to specific DNA sequences known as enhancers that can be near, in, or far from genes. page 382

Further levels of regulation after a gene is transcribed to mRNA include RNA processing, splicing, and editing. page 383

19.2 After an mRNA is transcribed and exported to the cytoplasm in eukaryotes, gene expression can be regulated at the level of mRNA stability, translation, and posttranslational modification of proteins.

Small regulatory RNAs, especially microRNA (miRNA) and small interfering RNA (siRNA), affect gene expression through their effects on translation or mRNA stability. page 384

Translational regulation controls the rate, timing, and location of protein synthesis. page 384

Translational regulation is determined by many features of an mRNA molecule, including the 5′ and 3′ UTR, the cap, and the poly(A) tail. page 385

Posttranslational modification comes into play after a protein is synthesized, and includes chemical modification of side groups of amino acids, affecting the structure and activity of a protein. page 385

Gene regulation is influenced by both genetic and environmental factors. page 386

19.3 Transcriptional regulation is illustrated in bacteria by the control of the production of proteins needed for the utilization of lactose, and in viruses by the control of the lytic and lysogenic pathways.

Transcriptional regulation can be positive, in which a gene is usually off and is turned on in response to the binding to DNA of a regulatory protein called an activator, or negative, in which a gene is usually on and is turned off in response to the binding to DNA of a regulatory protein called a repressor. page 387

Jacob and Monod studied the lactose operon in *E. coli* as a model for bacterial gene regulation. page 388

When lactose is added to culture of bacteria, the genes for the uptake of lactose (permease, encoded by *lacY*) and cleavage of lactose (β-galactosidase, encoded by *lacZ*) are expressed. page 388

The lactose operon is negatively regulated by the repressor protein (encoded by *lacI*), which binds to a DNA sequence known as the operator. page 389

When lactose is added to the medium, it induces an allosteric change in the repressor protein, preventing it from binding to the operator and allowing transcription of *lacY* and *lacZ*. In this way, lactose acts as an inducer of the lactose operon. page 390

An additional level of regulation of the lactose operon is provided by the CRP–cAMP complex, a positive activator of transcription. page 391

The lytic and lysogenic pathways of bacteriophage λ have also been well studied as a model of gene regulation. page 392

When bacteriophage λ infects *E. coli*, it can lyse the cell (the lytic pathway) or its DNA can become integrated into the bacterial genome (the lysogenic pathway). page 392

In infection of *E. coli* cells by bacteriophage λ, predominance of cro protein results in the lytic pathway, whereas predominance of the cI protein results in the lysogenic pathway. page 392

Self-Assessment

1. Distinguish between gene expression and gene regulation.
2. Explain what is meant by different "levels" of gene regulation and give some examples.
3. Give two examples of how DNA bases and chromatin can be modified to regulate gene expression, and explain

why these kinds of modifications result in increased or decreased gene expression.

4. Explain how *X*-inactivation in female mammals results in patchy coat color in calico cats.

5. Explain how one protein-coding gene can code for more than one polypeptide chain.

6. Name and describe three ways in which gene expression can be influenced after mRNA is processed and leaves the nucleus.

7. Diagram the lactose operon in *E. coli* with the proper order of the elements *lacI, lacO, lacY,* and *lacZ,* and explain how expression is controlled in the presence and in the absence of lactose.

8. Describe the role of the CRP–cAMP complex in positive regulation of the lactose operon in *E. coli.*

9. Describe what is meant by lysis and lysogeny, and explain how gene regulation controls these two pathways.

Log in to LaunchPad to check your answers to the Self-Assessment questions, and to access additional learning tools.

Perrennou Nuridsany/Science Source.

CHAPTER 20

Genes and Development

Core Concepts

20.1 In the development of humans and other animals, stem cells become progressively more restricted in their possible pathways of cellular differentiation.

20.2 The genetic control of development is a hierarchy in which genes are activated in groups that in turn regulate the next set of genes.

20.3 Many proteins that play key roles in development are evolutionarily conserved but can have dramatically different effects in different organisms.

20.4 Combinatorial control is a developmental strategy in which cellular differentiation depends on the particular combination of transcription factors present in a cell.

20.5 Ligand–receptor interactions activate signal transduction pathways that converge on transcription factors and genes that determine cell fate.

Altogether, the human body contains about 200 different types of cell, all of which derive from a single cell, the zygote. Some cells derived from the zygote become muscle cells, others nerve cells, still others connective tissue. Almost all of these cells have exactly the same genome: They differ not in their content of genes, but instead in the groups of genes that are expressed or repressed. In other words, these cell types differ as a result of gene regulation, discussed in the previous chapter.

Gene regulation is especially important in multicellular organisms because it underlies **development,** the process in which a fertilized egg undergoes multiple rounds of cell division to become an embryo with specialized tissues and organs. During development, cells undergo changes in gene expression as genes are turned on and off at specific times and places. Gene regulation causes cells to become progressively more specialized, a process known as **differentiation.**

In this chapter, we focus on the general principles by which genes control development. We will see that, as cells differentiate along one pathway, they progressively lose their ability to differentiate along other pathways. Yet gene expression can sometimes be reprogrammed to reopen pathways of differentiation that had previously been shut off, a process that has important implications for therapeutic replacement of diseased or damaged tissue. From a broader perspective, the study of evolutionary changes in developmental processes constitute the field of evolutionary developmental biology, often called **evo-devo.** We will see that, while some of the key molecular mechanisms of development are used over and over again in different organisms, they have evolved to yield such differences in shape and form as to conceal the underlying similarity in mechanism.

20.1 GENETIC PROGRAMS OF DEVELOPMENT

Genetic programs and computer programs have some features in common. Computer code is written as a linear string of letters, analogous to the sequence of nucleotides in genomic DNA. Once a computer program is initiated, it automatically runs and performs its coded task. Small mistakes in the code, like mutations, can have big consequences and even cause the program to crash.

The analogy between genetic programs and computer programs has an important limitation. Computer programs, designed by humans, are consciously written, whereas genetic programs evolve. The genetically encoded developmental programs of all living organisms emerged over billions of years through mutation and natural selection. These developmental programs changed through time, persisting only if they produced organisms that could successfully survive and reproduce in the existing environment. Here, we explore the genetic program of development—that is, the genetic instructions that lead a single fertilized egg to become a complex multicellular organism.

The fertilized egg is a totipotent cell.

In all sexually reproducing organisms, the fertilized egg is special because of its developmental potential. The fertilized egg is said to be **totipotent,** which means that it can give rise to a complete organism. In mammals, the egg also forms the membranes that surround and support the developing embryo (Chapter 42).

After fertilization, the fertilized egg, or **zygote,** undergoes successive mitotic cell divisions as it moves along a fallopian tube. One cell becomes two, two become four, four become eight, eight become sixteen, and so on, with all the cells contained within the egg's outer membrane (**Fig. 20.1**). Within 4–5 days after

FIG. 20.1 Early development of a human embryo. The zygote is a totipotent cell because its daughter cells can develop into any cell type and eventually into a complete organism.

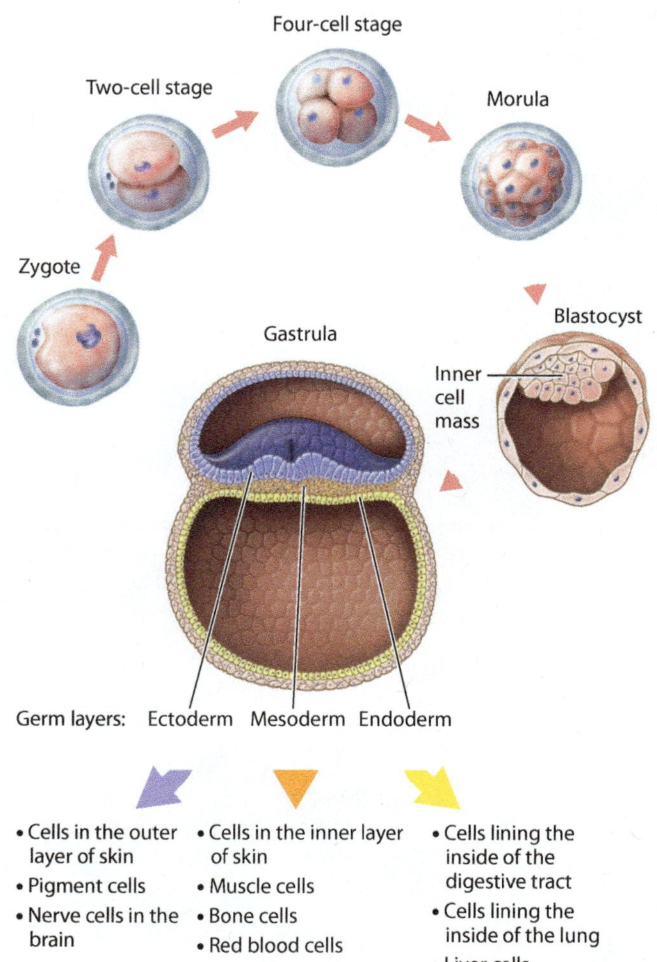

fertilization, the zygote has turned into a ball of cells called the **morula** and has traveled from the site of fertilization in one of the fallopian tubes to the uterus.

These early cell divisions are different from the mitotic cell divisions that occur later in life because the cells do not grow between divisions; they merely replicate their chromosomes and divide again. The result is that the cytoplasm of the egg is partitioned into smaller and smaller packages, with the new cells all bunched together inside the membrane that covers the developing embryo.

Cell division continues in the morula until there are a few thousand cells. The cells then begin to move relative to one another, pushing against and expanding the membrane that encloses them and rearranging themselves to form a hollow sphere called a **blastocyst** (Fig. 20.1). In one region of the inner wall of the blastocyst, there is a group of cells known as the **inner cell mass,** from which the body of the embryo develops. The wall of the blastocyst forms several membranes that envelop and support the developing embryo. Once the blastocyst forms, it implants in the uterine wall.

Once implanted in the uterine wall, the multiplying cells of the inner cell mass reorganize to form a **gastrula,** in which the blastula becomes organized into three **germ layers** (Fig. 20.1). Germ layers are sheets of cells that include the ectoderm, mesoderm, and endoderm and that differentiate further into specialized cells. Those formed from the **ectoderm** include the outer layer of the skin and nerve cells in the brain; cells from the **mesoderm** include cells that make up the inner layer of the skin, muscle cells, and red blood cells; and cells formed from the **endoderm** include cells of the lining of the digestive tract and lung, as well as liver cells and pancreas cells (Chapter 42).

Cellular differentiation increasingly restricts alternative fates.

At each successive stage in development in which the cells differentiate, they lose the potential to develop into any kind of cell. The fertilized egg is totipotent because it can differentiate into both the inner cell mass and supporting membranes, and eventually into an entire organism. The cells of the inner cell mass, called embryonic stem cells, are **pluripotent** because they can give rise to any of the three germ layers, and therefore to any cell of the body. However, pluripotent cells cannot on their own give rise to an entire organism, as a totipotent cell can. Cells further along in differentiation are **multipotent;** these cells can form a limited number of types of specialized cell. Cells of the germ layers are multipotent because they can give rise only to the cell types specified for each germ layer in Figure 20.1. Totipotent, pluripotent, and multipotent cells are all **stem cells,** cells that are capable of differentiating into different cell types.

→ **Quick Check 1** From what you know about embryonic development, do you think that a cell from the inner cell mass or one from the ectoderm has more developmental potential?

Why do differentiating cells increasingly lose their developmental potential? One hypothesis focuses on gene regulation. When cells become committed to a particular developmental pathway, genes no longer needed are turned off (that is, repressed) and are difficult to turn on again. Another hypothesis is genome reduction: As cells become differentiated, they delete the DNA for genes they no longer need.

These hypotheses can be distinguished by an experiment in which differentiated cells are reprogrammed to mimic earlier states. If loss of developmental potential is due to gene regulation, then differentiated cells could be reprogrammed to become pluripotent or multipotent. If loss of developmental potential is due to genome reduction, then differentiated cells could not be reprogrammed to become pluripotent or multipotent.

British developmental biologist John Gurdon carried out such experiments in the early 1960s (**Fig. 20.2**). Gurdon used a procedure called **nuclear transfer,** in which a hollow glass needle is used to insert the nucleus of a cell into the cytoplasm of an egg whose own nucleus has been destroyed or removed. Previous nuclear transfer experiments had been carried out in the leopard frog, *Rana pipiens.* Whereas nuclei from pluripotent or multipotent cells could often be reprogrammed to develop into normal tadpoles, attempts with nuclei from fully differentiated cells failed.

Gurdon tried the experiments in a different organism, the clawed toad *Xenopus laevis,* and demonstrated that nuclei from fully differentiated intestinal cells could be reprogrammed to support normal development of the tadpole (Fig. 20.2). Only 10 of 726 experiments succeeded, but this was sufficient to show that intestinal cell nuclei still contained a complete *Xenopus* genome. In other words, his findings supported the first hypothesis—all of the same genes are present in intestinal cells as in early embryonic cells, but some of the genes are turned off, or repressed, during development.

Fig. 20.3 summarizes the results of many nuclear transfer experiments carried out in mammals and amphibians. The percentage of reprogramming experiments that fail increases as cells differentiate. The best chance of success is to use pluripotent nuclei from cells in the blastocyst (or its amphibian equivalent, the blastula). However, even some experiments using nuclei from fully differentiated cells have been successful.

When nuclear transfer succeeds, the result is a **clone**—an individual that carries an exact copy of the nuclear genome of another individual. In this case, the new individual shares the same genome as that of the individual from which the donor nucleus was obtained. (The mitochondrial DNA is not from the nuclear donor, but from the donor of the egg cytoplasm.) The

HOW DO WE KNOW?

FIG. 20.2

How do stem cells lose their ability to differentiate into any cell type?

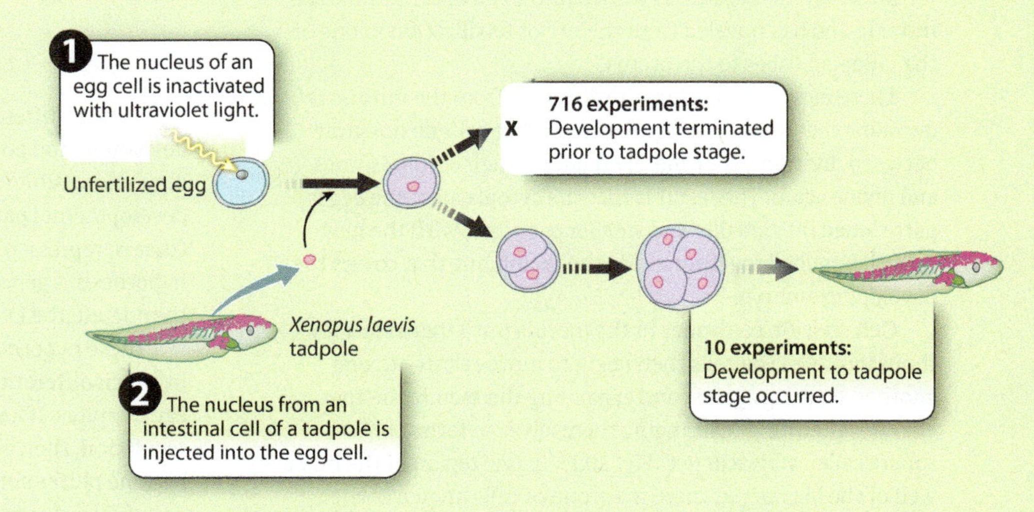

BACKGROUND During differentiation, cells become progressively more specialized and restricted in their fates. Early studies left the mechanisms of differentiation unclear.

HYPOTHESIS One hypothesis is that differentiation occurs as a result of changes in gene expression. A second hypothesis is that differentiation occurs as a result of genome reduction, in which genes that are not needed are deleted.

EXPERIMENT John Gurdon carried out experiments in the amphibian *Xenopus laevis* to test these hypotheses. He transferred nuclei from differentiated cells into unfertilized eggs whose nuclei had been inactivated with ultraviolet light. If differentiation is due to changes in gene expression, then the differentiated nucleus should be able to reprogram itself in the egg cytoplasm and differentiate again into all the cells of a tadpole. If differentiation is accompanied by loss of genes, then differentiation is irreversible and development will not proceed.

RESULTS The experiment was carried out 726 times. In 716 cases, development did not occur; in 10, development proceeded normally.

CONCLUSION Although the experiment succeeded in only 10 of 726 attempts, it showed that the nucleus of an intestinal cell and the cytoplasm of the unfertilized egg are able to support complete development of a normal animal. This result allows us to reject the hypothesis that differentiation occurs by the loss of genes. The first hypothesis—that cells become differentiated as a result of changes in gene expression—was supported. But, because of the small number of successes in reprogramming, additional experiments were needed to validate the conclusions.

FOLLOW-UP WORK Gurdon's work was controversial. Some critics argued that the successful experiments resulted from a small number of undifferentiated cells present in intestinal epithelium. Others accepted the conclusion but expressed misgivings about possible applications to humans. Later experiments that succeeded in cloning mammals from fully differentiated cells confirmed the original conclusion.

SOURCE Gurdon, J. B. 1962. "The Developmental Capacity of Nuclei Taken from Intestinal Epithelium Cells of Feeding Tadpoles." *Journal of Embryology & Experimental Morphology* 10:622–640.

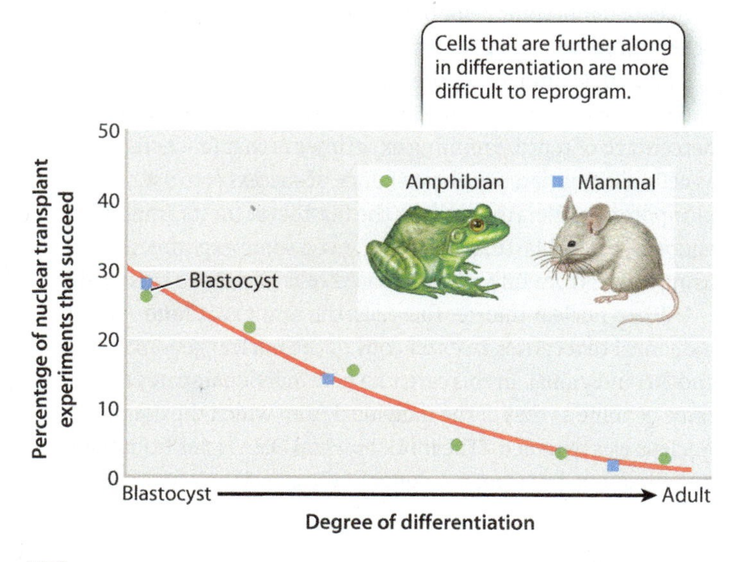

FIG. 20.3 Results of nuclear transfer of differentiated cells. *Data from: J. B. Gurdon and D. A. Melton, "Nuclear Reprogramming in Cells," 2008, Science 322:1811–1815.*

first mammalian clone was a lamb called Dolly (**Fig. 20.4a**), born in 1996. She was produced from the transfer of the nucleus of a cell in the mammary gland of a sheep to an egg cell with no nucleus, and was the only successful birth among 277 nuclear transfers. Successful cloning in sheep soon led to cloning in cattle, pigs, and goats.

The first household pet to be cloned was a kitten named CopyCat (**Fig. 20.4b**), born in 2001 and derived from the nucleus of a differentiated ovarian cell. CopyCat was the only success among 87 attempts. As shown in Fig. 20.4b, the cat from which the donor nucleus was obtained was a calico, but CopyCat herself was not,

FIG. 20.4 Celebrity clones and their genetic mothers. (a) Dolly; (b) CopyCat. *Sources: a. AP Photo/John Chadwick; b. AP Photo/Pat Sullivan.*

even though the two cats are clones of each other. The reason for their different appearance has to do with X-inactivation, discussed in Chapter 19. Recall that the mottled orange and black calico pattern results from random inactivation of one of the two X chromosomes during development. The lack of a calico pattern in CopyCat implies that the X chromosomes in the transferred nucleus did not "reset" as they do in normal embryos. Instead, the inactive X in the donor nucleus remained inactive in all the cells in the clone. Hence, while CopyCat and her mother share the same nuclear genome, the genes were not expressed in the same way because of irreversible epigenetic regulation in the donor nucleus.

→ **Quick Check 2** X-inactivation can result in two clones of a cell that differ in the genes they express. Can you think of other reasons why two genetically identical individuals might look different from each other?

? CASE 3 YOU, FROM A TO T: YOUR PERSONAL GENOME

Can cells with your personal genome be reprogrammed for new therapies?

Stem cells play a prominent role in **regenerative medicine,** which aims to use the natural processes of cell growth and development to replace diseased or damaged tissues. Stem cells are already used in bone marrow transplantation and may someday be used to treat Parkinson's disease, Alzheimer's disease, heart failure, certain types of diabetes, severe burns and wounds, and spinal cord injury.

At first, it seemed as though the use of embryonic stem cells gave the greatest promise for regenerative medicine because of their pluripotency. This approach proved ethically controversial because obtaining embryonic stem cells requires the destruction of human blastocysts—that is, early-stage embryos. A major breakthrough occurred in 2006 when Japanese scientists demonstrated that adult cells can be reprogrammed by activation of just a handful of genes, most of them encoding transcription factors or chromatin proteins. The reprogrammed cells were pluripotent and were therefore called **induced pluripotent stem cells (iPS cells).**

The success rate was only about one iPS cell per thousand, and the genetic engineering technique required the use of viruses that can sometimes cause cancer. Nevertheless, the result was regarded as spectacular. Other researchers soon found other genes that could be used to reprogram adult cells into pluripotent or multipotent stem cells, and still other investigators developed virus-free methods for delivering the genes. In recent years, researchers have even discovered small organic molecules that can reprogram adult cells.

This kind of reprogramming opens the door to personalized stem cell therapies. The goal is to create stem cells derived from the adult cells of the individual patient. Since these cells contain the patient's own genome, problems with tissue rejection are minimized or eliminated (Chapter 43). There remains much to learn before therapeutic use of induced stem cells becomes routine. Researchers will face challenges such as increasing the efficiency of reprogramming, verifying that reprogramming is complete, making sure that the reprogrammed cells are not prone to cancer, and demonstrating that the reprogrammed cells differentiate as they should. Nevertheless, researchers hope that someday soon your own cells containing your personal genome could be reprogrammed to restore cells or organs damaged by disease or accident.

20.2 HIERARCHICAL CONTROL OF DEVELOPMENT

During development of a complex multicellular organism, many genes are activated and repressed at different times, thus restricting cell fates. One of the key principles of development is that genes expressed early in an organism's development control the activation of other groups of genes that act later in development. Gene regulation during development is therefore **hierarchical** in the sense that genes expressed at each stage in the process control the expression of genes that act later.

Drosophila development proceeds through egg, larval, and adult stages.

The fruit fly *Drosophila melanogaster* has played a prominent role in our understanding of the genetic control of early development, and in particular the hierarchical control of development. Researchers have isolated and analyzed a large number of mutant genes that lead to a variety of defects at different stages in development. These studies have revealed many of the key genes and processes in development, which are the focus of the following sections.

The major events in *Drosophila* development are illustrated in **Fig. 20.5.** DNA replication and nuclear division begin soon after the egg and sperm nuclei fuse (Fig. 20.5a). Unlike in mammalian development, the early nuclear divisions in the *Drosophila* embryo occur without cell division, and therefore the embryo consists of a single cell with many nuclei in the center (Fig. 20.5b). When there are roughly 5000 nuclei, they migrate to the periphery (Fig. 20.5c), where each nucleus becomes enclosed in its own cell membrane, and together they form the **cellular blastoderm** (Fig. 20.5d).

Then begins the process of **gastrulation,** in which the cells of the blastoderm migrate inward, creating layers of cells within the embryo. As in humans and most other animals (section 20.1 and Chapter 42), gastrulation forms the three germ layers (ectoderm, mesoderm, and endoderm) that differentiate into different types of cell. A *Drosophila* embryo during gastrulation is shown in Fig. 20.5e. At this stage, the embryo already shows an organization into discrete parts or segments, the formation of which is known as **segmentation.** There are three cephalic segments, C1–C3 (the term "cephalic" refers to the head); three thoracic segments, T1–T3 (the thorax is the middle region of an insect); and eight abdominal segments (A1–A8). Each of these segments has a different fate in development.

About one day after fertilization, the embryo hatches from the egg as a larva (Fig. 20.5f). Over the next eight days, the larva grows and replaces its rigid outer shell, or cuticle, twice (Fig. 20.5g and 20.5h). After a week of further growth, the cuticle forms a casing—called the pupa (Fig. 20.5i)—in which the larva undergoes

FIG. 20.5 Life cycle of the fruit fly *Drosophila melanogaster*. The life cycle begins with (a) a fertilized egg, followed by (b–e) a developing embryo, (f–h) larval stages, (i) pupa, and (j) adult.

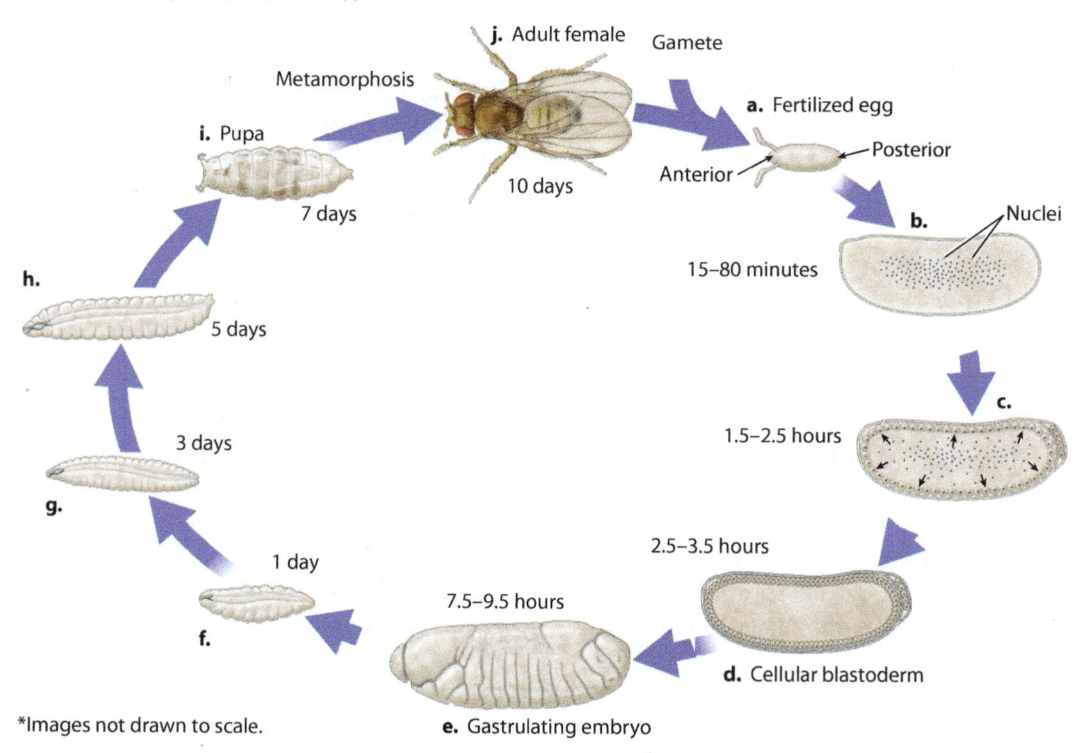

FIG. 20.6 Normal and mutant *Drosophila* larvae. (a) Nonmutant larva have anterior, middle, and posterior segments. (b) *Bicoid* mutant larva lack anterior segments. (c) *Nanos* mutant larva lack posterior structures.

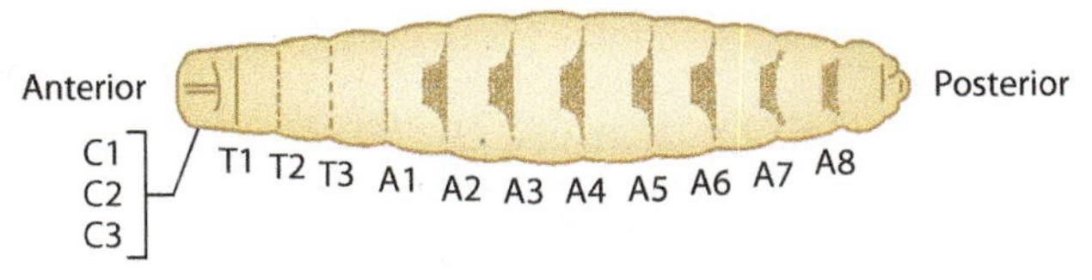

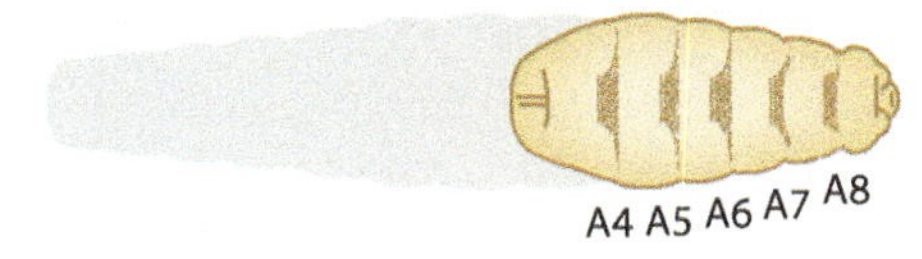

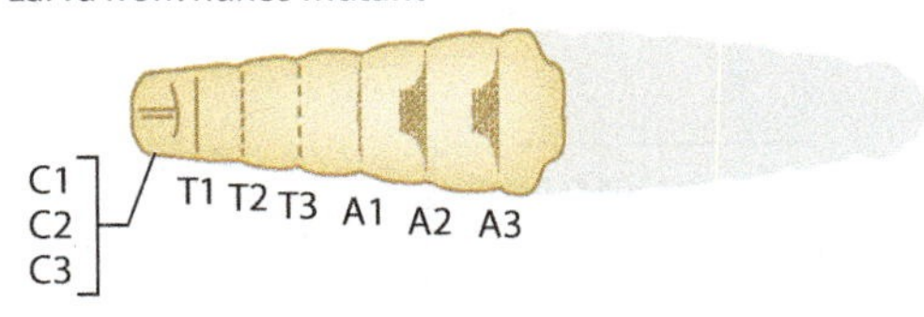

the dramatic developmental changes known as metamorphosis that give rise to the adult fruit fly.

The egg is a highly polarized cell.

How genes control development in *Drosophila* was inferred from systematic studies of mutants by Christiane Nusslein-Volhard and Eric F. Wieschaus, work for which they were awarded the 1995 Nobel Prize in Physiology or Medicine. One of their findings was that development starts even before a zygote is formed, in the maturation of the **oocyte,** the unfertilized egg cell produced by the mother. The oocyte, which matures under control of the mother's genes, is also important for normal embryonic development. This finding applies not only to insects like *Drosophila,* but also to many multicellular animals.

Among the striking mutants Nusslein-Volhard and Wieschaus generated and investigated were ones that significantly affected early development (**Fig. 20.6**). These defects in very early development can be easily seen by the time the mutants reach the larval stage. In one class of mutants, called *bicoid,* larvae are missing segments at the anterior end. In another class of mutants, called *nanos,* larvae are missing segments at the posterior end. The mutant larvae are grossly abnormal and do not survive. Nusslein-Volhard and Wieschaus were able to identify each segment that was missing based on each segment's distinctive pattern of hairlike projections. In Fig. 20.6, the patterns are shown as dark shapes on each segment.

A distinguishing feature of *bicoid* and *nanos* mutants is that the abnormalities in the embryo depend on the genotype of the mother, not the genotype of the embryo. The reason the genotype of the mother can affect the phenotype of the developing embryo is that successful development requires a functioning oocyte. In *Drosophila* and many other organisms, the composition of the egg includes macromolecules (such as RNA and protein) synthesized by cells in the mother and transported into the egg. If the mother carries mutations in genes involved in the development of the oocyte, the offspring can be abnormal. Genes such as *bicoid* and *nanos* that are expressed by the mother but affect the phenotype of the offspring (in this case the developing embryo and larva) are called **maternal-effect genes.**

A normal *Drosophila* oocyte is highly polarized, meaning that one end is distinctly different from the other. For example, there are gradients of macromolecules that define the anterior–posterior (head-to-tail) axis of the embryo as well as the dorsal–ventral (back-to-belly) axis. The best known of these gradients are those of messenger RNAs that are transcribed from the maternal-effect genes *bicoid* and *nanos.* **Fig. 20.7a** shows the gradient of *bicoid* mRNA across the oocyte. The bulk of *bicoid* mRNA comes from the mother, and is localized in the anterior of the egg by proteins that attach them to the cytoskeleton.

The mRNA corresponding to the maternal-effect gene *nanos* is also present in a gradient, but most of it is at the posterior end (**Fig. 20.7b**). Like *bicoid,* mRNA for *nanos* is synthesized by the mother's cells and then imported into the oocyte. After fertilization, the zygote produces Bicoid and Nanos proteins from the localized mRNAs, and they have concentration gradients resembling those of the mRNAs in the oocyte (Fig. 20.7).

FIG. 20.7 Gradients of *bicoid* and *nanos* mRNA and protein in the developing embryo. (a) The mRNA and protein for *bicoid* are localized in the anterior end of the egg. (b) The mRNA and protein for *nanos* are localized in the posterior end of the egg.

a. Distribution of *bicoid* mRNA and protein in the egg

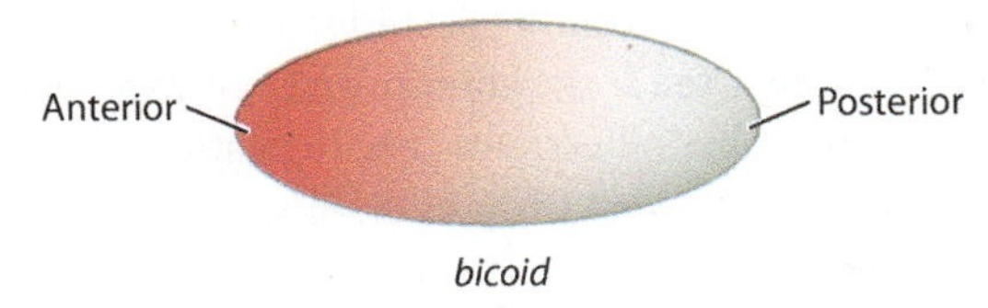

b. Distribution of *nanos* mRNA and protein in the egg

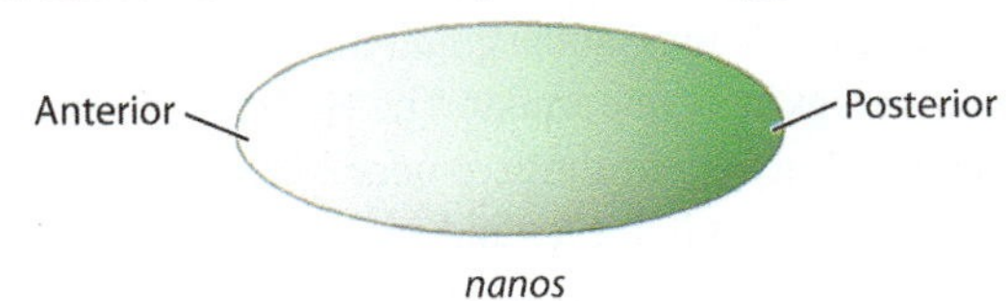

FIG. 20.8 Caudal and Hunchback gradients in the developing embryo. (a) mRNA levels of *hunchback* and *caudal* are uniform across the embryo. (b) Hunchback and Caudal protein levels are localized to the anterior and posterior ends of the embryo, respectively, because Bicoid and Nanos control the translation of *hunchback* and *caudal* mRNA.

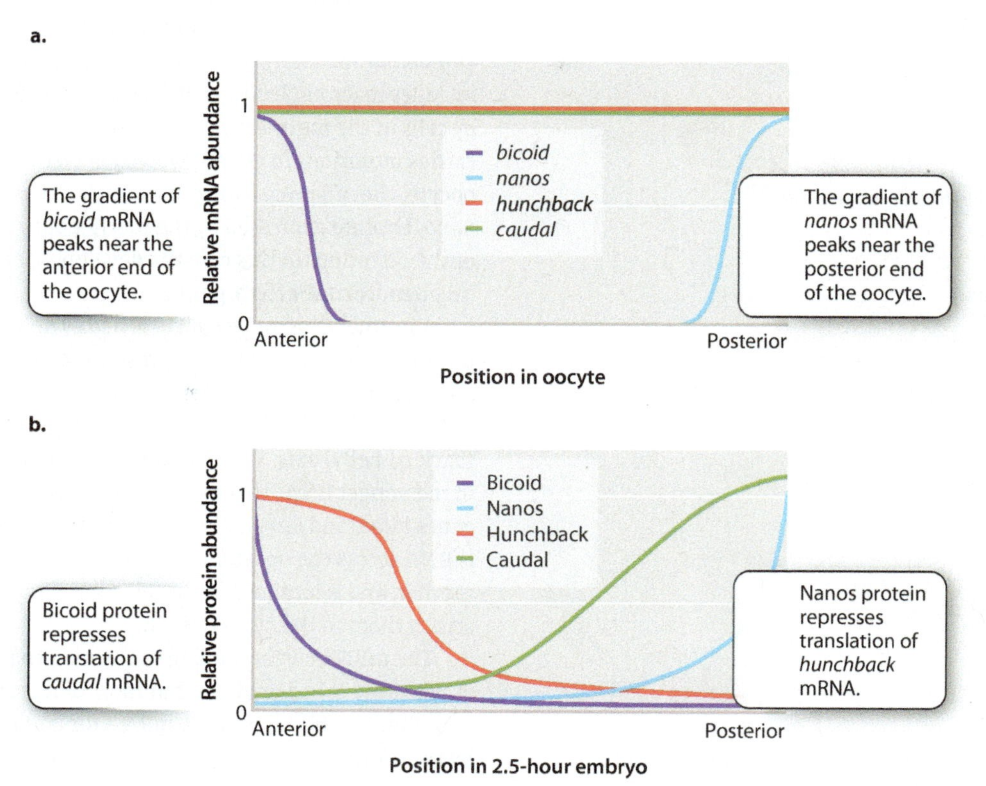

The anterior–posterior axis set up by the gradients of Bicoid and Nanos proteins is reinforced by gradients of two transcription factors called Caudal and Hunchback (**Fig. 20.8**). Like the mRNAs for Bicoid and Nanos, the mRNAs for Caudal and Hunchback are transcribed from the mother's genome and transported into the egg. As shown in Fig. 20.8a, the mRNAs for *caudal* and *hunchback* are spread uniformly in the cytoplasm of the fertilized egg. However, the mRNAs are not translated uniformly in the egg. Bicoid protein represses translation of *caudal*, and Nanos protein represses translation of *hunchback* (Fig. 20.8b). Caudal protein is therefore concentrated at the posterior end and Hunchback protein is concentrated at the anterior end. The expression of Caudal and Hunchback illustrates gene regulation at the level of translation, discussed in Chapter 19. Bicoid protein is also a transcription factor that promotes transcription of the *hunchback* gene from zygotic nuclei, which reinforces the localization of Hunchback protein at the anterior end.

The Hunchback and Caudal gradients set the stage for the subsequent steps in development. The Hunchback transcription factor targets genes of the embryo needed for the development of anterior structures like eyes and antennae, and Caudal targets genes of the embryo needed for the development of posterior structures like genitalia. In this way, maternal genes expressed early in development influence the expression of genes of the embryo that are important in later development. Because the products of the *bicoid* and *nanos* mRNA are the ones initially responsible for organizing the anterior and posterior ends of the embryo, respectively, mothers that are mutant for *bicoid* have larvae that lack anterior structures, and mothers that are mutant for *nanos* have larvae that lack posterior structures.

→ **Quick Check 3** Would development happen normally if the mother has normal *bicoid* function, but the embryo does not? Why or why not?

Development proceeds by progressive regionalization and specification.

Nusslein-Volhard and Wieschaus also discovered mutants of the embryo's developmental genes. They discovered three classes of such mutants, and their analysis showed that genes in the embryo controlling its development are turned on in groups, and that each successive group acts to refine and narrow the pattern of

FIG. 20.9 Normal gap-gene expression pattern and mutant phenotype. *Source: James Langeland, Steve Paddock and Sean Carroll, HHMI, University of Wisconsin–Madison.*

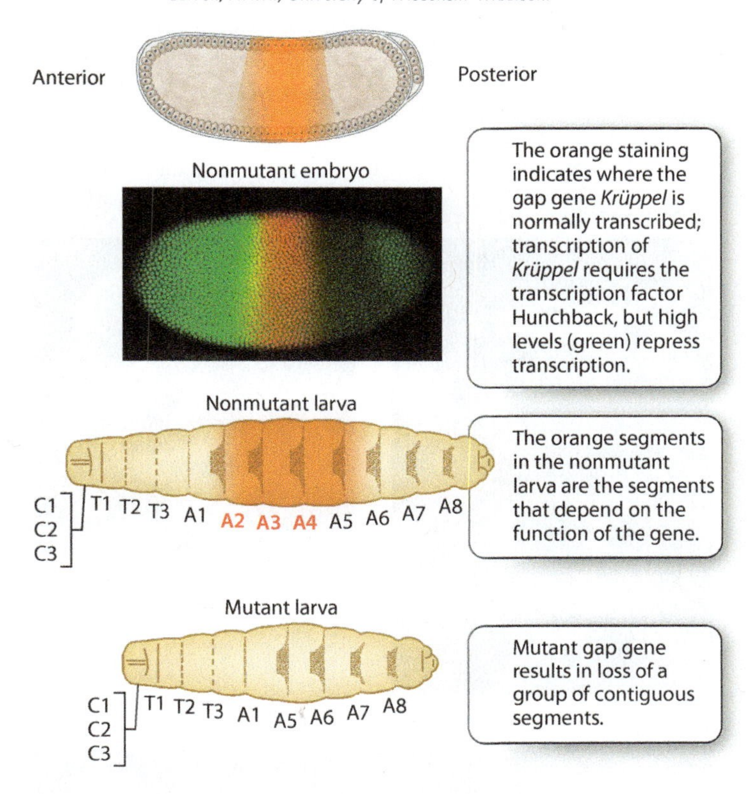

FIG. 20.10 Normal pair-rule gene expression pattern and mutant phenotype. *Source: James Langeland, Steve Paddock and Sean Carroll, HHMI, University of Wisconsin–Madison.*

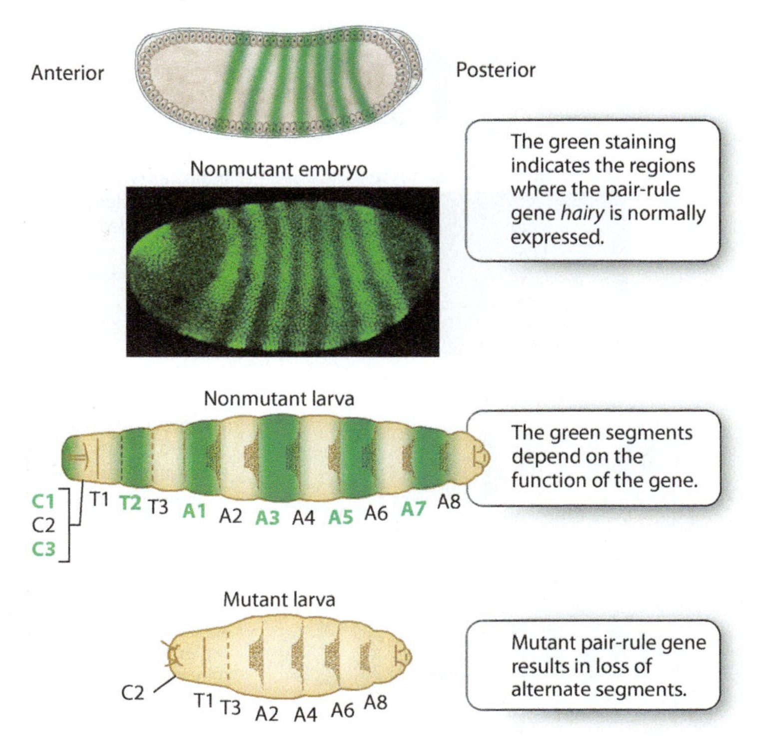

differentiation generated by previous groups. This, too, is a general principle of development in many multicellular organisms.

The anterior–posterior gradient set up by the maternal-effect genes is first narrowed by genes called gap genes (**Fig. 20.9**), each of which is expressed in a broad region of the embryo. The name "gap gene" derives from the phenotype of mutant embryos, which are missing groups of adjoining segments, leaving a gap in the pattern of segments. Fig. 20.9 shows the expression pattern of the gap gene *Krüppel,* which is expressed in the middle region of the embryo. Mutants of *Krüppel* lack some thoracic and abdominal segments when they reach the larval stage. *Krüppel* is expressed in the pattern shown in Fig. 20.9 because it is under the control of the transcription factor Hunchback, which in this embryo is stained in green. High concentrations of Hunchback repress *Krüppel* transcription entirely, and low concentrations fail to induce *Krüppel* transcription. Because Hunchback is present in an anterior–posterior gradient (see Fig. 20.8), the pattern of Hunchback expression means that *Krüppel* is transcribed only in the middle region of the embryo where Hunchback is present but not too abundant.

The gap genes encode transcription factors that control genes in the next level of the regulatory hierarchy, which consists of pair-rule genes (**Fig. 20.10**). Pair-rule genes receive their name because larvae with these mutations lack alternate body segments. The example in Fig. 20.10 is *hairy,* whose mutants lack the odd-numbered thoracic segments and the even-numbered abdominal segments. The pair-rule genes help to establish the uniqueness of each of seven broad stripes across the anterior–posterior axis.

The pair-rule genes in turn help to regulate the next level in the segmentation hierarchy, which consists of segment-polarity genes (**Fig. 20.11**). The segment-polarity genes refine the 7-striped pattern still further into a 14-striped pattern. Each of the 14 stripes has distinct anterior and posterior ends determined by the segment-polarity genes. Embryos with mutations in segment-polarity genes lose this anterior–posterior differentiation, with the result that the anterior and the posterior halves of each segment are mirror images. The example in Fig. 20.11 is *engrailed,* which eliminates the posterior pattern element in each stripe and replaces it with a mirror image of the anterior pattern element.

→ **Quick Check 4** Would the pattern of segment-polarity gene expression be normal if one or more gap genes were not expressed properly? Why or why not?

FIG. 20.11 Normal segment-polarity gene expression pattern and mutant phenotype. *Source: James Langeland, Steve Paddock and Sean Carroll, HHMI, University of Wisconsin–Madison.*

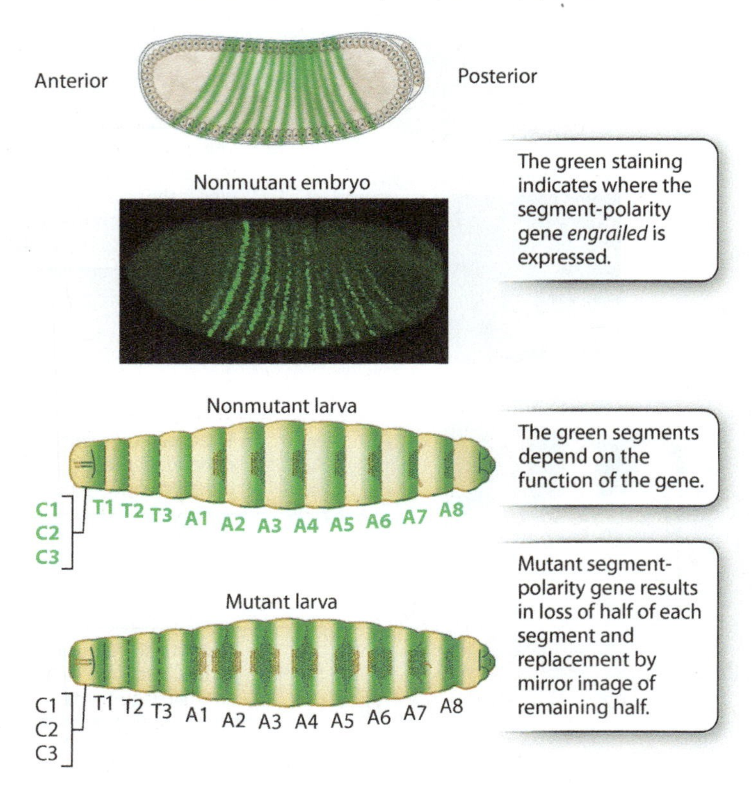

Homeotic genes determine where different body parts develop in the organism.

Together, the segment-polarity genes and other genes expressed earlier in the hierarchy control the pattern of expression of another set of genes called **homeotic (*Hox*) genes.** Originally discovered in *Drosophila,* homeotic genes encode some of the most important transcription factors in animal development. A homeotic gene is a gene that specifies the identity of a body part or segment during embryonic development. For example, homeotic genes instruct the three thoracic segments (T1, T2, and T3) each to develop a set of legs, and the second thoracic segment (T2) also to develop wings.

Two classic examples of the consequences of mutations in homeotic genes in *Drosophila* are shown in **Fig. 20.12.** Fig. 20.12a shows what happens when the homeotic gene *Antennapedia,* which specifies the development of the leg, is inappropriately expressed in anterior segments. The mutation causes legs to grow where antennae usually would. Similarly, a mutation in the homeotic gene *Bithorax* results in the transformation of thoracic segment 3 (T3) into thoracic segment 2 (T2), so that the fruit fly has two T2 segments in a row. As shown in the Fig. 20.12b, the result is a fruit fly with two complete sets of wings.

In *Drosophila,* the adult body parts like legs, antennae, and wings are formed from organized collections of tissue located throughout the larval body (**Fig. 20.13**). The development of these tissues and their metamorphosis into the adult body parts are regulated by the homeotic genes. First expressed at about the same time as the segment-polarity genes (see Fig. 20.11), the homeotic genes continue to be expressed even after the genes that regulate early development have shut down. Their continuing activity is due to the presence of chromatin remodeling proteins that keep the chromatin physically accessible to the transcription complex (Chapter 19).

Homeotic genes encode transcription factors. The DNA-binding domain in the homeotic proteins is a sequence of 60 amino acids called a **homeodomain,** whose sequences are very similar from one homeotic protein to the next across different species.

FIG. 20.12 Homeotic genes and segment identity. (a) Normal antennae are transformed into legs in an *Antennapedia* mutant. (b) Normal structures in the third thoracic segment are transformed into wings in a *Bithorax* mutant. *Sources: a. (left and right) F. Rudolf Turner, Ph.D., Indiana University; b. (left) Thomas Deerinck, NCMIR/Science Source; (right) David Scharf/Science Source.*

Normal | Mutant

a. *Antennapedia*

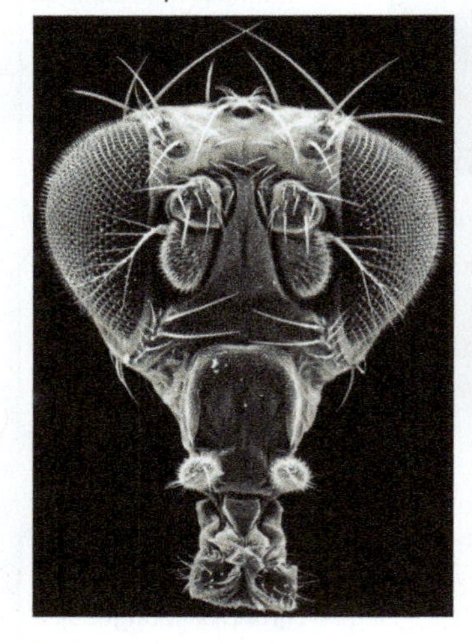

b. *Bithorax*

FIG. 20.13 Tissues controlled by homeotic genes. Homeotic genes specify the fate of clumps of tissue in the *Drosophila* larva.

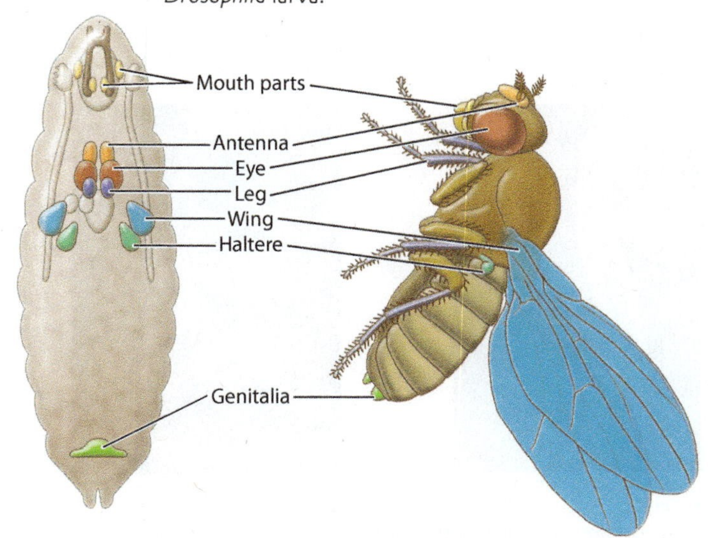

FIG. 20.14 Organization of the *Hox* gene clusters in *Drosophila* and the body parts that they affect. The order of genes along the chromosome corresponds to their positions along the anterior–posterior axis in the developing embryo.

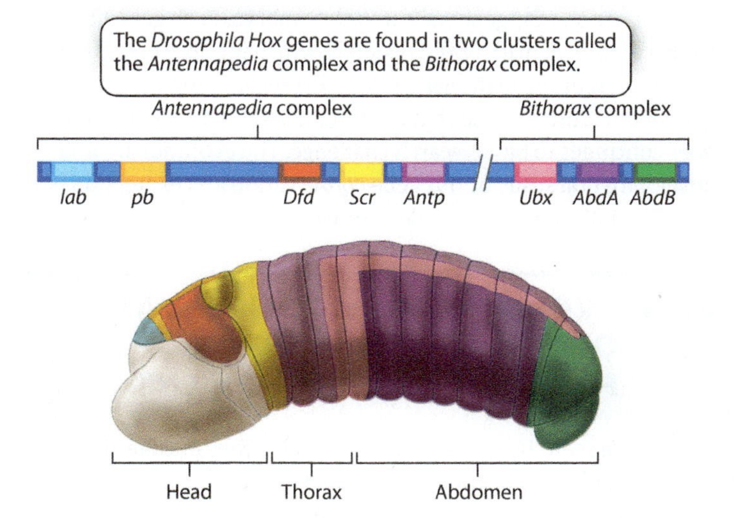

The *Drosophila* genome contains eight *Hox* genes comprising two distinct clusters, the *Antennapedia* complex and the *Bithorax* complex (**Fig. 20.14**). The genes are arranged along the chromosome in the same order as their products function in anterior–posterior segments along the embryo. In addition, the timing of their expression corresponds to their order along the chromosome and location of expression, with genes that are expressed closer to the anterior end turned on earlier than genes that are expressed closer to the posterior end. The correlation among linear order along the chromosome, anterior–posterior position in the embryo, and timing of expression is observed in *Hox* clusters in almost all organisms studied.

Because the amino acid sequences of the homeodomains of *Hox* gene products are very similar from one organism to the next, *Hox* gene clusters have been identified in a wide variety of animals with bilateral symmetry (organisms in which both sides of the midline are mirror images), from insects to mammals. Comparison of the number and types of *Hox* genes in different species supports the hypothesis that the ancestral *Hox* gene cluster had an organization very similar to what we now see in most organisms with *Hox* gene clusters. In its evolutionary history, the vertebrate genome underwent two whole-genome duplications; hence, vertebrates have four copies of the *Hox* gene cluster (**Fig. 20.15**).

FIG. 20.15 Organization and content of *Hox* gene clusters in the mouse and the regions of the embryo in which they are expressed.

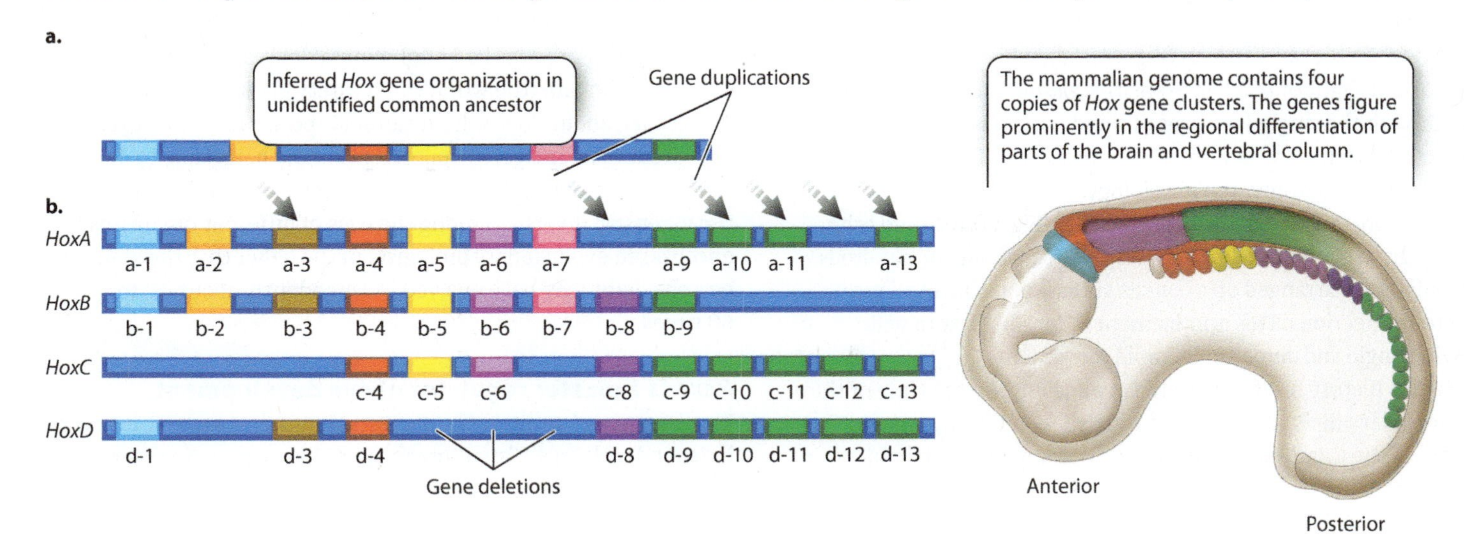

Unlike the *Hox* genes in *Drosophila,* mammalian *Hox* genes do not specify limbs but are important in the embryonic development of structures that become parts of the hindbrain, spinal cord, and vertebral column (Fig. 20.15). As in *Drosophila,* the genes in each cluster are expressed according to their linear order along the chromosome, which coincides with the linear order of regions the genes affect in the embryo. Each gene helps to specify the identity of the region in which it is expressed. Many of the genes in the mammalian *Hox* clusters have redundant or overlapping functions so that learning exactly what each gene does continues to be a research challenge. The evolutionary and developmental study of *Hox* gene conservation and expression is a good example of recent evo-devo research.

20.3 EVOLUTIONARY CONSERVATION OF KEY TRANSCRIPTION FACTORS IN DEVELOPMENT

As we have seen, *Hox* genes and the proteins they encode are very similar in a wide range of organisms. As a result, they can be identified by their similarity in DNA or amino acid sequence. Molecules that are similar in sequence among distantly related organisms are said to be evolutionarily conserved. Their similarity in sequence suggests that they were present in the most recent common ancestor and have changed very little over time because they carry out a vital function. Many transcription factors important in development are evolutionarily conserved. An impressive example is found in the development of animal eyes.

Animals have evolved a wide variety of eyes.

Animal eyes show an amazing diversity in their development and anatomy (**Fig. 20.16**). Among the simplest eyes are those of the planarian flatworm (Fig. 20.16a), which are only pit-shaped cells containing light-sensitive photoreceptors. The planarian eye has no lens to focus the light, but the animal can perceive differences in light intensity. The incorporation of a spherical lens, as in some jellyfish (Fig. 20.16b), improves the image.

FIG. 20.16 **Diversity of animal eyes.** *Sources: a. David M. Dennis/age footstock; b. Masa Ushioda/SeaPics.com; c. Reinhard Dirscherl/age fotostock; d. Andrey Armyagov/iStockphoto; e. Julian Brooks/age fotostock; f. Walter Geiersperger/age footstock.*

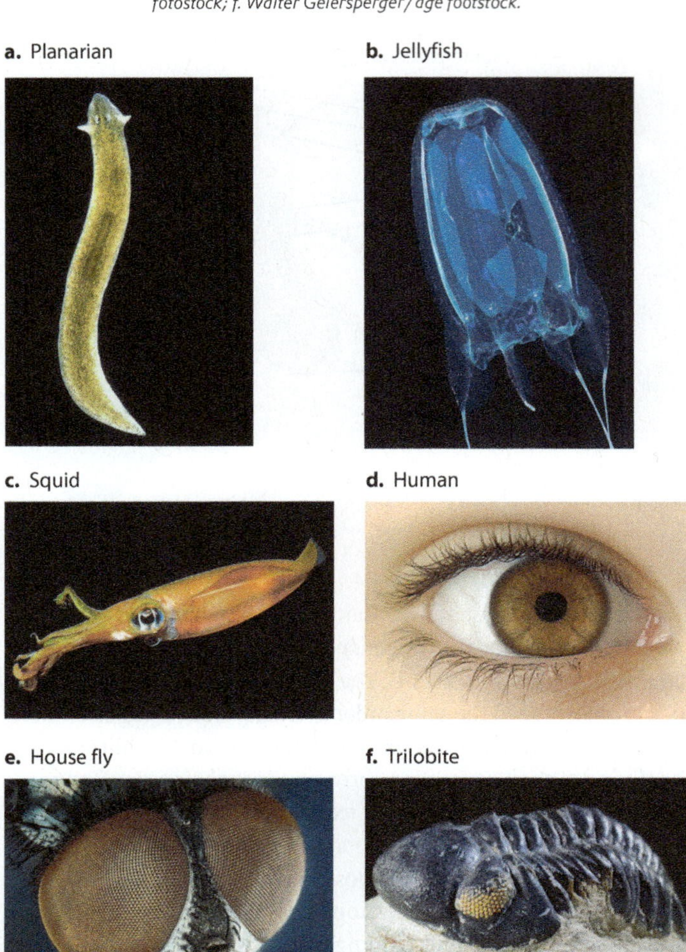

Among the most complex eyes are the camera-type eyes of the squid (Fig. 20.16c) and human (Fig. 20.16d), which have a single lens to focus light onto a light-sensitive tissue, the retina (Chapter 36). Though the single-lens eyes of squids and vertebrates are similar in external appearance, they are vastly different in their development, anatomy, and physiology.

Some organisms, such as the house fly (Fig. 20.16e) and other insects, have compound eyes, that is, eyes consisting of hundreds of small lenses arranged on a convex surface pointing in slightly different directions. This arrangement of lenses allows a wide viewing angle and detection of rapid movement.

In the history of life, eyes are very ancient. By the time of the extraordinary diversification of animals 542 million years ago known as the Cambrian explosion, organisms such as trilobites (Fig. 20.16f) already had well-formed compound eyes. These differed greatly from the eyes in modern insects, however. For example, trilobite eyes had hard mineral lenses composed of calcium carbonate (the world's first safety glasses, so to speak). Altogether, about 96% of living animal species have true eyes that produce an image, as opposed to simply being able to detect differences in light intensity. Despite this common feature, the extensive diversity among the eyes of different organisms encouraged evolutionary biologists in the belief that the ability to perceive light may have evolved independently about 40 to 60 times.

Pax6 is a master regulator of eye development.

The diversity of animal eyes suggests that they may have evolved independently in different organisms. However, more recent evidence suggests an alternative hypothesis—that they evolved once, very early in evolution, and subsequently diverged over time. One piece of evidence that supports the hypothesis of a single origin for light perception is the observation that the light-sensitive molecule in all light-detecting cells is the same, a derivative of vitamin A in association with a protein called opsin (Chapter 36). The presence of the same light-sensitive molecule in diverse eyes argues that it may have been present in the common ancestor of all animals with eyes and has been retained over time.

Another argument against multiple independent origins of eyes came from studies of eye development. Researchers identified *eyeless,* a gene in the fruit fly *Drosophila.* As its name implies, the phenotype of *eyeless* mutants is abnormal eye development (**Fig. 20.17a**). When the protein product of the *eyeless* gene was identified, it was found to be a transcription factor called Pax6. Mutant forms of a *Pax6* gene were already known to cause small eyes in the mouse (**Fig. 20.17b**) and aniridia (absence of the iris) in humans.

The mutations in the *Pax6* gene that cause the development defects in the eye in fruit flies, mice, and humans are **loss-of-function mutations.** A loss-of-function mutation is just that: a mutation that inactivates the normal function of a gene. In this case, loss-of-function mutations in *Pax6* make a defective version of the Pax6 transcription factor that is not able to carry out its function.

FIG. 20.17 Effect of *Pax6* mutations on eye development. (a) Normal and *eyeless* mutant in *Drosophila;* (b) normal and *small eye* mutant in mouse. *Sources: a. (left and right) David Scharf/Science Source; b. (left) INSADCO Photography/Alamy; (right) Jennifer L. Torrance, Photographer, The Jackson Laboratory.*

Normal

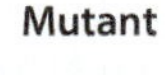

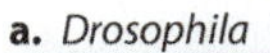

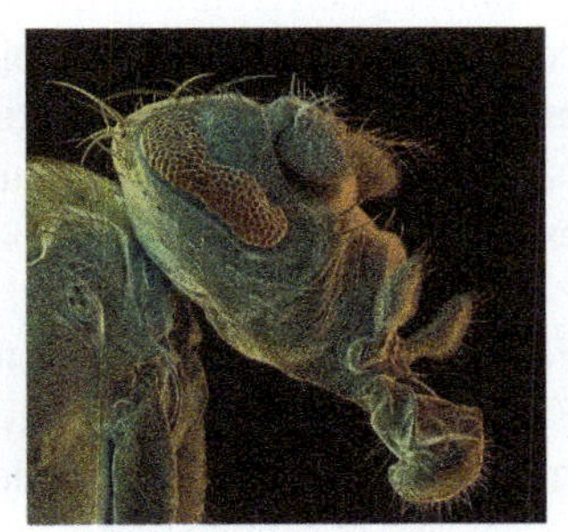

b. Mouse

FIG. 20.18 *Pax6,* a master switch controlling eye development. (a) Normal fly antenna; (b) eye tissue induced by expression of fruit fly *Pax6* in the antenna; (c) eye tissue induced by expression of mouse *Pax6* in the antenna. *Sources: a. Cheryl Power/Science Source; b. Eye of Science/Science Source; c. Prof. Walter Gehring/Science Source.*

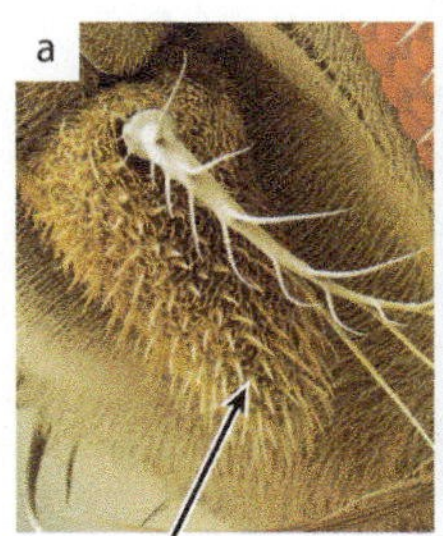

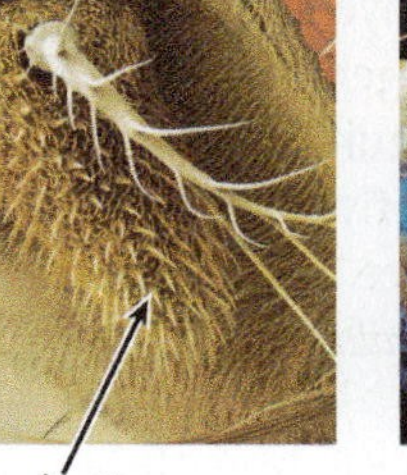

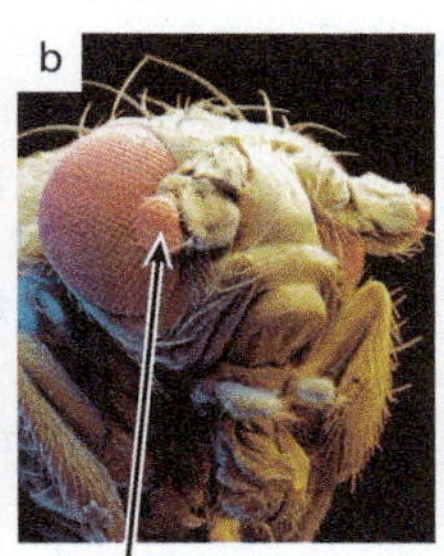

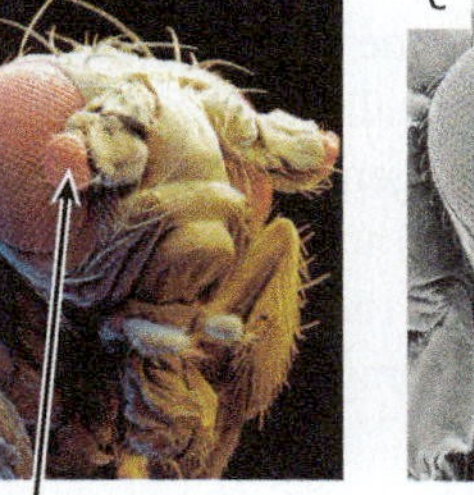

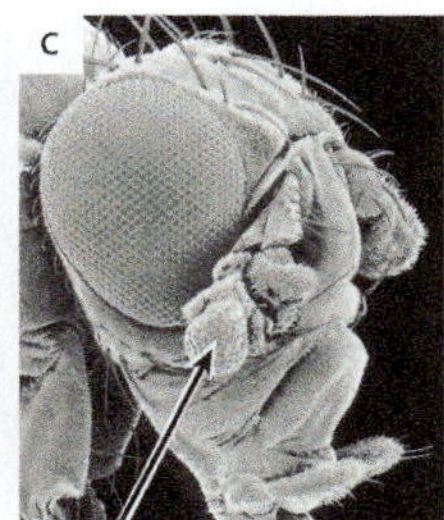

a Normal antenna

b Antennal eye induced by *Drosophila Pax6* gene

c Antennal eye induced by mouse *Pax6* gene

The similarity in phenotypes of *Pax6* mutants, along with the strong conservation of Pax6 in *Drosophila* and mouse, led Swiss developmental biologist Walter Gehring to hypothesize that Pax6 might be a master regulator of eye development. In other words, he hypothesized that Pax6 binds to regulatory regions of a set of genes that turns on a developmental program that induces eye development. In theory, this means that Pax6 can induce the development of an eye in any tissue in which it is expressed.

To test this hypothesis, Gehring and collaborators genetically engineered fruit flies that would produce the normal Pax6 transcription factor in the antenna, where it is not normally expressed (**Fig. 20.18a**). Any mutation in which a gene is expressed in the wrong place or at the wrong time is known as a **gain-of-function mutation.** (We saw an example of gain-of-function mutations when the *Hox* genes were expressed in the wrong segments of the fruit fly, producing legs where antennae normally develop and thus resulting in four-winged fruit flies.) For genes that control a developmental pathway, loss-of-function mutations and gain-of-function mutations often have opposite effects on phenotype.

→ **Quick Check 5** For genes that control pathways of development, loss-of-function mutations are usually recessive whereas gain-of-function mutations are usually dominant. Can you suggest a reason why?

Since loss-of-function mutations in *Pax6* result in an eyeless phenotype, a gain-of-function mutation should result in eyes developing in whatever tissue *Pax6* is expressed. In the gain-of-function mutation, the antenna developed into a miniature compound eye, and electrical recordings demonstrated that some of these antennal eyes were functional (**Fig. 20.18b**). The researchers also created other gain-of-function mutations that led to eyes on the legs, wings, and other tissues, which the *New York Times* publicized in an article headlined "With New Fly, Science Outdoes Hollywood."

Gehring and his group then went one step further. They took the *Pax6* gene from mice and expressed it in fruit flies to see whether the mouse *Pax6* gene is similar enough to the fruit fly version of the gene that it could induce eye development in the fruit fly. Specifically, they created transgenic fruit flies that expressed the *Pax6* gene from mice in the fruit fly antenna. The mouse gene induced a miniature eye in the fly (**Fig. 20.18c**). Note, however, that the *Pax6* gene from mice induced the development of a compound eye of *Drosophila,* not the single-lens eye of mouse.

The ability of mouse Pax6 to make an eye in fruit flies suggests that mouse and fruit fly *Pax6* are not only similar in DNA and amino acid sequence, but also similar in function, and indeed act as a master switch that can turn on a developmental program leading to the formation of an eye. But these observations lead to another question: Why did the mouse *Pax6* gene produce fruit fly eyes instead of mouse eyes?

The answer is that the fruit fly genome does not include the genes needed to make mouse eyes. Mouse Pax6 protein induces fly eyes because of the **downstream genes** affected by Pax6, those that function later in the process of eye development. Transcription factors like Pax6 interact with their target genes by binding with short DNA sequences adjacent to the gene, usually at the 5′ end, called **cis-regulatory elements.** These regulatory elements are located in the promoter and help determine whether the adjacent DNA is transcribed. When bound to cis-regulatory elements, some transcription factors act as repressors that prevent transcription of the target gene, and others serve as activators by recruiting the transcriptional machinery to the target gene (Chapter 19). A transcription factor can even repress some of its target genes and activate others.

In the fruit fly, Pax6 binds to cis-regulatory elements in many genes, turning some genes on and others off. The products of these downstream genes in turn affect the expression of further downstream genes. The total number of genes that are downstream of Pax6 and that are needed for eye development is estimated at about 2000. Most are not direct targets of Pax6 but are activated indirectly through other transcription factors downstream of Pax6. When mouse Pax6 is expressed in fruit flies, it is similar enough in sequence to activate the genes involved in fruit fly eye development, so it makes sense that mouse Pax6 leads to the development of a fruit fly eye, not a mouse eye.

One scenario of how Pax6 became a master switch for eye development in a wide range of organisms, but produces a diversity of eyes in these organisms, is that Pax6 evolved early in the history of life as a transcription factor able to bind to and regulate genes involved in early eye development. Over time, different genes in different organisms acquired new Pax6-binding cis-regulatory elements by mutation, and if these were beneficial they persisted. The downstream genes that are targets of Pax6 therefore are different in different organisms, but they share two features—they are regulated by Pax6 and they are involved in eye development. So, the early steps are conserved, but the later ones are not.

The *Pax6* gene and the *Hox* genes reveal an important principle in evo-devo, which is that master regulatory genes that control development are often evolutionarily conserved, whereas the downstream genes that they regulate may not be. Downstream genes can evolve new functions, or genes not originally controlled by the master regulator may evolve to come under its influence, or genes formerly controlled by the master regulator may evolve to be unresponsive. In this way, a conserved master regulatory mechanism may result in distinct developmental outcomes in different organisms. In the case of *Pax6,* for example, the master regulatory mechanism for eye development evolved early and is shared among a wide range of animals, even though the eyes that are produced are quite diverse.

20.4 COMBINATORIAL CONTROL IN DEVELOPMENT

Most cis-regulatory elements are located near one or more binding sites for transcription factors, some of which are activators of transcription and others repressors of transcription. The rate of transcription of any gene in any type of cell is therefore determined by the combination of transcription factors that are present in the cell and by the relative balance of activators and repressors. Regulation of gene transcription according to the mix of transcription factors in the cell is known as **combinatorial control.** Combinatorial control of transcription is another general principle often seen in all multicellular organisms at many stages of development. Here, we discuss flower development as an example of this general principle.

Floral differentiation is a model for plant development.

The plant *Arabidopsis thaliana,* a weed commonly called mouse-ear cress, is a model organism for developmental studies and presents a clear case of combinatorial control. As in all plants, *Arabidopsis* has regions of undifferentiated cells, called meristems, where growth can take place (Chapter 31). Meristem cells are similar to stem cells in animals. They are the growing points where shoots, roots, and flowers are formed. Floral meristems consist of multipotent cells that can differentiate into different structures in the flower. In floral meristems of *Arabidopsis,* the flowers develop from a pattern of four concentric circles of cells, or whorls, each of which differentiates into a distinct type of floral structure.

FIG. 20.19 Development of the *Arabidopsis* floral organs from concentric whorls of cells in the floral meristem. *Photo source: Juergen Berger/Max Planck Institute for Developmental Biology, Tuebingen, Germany.*

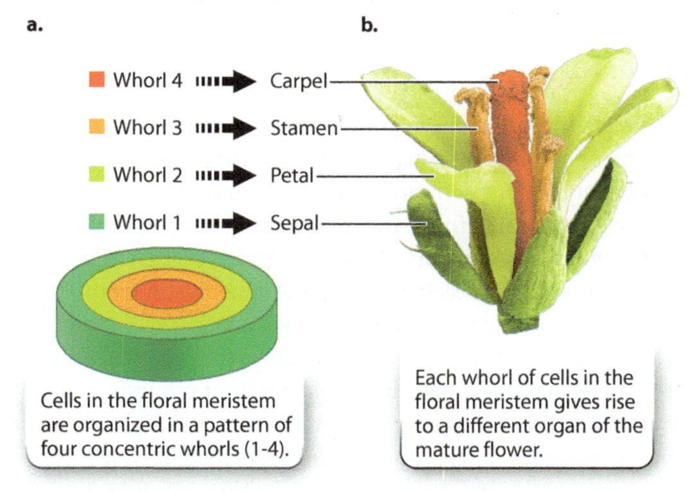

From the outermost whorl (whorl 1) of cells to the innermost whorl (whorl 4), the floral organs are formed as shown in **Fig. 20.19:**

- Cells in whorl 1 form the green sepals, which are modified leaves forming a protective sheath around the petals in the flower bud that unfold like petals when the flower opens.
- Cells in whorl 2 form the petals.
- Cells in whorl 3 form the stamens, the male sexual structures in which pollen is produced. (The small granules on the stamens in Fig. 20.19b are pollen grains.)
- Cells in whorl 4 form the carpels, the female reproductive structures that receive pollen and contain the ovaries.

The identity of the floral organs is determined by combinatorial control.

The genetic control of flower development in *Arabidopsis* was discovered by the analysis of mutant plants, an investigation similar in principle to the way that genetic control of development in fruit flies was determined. The plant mutants fell into three classes showing characteristic floral abnormalities (**Fig. 20.20**). Mutations in the gene *apetala-2* inactivate the gene's function and affect whorls 1 and 2 (Fig. 20.20b), those in *apetala-3* or *pistillata* affect whorls 2 and 3 (Fig. 20.20c), and those in *agamous* affect whorls 3 and 4 (Fig. 20.20d). (There are different standards for naming genes and proteins in different organisms. In *Arabidopsis*, wild-type alleles are written in capital letters with italics, and wild-type proteins in capital letters without italics. Mutant alleles and proteins follow the same rules but are written in lower-case letters.)

The observation that the mutant phenotypes affect development of pairs of adjacent whorls already hints at combinatorial control, which researchers incorporated into a model of floral development called the **ABC model.** The model invokes three activities arbitrarily called A, B, and C. These activities represent the function of one or more different proteins that

FIG. 20.20 Phenotypes of the normal flower and floral mutants of *Arabidopsis.* *Source: Courtesy of John Bowman.*

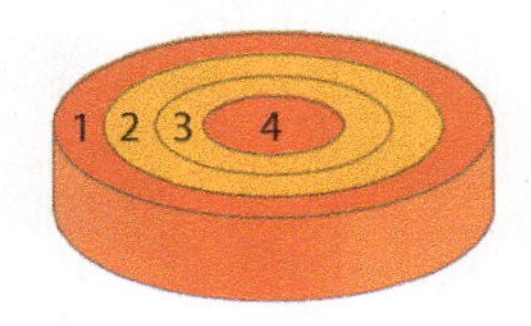

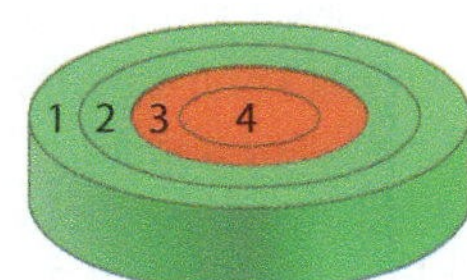

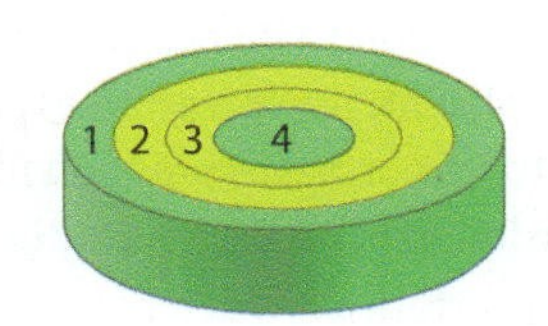

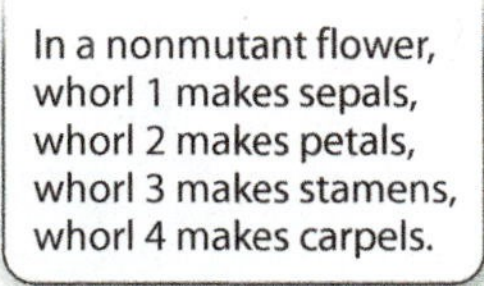

In a nonmutant flower, whorl 1 makes sepals, whorl 2 makes petals, whorl 3 makes stamens, whorl 4 makes carpels.

In the mutant *APETALA-2*, whorl 1 makes carpels, whorl 2 makes stamens, whorl 3 makes stamens, whorl 4 makes carpels.

In the mutant *APETALA-3* or the mutant *PISTILLATA*, whorl 1 makes sepals, whorl 2 makes sepals, whorl 3 makes carpels, whorl 4 makes carpels.

In the mutant *AGAMOUS*, whorl 1 makes sepals, whorl 2 makes petals, whorl 3 makes petals, whorl 4 makes sepals.

FIG. 20.21 **The ABC model for flower development in *Arabidopsis*.**

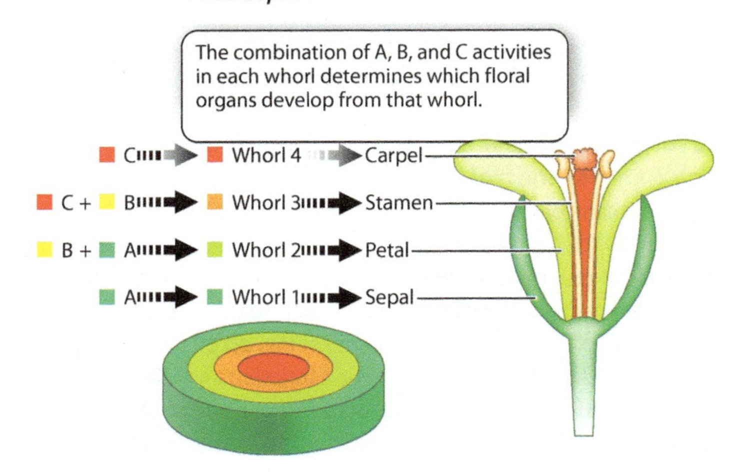

are hypothesized to be present in cells of each whorl as shown in **Fig. 20.21.** In the course of floral differentiation, the cells of each whorl become different from those in the other whorls, and different combinations of genes are activated or repressed in each whorl. Activity A is present in whorls 1 and 2, activity B in whorls 2 and 3, and activity C in whorls 3 and 4. According to the ABC model, activity A alone results in the formation of sepals, A and B together result in petals, B and C together result in stamens, and C alone results in carpels. Like any good hypothesis (Chapter 1), this one makes specific predictions. Mutants lacking A will have defects in whorls 1 and 2, mutants lacking B will have defects in whorls 2 and 3, and mutants lacking C will have defects in whorls 3 and 4.

Matching these predictions with the mutants in Fig. 20.20 yields the following correspondence:

- The A activity is encoded by the gene *APETALA-2*.
- The B activity is encoded jointly by the genes *APETALA-3* and *PISTILLATA*.
- The C activity is encoded by the gene *AGAMOUS*.

To account for the mutant phenotypes in Fig. 20.20, you need only to postulate that when A is absent, C is expressed in whorls 1 and 2 in addition to its normal expression in whorls 3 and 4; and when C is absent, A expression expands into whorls 3 and 4 in addition to its normal expression in whorls 1 and 2. The domains of expression expand in this way because, in addition to its role in activating certain downstream genes, the product of *APETALA-2* represses the expression of *AGAMOUS* in whorls 1 and 2. As a result, *AGAMOUS* is normally not expressed in whorls 1 and 2, but when *APETALA-2* is mutant, *AGAMOUS* is expressed wherever *APETALA-2* normally represses it. The result is activity C in whorls 1 and 2. Similarly, in addition to its role in activating other genes, the product of *AGAMOUS* represses the expression of *APETALA-2* in whorls 3 and 4, so *APETALA-2* is normally not expressed in whorls 3 and 4.

→ **Quick Check 6** Predict the phenotype of a flower in which *APETALA-2* (activity A) and *AGAMOUS* (activity C) are expressed normally, but *APETALA-3* and *PISTILLATA* (activity B) are expressed in all four whorls.

The identification of the products of these genes demonstrates combinatorial control and explains the mutant phenotypes. All of the genes encode transcription factors, which are called AP2, AP3, PI, and AG, corresponding to the four genes. The transcription factors bind to cis-regulatory elements of genes that encode proteins necessary for each whorl's development, similar to what we saw earlier for Pax6 and eye development. AP3 and PI are both necessary for the B activity because the proteins form a heterodimer, a protein made up of two different subunits.

Although other transcription factors associated with other activities that also contribute to flower development were discovered later, the original ABC model is still valid and serves as

FIG. 20.22 **Floral diversity.** (a) Colorado blue columbine (*Aquilegia coaerulea*); (b) slipper orchid (*Paphiopedilum holdenii*); (c) ginger flower (*Smithatris supraneanae*); (d) garden rose (hybrid). *Sources: a. ljh images/Shutterstock; b. Cesar Chavez Photography/Big Stock; c. Nonn Panitvong; d. Maria Mosolova/age fotostock.*

an elegant example of combinatorial control. As in the case of the *Pax6* gene in the development of animals' eyes, evolution of the cis-regulatory sequences in downstream genes has resulted in a wide variety of floral morphologies in different plant lineages, some of which are shown in **Fig. 20.22.**

20.5 CELL SIGNALING IN DEVELOPMENT

As we have seen in the discussion of stem cells, differentiated cells can be reprogrammed by the action of only a few key genes or small organic molecules. In some cases, the processes that push differentiation in a forward direction are also quite simple. An important example is **signal transduction,** in which an extracellular molecule acts as a signal to activate a membrane protein that in turn activates molecules inside the cell that control differentiation (Chapter 9). The signaling molecule is called the **ligand** and the membrane protein that it activates is called the **receptor.** The following example shows how a simple ligand–receptor pair can have profound effects not only on the differentiation of the cell that carries the receptor, but also on its neighbors.

A signaling molecule can cause multiple responses in the cell.

Pioneering experiments on signal transduction in development were carried out in the soil nematode *Caenorhabditis elegans* (**Fig. 20.23**). This worm is a useful model organism for developmental studies because of its small size (the adult is about 1 mm long), short generation time (2½ days from egg to sexually mature adult), and stereotyped process of development. (A stereotyped process is one that is always the same in each individual.) The adult animal consists of exactly 959 somatic (body) cells, and the cells' patterns of division, migration, differentiation, and morphogenesis are identical in each individual *C. elegans.* The adult is typically a hermaphrodite that produces both male and female reproductive cells. The eggs are fertilized internally by sperm, and after undergoing five or six cell divisions the eggs are laid through an organ known as the vulva (Fig. 20.23).

Vulva development is relatively simple and therefore amenable to experimental studies. The entire structure arises from only three multipotent cells that undergo either of two types of differentiation and divide to produce a vulva consisting of exactly 22 cells. Cells resulting from type 1 differentiation form the opening of the vulva, and cells resulting from type 2 differentiation form a supporting structure around the vulval opening. In each of the two types of differentiation, different genes are activated or repressed, resulting in differentiated cells with different functions.

How each of the three original cells, called progenitor cells, differentiates is determined by its position in the developing worm (**Fig. 20.24**). The fate of the progenitor cells is determined by their proximity to another cell, called the anchor cell ("AC" in

FIG. 20.23 The nematode worm *Caenorhabditis elegans.*

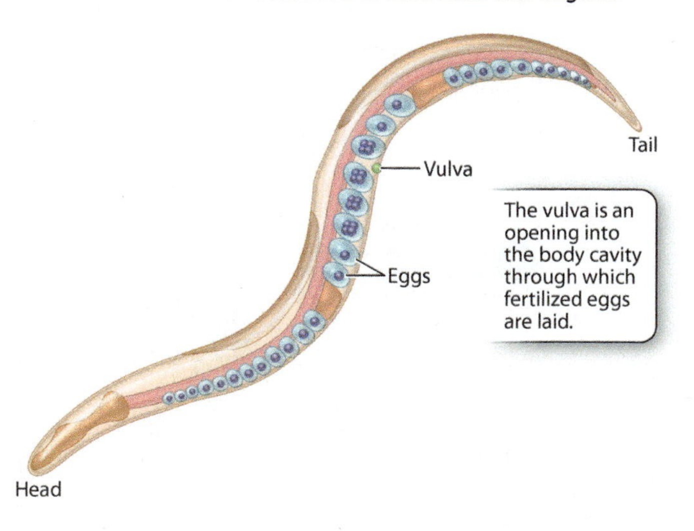

FIG. 20.24 Vulval development in *Caenorhabditis elegans.* The cells of the vulva differentiate in response to molecular signals from other cells.

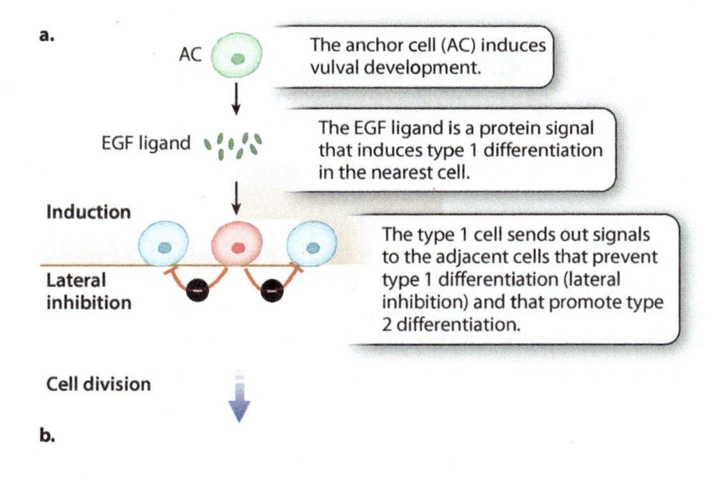

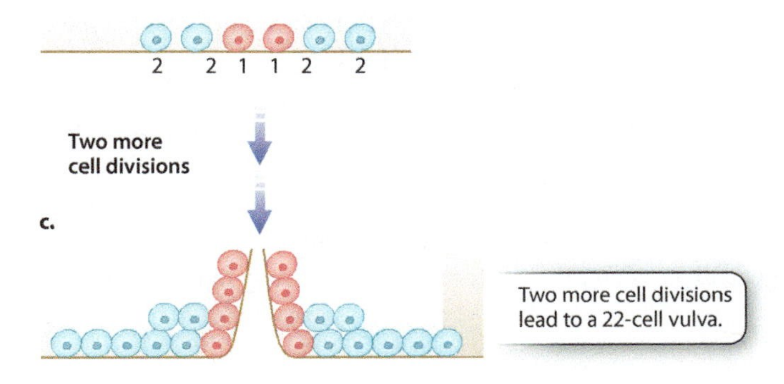

FIG. 20.25 Signal transduction in vulval development in *Caenorhabditis elegans.*

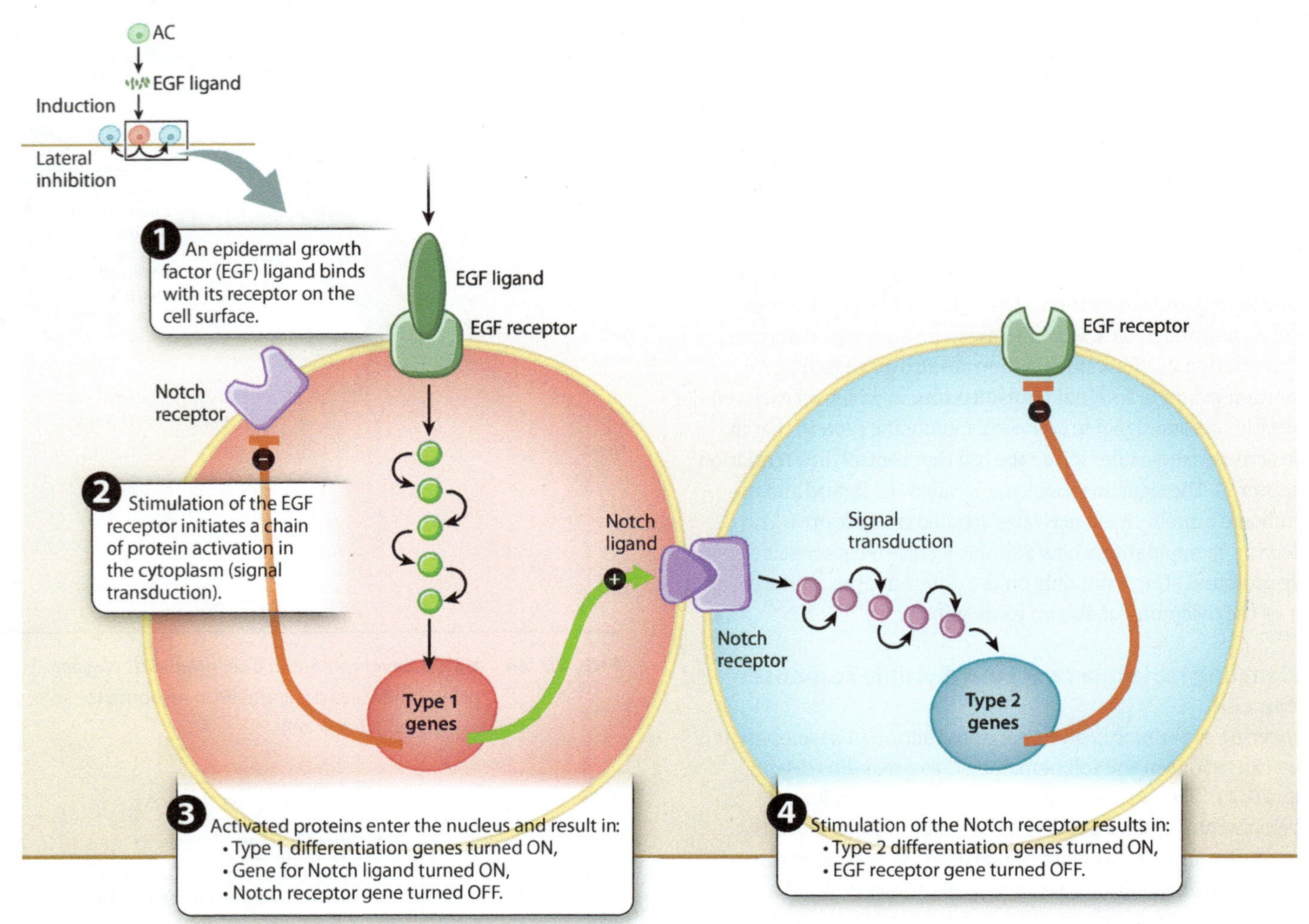

Fig. 20.24a), which secretes a protein called epidermal growth factor (EGF) that binds to and activates a transmembrane EGF receptor (Chapter 9). The progenitor cell closest to the anchor cell receives the most amount of signal, and upon activation of its EGF receptor the progenitor cell carries out three functions:

- It activates the genes for differentiation into a type 1 cell (Fig. 20.24a).
- It prevents the adjacent cells from differentiating as type 1 (a process called **lateral inhibition**).
- It induces the adjacent cells to differentiate as type 2 cells.

Developmental signals are amplified and expanded.

How can a single ligand–receptor pair cause so many changes in gene expression that it determines the pathway of differentiation not only of the cell itself but also of its neighbors? **Fig. 20.25** shows the mechanisms in simplified form. On the left is the progenitor cell nearest the anchor cell, which receives the strongest EGF signal. Activation of the EGF receptor by the ligand initiates a process of signal transduction in the cytoplasm, in which the signal is transmitted from one protein to the next by means of proteins at each stage phosphorylating several others, which amplifies the signal at each stage (Chapter 9). The result of signal transduction is that a set of transcription factors is activated.

In the nucleus, the transcription factors activate transcription of genes for type 1 differentiation. The transcription factors also activate transcription of genes that prevent type 1 differentiation in neighboring cells, including the genes that produce another type of protein ligand, called Notch. The Notch ligand is a transmembrane protein that activates Notch receptors in the neighboring cells. Activation of the Notch receptor in these cells activates a signal transduction cascade in these cells, which results in transcription of the genes for type 2 differentiation. The cascade started by the binding of Notch also activates transcription of other genes whose products inhibit the EGF receptor. Inhibiting the EGF receptor in type 2 cells prevents EGF from eliciting a type 1 response in the type 2 cell. In addition, at the same time that the type 1 cell produces the Notch ligand, it produces proteins that inhibit its own Notch receptors, and this prevents Notch from initiating a type 2 response in the type 1 cell.

While vulva development in nematodes is a fairly simple example of the importance of ligand–receptor signaling in development, EGF and Notch ligands and their receptors are found in virtually all animals. They are among dozens of ligand–receptor pairs that have evolved as signaling mechanisms to regulate processes in cellular metabolism and development. Humans have several distinct but related gene families of EGF and Notch ligands and receptors. Human EGF is important in cell survival, proliferation, and differentiation. EGF functions in the differentiation and repair of multiple types of cells in the skin; it is present in all body fluids and helps regulate rapid metabolic responses to changing conditions. Human Notch ligands are involved in development of the nervous and immune systems as well as heart, pancreas, and bone. Abnormalities in EGF or Notch signaling are associated with many different types of cancer.

Therefore, just as we saw in our discussion of eye and flower development, the molecular players involved in development are often evolutionarily conserved across a wide range of organisms. This is true even of genes that we typically associate with disease, such as *BRCA1* and *BRCA2* and their link with cancer. These genes not only play a role in cell cycle control in the adult, but also in early development in many organisms. In fact, although heterozygous mutations in these genes predispose individuals to breast and ovarian cancers in humans, homozygous mutations are lethal in early embryonic stages. In **Fig. 20.26,** we focus on the *BRCA1* gene to summarize key concepts about DNA replication, mutation, genetic variation, inheritance, gene regulation, and development. ■

VISUAL SYNTHESIS **Genetic Variation and Inheritance**

FIG. 20.26 **Integrating concepts from Chapters 12–20**

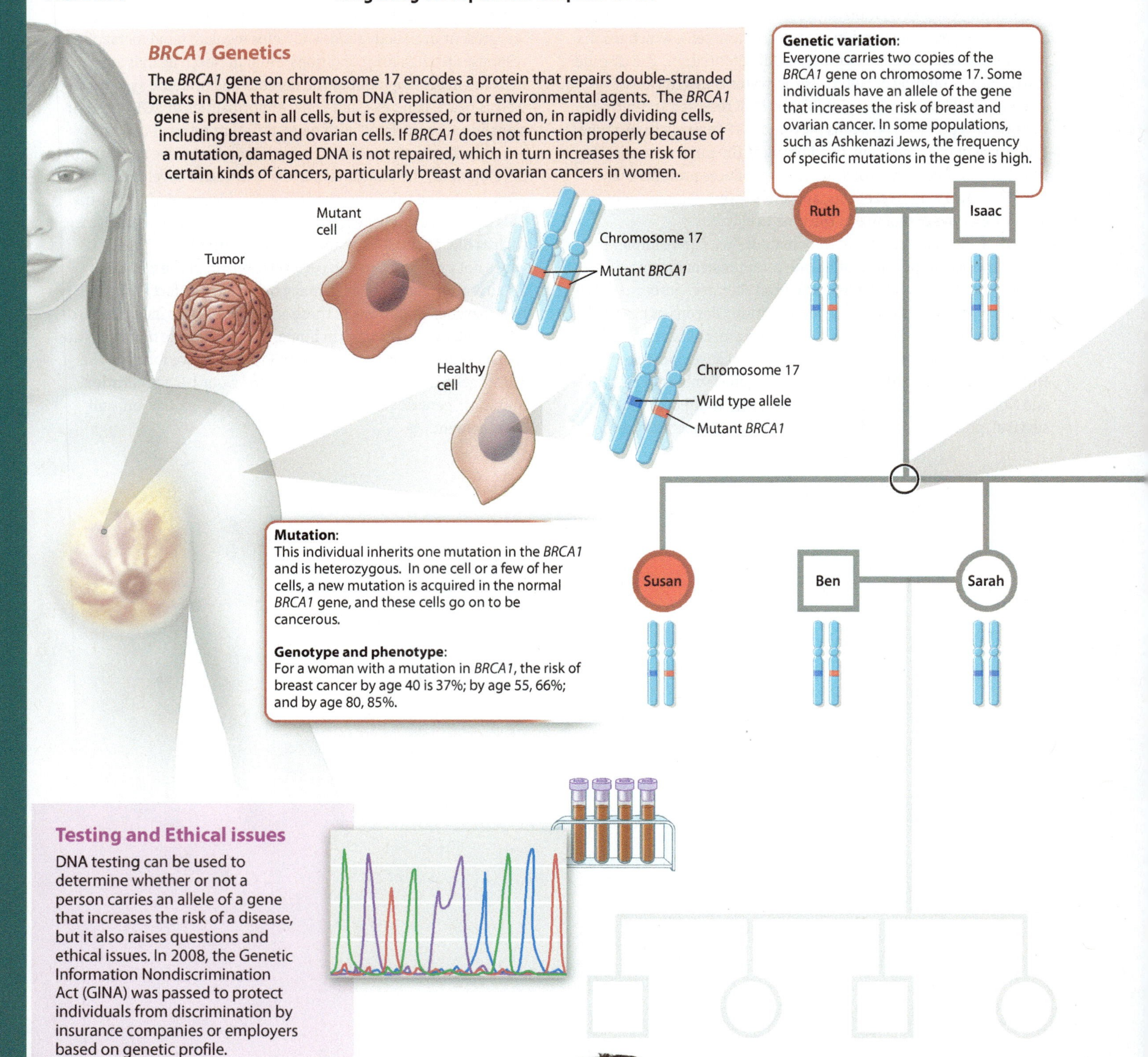

DNA testing can be used to determine whether any of these individuals carry a mutation in the *BRCA1* gene.
If any individual in this family gets a positive result, everyone else can figure out their risk of carrying the mutation.
If an individual tests positive, what should he or she do?
What happens if an individual finds a mutation in the *BRCA1* gene which is not known to be associated with cancer?

? ? ? ?

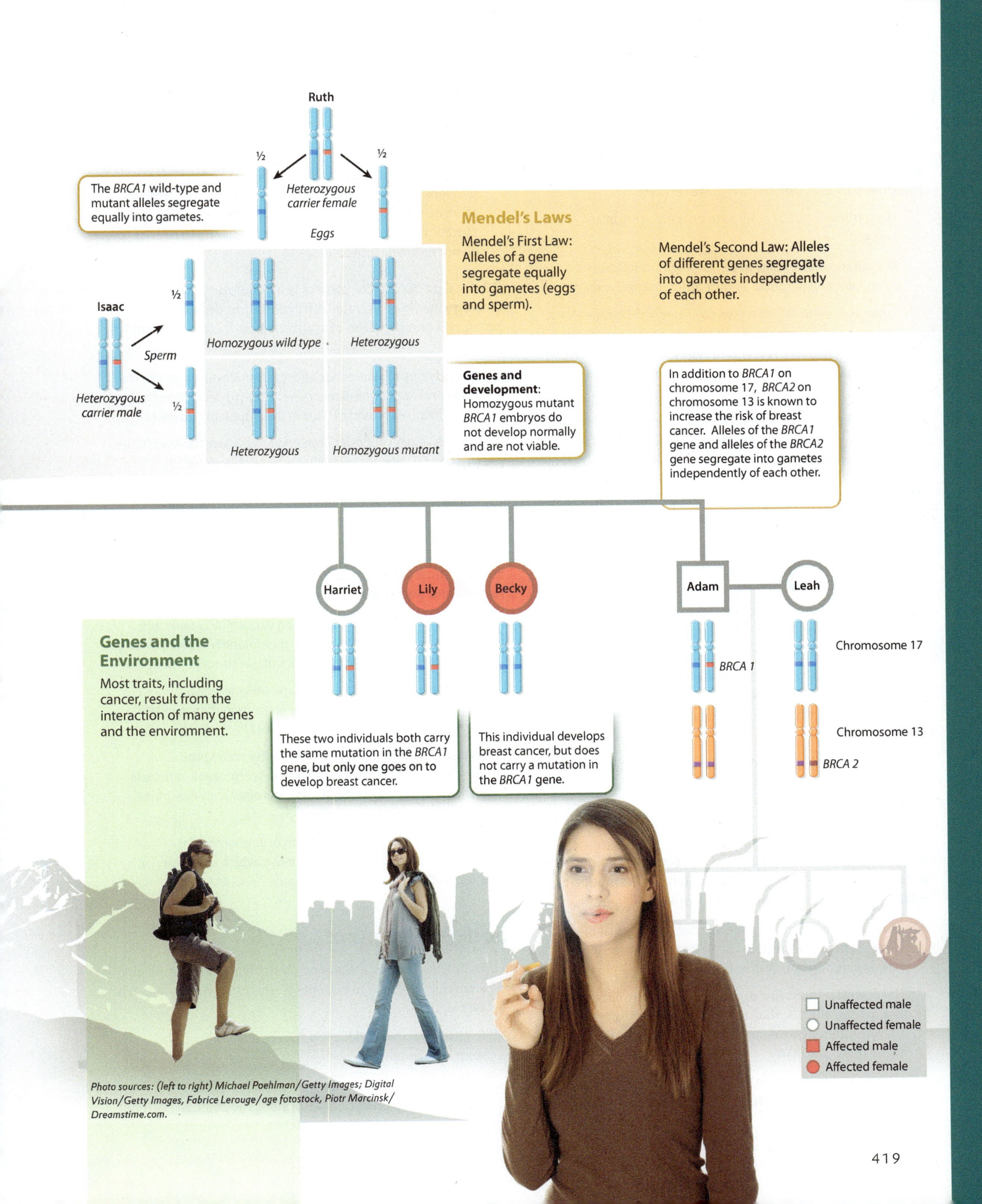

Photo sources: (left to right) Michael Poehlman/Getty Images; Digital Vision/Getty Images, Fabrice Lerouge/age fotostock, Piotr Marcinsk/Dreamstime.com.

Core Concepts Summary

20.1 In the development of humans and other animals, stem cells become progressively more restricted in their possible pathways of cellular differentiation.

The fertilized egg can give rise to a complete organism. page 400

At each successive stage in development, cells lose developmental potential as they differentiate. page 401

Embryonic stem cells can give rise to any of the three germ layers, those further along in differentiation can form only a limited number of specialized cell types, and those still further along can form only one cell type. page 401

Stem cells play a prominent role in regenerative medicine, in which stem cells—in some cases reprogrammed cells from the patient's own body—are used to replace diseased or damaged tissues. page 403

20.2 The genetic control of development is a hierarchy in which genes are activated in groups that in turn regulate the next set of genes.

Hierarchical gene control can be seen in fruit fly (*Drosophila*) development. page 404

The oocyte of a fruit fly is highly polarized, with gradients of maternal mRNA that set up anterior–posterior and dorsal–ventral axes. page 405

These gradients in turn affect the expression of segmentation genes in the zygote, including the gap, pair-rule, and segment-polarity genes, which define specific regions in the developing embryo. page 407

The segmentation genes direct the expression of homeotic genes, key transcription factors that specify the identity of each segment of the fly and that are conserved in animal development. page 408

20.3 Many proteins that play key roles in development are evolutionarily conserved but can have dramatically different effects in different organisms.

Many proteins important in development are similar in sequence from one organism to the next. Such proteins are said to be evolutionarily conserved. page 410

The downstream targets of homeotic genes are different in different animals, allowing homeotic genes to activate different developmental pathways in different organisms. page 410

Although animals exhibit an enormous diversity in eye morphology, the observation that the proteins involved in light perception are evolutionarily conserved suggests that the ability to perceive light may have evolved once, early in the evolution of animals. page 410

The Pax6 transcription factor is a master regulator of eye development. Loss-of-function mutations in *Pax6* result in abnormalities in eye development, whereas gain-of-function mutations result in eye development in tissues in which eyes do not normally form. page 410

20.4 Combinatorial control is a developmental strategy in which cellular differentiation depends on the particular combination of transcription factors present in a cell.

By analyzing mutants that affect flower development in the plant *Arabidopsis*, researchers were able to determine the genes involved in normal flower development. page 412

The ABC model of flower development invokes three activities (A, B, and C) present in circular regions (whorls) of the developing flower, with the specific combination of factors determining the developmental pathway in each whorl. page 413

20.5 Ligand–receptor interactions activate signal transduction pathways that converge on transcription factors and genes that determine cell fate.

Cell signaling involves a ligand, an extracellular molecule that acts as a signal to activate a membrane receptor protein, which in turn activates molecules inside the cell. page 415

Activation of a receptor sets off a pathway of signal transduction, in which a series of proteins in the cytoplasm become sequentially activated. page 416

Because signal transduction can amplify and expand a developmental signal, a single ligand–receptor pair can cause major changes in gene expression and ultimately determine the pathway of differentiation. page 416

An example of signal transduction in development is differentiation of the nematode vulva, which is determined by means of an EGF ligand and receptor. page 416

Self-Assessment

1. Distinguish among totipotent, pluripotent, and multipotent stem cells, and give an example of where you would find each type of cell.

2. Explain how an individual's own cells might be used in stem cell therapy.

3. Draw a diagram to illustrate how a concentration gradient of a transcription factor along the anterior–posterior

axis of a *Drosophila* embryo can create a region in the middle in which transcription of a target gene takes place without being transcribed in either the anterior or posterior region.

4. Expression of a homeotic gene in the wrong tissue in *Drosophila* results in the development of an inappropriate body part from that tissue. Explain why this happens and how it shows that homeotic genes are positive regulators of developmental pathways.

5. Explain why master regulatory genes tend to be more strongly conserved in evolution than are the downstream genes they regulate.

6. Define combinatorial control in the context of the ABC model of floral development.

7. Diagram a pathway of signal transduction including a ligand, receptor, and ultimately a transcription factor that activates a gene that inhibits the receptor.

Log in to LaunchPad to check your answers to the Self-Assessment questions, and to access additional learning tools.

CASE 4

Malaria

Coevolution of Humans and a Parasite

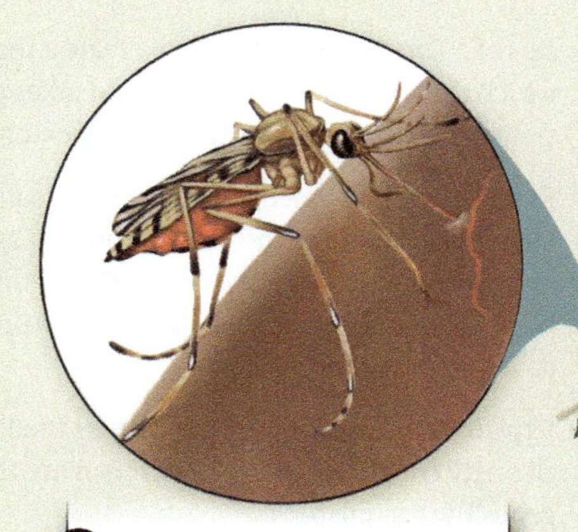

1 As it feeds, the mosquito injects *Plasmodium* parasites along with its saliva into the host. The parasites migrate through the bloodstream to the host's liver.

Most people would say the world has more than enough mosquitoes, but in 2010 scientists at the University of Arizona conjured up a new variety. The high-tech bloodsuckers were genetically engineered to resist *Plasmodium,* the single-celled eukaryote that causes malaria.

Normally, the parasite grows in the mosquito's gut and is spread to humans by the insect's bite. By altering a single gene in the mosquito's genome, the researchers had made the insects immune to the malaria parasite. The accomplishment is a noteworthy advance, the latest in a long line of efforts to stop *Plasmodium* in its tracks.

As humans have tried a succession of weapons to defeat P. falciparum, the parasite has evolved, thwarting their efforts.

Malaria is one of the most devastating diseases on the planet. The World Health Organization estimates that 500 million people contract malaria annually, primarily in tropical regions. The disease is thought to claim about a million lives each year. Of those deaths, 85% to 90% occur in sub-Saharan Africa, mostly among children under 5.

Five species of *Plasmodium* can cause malaria in humans. One of them, *P. falciparum,* is particularly dangerous, accounting for the vast majority of malaria fatalities. Over thousands of years, this parasite and its human host have played a deadly game of tug-of-war. As humans have tried a succession of weapons to defeat *P. falciparum,* the parasite has evolved, thwarting their efforts. And in turn, the tiny organism has helped to shape human evolution.

Plasmodium is a wily and complicated parasite, requiring both humans and mosquitoes to complete its life cycle. The human part of the cycle begins with a single bite from an infected mosquito. As the insect draws blood, it releases *Plasmodium*-laden saliva into the bloodstream. Once inside their human host, the *Plasmodium* parasites invade the liver cells. There they undergo cell division for several days, their numbers increasing. Eventually, the parasites infect red blood cells, where they continue to grow and multiply. The mature parasites burst from the red blood cells at regular intervals, triggering malaria's telltale cycle of fever and chills.

Some of the freed *Plasmodium* parasites go on to infect new red blood cells; others divide to form gametocytes, precursor cells to male and female gametes, which travel through the victim's blood vessels. When another mosquito bites the infected individual, it takes up the gametocytes with its blood meal. Inside the insect, the parasite completes its life cycle. The gametocytes fuse to form zygotes. Those zygotes bore into the mosquito's stomach, where they form cells called oocysts that give rise to a new generation of parasites. When the infected mosquito sets out to feed, the cycle begins again.

The battle between malaria and humankind has raged through the ages. Scientists have recovered *Plasmodium* DNA from the bodies of 3500-year-old Egyptian mummies—evidence that those ancient humans were infected with the malaria parasite. The close connection between humans, mosquitoes, and the malaria parasite almost certainly extends back much further.

In fact, people who hail from regions where malaria is endemic are more likely than others to have certain genetic signatures that offer some degree of protection from the parasite. That indicates that *Plasmodium* has been exerting evolutionary pressure on humankind for quite some time.

Meanwhile, we've done our best to fight back. For centuries, humans fought the infection with quinine, a chemical found in the bark of the South American cinchona tree. In the 1940s, scientists developed a more sophisticated drug based on the cinchona compound. That drug, chloroquine, was effective, inexpensive, easy to administer, and caused few side effects. As a result, it was widely used, and in the late 1950s, *P. falciparum* began showing signs of resistance to chloroquine. Within 20 years, resistance had spread to Africa, and today most strains of *P. falciparum* have evolved resistance to the once-potent medication.

Just as bacteria develop resistance to antibiotics, *Plasmodium* evolves resistance to the antiparasitic drugs designed to fight it. In poor, rural areas where malaria is prevalent, people often can't follow the recommended protocols for antimalarial treatment. Sick individuals may be able to afford only a few pills rather than the full recommended dose. The strength and quality of those pills may be questionable, and the drugs are often taken without oversight from a medical professional.

Unfortunately, inadequate use of the drugs fuels resistance. When the pills are altered or the course of treatment is abbreviated, not all *Plasmodium* parasites are

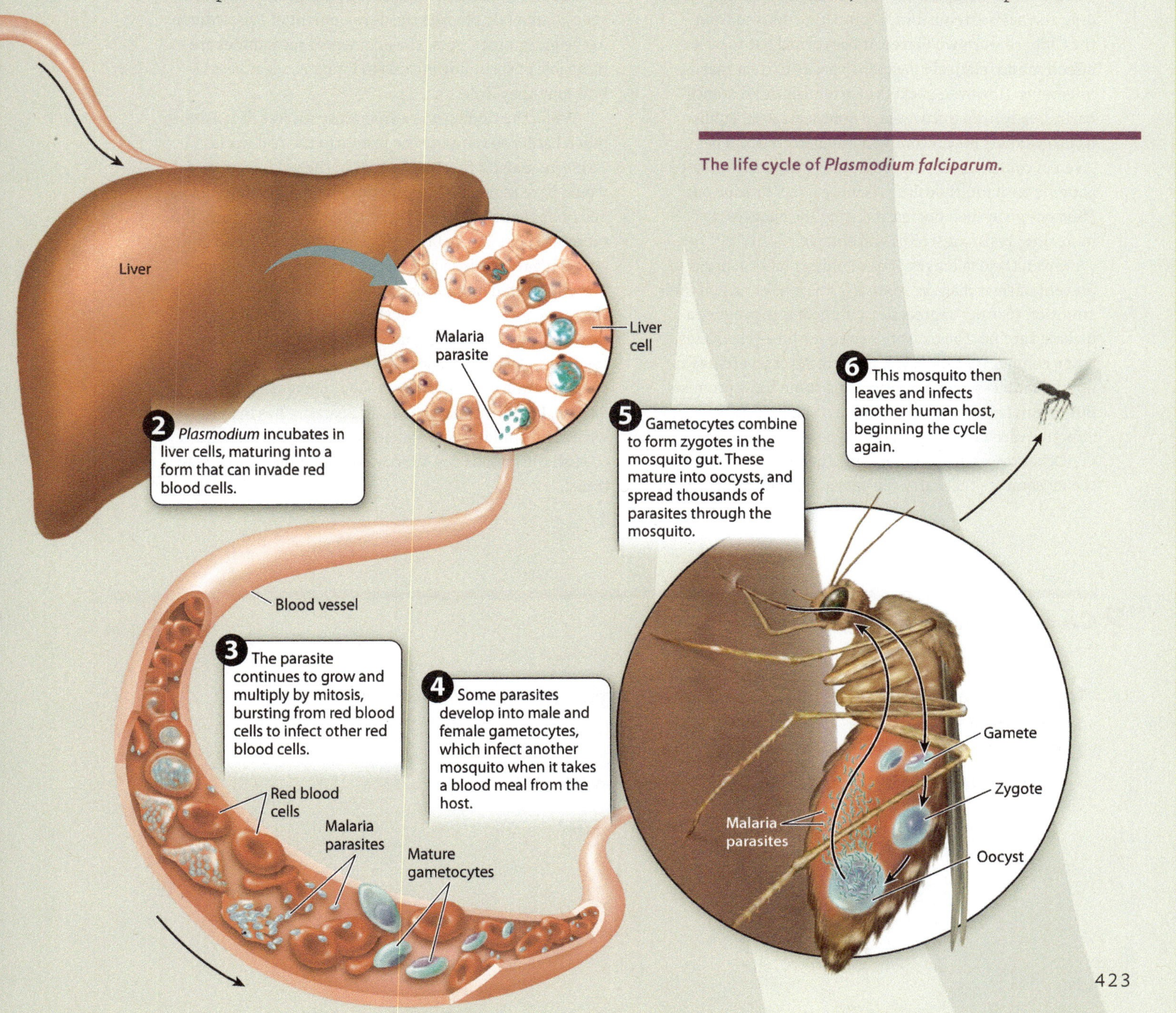

The life cycle of *Plasmodium falciparum*.

wiped out. Those that survive in the presence of the drug are likely to evolve resistance to the drug. Because the resistant parasites have a survival advantage, the genes for drug resistance spread quickly through the population.

Since chloroquine resistance emerged, pharmaceutical researchers have developed a variety of new medications to prevent or treat malaria. Most, however, are far too expensive for people in the poverty-stricken regions where malaria is rampant. And just as was the case with chloroquine, almost as quickly as new drugs are developed, *Plasmodium* begins evolving resistance.

One of the weapons more recently added to the drug arsenal is artemisinin, a compound derived from the Chinese wormwood tree. It has turned out to be an effective and relatively inexpensive way to treat malaria infections. However, pockets of artemisinin resistance have already been uncovered in Southeast Asia. Public health workers now recommend that artemisinin be given in combination with other drugs. Treating infected patients with multiple drugs is more likely to wipe out *Plasmodium* in their bodies, reducing the chances that more drug-resistant strains will emerge.

Given the challenges of developing practical drugs to prevent or treat malaria, some scientists have turned their attention to other approaches. One goal is to produce a malaria vaccine. The parasite's complex life cycle makes that a complicated endeavor, though. Several vaccines are now in various stages of testing, and some show promise. But researchers expect it will be years before a safe, effective vaccine for malaria could be available.

Other researchers are focusing their efforts on the mosquitoes that carry the parasite, rather than on *Plasmodium* itself. The genetically engineered insects created by the team in Arizona are a promising step in that direction.

The Arizona researchers set out to alter a cellular signaling gene that plays a role in the mosquito's life cycle. Mosquitoes normally live 2 to 3 weeks, and *Plasmodium* takes about 2 weeks to mature in the mosquito's gut. The researchers hoped to create mosquitoes that would die prematurely, before the parasite is mature. The genetic modification worked as planned. The engineered mosquitoes' life-spans were shortened by 18% to 20%. The genetic tweak also had a surprising side effect. The altered gene completely blocked the development of *Plasmodium* in the mosquitoes' guts. The engineered mosquitoes are incapable of spreading malaria to humans, regardless of how long they live.

While the finding was a laboratory success, it will be much harder to translate the results to the real world. To create malaria-free mosquitoes in the wild, scientists would have to release the genetically modified mosquitoes and hope that their altered gene spreads through the wild mosquito population. But that gene would be passed on only if it gave the insects a distinct evolutionary advantage. The engineered mosquitoes may be malaria free, but so far they have no advantage over their wild counterparts.

It's clear that slashing malaria rates will not be an easy task. Despite decades of research, insecticide-treated bed nets are still the best method for preventing the disease. For millennia, the malaria parasite has managed to withstand our efforts to squelch it, yet science continues to push the boundaries. Who will emerge the victor? Stay tuned.

? CASE 4 QUESTIONS

Special sections in Chapters 21–24 discuss the following questions related to Case 4.

1. **What genetic differences make some individuals more and some less susceptible to malaria?** See page 434.
2. **How did malaria come to infect humans?** See page 455.
3. **What human genes are under selection for resistance to malaria?** See page 501.

Roc Canals Photography/Getty Images.

CHAPTER 21

Evolution

How Genotypes and Phenotypes Change over Time

Core Concepts

21.1 Genetic variation refers to differences in DNA sequences.

21.2 Patterns of genetic variation can be described by allele frequencies.

21.3 Evolution is a change in the frequency of alleles or genotypes over time.

21.4 Natural selection leads to adaptations, which enhance the fit between an organism and its environment.

21.5 Migration, mutation, genetic drift, and non-random mating are non-adaptive mechanisms of evolution.

21.6 Molecular evolution is a change in DNA or amino acid sequences over time.

Variation is a fact of nature. A walk down any street reveals how variable our species is: Skin color and hair color, for example, vary from person to person. Until the publication in 1859 of Charles Darwin's *On the Origin of Species,* scientists tended to view all the variation we see in humans and other species as biologically unimportant. According to the traditional view, not only were species individually created in their modern forms by a divine Creator, but, because the Creator had a specific design in mind for each species, they were fixed and unchanging. Departures or variations from this divinely ordained type were therefore ignored.

Since Darwin, however, we have appreciated that a species does not conform to a type. Rather, a species consists of a range of variants. In our own species, people may be tall, short, dark-haired, fair-haired, and so on. Furthermore, variation is an essential ingredient of Darwin's theory because the mechanism he proposed for evolution, natural selection, depends on the differential success—in terms of surviving and reproducing—of variants. Darwin changed how we view variation. Before Darwin, variation was irrelevant, something to be ignored; after Darwin, it was recognized as the key to the evolutionary process.

21.1 GENETIC VARIATION

Variation is a major feature of the natural world. We humans are particularly good at noticing phenotypic variation among individuals of our own species. As we discussed in Chapter 16, a phenotype is an observable trait, such as human height or wing color in butterflies. As we saw in Chapter 18, two factors contribute to phenotype: an individual's genotype, which is the set of alleles possessed by the individual, and the environment in which the individual lives. We can take the environment out of the equation by looking directly at genotypic differences through sequencing DNA regions in multiple individuals. We now explore genetic variation directly, in terms of differences at the DNA sequence level.

Population genetics is the study of patterns of genetic variation.

Remarkably, in spite of a high degree of phenotypic variation, humans actually rank low in terms of overall genetic variation compared with other species. Any two randomly selected humans differ from each other on average by one DNA base per thousand (the two genomes are 99.9% identical), while two fruit flies differ by ten bases per thousand (the two genomes are 99% identical). Even one of the most seemingly uniform species on the planet, the Adélie penguins seen in **Fig. 21.1,** is two to three times more genetically variable than we are.

As we discuss in Chapter 22, a **species** consists of individuals that can exchange genetic material through interbreeding. From a genetic perspective, a species is therefore a group of individuals capable, through reproduction, of sharing alleles with one another. Individuals represent different combinations of alleles drawn from the species' **gene pool,** that is, from all the alleles present in all individuals in the species. The human gene pool includes alleles that cause differences in skin color, hair type, eye color, and so on. Each one of us has a different set of those alleles—alleles that cause brown hair and brown eyes, for example, or black hair and blue eyes—drawn from that gene pool.

Population genetics is the study of genetic variation in natural **populations,** which are interbreeding groups of organisms of the same species living in the same geographical area. What factors determine the amount of variation in a population and in a species? Why are humans genetically less variable than penguins? What factors affect the distribution of particular variations? Population genetics addresses detailed questions about patterns of variation. And small differences, given enough time, can lead to the major differences we see among organisms today.

FIG. 21.1 Genetic diversity in Adélie penguins. Adélie penguins are uniform in appearance but are actually more genetically diverse than humans. *Source: Tim Davis/Corbis.*

Mutation and recombination are the two sources of genetic variation.

Genetic variation has two sources: Mutation generates new variation, and recombination followed by segregation of homologous chromosomes during meiotic cell division shuffles mutations to create new combinations. In both cases, new alleles are formed, as shown in **Fig. 21.2.**

As we saw in Chapter 14, mutations can be **somatic,** occurring in the body's tissues, or **germline,** occurring in the reproductive cells and therefore passed on to the next generation. From an evolutionary viewpoint, we are primarily interested in germ-line mutations. A somatic mutation affects only the cells descended from the one cell in which the mutation originally arose, and thus affects only that one individual. However, a germ-line mutation appears in every cell of an individual derived from the fertilization involving the mutation-bearing gamete, and thus appears in its descendants.

Mutations can also be classified by their effects on an organism (Chapter 15). Mutations occur randomly throughout the genome, and, because most of the genome consists of noncoding DNA, most mutations are **neutral,** having little or no effect on the organism. Most mutations that do occur in protein-coding regions of the genome, however, have a **deleterious,** or harmful, effect on an organism. Rarely, a mutation occurs that has a beneficial effect. Mutations like these are **advantageous** if they improve their carriers' chances of survival or reproduction.

FIG. 21.2 Mutation and recombination. The formation of new alleles occurs by mutation and recombination.

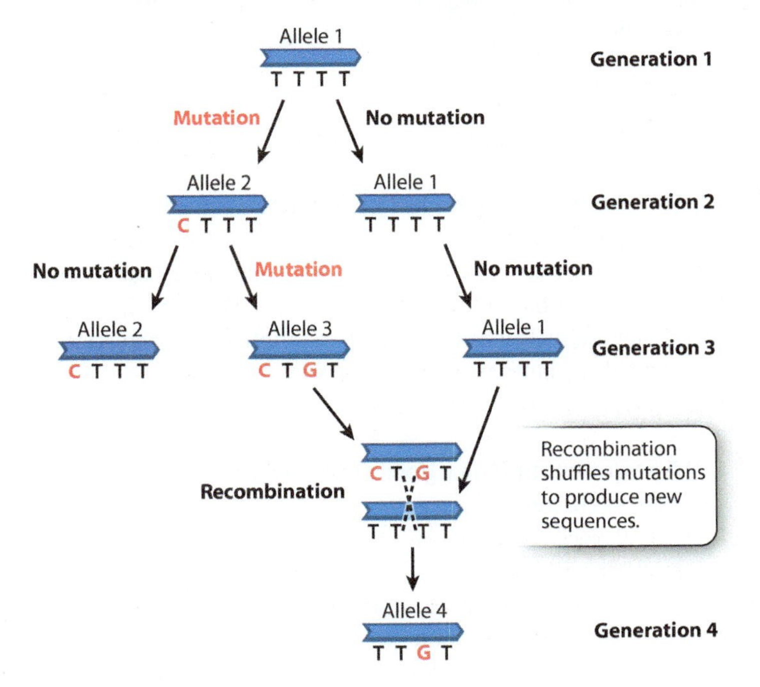

Advantageous mutations, as we will see, can increase in frequency in a population until eventually they are carried by every member of a species. These mutations are the ones that result in a species that is adapted to its environment—better able to survive and reproduce in that environment.

21.2 MEASURING GENETIC VARIATION

Mutations, whether deleterious, neutral, or advantageous, are sources of genetic variation. The goal of population genetics is to make inferences about the evolutionary process from patterns of genetic variation in nature. The raw information for this comes from the rates of occurrence of alleles in populations, or **allele frequencies.**

To understand patterns of genetic variation, we require information about allele frequencies.

The allele frequency of an allele x is simply the number of x's present in the population divided by the total number of alleles. Consider, for example, pea color in Mendel's pea plants. In Chapter 16, we discussed how pea color (yellow or green) results from variation at a single gene. Two alleles of this gene are the dominant *A* (yellow) allele and the recessive *a* (green) allele. *AA* homozygotes and *Aa* heterozygotes produce yellow peas, and *aa* homozygotes produce green peas. Imagine that in a population every pea plant produces green peas, meaning that only one allele, *a*, is present: The allele frequency of *a* is 100%, and the allele frequency of *A* is 0%. When a population exhibits only one allele at a particular gene, we say that the population is **fixed** for that allele.

Now consider another population of 100 pea plants with genotype frequencies of 50% *aa*, 25% *Aa*, and 25% *AA*. A genotype frequency is the proportion in a population of each genotype at a particular gene or set of genes. These genotype frequencies give us 50 green-pea pea plants (*aa*), 25 yellow-pea heterozygotes (*Aa*), and 25 yellow-pea homozygotes (*AA*). What is the allele frequency of *a* in this population? Each of the 50 *aa* homozygotes has two *a* alleles and each of the 25 heterozygotes has one *a* allele. Of course, there are no *a* alleles in *AA* homozygotes. The total number of *a* alleles is thus $(2 \times 50) + 25 = 125$. To determine the allele frequency of *a*, we divide the number of *a* alleles by the total number of alleles in the population, 200 (because each pea plant is diploid, meaning that it has two alleles): $125/200 = 62.5\%$. Because we are dealing with only two alleles in this example, the allele frequency of *A* is $100\% - 62.5\% = 37.5\%$

Thus, the allele frequencies of *A* and *a* provide a measure of genetic variation at one gene in a given population. In this example, we were given the genotype frequencies, and from this information we determined the allele frequencies. But how are genotype and allele frequencies measured? We consider three ways to measure genotype and allele frequencies in populations: observable traits, gel electrophoresis, and DNA sequencing.

→ **Quick Check 1** Using the example of pea color in Mendel's pea plants, can you devise equations to determine the allele frequencies of *A* and *a* from the genotype frequencies of *aa*, *Aa*, and *AA*?

Early population geneticists relied on observable traits and gel electrophoresis to measure variation.

It would be a simple matter to measure genetic variation in a population if we could use observable traits. Then we could simply count the individuals displaying variant forms of a trait and have a measure of the variation of that trait's gene. However, as we saw in Chapter 18, this approach can work only rarely, for two important reasons. First, many traits are encoded by a large number of genes. In these cases, it is difficult, if not impossible, to make direct inferences from a phenotype to the underlying genotype. Even traits that seem to have a simple set of phenotypes often prove to have a complicated genetic basis. For instance, human skin color is determined by at least six different genes. Second, the phenotype is a product of both the genotype and the environment.

Until the 1960s, there was only one workable solution: to limit population genetics to the study of phenotypes that are encoded by a single gene. As these are few, the number of genes that population geneticists could study was extremely small. Human blood groups, including the ABO system, provided an early example of a trait encoded by a single gene with multiple alleles. At this gene, there are three alleles in the population—*A*, *B*, and *O*—and therefore six possible genotypes, which result in four different phenotypes (**Table 21.1**).

TABLE 21.1 The ABO blood system.

PHENOTYPE	GENOTYPE
A	*AA* or *AO*
B	*BB* or *BO*
AB	*AB*
O	*OO*

Other instances in which phenotypic variation can be readily correlated with genotype include certain markings in invertebrates. For example, the coloring of the two-spot ladybug *Adalia bipunctata* is controlled by a single gene (**Fig. 21.3**). However, the genetic basis of most traits is not so simple.

Single-gene variation became much easier to detect in the 1960s with the application of gel electrophoresis. In Chapter 12, we saw how gel electrophoresis separates segments of DNA according to their size. Before DNA technologies were developed, the same basic process was applied to proteins to separate them according to their electrical charge and their size. In gel electrophoresis, the proteins being studied migrate through a gel when an electrical charge is applied. The rate at which the proteins

FIG. 21.3 **Single-gene variation.** A genetic difference in color in the two-spot ladybug, *Adalia bipunctata*, results from variation in a single gene. *Sources: (top) © Biopix: G. Drange; (bottom) Howard Marsh/Shutterstock.*

move from one end of the gel to the other is determined by their charge and their size.

Early studies of protein electrophoresis focused on enzymes that catalyze reactions that can be induced to produce a dye when the substrate for the enzyme is added. If we add the substrate, we can see the locations of the proteins in the gel. The bands in the gel provide a visual picture of genetic variation in the population, revealing what alleles are present and what their frequencies are. **Fig. 21.4** shows this sort of experiment.

DNA sequencing is the gold standard for measuring genetic variation.

Protein gel electrophoresis was a leap forward in our ability to detect genetic variation, but this technique had significant limitations. Researchers could study only enzymes because they needed to be able to stain specifically for enzyme activity and could detect only mutations that resulted in amino acid

HOW DO WE KNOW?

FIG. 21.4

How is genetic variation measured?

BACKGROUND The introduction of protein gel electrophoresis in 1966 gave researchers the opportunity to identify differences in amino acid sequence in proteins both among individuals and, in the case of heterozygotes, within individuals. Proteins with different amino acid sequences run at different rates through a gel in an electric field. Often, a single amino acid difference is enough to affect the mobility of a protein in a gel.

METHOD Starting with crude tissue—in this case, the ground-up whole body of a fruit fly—we load the material on a gel and turn on the current. The rate at which a protein migrates depends on its size and charge, both of which may be affected by its amino acid sequence. To visualize the protein at the end of the gel run, we use a biochemical indicator that produces a stain when the protein of interest is active. Here, we use an indicator that detects the activity of alcohol dehydrogenase (Adh). The result is a series of bands on the gel showing where the Adh proteins in each lane migrated.

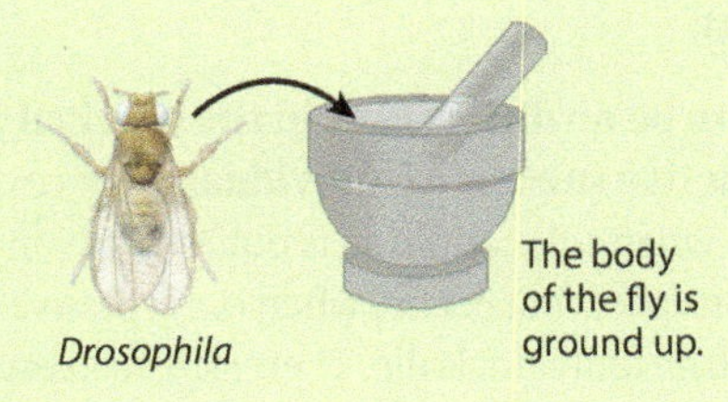

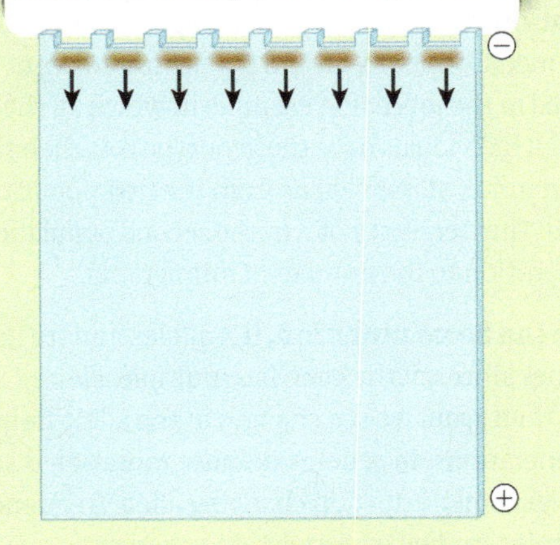

RESULTS The *Adh* gene has two common alleles, distinguished by a single amino acid difference that changes the charge of the protein. One allele, *Fast* (*F*), accordingly runs faster than the other, *Slow* (*S*). Four individuals are *S* homozygotes; two are *F* homozygotes; and two are *FS* heterozygotes. Note that the heterozygotes do not stain as strongly on the gel because each band has half the intensity of the single band in the homozygote. We can measure the allele frequencies simply by counting the alleles. Each homozygote has two of the same allele, and each heterozygote has one of each.

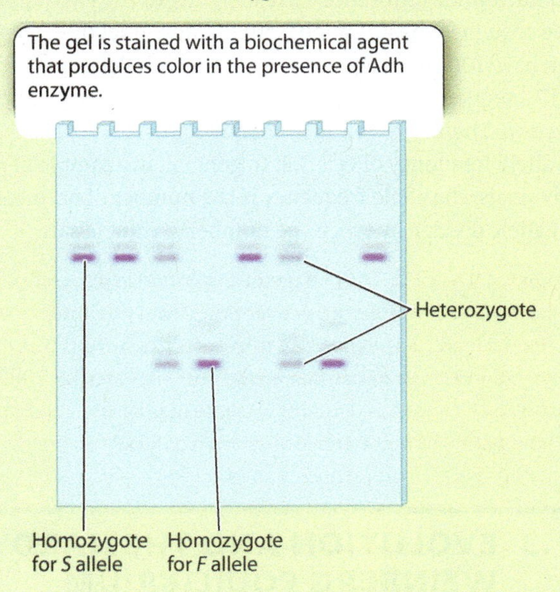

Total number of alleles in the population of 8 individuals $= 8 \times 2 = 16$

Number of *S* in the population $= 2 \times$ (number of *S* homozygotes) + (number of heterozygotes) $= 8 + 2 = 10$

Frequency of $S = \frac{10}{16} = \frac{5}{8}$

Number of *F* in the population $= 2 \times$ (number of *F* homozygotes) + (number of heterozygotes) $= 4 + 2 = 6$

Frequency of $F = \frac{6}{16} = \frac{3}{8}$

Note that the two allele frequencies add to 1.

CONCLUSION We now have a profile of genetic variation at this gene for these individuals. Population genetics involves comparing data such as these with data collected from other populations to determine the forces shaping patterns of genetic variation.

FOLLOW-UP WORK This technique is seldom used these days because it is easy now to recover much more detailed genetic information about genetic variation from DNA sequencing.

SOURCE Lewontin, R. C., and J. L. Hubby. 1966. "A Molecular Approach to the Study of Genic Heterozygosity in Natural Populations. II. Amount of Variation and Degree of Heterozygosity in Natural Populations of *Drosophila pseudoobscura*." *Genetics* 54:595–609.

substitutions that changed a protein's mobility in the gel. Only with DNA sequencing did researchers finally have an unambiguous means of detecting all genetic variation in a stretch of DNA, whether in a coding region or not. The variations studied by modern population geneticists are differences in DNA sequence, such as a T rather than a G at a specified nucleotide position in a particular gene.

Calculating allele frequencies, then, simply involves collecting a population sample and counting the number of occurrences of a given mutation. We can look even closer at the example of the *Drosophila Adh* gene from Fig. 21.4 to focus not on the amino acid difference between the Fast and Slow phenotypes, but on the A or G nucleotide difference corresponding to the two phenotypes. If we sequence the *Adh* gene from 50 individual flies, we will then have 100 gene sequences from these diploid individuals. We find 70 sequences have an A and 30 have a G at the position in question. Therefore, the allele frequency of A is $70/100 = 0.7$ and the allele frequency of G is 0.3. In general, in a sample of *n* diploid individuals, the allele frequency is the number of occurrences of that allele divided by twice the number of individuals.

→ **Quick Check 2** Data on genetic variation in populations have become ever more precise over time, from phenotypes that are determined by a single gene to gel electrophoresis that looks at variation among genes that encode for enzymes, to analysis of the DNA sequence. Has this increase in precision resulted in the uncovering of more genetic variation or less?

21.3 EVOLUTION AND THE HARDY–WEINBERG EQUILIBRIUM

Determining allele frequencies gives us information about genetic variation. Following and measuring change in that variation over time is key to understanding the genetic basis of evolution.

Evolution is a change in allele or genotype frequency over time.

At the genetic level, evolution is simply a change in the frequency of an allele or a genotype from one generation to the next. For example, if there are 200 copies of an allele that causes blue eye color in a population in generation 1 and there are 300 copies of that allele in a population of the same size in generation 2, evolution has occurred. In principle, evolution may occur without allele frequencies changing. For instance, even if, in our fruit fly example, the A/G allele frequencies stay the same from one generation to the next, the frequencies of the different genotypes (that is, of AA, AG, and GG) may change. This would be evolution *without* allele frequency change.

Evolution is therefore a change in the genetic makeup of a population over time. Note an important and often misunderstood aspect of this definition: *Populations* evolve, not *individuals.* Note, too, that this definition does not specify a mechanism for this change. As we will see, many mechanisms can cause allele or genotype frequencies to change. Regardless of which mechanisms are involved, any change in allele frequencies, genotype frequencies, or both constitutes evolution.

The Hardy–Weinberg equilibrium describes situations in which allele and genotype frequencies do not change.

Allele and genotype frequencies change over time only if specific forces act on the population. This principle was demonstrated independently in 1908 by the English mathematician G. H. Hardy and the German physician Wilhelm Weinberg, and has become known as the **Hardy–Weinberg equilibrium.** In essence, the Hardy–Weinberg equilibrium describes the situation in which evolution does *not* occur. In the absence of evolutionary forces (such as natural selection), allele and genotype frequencies do not change.

To determine whether or not evolutionary forces are at work, we need to determine whether or not a population is in Hardy–Weinberg equilibrium. The Hardy–Weinberg equilibrium specifies the relationship between allele frequencies and genotype frequencies when a number of key conditions are met. In these cases, we can conclude that evolutionary forces are not acting on the gene in the population we are studying. In many ways, then, the Hardy–Weinberg equilibrium is most interesting when we find instances in which allele or genotype frequencies *depart* from expectations. This finding implies that one or more of the conditions are *not* met and that evolutionary mechanisms are at work.

A population that is in Hardy–Weinberg equilibrium meets these conditions:

1. **There can be no differences in the survival and reproductive success of individuals.** Let's examine what happens when this condition is not met. Given two alleles, *A* and *a*, consider what occurs when *a*, a recessive mutation, is lethal. All *aa* individuals die. Therefore, in every generation, there is a selective elimination of *a* alleles, meaning that the frequency of *a* will gradually decline (and the frequency of *A* correspondingly increase) over the generations. As we discuss below, we call this differential success of alleles **selection.**

2. **Populations must not be added to or subtracted from by migration.** Again, let's see what happens when this condition is not met. Consider a second population adjacent to the one we used in the preceding example in which all the alleles are *A* and all individuals have the genotype *AA*. Then there is a sudden influx of individuals from the first population into the second. The frequency of *A* in the second population changes in proportion to the number of immigrants.

3. **There can be no mutation.** If *A* alleles mutate into *a* alleles (or other alleles, if the gene has multiple alleles), and vice versa, then again we see changes in the allele frequencies over the generations. In general, because mutation is so rare, it has a very small effect on changing allele frequencies on the timescales studied by population geneticists.

4. **The population must be sufficiently large to prevent sampling errors.** Small samples are likely to be more misleading than large ones. Campus-wide, a college's sex ratio may be close to 50:50, but in a small class of 8 individuals it is not improbable that we would have 6 women and 2 men (a 75:25 ratio). Sample size, in the form of population size, also affects the Hardy–Weinberg equilibrium such that it technically holds only for infinitely large populations. A change in the frequency of an allele due to the random effects of limited population size is called **genetic drift.**

5. **Individuals must mate at random.** For the Hardy–Weinberg equilibrium to hold, mate choice must be made without regard to genotype. For example, an *AA* homozygote when offered a choice of mate from among *AA*, *Aa*, or *aa* individuals should choose at random. In contrast, **non-random mating** occurs when individuals do not mate randomly. For example, *AA* homozygotes might preferentially mate with other *AA* homozygotes. Non-random mating affects genotype frequencies from generation to generation, but does not affect allele frequencies.

The Hardy–Weinberg equilibrium relates allele frequencies and genotype frequencies.

Now that we have established the conditions required for a population to be in Hardy–Weinberg equilibrium, let us explore the idea in detail. In the *Drosophila* example we looked at earlier, we know the frequency of the two alleles, one with A and the other with G in the *Adh* gene. What are the genotype frequencies? That is, how many AA homozygotes, AG heterozygotes, and GG homozygotes do we see? The Hardy–Weinberg equilibrium predicts the expected genotype frequencies from allele frequencies.

The logic is simple. Random mating is the equivalent of putting all the population's gametes into a single pot and drawing out pairs of them at random to form a zygote, which is the same principle we saw in action in the discussion of independent assortment in Chapter 16. We therefore put in our 70 A alleles and 30 G alleles and pick pairs at random. What is the probability of picking an AA homozygote (that is, what is the probability of picking an A allele followed by another A allele)? The probability of picking an A allele is its frequency in the population, so the probability of picking the first A is 0.7. What is the probability of picking the second A? Also 0.7. What then is the probability of picking an A followed by another A? It is the product of the two probabilities: $0.7 \times 0.7 = 0.49$. Thus, the frequency of an AA genotype is 0.49. We take the same approach to determine the genotype frequency for the GG genotype: Its frequency is 0.3×0.3, or 0.09.

What about the frequency of the heterozygote, AG? This is the probability of drawing G followed by A, or A followed by G. There are thus two separate ways in which we can generate the heterozygote. Its frequency is therefore $(0.7 \times 0.3) + (0.3 \times 0.7) = 0.42$.

As seen in the Punnett square on the left in **Fig. 21.5,** we can generalize these calculations algebraically by substituting letters for the numbers we have computed to derive the relation defined by the Hardy–Weinberg equilibrium. If the allele frequency of one allele, A, is p, and the other, G, is q, then $p + q = 1$ (because there are no other alleles at this site).

Genotypes	AA	AG	GG
Frequencies	p^2	$2pq$	q^2

In the graph on the right hand side of Fig 21.5, we have replaced the p's and q's with numbers. If no *a* alleles are present, then $q = 0$ and $p = 1$ and the frequency of *AA* is 1. In this case, all the genotypes in the population are *AA* and, accordingly, the blue line representing the frequency of *AA* in the population is at 1, and the red line representing the frequency of *aa* and the purple line representing the frequency of *Aa* are both at 0. When no A alleles are present, all individuals have genotype *aa*, and the red line is at 1 and the others at 0.

Not only does the Hardy–Weinberg relation predict genotype

FIG. 21.5 Hardy–Weinberg relation. The Hardy–Weinberg relation predicts genotype frequencies from allele frequencies, and vice versa.

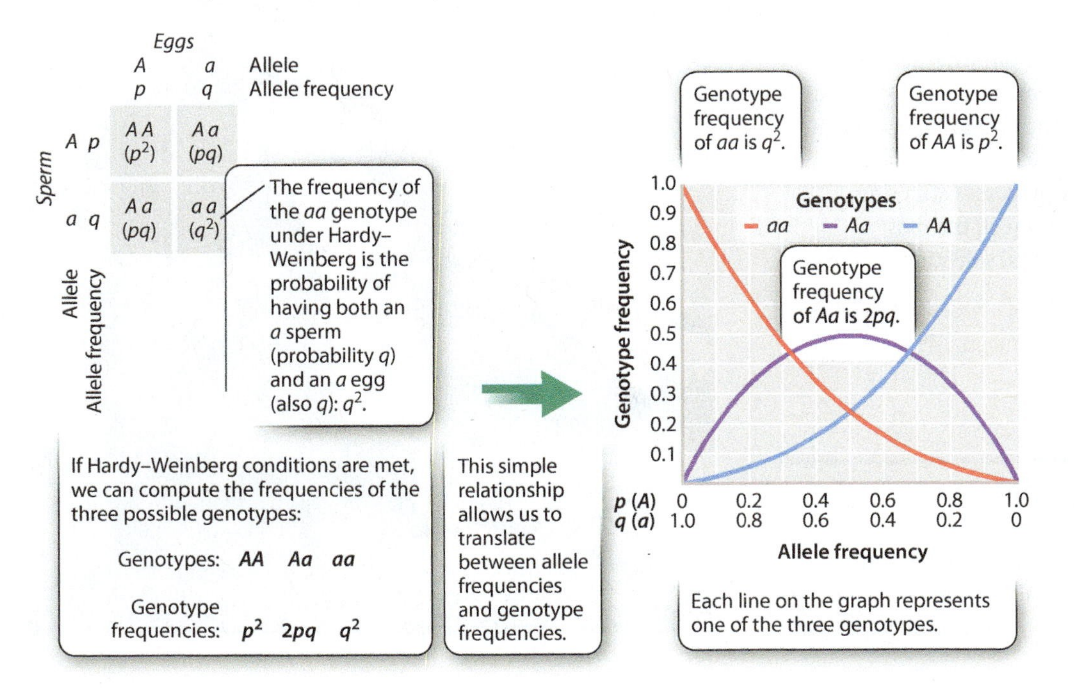

frequency from allele frequencies, but it works in reverse, too: Genotype frequencies predict allele frequencies.

The graph in Fig. 21.5 also shows how allele and genotype frequencies are related. We can use the graph to determine allele frequencies for given genotype frequencies and genotype frequencies for given allele frequencies. For example, if the population we are examining has a 0.5 frequency of heterozygotes, *Aa*, the purple line in the graph indicates that both p and q are 0.5. If we know that $p = 0.5$ (and therefore $q = 0.5$), we can look at the lines to infer that the frequency of heterozygotes is 0.5 and the frequency of both homozygotes, *AA* and *aa*, is the same, 0.25.

We can do this mathematically as well. Knowing the genotype frequency of *AA*, for example, permits us to calculate allele frequencies: if, as in our *Adh* example, p^2 is 0.49 (that is, 49% of the population has genotype *AA*), then p, the allele frequency of *A*, is $\sqrt{0.49} = 0.7$. Because $p + q = 1$, then q, the allele frequency of *a*, is $1 - 0.7 = 0.3$.

Note that these relationships hold only if the Hardy–Weinberg conditions are met. If not, then allele frequencies can be determined only from genotype frequencies, as described earlier in section 21.2.

The Hardy–Weinberg equilibrium is the starting point for population genetic analysis.

Recall our definition of evolution: a change in allele or genotype frequency from one generation to the next. Given this definition, it might seem odd to be discussing factors necessary for allele frequencies to stay the same. The Hardy–Weinberg equilibrium not only provides a means of converting between allele and genotype frequencies, but, critically, it also serves as an indicator that something interesting is happening in a population when it is not upheld.

If we find a population whose allele or genotype frequencies are not in Hardy–Weinberg equilibrium, we can infer that evolution has occurred. With further study of the population, we can then consider, for the gene in question, whether the population is subject to selection, migration, mutation, genetic drift, or non-random mating. These are the primary mechanisms of evolution. The Hardy–Weinberg equilibrium gives us a baseline from which to explore the evolutionary processes affecting populations. We will start by considering one of the most important evolutionary mechanisms: natural selection.

→ **Quick Check 3** When we find a population whose allele frequencies are *not* in Hardy–Weinberg equilibrium, what can and can't we conclude about that population?

21.4 NATURAL SELECTION

Natural selection results in allele frequencies changing from generation to generation according to the allele's impact on the survival and reproduction of individuals. New mutations that are deleterious and eliminated by natural selection have no long-term evolutionary impact; ones that are beneficial, however, can result in adaptation to the environment over time.

Natural selection brings about adaptations.

The adaptations we see in the natural world—the exquisite fit of organisms to their environment—were typically taken by pre-Darwinian biologists as evidence of a divine Creator's existence. Each species, they argued, was so well adapted—the desert plant so physiologically adept at coping with minimal levels of rainfall and the fast-swimming fish so hydrodynamically streamlined—that it must have been designed by a Creator.

With the publication of *On the Origin of Species* in 1859, Darwin, pictured in **Fig. 21.6,** overturned the biological convention of his day on two fronts. First, he showed that species are not unchanging; they have evolved over time. Second, he suggested a mechanism, natural selection, that brings about adaptation. Natural selection was a brilliant solution to the central problem of biology: how organisms come to fit so well in their environments. From where does the woodpecker get its powerful chisel of a bill? And the hummingbird its long delicate bill for probing the nectar stores

FIG. 21.6 **Charles Darwin.** This photograph was taken at about the time Darwin was writing *On the Origin of Species*. *Source: akg-images.*

FIG. 21.7 Alfred Russel Wallace. This photograph was taken in Singapore during Wallace's expedition to Southeast Asia. *Source: A. R. Wallace Memorial Fund & G. W. Beccaloni.*

in flowers? Darwin showed how a simple mechanism, without foresight or intentionality, could result in the extraordinary range of adaptations that all of life is testimony to.

For 20 years after first conceiving the essence of his theory, Darwin collected supporting evidence. In 1858, however, he was spurred to begin writing *On the Origin of Species* by a letter from a little-known naturalist collecting specimens in what is today Indonesia. By a remarkable coincidence, Alfred Russel Wallace, shown in **Fig. 21.7,** had also developed the theory of evolution by natural selection. Aware that Darwin was interested in the problem but having no idea that Darwin was working on the same theory, Wallace wrote to Darwin in 1858 to see what he thought of his idea.

Suddenly, Darwin was confronted with the prospect of losing his claim on the theory that he had been quietly nurturing for 20 years. But all was not lost. Darwin's colleagues arranged for the publication of a joint paper by Wallace and Darwin in 1858. This was done without consulting Wallace, who nonetheless never resented Darwin and afterward was careful to insist that the idea rightly belonged to the older man. It was Darwin's publication of *On the Origin of Species* in 1859 that brought both evolution and natural selection, its underlying mechanism of adaptation, to public attention. Wallace is only fleetingly mentioned in Darwin's great work, and Darwin, not Wallace, is now the name associated with the discovery.

Both Darwin and Wallace recognized their debt to the writings of a British clergyman, Thomas Malthus. In his *Essay on the Principle of Population,* first published in 1798, Malthus pointed out that natural populations have the potential to increase in size geometrically, meaning that populations get larger at an ever-increasing rate. Imagine that human couples can have just four children (two males and two females), so the population doubles every generation. Starting with a single couple, by the twentieth generation, the population will have grown to over a million—1,048,576, to be precise.

However, this geometric expansion of populations does not occur. In fact, population sizes are typically stable from generation to generation. This is because the resources upon which populations are dependent—food, water, places to live—are limited. In each generation many fail to survive or reproduce; there simply is not enough food and other resources to go around. This implies in turn that individuals within a population must compete for resources.

Which individuals will win the competition? Darwin and Wallace suggested that those that are best adapted would most likely survive and leave more offspring. Genetic variation among individuals results in some that are more likely than others to survive and reproduce, passing their genetic material to the next generation. As a result, the next generation will have a higher proportion of these same advantageous alleles. Darwin used the term "natural selection" for the filtering process that acts against deleterious alleles and in favor of advantageous ones.

Competitive advantage is a function of how well an organism is adapted to its environment. A desert plant that is more efficient at minimizing water loss than another plant is better adapted to the desert environment. An organism that is better adapted to its environment is more fit. **Fitness,** in this context, is a measure of the extent to which the individual's genotype is represented in the next generation. We say that the first plant's fitness is higher than the second's if it leaves more surviving offspring either because the plant itself survives for longer, giving it greater opportunity to reproduce, or because it has some other reproductive advantage, such as the ability to produce more seeds. Assuming that the trait maintains its advantage, natural selection then acts over generations to increase the overall fitness of a population. A plant population newly arrived in a desert may be poorly adapted to its environment, but, over time, alleles that minimize water loss increase under natural selection, resulting in a population that is better adapted to the desert.

Such changes in populations take time. Borrowing from the geologists of his day, Darwin recognized that time was a critical ingredient of his theory. Geologists had put forward a view of Earth's history that argued that large geological changes—like the carving of the Grand Canyon—can be explained by simple day-to-day processes operating over vast timescales. Darwin applied this worldview to biology. He recognized that small changes, like subtle shifts in the frequencies of alleles, could add up to major changes given long enough time periods. What might seem to us to be a trivial change over the short term can, over the long term, result in substantial differences among populations.

The Modern Synthesis combines Mendelian genetics and Darwinian evolution.

Darwinian evolution involves the change over time of the genetic composition of populations and is thus a genetic theory. Although Mendel published his genetic studies of pea plants in 1866, not long after *The Origin,* Darwin never saw them, so a key component of the theory was missing. The rediscovery of Mendel's work in 1900 unexpectedly provoked a major controversy among evolutionary biologists. Some argued that Mendel's discoveries did not apply to most genetic variation because the traits studied by Mendel were discrete, meaning that they had clear alternative states, such as either yellow or green color in peas. Most of the variation we see in natural populations, in contrast, is continuous, meaning that variation occurs across a spectrum (Chapter 18). Human height, for example, does not come in discrete classes. People are not either 5 feet tall or 6 feet tall and of no height in between. Instead, they may be any height within a certain range.

How could the factors that controlled Mendel's discrete traits account for the continuous variation seen in natural populations? This question was answered by the English theoretician Ronald Fisher, who realized that, instead of a single gene contributing to a trait like human height, there could be several genes that contribute to the trait. He argued that extending Mendel's theory to include multiple genes per trait could account for patterns of continuous variation that we see all around us.

Fisher's insight formed the basis of a synthesis between Darwin's theory of natural selection and Mendelian genetics that was forged during the middle part of the twentieth century. The product of this **Modern Synthesis** is our current theory of evolution.

Natural selection increases the frequency of advantageous mutations and decreases the frequency of deleterious mutations.

Natural selection increases the frequency of advantageous alleles, resulting in adaptation. In some cases, it can promote the fixation of advantageous alleles, meaning the allele has a frequency of 1. To start with, a new advantageous allele will exist as a single copy in a single individual (that is, as a heterozygote), but, under the influence of natural selection, the advantageous allele can eventually replace all the other alleles in the population. Natural selection that increases the frequency of a favorable allele is called **positive selection.**

As we have seen, most mutations to functional genes are deleterious. In extreme cases, they are lethal to the individuals carrying them and are thus eliminated from the population.

Sometimes, however, natural selection is inefficient in getting rid of a deleterious allele. Consider a recessive lethal mutation, *b* (that is, one that is lethal only as a homozygote, *bb,* and has no effect as a heterozygote, *Bb*). When it first arises, all the other alleles in the population are *B,* which means that the first *b* allele that appears in the population must be paired with a *B* allele, resulting in a *Bb* heterozygote. Because natural selection does not act against heterozygotes in this case, the *b* allele may increase in frequency by chance alone (we discuss below how this happens). Only when two *b* alleles come together to form a *bb* homozygote does natural selection act to rid the population of the allele. Natural selection that decreases the frequency of a deleterious allele is called **negative selection.**

Many human genetic diseases show this pattern: The deleterious allele is rare and recessive. Because it is rare, homozygotes for it are formed only infrequently. Remember that the expected frequency of homozygotes in a population under the Hardy–Weinberg equilibrium is the square of the frequency of the allele in the population. Therefore, if the allele frequency is 0.01, we expect 0.01×0.01, or 1 in every 10,000 individuals, to be homozygous for it. Thus, the genetic disease occurs rarely, and the allele remains in the population because it is recessive and not expressed as a heterozygote.

? CASE 4 MALARIA: CO-EVOLUTION OF HUMANS AND A PARASITE

What genetic differences have made some individuals more and some less susceptible to malaria?

In addition to allowing alleles to be either eliminated or fixed, natural selection can also maintain an allele at some intermediate frequency between 0% and 100%. This form of natural selection is called **balancing selection,** and it acts to maintain two or more alleles in a population. A simple case is members of a species that face different conditions depending upon where they live. One allele might be favored by natural selection in a dry area, but a different one favored in a wet area. Taking the species as a whole, these alleles are maintained by natural selection at intermediate frequencies.

Another example of balancing selection occurs when the heterozygote's fitness is higher than that of either of the homozygotes, resulting in selection that ensures that both alleles remain in the population at intermediate frequencies. This form of balancing selection is called **heterozygote advantage,** and it is exemplified by human populations in Africa, where malaria has been a long-standing disease. Because the malaria parasite spends part of its life cycle in human red blood cells, mutations in the hemoglobin molecule that affect the structure of the red blood cells have a negative impact on the parasite and can reduce the severity of malarial attacks.

Two alleles of the gene for one of the subunits of hemoglobin are *A* and *S* (Chapter 15). The *A* allele codes for normal hemoglobin, resulting in fully functional, round red blood cells. The *S* allele encodes a polypeptide that differs from the *A* allele's product in just a single amino acid, which is enough to cause the molecules to aggregate end to end, so the red blood cell is distorted into a sickle.

In regions of the world with malaria, heterozygous individuals (*SA*) have an advantage over homozygous individuals (*SS* and *AA*). *SS* homozygotes are protected against malaria, but they are burdened with severe sickling disease. Sickle-shaped red blood cells can block capillaries, and therefore people with the *SS* genotype are prone to debilitating, painful, and sometimes fatal episodes resulting from capillary blockage. *AA* homozygotes lack sickling disease but are vulnerable to malaria. *SA* heterozygotes, however, do not have severe sickling disease and have some protection from malaria. As a result, natural selection maintains both the *S* and *A* alleles in the population at intermediate frequencies.

In areas where there is no malaria, this balance is shifted. Many African-Americans, descended from Africans upon whom natural selection operated in favor of the heterozygote, still carry the *S* allele, even though the allele is no longer useful to them in their malaria-free environment. If natural selection were to run its course among African-Americans, the *S* allele would gradually be eliminated. However, this is a slow process, and many more people will continue to suffer from sickle-cell anemia before it is complete.

Natural selection can be stabilizing, directional, or disruptive.

Up to this point, we have followed the fate of individual mutations, which can increase, decrease, or be maintained at an intermediate frequency under the influence of natural selection. We can also look at the consequence of natural selection from a different perspective. Instead of following individual mutations, we can look at changes over time of a particular trait of an organism. For example, we might track the evolution of height in a population, despite not knowing the specifics of the genetic basis of height differences. When we look at natural selection from this perspective, we see three types of patterns: stabilizing, directional, and disruptive.

FIG. 21.9 Directional selection. Directional selection in a population of Galápagos finches caused a shift toward larger bills in a single drought year, 1977. *Data from Freeman & Herron Evolutionary Analysis 2004.*

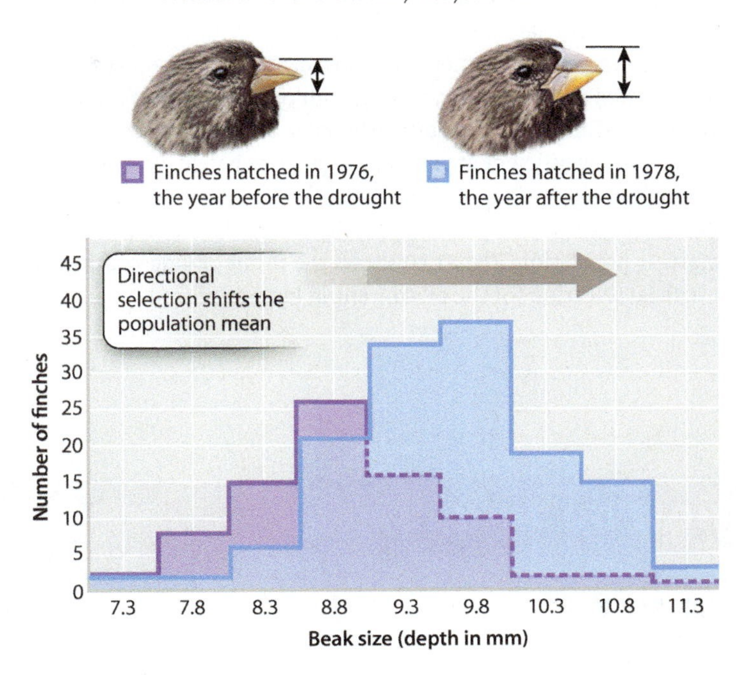

FIG. 21.8 Stabilizing selection. Stabilizing selection on human birth weight results in selection against babies that are either too small or too large. *Source: Data from L. L. Cavalli-Sforza and W. F. Bodmer, 1971, The Genetics of Human Populations, San Francisco: W. H. Freeman, p. 613.*

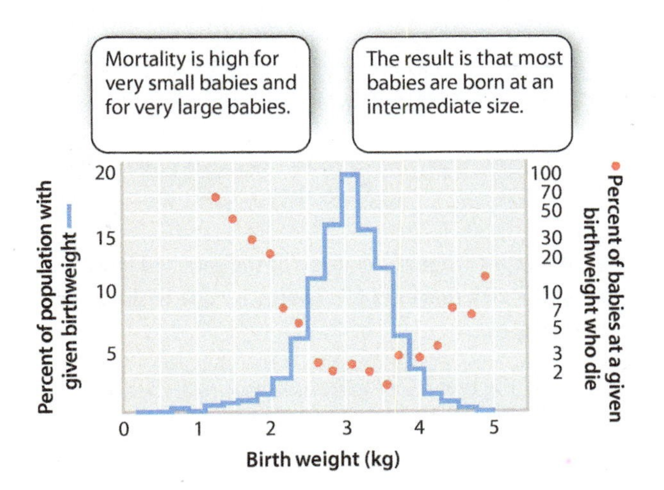

Stabilizing selection maintains the status quo and acts against extremes. A good example is provided by human birth weight, a trait affected by a number of factors, including many fetal genes (**Fig. 21.8**). If a baby is too small, then its chances of survival after birth are low. However, if it is too big, there may be complications during delivery that endanger both mother and baby. Thus, the optimum birth weight is between these two extremes. In this case, natural selection acts against the extremes. The vast majority of natural selection is of this kind as deleterious mutations that cause a departure from the optimal phenotype are selected against.

Whereas stabilizing selection keeps a trait the same over time, **directional selection** leads to a change in a trait over time. A well-documented case of directional selection is found in Darwin's finches on the Galápagos Islands (**Fig. 21.9**). A severe drought in 1977 killed a significant amount of the vegetation that provided food for one island's population of seed-eating ground finches, *Geospiza fortis*, reducing that population's numbers from 750 to 90. Because plant species that produce big seeds fared better in drought conditions than other plant species did, the average size of seeds available to the remaining birds increased. Birds with bigger bills were better at handling the big seeds and therefore had higher survival rates than birds with smaller bills. Because bill size is genetically determined, the drought resulted in directional selection for increased bill size.

Artificial selection, which has been practiced by humans since at least the dawn of agriculture, is a form of directional selection. Artificial selection is analogous to natural selection, but the competitive element is removed. Successful genotypes are selected by the breeder, not through competition. Because it can be carefully controlled by the breeder, artificial selection is astonishingly efficient at generating genetic change. Practiced over many generations, artificial selection can create a population in which the selected phenotype is far removed from that of the starting population. **Fig. 21.10** shows the result of long-continued artificial selection for the oil content in kernels of corn.

A third mode of selection, known as **disruptive selection,** operates in favor of extremes and against intermediate forms. Apple maggot flies of North America, *Rhagotletis pomonella,* provide an example (**Fig. 21.11**). The larvae of these flies feed on the fruit of hawthorn trees. However, with the introduction of apples from Europe about 150 years ago, these flies have become pests of apples. Apple trees flower and produce fruit earlier every summer than hawthorns, so disruptive selection has resulted in the production of two genetically distinct groups of flies, one specializing on apple trees and the other on hawthorn trees. Disruptive selection acts against intermediates between the two groups, which miss the peaks of both the apple and hawthorn seasons. We explore this mechanism, which can lead to the evolution of new species, in more detail in the next chapter.

HOW DO WE KNOW?

FIG. 21.10

How far can artificial selection be taken?

BACKGROUND Begun in the 1890s and continuing to this day at the University of Illinois as an attempt to manipulate the properties of corn, this experiment has become one of the longest-running biological experiments in history.

HYPOTHESIS Researchers hypothesized that there is a limit to the extent to which a population can respond to continued directional selection.

EXPERIMENT Corn was artificially selected for either high oil content or low oil content: Every generation, researchers bred together just the plants that produced corn with the highest oil content, and did the same for the plants that produced corn with the lowest oil content. Every generation, kernels showed a range of oil levels, but only the 12 kernels with the highest or the lowest oil content were used for the next generation.

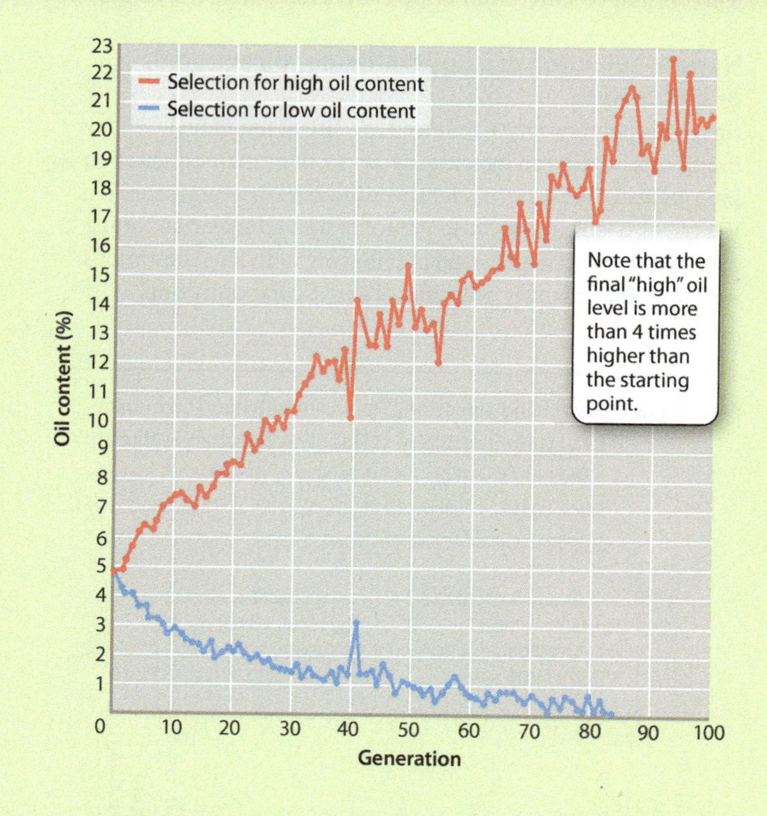

RESULTS In the line selected for high oil content, the percentage of oil more than quadrupled, from about 5% to more than 20%. In the line selected for low oil content, the oil content fell so close to zero that it could no longer be measured accurately, and the selection was terminated. Both selected lines are completely outside the range of any phenotype observed at the beginning of the experiment.

FOLLOW-UP WORK Genetic analysis of the selected lines indicates that the differences in oil content are due to the effects of at least 50 genes.

SOURCE Moose, S. P., J. W. Dudley, and T. R. Rocheford. 2004. "Maize Selection Passes the Century Mark: A Unique Resource for 21st Century Genomics." *Trends in Plant Science* 9:358–364.

FIG. 21.11 **Disruptive selection.** Disruptive selection has produced two genetically distinct populations of apple maggot fly, each one coordinated with fruiting times of two different species of tree. *Sources: (photo) Rob Oakleaf (National Science Foundation); Data from Filchak et al. 2000 Nature 407:739–42.*

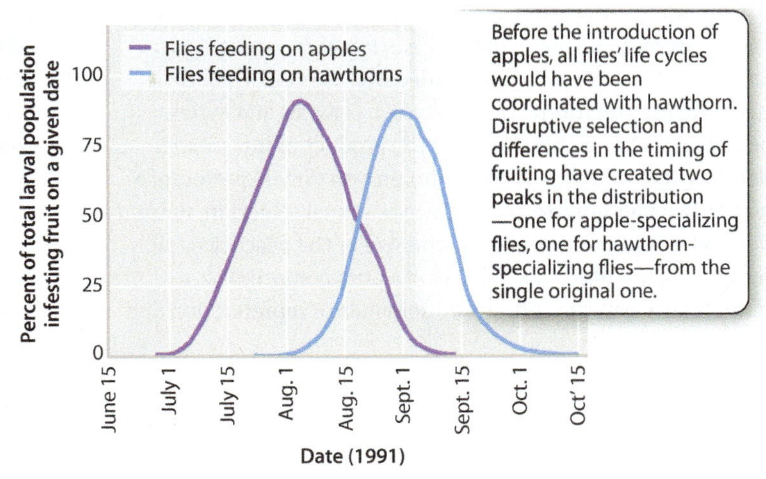

Sexual selection increases an individual's reproductive success.

Initially, Darwin was puzzled by features of organisms that seemed to reduce an individual's chances of survival. In a letter a few months after the publication of *The Origin,* he wrote, "The sight of a feather in a peacock's tail, whenever I gaze at it, makes me sick!" The tail is metabolically expensive to produce; it is an advertisement to potential predators; and it is an encumbrance in any attempt to escape a predator. How could such a feature evolve under natural selection?

In his 1871 book, *The Descent of Man, and Selection in Relation to Sex,* Darwin introduced a solution to this problem. Natural selection is indeed acting to reduce the showiness and size of the peacock's tail, but another form of selection, **sexual selection,** is acting in the opposite direction. Sexual selection promotes traits that increase an individual's access to reproductive opportunities.

Darwin recognized that this could occur in two different ways (**Fig. 21.12**). In one form of sexual selection, members of one sex (usually the males) compete with one another for access to the other sex (usually the females). This form is called **intrasexual selection** since it focuses on interactions between individuals of one sex. Because competition typically occurs among males, it is in males that we see physical traits such as large size and horns and other elaborate weaponry, as well as fighting ability. Larger, more

FIG. 21.12 **Sexual selection.** (a) Intrasexual selection often involves competition between males, as in this battle between two male elk. (b) Intersexual selection often involves bright colors and displays by males to attract females, as shown by this male and female Japanese Red-crowned Crane. *Source: (a) Kelly Funk/All Canada Photos/Corbis; (b) Steven Kaufman/Getty Images.*

powerful males tend to win more fights, hold larger territories, and have access to more females.

Darwin also recognized a second form of sexual selection. Here, males (typically) do not fight with one another, but instead compete for the attention of the female with bright colors or advertisement displays. In this case, females choose their mates. This form of selection is called **intersexual selection** since it focuses on interactions between females and males. The peacock's tail is thought to be the product of intersexual selection: Its evolution has been driven by a female preference for ever-showier tails. In the absence of sexual selection, natural selection would act to minimize the size of the peacock's tail. Presumably, the peacocks' tails we see are a compromise, a trade-off between the conflicting demands of reproduction and survival.

→ **Quick Check 4** Sexual selection tends to cause bigger size, more elaborate weaponry, or brighter colors in males. Is this an example of stabilizing, directional, or disruptive selection?

21.5 MIGRATION, MUTATION, GENETIC DRIFT, AND NON-RANDOM MATING

Selection is evolution's major driving force, enriching each new generation for the mutations that best fit organisms to their environments. However, as we have seen, it is not the only evolutionary mechanism. There are other mechanisms that can cause allele and genotype frequencies to change. These are migration, mutation, genetic drift, and non-random mating. Like natural selection, these mechanisms can cause allele frequencies to change. Unlike natural selection, they do not lead to adaptations.

Migration reduces genetic variation between populations.

Migration is the movement of individuals from one population to another, resulting in **gene flow,** the movement of alleles from one population to another. It is relatively simple to see how movements of individuals and alleles can lead to changes in allele frequencies. Consider two isolated island populations of rabbits, one white and the other black. Now imagine that the isolation breaks down—a bridge is built between the islands—and migration occurs. Over time, black alleles enter the white population and vice versa, and the allele frequencies of the two populations gradually become the same.

The consequence of migration is therefore the homogenizing of populations, making them more similar to each other and reducing genetic differences between them. Because populations are often adapted to their particular local conditions (think of dark-skinned humans in regions of high sunlight versus fair-skinned humans in regions of low sunlight), migration may be worse than merely non-adaptive—it may be maladaptive, in that it causes a decrease in a population's average fitness. Fair-skinned people arriving in an equatorial region, for example, are at risk of sunburn and skin cancer.

Mutation increases genetic variation.

As we saw earlier in this chapter, mutation is a rare event. This means that it is generally not important as an evolutionary mechanism that leads allele frequencies to change. However, as we have also seen, it is the source of new alleles and the raw material on which the other forces act. Without mutation, there would be no genetic variation and no evolution.

Genetic drift has a large effect in small populations.

Genetic drift is the random change in allele frequencies from generation to generation. By "random," we mean that frequencies can either go up or down simply by chance. An extreme case is a population **bottleneck,** which occurs when an originally large population falls to just a few individuals.

Consider a rare allele, *A*, with a frequency of 1/1000. Habitat destruction then reduces the population to just one pair of individuals, one of which is carrying *A*. The frequency of *A* in this new population is 1/4 because each individual has two alleles, giving a total of four alleles. In other words, the bottleneck resulted in a dramatic change in allele frequencies. It also caused a loss of genetic variation as much of the variation present in the original population was not present in the surviving pair. That is, the surviving pair carries only a few of the alleles that were present in their original population. A population of Galápagos tortoises that has very low levels of genetic diversity probably went through just such a bottleneck about 100,000 years ago when a volcanic eruption eliminated most of the tortoises' habitat.

Genetic drift also occurs when a few individuals start a new population, in what is called a **founder event.** Such events occur, for example, when a small number of individuals arrive on an island and colonize it. Once again, relative to the parent population, allele frequencies are changed and genetic variation is lost.

Earlier, we considered the fate of beneficial and harmful mutations under the influence of natural selection. What about neutral mutations? Natural selection, by definition, does not govern the fate of neutral mutations. Consider a neutral mutation, *m*, which occurs in a noncoding region of DNA and therefore has no effect on fitness. At first, it is in just a single heterozygous individual. What happens if that individual fails to reproduce (for reasons unrelated to *m*)? In this case, *m* is lost from the population, but not by natural selection (which does not select against *m*). Alternatively, the *m*-bearing individual might by chance leave many offspring (again for reasons unrelated to *m*), in which case the frequency of *m* increases. In principle, it is possible over a long period of time for *m* to become fixed in the population. At the end of the process, every member of the population is homozygous *mm*.

Like natural selection, genetic drift leads to allele frequency changes and therefore to evolution. Unlike natural selection, however, it does not lead to adaptations since the alleles whose frequencies are changing as a result of drift do not affect an individual's ability to survive or reproduce.

The impact of genetic drift depends on population size (**Fig. 21.13**). If *m* arises in a very small population, its frequency will change rapidly, as shown in Figs. 21.13a and 21.13b. Imagine *m* arising in a population of just six individuals (or three pairs). Its initial frequency is 1 in 12, or about 8% (there are a total of 12 alleles because each individual is diploid). If, by chance, one pair fails to breed and the other two (including the one who is an *Mm* heterozygote) each produce three offspring, and all three of the *Mm* individual's offspring happen to inherit the *m* allele, then the frequency of *m* will increase to 3 in 12 (25%) in a single generation. In effect, genetic drift is equivalent to a sampling error. In a small sample, extreme departures from the expected outcome are common.

On the other hand, if the population is large, as in Figs. 21.13c and 21.13d, then changes in allele frequency from generation to generation are much smaller, typically less than 1%. A large population is analogous to a large sample size, in which we tend not to see large departures from expectation. Toss a coin 1000 times, and you will end up with approximately 500 heads. Toss a coin 5 times, and you might well end up with zero heads. In other words, in a small sample of coin tosses, we are much more likely to see departures from our 50 : 50 expectation than in a large sample. The same is true of genetic drift. It is likely to be much more significant in small populations than in large ones.

→ **Quick Check 5** Why, of all the evolutionary mechanisms, is selection the only one that can result in adaptation?

FIG. 21.13 **Genetic drift.** The fate of neutral mutations is governed by genetic drift, the effect of which is more extreme in small populations than in large populations.

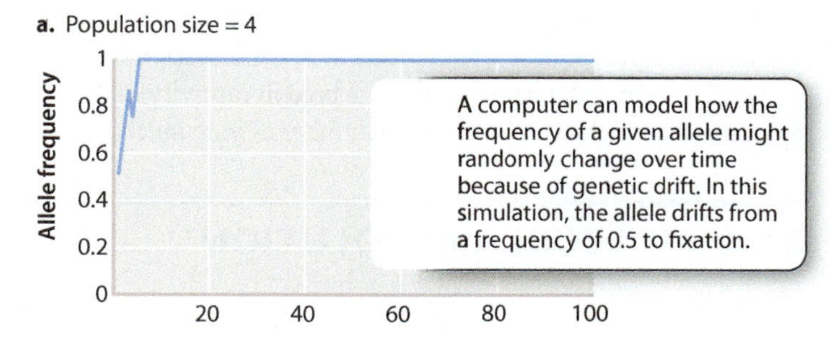

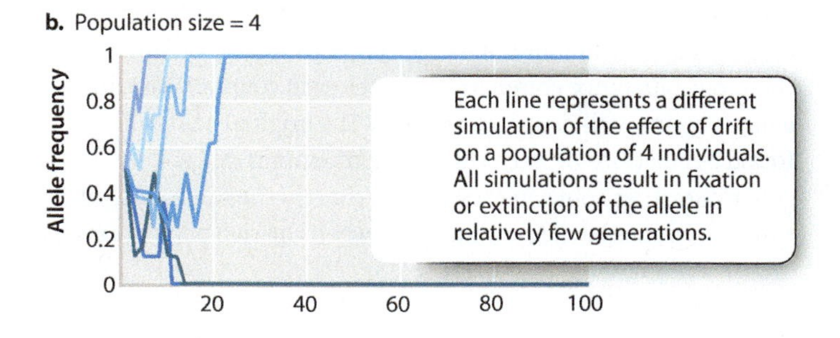

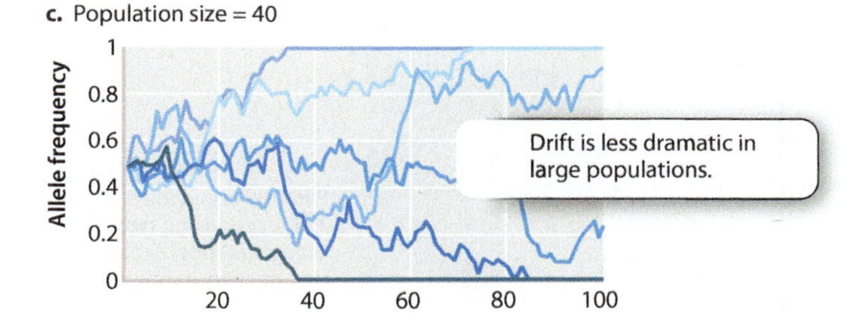

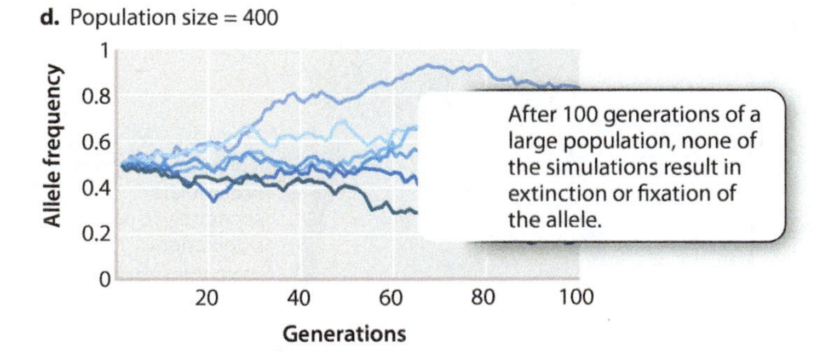

Non-random mating alters genotype frequencies without affecting allele frequencies.

As we saw earlier, another way that a population can evolve is through non-random mating. In random mating, individuals select mates without regard for genotype. In non-random mating, by contrast, individuals preferentially choose mates according to their genotypes. The result is that certain phenotypes increase and others decrease. Because non-random mating just rearranges alleles already in the gene pool and, unlike migration or mutation, does not add new alleles to the population, the genotype frequencies change, whereas the allele frequencies do not.

Probably the most evolutionarily significant form of non-random mating is inbreeding, in which mating occurs between close relatives. Inbreeding increases the frequency of homozygotes and decreases the number of heterozygotes in a population without affecting allele frequencies.

Consider a rare allele *b* that is at frequency 0.001 in a population. According to the Hardy–Weinberg equilibrium, the expected frequency of *bb* homozygotes is $0.001^2 = 0.000001$. Now let's see what happens to the frequency of *bb* homozygotes when there is inbreeding. A father with one *b* allele has a son and daughter who mate and have offspring. We know that the probability that the brother inherited the allele from his father is 0.5, and the same is true for his sister. The probability that the siblings' child is homozygous *bb* is the probability each sibling inherited a copy from their father (0.5×0.5) multiplied by the probability that their child inherited *b* from both of them (0.5 × 0.5), or $(0.5)^4 = 0.0625$. Needless to say, 0.0625 is a considerably higher probability than 0.000001.

If *b* is a deleterious recessive mutation, it may contribute to **inbreeding depression** in the child, a reduction in the child's fitness caused by homozygosity of deleterious recessive mutations. Inbreeding depression is a major problem in conservation biology, especially when endangered species are bred in captivity in programs starting with a just a small number of individuals.

21.6 MOLECULAR EVOLUTION

How do DNA sequence differences arise among species? Imagine starting with two pairs of identical twins, one pair male and the other female. Now we place one member of each pair together on either side of a mountain range (**Fig. 21.14**). Let's assume the mountain range completely isolates each couple. What, in genetic terms, will happen over time? The original pairs will found populations on each side of the mountain range. The genetic starting point, in each case, is exactly the same, but, over time, differences will accumulate between the two populations.

FIG. 21.14 Genetic divergence in isolated populations.

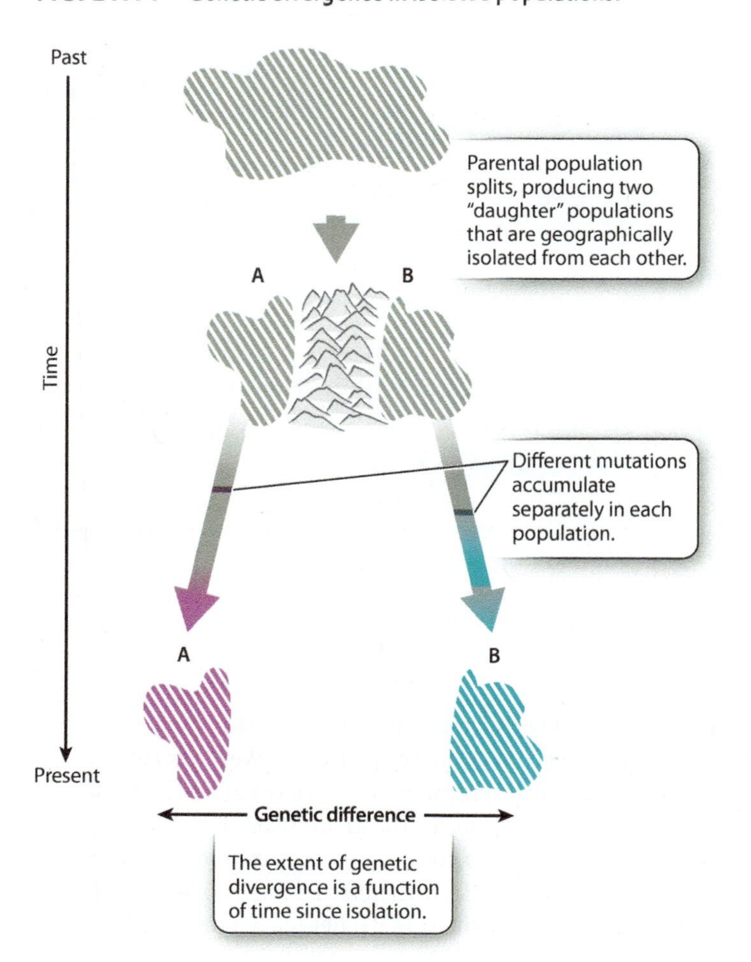

Mutations will occur in one population that will not have arisen in the other population, and vice versa.

A mutation in either population has one of three fates: It goes to fixation (either through genetic drift or through positive selection); it is maintained at intermediate frequencies (by balancing selection); or it is eliminated (either through natural selection or genetic drift). Different mutations will be fixed in each population. When we come back thousands of generations later and sequence the DNA of our original identical individuals' descendants, we will find that many differences have accumulated. The populations have diverged genetically. What we are seeing is evidence of **molecular evolution.**

Species are the biological equivalents of islands because they, too, are isolated. They are *genetically* isolated because, by definition, members of one species cannot exchange genetic material with members of another (Chapter 22). The amount of time that two species have been isolated from each other is the time since their most recent common ancestor. Thus, humans and chimpanzees, whose most recent common ancestor lived about 6–7 million years ago, have been isolated from each other for about 6–7 million years. Mutations arose and were fixed in the human lineage over that period; mutations, usually different ones, also arose and were fixed in the chimpanzee lineage over the same period. The result is the genetic difference between humans and chimpanzees.

The molecular clock relates the amount of sequence difference between species and the time since the species diverged.

The extent of genetic difference, or genetic divergence, between two species is a function of the time they have been genetically isolated from each other. The longer they have been apart, the greater the opportunity for mutation and fixation to occur in each population. This correlation between the time two species have been evolutionarily separated and the amount of genetic divergence between them is known as the **molecular clock.**

For a clock to function properly, it not only needs to keep time, but it also needs to be set. We set the clock using dates from the fossil record. For example, in a 1967 study, Vince Sarich and Allan Wilson determined from fossils that the lineages that gave rise to the Old and New World monkeys separated about 30 million years ago. Finding that the amount of genetic divergence between humans and chimpanzees was about one-fifth of that between Old and New World monkeys, they concluded that humans and chimpanzees had been separated one-fifth as long, or about 6 million years. Although this is the generally accepted number today, Sarich and Wilson's result was revolutionary at the time, when it was thought that the two species had been separated for as long as 25 million years.

The rate of the molecular clock varies.

Molecular clocks can be useful for dating evolutionary events like the separation of humans and chimpanzees. However, because

FIG. 21.15 The molecular clock. Different genes evolve at different rates because of differences in the intensity of negative selection. *After Fig. 20-3, p. 733, in A. J. F. Griffiths, S. R. Wessler, S. B. Carroll, and J. Doebley, 2012, Introduction to Genetic Analysis, 10th ed., New York: W. H. Freeman.*

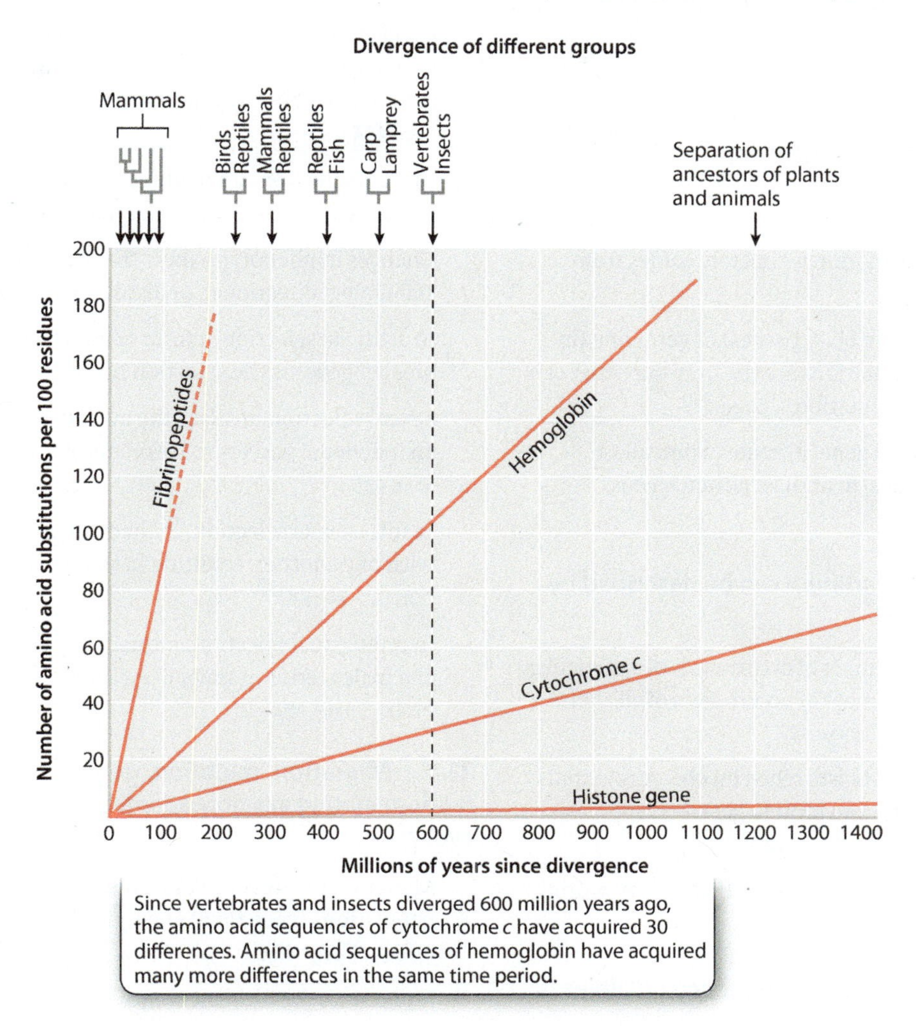

the rates of molecular clocks vary from gene to gene, clock data should be interpreted cautiously. These rate differences can be attributed largely to differences in intensity of negative selection (which results in the elimination of harmful mutations) among different genes. The slowest molecular clock on record belongs to the histone genes, which encode the proteins around which DNA is wrapped to form chromatin (Chapters 3 and 13). These proteins are exceptionally similar in all organisms; only 2 amino acids (in a chain of about 100) distinguish plant and animal histones. Plants and animals last shared a common ancestor more than 1 billion years ago, which means, because each evolutionary lineage is separate, that there have been at least 2 billion years of evolution since they were in genetic contact. And yet the histones have hardly changed at all. Almost any amino acid change fatally disrupts the histone protein, preventing it from carrying out its proper function. Negative selection has thus been extremely effective in eliminating just about every amino acid–changing histone mutation over 2 billion years of evolution. The histone molecular clock is breathtakingly slow.

Other proteins are less subject to such rigorous negative selection. Occasional mutations may therefore become fixed, either through drift (if they are neutral) or selection (if beneficial). The extreme case of a fast molecular clock is that derived from a **pseudogene,** a gene that is no longer functional. Because *all* mutations in a pseudogene are by definition neutral—there is no function for a mutation to disrupt, so a mutation is neither deleterious nor beneficial—we expect to see a pseudogene's molecular clock tick at a very fast rate. In the histone genes, virtually all mutations are selected against, constraining the rate of evolution; in pseudogenes, none is. **Fig. 21.15** shows the varying rates of the molecular clock for different genes.

In the next chapter, we examine how genetic divergence between populations can lead to the evolution of new species. ■

Core Concepts Summary

21.1 Genetic variation refers to differences in DNA sequences.

Visible differences among members of a species (phenotypic variation) are the result of differences at the DNA level (genetic variation) as well as the influence of the environment. page 426

Mutation and recombination are the two sources of genetic variation, but all genetic variation ultimately comes from mutation. page 427

Mutations can be somatic (in body tissues) or germ line (in gametes), but germ-line mutations are the only ones that can be passed on to the next generation. page 427

When a mutation occurs in a gene, it creates a new allele. Mutations can be deleterious, neutral, or advantageous. page 427

21.2 Patterns of genetic variation can be described by allele frequencies.

An allele frequency is the number of occurrences of a particular allele divided by the total number of occurrences of all alleles of that gene in a population. page 427

In the past, population geneticists relied on observable traits determined by a single gene and protein gel electrophoresis to measure genetic variation. page 428

DNA sequencing is now the standard technique for measuring genetic variation. page 428

21.3 Evolution is a change in the frequency of alleles or genotypes over time.

The Hardy–Weinberg equilibrium describes situations in which allele frequencies do not change. By seeing if a population is in Hardy–Weinberg equilibrium, we can determine whether or not evolution is occurring in a population. page 430

The Hardy–Weinberg equilibrium makes five assumptions. These assumptions are that the population experiences no selection, no migration, no mutation, no sampling error due to small population size, and random mating. page 430

The Hardy–Weinberg equilibrium allows allele frequencies and genotype frequencies to be calculated from each other. page 431

21.4 Natural selection leads to adaptation, which enhances the fit between an organism and its environment.

Independently conceived by Charles Darwin and Alfred Russel Wallace, natural selection is the differential reproductive success of genetic variants. page 432

Under natural selection, a harmful allele decreases in frequency, and a beneficial one increases in frequency. Natural selection does not affect the frequency of neutral mutations. page 434

Natural selection can maintain alleles at intermediate frequencies by balancing selection. page 434

Changes in phenotype show that natural selection can be stabilizing, directional, or disruptive. page 435

In artificial selection, a form of directional selection, a breeder governs the selection process. page 436

Sexual selection involves the evolution of traits that increase an individual's access to members of the opposite sex. page 436

In intrasexual selection, individuals of the same sex compete with one another, resulting in traits like large size and horns. page 437

In intersexual selection, interactions between females and males result in traits like elaborate plumage in male birds. page 438

21.5 Migration, mutation, genetic drift, and non-random mating are non-adaptive mechanisms of evolution.

Migration involves the movement of alleles between populations (gene flow) and tends to have a homogenizing effect. page 438

Mutation is the ultimate source of variation, but it also can change allele frequencies on its own. page 438

Genetic drift is a kind of sampling error, which acts more strongly in small populations than in large ones. page 438

Non-random mating, such as inbreeding, results in an increase in homozygotes and a decrease in heterozygotes, but does not change allele frequencies. page 439

21.6 Molecular evolution is a change in DNA or amino acid sequences over time.

The extent of sequence difference between two species is a function of the time they have been genetically isolated from each other. page 440

Correlation between sequence differences among species and time since common ancestry of those species is known as the molecular clock. page 440

The rate of the molecular clock varies among genes because some genes are more selectively constrained than others. page 441

Self-Assessment

1. Why does the relation between phenotype and genotype matter in evolution?
2. What is a neutral mutation and what evolutionary mechanism causes it to change in frequency over time?
3. Define genetic variation and explain how it can be measured.
4. How would you calculate the allele frequencies for a two-allele trait in a population if given the genotype frequencies?
5. Can evolution occur without allele frequency changes? If not, why not? If so, how?
6. Describe what happens to allele and genotype frequencies under the Hardy–Weinberg equilibrium.
7. Name the five assumptions of the Hardy–Weinberg equilibrium and, for each one, explain what happens in a population in which that condition is not met.
8. How would you calculate genotype frequencies of a population in Hardy–Weinberg equilibrium, given the allele frequencies of that trait?
9. Define natural selection and explain how it is different from other mechanisms of evolution.
10. Explain how a molecular clock can be used to determine the time of divergence of two species.

Log in to LaunchPad to check your answers to the Self-Assessment questions, and to access additional learning tools.

Rangzen/Shutterstock.

CHAPTER 22

Species and Speciation

Core Concepts

22.1 Reproductive isolation is the key to the biological species concept.

22.2 Reproductive isolation is caused by barriers to reproduction before or after egg fertilization.

22.3 Speciation underlies the diversity of life on Earth.

22.4 Speciation can occur with or without natural selection.

Imagine for a moment a world without **speciation,** the process that produces new and distinct forms of life. Life would have originated and natural selection would have done its job of winnowing advantageous mutations from disadvantageous ones, but the planet would be inhabited by a single kind of generally adapted organism. Instead of the staggering biological diversity we see around us—by current estimates, between 10 and 100 million species call Earth home—there would be just a single life-form. From a biological perspective, the planet would be a decidedly dull place. Evolution is as much about speciation, the engine that generates this breathtaking biodiversity, as it is about adaptation, the result of natural selection.

22.1 THE BIOLOGICAL SPECIES CONCEPT

The definition of **species** has been a long-standing problem in biology. Many biologists respond to the problem in the same way that Darwin himself did. In *On the Origin of Species,* Darwin wrote, "No one definition has as yet satisfied all naturalists; yet every naturalist knows vaguely what he means when he speaks of a species." The difficulty of defining species has come to be called the species problem.

Here is the problem in a nutshell: The species, as an evolutionary unit, must by definition be fluid and capable of changing, giving rise through evolution to new species. The whole point of the Darwinian revolution is that species are not fixed. How, then, can we define something that changes over time, and, by the process of speciation, even gives rise to two species from one?

Species are reproductively isolated from other species.

We can plainly see biodiversity, but are what we call "species" real biological entities? Or was the term coined by biologists to simplify their description of the natural world? To test whether or not species are real, we can examine the natural world, measure some characteristic of the different living organisms we see, and then plot these measurements on a graph. **Fig. 22.1** shows such a plot, graphing antenna length and wing length of three different types of butterfly. Note that the dots, representing individual organisms, fall into non-overlapping clusters. Today, we can add a molecular dimension to this kind of analysis. When we compare genomes of multiple organisms, they, too, cluster on the basis of similarity. Each cluster is a species, and the fact that the clusters are distinct implies that species are biologically real.

The distances we see between the dots within a cluster reflect variation from one individual to the next within a species (Chapter 21). Humans are highly variable, but overall we are more similar to one another than to our most humanlike relative, the chimpanzee. We form a messy cluster, but that cluster does not overlap with the chimpanzee cluster.

Species, then, are real biological entities, not just a convenient way to group organisms. Whether or not two individuals are

FIG. 22.1 Species clusters on the basis of two characteristics.

Source: Museum of Comparative Zoology, Harvard University.

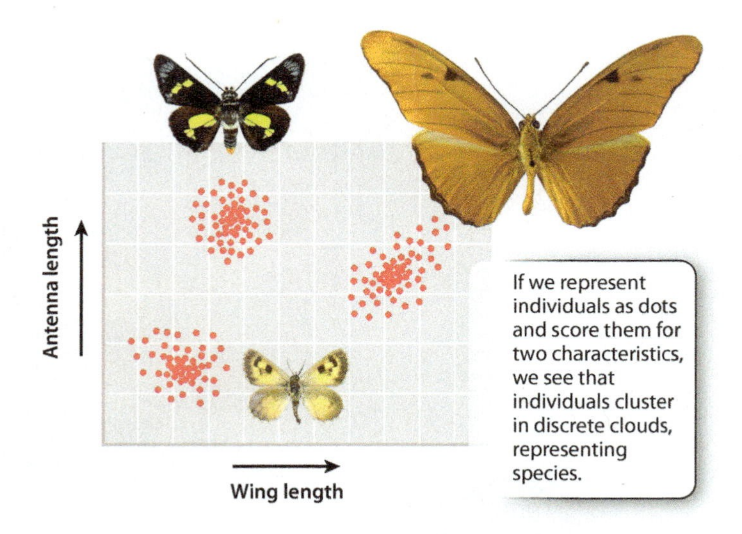

members of the same species is not a matter of judgment, but rather a reflection of their ability (or inability) to exchange genetic material by producing fertile offspring. Consider a new advantageous mutation that appears initially in a single individual. That individual and its offspring inheriting the mutation will have a competitive advantage over other members of the population, and the mutation will increase in frequency until it reaches 100%. Migration among populations causes the mutation to spread further until all individuals within the species have it. The mutation spreads *within* the species but, with some exceptions, cannot spread beyond it.

Therefore, a species represents a closed gene pool, with alleles being shared among members of that species but usually not with members of others. As a result, it is species that become extinct and it is species that, through genetic divergence, give rise to new species.

The definition of "species" continues to be debated to this day. The most widely used and generally accepted definition of a species is known as the **biological species concept (BSC).** The BSC was described by the great evolutionary biologist Ernst Mayr (1904–2005) as follows:

> Species are groups of actually or potentially interbreeding populations that are reproductively isolated from other such groups.

Let us look at this definition closely. At its heart is the idea of reproductive compatibility. Members of the same species are capable of producing offspring together, whereas members of different species are incapable of producing offspring together. In other words, members of different species are **reproductively isolated** from one another.

FIG. 22.2 ***Elephas maximus*** **in Sri Lanka and in India.** *Sources: (top) Louise Morgan/Getty Images; (bottom) Mike Smith/age footstock.*

As Mayr and many others realized, however, reproductive compatibility entails more than just the ability to produce offspring. The offspring must be fertile and therefore capable of passing their genes on to their own offspring. For example, although a horse and a donkey, two different species, can mate to produce a mule, the mule is infertile. If a hybrid offspring is infertile, it is a genetic dead end.

Note, too, the "actually or potentially" part of Mayr's definition. An Asian elephant living on the island of Sri Lanka and one living in nearby India are considered members of the same species, *Elephas maximus*, even though Sri Lankan and Indian Asian elephants never have a chance to mate with each other in nature because they are geographically separated (**Fig. 22.2**).

The BSC is more useful in theory than in practice.

Species, then, are real biological entities and the BSC is the most useful way to define them. However, the BSC has shortcomings, the most important of which is that it can be difficult to apply. Imagine you are on a field expedition in a rain forest and you find two insects that look reasonably alike. Are they members of the same species or not? To use the BSC, you would need to test whether or not they are capable of producing fertile offspring. However, in practice, you probably will not have the time and resources to perform such a test. And even if you do, it is possible that members of the same species will not reproduce when you place them together in a laboratory setting or in your field camp. The conditions may be too unnatural for the insects to behave in a normal way. And even if they do mate, you may not have time to determine whether or not the offspring are fertile.

Thus, we should consider the BSC as a valid framework for thinking about species, but one that is difficult to test. On a day-to-day basis, therefore, biologists often use a rule of thumb called the **morphospecies concept.** Stated simply, the morphospecies concept holds that members of the same species usually look alike. For example, the shape, size, and coloration of the bald eagle make it easy to determine whether or not a bird observed in the wild is a member of the bald eagle species.

Today, the morphospecies concept has been extended to the molecular level: Members of the same species usually have similar DNA sequences that are distinct from those of other species. A remarkable project called the Barcode of Life has established a database linking DNA sequences to species. Scientists can therefore identify species from DNA alone by sequencing the relevant segment of DNA and finding the matching sequence (and therefore the species) in the Barcode of Life database. Here, at both morphological and molecular levels, we are returning to those biological clusters of Fig. 22.1 in which members of a species fall close to one another within a single, discrete group.

Although the morphospecies concept is useful and generally applicable, it is not infallible. Members of a species may not always look alike, but instead show different phenotypes called polymorphisms (Chapter 15)—for example, color difference in some species of birds. Sometimes, males of a species may look different from females—think of the showy peacock, which differs dramatically from the relatively drab peahen. Young can be very different from old: Caterpillars that mature into butterflies are a striking example.

So members of the *same* species can look quite different from one another. It is also true that members of *different* species can look quite similar. Some species of butterfly, for example, appear similar but have chromosomal differences that can be observed only with the aid of a microscope (**Fig. 22.3**). These chromosomal differences prevent the formation of interspecies hybrids, so those butterflies meet the BSC criterion—they are different species—but they do not meet the morphospecies criterion because they look so much alike.

With the application of ever more powerful DNA-based approaches to studies of natural populations, scientists frequently uncover what are called cryptic species. These are composed of organisms that had been traditionally considered as belonging to one species because they look similar, but turn out to belong to two species because of a distinction at the DNA sequence level.

The BSC does not apply to asexual or extinct organisms.

In addition to the difficulties of putting the BSC into practice, a second problem is that it overlooks some organisms. For example, because it is based on the sexual exchange of genetic information, the BSC cannot apply to species that reproduce asexually, such as bacteria. Although some asexual species do occasionally exchange genetic information in a process known as conjugation (Chapter 26), true asexual organisms do not fit the BSC.

Furthermore, because it depends on reproduction, the BSC obviously cannot be applied to species that have become extinct

FIG. 22.3 Three species of *Agrodiaetus* butterflies. These similar-appearing butterflies are identifiable only by differences in their chromosome numbers. *Source: Vladimir Lukhtanov, Zoological Institute of Russian Academy of Sciences, St. Petersburg, Russia.*

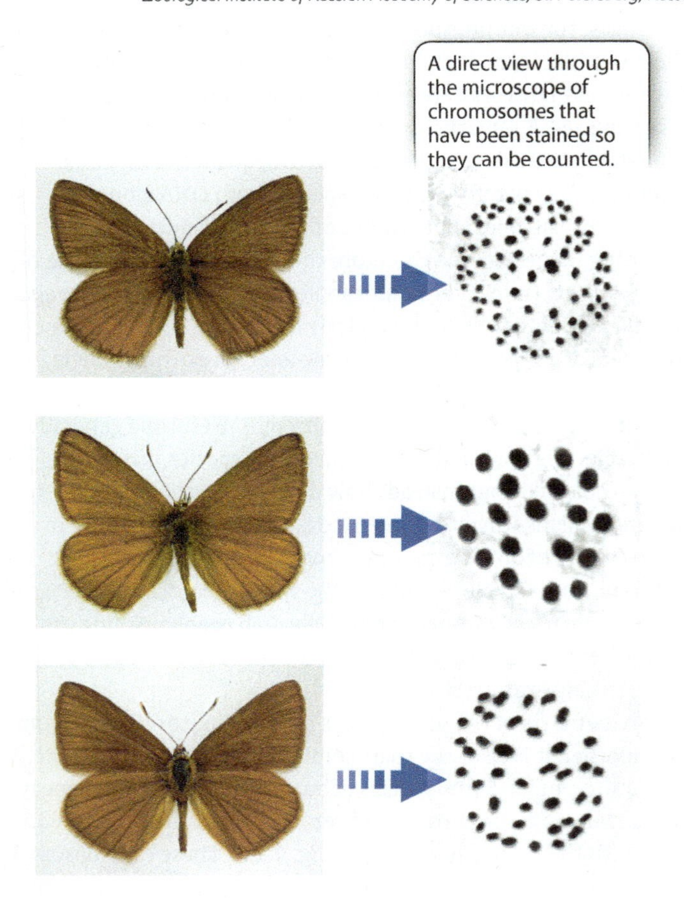

and are known only through the fossil record, like different "species" of trilobites and dinosaurs.

Ring species and hybridization complicate the BSC.

An unusual but interesting geographic pattern shown by **ring species,** species with populations that are reproductively but not genetically isolated, highlights another shortcoming of the BSC. Here, we find that some populations within a species are reproductively isolated from each other, but others are not.

For example, populations of the greenish warbler, *Phylloscopus trochiloides,* are distributed in a large geographic loop around the Himalayas (**Fig. 22.4**). In Russia to the north, members of two neighboring populations do not interbreed. Therefore, according to the BSC, they are different species. However, the more western of the two Russian populations is capable of reproducing with the population to the south of it, and that population is capable of reproducing with the population to the east of it, and so on. Eventually, the loop of genetically exchanging populations comes all the way round to the more eastern of the two Russian populations. Thus, though members of the two Russian populations cannot exchange genes directly, they can do so indirectly, with the genetic material passing through many intermediate populations. This situation is more complicated than anything predicted by the BSC. The two Russian populations are reproductively isolated from each other but they are not genetically isolated from each other because of gene flow around the ring.

We also see a complicated situation in some groups of closely related species of plants. Despite apparently being good morphospecies that can be distinguished by appearance alone, many different species of willow (*Salix*), oak (*Quercus*), and dandelion (*Taraxacum*) are still capable of exchanging genes with other species in their genera through **hybridization,** or interbreeding, between species. By the BSC, these different forms should be considered one large species because they are able to reproduce and produce fertile offspring. However, because they maintain their distinct appearances, natural selection must work against the hybrid offspring.

This unusual phenomenon seems to occur mainly in plants, but, with the application of powerful comparative genomic approaches, we are discovering that the boundaries between closely related animal species are also not as strictly drawn as traditionally supposed. As we will see in Chapter 24, for example, we now know that our own species, *Homo sapiens,* interbred in the past with another species, *Homo neanderthalensis.*

Ecology and evolution can extend the BSC.

Because of these problems, much work has been put into modifying and improving the BSC and many alternative definitions of species have been suggested. In general, these efforts highlight how difficult it is to make all species fit easily into one definition. The natural world truly defies neat categorization!

Several useful ideas, however, have come out of this literature. One of these is the notion that a species can sometimes be characterized by its **ecological niche,** which, as we will discuss further in Chapter 47, is a complete description of the role the species plays in its environment—its habitat requirements, its nutritional and water needs, and the like. It turns out that it is impossible for two species to coexist in the same location if their niches are too similar because competition between them for resources will inevitably lead to the extinction of one of them. This observation has given rise to the **ecological species concept (ESC),** the idea that there is a one-to-one correspondence between a species and its niche. Thus, we can determine whether or not asexual bacterial lineages are distinct species on the basis of differences or similarities in their ecological requirements. If two lineages have very different nutritional needs, for example, we can infer on ecological grounds that they are separate species.

Another species concept is the **phylogenetic species concept (PSC),** which is the idea that members of a species all

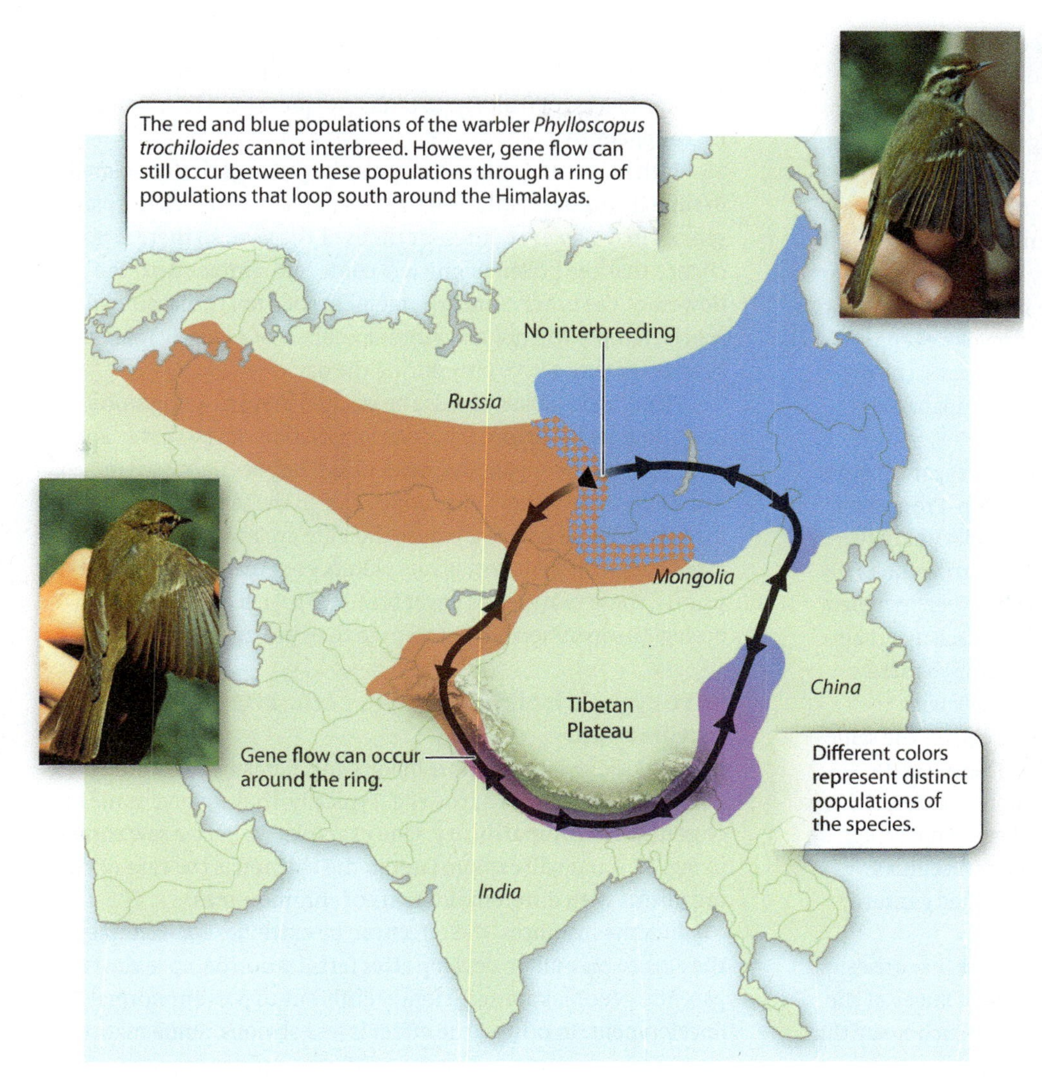

FIG. 22.4 Ring species. Ring species contain populations that are reproductively isolated from each other but can still exchange genetic material through other linking populations. *Source: Courtesy of Darren E. Irwin.*

share a common ancestry and a common fate. It is, after all, species rather than individuals that become extinct. The PSC requires that all members of a species are descended from a single common ancestor. It does not specify, however, on what scale this idea should be applied. All mammals derive from a single common ancestor that lived about 200 million years ago, but there are thousands of what we recognize as species of mammals that have evolved since that long-ago common ancestor. But, under a strict application of the PSC, would we consider all mammals to be a single species? Similarly, siblings and cousins are all descended from a common ancestor, a grandmother, but that is surely not sufficient grounds to classify more-distant relatives as a distinct species. The PSC can be useful when thinking about asexual species, but, given the arbitrariness of the decisions involved in assessing whether or not the descendants of a single ancestor warrant the term "species," its utility is limited.

It is worth bearing these ecological and phylogenetic considerations in mind when thinking about species. Although the BSC remains our most useful definition of species, the ESC and PSC broaden and generalize the concept. For example, in the case of a normally asexual species that is difficult to study because conjugation is so infrequent, we can jointly apply the ESC and PSC. We can use the ESC to loosely define the species in terms of its ecological characteristics (for example, its nutritional requirements), and we can refine that definition by using genetic analyses to determine whether the group is indeed a species by the PSC's standard (that is, that all its members derive from a single common ancestor).

Despite the shortcomings of the BSC and the usefulness of alternative ideas, we stress that the BSC is the most constructive way to think about species. In particular, by focusing on reproductive isolation—the inability of two different species to produce viable, fertile offspring—the BSC gives us a means of studying and understanding speciation, the process by which two populations, originally members of the same species, become distinct.

→ **Quick Check 1** Why haven't we been able to come up with a single, comprehensive, and agreed-upon species concept?

22.2 REPRODUCTIVE ISOLATION

Factors that cause reproductive isolation are generally divided into two categories, depending on when they act. **Pre-zygotic** isolating factors act before the fertilization of an egg, and **post-zygotic** factors come into play after fertilization. In other words, pre-zygotic factors prevent fertilization from taking place, whereas post-zygotic factors result in the failure of the fertilized egg to develop into a fertile individual.

Pre-zygotic isolating factors occur before egg fertilization.

Most species are reproductively isolated by pre-zygotic isolating factors, which can take many forms. Among animals, species are often **behaviorally isolated,** meaning that individuals mate only with other individuals based on specific courtship rituals, songs, or other behaviors. Chimpanzees may be our closest relative—and therefore the species we are most likely to confuse with our own—but a chimpanzee of the appropriate sex, however attractive to a chimpanzee of the opposite sex, fails to provoke even the faintest reproductive impulse in a human. In this case, the pre-zygotic reproductive isolation of humans and chimpanzees is behavioral.

Behavior does not play a role in plants, but pre-zygotic factors can still be important in their reproductive isolation. Pre-zygotic isolation in plants can take the form of incompatibility between the incoming pollen and the receiving flower, so fertilization fails to take place. We see similar forms of isolation between members of marine species, such as abalone, which simply discharge their gametes into the water. In these cases, membrane-associated proteins on the surface of sperm interact specifically with membrane-associated proteins on the surface of eggs of the same species but not with those of different species. These specific interactions ensure that a sperm from one abalone species, *Haliotis rufescens,* fertilizes only an egg of its own species and not an egg from *H. corrugata,* a closely related species. Incompatibilities between the gametes of two different species is called **gametic isolation.**

In some animals, especially insects, incompatibility arises earlier in the reproductive process. The genitalia of males of the fruit fly *Drosophila melanogaster* are configured in such a way that they fit only with the genitalia of females of the same species. Attempts by males of *D. melanogaster* to copulate with females of another species of fruit fly, *D. virilis,* are prevented by **mechanical incompatibility.**

Both plants and animals may also be pre-zygotically isolated in time (**temporal isolation**). For example, closely related plant species may flower at different times of the year, so there is no chance that the pollen of one will come into contact with the flowers of the other. Similarly, members of a nocturnal animal species simply will not encounter members of a closely related species that are active only during the day.

Plants and animals can also be isolated in space (**geographic** or **ecological isolation**). This type of isolation can be subtle. For example, the two Japanese species of ladybug beetle shown in **Fig. 22.5** can be found living side by side in the same field, but they feed on different plants. Because their life cycles are so intimately associated with their host plants (adults even mate on their host plants), these two species never breed with each other. This ecological separation is what leads to their pre-zygotic isolation.

Post-zygotic isolating factors occur after egg fertilization.

Post-zygotic isolating factors involve mechanisms that come into play after fertilization of the egg. Typically, they involve some kind of **genetic incompatibility.** One example, which we saw earlier in Fig. 22.3 and will explore later in the chapter, is the case of two organisms with different numbers of chromosomes.

In some instances, the effect can be extreme. For example, the zygote may fail to develop after fertilization because the two parental genomes are sufficiently different to prevent normal development. In others, the effect is less obvious. Some matings between different species produce perfectly viable adults, as in the case of the horse–donkey hybrid, the mule. As we have seen, though, all is not well with the mule from an evolutionary perspective. The horse and donkey genomes are different enough to cause the mule to be infertile. As a general rule, the more closely related—and therefore genetically similar (Chapter 21)—a pair of species, the less extreme the genetic incompatibility between their genomes.

FIG. 22.5 Ecological isolation. The ladybugs (a) *Henosepilachna yasutomii* and (b) *H. niponica* are reproductively isolated from each other because they feed and mate on different host plants. *Source: Courtesy Dr. Haruo Katakura.*

a.

b.

22.3 SPECIATION

Recognizing that species are groups of individuals that are reproductively isolated from other such groups, we are now in a position to recast the key question: How does speciation occur?

FIG. 22.6 Speciation. Speciation occurs when two populations that are genetically diverging become reproductively isolated from each other.

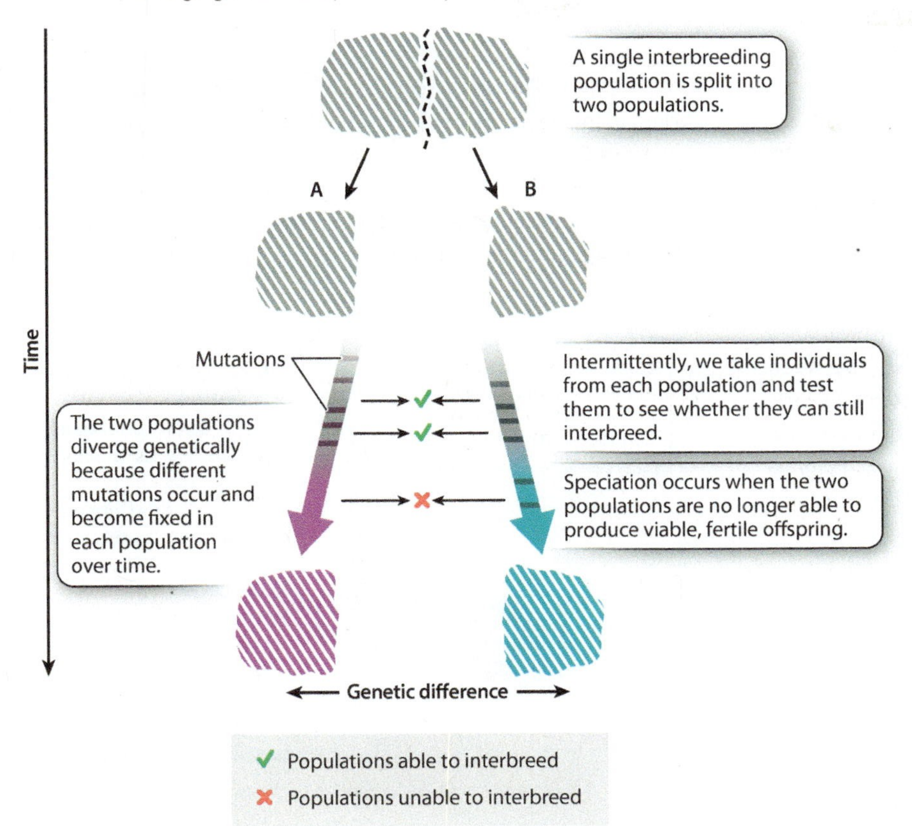

Instead of asking, "How do new species arise?" we can ask, "How does reproductive isolation arise between populations?"

Speciation is a by-product of the genetic divergence of separated populations.

The key to speciation is the fundamental evolutionary process of genetic divergence between genetically separated populations. As we saw in Chapter 21, if a single population is split into two populations that are unable to interbreed, different mutations will appear by chance in the two populations. Like all mutations, these will be subject to genetic drift or natural selection (or both), resulting over time in the genetic divergence of the two populations. Two separate populations that are initially identical will, over long periods of time, gradually become distinct as different mutations are introduced and propagated in each population. At some stage in the course of divergence, changes occur in one population that lead to its members' being reproductively isolated from members of the other population (**Fig. 22.6**). It is this process that results in speciation.

Speciation—the development of reproductive isolation between populations—is, therefore, typically just a by-product of the genetic divergence of separated populations.

As Fig. 22.6 shows, speciation is typically a gradual process. If we try to cross members of two populations that have genetically diverged but not diverged far enough for full reproductive isolation (that is, speciation) to have arisen, we may find that the populations are **partially reproductively isolated.** They are not yet separate species, but the genetic differences between them are extensive enough that the hybrid offspring they produce have reduced fertility or viability compared to offspring produced by crosses between individuals within each population.

Allopatric speciation is speciation that results from the geographical separation of populations.

Speciation is the process by which two groups of organisms become reproductively isolated from each other. Because this process requires genetic isolation between the diverging populations and because geography is the easiest way to ensure physical and therefore genetic isolation, many models of speciation focus on geography.

The process usually begins with the creation of **allopatric** (literally, "different place") populations, populations that are geographically separated from each other. Clearly, physical separation will not immediately cause reproductive isolation. If a single population is split in two by a geographic barrier, and, after a couple of generations, individuals from each population are allowed to interbreed again, they will still be capable of producing fertile offspring and therefore still be members of the same species. What's also required for speciation to occur is time. Time allows for different mutations to become fixed in the two separated populations so that eventually they become reproductively isolated from each other.

Because genetic divergence is typically gradual, we often find allopatric populations that have yet to evolve even partial reproductive isolation but which have accumulated a few population-specific traits. This genetic distinctness is sometimes recognized by taxonomists, who call each geographic form a **subspecies** by adding a further designation after its species name. For example, Sri Lankan Asian elephants, subspecies *Elephas maximus maximus*, are generally larger and darker than Indian ones, subspecies *Elephas maximus indicus* (see Fig. 22.2).

Dispersal and vicariance can isolate populations from each other.

How do populations become allopatric? There are two ways. The first is by **dispersal,** in which some individuals colonize a distant

FIG. 22.7

Can vicariance cause speciation?

BACKGROUND Three and a half million years ago, the Isthmus of Panama was not completely formed. Several marine corridors remained open, allowing interbreeding between marine populations in the Caribbean and the eastern Pacific. Subsequently, the gaps in the isthmus were plugged, separating the Caribbean and eastern Pacific populations and preventing gene flow between them.

HYPOTHESIS Nancy Knowlton and her colleagues hypothesized that patterns of speciation would reflect the impact of the vicariance resulting from the closing of the direct marine connections between the Pacific and the Caribbean. Specifically, they predicted that each ancestor species (from the time before the formation of the isthmus) split into two "daughter" species, one in the Caribbean and the other in the Pacific. The closest relative of each current Pacific species, then, is predicted to be a Caribbean species (and vice versa).

Interbreeding between eastern Pacific and Caribbean populations of *Alpheus* was possible via the corridors that existed before the final formation of the Isthmus of Panama.

3.5 million years ago

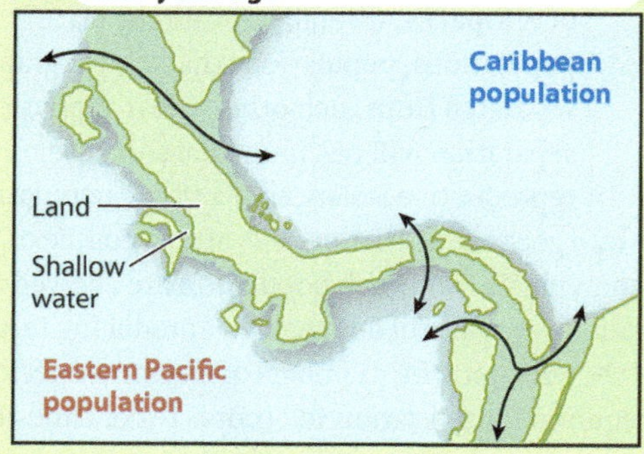

Interbreeding between eastern Pacific and Caribbean populations is no longer possible because of the geographic barrier.

Today

Nicaragua
Caribbean population
Costa Rica
Caribbean Sea
Panama
Eastern Pacific population
Pacific Ocean
Colombia

EXPERIMENT This study focused on 17 species of snapping shrimp in the genus *Alpheus,* a group that is distributed on both sides of the isthmus. The first step was to sequence the same segment of DNA from each species. The next step was to compare those sequences in order to reconstruct the phylogenetic relationships among the species.

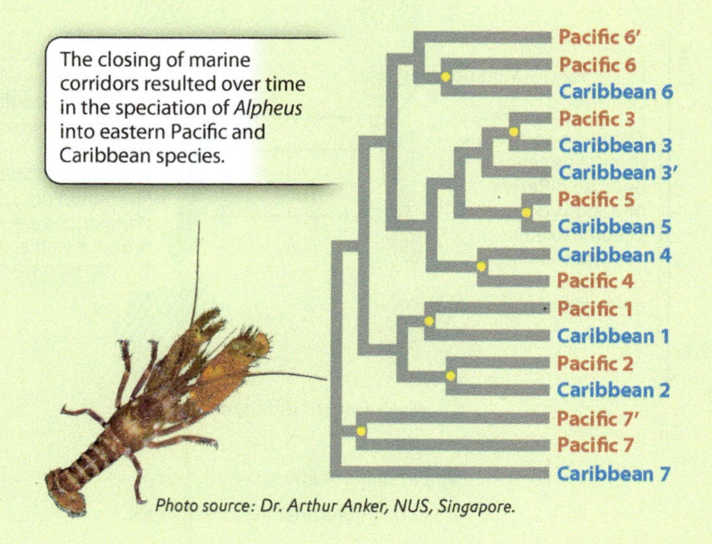

Photo source: Dr. Arthur Anker, NUS, Singapore.

RESULTS The phylogeny reveals that species show a distinctly paired pattern of relatedness: The closest relative of each species is one from the other side of the isthmus.

CONCLUSION That we see these consistent Pacific/Caribbean sister species pairings strongly supports the hypothesis that the vicariance caused by the formation of the isthmus has driven speciation in *Alpheus.* Each Pacific/Caribbean pairing is derived from a single ancestral species (indicated by a yellow dot) whose continuous distribution between the Caribbean and eastern Pacific was disrupted by the formation of the isthmus. Here, we see striking evidence of the role of vicariance in multiple speciation events.

FOLLOW-UP WORK Speciation is about more than just genetic differences between isolated populations, as shown here. Knowlton and colleagues also tested the different species for reproductive isolation and found that there were high levels of isolation between Caribbean/Pacific pairs. Given that we know that each species pair has been separated for at least 3 million years, we can then calibrate the rate at which new species are produced.

SOURCE Knowlton, N., et al. 1993. "Divergence in Proteins, Mitochondrial DNA, and Reproductive Compatibility Across the Isthmus of Panama." *Science* 260 (5114): 1629–1632.

place, such as an island, far from the main source population. The second is by **vicariance,** in which a geographic barrier arises within a single population, separating it into two or more isolated populations. For example, when sea levels rose at the end of the most recent ice age, new islands formed along the coastline as the low-lying land around them was flooded. The populations on those new islands suddenly found themselves isolated from other populations of their species. This kind of island formation is a vicariance event.

Regardless of how the allopatric populations came about—whether through dispersal or vicariance—the outcome is the same. The two separated populations will diverge genetically until speciation occurs.

Often, vicariance-derived speciation is the easier to study because we can date the time at which the populations were separated if we know when the vicariance occurred. One such event whose history is well known is the formation of the Isthmus of Panama between Central and South America, shown in **Fig. 22.7.** This event took place about 3.5 million years ago. As a result, populations of marine organisms in the western Caribbean and eastern Pacific that had formerly been able to interbreed freely were separated from each other. After a period of time, the result was the formation of many distinct species, with each one's closest relative being on the other side of the isthmus.

Dispersal is important in a specific kind of allopatric speciation known as **peripatric speciation** (that is, in a peripheral place). In this model, a few individuals from a **mainland population** (the central population of a species) disperse to a new location remote from the original population and evolve separately. This may be an intentional act of dispersal, such as young mammals migrating away from where they were raised, or it could be an accident brought about by, for example, an unusual storm that blows migrating birds off their normal route. The result is a distant, isolated **island population.** "Island" in this case may refer to a true island—like Hawaii—or may simply refer to a patch of habitat on the mainland that is appropriate for the species but is geographically remote from the mainland population's habitat area. For a species adapted to life on mountaintops, a new island might be another previously uninhabited mountaintop. For a rain forest tree species, that new island might be a patch of lowland forest on the far side of a range of mountains that separates it from its mainland forest population.

The island population is classically small and often in an environment that is slightly different from that of the mainland population. The peripatric speciation model suggests that change accumulates faster in these peripheral isolated populations than in the large mainland populations, both because genetic drift is more pronounced in smaller populations than in larger ones and because the environment may differ between the mainland and island in a way that results in natural selection driving differences between the two populations. These mechanisms cause genetic divergence of the island population from the mainland one, ultimately leading to speciation.

It is possible to glimpse peripatric differentiation in action (**Fig. 22.8**). Studies of a kingfisher, *Tanysiptera galatea,* in New

FIG. 22.8 Peripatric speciation in action among populations of New Guinea kingfishers. *Sources: Data from D. J. Futuyma, 2009, Evolution, 2nd ed., Sunderland, MA: Sinauer Associates, Fig. 18.7, p. 484; Photo from C. H. Greenewalt/VIREO.*

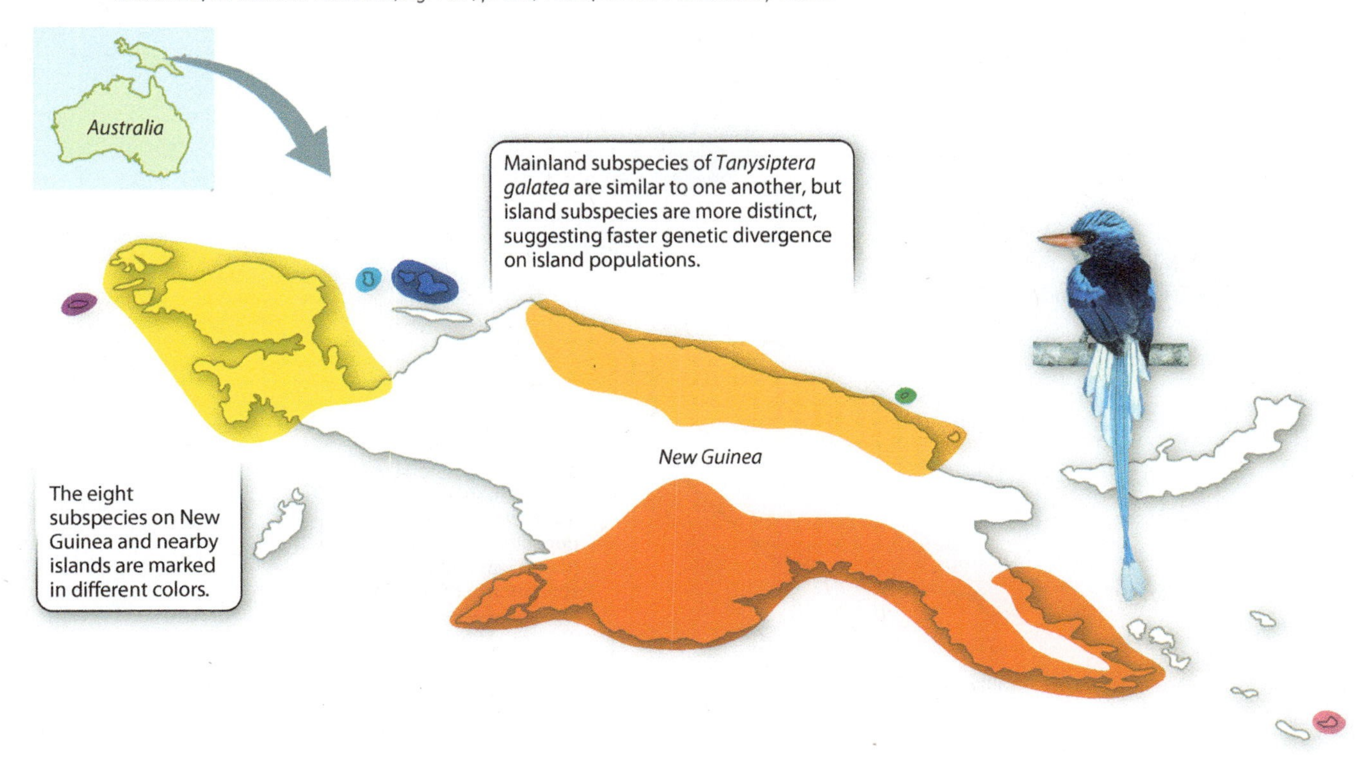

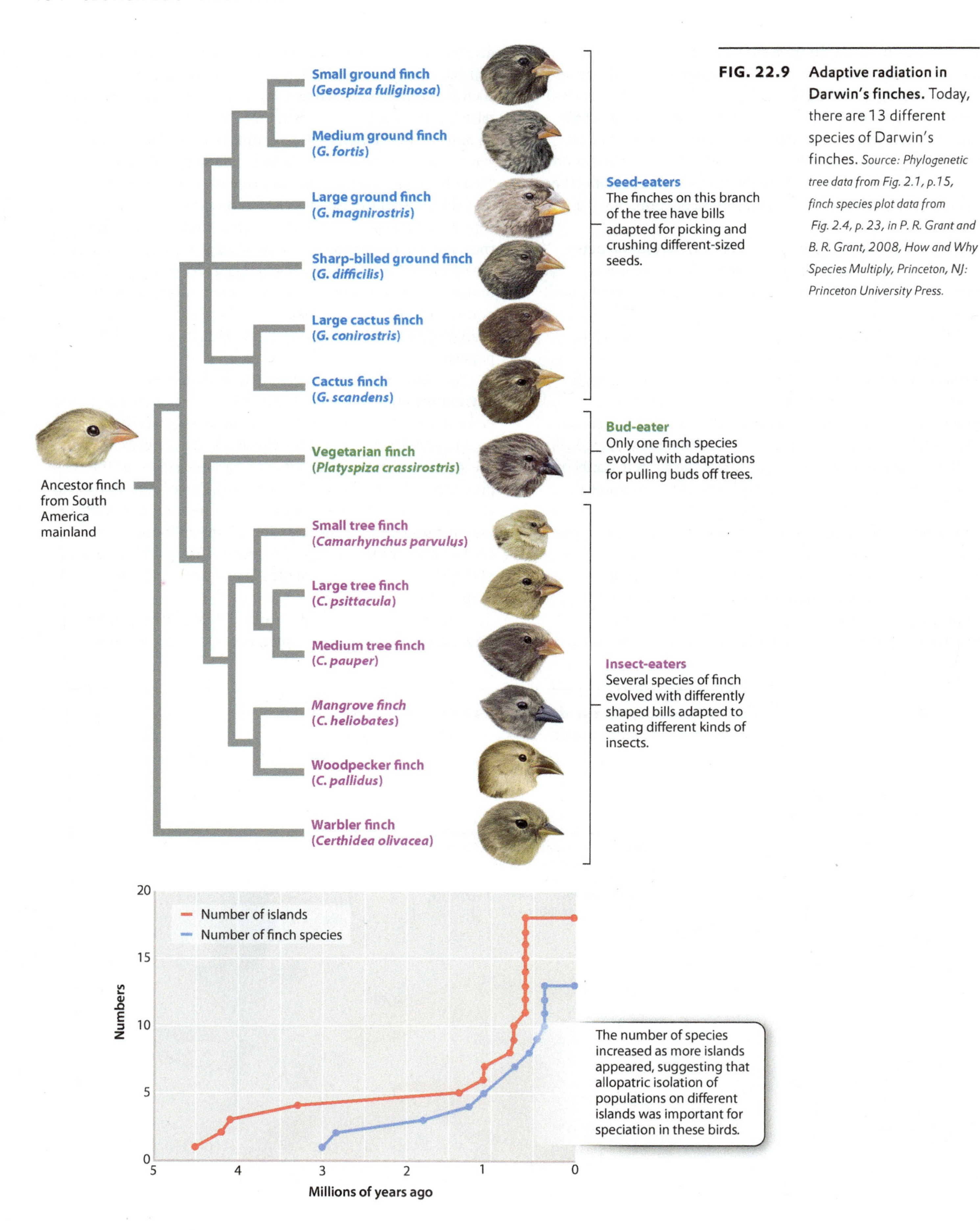

FIG. 22.9 Adaptive radiation in Darwin's finches. Today, there are 13 different species of Darwin's finches. *Source: Phylogenetic tree data from Fig. 2.1, p.15, finch species plot data from Fig. 2.4, p. 23, in P. R. Grant and B. R. Grant, 2008, How and Why Species Multiply, Princeton, NJ: Princeton University Press.*

Guinea and nearby islands show the process under way. There are eight recognized subspecies of *T. galatea,* three on mainland New Guinea (where they exist in large populations separated by mountain ranges) and five on nearby islands (where, because the islands are small, the populations are correspondingly small). The mainland subspecies are still quite similar to one another, but the island subspecies are much more distinct, suggesting that genetic divergence is occurring faster in the small island populations. If we wait long enough, these subspecies will probably diverge into new species.

Because dating such dispersal events is tricky—newcomers on an island tend not to leave a record of when they arrived—we can sometimes use vicariance information to study the timing of peripatric speciation. For instance, we know that the oldest of the Galápagos Islands were formed 4–5 million years ago by volcanic action. Sometime early in the history of the Galápagos, individuals of a small South American finch species arrived there. Conditions on the Galápagos are very different from those on the South American continent, where the mainland population of this ancestral finch lived, and so the isolated island population evolved to become distinct from its mainland ancestor and eventually became a new species.

The finches' subsequent dispersal among the other islands of the Galápagos has promoted further peripatric speciation (**Fig. 22.9**). In the graph at the bottom of the figure, we see that the number of finch species is correlated with the number of islands in the archipelago. This is clear indication of the importance of geographic separation (allopatry) in speciation: The availability of islands provided opportunities for populations to become isolated from one another, allowing speciation. The result was the evolution of 13 different species of finches, collectively known today as Darwin's finches.

The Galápagos finches and their frenzy of speciation illustrate the important evolutionary idea of **adaptive radiation,** a bout of unusually rapid evolutionary diversification in which natural selection accelerates the rates of both speciation and adaptation. Adaptive radiation occurs when there are many ecological opportunities available for exploitation. Consider the ancestral finch immigrants arriving on the Galápagos. A wealth of ecological opportunities was open and available. Until the arrival of the colonizing finches, there were no birds on the islands to eat the plant seeds, or to eat the insects on the plants, and so on. Suppose that the ancestral finches fed specifically on medium-sized seeds on the South American mainland (that is, they were medium-seed specialists, with bills that are the right size for handling medium-sized seeds). On the mainland, they were constrained to that size of seed because any attempt to eat larger or smaller ones brought them into conflict with other species—a large-seed specialist and a small-seed specialist—that already used these resources. In effect, stabilizing selection (Chapter 21) was operating on the mainland to eliminate the extremes of the bill-size spectrum in the medium-seed specialists.

When the medium-billed immigrant finches first arrived on the Galápagos, however, no such competition existed. Therefore, natural selection promoted the formation of new species of small- and large-seed specialists from the original medium-seed-eating ancestral stock. With the elimination of the stabilizing selection that kept the medium-billed finches medium-billed on the mainland, selection actually operated in the opposite direction, favoring the large and small extremes of the bill-size spectrum because these individuals could take advantage of the abundance of unused resources, small and large seeds. It is this combination of emptiness—the availability of ecological opportunity—and the potential for allopatric speciation that results in adaptive radiation.

→ **Quick Check 2** Why do we see so many wonderful examples of adaptive radiation on mid-ocean volcanic archipelagos like the Galápagos?

Co-speciation is speciation that occurs in response to speciation in another species.

As we have seen, physical separation is often a critical ingredient in speciation. Two populations that are not fully separated from each other—that is, there is gene flow between them—will typically not diverge from each other genetically, because genetic exchange homogenizes them. This is why most speciation is allopatric. Allopatric speciation brings to mind populations separated from each other by stretches of ocean or deserts or mountain ranges. However, separation can be just as complete even in the absence of geographic barriers.

Consider an organism that parasitizes a single host species. Suppose that the host undergoes speciation, producing two daughter species. The original parasite population will also be split into two populations, one for each host species. Thus, the two new parasite populations are physically separated from each other and will diverge genetically, ultimately undergoing speciation. This divergence results in a pattern of coordinated host–parasite speciation called **co-speciation,** a process in which two groups of organisms speciate in response to each other and at the same time.

Phylogenetic analysis of lineages of parasites and their hosts that undergo co-speciation reveals trees that are similar for each group. Each time a branching event—that is, speciation—has occurred in one lineage, a corresponding branching event occurred in the other (**Fig. 22.10**).

? CASE 4 MALARIA: CO-EVOLUTION OF HUMANS AND A PARASITE

How did malaria come to infect humans?

Now let's look at a human parasite, *Plasmodium falciparum,* the single-celled eukaryote that causes malaria. It had been suggested that *P. falciparum*'s closest relative is another *Plasmodium* species, *P. reichenowi,* found in chimpanzees, our closest living relative. Were *P. falciparum* and *P. reichenowi* the products of co-speciation? When the ancestral population split millions of years ago to give

FIG. 22.10 Co-speciation. Parasites and their hosts often evolve together, and the result is similar phylogenies. *Sources: Data from J. P. Huelsenbeck and B. Rannala, 1997, "Phylogenetic Methods Come of Age: Testing Hypotheses in an Evolutionary Context," Science 276:227, doi: 10.1126/science.276.5310.227; Louse photo by Alex Popinga, courtesy of James Demastes; pocket gopher photo by Richard Ditch @ richditch.com.*

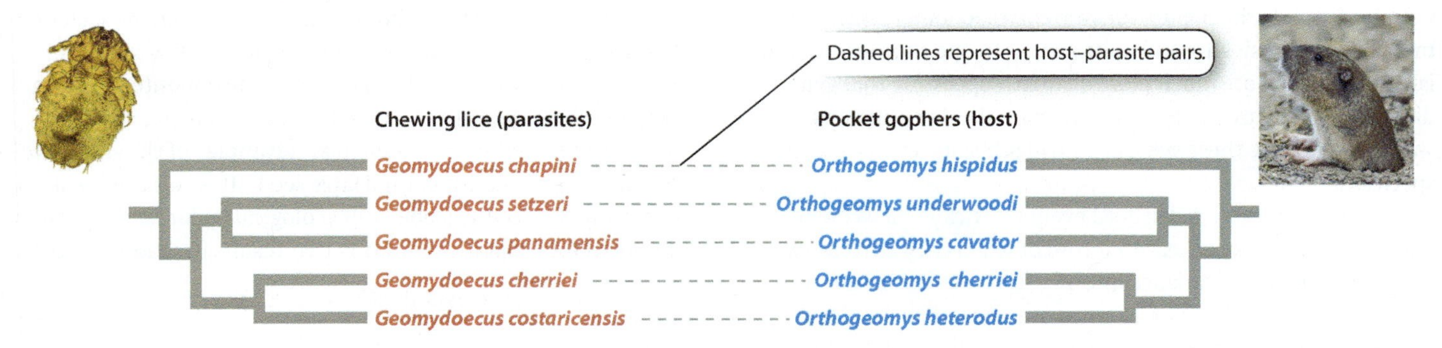

rise to human and chimpanzee lineages, that population's parasitic *Plasmodium* population could also have been split, ultimately yielding *P. falciparum* and *P. reichenowi.*

Recent studies, however, have disproved this hypothesis. We now know that *P. falciparum* was introduced to humans relatively recently from gorillas. Why doesn't the evolutionary history of *Plasmodium* follow the classical host–parasite co-speciation pattern? We know that this history is complex and is still being unraveled. However, we also know that the mosquito-borne phase of its life cycle facilitates transfer to new hosts. Malaria parasites are thus not as inextricably tied to their hosts as the pocket gopher lice in Fig 22.10 and so their evolutionary history is not as parallel to their hosts.

Sympatric populations—those not geographically separated—may undergo speciation.

Can speciation occur *without* complete physical separation of populations? Yes, though evolutionary biologists are still exploring how common this phenomenon is. Recall how separated populations inevitably diverge genetically over time (see Fig. 22.6). If a mutation arises in population A after it has separated from population B, that mutation is present only in population A and may eventually become fixed (100% frequency) in that population, either through natural selection, if it is advantageous, or through drift, if it is neutral. Once the mutation is fixed in population A, it represents a genetic difference between populations A and B. Repeated independent fixations of different mutations in the two populations result over time in the genetic divergence of separated populations.

Now imagine that populations A and B are not completely separated and there is some gene flow between them. The mutation that arose in population A can, in principle, appear in population B as members of the two populations interbreed. The new mutation may indeed have become fixed in population A, but a migrant from population A to population B may introduce the mutation to population B as well. Gene flow effectively negates the genetic divergence of populations. If there is gene flow, a pair of populations may change over time, but they do so together. How, then, can speciation occur if gene flow exists? The term we use to describe populations that are in the same geographic location is **sympatric** (literally, "same place"). So we can rephrase the question in technical terms: How can speciation occur sympatrically?

FIG. 22.11 Sympatric speciation by disruptive selection. Natural selection eliminates individuals in the middle of the spectrum.

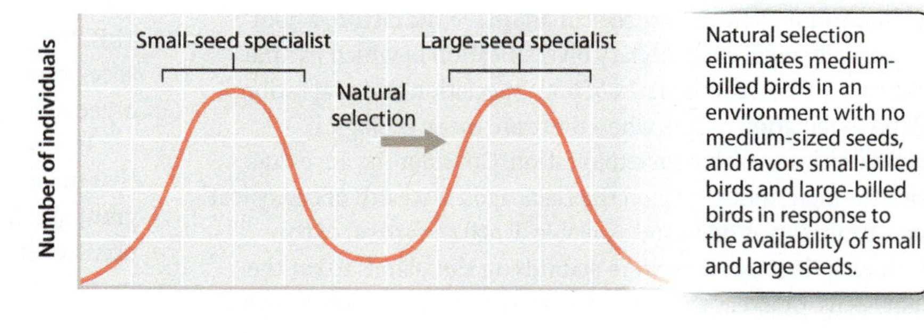

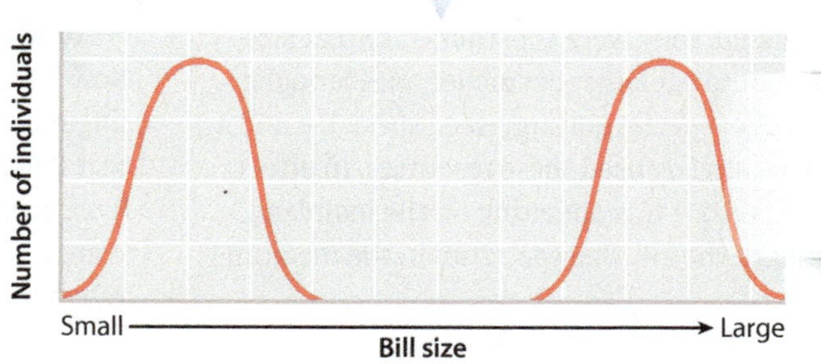

For speciation to occur sympatrically, natural selection must act strongly to counteract the homogenizing effect of gene flow. Consider two sympatric populations of finch-like birds, represented in the graph in **Fig. 22.11.** One population begins to specialize on small seeds and the other on large seeds. If the two populations freely interbreed, no genetic differences between the two will occur and speciation will not take place. Now suppose that the offspring produced by the pairing of a big-seed specialist with a small-seed specialist is an individual best adapted to eat medium seeds, and there are no medium-sized seeds available in the environment. Natural selection will act against the hybrids, which will starve to death because there are no medium-sized seeds for them to eat and they are not well adapted to compete with the big- or small-seed specialists. Natural selection would then, in effect, eliminate the products of gene flow. So, although gene flow is occurring, it does not affect the divergence of the two populations because the hybrid individuals do not survive to reproduce. As discussed in Chapter 21, this form of natural selection, which operates against the middle of a spectrum of variation, is called disruptive selection.

However, it turns out to be difficult to find evidence of sympatric speciation in nature, though it may not be especially rare in plants, as we will see. We might find two very closely related species in the same location and argue that they must have arisen through sympatric speciation. However, there is an alternative explanation. One species could have arisen elsewhere by, for example, peripatric speciation and subsequently moved into the environment of the other one. In other words, the speciation occurred in the past by allopatry, and the two species are only currently sympatric because migration after speciation occurred.

However, recent studies of plants on an isolated island have provided strong, if not definite, evidence of sympatric speciation. Lord Howe Island is a tiny island (about 10 km long and 2 km across at its widest) about 600 km east of Australia. Two species of palm tree that are only found on the island are each other's closest relatives. Because the island is so small, there is little chance that the two species could be geographically separated from each other, meaning that the species are and were sympatric, and, because of the distance of Lord Howe from Australia (or other islands), it is extremely likely that they evolved and speciated from a common ancestor on the island.

This and other demonstrations show that sympatric speciation can occur. We must recognize, however, that we do not yet know just how much of all speciation is sympatric and how much is allopatric. **Fig. 22.12** summarizes modes of speciation based on geography.

FIG. 22.12 Modes of speciation.

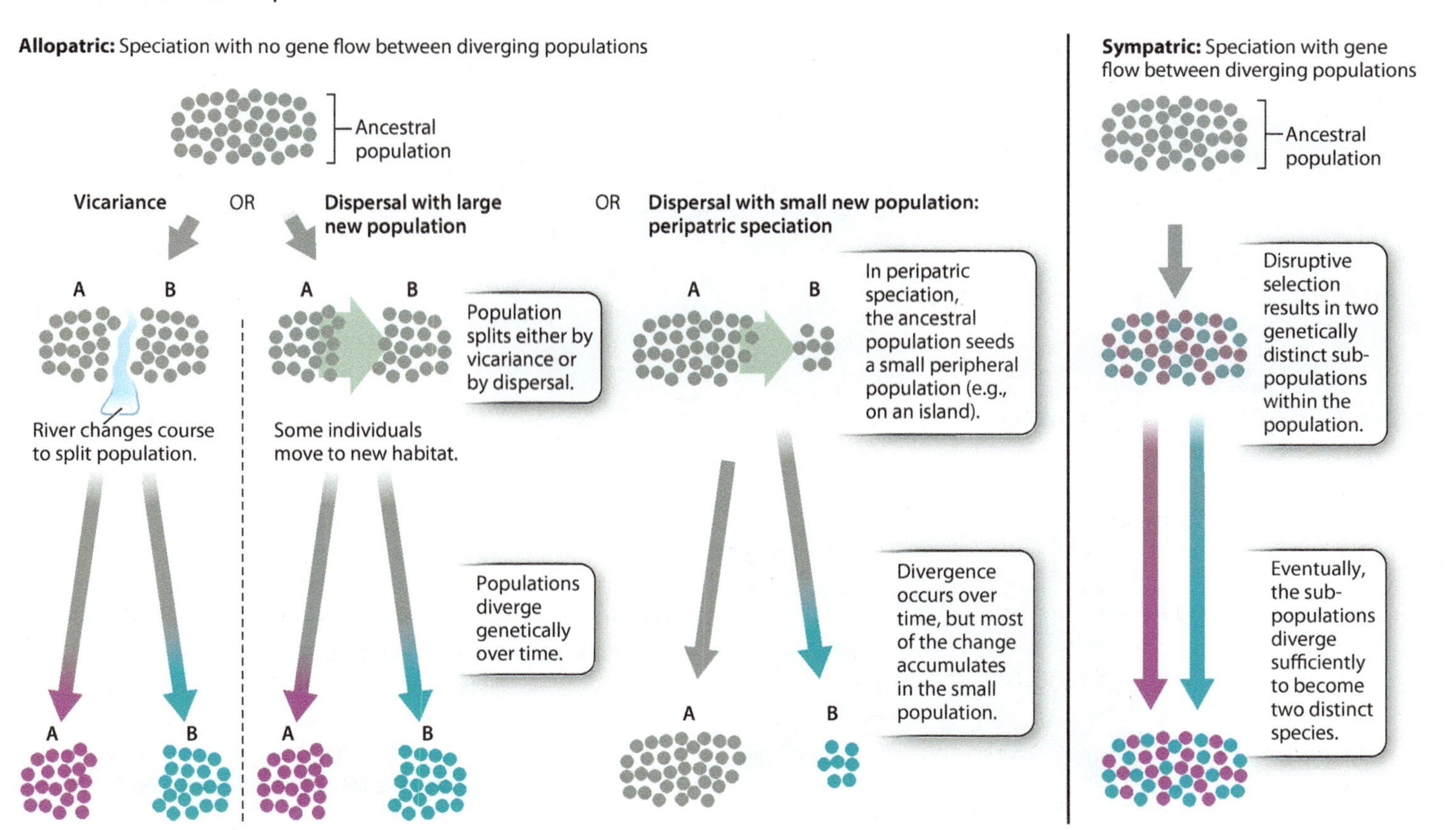

→ **Quick Check 3** There are hundreds of species of cichlid fish in Lake Victoria in Africa. Some scientists argue that they evolved sympatrically, but recent studies of the lake suggest that it periodically dried out, leaving a series of small ponds. Why is this observation relevant to evaluating the hypothesis that these species arose by sympatric speciation?

Speciation can occur instantaneously.

Although speciation is typically a lengthy process, it can occasionally occur in a single generation, making it sympatric by definition. Typically, cases of such **instantaneous speciation** are caused by hybridization between two species in which the offspring are reproductively isolated from both parents.

For example, hybridization in the past between two sunflower species, *Helianthus annuus* and *H. petiolaris* (the ancestor of the cultivated sunflower) has apparently given rise to three new sunflower species, *H. anomalus, H. paradoxus,* and *H. deserticola* (**Fig. 22.13**). *H. petiolaris* and *H. annuus* have probably formed innumerable hybrids in nature, virtually all of them inviable. However, a few—the ones with a workable genetic complement—are the ones that survived to yield these three daughter species. In this case, each one of these new species has acquired a different mix of parental chromosomes. It is this species-specific chromosome complement that makes all three distinct and reproductively isolated from the parent species and from one another.

In many cases of hybridization, chromosome numbers may change. Two diploid parent species with 5 pairs of chromosomes, for a total of 10 chromosomes each, may produce a hybrid with double the number of chromosomes (that is, the hybrid inherits a full paired set of chromosomes from each parental species)—a total of 20. In this case, the hybrid has four genomes rather than the diploid number of two. We call such a double diploid a tetraploid. In general, animals cannot sustain this kind of expansion in chromosome complement, but plants are more likely to do so. As a result, the formation of new species through **polyploidy**—multiple chromosome sets (Chapter 13)—has been relatively common in plants.

Polyploids may be allopolyploids, meaning that they are produced from hybridization of two different species. For example, related species of *Chrysanthemum* appear, on the basis of their chromosome numbers, to be allopolyploids. Alternatively, polyploids may be autopolyploids, meaning that they are derived from an unusual reproductive event between members of a single species. In this case, through an error of meiosis in one or both parents in which homologous chromosomes fail to separate, a gamete may be produced that is not haploid. For example, *Anemone rivularis,* a plant in the buttercup family, has 16 chromosomes, and its close relative *A. quinquefolia* has 32. *Anemone quinquefolia* appears to be an autopolyploid derived from the joining of *A. rivularis* gametes with two full sets of chromosomes each.

So rampant is speciation by polyploidy in plants that it affects the pattern of chromosome numbers across all plants. In **Fig. 22.14,** we see the haploid chromosome numbers of thousands of plant species plotted on a graph. Note that as the numbers get higher, even numbers tend to predominate, suggesting that the doubling of the total number of chromosomes (a form

FIG. 22.13 Speciation by hybridization. In sunflowers, *Helianthus anomalus* (bottom) is the product of natural hybridization between *H. annuus* (top left) and *H. petiolaris* (top right). *Source: (top left) Gary A. Monroe @ USDA-NRCS PLANTS Database, (top right) Jason Rick, (bottom) Gerald J. Seiler, USDA-ARS.*

FIG. 22.14 **Plant chromosome numbers.** Plant chromosome numbers suggest that polyploidy has played an important role in plant evolution. *Data from Fig. 20-18, p. 752, in A. J. F. Griffiths, S. R. Wessler, S. B. Carroll, and J. Doebley, 2012, Introduction to Genetic Analysis, 10th ed., New York: W. H. Freeman.*

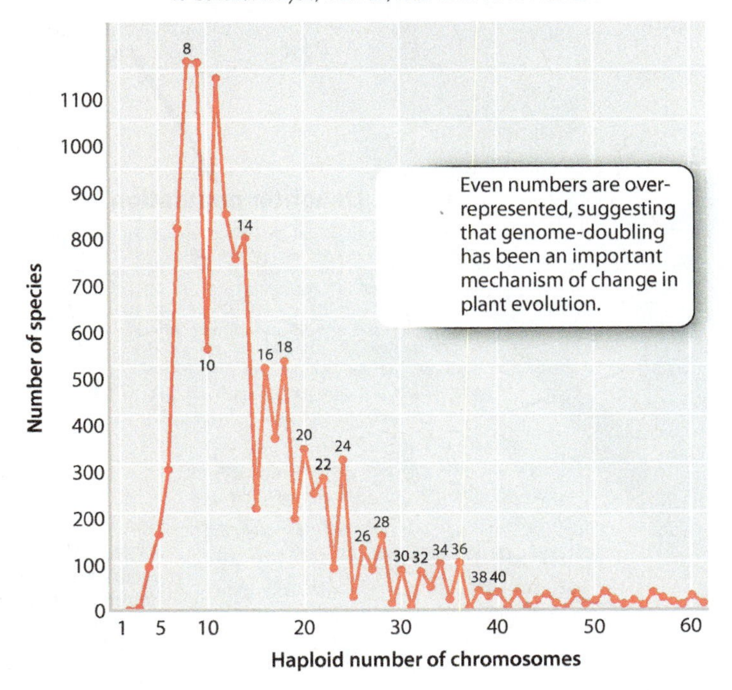

of polyploidy that will always result in an even number of chromosomes) is an important factor in plant evolution.

22.4 SPECIATION AND SELECTION

The association of both the origin of species and natural selection with Charles Darwin may make it seem that one cannot occur without the other. However, speciation can occur in the presence or absence of natural selection, and natural selection does not always lead to speciation.

Speciation can occur with or without natural selection.

Natural selection may or may not play a role in speciation. The genetic divergence of two populations can be entirely due to genetic drift, for example, with no role for natural selection.

We have also seen two ways in which natural selection can be involved in speciation. First, sympatric speciation requires some form of disruptive natural selection, as when hybrid offspring are competitively inferior. Second, allopatric speciation (and adaptive radiation) may be facilitated by natural selection. For example, when a peripheral population is in a new environment, natural selection will act to promote its adaptation to the new conditions, accelerating in the process the rate of genetic divergence between it and its parent population.

Natural selection can enhance reproductive isolation.

There is a third way in which natural selection may play an important role in speciation. In this case, natural selection contributes directly to the process of speciation (rather than contributing indirectly by causing differences among the diverging populations), when individuals better at choosing mates from their own group are selectively favored over those that frequently mate with members of the "wrong" group.

Recall the two hypothetical bird populations, one with large bills and the other with small bills, whose medium-billed hybrid offspring are at a disadvantage because of a lack of medium-sized seeds in their environment. If a large-billed individual cannot distinguish between large- and small-billed individuals as potential mates (that is, there is a lack of pre-zygotic isolation between them), it will frequently make the "wrong" mate choice, picking a small-billed individual and paying a considerable evolutionary cost of producing poorly adapted hybrid offspring.

Now imagine a new mutation in the large-billed group that permits individuals carrying it to distinguish between the two groups of birds and to mate only with other large-billed individuals. Such a mutation would spread under natural selection because it would prevent the wasted reproductive effort of producing disadvantaged hybrid offspring. This is an example of **reinforcement of reproductive isolation,** or **reinforcement** for short. Reinforcement is the process by which diverging populations undergo natural selection in favor of traits that enhance pre-zygotic isolation, thereby preventing the production of less fit hybrid offspring. In this case, enhanced pre-zygotic isolation takes the form of mating discrimination, an increased ability to recognize and mate with members of one's own population.

The best evidence in support of reinforcement comes from a study of related fruit-fly species living either in allopatry (geographically separated) or sympatry (not geographically separated). Sympatric species evolve pre-zygotic isolating mechanisms more rapidly than allopatric species. Why would this be? When the two populations are geographically separated, the formation of less fit hybrids is impossible since the two populations do not interbreed, so a mutation that increases mating discrimination between the two groups would not be favored by natural selection. In sympatry, however, where the production of less fit hybrids is a problem, such a mutation provides a fitness benefit, so natural selection favors its spread in the population and reinforces reproductive isolation between the two groups.

Fig. 22.15 summarizes the evolutionary mechanisms that lead to speciation. Speciation is caused by the accumulation of genetic differences between populations. Mutation is therefore a key component of the process, but so, too, is the fixation process whereby mutations go to 100% in a population (Chapter 21). A mutation can go to fixation by selection if it is advantageous and by genetic drift. The differences that accumulate between diverging populations are therefore produced by multiple forces acting over long periods on each population. ■

VISUAL SYNTHESIS Speciation

FIG. 22.15

Integrating concepts from Chapters 21 and 22

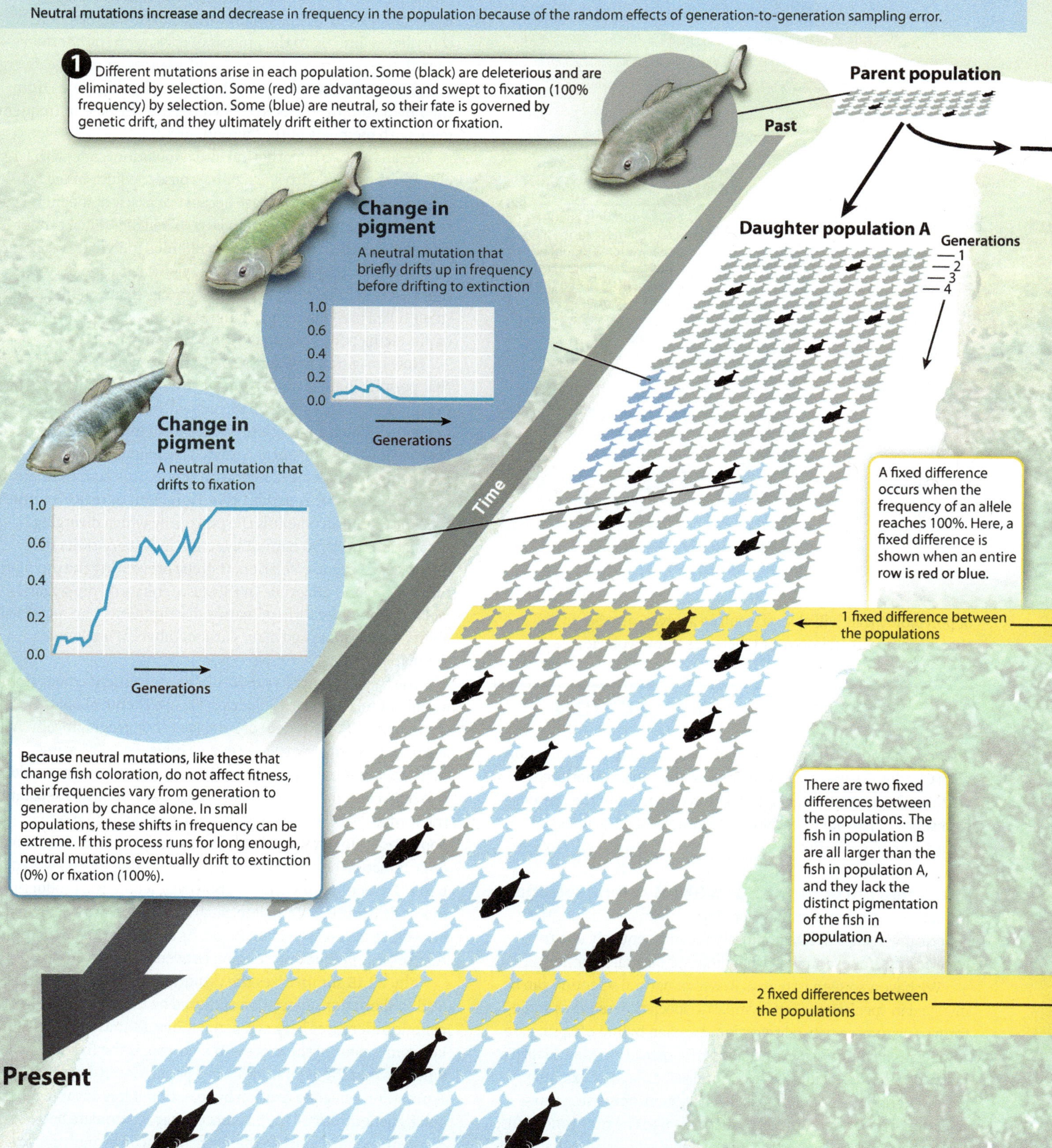

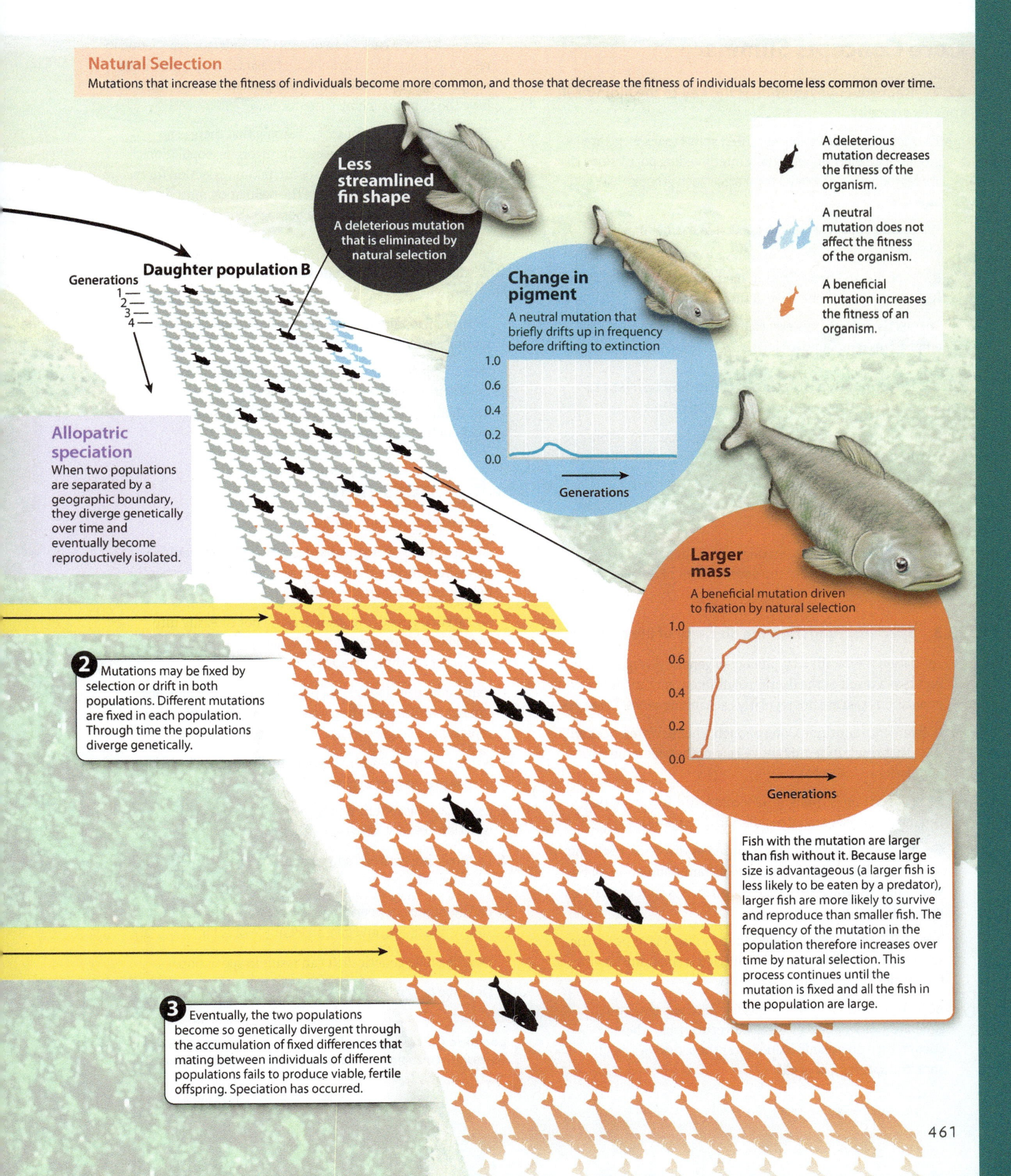
Natural Selection
Mutations that increase the fitness of individuals become more common, and those that decrease the fitness of individuals become less common over time.
Less streamlined fin shape
A deleterious mutation that is eliminated by natural selection
A deleterious mutation decreases the fitness of the organism.
A neutral mutation does not affect the fitness of the organism.
A beneficial mutation increases the fitness of an organism.
Daughter population B
Generations
1
2
3
4
Change in pigment
A neutral mutation that briefly drifts up in frequency before drifting to extinction
1.0
0.6
0.4
0.2
0.0
Generations
Allopatric speciation
When two populations are separated by a geographic boundary, they diverge genetically over time and eventually become reproductively isolated.
Larger mass
A beneficial mutation driven to fixation by natural selection
1.0
0.6
0.4
0.2
0.0
Generations
2 Mutations may be fixed by selection or drift in both populations. Different mutations are fixed in each population. Through time the populations diverge genetically.
Fish with the mutation are larger than fish without it. Because large size is advantageous (a larger fish is less likely to be eaten by a predator), larger fish are more likely to survive and reproduce than smaller fish. The frequency of the mutation in the population therefore increases over time by natural selection. This process continues until the mutation is fixed and all the fish in the population are large.
3 Eventually, the two populations become so genetically divergent through the accumulation of fixed differences that mating between individuals of different populations fails to produce viable, fertile offspring. Speciation has occurred.

Core Concepts Summary

22.1 Reproductive isolation is the key to the biological species concept.

The biological species concept (BSC) states that species are groups of actually or potentially interbreeding populations that are reproductively isolated from other such groups. page 446

We cannot apply the BSC to asexual or extinct organisms. page 447

Ring species and hybridization further demonstrate that the BSC is not a comprehensive definition of species. page 448

The BSC is nevertheless especially useful because it emphasizes reproductive isolation. page 449

22.2 Reproductive isolation is caused by barriers to reproduction before or after egg fertilization.

Reproductive barriers can be pre-zygotic, occurring before egg fertilization, or post-zygotic, occurring after egg fertilization. page 449

Pre-zygotic isolation may be behavioral, gametic, temporal, or ecological. page 450

In post-zygotic isolation, mating occurs but genetic incompatibilities prevent the development of a viable, fertile offspring. page 450

22.3 Speciation underlies the diversity of life on Earth.

Speciation is typically a by-product of genetic divergence that occurs as a result of the fixation of different mutations in two populations that are not regularly exchanging genes. page 451

If divergence continues long enough, chance differences will arise that result in reproductive barriers between the two populations. page 451

Most speciation is thought to be allopatric, involving two geographically separated populations. page 451

Geographic separation may be caused by dispersal, resulting in the establishment of a new and distant population, or by vicariance, in which the range of a species is split by a change in the environment. page 451

A special case of allopatric speciation by dispersal is peripatric speciation, in which the new population is small and outside the species' original range. page 453

Adaptive radiation, in which speciation occurs rapidly to generate a variety of ecologically diverse forms, is best documented on oceanic islands after the arrival of a single ancestral species. page 455

Co-speciation occurs when one species undergoes speciation in response to speciation in another. In parasites and their hosts, co-speciation can result in host and parasite phylogenies that have the same branching patterns. page 455

Speciation may be sympatric, meaning that there is no geographic separation between the diverging populations. For this type of speciation to occur, natural selection for two or more different types within the population must act so strongly that it overcomes the homogenizing effect of gene flow. page 456

22.4 Speciation can occur with or without natural selection.

Separated populations can diverge as a result of genetic drift, natural selection, or both. page 459

Natural selection may act on mutations that allow individuals to identify and mate with individuals that are more like themselves. This process is called reinforcement of reproductive isolation. page 459

Self-Assessment

1. Define the term "species."
2. Given a group of organisms, describe how you would test whether they all belong to one species or whether they belong to two separate species.
3. Name two types of organism that do not fit easily into the biological species concept. What species concept would work best for these organisms?
4. Explain how ecological and phylogenetic considerations can help inform whether or not a group of organisms represents a single species.
5. Name four reproductive barriers and indicate whether each is pre- or post-zygotic.
6. Describe how genetic divergence and reproductive isolation are related to each other.
7. Differentiate between allopatric and sympatric speciation, and state which is thought to be more common and why.
8. Differentiate between allopatric speciation by dispersal and by vicariance, and give an example of each.
9. Describe how genetic drift can result in speciation.
10. Describe how natural selection can result in speciation.

Log in to LaunchPad to check your answers to the Self-Assessment questions, and to access additional learning tools.

Dudley Museums Service.

CHAPTER 23

Evolutionary Patterns

Phylogeny and Fossils

Core Concepts

23.1 A phylogenetic tree is a reasoned hypothesis of the evolutionary relationships among organisms.

23.2 A phylogenetic tree is built on the basis of shared derived characters.

23.3 The fossil record provides direct evidence of evolutionary history.

23.4 Phylogeny and fossils provide independent and corroborating evidence of evolution.

All around us, nature displays nested patterns of similarity among species. For example, as noted in Chapter 1, humans are more similar to chimpanzees than either humans or chimpanzees are to monkeys. Humans, chimpanzees, and monkeys, in turn, are more similar to one another than any one of them is to a mouse. And humans, chimpanzees, monkeys, and mice are more similar to one another than any of them is to a catfish. This pattern of nested similarity was recognized more than 200 years ago and used by the Swedish naturalist Carolus Linnaeus to classify biological diversity. A century later, Charles Darwin recognized this pattern as the expected outcome of a process of "descent with modification," or evolution.

Evolution produces two distinct but related patterns, both evident in nature. First is the nested pattern of similarities found among species on present-day Earth. The second is the historical pattern of evolution recorded by fossils. Life, in its simplest form, originated more than 3.5 billion years ago. Today, an estimated 10 million species inhabit the planet. Short of inventing a time machine, how can we reconstruct those 3.5 billion years of evolutionary history in order to understand the extraordinary events that have ultimately resulted in the biological diversity we see around us today? These two great patterns provide the answer.

Darwin recognized that the species he observed must be the modified descendants of earlier ones. Distinct populations of an ancestral species separate and diverge through time, again and again, giving rise to multiple descendant species. The result is the pattern of nested similarities observed in nature (see Fig. 1.17). This history of descent with branching is called **phylogeny,** and is much like the genealogy that records our own family histories.

The evolutionary changes inferred from patterns of relatedness among present-day species make predictions about the historical pattern of evolution we should see in the fossil record. For example, groups with features that we infer to have evolved earlier than others should appear earlier in time as fossils. Paleontological research reveals that the history of life is indeed laid out in the chronological order predicted on the basis of comparative biology. How do we reconstruct the history of life from evolution's two great patterns, and how do they compare?

23.1 READING A PHYLOGENETIC TREE

Chapter 22 introduced the concept of speciation, the set of processes by which physically, physiologically, or ecologically isolated populations diverge from one another to the point where they can no longer produce fertile offspring. As illustrated in **Fig. 23.1,** speciation can be thought of as a process of branching. Now consider what happens as this process occurs over and over in a group through time. As species proliferate, their evolutionary relationships to one another unfold in a treelike pattern, with individual species at the twig tips and their closest relatives connected to them at the nearest fork in the branch, called a **node.** A node thus represents the most recent common ancestor of two descendant species.

FIG. 23.1 The relationship between speciation and a phylogenetic tree. The phylogenetic tree, on the right, depicts the evolutionary relationships that result from the two successive speciation events diagrammed on the left.

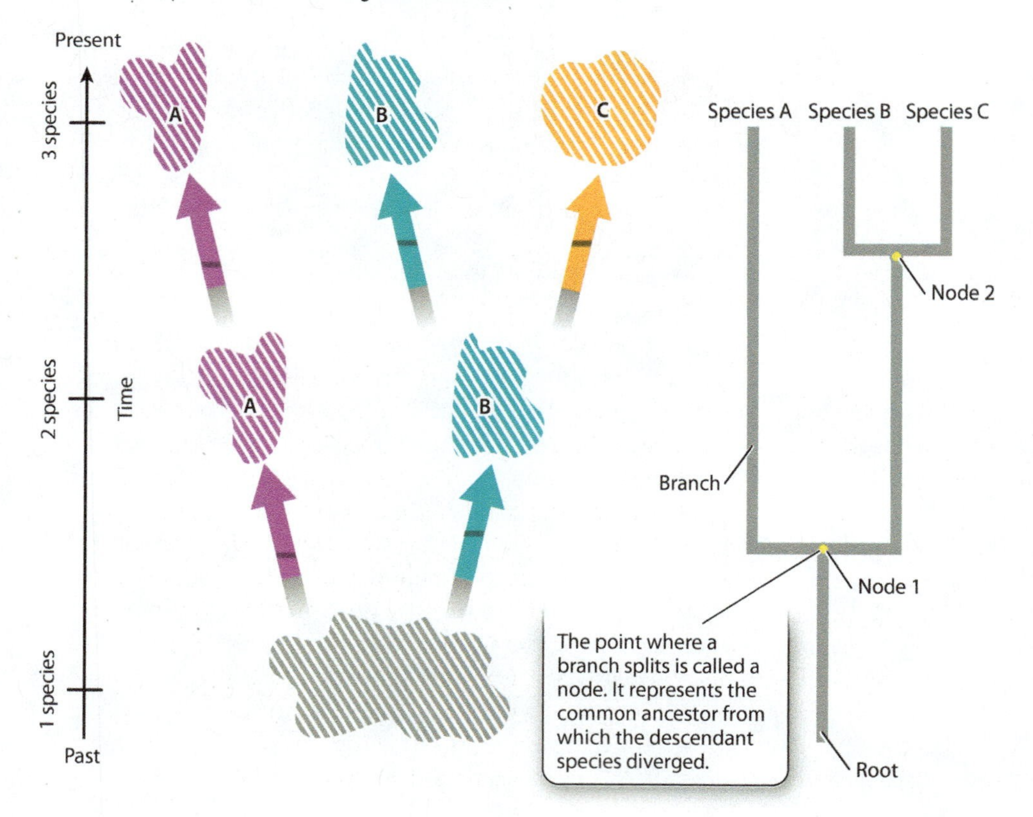

Phylogenetic trees provide hypotheses of evolutionary relationships.

Phylogenetics is one of two related disciplines within systematics, the study of evolutionary relationships among organisms. The other is taxonomy, the classification of organisms.

The aim of taxonomy is to recognize and name groups of individuals as species, and, subsequently, to group closely related species into the more inclusive taxonomic group of the genus, and so on up through the taxonomic ranks—species, genus, family, order, class, phylum, kingdom, domain. Taxonomy, then, provides us with a hierarchical classification of species in groups that are more and more inclusive, giving us a convenient way to communicate information about the features each group possesses. So, if we want to tell someone about a small animal we have seen with fur, mammary glands, and extended finger bones that permit it to fly, we can give them this long description, or we can just say we saw a bat, or a member of Order Chiroptera. All the rest is understood (or can be looked up in a reference).

Phylogenetics, on the other hand, aims to discover the pattern of evolutionary relatedness among groups of species or other groups by comparing their anatomical or molecular features, and to depict these relationships as a **phylogenetic tree.** A phylogenetic tree is a hypothesis about the evolutionary history, or phylogeny, of the species. Phylogenetic trees are hypotheses because they represent the best model, or explanation, of the relatedness of organisms on the basis of all the existing data. As with any model or hypothesis, new data may provide evidence for alternative relationships, leading to changes in the hypothesized pattern of branching on the tree.

Many phylogenetic trees explore the relatedness of particular groups of individuals, populations, or species. We may, for example, want to understand how wheat is related to other, non-commercial grasses, or how disease-causing populations of *Escherichia coli* relate to more benign strains of the bacterium. At a much larger scale, universal similarities of molecular biology indicate that all living organisms are descended from a single common ancestor. This insight inspires the goal of reconstructing phylogenetic relationships for all species in order to understand how biological diversity has evolved since life originated. This universal tree is commonly referred to as the tree of life (Chapter 1). In Part 2 of this book, we will make use of the tree of life and many smaller-scale phylogenetic trees to understand our planet's biological diversity.

FIG. 23.2 A phylogeny of vertebrate animals. The branching order constitutes a hypothesis of evolutionary relationships within the group.

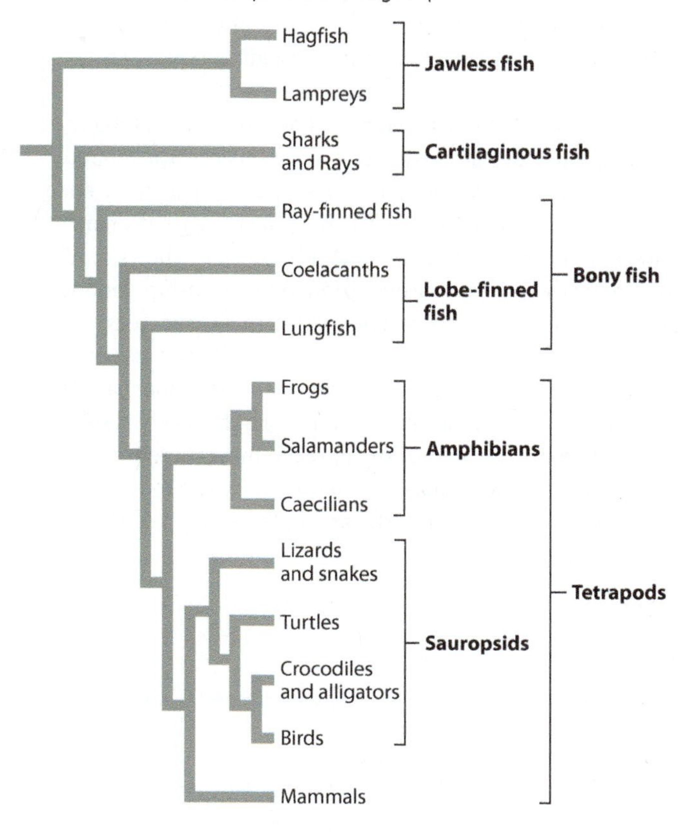

Fig. 23.2 shows a phylogenetic tree for vertebrate animals. The informal name at the end of each branch represents a group of organisms, many of them familiar. We sometimes find it useful to refer to groups of species this way (for example, "frogs," or "Class Anura") rather than name all the individual species or list the characteristics they have in common. It is important, however, to remember that such named groups represent a number of member species. If, for example, we were able to zoom in on the branch labeled "Frogs," we would see that it consists of many smaller branches, each representing a distinct species of frog, either living or extinct.

This tree provides information about evolutionary relationships among vertebrates. For example, it proposes that the closest living relatives of birds are crocodiles and alligators. The tree also proposes that the closest relatives of all tetrapod (four-legged) vertebrates are lungfish, which are fish with lobed limbs and the ability to breathe air. Phylogenetic trees are built from careful analyses of the morphological and molecular attributes of the species or other groups under study. A tree is therefore a hypothesis about the order of branching events in evolution, and it can be tested by gathering more information about anatomical and molecular traits.

A phylogenetic tree does not in any way imply that more recently evolved groups are more advanced than groups that arose earlier. A modern lungfish, for example, is not more primitive or "less evolved" than an alligator, even though its group branches off the trunk of the vertebrate tree earlier than the alligator group does. After all, both species are the end products of the same interval of evolution since their divergence from a common ancestor more than 370 million years ago.

The search for sister groups lies at the heart of phylogenetics.

Two species, or groups of species, are considered to be closest relatives if they share a common ancestor not shared by any other species or group. In Fig. 23.2, for example, we see that frogs are more closely related to salamanders than to any other group of organisms because frogs and salamanders share a common amphibian ancestor not shared by any other group. Similarly, lungfish are more closely related to tetrapods than to any other group. A lungfish may look more like a fish than it does an amphibian, but lungfish are more closely related to amphibians than they are to other fish because lungfish share a common ancestor with amphibians (and other tetrapods) that was more recent than their common ancestor with other fish (Fig. 23.2).

Groups that are more closely related to each other than either of them is to any other group, like lungfish and tetrapods, are called **sister groups.** Simply put, phylogenetic hypotheses amount to determining sister-group relationships because the simplest phylogenetic question we can ask is which two of any three species (or other groups) are more closely related to each other than either is to the third. In this light, we can see that a phylogenetic tree is simply a set of sister-group relationships; adding a species to the tree entails finding its sister group in the tree.

Closeness of relationship is then determined by looking to see how recently two groups share a common ancestor. Shared ancestry is indicated by a node, or branch point, on a phylogenetic tree. Nodes can be rotated without changing the evolutionary relationships of the groups. **Fig. 23.3,** for example, shows four phylogenetic trees depicting evolutionary relationships among birds, crocodiles and alligators, and turtles. In all four trees, birds are a sister group to crocodiles and alligators because birds, crocodiles, and alligators share a common ancestor not shared by turtles. The more recent a common ancestor, the more closely related two groups are. Evolutionary relatedness therefore is determined by following nodes from the tips to the root of the tree, and is not determined by the order of the tips from the top to bottom of a page.

→ **Quick Check 1** Does either of these two phylogenetic trees indicate that humans are more closely related to lizards than to mice?

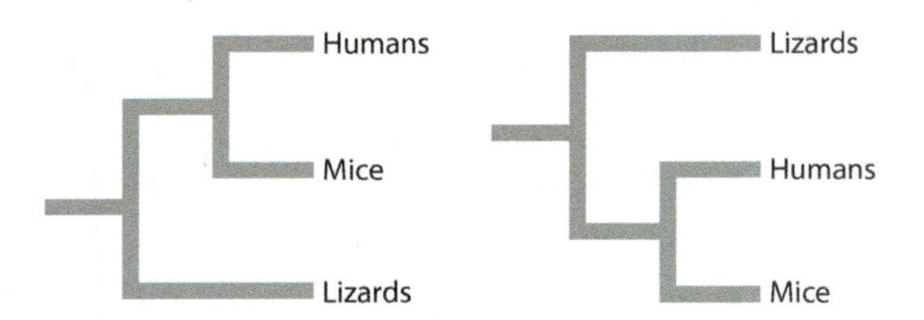

A monophyletic group consists of a common ancestor and all its descendants.

Up to this point, we have used the word "group" to mean all the species in some taxonomic entity under discussion. A more technical word is **taxon** (plural, **taxa**), with taxonomy providing a formal means of naming groups. Recently, biologists have worked to integrate evolutionary history with taxonomic classification. The resulting classification emphasizes groups that are **monophyletic,** meaning that all members share a single common ancestor not shared with any other species or group of species. In Fig. 23.2, the tetrapods are monophyletic because they all share a common ancestor not shared by any other taxa. Similarly, amphibians are monophyletic.

In contrast, consider the group of animals traditionally recognized as reptiles, which includes turtles, snakes, lizards, crocodiles, and alligators (Fig. 23.2). The group "reptiles" excludes birds, although they share a common ancestor with the included animals. Such a group is **paraphyletic.** A paraphyletic group includes some, but not all, of the descendants of a common ancestor. Early zoologists separated birds from reptiles because they are so distinctive. However, many features of skeletal anatomy and DNA sequence strongly support the placement of birds as a sister group to the crocodiles and alligators.

There is a simple way to distinguish between monophyletic and paraphyletic groups, illustrated in **Fig. 23.4.** If in order to separate a group from the rest of the phylogenetic tree you need only to make one cut, the group is monophyletic. If you need a second cut to trim away part of the separated branch, the group is paraphyletic.

Groupings that do not include the last common ancestor of all members are called **polyphyletic.** For example, clustering bats and birds together as flying tetrapods results in a polyphyletic group (Fig. 23.4).

Identifying monophyletic groups is a main goal of phylogenetics because monophyletic groups include *all* descendants of a common ancestor and *only* the descendants of that common ancestor. This means that monophyletic groups alone show the evolutionary path a given group has taken since its origin. Omitting some members of a group, as in the case of reptiles and other paraphyletic groups, can provide a misleading sense of evolutionary history. By using monophyletic groups in taxonomic classification, we effectively convey our knowledge of their evolutionary history.

→ **Quick Check 2** Look at Fig. 23.4. Are fish a monophyletic group?

FIG. 23.3 Sister groups. The four trees illustrate the same set of sister-group relationships.

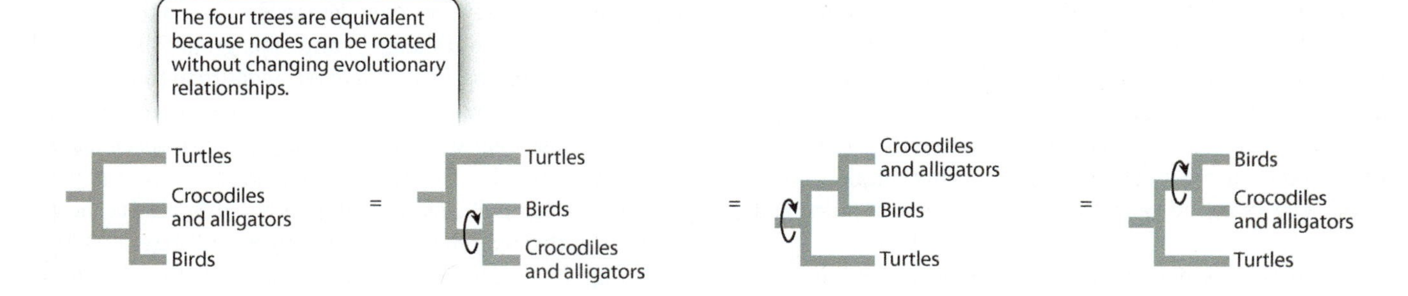

FIG. 23.4 Monophyletic, paraphyletic, and polyphyletic groups. Only monophyletic groups reflect evolutionary relationships because only they include all the descendants of a common ancestor.

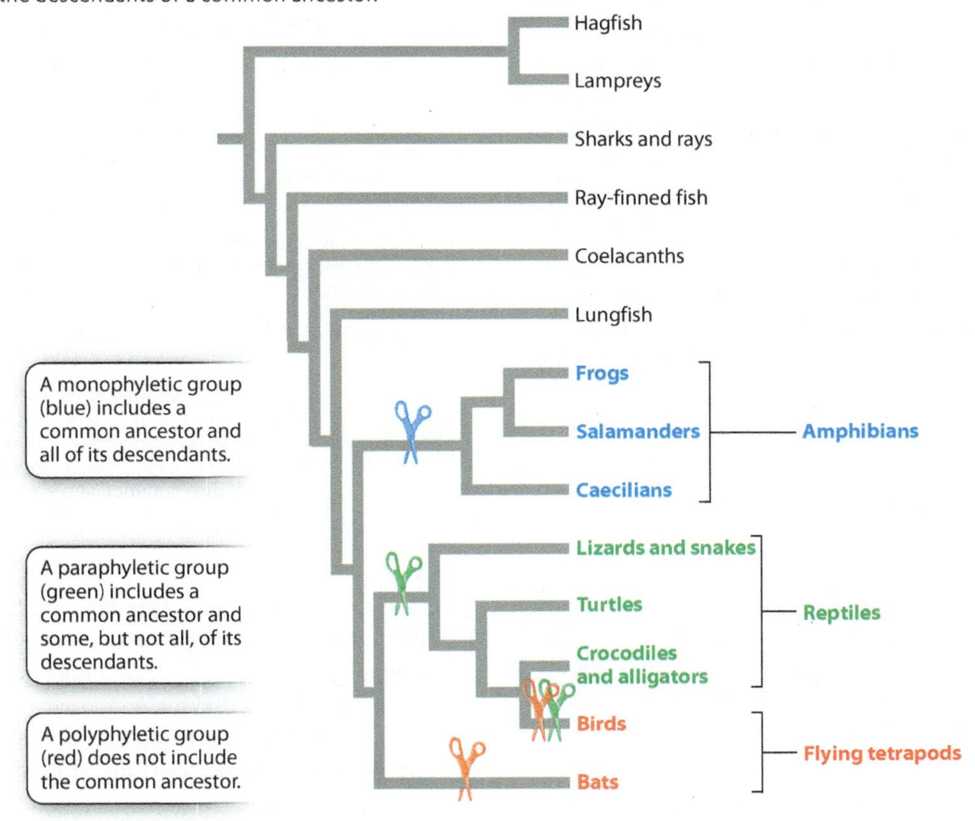

Taxonomic classifications are information storage and retrieval systems.

The nested pattern of similarities among species has been recognized by naturalists for centuries. In the vocabulary of formal classification, closely related species are grouped into a **genus** (plural, **genera**). Closely related genera, in turn, belong to a larger, more inclusive branch of the tree, as a **family.** Closely related families, in turn, form an **order,** orders form a **class,** classes form a **phylum** (plural, **phyla**), and phyla form a **kingdom,** each more inclusive taxonomic level occupying a successively larger limb on the tree (**Fig. 23.5**). Biologists today commonly refer to the three largest limbs of the entire tree of life as **domains** (Eukarya, or eukaryotes; Bacteria; and Archaea).

The ranks of classification form a nested hierarchy, but the boundaries of ranks above the species level are arbitrary

FIG. 23.5 Classification. Classification reflects our understanding of phylogenetic relationships, and the taxonomic hierarchy reflects the order of branching.

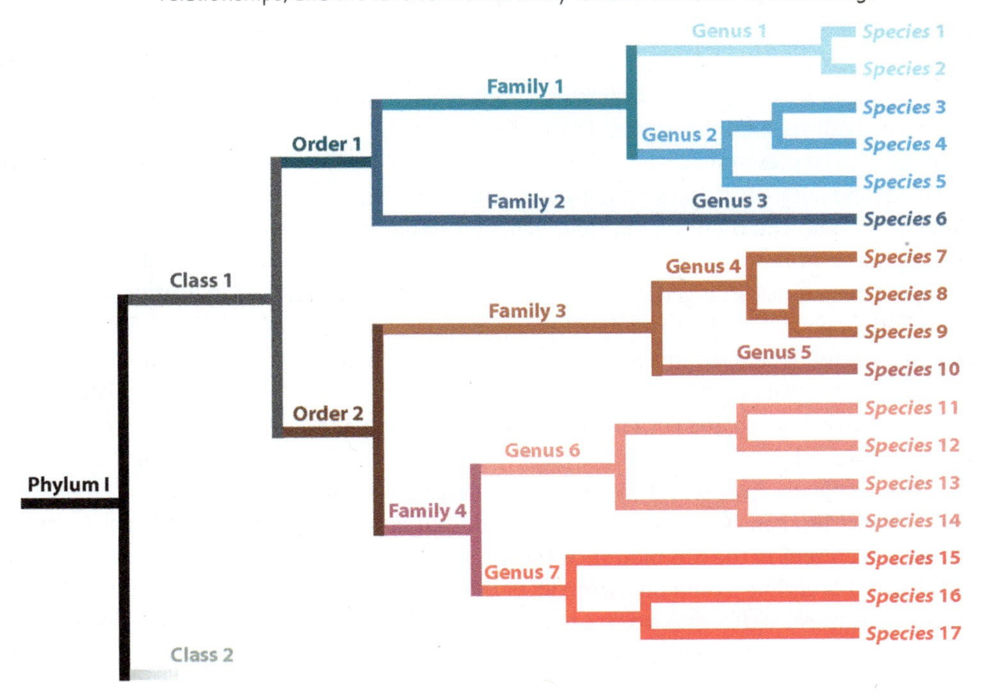

in that there is nothing particular about a group that makes it, for example, a class rather than an order. A taxonomist examining the 17 species included in Fig. 23.5 might decide that species 3 and 4 are sufficiently distinct from species 5 to warrant placing them into a distinct genus. The same taxonomist might then decide that the grouping of the two new genera together should rank as a family. For this reason, it is not necessarily true that orders or classes of, for example, birds and ferns are equivalent in any meaningful way. In contrast, sister groups are equivalent in several ways—notably, in that they diverged from a single ancestor at a single point in time. Therefore, if one branch is a sister group that contains 500 species and another has 6, the branches have experienced different rates of speciation, extinction, or both since they diverged. In Fig 23.5, Families 1 and 2 are sister groups, but Family 1 has five species while Family 2 has just one.

FIG. 23.6 Homology and analogy. A homology is a similarity that results from shared ancestry, whereas an analogy is a similarity that results from convergent evolution.

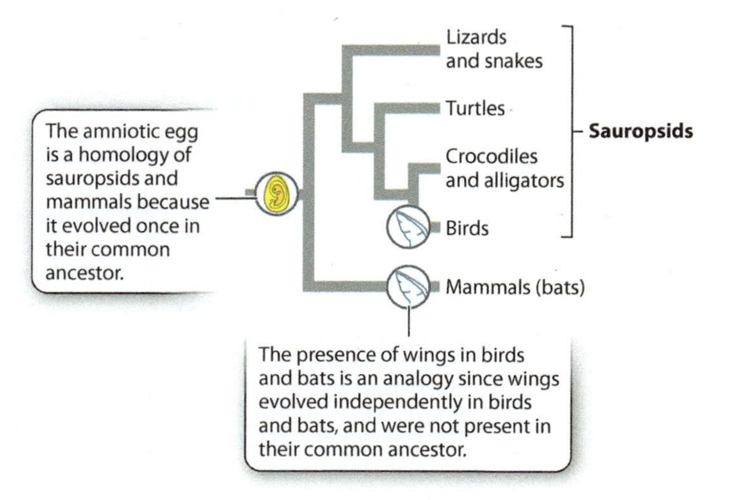

23.2 BUILDING A PHYLOGENETIC TREE

Up to this point, we have focused on how to interpret a phylogenetic tree, a diagram that depicts the evolutionary history of organisms. But how do we infer evolutionary history from a group of organisms? That is, how do we actually construct a phylogenetic tree? Biologists use characteristics of organisms to figure out their relationships. Similarities among organisms are particularly important in that similarities sometimes suggest shared ancestry. However, a key principle of constructing trees is that only *some* similarities are actually useful. Others can in fact be misleading.

Homology is similarity by common descent.

Phylogenetic trees are constructed by comparison of character states shared among different groups of organisms. **Characters** are the anatomical, physiological, or molecular features that make up organisms. In general, characters have several observed conditions, called **character states.** In the simplest case, a character can be present or absent—lungs are present in tetrapods and lungfish, but absent in other vertebrate animals. Commonly, however, there are multiple character states. Petals are a character of flowers, for example, and each observed arrangement—petals arranged in a helical pattern, petals arranged in a whorl, or petals fused into a tube—can be considered a state of the character of petal arrangement. All species contain some character states that are shared with other members of their group, some that are shared with members of other groups, and some that are unique.

Character states in different species can be similar for one of two reasons: The character state (for example, helically arranged petals) was present in the common ancestor of the two groups and retained over time (common ancestry), or the character state independently evolved in the two groups as an adaptation to similar environments (convergent evolution).

Consider two examples. Mammals and birds both produce amniotic eggs. Amniotic eggs occur only in groups descended from the common ancestor at the node connecting the mammal and sauropsid branches of the tree, and so we reason that birds and mammals each inherited this character from a common ancestor in which the trait first evolved (**Fig. 23.6**). Characters that are similar because of descent from a common ancestor are said to be **homologous.**

Not all similarities arise in this way, however. Think of wings, a character exhibited by both birds and bats. Much evidence supports the view that wings in these two groups do not reflect descent from a common, winged ancestor but rather evolved independently in the two groups. Similarities due to independent adaptation by different species are said to be **analogous.** They are the result of convergent evolution.

Innumerable examples of convergent evolution less dramatic than wings are known. In some, we even understand the genetic basis of the convergence. For example, echolocation has evolved in bats and in dolphins, but not in other mammals. Prestin is a protein in the hair cells of mammalian ears that is involved in hearing ultrasonic frequencies. Both bats and dolphins independently evolved similar changes in their *Prestin* genes, apparently convergent adaptations for echolocation. Similarly, unrelated fish that live in freezing water at the poles, Arctic and Antarctic, have evolved similar glycoproteins that act as molecular "antifreeze," preventing the formation of ice in their tissues.

In principle, two characters or character states are homologous if they are similar because of descent from a common ancestor with the same character or character state; they are analogous if they arose independently because of similar selective pressures. In practice, to determine if characters observed in two organisms are homologous or analogous, we can weigh evidence from where other traits place the two organisms on a phylogenetic tree, we can look at where on the organisms the trait occurs, and we can look at the anatomical or genetic details of how the trait is constructed. Wings in birds and bats are similar in morphological position (both are modified forelimbs) but differ in details of construction (the bat wing is supported by long fingers), and all other traits of birds and bats place them at the tips of different lineages, with many nonwinged species between them and their most recent common ancestor.

➔ **Quick Check 3** Fish and dolphins have many traits in common, including a streamlined body and fins. Are these traits homologous or analogous?

Shared derived characters enable biologists to reconstruct evolutionary history.

Because homologies result from shared ancestry, only homologies, and not analogies, are useful in constructing phylogenetic trees. However, it turns out that only some homologies are useful. For example, character states that are unique to a given species or other monophyletic group can't tell us anything about its sister group. They evolved after the divergence of the group from its sister group and so can be used to characterize a group but not to relate it to other groups. Similarly, homologies formed in the common ancestor of the entire group and therefore present in all its descendants do not help to identify sister-group relationships among the descendants of that common ancestor.

What we need to build phylogenetic trees are homologies that are shared by some, but not all, of the members of the group under consideration. These shared derived characters are called **synapomorphies.** A derived character state is an evolutionary innovation (for example, the change from five toes to a single toe—the hoof—in the ancestor of horses and donkeys). When such a novelty arises in the common ancestor of two taxa, it is shared by both (thus, the hoof is a synapomorphy defining horses and donkeys as sister groups).

In **Fig. 23.7,** we indicate the major synapomorphies that have helped us construct the phylogeny of vertebrates. For example, the lung is a character present in lungfish and tetrapods, but absent in other vertebrates. Thus, the presence of lungs provides one piece of evidence that lungfish are the sister group of tetrapods. Phylogenetic reconstruction on the basis of synapomorphies is called **cladistics.**

The simplest tree is often favored among multiple possible trees.

To show how synapomorphies help us chart out evolutionary relationships, let's consider the simple example in **Fig. 23.8.** We begin with four species of animals (labeled "A" through "D") in a group we wish to study that we will call our ingroup; we believe the species to be closely related to each other. For comparison, we have a species that we believe is outside this ingroup—that is, it falls on a branch that splits off nearer the root of the tree—and so is called an outgroup (labeled "OG" in Fig. 23.8). Each species in the ingroup and the outgroup has a different combination of characters, such as leg number, presence or absence of wings, and whether development of young to adult is direct or goes through a pupal stage (Fig. 23.8a).

We are interested in the relationships among species A–D and so focus on potential synapomorphies, character states shared by some but not all species within the group. For example, only C and D have

FIG. 23.7 **Synapomorphies, or shared derived characters.** Homologies that are present in some, but not all, members of a group help us to construct phylogenetic trees.

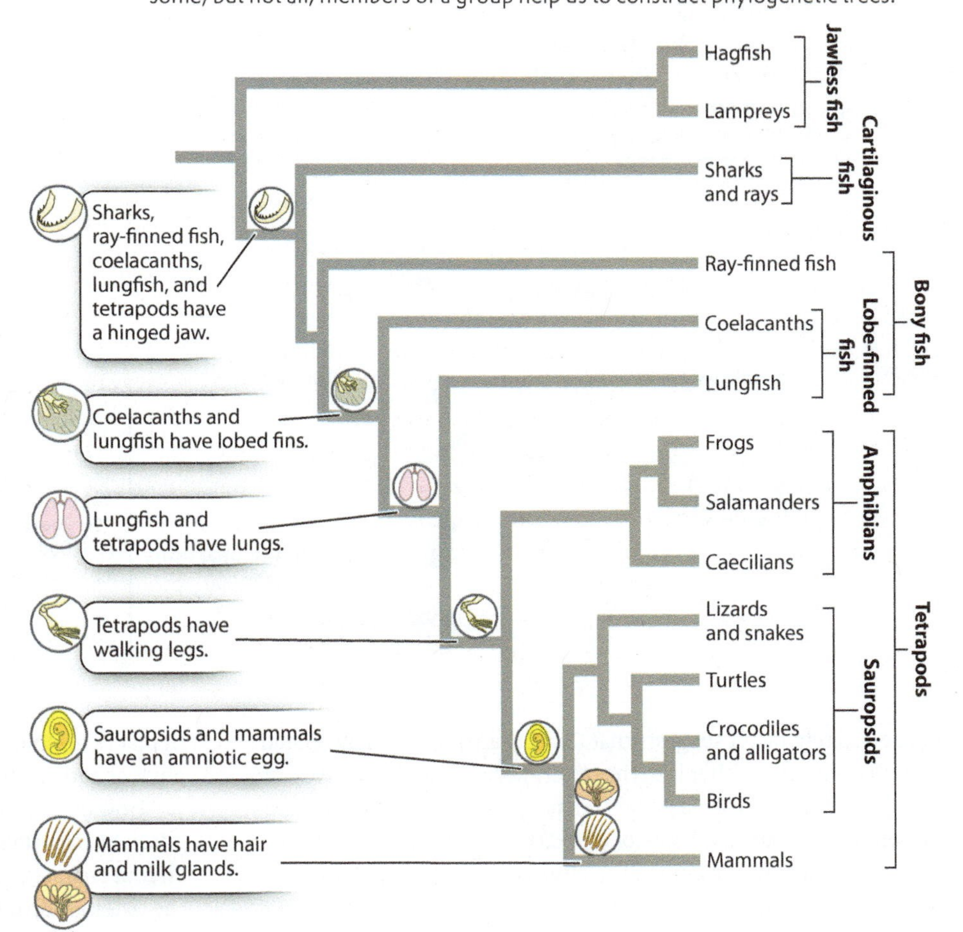

FIG. 23.8 Constructing a phylogenetic tree from shared derived traits. The strongest hypothesis of evolutionary relationships overall is the tree with the fewest number of changes because it minimizes the total number of independent origins of character states.

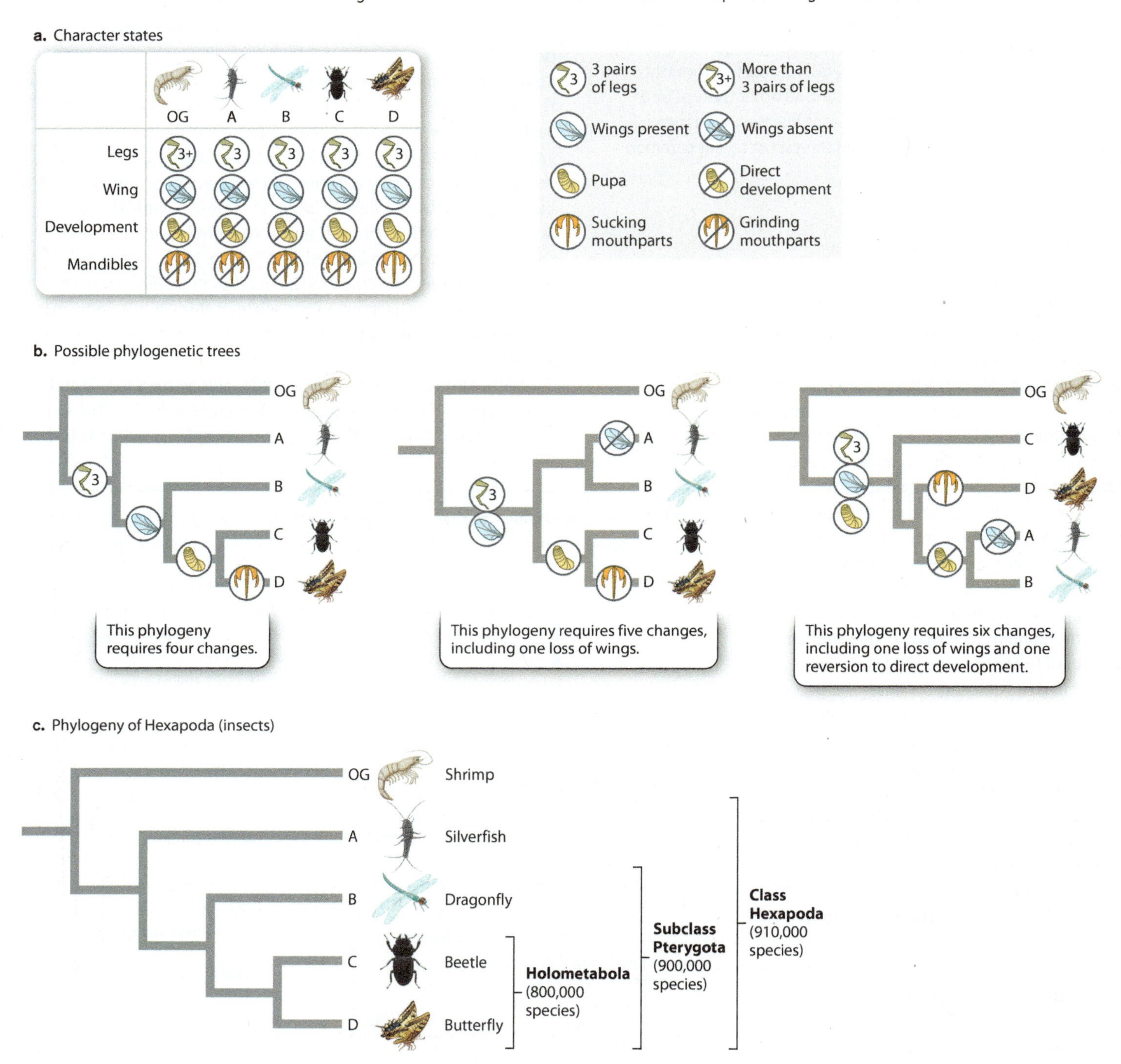

pupae. This character suggests that C and D are more closely related to each other than either is to the other species. The alternatives are that C and D each evolved pupae independently or that pupae were present in the common ancestor of A–D but were lost in A and B.

How do we choose among the alternatives? Studies of the outgroup show that it does not form pupae, supporting the hypothesis that pupal development evolved within the ingroup. In practice, biologists examine multiple characters and choose the phylogenetic hypothesis that best fits all of the data.

How do we determine "best fit"? Fig. 23.8b illustrates three different hypotheses for the relationships among the species based on four characters and their various character states. Each reflects the sister-group relationship between C and D proposed earlier. The leftmost tree requires exactly four character-state changes during the evolution of these species: reduction of leg pairs to three in the group ABCD, wings in the group BCD, and pupae in group CD, plus a change of form of the mandible, or jaw, in species D.

Now consider the middle tree in Fig. 23.8b. It groups A and B together, and so differs from the tree on the left in requiring either a loss of wings in A, or an additional origin of wings in B, independent of that in the common ancestor of C and D—five changes in all. The tree on the right groups A and B together, and requires two extra steps for a total of six steps. No tree that we can construct from species A–D requires fewer than four evolutionary changes, so the left-hand version in Fig. 23.8b is the best available hypothesis of evolutionary relatedness. In fact, this is the phylogeny for a sample of species from the largest group of animals on Earth, the Hexapoda—insects and their closest relatives (Fig. 23.8c).

In general, trees with fewer character changes are preferred to ones that require more because they provide the simplest explanation of the data. This approach is an example of **parsimony,** that is, choosing the simpler of two or more hypotheses to account for a given set of observations. When we use parsimony in phylogenetic reconstruction, we make the implicit assumption that evolutionary change is typically rare. Over time, most features of organisms stay the same—we have the same number of ears as our ancestors, the same number of fingers, and so on. Thus, biologists commonly prefer the phylogenetic tree requiring the fewest evolutionary steps.

In systematics, parsimony suggests counting character changes on a phylogenetic tree to find the simplest tree for the data (the one with the fewest number of changes). Each change corresponds to a mutation (or mutations) in an ancestral species, and the more changes or steps we propose, the more independent mutations we must also hypothesize.

Note also that it isn't necessary to make decisions in advance about which characters are homologies and which are analogies. We can construct all possible trees and then choose the one requiring the fewest evolutionary changes. This is a simple matter for the example in Fig. 23.8 because four species can be arranged into only 15 possible different trees. As the number of groups increases, however, the number of possible trees connecting them increases as well, and dramatically so. There are 105 trees for 5 groups and 945 trees for 6 groups, and there are nearly 2 million possible trees for 10. For 50 groups the possibilities balloon to 3×10^{76}! Clearly, computers are required to sort through all the possibilities.

As is true for all hypotheses, phylogenetic hypotheses can be supported strongly or weakly. Biologists use statistical methods to evaluate a given phylogenetic hypothesis. Available character data may not strongly favor any hypothesis. When support for a specific branching pattern is weak, biologists commonly depict the relationships as unresolved and show multiple groups diverging from one node, rather than just two. Such branching patterns are not meant to suggest that multiple species diverged simultaneously, but rather to indicate that we lack the data to choose unequivocally among several different hypotheses of relationship. In Part 2, we show unresolved branches in a number of groups. These shouldn't be read as admissions of defeat, but instead as problems awaiting resolution and opportunities for future research.

Molecular data complement comparative morphology in reconstructing phylogenetic history.

Trees can be built using anatomical features, but increasingly tree construction relies on molecular data. The amino acids at particular positions in the primary structure of a protein can be used, as can the nucleotides at specific positions along a strand of DNA.

From genealogy to phylogeny, tracing mutations in DNA or RNA sequences has revolutionized the reconstruction of historical genetic connections. Whether we are tracing the paternity of the children of Sally Hemings, mistress of Thomas Jefferson, identifying the origin of a recent cholera epidemic in Haiti, or placing baleen whales near the hippopotamus family in a phylogenetic tree of mammals, molecular data are a rich source of phylogenetic insight.

There is nothing about molecular data that provides a better record of history than does anatomical data; molecular data simply provide more details because there are more characters that can vary among the species. A sequence of DNA with hundreds or thousands of nucleotides can represent that many characters, as opposed to the tens of characters usually visible in morphological studies. Indeed, for microbes and viruses there is very little morphology available, so molecular information is critical for phylogenetic reconstruction. Once a gene or other stretch of DNA or RNA is identified that seems likely (based on previous studies of other species) to vary among the species to be studied, sequences are obtained and aligned to identify homologous nucleotide sites. Analyses of this kind commonly involve comparisons of sequences of about 1000 nucleotides from one or more genes. Increasingly, though, the availability of whole-genome sequences is changing the way we do molecular phylogenetics. Rather than comparing the sequences of a few genes, we compare the sequences of entire genomes.

The process of using molecular data is conceptually similar to the process described earlier for morphological data. Through comparison to an outgroup, we can identify derived and ancestral molecular characters (whether DNA nucleotides or amino acids in proteins) and generate the phylogeny on the basis of synapomorphies as before.

An alternative method of reconstruction is based on overall similarity rather than synapomorphies. Here, the premise is simple: The descendants of a recent common ancestor will have had relatively little time to evolve differences, whereas the descendants of an ancient common ancestor have had a lot of time to evolve differences. Thus, the extent of similarity (or distance) indicates how recently two groups shared a common ancestor.

Underpinning this approach is the assumption that the rate of evolution is constant. (Otherwise, a pair of taxa with a recent common ancestor could be more different than expected because of an unusually fast rate of evolution.) This rate-constancy assumption is less likely to be violated when we are using molecular data than when we are using morphological data. Recall from Chapter 21

FIG. 23.9 Phylogenetic trees of DNA sequences based on (a) synapomorphies and (b) overall similarity.

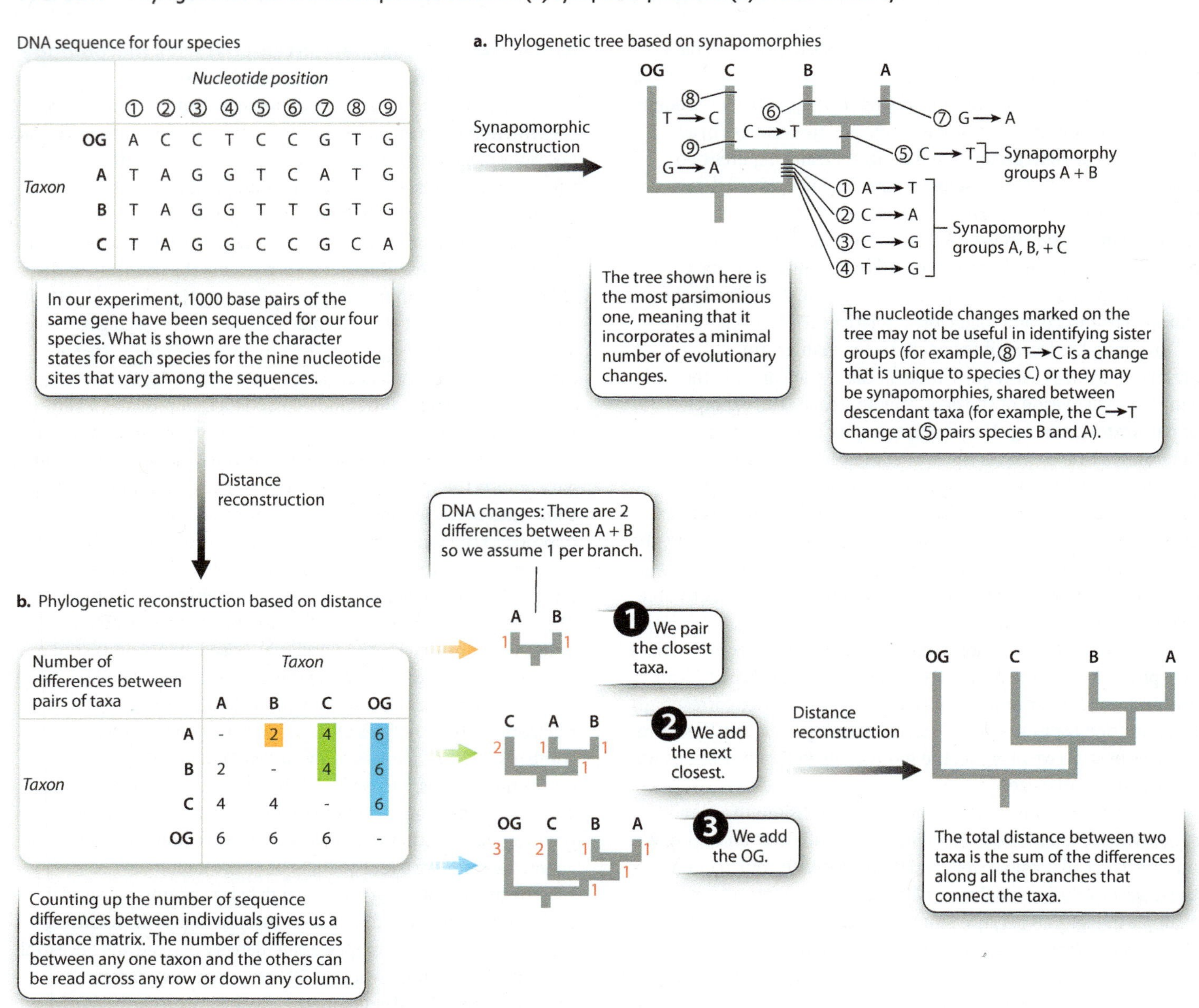

	Nucleotide position								
Taxon	①	②	③	④	⑤	⑥	⑦	⑧	⑨
OG	A	C	C	T	C	C	G	T	G
A	T	A	G	G	T	C	A	T	G
B	T	A	G	G	T	T	G	T	G
C	T	A	G	G	C	C	G	C	A

Number of differences between pairs of taxa	A	B	C	OG
A	-	2	4	6
B	2	-	4	6
C	4	4	-	6
OG	6	6	6	-

that the molecular clock is based on the observation of constant accumulation of genetic divergence through time. **Fig. 23.9** shows a simple DNA sequence dataset that we can analyze either on the basis of synapomorphies (Fig. 23.9a) or on the basis of distance (Fig. 23.9b). Note that both give the same result in this case.

Molecular data are often combined with morphological data, and each can also serve as an independent assessment of the other. Not surprisingly, results from analyses of each kind of data are commonly compatible, at least for plants and animals rich in morphological characters.

The single largest library of taxonomic information is GenBank, the National Institutes of Health's genetic data storage facility. As of this writing, GenBank gives users access to more than 100 billion observations (mostly nucleotides) collected under more than 430,000 taxonomic names. A growing internet resource is the Encyclopedia of Life, which is gathering additional biological information about species, including ecology, geographic distributions, photographs, and sounds in pages for individual species that are easy to navigate. Another web resource, the Tree of Life, provides information on phylogenetic trees for many groups of organisms.

Phylogenetic trees can help solve practical problems.

The sequence of changes on a tree from its root to its tips documents evolutionary changes that have accumulated through time. Trees suggest which groups are older than others, and which

HOW DO WE KNOW?

FIG. 23.10

Did an HIV-positive dentist spread the AIDS virus to his patients?

BACKGROUND In the late 1980s, several patients of a Florida dentist contracted AIDS. Molecular analysis showed that the dentist was HIV-positive.

HYPOTHESIS It was hypothesized that the patients acquired HIV during dental procedures carried out by the infected dentist.

METHOD Researchers obtained two HIV samples each (denoted 1 and 2 in the figure) from several people, including the dentist (Dentist 1 and Dentist 2), several of his patients (Patients A through G), and other HIV-positive individuals chosen at random from the local population (LP). In addition, a strain of HIV from Africa (HIVELI) was included in the analysis.

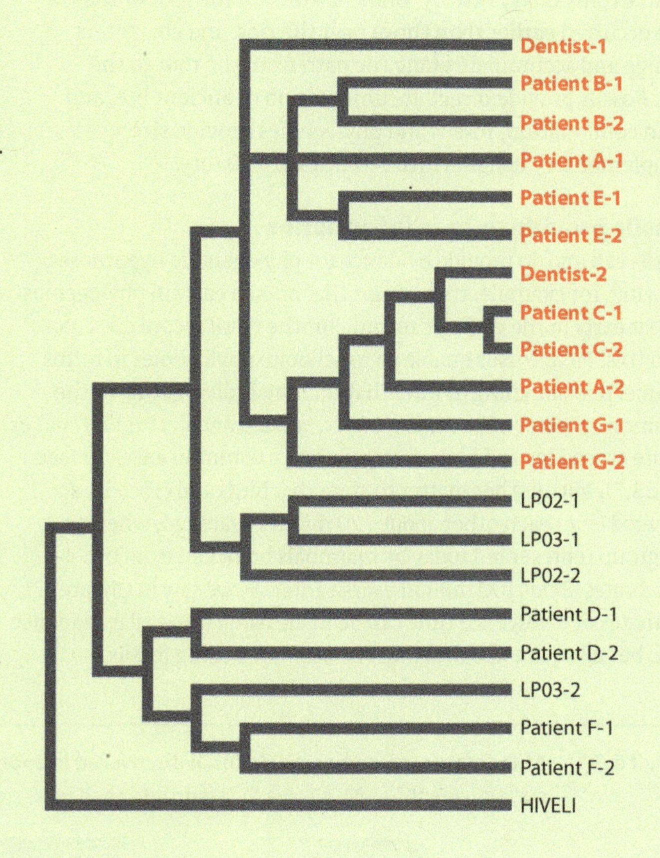

RESULTS Biologists constructed a phylogeny based on the nucleotide sequence of a rapidly evolving gene in the genome of HIV. Because the gene evolves so quickly, its mutations preserve a record of evolutionary relatedness on a very fine scale. HIV in some of the infected patients—patients A, B, C, E, and G—were more similar to the dentist's HIV than they were to samples from other infected individuals. Some patients' sequences, however, did not cluster with the dentist's, suggesting that these patients, D and F, had acquired their HIV infections from other sources.

CONCLUSION HIV phylogeny makes it highly likely that the dentist infected several of his patients. The details of how the patients were infected remain unknown, but rigidly observed safety practices make it unlikely that such a tragedy could occur again.

FOLLOW-UP WORK Phylogenies based on molecular sequence characters are now routinely used to study the origin and spread of infectious diseases, such as swine flu and Ebola.

SOURCE Hillis, D. M., J. P. Huelsenbeck, and C. W. Cunningham. 1994. "Application and Accuracy of Molecular Phylogenies." *Science* 264: 671–677.

traits came first and which followed later. Proper phylogenetic placement thus reveals a great deal about evolutionary history, and it can have practical consequences as well. For example, oomycetes, microorganisms responsible for potato blight and other important diseases of food crops, were long thought to be fungi because they look like some fungal species. The discovery, using molecular characters, that oomycetes belong to a very different group of eukaryotic organisms, has opened up new possibilities for understanding and controlling these plant pathogens. Similarly, in 2006, researchers used DNA sequences to identify the Malaysian parent population of a species of butterfly called lime swallowtails that had become an invasive species in the Dominican Republic, pinpointing the source populations from which natural enemies of this pest can be sought.

Phylogenetics solved a famous case in which an HIV-positive dentist in Florida was accused of infecting his patients (**Fig. 23.10**). HIV nucleotide sequences evolve so rapidly that biologists can build phylogenetic trees that trace the spread of specific strains from one individual to the next. Phylogenetic study of HIV present in samples from several infected patients, the dentist, and other individuals provided evidence that the dentist had, indeed, infected his patients.

Similarly, phylogenetic studies of influenza virus strains show their origins and subsequent movements among geographic regions and individual patients. Today, there is a growing effort to use specific DNA sequences as a kind of fingerprint or barcode for tracking biological material. Such information could quickly identify samples of shipments of meat as being from endangered species, or track newly emerging pests. The Consortium for the Barcode of Life has already accumulated species-specific DNA barcodes for more than 100,000 species. Phylogenetic evidence provides a powerful tool for evolutionary analysis and is useful across timescales ranging from months to the entire history of life, from the rise of epidemics to the origins of metabolic diversity.

23.3 THE FOSSIL RECORD

Phylogenies based on living organisms provide hypotheses about evolutionary history. Branches toward the root of the tree occurred earlier than those near the tips, and characters change and accumulate along the path from the root to the tips. Fossils provide direct documentation of ancient life, and so, in combination, fossils and phylogenies provide strong complementary insights into evolutionary history.

Fossils provide unique information.

Fossils can and do provide evidence for phylogenetic hypotheses, showing, for example, that groups that branch early in phylogenies appear early in the geologic record. But the fossil record does more than this. First, fossils enable us to calibrate phylogenies in terms of time. It is one thing to infer that mammals diverged from the common ancestor of birds, crocodiles, turtles, and lizards and snakes before crocodiles and birds diverged from a common ancestor (see Fig. 23.7), but another matter to state that birds and crocodiles diverged from each other about 220 million years ago, whereas the group represented today by mammals branched from other vertebrates about 100 million years earlier. As we saw in Chapter 21, estimates of divergence time can be made using molecular sequence data, but all such estimates must be calibrated using fossils.

The evolutionary relationship between birds and crocodiles highlights a second kind of information provided by fossils. Not only do fossils record past life, they also provide our only record of extinct species. The phylogeny in Fig. 23.7 contains a great deal of information, but it is silent about dinosaurs. Fossils demonstrate that dinosaurs once roamed Earth, and details of skeletal structure place birds among the dinosaurs in the vertebrate tree. Indeed, some remarkable fossils from China show that the dinosaurs most closely related to birds had feathers (**Fig. 23.11**).

A third, and also unique, contribution of fossils is that they place evolutionary events in the context of Earth's dynamic environmental history. Again, dinosaurs illustrate the point. As discussed in Chapter 1, geologic evidence from several continents suggests that a large meteorite triggered drastic changes in the global environment 66 million years ago, leading to the extinction of dinosaurs (other than birds). In fact, at five times in the past, large environmental disturbances sharply decreased Earth's biological diversity. These events, called mass extinctions, have played a major role in shaping the course of evolution.

Fossils provide a selective record of past life.

Fossils are the remains of once-living organisms, preserved through time in sedimentary rocks. If we wish to use fossils to complement phylogenies based on modern organisms, we must understand how fossils form and how the processes of formation govern what is and is not preserved.

For all its merits, the fossil record should not be thought of as a complete dictionary of everything that ever walked, crawled, or swam across our planet's surface. Fossilization requires burial, as when a clam dies on the seafloor and is quickly covered by sand, or a leaf falls to the forest floor and ensuing floods cover it in mud. Through time, accumulating sediments harden into sedimentary

FIG. 23.11 ***Microraptor gui,* a remarkable fossil discovered in approximately 125-million-year-old rocks from China.** The structure of its skeleton identifies *M. gui* as a dinosaur, yet it had feathers on its arms, tail, and legs. *Source: Nature/Xing Xu/Getty Images.*

FIG. 23.12 The Grand Canyon. Erosion has exposed layers of sedimentary rock that record Earth history. *Source: National Park Service.*

rocks such as those exposed so dramatically in the walls of the Grand Canyon (**Fig. 23.12**). If they are not buried, the remains of organisms are eventually recycled by biological and physical processes, and no fossil forms. In general, the fossil record of marine life is more completely sampled than that for land-dwelling creatures because marine habitats are more likely than those on land to be places where sediments accumulate and become rock. Thus, trees and elks living high in the Rocky Mountains have a low probability of fossilization, whereas clams and corals on the shallow seafloor are commonly buried and become fossils.

Biological factors contribute to the incompleteness of the fossil record. Most fossils preserve the hard parts of organisms, those features that resist decay after death. For animals, this usually means mineralized skeletons. Clams and snails that secrete shells of calcium carbonate have excellent fossil records. More than 80% of the clam species found today along California's coast also occur as fossils in sediments deposited during the past million years. In contrast, nematodes, tiny worms that may be the most abundant animals on Earth, have no mineralized skeletons and almost no fossil record. The wood and pollen of plants, which are made in part of decay-resistant organic compounds, enter the fossil record far more commonly than do flowers. And, among unicellular organisms, the skeleton-forming diatoms, radiolarians, and foraminiferans have exceptionally good fossil records, whereas most amoebas are unrepresented.

Together, then, the properties of organisms (do they make skeletons or other features that resist decay after death?) and environment (did the organisms live in a place where burial was likely?) determine the probability that an ancient species will be represented in the fossil record. Fortunately, for species that make mineralized skeletons and live on the shallow seafloor, the fossil record is very good, preserving a detailed history of evolution through time.

Organisms that lack hard parts can leave a fossil record in two other distinctive ways. Many animals leave tracks and trails as they move about or burrow into sediments. These **trace fossils,** from dinosaur tracks to the feeding trails of snails and trilobites, preserve a record of both anatomy and behavior (**Fig. 23.13**).

Organisms can also contribute **molecular fossils** to the rocks. Most biomolecules decay quickly after death. Proteins and DNA, for example, generally break down before they can be preserved, although, remarkably, a sizable fraction of the Neanderthal genome has been pieced together from DNA in 40,000-year-old

FIG. 23.13 Trace fossils. These footprints in 150-million-year-old rocks record both the structure and behavior of the dinosaurs that made them. *Source: José Antonio Hernaiz/age fotostock.*

bones. Other molecules, especially lipids like cholesterol, are more resistant to decomposition. Sterols, bacterial lipids, and some pigment molecules can accumulate in sedimentary rocks, documenting organisms that rarely form conventional fossils, especially bacteria and single-celled eukaryotes.

Rarely, unusual conditions preserve fossils of unexpected quality, including animals without shells, delicate flowers or mushrooms, fragile seaweeds or bacteria, even the embryos of plants and animals. For example, 505 million years ago, during the Cambrian Period, a sedimentary rock formation called the **Burgess Shale** accumulated on a relatively deep seafloor covering what is now British Columbia. Waters just above the basin floor contained little or no oxygen, so that when mud swept into the basin, entombed animals were sealed off from scavengers, disruptive burrowing activity, and even bacterial decay. For this reason, Burgess rocks preserve a remarkable sampling of marine life during the initial diversification of animals (**Fig. 23.14**).

FIG. 23.14 **A Burgess Shale fossil.** This fossil is *Opabinia regalis,* an extinct early relative of the arthropods, from the 505-million-year-old Burgess Shale. *Source: Courtesy of Smithsonian Institution. Photo by Chip Clark.*

The **Messel Shale** formed more recently—about 50 million years ago—in a lake in what is now Germany. Release of toxic gases from deep within the Messel lake suffocated local animals, and their carcasses settled into oxygen-poor muds on the lake floor. Fish, birds, mammals, and reptiles are preserved as complete and articulated skeletons, and mammals retain impressions of fur and color patterning (**Fig. 23.15**). Plants were also beautifully preserved, as were insects, some with a striking iridescence still intact. Messel rocks provide a truly outstanding snapshot of life on land as the age of mammals began.

In general, the fossil record preserves some aspects of biological history well and others poorly. Fossils provide a good sense of how the forms, functions, and diversity of skeletonized animals have changed over the past 500 million years. The same is true for land plants and unicellular organisms that form mineralized skeletons. These fossils shed light on major patterns of morphological evolution and diversity change through time; their geographic distributions record the movements of continents over millions of years; and the radiations and extinctions they document show how life responds to environmental change, both gradual and catastrophic.

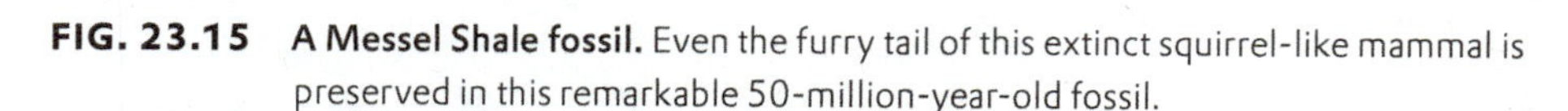

FIG. 23.15 **A Messel Shale fossil.** Even the furry tail of this extinct squirrel-like mammal is preserved in this remarkable 50-million-year-old fossil.

Geological data indicate the age and environmental setting of fossils.

How do we know the age of a fossil? Beginning in the nineteenth century, geologists recognized that groups of fossils change systematically from the bottom of a sedimentary rock formation to its top. As more of Earth's surface was mapped and studied, it became clear that certain fossils always occur in layers that lie beneath (and so are older than) layers that contain other species. From these patterns, geologists concluded that fossils mark time in Earth history. At first, geologists didn't know why fossils changed from one bed to the next, but after Darwin the reason became apparent:

Fossils record the evolution of life on Earth. They eventually mapped out the **geologic timescale,** the series of time divisions that mark Earth's long history (**Fig. 23.16**).

The layers of fossils in sedimentary rocks can tell us that some rocks are older than others, but they cannot by themselves provide an absolute age. Calibration of the timescale became possible with the discovery of radioactive decay. In Chapter 2, we discussed isotopes, variants of an element that differ from one another in the number of neutrons they contain. Many isotopes are unstable and spontaneously break down to form other, more stable isotopes. In the laboratory, scientists can measure how fast unstable isotopes decay. Then, by measuring the amounts of the

FIG. 23.16 The geologic timescale, showing major events in the history of life on Earth. *Sources: (top, left to right) Scott Orr/iStockphoto; Hans Steur, The Netherlands; T. Daeschler/VIREO; Reconstruction artwork: Mark A. Klingler/Carnegie Museum of Natural History, from the cover of Science, 25 MAY 2001 VOL 292, ISSUE 5521, Reprinted with permission from AAAS; dimair/Shutterstock; DEA/G. Cigolini/Getty Images; (bottom, left to right) Eye of Science/Science Source; Andrew Knoll, Harvard University; Antonio Guillén, Proyecto Agua, Spain; Andrew Knoll, Harvard University.*

Marine animals diversify.
First land plant fossils
First land vertebrates
First mammals
Age of the dinosaurs
Homo sapiens
Cambrian
Ordovician
Silurian
Devonian
Carboniferous
Permian
Triassic
Jurassic
Cretaceous
Paleogene
Neogene
Quaternary
Period
Paleozoic
Mesozoic
Cenozoic
Era
Phanerozoic
Eon
541 485 443 419 359 299 252 201 145 66 23 2.5 Present
Millions of years ago

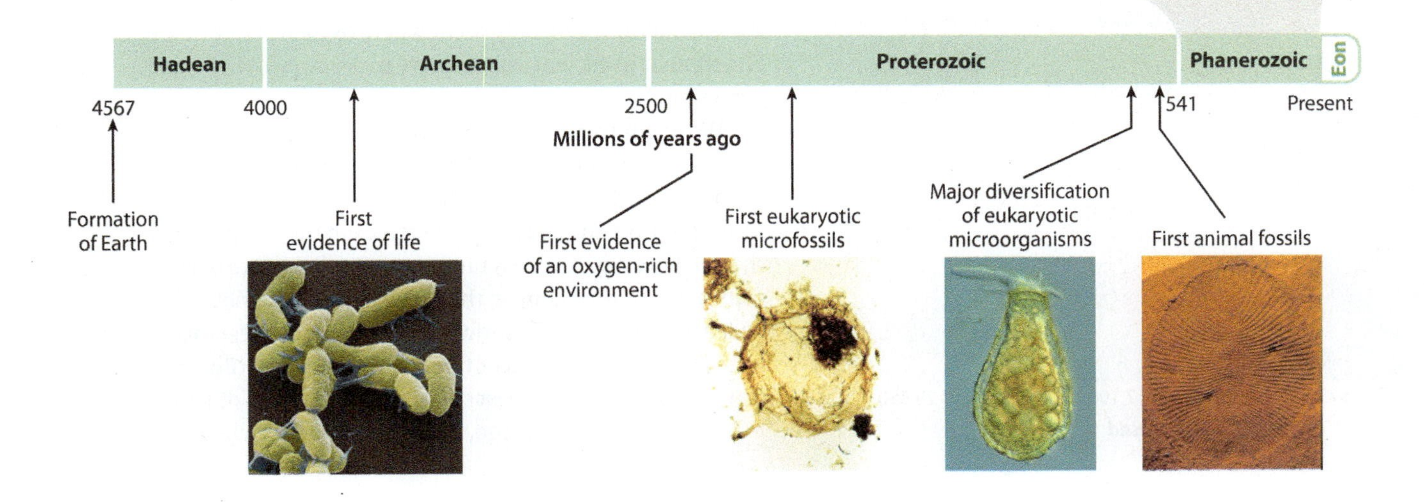

FIG. 23.17 **^{14}C decay.** Scientists can determine the age of relatively young materials such as wood and bone from the amount of ^{14}C they contain.

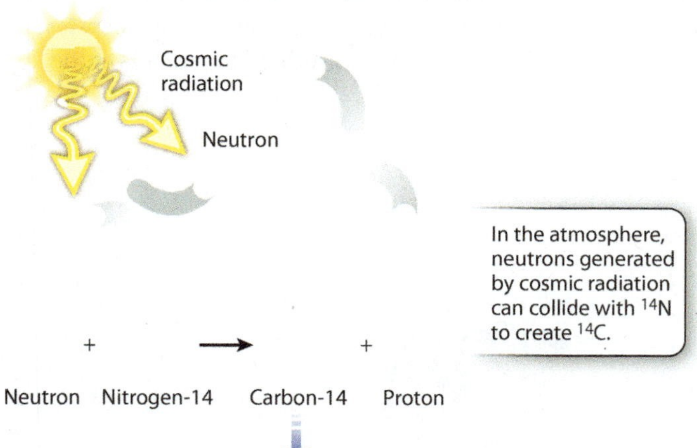

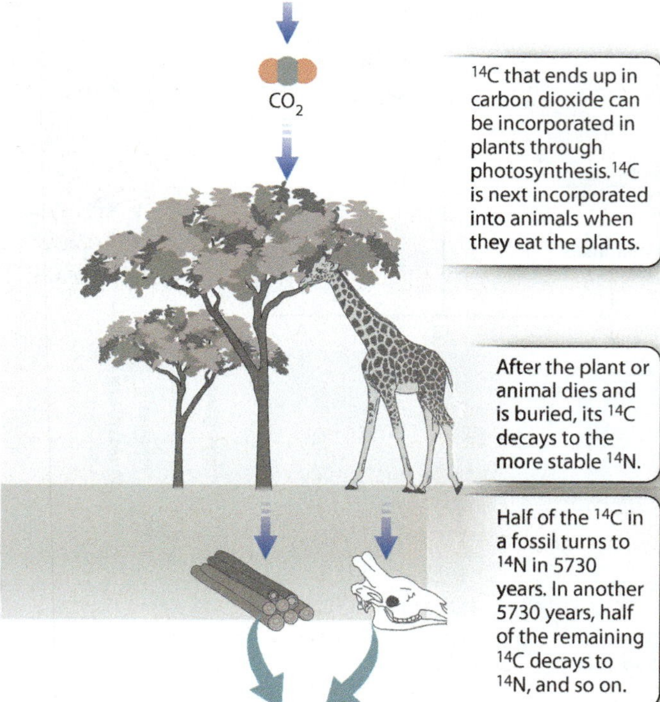

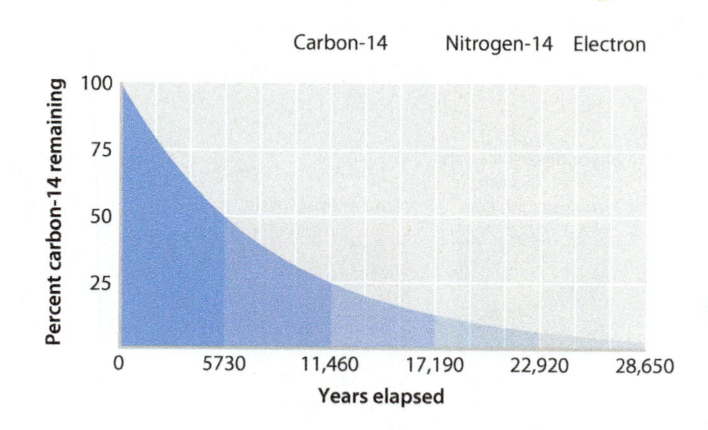

unstable isotope and its stable daughter inside a mineral, they can determine when the mineral formed.

Archaeologists commonly use the radioactive decay of the isotope carbon-14, or ^{14}C, to date wood and bone, a process called **radiometric dating.** As shown in **Fig. 23.17,** cosmic rays continually generate ^{14}C in the atmosphere, much of which is incorporated into atmospheric carbon dioxide. Through photosynthesis, carbon dioxide that contains ^{14}C is incorporated into wood, and animals incorporate small amounts of ^{14}C into their tissues when they eat plant material. After the organism's death, the unstable ^{14}C in these tissues begins to break down, losing an electron to form ^{14}N, a stable isotope of nitrogen. Laboratory measurements indicate that half of the ^{14}C in a given sample will decay to nitrogen in 5730 years, a period called its **half-life** (Fig. 23.17). Armed with this information, scientists can measure the amount of ^{14}C in an archaeological sample and, by comparing it to the amount of ^{14}C in a sample of known age—annual rings in trees, for example, or yearly growth of coral skeletons—determine the age of the sample.

Because its half-life is so short (by geological standards), ^{14}C is useful only in dating materials younger than 50,000 to 60,000 years. Beyond that, there is too little ^{14}C left to measure accurately. Older geological materials are commonly dated using the radioactive decay of uranium (U) to lead (Pb): ^{238}U, incorporated in trace amounts into the minerals of volcanic rocks, breaks down to ^{206}Pb with a half-life of 4.47 billion years; ^{235}U decays to ^{207}Pb with a half-life of 704 million years. Calibration of the geologic timescale is based mostly on the ages of volcanic ash interbedded with sedimentary rocks that contain key fossils, as well as volcanic rocks that intrude into (and so are younger than) layers of rock containing fossils. In turn, the ages of fossils provide calibration points for phylogenies.

The sedimentary rocks that contain fossils also preserve, encrypted in their physical features and chemical composition, information about the environment in which they formed. Sandstone beds, for example, may have rippled surfaces, like the ripples produced by currents that we see today in the sand of a seashore or lake margin. Pyrite (FeS_2), or fool's gold, forms when H_2S generated by anaerobic bacteria reacts with iron. As these conditions generally occur where oxygen is absent, pyrite enrichment in ancient sedimentary rocks can signal oxygen depletion.

We might think our moment in geologic time is representative of Earth as it has always existed, but nothing could be further from the truth. In the location and sizes of its continents, ocean chemistry, and atmospheric composition, the Earth we experience is unlike any previous state of the planet. Today, for example, the continents are distributed widely over the planet's surface, but 290 million years ago they were clustered in a supercontinent called Pangaea (**Fig. 23.18**). Oxygen gas permeates most surface environments of Earth today, but 3 billion years ago, there was no O_2 anywhere. And,

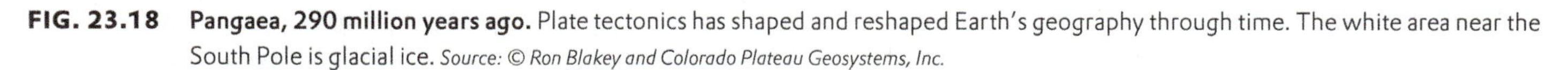

FIG. 23.18 Pangaea, 290 million years ago. Plate tectonics has shaped and reshaped Earth's geography through time. The white area near the South Pole is glacial ice. *Source: © Ron Blakey and Colorado Plateau Geosystems, Inc.*

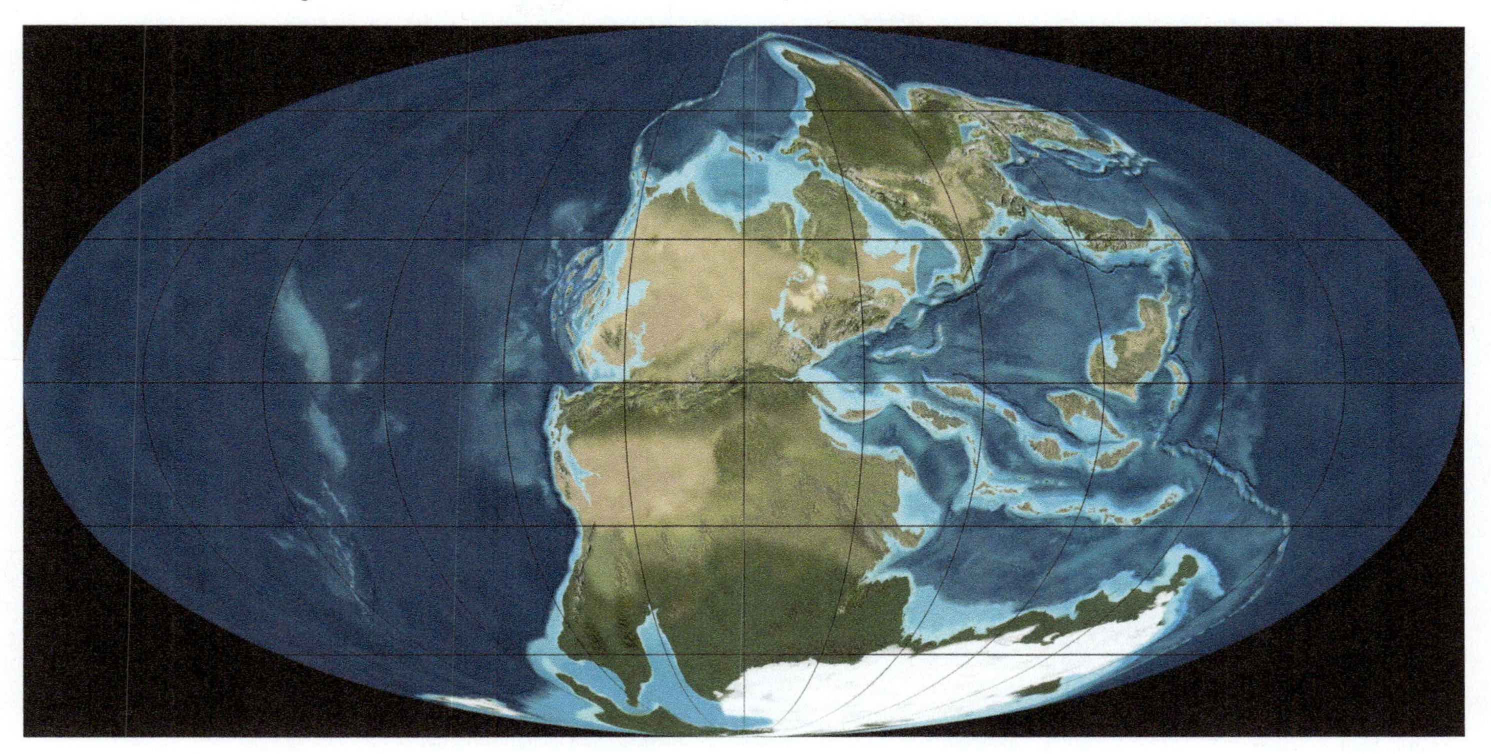

just 20,000 years ago, 2 km of glacial ice stood where Boston lies today. Sedimentary rocks record the changing state of Earth's surface over billions of years and show that life and environment have changed together through time, each influencing the other.

Fossils can contain unique combinations of characters.

Phylogenies hypothesize impressive morphological and physiological shifts through time—amphibians from fish, for example, or land plants from green algae. Do fossils capture a record of these transitions as they took place?

Let's begin with an example introduced earlier in this chapter. Phylogenies based on living organisms generally place birds as the sister group to crocodiles and alligators, but birds and crocodiles are decidedly different from each other in structure—birds have wings, feathers, toothless bills, and a number of other skeletal features distinct from those of crocodiles. In 1861, just two years after publication of *On the Origin of Species,* German quarry workers discovered a remarkable fossil that remains paleontology's most famous example of a transitional form. *Archaeopteryx lithographica,* now known from 11 specimens splayed for all time in fine-grained limestone, lived 150 million years ago. Its skeleton shares many characters with dromaeosaurs, a group of small, agile dinosaurs, but several features—its pelvis, its braincase, and, especially, its winglike forearms—are distinctly birdlike. Spectacularly, the fossils preserve evidence of feathers. *Archaeopteryx* clearly suggests a close relationship between birds and dinosaurs, and phylogenetic reconstructions that include information from fossils show that many of the characters found today in birds accumulated through time in their dinosaur ancestors. As noted earlier, even feathers first evolved in dinosaurs (**Fig. 23.19**).

Tiktaalik roseae and other skeletons in rocks deposited 375 to 362 million years ago record an earlier but equally fundamental transition: the colonization of land by vertebrates. Phylogenies show that all land vertebrates, from amphibians to mammals, are descended from fish. As seen in **Fig. 23.20,** *Tiktaalik* had fins, gills, and scales like other fish of its day, but its skull was flattened, more like that of a crocodile than a fish, and it had a functional neck and ribs that could support its body—features today found only in tetrapods. Along with other fossils, *Tiktaalik* captures key moments in the evolutionary transition from water to land, confirming the predictions of phylogeny.

→ **Quick Check 4** You have just found a novel vertebrate skeleton in 200-million-year-old rocks. How would you integrate this new fossil species into the phylogenetic tree depicted in Fig. 23.7?

FIG. 23.19 **Dinosaurs and birds.** A number of dinosaur fossils link birds phylogenetically to their closest living relatives, the crocodiles. The fossil at the bottom is *Archaeopteryx*. *Data from C. Zimmer, 2009, The Tangled Bank, Greenwood Village, CO: Roberts and Company; (photo) Jason Edwards/Getty Images.*

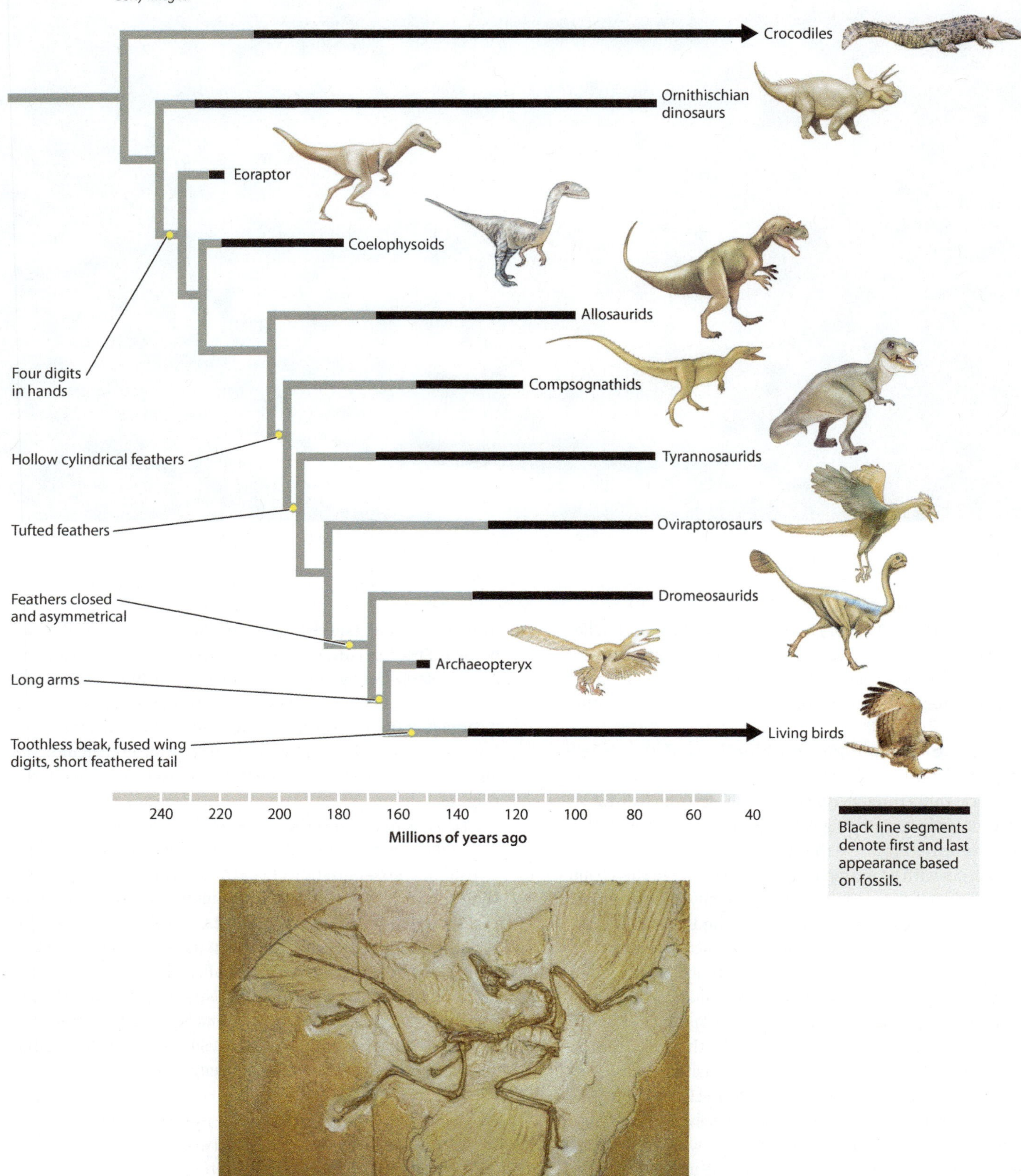

FIG. 23.20

Can fossils bridge the evolutionary gap between fish and tetrapod vertebrates?

BACKGROUND Phylogenies based on both morphological and molecular characters indicate that fish are the closest relatives of four-legged land vertebrates.

HYPOTHESIS Land vertebrates evolved from fish by modifications of the skeleton and internal organs that made it possible for them to live on land.

OBSERVATION Fossil skeletons 390 to 360 million years old show a mix of features seen in living fish and amphibians. Older fossils have fins, fishlike heads, and gills, and younger fossils have weight-bearing legs, skulls with jaws able to grab prey, and ribs that help ventilate lungs. Paleontologists predicted that key intermediate fossils would be preserved in 380–370-million-year-old rocks. In 2004, Edward Daeschler, Neil Shubin, and Farish Jenkins discovered *Tiktaalik,* a remarkable fossil that has fins, scales, and gills like fish, but wrist bones and fingers, an amphibian-like skull, and a true neck (which fish lack).

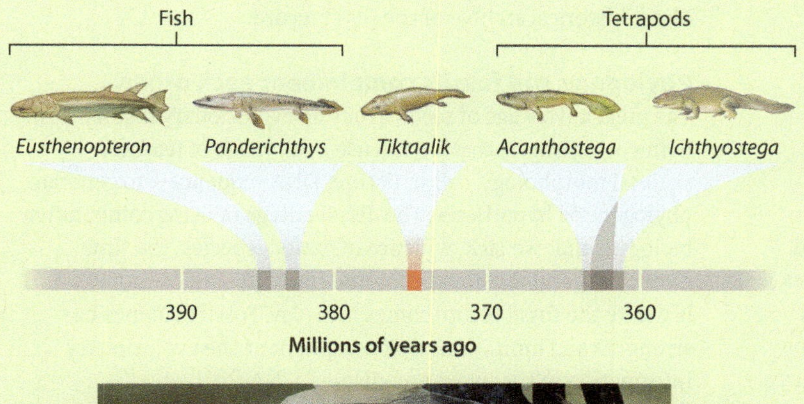

Photo source: Shubin Lab/ University of Chicago.

CONCLUSION Fossils confirm the phylogenetic prediction that tetrapod vertebrates evolved from fish by the developmental modification of limbs, skulls, and other features.

FOLLOW-UP WORK Research into the genetics of vertebrate development shows that the limbs of fish and amphibians are shaped by similar patterns of gene expression, providing further support for the phylogenetic connection between the two groups.

SOURCE Daeschler, E. B., N. H. Shubin, and F. A. Jenkins, Jr. 2006. "A Devonian Tetrapod-like Fish and the Evolution of the Tetrapod Body Plan." *Nature* 440:757–763.

Rare mass extinctions have altered the course of evolution.

Beginning in the 1970s, American paleontologist Jack Sepkoski scoured the paleontological literature, recording the first and last appearances in the geologic record for every genus of marine animals he could find. The results of this monumental effort are shown in **Fig. 23.21.** Sepkoski's diagram shows that animal diversity has increased over the past 540 million years: The biological diversity of animals in today's oceans may well be higher than it has ever been in the past. But it is also clear that animal evolution is not simply a history of gradual accumulation.

Repeatedly during the past 500 million years, animal diversity in the oceans dropped both rapidly and substantially, and extinctions also occurred on land. Known as **mass extinctions,** these events eliminated ecologically important taxa and thereby provided evolutionary opportunities for the survivors. In Chapter 22, we explored how species often radiate on islands where there is little or no competition or predation. For survivors, mass extinctions provided ecological opportunities on a grand scale.

The best-known mass extinction occurred 66 million years ago, at the end of the Cretaceous Period (Fig. 23.21). On land, dinosaurs disappeared abruptly, following more than 150 million years of dominance in terrestrial ecosystems. In the oceans, ammonites, cephalopod mollusks that had long been abundant predators, also became extinct, and most skeleton-forming microorganisms in the oceans disappeared as well. As discussed in Chapter 1, a large body of geologic evidence supports the hypothesis that this biological catastrophe was caused by the impact of a giant meteorite. By eliminating dinosaurs and much more of the biological diversity that had built up over millions of years on land, the mass extinction at the end of the Cretaceous Period also had the closely related effect of generating new evolutionary possibilities for terrestrial animals that survived the event, including mammals.

Before dinosaurs ever walked on Earth, the greatest of all mass extinctions occurred 252 million years ago, at the end of the Permian Period, when environmental catastrophe eliminated half of all families in the oceans and about 80% of all genera. Estimates of marine species loss run as high as 90%. Geologists have hypothesized that this mass extinction resulted from the catastrophic effects

FIG. 23.21 Mass extinctions. Though they eliminate much of the life found on Earth at the time, mass extinctions allow new groups to proliferate and diversify. *Source: Sepkoski's Online Genus Database. http://strata.geology.wisc.edu/jack/. Data from J. J. Sepkoski, Jr., 2002, "A Compendium of Fossil Marine Animal Genera," David Jablonski and Michael Foote (eds.), Bulletin of American Paleontology 363:1–560.*

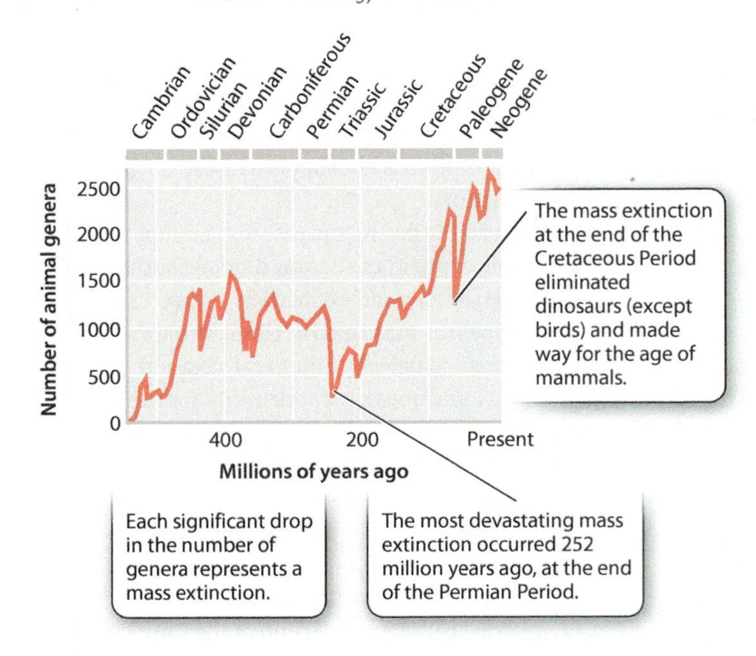

of massive volcanic eruptions. At the end of the Permian Period, most continents were gathered into the supercontinent Pangaea (see Fig. 23.18), and a huge ocean covered more than half of the Earth. Levels of oxygen in the deep waters of this ocean were low, the result of sluggish circulation and warm seawater temperatures. Then, a massive outpouring of ash and lava—a million times larger than any volcanic eruption experienced by humans—erupted across what is now Siberia. Enormous emissions of carbon dioxide and methane from the volcanoes caused global warming (so even less oxygen reached deep oceans) and ocean acidification (making it difficult for animals and algae to secrete calcium carbonate skeletons).

The three-way insult of lack of oxygen, ocean acidification, and global warming doomed many species on land and in the seas. Seascapes dominated for 200 million years by corals and shelled invertebrates called brachiopods disappeared. As ecosystems recovered from this mass extinction, they came to be dominated by new groups descended from survivors of the extinction. Bivalves and gastropods diversified, new groups of arthropods radiated, including the ancestors of the crabs and shrimps we see today, and surviving sea anemones evolved a new capacity to make skeletons of calcium carbonate, resulting in the corals that build modern reefs. In short, mass extinction reset the course of evolution, much as it did and the end of the Cretaceous Period.

When you stroll through a zoo or snorkel above a coral reef, you are not simply seeing the products of natural selection played out over Earth history. You are encountering the descendants of Earth's biological survivors. Current biological diversity reflects the interplay through time of natural selection and rare massive perturbations to ecosystems on land and in the sea. The fossil record provides a silent witness to our planet's long and complex evolutionary history.

23.4 COMPARING EVOLUTION'S TWO GREAT PATTERNS

The diversity of life we see today is the result of evolutionary processes playing out over geologic time. Evolutionary process can be studied by experiment, both in the field and in the laboratory, but evolutionary history is another matter. There is no experiment we can do to determine why the dinosaurs became extinct—we cannot rerun the events of 66 million years ago, this time without the meteorite impact. The history of life must be reconstructed from evolution's two great patterns: the nested similarity observed in the forms and molecular sequences of living organisms, and the direct historical archive of the fossil record.

Phylogeny and fossils complement each other.

The great advantage of reconstructing evolutionary history from living organisms is that we can use a full range of features—skeletal morphology, cell structure, DNA sequence—to generate phylogenetic hypotheses. The disadvantage of using comparative biology is that we lack evidence of extinct species, the time dimension, and the environmental context. This, of course, is where the fossil record comes into play. Fossil evidence has strengths and limitations that complement the evolutionary information in the living organisms.

We can use phylogenetic methods based on DNA sequences to infer that birds and crocodiles are closely related, but only fossils can show that the evolutionary link between birds and crocodiles runs through dinosaurs. And only the geologic record can show that mass extinction removed the dinosaurs, paving the way for the emergence of modern mammals. Paleontologists and biologists work together to understand evolutionary history. Biology provides a functional and phylogenetic framework for the interpretation of fossils, and fossils provide a record of life's history in the context of continual planetary change.

Agreement between phylogenies and the fossil record provides strong evidence of evolution.

Phylogenies based on morphological or molecular comparisons of living organisms make hypotheses about the timing of evolutionary changes through Earth history. We humans are a case in point. As we discuss in Chapter 24, comparisons of DNA sequences suggest that chimpanzees are our closest living relatives. This is hardly surprising, as simple observation shows that chimpanzees and humans share many features. However, among other differences, chimpanzees have smaller stature, smaller brains, long arms that facilitate knuckle-walking (the arms help support the body as the chimpanzee moves forward), a more prominent snout, and larger teeth.

How did these differences accumulate over the 7 million years or so since the human and chimpanzee lineages split? Fossils painstakingly unearthed over the past century show, for example, that the ability to walk upright evolved before the large human brain. Lucy, a famous specimen of *Australopithecus afarensis* dating to about 3.2 million years ago, was fully bipedal but had a brain the size of a chimpanzee's. Fossils like Lucy that are a mix of ancestral (brain size) and derived (bipedalism) characters provide strong support for our phylogenetic tree that shows that humans and chimpanzees are sister taxa.

The agreement of comparative biology and the fossil record can be seen at all scales of observation. Humans form one tip of a larger branch that contains all members of the primate family. Primates, in turn, are nested within a larger branch occupied by mammals, and mammals nest within a still larger branch containing all vertebrate animals, which include fish. This arrangement predicts that the earliest fossil fish should be older than the earliest fossil mammals, the earliest fossil mammals older than the earliest primates, and the earliest primates older than the earliest humans. This is precisely what the fossil record shows (**Fig. 23.22**).

The agreement between fossils and phylogenies can be seen again and again when we examine different branches of the tree of life or, for that matter, the tree as a whole. All phylogenies indicate that microorganisms diverged early in evolutionary history, and mammals, flowering plants, and other large complex organisms diverged more recently. The tree's shape implies that diversity has accumulated through time, beginning with simple organisms and later adding complex macroscopic forms.

The geologic record shows the same pattern. For nearly 3 billion years of Earth history, microorganisms dominate the fossil record, with the earliest animals appearing about 600 million years ago, the earliest vertebrate animals 520 million years ago, the earliest tetrapod vertebrates about 360 million years ago, the earliest mammals 210 million years ago, the earliest primates perhaps 55 million years ago, the earliest fossils of our own species a mere 200,000 years ago. Similarly, if we focus on photosynthetic organisms, we find a record of photosynthetic bacteria beginning at least 3500 million years ago, algae 1200 million years ago, simple land plants 470 million years ago, seed plants 370 million years ago, flowering plants about 140 million years ago, and the earliest grasses 70 million years ago.

FIG. 23.22 Phylogeny and fossils of vertebrates. The branching order of the phylogeny corresponds to the order of appearance of each group in the fossil record.

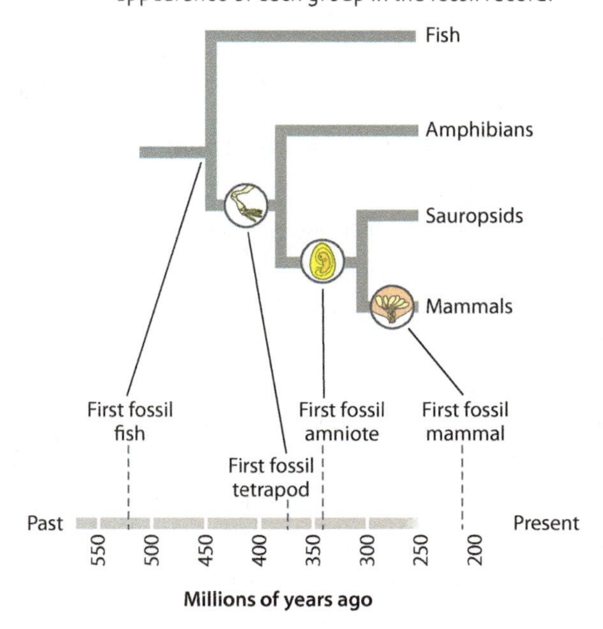

In Part 2, we explore the evolutionary history of life in some detail. Here, it is sufficient to draw the key general conclusion: The fact that comparative biology and fossils, two complementary but independent approaches to reconstructing the evolutionary past, yield the same history is powerful evidence of evolution. ■

Core Concepts Summary

23.1 A phylogenetic tree is a reasoned hypothesis of the evolutionary relationships among organisms.

The nested pattern of similarities seen among organisms is a result of descent with modification and can be represented as a phylogenetic tree. page 465

The order of branches on a phylogenetic tree indicates the sequence of events in time. page 465

Sister groups are more closely related to one another than they are to any other group. page 465

A node is a branching point on a tree, and it can be rotated without changing evolutionary relationships. page 466

A monophyletic group includes all the descendants of a common ancestor, and it is considered a natural grouping of organisms based on shared ancestry. page 466

A paraphyletic group includes some, but not all, of the descendants of a common ancestor. page 466

A polyphyletic group includes organisms from distinct groups based on shared characters, but it does not include a common ancestor. page 466

Organisms are classified into domain, kingdom, phylum, class, order, family, genus, and species. page 467

23.2 A phylogenetic tree is built on the basis of shared derived characters.

Characters, or traits, existing in different states are used to build phylogenetic trees. page 468

Homologies are similarities based on shared ancestry, while analogies are similarities based on independent adaptations. page 468

Homologies can be ancestral, unique to a particular group, or present in some, but not all, of the descendants of a common ancestor (shared derived characters). page 468

Only shared derived characters, or synapomorphies, are useful in constructing a phylogenetic tree. page 469

Molecular data provide a wealth of characters that complement other types of information in building phylogenetic trees. page 471

Phylogenetic trees can be used to understand evolutionary relationships of organisms and solve practical problems, such as how viruses evolve over time. page 473

23.3 The fossil record provides direct evidence of evolutionary history.

Fossils are the remains of organisms preserved in sedimentary rocks. page 474

The fossil record is imperfect because fossilization requires burial in sediment, sediments accumulate episodically and discontinuously, and fossils typically preserve only the hard parts of organisms. page 474

Radioactive decay of certain isotopes of elements provides a means of dating rocks. page 477

Archaeopteryx and *Tiktaalik* are two fossil organisms that document, respectively, the bird–dinosaur transition and the fish–tetrapod transition. page 479

The history of life is characterized by five mass extinctions that changed the course of evolution. page 481

The extinction at the end of the Cretaceous Period 66 million years ago led to the extinction of the dinosaurs (other than birds). page 481

The extinction at the end of the Permian Period 252 million years ago is the largest documented mass extinction in the history of Earth. page 481

23.4 Phylogeny and fossils provide independent and corroborating evidence of evolution.

Phylogeny makes use of living organisms, and the fossil record supplies a record of species that no longer exist, absolute dates, and environmental context. page 482

Data from phylogeny and fossils are often in agreement, providing strong evidence for evolution. page 482

Self-Assessment

1. Draw a phylogenetic tree of three groups of organisms and explain how a nested pattern of similarity can be seen in the tree and how it might arise.
2. Distinguish among monophyletic, paraphyletic, and polyphyletic groups, and give an example of each.
3. List the levels of classification, from the least inclusive (species) to the most inclusive (domain).
4. Define "homology" and "analogy" and describe two traits that are homologous and two that are analogous.
5. Name a type of homology that is useful in building phylogenetic trees and explain why this kind of homology, and not others, is useful.
6. Describe three ways that an organism can leave a record in sedimentary rocks and explain why this means that there are gaps in the fossil record.
7. Explain how the fossil record can be used to determine both the relative and the absolute timescales of past events.
8. Describe the significance of *Archeopteryx* and *Tiktaalik*.
9. Describe how mass extinctions have shaped the ecological landscape.

CHAPTER 24

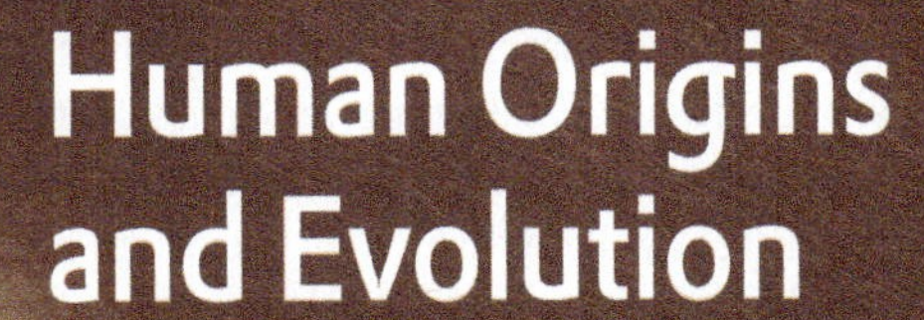

Human Origins and Evolution

Core Concepts

24.1 Anatomical, molecular, and fossil evidence shows that the human lineage branches off the great apes tree.

24.2 Phylogenetic analysis of mitochondrial DNA and the Y chromosome shows that our species arose in Africa.

24.3 During the 5–7 million years since the most recent common ancestor of humans and chimpanzees, our lineage acquired a number of distinctive features.

24.4 Human history has had an important impact on patterns of genetic variation in our species.

24.5 Culture is a potent force for change in modern humans.

Charles Darwin carefully avoided discussing the evolution of our own species in *On the Origin of Species*. Instead, he wrote only that he saw "open fields for far more important researches," and that "[l]ight will be thrown on the origin of man and his history." Darwin, an instinctively cautious man, realized that the ideas presented in *On the Origin of Species* were controversial enough without his adding humans to the mix. He presented his ideas on human evolution to the public only when he published *The Descent of Man* 12 years later, in 1871.

As it turned out, Darwin's delicate sidestepping of human origins had little effect. The initial print run of *The Origin* sold out on the day of publication, and the public was perfectly capable of reading between the lines. The Victorians found themselves wrestling with the book's revolutionary message: that humans are a species of ape.

Darwin's conclusions remain controversial to this day among the general public, but they are not controversial among scientists. The evidence that humans are descended from a line of apes whose modern-day representatives include gorillas and chimpanzees is compelling. We know now that about 5–7 million years ago the family tree of the great apes split, one branch ultimately giving rise to chimpanzees and the other to our species. It is those 5–7 million years that hold the key to our humanity. It was over this period—brief by evolutionary standards—that the attributes that make our species so remarkable arose. This chapter discusses what happened over those 5–7 million years and how we came to be the way we are.

24.1 THE GREAT APES

We can approach the question of our place in the tree of life in three different ways: through comparative anatomy, through molecular analysis, and through the fossil record. In this section, we use data from all three sources as we apply the standard methods of phylogenetic reconstruction (Chapter 23) to figure out the evolutionary relationships between humans and other mammals.

Comparative anatomy shows that the human lineage branches off the great apes tree.

There are about 400 species of **primates,** which include prosimians (lemurs, bushbabies), monkeys, and apes (**Fig. 24.1**). All primates share a number of general features that distinguish them from other mammals: nails rather than claws together with a versatile thumb allow objects to be manipulated more dexterously, and eyes on the front of the face instead of the side allow stereoscopic (three-dimensional) vision.

Prosimians are thought to represent a separate primate group from the one that gave rise to humans. Lemurs, which today are confined to the island of Madagascar, are thus only distantly related to humans. Monkeys underwent independent bouts of evolutionary change in the Americas and in Africa and Eurasia, so the family tree is split along geographic lines into New World and Old World monkeys. Though both groups have evolved similar habits, there are basic distinctions. There are differences between the teeth of the two groups, and in New World monkeys the nostrils tend to be widely spaced, whereas in Old World species they are closer together.

One line of Old World monkeys gave rise to the apes, which lack a tail and show more sophisticated behaviors than other monkeys. The apes are split into two groups, the lesser and the great apes (**Fig. 24.2**). Lesser apes include the fourteen species of gibbon, all of which are found in Southeast Asia. The great apes include orangutans, gorillas, chimpanzees, and humans. Taxonomists classify all the descendants of a specified common

FIG. 24.1 The primate family tree. This tree shows the evolutionary relationships of prosimians, monkeys, and apes. *Sources: (left to right) George Holton/Science Source; Penelope Dearman/Getty Images; Daily Mail/Rex/Alamy; Patrick Shyu/Getty Images; Yellow Dog Productions/Getty Images.*

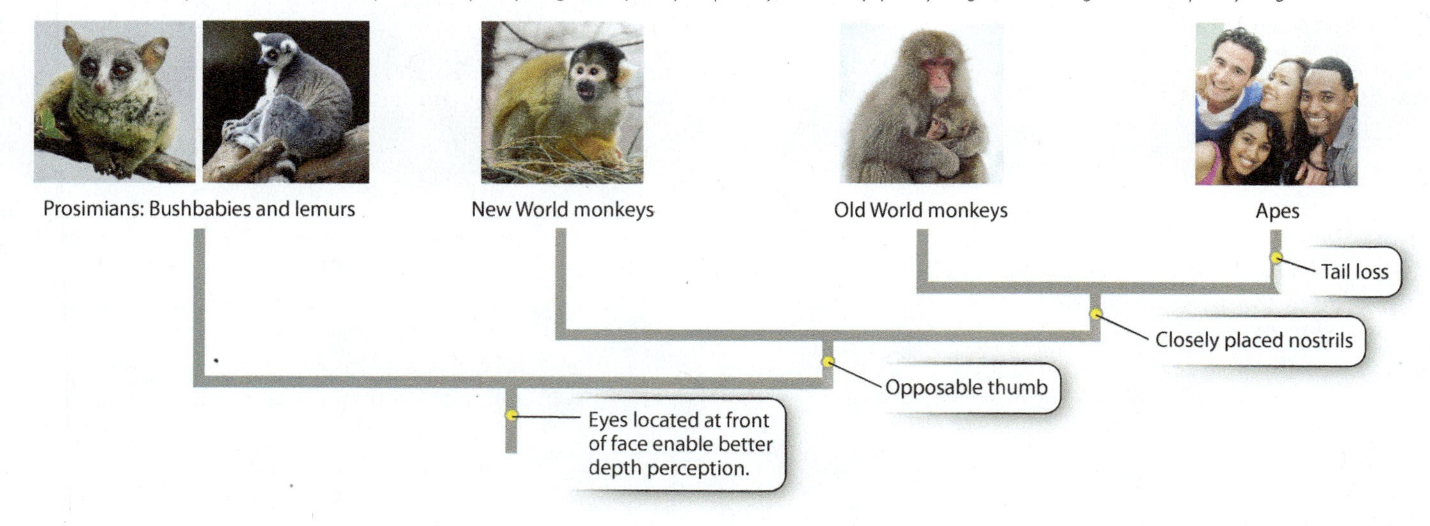

FIG. 24.2 The family tree of the lesser and great apes. The apes consist of two major groups, the lesser apes and great apes. The great apes group includes humans. *Sources: (left to right) Zoonar/K. Jorgensen/age fotostock; S Sailer/A Sailer/age fotostock; J & C Sohns/age fotostock; Michael Dick/Animals Animals–Earth Scenes; FLPA/Jurgen & Christi/age fotostock; Yellow Dog Productions/Getty Images.*

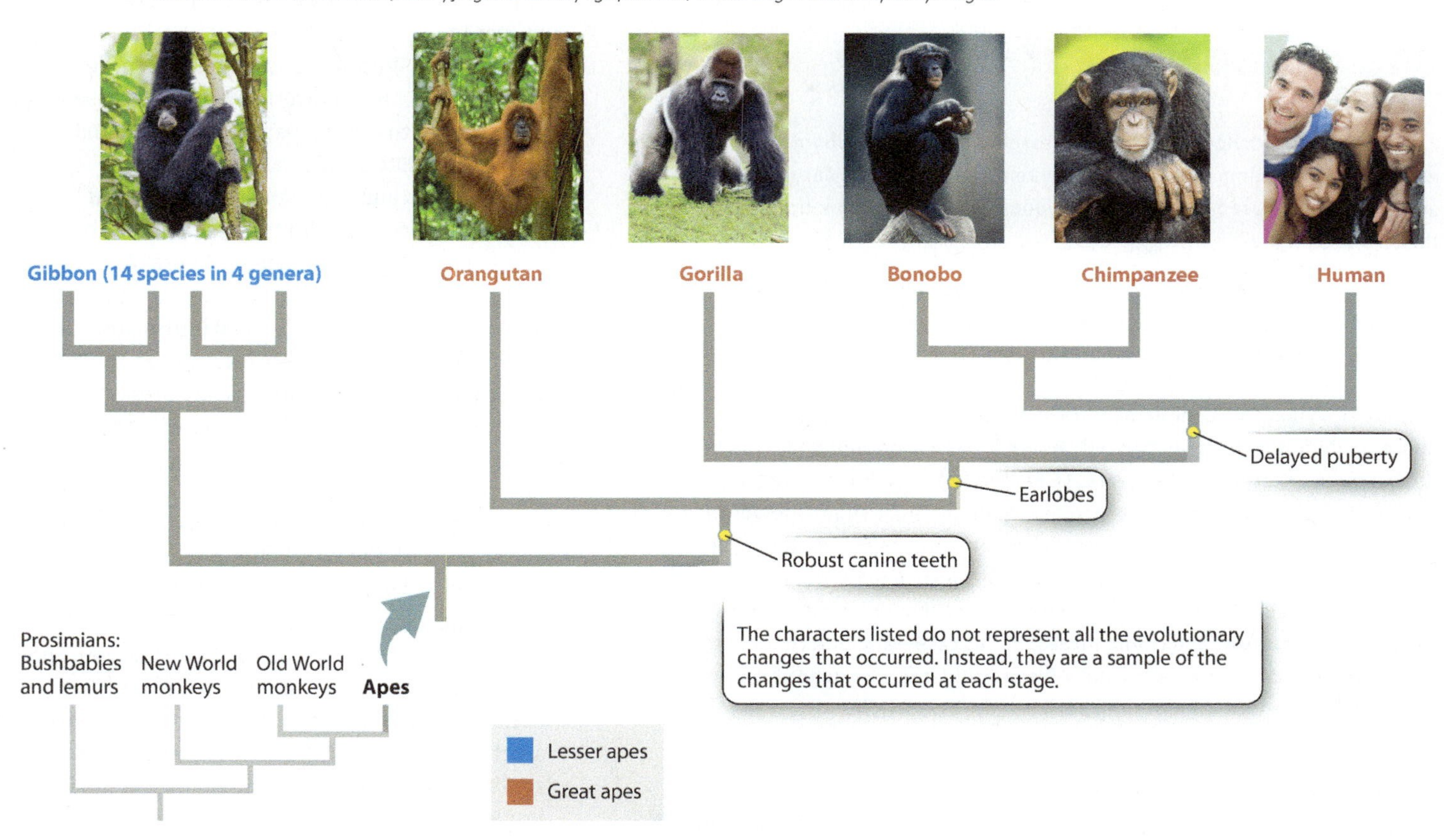

ancestor as belonging to a monophyletic group (Chapter 23). Thus, humans, blue whales, and hedgehogs are all mammals because all three are descended from the first mammal, the original common ancestor of all mammals. Because human ancestry can be traced to the common ancestor of orangutans, gorillas, and chimpanzees, humans, too, are a member of the monophyletic great ape group.

Molecular analysis reveals that the human lineage split from the chimpanzee lineage about 5–7 million years ago.

Which great ape is most closely related to humans? That is, which is our sister group? Traditional approaches of reconstructing evolutionary history by comparing anatomical features failed to determine which of two candidates, gorillas or chimpanzees, is the sister group to humans. It was only with the introduction of molecular methods of assessing evolutionary relationships—through the comparison of DNA and amino acid sequences from the different species—that we had the answer. Our closest relative is the chimpanzee (Fig. 24.2), or, more accurately, the chimpanzees, plural, because there are two closely related chimpanzee species, the smaller of which is often called the bonobo.

Just how closely are humans and chimpanzees related? To answer this question, we need to know the timing of the evolutionary split that led to chimpanzees along one fork and to humans along the other. As we saw in Chapter 21, DNA sequence differences accumulate between isolated populations or species, and they do so at a more or less constant rate. As a result, the extent of sequence difference between two species is a good indication of the amount of time they have been separate, that is, the amount of time since their last common ancestor.

The first thorough comparison of DNA molecules between humans and chimpanzees was carried out before the advent of DNA sequencing methods by Mary-Claire King and Allan Wilson at the University of California at Berkeley (**Fig. 24.3**). One of their methods of measuring molecular differences between species relied on DNA–DNA hybridization (Chapter 12). Two complementary strands of DNA in a double helix can be separated by heating the sample. If the two strands are not perfectly complementary, as is the case if there is a base-pair mismatch (for example, a G paired with a T rather than with a C), less heat is required to separate the strands.

King and Wilson used this observation to examine the differences between a human strand and the corresponding

HOW DO WE KNOW?

FIG. 24.3

How closely related are humans and chimpanzees?

BACKGROUND Phylogenetic analysis based on anatomical characteristics had established that chimpanzees are closely related to humans. Mary-Claire King and Allan Wilson used molecular techniques to determine *how* closely related the two species are.

HYPOTHESIS Despite marked anatomical and behavioral differences between the two species, the genetic distance between the two is small, implying a relatively recent common ancestor.

METHOD When two complementary strands of DNA are heated, the hydrogen bonds pairing the two helices are broken at around 95°C and the double helix denatures, or separates (Chapter 12). Two complementary strands with a few mismatches separate at a temperature slightly lower than 95°C because fewer hydrogen bonds are holding the helix together. More mismatches between the two sequences results in an even lower denaturation temperature. Using hybrid DNA double helices with one strand contributed by each species—humans and chimpanzees, in this case—and determining their denaturation temperature, King and Wilson could infer the genetic distance (the extent of genetic divergence) between the two species.

Perfect complementarity: 95°C denaturation

Some mismatch: 93°C denaturation

More mismatch: 91°C denaturation

RESULTS King and Wilson found that human–chimpanzee DNA molecules separated at a temperature approximately 1°C lower than the temperature at which human–human DNA molecules separate. This difference could be calibrated on the basis of studies of other species whose genetic distances were known from other methods. The DNA of humans differs from that of chimpanzees by about 1%.

CONCLUSION AND INTERPRETATION King and Wilson noted the discrepancy between the extent of genetic divergence (small) and the extent of anatomical and behavioral divergence (large) between humans and chimpanzees. They suggested that one way in which relatively little genetic change could produce extensive phenotypic change is through differences in gene regulation (Chapter 19). A small genetic change in a control region responsible for switching a gene on and off might have major consequences for the organism.

FOLLOW-UP WORK The sequences of the chimpanzee and human genomes allow us to compare the two sequences directly, and these data confirm King and Wilson's observations.

SOURCE King, M.-C., and A. C. Wilson. 1975. "Evolution at Two Levels in Humans and Chimpanzees." *Science* 188:107–116.

chimpanzee strand. They inferred the extent of DNA sequence divergence from the melting temperature and made a striking discovery: Human and chimpanzee DNA differ in sequence by just 1%.

King and Wilson's conclusions have proved robust as ever more powerful techniques have been applied to the comparison of human and chimpanzee DNA. Today, after sequencing entire human and chimpanzee genomes and literally counting the differences between the two sequences, we find that the figure of 1% still approximately holds. Genome sequencing studies give us superbly detailed information on the differences and similarities between the two genomes, revealing regions that are present in one species but not the other, and which parts of the genomes have evolved more rapidly than others.

Because the amount of sequence difference is correlated with the length of time the two species have been isolated, we can convert the divergence results into an estimate of the timing of the split between the human and chimpanzee lineages. That split occurred about 5–7 million years ago. All the extraordinary characteristics that set our species apart from the rest of the natural world—those attributes that are ours and ours alone—arose in just 5–7 million years.

→ **Quick Check 1** Did humans evolve from chimpanzees? Explain.

The fossil record gives us direct information about our evolutionary history.

Molecular analysis is a powerful tool for comparing species and populations within species. It allows us to compare humans and chimpanzees and look at differences among groups of humans or groups of chimpanzees. However, for a full picture of human evolution, we must turn to fossils (Chapter 23).

For the first several million years of human evolution, all the fossils from the human lineage are found in Africa. This is not surprising. Charles Darwin himself noted that it was likely that the human lineage originated there, as humans' two closest relatives, chimpanzees and gorillas, live only in Africa. The fossil material varies in quality, and a great deal of ingenuity is often required to reconstruct the appearance and attributes of an individual from fragmentary fossil material. It's hard to determine whether two fossil specimens with slight differences belong to the same or different species. And, of course, interbreeding, the criterion that researchers most often use to define a species (Chapter 22), cannot readily be applied to fossils. As a result, experts often disagree over the details of the human fossil record—whether or not, for example, a specimen belongs to a particular species. Regardless, several robust general conclusions may be drawn.

The many different species, including modern humans, that have arisen on the human side of the split since the human/chimpanzee common ancestor are called **hominins.** Let's start with the earliest known hominin, *Sahelanthropus tchadensis* (**Fig. 24.4**). Discovered in Chad in 2002, the skull of *S. tchadensis* combines both modern (human) and ancestral features. *S. tchadensis,* which has been dated to about 7 million years ago and has a chimpanzee-sized brain but hominin-type brow ridges, probably lived shortly after the split between the hominin and chimpanzee lineages, although this conclusion is controversial because of difficulties in interpreting whether a fossil is part of a newly diverged lineage or just another specimen of the ancestral lineage.

An important early hominin, dating from about 4.4 million years ago, is a specimen of *Ardipithecus ramidus* from Ethiopia. This individual, known as **Ardi,** was capable of walking upright, using two legs on the ground but all four limbs in the trees.

An unusually complete early hominin fossil, **Lucy,** found in 1974 at Hadar, Ethiopia, represents the next step in the evolution of hominin gait: She was fully **bipedal,** habitually walking upright (**Fig. 24.5**). Her name comes from the Beatles' song "Lucy in the Sky with Diamonds," which was playing in the paleontologists' field camp when she was unearthed. This fossil dates from around 3.2 million years ago. Lucy was a member of the species *Australopithecus afarensis* and was much smaller than modern humans, less than 4 feet tall, and had a considerably smaller brain (even when the difference in body size is taken into consideration). In many ways, then, Lucy was similar to the common ancestor of humans and chimpanzees, except that she was bipedal.

FIG. 24.4 The skull of *Sahelanthropus tchadensis.* This skull, which is about 7 million years old and was found in Chad, has both human and chimpanzee features. *Source: Richard T. Nowitz/Science Source.*

FIG. 24.5 Lucy, a specimen of *Australopithecus afarensis.* Lucy was fully bipedal, establishing that the ability to walk upright evolved at least 3.2 million years ago. *Source: The Natural History Museum, London/The Image Works.*

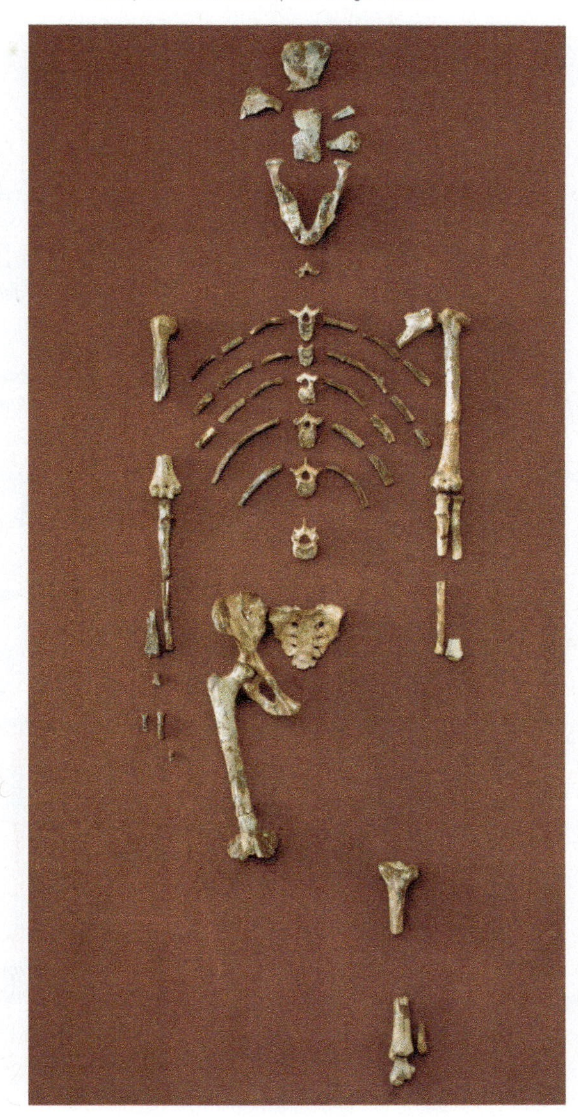

FIG. 24.6 Hominin lineages. Three main lineages, *Ardipithecus, Australopithecus,* and *Homo* existed at various and sometimes overlapping times in history. The *Homo* lineage led to modern humans, *Homo sapiens*. *Sources: Data from R. G. Klein, 2009, The Human Career, Chicago: University of Chicago Press, p. 244; Homo floresiensis skull from R. D. Martin, A. M. MacLarnon, J. L. Phillips, L. Dussubieux, P. R. Williams and W. B. Dobyns, 2006, Comment on "The Brain of LB1, Homo floresiensis," Science 312:999.*

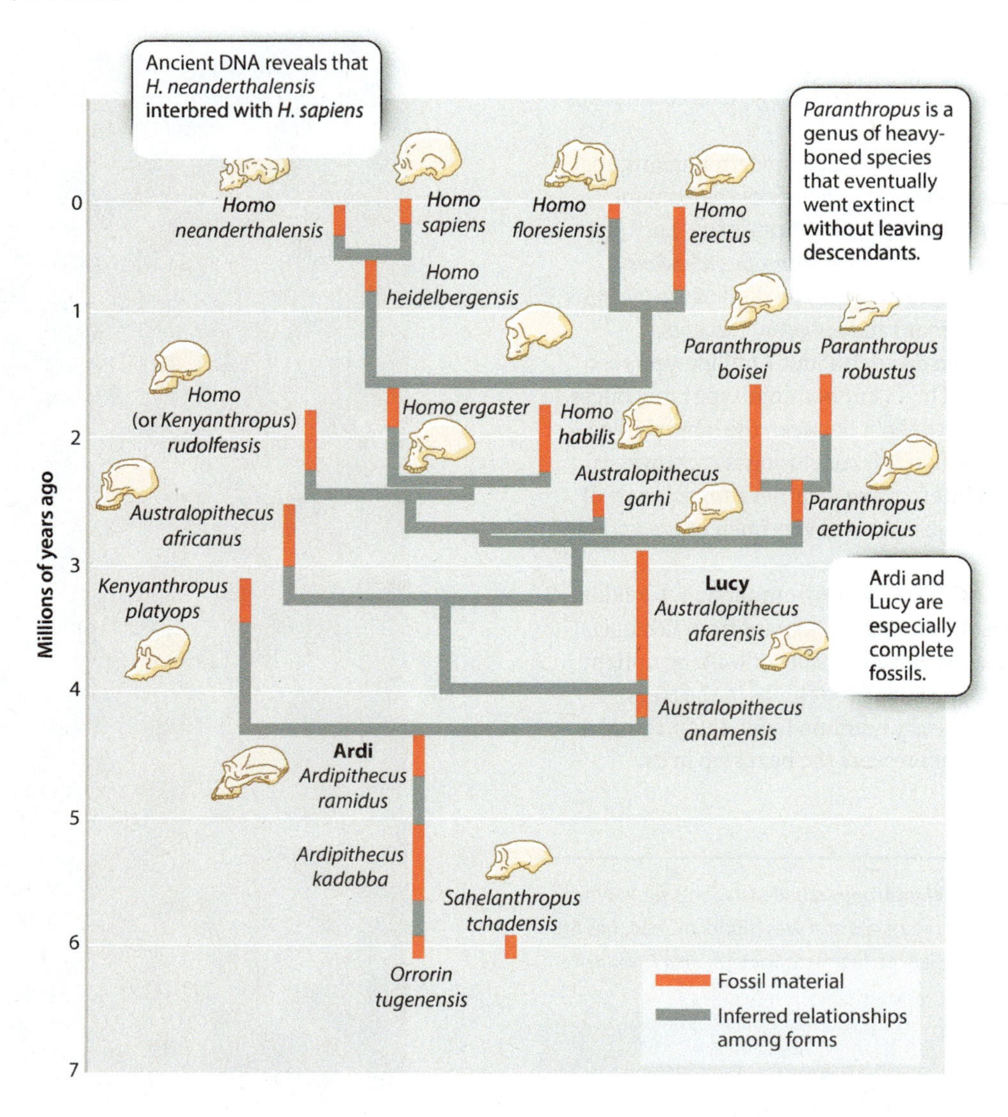

Remarkably, there were many different hominin species living in Africa at the same time. **Fig. 24.6** shows the main hominin species and the longevity of each in the fossil record, along with suggested evolutionary relationships among the species. Note that all hominins have a common ancestor, but not all of these groups lead to modern humans, producing instead other branches of the hominin tree that ultimately went extinct.

A number of trends can be seen when we look over the entire record. Body size increased. Most striking is the increase in size of the cranium and therefore, by inference, of the brain, as shown in **Fig. 24.7.**

The earliest hominin fossils found in Asia are about 2 million years old, indicating that at least one group of hominins ventured out of Africa then. The individuals that first left Africa were members of a species that is sometimes called *Homo ergaster* and sometimes *Homo erectus*. The confusion stems from the controversies that surround the naming of fossil species. Some researchers contend that *H. ergaster* is merely an early form of *H. erectus*. Here, we designate this first hominin *Homo ergaster* and a later descendant species *H. erectus* (see Fig. 24.6). The naming details, however, are relatively unimportant. What matters is that some hominins first colonized areas beyond Africa about 2 million years ago.

FIG. 24.7 Increase in brain size (as inferred from fossil cranium volume) over 3 million years. *Data from T. Deacon, "The Human Brain," pp. 115–123, in J. Jones, R. Martin, and D. Pilbeam (eds.), 1992, The Cambridge Encyclopedia of Human Evolution, Cambridge, England: Cambridge University Press.*

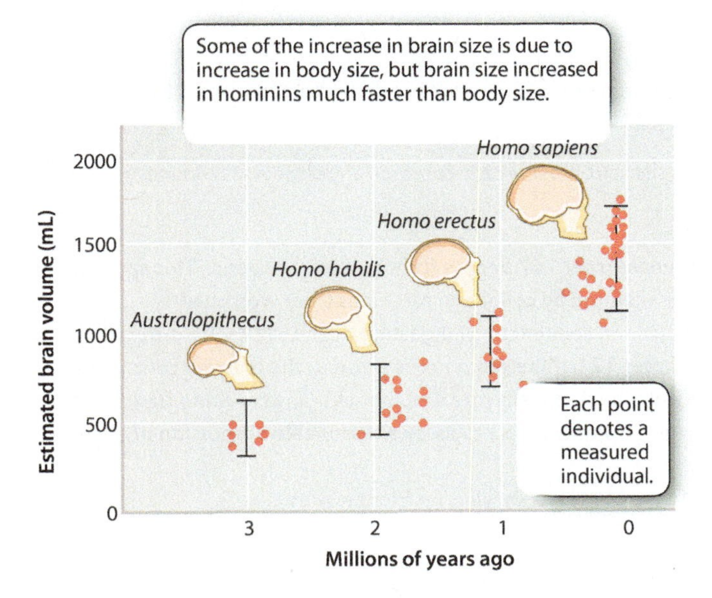

Another hominin species was *Homo neanderthalensis,* whose fossils appear in Europe and the Middle East. **Neanderthals** represent a second hominin exodus from Africa dating from around 600,000 years ago. Thicker boned than us, and with flatter heads that contained brains about the same size as, or slightly larger than, ours, Neanderthals disappeared around 30,000 years ago. As we will see, genetic analysis suggests that this group likely interbred with our own group, *Homo sapiens,* so this disappearance was perhaps not as complete as the fossil record suggests.

Another hominin species became extinct only about 12,000 years ago. This was *H. floresiensis,* known popularly as the Hobbit. *H. floresiensis* is peculiar. Limited to the Indonesian island of Flores, adults were only just over 3 feet tall. Some have suggested that *H. floresiensis* is not a genuinely distinct species, but, rather, is an aberrant *H. sapiens.* Plenty of morphological evidence, however, suggests that *H. floresiensis* is a distinct species derived from an archaic *Homo* species, probably *H. erectus.* Mammals often evolve small body size on islands because of the limited availability of food.

24.2 AFRICAN ORIGINS

From the wealth of fossil evidence, it's clear that modern humans first evolved in Africa. But for a while the timing of this event was unclear. Can all modern humans, *Homo sapiens,* date their ancestry to early hominins in Africa that migrated out and spread around the world about 2 million years ago? Or did the groups that left Africa ultimately go extinct, so that modern humans evolved from a group of hominins in Africa that migrated out much more recently? Molecular studies helped to answer this question.

Studies of mitochondrial DNA reveal that modern humans evolved in Africa relatively recently.

For a long time, it was argued that modern humans derive from the *Homo ergaster* (or *H. erectus*) populations that spread around the world starting from around the time of the early emigration from Africa 2 million years ago. This idea is called the **multiregional hypothesis** of human origins because it implies that different *Homo ergaster* populations throughout Africa and Eurasia evolved in parallel, with some limited gene flow among them, each producing modern *H. sapiens* populations. In short, racial differences among humans would have evolved over 2 million years in different geographic locations.

This idea was overturned in 1987 by another study from Allan Wilson's laboratory, which instead suggested that modern humans arose much more recently from *Homo ergaster* descendants that remained in Africa (sometimes called *Homo heidelbergensis;* see Fig 24.6), and are all descended from an African common ancestor dating from around 200,000 years ago. This newer idea is the **out-of-Africa hypothesis** of human origins.

To test these two hypotheses about human origins, Rebecca Cann, a student in Allan Wilson's laboratory, analyzed DNA sequences to reconstruct the human family tree. Specifically, she studied sequences of a segment of mitochondrial DNA from people living around the world (**Fig. 24.8**).

Mitochondrial DNA (mtDNA) is a small circle of DNA, about 17,000 base pairs long, found in every mitochondrion (Chapter 17). Cann chose to study mtDNA for several reasons. Although a typical cell contains a single nucleus with just two copies of nuclear DNA, each cell has many mitochondria, each carrying multiple copies of mtDNA, making mtDNA much more abundant than nuclear DNA and therefore easier to extract. More important, however, is its mode of inheritance. All your mtDNA is inherited from your mother in the egg she produces because sperm do not contribute mitochondria to the zygote. This means that there is no opportunity for genetic recombination between different mtDNA molecules, so the only way in which sequence variation can arise is through mutation. In nuclear DNA, by contrast, differences between two sequences can be introduced through both mutation and recombination. Recombination obscures genealogical relationships because it mixes segments of DNA with different evolutionary histories.

HOW DO WE KNOW?

FIG. 24.8

When and where did the most recent common ancestor of all living humans live?

BACKGROUND With the availability of DNA sequence analysis tools, it became possible to investigate human prehistory by comparing the DNA of living people from different populations.

HYPOTHESIS The multiregional hypothesis of human origins suggests that our most recent common ancestor was living at the time *Homo ergaster* populations first left Africa, about 2 million years ago.

METHOD Rebecca Cann compared mitochondrial DNA (mtDNA) sequences from 147 people from around the world. This approach required a substantial amount of mtDNA from each individual, which she acquired by collecting placentas from women after childbirth. Instead of sequencing the mtDNA, Cann inferred differences among sequences by digesting the mtDNA with different restriction enzymes, each of which cuts DNA at a specific sequence (Chapter 12). If the sequence is present, the enzyme cuts. If any base in the recognition site of the restriction enzyme has changed in an individual, the enzyme does not cut. The resulting fragments were then separated by gel electrophoresis. By using 12 different enzymes, Cann was able to assay a reasonable proportion of all the mtDNA sequence variation present in the sample.

Here, we see a sample dataset for four people and a chimpanzee. There are seven varying restriction sites:

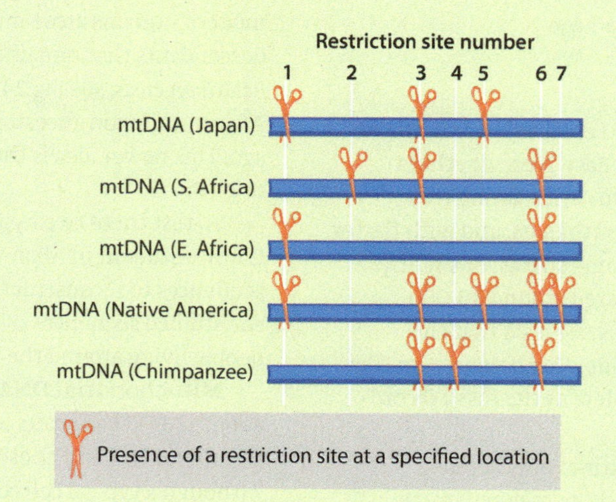

ANALYSIS The data were converted into a family tree by using shared derived characters, described in Chapter 23. By mapping the pattern of changes in this way, the phylogenetic relationships of the mtDNA sequences can be reconstructed. For example, the derived "cut" state at restriction site 1 shared by the East African, Japanese, and Native American sequences implies that the three groups had a common ancestor in which the mutation that created the new restriction site occurred.

Mitochondrial DNA sequences offered Cann a clear advantage. With sequence information from the mtDNA of 147 people from around the world, Cann reconstructed the human family tree (Fig. 24.8). The tree contained two major surprises. First, the two deepest branches—the ones that come off the tree earliest in time—are African. All non-Africans are branches off the African tree. The implication is that *Homo sapiens* evolved in Africa and only afterward did populations migrate out of Africa and become established elsewhere. This finding contradicted the expectations of the multiregional theory, which predicted that *Homo sapiens*

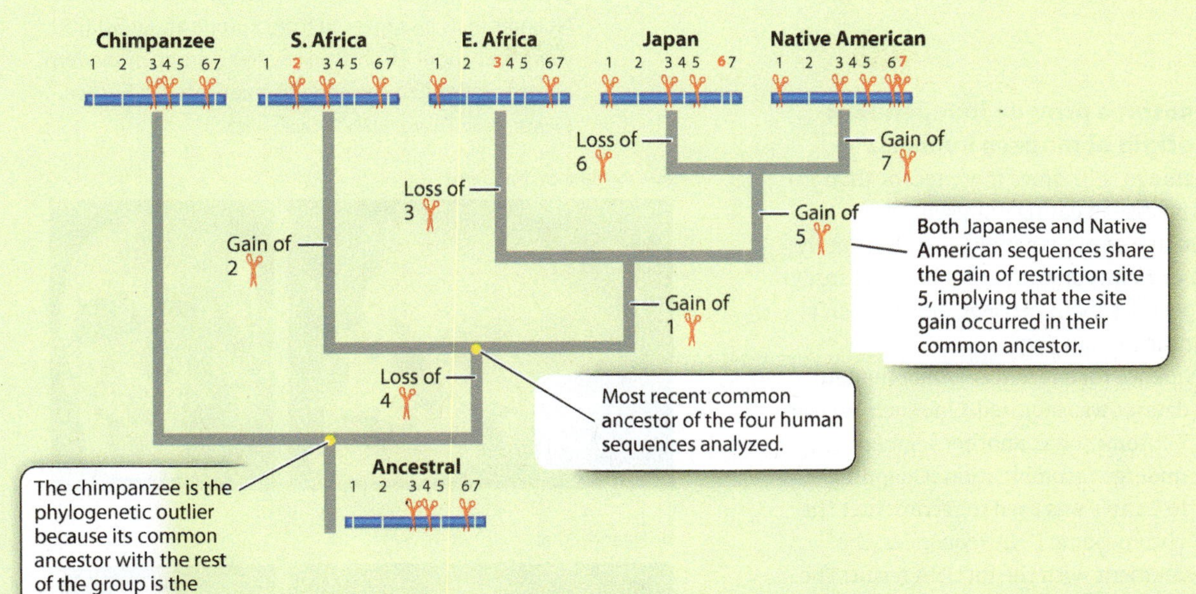

RESULTS AND CONCLUSION A simplified version of Cann's phylogenetic tree based on mtDNA is shown below. Modern humans arose relatively recently in Africa, and their most common ancestor lived about 200,000 years ago. The multiregional hypothesis is rejected. Also, the data revealed that all non-Africans derive from within the African family tree, implying that *H. sapiens* left Africa relatively recently. The Cann study gave rise to the out-of-Africa theory of human origins.

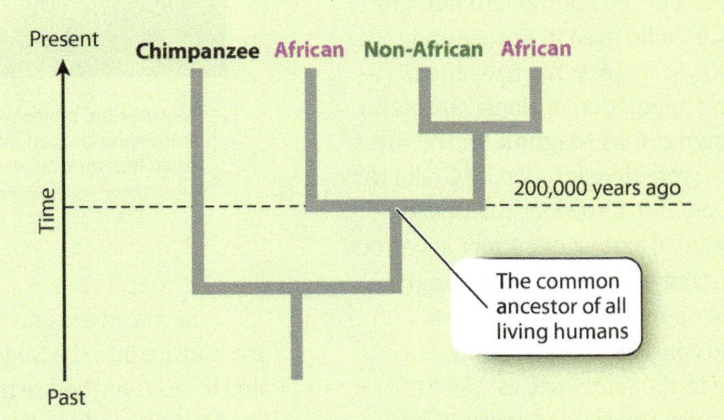

FOLLOW-UP WORK This study set the stage for an explosion in genetic studies of human prehistory that used the same approach of comparing sequences in order to identify migration patterns. Today, with molecular tools vastly more powerful than those available to Cann, studies of the distribution of genetic variation are giving us a detailed pattern, even on relatively local scales, of demographic events.

SOURCE Cann, R. L., M. Stoneking, and A. C. Wilson. 1987. "Mitochondrial DNA and Human Evolution." *Nature* 325:31–36.

evolved independently in different locations throughout the Old World.

Second, the tree is remarkably shallow. That is, even the most distantly related modern humans have a relatively recent common ancestor. By calibrating the rate at which mutations occur in mtDNA, Cann was able to estimate the time back to the common ancestor of all modern humans as about 200,000 years. Subsequent analyses have somewhat refined this estimate, but the key message has not changed: The data contradict the multiregional hypothesis, which predicted a number closer to 2 million years.

→ **Quick Check 2** What do the multiregional and out-of-Africa hypotheses predict about the age of the common ancestor of all humans living today?

Studies of the *Y* chromosome provide independent evidence for a recent origin of modern humans.

Two hundred thousand years ago is 10 times more recent than 2 million years ago, so this change in dating represents a revolution in our understanding of human prehistory. But surely it is risky to make such major claims on the basis of a single study. Maybe there's something peculiar about mtDNA, and it doesn't offer an accurate picture of the human past.

To find independent evidence for a recent origin of modern humans in Africa, another dataset was required. One such dataset comes from studies of the Y chromosome, another segment of human DNA that does not undergo recombination (Chapter 17). When an approach similar to Cann's was used to reconstruct the human family tree using Y chromosome DNA sequences, the result was completely in agreement with the mtDNA result: The human family is young, and it arose in Africa.

Neanderthals disappear from the fossil record as modern humans appear, but have contributed to the modern human gene pool.

If we exclude *H. floresiensis,* the last of the nonmodern humans were the Neanderthals (**Fig. 24.9**), who lived in Europe and Western Asia until about 30,000 years ago. What happened? Were they eliminated by the first population of *Homo sapiens* to arrive in Europe (a group known as **Cro-Magnon** for the site in France from which specimens were first described)? Or did they interbreed with the Cro-Magnons, and, if so, does that mean that Neanderthals should be included among our direct ancestors? Indeed, if modern humans interbred with Neanderthals and our definition of species is based on reproductive isolation (Chapter 22), perhaps we should consider *H. sapiens* and *H. neanderthalensis* as belonging to the same species.

In 1997, we thought we had a clear answer. Matthias Krings, working in Svante Pääbo's lab in Germany, extracted intact stretches of mtDNA from 30,000-year-old Neanderthal bone. If Neanderthals had interbred with our ancestors, we would expect to see evidence of their genetic input in the form of Neanderthal mtDNA in the modern human gene pool. However, the Neanderthal sequence was strikingly different from that of modern humans. That we see nothing even close to the Neanderthal mtDNA sequence in modern humans strongly suggested that Neanderthals did not interbreed with our ancestors (**Fig. 24.10a**).

This conclusion, however, has been reversed following remarkable technological advances that permitted Pääbo and others to sequence the entire Neanderthal genome (rather than just its mtDNA). Careful population genetic analysis revealed that our ancestors did interbreed with Neanderthals, and that 1% to 4% of the genome of every non-African is Neanderthal-derived.

FIG. 24.9 (a) ***Homo sapiens*** and (b) ***Homo neanderthalensis.*** Neanderthals disappeared from Europe about 30,000 years ago, about 10,000 years after a group of modern humans called Cro-Magnon first appeared in Europe. *Sources: Javier Trueba/MSF/Science Source.*

a. Cro-Magnon (modern)

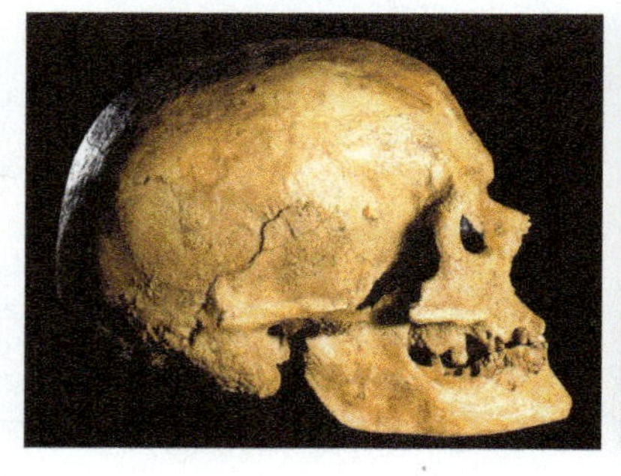

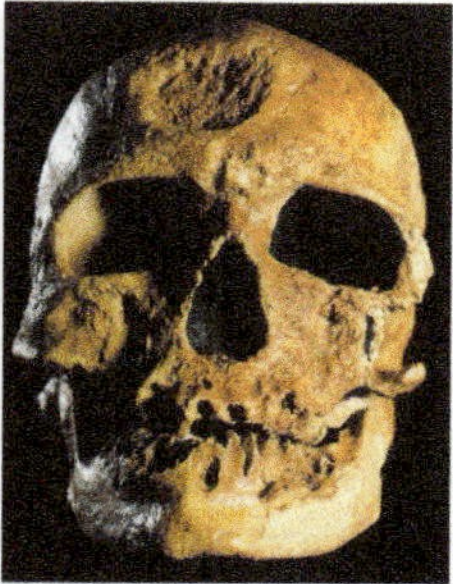

b. Neanderthal

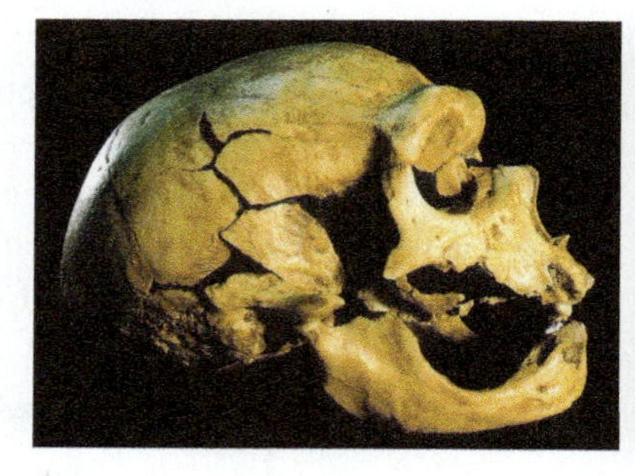

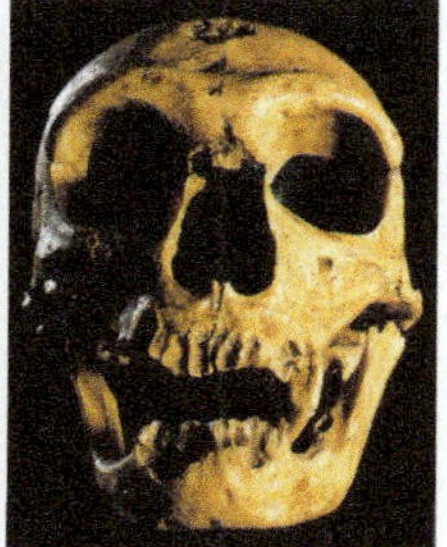

Neanderthal brains were slightly larger than ours, but their skulls were shaped differently. The Neanderthal forehead was much less pronounced than ours, and the skull in general lower.

A new scenario emerged: As populations of our ancestors headed from Africa into the Middle East, they encountered Neanderthals, and it was then, before modern humans had spread out across the world, that interbreeding took place (**Fig. 24.10b**). Remarkably, ancient DNA analysis has further revealed that early non-African *Homo sapiens* did not interbreed only with Neanderthals. Another group of Eurasian hominins, relatives of Neanderthals known as Denisovans after the Siberian cave in which their remains were found, also contributed genetically to some modern human populations, especially indigenous people from Australasia.

How can we reconcile the two apparently contradictory results about the contributions of Neanderthals to the modern human gene pool? The mtDNA study indicates there was no genetic input from Neanderthals in modern humans, but the whole-genome study suggests that in fact there was. One possible solution lies in the difference in patterns of transmission between mtDNA and genomic material (Chapter 17). Recall that mtDNA is maternally inherited because sperm do not contribute mitochondrial material. Your mtDNA, whether you are male or female, is derived solely

FIG. 24.10 Evidence for and against early interbreeding between Neanderthals and early humans. (a) Studies using only mtDNA did not support the idea of interbreeding, but (b) later studies using whole genomes did.

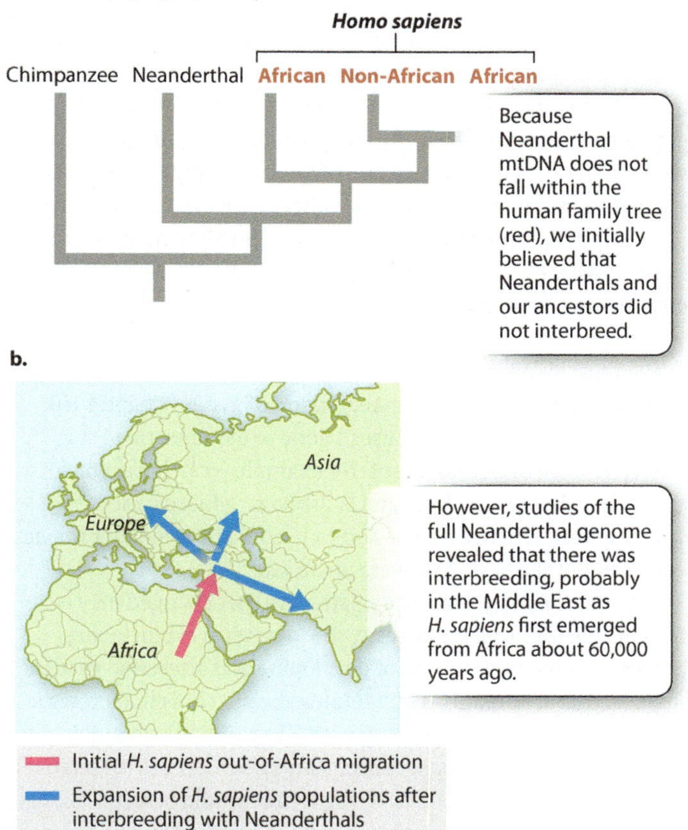

from your mother. Therefore, it is possible that the Neanderthal mtDNA lineage has been lost through genetic drift (Chapter 21).

An alternative is that the discrepancy in the ancient DNA stems from a sex-based difference in interbreeding. If Neanderthal females did not interbreed with our ancestors, Neanderthal mtDNA would not have entered the modern human gene pool. If only male Neanderthals interbred with our ancestors, we would expect exactly the pattern we observe: no Neanderthal mtDNA in modern humans, but Neanderthal genomic DNA in the modern human gene pool.

Regardless of the details of what happened when modern humans, Neanderthals, and Denisovans encountered each other, one thing is clear: We are forced to reimagine the ancestry of a large portion of the human population. Every non-African on the planet is part Neanderthal, albeit a very small part, and many non-Africans can count Denisovans among their ancestors as well.

→ **Quick Check 3** Explain how genetic data are consistent with the possibility that only male Neanderthals interbred with our ancestors.

24.3 DISTINCT FEATURES OF OUR SPECIES

Many extraordinary changes in anatomy and behavior occurred in the 5–7 million years since our lineage split from the lineage that gave rise to the chimpanzees. Fossils tell us a great deal about those changes, especially when high-quality material such as Lucy or Ardi is available, but in general this is an area in which fossils are hard to come by and there is a lot of speculation. Speculation is especially common when we try to explain the reasons behind the evolution of a particular trait. Why, for example, did language evolve? It is easy enough to think of a scenario in which natural selection favors some ability to communicate—maybe language arose to facilitate group hunting. There are plenty of plausible ideas on the subject, but, in most cases, no evidence, so it is impossible to distinguish among competing hypotheses. We can, however, be confident that the events that produced language occurred in Africa, and, through paleontological studies of past environments, we can conclude that humans evolved in an environment similar in many ways to today's East African savanna.

Bipedalism was a key innovation.

The shift from walking on four legs to walking on two was probably one of the first changes in our lineage. Many primates are partially bipedal—chimpanzees, for example, may be observed knuckle-walking—but humans are the only living species that is wholly bipedal. As we've seen, Lucy, dating from about 3.2 million years ago, was already bipedal, and 4.4 million years ago, Ardi was partially bipedal. We can further refine our estimate of when full bipedalism arose from the evidence of a trace fossil. A set of 3.5-million-year-old fossil footprints discovered in Laetoli, Tanzania, reveal a truly upright posture.

Becoming bipedal is not simply a matter of standing up on hind legs. The change required substantial shifts in a number of basic anatomical characteristics, described in **Fig. 24.11.**

Why did hominins become bipedal? Maybe it gave them access to berries and nuts located high on a bush. Maybe it allowed them to scan the vicinity for predators. Maybe it made long-distance travel easier. Whatever the reason, bipedalism freed our ancestors' hands. No longer did they need their hands for locomotion, so for the first time there arose the possibility that specialized hand function could evolve. Most primates have some kind of opposable thumb, but the human version is much more dexterous. The human thumb has three muscles that are not present in the thumb of chimpanzees, and these allow much finer motor control of the thumb. Tool use, present but crude in chimpanzees, can be much more subtle and sophisticated with a human hand.

Bipedalism also made it possible to carry material over long distances. Hominins could then set up complex foraging strategies whereby some individuals supplied others with resources. Exactly when sophisticated tool use arose in our ancestors is controversial, but it is indisputable that bipedalism contributed. Similarly, the

FIG. 24.11 The shift from four to two legs. Anatomical changes were required for the shift, especially in the structure of the skull, spine, legs, and feet.

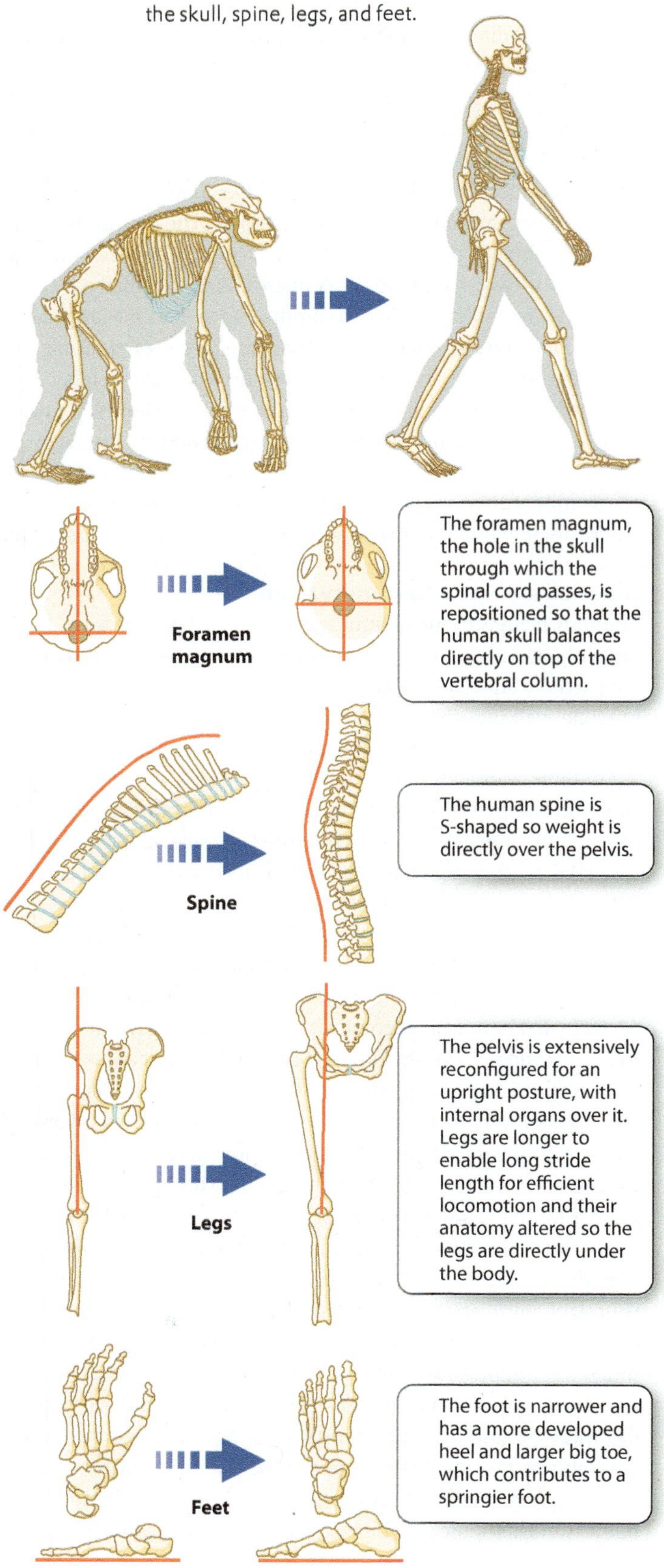

Foramen magnum

The foramen magnum, the hole in the skull through which the spinal cord passes, is repositioned so that the human skull balances directly on top of the vertebral column.

Spine

The human spine is S-shaped so weight is directly over the pelvis.

Legs

The pelvis is extensively reconfigured for an upright posture, with internal organs over it. Legs are longer to enable long stride length for efficient locomotion and their anatomy altered so the legs are directly under the body.

Feet

The foot is narrower and has a more developed heel and larger big toe, which contributes to a springier foot.

ability to manipulate food with the hands, and to carry material using the hands rather than the mouth, likely permitted the evolution of the human jaw—indeed, the entire facial structure—in such a way that language became a possibility.

Adult humans share many features with juvenile chimpanzees.

King and Wilson's 1975 discovery that the DNA of humans and of chimpanzees are 99% identical has recently been confirmed by DNA sequence comparison of the human and chimpanzee genomes. It turns out that the two genomes are extraordinarily similar: All our genes are also present in the chimpanzee and their sequences in humans and in chimpanzees are extremely similar, suggesting that the functions of the proteins they code for are the same. If we are so similar, how can we account for the extensive differences between the two species?

In their original paper, King and Wilson suggested that much of the most significant evolution along the hominin lineage came about through changes in the regulation of genes (Chapter 19). A small change—one that causes a gene to be transcribed at a different stage of development, for example—can have a major effect. In other words, King and Wilson introduced a model whereby small changes in the software could have a major impact even though the basic hardware is the same.

One of the gene-regulatory pathways that changed may be responsible for human **neoteny,** the long-term evolutionary process in which the timing of development is altered so that a sexually mature organism still retains the physical characteristics of the juvenile form. In keeping with King and Wilson's idea, this shift in development could conceivably be achieved with relatively little genetic change. All that it might take would be a few changes to the regulatory switches that control the timing of development.

As early as 1836, French naturalist Étienne Geoffroy Saint-Hilaire noted that a young orangutan on exhibit in Paris looked considerably more like a human than the adult of its own species (**Fig. 24.12**). Several human attributes support this model, including heads that are large relative to our bodies (and our correspondingly large brains), a feature of juvenile great apes. A second human attribute is our lack of hair. The juveniles of other great apes are not as hairless as humans, but they're considerably less hairy than adult apes. A third attribute is the position of the foramen magnum at the base of the skull. In primate development, the foramen magnum starts off in the position it occupies in the adult human and then, in nonhuman great apes, migrates toward the back of the skull. Adult humans have retained the juvenile great ape foramen magnum position. Finally, it has been suggested that our mentality, with its questioning and playfulness, is equivalent in many ways to that of a juvenile ape rather than to that of the comparatively inflexible adult ape.

Humans have large brains relative to body size.

In mammals, brain size is typically correlated with body size. Humans are relatively large-bodied mammals, but our brains

FIG. 24.12 A juvenile and an adult orangutan. Humans look more like juvenile orangutans than like adult orangutans, suggesting that humans may be neotenous great apes. *Sources: (top to bottom) McPHOTO/W. Layer/age fotostock; Russell Watkins/Shutterstock.*

are large even for our body size (**Fig. 24.13**). It is our large brains that have allowed our species' success, extraordinary technological achievements, and at times destructive dominion over the planet.

What factors acted as selective pressure for the evolution of the large human brain? Again, speculation is common, but because a large brain is metabolically expensive to produce and to maintain, we can conclude that the brain has adaptive features that natural selection acted in favor of. What are the selective factors? Here are some possibilities:

- *Tool use.* Bipedalism permitted the evolution of manual dexterity, which in turn requires a complex nervous organization if hands are to be useful.
- *Social living.* Groups require coordination, and coordination requires some form of communication and the means of integrating and acting upon the information conveyed. One scenario, for example, sees group hunting as critical in the evolution of the brain: Natural selection favored those individuals who cooperated best as they pursued large prey.
- *Language.* Did the evolution of language drive the evolution of large brains? Or did language arise as a result of having large brains? Again, we will probably never know, but it is tempting to speculate that the brain and our extraordinary powers of communication evolved in concert.

Probably, as with bipedalism, there was no single factor but a mix of elements that worked together to result in the evolution of large brain size. We tend to focus on brain size because it is a convenient stand-in for mental power and because we can measure it in the fossil record by making the reasonable assumption that the volume of a fossil's cranium reflects the size of its brain. However, it should be emphasized that our minds are not the products solely

FIG. 24.13 Brain size plotted against body size for different species. Humans have large brains for their body size. *Data from Fig. 2.4, p. 44, in H. J. Jerison, 1973, Evolution of the Brain and Intelligence, New York: Academic Press.*

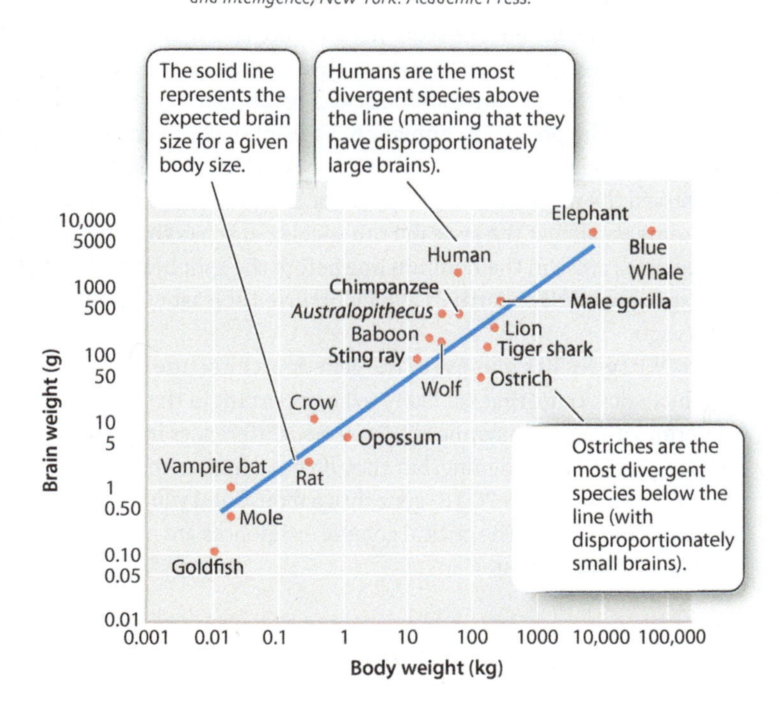

of larger brains. Rather, the reorganization of existing structures and pathways is the more important product of the evolution of the brain. Not only was the brain expanding over those 7 million years of human evolution, it was also being rewired.

The human brain evolved through natural selection. What we have today is a learning machine capable of generating many more skills and abilities than just those that directly enhance evolutionary fitness. For example, let's assume that the brain evolved to make group hunting more efficient. The result is that the brain does indeed enhance our ability to hunt together, but there are numerous by-products of this brain that take its abilities beyond just hunting. Making art, for instance, has nothing to do with group hunting, and yet the brain allows us to do it. A brain evolved for an essential but mundane task like group hunting lies at the heart of much that is wonderful about humanity.

The human and chimpanzee genomes help us identify genes that make us human.

The key genetic differences between humans and chimpanzees must lie in the approximately 1% of protein-coding DNA sequence that differs between the two genomes. What do we see when we compare human and chimpanzee genomes?

A gene that has attracted a lot of interest is *FOXP2*, a member of a large family of evolutionarily conserved genes that encode transcription factors that play important roles in development. Individuals with mutations in *FOXP2* often have difficulty with speech and language. Interestingly, laboratory animals whose *FOXP2* has been knocked out also have communication impairments. Songbirds are less capable of learning new songs, and the high-frequency "songs" of mice are disrupted. In addition to its effects on the brain, *FOXP2* plays a role in the development of many tissues.

Studies of the gene's amino acid sequence reveal a pattern of extreme conservation. The sequences in mice and chimpanzees, whose most recent common ancestor lived about 75 million years ago, differ by a single amino acid (**Fig. 24.14**). However, two amino acids are present in humans but absent in chimpanzees. Studies of Neanderthal DNA have shown that Neanderthals possessed the modern human version of *FOXP2*. The presence of the same version of the gene in both species suggests that the differences arose in the hominin line before the split between our species and Neanderthals, which occurred at least 600,000 years ago.

FOXP2 gives us a glimpse of the genetic architecture of the biological traits that are likely to be important in the determination of "humanness." Those two differences in amino acid sequence are intriguing, but they do not a human make. Substitute the human *FOXP2* gene into a mouse and you will not get a talking mouse. The critical genetic differences are many, subtle, and interacting.

→ **Quick Check 4** The *FOXP2* gene is sometimes called the "language gene." Why is this name inaccurate?

FIG. 24.14 A family tree of mammals showing amino acid changes in *FOXP2*. The gene is highly conserved, but two changes are seen in the Neanderthals and modern humans.

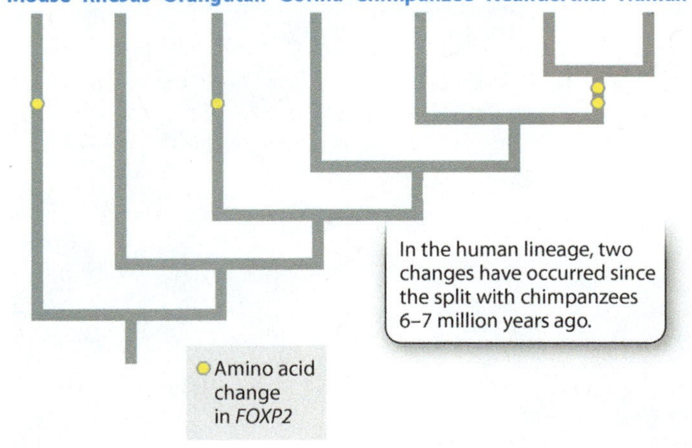

24.4 HUMAN GENETIC VARIATION

So far, we have treated humans as all alike, and in many ways, of course, we are. But there are also many differences from one person to the next. Those differences ultimately have two sources: genetic variation and differences in environment (Chapter 18). A person may be born with dark skin, or a person born with pale skin may acquire darker skin—a tan—in response to exposure to sun.

The differences we see from one person to the next are deceptive. Despite appearances, ours is not the most genetically variable species on Earth. While it is certainly true that everyone alive today (except for identical twins) is genetically unique, our species is actually rather low in overall amounts of genetic variation. Modern estimates based on comparisons of many human DNA sequences indicate that, on average, about 1 in every 1000 base pairs differs among individuals (that is, our level of DNA variation is 0.1%). That's about 10 times less genetic variation than in fruit flies (which nevertheless all look the same to us) and about two to three times less than in Adélie penguins, which look strikingly similar to one another (see Fig. 21.1).

Why, then, are we all so phenotypically different if there is so little genetic variation in our species? Given the large size of our genome, a level of variation of 0.1% translates into a great many genetic differences. Our genome consists of approximately

3 billion base pairs, so 0.1% variation means that 3 million bp differ between any two people chosen at random. Many of those differences are in noncoding DNA, but some fall in regions of DNA that encode proteins and therefore influence the phenotype. When those mutations are reshuffled by recombination, we get the vast array of genetic combinations present in the human population.

The prehistory of our species has had an impact on the distribution of genetic variation.

The reasons for our species' relative lack of genetic variation compared to other species lie in prehistory, and factors affecting the geographical distribution of that variation also lie in the past. Studies of the human family tree initiated by Rebecca Cann's original mtDNA analysis are giving a detailed picture of how our ancestors colonized the planet and how that process affected the distribution of genetic variation across populations today. Detailed analyses of different populations, often using mtDNA or Y chromosomes, allow us to reconstruct the history of human population movements.

As we have seen, *Homo sapiens* arose in Africa. Genetic analyses indicate that perhaps 60,000 years ago, populations started to venture out through the Horn of Africa and into the Middle East (**Fig. 24.15**). The first phase of colonization took our ancestors through Asia and into Australia by about 50,000 years ago. Archaeological evidence indicates that it was not until about 15,000 years ago that the first modern humans crossed from Siberia to North America to populate the New World.

Genetic analyses also indicate that other colonizations were even later. Despite its closeness to the African mainland, the first humans arrived in Madagascar only about 2000 years ago, and the colonists came from Southeast Asia, not Africa. Madagascar populations to this day bear the genetic imprint of this surprising Asian input. The Pacific Islands were among the last habitable places on Earth to be colonized during the Polynesians' extraordinary seaborne odyssey from Samoa, which began about 2000 years ago. Hawaii was colonized about 1500 years ago, and New Zealand only 1000 years ago.

By evolutionary standards, the beginning of the spread of modern humans out of Africa about 60,000 years ago is very recent. There has therefore been relatively little time for differences to accumulate among regional populations, and most of the variation in human populations today arose in ancestral populations before any humans left Africa. When we compare levels of variation in a contemporary African population with that in a non-African population, like Europeans or Asians, we

FIG. 24.15 Human migratory routes. Tracking the spread of mitochondrial DNA mutations around the globe allows us to reconstruct the colonization history of our species. *Data from Fig. 6.18, p. 149, in D. J. Futuyma, 2009, Evolution, 2nd ed., Sunderland, MA: Sinauer Associates.*

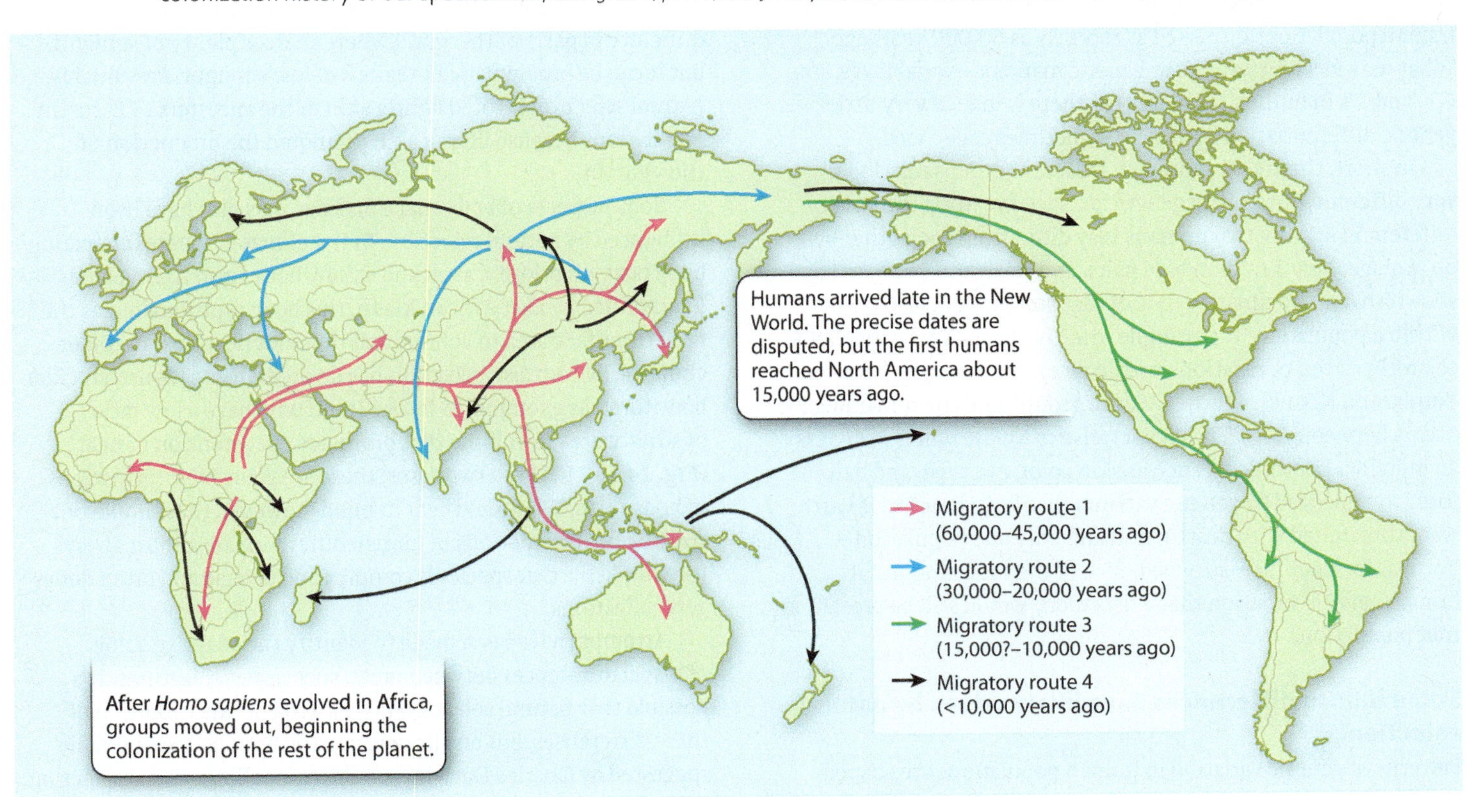

find there is more variation in the African population. This is because the individuals that left Africa 60,000 years ago to found populations in the Middle East and beyond were a relatively small sample of the total amount of genetic variation then present in the human population. Non-African populations therefore began with less genetic variation, and that initial lower variation is reflected in their genetic profiles today.

The recent spread of modern humans means that there are few genetic differences between groups.

Because the out-of-Africa migration was so recent, the genetic differences we see among geographical groups—sometimes called races—are minor. This fact is highly counterintuitive. We see many superficial differences between an African and a Caucasian, such as skin color, facial form, and hair type, and assume that these superficial differences must reflect extensive genetic differences. This assumption made sense when the standard theory about the origin of modern humans was the multiregional one. If European and African populations really had been geographically isolated from each other for as long as 2 million years, then we would expect significant genetic differences among populations.

We expect isolated populations to diverge genetically over time as different mutations occur and are fixed in each population. The longer two populations have been isolated from each other, the more genetic differences between them we expect to see (Chapter 21). Isolation lasting 2 million years implies that the differences are extensive, but isolation of just 60,000 years suggests they are relatively few. Patterns of genetic variation among different human populations support the hypothesis that human populations dispersed as recently as 60,000 years ago. What we see when we look at genetic markers—variable A's, T's, G's, and C's in human DNA—is that there is indeed very little genetic differentiation by what is sometimes called race.

In short, there's a disconnect: Different groups may look very different, but, from a genetic perspective, they're not very different at all. Any two humans may differ from each other by, on average, only 3 million base pairs, and statistical analyses have shown that approximately 85% of that genetic variation occurs within a population (for example, the Yoruba in West Africa); 8% occurs between populations within races (for example, between Yoruba and Kikuyu, another African group); and the remaining 7% occurs between races. The characteristics we use when we assess an individual's ethnicity, such as skin color, eye type, and hair form, are encoded by genetic variants that lie in that 7%. If Earth were threatened with destruction and only one population—Yoruba, for example—survived, 85% of the total amount of human genetic variation that exists today would still be present in that population.

Some human differences have likely arisen by natural selection.

Patterns of genetic variation in human populations are shaped by the set of evolutionary forces discussed in Chapter 21. Neutral variants, for example, are subject to genetic drift, and, given the stepwise global colonization history of our species, it is likely that founder events have played a role as well.

Selection, too, has been important. It is apparent, for example, that the genetic variants affecting traits we can easily see are an especially prominent feature of the 7% of human genetic variation that occurs between so-called races. If we look at other genetic variants, ones that don't affect traits that we can see, there is little or no racial pattern: An African is as likely to have a particular base-pair mutation in a randomly chosen gene as a European. So why are visible traits so markedly different among races and other traits are not? Given the short amount of time (by evolutionary standards) since all *Homo sapiens* were in Africa, it is likely that the differences we see between groups are the product of selection.

People with dark skin tend to originate from areas in lower latitudes with high levels of solar radiation, and people with light skin tend to originate from areas in higher latitudes with low levels of solar radiation. It is likely that natural selection is responsible for the physical differences between these populations. Assuming that the ancestors of non-African populations were relatively dark-skinned, what selective factors can account for the loss of pigmentation?

A likely factor is an essential vitamin, vitamin D, which is particularly important in childhood because it is needed for the production of bone. A deficiency of vitamin D can result in the skeletal malformation known as rickets. The body can synthesize vitamin D, but the process requires ultraviolet radiation. Heavily pigmented skin limits the entry of UV radiation into cells and so limits the production of vitamin D. This does not present a difficulty in parts of the world where there is plenty of sunlight, but it can be problematic in regions of low sunlight. Presumably, natural selection favored lighter skin in the ancestors of Eurasian populations because lighter skin promoted the production of the vitamin.

Some aspects of body shape and size may also have been influenced by natural selection. In hot climates, where dissipating body heat is a priority, a tall and skinny body form has evolved. Exemplified by East African Masai, this body type maximizes the ratio of surface area to volume and thus aids heat loss. In colder climates, by contrast, selection has favored a more robust, stockier body form, as exemplified by the Inuit, who have a low ratio of surface area to volume that promotes the retention of heat (**Fig. 24.16**). In these two cases, these are plausible explanations of body form. We should bear in mind, however, that simple one-size-fits-all explanations of human difference are almost always too simplistic. Our species is complex and diverse and often defies generalizations.

Attempts have been made to identify the adaptive value of visual differences between races, such as facial features. It's possible that natural selection played a role in the evolution of these differences, but an alternative explanation, one originally suggested by Charles Darwin, is more compelling: sexual selection (Chapters 21 and 45).

FIG. 24.16 Evolutionary responses of body shape to climate. (a) A heat-adapted Masai in Kenya and (b) a cold-adapted Inuit in Greenland.
Sources: a. Ryan Heffernan/Aurora/Getty Images; b. B&C Alexander/ArcticPhoto.

As we have seen, there is an apparent mismatch between the extent of difference among groups in visible characters, such as facial features, and the overall level of genetic difference between human groups. Sexual selection can account for this mismatch because it operates solely on characteristics that can readily be seen—think of the peacock's tail. As we learn more about the genetic underpinnings of the traits in question, we will be able to investigate directly the factors responsible for the differences we see among groups.

? CASE 4 MALARIA: CO-EVOLUTION OF HUMANS AND A PARASITE

What human genes are under selection for resistance to malaria?

We see evidence of regional genetic variation in response to local challenges, especially those posed by disease. Malaria, for example, is largely limited to warm climates because it is transmitted by a species of mosquito that can survive only in these regions. Historically, the disease has been devastating in Africa and the Mediterranean. As we saw in Chapter 21, the sickle allele of hemoglobin, *S*, has evolved to be present at high frequencies in these regions because in heterozygotes it confers some protection against the disease. But in homozygotes, the *S* allele is highly detrimental because it causes sickle-cell anemia. Homozygotes for the allele encoding normal hemoglobin are also at a disadvantage because they are entirely unprotected from the parasite.

The *S* allele is beneficial only in the presence of malaria. If there is no malaria in an area, the *S* allele is disadvantageous, so natural selection presumably acted rapidly to eliminate it in the ancestors of Europeans when they arrived in malaria-free regions. The continued high frequency of the *S* allele in Africans, some Mediterranean populations, and in populations descended recently from Africans (such as African-Americans) is, however, a reflection of the response of natural selection to a regional disease.

The hemoglobin genes are not the only genes that are under selection for resistance to malaria. *Glucose-6-phosphate dehydrogenase* (*G6PD*), a gene involved in glucose metabolism, is one of several other genes implicated. People who are heterozygotes for a mutation in the *G6PD* gene—and therefore have a G6PD enzyme deficiency—can develop severe anemia when they eat certain foods (most notably fava beans; hence, the condition is called favism). People who are heterozygotes for a mutation in the *G6PD* gene, however, also have increased resistance to malaria, apparently because they are better at clearing infected red blood cells from their bloodstream. In areas where malaria is common, the advantage of malaria resistance offsets the disadvantage of favism.

Detailed evolutionary analysis of mutations in *G6PD* shows that favism has arisen multiple times, each time selectively favored because of its role in the body's response to the malaria parasite. As expected, favism, like sickle-cell anemia, is mainly a feature of populations in malarial areas or of populations whose evolutionary roots lie in these areas.

24.5 CULTURE, LANGUAGE, AND CONSCIOUSNESS

The most remarkable outcome of the evolutionary process described in this chapter is the human brain. This allows us to do extraordinary things, like appreciate Bach's music, read books, and build skyscrapers. But does this wonderful brain make us qualitatively different from other organisms? Does it in some way take us out of nature? Traditionally, the answer to these questions would have been a resounding yes. However, research into the capabilities of other species is questioning this conclusion: The human brain is certainly remarkable, but, in essence, what we can do is merely an extension of what other animals can do.

Culture changes rapidly.

Culture is generally defined as a body of learned behavior that is socially transmitted among individuals and passed down from one generation to the next. Culture has permitted us in part to transcend our biological limits. To take a simple example, clothing and ingeniously constructed shelters have enabled us to live in extraordinarily inhospitable parts of the planet, like the Arctic (**Fig. 24.17**). The capacity to innovate coupled with the ability to transmit culture is the key to the success of humans.

Culture, of course, changes over time. In many ways, cultural change is responsible for our species' extraordinary achievements. Genetic evolution is slow because it involves mutation followed by changes in allele frequencies that take place over many generations. Cultural change, on the other hand, can occur much more rapidly. Ten years ago, nobody had heard of smartphones, but today, millions of people own them. Or think of the speed at which a change in clothing style—a shift from flared to straight-leg jeans, for example—spreads through a population. Today's human population as a whole is genetically almost identical to the population when your grandparents were young. But think of the cultural changes that have occurred in the 50 years or so between your grandparents' youth and your own.

Despite this clear contrast between biological evolution and cultural change, we should not necessarily think of the two processes as independent of each other. Sometimes cultural change drives biological evolution.

A good example of the interaction between cultural change and biological evolution is the evolution of lactose tolerance in populations for which domesticated animals became an important source of dairy product. Most humans are lactose intolerant. Lactose, a sugar, is a major component of mammalian milk, including human breast milk. We have an enzyme, lactase, that breaks down lactose in the gut, but, typically, the enzyme is produced only in the first years of life, when we are breast-feeding. Once a child is weaned, lactase production is turned off. Lactase, however, is clearly a useful enzyme to have if there is a major dairy component to your diet.

FIG. 24.17 **The power of culture.** Inventions (such as clothing) have allowed our species to expand its geographic range. *Source: B&C Alexander/ArcticPhoto.*

Archaeological and genetic analyses indicate that cattle were domesticated probably three separate times in the past 10,000 years, in three separate places: in the Middle East, in East Africa, and in the Indus Valley. In at least two of these cases, there has been subsequent human biological evolution in favor of lactose tolerance, that is, continued lactase production throughout life. Analysis of the gene region involved in switching lactase production on and off has revealed mutations in European lactose-tolerant people that are different from those in African lactose-tolerant people, implying independent, convergent evolution of this trait in the two populations. Furthermore, we see evidence that these changes have evolved very recently, implying that they arose as a response to the domestication of cattle. Here, we see the interaction between cultural change and biological evolution. Biological

FIG. 24.18 Nonhuman culture. (a) English blue tits steal cream from a milk bottle on the doorstep. (b) Adult meerkats teach their young how to handle their prey. (c) Chimpanzees in different populations have devised different ways of using tools to hunt insects. *Sources: a. Colin F. Sargent; b. Dr. Alex Thornton, University of Exeter; c. FLPA/Peter Davey/age fotostock.*

evolution of continued lactase production has resulted from the change in a cultural practice, namely cattle domestication.

Is culture uniquely human?

Even nonprimates are known to be capable of learning from another member of their own species—culture is not uniquely human. Bird songs, for instance, may vary regionally because juveniles learn the song from local adults. In a famous case, shown in **Fig. 24.18a,** small birds called blue tits (*Parus caeruleus*) in Britain learned from each other how to peck through the aluminum caps of milk bottles left on doorsteps by milkmen to reach the rich cream at the top of the milk. Presumably, one individual discovered accidentally how to peck through a milk bottle cap and the others then imitated the action.

Nor is teaching, one way in which culture is transmitted, limited to humans. In teaching, one individual tailors the information available to another in order to facilitate learning. Teaching and imitation together make learning highly efficient. Adult meerkats teach young meerkats how to handle prey by giving them the opportunity to interact with live prey (**Fig. 24.18b**).

The most sophisticated example of nonhuman culture is shown by chimpanzees. Detailed studies of several geographically isolated populations have revealed 39 culturally transmitted behaviors, such as ways of using tools to catch insects, which are specific to a particular population (**Fig. 24.18c**). West African Chimpanzees in Guinea and the Ivory Coast both use tools to help them harvest the same species of army ants for food (and avoid the painful defensive bites of the ants), but each population has its own distinct way of doing this. Chimpanzees, then, are like us in having regional variation in culture.

Despite the cultural achievements of other species, it is clear that culture in *Homo sapiens* is very far removed from anything seen in the natural world. We are amazingly adept at imitation, learning, and acquiring culture. Teaching, learning, and cultural transmission all benefit from another extraordinary human attribute, language.

Is language uniquely human?

Language is a complex form of communication involving the transfer of information encoded in specific ways, such as speech, facial expressions, posture, and the like. Despite previous claims that language is unique to humans, it is not. The waggle dance of a worker bee on its return to the hive communicates information on the direction, distance, and nature of a food resource (Chapter 45). Vervet monkeys use warning vocalizations to specify the identity of a potential predator: They have one call for "leopard," for example, and another for "snake." But nonhuman animal languages are limited. Attempts to teach even our closest relatives, chimpanzees, to communicate using sign language have met with only limited success (**Fig. 24.19**).

Grammar provides a set of rules that allow the combination of words into a virtually infinite array of meanings. Noam Chomsky, father of modern linguistics, has pointed out that all human languages are basically similar from a grammatical viewpoint. Humans have what he has called a "universal grammar" that would lead a visiting linguist from another planet to conclude that all Earth's languages are dialects of the same basic language. That universal grammar is in some way hard wired into the human brain in such a way that every human infant spontaneously strives to acquire language. The specific attributes of the language depend on the baby's environment—a baby in France will learn French, and one in Japan will learn Japanese—but the basic process is similar in every case. Chimpanzees, and the entire natural world, lack the drive toward the acquisition of a grammatical language.

Is consciousness uniquely human?

Descartes famously wrote, "Cogito, ergo sum"—"I think, therefore I am." Can we legitimately rewrite his statement to declare, "Animals think, therefore they are"? With the growth

FIG. 24.19 Dr. Susan Savage Rumbaugh with Panbanisha, a female bonobo who learned to communicate using sign language. Chimpanzees and bonobos are able to learn and use sign language to express words and simple sentences. *Source: Anna Clopet/CORBIS.*

of the animal rights movement, particularly in reference to the treatment of animals in factory farms, this question is of more than academic interest. We now have many examples of animal thinking from a range of species, including, not surprisingly, chimpanzees and gorillas.

Perhaps more remarkable are the examples that come from animal species that are not closely related to us. In experiments carried out by Alex Kacelnik in Oxford, England, a pair of New Caledonian Crows was presented with two pieces of wire, one straight and the other hooked, and offered a food reward that could be obtained only by using the hooked wire. One member of the pair, the male, disregarded the experiment and flew off with the hooked wire. The female, however, having discovered that she could not get the food reward with the straight wire, went to some considerable trouble to bend a hook into the straight wire. She succeeded in getting the food. It is difficult to deny that the crow thought about the problem and was able to solve it, perhaps in the same way as we would. Definitions of consciousness are contested, but, as with language and culture, it seems clear that other species are capable of some form of conscious thought.

The evolutionary biologist Theodosius Dobzhansky once said, "All species are unique, but humans are uniquest." Our "uniquest" status is not derived from having attributes absent in other species, but from the extent to which those attributes are developed in us. Human language, culture, and consciousness are extraordinary products of our extraordinary brains. Nevertheless, as Darwin taught us and as we should never forget, we are fully a part of the natural world. ■

Core Concepts Summary

24.1 Anatomical, molecular, and fossil evidence shows that the human lineage branches off the great apes tree.

Anatomical features indicate that primates are a monophyletic group that includes prosimians, monkeys, and apes. The apes in turn include the lesser and great apes. page 486

The great apes include orangutans, gorillas, chimpanzees, and humans. page 487

Analysis of sequence differences between humans and our closest relatives, chimpanzees, indicate that our lineage split from chimpanzees 5–7 million years ago. page 487

Lucy, an unusually complete specimen of *Australopithecus afarensis,* demonstrates that our ancestors were bipedal by about 3.2 million years ago. page 489

Hominin fossils occur only in Africa until about 2 million years ago, when *Homo ergaster* migrated out of Africa to colonize the Old World. page 490

24.2 Phylogenetic analysis of mitochondrial DNA and the *Y* chromosome shows that our species arose in Africa.

Studies of mitochondrial DNA (mtDNA) suggest that the time back to the common ancestor of modern humans is about 200,000 years, implying that modern humans (*Homo sapiens*) arose in Africa (the out-of-Africa theory). page 491

The mtDNA out-of-Africa pattern is supported by Y chromosome analysis, which also shows a recent African origin of modern humans. page 494

Analysis of Neanderthal DNA from 30,000-year-old material indicates that, as the ancestors of non-African humans emigrated from Africa, they interbred with the Neanderthals. page 494

Our species originated in Africa and subsequently colonized the rest of the planet, starting about 60,000 years ago. page 495

24.3 During the 5–7 million years since the most recent common ancestor of humans and chimpanzees, our lineage acquired a number of distinctive features.

The development of bipedalism involved a wholesale restructuring of anatomy. page 495

Neoteny is the process in which the timing of development is altered over evolution so that a sexually mature organism retains the physical characteristics of the juvenile form; humans are neotenous, exhibiting many traits as adults that chimpanzees exhibit as juveniles. page 496

There are many possible selective factors that explain the evolution of our large brain, including tool use, social living, and language. page 497

FOXP2, a transcription factor involved in brain development, may be important in language, as mutations in the gene that encodes FOXP2 are implicated in speech pathologies. page 498

24.4 Human history has had an important impact on patterns of genetic variation in our species.

Because our ancestors left Africa very recently in evolutionary terms, there has been little chance for genetic differences to accumulate among geographically separated populations. page 499

Humans have very little genetic variation, with only about 1 in every 1000 base pairs varying among individuals. page 499

Most of the variation among humans occurs within populations. As much as 85% of the total amount of genetic variation in humans can be found within a single population. Only about 7% of human genetic variation segregates between groups that are commonly called races. page 500

Some racial differences, such as skin color and resistance to malaria, have probably arisen by natural selection. page 500

Other racial differences have probably arisen by sexual selection. page 501

24.5 Culture is a potent force for change in modern humans.

Cultural evolution and biological evolution may interact, as in the case of the evolution of lactose tolerance in regions where cattle were domesticated. page 502

Other animals possess simple versions of culture, language, and even consciousness, but the capabilities of our species in all three are truly exceptional. page 503

Self-Assessment

1. Describe the evidence suggesting that chimpanzees are the closest living relatives of humans.
2. Explain the out-of-Africa theory of human origins and how studies of mitochondrial DNA and the Y chromosome support it.
3. List four anatomical differences between chimpanzees and humans, and explain how these changes facilitated walking upright.
4. Given the high genetic similarity of humans and chimpanzees, how can we account for the differences we see between the two species?
5. Describe three possible selective factors underlying the evolution of large brains in our ancestors.
6. Explain how differences among different human populations arose by natural and sexual selection.
7. Provide arguments for and against the idea that culture, language, and consciousness are uniquely human.

Log in to LaunchPad to check your answers to the Self-Assessment questions, and to access additional learning tools.

QUICK CHECK ANSWERS

Chapter 1

1. One hypothesis is that their tan fur protects the mice from predators by allowing them to blend in with their surroundings.
2. One experiment would be to take two large populations of laboratory mice and introduce cigarette smoke into the cages of one population but not the other. Do mice subjected to smoke develop cancer at rates that are significantly higher than those of the control group? We can also make observations of human populations: Do smokers develop lung cancer at rates significantly higher than those of nonsmokers?
3. DNA directs the formation of proteins that do the cell's work. Mutations in DNA can be transcribed and translated into proteins with altered structure and, therefore, different functions. Changes in protein function can cause the cell to work improperly, or fail altogether, resulting in some cases in disease. In some cases, however, altered proteins improve the function of the cell.
4. The use of antibiotics selects for antibiotic resistance among bacteria. Bacteria do not become antibiotic resistant out of "need" or because resistance would be advantageous for them. Instead, before the application of antibiotics, bacteria with antibiotic resistance exist in low numbers. The application of antibiotics allows these bacteria to grow and reproduce more successfully than those that are susceptible to antibiotics. Mutations, such as those responsible for antibiotic resistance, are not influenced by whether or not the organism is in an environment in which that mutation would be advantageous.

Chapter 2

1. One conclusion is that atoms consist mainly of empty space, and hence most positively charged particles passing through the gold foil do not come close enough to any other positive charge to be deflected. Another conclusion is that the positively charged protons in the nucleus must be small and densely packed.
2. Hydrogen and lithium are in the same column, or group, in the periodic table. Each has one valence electron in their outer orbital. As a result, one atom of lithium combines with one atom of hydrogen to make lithium hydride, with a full complement of two electrons in the single molecular orbital.
3. Ice is less dense than liquid water. As a result, when water freezes, it expands in volume and can burst closed containers, such as cans of soda or water pipes in houses. This property is unusual. For most substances, the solid phase is more dense than the liquid phase.
4. Glucose and galactose differ only in the orientation of the –OH and –H groups attached to carbon 4.

Chapter 3

1. Because R = A or G, then %R = %A + %G; and because %A = %T and %G = %C, we can write %R = %A + %G = %T + %C. But T or C = Y, and so %T + %C = %Y. It follows that %R = %Y.
2. The RNA transcript has the sequence 5′-AUCGCUGAAAGU-3′.
3. The incorporation of a nucleotide with a 3′ H rather than a 3′ OH will stop subsequent elongation because the 3′ OH is necessary to attack the high energy phosphate bond of the incoming nucleoside triphosphate. The incorporation of a nucleotide with a 2′ H rather than a 2′ OH will have no effect on elongation, as this group is not involved in the polymerization reaction.
4. The eukaryotic DNA sequence contains introns, which the bacterial cell cannot splice out properly, and so the correct protein is not produced from the information in the bacterial RNA transcript.

Chapter 4

1. The sequence of amino acids in a protein determines how a protein folds, so a change in even a single amino acid can affect the way the protein folds and can disrupt its function. For example, if the hydrophobic R groups of two amino acids must aggregate for proper structure and function, then a mutation that changes one of the hydrophobic amino acids for an acidic or a basic amino acid will prevent this aggregation and disrupt structure and function. Similarly, if proper folding requires interaction between the R groups of an acidic and a basic amino acid, then, if either one of them is changed to a hydrophobic amino acid, proper folding will not take place.
2. The three reading frames are:

   UUU GGG UUU GGG..., which codes for repeating Phe–Gly–Phe–Gly...

   UUG GGU UUG GGU..., which codes for repeating Leu–Gly–Leu–Gly...

   UGG GUU UGG GUU..., which codes for repeating Trp–Val–Trp–Val...
3. With proper eukaryotic processing, the RNA transcript from the bacterial DNA will be capped at the 5′ end. The initiation complex will form at the 5′ cap and move along the mRNA until the first AUG codon is encountered, and then translation begins. When one of the termination codons is encountered, the polypeptide is released. Translation of the downstream polypeptides cannot take place because the Shine–Dalgarno sequences preceding them are not recognized by the eukaryotic translational machinery.
4. A mutation that decreases survival or reproduction will likely decrease in number in each generation because nonmutant forms leave more offspring. Eventually, the harmful mutation may disappear from the population because its final carriers failed to survive or reproduce. In the extreme case when the harmful mutation causes death or sterility, it can disappear in only one generation.

Chapter 5

1. Saturated fatty acids are less mobile within the membrane compared to unsaturated fatty acids. As a result, saturated fatty acids tend to be solid at room temperature, whereas unsaturated fatty acids tend to be liquid. Margarine and many other animal fats

contain saturated fatty acids and are solid, whereas many plant and fish oils contain unsaturated fatty acids and are liquid at room temperature.

2. Water molecules move in both directions, but the *net* movement of water molecules is from side A to side B. Water moves from regions of higher water concentration to regions of lower water concentration. Likewise, sodium and chloride ions move in both directions, but the *net* movement of sodium and chloride ions is from side B to side A. Movement of water and ions results from diffusion, the random motion of substances. Even when the concentration of all molecules is the same on the two sides, diffusion still occurs, but there is no net movement of water molecules or ions.
3. If the sodium-potassium pump is made inactive by poison, the cell will swell and even burst, as the intracellular fluid becomes hypertonic relative to the outside of the cell and water moves into the cell by osmosis.

Chapter 6

1. No, because the second law of thermodynamics applies to the universe as a whole. This means that we have to consider not just the air in the room but the heat released to the outdoors as well. An air conditioner produces more hot air than cold air, and therefore total entropy increases, as described by the second law of thermodynamics.
2. Increasing the temperature increases the value of $T\Delta S$, which decreases ΔG, since $\Delta G = \Delta H - T\Delta S$. As a result, an increase in temperature makes it more likely that a reaction will proceed without a net input of energy.
3. Enzymes increase the reaction rate and decrease the activation energy. The other parameters are not changed by enzymes.

Chapter 7

1. The reduced molecules are NADH, $FADH_2$, and $C_6H_{12}O_6$, and the oxidized molecules are NAD^+, FAD, and CO_2. The reduced forms have more chemical energy than their corresponding oxidized forms.
2. At the end of glycolysis, the energy in the original glucose molecule is contained in pyruvate, ATP, and NADH.
3. At the end of the citric acid cycle, the energy in the original glucose molecule is held in ATP, NADH, and $FADH_2$.
4. Oxygen is consumed in cellular respiration. Oxygen is the final electron acceptor in the electron transport chain and is converted to water.
5. Uncoupling agents decrease the proton gradient and therefore decrease levels of ATP. The energy of the proton gradient is not used for oxidative phosphorylation, but instead is dissipated as heat. Uncoupling agents are found naturally in certain tissues, such as fat, for heat generation. They can also act as poisons.
6. Yeast cells are eukaryotes. In bread making, yeast can use sugar as a food source for ethanol fermentation. The carbon dioxide produced in the process causes the bread to rise. The ethanol is removed in the baking process.

Chapter 8

1. You should label the oxygen in CO_2 (that is, inject $C^{18}O_2$) because the entire CO_2 molecule is used in synthesizing carbohydrates, whereas H_2O donates only the electron needed for the reduction step of the Calvin cycle. The extraction of electrons from water releases O_2 as a by-product.
2. NADPH supplies the major input of energy that is used to synthesize carbohydrates in the Calvin cycle.
3. Antenna chlorophyll molecules transfer absorbed energy from one antenna chlorophyll molecule to another, and ultimately to the reaction center. Reaction center chlorophylls transfer electrons to an electron acceptor, resulting in the oxidation of reaction center chlorophyll molecules.
4. One photosystem is needed to pull electrons from water, and a second photosystem is needed to raise the energy of these electrons enough that they can reduce $NADP^+$.
5. Like cellular respiration, photorespiration consumes O_2 and releases CO_2. Unlike cellular respiration, it consumes rather than produces ATP.
6. Rubisco faces a fundamental trade-off between selectivity and speed because it can use both CO_2 and O_2 as substrates. High selectivity of CO_2 over O_2 requires that the reaction have a high energy barrier, leading to a lower catalytic rate.

Chapter 9

1. Only cells that have receptors for the hormone respond to the signal. Therefore, signaling can be specific for particular cells.
2. No. A G protein-coupled receptor is a transmembrane receptor that interacts with a G protein located inside the cell on the cytoplasmic side of the plasma membrane.
3. The signal to increase heart rate carried by adrenaline can be reversed in at least four ways: (1) by decreasing the amounts of the signaling molecule available to bind and activate the G protein-coupled receptor; (2) by inactivating the G protein; (3) by decreasing the amount of the second messenger cAMP; and (4) by dephosphorylating the target proteins that cause the increased rate of contraction of the muscle cells.

Chapter 10

1. A defect in dynein would cause the melanin granules to remain dispersed because dynein transports the granules back toward the minus end of the microtubule during granule aggregation.
2. Adherens junctions form a belt around cells, whereas desmosomes are button-like attachments. In addition, adherens junctions connect to actin microfilaments, whereas desmosomes connect to intermediate filaments.

3. Tight junctions prevent the passage of materials through the space between cells. Adherens junctions and desmosomes attach cells to one another.
4. Integrins are responsible for the production of milk proteins by mammary cells in response to the extracellular matrix because integrins bind extracellular matrix proteins. By contrast, cadherins bind to other cadherins when cells adhere to other cells.

Chapter 11

1. A mutation that disrupts the function of the FtsZ protein will block cell division.
2. Sister chromatids are the result of DNA replication during S phase, and so they have identical DNA sequences (with the exception of a few changes due to rare mutations). The two homologous chromosomes are inherited from two different parents. The DNA sequences of these chromosomes are therefore similar, but not identical.
3. A cell that undergoes mitosis but not cytokinesis will become a single cell with two nuclei (and therefore with twice the normal amount of DNA); this type of cell is called a multinucleate cell.
4. In human cells at the end of prophase I, there are 92 chromatids, 46 centromeres, and 23 bivalents.
5. In meiosis I, homologous chromosomes pair, undergo crossing over, and segregate from each other. These events do not occur in mitosis. In mitosis, centromeres divide and sister chromatids separate, events that do not take place in meiosis I.
6. The products of meiosis are different from each other as a result of two key processes: (1) crossing over, which occurs at essentially random positions along the chromosomes and creates unique combinations of genetic differences that may be present in the maternal and paternal chromosomes, and (2) random orientation of the homologous chromosomes on the spindle in metaphase I, so each nucleus receives a random combination of maternal and paternal homologs.
7. The function of the p53 protein can be disrupted by a mutation in the *p53* gene. Alternatively, certain viral proteins, such as the E6 protein of HPV discussed in Case 2: Cancer, can interfere with the function of the p53 protein.
8. An oncogene causes cancer by producing an excess of protein activity that pushes the cell to divide. A tumor suppressor like *p53* functions oppositely: Its normal function is to prevent cell division and its absence is what allows the cell to divide uncontrollably.

Chapter 12

1. After one round of replication, you would predict only $^{14}N/^{15}N$ hybrid DNA, which has a density of 1.715 gm/cm^3. After two rounds of replication, you would predict half the molecules to be $^{14}N/^{15}N$ hybrid DNA (density 1.715 gm/cm^3) and half to be $^{15}N/^{15}N$ heavy DNA (density 1.722 gm/cm^3).
2. The most likely reason for observing an unexpected PCR product is that the primers annealed not only to sites flanking the target region but also annealed in the correct orientation to sites flanking a different, nontarget region of the genome. The size of the nontarget region is completely unpredictable, so there is nothing significant in the size of the unexpected product being 550 bp.
3. The sequence of the template strand is antiparallel to the synthesized strand and inferred from complementary base pairing as 5′-ACTCGGTAGT-3′.
4. The value of sticky ends is that they give the researcher greater control over which restriction fragments can come together and be attached. Sticky ends can pair only with other sticky ends that have complementary 3′ and 5′ overhangs. Hence, restriction fragments produced by *Bam*HI can combine only with other fragments produced by *Bam*HI, and not with fragments produced by, for example, *Hin*dIII. However, any blunt end can be attached to any other blunt end, for example a *Hpa*I end to a *Sma*I end.

Chapter 13

1. DNA sequencing technology is limited to DNA molecules that are much smaller than the size of a chromosome, so the challenge of genome sequencing is to piece together smaller sequenced DNA fragments.
2. Even when short repeated sequences can be sequenced, there is an assembly problem similar to that for longer repeats in which the researcher has no way of knowing where in the repeat any particular sequenced fragment should be assigned. The result is that the total number of repeats remains unresolved. It could be in the hundreds, thousands, tens of thousands, hundreds of thousands, or even millions.
3. The observation suggests that lentiviruses and their hosts evolve together and are closely matched. This in turn suggests that lentiviruses are usually unable to infect hosts that are not evolutionarily very closely related to their natural host.
4. You can't tell. The C-value paradox means that you cannot predict genome size from the complexity of the organism. In fact, the genome size of the two-toed salamander is 30 times larger than that of the human genome.

Chapter 14

1. Mutant genes that are harmful or neutral are much less likely to persist in a population than ones that result in increased survival and reproduction because the latter mutations would lead to greater fitness.
2. The number of nucleotides that are inserted or deleted is almost always an exact multiple of 3 because each codon in the genetic code consists of three nucleotides. Any insertion or deletion that includes a number of nucleotides that is not an exact multiple of 3 shifts the reading frame, and the resulting sequencing will most likely code for a nonfunctional protein.
3. Mutations are random; they are not directed by the environment. This does not mean that the environment cannot affect the *rate* of mutation. Mutagens increase the rate of the mutation, but they

cannot induce specific mutations that would be beneficial to the organism in response to the environment.

Chapter 15

1. Mutations that cause antibiotic resistance are clearly beneficial to bacteria when antibiotic is present, but are neutral or can even be harmful in the absence of antibiotic. The effect of a mutation on an organism often depends on the environment.
2. Only a small fraction of the human genome codes for proteins or other functional elements (Chapter 13), so most mutations are neutral.
3. If the suspect is the actual source of a sample, then the DNA fingerprints must match exactly, and any mismatch rules out the suspect as the source of the sample. When there is a match, however, there is always a small chance (typically very small) that the sample came from another person with the same DNA fingerprint.
4. In a VNTR, the restriction fragments are different lengths because of a variable number of repeats between restriction sites (one, two, three, four, or more). In a RFLP, the restriction fragments are different lengths because a restriction site is removed in one DNA sequence and not in another, making the distances between restriction sites different.
5. A point mutation is a change in a single nucleotide in an individual cell, such as C to G. An SNP results from a point mutation that occurred at some time in the past so that now, in the population, there are two or more different single nucleotides at a given position. For example, some people might have C at a certain position and others might have G at that position.
6. Both *Y* chromosomes in the *XYY* baby must come from the father, so nondisjunction took place in the father. In normal meiosis, the first meiotic division separates the *X* chromosome from the *Y* chromosome, and the second meiotic division separates the sister chromatids of the *X* chromosome and the sister chromatids of the *Y* chromosome. For the *Y* chromosomes to remain together and be included in the same sperm, nondisjunction must take place in the second meiotic division (see Fig. 15.10c).

Chapter 16

1. For the true-breeding plants with yellow seeds, the phenotype is yellow seeds, and the genotype is *AA*. For the true-breeding plants with green seeds, the phenotype is green seeds, and the genotype is *aa*.
2. Yes in both cases. In the case of simple dominance, the homozygous *AA* and heterozygous *Aa* genotypes have the same phenotype but different genotypes. In traits that are influenced by the environment, individuals with the same genotype can have different phenotypes because they have different environments. An example discussed in Chapter 15 is how tobacco smoking increases the severity of emphysema resulting from the *PiZ* allele of the α*-1 antitrypsin* gene.
3. Half the progeny will have the *AA* genotype and half will have the *Aa* genotype. All the plants will have yellow seeds.
4. The probability that the first pea is green is ¼; the probability that the second pea is green is also ¼; the probability that the third pea is yellow is ¾; *and* the probability that the fourth pea is yellow is ¾. We use the multiplication rule (because these are mutually exclusive events) to figure out that the probability of this configuration is $1/4 \times 1/4 \times 3/4 \times 3/4 = 9/256$. There are six different ways that a pod of four seeds can have two green seeds and two yellow seeds (the green seeds can be in positions 1 and 2; 1 and 3; 1 and 4; 2 and 3; 2 and 4; or 3 and 4). Each of these configurations occurs with a probability of 9/256 (as previously described). The probability of *any* of these occurring (*either* one configuration *or* another *or* another) is given by the addition rule, or $9/256 + 9/256 + 9/256 + 9/256 + 9/256 + 9/256 = (9/256) \times 6 = 54/256$, or about 21%.
5. The F_1 chickens have white feathers because the dominant inhibitor allele *I* inhibits expression of the pigment allele *C*.
6. In the population as a whole, there are many copies of each chromosome, so any gene can have multiple alleles present in the different copies. Any one individual can have only two copies of any chromosome, and so any individual can have no more than two different alleles.

Chapter 17

1. A woman whose father is color blind must be heterozygous for the mutant allele. If she has children with a man who is color blind, then half of the female offspring are expected to be homozygous mutant and therefore color blind.
2. Even when one (or more) crossover occurs between the genes, only two of the four products of meiosis are recombinant because crossing over takes place at the four-strand stage of meiosis (Fig. 17.10a). With two strands that are recombinant and two that are nonrecombinant, the frequency of recombination is $2/4 = 50\%$, and so this is the maximum.
3. Independent assortment means that a doubly heterozygous genotype like *AB ab* produces gametes in the ratio ¼ *AB* : ¼ *Ab* : ¼ *aB* : ¼ *ab*. The first and last are nonrecombinant gametes, and the second and third are recombinant gametes. The frequency of recombination is therefore $1/4 + 1/4 = 1/2$, or 50%. This means that independent assortment is observed for genes that are far apart in the same chromosome as well as for genes in different chromosomes.

Chapter 18

1. No. Regression toward the mean simply indicates that, in any population, parents with extreme phenotypes (very tall or very short, for example) will tend to have offspring that are closer to the average height of the population.
2. "Heritability" refers to the relative magnitude of variation in a trait among individuals that can be attributed to differences in genotype

or to differences in environment. A high heritability means that genetic differences account for relatively more variation in the trait than environmental differences. In this case, the 90% heritability means that only 10% of the variation in fingerprint ridge count can be attributed to environment; 90% of the variation is due to differences in genotype.

3. The few genes with large effects are the easier to detect. The magnitude of gene effects that can be detected depends on the number of individuals studied. The larger the number of individuals, the smaller the effects that can be detected. Whatever the size of the study, however, the genes with the largest effects are always easiest to detect, even though there may be few of them.

Chapter 19

1. Combinatorial control requires far fewer genes because there is not a one-to-one correspondence between the genes that are regulated and those that do the regulating. The transcription factors produced by a relatively small number of regulatory genes can, in various combinations, control the expression of a far larger number of target genes. As an analogy, consider the virtually limitless combinations of sounds that can be played on a standard piano that has only 88 keys (36 black and 52 white).

2. Small regulatory RNAs are not translated into proteins (Chapter 3). As a result, they are often called noncoding RNAs since they do not encode for proteins. Their function is often to regulate the expression of other genes. By contrast, mRNAs are translated into proteins.

3. Binding with the small molecule results in a conformational change in one part of the protein that initiates a chain of interactions that propagate and alter its structure at sites far removed from the small molecule. These changes may conceal amino acid side chains responsible for the original DNA-binding specificity and expose a new combination of amino acid side chains that have a different DNA-binding specificity.

4. A mutation in the repressor gene that does not allow the repressor protein to bind allactose means that the repressor will never be blocked from binding the lactose operon promoter. This will lead to a cell that is not able to produce β-galactosidase in the presence or absence of lactose. In other words, the lactose operon will not be inducible.

Chapter 20

1. A cell from the inner cell mass of a blastocyst has more developmental potential than a cell from the ectoderm, which is one of the three germ layers, because the blastocyst occurs earlier in development than the formation of the three germ layers.

2. In Chapter 19, we saw that many common traits result from an interaction of genes and the environment, and so even clones can look different from each other because of environmental influences. Furthermore, if the clone is produced by nuclear transplantation, as in the cases of Dolly and CopyCat, the nuclear genomes of the clone and parent are identical, but their mitochondrial genomes are different.

3. Yes. The embryo would develop normally, even though the embryo does not have normal *bicoid* function, because *bicoid* is a maternal-effect gene, in which the mother's genotype affects the phenotype of the offspring.

4. The gap genes control the expression of the pair-rule genes, which in turn control the expression of segment-polarity genes. Therefore, if gap-gene expression is not normal, you would predict that segment-polarity gene expression would also not be normal.

5. Sexually reproducing organisms carry two copies of most genes, one from the mother and the other from the father. If either of these is knocked out by a loss-of-function mutation, the other is still present and compensates for the mutant. Hence, a loss-of-function mutation is expected to be recessive. In a gain-of-function mutation, one of the gene copies is expressed in the wrong amount, or the wrong tissue, or at the wrong time, and if such a gene turns on a developmental pathway, expression of only one copy of the gene is sufficient to turn the pathway on. Hence, a gain-of-function mutation in a gene that controls a developmental pathway is expected to be dominant.

6. Because A and B together result in petals, and B and C together result in stamens, a flower expressing B activity in all four whorls would have, from outside to inside, petals, petals, stamens, and stamens.

Chapter 21

1. The allele frequency of *a* was calculated as follows:

$$\text{Frequency}(a) = [2 \times (\text{number } aa) + 1 \times (\text{number } Aa)]/[2 \times (\text{total number of individuals})]$$

This equation can be rewritten as

$$\text{Frequency}(a) = [(\text{number } aa) + \tfrac{1}{2} \times (\text{number } Aa)]/(\text{total number of individuals})$$

Note that

$$\text{Number } aa/\text{total number of individuals} = \text{frequency}(aa)$$

and

$$\tfrac{1}{2} \times (\text{number } Aa)/\text{total number of individuals} = \tfrac{1}{2}\,\text{frequency}(Aa)$$

Therefore,

$$\text{Frequency}(a) = \text{frequency}(aa) + \tfrac{1}{2}\,\text{frequency}(Aa)$$

Stated in words, the frequency of allele *a* equals the frequency of *aa* homozygotes plus half the frequency of *Aa* heterozygotes.

By similar logic,

$$\text{Frequency}(A) = \text{frequency}(AA) + \tfrac{1}{2}\,\text{frequency}(Aa)$$

These equations are very useful for determining allele frequencies directly from genotype frequencies.

2. More. Take, for example, a protein variant that can be identified through protein gel electrophoresis. Now that we can sequence the DNA of multiple copies of that allele, we may find that there are several other amino acid differences that do not affect the mobility of the protein on a gel.

3. We can conclude that the population is evolving. What we cannot tell is what mechanism—selection, migration, mutation, genetic drift, or non-random mating—is causing it to evolve. To determine what mechanism(s) is (are) driving the process requires more-detailed population genetics analysis.
4. Sexual selection for bigger size, more elaborate weaponry, or brighter colors in males is an example of directional selection.
5. Adaptation is the fit between an organism and its environment. Of all the evolutionary mechanisms, only selection causes allele frequencies to change based on how they contribute to the success of an individual in terms of survival and reproduction. This means that allele frequencies in the next generation are ultimately governed by the environment. Because phenotype is in part determined by genotype, organisms become adapted to their environment under the influence of selection over time.

Chapter 22

1. Species change over time, making it difficult to craft a single definition that can be applied in all cases. Also, a species concept has to apply to such an astonishing variety of living and dead biological forms—everything from a living microbe to a long-extinct dinosaur—that it seems impossible to find an ideal one-rule-fits-all definition. Thus, though the BSC is useful in many cases, it is not applicable to asexual organisms and organisms known only from fossils.
2. Volcanic island chains offer many examples of adaptive radiation because they are dependent on the serendipitous process of colonization. This means that only some plants and animals are able to get there. In addition, because of the absence of competitors on these islands, there are often many available ecological opportunities for the colonizers. For example, there may be no insect-eating birds on the islands, providing an ecological niche to be filled.
3. If populations of fish became separated from other populations of fish in different ponds during dry periods, this scenario suggests that speciation was more likely to be allopatric rather than sympatric. The allopatrically evolved species became sympatric only when the lake flooded again, combining all the separate ponds into a single body of water.

Chapter 23

1. No. The two trees are equivalent. The node shared by humans and mice is simply rotated in one tree. In both trees, the closest relative of humans is mice because they share a node (where they split) not shared with lizards.
2. A group called "fish" is paraphyletic because it includes some, but not all, of the descendants of a common ancestor. The descendants of the common ancestor of fish also include amphibians, sauropsids, and mammals.
3. The traits are analogous, the result of convergent evolution. Both animals are adapted for swimming in water and converged independently on similar traits, including a streamlined body and fins. However, they are only distantly related, as one is a fish and the other is a mammal.
4. Character traits can be measured for fossils as well as living organisms, so you can examine the features of the skeleton and then construct a phylogenetic tree that takes into account the skeletal traits shared with known vertebrates.

Chapter 24

1. No. Modern humans and modern chimpanzees share a common ancestor. Changes have occurred along both lineages: from the common ancestor to humans, and from the common ancestor to chimpanzees.
2. The multiregional hypothesis suggests that the most recent common ancestor of all living humans lived about 2 million years ago, when *H. ergaster* first migrated from Africa to colonize Europe and Asia. The out-of-Africa hypothesis suggests that *H. sapiens* evolved much more recently in Africa and then moved out of Africa, replacing the remnants of populations originally established by *H. ergaster* when it left Africa. The out-of-Africa hypothesis suggests a much more recent common ancestor for all present-day humans, one that lived about 200,000 years ago.
3. Studies of mtDNA, which is maternally transmitted, indicated that there was no input of Neanderthal mtDNA into the modern human population. In contrast, studies of Neanderthal genomic DNA showed that 1% to 4% of non-Africans' genomes are composed of Neanderthal DNA, implying that there was interbreeding between the ancestors of non-Africans and Neanderthals, presumably when the *H. sapiens* population first left Africa. If there is evidence of genetic input but no evidence of a female-based genetic contribution, we can hypothesize that male Neanderthals were the key contributors.
4. It is inaccurate to label *FOXP2* the "language gene" for several reasons. First, the gene is expressed in many tissues, so its effects are not limited to speech and language. Second, many genes are required for language, not just *FOXP2*. Nevertheless, its association with vocal communication in multiple species is striking.

GLOSSARY

10-nm fiber A relaxed 30-nm chromatin fiber, the state of the chromatin fiber in regions of the nucleus where transcription is currently taking place.

3′ end The end of a nucleic acid strand that carries a free 3′ hydroxyl.

30-nm fiber A chromosomal conformation created by the folding of the nucleosome fiber of DNA and histones.

3-phosphoglycerate (3-PGA) A 3-carbon molecule; two molecules of 3-PGA are the first stable products of the Calvin cycle.

5′ cap The modification of the 5′ end of the primary transcript by the addition of a special nucleotide attached in an unusual chemical linkage.

5′ end The end of a nucleic acid strand containing a free 5′ phosphate group.

ABC model A model of floral development that invokes three activities, A, B, and C, each of which represents the function of a protein or proteins that specify organ identity in each whorl.

abiotic factors Aspects of the physical environment, such as temperature, water availability, or wind, that affect an organism.

abomasum The fourth chamber in the stomach of ruminants, where protein digestion begins.

abscisic acid A plant hormone that triggers stomatal closure, stimulates root elongation, and maintains seed dormancy.

absolute temperature (T) Temperature measured on the Kelvin scale.

absorption The direct uptake of molecules by organisms, commonly to obtain food. In vertebrate digestion, it is the process by which breakdown products are taken up into the bloodstream.

accessory pigment A light-absorbing pigment other than chlorophyll in the photosynthetic membrane; carotenoids are important accessory pigments.

acclimatization An adaptive change in body function to a new environment.

acidic Describes a solution in which the concentration of protons is higher than that of hydroxide ions (the pH is lower than 7).

acoelomate A bilaterian without a body cavity.

acrosome An organelle that surrounds the head of the sperm containing enzymes that enable sperm to transverse the outer coating of the egg.

actin A protein subunit that makes up microfilaments; used by both striated and smooth muscles to contract and generate force.

action potential A brief electrical signal transmitted from the nerve cell body along one or more axon branches.

activation energy (E_A) The energy input necessary to reach the transition state.

activator A synthesized compound that increases the activity of an enzyme.

active site The portion of the enzyme that binds substrate and converts it to product.

active transport The "uphill" movement of substances against a concentration gradient requiring an input of energy.

adaptation In an evolutionary context, the fit between an organism and its environment that results from evolution by natural selection. In sensory reception, the process in which sensory receptors reduce their firing rate when a stimulus continues over a period of time.

adaptive (acquired) immunity The part of the immune system that is specific to given pathogens.

adaptive radiation A period of unusually rapid evolutionary diversification in which natural selection accelerates the rates of both speciation and adaptation in a single lineage.

addition rule The principle that the probability of either of two mutually exclusive outcomes occurring is given by the sum of their individual probabilities.

adenine (A) A purine base.

adenosine triphosphate (ATP) The molecule that provides energy in a form that all cells can readily use to perform the work of the cell. ATP is the universal energy currency for all cells.

adherens junction A beltlike junctional complex composed of cadherins that attaches a band of actin to the plasma membrane.

adrenal glands Paired glands located adjacent to the kidneys that secrete cortisol in times of stress.

adrenal medulla The inner part of the adrenal gland, which is stimulated by the sympathetic nervous system.

adrenaline (epinephrine) A hormone released by the adrenal gland that causes alertness and arousal.

advantageous mutations Genetic changes that improve their carriers' chances of survival or reproduction.

advertisement display Behavior by which individuals draw attention to themselves or their status.

aerobic Utilizing oxygen.

aerobic metabolism Energy metabolism that uses oxygen gas to oxidize organic compounds, generating ATP; found in Bacteria, Archaea, and the mitochondria of eukaryotic organisms.

age structure The number of individuals within each age group of a population.

agonist muscles Muscle pairs that combine to produce similar motions.

aldosterone A hormone produced by the adrenal glands that stimulates the distal convoluted tubule and collecting ducts to take up more salt and water.

alga (plural, algae) A photosynthetic protist.

alkaloid Any one of a group of nitrogen-bearing compounds that damages the nervous system of animals, produced by some plants as a defensive mechanism.

allantois In the amniotic egg, a membrane that encloses a space where metabolic wastes collect.

allele frequency Among all the alleles of a gene in a population, the proportion that are of a specified allele.

alleles The different forms of a gene, corresponding to different DNA sequences in each different form.

allopatric Describes populations that are geographically separated from each other.

allosteric effect A change in the activity or affinity of a protein as the result of binding of a molecule to a site other than the active site.

allosteric enzyme An enzyme whose activity is affected by binding a molecule at a site other than the active site. Typically, allosteric enzymes change their shape on binding an activator or inhibitor.

alpha (α) carbon The central carbon atom of each amino acid.

alpha (α) helix One of the two principal types of secondary structure found in proteins.

alternation of generations The life cycle in which a haploid phase, the gametophyte, and a diploid phase, the sporophyte, follow one after the other.

alternative splicing A process in which primary transcripts from the same gene can be spliced in different ways to yield different mRNAs and therefore different protein products.

altruism Self-sacrificial behavior in which an individual's actions decrease its own fitness while increasing that of another individual.

Alveolata A eukaryotic superkingdom, defined by the presence of cortical alveoli, small vesicles that, in some species, store calcium ions.

alveoli (singular, alveolus) Clusters of tiny thin-walled sacs where gas exchange by diffusion takes place; found at the ends of very fine bronchioles.

amacrine cell A type of interneuron in the retina that communicates between neighboring bipolar cells and ganglion cells, enhancing motion detection and adjusting for changes in illumination of the visual scene.

amine hormone A hormone that is derived from a single aromatic amino acid, such as tyrosine.

amino acid An organic molecule containing a central carbon atom, a carboxyl group, an amino group, a hydrogen atom, and a side chain. Amino acids are the building blocks of proteins.

amino acid replacement A change in the identity of an amino acid at a particular site in a protein resulting from a mutation in the gene.

amino end The end of a polypeptide chain that has a free amino group.

amino group NH_2; a nitrogen atom bonded to two hydrogen atoms, covalently linked to the central carbon atom of an amino acid.

aminoacyl (A) site One of three binding sites for tRNA on the large subunit of a ribosome.

aminoacyl tRNA synthetase An enzyme that attaches a specific amino acid to a specific tRNA molecule.

amnion In the amniotic egg, a membrane surrounding a fluid-filled cavity that allows the embryo to develop in a watery environment.

amniotes The group of vertebrate animals that produces amniotic eggs; this group includes lizards, snakes, turtles, and crocodilians.

amniotic egg An egg that can exchange gases while retaining water, permitting amniotes to reproduce in dry terrestrial habitats that amphibian eggs cannot tolerate.

Amoebozoa A superkingdom of eukaryotes with amoeba-like cells that move and gather food by means of pseudopodia.

Amphibia A monophyletic group of vertebrates, including frogs and salamanders, with an aquatic larval form with gills and an adult terrestrial form that usually has lungs.

amphipathic Having both hydrophilic and hydrophobic regions.

amplified In PCR technology, an alternative term for "replicated."

amylase An enzyme that breaks down starch into smaller subunits.

anabolism The set of chemical reactions that build molecules from smaller units utilizing an input of energy, usually in the form of ATP. Anabolic reactions result in net energy storage within cells and the organism.

anaerobic metabolism Energy metabolism in the absence of oxygen.

analogous Describes similar characters that evolved independently in different organisms as a result of adaptation to similar environments.

anammox Anaerobic ammonia oxidation; energy metabolism in which ammonium ion is oxidized by nitrite, yielding nitrogen gas as a by-product.

anaphase The stage of mitosis in which the sister chromatids separate.

anaphase I The stage of meiosis I in which the two homologous chromosomes of each bivalent separate as they are pulled in opposite directions, but the sister chromatids remained joined at the centromere.

anaphase II The stage of meiosis II in which the centromere of each chromosome splits and the separated chromatids are pulled toward opposite poles of the spindle.

anchor A membrane protein that attaches to other proteins and helps to maintain cell structure and shape.

angiosperms The flowering plants; angiosperms are a monophyletic group of seed plants characterized by flowers, double fertilization, and fruits.

angiotensin II A hormone that causes the smooth muscles of arterioles throughout the body to constrict, which increases blood pressure as well as stimulates the release of the hormone aldosterone, increasing reabsorption of electrolytes and water by the kidneys.

annealing The coming together of complementary strands of single-stranded nucleic acids by base pairing.

annelid worms A phylum of worms (Annelida) that have a cylindrical body with distinct segments and a bilaterian body plan.

annual clock A biological clock that corresponds closely to a solar year.

anoxygenic Not producing oxygen; anoxygenic photosynthetic bacteria do not gain electrons from water and so do not generate oxygen gas.

antagonist muscles Muscle pairs that pull in opposite directions to produce opposing motions.

anterior pituitary gland The region of the pituitary gland that forms from epithelial cells that develop and push up from the roof of the mouth; it receives hormones from the hypothalamus that stimulate it to release hormones in turn.

anther In flowering plants, a structure at the top of a stamen consisting of several sporangia in which pollen is produced.

Anthropocene Period The modern era, so named to reflect the dominant impact of humans on Earth.

antibody A large protein produced by plasma cells that binds to molecules called antigens.

anticodon The sequence of three nucleotides in a tRNA molecule that base pairs with the corresponding codon in an mRNA molecule.

antidiuretic hormone (ADH) A posterior pituitary gland hormone that acts on the kidneys and controls the water permeability of the collecting ducts, thus regulating the concentration of urine that an animal excretes; also known as *vasopressin*.

antigen Any molecule that leads to the production of antibodies.

antigenic drift The gradual process in which a high rate of mutation leads to changes in the amino acid sequences of antigens, thus allowing a population of viruses to evolve over time and evade memory T and B cells.

antigenic shift Reassortment of RNA strands in the viral genome, leading to sudden changes in cell-surface proteins, thus making it difficult to predict from year to year which virus strains will be most prevalent and therefore what vaccine will be most effective.

antigenic variation The encoding of a protein at different times by any one of a number of different genes.

antigen-presenting cell A type of cell (including macrophages, dendritic cells, and B cells) that takes up an antigen and returns portions of it to the cell surface bound to MHC class II proteins.

antiparallel Oriented in opposite directions; the strands in a DNA duplex are antiparallel.

aorta A large artery through which oxygenated blood flows from the left ventricle to the head and rest of the body.

aortic body A sensory structure of the vertebrate aorta that monitors the levels of oxygen and carbon dioxide and the pressure of blood moving through the body.

aortic valve A valve beween the left ventricle and the aorta.

apical dominance The suppression of growth of axillary buds by the shoot apical meristem.

apomixis A process in some species of flowering plants in which seeds can develop even in the absence of fertilization.

appendicular Describes the part of the vertebrate skeleton that consists of the bones of the limbs, including the shoulder and pelvis.

appendix A narrow, tubelike structure that extends from the cecum. The appendix is a vestigial structure that has no clear function in nonherbivorous animals.

aquaporin A protein channel that allows water to flow through the plasma membrane more readily by facilitated diffusion.

aqueous Watery.

aqueous humor A clear watery liquid that fills the interior region in front of the lens of the vertebrate eye.

Archaea One of the three domains of life, consisting of single-celled organisms with a single circular chromosome and no true nucleus that divide by binary fission and differ from bacteria in many aspects of their cell and molecular biology.

Archaeplastida A eukaryotic superkingdom of photosynthetic organisms; includes the land plants.

Ardi A specimen of *Ardipethecus ramidus*, an important early hominin, dating from about 4.4 million years ago.

arteriole A small branch of an artery.

artery A large, high-pressure vessel that moves blood flow away from the heart to the tissues.

Arthropoda A monophyletic group of animals that includes insects and contains more than half of all known animal species; distinguished by their segmented bodies and jointed legs.

artificial selection A form of directional selection analogous to natural selection, but without the competitive element; successful genotypes are selected by the breeder, not by competition.

ascomycetes A monophyletic fungal subgroup of the Dikarya, making up 64% of all fungal species, in which nuclear fusion and meiosis take place in an elongated saclike cell called an ascus; also called *sac fungi*.

asexual reproduction The reproduction of cells or single-celled organisms by cell division; offspring are clones of the parent.

assimilation The process by which organisms take up nutrients from the environment.

associative learning (conditioning) Learning that two events are correlated.

astrocyte A type of star-shaped glial cell that contributes to the blood–brain barrier by surrounding blood vessels in the brain and thus limiting the size of compounds that can diffuse from the blood into the brain.

atom The basic unit of matter.

atomic mass The mass of the atom determined by the number of protons and neutrons.

ATP synthase An enzyme that couples the movement of protons through the enzyme with the synthesis of ATP.

atrioventricular (AV) node A specialized region of the heart containing pacemaker cells that transmit action potentials from the sinoatrial nodes to the ventricles of the heart.

atrioventricular (AV) valve A valve between the right atrium and the right ventricle and between the left ventricle and left atrium.

atrium (plural, atria) A heart chamber that receives blood from the lungs or the rest of the body.

auditory cortex The area of the brain that processes sound.

autocrine signaling Signaling between different parts of a cell; the signaling cell and the responding cell are one and the same.

autoimmune disease A disease in which tolerance is lost and the immune system becomes active against antigens of the host.

autonomic nervous system The involuntary component of the peripheral nervous system, which controls internal functions of the body such as heart rate, blood flow, digestion, excretion, and temperature.

autosome Any chromosome other than the sex chromosomes.

autotroph Any organism that is able to convert carbon dioxide into glucose, thus providing its own organic source of carbon.

auxin A plant hormone that causes shoots to elongate and guides vascular differentiation.

avirulent In plants, describes pathogens that damage only a small part of the plant because the host plant is able to contain the infection. In animals, describes nonpathogenic microorganisms.

axial Describes the part of the vertebrate skeleton that consists of the skull and jaws of the head, the vertebrae of the spinal column, and the ribs.

axillary bud A meristem that forms at the base of each leaf.

axon The fiberlike extension from the cell body of a neuron that transmits signals away from the nerve's cell body; the output end of a nerve cell.

axon hillock The junction of the nerve cell body and its axon.

B lymphocyte (B cell) A cell type that matures in the bone marrow of humans and produces antibodies.

Bacteria One of the three monophyletic domains of life, consisting of single-celled organisms with a single circular chromosome but no nucleus that divide by binary fission and differ from archaeons in many aspects of their cell and molecular biology.

bacteriochlorophyll A light-harvesting pigment closely related to the chlorophyll found in plants, algae, and cyanobacteria.

bacteriophage Virus that infects bacterial cells.

balancing selection Natural selection that acts to maintain two or more alleles of a given gene in a population.

ball-and-socket joint A joint that allows rotation in three axes, like the hip and shoulder.

basal lamina A specialized form of extracellular matrix that underlies and supports all epithelial tissues.

base A nitrogen-containing compound that makes up part of a nucleotide.

base excision pair A specialized repair system in which an incorrect DNA base and its sugar are both removed and the resulting gap is repaired.

base stacking Stabilizing hydrophobic interactions between bases in the same strand of DNA.

basic Describes a solution in which the concentration of protons is lower than that of hydroxide ions (the pH is higher than 7).

basidiomycetes A monophyletic fungal subgroup of the Dikarya, including smuts, rusts, and mushrooms, in which nuclear fusion and meiosis take place in a club-shaped cell called a basidium; also called club fungi.

basilar membrane The membrane that, with the cochlear duct, separates the upper and lower canal of the cochlea.

basophil A type of granulocyte that, along with eosinphils, defends against parasitic infections but also contributes to allergies.

behaviorally isolated Describes individuals that only mate with other individuals on the basis of specific courtship rituals, songs, and other behaviors.

beta (β) sheet One of the two principal types of secondary structure found in proteins.

beta-(β-)oxidation The process of shortening fatty acids by a series of reactions that sequentially remove two carbon units from their ends.

bilateral symmetry Symmetry on both sides of a midline; animals with bilateral symmetry have a distinct head and tail, marking front and back, with a single plane of symmetry running between them at the midline.

Bilateria The monophyletic group of animals with bilateral symmetry.

bilayer A two-layered structure of the cell membrane with hydrophilic "heads" pointing outward toward the aqueous environment and hydrophobic "tails" oriented inward, away from water.

bile A fluid produced by the liver that aids in fat digestion by breaking large clusters of fats into smaller lipid droplets.

binary fission The process by which cells of bacteria or archaeons divide.

binding affinity The tightness of the binding between the receptor and the signaling molecule.

biodiversity Biological diversity; the aggregate number of species, or, more broadly, also the diversity of genetic sequences, cell types, metabolism, life history, phylogenetic groups, communities, and ecosystems.

biodiversity hotspots Relatively small areas that have unusually high numbers of endemic species and that are under threat from human activities.

biogeochemical cycles The cycling of carbon and other biologically important elements through the biosphere.

biological species concept (BSC) As described by Ernst Mayr, the concept that "species are groups of actually or potentially interbreeding populations that are reproductively isolated from other such groups." The BSC is the most widely used and accepted definition of a species, but cannot be applied to Bacteria or Archaea.

biologist A scientist who studies life.

biology The science of life and how it works.

biome The distinctive and stable assemblage of species found over a broad region of Earth; terrestrial biomes are each recognized by their distinctive vegetation.

biomineralization The precipitation of minerals by organisms, as in the formation of skeletons.

biotrophic pathogen A plant pathogen that obtains resources from living cells.

biparental inheritance A type of inheritance in which the organelles in the offspring cells derive from those in both parents.

bipedal Moving by two feet and habitually walking upright.

bipolar cell A type of interneuron in the retina that adjusts its release of neurotransmitter in response to input from rod and cone cells.

biotic factors Any aspect of organisms that affects another organism.

bivalent The four-stranded structure consisting of two pairs of sister chromatids aligned along their length and held together by chiasmata.

bivalves A group of mollusks that includes clams, oysters, and mussels; they have an enclosing skeleton in which two hard shells are connected by a flexible hinge.

bladder A hollow organ in mammals and fishes for the storage and elimination of urine.

blastocyst A hollow sphere produced by cells in the morula that move in relation to one another, pushing against and expanding the membrane that encloses them. A blastocyst forms from the blastula, has an inner cell mass, and occurs only in mammals.

blastula A fluid-filled ball of undifferentiated cells formed after the fertilized egg has undergone several rounds of mitotic cell division following the morula stage.

blending inheritance The now-discredited model in which heredity factors transmitted by the parents become intermingled in the offspring instead of retaining their individual genetic identities.

blood The circulatory fluid in vertebrates.

bone marrow A fatty tissue between trabeculae and within the central cavity of a bone that contains many important cell populations.

bottleneck An extreme, usually temporary, reduction in population size that often results in marked genetic drift.

Bowman's capsule A membranous sac that encases the glomerulus.

brain The centralized concentration of neurons in an organ that processes complex sensory stimuli from the environment or from anywhere in the body.

brainstem Part of the vertebrate brain, formed from the midbrain, which activates the forebrain by relaying information from lower spinal levels.

bronchiole Any one of the fine branches of secondary bronchi.

bryophytes A paraphyletic group of nonvascular plants that includes the mosses, liverworts, and hornworts.

bud scale One of many like structures formed from leaf primordia that, together, protect shoot apical meristems from desiccation and damage due to cold.

budding A form of asexual reproduction in fungi, plants, and some animals in which a bud forms on the organism and eventually breaks off to form a new organism that is smaller than its parent.

bulbourethral glands Glands below the prostate gland that produce a clear fluid that lubricates the urethra for passage of the sperm.

bulk flow The movement of molecules through organisms due to pressure differences at rates beyond those possible by diffusion across a concentration gradient.

bundle sheath A cylinder of cells that surrounds each vein in C_4 plants in which carbon dioxide is concentrated in bundle sheath cells, suppressing photorespiration.

Burgess Shale A sedimentary rock formation in British Columbia, Canada, that preserves a remarkable sampling of marine life during the initial diversification of animals.

C_3 plant A plant that does not use 4-carbon organic acids to supply the Calvin cycle with carbon dioxide.

C_4 plant A plant in which carbon dioxide is incorporated into 4-carbon organic acids that are then used to supply the Calvin cycle.

cadherin A calcium-dependent adherence protein, important in the adhesion of cells to other cells.

calmodulin A protein that binds with Ca^{21} and activates the enzyme myosin kinase.

Calvin cycle The process in which carbon dioxide is reduced to synthesize carbohydrates, with ATP and NADPH as the energy sources.

Cambrian explosion A transition period in geologic time during which the body plans characteristic of most bilaterian phyla developed.

canine One of the teeth in carnivores specialized for piercing the body of prey.

capacitation A series of physiological changes that allow the sperm to fertilize the egg.

capillary A very small blood vessel, arranged in finely branched networks connected to arterioles or venules, where gases are exchanged by diffusion with surrounding tissues.

capsid The protein coat that surrounds the nucleic acid of a virus.

carbohydrate An organic molecule containing C, H, and O atoms that provides a source of energy for metabolism and that forms the starting point for the synthesis of all other organic molecules.

carbon cycle The intricately linked network of biological and physical processes that shuttles carbon among rocks, soil, oceans, air, and organisms.

carboxyl end The end of a polypeptide chain that has a free carboxyl group.

carboxyl group COOH; a carbon atom with a double bond to oxygen and a single bond to a hydroxyl group.

carboxylation The first step of the Calvin cycle, in which carbon dioxide absorbed from the air is added to a 5-carbon molecule.

cardiac cycle The contraction of the two atria of the heart followed by contraction of the two ventricles.

cardiac muscle Muscle cells that make up the walls of the atria and ventricles and contract to pump blood through the heart.

cardiac output (CO) The volume of blood pumped by the heart over a given interval of time, the key measure of heart function.

carnivores A monophyletic group of animals, including cats, dogs, seals, and their relatives, that consume other animals.

carotid body A chemosensory structure of the carotid artery that monitors the levels of oxygen and carbon dioxide and the pH of blood moving to the brain.

carpel An ovule-producing floral organ in the center whorl.

carrier A transporter that facilitates movement of molecules.

carrying capacity (K) The maximum number of individuals a habitat can support.

cartilage A type of connective tissue found, for example, in the walls of intervertebral discs and the joint surfaces between adjacent bones.

Casparian strip A thin band of hydrophobic material that encircles each cell of the endodermis of a root, controlling which materials enter the xylem.

catabolism The set of chemical reactions that break down molecules into smaller units and, in the process, produces ATP to meet the energy needs of the cell.

causation A relationship in which one event leads to another.

cavitation The abrupt replacement of the water in a conduit by water vapor, which blocks water flow in xylem.

cecum A chamber that branches off the large intestine; along with the colon, the site of hindgut fermentation.

cell The simplest self-replicating entity that can exist as an independent unit of life.

cell adhesion molecule A cell-surface protein that attaches cells to one another and to the extracellular matrix.

cell cycle The collective name for the steps that make up eukaryotic cell division.

cell division The process by which cells make more cells.

cell junction A complex of proteins in the plasma membrane where a cell makes contact with another cell or the extracellular matrix.

cell theory The theory that the cell is the fundamental unit of life in all organisms and that cells come only from preexisting cells.

cell wall A defining boundary in many organism, external to the cell membrane, that helps maintain the shape and internal composition of the cell.

cell-mediated immunity The ability of T cells, which do not secrete antibodies, to recognize and act against pathogens directly.

cellular blastoderm In *Drosophila* development, the structure formed by the nuclei in the single-cell embryo when they migrate to the periphery of the embryo and each nucleus becomes enclosed in its own cell membrane.

cellular respiration A series of chemical reactions that convert the energy stored in nutrients into a chemical form that can be readily used by cells.

cellulase The enzyme that breaks down cellulose.

central dogma The theory that information transfer in a cell usually goes from DNA to RNA to protein.

central nervous system (CNS) In vertebrates, the brain and spinal cord; in invertebrates, centralized information-processing ganglia.

centromere A constriction that physically holds sister chromatids together; the site of the attachment of the spindle fibers that move the chromosome in cell division.

centrosome A compact structure that is the microtubule organizing center for animal cells.

cephalization The concentration of nervous system components at one end of the body.

cephalochordates A subphylum of Chordata that shares key features of body organization with vertebrates but lacks a well-developed brain and eyes, has no lateral appendages, and does not have a mineralized skeleton.

cephalopods A monophyletic group of mollusks, including squid, cuttlefish, octopus, and chambered nautilus, with distinctive adaptations such as well-developed eyes and muscular tentacles that capture prey and sense the environment.

cerebellum Part of the vertebrate brain, formed from the hindbrain, which coordinates complex motor tasks by integrating motor and sensory information.

cerebral cortex Part of the vertebrate brain formed from a portion of the forebrain that is greatly expanded in mammals, particularly primates.

cerebrum The outer left and right hemispheres of the cerebral cortex.

cervix The end, or neck, of the uterus.

chain terminator A term for a dideoxynucleotide, which if incorporated into a growing daughter strand stops strand growth because there is no hydroxyl group to attack the incoming nucleotide.

channel A transporter with a passage that allows the movement of molecules through it.

chaperone A protein that helps shield a slow-folding protein until it can attain its proper three-dimensional structure.

character In the discipline of systematics, an anatomical, physiological, or molecular feature of an organism that varies among closely related species.

character state The observed condition of a character, such as presence or absence of lungs or arrangement of petals.

checkpoint One of multiple regulatory mechanism that coordinate the temporal sequence of events in the cell cycle.

chelicerates One of the four main groups of arthropods, including spiders and scorpions, chelicerates have pincer-like claws and are the only arthropods that lack antennae.

chemical bond Any form of attraction between atoms that holds them together.

chemical energy A form of potential energy held in the chemical bonds between pairs of atoms in a molecule.

chemical reaction The process by which molecules are transformed into different molecules.

chemoautotroph A microorganism that obtains energy from chemical compounds, not from sunlight.

chemoreceptor A receptor that responds to molecules that bind to specific protein receptors on the cell membrane of the sensory receptor.

chemiosmotic hypothesis The hypothesis that the gradient of protons across a membrane provides a source of potential energy that is converted into chemical energy stored in ATP.

chemotroph An organism that derives its energy directly from organic molecules such as glucose.

chiasma (plural, chiasmata) A crosslike structure within a bivalent constituting a physical manifestation of crossing over.

chitin A modified polysaccharide containing nitrogen that makes up the cell walls of fungi and the hard exoskeletons of arthropods.

chloride cell A type of specialized cell in the gills of marine bony fishes that counters the ingestion and diffusion of excess electrolytes into the animal by pumping chloride ions into the surrounding seawater; chloride cells in freshwater fishes have opposite polarity.

chlorophyll The major photosynthetic pigment contained in the thylakoid membrane; it plays a key role in the chloroplast's ability to capture energy from sunlight. Chlorophyll appears green because it is poor at absorbing green wavelengths.

chloroplast An organelle that converts energy of sunlight into chemical energy by synthesizing simple sugars.

chloroplast genome In photosynthetic eukaryotes, the genome of the chloroplast.

choanocyte A type of cell that lines the interior surface of a sponge; choanocytes have flagella and function in nutrition and gas exchange.

choanoflagellate One of a group of mostly unicellular protists characterized by a ring of microvilli around the cell's single flagellum.

cholecystokinin (CCK) A peptide hormone that causes the gallbladder to contract and thus release bile into the duodenum.

cholesterol An amphipathic lipid that is a major component of animal cell membranes.

Chondrichthyes Cartilaginous fish, a monophyletic group that includes about 800 species of sharks, rays, and chimaeras.

Chordata One of the three major phyla of deuterostomes, this group includes vertebrates and closely related invertebrate animals such as sea squirts.

chorion In the amniotic egg, a membrane that surrounds the entire embryo along with its yolk and allantoic sac.

chromatin A complex of DNA, RNA, and proteins that gives chromosomes their structure; chromatin fibers are either 30 nm in diameter or, in a relaxed state, 10 nm.

chromatin remodeling The process in which the nucleosomes are repositioned to expose different stretches of DNA to the nuclear environment.

chromosome In eukaryotes, the physical structure in which DNA in the nucleus is packaged; used more loosely to refer to the DNA in bacterial cells or archaeons.

chromosome condensation The progressive coiling of the chromatin fiber, an active, energy-consuming process requiring the participation of several types of proteins.

chytrid A single-celled aquatic fungus with chitin walls that attaches to decomposing organic matter.

cilium (plural, cilia) A hairlike organelle that propels the movement of cells or of substances within cells or out of the body; shorter than a flagellum.

circadian clock A biological clock, on a near-daily cycle, that can be set by external cues and regulates many daily rhythms. In animals, the circadian clock affects feeding, sleeping, hormone production, and core body temperature.

circular muscle Smooth muscle that encircles the body or an organ; in the digestive tract, a circular muscle layer contracts to reduce the size of the lumen. A circular muscle layer contracts alternately with longitudinal muscle to move contents through the digestive tract and to enable locomotion in animals with hydrostatic skeletons.

circulation The movement of a specialized body fluid that carries oxygen and carbon dioxide, and nutrients and waste products, through the body.

cis-regulatory element A short DNA sequence adjacent to a gene, usually at the 54′ end, that interacts with transcription factors.

cisternae (singular, cisterna) The series of flattened membrane sacs that make up the Golgi apparatus.

citric acid cycle The third stage of cellular respiration, in which acetyl-CoA is broken down and more carbon dioxide is released.

cladistics Phylogenetic reconstruction on the basis of shared evolutionary changes in characters, often called synapomorphies.

class In the Linnaean system of classification, a group of closely related orders.

classical conditioning Associative learning in which two stimuli are paired.

cleavage The successive mitotic divisions of the zygote after fertilization, in which the single large egg is divided into many smaller cells.

climax community A mature assembly, a final stage in succession, in which there is little further change in species composition.

clitoris The female homolog of the glans penis.

clonal selection A hypothesis proposing that the antigen instructs the antibody to fold in a particular way so that the two interact in a specific manner; now a central principle of immunology.

clone An individual that carries an exact copy of the nuclear genome of another individual; clones are genetically identical cells or individuals.

closed circulatory system A circulatory system made up of a set of internal vessels and a heart that functions as a pump to move blood to different regions of the body.

Cnidaria A phylum characterized by radial symmetry, two germ tissues in the embryo, a closed internal gastric cavity, and well-developed tissues but not organs; includes jellyfish, sea anemones, and corals.

cochlea A coiled chamber within the skull containing hair cells that convert pressure waves into an electrical impulse that is sent to the brain.

cochlear duct A fluid-filled cavity in the cochlea, next to the upper canal, that houses the organ of Corti.

codon A group of three adjacent nucleotides in RNA that specifies an amino acid in a protein or that terminates polypeptide synthesis.

coelacanth A genus of two species of lobe-finned fish found off the coast of Africa and thought to have been extinct for 80 million years but which is still living today; along with lungfish, the nearest relative of tetrapods.

coelom A body cavity surrounding the gut.

coenocytic Containing many nuclei within one giant cell; the nucleus divides multiple times, but the nuclei are not partitioned into individual cells.

coenzyme Q (CoQ) In respiration, a mobile electron acceptor that transports electrons from complexes I and II to complex III in the electron transport chain and moves protons from the mitochondrial matrix to the intermembrane space.

coevolution The process in which species evolve together, each responding to selective pressures from the other.

cofactor A substance that associates with an enzyme and plays a key role in its function.

cognition The ability of the brain to process and integrate complex sources of information, interpret and remember past events, solve problems, reason, and form ideas.

cohesion Attraction between molecules; one consequence of cohesion is high surface tension.

cohort A group of the individuals born at a given time.

collagen A strong protein fiber found in bone, artery walls, and other connective tissues.

collecting duct A type of duct in the vertebrate kidney where urine collects.

colon Part of the hindgut and the site of reabsorption of water and minerals; also known as the *large intestine*.

combinatorial control Regulation of gene transcription by means of multiple transcription factors acting together.

commensalism An interaction between species in which one partner benefits with no apparent effect on the other.

communication The transfer of information between two individuals, the sender and the receiver.

community The set of all populations found in a given place.

compact bone Dense, mineralized bone tissue that forms the walls of a bone's shaft.

companion cell In angiosperms, a cell that carries out cellular functions such as protein synthesis for sieve tubes.

comparative genomics The analysis of the similarities and differences in protein-coding genes and other types of sequence in the genomes of different species.

competition An interaction in which the use of a mutually needed resource by one individual or group of individuals lowers the availability of the resource for another individual or group.

competitive exclusion The result of an antagonistic interaction in which one species is prevented from occupying a particular habitat or niche.

complement system The collective name for certain proteins circulating in the blood that participate in innate immune function and thus complement other parts of the immune system.

complementary Describes the relationship of purine and pyrimidine bases, in which the base A pairs only with T and G pairs only with C.

complex carbohydrate A long, branched chain of monosaccharides.

complex trait A trait that is influenced by multiple genes as well as by the environment.

compound eye An eye structure found in insects and crustaceans that consists of a number of ommatidia, individual light-focusing elements.

concordance The percentage of cases in which both members of a pair of twins show the trait when it is known that at least one member shows it.

cone cell A type of photoreceptor cell on the retina that detects color.

conjugation The direct cell-to-cell transfer of DNA, usually in the form of a plasmid.

conservation biology The efforts by biologists and policymakers to address the challenge of sustaining biodiversity in a changing world crowded with people.

conserved Describes sequences that are similar in different organisms.

constant (C) Describes an unchanging region of the H and L chains.

constitutive Describes expression of a gene that occurs continuously.

constitutive defense A defense that is always active.

consumer An organism that obtains the carbon it needs for growth and reproduction from the foods it eats and gains energy by respiring food molecules; heterotrophic organisms of all kinds that directly consume primary producers or consume those that do.

contractile ring In animal cells, a ring of actin filaments that forms at the equator of the cell perpendicular to the axis of what was the spindle at the beginning of cytokinesis.

contractile vacuole A type of cellular compartment that takes up excess water and waste products from inside the cell and expels them into the external environment.

control group The group that is not exposed to the variable in an experiment.

cooperative binding The increase in binding affinity with additional binding of O_2.

copy-number variation (CNV) Differences among individuals in the number of copies of a region of the genome.

Coriolis effect The phenomenon in which, because of Earth's counterclockwise rotation about its axis, winds and oceanic currents in the Northern Hemisphere deflect to the right, and those in the Southern Hemisphere deflect to the left.

cork cambium Lateral meristem that renews and maintains an outer layer that protects the stem against herbivores, mechanical damage, desiccation, and fire.

cornea The transparent portion of the sclera in the front of the vertebrate eye.

corpus luteum A temporary endocrine structure that secretes progesterone.

correlation The co-occurrence of two events or processes; correlation does not imply causation.

corridors Areas that provide species with routes for migration from one reserve to another.

cortex In a stem, the region between the epidermis and the vascular bundles, composed of parenchyma cells. In the mammalian brain, the highly folded outer layer of gray matter, about 4 mm thick, made up of densely packed neuron cell bodies and their dendrites. In the mammalian renal system, the outer layer of the kidney.

co-speciation A process in which two groups of organisms speciate in response to each other and at the same time, producing matching phylogenies.

countercurrent exchange A mechanism in which two fluids flow in opposite directions, exchanging properties.

countercurrent multiplier A system that generates a concentration gradient as two fluids move in parallel but in opposite directions.

covalent bond A chemical bond formed by a shared pair of electrons holding two different atoms together.

CpG island A cluster of CpG sites on a DNA strand where cytosine (C) is adjacent to guanosine (G); the "p" represents the phosphate in the backbone.

cranial nerve In vertebrates, a nerve that links specialized sensory organs to the brain; most contain axons of both sensory and motor neurons.

craniates A subphylum of Chordata, distinguished by a bony cranium that protects the brain; also known as *vertebrates.*

crassulacean acid metabolism (CAM) A mechanism in plants that helps balance carbon dioxide gain and water loss by capturing carbon dioxide into 4-carbon organic acids at night, when transpiration rates are low, and then using it to supply the Calvin cycle during the day while stomata remain closed.

Crenarchaeota One of the three major divisions of Archaea; includes acid-loving microorganisms.

CRISPR (clustered regularly interspaced short palindromic repeats) A method of DNA editing in which any sequence in the genome can be replaced with any other sequence; the phrase describes the organization of the viral DNA segments in certain bacterial genomes from which the DNA editing method derives.

crisscross inheritance A pattern in which an X chromosome present in a male in one generation is transmitted to a female in the next generation, and in the generation after that can be transmitted back to a male.

Cro-Magnon The first known population of *Homo sapiens* in Europe, named for the site in France where specimens were first described.

crop The last part of the foregut, which serves as an initial storage and digestive chamber; also known as the *stomach.*

cross-bridge The binding of the head of a myosin molecule to actin at a specific site between the myosin and actin filaments.

cross-bridge cycle Repeated sequential interactions between myosin and actin filaments at cross-bridges that cause a muscle fiber to contract.

crosscurrent Running across another current at 90 degrees.

crossover The physical breakage, exchange of parts, and reunion between non-sister chromatids.

crustaceans One of the four main groups of arthropods, including lobsters, shrimp, and crabs; distinguished by two pairs of antennae and their branched legs or other appendages.

ctenophores Comb-jellies; species in this phylum have a radial body plan but a flow-through gut; they propel themselves by cilia arranged like a comb along the long axis of the body.

cultural transmission The transfer of information among individuals through learning or imitation.

cuticle In leaves, a protective layer of a waxy substance secreted by epidermal cells that limits water loss; also, an exoskeleton that covers the bodies of invertebrates such as nematodes and arthropods.

C-value paradox The disconnect between genome size and organismal complexity (the C-value is the amount of DNA in a reproductive cell).

cyanobacteria A monophyletic group of bacteria capable of oxygenic photosynthesis.

cyclic electron transport An alternative pathway for electrons during the Calvin cycle that increases the production of ATP.

cyclin A regulatory protein whose levels rise and fall with each round of the cell cycle.

cyclin-dependent kinase (CDK) A kinase that is always present within the cell but active only when bound to the appropriate cyclin.

cytochrome-*b6f* complex Part of the photosynthetic electron transport chain, through which electrons pass between photosystem II and photosystem I.

cytochrome c The enzyme to which electrons are transferred in complex III of the electron transport chain.

cytokine A chemical messenger released by phagocytes that recruits other immune cells to the site of injury or infection.

cytokinesis In eukaryotic cells, the division of the cytoplasm into two separate cells.

cytokinin A plant hormone that stimulates cell division.

cytoplasm The contents of the cell other than the nucleus.

cytosine (C) A pyrimidine base.

cytoskeleton In eukaryotes, an internal protein scaffold that helps cells to maintain their shape and serves as a network of tracks for the movement of substances within cells.

cytosol The region of the cell inside the plasma membrane but outside the organelles; the jelly-like internal environment that surrounds the organelles.

cytotoxic T cell One of a subpopulation of T cells, activated by cytokines released from helper T cells, that can kill other cells.

daughter strand In DNA replication, the strand synthesized from a parental template strand.

day-neutral plant A plant that flowers independently of any change in day length.

decomposer An organism that breaks down dead tissues, feeding on the dead cells or bodies of other organisms.

delayed hypersensitivity reaction Reactions initiated by helper T cells, which release cytokines that attract macrophages to the site of exposure, which is typically the skin.

deleterious mutations Genetic changes that are harmful to an organism.

deletion A missing region of a gene or chromosome.

demography The study of the size, structure, and distribution of populations over time, including changes in response to birth, aging, migration, and death.

denaturation The unfolding of proteins by chemical treatment or high temperature; the separation of paired, complementary strands of nucleid acid.

dendrite A fiberlike extension from the cell body of a neuron that receives signals from other nerve cells or from specialized sensory endings; the input end of a nerve cell.

dendritic cell A type of cell with long cellular projections that is typically part of the natural defenses found in the skin and mucous membranes.

denitrification The process in which some bacteria use nitrate as an electron acceptor in respiration.

density-dependent Describes processes affecting populations that are influenced by the number of individual organisms, such as the use of resources or susceptibility to predation or parasitism.

density-independent Describes factors such as severe drought that influence population size without regard for the density of the population.

deoxyribonucleic acid (DNA) A linear polymer of four subunits; the information archive in all organisms.

deoxyribose The sugar in DNA.

depolarization An increase in membrane potential

dermis The layer of skin beneath the epidermis, consisting of connective tissue, hair follicles, blood and lymphatic vessels, and glands. It supports the epidermis both physically and by supplying it with nutrients and provides a cushion surrounding the body.

desiccation Excessive water loss; drying out.

desiccation tolerance A suite of biochemical traits that allows cells to survive extreme dehydration without damage to membranes or macromolecules.

desmosome A buttonlike point of adhesion that holds the plasma membranes of adjacent cells together.

deuterostome A bilaterian in which the blastopore, the first opening to the internal cavity of the developing embryo, becomes the anus. The taxonomic name is Deuterostomia and includes humans and other chordates.

development The process in which a fertilized egg undergoes multiple rounds of cell division to become an embryo with specialized tissues and organs.

diabetes mellitus A disease that results when the control of blood-glucose levels by insulin fails.

diaphragm A domed sheet of muscle at the base of the lungs that separates the thoracic and abdominal cavities and contracts to drive inhalation.

diaphysis The middle region of a bone; blood vessels invading the diaphysis and epiphysis trigger the transformation of cartilage into bone.

diastole The relaxation of the ventricles, allowing the heart to fill with blood.

dideoxynucleotide A nucleotide lacking both the 2′ and 3′ hydroxyl groups on the sugar ring.

dietary mineral A chemical element other than carbon, hydrogen, oxygen, or nitrogen that is required in the diet and must be obtained in food; see also *minerals.*

differentiation The process in which cells become progressively more specialized as a result of gene regulation.

diffusion The random motion of individual molecules, with net movement occurring where there are areas of higher and lower concentration of the molecules.

digestive tract Collectively, the passages that connect the mouth, digestive organs, and anus; also known as the *gut.*

Dikarya A vast fungal group that includes about 98% of all described fungal species and in which dikaryotic cells are formed.

dikaryotic ($n + n$) Having two haploid nuclei, one from each parent, in each cell.

diploid Describes a cell with two complete sets of chromosomes.

directional selection A form of selection that selects against one of two extremes and leads over time to a change in a trait.

dispersal The process in which some individuals colonize a distant place far from the main source population.

display A pattern of behavior that is species specific and tends to be highly repeatable and similar from one individual to the next.

disruptive selection A form of selection that operates in favor of extremes and against intermediate forms, selecting against the mean.

distal convoluted tubule The third portion of the renal tubule, in which urea is the principal solute and into which other wastes from the bloodstream are secreted.

disturbance A severe physical impact on a habitat that has density-independent effects on populations of interacting species.

divergence The slow accumulation of differences between duplicate copies of a gene that occurs over time.

DNA fingerprinting The analysis of a small quantity of DNA to uniquely identify an individual; also known as *DNA typing.*

DNA ligase An enzyme that uses the energy in ATP to close a nick in a DNA strand, joining the 3′ hydroxyl of one end to the 5′ phosphate of the other end.

DNA polymerase An enzyme that is a critical component of a large protein complex that carries out DNA replication.

DNA replication The process of duplicating a DNA molecule, during which the parental strands separate and new partner strands are made.

DNA transposons Repeated DNA sequences that replicate and can move from one location to another in the genome by DNA replication and repair.

DNA typing The analysis of a small quantity of DNA to uniquely identify an individual; also known as *DNA fingerprinting.*

domain One of the three largest limbs of the tree of life: Eukarya, Bacteria, or Archaea.

dominant The trait that appears in the heterozygous offspring of a cross between homozygous genotypes.

donor In recombinant DNA technology, the source of the DNA fragment that is inserted into a cell of another organism.

dormancy Describes a time period in the life of an organism when growth, development, and metabolism slow down or stop; in plants, a state in which seeds are prevented from germinating.

dorsal nerve cord A nerve cord that develops in a location dorsal to the notochord; this embryonic feature is unique to chordates.

dosage The number of copies of each gene in a chromosome.

dosage compensation The differential regulation of X-chromosomal genes in females and in males.

double bond A covalent bond in which covalently joined atoms share two pairs of electrons.

double fertilization In angiosperms, the process in which two sperm from a single pollen tube fuse with (1) the egg and (2) the two haploid nuclei of the central cell.

double helix The structure formed by two strands of complementary nucleotides that coil around each other.

Down syndrome A condition resulting from the presence of an extra copy of chromosome 21; also known as *trisomy 21.*

downstream gene A gene that functions later than another in development.

duodenum The initial section of the small intestine, into which food enters from the stomach.

duplication A region of a chromosome that is present twice instead of once.

duplication and divergence The process of creating new genes by duplication followed by change in sequence over evolutionary time.

dynamic instability Cycles of shrinkage and growth in microtubules.

dynein A motor protein that carries cargo away from the plasma membrane toward the minus ends of microtubules.

eardrum In mammals, another name for the *tympanic membrane,* which transmits airborne sounds into the ear.

Echinodermata One of the three major phyla of deuterosomes, defined by five-part symmetry; this group includes sea urchins and sea stars.

echolocation Using sound waves to locate an object; bats find insect prey by emitting short bursts of high-frequency sound that bounce off surrounding objects and are reflected to the bat's ears.

ecological footprint The quantification of individual human claims on global resources by adding up all the energy, food, materials, and services used and estimating how much land is required to provide those resources.

ecological isolation Pre-zygotic isolation between individuals that specialize ecologically in different ways.

ecological niche A complete description of the role a species plays in its environment.

ecological species concept (ESC) The concept that there is a one-to-one correspondence between a species and its niche.

ecology The study of how organisms interact with one another and with their physical environment in nature.

ecosystem A community of organisms and the physical environment it occupies.

ecosystem services Benefits to humans provided by biodiversity, such as cleaner air and water, greater primary productivity, improved resilience to environmental disruption, and untapped sources of food and molecular compounds for use in medicine and agriculture.

ectoderm The outer germ layer, which differentiates into epithelial cells, pigment cells in the skin, nerve cells in the brain, and the cornea and lens of the eye.

ectomycorrhizae One of the two main types of mycorrhizae; ectomycorrhizae produce a thick sheath of fungal cells (hyphae) that surround, but do not penetrate, root cells.

ectotherm An animal that obtains most of its heat from the environment.

ejaculatory duct The duct through which sperm travel from the vas deferens to the urethra.

elastin A protein fiber found in artery walls that provides elasticity.

electrochemical gradient A gradient that combines the charge gradient and the chemical gradient of protons and other ions.

electromagnetic receptor A receptor that responds to electrical, magnetic, or light stimuli.

electron A negatively charged particle that moves around the atomic nucleus.

electron carrier A molecule that carries electrons (and energy) from one set of reactions to another.

electron donor A molecule that loses electrons.

electron transport chain The system that transfers electrons along a series of membrane-associated proteins to a final electron acceptor, using the energy released as electrons move down the chain to produce ATP.

electronegativity The ability of atoms to attract electrons.

electroreceptor A sensory receptor found in some fish that enables them to detect weak electrical signals emitted by other organisms.

element A pure substance, such as oxygen, copper, gold, or sodium, that cannot be further broken down by the methods of chemistry.

elongation The process in protein translation in which successive amino acids are added one by one to the growing polypeptide chain.

elongation factor A protein that breaks the high-energy bonds of the molecule GTP to provide energy for ribosome movement and elongation of a growing polypeptide chain.

embryo An early stage of multicellular development that results from successive mitotic divisions of the zygote.

endemic species Species found in one place in the world and nowhere else.

endergonic Describes reactions with a positive □G that are not spontaneous and so require an input of energy.

endocrine signaling Signaling by molecules that travel through the bloodstream.

endocrine system A system of cells and glands that secretes hormones and works with the nervous system to regulate an animal's internal physiological functions.

endocytosis The process in which a vesicle buds off from the plasma membrane, bringing material from outside the cell into that vesicle, which can then fuse with other organelles.

endoderm The germ layer that differentiates into cells of the lining of the digestive tract and lung, liver cells, pancreas cells, and gallbladder cells.

endodermis In plants, a layer of cells surrounding the xylem and phloem at the center of the root that controls the movement of nutrients into the xylem. Also, the inner lining of the cnidarian body.

endomembrane system A cellular system that includes the nuclear envelope, the endoplasmic reticulum, the Golgi apparatus, lysosomes, the plasma membrane, and the vesicles that move between them.

endomycorrhizae One of the two main types of mycorrhizae; endomycorrhizal hyphae penetrate into root cells, where they produce highly branched structures (arbuscules) that provide a large surface area for nutrient exchange.

endophyte A fungus that lives within leaves and that may help the host plant by producing chemicals that deter pathogens and herbivores.

endoplasmic reticulum (ER) The organelle involved in the synthesis of proteins and lipids.

endoskeleton The bony skeletal system of vertebrate animals, which lies internal to most of the animal's soft tissues.

endosperm A tissue formed by many mitotic divisions of a triploid cell, it supplies nutrition to the angiosperm embryo.

endosymbiosis A symbiosis in which one partner lives within the other.

endotherm An animal that produces most of its own heat as by-products of metabolic reactions.

energetic coupling The driving of a non-spontaneous reaction by a spontaneous reaction.

energy A property of objects that can be transferred from one object to another, and that cannot be created or destroyed

energy balance A form of homeostasis in which the amount of energy calories from food taken in equals the amount of calories used over time to meet metabolic needs.

energy intake Sources of energy.

energy use The ways in which energy is expended.

enhancer A specific DNA sequence necessary for transcription.

enthalpy (H) The total amount of energy in a system.

entropy (S) The degree of disorder in a system.

envelope A lipid structure that surrounds the capsids of some viruses.

environmental risk factor A characteristic in a person's surroundings that increases the likelihood of developing a particular disease.

environmental variation Variation among individuals that is due to differences in the environment.

enzyme A protein that functions as a catalyst to accelerate the rate of a chemical reaction; enzymes are critical in determining which chemical reactions take place in a cell.

eosinophil A type of granulocyte that, along with basophils, defends against parasitic infections but also contributes to allergies.

epidermis In mammals, the outer layer of skin, which serves as a water-resistant, protective barrier. In plants, the outermost layer of cells in leaves, young stems (lacking secondary growth), and roots. Also, the outer layer of the cnidarian body.

epididymis An organ that lies above the testes where sperm become motile and are stored prior to ejaculation.

epigenetic Describes effects on gene expression due to differences in DNA packaging, such as modifications in histones or chromatin structure.

epiglottis A flap of tissue at the bottom of the pharynx that prevents food from entering the trachea and lungs.

epiphysis The end region of a bone; blood vessels invading the epiphysis and diaphysis trigger the transformation of cartilage into bone.

epiphyte A plant that grows high in the canopy of other plants, or on branches or trunks of trees, without contact with the soil.

epistasis Interaction between genes that modifies the phenotypic expression of genotypes.

equational division Another name for meiosis II because cells in meiosis II have the same number of chromosomes at the beginning and at the end of the process.

esophagus Part of the foregut; the passage from the mouth to the stomach.

essential amino acid An amino acid that cannot be synthesized by cellular biochemical pathways and instead must be ingested.

estrogen A hormone secreted by the ovaries that stimulates the development of female secondary sexual characteristics.

estrus cycle A cycle in placental mammals other than humans and chimpanzees characterized by phases in which females are sexually receptive.

ethanol fermentation The fermentation pathway in plants and fungi during which pyruvate releases carbon dioxide to form acetaldehyde and electrons from NADH are transferred to acetaldehyde to produce ethanol and NAD^+.

Eukarya The eukaryotes; one of the three domains of life, in which cells have a true nucleus and divide by mitosis.

eukaryote An organism whose cells have a true nucleus.

Eumetazoa The monophyletic group of all animals other than sponges.

Euryarchaeota One of the three major divisions of Archaea; includes acid-loving, heat-loving, methane-producing, and salt-loving microorganisms.

eusocial Describes behavior most commonly observed in species of Hymenopteran insects, in which they have overlapping generations in a nest, cooperative care of the young, and clear and consistent division of labor between reproducers (the queen of a honeybee colony) and nonreproducers (the workers).

eutrophication The process in which added nutrients lead to a great increase in the populations of algae and cyanobacteria.

evaporation The amount of water evaporated from the Earth's surface, including ponds, rivers, and soil.

evo-devo The short name for evolutionary-developmental biology, a field of study that compares the genetic programs for growth and development in species on different branches of phylogenetic reconstructions.

evolution Changes in the genetic make-up of populations over time, sometimes resulting in adaptation to the environment and the origin of new species.

evolutionarily conserved Characteristics that persist relatively unchanged through diversification of a group of organisms and therefore remain similar in related species.

evolutionarily stable strategy A type of behavior that cannot readily be driven to extinction by an alternative strategy.

excitation–contraction coupling The process that produces muscle force and movement, by excitation of the muscle cell coupled to contraction of the muscle.

excitatory postsynaptic potential (EPSP) A positive change in the postsynaptic membrane potential.

excretion The elimination of waste generated by metabolism.

excretory tubule In renal systems, a type of tube that drains waste products and connects to the outside of the body.

exergonic Describes reactions with a negative ΔG that release energy and proceed spontaneously.

exhalation The expelling of oxygen-poor air by the elastic recoil of the lungs and chest wall.

exit (E) site One of three binding sites for tRNA on the large subunit of a ribosome.

exocytosis The process in which a vesicle fuses with the plasma membrane and empties its contents into the extracellular space or delivers proteins to the plasma membrane.

exon A sequence that is left intact in mRNA after RNA splicing.

exoskeleton A rigid skeletal system that lies external to the animal's soft tissues.

experimentation A disciplined and controlled way of learning about the world and testing hypotheses in an unbiased manner.

exponential growth The pattern of population increase that results when r (the per capita growth rate) is constant through time.

expressed Turned on or activated, as a gene or protein.

extension (PCR) A step in the polymerase chain reaction (PCR) for producing new DNA fragments in which the reaction mixture is heated to the optimal temperature for DNA polymerase, and each primer is elongated by means of deoxynucleoside triphophosphates.

extension (joint) The joint motion in which bone segments move apart.

external fertilization Fertilization that takes place outside the body of the female; in aquatic organisms, for example, eggs and sperm are released into the water.

extracellular digestion The process in most animals in which food is isolated and broken down outside a cell, in a body compartment.

extracellular matrix A meshwork of proteins and polysaccharides outside the cell; the main constituent of connective tissue.

extraembryonic membrane In the amniotic egg, one of several sheets of cells that extend out from the developing embryo and form the yolk sac, amnion, allantois, and chorion.

extravasation The process in which phagocytes that travel in the blood move from a blood vessel to the site of infection.

eyecup An eye structure found in flatworms that contains photoreceptors that point up and to the left or right.

F_1 generation The first filial, or offspring, generation.

F_2 generation The second filial generation; the offspring of the F_1 generation.

facilitated diffusion Diffusion through a membrane protein, bypassing the lipid bilayer.

facilitation A beneficial indirect interaction between organisms that are independently interacting directly with a third, as when two different organisms attack another organism, and the attack of each is aided as a result.

facultative In mutualisms, describes one in which one or both sides can survive without the other. In metabolism, describes a means of obtaining energy that is sometimes but not always used.

fallopian tube (oviduct) A tube from each ovary, through one of which a released oocyte passes.

family A group of closely related genera.

fast-twitch Describes muscle fibers that generate force quickly, producing rapid movements, but consume much more ATP than do slow-twitch fibers.

fatty acid A long chain of carbons attached to a carboxyl group; three fatty acid chains attached to glycerol form a triacylglycerol, a lipid used for energy storage.

feature detector A specialized sensory receptor or group of sensory receptors that respond to important signals in the environment.

fermentation A process of breaking down pyruvate through a wide variety of metabolic pathways that extract energy from fuel molecules such as glucose; the partial oxidation of complex carbon molecules to molecules that are less oxidized than carbon dioxide.

ferns and horsetails A monophyletic group of vascular plants that have leaves and disperse by spores.

fertilization The union of gametes to produce a diploid zygote.

fetus In humans, the embryo toward the end of the first trimester.

fiber In angiosperms, a narrow cell with thick walls that provides mechanical support in wood. In animals, a term for a muscle cell, which produces forces within an animal's body and exerts forces on the environment.

filament In animals, a thin thread of proteins that interacts with other filaments to cause muscles to shorten. In plants, the part of the stamen that supports the anther.

filial imprinting Imprinting in which newborn offspring rapidly learn to treat any animal they see shortly after birth as their mother.

filtration The separation of solids from fluids, as when circulatory pressure pushes fluid containing wastes through specialized filters into an extracellular space.

firing rate The number of action potentials fired over a given period of time.

first law of thermodynamics The law of conservation of energy: Energy can neither be created nor destroyed—it can only be transformed from one form into another.

first-division nondisjunction Failure of chromosome separation in meiosis I.

fitness A measure of the extent to which an individual's genotype is represented in the next generation

fixed Describes a population that exhibits only one allele at a particular gene.

fixed action pattern (FAP) A sequence of behaviors that, once triggered, is followed through to completion.

flagellum (plural, flagella) An organelle that propels the movement of cells or of substances within cells; longer than a cilium.

flexion The joint motion in which bone segments rotate closer together.

fluid mosaic model A model that proposes that the lipid bilayer is a fluid structure that allows molecules to move laterally within the membrane and is a mosaic of two types of molecules, lipids and proteins.

flux The rate at which a substance, for example carbon, flows from one reservoir to another.

folding domain A region of a protein that folds in a similar way across a protein family relatively independently of the rest of the protein.

follicle A type of cell that makes up the shell of cells surrounding an oocyte that supports the developing oocyte.

follicle-stimulating hormone (FSH) A hormone secreted by the anterior pituitary gland that stimulates the male and female gonads to secrete testosterone in males and estrogen and progesterone in females.

follicular phase The phase of the menstrual cycle during which FSH acts on granulosa cells, resulting in the maturation of several oocytes, of which, usually, only one becomes completely mature.

food chain The linear transfer of carbon from one organism to another.

food web A map of the interactions that connect consumer and producer organisms within the carbon cycle; the movement of carbon through an ecosystem.

force An interaction that changes the movement of an object, such as a push or pull by one object interacting with another object.

forebrain The region of the vertebrate brain that governs cognitive functions.

foregut The first part of an animal's digestive tract, including the mouth, esophagus, and stomach.

fossil The remains, impressions, or evidence of a once-living organism, preserved through time, most often in sedimentary rocks or tree resins.

founder event A type of bottleneck that occurs when only a few individuals establish a new population.

fovea The center of the visual field of most vertebrates, where cone cells are most concentrated; the region of greatest acuity.

fragmentation A form of asexual reproduction in which new individuals arise by the splitting of one organism into pieces, each of which develops into a new individual.

frameshift mutation A mutation in which an insertion or deletion of some number of nucleotides that is not a multiple of three causes a shift in the reading frame of the mRNA, changing all following codons.

fraternal (dizygotic) twins Twins that arise when two separate eggs, produced by double ovulation, are fertilized by two different sperm.

free energy The capacity to do work.

frequency of recombination The proportion of recombinant chromosomes among the total number of chromosomes observed.

frontal lobe The region of the brain located in the anterior region of the cerebral cortex, important in decision making and planning.

fruit In angiosperms, the structure that develops from the ovary, sometimes united with adjacent tissues, and serves to protect immature seeds and enhance dispersal once the seeds are mature.

fruiting body A multicellular structure in some fungi that facilitates the dispersal of sexually produced spores.

functional groups Groups of one or more atoms that have particular chemical properties of their own, regardless of what they are attached to.

fundamental niche The full range of climate conditions and food resources that permit the individuals in a species to live.

fungi An abundant and diverse group of heterotrophic eukaryotic organisms, principally responsible for the decomposition of plant and animal tissues.

G protein A protein that binds to the guanine nucleotides GTP and GDP.

G protein-coupled receptor A receptor that couples to G proteins, which bind to the guanine nucleotides GTP and GDP.

G_0 phase The gap phase in which cells pause in the cell cycle between M phase and S phase; may last for periods ranging from days to more than a year.

G_1 phase The gap phase in which the size and protein content of the cell increase and specific regulatory proteins are made and activated in preparation for S-phase DNA synthesis.

G_2 phase The gap phase in which the size and protein content of the cell increase in preparation for M-phase mitosis and cytokinesis.

gain-of-function mutation Any mutation in which a gene is expressed in the wrong place or at the wrong time.

gallbladder The organ in which bile produced by the liver is stored.

gamete A reproductive haploid cell; gametes fuse in pairs to form a diploid zygote. In many species, there are two types of gametes: eggs in females, sperm in males.

gametic isolation Incompatibility between the gametes of two different species.

gametogenesis The formation of gametes.

gametophyte In alternation of generation, describes the haploid multicellular generation that gives rise to gametes.

ganglion (plural, ganglia) A group of nerve cell bodies that processes sensory information received from a local, nearby region, resulting in a signal to motor neurons that control some physiological function of the animal.

ganglion cell A type of interneuron in the retina that synapses with bipolar cells and, if activated, transmits action potentials along the optic nerve to the visual cortex in the brain.

gap junction A type of connection between the plasma membranes of adjacent animal cells that permits materials to pass directly from the cytoplasm of one cell to the cytoplasm of another.

gas exchange The transport of oxygen and carbon dioxide between an organism and its environment.

gastric cavity In cnidarians, a closed internal site where extracellular digestion and excretion take place.

gastrin A peptide hormone produced in the stomach that stimulates cells lining the stomach to increase their production of HCl.

gastropods A group of mollusks consisting of snails and slugs.

gastrula A layered structure formed when the inner cell mass cells of the blastocyst migrate and reorganize.

gastrulation A highly coordinated set of cell movements in which the cells of the blastoderm migrate inward, creating germ layers of cells within the embryo.

gel electrophoresis A procedure to determine the size of a DNA fragment, in which DNA samples are inserted into slots or wells in a gel and a current passed through. Fragments move toward the positive pole according to size.

gene The unit of heredity; the stretch of DNA that affects one or more traits in an organism, usually through an encoded protein or noncoding RNA.

gene expression The production of a functional gene product.

gene family A group of genes with related functions, usually resulting from multiple rounds of duplication and divergence.

gene flow The movement of alleles from one population to another through interbreeding of some of their respective members.

gene pool All the alleles present in all individuals in a population or species.

gene regulation The various ways in which cells control gene expression.

general transcription factors A set of proteins that bind to the promoter of a gene whose combined action is necessary for transcription.

genetic code The correspondence between codons and amino acids, in which 20 amino acids are specified by 64 codons.

genetic drift A change in the frequency of an allele due to the random effects of finite population size.

genetic incompatibility Genetic dissimilarity between two organisms, such as different numbers of chromosomes, that is sufficient to act as a post-zygotic isolating factor.

genetic information Information carried in DNA, organized in the form of genes.

genetic map A diagram showing the relative positions of genes along a chromosome.

genetic risk factor Any mutation that increases the risk of a given disease in an individual.

genetic test A method of identifying the genotype of an individual.

genetic variation Differences in genotype among individuals in a population.

genetically modified organism (GMO) An organism that has been genetically engineered, such as modified viruses and bacteria, laboratory organisms, agricultural crops, and domestic animals; also known as a *transgenic organism.*

genome The genetic material transmitted from a parental cell or organism to its offspring.

genome annotation The process by which researchers identify the various types of sequence present in genomes.

genomic rearrangement The process of joining different gene segments as a B cell differentiates to produce a specific antibody.

genotype The genetic makeup of a cell or organism; the particular combination of alleles present in an individual.

genotype-by-environment interaction Unequaleffects of the environment on different genotypes, resulting in different phenotypes.

genus (plural, genera) A group of closely related species.

geographic isolation Spatial segregation of individuals.

geologic timescale The series of time divisions that mark Earth's long history.

germ cells The reproductive cells that produce sperm or eggs and the cells that give rise to them.

germ layers Three sheets of cells, the ectoderm, mesoderm, and endoderm, formed by migrating cells of the gastrula that differentiate further into specialized cells.

germ-line mutation A mutation that occurs in eggs and sperm or in the cells that give rise to these reproductive cells and therefore is passed on to the next generation.

gibberellic acid A plant hormone that stimulates the elongation of stems.

Gibbs free energy (G) The amount of energy available to do work.

gills Highly folded delicate structures in aquatic animals that facilitate gas exchange with the surrounding water.

gizzard In birds, alligators, crocodiles, and earthworms, a compartment with thick muscular walls in the digestive tract where food mixed with ingested rock or sediment is broken down into smaller pieces.

glans penis The head of the human penis.

glial cell A type of cell that surrounds neurons and provides them with nutrition and physical support.

glomeromycetes A monophyletic fungal group of apparently low diversity but tremendous ecological importance that occurs in association with plant roots.

glomerulus A tufted loop of porous capillaries in the vertebrate kidney that filters blood.

glycerol A 3-carbon molecule with OH groups attached to each carbon.

glycogen The form in which glucose is stored in animals.

glycolysis The breakdown of glucose to pyruvate; the first stage of cellular respiration.

glycosidic bond A covalent bond that attaches one monosaccharide to another.

Golgi apparatus The organelle that modifies proteins and lipids produced by the ER and acts as a sorting station as they move to their final destinations.

gonad In mammals, the part of the reproductive system where haploid gametes are produced. Male gonads are testes, where sperm are produced. Female gonads are ovaries, where eggs are produced.

gonadotropin-releasing hormone (GnRH) A hormone released by the hypothalamus that stimulates the anterior pituitary gland to secrete luteinizing hormone and follicle-stimulating hormone.

gram-positive bacteria Bacteria that retain, in their thick peptidoglycan walls, the diagnostic dye developed by Hans Christian Gram. (Bacteria with thin walls, which do not retain the dye, are said to be gram negative.)

grana (singular, granum) Interlinked structures that form the thylakoid membrane.

granulocyte A type of phagocytic cell that contains granules in its cytoplasm.

granuloma A structure formed by lymphocytes surrounding infected macrophages that helps to prevent the spread of an infection and aids in killing infected cells.

gravitropic Bending in response to gravity. A negative gravitropic response, as in stems, is growth upward against the force of gravity; a positive gravitropic response, as in roots, is with the force of gravity.

gray matter Densely packed neuron cell bodies and dendrites that make up the cortex, a highly folded outer layer of the mammalian brain about 4 mm thick.

greenhouse gas A gas in the atmosphere that allows incoming solar radiation to reach the Earth's surface, but absorbs radiation re-emitted as heat, trapping it in the atmosphere and causing the temperature to rise.

group selection Selection caused by the differential success of groups rather than individuals.

growth factor Any one of a group of small, soluble molecules, usually the signal in paracrine signaling, that affect cell growth, cell division, and changes in gene expression.

growth plate A region of cartilage near the end of a bone where growth in bone length occurs.

growth ring One of the many rings apparent in the cross section of the trunk of a tree, produced by decreases in the size of secondary xylem cells at the end of the growing season, that make it possible to determine the tree's age.

guanine (G) A purine base.

guard cell One of two cells surrounding the central pore of a stoma.

gustation The sense of taste.

gut Collectively, the passages that connect the mouth, digestive organs, and anus; also known as the *digestive tract.*

gymnosperms Seed plants whose ovules are not enclosed in a carpel; gymnosperms include pine trees and other conifers.

habituation The reduction or elimination of a behavioral response to a repeatedly presented stimulus.

hagfish One of the earliest-branching craniates, with a cranium built of cartilage but no jaws; hagfish feed on marine worms and dead and dying sea animals.

hair cell A specialized mechanoreceptor that senses movement and vibration.

half-life The time it takes for an amount of a substance to reach half its original value. Radioactive half-life is the time it takes for half of the atoms in a given sample of a substance to decay.

haploid Describes a cell with one complete set of chromosomes.

haplotype The particular combination of alleles present in any defined region of a chromosome.

Hardy–Weinberg equilibrium A state in which particular allele and genotype frequencies do not change over time, implying the absence of evolutionary forces. It also specifies a mathematical relationship between allele frequencies and genotype frequencies.

heart The pump of the circulatory system, which moves blood to different regions of the body.

heart rate (HR) The number of heartbeats per unit time.

heartwood The center of the stem in long-lived trees, which does not conduct water.

heavy (H) chains Two of the four polypeptide chains that make up the simplest antibody molecule.

helicase A protein that unwinds the parental double helix at the replication fork.

helper T cell One of a subpopulation of T cells that help other cells of the immune system by secreting cytokines, thus activating B cells to secrete antibodies.

hematocrit The fraction of red blood cells within the blood of vertebrates.

Hemichordata One of the three major phyla of deuterosomes, this group includes acorn worms and pterobranchs.

hemidesmosome A type of desmosome in which integrins are the prominent cell adhesion molecules.

hemoglobin An iron-containing molecule specialized for oxygen transport.

hemolymph The circulatory fluid in invertebrates.

hemophilia A trait characterized by excessive bleeding that results from a recessive mutation in a gene encoding a protein necessary for blood clotting.

herbivory The consumption of plant parts.

heritability In a population, the proportion of the total variation in a trait that is due to genetic differences among individuals.

heterokaryotic Describes a stage in the life cycle of some fungi, in which plasmogamy is not followed immediately by karyogamy and the cells have unfused haploid nuclei from both parents.

heterotroph An organism that obtains its carbon from organic molecules synthesized by other organisms.

heterozygote advantage A form of balancing selection in which the heterozygote's fitness is higher than that of either of the homozygotes, resulting in selection that ensures that both alleles remain in the population at intermediate frequencies.

heterozygous inherits different types of alleles from the parents, or genotypes in which the two alleles for a given gene are different.

hierarchical Describes gene regulation during development, in which the genes expressed at each stage in the process control the expression of genes that act later.

hindbrain Along with the midbrain, the region of the vertebrate brain that controls basic body functions and behaviors.

hindgut The last part of an animal's digestive tract, including the large intestine and rectum.

hinge joint A simple joint that allows one axis of rotation, like the elbow and knee.

hippocampus A posterior region of the limbic system involved in long-term memory formation.

histamine A chemical messenger released by mast cells and basophils; an important contributor to allergic reactions and inflammation.

histone code The pattern of modifications of the histone tails that affects the chromatin structure and gene transcription.

histone tail A string of amino acids that protrudes from a histone protein in the nucleosome.

homeodomain The DNA-binding domain in homeotic proteins, whose sequences are very similar from one homeotic protein to the next.

homeostasis The active regulation and maintenance, in animals, organs, or cells, of a stable internal physiological state in the face of a changing external environment.

homeotic gene A gene that specifies the identity of a body part or segment during embryonic development. One type of homeotic gene is known as a *Hox* gene.

hominins A member of one of the different species in the group leading to humans.

homologous Describes characters that are similar in different species because of descent from a common ancestor.

homologous chromosomes Pairs of chromosomes, matching in size and appearance, that carry the same set of genes; one of each pair was received from the mother, the other from the father.

homozygous Describes an individual who inherits an allele of the same type from each parent, or a genotype in which both alleles for a given gene are of the same type.

horizontal cell A type of interneuron in the retina that communicates between neighboring pairs of photoreceptors and bipolar cells, enhancing contrast through lateral inhibition to sharpen the image.

horizontal gene transfer The transfer of genetic material between organisms that are not parent and offspring.

hormone A chemical signal that influences physiology and development in both plants and animals; in animals hormones are released into the bloodstream and circulate throughout the body.

host cell A cell in which viral reproduction occurs.

host plant A plant species that can be infected by a given pathogen.

hotspot A site in the genome that is especially mutable.

housekeeping gene A gene that is transcribed continually because its product is needed at all times and in all cells.

human chorionic gonadotropin (hCG) A hormone released by the developing embryo that maintains the corpus luteum.

human genome The DNA in the chromosomes present in a human sperm or egg. The term is often used informally to mean all of the genetic material in a human cell or organism.

hybridization Interbreeding between two different varieties or species.

hydrogen bond A weak bond between a hydrogen atom in one molecule and an electronegative atom in another molecule.

hydrophilic "Water loving"; describes a class of molecules with which water can undergo hydrogen bonding.

hydrophobic "Water fearing"; describes a class of molecules poorly able to undergo hydrogen bonding with water.

hydrophobic effect The exclusion of nonpolar molecules by polar molecules, which drives biological processes such as the formation of cell membranes and the folding of proteins.

hydrostatic skeleton A skeletal system in which fluid contained within a body cavity is the supporting element.

hydroxyapatite The calcium phosphate mineral found in bone.

hypersensitive response A type of plant defense against infection in which uninfected cells surrounding the site of infection rapidly produce large numbers of reactive oxygen species, triggering cell wall reinforcement and causing the cells to die, thus creating a barrier of dead tissue.

hyperthermophile An organism that requires an environment with high temperature.

hyphae In fungi, highly branched filaments that provide a large surface area for absorbing nutrients.

hypothalamus The underlying brain region of the forebrain, which interacts with the autonomic and endocrine systems to regulate the general physiological state of the body.

hypothesis A tentative explanation for one or more observations that makes predictions that can be tested by experiments or additional observations.

identical (monozygotic) twins Twins that arise from a single fertilized egg, which after several rounds of cell division, separates into two distinct, but genetically identical, embryos.

IgA One of the five antibody classes, IgA is typically a dimer and the major antibody on mucosal surfaces.

IgD One of the five antibody classes, IgD is a monomer and is typically found on the surface of B cells; it helps initiate inflammation.

IgE One of the five antibody classes, IgE is a monomer that plays a central role in allergies, asthma, and other immediate hypersensitivity reactions.

IgG The most abundant of the five antibody classes, IgG is a monomer that circulates in the blood and is particularly effective against bacteria and viruses.

IgM One of the five antibody classes, IgM is a pentamer in mammals and a tetramer in fish and is particularly important in the early response to infection, activating the complement system and stimulating an immune response.

ileum A section of the small intestine that, with the jejunum, carries out most nutrient absorption.

imitation Observing and copying the behavior of another.

immediate hypersensitivity reaction A reaction characterized by a heightened or an inappropriate immune response to common antigens.

immunodeficiency Any disease in which part of the immune system does not function properly.

imprinting A form of learning typically seen in young animals in which a certain behavior is acquired in response to key experiences during a critical period of development.

in vitro fertilization (IVF) A process in which eggs and sperm are brought together in a petri dish, where fertilization and early cell divisions occur.

inbred line A true-breeding, homozygous strain.

inbreeding depression A reduction in fitness resulting from breeding among relatives causing homozygosity of deleterious recessive mutations.

incisor One of the teeth in the front of the mouth, used for biting.

incomplete dominance Describes inheritance in which the phenotype of the heterozygous genotype is intermediate between those of homozygous genotypes.

incomplete penetrance The phenomenon in which some individuals with a genotype corresponding to a trait do not show the phenotype, either because of environmental effects or because of interactions with other genes.

incus A small bone in the vertebrate middle ear that helps amplify the sound waves that strike the tympanic membrane.

induced pluripotent stem cell (iPS cell) A cell that has been reprogrammed to become pluripotent by activation of certain genes, most of them encoding transcription factors or chromatin proteins.

inducer A small molecule that elicits gene expression.

inflammation A physiological response of the body to injury that removes the inciting agent if present and begins the healing process.

inhalation The drawing of oxygen-rich blood into the lungs by the expansion of the thoracic cavity.

inhibitor A synthesized compound that decreases the activity of an enzyme.

inhibitory postsynaptic potential (IPSP) A negative change in the postsynaptic membrane potential.

initiation The stage of translation in which methionine is established as the first amino acid in a new polypeptide chain.

initiation factor A protein that binds to mRNA to initiate translation.

innate Describes behaviors that are instinctive and carried out regardless of earlier experience.

innate (natural) immunity The part of the immune system that provides protection in a nonspecific manner against all kinds of infection; it does not depend on exposure to a pathogen.

inner cell mass A mass of cells in one region of the inner wall of the blastocyst from which the body of the embryo develops.

inner ear The part of the vertebrate ear that includes the cochlea and semicircular canals.

insects The most diverse of the four main groups of arthropods.

instantaneous speciation Speciation that occurs in a single generation.

integral membrane protein A protein that is permanently associated with the cell membrane and cannot be separated from the membrane experimentally without destroying the membrane itself.

integrin A transmembrane protein, present on the surface of virtually every animal cell, that enables cells to adhere to the extracellular matrix.

intercostal muscles Muscles of vertebrates attached to adjacent pairs of ribs that assist the diaphragm by elevating the ribs on inhalation and depressing them during exhalation.

intermediate filament A polymer of proteins, which vary according to cell type, that combine to form strong, cable-like filaments that provide animal cells with mechanical strength.

intermembrane space The space between the inner and outer mitochondrial membranes.

internal fertilization Fertilization that takes place inside the body of the female.

interneuron A neuron that processes information received by sensory neurons and transmits it to motor neurons in different body regions.

internode The segment between two nodes on a shoot.

interphase The time between two successive M phases.

intersexual selection A form of sexual selection involving interaction between males and females, as when females choose from among males.

interspecific competition Competition between individuals of different species.

intervertebral disc A fluid-filled support structure found between the bony vertebrae of the backbone that enables flexibility and provides cushioning of loads.

intracellular digestion The process in single-celled protists in which food is broken down within cells.

intrasexual selection A form of sexual selection involving interactions between individuals of one sex, as when members of one sex compete with one another for access to the other sex.

intraspecific competition Competition within species.

intrinsic growth rate The per capita growth rate; the maximum rate of growth when no environmental factors limit population increase.

intron A sequence that is excised from the primary transcript and degraded during RNA splicing.

invasive species Non-native species; since they are removed from natural constraints on population growth, invasive species can expand dramatically when introduced into new areas, sometimes with devastating consequences for native species and ecosystems.

inversion The reversal of the normal order of a block of genes.

involuntary Describes the component of the nervous system that regulates internal bodily functions.

ion An electrically charged atom or molecule.

ion channels Cell-surface receptors that open and close, thereby altering the flow of ions across the plasma membrane.

iris A structure found at the front of the vertebrate eye, surrounding the pupil, that opens and closes to adjust the amount of light that enters the eye.

island population An isolated population.

isomers Molecules that have the same chemical formula but different structures.

isometric The generation of force without muscle movement.

isotopes Atoms of the same element that have different numbers of neutrons.

jejunum A section of the vertebrate small intestine that, with the ileum, carries out most nutrient absorption.

juxtaglomerular apparatus The structure formed by specialized cells of the efferent arteriole leaving the glomerulus of each nephron, which secretes the hormone rennin into the bloodstream.

karyogamy The fusion of two nuclei following plasmogamy.

karyotype A standard arrangement of chromosomes, showing the number and shapes of the chromosomes representative of a species.

key stimulus A stimulus that initiates a fixed action pattern.

keystone species Pivotal populations that affect other members of the community in ways that are disproportionate to their abundance or biomass.

kidneys In vertebrates, paired organs of the renal system that remove waste products and excess fluid; their action contributes to homeostasis.

kin selection A form of natural selection that favors the spread of alleles promoting behaviors that help close relatives.

kinesin A motor protein, similar in structure to myosin, that transports cargo toward the plus end of microtubules.

kinesis (plural, kineses) A random, undirected movement in response to a stimulus.

kinetic energy The energy of motion.

kinetochore The protein complexes on a chromatid where spindle fibers attach.

kingdom A group of closely related phyla.

Klinefelter syndrome A sex-chromosomal abnormality in which an individual has 47 chromosomes, including two X chromosomes and one Y chromosome.

knee-extension reflex A reflex commonly tested by physicians to evaluate peripheral nervous and muscular system function.

K-strategist A species that produces relatively few young but invests considerable resources into their support

labia majora Outer folds of skin in the vulva.

labia minora Inner folds of skin in the vulva that meet at the clitoris.

lactic acid fermentation The fermentation pathway in animals and bacteria during which electrons from NADH are transferred to pyruvate to produce lactic acid and NAD^1.

lagging strand A daughter strand that has its 5⊠ end pointed toward the replication fork, so as the parental double helix unwinds, a new DNA piece is initiated at intervals, and each new piece is elongated at its 3⊠ end until it reaches the piece in front of it.

lamellae (singular, lamella) The many thin, sheetlike structures spread along the length of each gill filament, giving gills an enormous surface area relative to their size.

lamprey One of the earliest-branching craniates, with a cranium and vertebral column built of cartilage but no jaws; many lampreys live parasitically, sucking body fluids from fish prey.

large intestine Part of the hindgut and the site of reabsorption of water and minerals; also known as the colon.

lariat A loop and tail of RNA formed after RNA splicing.

larynx The structure, above the trachea, that contains the vocal cords.

lateral inhibition Inhibition of a process in cells adjacent to the cell receiving a signal inducing that process, enhancing the strength of a signal locally but diminishing it peripherally.

lateral line system In fish and sharks, a sensory organ along both sides of the body that uses hair cells to detect movement of the surrounding water.

lateral meristem The source of new cells that allows plants to grow in diameter.

latex A white sticky liquid produced in some plants.

latitudinal diversity gradient The increase in species diversity from the poles to the equator.

leading strand A daughter strand that has its 3⊠ end pointed toward the replication fork, so as the parental double helix unwinds, this daughter strand can be synthesized as one long, continuous polymer.

leaf The principal site of photosynthesis in vascular plants.

leaf primordia A lateral outgrowth of the apical meristem that will eventually develop into a leaf.

learned Describes a behavior that depends on an individual's experience.

learning The process in which experience leads to changes in behavior.

lengthening contraction The contraction of a muscle against a load greater than the muscle's force output, leading to a lengthening of the muscle.

lens A flexible structure in the vertebrate eye through which light passes after entering through the pupil; it is controlled by ciliary muscles that contract or relax to the adjust the shape of the lens to focus light images.

lenticel A region of less tightly packed cells in the outer bark that allows oxygen to diffuse into the stem.

Leydig cell A type of cell in the testes that secretes testosterone.

Liebig's Law of the Minimum The principle that primary production is limited by the nutrient that is least available relative to its use by primary producers.

lichens Stable associations between a fungus and a photosynthetic microorganism, usually a green alga but sometimes a cyanobacterium.

life history The typical pattern of resource investment in each stage of a given species' lifetime.

life table A table that presents information about how many individuals of a cohort are alive at different points in time.

ligand Alternative term for a signaling molecule that binds with a receptor, usually a protein.

ligand-binding site The specific location on the receptor protein where a signaling molecule binds.

ligand-gated ion channel A receptor that alters the flow of ions across the plasma membrane when bound by its ligand.

light (L) chains Two of the four polypeptide chains that make up the simplest antibody molecule.

limbic system Inner components of the forebrain that control physiological drives, instincts, emotions, motivation, and the sense of reward.

linked Describes genes that are sufficiently close together in the same chromosome that they do not assort independently.

lipase A type of enzyme produced by the pancreas that breaks apart lipids, thus enabling their more effective digestion.

lipid An organic molecule that stores energy, acts as a signaling molecule, and is a component of cell membranes.

lipid raft Lipids assembled in a defined patch in the cell membrane.

liposome An enclosed bilayer structure spontaneously formed by phospholipids in environments with neutral pH, like water.

liver A vertebrate organ that aids in the digestion of proteins, carbohydrates, and fats in the duodenum by producing bile, which breaks down fat.

lobe-finned fish Species of fish with paired pectoral and pelvic fins that have a bone structure similar to that of tetrapod limbs.

logistic growth The pattern of population growth that results as growth potential slows down as the population size approaches *K*, its maximum sustainable size.

long-day plant A plant that flowers only when the light period exceeds a critical value.

longitudinal muscle Smooth muscle that runs lengthwise along a body or organ; in the digestive tract, a longitudinal muscle layer contracts to shorten small sections of the gut. A longitudinal muscle layer contracts alternately with circular muscle to move contents through the digestive tract and to enable locomotion in animals with hydrostatic skeletons.

loop of Henle The middle portion of the vertebrate renal tubule, which creates a concentration gradient that allows water passing through the collecting duct to be reabsorbed.

loss-of-function mutation A genetic change that inactivates the normal function of a gene.

Lucy An unusually complete early hominin fossil, *Australopithecus afarensis,* found in 1974 in Ethiopia.

lumen In eukaryotes, the continuous interior of the endoplasmic reticulum; in plants, a fluid-filled compartment enclosed by the thylakoid membrane; generally, the interior of any tubelike structure.

lunar clock A moon-based biological clock that times activities in some species, especially those living in habitats where tides are important.

lungfish Several species of lobe-finned fish that use a simple lung to survive periods when their watery habitat dries by burying themselves in moist mud and breathing air; along with coelacanths, the nearest relative of tetrapods.

lungs The internal organs for gas exchange in many terrestrial animals.

luteal phase The phase of the menstrual cycle beginning with ovulation.

luteinizing hormone (LH) A hormone secreted by the anterior pituitary gland that stimulates the male and female gonads to secrete testosterone in males and estrogen and progesterone in females.

lycophytes A monophyletic major group of spore-dispersing vascular plants that are the sister group to all other vascular plants.

lymph The fluid in the lymphatic system in which T and B cells ciculate.

lymphatic system A network of vessels distributed through the body with important functions in the immune system.

lysis The breakage, or bursting, of a cell.

lysogenic pathway The alternative to the lytic pathway; in the lysogenic pathway, a virus integrates its DNA into the host cell's DNA, which is then transmitted to offspring cells.

lysogeny The integration of viral DNA into a host cell's DNA, which is then transmitted to offspring cells.

lysosome A vesicle derived from the Golgi apparatus that contains enzymes that break down macromolecules such as proteins, nucleic acids, lipids, and complex carbohydrates.

lytic pathway The alternative to the lysogenic pathway; in the lytic pathway, a virus bursts, or lyses, the cell it infects, releasing new virus particles.

M phase The stage of the cell cycle consisting of mitosis and cytokinesis, in which the parent cell divides into two daughter cells.

macrophage A type of large phagocytic cell that is able to engulf other cells.

mainland population The central population of a species.

major groove The larger of two uneven grooves on the outside of a DNA duplex.

major histocompatibility complex (MHC) A group of proteins that appear on the surface of most mammalian cells; only the antigen associated with MHC proteins is recognized by T cell receptors.

malleus A small bone in the middle ear that helps amplify the waves that strike the tympanic membrane.

Malpighian tubule One of the tubes in the main body cavity of insects and other terrestrial arthropods through which fluid passes and which empties into the hindgut.

mammals A class of vertebrates distinguished by body hair and mammary glands from which they feed their young.

map information The knowledge of where an individual is in respect to the goal.

map unit A unit of distance in a genetic map equal to the distance between genes resulting in 1% recombination.

mark-and-recapture A method in which individuals are captured, marked in way that doesn't affect their function or behavior, and then released. The percentage of marked individuals in a later exercise of capture enables ecologists to estimate population size.

marsupials A group of mammals that includes kangaroos, koalas, and opossums; their young are born at an early stage of development and must crawl to a pouch where mammary glands equipped with nipples provide them with milk.

mass extinction A catastrophic drop in recorded diversity, which has occurred five or more times in the past 541 million years.

mast cell A cell that releases histamine.

maternal inheritance A type of inheritance in which the organelles in the offspring cells derive from those in the mother.

maternal-effect gene A gene that is expressed by the mother that affects the phenotype of the offspring, typically through the composition or organization of the oocyte.

mating types Genetically distinct forms of individuals of a fungus species that, by enabling fertilization only between different types, prevent self-fertilization and promote out-crossing.

mechanical incompatibility Structural configuration of the genitalia that prevents mating with another species.

mechanoreceptor A sensory receptor that responds to physical deformations of its membrane produced by touch, stretch, pressure, motion, or sound.

mediator complex A complex of proteins that interacts with the Pol II complex and allows transcription to begin.

medulla A part of the brainstem; also, the inner layer of the mammalian kidney.

meiosis I Reductional division, the first stage of meiotic cell division, in which the number of chromosomes is halved.

meiosis II Equational division, the second stage of meiotic cell division, in which the number of chromosomes is unchanged.

meiotic cell division A form of cell division that includes only one round of DNA replication but two rounds of nuclear division; meiotic cell division makes sexual reproduction possible.

membrane attack complex (MAC) A complex of complement proteins that makes holes in bacterial cells, leading to cell lysis.

membrane potential A difference in electrical charge across the plasma membrane.

memory cell A type of long-lived cell that contains membrane-bound antibodies having the same specificity as the parent cell.

menopause The cessation of menstrual cycles resulting from decreased production of estradiol and progesterone by the ovaries.

menstrual cycle A monthly cycle in females in which oocytes mature and are released from the ovary under the influence of hormones.

menstruation The monthly shedding of the uterine lining.

meristem A discrete population of actively dividing, totipotent cells; apical meristems are located at the tip of stems and roots and produce cells that allow plants to grow in length, while lateral meristems surround stems and roots and produce cells that allow growth in diameter.

meristem identity gene A gene that contributes to meristem stability and function.

mesentery A membrane in the abdominal cavity through which blood vessels, nerves, and lymph travel to supply the gut.

mesoderm The intermediate germ layer, which differentiates into cells that make up connective tissue, muscle cells, red blood cells, bone cells, kidney cells, and gonad cells.

mesoglea In cnidarians, a gelatinous mass enclosed in by the epidermis and endodermis.

mesohyl A gelatinous mass that lies between the interior and exterior cell layers of a sponge that contains some amoeba-like cells that function in skeleton formation and the dispersal of nutrients.

mesophyll A leaf tissue of loosely packed photosynthetic cells.

Messel Shale A sedimentary rock formation in Germany, preserving fossils that document fish, birds, mammals, and reptiles from the beginning of the age of mammals.

messenger RNA (mRNA) The RNA molecule that combines with a ribosome to direct protein synthesis; it carries the genetic "message" from the DNA to the ribosome.

metabolic rate An animal's overall rate of energy use.

metabolism The chemical reactions occurring within cells that convert molecules into other molecules and transfer energy in living organisms.

metamorphosis The process in some animals in which the body changes dramatically at key stages in development.

metanephridia A pair of excretory organs in each body segment of annelid worms that filters the body fluid.

metaphase The stage of mitosis in which the chromosomes are aligned in the middle of the dividing cell.

metaphase I The stage of meiosis I in which the meiotic spindle is completed and the bivalents move to lie on an imaginary plane cutting transverely across the spindle.

metaphase II The stage of meiosis II in which the chromosomes line up so that their centromeres lie on an imaginary plane cutting across the spindle.

metapopulation A large population made up of smaller populations linked by migration.

methanogens Euryarchaeotes that generate natural gas (methane, CH_4).

MHC class I (genes and proteins) MHC genes and proteins in vertebrates that are expressed on the surface of all nucleated cells.

MHC class II (genes and proteins) MHC genes and proteins in vertebrates that are expressed on the surface of macrophages, dendritic cells, and B cells.

MHC class III (genes and proteins) MHC genes and proteins in vertebrates that encode several proteins of the complement system and proteins involved in inflammation.

micelle A spherical structure in which lipids with bulky heads and a single hydrophobic tail are packed.

microfilament A helical polymer of actin monomers, present in various locations in the cytoplasm, that helps make up the cytoskeleton.

microfossil A microscopic fossil, including fossils of bacteria and protists.

microRNA (miRNA) Small, regulatory RNA molecules that can cleave or destabilize RNA or inhibit its translation.

microtubule A hollow, tubelike polymer of tubulin dimers that helps make up the cytoskeleton.

microvilli Highly folded surfaces of villi, formed by fingerlike projections on the surfaces of epithelial cells.

midbrain Along with the hindbrain, the region of the vertebrate brain that controls basic body functions and behaviors; it is a part of the brainstem.

middle ear The part of the mammalian ear containing three small bones, the malleus, incus, and stapes, which amplify the waves that strike the tympanic membrane.

midgut The middle part of an animal's digestive tract, including the small intestine.

migration The movement of individuals from one population to another.

mineral nutrients Elements that come from the soil or water that play essential roles in all metabolic and structural processes in plants and algae; see also *dietary minerals.*

minor groove The smaller of two unequal grooves on the outside of a DNA duplex.

mitochondria (singular, mitochondrion) Specialized organelles that harness energy for the cell from chemical compounds like sugars and convert it into ATP.

mitochondrial DNA (mtDNA) A small circle of DNA, about 17,000 base pairs long in humans, found in every mitochondrion.

mitochondrial genome In eukaryotic cells, the DNA in the mitochondria.

mitochondrial matrix The space enclosed by the inner membrane of the mitochondria.

mitosis In eukaryotic cells, the division of the nucleus, in which the chromosomes are separated into two nuclei.

mitotic spindle A structure in the cytosol made up predominantly of microtubules that pull the chromosomes into separate daughter cells.

Modern Synthesis The current theory of evolution, which combines Darwin's theory of natural selection and Mendelian genetics.

molar One of the teeth in the back of the mouth of mammals specialized for crushing and shredding tough foods such as meat and fibrous plant material.

molecular clock Estimates of the time when different taxa diverged, based on the amount of genetic divergence between them.

molecular evolution Evolution at the level of DNA, which in time results in the genetic divergence of populations.

molecular fossils Sterols, bacterial lipids, and some pigment molecules, which are relatively resistant to decomposition, that accumulate in sedimentary rocks and document organisms that rarely form conventional fossils.

molecular orbital A merged orbital traversed by a pair of shared electrons.

molecular self-assembly The process by which, when conditions and relative amounts are suitable, viral components spontaneously interact and assemble into mature virus particles.

molecule A substance made up of two or more atoms.

mollusks A monophyletic group distinguished by a mantle, which plays a major role in movement, skeleton-building, breathing, and excretion; includes clams, snails, and squid.

molting Periodic shedding, as of an exoskeleton.

monophyletic Describes groupings in which all members share a single common ancestor not shared with any other species or group of species.

monosaccharide A simple sugar.

morphospecies concept The idea that members of the same species usually look like each other more than like other species.

morula The solid ball of cells resulting from early cell divisions of the fertilized egg.

motor endplate The region on a muscle cell where acetylcholine binds with receptors.

motor neuron A neuron that, on receiving information from interneurons, effects a response in the body.

motor protein Any of various proteins that are involved in intracellular transport or cause muscle contraction by moving the actin microfilaments inside muscle cells.

motor unit A vertebrate motor neuron and the population of muscle fibers that it innervates.

mouth The first part of the foregut, which receives food.

mucosa An inner tissue layer with secretory and absorptive functions surrounding the lumen of the digestive tract.

multiple alleles Two or more different alleles of the same gene, occurring in a population of organisms.

multiplication rule The principle that the probability of two independent events occurring together is the product of their respective probabilities.

multipotent Describes cells that can form a limited number of types of specialized cell.

multiregional hypothesis The idea that modern humans derive from the *Homo ergaster* populations that spread around the world starting 2 million years ago.

mutagen An agent that increases the probability of mutation.

mutation Any heritable change in the genetic material, usually a change in the nucleotide sequence of a gene.

mutualism An interaction between two or more species that benefits all.

mycelium A network of branching hyphae.

mycorrhizae Symbioses between roots and fungi that enhance nutrient uptake.

myelin Lipid-rich layers or sheaths formed by glial cells that wrap around the axons of vertebrate neurons and provide electrical insulation.

myofibril A long rodlike structure in muscle fibers that contains parallel arrays of the actin and myosin filaments.

myoglobin An oxygen-binding protein in the cells of vertebrate muscles, related to hemoglobin, that facilitates oxygen delivery to mitochondria.

myosin A motor protein found in cells that carries cargo to the plus ends of microfilaments and is also used by both striated and smooth muscles to contract and generate force.

myotome In chordates, any one of a series of segments that organizes the body musculature.

myriapods One of the four main groups of arthropods, including centipedes and millipedes; distinguished by their many pairs of legs.

natural killer cell A cell type of the innate immune system that does not recognize foreign cells but instead recognizes and kills host cells that are infected by a virus or have become cancerous or otherwise abnormal.

natural selection The process in which, when there is inherited variation in a population of organisms, the variants best suited for growth and reproduction in a given environment contribute disproportionately to future generations. Of all the evolutionary mechanisms, natural selection is the only one that leads to adaptations.

Neanderthal *Homo neanderthalensis,* a species similar to humans, but with thicker bones and flatter heads that contained brains about the same size as humans'; present in the fossil record 600,000–30,000 years ago.

necrotrophic pathogen A plant pathogen that kills cells before drawing resources from them.

negative feedback Describes the effect in which the final product of a biochemical pathway inhibits the first step; the process in which a stimulus acts on a sensor that communicates with an effector, producing a response that opposes the initial stimulus. Negative feedback is used to maintain steady conditions, or homeostasis.

negative regulation The process in which a regulatory molecule must bind to the DNA at a site near the gene to prevent transcription.

negative selection Natural selection that reduces the frequency of a deleterious allele.

negatively selected Describes T cells that react too strongly to self antigens in association with MHC and are eliminated through cell death.

nematodes Roundworms, the most numerous of all animals; a phylum of the Ecdysozoa.

neoteny The process in which the timing of development is altered so that a sexually mature organism retains the physical characteristics of the juvenile form.

nephron The functional unit of the kidney, consisting of the glomerulus, capsule, renal tubules, and collecting ducts.

nerve A bundle of long fiberlike extensions from multiple nerve cells.

nerve cord A bundle of long fiberlike extensions from multiple nerve cells that serves as the central nervous system of invertebrates such as flatworms and earthworms.

nervous system A network of many interconnected nerve cells.

neural tube In chordates, a cylinder of embryological tissue that develops into a dorsal nerve cord.

neuron Nerve cell; the basic fundamental unit of nervous systems.

neurosecretory cell A neuron in the hypothalamus that secretes hormones into the bloodstream.

neurotransmitter A molecule that conveys a signal from the end of the axon to the postsynaptic target cell.

neutral mutations Genetic changes that have no effect or negligible effects on the organism, or whose effects are not associated with differences in survival or reproduction.

neutron An electrically neutral particle in the atomic nucleus.

neutrophil A type of phagocytic cell that is very abundant in the blood and is often one of the first cells to respond to infection.

niche The combination of traits and habitat in which a species exists.

nicotinamide adenine dinucleotide phosphate (NADPH) An important cofactor in many biosynthetic reactions; the reducing agent used in the Calvin cycle.

nitrification The process by which chemoautotrophic bacteria oxidize ammonia (NH_3) to nitrite (NO_2^-) and then nitrate (NO_3^-).

nitrogen fixation The process in which nitrogen gas (N_2) is converted into ammonia (NH_3), a form biologically useful to primary producers.

nitrogenous waste Waste in the form of ammonia, urea, and uric acid, which are toxic to organisms in varying degrees.

nociceptor A type of nerve cell with dendrites in the skin and connective tissues of the body that responds to excessive mechanical, thermal, or chemical stimuli by withdrawal from the stimulus and by the sensation of pain.

node In phylogenetic trees, the point where a branch splits, representing the common ancestor from which the descendant species diverged. In plants, the point on a shoot where one or more leaves are attached.

nodes of Ranvier Sites on an axon that lie between adjacent myelin-wrapped segments, where the axon membrane is exposed.

non-associative learning Learning that occurs in the absence of any particular outcome, such as a reward or punishment.

nondisjunction The failure of a pair of chromosomes to separate normally during anaphase of cell division.

nonpolar Describes compounds that do not have regions of positive and negative charge.

nonpolar covalent bond A covalent bond between atoms that have the same, or nearly the same, electronegativity.

non-random mating Mate selection biased by genotype or relatedness.

nonrecombinants Progeny in which the alleles are present in the same combination as that present in a parent.

nonself From the point of view of a given organism, describes the molecules and cells of another organism.

nonsense mutation A mutation that creates a stop codon, terminating translation.

non-sister chromatids Chromatids of different members of a pair of homologous chromosomes; although they carry the same complement of genes, they are not genetically identical.

nonsynonymous (missense) mutation A point mutation (nucleotide substitution) that causes an amino acid replacement.

nontemplate strand The untranscribed partner of the template strand of DNA used in transcription.

norm of reaction A graphical depiction of the change in phenotype across a range of environments.

normal distribution A distribution whose plot is a bell-shaped curve.

notochord In chordates, a stiff rod of collagen and other proteins that runs along the back and provides support for the axis of the body.

nuclear envelope The cell structure, composed of two membranes, inner and outer, that defines the boundary of the nucleus.

nuclear genome In eukaryotic cells, the DNA in the chromosomes.

nuclear localization signal The signal sequence for the nucleus that enables proteins to move through pores in the nuclear envelope.

nuclear pore One of many protein channels in the nuclear envelope that act as gateways that allow molecules to move into and out of the nucleus and are thus essential for the nucleus to communicate with the rest of the cell.

nuclear transfer A procedure in which a hollow glass needle is used to insert the nucleus of a cell into the cytoplasm of an egg whose own nucleus has been destroyed or removed.

nucleic acid A polymer of nucleotides that encodes and transmits genetic information.

nucleoid In prokaryotes, a cell structure with multiple loops formed from supercoils of DNA.

nucleolus A distinct, dense, non–membrane-bound spherical structure within the nucleus that contains the genes and transcripts for ribosomal RNA.

nucleoside A molecule consisting of a 5-carbon sugar and a base.

nucleosome A beadlike repeating unit of histone proteins wrapped with DNA making up the 10-nm chromatin fiber.

nucleotide A constituent of nucleic acids, consisting of a 5-carbon sugar, a nitrogen-containing base, and one or more phosphate groups.

nucleotide excision repair The repair of multiple mismatched or damaged bases across a region; a process similar to mismatch repair, but over a much longer piece of DNA, sometimes thousands of nucleotides.

nucleotide substitution A mutation in which a base pair is replaced by a different base pair; this is the most frequent type of mutation; also known as a *point mutation.*

nucleus (of an atom) The dense central part of an atom containing protons and neutrons.

nucleus (of a cell) The compartment of the cell that houses the DNA in chromosomes.

obligate Describes a mutualism in which one or both sides cannot survive without the other.

observation The act of viewing the world around us.

occipital lobe The region of the brain that processes visual information from the eyes.

ocean acidification An increase in the abundance of carbon dioxide in the oceans that causes the pH of seawater to go down.

Okazaki fragment In DNA replication, any of the many short DNA pieces in the lagging strand.

olfaction The sense of smell.

oligodendrocyte A type of glial cell that insulates cells in the brain and spinal cord by forming a myelin sheath.

omasum The third chamber in the stomach of ruminants, into which the mixture of food and bacteria passes and where water is reabsorbed.

ommatidia (singular, ommatidium) Individual light-focusing elements that make up the compound eyes of insects and crustaceans; the number of ommatidia determines the resolution of the image.

oncogene A cancer-causing gene.

oocyte The unfertilized egg cell produced by the mother; the developing female gamete.

oogenesis The formation of ova or eggs.

open circulatory system A circulatory system found in many smaller animals that contains few blood vessels and in which most of the circulating fluid is contained within the animal's body cavity.

open reading frame (ORF) A stretch of DNA or RNA consisting of codons for amino acids uninterrupted by a stop codon. In genome annotation, this sequence motif identifies the region as potentially protein coding.

operant conditioning Associative learning in which a novel behavior that was initially undirected has become paired with a particular stimulus through reinforcement.

operator The binding site for a repressor protein.

operon A group of functionally related genes located in tandem along the DNA and transcribed as a single unit from one promoter; the region of DNA consisting of the promoter, the operator, and the coding sequence for the structural genes.

Opisthokonta A superkingdom that encompasses animals, fungi, and related protists.

opsin A photosensitive protein that converts the energy of light photons into electrical signals in the receptor cell.

opsonization The binding of a molecule to a pathogen to facilitate uptake by a phagocyte.

optic nerve A cranial nerve that transmits action potentials from ganglion cells in the retina to the visual cortex of the brain.

orbital A region in space where an electron is present most of the time.

order A group of closely related families.

organ Two or more tissues that combine and function together.

organ of Corti A structure in the cochlear duct, supported by the basilar membrane, with specialized hair cells with stereocilia, that functions to convert mechanical vibrations to electrical impulses.

organelle Any one of several compartments in eukaryotes that divide the cell contents into smaller spaces specialized for different functions.

organic molecule A carbon-containing molecule.

organogenesis The transformation of the three germ layers into all the organ systems of the body.

origin of replication Any point on a DNA molecule at which DNA synthesis is initiated.

osmoconformer An animal that matches its internal osmotic pressure to that of its external environment.

osmoregulation The regulation of water and solute levels to control osmotic pressure.

osmoregulator An animal that maintains internal solute concentrations that differ from that of its environment.

osmosis The net movement of a solvent, such as water, across a selectively permeable membrane toward the side of higher solute concentration.

osmotic pressure The pressure needed to prevent water from moving from one solution into another by osmosis.

Osteichthyes Bony fish that have a cranium, jaws, and mineralized bones; there are about 20,000–25,000 species.

osteoblast A type of cell that forms bone tissue.

osteoclast A type of cell that secretes digestive enzymes and acid that dissolves the calcium mineral and collagen in bone.

outer ear The part of the human ear that includes the pinna, the ear canal, and tympanic membrane.

out-of-Africa hypothesis The idea that modern humans arose from *Homo ergaster* descendants in Africa before dispersing beyond Africa around 60,000 years ago.

oval window The thin membrane between the stapes of the middle ear and the cochlea of the inner ear.

ovary In plants, a hollow structure at the base of the carpel in which the ovules develop and which protects the ovules from being eaten or damaged by animals; in animals, the female gonad where eggs are produced.

oviparity Laying eggs.

ovoviviparity Giving birth to live young, with nutritional support of the embryo from the yolk.

ovule A reproductive structure in seed plants consisting of the female gametophyte developing within a sporangium and surrounded by outer protective layers. Ovules, when fertilized, develop into seeds.

ovule cone A reproductive shoot in gymnosperms that produces ovules.

ovum (egg) (plural, ova) The larger, female gamete.

oxidation reaction Describes a reaction in which a molecule loses electrons.

oxidation–reduction reaction A reaction involving the loss and gain of electrons between reactants. In biological systems these reactions are often used to store or release chemical energy.

oxidative phosphorylation A set of metabolic reactions that occurs by passing electrons along an electron transport chain to the final electron acceptor, oxygen, pumping protons across a membrane, and using the proton electrochemical gradient to drive synthesis of ATP.

oxygen dissociation curve The curve that results when blood pO_2 is plotted against the percentage of O_2 bound to hemoglobin.

oxygenic Producing oxygen.

oxytocin A posterior pituitary gland hormone that causes uterine contraction during labor and stimulates the release of milk during breastfeeding.

P_1 generation The parental generation in a series of crosses.

pacemaker Describes cardiac muscle cells that function as a regulator of heart rhythm.

palindromic Reading the same in both directions; describes sequence identity along the paired strands of a duplex DNA molecule; a symmetry typical of restriction sites.

pancreas A secretory gland that has both endocrine function, secreting hormones, including insulin, directly into the blood, and exocrine function, aiding the digestion of proteins, carbohydrates, and fats by secreting digestive enzymes into ducts that connect to the duodenum.

paracrine signaling Signaling by a molecule that travels a short distance to the nearest neighboring cell to bind its receptor and deliver its message.

paraphyletic Describes groupings that include some, but not all, the descendants of a common ancestor.

parasexual Describes asexual species that generate genetic diversity by the crossing over of DNA during mitosis.

parasites Organisms that live in close association with another species, consuming or gaining nutrition from their hosts' tissues, generally without killing them.

parasympathetic division The division of the autonomic nervous system that slows the heart and stimulates digestion and metabolic processes that store energy, enabling the body to "rest and digest."

parathyroid gland A gland adjacent to the thyroid gland that secretes parathyroid hormone (PTH), which, with calcitonin, regulates the actions of bone cells.

parenchyma In plants, describes thin-walled, undifferentiated cells.

parietal lobe The brain region, posterior to the frontal lobe, that controls body awareness and the ability to perform complex tasks.

parsimony Choosing the simplest hypothesis to account for a given set of observations.

parthenogenesis A form of asexual reproduction in which females produce eggs that are not fertilized by males but divide by mitosis and develop into new individuals.

partial pressure (p) The fractional concentration of a gas relative to other gases present multiplied by the atmospheric pressure exerted on the gases.

partially reproductively isolated Describes populations that have not yet diverged as separate species but whose genetic differences are extensive enough that the hybrid offspring they produce have reduced fertility or viability compared with offspring produced by crosses between individuals within each population.

patch A bit of habitat that is separated from other bits by inhospitable environments that are difficult or risky for individuals to cross.

paternal inheritance A type of inheritance in which the organelles in the offspring cells derive from those in the father.

pathogens Organisms and other agents, such as viruses, that cause disease.

peat bog Wetland in which dead organic matter accumulates.

pedigree A diagram of family history that summarizes the record of the ancestor-descendent relationships among individuals.

penis The male copulatory organ.

pepsin An enzyme produced in the stomach that breaks down proteins.

peptide bond A covalent bond that links the carbon atom in the carboxyl group of one amino acid to the nitrogen atom in the amino group of another amino acid.

peptide hormone A hormone that is a short chain of linked amino acids.

peptidoglycan A complex polymer of sugars and amino acids that makes up the cell wall.

peptidyl (P) site One of three binding sites for tRNA on the large subunit of a ribosome.

pericycle In roots, a single layer of cells just to the inside of the endodermis from which new root apical meristems are formed, allowing roots to branch.

periodic selection The episodic loss of diversity as a successful variant outcompetes others.

periodic table of the elements The arrangement of the chemical elements in tabular form, organized by their chemical properties.

peripatric speciation A specific kind of allopatric speciation in which a few individuals from a mainland population disperse to a new location remote from the original population and evolve separately.

peripheral membrane protein A protein that is temporarily associated with the lipid bilayer or with integral membrane proteins through weak noncovalent interactions.

peripheral nervous system (PNS) Collectively, the sensory and motor nerves, including the cranial and spinal nerves, and interneurons and ganglia.

peristalsis Waves of muscular contraction that move food toward the base of the stomach.

personalized medicine An approach in which the treatment is matched to the patient, not the disease; examination of an individual's genome sequence, by revealing his or her disease susceptibilities and drug sensitivities, allows treatments to be tailored to that individual.

peroxisomes Organelles in eukaryotic cells that contain many different enzymes and are involved in metabolic reactions.

petal A structure, often brightly colored and distinctively shaped, occurring in the next-to-outermost whorl of a flower; petals attract and orient animal pollinators.

phagocyte A type of immune cell that engulfs and destroys foreign cells or particles.

phagocytosis The engulfing of a cell or particle by another cell.

pharyngeal slit A vertical opening separated from other slits by stiff rods of protein in the pharynx of hemichordates.

pharynx The region of the throat that connects the nasal and mouth cavities; in hemichordates, a tube that connects the mouth and the digestive tract.

phenol Any one of a class of compounds, produced by some plants as a defensive mechanism.

phenotype The expression of a physical, behavioral, or biochemical trait; an individual's observable phenotypes include height, weight, eye color, and so forth.

pheromone A water- or airborne chemical compound released by animals into the environment that signals and influences the behavior of other members of their species.

phloem The vascular tissue that transports carbohydrates from leaves to the rest of the plant body.

phloem sap The sugar-rich solution that flows through both the lumen of the sieve tubes and the sieve plate pores.

phosphatase An enzyme that removes a phosphate group from another molecule.

phosphate group A chemical group consisting of a phosphorus atom bonded to four oxygen atoms.

phosphodiester bond A bond that forms when a phosphate group in one nucleotide is covalently joined to the sugar unit in another nucleotide. Phosophodiester bonds are relatively stable and form the backbone of a DNA strand.

phospholipid A type of lipid and a major component of the cell membrane.

photic zone The surface layer of the ocean through which enough sunlight penetrates to enable photosynthesis.

photoheterotroph An organism that uses the energy from sunlight to make ATP and relies on organic molecules obtained from the environment as the source of carbon for growth and other vital functions.

photoperiod Day length.

photoperiodism The effect of the photoperiod, or day length, on flowering.

photoreceptor A molecule whose chemical properties are altered when it absorbs light; also, photoreceptors are the sensory receptors in the eye.

photorespiration A process in which rubisco acts as an oxygenase, resulting in a release of CO_2 and a net loss of energy.

photosynthesis The biochemical process in which carbohydrates are built from carbon dioxide and the energy of sunlight.

photosynthetic electron transport chain A series of redox reactions in which light energy absorbed by chlorophyll is used to power the movement of electrons; in oxygenic photosynthesis, the electrons ultimately come from water and the terminal electron acceptor is $NADP^+$.

photosystem A protein-pigment complex that absorbs light energy to drive redox reactions and thereby sets the photosynthetic electron transport chain in motion.

photosystem I The photosystem that energizes electrons with a second input of light energy so they have enough energy to reduce $NADP^+$.

photosystem II The photosystem that supplies electrons to the beginning of the electron transport chain. When photosystem II loses an electron, it can pull electrons from water.

phototroph An organism that captures energy from sunlight.

phototropic Bending in response to light. A positive phototropic response, as in stems, is toward light; a negative phototropic response, as in roots, is away from light.

phragmoplast In dividing plant cells, a structure formed by overlapping microtubules that guide vesicles containing cell wall components to the middle of the cell.

phylogenetic niche conservatism The observed similarity in closely related species of some aspect of their niches, indicating its presence in their common ancestor.

phylogenetic species concept (PSC) The idea that members of a species all share a common ancestry and a common fate.

phylogenetic tree A diagrammed hypothesis about the evolutionary history, or phylogeny, of a species.

phylogeny The history of descent with modification and the accumulation of change over time.

phylum (plural, phyla) A group of closely related classes, defined by having one of a number of distinctive body plans.

phytochrome A photoreceptor for red and far-red light that switches back and forth between two stable forms, active and inactive, depending on its exposure to light.

pilus (plural, pili) A threadlike that connects bacteria, allowing plasmids to be transferred between them.

pineal gland A gland located in the thalamic region of the brain that responds to autonomic nervous system input by secreting melatonin, which controls wakefulness.

pinna (plural, pinnae) Small unit of the photosynthetic surface on fern leaves; also, the external structure of mammalian ears that enhance the reception of sound waves contacting the ear.

pit A circular or ovoid region in the walls of xylem cells where the lignified cell wall layer is not produced.

pith In a stem, the region inside the ring of vascular bundles.

pituitary gland A gland beneath the brain that produces a number of hormones, including growth hormone.

placenta In placental mammals, an organ formed by the fusion of the chorion and allantois that allows the embryo to obtain nutrients directly from the mother.

placental mammal A mammal that provides nutrition to the embryo through the placenta, a temporary organ that develops in the uterus; placental mammals include carnivores, primates, hooved mammals, and whales.

placozoans Possibly the simplest of all animals; each contains only a few thousand cells arranged into upper and lower epithelia that sandwich an interior fluid crisscrossed by a network of multinucleate fiber cells.

plasma (effector) cell A cell that secretes antibodies.

plasma membrane The membrane that defines the space of the cell, separating the living material within the cell from the nonliving environment around it.

plasmid In bacteria, a small circular molecule of DNA carrying a small number of genes that can replicate independently of the bacterial genomic DNA.

plasmodesmata (singular, plasmodesma) Connections between the plasma membranes of adjacent plant cells that permit molecules to pass directly from the cytoplasm of one cell to the cytoplasm of another.

plasmogamy The cytoplasmic union of two cells.

plate tectonics The dynamic movement of Earth's crust, the outer layer of Earth.

pleiotropy The phenomenon in which a single gene has multiple effects on seemingly unrelated traits.

pluripotent Describes embryonic stem cells (cells of the inner mass), which can give rise to any of the three germ layers and therefore to any cell of the body.

podocytes Cells with footlike processes that make up one of the three layers of the filtration barrier of the glomerulus.

point mutation A mutation in which a base pair is replaced by a different base pair; this is the most frequent type of mutation; also known as a *nucleotide substitution.*

Pol II The RNA polymerase complex responsible for transcription of protein-coding genes.

polar (molecule) A molecule that has regions of positive and negative charge.

polar body A small cell produced by the asymmetric first meiotic division of the primary oocyte.

polar covalent bond Bonds that do not share electrons equally.

polar transport The coordinated movement of auxin across many cells.

polarity An asymmetry such that one end of a structure differs from the other.

polarized Having opposite properties in opposite parts; describes a resting membrane potential in which there is a buildup of negatively charged ions on the inside surface of the cell's plasma membrane and positively charged ions on its outer surface.

pollen In seed plants, the multicellular male gametophyte surrounded by a sporopollenin-containing outer wall.

pollen cone A reproductive shoot in gymnosperms that produces pollen.

pollen tube A structure produced by the male gametophyte that grows through an opening in the sporopollenin-containing pollen wall and eventually conveys the sperm to the ovule.

pollination The process in which pollen is carried to an ovule.

poly(A) tail The nucleotides added to the 3′ end of the primary transcript by polyadenylation.

polyadenylation The addition of a long string of consecutive A-bearing ribonucelotideas to the 3′ end of the primary transcript.

polycistronic mRNA A single molecule of messenger RNA that is formed by the transcription of a group of functionally related genes located next to one another along bacterial DNA.

polymer A complex organic molecule made up of repeated simpler units connected by covalent bonds.

polymerase chain reaction (PCR) A selective and highly sensitive method for making copies of a piece of DNA, which allows a targeted region of a DNA molecule to be replicated into as many copies as desired.

polymorphism Any genetic difference among individuals sufficiently common that it is likely to be present in a group of 50 randomly chosen individuals.

polypeptide A polymer of amino acids connected by peptide bonds.

polyphyletic Describes groupings that do not include the last common ancestor of all members.

polyploidy The condition of having more than two complete sets of chromosomes in the genome.

polysaccharide A polymer of simple sugars. Polysaccharides provide long-term energy storage or structural support.

polyspermy In animals, fertilization by more than one sperm.

pons A part of the brainstem.

population All the individuals of a given species that live and reproduce in a particular place; one of several interbreeding groups of organisms of the same species living in the same geographical area.

population density The size of a population divided by its range.

population size The number of individuals of all ages alive at a particular time in a particular place.

positive feedback In the nervous system, the type of feedback in which a stimulus causes a response that leads to an enhancement of the original stimulus that leads to a larger response. In the endocrine system, the type of feedback in which a stimulus causes a response, and that response causes a further response in the same direction. In both cases, the process reinforces itself until interrupted.

positive regulation The process in which a regulatory molecule must bind to the DNA at a site near the gene in order for transcription to take place.

positive selection Natural selection that increases the frequency of a favorable allele.

positively selected Describes T cells that recognize self MHC molecules on epithelium cells and continue to mature.

posterior pituitary gland The region of the pituitary gland that develops from neural tissue at the base of the brain and into which neurosecretory cells of the hypothalamus extend that secrete releasing factors.

postreplication mismatch repair The correction of a mismatched base in a DNA strand by cleaving one of the strand backbones, degrading the sequence with the mismatch, and resynthesizing from the intact DNA strand.

posttranslational modification The modification, after translation, of proteins in ways that regulate their structure and function.

post-zygotic isolation Describes factors that cause the failure of the fertilized egg to develop into a fertile individual.

potential energy Stored energy that is released by a change in an object's structure or position.

potential evapotranspiration The amount of evapotranspiration that temperature, humidity, and wind would cause if water supply weren't limiting; the demand on the water resources of an ecosystem.

potential evapotranspiration ratio In an ecosystem, the ratio of water demand to supply.

power stroke The stage in the cross-bridge cycle in which the myosin head pivots and generates a force, causing the myosin and actin filaments to slide relative to each other.

predation An interaction between organisms in which one (the predator) consumes the other (the prey).

pregnancy The carrying of one or more embryos in the mammalian uterus.

premolar One of the teeth of mammals between the canines and molars that are specialized for shearing tough foods.

pre-zygotic isolation Describes factors that that prevent the fertilization of an egg.

primary active transport Active transport that uses the energy of ATP directly.

primary bronchi The two divided airways from the trachea, supported by cartilage rings, each airway leading to a lung.

primary consumers Herbivores, which consume primary (plant or algae) producers. Sometimes called grazers.

primary growth The increase in plant length made possible by apical meristems.

primary motor cortex The part of the frontal lobe of the brain that produces complex coordinated behaviors by controlling skeletal muscle movements.

primary oocyte A diploid cell formed by mitotic division of oogonia during fetal development.

primary producer An organism that takes up inorganic carbon, nitrogen, phosphorus, and other compounds from the environment and converts them into organic compounds that will provide food for other organisms in the local environment.

primary response The response to the first encounter with an antigen, during which there is a short lag before antibody is produced.

primary somatosensory cortex The part of the parietal lobe that integrates tactile information from specific body regions and relays it to the motor cortex.

primary spermatocyte A diploid cell formed by mitotic division of spermatogonia at the beginning of spermatogenesis.

primary structure The sequence of amino acids in a protein.

primary transcript The initial RNA transcript that comes off the template DNA strand.

primate A member of an order of mammals that share a number of general features that distinguish them from other mammals, including nails rather than claws, front-facing eyes, and an opposable thumb.

primer A short stretch of RNA at the beginning of each new DNA strand that serves as a starter for DNA synthesis; an oligonucleotide that serves as a starter in the polymerase chain reaction.

primordia (singular, primordium) An organ in its earliest stage of development; leaf primordia form near the tips of shoot apical meristems and develop into leaves.

principle of independent assortment The principle that segregation of one set of alleles of a gene pair is independent of the segregation of another set of alleles of a different gene pair.

principle of segregation The principle by which half the gametes receive one allele of a gene and half receive the other allele.

probability Among a very large number of observations, the expected proportion of observations that are of a specified type.

probe A labeled DNA fragment that can be tracked in a procedure such as a Southern blot.

procambial cell A cell that retains the capacity for cell division and gives rise to both xylem and phloem.

product Any one of the transformed molecules that result from a chemical reaction.

progesterone A hormone secreted by the vertebrate ovaries that maintains the thickened and vascularized uterine lining.

prokaryote An organism whose cell or cells does not have a nucleus. Often used to refer collectively to archaeons and bacteria.

prometaphase The stage of mitosis in which the nuclear envelope breaks down and the microtubules of the mitotic spindle attach to chromosomes.

prometaphase I The stage of meiosis I in which the nuclear envelope breaks down and the meiotic spindles attach to kinetochores on chromosomes.

prometaphase II The stage of meiosis II in which the meiotic spindles attach to kinetochores on chromosomes.

promoter A regulatory region where RNA polymerase and associated proteins bind to the DNA duplex.

proofreading The process in which DNA polymerases can immediately correct their own errors by excising and replacing a mismatched base.

prophase The stage of mitosis characterized by the appearance of visible chromosomes.

prophase I The beginning of meiosis I, marked by the visible manifestation of chromosome condensation.

prophase II The stage of meiosis II in which the chromosomes in the now-haploid nuclei recondense to their maximum extent.

prostate gland An exocrine gland that produces a thin, slightly alkaline fluid that helps maintain sperm motility and counteracts the acidity of the female reproductive tract.

protease inhibitor An antidigestive protein that binds to the active site of enzymes that break down proteins in a herbivore's digestive system.

protein family A group of proteins that are structurally and functionally related.

protein sorting The process by which proteins end up where they need to be in the cell to perform their function.

proteins The key structural and functional molecules that do the work of the cell, providing structural support and catalyzing chemical reactions. The term "protein" is often used as a synonym for "polypeptide."

proteobacteria The most diverse bacterial group, defined largely by similarities in rRNA gene sequences; it includes many of the organisms that populate the expanded carbon cycle and other biogeochemical cycles.

protist An organism having a nucleus but lacking other features specific to plants, animals, or fungi.

proton A positively charged particle in the atomic nucleus.

protonephridia Excretory organs in flatworms that isolate waste from the body cavity.

proto-oncogene The normal cellular gene counterpart to an oncogene, which is similar to a viral oncogene but can cause cancer only when mutated.

protostome A bilaterians in which the blastopore, the first opening to the internal cavity of the developing embryo, becomes the mouth.

protozoan (plural, protozoa) A heterotrophic protist.

proximal convoluted tubule The first portion of the renal tubule from which electrolytes and other nutrients are reabsorbed into the blood.

pseudogene A gene that is no longer functional.

pulmonary artery One of two arteries, left and right, that carries deoxygenated blood from the right ventricle to the lungs.

pulmonary capillary A small blood vessel that supplies the alveolar wall.

pulmonary circulation Circulation of the blood to the lungs.

pulmonary valve A valve between the right ventrical and the pulmonary trunk, which divides into the pulmonary arteries.

pulmonary vein A vein that returns oxygenated blood from the lungs to the left ventricle.

Punnett square A worksheet in the form of a checkerboard used to predict the consequences of a random union of gametes.

pupa The quiescent stage of metamorphosis in insects, during which the body tissues undergo a transformation from larva to an adult.

pupil An opening in the vertebrate eye through which light enters.

purine In nucleic acids, either of the bases adenine and gunanine, which have a double-ring structure.

pyloric sphincter A band of muscle at the base of the stomach that opens to allow small amounts of digested food to enter the small intestine.

pyrimidine In nucleic acids, any of the bases thymine, cytosine, and uracil, which have a single-ring structure.

quantitative trait A complex trait in which the phenotype is measured along a continuum with only small intervals between similar individuals.

quaternary structure The structure that results from the interactions of several polypeptide chains.

***R* gene** Any one of the group of genes that express the R proteins in plants.

R group A chemical group attached to the central carbon atom of an amino acid, whose structure and composition determine the identity of the amino acid; also known as a *side chain*.

R protein Any one of a group of receptors in plant cells, each expressed by a different gene, that function as part of the plant's immune system by each binding to a specific pathogen-derived protein.

radiometric dating Dating by using the decay of radioisotopes as a yardstick, including (for time intervals up to a few tens of thousands of years) the decay of radioactive ^{14}C to nitrogen and (for most of Earth history) the decay of radioactive uranium to lead.

rain shadow The area on the lee side of mountains, where air masses descend, warming and taking up water vapor; as a result, lands in the rain shadow are arid.

range The extent of the geographic area over which the populations of a species are distributed.

reabsorption In renal systems, an active or passive process in which substances that are important for an animal to retain are taken up by cells of the excretory tubule and returned to the bloodstream.

reactant Any of the starting molecules in a chemical reaction.

reaction center Specially configured chlorophyll molecules where light energy is converted into electron transport.

reactive oxygen species Highly reactive forms of oxygen.

reading frame Following a start codon, a consecutive sequence of codons for amino acids.

realized niche The actual range of habitats occupied by a species.

receiver The individual who, during communication, receives from the sender a signal that elicits a response.

receptor A molecule on cell membranes that detects critical features of the environment. Receptors detecting signals that easily cross the cell membrane are sometimes found in the cytoplasm.

receptor activation The "turning on" of a receptor, which often occurs when a signaling molecule binds to a receptor on a responding cell.

receptor kinase A receptor that is an enzyme that adds a phosphate group to another molecule.

receptor protein The molecule on the responding cell that binds to the signaling molecule.

recessive The trait that fails to appear in heterozygous genotypes from a cross between the corresponding homozygous genotypes.

reciprocal altruism The exchange of favors between individuals.

reciprocal cross A cross in which the female and male parents are interchanged.

reciprocal inhibition The activation of opposing sets of muscles so that one set is inhibited as the other is activated, allowing the movement of joints such as the knee.

reciprocal translocation Interchange of parts between nonhomologous chromosomes.

recombinant An offspring with a different combination of alleles from that of either parent, resulting from one or more crossovers in prophase I of meiosis.

recombinant DNA DNA molecules from two (or more) different sources combined into a single molecule.

recovery metabolism An animal's elevated consumption of oxygen following activity.

rectum The part of the hindgut where feces are stored until elimination.

Redfield ratio On average, cells contain atoms in the ratio 1 P:16 N:100 C, a correspondence first described by the American biologist Alfred Redfield.

reduction A reaction in which a molecule gains electrons.

reductional division An alternative name for meiosis I, since this division reduces the number of chromosomes by half.

refractory period The period following an action potential during which the inner membrane voltage falls below and then returns to the resting potential.

regeneration In the context of photosynthesis, the third step of the Calvin cycle, in which the 5-carbon molecule needed for carboxylation is produced.

regenerative medicine A discipline that aims to use the natural processes of cell growth and development to replace diseased or damaged tissues.

regression toward the mean With regard to complex traits, the principle that offspring exhibit an average phenotype that is intermediate between that of the parents and that of the population as a whole.

regulatory transcription factor A protein that recruits the components of the transcription complex to the gene.

reinforcement of reproductive isolation (reinforcement) The process by which diverging populations undergo natural selection in favor of enhanced pre-zygotic isolation to prevent the production of inferior hybrid offspring.

release factor A protein that causes a finished polypeptide chain to be freed from the ribosome.

releasing factor A peptide hormone that signals to the anterior pituitary gland through blood vessels, leading to a much larger release of associated hormones from that organ.

renal tubules Tubes in the vertebrate kidney that process the filtrate from the glomerulus by reabsorption and secretion.

renaturation The base pairing of complementary single-stranded nucleic acids to form a duplex; also known as *hybridization*, it is the opposite of denaturation.

renin An enzyme produced by the kidneys that converts angiotensinogen to angiotensin I and is involved in the regulation of blood volume and pressure.

repetitive DNA The collective term for repeated sequences of various types in eukaryotic genomes.

replication The process of copying DNA so genetic information can be passed from cell to cell or from an organism to its progeny.

replication bubble A region formed by the opening of a DNA duplex at an origin of replication, which has a replication fork at each end.

replication fork The site where the parental DNA strands separate as the DNA duplex unwinds.

repressor A protein that, when bound with a sequence in DNA, can inhibit transcription.

reservoir A supply or source of a substance. Reservoirs of carbon, for example, include organisms, the atmosphere, soil, the oceans, and sedimentary rocks.

residue In the context of protein synthesis, any of the amino acids that is incorporated into a protein.

resource partitioning A pattern in which species whose niches overlap may diverge to minimize the overlap.

responding cell The cell that receives information from the signaling molecule.

response A change in cellular behavior, such as activation of enzymes or genes, following a signal.

resting membrane potential The negative voltage across the membrane at rest.

restriction enzyme Any one of a class of enzymes that recognizes specific, short nucleotide sequences in double-stranded DNA and cleaves DNA at or near these sites.

restriction fragment length polymorphism (RFLP) A polymorphism in which the length of the restriction fragments is different in the two alleles.

restriction site A recognition sequence in DNA cutting, which is typically four or six base pairs long; most restriction enzymes cleave double-stranded DNA at or near these restriction sites.

reticulum The second chamber in the stomach of ruminants, which, along with the rumen, harbors large populations of anaerobic bacteria that break down cellulose.

retina A thin tissue in the posterior of the vertebrate eye that contains the photoreceptors and other nerve cells that sense and initially process light stimuli.

retinal A derivative of vitamin A that absorbs light and binds to rhodopsin, a transmembrane protein in the photosensitive cells of vertebrates.

retrotransposons Transposable elements in DNA sequences in which RNA is used as a template to synthesize complementary strands of DNA, a reversal of the usual flow of genetic information from DNA into RNA.

reverse transcriptase An RNA-dependent DNA polymerase that uses a single-stranded RNA as a template to synthesize a DNA strand that is complementary in sequence to the RNA.

rhizosphere The soil layer that surrounds actively growing roots.

ribonucleic acid (RNA) A molecule chemically related to DNA that is synthesized by proteins from a DNA template.

ribose The sugar in RNA.

ribosomal RNA (rRNA) Noncoding RNA found in all ribosomes that aid in translation.

ribosome A complex structure of RNA and protein, bound to the cytosolic face of the RER in the cytoplasm, on which proteins are synthesized.

ribulose bisphosphate carboxylase oxygenase (rubisco) The enzyme that catalyzes the carboxylation reaction in the Calvin cycle.

ribulose-1,5-bisphosphate (RuBP) The 5-carbon sugar to which carbon dioxide is added by the enzyme rubisco.

ring species Species that contain populations that are reproductively isolated from each other but can exchange genetic material through other, linking populations.

RISC (RNA-induced silencing complex) A protein complex that is targeted to specific mRNA molecules by base pairing with short regions on the target mRNA, inhibiting translation or degrading the RNA.

ritualization The process of co-opting and modifying behaviors used in another context by increasing the conspicuousness of the behavior, reducing the amount of variation in the behavior so that it can be immediately recognized, and increasing its separation from the original function.

RNA editing The process in which some RNA molecules become a substrate for enzymes that modify particular bases in the RNA, thereby changing its sequence and sometimes what it codes for.

RNA polymerase The enzyme that carries out polymerization of ribonucleoside triphosphates from a DNA template to produce an RNA transcript.

RNA polymerase complex An aggregate of proteins that synthesizes the RNA transcript complementary to the template strand of DNA.

RNA primase An RNA polymerase that synthesizes a short piece of RNA complementary to the DNA template and does not require a primer.

RNA processing Chemical modification that converts the primary transcript into finished mRNA, enabling the RNA molecule to be transported to the cytoplasm and recognized by the translational machinery.

RNA splicing The process of intron removal from the primary transcript.

RNA transcript The RNA sequence synthesized from a DNA template.

RNA world hypothesis The hypothesis that the earliest organisms relied on RNA for both catalysis and information storage.

rod cell A type of photoreceptor cell on the retina that detects light and shades ranging from white to shades of gray and black, but not color.

root apical meristem A group of totipotent cells near the tip of a root that is the source of new root cells.

root cap A structure that covers and protects the root apical meristem as it grows through the soil.

root hair A slender outgrowth produced by epidermal cells that greatly increases the surface area of the root.

root nodule A structure, formed by dividing root cells, in which nitrogen-fixing bacteria live.

roots A major organ system of vascular plants, generally belowground.

rough endoplasmic reticulum (RER) The part of the endoplasmic reticulum with attached ribosomes.

***r*-strategist** A species that produces large numbers of offspring but provides few resources for their support.

rumen The first chamber in the stomach of ruminants, which, along with the reticulum, harbors large populations of anaerobic bacteria that break down cellulose.

S phase The phase of interphase in which the entire DNA content of the nucleus is replicated.

saccharide The simplest carbohydrate molecule, also called a *sugar.*

saltatory propagation The movement of an action potential along a myelinated axon, "jumping" from node to node.

Sanger sequencing A procedure in which chemical termination of daughter strands help in determining the DNA sequence.

sapwood In long-lived trees, the layer adjacent to the vascular cambium that contains the functional xylem.

sarcomere The region from one Z disc to the next, the basic contractile unit of a muscle.

sarcoplasmic reticulum (SR) A modified form of the endoplasmic reticulum surrounding the myofibrils of muscle cells.

saturated Describes fatty acids that do not contain double bonds; the maximum number of hydrogen atoms is attached to each carbon atom, "saturating" the carbons with hydrogen atoms.

scaffold A supporting protein structure in a metaphase chromosome.

scientific method A deliberate, careful, and unbiased way of learning about the natural world.

sclera A tough, white outer layer surrounding the vertebrate eye.

scrotum A sac outside the abdominal cavity of the male that holds the testes.

second law of thermodynamics The principle that the transformation of energy is associated with an increase in the degree of disorder in the universe.

secondary consumers Predators or scavengers that feed on primary consumers.

second messenger An intermediate cytosolic signaling molecule that transmits signals from a receptor to a target within the cell. (First messengers transmit signals from outside the cell to a receptor.)

secondary active transport Active transport that uses the energy of an electrochemical gradient to drive the movement of molecules.

secondary growth The increase in plant diameter resulting from meristems that surround stems and roots.

secondary oocyte A large cell produced by the asymmetric first meiotic division of the primary oocyte.

secondary phloem New phloem cells produced by the vascular cambium, which are located to the outside of the vascular cambium.

secondary response The response to re-exposure to an antigen, which is quicker, stronger, and longer than the primary response.

secondary sexual characteristic A trait that characterizes and differentiates the two sexes but that does not relate directly to reproduction.

secondary spermatocyte A diploid cell formed during the first meiotic division of the primary spermatocyte.

secondary structure The structure formed by interactions between stretches of amino acids in a protein.

secondary xylem New xylem cells produced by vascular cambium, which are located to the inside of the vascular cambium.

second-division nondisjunction Disjunction in the second meiotic division.

secretin A hormone released by cells lining the duodenum in response to the acidic pH of the stomach contents entering the small intestine and that stimulates the pancreas to secrete bicarbonate ions.

secretion In renal systems, an active process that eliminates substances that were not previously filtered from the blood.

seed A fertilized ovule; seeds are multicellular structures that allow offspring to disperse away from the parent plant.

seed coat A protective outer structure surrounding the seed.

segmentation The formation of discrete parts or segments in the insect embryo.

segregate Separate; applies to chromosomes or members of a gene pair moving into different gametes.

selection The retention or elimination of mutations in a population of organisms.

selectively permeable Describes the properties of some membranes, including the plasma membrane, which lets some molecules in and out freely, lets others in and out only under certain conditions, and prevents other molecules from passing through at all.

self Describes an organism's own molecules and cells.

self-compatible Describes species in which pollen and ovules produced by flowers on the same plant can produce viable offspring.

self-incompatible Describes species in which pollination by the same or a closely related individual does not lead to fertilization.

self-propagating Continuing without input from an outside source; action potentials are self-propagating in that they move along axons by sequentially opening and closing adjacent ion channels.

semen A fluid that nourishes and sustains sperm as they travel in the male and then the female reproductive tracts.

semicircular canal One of three connected fluid-filled tubes in the mammalian inner ear that contains hair cells that sense angular motions of the head in three perpendicular planes.

semiconservative replication The mechanism of DNA replication in which each strand of a parental DNA duplex serves as a template for the synthesis of a new daughter strand.

seminal vesicles Two glands at the junction of the vas deferens and the prostate gland that secrete a protein- and sugar-rich fluid that makes up most of the semen and provides energy for sperm motility.

seminiferous tubules A series of tubes in the testes where sperm are produced.

sender The indivdual who, during communication, supplies a signal that elicits a response from the receiver.

sensitization The enhancement of a response to a stimulus that is achieved by first presenting a strong or novel stimulus.

sensory neuron A neuron that receives and transmits information about an animal's environment or its internal physiological state.

sensory organ A group of sensory receptors that converts particular physical and chemical stimuli into nerve impulses that are processed by a nervous system and sent to a brain.

sensory receptor cell A sensory neuron with specialized membranes in which receptor proteins are embedded.

sensory transduction The conversion of physical or chemical stimuli into nerve impulses.

sepal A structure, often green, that forms the outermost whorl of a flower with other sepals and encases and protects the flower during its development.

septum (plural, septa) In fungi, a wall that partially divides the cytoplasm into separate cells in hyphae.

sequence assembly The process in which short nucleotide sequences of a long DNA molecule are arranged in the correct order to generate the complete sequence.

sequence motif Any of a number of sequences or sequence arrangements that indicate the likely function of a segment of DNA.

serosa An outer layer of cells and connective tissue that covers and protects the gut.

Sertoli cell A type of cell in the seminiferous tubules that supports sperm production.

sex chromosome Any of the chromosomes associated with sex.

sex determination The factors that trigger development of male or female characteristics; in humans, sex determination occurs because of the Y chromosome.

sexual reproduction The process of producing offspring that receive genetic material from two parents; in eukaryotes, the process occurs through meiosis and fertilization.

sexual selection A form of selection that promotes traits that increase an individual's access to reproductive opportunities.

shell (of an atom) An energy level.

shoot The collective name for the leaves, stems, and reproductive organs, the major aboveground organ systems of vascular plants.

shoot apical meristem A group of totipotent cells near the tip of a stem or branch that gives rise to new shoot tissues in plants.

short-day plant A plant that flowers only when the day length is less than a critical value.

shotgun sequencing DNA sequencing method in which the sequenced fragments do not originate from a particular gene or region, but from sites scattered randomly across the molecule.

sickle-cell anemia A condition in which hemoglobin molecules tend to crystallize when exposed to lower-than-normal levels of oxygen, causing the red blood cells to collapse and block capillary blood vessels.

side chain A chemical group attached to the central carbon atom of an amino acid, whose structure and composition determine the identity of the amino acid; also known as an *R group.*

sieve plate A modified end wall with large pores that links sieve elements.

sieve tube A multicellular unit composed of sieve elements that are connected end to end, through which phloem transport takes place.

sigma factor A protein that associates with RNA polymerase that facilitates its binding to specific promoters.

signal sequence An amino acid sequence that directs a protein to its proper cellular compartment.

signal transduction The process in which an extracellular molecule acts as a signal to activate a receptor, which transmits information through the cytoplasm.

signal-anchor sequence In protein sorting, an amino acid sequence in a polypeptide chain that embeds the chain in the membrane.

signaling cell The source of the signaling molecule.

signaling molecule The carrier of information transmitted when the signaling molecule binds to a receptor; also referred to as a *ligand.*

signal-recognition particle (SRP) An RNA–protein complex that binds with part of a polypeptide chain and marks the molecule for incorporation into the endoplasmic reticulum (eukaryotes) or the plasma membrane (prokaryotes).

silencers DNA sequences that bind with regulatory transcription factors and repress transcription.

single-gene trait A trait determined by Mendelian alleles of a single gene with little influence from the environment.

single-lens eye An eye structure found in vertebrates and cephalopod mollusks that works like a camera to produce a sharply defined image of the animal's visual field.

single-nucleotide polymorphism (SNP) A site in the genome where the base pair that is present differs among individuals in a population.

single-strand binding protein A protein that binds single-stranded nucleic acids.

sink In plants, any portion of the plant that needs carbohydrates to fuel growth and respiration, such as a root, young leaf, or developing fruit.

sinoatrial (SA) node A specialized region of the heart containing pacemaker cells where the heartbeat is initiated.

sister chromatids The two copies of a chromosome produced by DNA replication.

sister groups Groups that are more closely related to each other than either of them is to any other group.

skeletal muscle Muscle that connects to the body skeleton to move an animal's limbs and torso.

sliding filament model The hypothesis that muscles produce force and change length by the sliding of actin filaments relative to myosin filaments.

slow-twitch Describes muscle fibers that contract slowly and consume less ATP than do fast-twitch fibers to produce force.

small interfering RNA (siRNA) A type of small double-stranded regulatory RNA that becomes part of a complex able to cleave and destroy single-stranded RNA with a complementary sequence.

small intestine Part of the midgut; the site of the last part of digestion and most nutrient absorption.

small nuclear RNA (snRNA) Noncoding RNA found in eukaryotes and involved in splicing, polyadenylation, and other processes in the nucleus.

small regulatory RNA A short RNA molecule that can block transcription, cleave or destabilize RNAA, or inhibit mRNA translation.

smooth endoplasmic reticulum (SER) The portion of the endoplasmic reticulum that lacks ribosomes.

smooth muscle The muscle in the walls of arteries, the respiratory system, and the digestive and excretory systems; smooth muscle appears uniform under the light microscope.

solubility The ability of a substance to dissolve.

solute A dissolved molecule such as the electrolytes, amino acids, and sugars often found in water, a solvent.

solvent A liquid capable of dissolving a substance.

somatic cell A nonreproductive cell, the most common type of cell in body.

somatic mutation A mutation that occurs in somatic cells.

somatic nervous system The voluntary component of the peripheral nervous system, which is made up of sensory neurons that respond to external stimuli and motor neurons that synapse with voluntary muscles.

source In plants, a region that supplies carbohydrates to other parts of the plant.

Southern blot A method for determining the size and number of copies of a DNA sequence of interest by means of a labeled probe hybridized to DNA fragments separated by size by means of electrophoresis.

spatial summation The converging of multiple receptors onto a neighboring neuron, increasing its firing rate proportionally to the number of signals received.

speciation The process whereby new species are produced.

species A group of individuals that can exchange genetic material through interbreeding to produce fertile offspring.

species–area relationship The relationship between island size and equilibrium species diversity.

spermatogenesis The formation of sperm.

spermatozoa (sperm) The smaller, male gametes.

spinal cord In vertebrates, a central tract of neurons that passes through the vertebrae to transmit information between the brain and the periphery of the body.

spinal nerve In vertebrates, a nerve running from the spinal cord to the periphery containing axons of both sensory and motor neurons.

spiracle An opening in the exoskeleton on either side of an insect's abdomen through which gases are exchanged.

spliceosome A complex of RNA and protein that catalyzes RNA splicing.

spongy bone Vertebrate bone tissue consisting of trabeculae, and thus lighter than compact bone, found in the ends of limb bones and within vertebrae.

spontaneous Occurring in the absence of any assignable cause; most mutations are spontaneous.

sporangium In plants, a multicellular structure in which haploid spores are formed by meiosis.

spore A reproductive cell capable of growing into a new individual without fusion with another cell.

sporophyte Describes the diploid multicellular generation in plants that produces spores.

sporopollenin A complex mixture of polymers that is remarkably resistant to environmental stresses such as ultraviolet radiation and desiccation.

stabilizing selection A form of selection that selects against extremes and so maintains the status quo.

stamen A pollen-producing floral organ.

stapes A small bone in the middle ear that helps amplify the waves that strike the tympanic membrane; the stapes connects to the oval window of the cochlea.

starch The form in which glucose is stored in plants.

Starling's Law The correspondence between change in stroke volume and change in the volume of blood filling the heart.

statocyst A type of gravity-sensing organ found in most invertebrates.

statolith In plants, a large starch-filled organelle in the root cap that senses gravity; in animals, a dense particle that moves freely within a statocyst, enabling it to sense gravity.

stem cell An undifferentiated cell that can undergo an unlimited number of mitotic divisions and differentiate into any of a large number of specialized cell types.

stereocilia Nonmotile cell-surface projections on hair cells whose movement causes a depolarization of the cell's membrane.

steroid A type of lipid.

steroid hormone A hormone that is derived from cholesterol.

stigma The surface at the top of the carpel, to which pollen adheres.

stomach The last part of the foregut, which serves as a storage and digestive chamber; also known as the *crop*.

stomata (singular, stoma) Pores in the epidermis of a leaf that regulate the diffusion of gases between the interior of the leaf and the atmosphere.

Stramenopila A eukaryotic superkingdom including unicellular organisms, giant kelps, algae, protozoa, free-living cells, and parasites; distinguished by a flagellum with two rows of stiff hairs and, usually, a second, smooth flagellum.

striated muscle Skeletal muscle and cardiac muscle, which appear striped under a light microscope.

strigolactone A hormone, produced in roots and transported upward in the xylem, that inhibits the outgrowth of axillary buds.

stroke volume (SV) The volume of blood pumped during each heart beat.

stroma The region surrounding the thylakoid, where carbohydrate synthesis takes place.

stromatolite A layered structure that records sediment accumulation by microbial communities.

structural gene A gene that codes for the sequence of amino acids in a polypeptide chain.

style A cylindrical stalk between the ovary and the stigma, through which the pollen tube grows.

suberin A waxy compound coating cork cells that protects against mechanical damage, the entry of pathogens, and water loss.

submucosa A tissue layer surrounding the mucosa that contains blood vessels, lymph vessels, and nerves.

subspecies Allopatric populations that have yet to evolve even partial reproductive isolation but which have acquired population-specific traits.

substrate (S) A molecule acted upon by an enzyme.

substrate-level phosphorylation A way of generating ATP in which a phosphate group is transferred to ADP from an organic molecule, which acts as a phosphate donor or substrate.

succession The replacement of species by other species over time.

sugar The simplest carbohydrate molecule; also called a *saccharide*.

sulci Deep crevices in the brain that separate the lobes of the cerebral hemispheres.

supercoil A coil of coils; a circular molecule of DNA can coil upon itself to form a supercoil.

superkingdom One of seven major groups of eukaryotic organisms, classified by molecular sequence comparisons.

supernormal stimulus An exaggerated stimulus that elicits a response more strongly than the normal stimulus.

surfactant A compound that reduces the surface tension of a fluid film.

survivorship The proportion of individuals from an initial cohort that survive to each successive stage of the life cycle.

suspension filter feeding The most common form of food capture by animals, in which water with food suspended in it passes through a sievelike structure.

sustainable development Use of natural resources at rates no higher than the rate at which they can be replenished.

symbiont An organism that lives in closely evolved association with another species.

symbiosis (plural, symbioses) Close interaction between species that live together, often interdependently.

sympathetic division The division of the autonomic nervous system that generally produces arousal and increased activity; active in the fight-or-flight response.

sympatric Describes populations that are in the same geographic location.

synapomorphy A shared derived character; a homology shared by some, but not all, members of a group.

synapse A junction through which the axon terminal communicates with a neighboring cell.

synapsis The gene-for-gene pairing of homologous chromosomes in prophase I of meiosis.

synaptic cleft The space between the axon of the presynaptic cell and the neighboring postsynaptic cell.

synaptic plasticity The ability to adjust synaptic connections between neurons.

synonymous (silent) mutation A mutation resulting in a codon that does not alter the corresponding amino acid in the polypeptide.

systemic acquired resistance (SAR) The ability of a plant to resist future infections, occurring in response to a wide range of pathogens.

systemic circulation Circulation of the blood to the body, excluding the lungs.

systole The phase of the vertebrate cardiac cycle when the ventricles contract to eject blood from the heart.

T cell receptor (TCR) A protein receptor on a T cell that recognizes and binds to an antigen.

T lymphocyte (T cell) A cell type that matures in the thymus and includes helper and cytotoxic cells.

tannin Any one of a group of phenols found widely in plant tissues that bind with proteins and reduce their digestibility.

taste bud One of the sensory organs for taste.

TATA box A DNA sequence present in many promoters in eukaryotes and archaeons that serves as a protein-binding site for a key general transcription factor.

taxis (plural, taxes) Movement in a specific direction in response to a stimulus.

taxon (plural, taxa) All the species in a taxonomic entity such as family or genus.

tectorial membrane A rigid membrane in the cochlear duct, against which the stereocilia of hair cells in the organ of Corti bend when stimulated by vibration, setting off an action potential.

telomerase An enzyme containing an RNA template from which complementary telomere repeats are synthesized.

telomere A repeating sequence at each end of a eukaryotic chromosome.

telophase The stage of mitosis in which the nuclei of the daughter cells are formed and the chromosomes uncoil to their original state.

telophase I The stage of meiosis I in which the chromosomes uncoil slightly, a nuclear envelope briefly reappears, and in many species the cytoplasm divides, producing two separate cells.

telophase II The stage of meiosis II in which the chromosomes uncoil and become diffuse, a nuclear envelope forms around each set of chromosomes, and the cytoplasm divides by cytokinesis.

template A strand of DNA or RNA whose squence of nucleotides is used to synthesize a compementary strand.

template strand In DNA replication, the parental strand whose sequence is used to synthesize a complementary daughter strand.

temporal isolation Pre-zygotic isolation between individuals that are reproductively active at different times.

temporal lobe The region of the brain involved in the processing of sound, language and reading, and object identification and naming.

temporomandibular joint A specialized jaw joint in mammals that allows the teeth of the lower and upper jaws to fit together precisely.

tendon A collagen structure that attaches muscles to the skeleton and transmits muscle forces over a wide range of joint motion.

termination In protein translation, the time at which the addition of amino acids stops and the completed polypeptide chain is released from the ribosome. In cell communication, the stopping of a signal.

terminator A DNA sequence at which transcription stops and the transcript is released.

terpene Any one of a group of compounds that do not contain nitrogen and are produced by some plants as a defensive mechanism.

tertiary structure The overall three-dimensional shape of a protein, formed by interactions between secondary structures.

test (of a protist) A "house" constructed of organic molecules that shelters a protist.

test group The experimental group that is exposed to the variable in an experiment.

testcross Any cross of an unknown genotype with a homozygous recessive genotype.

testis (plural, testes) The male gonad, where sperm are produced.

testosterone A steroid hormone, secreted by the testes, that plays key roles in male growth, development, and reproduction.

tetanus A muscle contraction of sustained force.

tetraploid A cell or organism with four complete sets of chromosomes; a double diploid.

Tetrapoda A monophyletic group of animals whose last common ancestor had four limbs; this group includes amphibians, lizards, turtles, crocodilians, birds, and mammals (some tetrapods, like snakes, have lost their legs in the course of evolution).

thalamus The inner brain region of the forebrain, which acts as a relay station for sensory information sent to the cerebrum.

thallus A flattened photosynthetic structure produced by some bryophytes.

Thaumarchaeota One of the three major divisions of Archaea; thaumarchaeota are chemotrophs, deriving energy from the oxidation of ammonia.

theory A general explanation of a natural phenomenon supported by a large body of experiments and observations.

theory of island biogeography A theory that states that the number of species that can occupy a habitat island depends on two factors: the size of the island and the distance of the island from a source of colonists.

thermoreceptor A sensory receptor in the skin and in specialized regions of the central nervous system that responds to heat and cold.

thick filament A parallel grouping of myosin molecules that makes up the myosin filament.

thin filament Two helically arranged actin filaments twisted together that make up the actin filament.

threshold potential The critical depolarization voltage of −50 mV required for an action potential.

thylakoid membrane A highly folded membrane in the center of the chloroplast that contains light-collecting pigments and that is the site of the photosynthetic electron transport chain.

thymine (T) A pyrimidine base.

thyroid gland A gland located in the front of the neck that leads to the release of two peptide hormones, thyroxine and triiodothyorine.

Ti plasmid A small circular DNA molecule in virulent strains of *R. radiobacter* containing genes that can be integrated into the host cell's genome, as well as the genes needed to make this transfer.

tidal ventilation A breathing technique in most land vertebrates in which air is drawn into the lungs during inhalation and moved out during exhalation.

tidal volume The amount of air inhaled and exhaled in a cycle; in humans, tidal volume is 0.5 liter when breathing at rest.

tight junction A junctional complex that establishes a seal between cells so that the only way a substance can travel from one side of a sheet of epithelial cells to the other is by moving through the cells by a cellular transport mechanism.

tissue A collection of cells that work together to perform a specific function.

tolerance The ability of T cells not to respond to self antigens even though the immune system functions normally otherwise.

toll-like receptors (TLRs) A family of transmembrane receptors on phagocytes that recognize and bind to molecules on the surface of microorganisms, providing an early signal that an infection is present.

topography The physical features of Earth.

topoisomerase Any one of a class of enzymes that regulates the supercoiling of DNA by cleaving one or both strands of the DNA double helix, and later repairing the break.

topoisomerase II An enzyme that breaks a DNA double helix, rotates the ends, and seals the break.

totipotent Describes cells that have the potential to give rise to a complete organism; a fertilized egg is a totipotent cell.

trabeculae Small plates and rods with spaces between them, found in spongy bone.

trace fossil A track or trail, such as a dinosaur track or the feeding trails of snails and trilobites, left by an animal as it moves about or burrows into sediments.

trachea The central airway leading to the lungs, supported by cartilage rings.

tracheae An internal system of tubes in insects that branch from openings along the abdominal surface into smaller airways, directing oxygen to and removing carbon dioxide from respiring tissues.

tracheid A unicellular xylem conduit.

trade-off An exchange in which something is gained at the expense of something lost.

trait A characteristic of an individual.

transcription The synthesis of RNA from a DNA template.

transcriptional activator protein A protein that binds to a sequence in DNA to enable transcription to begin.

transcriptional regulation The mechanisms that collectively regulate whether or not transcription occurs.

transduction Horizontal gene transfer by means of viruses.

transfer RNA (tRNA) Noncoding RNA that carries individual amino acids for use in translation.

transformation The conversion of cells from one state to another, as from nonvirulent to virulent, when DNA released to the environment by cell breakdown is taken up by recipient cells. In recombinant DNA technology, the introduction of recombinant DNA into a recipient cell.

transgenic organisms An alternative term for genetically modified organisms.

transition state The brief time in a chemical reaction in which chemical bonds in the reactants are broken and new bonds in the product are formed.

translation Synthesis of a polypeptide chain corresponding to the coding sequence present in a molecule of messanger RNA.

transmembrane proteins Proteins that span the entire lipid bilayer; most integral membrane proteins are transmembrane proteins

transmission genetics The discipline that deals with the manner in which genetic material is passed from generation to generation.

transpiration The loss of water vapor from leaves.

transporters Membrane proteins that move ions or other molecules across the cell membrane.

transposable element (TE) A DNA sequence that can replicate and move from one location to another in a DNA molecule; also known as *transposon.*

transposase The enzyme that cleaves a transposon from its original location in the genome and inserts it into a different position.

transposition The movement of a transposable element.

transposon A DNA sequence that can replicate and move from one location to another in a DNA molecule; also known as *transposable element (TE).*

triacylglycerol A lipid composed of a glycerol backbone and three fatty acids.

trimesters The three periods of pregnancy, each lasting about 3 months.

triose phosphate A 3-carbon carbohydrate molecule, produced by the Calvin cycle and exported from the chloroplast.

triploid A cell or organism with three complete sets of chromosomes.

trisomy 21 A condition resulting from the presence of three, rather than two, copies of chromosome 21; also known as *Down syndrome.*

trophic level An organism's typical place in a food web as a producer or consumer.

trophic pyramid A diagram that traces the flow of energy through communities, showing the amount of energy available at each level to feed the next. The pyramid shape results because biomass and the energy it represents generally decrease from one trophic level to the next.

tropic hormone A hormone that controls the release of other hormones.

tropism The bending or turning of an organism in response to an external signal such as light or gravity.

tropomyosin A protein that runs in the grooves formed by the actin helices and blocks the myosin-binding sites.

troponin A protein that moves tropomyosin away from myosin-binding sites, allowing cross-bridges between actin and myosin to form and the muscle to contract.

true breeding Describes a trait whose physical appearance in each successive generation is identical to that in the previous one.

trypsin A digestive enzyme produced by the pancreas that breaks down proteins.

tube feet Small projections of the water vascular system that extend outward from the body surface and facilitate locomotion, sensory perception, food capture, and gas exchange in echinoderms.

tubulin Dimers (composed of an α tubulin and a β tubulin) that assemble into microfilaments.

tumor suppressors A family of genes that encode proteins whose normal activities inhibit cell division.

tunicates A subphylum of Chordata that includes about 3000 species of filter-feeding marine animals, such as sea squirts and salps.

turgor pressure Pressure within a cell resulting from the movement of water into the cell by osmosis and the tendency of the cell wall to resist deformation.

Turner syndrome A sex-chromosomal abnormality in which an individual has 45 chromosomes, including only one X chromosome.

twitch A muscle contraction that results from a single action potential.

twofold cost of sex Population size can increase more rapidly in asexually reproducing organisms than in sexually reproducing organisms because only female produce offspring, and sexual females have only half the fitness of asexual parents.

tympanic membrane A thin sheet of tissue at the surface of the ear that vibrates in response to sound waves, amplifying airborne vibrations; in mammals, also known as the *eardrum.*

unbalanced translocation Translocation in which only part of a reciprocal translocation (and one of the nontranslocated chromosomes) is inherited from one of the parents.

unsaturated Describes fatty acids that contain carbon–carbon double bonds.

uracil (U) A pyrimidine base in RNA, where it replaces the thymine found in DNA.

urea A waste product of protein metabolism that many animals excrete.

ureter A large tube in the vertebrate kidney that brings urine from the kidneys to the bladder.

urethra A tube from the bladder that in males carries semen as well as urine from the body.

uterus A hollow organ within the reproductive tract of female mammals with thick, muscular walls that is adapted to support the developing embryo if fertilization occurs and to deliver the baby during birth.

vaccination The deliberate delivery in a vaccine of an antigen from a pathogen to induce a primary response but not the disease, thereby providing future protection from infection.

vacuole A cell structure that absorbs water and contributes to turgor pressure.

vagina A tubular channel connecting the uterus to the exterior of the body; also known as the birth canal.

valence electrons The electrons farthest from the nucleus, which are at the highest energy level.

van der Waals interactions The binding of temporarily polarized molecules because of the attraction of opposite charges.

variable (in experimentation) The feature of an experiment that is changed by the experimenter from one treatment to the next.

variable (V) region A region of the heavy (H) and light (L) chains of an antibody; the variable region distinguishes a given antibody from all others.

variable expressivity The phenomenon in which a particular phenotype is expressed with a different degree of severity in different individuals.

variable number tandem repeat (VNTR) A genetic difference in which the number of short repeated sequences of DNA differs from one chromosome to the next.

vas deferens A long, muscular tube from the scrotum, through the abdominal cavity, along the bladder, and connecting with the ejaculatory duct.

vasa recta The blood vessels in the kidneys.

vascular cambium Lateral meristem that is the source of new xylem and phloem.

vascular plant A plant that produces both xylem and phloem.

vasoconstriction The process in the supply of blood to the limbs is reduced by constriction of the arteries that supply the limbs.

vasodilation The process in which resistance in the arteries is decreased and blood flow increased following relaxation of the smooth arterial muscles.

vasopressin A posterior pituitary gland hormone that acts on the kidneys and controls the water permeability of the collecting ducts, thus regulating the concentration of urine that an animal excretes; also known as *antidiuretic hormone (ADH)*.

vector In recombinant DNA, a carrier of the donor fragment, usually a plasmid.

vegetative reproduction Reproduction by growth and fragmentation. In plants, the production of upright shoots from horizontal stems, permitting new plants to be produced at a distance from the site where the parent plant originally germinated.

veins In plants, the system of vascular conduits within the leaf; in animals, the large, low-pressure vessels that return blood to the heart.

vena cava (plural, venae cavae) One of two large veins in the body that drain blood from the head and body into the heart.

ventilation The movement of an animal's respiratory medium—water or air—past a specialized respiratory surface.

ventricle A heart chamber that pumps blood to the lungs or the rest of the body.

venule A blood vessel into which capillaries drain as blood is returned to the heart.

vernalization A prolonged period of exposure to cold temperatures necessary to induce flowering.

vertebrae (singular, vertebra) The series of hard bony segments making up the jointed skeleton that runs along the main axis of the body in vertebrates.

vertebral column A skeletal structure in vertebrates that functionally replaces the embryonic notochord that supports the body.

vertebrates A subphylum of Chordata, distinguished by a bony cranium that protects the brain and (unless lost through evolution), a vertebral column; also known as *craniates*.

vesicle A small membrane-enclosed sac that transports substances within the cell.

vessel A multicellular xylem conduit.

vessel element An individual cell that is part of a xylem vessel for water transport in plants.

vestibular system A system in the mammalian inner ear made up of two statocyst chambers and three semicircular canals.

vestigial structure A structure that has lost its original function over time and is now much reduced in size.

vicariance The process in which a geographic barrier arises within a single population, separating it into two or more isolated populations.

villi Highly folded inner surfaces of the jejunum and ileum of the small intestine.

virulent Describes pathogens that are able to overcome a host's defenses and lead to disease.

virus A small infectious agent that contains a nucleic acid genome packaged inside a protein coat called a capsid.

visible light The portion of the electromagnetic spectrum apparent to our eyes.

visual cortex The part of the brain that processes visual images.

vitamin An organic molecule that is required in very small amounts in the diet.

vitreous humor A gel-like substance filling the large cavity behind the lens that makes up most of the volume of the vertebrate eye.

viviparity Giving birth to live young, with nutritional support of the embryo from the mother.

vocal cords Twin organs in the larynx that vibrate as air passes over them, enabling speech, song, and sound production.

voltage-gated channels Ion membrane channels that open and close in response to changes in membrane potential.

voluntary Describes the component of the nervous system that handles sensing and responding to external stimuli.

vulva The external genitalia of the female.

water vascular system A series of fluid-filled canals that permit bulk transport of oxygen and nutrients in echinoderms.

white blood cell (leukocyte) A type of cell in the immune system that arises by differentiation from stem cells in the bone marrow.

white matter Collectively, the axons of cortical neurons in the interior of the brain; it is the fatty myelin produced by glial cells surrounding the axons that makes this region of the brain white.

wild type The most common allele, genotype, or phenotype present in a population; nonmutant.

***X* chromosome** One of the sex chromosomes; a normal human female has two copies of the X chromosome; a normal male has one X and one Y chromosome.

xanthophyll Any one of several yellow-orange pigments that slow the formation of reactive oxygen species by reducing excess light energy; these pigments accept absorbed light energy directly from chlorophyll and convert this energy to heat.

***X*-inactivation** The process in mammals in which dosage compensation occurs through the inactivation of one X chromosome in each cell in females.

***X*-linked gene** A gene in the X chromosome.

xylem Vascular tissue consisting of lignified conduits that transports water and nutrients from the roots to the leaves.

***Y* chromosome** One of the sex chromosomes; a normal human male has one X and one Y chromosome.

yeast A single-celled fungus found in moist, nutrient-rich environments.

***Y*-linked gene** A gene that is present in the region of the Y chromosome that shares no homology with the X chromosome.

yolk A substance in the eggs of animals with external fertilization that provides all the nutrients that the developing embryo needs until it hatches.

Z disc A protein backbone found regularly spaced along the length of a myofibril.

Z scheme Another name for the photosynthetic electron transport chain, so called because the overall energy trajectory resembles a "Z."

zygomycetes Fungi groups that produce hyphae undivided by septa and do not form multicellular fruiting bodies; they make up less than 1% of known fungal diversity.

zygote The diploid cell formed by the fusion of two gametes.

Index

bold face indicates a definition
italics indicate a figure
t indicates a table

D

I

J

K

L

Q

R